Wörterbuch der Tiernamen

Theodor C. H. Cole

Wörterbuch der Tiernamen

Latein-Deutsch-Englisch
Deutsch-Latein-Englisch

2. Auflage

Theodor C. H. Cole
Heidelberg, Deutschland

ISBN 978-3-662-44241-8   ISBN 978-3-662-44242-5 (eBook)
DOI 10.1007/978-3-662-44242-5

Die Deutsche Nationalbibliothek verzeichnet diese Publikation in der Deutschen Nationalbibliografie; detaillierte bibliografische Daten sind im Internet über http://dnb.d-nb.de abrufbar.

Springer Spektrum
© Springer-Verlag Berlin Heidelberg 2000, Softcover 2015
Das Werk einschließlich aller seiner Teile ist urheberrechtlich geschützt. Jede Verwertung, die nicht ausdrücklich vom Urheberrechtsgesetz zugelassen ist, bedarf der vorherigen Zustimmung des Verlags. Das gilt insbesondere für Vervielfältigungen, Bearbeitungen, Übersetzungen, Mikroverfilmungen und die Einspeicherung und Verarbeitung in elektronischen Systemen.
Die Wiedergabe von Gebrauchsnamen, Handelsnamen, Warenbezeichnungen usw. in diesem Werk berechtigt auch ohne besondere Kennzeichnung nicht zu der Annahme, dass solche Namen im Sinne der Warenzeichen- und Markenschutz-Gesetzgebung als frei zu betrachten wären und daher von jedermann benutzt werden dürften. Der Verlag, die Autoren und die Herausgeber gehen davon aus, dass die Angaben und Informationen in diesem Werk zum Zeitpunkt der Veröffentlichung vollständig und korrekt sind. Weder der Verlag noch die Autoren oder die Herausgeber übernehmen, ausdrücklich oder implizit, Gewähr für den Inhalt des Werkes, etwaige Fehler oder Äußerungen.

Planung und Lektorat: Merlet Behncke-Braunbeck, Bettina Saglio

Gedruckt auf säurefreiem und chlorfrei gebleichtem Papier

Springer-Verlag GmbH Berlin Heidelberg ist Teil der Fachverlagsgruppe Springer Science+Business Media
(www.springer.com)

To

Jane Goodall

and a life committed to promoting
the values of biodiversity and bioethics

Vorwort

"Man gave Names to all the Animals"

Dieses *Wörterbuch der Tiernamen* ist ein Referenz- und Nachschlagewerk für Wissenschaftler, Übersetzer und Laien, die sich auf verschiedenste Weise mit Tieren beschäftigen, zum Beispiel in den Bereichen:

- Ökologie / Biodiversität
- Artenschutz / Naturschutz
- Umweltbestandsaufnahmen
- Parasitenkunde
- Schädlinge und Nützlinge

- Zoologische Gärten
- Tier- und Wildparks
- Naturkundemuseen
- Aquaristik / Terraristik
- Meeresbiologie / Tauchen

- Tierhandel
- Pelzhandel
- Lebensmittelsektor
- Hobby (Vögel/Muscheln)
- Dokumentation / Archiv

Die Auswahl von annähernd 16.000 Tiernamen - als Querschnitt des gesamten Tierreiches - erfolgte nach Häufigkeit, Bekanntheitsgrad sowie ökologischer und wirtschaftlicher Bedeutung. Selbst bei der Zielsetzung, die Tierwelt möglichst breit zu erfassen (gemäß *Grzimeks Tierleben*), liegt der Schwerpunkt dieses Buches aus Platzgründen auf der europäischen Fauna. Vertreter der meisten Tierstämme sind aufgeführt, circa zwei Drittel Wirbeltiere und ein Drittel Wirbellose. Zu Grunde liegt eine umfangreiche Literaturbearbeitung enzyklopädischer Werke, Faunen, Naturführer, Roter Listen, und Monographien einzelner Tiergruppen (siehe Anhang).

In der Natur, in Zoos, Wildparks oder im Naturkundemuseum begegnen wir Tieren über deren Namen wir erste Beziehungen knüpfen. Tiere werden gehandelt, verarbeitet, gegessen - gejagt, gesammelt oder geschützt! Deutsche Tiernamen werden in der Landwirtschaft und im Lebensmittelsektor ebenso verwendet wie im Tierhandel, bei Aquarianern, Sammlern, Tauchern, Kleintierzüchtern, Pelzhändlern oder Jägern. Dies gilt entsprechend für den englischen Sprachraum. Die Vielfalt der Tiernamen in den Sprachen führt aber leider oft zu babylonischer Sprachverwirrung. Die Wissenschaft schuf deshalb ein einheitliches System der Namensgebung in lateinischer Sprache mit universeller Geltung. Hier zu vermitteln ist Ziel dieses Wörterbuchs.

Neben Artnamen werden auch höhere Taxa erfasst, vor allem auf der Ebene der Ordnung und Familie. Subspezies (Unterarten) erscheinen nur in Ausnahmefällen.

Tiere, für die in der Literatur keine entsprechenden Populärnamen (volkstümliche Namen, Trivialnamen, Vulgärnamen oder Handelsbezeichnungen „Hbz.") erwähnt sind, erhalten in der vorliegenden Liste eine Bezeichnung (jeweils mit Sternchen * gekennzeichnet), die sich von dem wissenschaftlichen Namen, der geographischen Verbreitung oder anderen prägnanten Merkmalen ableitet. Populärnamen sollten allerdings möglichst nur in Verbindung mit dem validen (derzeitig gültigen) wissenschaftlichen Namen verwendet werden.

Kritisch zu beurteilen ist die zunehmende Patronymisierung bei der biologischen Namensgebung. Die ursprüngliche Absicht Linnés war eine Kennzeichnung anhand prägnanter unterscheidender Merkmale bzw. des geographischen Vorkommens der Lebewesen. Deskriptive Tiernamen tragen allemal einen höheren Informationsgehalt als Personennamen. „Grabstein-Taxonomie" hingegen ist Ausdruck anthropozentrischer Naturphilosophie. Auch bei der Namensgebung haben Biologen die Aufgabe für die Achtung und Würde der Tiere einzutreten, die sie schätzen und schützen.

Da Gattungsnamen sich durch Revision ändern können, wird empfohlen, falls eine Tierart nicht über den Gattungsnamen auffindbar sein sollte, anhand der CD-ROM den Artnamen zu suchen (Achtung: sprachlicher Genus!). Nach allgemein gültigen Nomenklaturregeln wird meist der Artnamen (Art-Epitheton) der Erstbestimmung beibehalten. Mit der CD-ROM können die Begriffe zusätzlich nach Wortfeldern geordnet werden, wodurch der Benützer eigene Listen spezieller Tiergruppen erstellen kann.

Die neue deutsche Rechtschreibung wurde berücksichtigt (z.B. Blässhuhn, Känguru, Gämse, Rauzahndelphin). Teilweise wurden alte Schreibweisen mit aufgenommen. Verzichtet wurde auf die optionalen Schreibweisen: Panter, Delfin, und Tunfisch.

Danksagungen. Besondere Anerkennung und Dank für die Durchsicht und Bearbeitung einzelner Tiergruppen sowie fachlicher Hinweise und kritischer Kommentare gebührt den Fachreferenten und Beratern:

Prof. Dr. Gerd Alberti (Universität Greifswald)
Prof. Dr. Rolf Beiderbeck (Universität Heidelberg)
Dr. Heiko Bellmann (Universität Ulm)
Dr. Paul Cornelius (British Museum, Natural History, London)
Ms. Ann Datta, Zoology Librarian (British Museum, Natural History, London)
Dr. Manfred Grasshoff (Senckenberg, Frankfurt/M.)
Dr. Sheila Halsey (British Museum, Natural History, London)
Ms. Julie M.V. Harvey, Entomology Librarian (British Museum, Natural History, London)
Dr. Ingrid Haußer (Universität Heidelberg)
Dr. Klaus Honomichl (Universität Mainz)
Dr. Raymond Ingle (British Museum, Natural History, London)
Frau Kornelia Jentoch[† (12.5.1999)] (Senckenberg, Frankfurt/M.)
Prof. Dr. Masahiro Kagami (Tokyo Gakugei University)
Prof. Dr. Wolfgang Klausewitz (Senckenberg, Frankfurt/M.)
Dr. Dieter Kock (Senckenberg, Frankfurt/M.)
Dr. Damir Kovac (Senckenberg, Frankfurt/M.)
Dr. Judith Marshall (British Museum, Natural History, London)
Herrn Curt Rambow (Universität Heidelberg)
Dr. Dietrich Schulz (Umweltbundesamt, UBA, Berlin)
Ms. Gail Skidmore, Librarian (Monterey Bay Aquarium, California)
Ms. Mary E. Spencer Jones (British Museum, Natural History, London)
Dr. Matthias Stehmann (Bundesforschungsanstalt für Fischerei, Hamburg)
Prof. Dr. Volker Storch (Universität Heidelberg)
Dr. Michael Türkay (Senckenberg, Frankfurt/M.)
Ms. Clare Valentine (British Museum, Natural History, London)
Herrn Uwe Zajonz, Dipl. biol. (Senckenberg, Frankfurt/M.)
Dr. Bernhard Ziegler (Englisches Institut, Heidelberg)

Merlet Behncke-Braunbeck, Lektorin für Biologie bei Spektrum Akademischer Verlag, danke ich für Determination, Inspiration und Geduld; für die positive Arbeitsatmosphäre meine besondere Wertschätzung!

Dankbar bin ich vor allem meiner Familie, die mich mit Rat und Tat durch die Höhen und Tiefen dieser „Expedition ins Tierreich" begleitete, nunmehr mein viertes und größtes Buchprojekt. Für Erika, Cynthia, Celline und mich waren die Tiere und ihre Geschichten: *Inspiration*!

Heidelberg, im Jahr 2000   Theodor C.H. Cole

Preface

" Man gave Names to all the Animals "

This *Dictionary of Animal Names* is designed as a ready reference for scientists, translators, and laypersons involved in the fields of:

- Ecology - Biodiversity
- Nature - Environment
- Endangered Species
- Parasites - Pest Control
- Agriculture
- Zoos - Wildlife Parks
- Wildlife Management
- Natural History Museums
- Aquaristics - Terraristics
- Marine Biology - Diving
- Animal Trade
- Fur Trade
- Food - Nutrition
- Hobby (Birding - Shells..)
- Documentation - Archives

Some 16,000 animal names were selected on the basis of their distribution, popularity, economic and medical importance, and ecological significance. Emphasis was placed on the European fauna, while also including a broad selection of globally important animals. Representatives of most animal phyla are listed - approximately two-thirds vertebrates and one-third invertebrates. Names were obtained from encyclopedic sources, faunas, field guides, red data lists, and monographs on individual animal groups.

Animals are encountered in nature, in zoos, wildlife parks, natural history museums, pet shops and sold for food at most grocery shops. Their names help us establish first relationships and allow us to communicate with other 'conspecifics' about these animals in an abstract and intellectual context. Animals are domesticated, traded, processed, and eaten - hunted, gathered, collected, and beaten - or protected! English names are used the world over in agriculture, trade, and the food sector as much as among aquarists, collectors, divers, breeders, and hunters - and commonly listed in governmental regulations and red data lists.

The diversity of names has led to utmost confusion. One animal may be known by several English names with several local geographical variations. One name may be used by different people to designate different animals. Besides, names and their usage may change over time. International communication under such premises would scarcely be efficient. Biologists thus introduced the Latin binominal system in order to establish a universally accepted nomenclature. The goal of this dictionary is to establish a link between science and the general public - between the languages and scientific animal names.

How common must a common name be in order to be considered common? This listing is a compilation of common names, rare names, colloquial names, vernacular names, trivial names, trade names, sometimes even vulgar names without claiming the authority of fixing any one name as the only acceptable English name, while such a ranking may be desirable.

Attempts have been made by the scientific community to establish uniformity in common names for individual groups of animals, mainly those of foremost economic importance, esp. insects, birds, fishes.

As stated by the Committee on Names of Aquatic Invertebrates (CNAI) of the American Fisheries Society (AFS): "Common names have the potential to be stabilized by general agreement; scientific names, on the other hand, often change with advancing knowledge." This alludes to the volatility of genus names; species/specific names (epithets) are more stable according to international binominal rules, and thus rarely change.

The use of patronymic names has been critically referred to as 'graveyard taxonomy', a questionable form of commemorating honorable scholars. It is argued that such anthropocentric names are less useful than descriptive names which allude to an animal's pertinent distinguishing characteristics or geographical distribution. Also, the question is raised whether people's names for animals are ethically justified: the name of an animal should reflect an appreciation for its uniqueness and respect of its dignity. Ethically there may be a difference between dedicating a scholarly work (as a book) or an animal to the commemoration of an honorable person. Thus the use of commemorative personal names in biological taxonomy should be discouraged.

Animals for which common names have not been identified in the literature have been assigned a suggested English/German name (marked by asterisk *); these names are derived from their scientific names, geographic occurrance, or other pertinent characteristics. May some of these names eventually become "common"!

Common names should only be used in conjunction with currently accepted scientific names.

Acknowledgements. Sincere thanks are expressed to the following colleagues who reviewed individual animal groups, made suggestions, and gave valuable comments and advice:

Prof. Dr. Gerd Alberti (University of Greifswald)
Prof. Dr. Rolf Beiderbeck (University of Heidelberg)
Dr. Heiko Bellmann (Unversity of Ulm)
Dr. Paul Cornelius (British Museum, Natural History, London)
Ms. Ann Datta, Zoology Librarian (British Museum, Natural History, London)
Dr. Sheila Halsey (British Museum, Natural History, London)
Ms. Julie MV Harvey, Entomology Librarian (British Museum, Natural History, London)
Dr. Ingrid Haußer (University of Heidelberg)
Dr. Klaus Honomichl (University of Mainz)
Dr. Raymond Ingle (British Museum, Natural History, London)
Ms. Kornelia Jentoch[† (12 May 1999)] (Senckenberg, Frankfurt/M.)
Prof. Dr. Masahiro Kagami (Tokyo Gakugei University)
Prof. Dr. Wolfgang Klausewitz (Senckenberg, Frankfurt/M.)
Dr. Dieter Kock (Senckenberg, Frankfurt/M.)
Dr. Damir Kovac (Senckenberg, Frankfurt/M.)
Dr. Judith Marshall (British Museum, Natural History, London)
Mr. Curt Rambow (University of Heidelberg)
Dr. Dietrich Schulz (German Environmental Agency, UBA, Berlin)
Ms. Gail Skidmore, Marine Biology Librarian (Monterey Bay Aquarium, California)
Ms. Mary E. Spencer Jones (British Museum, Natural History, London)
Dr. Matthias Stehmann (Bundesforschungsanstalt für Fischerei, Hamburg)
Prof. Dr. Volker Storch (University of Heidelberg)
Dr. Michael Türkay (Senckenberg, Frankfurt/M.)
Ms. Clare Valentine (British Museum, Natural History, London)
Mr. Uwe Zajonz, Dipl.biol. (Senckenberg, Frankfurt/M.)
Dr. Bernhard Ziegler (Englisches Institut, Heidelberg)

The libraries of the Senckenberg Research Institution (Frankfurt), the University of Heidelberg, the University of Maryland, the British Museum (Natural History), and Monterey Bay Aquarium (California) have been most valuable in researching the content of this book.

Special compliments to Merlet Behncke-Braunbeck of Spektrum Publishers, for intuitive judgement, professional guidance, and personal encouragement which decisively motivated this project!

Erika Siebert-Cole (M.A.) accompanied me through this fourth book project with her expertise. Without her support and inspiration this project would never have been accomplished.
Cindy and Celline have helped in many ways - and learned. With them we are grateful for the inspiration afforded by the uniqueness and beauty inherent to each of the encountered animals.

Heidelberg, Germany, in the year 2000   Theodor C.H. Cole

Abkürzungen

u.a. - hier: „unter anderen"
(der Name wird auch für andere ähnliche Arten/Gattungen verwendet)

a.o. - hier: „among others"
(der Name wird auch für andere ähnliche Arten/Gattungen verwendet)

spp. - „Spezies" *pl.* = Arten
(aufgeführte Namen werden für die gesamte Artgruppe/Gattung verwendet)

kl. - „Klepton" (z.B. *Rana* kl. *esculenta*)

Hbz. - Handelsbezeichnung

Austr. - Australien (Australischer Name)

Br. - Britisch (Britischer Name)

U.S. - United States/Vereinigte Staaten (Amerikanischer Name)

FAO - Food and Agricultural Organization of the United Nations
(so gekennzeichnete englische Namen sind FAO-akkreditiert,
u.a. gemäß FISHBASE der ICLARM, Manila - *siehe Anhang*)

IUCN - International Union for Conservation of Nature and Natural Resources
(so gekennzeichnete englische Namen sind IUCN-akkreditiert,
gemäß IUCN Publikationen - *siehe Anhang*)

Bedeutung der lateinischen Suffxe:

- acea → Klasse
- formes → Ordnung
- oidea → Überfamilie
- idae → Familie
- inae → Unterfamilie
- ini → Tribus

Abacarus hystrix	Getreiderostmilbe	grain rust mite, cereal rust mite
Abastor erythrogrammus/ Farancia erythrogramma	Regenbogennatter	rainbow snake
Abax parallelepipedus	Großer Breitkäfer	parallel-sided ground beetle
Abida secale	Roggenkornschnecke	large chrysalis snail, juniper chrysalis shell
Abietinaria abietina	Tannenmoos, Moostanne, Meertanne	sea fir
Abietinaria filicula	Farnmoos*	fern hydroid
Ablepharus kitaibelii	Johannisechse	juniper skink, snake-eyed skink
Ablepharus spp.	Schlangenaugen-Skinke, Natternaugen-Skinke	snake-eyed skinks, ocellated skinks
Abra aequalis	Gemeine Atlantische Pfeffermuschel	common Atlantic abra (common Atlantic furrow-shell)
Abra alba	Weiße Pfeffermuschel, Kleine Pfeffermuschel	white abra (white furrow-shell)
Abra nitida	Glänzende Pfeffermuschel	shiny abra (glossy furrow-shell)
Abra prismatica	Lange Pfeffermuschel	prismatic abra, elongate abra (elongate furrow-shell)
Abra spp.	Kleine Pfeffermuscheln	abras (lesser European abras)
Abra tenuis	Platte Pfeffermuschel	flat European abra (flat furrow-shell)
Abralia veranyi	Blitzaugenkalmar*	eye-flash squid
Abramis ballerus	Zope	zope (FAO), blue bream
Abramis bjoerkna/ Blicca bjoerkna/	Güster, Blicke, Pliete (Halbbrachsen)	silver bream, white bream (FAO)
Abramis brama	Blei, Brachsen, Brassen, Brasse	common bream, freshwater bream, carp bream (FAO)
Abramis sapa	Donau-Brachse, Zobel	Danube bream, Danubian bream, white-eye bream (FAO)
Abraxas grossulariata	Harlekin, Stachelbeerspanner	magpie moth, currant moth
Abrocoma spp.	Chinchillaratten	chinchilla rats, chinchillones
Abronia spp.	Baumschleichen	tree anguids, arboreal alligator lizards
Abrostola spp.	Nesseleulen, Brennnesseleulen	dark spectacles
Abudefduf saxatlis	Sergeant Major, Hauptfeldwebelfisch, Fünfstreifen-Zebrafisch	sergeant major
Abudefduf sexfasciatus	Zebra-Riffbarsch	striped damselfish, scissortail sergeant (FAO)
Abudefduf spp.	Riffbarsche, Jungfernfische	sergeant damselfishes
Acalitus essigi	Brombeermilbe	redberry mite, blackberry mite
Acalitus rudis	Birken-Hexenbesenmilbe	witches' broom mite (of birch)
Acalyptophis peronii	Peron-Seeschlange	Peron's sea snake
Acanthacaris caeca	Atlantischer Tiefseehummer	Atlantic deepsea lobster
Acantharchus pomotis	Schlammbarsch	mud sunfish
Acanthaster planci	Dornenkronenseestern	crown-of-thorns starfish

Acanthella acuta	Kaktusschwamm	cactus sponge
Acanthepeira stellata	Sternbauchspinne*	starbellied spider, star-bellied spider
Acanthina monodon	Einhornschnecke	unicorn, one-toothed thais
Acanthina spp.	Einhornschnecken	unicorns, unicorn snails
Acanthinula aculeata	Stachelschnecke	prickly snail
Acanthisitta chloris	Grenadier	rifleman
Acanthisittidae	Neuseeland-Grenadiere	New Zealand wrens
Acanthistius brasilianus	Argentinischer Zackenbarsch	Argentine seabass
Acanthizidae	Südseegrasmücken	thornbills, whitefaces
Acanthobdelliformes	Borstenegel	bristly leeches
Acanthocardia aculeata	Stachelige Herzmuschel, Große Herzmuschel	spiny cockle
Acanthocardia echinata	Dornige Herzmuschel	prickly cockle, thorny cockle
Acanthocardia tuberculata	Knotige Herzmuschel	tuberculate cockle, rough cockle
Acanthocarpus alexandri	Gladiatorkrabbe*	gladiator box crab
Acanthocephala	Kratzer	spiny-headed worms, thorny-headed worms, acanthocephalans
Acanthocephalus lucii	Fischkratzer	thorny-headed worm of fish
Acanthochelys macrocephala/ Platemys macrocephala	Großkopf-Plattschildkröte, Großkopf-Pantal-Sumpfschildkröte	Pantanal swamp turtle, big-headed Pantanal swamp turtle
Acanthochelys pallidipectoris/ Platemys pallidipectoris	Sporen-Plattschildkröte	Chaco side-necked turtle
Acanthochermes quercus	Sternwarzenzwerglaus	
Acanthochitona communis (Polyplacophora)	Gemeine Stachel-Käferschnecke	velvety mail chiton (velvety mail shell)
Acanthochitona crinita/ Acanthochitona fascicularis/ Chiton fascicularis (Polyplacophora)	Europäische Stachel-Käferschnecke	bristly mail chiton (bristly mail shell)
Acanthocinus aedilis	Zimmermannsbock	common timberman beetle, timberman
Acanthoclinidae	Stachelzwergbarsche	spiny basslets
Acanthocybium solandri	Wahoo	wahoo
Acanthodactylus erythrurus	Gewöhnlicher Fransenfinger, Europäischer Fransenfinger	spiny-footed lizard, fringe-fingered lizard
Acanthodactylus spp.	Fransenfinger	fringe-toed lacertids, fringe-fingered lizards
Acanthoderes clavipes	Scheckenbock	
Acanthodii	Stachelfische, „Dornhaie", Acanthodier	spiny fishes, acanthodians
Acanthodoras spinosissimus	Dornwels	talking catfish (FAO), spiny catfish, croaking catfish
Acantholabrus palloni	Augenschuppen-Lippfisch, Sattelfleck-Lippfisch	scale-rayed wrasse
Acantholyda erythrocephala	Stahlblaue Kiefernschonungs-Gespinstblattwespe	pine false webworm

Acantholyda posticalis	Große Kieferngespinstblattwespe	pine web-spinning sawfly
Acanthonyx lunulatus	Tangkrabbe	surf crab*
Acanthophilus helianthis	Saflorfliege	safflower fly
Acanthophis antarcticus	Todesotter	death adder
Acanthophthalmus kuhli/ Pangio kuhli	Dornauge	coolie loach
Acanthophthalmus shelfordi/ Pangio shelfordi	Borneo-Dornauge	Borneo loach
Acanthopleura granulata (Polyplacophora)	Struppige Käferschnecke	West Indian fuzzy chiton, granulated chiton
Acanthosaura lepidogaster	Brauner Nackenstachler	brown pricklenape
Acanthosaura spp.	Nackenstachler, Winkelkopfagamen u.a.	pricklenapes
Acanthoscelides obtectus	Speisebohnenkäfer	bean weevil
Acanthoscurria antillensis	Antillen-Vogelspinne	Antilles tarantula
Acanthosoma haemorrhoidale	Stachelwanze, Wipfelwanze	hawthorn shieldbug
Acanthosomatidae	Stachelwanzen	shieldbugs
Acanthostracion quadricornis/ Ostracion quadricornis	Vierhorn-Kofferfisch	scrawled cowfish
Acanthuridae	Doktorfische & Einhornfische	surgeonfishes & unicornfishes
Acanthurinae	Doktorfische	surgeonfishes
Acanthurus coerulens	Blauer Doktorfisch	blue tang surgeonfish
Acanthurus leucosternon	Weißkehlseebader, Weißbrust-Doktorfisch, Weißkehl-Doktorfisch	white-breasted surgeonfish, powder-blue tang, powderblue surgeonfish (FAO)
Acanthurus lineatus	Blaustreifen-Doktorfisch, Streifen-Doktorfisch	clown surgeonfish, lined surgeonfish (FAO), blue-lined surgeonfish
Acanthurus olivaceus	Gelber Doktorfisch	olive surgeonfish, orangespot surgeonfish (FAO)
Acanthurus pyroferus	Orangefleck-Doktorfisch	chocolate surgeonfish
Acanthurus sohal	Rotmeer-Doktorfisch	Red Sea surgeonfish, Red Sea clown surgeon, Sohal surgeonfish (FAO)
Acanthurus triostegus	Sträflings-Doktorfisch	convict tang, convict surgeonfish (FAO)
Acanthurus xanthopterus	Weißschwanz-Doktorfisch	purple surgeonfish, yellowfin surgeonfish (FAO)
Acantophiops lineatus	Gestreifter Blindskink	woodbush legless skink
Acar bailyi/Barbatia bailyi	Miniatur-Archenmuschel	miniature ark
Acarapis spp.	Tracheenmilben	tracheal mites
Acarapis woodi	Bienenmilbe, Tracheenmilbe	bee mite, honey bee mite
Acari/Acarina	Milben & Zecken	mites & ticks
Acaridae/Tyroglyphidae	Vorratsmilben	acarid mites, house dust mites
Acarnus erithacus	Roter Vulkanschwamm	red volcano sponge
Acarus siro/ Tyroglyphus farinae	Mehlmilbe (>Bäckerkrätze)	flour mite, grain mite
Accipiter brevipes	Kurzfangsperber	Levant sparrowhawk
Accipiter cooperii	Rundschwanzsperber	Cooper's hawk

Accipiter gentilis	Habicht	goshawk, northern goshawk
Accipiter nisus	Sperber	northern sparrowhawk, sparrow hawk
Accipiter novaehollandiae	Neuhollandhabicht	white goshawk
Accipiter striatus	Eckschwanzsperber	sharp-shinned hawk
Accipiter virgatus	Besrasperber	Besra sparrowhawk
Accipitrinae	Habichtartige	hawks & eagles & accipiters & kites
Aceria ficus/Aceria fici	Feigenmilbe	fig mite
Aceria fraxinivora	Klunkern-Gallmilbe	ash gall mite
Aceria lycopersici	Tomatenmilbe	tomato erineum mite
Aceria macrochelus	Ahornblatt-Gallmilbe	maple leaf solitary-gall mite
Aceria tristriatus	Walnussblatt-Gallmilbe	walnut leaf gall mite, Persian walnut leaf blister mite
Acetes japonicus	Akiami-Garnele	Akiami paste shrimp
Achaearanea tepidariorum	Gewächshausspinne, Amerikanische Hausspinne	house spider, domestic spider, American common house spider, common American house spider
Achaeus cranchii	Cranchs Seespinne	Cranch's spider crab
Achatina achatina	Echte Achatschnecke	common African snail
Achatina fulica	Große Achatschnecke, Gemeine Riesenschnecke, Afrikanische Riesenschnecke	giant African snail, giant African land snail
Achatinella spp.	Hawaiianische Baumschnecken	Hawaiian tree snails
Achatinellidae	Hawaiianische Baumschnecken	Hawaiian tree snails
Achatinidae	Afrikanische Riesenschnecken	giant African snails
Acherontia atropos	Totenkopfschwärmer, Totenkopf	death's-head hawkmoth
Acheta domestica	Heimchen, Hausgrille	house cricket (domestic cricket, domestic gray cricket)
Achiridae	Amerikanische Seezungen	American soles
Achiropsettidae	Südflundern	southern flounders
Achirus lineatus	Pazifische Seezunge	lined sole
Achroia grisella	Kleine Wachsmotte	lesser wax moth
Acicula fusca/Acme fusca	Braune Nadelschnecke	point snail, brown point snail (point shell)
Acicula lineata	Gestreifte Nadelschnecke	striped point snail
Aciculidae/Acmidae	Nadelschnecken	point snails
Acilius spp.	Furchenschwimmer	pond beetles
Acilius sulcatus	Gemeiner Furchenschwimmer	pond beetle, common pond beetle
Acinonyx jubatus	Gepard	cheetah
Acipenser baerii	Sibirischer Stör	Siberian sturgeon
Acipenser brevirostrum	Kurznasenstör	shortnose sturgeon
Acipenser dabryanus	Yangtze-Stör	Yangtze sturgeon
Acipenser fulvescens	Roter Stör	lake sturgeon
Acipenser gueldenstaedti	Waxdick	Danube sturgeon, Russian sturgeon (FAO), osetr
Acipenser medirostris	Grüner Stör	green sturgeon
Acipenser mikadoi	Sachalin-Stör	Sakhalin sturgeon

Acipenser naccarii	Adria-Stör, Adriatischer Stör	Adriatic sturgeon
Acipenser nudiventris	Glattdick	ship sturgeon
Acipenser oxyrhynchus	Atlantischer Stör	Atlantic sturgeon
Acipenser platorynchus/ Scaphirhynchus platorhynchus	Schaufelstör, Schaufelnasenstör	shovelnose sturgeon
Acipenser ruthenus	Sterlett, Sterlet	sterlet (FAO), Siberian sterlet
Acipenser sinensis	Chinesischer Stör	Chinese sturgeon
Acipenser stellatus	Sternhausen, Scherg	starry sturgeon, sevruga, stellate sturgeon (IUCN)
Acipenser sturio	Stör, Baltischer Stör	common sturgeon (IUCN), Atlantic sturgeon, European sturgeon
Acipenser transmontanus	Weißer Stör, Sacramento-Stör	white sturgeon
Acipenseridae	Eigentliche Störe	sturgeons
Acipenseriformes	Störe & Löffelstöre	sturgeons & sterlets & paddlefishes
Acleris variegana	Heidelbeerwickler	
Acmaea mitra	Pazifische Weiße Schildkrötenschnecke	whitecap limpet, Pacific white tortoiseshell limpet, Pacific white cap limpet
Acmaea testudinalis/ Tectura tessulata/ Collisella tessulata	Schildkröten-Napfschnecke, Klippkleber	tortoiseshell limpet, common tortoiseshell limpet
Acmaea virginea/ Tectura virginea	Weiße Schildkröten-Napfschnecke, Klippkleber, Jungfräuliche Napfschnecke	white tortoiseshell limpet
Acmaeidae	Schildkrötenschnecken	tortoiseshell limpets
Acmaeodera degener	Gelbgefleckter Prachtkäfer	
Acmaeops collaris	Blauschwarzer Kugelhalsbock	
Acmidae/Aciculidae	Nadelschnecken	point snails
Acomys cahirinus	Ägyptische Stachelmaus	Egyptian spiny mouse
Acomys minous	Kreta-Stachelmaus	Cretan spiny mouse
Acomys spp.	Stachelmäuse	spiny mice
Aconaemys fuscus	Südamerikanische Felsenratte	rock rat
Acontias spp.	Lanzenskinke	lance skinks
Acosmetia caliginosa	Färberscharteneule	reddish buff (moth)
Acrania	Schädellose	acranians
Acrantophis madagascariensis	Madagaskar-Boa	Madagascan boa
Acrantophis dumerili	Dumerils Boa	Dumeril's boa
Acrida hungarica/ Acrida ungarica	Nasenschrecke, Turmschrecke, Gewöhnliche Nasenschrecke	snouted grasshopper, long-headed grasshopper
Acrididae/Acridoidea	Feldheuschrecken (Grashüpfer/Heuhüpfer)	grasshoppers (short-horned grasshoppers)
Acridotheres albocinctus	Halsbandmaina	white-collared mynah
Acridotheres cristatellus	Haubenmaina	crested mynah, Chinese jungle mynah
Acridotheres fuscus	Dschungelmaina, Braunmaina	Indian jungle mynah
Acridotheres ginginianus	Ufermaina	bank mynah
Acridotheres grandis	Langschopfmaina	great mynah

Acridotheres spp.	Mainas	mynahs, mynas
Acridotheres tristis	Hirtenmaina	common mynah, common myna
Acris crepitans crepitans	Nördlicher Grillenfrosch (Nördlicher Heuschreckenfrosch)	northern cricket frog
Acris gryllus gryllus	Südlicher Grillenfrosch (Südlicher Heuschreckenfrosch)	southern cricket frog
Acris spp.	Grillenfrösche (Heuschreckenfrösche)	cricket frogs
Acrobates pygmaeus	Zwerggleitbeutler, Mausgleitbeutler, Mausflugbeutler	pygmy gliding possum, feathertail glider
Acrocephalinae	Rohrsänger	leaf-warblers
Acrocephalus aedon	Dickschnabel-Rohrsänger	thick-billed warbler
Acrocephalus agricola	Feldrohrsänger	paddyfield warbler
Acrocephalus arundinaceus	Drosselrohrsänger	great reed warbler
Acrocephalus dumetorum	Buschrohrsänger	Blyth's reed warbler
Acrocephalus melanopogon	Mariskensänger	moustached warbler
Acrocephalus paludicola	Seggenrohrsänger	aquatic warbler
Acrocephalus palustris	Sumpfrohrsänger	marsh warbler
Acrocephalus schoenobaenus	Schilfrohrsänger	sedge warbler
Acrocephalus scirpaceus	Teichrohrsänger	reed warbler
Acrocephalus stentoreus	Stentorrohrsänger	clamorous reed warbler
Acroceridae	Kugelfliegen, Spinnenfliegen	small-headed flies
Acrochordidae	Warzenschlangen	wart snakes, file snakes
Acrochordus arafurae	Arafura-Warzenschlange	Arafura file snake
Acrochordus granulatus/ Chersydrus granulatus	Kleine Warzenschlange, Indische Warzenschlange	leeser wart snake, Indian wart snake
Acrochordus javanicus	Javanische Warzenschlange	elephant-trunk snake
Acrocnida brachiata	Langarmiger Schlangenstern	longarm brittlestar
Acrolepia assectella	Lauchmotte	leek moth
Acrolepiidae	Halbmotten	smudges
Acroloxidae	Teichnapfschnecken	shield snails, lake limpets
Acroloxus lacustris/ Ancylus lacustris	Teichnapfschnecke, Seenapfschnecke	lake limpet, shield snail
Acrometopa servillea	Langbeinschrecke*	longlegged bushcricket
Acronicta aceris	Ahorneule, Rosskastanieneule	sycamore moth
Acronicta alni	Erleneule	alder moth
Acronicta leporina	Wolleule	miller
Acronicta megacephala	Großkopf	poplar grey (moth)
Acronicta psi	Pfeileule	grey dagger
Acronicta rumicis	Ampfereule	knot grass (moth)
Acronicta tridens	Aprikoseneule	dark dagger
Acropomatidae	Indo-Pazifische Leuchtfische	temperate ocean-basses, lanternfishes
Acropora cervicornis	Hirschgeweihkoralle	staghorn coral
Acropora clathrata	Plattenkoralle	plate coral*
Acropora palmata	Elchhornkoralle, Elchgeweihkoralle	elkhorn coral
Acropora prolifera	Fächergeweihkoralle*	fused staghorn coral

Acropora spp.	Geweihkorallen, Baumkorallen	horn corals
Acryllium vulturinum	Geierperlhuhn	vulturine guineafowl
Acteon eloisae	Eloisenschnecke	Eloise's acteon
Acteon tornatilis	Drechselschnecke, Europäische Drechselschnecke	European acteon, lathe acteon
Acteon virgatus	Gestreifte Drechselschnecke	striped acteon
Acteonidae	Drechselschnecken	barrel snails, baby bubbles (barrel shells, small bubble shells)
Actias luna	Mondspinner	luna moth
Actias spp.	Mondspinner	luna moths
Actinauge verrillii	Netzanemone*	reticulate anemone
Actinia cari	Gürtelrose, Ringelrose	green sea anemone
Actinia equina	Purpurrose, Pferdeaktinie	beadlet anemone, red sea anemone, plum anemone
Actinia fragacea	Erdbeerrose	strawberry anemone
Actiniaria	Seeanemonen	sea anemones
Actinodendron arboreum	Bäumchen-Anemone	tree anemone
Actinodendron plumosum	Gefiederte Anemone	hell's fire sea anemone, hell's fire anemone, pinnate anemone
Actinopterygii	Strahlenflosser	ray-finned bony fishes, rayfin fishes, actinopterygians, actinopts
Actinothoe clavata	Schlangenhaarrose	snakehair anemone*
Actinothoe sphyrodeta	Felsrose	cliff anemone*
Actinotrichida	Actinotriche Milben	actinotrichid mites
Actitis hypoleucos/ Tringa hypoleucos	Flussuferläufer	common sandpiper
Actitis macularia	Drosseluferläufer	spotted sandpiper
Actornis l-nigrum	Schwarzes L	
Aculepeira ceropegia	Eichenblatt-Radspinne	oakleaf orbweaver*
Aculifera/Amphineura	Stachelweichtiere	aculiferans, amphineurans
Aculops acericola	Platanengallmilbe	sycamore gall mite
Aculops lycopersici	Tomatenrostmilbe*	tomato russet mite
Aculus fockeui/ Aculus cornutus	Pflaumenrostmilbe	plum rust mite, peach silver mite
Aculus schlechtendali	Apfelrostmilbe	apple rust mite, apple leaf and bud mite
Acyrthosiphon pisum	Grüne Erbsenblattlaus, Grüne Erbsenlaus	pea aphid
Adalia bipunctata	Zweipunkt	two-spot ladybird, 2-spot ladybird
Adalia decimpunctata	Zehnpunkt-Marienkäfer	ten-spot ladybird
Adamsia palliata/ Adamsia carciniopados	Mantelaktinie	cloak anemone (hermit crab anemone a.o.)
Addax nasomaculatus	Mendesantilope	addax
Adelges laricis	Rote Fichtengallenlaus, Frühe Fichten-Kleingallenlaus	red larch gall adelgid, larch adelges, larch woolly aphid
Adelgidae	Tannengallläuse	conifer aphids (pine and spruce aphids)
Adelidae/Incurvariidae	Langhornmotten, Langfühlermotten	longhorn moths, bright moths

*A*delocera

Adelocera murina	Mausgrauer Schnellkäfer	
Adelotus brevis	Rüsselfrosch, Australischer Rüsselfrosch	tusked frog
Adenorhinos barbouri	Ostafrikanische Erdviper	worm-eating viper
Aderidae/Hylophilidae	Mulmkäfer	aderid beetles, hylophilid beetles
Aderus nigrinus	Gemeiner Baummulmkäfer	
Adioryx spinifer	Großer Soldatenfisch	scarlet-fin soldierfish, spiny squirrelfish, sabre squirrelfish (FAO)
Adocia cinerea	Wattschwamm	mudflat sponge*
Adoxophyes orana/ Capua reticulana	Apfelschalenwickler, Fruchtschalenwickler	summer fruit tortrix
Adrianichthyidae	Schaufelkärpflinge	adrianichthyids (killifish-like medakas)
Adscita chloros/ Jordanita chloros	Kupferglanz-Grünwidderchen	copper forester*
Adscita statices/ Procris statices	Grünwidderchen, Grasnelken-Widderchen	forester, common forester
Aechmophorus occidentalis	Renntaucher	western grebe
Aedes aegypti/ Stegomyia aegypti	Gelbfiebermücke	yellow fever mosquito
Aedes maculatus	Waldstechmücke	forest mosquito*
Aega psora	Fischassel (Peterassel)	fish louse
Aegeria apiformis/ Sesia apiformis	Hornissenschwärmer, Bienenglasflügler	poplar hornet clearwing, hornet moth
Aegeriidae/Sesiidae	Glasflügler (Glasschwärmer)	clearwing moths, clear-winged moths
Aegithalidae	Schwanzmeisen	long-tailed tits, bushtits
Aegithalos caudatus	Schwanzmeise	long-tailed tit
Aegithininae	Ioras	ioras
Aegolius acadicus	Sägekauz	saw-whet owl, northern saw-whet owl
Aegolius funereus	Rauhfußkauz	Tengmalm's owl, boreal owl, Richardson's owl (N.Am.)
Aegopinella nitidula	Rötliche Glanzschnecke	waxy glass snail, smooth glass snail, dull glass snail
Aegopinella pura	Kleine Glanzschnecke	clear glass snail, delicate glass snail
Aegopis verticillus	Riesenglanzschnecke, Wirtelschnecke	large glass snail, giant glass snail
Aegothelidae	Höhlenschwalme, Zwergschwalme	owlet-nightjars
Aegypius monachus	Mönchsgeier	black vulture
Aegypius tracheliotus	Ohrengeier	lappet-faced vulture
Aelia acuminata	Spitzling, Getreidespitzwanze	bishop's mitre, bishop's mitre bug
Aeluroscalabotes felinus	Katzengecko	cat gecko
Aeolidia papillosa	Breitwarzige Fadenschnecke	maned nudibranch, plumed sea slug, grey sea slug
Aeolidiacea/Eolidiacea	Fadenschnecken	aeolidacean snails, aeolidaceans

Aeoliscus strigatus	Rasiermesserfisch, Schnepfenmesserfisch	shrimpfish, razorfish (FAO)
Aeolothripidae	Bänder-Fransenflügler*, Rennthripse	banded thrips
Aepyceros melampus	Schwarzfersenantilope, Impala	impala
Aepyornithiformes	Elefantenvögel, Madagaskarstrauße	elephant birds
Aepyprymnus rufescens	Rotes Rattenkänguru(h), Großes Rattenkänguru(h)	rufous "rat"-kangaroo
Aequidens pulcher	Blaupunktbuntbarsch, Blaupunktbarsch	blue acara
Aequidens rivulatus	Goldsaum-Buntbarsch	green terror
Aequipecten lineolaris	Wellenlinien-Pilgermuschel	wavy-lined scallop
Aequipecten opercularis	Kleine Pilgermuschel, Bunte Kammmuschel	queen scallop
Aequorea aequorea	Gerippte Hydromeduse	many-ribbed hydromedusa
Aequorea victoria	Wässrige Hydromeduse*	water jellyfish, victoria hydromedusa
Aeretes melanopterus	Spaltzahn-Flughörnchen	groove-toothed flying squirrel
Aeromys spp.	Große Schwarze Gleithörnchen	large black flying squirrels
Aeshna affinis	Südliche Mosaikjungfer	southern European hawker
Aeshna coerulea	Alpen-Mosaikjungfer	blue aeshna, azure hawker
Aeshna cyanea	Blaugrüne Mosaikjungfer	blue-green darner, southern aeshna, southern hawker
Aeshna grandis	Braune Mosaikjungfer	brown aeshna, brown hawker, great dragonfly
Aeshna isosceles	Keilflecklibelle	Norfolk aeshna, Norfolk hawker (Br.)
Aeshna juncea	Torf-Mosaikjungfer	common aeshna, common hawker
Aeshna mixta	Herbst-Mosaikjungfer	scarce aeshna, migrant hawker
Aeshna subarctica	Hochmoor-Mosaikjungfer	subarctic peat-moor hawker*
Aeshna viridis	Grüne Mosaikjungfer	green hawker
Aeshnidae	Edellibellen, Teufelsnadeln	darners (U.S.), large dragonflies
Aethalops alecto	Zwergflughund	pygmy fruit bat
Aethia cristatella	Schopfalk	crested auklet
Aethia pusilla	Zwergalk	least auklet
Aethia pygmaea	Bartalk	whiskered auklet
Aethomys spp.	Busch-Felsratten	bush rats, rock rats
Aethopyga gouldiae	Gouldnektarvogel	Mrs. Gould's sunbird
Aethopyga siparaja	Scharlachnektarvogel	yellow-backed sunbird
Aethus flavicornis	Braune Erdwanze	brown groundbug*
Aethus nigritus	Schwarze Erdwanze	black groundbug*
Aetobatus narinari	Gefleckter Adlerrochen, Entenschnabelrochen	spotted eagle ray
Afrixalus aureus	Goldener Bananenfrosch	golden dwarf reed frog (IUCN), golden banana frog
Afrixalus dorsalis	Brauner Bananenfrosch	brown banana frog
Afrixalus spp.	Bananenfrösche, Dornenriedfrösche	banana frogs

Afropavo congensis	Kongopfau	Congo peafowl
Agabus bipustulatus/ Gaurodytes bipustulatus	Gemeiner Schnellschwimmer, Zweipunkt-Schnellschwimmer	
Agalenatea redii/ Araneus redii	Strauchradspinne, Körbchenspinne	bush orbweaver*, basketweaver*
Agalychnis callidryas	Rotaugenfrosch, Rotaugen-Laubfrosch	red-eyed treefrog
Agalychnis saltator	Kleiner Rotaugenfrosch	misfit leaf frog
Agalychnis spp.	Tropisch-Afrikanische Laubfrösche	leaf frogs
Agama aculeata	Erd-Agame*	ground agama
Agama agama	Siedleragame	common agama
Agama atra	Bergagame, Felsenagame	southern rock agama, South African rock agama
Agama atricollis/ Stellio atricollis	Blaukehlagame	blue-throated agama
Agama bibroni	Atlasagame	Bibron's agama
Agama caucasica/ Stellio caucasica	Kaukasusagame	northern rock agama, Caucasian agama
Agama kirkii	Kirks Agame	Kirk's rock agama
Agama mutabilis/ Trapelus mutabilis	Wüstenagame	desert agama
Agama paragama	Falsche Siedleragame	false agama
Agama planiceps	Stachelagame	Namib rock agama
Agama stellio/Stellio stellio	Hardun	roughtail rock agama, hardun
Agamidae	Agamen	agamas, chisel-teeth lizards
Agaonidae	Feigenwespen	fig wasps
Agapanthia cardui	Weißstreifiger Distelbock	whitestripe thistle longhorn beetle*
Agapanthia villosoviridescens	Scheckhorn-Distelbock	thistle longhorn beetle*
Agapanthia violacea	Metallfarbener Distelbock	metallic thistle longhorn beetle*
Agapornis fischeri	Erdbeerköpfchen, Fischers Unzertrennlicher	Fischer's lovebird
Agapornis nigrigenis	Rußköpfchen	black-cheeked lovebird
Agapornis personatus	Schwarzköpfchen, Maskenköpfchen	masked lovebird
Agapornis roseicollis	Rosenköpfchen	peach-faced lovebird
Agaricia agaricites	Salatkoralle, Lärchenpilzkoralle	lettuce coral
Agaricia fragilis	Zerbrechliche Tellerkoralle	fragile saucer coral
Agaricia lamarcki	Lamarcks Salatblattkoralle	Lamarck's sheet coral
Agaricia spp.	Salatblattkorallen	lettuce corals
Agaricia tenuifolia	Salatblattkoralle, Folienkoralle	lettuce-leaf coral*
Agaricia undata	Schnörkelkoralle*	scroll coral
Agaristinae (Noctuidae)	Förstermotten*	forester moths
Agaronia acuminata	Spitze Scheinolive	pointed ancilla
Agaronia contortuplicata	Gedrehte Scheinolive	twisted ancilla
Agaronia hiatula	Graue Scheinolive	olive-gray ancilla
Agaronia nebulosa	Gefleckte Scheinolive	blotchy ancilla
Agaronia propatula	Weitgeöffnete Scheinolive*	open-mouthed ancilla
Agelaius phoeniceus	Rotschulterstärling	red-winged blackbird
Agelas clathrodes	Elefantenohrschwamm	orange wall sponge
Agelas conifera	Brauner Röhrenschwamm	Pan Pipe sponge

Agelas oroides	Zäher Orangefarbener Hornschwamm, Zäher Goldschwamm	orange crater sponge, leathery gold sponge
Agelastes meleagrides	Weißbrust-Perlhuhn	white-breasted guineafowl
Agelastes niger	Schwarzperlhuhn	black guineafowl
Agelastica alni	Blauer Erlenblattkäfer	alder leaf-beetle
Agelena labyrinthica	Labyrinthspinne	grass funnel-weaver, maze spider*
Agelenidae	Trichterspinnen, Trichternetzspinnen	funnel web weavers, funnel-weavers
Ageneiosidae	Bartellose Welse	bottlenose catfishes, barbelless catfishes
Agkistrodon acutus/ Deinagkistrodon acutus	Chinesische Nasenotter	sharp-nosed pit viper, sharp-nosed viper
Agkistrodon bilineatus	Mexikanische Mokassinschlange	cantil, mocassin
Agkistrodon blomhoffi/ Gloydius blomhoffi	Mamushi, Japanische Grubenotter	Japanese mamushi
Agkistrodon contortrix	Kupferkopf, Kupferkopfotter	copperhead
Agkistrodon halys/ Gloydius halys	Halysschlange, Halys-Grubenotter, Halysotter	Asiatic pit viper, Pallas' viper, Halys' viper, mamushi
Agkistrodon himalayanus	Himalaya-Grubenotter	Himalaya pit viper
Agkistrodon hypnale/ Hypnale hypnale	Indische Nasenotter	Indian humpnose viper, hump-nosed viper
Agkistrodon piscivorus	Wassermokassinschlange, Wassermokassinotter	cottonmouth, water moccasin
Agkistrodon rhodostoma/ Calloselasma rhodostoma	Malaien-Mokassinschlange, Malaiische Grubenotter	Malayan pit viper
Agkistrodon saxatilis/ Gloydius saxatilis	Amurotter	brown mamushi
Aglaeactis aliciae	Purpurrückenkolibri	purple-backed sunbeam
Aglaiocercus kingi	Himmelssylphe	long-tailed sylph
Aglais urticae	Kleiner Fuchs	small tortoiseshell
Aglantha digitalis	Rosa Fingerhutqualle	pink helmet
Aglaophenia cupressina	Zypressen-Federpolyp, Zypressen-Nesselfarn	cypress plume hydroid, sea cypress
Aglaophenia spp.	Federpolypen, Nesselpolypen, Nesselfarne	plume hydroids, feather hydroids
Aglaophenia struthionides	Straußen-Federpolyp, Straußen-Nesselfarn	ostrich plume hydroid
Aglaophenia tubulifera	Federpolyp, Nesselpolyp, Nesselfarn	plume hydroid, plumed hydroid
Aglia tau	Nagelfleck	tau emperor
Aglossa pinguinalis	Fettzünsler	large stable tabby, tabby moth, grease moth
Agnatha	Kieferlose, Agnathen	jawless fishes, agnathans
Agnathus decoratus	Schmarotzer-Wollkäfer	
Agonidae	Panzergroppen	poachers
Agonostomus monticola	Amerikanische Bergmeeräsche	mountain mullet
Agonum marginatum	Gelbrand-Glanzflachläufer	yellowside ground beetle*

Agonum sexpunctatum	Sechspunkt-Glanzflachläufer, Sechspunktiger Putzkäfer	six-point ground beetle*
Agonus cataphractus	Steinpicker	pogge, hooknose (FAO), armed bullhead
Agouti spp./*Cuniculus* spp.	Pakas	pacas
Agouti taczanowskii	Bergpaka	mountain paca
Agrilus angustulus	Schmaler Prachtkäfer	slender oak borer
Agrilus biguttatus	Zweipunktiger Eichenprachtkäfer, Zweifleckiger Eichenprachtkäfer, Gefleckter Eichenprachtkäfer	twospotted oak borer
Agrilus sinuatus	Birnenprachtkäfer	rusty peartree borer*
Agrilus viridis	Buchenprachtkäfer	beech borer, flat-headed wood borer
Agriolimacidae	Ackerschnecken	field slugs
Agriolimax caruanae/ Limax brunneus/ Deroceras panormitanum	Brauner Uferschnegel	brown slug
Agrionemys horsfieldi/ Testudo horsfieldii	Steppenschildkröte, Vierzehenschildkröte	Horsfield's tortoise, four-toed tortoise, Central Asian tortoise
Agrionemys spp.	Vierzehenschildkröten	four-toed tortoises
Agriopis leucophaearia	Weißgrauer Breitflügelspanner	spring umber, spring usher
Agriopis marginaria	Braunrandiger Frostspanner	dotted border moth
Agriotes lineatus	Saatschnellkäfer	lined click beetle, striped click beetle
Agriotes obscurus	Düsterer Humusschnellkäfer	dusky click beetle
Agrius convolvuli/ Herse convolvuli/ Sphinx convolvuli	Windenschwärmer	convolvulus hawkmoth, morning glory sphinx moth
Agrobates galactotes/ Cercotrichas galactotes	Heckensänger	rufous scrub robin, rufous-tailed scrub robin, rufous warbler
Agrochola circellaris	Ulmen-Herbstfalter	
Agrochola helvola	Weiden-Herbsteule	
Agrochola lychnidis	Gelber Mönch	
Agroeca brunnea	Braune Feenlämpchenspinne	
Agromyzidae	Minierfliegen	leaf miner flies
Agrotis cinerea	Trockenrasen-Erdeule	light feathered rustic (moth)
Agrotis exclamationis	Ausrufungszeichen, Ausrufezeichen, Gemeine Graseule	heart and dart moth
Agrotis ipsilon/ Agrotis ypsilon/ Scotia ypsilon	Ypsiloneule	dark dart moth, dark sword-grass moth; black cutworm
Agrotis segetum	Saateule, Erdeule, Wintersaateule, Graswurzeleule	turnip moth (common cutworm)
Agrotis vestigialis	Kiefernsaateule	archer's dart moth
Agrypnus murinus	Mausgrauer Sandschnellkäfer	
Ahaetulla prasina	Baumschnüffler, Grüne Peitschennatter	long-nosed tree snake, long-nosed whipsnake, Oriental whipsnake
Ahaetulla spp.	Asiatische Peitschennattern	Asian longnose whipsnake
Ahasverus advena	Schimmel-Getreideblattkäfer	foreign grain beetle

Ailuroedus melanotis	Schwarzohr-Laubenvogel	spotted catbird
Ailuropoda melanoleuca	Bambusbär, Riesenpanda, Großer Panda	giant panda
Ailurus fulgens	Katzenbär	lesser panda, red panda
Aiolopus strepens	Braune Strandschrecke	southern longwinged grasshopper
Aiolopus thalassinus/ Epacromia thalassina	Grüne Strandschrecke	longwinged grasshopper
Aiptasia mutabilis	Glasrose, Siebanemone	trumpet anemone, glassrose anemone
Aiptasia pallida	Blasse Glasrose	pale anemone
Aipysurus duboisii	Dubois' Seeschlange	Dubois's sea snake
Aipysurus laevis	Olivbraune Seeschlange	olive-brown sea snake
Aix galericulata	Mandarinente	mandarin duck, mandarin
Aix sponsa	Brautente	wood duck
Ajaja ajaja	Rosalöffler	roseate spoonbill
Akera bullata	Kugelschnecke, Gemeine Kugelschnecke	common bubble snail
Akera soluta	Papier-Kugelschnecke*	papery bubble snail
Akimerus schaefferi/ Acimerus schaefferi	Breitschulterbock	broad-shouldered longhorn beetle
Akodon spp.	Südamerikanische Feldmäuse	South American field mice, grass mice
Akysidae	Flusswelse	stream catfishes
Alactagulus pumilio	Erdhase, Erdhäschen, Kleine Fünfzehenspringmaus, Zwergpferdespringer	lesser five-toed jerboa, little earth hare
Alaemon alaudipes	Wüstenläuferlerche	hoopoe lark, bifasciated lark
Alauda arvensis	Feldlerche	skylark
Alauda gulgula	Kleine Feldlerche	Oriental skylark
Alaudidae	Lerchen	larks
Alaus oculatus	Augenfleck-Schnellkäfer*	eyed elater, eyed click beetle
Albula vulpes	Grätenfisch	bonefish
Albulidae	Grätenfische	bonefishes
Alburnoides bipunctatus	Schneider	riffle minnow, schneider
Alburnus albidus	Weißer Ukelei, Laube	Italian bleak
Alburnus alburnus	Ukelei	bleak
Alca torda	Tordalk	razorbill
Alcedinidae	Eisvögel, Fischer, Lieste	kingfishers
Alcedo atthis	Eisvogel	kingfisher
Alcelaphus buselaphus	Kuhantilope	red hartebeestY
Alcelaphus lichtensteini/ Sigmoceros lichtensteini	Lichtensteins Hartebeest	Lichtenstein's hartebeest
Alces alces	Elch	elk, moose (U.S.)
Alcidae	Alken	auks & murres & puffins
Alcyonacea/Alcyonaria	Weichkorallen, Lederkorallen	soft corals, alcyonaceans
Alcyonidium gelatinosum	Gallert-Moostierchen	gelatinous bryozoan

Alcyoniidae	Weichkorallen, Lederkorallen	soft corals, alcyoniids
Alcyonium digitatum	Tote Manneshand, Meerhand, Nordische Korkkoralle (Lederkoralle)	dead-man's fingers, sea-fingers
Alcyonium glomeratum	Rote Korkkoralle	red sea-fingers, red fingers
Alcyonium palmatum	Seemannshand, Diebeshand, Mittelmeer-Korkkoralle	Mediterranean sea-fingers
Alectis alexandrinus	Alexandria-Fadenmakrele	Alexandria pompano, African threadfish
Alectis ciliaris/ Alectis crinitus	Afrikanische Fadenmakrele	pennant trevally, threadfin mirrorfish, African pompano (FAO)
Alectis indicus	Indische Fadenmakrele	Indian mirrorfish, Indian threadfish (FAO)
Alectis spp.	Fadenmakrelen	pompanos, threadfishes
Alectoris barbara	Felsenhuhn	barbary partridge
Alectoris chukar	Chukarsteinhuhn	chukar partridge
Alectoris chukar	Chukarhuhn	chukar
Alectoris graeca	Steinhuhn	rock partridge
Alectoris rufa	Rothuhn	red-legged partridge
Alectura lathami	Buschhuhn	brush turkey
Aleochara bipustulata	Zweipunkt-Tagkurzflügler	egg-eating rove beetle
Aleochara curtula	Schwarzer Tagkurzflügler	black diurnal rove beetle*
Aleochara spp.	Tagkurzflügler	small-headed rove beetles
Alepes djedaba	Goldene Stachelmakrele	shrimp scad (FAO), golden scad
Alepisauridae	Lanzenfische	lancetfishes
Alepisaurus ferox	Lanzenfisch, Langnasen-Lanzenfisch	lancetfish, longnose lancetfish (FAO)
Alepocephalidae	Glattkopffische, Schwarzköpfe	slickheads
Alepocephalus bairdii	Glattkopf, Bairds Glattkopffisch	Baird's smooth-head
Alepocephalus rostratus	Rissos Glattkopffisch	Risso's smooth-head
Alestes chaperi	Grüner Salmler	Chaper's characin
Alestidae/Alestiinae	Großaugensalmler (Echte Afrikanische Salmler)	African tetras
Aleyrodes fragariae/ Aleyrodes lonicerae	Erdbeer-Mottenschildlaus, Geißblatt-Mottenschildlaus	strawberry whitefly, honeysuckle whitefly
Aleyrodes proletella	Kohl-Mottenschildlaus	cabbage whitefly
Aleyrodidae	Mottenschildläuse (Mottenläuse, Weiße Fliegen)	whiteflies
Algyroides fitzingeri	Zwerg-Kieleidechse	Fitzinger's algyroides
Algyroides marchi	Spanische Kieleidechse	Spanish keeled lizard
Algyroides moreoticus	Peloponnesische Kieleidechse	Greek keeled lizard
Algyroides nigropunctatus	Pracht-Kieleidechse	blue-throated keeled lizard
Alisterus scapularis	Königssittich	Australian king parrot
Alkmaria romijni		tentacled lagoonworm
Allactaga spp.	Pferdespringer	four- & five-toed jerboas
Allantus cinctus	Gürtel-Rosenblattwespe, Gebänderte Rosenblattwespe	banded rose sawfly

Alle alle	Krabbentaucher	little auk, dovekie
Alleculidae	Pflanzenkäfer	comb-clawed beetles, comb-clawed bark beetles
Allenopithecus nigroviridis	Sumpfmeerkatze, Schwarzgrüne Meerkatze	Allen's monkey, Allen's swamp monkey (blackish-green guenon)
Alligator mississippiensis	Mississippi-Alligator	American alligator
Alligator sinensis	China-Alligator	Chinese alligator
Allocebus trichotis	Büschelohriger Katzenmaki, Kleiner Katzenmaki	hairy-eared dwarf lemur
Allodermanyssus sanguineus/ Liponyssoides sanguineus	Hausmausmilbe	house-mouse mite
Allopeas clavulinum	Stachlige Turmschnecke*	spike awlsnail
Allopeas gracile	Anmutige Turmschnecke*	graceful awlsnail
Allophyes oxyacanthae	Weißdorneule	green-brindled crescent
Allorossia glaucopsis	Blauäugige Rossie	blue-eyed bob-tailed squid
Alloteuthis media	Marmorierter Zwergkalmar, Mittelländischer Zwergkalmar	marbled little squid, midsize squid (FAO)
Alloteuthis subulata	Gepfriemter Zwergkalmar	little squid, European common squid (FAO)
Allothrombium fulginosum	Bräunlichrote Samtmilbe	red velvet mite
Allothunnus fallai	Schlankthun	slender tuna
Alopex lagopus	Eisfuchs, Polarfuchs (Hbz. Weißfuchs/Blaufuchs)	Arctic fox
Alopias pelagicus	Indopazifischer Fuchshai	pelagic thresher
Alopias superciliosus	Großaugen-Fuchshai, Großäugiger Fuchshai	bigeye thresher
Alopias vulpinus	Fuchshai, Drescher	thresher shark, thintail thresher (FAO), thintail thresher shark, fox shark
Alopiidae	Fuchshaie, Drescherhaie	thresher sharks
Alopochon aegyptiacus	Nilgans	Egyptian goose
Alosa aestivalis	Blaurückenhering, Kanadische Alse	blueback herring, blueback shad (FAO)
Alosa alosa	Maifisch, Alse, Gewöhnliche Alse	allis shad
Alosa fallax/Alosa finta	Finte	twaite shad (FAO), finta shad
Alosa mediocris	Westatlantische Alse	hickory shad
Alosa pontica	Pontische Alse, Donauhering	Pontic shad (FAO), Black Sea shad
Alosa pseudoharengus	Nordamerikanischer Flusshering	alewife (FAO), river herring
Alosa sapidissima	Amerikanische Alse, Amerikanischer Maifisch	American shad
Alouatta belzebul	Rothandbrüllaffe	black-and-red howler
Alouatta caraya	Schwarzer Brüllaffe	black howler
Alouatta fusca	Brauner Brüllaffe	brown howler
Alouatta palliata	Mantelbrüllaffe	mantled howler
Alouatta seniculus	Roter Brüllaffe	red howler
Alouatta spp.	Brüllaffen	howler monkeys
Alouatta villosa	Guatemala-Brüllaffe	Guatemalan howler

Alpheidae	Knallkrebschen, Pistolenkrebse	snapping shrimps, snapping prawns, cracker shrimps
Alpheus armatus	Braunes Knallkrebschen, Brauner Pistolenkrebs	brown snapping prawn, brown pistol shrimp
Alpheus dentipes	Mittelmeer-Knallgarnele, Mittelmeer-Pistolenkrebs, Knallkrebschen	Mediterranean snapping prawn
Alpheus macrocheles	Großscheren-Knallkrebschen, Großscheren-Pistolenkrebs	snapping prawn, snapping shrimp, big-claw snapping prawn*
Alpheus spp.	Knallkrebschen, Pistolenkrebse	snapping prawns, snapping shrimps, pistol shrimps
Alphitobius diaperinus	Glänzender Getreideschimmelkäfer	lesser mealworm
Alphitobius laevigatus	Mattschwarzer Getreideschimmelkäfer	black fungus beetle (lesser mealworm beetle)
Alphitophagus bifasciatus	Zweibindiger Pilzschwarzkäfer	twobanded fungus beetle
Alsophila aescularia	Kreuzflügel, Rosskastanien-Frostspanner	March moth
Alsophis ater	Schwarze Rennnatter	black racer
Alsophis vudii	Bahama-Rennnatter	West Indian racer, Bahamas racer
Alsophylax pipiens	Kaspischer Geradfingergecko	Caspian even-fingered gecko, Caspian straight-fingered gecko
Altica chalybea	Amerikanischer Traubenerdfloh	grape flea beetle
Altica ericeti	Heidekrauterdfloh	heather flea beetle
Altica lythri	Großer Blauer Erdfloh	large blue flea beetle
Altica oleracea	Falscher Kohlerdfloh	false turnip flea beetle
Altica quercetorum	Eichenerdfloh	oak flea beetle
Alticola spp.	Gebirgswühlmäuse	high mountain voles
Alucita hexadactyla	Geißblattfedermotte	many-plumed moth, twenty plumed moth
Alucitidae	Federmotten, Geistchen	plumed moths, many-plume moths, alucitids
Alutera monoceros	Einhorn-Lederjacke	unicorn leatherjacket
Aluteridae	Lederjacken, Einstachler	leatherjackets
Aluterus schoepfi	Orangeroter Feilenfisch	orange filefish
Aluterus scriptus	Gezeichneter Feilenfisch	long-tailed filefish, scribbled filefish, scrawled filefish (FAO)
Alydidae	Krummfühlerwanzen	broad-headed bugs
Alydus calcaratus	Rotrückiger Irrwisch	redbacked bug, redbacked broadheaded bug
Alytes cisternasii	Iberische Geburtshelferkröte, Spanische Geburtshelferkröte	Iberian midwife toad
Alytes muletensis	Balearenkröte	Mallorcan midwife toad, ferreret
Alytes obstetricans	Geburtshelferkröte (Glockenfrosch, Fresser, Steinkröte)	midwife toad
Amaea magnifica	Prächtige Wendeltreppe	magnificient wentletrap
Amalda marginata	Geränderte Scheinolive*	margin ancilla
Amandava amandava	Tigerfink	red avadavat
Amandava formosa	Olivastrild	green munia

Amara aenea	Erz-Kanalkäfer, Erzfarbener Kanalkäfer	brazen sun beetle
Amara apricaria	Dunkler Kanalkäfer	dusky sun beetle*
Amara aulica	Prächtiger Kanalkäfer	magnificent sun beetle
Amara familiaris	Gelbbeiniger Kanalkäfer, Geselliger Kanalkäfer	social sun beetle
Amara plebeja	Gemeiner Kanalkäfer, Gewöhnlicher Kanalkäfer	common sun beetle
Amara spp.	Kanalkäfer, Kamelläufer	sun beetles
Amarsipidae	Amarsipiden	amarsipida
Amaurobiidae	Finsterspinnen	white-eyed spiders
Amaurobius fenestralis	Weißaugen-Finsterspinne	white-eyed spider, window lace-weaver
Amaurobius ferox	Schwarze Finsterspinne	black lace-weaver
Amauropsis islandica/ Bulbus islandicus	Isländische Mondschnecke	Iceland moonsnail
Amazona aestiva	Rotbugamazone, Blaustirnamazone	blue-fronted amazon
Amazona albifrons	Weißstirnamazone	white-fronted amazon
Amazona amazonica	Venezuela-Amazone	orange-winged parrot, orange-winged amazon
Amazona guildingii	Königsamazone	St. Vincent parrot, St. Vincent amazon
Amazona imperialis	Kaiseramazone	imperial parrot, imperial amazon
Amazona ochrocephala	Gelbscheitelamazone	yellow-crowned parrot, yellow-headed parrot, yellow-crowned amazon
Amazona versicolor	Blaumaskenamazone	St. Lucia parrot, St. Lucia amazon
Amazona viridigenalis	Grünwangenamazone	red-crowned parrot (Mexican red-head amazon)
Amazona vittata	Puerto Rico-Amazone	Puerto Rican parrot, Puerto Rican amazon
Ambassidae	Glasbarsche	glassies
Ambloplites rupestris	Steinbarsch	rock bass (red-eye)
Amblycipitidae	Schlankwelse	torrent catfishes
Amblycirrhitus pinos	Karibischer Büschelbarsch, Karibischer Korallenwächter	redspotted hawkfish
Amblyeleotris aurora	Rotbinden-Schläfergrundel	pinkbar goby
Amblyeleotris wheeleri	Wheeler's Partnergrundel	gorgeous prawn-goby (FAO)
Amblyomma americanum	Amerikanische Buntzecke, Amerikanische Sternzecke*	lone star tick
Amblyomma cajennense	Cayenne-Zecke	Cayenne tick
Amblyomma hebraeum	Buntzecke	bont tick
Amblyomma maculatum	Amerikanische Golfküsten-Buntzecke	Gulf Coast tick
Amblyomma variegatum	Tropische Buntzecke	tropical bont tick
Amblyopsidae	Blindfische	cavefishes
Amblyornis inornatus	Hüttengärtner	vogelkop gardener bowerbird
Amblyospiza albifrons	Weißstirnweber	grosbeak weaver
Amblypygi	Geißelspinnen	tailless whipscorpions
Ambylyrhynchus cristatus	Meerechse	marine iguana, Galapagos marine iguana

Ambystoma annulatum	Ringelquerzahnmolch	ringed salamander
Ambystoma californiense	Kalifornischer Querzahnmolch	California tiger salamander
Ambystoma cingulatum	Genetzter Querzahnmolch	reticulated salamander, flatwood salamander
Ambystoma gracile	Nordwestamerikanischer Querzahnmolch*	northwestern salamander
Ambystoma laterale	Baufleck-Querzahnmolch	blue-spotted salamander
Ambystoma macrodactylum	Langzehen-Querzahnmolch	long-toed salamander
Ambystoma maculatum	Fleckenquerzahnmolch	spotted salamander
Ambystoma mexicanum	Axolotl	axolotl
Ambystoma opacum	Marmorquerzahnmolch	marbled salamander
Ambystoma platineum	Silberquerzahnmolch	silvery salamander
Ambystoma rosaceum	Chihuahua-Querzahnmolch	Chihuahuan mole salamanders
Ambystoma spp.	Amerikanische Querzahnmolche	American mole salamanders
Ambystoma talpoideum	Maulwurf-Querzahnmolch	mole salamander
Ambystoma texanum	Schmalkopf-Querzahnmolch	small-mouthed salamander, smallmouth salamander
Ambystoma tigrinum	Tigerquerzahnmolch	tiger salamander
Ambystoma velascoi	Hochland-Querzahnmolch	plateau tiger salamander
Ambystomatidae	Querzahnmolche	ambystomids, mole salamanders
Ameiurus nebulosus/ Ictalurus nebulosus	Brauner Zwergwels, Langschwänziger Katzenwels	brown bullhead (FAO), "speckled catfish", horned pout, American catfish
Ameiva ameiva	Ameive	jungle runner
Ameiva spp.	Ameiven	jungle runners
Americardia media	Atlantische Erdbeer-Herzmuschel*	Atlantic strawberry-cockle
Amia calva	Schlammfisch, Kahlhecht, Amerikanischer Schlammfisch	bowfin
Amicula vestita (Polyplacophora)	Arktische Löcher-Käferschnecke	concealed Arctic chiton
Amiidae	Kahlhechte, Schlammfische	bowfins
Amiiformes	Kahlhechte, Schlammfische	modern bowfins
Ammobiota festiva/ Ammobiota hebe	Englischer Bär, Wolfsmilchbär	
Ammodillus imbellis	Somalia-Rennmaus	walo, Somali gerbil
Ammodorcas clarkei	Stelzengazelle, Lamagazelle, Dibatag	dibatag
Ammodytes americanus	Amerikanischer Sandaal	American sandlance
Ammodytes cicerelus/ Ammodytes cicerellus/ Gymnammodytes cicerelus	Nacktsandaal, Mittelmeersandaal, Mittelmeer-Nacktsandaal	sandlance, smooth sandlance, Mediterranean sandeel (FAO)
Ammodytes dubius	Westatlantischer Sandaal	northern sandlance
Ammodytes hexapterus	Pazifischer Sandaal	Pacific sandlance
Ammodytes lanceolatus/ Hyperoplus lanceolatus	Großer Sandaal, Großer Sandspierling	greater sandeel, lance, great sandeel (FAO)
Ammodytes marinus	Nordischer Sandaal	lesser sandeel (FAO)
Ammodytes tobianus/ Ammodytes lancea	Sandaal, Kleiner Sandaal, Tobiasfisch	sand eel, lesser sandeel, small sandeel (FAO)
Ammodytidae	Sandaale	sand lances, sandlances, sandeels

Ammomanes cincturus	Sandlerche	bar-tailed desert lark
Ammomanes deserti	Steinlerche	desert lark
Ammomanes phoenicurus	Rotschwanzlerche	rufous-tailed desert lark, rufous-tailed lark
Ammonoidea	Ammoniten	ammonites
Ammoperdis griseogularis	Persisches Wüstenhuhn	see-see partridge
Ammospermophilus spp.	Antilopen-Erdhörnchen	antelope ground squirrels, antelope squirrels
Ammotragus lervia	Mähnenspringer	barbary sheep, aoudad
Amoebozoa	Amöben, Wechseltierchen, Wurzeltierchen, Rhizopoden	amebas, amoebas
Amolops spp.	Kaskadenfrösche	sucker frogs
Amoria damonii	Damons Walzenschnecke	Damon's volute
Amoria ellioti	Elliots Walzenschnecke	Elliot's volute
Amoria zebra	Zebra-Walzenschnecke	zebra volute
Ampedus sanguineus	Blutroter Schnellkäfer	bloodred click beetle*
Amphibia	Lurche, Amphibien	amphibians
Amphibolurus barbatus/ Pogona barbatus	Bartagame	bearded dragon
Amphibolurus diemensis	Australische Bergagame	mountain dragon
Amphibolurus maculatus/ Ctenophorus maculatus	Gefleckte Bodenagame	spotted dragon, spotted bearded dragon
Amphibolurus maculosus/ Ctenophorus maculosus	Lake-Eyre-Agame	Lake Eyre dragon
Amphibolurus muricatus	Australischer Blutsauger (Bodenagame)	Australian "bloodsucker", jacky lizard, jacky
Amphibolurus spp.	Australische Bodenagamen	bearded dragons, Australian dragon lizards
Amphicoela	Urfrösche	amphicoelans
Amphictene koreni/ Lagis koreni/ Pectinaria koreni	Köcherwurm	quiver worm*
Amphiliidae	Kaulquappenwelse	loach catfishes
Amphimallon solstitialis/ Rhizotragus solstitialis	Junikäfer (Gemeiner Brachkäfer, Sonnwendkäfer)	summer chafer
Amphinome spp.	Feuerwürmer u.a.	firewoms a.o.
Amphinomidae	Feuerwürmer	firewoms, amphinomids
Amphiodia occidentalis	Grabender Schlangenstern*	burrowing brittlestar
Amphipholus squamata	Schuppiger Schlangenstern	scaly brittlestar
Amphipoda	Flohkrebse, Flachkrebse	beach hoppers, sand hoppers, scuds and relatives
Amphiporus lactifloreus (Nemertini)	Heller Schnurwurm, Helle Nemertine	
Amphiprion akallopisos	Indischer Weißrücken-Anemonenfisch, Weißrücken-Clownfisch	skunk clownfish (FAO), skunk anemonefish
Amphiprion akindynos	Barriere-Riff-Anemonenfisch	Barrier Reef anemonefish
Amphiprion bicinctus	Zweibinden-Anemonenfisch, Zweiband-Anemonenfisch, Orangeringelfisch, Spitzbinden-Anemonenfisch, Rotmeer-Clownfisch	two-banded anemonefish

Amphiprion chrysopterus	Orangenflossen-Anemonenfisch	orange-fin anemonefish, orange-finned anemonefish
Amphiprion clarki	Dreistreifen-Anemonenfisch, Clarks Anemonenfisch	Clark's anemonefish, goldbelly
Amphiprion ephippium	Glühkohlenfisch, Glühkohlen-Anemonenfisch	saddle anemonefish (FAO), red saddleback anemonefish
Amphiprion frenatus	Halsband-Anemonenfisch, Weißbinden-Glühkohlen-Anemonenfisch, Roter Clownfisch, Unechter Glühkohlenfisch	red clownfish, tomato anemonefish, tomato clownfish (FAO)
Amphiprion laticlavius/ Amphiprion polymnus	Sattel-Anemonenfisch, Sattelfleck-Anemonenfisch	saddleback anemonefish
Amphiprion melanopus	Schwarzer Anemonenfisch, Schwarzroter Anemonenfisch	black anemonefish, dusky anemonefish, fire clownfish (FAO)
Amphiprion nigripes	Schwarzflossen-Anemonenfisch, Malediven-Anemonenfisch	black-finned anemonefish, Maldive anemonefish (FAO)
Amphiprion ocellaris	Orangenringelfisch, Orangeringelfisch, Falscher Clown-Anemonenfisch	false clown anemonefish
Amphiprion percula	Orange-Anemonenfisch, Clown-Anemonenfisch, Schwarzgeränderter Orangenringelfisch	orange clownfish (FAO), clown anemonefish
Amphiprion perideraion	Rosa Anemonenfisch, Halsband-Anemonenfisch, Falscher Weißrücken-Clownfisch	pink anemonefish (FAO), false skunk-striped anemonefish
Amphiprion polymnus/ Amphiprion laticlavius	Sattel-Anemonenfisch, Sattelfleck-Anemonenfisch	saddleback anemonefish
Amphiprion sandaracinos	Philippinischer Rückenstreifen-Anemonenfisch, Philippinen-Weißrücken-Clownfisch	white-backed anemonefish
Amphiprion sebae	Dreibinden-Anemonenfisch, Goldflößchen, Gelbschwanzringelfisch, Sebaes Anemonenfisch	Sebae's anemonefish
Amphiprion spp.	Anemonenfische, Clownfische	anemonefishes, clownfishes
Amphiprion tricinctus	Dreiband-Anemonenfisch	three-band anemonefish
Amphiprioninae	Anemonenfische	clownfishes
Amphipyra pyramidea	Pyramideneule	copper underwing
Amphisbaena alba	Rote Wurmschleiche	red worm lizard
Amphisbaena darwini	Darwins Wurmschleiche	Darwin's ringed worm lizard
Amphisbaena fuliginosa	Gefleckte Wurmschleiche	speckled worm lizard
Amphisbaenia	Doppelschleichen, Wurmschleichen	worm lizards, amphisb(a)enids, amphisbenians
Amphisbaenidae	Eigentliche Wurmschleichen	tropical worm lizards
Amphitrite ornata	Zierwurm*	ornate worm
Amphiuma means	Zweizehen-Aalmolch	two-toed amphiuma
Amphiuma pholeter	Einzehen-Aalmolch	one-toed amphiuma
Amphiuma tridactylum	Dreizehen-Aalmolch	three-toed amphiuma
Amphiumidae	Aalmolche	amphiumas, "Congo eels"
Amphiura filiformis	Fadenstern, Fadenförmiger Schlangenstern, Fadenarm-Schlangenstern	filiform burrowing brittlestar

Amphiura spp.	Fadensterne, Fadenförmige Schlangensterne	filiform burrowing brittlestars
Amphizoidae	Amphizoiden	trout-stream beetles
Amplexidiscus fenestrafer	Großes Elefantenohr	giant elephant's ear*
Ampulla priamus	Gefleckte Walzenschnecke, Nordostatlantische Fleckenwalze*	spotted flask
Ampullariidae	Kugelschnecken, Blasenschnecken	bubble snails (bubble shells)
Ampullarius gigas	Große Kugelschnecke	giant bubble snail (giant bubble shell)
Ampullarius scalaris	Apfelschnecke	scaled bubble snail*
Amusium japonicum	Japanische „Sonne und Mond"-Muschel	Japanese sun-and-moon scallop
Amusium papyraceum	Papier-Kammmuschel	paper scallop
Amusium pleuronectes	Kompassmuschel	Asian moon scallop
Amyda cartilaginea/ Trionyx cartilagineus	Asiatische Weichschildkröte, Knorpel-Weichschildkröte	Asiatic softshell turtle, black-rayed softshell turtle
Amygdalum papyrium	Atlantische Papiermuschel	Atlantic papermussel
Amygdalum phaseolinum	Bohnen-Papiermuschel	kidney-bean horse mussel, kidney-bean papermussel
Amytornithinae	Grasschlüpfer	grasswrens
Anabantidae	Kletterfische & Buschfische	climbing gouramies
Anabas testudineus	Kletterfisch	climbing perch (FAO), climbing gourami, walking fish
Anablepidae	Vieraugenfische	four-eyed fishes
Anableps anableps	Vierauge	foureyes, largescale foureyes (FAO)
Anacanthobatidae	Fadenschwanzrochen	smooth skates, leg skates
Anachis obesa	Dickbauchige Täubchenschnecke*	fat dovesnail
Anacridium aegypticum	Ägyptische Heuschrecke, Ägyptische Wanderheuschrecke	Egyptian grasshopper
Anadara granosa	Genarbte Archenmuschel, Westpazifische Archenmuschel	granular ark (granular ark shell)
Anadara notabilis	Öhrchenmuschel*	eared ark (eared ark shell)
Anadara ovalis	Blutrote Archenmuschel	blood ark (blood ark shell)
Anadara subcrenata	Japanische Archenmuschel	mogal clam
Anadara uropygimelana	Verbrannte Archenmuschel	burnt-end ark (burnt-end ark shell)
Anaesthetis testacea	Kragenbock	
Anagasta kuehniella	Mehlmotte	Mediterranean flour moth
Anaglyptus mysticus	Dunkler Zierbock	grey-coated longhorn beetle
Anaitides maculata	Gefleckter Blattwurm	spotted leafworm*
Analgidae	Federmilben	feather mites
Anapagurus laevis	Glattscheriger Einsiedler, Gelber Einsiedler	yellow hermit crab
Anaplectoides prasina	Grüne Heidelbeereule	green arches
Anarhichadidae	Wolfsfische, Seewölfe	wolffishes
Anarhichas latifrons/ Anarhichas denticulatus	Blauer Katfisch, Wasserkatze	jelly cat, jelly wolffish, northern wolffish (FAO)
Anarhichas lupus	Seewolf, Gestreifter Seewolf, Kattfisch, Katfisch	Atlantic wolffish, wolffish (FAO), cat fish, catfish

Anarhichas minor	Gefleckter Seewolf, Gefleckter Katfisch	spotted wolffish (FAO), spotted sea-cat, spotted catfish, spotted cat
Anarrhichthys ocellatus	Pazifischer Seewolf	wolf-eel, Pacific wolf-eel
Anarsia lineatella	Pfirsichmotte	peach twig borer
Anarta cordigera	Moorbunteule	small dark yellow underwing
Anarta melanopa		broad-bordered white underwing
Anarta myrtilli	Heidekrauteule, Heidekrauteulchen	beautiful yellow underwing
Anas acuta	Spießente	northern pintail
Anas americana	Nordamerikanische Pfeifente	American wigeon, baldpate
Anas aucklandica	Aucklandente	New Zealand black duck
Anas clypeata	Löffelente	northern shoveler
Anas crecca	Krickente	teal, green-winged teal
Anas cyanoptera	Zimtente	cinnamon teal
Anas discors	Blauflügelente	blue-winged teal
Anas falcata	Sichelente	falcated duck, falcated teal
Anas formosa	Gluckente	Baikal teal
Anas penelope	Pfeifente	wigeon
Anas platyrhynchos	Stockente	mallard
Anas querquedula	Knäkente	garganey
Anas rubripes	Dunkelente	black duck, North American black duck
Anas strepera	Schnatterente	gadwall
Anaspis frontalis	Gemeiner Scheinstachelkäfer	
Anastomus lamelligerus	Schwarzklaffschnabel	African open-bill stork
Anathana ellioti	Elliots Tupaia	Indian tree shrew
Anatidae	Enten & Gänse & Schwäne	ducks & geese & swans
Anatina anatina	Glatte Ententrogmuschel	smooth duckclam
Anatis ocellata	Augenmarienkäfer	eyed ladybird, pine ladybird beetle
Anatoma crispata	Kräusel-Riss-Schnecke*	crispate scissurelle
Anax imperator	Große Königslibelle	emperor dragonfly
Anax junius	Grüne Königslibelle	green darner, green emperor dragonfly
Anax parthenope	Kleine Königslibelle	lesser emperor dragonfly
Anchoa hepsetus	Breitstreifensardelle	striped anchovy, broad-striped anchovy (FAO)
Anchoa mitchilli	Nordwestatlantische Sardelle	bay anchovy
Ancilla castanea	Kastanien-Scheinolive	chestnut ancilla
Ancilla cingulata	Goldstreifen-Scheinolive*	honey-banded ancilla
Ancilla glabrata	Goldene Scheinolive*	golden ancilla
Ancilla lienardi	Lienardos Scheinolive*	Lienardo's ancilla
Ancistroteuthis lichtensteini	Engelskalmar	angel squid
Ancistrus dolichopterus/ Xenocara dolichoptera	Blauer Antennenwels	blue-chin ancistrus, blue-chin xenocara, bushymouth catfish (FAO)
Ancistrus spp.	Antennenwelse	bristle-noses
Ancylidae	Flussnapfschnecken	river limpets, freshwater limpets
Ancylostoma caninum	Hunde-Hakenwurm	canine hookworm

Ancylostoma duodenale	Altwelt-Hakenwurm, Grubenwurm	Old World hookworm
Ancylus fluviatilis	Flussnapfschnecke, Gemeine Flussnapfschnecke	river limpet, common river limpet
Ancylus lacustris/ Acroloxus lacustris	Teichnapfschnecke, Seenapfschnecke	lake limpet, shield snail
Andigena laminirostris	Leistenschnabeltukan	plate-billed mountain toucan
Andigena nigrirostris	Schwarzschnabeltukan	black-billed mountain toucan
Andinomys edax	Andenmaus	Andean mouse
Andrena fulva/ Andrena armata	Rotpelzige Sandbiene	tawny burrowing bee
Andrena gravida	Gebänderte Sandbiene*	banded mining bee
Andrena spp.	Sandbienen	mining bees, burrowing bees
Andrenidae	Sandbienen	mining bees, burrowing bees
Andresia parthenope	Grabende Anemone	burrowing anemone
Andrias davidianus	Chinesischer Riesensalamander	Chinese giant salamander
Andrias japonicus	Japanischer Riesensalamander	Japanese giant salamander
Andrias spp.	Asiatische Riesensalamander	Asiatic giant salamanders
Andricus albopunctatus/ Andricus paradoxus	Weißfleckige Eichengallwespe	oak leaf gall wasp
Andricus curvator	Eichenkragengallwespe (>Kragengalle)	oak bud collared-gall wasp
Andricus fecundator/ Andricus foecundatrix	Eichenrosengallwespe	artichoke gall wasp, larch cone gall cynipid, hop gall wasp (>artichoke gall)
Andricus inflator	Grünkugelgallwespe (>Grüne Kugelgalle)	oak bud globular-gall wasp
Andricus kollari	Schwammkugelgallwespe (>Schwammkugelgalle)	marble gall wasp (>marble gall/oak nut)
Andricus lignicola	'Kolanuss'-Gallwespe*	'cola-nut' gall wasp
Andricus nudus	Spindelgallwespe (>Kleine Spindelgalle)	oak catkin gall wasp
Andricus ostreus	Eichenblatt-Austerngallwespe	oak leaf oyster-gall wasp
Andricus quercuscalicis	Eichenknopperngallwespe	acorn cup gall wasp (>knopper gall)
Andricus quercusradicis	Wurzelknötchengallwespe* (>Knötchengalle)	oak root truffle-gall wasp
Andricus solitarius	Solitär-Filzgallwespe* (Braune Filzgalle)	solitary oak leaf gall wasp
Andricus spp.	Eichengallwespen	oak gall wasps, oak gall cynipids
Andricus testaceipes	Eichenwurzelknoten-Gallwespe	oak red barnacle-gall wasp
Androctonus australis	Dickschwanzskorpion	fattailed scorpion, fat-tailed scorpion, African fat-tailed scorpion
Androlaelaps casalis	Vogelnestmilbe, Hühnernestmilbe	poultry litter mite, cosmopolitan nest mite
Androniscus dentiger	Kellerassel	cellar woodlouse, rosy woodlouse
Anechura bipunctata	Zweipunkt-Ohrwurm	two-spotted earwig
Aneides aeneus	Erzsalamander	green salamander
Aneides ferreus	Marmorierter Salamander	clouded salamander
Aneides flavipunctatus	Schwarzer Baumsalamander	black salamander
Aneides lugubris	Alligatorsalamander	arboreal salamander
Aneides spp.	Baumsalamander, Klettersalamander	arboreal salamanders, climbing salamanders

Anelytropsis papillosus	Mexikanische Schlangenechse	Mexican blind lizard
Anemonia sargassensis	Sargassum-Anemone	sargassum anemone
Anemonia sulcata/ Anemonia viridis	Wachsrose	snakelocks anemone, opelet anemone
Aneuridae	Aneuriden	barkbugs
Angaria delphinus	Delphinschnecke	dolphin snail (dolphin shell)
Angariidae	Delphinschnecken	dolphin snails (dolphin shells)
Angerona prunaria	Schlehenspanner, Pflaumenspanner	orange moth
Anguidae	Schleichen	lateral fold lizards, anguids
Anguilla anguilla	Europäischer Flussaal	eel, European eel (FAO), river eel
Anguilla australis	Australischer Aal	Australian eel, shortfin eel (FAO)
Anguilla japonica	Japanischer Aal	Japanese eel
Anguilla rostrata	Amerikanischer Aal	American eel
Anguillidae	Aale	freshwater eels
Anguilliformes	Aalfische, Aalartige	eels
Anguillula aceti/ Turbatrix aceti	Essigälchen	vinegar eel, vinegar nematode
Anguina tritici	Weizenälchen, Weizengallenälchen (Radekrankheit)	wheat-gall nematode, wheat nematode, wheat eelworm
Anguis fragilis	Blindschleiche	European slow worm (blindworm), slow worm
Angulus fabula/ Fabulina fabula/ Tellina fabula	Gerippte Tellmuschel	bean-like tellin, semi-striated tellin
Angulus incarnatus/ Tellina incarnatus	Rote Tellmuschel	red tellin
Angulus planatus	Mittelmeer-Tellmuschel	Mediterranean tellin
Angulus tenuis/Tellina tenuis	Platte Tellmuschel, Zarte Tellmuschel, Plattmuschel	thin tellin, plain tellin, petal tellin
Anhimidae	Wehrvögel	screamers
Anhinga anhinga	Amerikanischer Schlangenhalsvogel	American darter, anhinga
Anhingidae	Schlangenhalsvögel	anhingas, darters
Anhydrophryne rattrayi	Rattray-Frosch	Rattray's frog, hogsback frog (IUCN)
Aniculus maximus	Großer Schuppeneinsiedler	large hermit crab
Aniliidae	Rollschlangen	pipe snakes, pipesnakes
Anilius scytale	Korallen-Rollschlange	coral pipesnake, false coral snake
Anisodactylus binotatus	Schwarzer Schmuckkäfer	
Anisolabis annulipes	Südlicher Ohrwurm	ring-legged earwig, spotted earwig
Anisolabis maritima	Küsten-Ohrwurm*	seaside earwig
Anisomys imitator	Neu Guinea-Riesenratte*	powerful-toothed rat, New Guinea giant rat (squirrel-toothed rat), uneven-toothed rat
Anisopodidae	Pfriemenmücken	window midges, wood gnats

Anisoptera	Großlibellen	dragonflies; hawkers (Europe)
Anisotremus virginicus	Schweinsfisch, Schweinsgrunzer	porkfish
Anisus leucostoma	Weißmündige Tellerschnecke	white-lipped ram's horn snail, white-lipped ramshorn snail, button ram's horn
Anisus vortex	Scharfe Tellerschnecke, Spiralige Tellerschnecke	whirlpool ram's horn snail, whirlpool ramshorn snail
Anisus vorticulus	Zierliche Tellerschnecke	delicate ram's horn snail*, delicate ramshorn snail*
Annamemys annamensis	Annam-Schildkröte	Annam leaf turtle, Vietnamese leaf turtle
Annelida	Ringelwürmer, Gliederwürmer, Borstenfüßer, Anneliden	segmented worms, annelids
Anniella geronimensis	Geronimo-Ringelschleiche	Baja California legless lizard
Anniella pulchra	Kalifornische Ringelschleiche	California legless lizard
Anniella pulchra nigra	Schwarze Ringelschleiche	black legless lizard
Anobiidae	Klopfkäfer, Nagekäfer, Pochkäfer	furniture beetles, drug-store beetles, death-watch beetles
Anobium pertinax/ *Dendrobium pertinax*	Trotzkopf, Totenuhr	furniture beetle, furniture borer, common furniture beetle
Anobium punctatum	Gemeiner Nagekäfer, Gewöhnlicher Nagekäfer, Holzwurm	furniture beetle, woodworm (*larva*)
Anocentor nitens/ *Dermacentor nitens*	Tropische Pferdezecke	tropical horse tick
Anodonta anatina	Flache Teichmuschel, „Entenmuschel"	duck mussel
Anodonta anodonta	Gemeine Teichmuschel	common pond mussel*
Anodonta cygnea	Große Teichmuschel, Schwanenmuschel	swan mussel
Anodonta spp.	Teichmuscheln	pond mussels*, floaters
Anodontia alba	Butterblumen-Mondmuschel*	buttercup lucine
Anodontostoma chacunda	Chacunda	chacunda gizzard shad
Anodorhynchus hyacinthinus	Hyazinthara	hyacinth macaw
Anodorhynchus leari	Lear-Ara	Lear's macaw, indigo macaw
Anolis auratus	Grasanolis	grass anole
Anolis carolinensis	Rotkehl-Anolis, Rotkehlanolis	green anole
Anolis chlorocyanus	Laubanolis	Hispaniola green anole
Anolis cristatellus	Kamm-Anolis	crested anole
Anolis cybotes	Dickkopfanolis	large-headed anole
Anolis distichus	Zaunanolis	bark anole
Anolis equestris	Ritteranolis	knight anole
Anolis lineatopus	Strichfußanolis	stripefoot anole
Anolis lucius	Höhlenanolis	cave anole
Anolis ricordi	Haiti-Riesenanolis	Haitian green anole
Anolis roosevelti	Riesenanolis	giant anole, Culebra Island giant anole

Anolis sagrei	Braunanolis, Kuba-Anolis	brown anole, Cuban anole
Anolis vermiculatus	Wurmanolis	Vinales anole
Anomala dubia	Julikäfer	margined vine chafer
Anomalepidae	Amerikanische Schlankblindschlangen	dawn blind snakes
Anomalochromis thomasi	Afrikanischer Schmetterlingsbuntbarsch, Thomas-Prachtbuntbarsch	dwarf jewel fish
Anomalopidae	Laternenfische, Blitzlichtfische	flashlight fishes, lanterneye fishes
Anomaluridae	Dornschwanzhörnchen	scaly-tailed squirrels, scaly tails, anomalures
Anomalurus spp.	Dornschwanzhörnchen	scaly-tailed flying squirrels, anomalures
Anomia ephippium	Europäische Sattelauster, Sattelmuschel, Zwiebelmuschel, Zwiebelschale	European saddle oyster (European jingle shell)
Anomia simplex	Gewöhnliche Sattelmuschel	common saddle oyster (common jingle shell)
Anomia spp.	Sattelmuscheln	saddle oysters (jingle shells)
Anomiidae	Sattelmuscheln, Zwiebelmuscheln	saddle oysters (jingle shells)
Anomochilus weberi	Rohrschlange*, Südostasiatische Walzenschlange	pipe snake
Anonconotus alpinus	Kleine Alpenschrecke	small Alpine bushcricket
Anonymomys mindorensis	Mindoro-Ratte	Mindoro rat
Anopheles atroparvus	Gemeine Fiebermücke	common anopheles mosquito, common malaria anopheles mosquito
Anopheles plumbens	Astloch-Gabelmücke*	tree-hole mosquito
Anopheles spp.	Fiebermücken, Malariamücken, Gabelmücken	malaria mosquitoes
Anoplius viaticus/ Anoplius fuscus/ Pompilus viaticus	Frühlings-Wegwespe	black-banded spider wasp
Anoplocephala spp.	Pferdebandwürmer	horse tapeworms, equine tapeworms
Anoplogaster cornuta	Blattschupper	fangtooth
Anoplogasteridae	Blattschupper	fangtooths
Anoplopoma fimbria	Kohlenfisch	sablefish
Anoplopomatidae	Schwarzfische, Säbelfische	sablefishes
Anoplotrupes stercorosus	Waldmistkäfer	
Anoplura (Siphunculata)	Läuse (Tierläuse)	sucking lice
Anoptichthys jordani/ Astyanax fasciatus mexicanus	Blindhöhlenfisch	blind cavefish
Anostomidae	Kopfsteher, Engmaulsalmler	anostomids
Anostomus anostomus	Prachtkopfsteher	striped headstander (FAO), striped anostomus
Anostraca	Schalenlose, Kiemenfußkrebse, Kiemenfüße	fairy shrimps, anostracans
Anotheca spinosa	Kronenlaubfrosch	spiny-headed treefrog

Anotomys leander	Südamerikanische Fischratte	fish-eating rat, aquatic rat
Anotopteridae	Speerfische	daggertooths
Anotopterus pharao	Speerfisch	daggertooth
Anoura spp.	Geoffroys Langnasen-Fledermaus	Geoffroy's long-nosed bats
Anourosorex squamipes	Wühlspitzmaus, Stummelschwanzspitzmaus	mole shrew, Szechwan burrowing shrew
Anous stolidus	Noddiseeschwalbe	noddy tern, noddy
Anous tenuirostris	Schlankschnabelnoddi, Kleine Noddiseeschwalbe	lesser noddy
Anoxypristis cuspidata	Spitzkopf-Sägerochen	pointed sawfish
Anser albifrons	Blässgans	white-fronted goose
Anser anser	Graugans	graylag goose, grey lag goose
Anser brachyrhynchus	Kurzschnabelgans	pink-footed goose
Anser caerulescens	Schneegans	snow goose
Anser canagicus/ Philacte canagicus	Kaisergans	emperor goose
Anser erythropus	Zwerggans	lesser white-fronted goose
Anser fabalis	Saatgans	bean goose
Anser indicus	Streifengans	bar-headed goose
Anseranas semipalmata	Spaltfußgans	magpie goose
Anseranatidae	Spaltfußgänse	magpie geese
Anseriformes	Entenvögel, Gänsevögel	screamers & waterfowl (ducks & geese & swans)
Anseropoda placenta	Gänsefußstern	goose-foot starfish, goosefoot starfish
Ansonia longidigita	Zirpkröte	long-fingered stream toad
Ansonia spp.	Bachkröte	stream toads
Antalis dentalis (Scaphopoda)	Meerzahn	European tusk
Antalis longitrorsum (Scaphopoda)	Langer Elefantenzahn	elongate tusk
Antalis pretiosum/ Dentalium pretiosum (Scaphopoda)	Indischer Elefantenzahn, Kostbarer Elefantenzahn	Indian money tusk, wampum tuskshell
Antalis tarentinum/ Dentalium vulgare (Scaphopoda)	Gemeiner Elefantenzahn, Gemeine Zahnschnecke	common elephant's tusk, common tusk (common tuskshell)
Antaresia childreni/ Liasis childreni/ Morelia childreni	Childrens Python	Children's python, Children's rock python
Antaresia maculosa	Fleckenpython	spotted python
Antaresia perthensis/ Liasis perthensis/ Morelia perthensis	Zwergpython	Perth pigmy python, pygmy python
Antaxius hispanicus	Vielfleck-Bergschrecke*	mottled bushcricket
Antaxius pedestris	Atlantische Bergschrecke	Pyrenean bushcricket
Antaxius spinibrachius	Stachelbeinige Bergschrecke	spiny-legged bushcricket
Antechinomys laniger	Springbeutelmaus	kultarr
Antechinus spp.	Breitfußbeutelmäuse	broad-footed marsupial "mice"
Antedon bifida		feather star
Antedon mediterranea	Mittelmeer-Haarstern	orange-red feather star

Antennariidae	Fühlerfische, Krötenfische	anglers, frogfishes
Antennarius commersoni/ Antennarius moluccensis	Gelber Krötenfisch	Commerson's frogfish
Antennarius hispidus	Krötenfisch, Gestreifter Anglerfisch	anglerfish, fishing frog, shaggy angler (FAO)
Antennarius multiocellatus	Augenfleck-Anglerfisch, Karibischer Augenflecken-Angler	longlure frogfish
Antennarius nummifer	Gefleckter Anglerfisch	coin-bearing frogfish
Antennarius phymatodes/ Antennarius maculatus	Warzen-Anglerfisch	warty frogfish
Antennarius radiosus	Einfleck-Anglerfisch*	big-eyed frogfish, singlespot frogfish (FAO)
Antennarius striatus/ Antennarius scaber	Fühlerfisch, Streifen-Anglerfisch	striated frogfish (FAO), splitlure frogfish
Anthaxia nitidula	Zierlicher Prachtkäfer	
Anthaxia quadripunctata	Vierpunkt-Kiefernprachtkäfer	
Anthelia glauca	Winkende Hand*	waving hand polyp a.o., pulse coral a.o.
Antheraea polyphemus	Polyphemusmotte*	polyphemus moth
Anthias anthias	Roter Fahnenbarsch, Rötling	swallow-tail sea perch, marine goldfish, swallowtail seaperch (FAO)
Anthias squamipinnis/ Pseudanthias squamipinnis	Juwelen-Fahnenbarsch	orange sea perch, sea goldie (FAO), lyretail coralfish
Anthicidae	Blütenmulmkäfer	antlike flower beetles
Anthicus floralis	Gemeiner Blütenmulmkäfer	narrownecked grain beetle, narrow-necked harvest beetle
Anthidium manicatum	Wollbiene, Große Wollbiene	wool carder bee
Anthidium spp.	Wollbienen & Harzbienen	wool carder bees
Anthidium strigatum/ Anthidiellum strigatum	Harzbiene, Kleine Harzbiene	small anthid bee
Anthiidae	Fahnenbarsche, Rötlinge	fairy basslets
Anthocharis cardamines	Aurorafalter	orange-tip
Anthocomus coccineus	Herbst-Zipfelkäfer	
Anthocoridae	Blumenwanzen	minute pirate bugs
Anthomyiidae	Blumenfliegen	anthomyids, flower flies
Anthonomus grandis	Baumwollknospenstecher	cotton boll weevil
Anthonomus piri	Birnenknospenstecher	apple bud weevil
Anthonomus pomorum	Apfelblütenstecher	apple blossom weevil
Anthonomus rubi	Erdbeerblütenstecher, Himbeerblütenstecher, Beerenstecher	berry blossom weevil, strawberry blossom weevil
Anthophora acervorum	Gemeine Pelzbiene	common Central European flower bee
Anthophora retusa	Töpfer-Pelzbiene	potter flower bee
Anthophora spp.	Pelzbienen	flower bees
Anthopleura artemisia	Mondschein-Anemone*	moonglow anemone
Anthopleura elegantissima	Klon-Anemone*	clonal anemone, aggregating anemone
Anthopleura xanthogrammica	Grüne Riesenanemone*	giant green anemone, great green anemone

Anthops ornatus	Blumennasen-Fledermaus	flower-faced bat
Anthozoa	Blumentiere, Blumenpolypen, Anthozoen	flower animals, anthozoans
Anthracoceros coronatus	Malabarhornvogel	Malabar pied hornbill
Anthrax anthrax	Düsterer Trauerschweber*	dusky beefly*
Anthrenus flavipes/ Anthrenus vorax	Knochenmehlkäfer (Keratinkäfer)	furniture carpet beetle
Anthrenus museorum	Museumskäfer	museum beetle
Anthrenus scrophulariae	Teppichkäfer, Kabinettkäfer	carpet beetle
Anthrenus verbasci	Textilkäfer	varied carpet beetle
Anthrenus vorax	Polsterwarenkäfer	
Anthreptes platurus	Erznektarvogel	pygmy sunbird
Anthribidae	Breitfüßler, Breitrüssler, Maulkäfer	fungus weevils
Anthropoides virgo	Jungfernkranich	demoiselle crane
Anthus campestris	Brachpieper	tawny pipit
Anthus cervinus	Rotkehlpieper	red-throated pipit
Anthus godlewskii	Steppenpieper	Blyth's pipit
Anthus gustavi	Petschorapieper	pechora pipit
Anthus hodgsoni	Waldpieper	olive-backed pipit
Anthus novaeseelandiae	Australspornpieper	Richard's pipit
Anthus petrosus	Strandpieper	rock pipit
Anthus pratensis	Wiesenpieper	meadow pipit
Anthus richardi	Spornpieper	Richard's pipit
Anthus rubescens	Pazifischer Wasserpieper	buff-bellied pipit
Anthus similis	Langschnabelpieper	long-billed pipit
Anthus spinoletta	Bergpieper	rock pipet, water pipit
Anthus trivialis	Baumpieper	tree pipit
Antidorcas marsupialis	Springbock	springbuck, springbok
Antigonia capros	Beilbäuchiger Eberfisch*	deep-bodied boarfish
Antigonia rubescens	Indischer Eberfisch	Indian boarfish
Antilocapra americana	Gabelbock, Gabelhornantilope	pronghorn
Antilope cervicapra	Hirschziegenantilope	blackbuck
Antimora rostrata	Blauhecht	blue antimora
Antipatharia	Dornkorallen, Dörnchenkorallen, Schwarze Edelkorallen	thorny corals, black corals, antipatharians
Antipathes atlantica	Atlantische Schwarze Koralle	Atlantic black coral
Antipathes subpinnata	Schwarze Koralle	black coral
Antonogadus macrophthalmus	Großaugen-Seequappe*	big-eyed rockling
Antrodiaetidae	Falttürspinnen	folding trapdoor spiders
Antrozous spp.	Blasse Fledermäuse	pallid bats, desert bats
Anura/Salientia	Froschlurche (Frösche & Kröten)	frogs & toads, anurans
Anuraphis farfarae	Taschengallen-Birkenlaus	pear-coltsfoot aphid
Anurida maritima	Strandspringer	seashore springtail, marine springtail
Anurida maritima	Strand-Springschwanz*, Strandkollembole*	seashore springtail

Anyphaena accentuata	Brummspinne*	buzzing spider
Anyphaenidae	Zartspinnen	ghost spiders
Aombus lapidarius/ Bombus lapidarius	Steinhummel	red-tailed bumble bee
Aonidia lauri	Lorbeerschildlaus	laurel scale*
Aonidiella aurantii	Rote Zitrusschildlaus, Rote Citrusschildlaus	California red scale
Aonyx capensis	Kapotter, Weißwangenotter (Hbz. Afrikanischer Otter)	African clawless otter
Aonyx cinerea	Zwergotter	Oriental small-clawed otter
Aonyx congica	Kongo-Otter, Kongo-Fingerotter	Congo clawless otter, swamp otter
Aonyx spp.	Kapottern, Fingerottern	clawless otters
Aotus trivirgatus	Nachtaffe	douroucouli, night monkey
Apaloderminae	Afrikanische Trogons	African trogons
Apalone ferox/Trionyx ferox	Wilde Dreiklaue, Florida-Weichschildkröte	Florida softshell turtle
Apalone mutica/ Trionyx muticus	Glattrandige Weichschildkröte, Glattrand-Weichschildkröte	smooth softshell turtle
Apalone spinifera/ Trionyx spiniferus	Dornrand-Weichschildkröte	spiny softshell turtle
Apamea lithoxylea	Graswurzeleule, Trockenrasen-Graswurzeleule	common light arches
Apamea monoglypha	Getreidewurzeleule, Wurzelfresser	dark arches
Apamea secalis/ Mesapamea secalis	Getreidewurzeleule, Getreideeule, Getreidesaateule, Roggeneule	common rustic (moth)
Apamea unanimis	Ufer-Glanzgraseule	small clouded brindle
Apamea zeta	Treitschkes Alpen-Grasbüscheleule	the exile (moth)
Apanteles glomeratus	Kohlweißlingsraupenwespe	common apanteles, common parasitic wasp
Aparasphenodon spp.	Panzerkopf-Laubfrösche	casque-headed frogs
Apatura ilia	Kleiner Schillerfalter, Espen-Schillerfalter	lesser purple emperor
Apatura iris	Großer Schillerfalter	purple emperor
Apatura spp.	Schillerfalter	emperors
Aphaenogaster subterraneus	Unterirdische Knotenameise	
Aphaniotis spp.	Blaumaul-Agamen	earless agamas
Aphanius dispar	Perlmutterkärpfling	pearly killifish*
Aphanius fasciatus	Mittelmeerkärpfling, Salinenkärpfling (Zebrakärpfling)	Mediterranean toothcarp
Aphanius iberus	Spanienkärpfling	Spanish killifish, Spanish toothcarp (FAO), Iberian toothcarp
Aphanopus carbo	Kurzflossen-Haarschwanz, Espada	black scabbardfish (FAO), espada
Aphantopus hyperantus	Brauner Waldvogel	ringlet
Aphelenchoides fragariae	Erdbeerälchen	strawberry leaf nematode, spring crimp nematode

Aphelenchoides spp.	Blattnematoden	leaf nematodes
Aphelocoma coerulescens	Buschhäher	scrub jay
Aphia minuta	Glasküling, Glasgrundel	transparent goby
Aphididae	Röhrenläuse, Röhrenblattläuse	aphids, "plant lice"
Aphidiidae	Blattlausschlupfwespen	aphid parasites
Aphidina	Blattläuse	aphids
Aphidoidea	Blattläuse	aphids & greenflies etc.
Aphis fabae	Schwarze Bohnenlaus, Schwarze Rübenlaus, Schwarze Bohnenblattlaus, Schwarze Rübenblattlaus	black bean aphid, "blackfly"
Aphis gossypii	Baumwollblattlaus	cotton aphid, melon aphid
Aphis grossulariae	Kleine Stachelbeertriebblattlaus	gooseberry aphid
Aphis idaei	Kleine Himbeerblattlaus, Kleine Himbeerlaus	small raspberry aphid
Aphis pomi	Grüne Apfellaus	apple aphid, green apple aphid
Aphis ruborum	Kleine Brombeerschildlaus	permanent blackberry aphid
Aphis sambuci	Holunderlaus, Schwarze Holunderblattlaus	elder aphid
Aphis schneideri	Kleine Johannisbeertriebblattlaus, Kleine Johannisbeerlaus	permanent currant aphid
Aphomia gularis/ Paralispa gularis	Samenzünsler	stored nut moth
Aphonopelma chalcodes	Mexikanische Blond-Vogelspinne	Mexican blond tarantula
Aphonopelma saltator	Blaue Argentinische Vogelspinne, Weißnacken-Vogelspinne	Argentine blue tarantula*
Aphonopelma seemanni	Gestreifte Guatemala-Vogelspinne	Costa Rican zebra tarantula
Aphredoderus sayanus	Piratenbarsch	pirate perch
Aphrodes bicinctus	Erdbeerzikade	strawberry leafhopper
Aphrodita aculeata	Gemeine Seemaus, Filzwurm, Wollige Seemaus, Seeraupe	sea mouse, European sea mouse
Aphrodita spp.	Seemäuse, Filzwürmer	sea mice
Aphroditidae	Seemäuse, Seeraupen	sea mice (scaleworms a.o.)
Aphrophora alni	Erlenschaumzikade	alder froghopper
Aphrophora salicina	Weidenschaumzikade	willow froghopper
Aphthona euphorbiae	Dunkelgrauer Flachserdfloh	large flax flea beetle
Aphyocharax rubripinnis/ Aphyocharax anisitsi	Rotflossensalmler	bloodfin
Aphyocypris pooni/ Hemigrammocypris lini	Venusfisch	garnet minnow
Aphyonidae	Glasgrundeln	aphyonids
Aphyosemion australe	Bunter Prachtkärpfling, Fahnenhechtling, Kap Lopez	lyretail panchax (FAO), Cape Lopez lyretail
Aphyosemion filamentosum	Fadenprachtkärpfling	plumed lyretail
Aphyosemion gulare	Tigerkärpfling, Gelber Prachtkärpfling	yellow gularis
Aphyosemion sjostedti	Blauer Prachtkärpfling	blue gularis (FAO), red aphyosemion, golden pheasant

Apidae

Apidae	Bienen	hive bees (bumble bees, honey bees & orchid bees)
Apioceridae	Blumenfliegen*	flower-loving flies
Apion apricans	Rotklee-Spitzmausrüssler	red clover seed weevil
Apion frumentarium/ Apion haematodes	Sauerampfer-Gallrüssler	sheep's sorrel gall weevil
Apion pomonae	Metallblauer Spitzmausrüssler	vetch seed weevil, tare seed weevil
Apion radiolus	Stockrosen-Spitzmausrüssler	hollyhock weevil
Apion spp.	Spitzmausrüssler (Spitzmäuschen)	flower weevils
Apion vorax	Gemeiner Samenstecher	bean flower weevil
Apioninae	Spitzmausrüssler (Spitzmäuschen)	weevils, flower weevils
Apis mellifera syriaca	Syrische Honigbiene	Syrian honey bee
Apis mellifera carnica	Krainer Honigbiene	Carniola honey bee
Apis mellifera adansoni	Afrikanische Honigbiene	African honey bee, Africanized honey bee
Apis mellifera cecropia	Griechische Honigbiene	Greek honey bee
Apis mellifera indica	Indische Honigbiene	Indian honey bee
Apis mellifera fasciata	Ägyptische Honigbiene	Egyptian honey bee
Apis mellifera cypria	Zyprische Honigbiene	Cyprian honey bee, Cyprian bee
Apis mellifera ligustica	Italienische Honigbiene	Italian honey bee
Apis mellifera mellifera	Honigbiene, Europäische Honigbiene, Gemeine Honigbiene	honey bee, hive bee
Apistogramma spp.	Zwergbuntbarsche	dwarf cichlids
Apistogramma trifasciata	Dreistreifen-Zwergbuntbarsch	blue apistogramma
Aplacophora	Aplacophoren, Wurmmollusken, Wurmmolluscen	aplacophorans
Aplasta ononaria	Magerrasen-Hauhechelspanner	rest harrow
Apletodon dentatus/ Lepadogaster microcephalus	Kleinköpfiger Ansauger	small-headed clingfish (FAO), small-headed sucker
Aplexa hypnorum	Moosblasenschnecke	moss bladder snail
Aplidium conicum	Kegel-Seescheide	conical sea-squirt
Aplidium elegans	Elegante Seescheide	elegant sea-squirt
Aplidium proliferum	Spross-Seescheide, Spross-Synascidie	prolific sea-squirt*
Aplidium stellatum	Stern-Seescheide, Stern-Synascidie	sea pork (*of the tunic*)
Aploactinidae	Aploactiniden	velvetfishes
Aplocheilidae	Hechtlinge	rivulines
Aplocheilus dayi	Grüner Streifenhechtling, Ceylonhechtling	Ceylon killifish (FAO), Day's killifish
Aplocheilus lineatus	Streifenhechtling	lined panchax
Aplocheilus panchax	Gemeiner Hechtling	blue panchax
Aplodactylidae	Marmorfische	marblefishes
Aplodontia rufa	Bergbiber, Biberhörnchen, Stummelschwanzhörnchen	sewellel, mountain beaver
Aplonis metallica	Weberstar	shining starling
Aplousobranchia	Mehrteilige Seescheiden	aplousobranchs

Aplustrum amplustre/ *Hydatina amplustre*	Prächtige Papierblase	royal paperbubble
Aplysia dactylomela	Großfleckiger Seehase	large-spotted sea hare
Aplysia depilans	Gefleckter Seehase	spotted sea hare
Aplysia fasciata	Band-Seehase, Großer Brauner Seehase	banded sea hare
Aplysia juliana	Gestelzter Seehase*	walking sea hare
Aplysia rosea/ *Aplysia punctata*	Getupfter Seehase, Kleingepunkteter Seehase	small rosy sea hare*, common sea hare
Aplysia spp.	Seehasen	sea hares
Aplysiacea/Anaspidea	Seehasen, Breitfußschnecken	sea hares
Aplysina fistularis	Neptunsschwamm	yellow tube sponge, candle sponge, sulphur sponge (dead man's fingers)
Aplysina lacunosa	Zylinderschwamm	giant tube sponge (convoluted barrel sponge)
Apocheima hispidaria	Gelbfühleriger Spanner	small brindled beauty
Apocheima pilosaria/ *Phigalia pilosaria*	Grauer Wollrückenspanner, Schlehenfrostspinner	pale brindled beauty
Apoda limacodes/ *Cochlidion limacodes*/ *Apoda avellana*	Große Schildmotte, Kleiner Asselspinner	festoon
Apodemus agrarius	Brandmaus, Feldmaus	Old World field mouse, striped field mouse
Apodemus alpicola	Alpenwaldmaus	Alpine wood mouse
Apodemus flavicollis	Gelbhalsmaus	yellow-necked mouse
Apodemus microps	Zwergwaldmaus, Kleine Waldmaus	pygmy field mouse, herb field mouse
Apodemus mystacinus	Felsenmaus	rock mouse, broad-toothed field mouse
Apodemus spp.	Waldmäuse & Feldmäuse	Old World wood & field mice
Apodemus sylvaticus	Waldmaus, Feld-Waldmaus	wood mouse, long-tailed field mouse
Apoderus coryli	Haselblattroller, Dickkopfrüssler	hazel weevil
Apodidae	Segler	swifts
Apodiformes/ Micropodiformes	Schwirrflügler, Seglervögel, Seglerartige (Segler & Kolibris)	swifts & hummingbirds
Apogon binotatus	Zweistreifenbarsch	barred cardinalfish
Apogon guamensis/ *Apogon nubilus*	Guam-Kardinalfisch	pearl cardinalfish
Apogon imberbis	Meerbarbenkönig	cardinalfish, cardinal fish (FAO)
Apogon lachneri	Leuchtstern-Kardinalfisch	whitestar cardinalfish
Apogon maculatus	Flammenfisch	flamefish
Apogon multitaeniatus	Vielstreifen-Kardinalbarsch	smallscale cardinal
Apogon nematopterus	Pyjama-Kardinalbarsch, Wimpel-Kardinalbarsch	pyjama cardinal fish
Apogon spp.	Kardinalbarsche	cardinalfish, cardinals
Apogonidae	Kardinalbarsche, Kardinalfische	cardinal fishes
Apolemia uvaria	Fadenqualle*	string jelly
Apolemichthys trimaculatus	Dreifleckiger Engelfisch	threespot angelfish

Apomys spp.	Philippinische Ratten u.a.	Philippine rats a.o.
Aporia crataegi	Baumweißling	black-veined white
Aporosaura anchietae	Sandechse	Namib sanddiver
Aporrhiadae	Pelikansfüße	pelican's foot snails
Aporrhias occidentalis	Amerikanischer Pelikansfuß	American pelican's foot
Aporrhias pespelicani	Pelikanfuß	common pelican's foot
Appendicularia/Larvacea	Appendicularien, Geschwänzte Schwimm-Manteltiere	appendicularians
Aprion virescens	Königsschnapper	king snapper, green jobfish (FAO), streaker
Apristurus aphyodes	Heller Tiefenkatzenhai	pale catshark
Apristurus brunneus	Brauner Katzenhai	brown catshark
Apristurus laurussonii (*Apristurus maderensis*)	Island-Katzenhai, Island-Tiefen-Katzenhai (Madeira-Katzenhai)	Iceland catshark (FAO) (Madeira catshark)
Apristurus profundorum	Tiefsee-Katzenhai	deepwater catshark
Aptenodytes forsteri	Kaiserpinguin	emperor penguin
Aptenodytes patagonicus	Königspinguin	king penguin
Apterichthus anguiformis		slender finless eel
Apterichthus caecus	Europäischer Flossenloser Aal*	European finless eel
Apteronotidae	Schwanzflossen-Messeraale, Peitschenmesseraale	ghost knifefishes
Apteronotus albifrons	Weißstirn-Seekuhaal, Weißstirn-Messeraal	black ghost
Apteronotus spp.	Seekuhaale	apteronotid eels
Apterygida albipennis	Gebüsch-Ohrwurm	short-winged earwig, apterous earwig
Apterygidae	Kiwis	kiwis
Apterygiformes	Kiwis	kiwis
Apteryx australis	Streifenkiwi	brown kiwi
Apteryx haastii	Großer Fleckenkiwi	great spotted kiwi
Apteryx owenii	Zwergkiwi, Kleiner Fleckenkiwi	little spotted kiwi
Aptinothrips rufus	Roter Blasenfuß	red glass thrips
Aptinus bombarda	Alpen-Bombardierkäfer (Großer Bombardierkäfer)	Alpine bombardier beetle
Apus affinis	Haussegler	little swift
Apus apus	Mauersegler	swift
Apus caffer	Kaffernsegler	white-rumped swift
Apus melba	Alpensegler	Alpine swift
Apus pallidus	Fahlsegler	pallid swift
Aquila audax	Keilschwanzadler	wedge-tailed eagle
Aquila chrysaetos	Steinadler	golden eagle
Aquila clanga	Schelladler	spotted eagle
Aquila heliaca	Kaiseradler	imperial eagle
Aquila nipalensis	Steppenadler	steppe eagle
Aquila pomarina	Schreiadler	lesser spotted eagle
Aquila rapax	Savannenadler	tawny eagle
Aquila verreauxii	Kaffernadler	Verreaux's eagle
Ara ararauna	Ararauna	brown and yellow macaw, blue-and-gold macaw

Ara chloroptera	Grünflügelara, Dunkelroter Ara	green-winged macaw
Ara macao	Arakanga	scarlet macaw
Ara nobilis	Zwergara	red-shouldered macaw
Arabella iricolor	Opalwurm	opal worm, iridescent worm
Arachnanthus oligopodus	Gebänderte Zylinderrose	banded tube anemone*
Arachnida	Spinnentiere, Arachniden	arachnids
Arachnocephalus vestitus	Buschgrille	Mediterranean wingless shrub cricket
Arachnothera longirostra	Weißkehl-Spinnenjäger	little spiderhunter
Arachnothera robusta	Langschnabel-Spinnenjäger	long-billed spiderhunter
Aradidae	Rindenwanzen	flatbugs, flat bugs
Aradus cinnamomeus	Kiefernrindenwanze	pine flatbug
Araecerus fasciculatus	Kaffeebohnenkäfer	coffee bean weevil
Aramidae	Rallenkranich, Riesenralle	limpkin
Aramus guarauna	Rallenkranich, Riesenralle	limpkin
Araneae	Spinnen, Webspinnen	spiders
Araneidae	Radnetzspinnen, Kreuzspinnen	orbweavers, orb-weaving spiders (broad-bodied orbweavers)
Araneinae	Eigentliche Kreuzspinnen	typical orbweavers
Araneus adiantus/ Neoscona adianta	Heide-Radspinne, Heideradspinne	heathland orbweaver
Araneus alsiae	Erdbeerspinne	strawberry spider
Araneus angulatus	Gehörnte Kreuzspinne	horned orbweaver*
Araneus cavaticus	Amerikanische Scheunenspinne	barn orbweaver, barn spider
Araneus ceropegius/ Aculepeira ceropegia	Eichblatt-Radspinne, Eichenblatt-Radnetzspinne	oakleaf orbweaver*
Araneus diadematus	Gemeine Kreuzspinne, Gartenkreuzspinne	cross orbweaver, European garden spider, cross spider
Araneus marmoreus	Marmorierte Kreuzspinne	marbled orbweaver, marbled spider
Araneus quadratus	Vierfleck-Kreuzspinne	fourspotted orbweaver*
Araneus redii/ Agalenatea redii	Strauchradspinne, Körbchenspinne	bush orbweaver*, basketweaver*
Araneus spp.	Kreuzspinnen	orbweavers (angulate & roundshouldered orbweavers)
Araneus trifolium	Shamrock-Spinne*	shamrock orbweaver, shamrock spider
Araneus umbraticus/ Nuctenea umbratica	Spaltenkreuzspinne	crevice spider*
Araniella cucurbitina/ Araneus cucurbitinus	Kürbisspinne	gourd spider, pumpkin spider
Araniella displicata	Amerikanische Sechspunktspinne*	sixspotted orbweaver, six-spotted orb weaver
Arapaima gigas	Arapaima, Piracurú	pirarucu
Araschnia levana	Landkärtchen	map butterfly
Aratinga auricapilla	Goldhaubensittich	golden-capped parakeet, golden-capped conure
Aratinga canicularis/ Eupsittula canicularis	Elfenbeinsittich	orange-fronted parakeet, orange-fronted conure

Aratinga erythrogenys	Guayaquilsittich	red-masked conure, cherry head conure
Aratinga guarouba/ Guaruba guarouba	Goldsittich	golden parakeet, golden conure
Aratinga holochlora/ Psittacara holochlora	Grünsittich	green parakeet
Aratinga jandaya	Jendajasittich	Jandaya conure
Aratinga pertinax	Braunwangensittich	brown-throated conure
Aratinga solstitialis	Sonnensittich	sun conure
Aratus pisonii	Mangrovekrabbe, Mangroven-Baumkrabbe	mangrove tree crab, mangrove crab
Arbacia lixula	Schwarzer Seeigel	black urchin
Arbacia punctulata	Atlantischer Violetter Seeigel	Atlantic purple urchin
Arborophila torqueola	Hügelhuhn	common hill partridge
Arca imbricata	Schuppige Archenmuschel	mossy ark (mossy ark shell)
Arca noae	Arche Noah, Archenmuschel	Noah's ark (Noah's ark shell)
Arca nodulosa	Knotige Archenmuschel	nodular ark (nodular ark shell)
Arca tetragona	Vierkantige Archenmuschel	tetragonal ark (tetragonal ark shell)
Arca zebra	Truthahnflügel	turkey wing
Archachatina marginata	Tropische Riesenachatschnecke	tropical giant snail
Archaeogastropoda/ Diotocardia	Schildkiemer, Altschnecken	limpets and allies, archeogastropods
Archaeognatha	Felsenspringer	jumping bristletails
Archaeopsylla erinacei	Igelfloh	hedgehog flea
Archaeornithes	Urvögel	ancestral birds, "lizard birds", archaeornithes
Archanara algae	Teich-Röhrichteule	rush wainscot
Archanara geminipuncta	Zweipunktschilfeule	
Archanara neurica	Kleine Röhrichteule	white-mantled wainscot
Archanara sparganii	Rohrkolbeneule	Webb's wainscot
Archidoris pseudoargus/ Archidoris tuberculata	Meerzitrone, Warzige Sternschnecke	sea lemon
Archiearis notha	Mittleres Jungfernkind	light orange underwing
Archiearis parthenias	Großes Jungfernkind, Jungfernsohn	orange underwing moth
Archilochus alexandri	Schwarzkinnkolibri	black-chinned hummingbird
Archilochus colubris	Rubinkehlkolibri	ruby-throated hummingbird
Archips argyrospilus	Amerikanischer Obstbaumwickler	fruittree leafroller
Archips crataegana	Weißdornwickler, Obstbaumwickler	fruit tree tortrix (moth), brown oak tortrix, oak red-barred twist
Archips oporana	Kiefernnadelwickler, Nadelholzwickler	pineneedle tortrix moth
Archips podana	Großer Obstbaumwickler, Bräunlicher Obstbaumwickler	large fruittree tortrix moth, apple tortrix (moth)
Archips rosana	Rosenwickler	rose tortrix (moth), rose twist
Architectonia laevigata	Glattschalige Sonnenuhrschnecke	smooth sundial (snail)

Architectonica nobilis	Amerikanische Sonnenuhrschnecke	American sundial (snail), common American sundial, common sundial
Architectonica perspectiva	Perspektivschnecke, Europäische Sonnenuhrschnecke	clear sundial, European sundial snail
Architectonicidae/Solariidae	Sonnenschnecken, Sonnenuhrschnecken	sundials, sundial snails (sundial shells, sun shells)
Architeuthidae	Riesenkalmare	giant squids
Architeuthis dux	Atlantischer Riesenkalmar	Atlantic giant squid
Architeuthis japonica	Japanischer Riesenkalmar	Japanese giant squid
Architeuthis spp.	Riesenkalmare	giant squids
Archon apollinus	Griechischer Apoll	false apollo
Archosargus probatocephalus	Schafskopf-Brasse, Sträflings-Brasse	sheepshead
Archosargus rhomboidalis	Ohrfleck-Brasse	western Atlantic seabream
Arcidae	Archenmuscheln	arks (ark shells)
Arcinella arcinella	Stachel-Juwelendose*	spiny jewelbox
Arcopagia crassa/ Tellina crassa	Stumpfe Tellmuschel	blunt tellin
Arctia caja	Brauner Bär	garden tiger (moth)
Arctia villica	Schwarzer Bär	cream-spot tiger (moth)
Arctica islandica	Islandmuschel	Icelandic cyprine, Iceland cyprina, ocean quahog
Arctictis binturong	Binturong	binturong
Arctides regalis	Rotband-Bärenkrebs	red-banded slipper lobster
Arctiidae	Bärenspinner (Bären)	tigermoths & footman moths & ermine moths (caterpillars: woolly bears)
Arctocebus aureus	Gold-Bärenmaki	golden angwantibo
Arctocebus calabarensis	Calabar-Bärenmaki	Calabar angwantibo, golden potto
Arctocephalus philippii	Juan Fernández Seebär	Juan Fernández fur seal
Arctocephalus spp.	Südliche Seebären	southern fur seals
Arctocephalus townsendi	Guadalupe Seebär	Guadalupe fur seal
Arctogadus borisovi	Ostsibirischer Dorsch	toothed cod
Arctogadus glacialis	Arktischer Dorsch, Arktisdorsch (Polardorsch, Grönland-Dorsch)	Arctic cod (FAO/U.K.), polar cod (U.S./Canada)
Arctogalidia trivirgata	Streifenroller	small-toothed palm civet, three-striped palm civet
Arctonyx collaris	Schweinsdachs	hog badger
Arctoscopus japonicus	Japanischer Sandfisch	Japanese sandfish, sailfin sandfish (FAO)
Arcyptera fusca	Großer Band-Grashüpfer, Große Höckerschrecke	large banded grasshopper
Arcyptera microptera	Kleiner Band-Grashüpfer, Kleine Höckerschrecke	small banded grasshopper
Ardea cinerea	Graureiher	grey heron
Ardea herodias	Kanadareiher	great blue heron
Ardea purpurea	Purpurreiher	purple heron
Ardeidae	Reiher (inkl. Dommeln)	herons & egrets & bitterns
Ardeola ibis/ Bubulcus ibis	Kuhreiher	cattle egret, buff-backed heron

Ardeola ralloides	Rallenreiher	squacco heron
Ardeotis nigriceps	Hindutrappe, Indische Trappe	great Indian bustard
Ardops nichollsi	Baumfledermaus	tree bat
Arenaeus cribrarius	Flecken-Schwimmkrabbe*	speckled crab
Arenaria interpres	Steinwälzer	turnstone
Arenicola cristata	Amerikanischer Sandpierwurm	American lug worm
Arenicola defoliens	Schwarzer Pierwurm	black lug worm, black lug
Arenicola marina	Pierwurm, Sandpier, Sandpierwurm, Köderwurm	European lug worm, blow lug
Arenicolidae	Sandwürmer	lug worms, lugworms
Arenivaga bolliana	Wüstenschabe	desert cockroach
Arenomya arenaria/ Mya arenaria	Sandmuschel, Sandklaffmuschel, Strandauster, Große Sandklaffmuschel	sand gaper, soft-shelled clam, softshell clam, large-neck clam, steamer
Arethusana arethusa/ Satyrus arethusa	Rostbinden-Samtfalter	false grayling
Argas persicus	Persische Zecke	fowl tick, Persian poultry tick (bluebug, abode tick, tampan)
Argas reflexus	Taubenzecke	pigeon tick
Argas spp.	Geflügelzecken	fowl ticks
Argasidae	Saumzecken, Lederzecken	soft ticks, softbacked ticks
Arge ochropus	Gelbe Rosenbürsthornwespe	rose sawfly
Argentina sialis	Pazifisches Glasauge	North-Pacific argentine
Argentina silus	Großes Glasauge, Goldlachs	greater argentine
Argentina sphyraena	Glasauge, Kleines Glasauge	argentine (FAO), lesser argentine
Argentinidae	Goldlachse	argentines, herring smelts
Argidae	Bürstenhornblattwespen	stout-bodied sawflies, argid sawflies
Argiope argentata	Silber-Wespenspinne	silver argiope, silver garden spider
Argiope aurantia	Gold-Wespenspinne	black-and-yellow argiope, black-and-yellow garden spider, yellow garden spider, writing spider
Argiope bruennichi	Wespenspinne, Zebraspinne	black-and-yellow argiope, black-and-yellow garden spider
Argiope trifasciata	Amerikanische Wespenspinne	banded argiope, banded garden spider, American banded garden spider, whitebacked garden spider
Argiope versicolor	Vielfarben-Wespenspinne	multicolored argiope
Argiopinae	Zebraspinnen	garden spiders
Argis dentata	Arktische Garnele*	Arctic argid
Argonauta argo	Papierboot	greater argonaut (FAO), common paper nautilus
Argonauta hians	Braunes Papierboot	winged argonaut, brown paper nautilus
Argonauta nodosus	Warziges Papierboot	knobby argonaut

Argonauta spp.	Papierboote	paper nautiluses
Argopecten circularis	Runde Pilgermuschel	circular scallop
Argopecten flabellum	Afrikanische Fächermuschel*	African fan scallop
Argopecten gibbus	Calico-Pilgermuschel	Calico scallop, calicot scallop
Argopecten irradians	Karibik-Pilgermuschel	bay scallop, Atlantic bay scallop
Argopecten purpuratus	Violette Pilgermuschel	purple scallop
Argopecten solidulus	Dickschalige Pilgermuschel	solid scallop
Argulidae	Fischläuse	fish lice
Argulus foliaceus	Karpfenlaus	carp louse
Argulus spp.	Fischläuse	fish lice
Argusianus argus	Argusfasan	great argus pheasant
Argynnis adippe/ Fabriciana adippe	Feuriger Perlmuttfalter, Märzveilchenfalter, Bergmatten-Perlmutterfalter, Hundsveilchen-Perlmutterfalter	high brown fritillary
Argynnis aglaja/ Mesoacidalia aglaja	Großer Permutterfalter, Großer Permuttfalter	dark green fritillary
Argynnis lathonia/ Issoria lathonia	Kleiner Perlmutterfalter, Kleiner Perlmuttfalter	Queen of Spain fritillary
Argynnis niobe/ Fabriciana niobe	Stiefmütterchen-Perlmutterfalter, Mittlerer Perlmuttfalter	niobe fritillary
Argynnis pandora/ Pandoriana pandora	Kardinal, Pandorafalter	cardinal
Argynnis paphia	Kaisermantel, Silberstrich	silver-washed fritillary
Argyresthia conjugella	Apfelmotte	apple fruit moth
Argyresthiidae	Silbermotten	argents (cypress moths)
Argyroneta aquatica	Wasserspinne	European water spider
Argyronetidae	Wasserspinnen	water spiders
Argyropelecus affinis	Pazifischer Silberbeil	Pacific hatchetfish
Argyropelecus spp.	Faltbauchfische, Silberbeile	hatchetfishes
Argyrosomus hololepidotus/ Sciaena hololepidota	Afrikanischer Adlerfisch	kob, southern meagre (FAO)
Argyrosomus japonicus	Japanischer Adlerfisch	Japanese meagre
Argyrosomus regius/ Sciaena aquila	Adlerfisch (Adlerlachs)	meagre
Argyrozona argyrozona	Tischler-Seebrasse	carpenter, carpenter seabream (FAO)
Arianta arbustorum	Gefleckte Schnirkelschnecke, Baumschnecke	orchard snail, copse snail
Aricia agestis	Kleiner Sonnenröschen-Bläuling, Dunkelbrauner Bläuling	brown argus
Aricia artaxerxes	Großer Sonnenröschen-Bläuling	northern brown argus
Ariidae/Tachysuridae	Meereswelse, Kreuzwelse, Maulbrüterwelse	sea catfishes
Arilus cristatus	Amerikanische Radwanze, Sägekamm-Raubwanze	wheel bug
Ariolimax columbianus	Kolumbianische Wegschnecke	Pacific banana slug, giant yellow slug
Ariomma indica	Indische Schwebmakrele	Indian driftfish

Arion

Arion ater	Schwarze Wegschnecke, Große Schwarze Wegschnecke, Große Wegschnecke	large black slug, greater black slug, black arion
Arion brunneus	Moor-Wegschnecke	bog slug, bog arion
Arion circumscriptus	Graue Wegschnecke	white-soled slug, grey garden slug, brown-banded arion
Arion distinctus	Distinkte Wegschnecke*	darkface slug, darkface arion
Arion fasciatus	Gelbstreifige Wegschnecke	orange-banded slug, orange-banded arion, banded slug
Arion hortensis	Gartenschnecke, Garten-Wegschnecke, Gartenwegschnecke	garden slug, garden arion, common garden slug, black field slug
Arion intermedius	Igel-Wegschnecke, Igelschnecke, Kleine Wegschnecke	hedgehog slug, hedgehog arion
Arion lusitanicus	Spanische Wegschnecke, Portugiesische Wegschnecke	Spanish slug, Lusitanian slug
Arion rufus	Große Rote Wegschnecke	large red slug, greater red slug, chocolate arion
Arion silvaticus	Wald-Wegschnecke	forest slug, forest arion
Arion spp.	Wegschnecken	slugs, land slugs, roundback slugs
Arion subfuscus	Braune Wegschnecke	dusky slug, dusky arion
Arionidae	Wegschnecken	roundback slugs
Ariosoma balearica	Balearen-Muräne (Azoren-Muräne)	Balearic conger
Aristaeomorpha foliacea	Rote Garnele, Rote Tiefseegarnele	giant gamba prawn, giant red shrimp, royal red prawn
Aristeidae	Tiefseegarnelen	gamba prawns, aristeid shrimp
Aristeus antennatus	Blassrote Tiefseegarnele, Blaurote Garnele	blue-and-red shrimp
Aristeus antillensis	Violettköpfige Tiefseegarnele	purplehead gamba prawn
Aristichthys nobilis/ Hypophthalmichthys nobilis	Marmorkarpfen	bighead carp
Ariteus flavescens	Jamaica Feigen-Fledermaus	Jamaican fig-eating bat
Arius felis	Katzen-Kreuzwels*	hardhead sea catfish
Arius heudeloti	Glattmaul-Kreuzwels	smoothmouth sea catfish
Arius maculatus	Gefleckter Kreuzwels	spotted catfish
Arius proops	Kruzifixwels	crucifix sea catfish (FAO), salmon catfish
Arius seemanni/ Arius jordani	Westamerikanischer Kreuzwels, „Minihai"	Tete Sea catfish (FAO), shark catfish
Arius spp.	Kreuzwelse	sea catfishes, salmon catfishes
Arius thalassinus	Riesenmeereswels	giant catfish
Arizona elegans	Arizonanatter	glossy snake
Armadillidiidae	Rollasseln, Kugelasseln	woodlice, pillbugs, sow bugs

Armadillidium vulgare	Gemeine Rollassel, Kugelassel	common woodlouse, common pillbug, sow bug
Armandia cirrhosa	Lagunen-Sandwurm*	lagoon sandworm
Arnoglossus imperialis	Königs-Lammzunge, Königs-Lammbutt	imperial scaldfish
Arnoglossus laterna	Lammzunge, Lammbutt	scaldfish
Arnoglossus thori	Thors Lammzunge, Thors Lammbutt	Thor's scaldfish (FAO), Grohman's scaldfish
Arnoldichthys spilopterus	Arnolds Rotaugensalmler, Afrikanischer Großschuppensalmler	red-eyed characin
Aromia moschata	Moschusbock	musk beetle
Arothron hispidus	Weißflecken-Kugelfisch, Grauer Puffer	toby, blaasop, toadfish, whitespotted puffer (FAO)
Arothron nigropunctatus	Schwarzflecken-Kugelfisch, Schwarzgefleckter Kugelfisch	black-spotted blowfish, blackspotted puffer (FAO)
Arothron stellatus	Riesen-Kugelfisch	star puffer
Arripis georgiana	Georgia-Lachs	ruff
Arripis trutta	Australischer Lachs	Australian salmon (ruff)
Artamidae	Schwalbenstare	woodswallows, currawongs
Artamus personatus	Maskenschwalbenstar	masked wood swallow
Artediellus atlanticus	Atlantische Ohrhakengroppe*	Atlantic hookear sculpin (FAO), Atlantic hook-eared sculpin
Artediellus uncinatus	Arktische Ohrhakengroppe*	Arctic hookear sculpin (FAO), Arctic sculpin
Artedius corallinus	Korallengroppe*, Korallenseeskorpion*	coralline sculpin
Artemesia longinaris	Argentinische Stilett-Garnele	Argentine stiletto shrimp
Artemia gracilis	Amerikanisches Salinenkrebschen	American brine shrimp
Artemia salina	Salzkrebschen, Salinenkrebs, Salinenkrebschen	brine shrimp
Arthroleptidae	Schrillfrösche*	screeching frogs
Arthroleptis spp.	Schrillfrösche*	screeching frogs
Arthroleptis troglodytes	Südrhodesischer Höhlenfrosch, Süd-Zimbabwe-Höhlenfrosch	cave frog, South Zimbabwe screeching frog
Arthropoda	Arthropoden, Gliederfüßer	arthropods
Artibeus spp.	Neotropische Flughunde	neotropical fruit bats
Articulata	Gliedertiere, Articulaten	articulates, articulated animals
Artiodactyla	Paarhufer	even-toed ungulates, cloven-hoofed animals, artiodactyls
Arvicanthis spp.	Kusu-Grasratten	unstriped grass mice, kusu rats
Arvicola sapidus	Westschermaus	southwestern water vole, southern water vole
Arvicola spp.	Schermäuse	water voles, bank voles
Arvicola terrestris	Ostschermaus	European water vole, northern water vole
Asaphis deflorata	Karibische Sandmuschel*	gaudy asaphis
Asaphis violascens	Pazifische Sandmuschel	Pacific asaphis, violet asaphis

Ascalaphidae	Schmetterlingshafte	owlflies
Ascaphidae	Schwanzfrösche	tailed frogs
Ascaphus truei	Schwanzfrosch	tailed frog
Ascaris lumbricoides	Spulwurm	giant intestinal worm, common intestinal roundworm
Ascaris suum	Schweine-Spulwurm	pig roundworm
Aschelminthes/ Nemathelminthes	Aschelminthen, Nemathelminthen, Schlauchwürmer, Rundwürmer *sensu lato* (Pseudocölomaten)	aschelminths, nemathelminths, pseudocoelomates
Ascideacea	Seescheiden, Ascidien	sea-squirts, sea squirts, ascidians
Ascidia callosa	Schwielen-Seescheide	callused sea-squirt
Ascidia mentula	Stumpen-Seescheide, Stumpen-Ascidie	stumpy sea-squirt*
Ascidia virginea	Weiße Seescheide, Jungfern-Seescheide	virgin sea-squirt*, virgin tube tunicate
Ascidiella aspersa	Spritz-Seescheide, Spritz-Ascidie	rough sea-squirt*
Ascidiella scabra	Gemeine Seescheide, Gemeine Ascidie	common sea-squirt
Asellia spp.	Dreizack-Blattnasen	trident leaf-nosed bats
Aselliscus spp.	Tates Dreispitznäsige Fledermaus, Dreispitz-Fledermaus	Tate's trident-nosed bats
Asellus aquaticus	Wasserassel, Gemeine Wasserassel	water louse
Asemum striatum	Düsterbock	pine longhorn beetle
Asilidae	Raubfliegen (Mordfliegen)	robberflies & grass flies
Asilus crabroniformis	Hornissenraubfliege	robberfly (Central European sanddune robberfly)
Asio capensis	Kapohreule	marsh owl, Algerian marsh owl
Asio flammeus	Sumpfohreule	short-eared owl
Asio otus	Waldohreule	long-eared owl
Asiomorpha coarctata	Treibhaustausendfüßer*	hothouse millepede
Asiphum tremulae/ Pachypappa tremulae	Espenblattnest-Blattlaus	spruce root aphid
Aslia lefevrei	Lefevres Seegurke	brown sea cucumber
Aspidelaps lubricus	Südafrikanische Korallenschlange	South African coral snake
Aspidelaps scutatus	Schildkobra	shield-nose snake
Aspidelaps spp.	Schildkobras	shield-nose snakes
Aspideretes gangeticus/ Trionyx gangeticus	Ganges-Weichschildkröte	Indian softshell turtle, Ganges softshell turtle
Aspideretes hurum/ Trionyx hurum	Pfauenaugen-Weichschildkröte	Indian peacock softshell turtle, peacock softshell turtle
Aspideretes leithii/ Trionyx leithii	Vorderindische Weichschildkröte	Leith's softshell turtle
Aspideretes nigricans/ Trionyx nigricans	Schwarze Weichschildkröte, Dunkle Weichschildkröte, Tempel-Weichschildkröte	black softshell turtle, sacred black softshell turtle, Bostami turtle, dark softshell turtle

Aspidiotus hederae/ *Aspidiotus nerii*	Oleanderschildlaus	oleander scale
Aspidiotus juglansregiae	Walnuss-Schildlaus	English walnut scale
Aspidites spp.	Schwarzkopfpythons	black-headed pythons
Aspidomorphus spp.	Kronenschlangen	crowned snakes
Aspidontus taeniatus	Falscher Putzerfisch, Putzernachahmer	mimic blenny, false cleanerfish (FAO)
Aspitates gilvaria	Schafgarbenspanner	straw belle (moth)
Aspitrigla cuculus/ *Chelidonichthys cuculus*	Seekuckuck, Kuckucks-Knurrhahn	red gurnard, cuckoo gurnard
Aspitrigla obscura/ *Chelidonichthys obscurus*	Glänzender Knurrhahn, Rauer Knurrhahn	longfin gurnard (FAO), shining gurnard
Aspius aspius	Rapfen	asp
Aspredinidae	Banjowelse, Bratpfannenwelse	banjo catfishes
Assa darlingtoni	Darlingtons Beutelfrosch, Darlingtonfrosch	pouched frog
Assiminea grayana	Kegelige Strandschnecke, Marschenschnecke	dun sentinel
Assimineidae		sentinel snails
Astacidae	Flusskrebse	crayfishes, river crayfishes
Astacus astacus	Edelkrebs	noble crayfish
Astacus leptodactylus	Sumpfkrebs, Galizier	long-clawed crayfish
Astacus torrentium/ *Austropotamobius torrentium*/ *Potamobius torrentium*	Steinkrebs	stone crayfish, torrent crayfish
Astarte arctica	Arktische Astarte	Arctic astarte
Astarte borealis	Nördliche Astarte	boreal astarte
Astarte castanea	Kastanien-Astarte, Glatte Astarte	smooth astarte
Astarte crenata	Kerben-Astarte*	crenulate astarte
Astarte elliptica	Elliptische Astarte	elliptical astarte
Astarte montagui/ *Tridonta montagui*	Schmalband-Astarte*	narrow-hinge astarte, Montagu tridonta
Astartidae	Astartiden	astartes
Asteiidae	Feinfliegen	asteiids
Asterias rubens	Gemeiner Seestern	common starfish, common European seastar
Asterias vulgaris	Nordmeer-Seestern*	northern seastar
Asterina cepheus	Königsseestern	royal seastar*
Asterina gibbosa	Fünfeckstern, Kleiner Scheibenstern, Kleiner Buckelstern	cushion star, cushion starlet
Asterinidae	Scheibensterne	cushion stars
Asterochelys radiata/ *Geochelone radiata*/ *Testudo radiata*	Madagaskar-Strahlenschildkröte, Madagassische Strahlenschildkröte	radiated tortoise
Asterochelys yniphora/ *Geochelone yniphora*/ *Testudo yniphora*	Madagaskar-Spornschildkröte, Madagassische Schnabelbrustschildkröte	Madagascan spurred tortoise, angonoka
Asteroidea	Seesterne	seastars, starfishes
Asterolecaniidae	Pockenschildläuse, Pockenläuse	pit scales
Asteropsis carinifera	Gekielter Seestern	keeled seastar*

Asteropus sarassinorum	Sternporenschwamm	starpore sponge
Astiotes sponsa	Großer Eichenkarmin	dark crimson underwing, scarlet underwing moth
Astraea heliotropium	Sonnenschnecke	sunsnail (sun shell), sunburst star turban (sunburst star shell)
Astraea rugosa/ Astralium rugosum	Stachelschnecke, Roter Runzelstern	spiny topsnail (rough starshell)
Astralium phoebium	Langstachelschnecke*	longspine starsnail
Astrangia oculata	Nördliche Sternkoralle	northern star coral
Astrapia mayeri	Schmalschwanz-Paradieselster	ribbon-tailed bird-of-paradise
Astreopora gracilis	Sternkoralle	star coral a.o.*
Astroba nuda	Indopazifisches Gorgonenhaupt	Indopacific basket star
Astroblepidae	Kletterwelse	climbing catfishes
Astroconger myriaster	Weißflecken-Congeraal	white-spotted conger
Astroides calycularis	Kelchkoralle	chalice coral*
Astronesthidae	Kehlzähner	snaggletooths
Astronotus ocellatus	Pfauenaugenbuntbarsch, Roter Oskar, Oskar	oscar (FAO), oscar's cichlid, velvet cichlid
Astropecten armatus	Gepanzerter Kammstern	armored seastar, armored comb star
Astropecten articulatus	Randplatten-Kammstern	plated-margined seastar, plate-margined comb star
Astropecten aurantiacus	Roter Kammstern, Großer Kammstern, Mittelmeer-Kammstern	red sand star, red comb star
Astropecten bispinosus	Schlanker Kammstern	slender sand star*
Astropecten irregularis	Nordischer Kammstern	sand star
Astropecten spinulosus	Kletter-Kammseestern	spiny comb star
Astrophyton muricatum	Karibisches Gorgonenhaupt	Caribbean basket star
Astropyga magnifica	Prächtiger Diadem-Seeigel	magnificent hatpin urchin
Astropyga radiata	Strahlen-Diadem-Seeigel	radiating hatpin urchin
Astrospartus mediterraneus	Mittelmeer-Gorgonenhaupt	Mediterranean basket star
Astylosternus spp.	Nachtfrösche	night frogs
Astyris rosacea/ Mitrella rosacea	Rosa Täubchenschnecke	rosy northern dovesnail
Atelecyclidae	Rundkrabben	horse crabs, circular crabs
Atelecyclus rotundatus	Gemeine Rundkrabbe	circular crab
Atelecyclus septemdentatus	Runzelgesicht-Rundkrabbe*	old-man's face crab
Ateleopodidae	Tiefseequappen	tadpole fishes, jellynose fishes
Atelerix algirus/ Erinaceus algirus	Algerischer Igel, Mittelmeerigel	Algerian hedgehog
Atelerix albiventris/ Erinaceus albiventris	Weißbauchigel	four-toed hedgehog
Atelerix frontalis/ Erinaceus frontalis	Kap-Igel	South African hedgehog
Atelerix sclateri/ Erinaceus sclateri	Sclaters Igel, Somalia-Igel	Somali hedgehog
Ateles belzebuth	Goldstirn-Klammeraffe	long-haired spider monkey
Ateles fusciceps	Braunkopf-Klammeraffe	brown-headed spider monkey
Ateles geoffroyi	Geoffroy-Klammeraffe	black-headed spider monkey

Ateles paniscus	Schwarzer Klammeraffe	black spider monkey
Ateles spp.	Klammeraffen	spider monkeys
Atelocynus microtis	Kurzohrfuchs	small-eared dog
Atelopidae/Atelopodidae	Stummelfußfrösche	stubfoot frogs, atelopids
Atelopus spp.	Stummelfußfrösche	stubfoot frogs
Athalia rosae	Kohlrübenblattwespe, Rübenblattwespe	turnip sawfly
Athanas nitescens	Haubengarnele	
Athene cunicularia	Kaninchenkauz	burrowing owl
Athene noctua	Steinkauz	little owl
Atheresthes evermanni	Pfeilzahn-Heilbutt, Asiatischer Pfeilzahnheilbutt	Kamchatka halibut
Atheresthes stomias	Amerikanischer Pfeilzahnheilbutt	arrow-tooth flounder
Atherina boyeri	Boyers Ährenfisch, Kleiner Ährenfisch	big-scale sand smelt (FAO), big-scaled sandsmelt, Boyer's sand smelt
Atherina hepsetus	Großer Ährenfisch	sandsmelt, silverside, Mediterranean sand smelt (FAO)
Atherina presbyter	Ährenfisch, Sand-Ährenfisch	sandsmelt, sand smelt (FAO)
Atherinidae	Ährenfische	silversides
Atheriniformes	Hornhechtartige, Ährenfischverwandte	silversides & skippers & flying fishes & others
Atherinomorus lacunosus	Hartkopf-Ährenfisch	hardyhead silverside
Atheris chloroechis	Grüne Buschviper	plain green bush viper
Atheris hindii	Ostafrikanische Bergotter	East African mountain viper
Atheris hispidus	Rauschuppen-Buschviper	rough-scaled tree viper
Atheris nitschei	Schwarzgrüne Buschviper	sedge viper
Atheris spp.	Buschvipern	African bush vipers
Atheris squamiger	Blattgrüne Buschviper	green bush viper, bush viper, leaf viper
Atheris superciliaris	Afrikanische Tiefland-Buschviper, Afrikanische Tieflandviper	lowland swamp viper, African lowland viper
Atherurus spp.	Quastenstachler	brush-tailed porcupine
Atheta crassicornis	Gemeiner Pilzkurzflügler	
Athetis pallustris	Feuchtwiesen-Staubeule, Wiesen-Staubeule	marsh moth
Athous haemorrhoidalis/ Athous obscurus	Rotbauchiger Laubschnellkäfer	garden click beetle
Atilax paludinosus	Sumpfichneumon, Sumpfmanguste, Wassermanguste	marsh mongoose, water mongoose
Atlanta peroni		Peron's sea butterfly
Atlantidae	Rollschwimmschnecken	atlantas
Atlantoxerus getulus	Nordafrikanisches Erdhörnchen, Atlashörnchen	barbary ground squirrel, North African ground squirrel
Atlides halesus	Großer Purpur-Zipfelfalter	great purple hairstreak
Atomaria linearis	Moosknopfkäfer	pygmy mangold beetle
Atractaspis bibroni	Südliche Erdotter	Bibron's mole viper, Bibron's burrowing viper
Atractaspis corpulenta	Westliche Erdotter	western mole viper, fat mole viper

Atractaspis irregularis	Veränderliche Erdotter	variable mole viper
Atractaspis microlepidota	Arabische Erdotter, Kleinschuppen-Erdviper	northern mole viper, Arabian mole viper
Atractaspis micropholis	Gewöhnliche Erdotter	common mole viper
Atractaspis spp.	Erdottern, Erdvipern	mole vipers
Atractoscion aequidens	Afrikanischer Umberfisch	African weakfish
Atractosteus spatula/ Lepisosteus spatula	Mississippi-Knochenhecht	alligator gar
Atrax robustus	Australische Trichternetz-Vogelspinne	Australian funnelweb spider, Sydney funnelweb spider
Atretium schistosum	Kielschuppen-Wassernatter	olive keelback
Atrichornis clamosus	Braunbauch-Dickichtvogel	noisy scrub-bird
Atrichornithidae	Dickichtschlüpfer, Dickichtvögel	scrub-birds
Atrina fragilis/ Pinna fragilis	Zerbrechliche Steckmuschel	fragile penshell, fan mussel (fragile fanshell)
Atrina rigida	Steife Steckmuschel	stiff penshell
Atrina seminuda	Halbentblößte Steckmuschel*	half-naked penshell
Altrina serrata	Sägezahn-Steckmuschel	sawtooth penshell, saw-toothed penshell
Atrina vexillum	Flaggenmuschel, Flaggen-Steckmuschel	flag penshell
Atrobucca nibe	Schwarzmaul-Umberfisch	blackmouth croaker, black croaker, longfin kob
Atta spp.	Blattschneiderameisen	leafcutting ants
Attacus atlas	Atlasspinner	atlas moth
Attagenus fasciatus	Tropischer Pelzkäfer	wardrobe beetle, tropical carpet beetle
Attagenus megatoma	Dunkler Pelzkäfer	black carpet beetle
Attagenus pellio	Pelzkäfer, Gefleckter Pelzkäfer	fur beetle (black carpet beetle)
Attelabinae	Blattroller	leaf-rolling weevils
Attelabus nitens	Roter Eichenkugelrüssler, Eichenblattroller, Tönnchenwickler	oak leaf roller, red oak roller
Atticora fasciata	Weißbandschwalbe	white-banded swallow
Atule mate/Caranx mate	Gelbflossen-Makrele	yellowfin scad
Atyidae	Süßwassergarnelen	atyid shrimps (freshwater shrimps)
Atypidae	Tapezierspinnen	purse-web spiders
Atypus affinis	Tapezierspinne	purse-web spider
Atys naucum	Weißer Pazifik-Atys	white Pacific glassy-bubble
Auchenipterichthys thoracatus	Mitternachtswels, Zamorawels	Zamora catfish
Auchenipteridae	Falsche Dornwelse	driftwood catfishes
Auchenoglanis occidentalis	Langstirnwels	bubu (FAO), giraffe catfish
Auchenorrhyncha (Homoptera)	Zikaden (Zirpen) & Schaumzikaden	cicadas & hoppers (*see*: spittlebugs, froghoppers)
Auchmeromyia luteola		Congo floor-maggot fly
Aulacaspis rosae	Kleine Rosenschildlaus	rose scale
Aulacomya ater	Magellan-Miesmuschel	Magellan mussel, black-ribbed mussel

Aulacorhynchus prasinus	Laucharassari	emerald toucanet
Aulacorhynchus sulcatus	Blauzügelarassari	groove-billed toucanet
Aulacorthum circumflexum	Gefleckte Gewächshauslaus	mottled arum aphid
Aulacorthum solani	Gefleckte Kartoffellaus	glasshouse aphid, potato aphid
Aulacorthum speyeri	Maiglöckchenlaus	lily-of-the-valley aphid
Aulica aulica/ Cymbiola aulica	Ohr-Walzenschnecke	princely volute
Aulica nobilis/ Cymbiola nobilis	Edle Walzenschnecke	noble volute
Auliscomys spp.	Südamerikanische Blattohrmäuse	South American leaf-eared mice
Aulonocara hansbaenschi	Rotschulter-Kaiserbuntbarsch	red-shoulder Malawi peacock cichlid, aulonnocara Fort Maguire (FAO)
Aulonocara jacobfreibergi	Feen-Kaiserbuntbarsch	fairy cichlid (FAO), jacobfreibergi
Aulonocara nyassae	Kaiserbuntbarsch	emperor cichlid
Aulopiformes	Fadensegelfische	aulopiform fishes
Aulopodidae	Fadensegelfische	aulopus
Aulopus spp.	Fadensegelfische	flagfins
Aulorhynchidae	Röhrenschnäbler	tubesnouts
Aulorhynchus flavidus	Röhrenschnäbler	tubesnout
Aulostomidae/ Aulostomatidae	Trompetenfische	trumpetfishes
Aurelia aurita	Ohrenqualle	moon jelly, common jellyfish
Aurelia limbata	Braungebänderte Ohrenqualle	brown banded moon jelly
Auriparus flaviceps	Goldköpfchen	verdin
Auripasser luteus	Sudangoldsperling	Sudan golden sparrow
Austroginella muscaria	Fliegenschnecke	fly marginella
Austroglossus microlepis	Westküsten-Seezunge	West Coast sole (South Africa)
Austroglossus pectoralis	Ostküsten-Seezunge	East Coast sole (South Africa)
Austropotamobius pallipes	Dohlenkrebs	white-clawed crayfish, freshwater white-clawed crayfish, river crayfish
Austropotamobius torrentium	Steinkrebs	stone crayfish, torrent crayfish
Autographa gamma	Gammaeule	silver Y
Autographa jota	Jota-Eule	plain golden Y
Autographa pulchrina	Ziesteule	beautiful golden Y
Automeris io	Amerikanisches Tagpfauenauge	io moth
Auxis rochei	Melvera-Fregattmakrele, Unechter Bonito	bullet tuna (FAO), bullet mackerel
Auxis thazard	Fregattmakrele, Fregattenmakrele, Unechter Bonito	frigate tuna (FAO), frigate mackerel
Avahi laniger	Wollmaki	avahi, wooly lemur
Aviceda subcristata	Papuaweih	crested baza
Avicularia avicularia	Gemeine Vogelspinne	pinktoe tarantula
Avicularia metallica	Rotfußvogelspinne, Weißfußvogelspinne	whitetoe tarantula
Avicularia pulchra	Rote Zwerg-Avicularia	red dwarf tarantula*
Avicularia purpurea	Purpur-Vogelspinne	Ecuadorian purple tarantula
Avicularia versicolor	Martinique Baum-Vogelspinne	Antilles pinktoe tarantula

Axiidae		lobster shrimps
Axinella spp.	Geweihschwämme u.a.	antler sponges
Axiognathus squamatus	Schuppiger Zwerg-Schlangenstern	dwarf brittlestar
Axiothella spp.	Bambuswürmer u.a.	bamboo worms a.o.
Axis axis/Cervus axis	Axishirsch, Chital	spotted deer, axis deer, chital
Axis porcinus	Schweinshirsch	hog deer
Aythya affinis	Kleine Bergente	lesser scaup
Aythya americana	Rotkopfente	redhead
Aythya collaris	Ringschnabelente	ring-necked duck
Aythya ferina	Tafelente	pochard
Aythya fuligula	Reiherente	tufted duck
Aythya marila	Bergente	scaup, greater scaup
Aythya nyroca	Moorente	ferruginous duck, white-eyed pochard
Aythya valisineria	Riesentafelente	canvasback
Azeca goodalli/ Azeca menkeana	Bezahnte Achatschnecke	three-toothed moss snail, three-toothed snail, glossy trident shell
Azemiops feae	Feaviper, Fea-Viper	Fea's viper

Babelomurex babelis	Babylon-Latiaxis-Schnecke	babylon latiaxis
Babelomurex spinosus	Stachelige Latiaxis-Schnecke	spined latiaxis
Babylonia spp.	Babylonische Türme	babylon snails
Babyrousa babyrussa	Babirusa	babirusa
Bactrocera oleae	Olivenfliege	olivefruit fly
Badidae	Blaubarsche	chameleonfishes
Badis badis	Blaubarsch	chameleonfish, badis (FAO)
Baeopogon indicator	Weißschwanzbülbül	honeyguide greenbul
Baetis pumilus	Kleine Glashafte*	little claret spinner
Bagarius bagarius	Großwels	goonch
Bagridae	Stachelwelse	bagrid catfishes
Bagrus docmac	Nilwels	nilotic catfish
Baiomys spp.	Amerikanische Zwergmäuse	pygmy mice
Balaena glacialis (*Eubalaena glacialis*)	Nordkaper	northern right whale
Balaena mysticetus	Grönlandwal	bowhead whale, Greenland right whale, Arctic right whale
Balaeniceps rex	Schuhschnabel	whale-headed stork
Balaenicipitidae	Schuhschnäbel	whale-headed storks, shoebills
Balaenidae	Glattwale	right whales
Balaenoptera acutorostrata	Zwergwal, Zwergfurchenwal	minke whale, lesser rorqual
Balaenoptera borealis	Seiwal	sei whale
Balaenoptera edeni	Brydes Wal	Bryde's whale
Balaenoptera musculus	Blauwal	blue whale, sulphur bottom whale
Balaenoptera physalus	Finnwal	fin whale, common rorqual
Balaenopteridae	Furchenwale	rorquals
Balanoglossus clavigerus (Enteropneusta)	Keulen-Eichelwurm	Mediterranean acorn worm
Balanophyllia europaea	Warzenkoralle, Europäische Warzenkoralle	European star coral
Balanophyllia floridana	Porige Warzenkoralle	porous star coral, porous cupcoral
Balanophyllia regia	Königliche Warzenkoralle	royal star coral
Balanophyllia verrucaria	Warzenkoralle, Warzige Sternkoralle	warty star coral, warty cupcoral
Balantiocheilos melanopterus	Haibarbe	bala shark, silver shark, tricolor sharkminnow (FAO)
Balantiopteryx spp.	Klein-Flügeltaschen-Fledermaus	least sac-winged bats
Balanus amphitrite	Kleine Streifenseepocke	little striped barnacle
Balanus balanoides/ *Semibalanus balanoides*	Gemeine Seepocke	northern rock barnacle, common rock barnacle
Balanus balanus	Große Seepocke	rough barnacle
Balanus eburneus	Elfenbein-Seepocke	ivory barnacle
Balanus improvisus	Kleine Elfenbeinseepocke	bay barnacle, little ivory barnacle
Balanus nubilis	Riesen-Seepocke	giant acorn barnacle, giant barnacle
Balanus perforatus	Kerb-Seepocke	perforated barnacle

Balaustium murorum	Mauermilbe	wall mite*
Balea biplicata	Gemeine Schließmundschnecke	common door snail, Thames door snail (*Br.*)
Balea perversa	Zahnlose Schließmundschnecke	toothless door snail, tree snail
Balearica pavonina	Kronenkranich	crowned crane
Balearicinae	Kronenkraniche	crowned cranes
Balionycteris maculata	Gefleckflügliger Flughund, Fleckenflügel-Flughund	spotted-winged fruit bat
Balistapus undulatus	Gestreifter Drückerfisch, Grüner Drückerfisch	undulate triggerfish, orange-lined triggerfish (FAO)
Balistes capriscus/ Balistes carolinensis	Schweinsdrückerfisch, Mittelmeerdrücker	triggerfish, grey triggerfish (FAO), clown triggerfish
Balistes erythrodon/ Odonus erythrodon/ Xenodon niger> Odonus niger	Atlantischer Drückerfisch, Nördlicher Drückerfisch, Gabelschwanz-Drückerfisch, Blauer Drücker, Rotzahn, Rotzahn-Drückerfisch	redtooth triggerfish (FAO), red-toothed triggerfish, redfang
Balistes punctatus	Gefleckter Drückerfisch	spotted triggerfish
Balistes vetula	Königin-Drückerfisch	queen triggerfish (FAO), old-wife
Balistes vidua/ Melichtys vidua	Braune Witwe	pinktail triggerfish
Balistidae	Drückerfische	triggerfishes
Balistoides conspicillum	Leoparden-Drückerfisch	big-spotted triggerfish, clown triggerfish (FAO)
Balitoridae/Homalopteridae	Plattschmerlen	river loaches, hillstream loaches
Bandicota bengalensis	Kleine Bandikutratte	lesser bandicoot rat
Bandicota indica	Große Bandikutratte	greater bandicoot rat
Bandicota spp.	Bandikutratten	bandicoot rats
Barbastella barbastellus	Mopsfledermaus	barbastelle
Barbastella spp.	Mopsfledermäuse	barbastelles
Barbatia amygdalumtostum	Gebrannte Mandel	burnt-almond ark
Barbatia barbata	Bärtige Archenmuschel	bearded ark (bearded ark shell)
Barbatia cancellaria	Rotbraune Archenmuschel	red-brown ark (red-brown ark shell)
Barbatia candida	Weißbärtige Archenmuschel	white-beard ark (white-beard ark shell)
Barbatia foliata	Blättrige Archenmuschel*	leafy ark (leafy ark shell)
Barbatula barbatula/ Noemacheilus barbatulus/ Nemacheilus barbatulus	Bartgrundel, Bachschmerle, Schmerle	stone loach
Barbitistes constrictus	Nadelholz-Säbelschrecke	eastern sawtailed bushcricket
Barbitistes serricauda	Laubholz-Säbelschrecke	sawtailed bushcricket
Barbourisia rufa	Roter „Walfisch"	red whalefish
Barbus barbus	Barbe, Flussbarbe, Gewöhnliche Barbe	barbel
Barbus brevipinnis	Kurzflossenbarbe	shortfin barb
Barbus comizo	Iberische Barbe	Iberian barbel
Barbus conchonius/ Puntius conchonius	Prachtbarbe	rosy barb

Barbus fasciatus/ *Puntius eugrammus*	Bandbarbe, Glühkohlenbarbe	striped barb (FAO), zebra barb, African banded barb
Barbus lateristriga/ *Puntius lateristriga*	Schwarzbandbarbe, Seitenstrichbarbe	T-barb, spanner barb (FAO)
Barbus meridionalis	Hundsbarbe	Mediterranean barbel
Barbus meridionalis petenyi	Semling, Afterbarbe	southern barbel, Danubian barbel
Barbus nigrofasciatus/ *Puntius nigrofasciatus*	Purpurkopfbarbe	black ruby barb
Barbus oligolepis/ *Puntius oligolepis*	Eilandbarbe	checkered barb (FAO), checkerboard, island barb
Barbus pentazona pentazona	Fünfgürtelbarbe	five-banded barb, fiveband barb (FAO), tiger barb
Barbus plebejus	Italienische Barbe	Italian barbel
Barbus schuberti	Goldgelbe Messingbarbe	golden barb
Barbus semifasciolatus	Messingbarbe	green barb, China barb, Chinese barb (FAO)
Barbus terio/ *Puntius terio*	Goldfleckbarbe	onespot barb
Barbus tetrazona/ *Puntius tetrazona*	Sumatrabarbe, Viergürtelbarbe	Sumatra barb (FAO), tiger barb
Barbus titteya/ *Puntius titteya*	Bitterlingsbarbe	cherry barb
Barbus tor/Tor tor	Tormahseer	mahseer (FAO), tor mahseer
Barbus viviparus/ *Puntius viviparus*	Schneiderbarbe	bowstripe barb
Barleeia unifasciata	Rotspiralschnecke*	red spire snail
Barnea candida	Weiße Bohrmuschel	white piddock
Barnea truncata	Westatlantische Bohrmuschel*	Atlantic mud-piddock
Bartholomea annulata	Ringelanemone	ringed anemone, curley-cue anemone
Bartholomea lucida	Leuchtanemone	luminant anemone
Bartramia longicauda	Präriewläufer	upland sandpiper
Basiliscus basiliscus	Helmbasilisk	common basilisk
Basiliscus galeritus	West-Basilisk	western basilisk
Basiliscus plumifrons	Stirnlappen-Basilisk	green basilisk, plumed basilisk, double-crested basilisk
Basiliscus spp.	Basilisken	basilisks
Basiliscus vittatus	Streifenbasilisk	brown basilisk
Basommatophora (Pulmonata)	Wasserlungenschnecken	freshwater snails
Bassaricyon spp.	Makibären	olingos
Bassariscus astutus	Nordamerikanisches Katzenfrett	ringtail
Bassariscus sumichrasti	Mittelamerikanisches Katzenfrett	Central American cacomistle
Batagur baska	Batagur, Batagur-Schildkröte	batagur, river terrapin
Bathothauma lyromma	Tiefenwunder	deepsea squid a.o.
Bathyclupeidae	Tiefseeheringe	deep-sea herings
Bathydraconidae	Antarktis-Drachenfische	Antarctic dragonfishes

Bathyergidae	Sandgräber	African mole-rats
Bathyergus spp.	Strandgräber	dune mole-rats
Bathylagidae	Kleinmünder	deep-sea smelts
Bathymasteridae	Bathymasteriden	ronquils
Bathyomphalus contortus	Riementellerschnecke	twisted ramshorn
Bathypolypus arcticus	Arktischer Tiefenkrake	spoonarm octopus, offshore octopus, Arctic deepsea octopus, North Atlantic octopus (FAO)
Bathypteroidae	Spinnenfische	spiderfishes
Bathyraja spinicauda	Grönlandrochen	spiny-tail skate, spinetail ray (FAO)
Bathyraja spinosissima/ Psammobatis spinosissima	Weißrochen, Pazifischer Weißrochen, Stachliger Tiefenrochen	white skate (FAO), spiny deep-water skate
Bathysaurus spp.	Tiefseehechte	deep-sea lizardfishes
Bathysolea profundicola	Tiefseezunge	deepwater sole
Batoidea	Rochen	rays & skates
Batomys dentatus	Luzon Waldratte	Luzon forest rat
Batomys granti	Grants Waldratte	Grant's Luzon forest rat
Batomys salomonseni	Mindanao Waldratte	Mindanao forest rat
Batrachemys nasuta/ Phrynops nasutus	Gewöhnliche Froschkopf-Schildkröte, Gewöhnliche Krötenkopf-Schildkröte	common South American toad-headed turtle
Batrachemys spp.	Froschkopf-Schildkröten	toad-headed turtles
Batrachoididae	Froschfische, Krötenfische	toadfishes
Batrachoidiformes	Froschfische, Krötenfische	toadfishes
Batrachophrynus macrostomus	Anden-Glattfrosch, Junin-Frosch	Andes smooth frog
Batrachoseps aridus	Wüsten-Wurmsalamander	desert slender salamander
Batrachoseps attenuatus	Kalifornischer Wurmsalamander	California slender salamander
Batrachoseps spp.	Wurmsalamander	slender salamanders
Batrachostomidae	Asiatische Froschmäuler	Asian frogmouths
Batrachuperus spp.	Gebirgsseemolche	mountain salamanders
Bdellidae	Schnabelmilben	snout mites
Bdellostoma burgeri	Japanischer Inger	inshore hagfish
Bdeogale spp.	Hundemangusten	black-legged mongooses
Beamys hindei	Kleine Hamsterratte	long-tailed pouched rat, lesser pouched rat, lesser hamster-rat
Bebrornis sechellensis	Seychellenrohrsänger	Seychelles brush warbler
Bedotia geayi	Rotschwanz-Ährenfisch	Madagascar rainbowfish
Belemnitida	Belemniten	belemnites
Belomys pearsoni	Pearsons Flughörnchen	hairy-footed flying squirrel
Belone belone	Hornhecht	garfish, garpike
Belonesox belizanus	Hechtkärpfling	pike top minnow, pike killifish (FAO)
Belonidae	Hornhechte	needlefishes
Belontia hasselti	Wabenschwanz-Gurami, Wabenschwanz-Makropode	Java combtail
Belontia signata	Ceylonmakropode	Ceylonese combtail
Belontiidae/Polyacanthidae	Guramis, Labyrinthfische	gouramies

Belostomatidae	Riesenwanzen, Riesenwasserwanzen	giant water bugs (toe biters)
Bembecia hylaeiformis	Himbeer-Glasflügler	raspberry clearwing
Bembecia scopigera	Sechsstreifen-Glasflügler*	six-belted clearwing
Bembidion spp.	Ahlenläufer	brassy ground beetles
Bembix rostrata/ Epibembix rostrata	Geschnäbelte Kreiselwespe	rostrate bembix wasp*
Bembridae	Tiefseeflundern	deep-water flatfishes, deepwater flatheads
Bena prasinana/ Bena fagana/ Hylophila prasinana/ Pseudoips fagana	Kahnspinner, Kleiner Kahnspinner, Buchenwickler, Buchenkahneule, Jägerhütchen	green silverlines, scarce silverlines
Benthodesmus simonyi	Frostfisch*	frostfish
Berardius arnuxii	Südlicher Vierzahnwal	southern giant bottlenosed whale, Arnoux's beaked whale, four-toothed whale
Berardius bairdii	Baird-Wal	Pacific giant bottlenosed whale, Baird's beaked whale, four-toothed whale
Beroe cucumis	Melonenqualle, Gurkenqualle	melon jellyfish, melon comb jelly
Beroe mitrata	Mützenqualle	sea mitre
Beroe ovata	Mittelmeer-Melonenqualle, Mittelmeer-Gurkenqualle	ovate comb jelly, ovate beroid
Beroe spp.	Melonenquallen, Gurkenquallen	melon jellyfish
Beroida	Melonenquallen, Gurkenquallen	beroids
Berycidae	Schleimköpfe	berycids, alfonsinos
Beryciformes	Schleimköpfe, Schleimkopfartige	squirrel fishes (primitive acanthopterygians)
Berytidae	Stelzenwanzen	stilt bugs
Beryx decadactylus	Nordischer Schleimkopf, Kaiserbarsch, Alfonsino	beryx, alfonsino (FAO), red bream
Beryx splendens	Lowes Alfonsino	Lowe's alfonsino, splendid alfonsino (FAO)
Bethylidae	Ameisenwespchen	bethylid wasps
Betta imbellis	Kleiner Kampffisch	peaceful betta, crescent betta (FAO)
Betta pugnax	Maulbrütender Kampffisch	mouthbrooding fighting fish, Penang betta (FAO)
Betta splendens	Siamesischer Kampffisch	Siamese fighting fish
Bettongia lesueur	Lesueur-Bürstenkänguru(h)	boodie
Bettongia penicillata	Bürstenrattenkänguru(h)	woylie
Bettongia spp.	Bürstenkängurus	short-nosed "rat"-kangaroos
Bibimys spp.	Südamerikanische Rotnasenratten*	crimson-nosed rats
Bibio hortulanus	Gartenhaarmücke	garden march fly
Bibio marci	Markusfliege	St.Mark's fly
Bibionidae	Haarmücken	march flies, St.Mark's flies
Biomphalaria glabrata	Blutegel-Tellerschnecke*, Bilharziose-Tellerschnecke*	bloodfluke planorb
Biorrhiza pallida/ Biorhiza pallida	Schwammgallwespe, Eichenschwammgallwespe	oak-apple gall wasp (>oak apple)

Bipedidae	Zweifuß-Doppelschleichen	two-legged worm lizards
Bipes biporus	Fünfgliedrige Handwühle	five-toed worm lizard
Bipes canabiculatus	Viergliedrige Handwühle	four-toed worm lizard
Bipes spp.	Handwühlen	two-legged worm lizards
Bipes tridactylus	Dreigliedrige Handwühle	three-toed worm lizard
Biphyllidae	Pilzplattkäfer	false skin beetles
Biplex perca/Gyrineum perca	Ahornblatt-Triton	winged triton, maple leaf triton
Birgus latro	Palmendieb, Kokosnusskrebs	coconut crab, robber crab, palm crab, tree crab, purse crab
Biserramenia psammobionta (Aplacophora)	Sand-Furchenfuß	sand glistenworm*
Bison bison	Amerikanischer Bison	American bison, buffalo
Bison bonasus	Wisent	European bison, wisent
Bispira variegata	Vielfarbiger Fächerröhrenwurm	variegated fanworm, multicoloured fanworm*
Bispira voluticornis	Fächerröhrenwurm, Doppelköpfiger Palmfächerwurm	twin-fanworm
Biston betularia	Birkenspanner	peppered moth
Biston strataria	Pappelspanner	oak beauty, oak brindled beauty
Bithynia leachi	Bauchige Schnauzenschnecke, Runde Langfühlerschnecke	globose bithynia, Leach's bithynia
Bithynia tentaculata	Gemeine Schnauzenschnecke, Große Langfühlerschnecke	common bithynia, faucet snail
Bithyniidae	Schnauzenschnecken	faucet snails
Bitia hydroides	Wasserschuppenkopf	keel-bellied water snake
Bitis arietans/ Bitis lachesis	Puffotter, Gewöhnliche Puffotter	puff adder
Bitis atropos	Bergpuffotter, Atroposviper, Südafrikanische Bergotter	berg adder
Bitis caudalis	Gehörnte Puffotter	horned puff adder, Cape horned viper
Bitis cornuta	Büschelbrauenotter, Büschelbrauen-Puffotter	hornsman adder, many-horned adder
Bitis gabonica	Gabunviper	Gaboon viper
Bitis nasicornis	Nashornviper	river jack, rhinoceros viper
Bitis peringueyi	Zwergpuffotter, Peringuey-Wüstenotter	Peringuey's viper, desert sidewinding viper
Bitis schneideri	Namaqua-Zwergpuffotter	Namaqua dwarf adder
Bitis spp.	Puffottern	puff adders
Bitis xeropaga	Wüsten-Bergpuffotter	desert mountain adder
Bittacidae	Mückenhafte	hanging scorpionflies, hangingflies
Bittium reticulatum	Netzhörnchen, Hornschnecke, Mäusedreck, Kleine Gitterschnecke	needle whelk (needle shell)

Bivalvia/ Pelecypoda/ Lamellibranchiata	Muscheln	bivalves, pelecypods, "hatchet-footed animals" (clams: sedimentary, mussels: freely exposed)
Biziura lobata	Lappenente	musk duck
Blaberidae	Riesenschaben	giant cockroaches
Blaberus cranifer	Totenkopfschabe	death's head cockroach
Blaberus giganteus	Brasilianische Schabe	Brazilian cockroach
Blaniulidae	Tüpfeltausendfüßer	snake millepedes, blaniulid millepedes
Blaniulus guttulatus	Gemeiner Tüpfeltausendfüßer, Gefleckter Doppelfüßer, Getüpfelter Tausendfuß	spotted snake millepede, snake millepede
Blanus cinereus	Maurische Netzwühle, Maurische Ringelwühle (Ringelschleiche)	Mediterranean worm lizard
Blanus spp.	Ringelwühlen	Mediterranean worm lizards
Blanus strauchi	Türkische Netzwühle, Türkische Ringelwühle	Turkish worm lizard
Blaps mortisaga	Großer Totenkäfer	giant churchyard beetle, giant cellar beetle
Blaps mucronata	Stachelspitzer Totenkäfer	mucronate churchyard beetle, mucronate cellar beetle
Blaps spp.	Totenkäfer	churchyard beetles, cellar beetles
Blarina brevicauda	Kurzschwanzspitzmaus	short-tailed shrew
Blarinella quadraticauda	Asiatische Kurzschwanzspitzmaus	Asiatic short-tailed shrew
Blarinomys breviceps	Brasilianischer Wühlhamster	Brazilian shrew-mouse
Blastesthia turionella	Kiefernknospenwickler	pine bud moth
Blastocerus dichotomus/ *Odocoileus dichotomus*	Sumpfhirsch	marsh deer
Blastodacna atra/ *Spuleria atra*	Apfelmarkschabe, Apfeltriebmotte	pith moth, apple pith moth
Blastoida	Knospenstrahler	blastoids
Blastophaga psenes	Feigengallwespe	caprifig wasp
Blastophagus piniperda/ *Tomicus piniperda*	Großer Waldgärtner, Gefurchter Waldgärtner, Kiefernmarkkäfer, Markkäfer	pine beetle, larger pith borer, large pine shoot beetle
Blastophagus minor/ *Tomicus minor*	Kleiner Waldgärtner, Rotbrauner Waldgärtner	minor pith borer
Blatella asahinai	Asiatische Schabe	Asian cockroach
Blatella lituricollis	Unechte Deutsche Schabe	false German cockroach
Blatta orientalis	Küchenschabe, Bäckerschabe	Oriental cockroach, common cockroach
Blattella germanica	Deutsche Schabe	German cockroach (croton-bug, shiner, steamfly)
Blattellidae	Holzschaben	wood cockroaches
Blattidae	Schaben	Oriental and American cockroaches
Blattodea	Schaben	cockroaches
Blenniidae	Schleimfische	blennies, combtooth blennies
Blennius canevae/ *Lipophrys canevae*	Gelbkehl-Schleimfisch, Gelbwangen-Schleimfisch	yellow-throat blenny, Caneva's blenny

Blennius fluviatilis/ *Lipophrys fluviatilis*/ *Salaria fluviatilis*	Süßwasser-Schleimfisch, Fluss-Schleimfisch	river blenny, freshwater blenny
Blennius nigriceps/ *Lipophrys nigriceps*	Schwarzkopf-Schleimfisch	black-headed blenny
Blennius ocellaris	Seeschmetterling	butterfly blenny
Blennius pavo/ *Lipophrys pavo*	Pfauenschleimfisch	peacock blenny
Blennius zvonimiri/ *Parablennius zvonimiri*	Hirschschleimfisch	Black Sea blenny
Blennocampa phyllocolpa	Kleinste Rosenblattwespe	least leaf-rolling rose sawfly
Blennocampa pusilla	Rosenblattwespe	leaf-rolling rose sawfly
Blephariceridae/ Blepharoceridae/ Blepharoceratidae	Lidmücken, Netzmücken	net-winged midges
Blepharidopterus angulatus	Gimp	black-kneed capsid
Blicca bjoerkna/ *Abramis bjoerkna*	Güster, Blicke, Pliete (Halbbrachsen)	silver bream, white bream (FAO)
Blissus leucopterus	Amerikanische Getreidewanze	chinch bug, American chinch bug
Blithophaga opaca	Brauner Rübenaaskäfer	beet carrion beetle
Boa constrictor	Königsschlange, Abgottschlange, Abgottboa	boa constrictor
Boa dumerili	Dumerils Boa	Dumeril's boa
Boa madagascariensis	Madagaskar-Boa	Madagascan boa
Boaedon fuliginosus/ *Lamprophis fuliginosus*	Braune Hausschlange	common house snake, common brown house snake
Boaedon lineatus/ *Lamprophis lineatus*	Afrikanische Hausnatter, Gewöhnliche Hausschlange	African house snake, African brown house snake
Boaedon spp./ *Lamprophis* spp.	Boazähner, Hausschlangen, Hausnattern	house snakes
Boarmia roboraria	Steineichen-Baumspanner, Steineichenspanner	great oak beauty
Bodianus diana	Dianas Lippfisch	Diana's wrasse
Bodianus rufus	Spanischer Schweinsfisch	Spanish hogfish
Boettgerilla pallens	Wurmnacktschnecke	worm slug
Boettgerillidae	Wurmnacktschnecken	worm slugs
Bohadschia argus	Augenfleck-Seewalze	eyespot holothurian
Bohadschia graeffei	Graeffes Seewalze	Graeffe's holothurian
Bohadschia marmorata	Marmor-Seewalze	marbled holothurian
Bohadschia tenuissima	Unscheinbare Seewalze	inconspicuous holothurian*
Boidae	Boas (Riesenschlangen)	boas
Boiga dendrophila	Mangroven-Nachtbaumnatter	mangrove snake
Boiga irregularis	Braune Baumschlange	brown tree snake
Boiga spp.	Nachtbaumnattern	tree snakes
Bolbometopon muricatum	Buckelkopf-Papageifisch	green humphead parrotfish (FAO)
Boleophthalmus boddarti	Glotzauge	mud-hopper, goggle-eyed goby
Bolina hydatina	Lappenqualle u.a.	lobate comb jelly a.o.
Bolinopsis infundibulum	Glas-Lappenqualle	common northern comb jelly
Bolinopsis microptera	Kurzlappenqualle*	short-lobed comb jelly
Bolinus brandaris/ *Murex brandaris*	Brandhorn, Herkuleskeule	purple dye murex, dye murex

Bolinus cornutus	Horn-Stachelschnecke	horned murex
Bolitoglossa spp.	Echte Pilzzungensalamander	tropical lungless salamanders
Boloria dia/Clossiana dia	Magerrasen-Perlmuttfalter, Hainveilchen-Perlmutterfalter, Kleinster Perlmutterfalter	violet fritillary, Weaver's fritillary
Boloria euphrosyne/ Clossiana euphrosyne	Silberfleck-Perlmuttfalter, Früher Perlmuttfalter, Veilchen-Perlmutterfalter	pearl-bordered fritillary
Boloria selene/ Clossiana selene	Braunscheckiger Perlmutterfalter, Braunfleckiger Perlmutterfalter	small pearl-bordered fritillary
Boltenia echinata	Kaktus-Seescheide	cactus sea-squirt
Bolyeria multicarinata	Bolyerschlange	Round Island burrowing boa
Bombidae/Bombinae	Hummeln	bumble bees
Bombina bombina	Rotbauchunke, Tieflandunke	fire-bellied toad
Bombina maxima	Riesenunke	Yunnan firebelly toad
Bombina orientalis	Chinesische Rotbauchunke	Oriental fire-bellied toad
Bombina spp.	Unken, Feuerkröten	fire-bellied toads, firebelly toads
Bombina variegata	Gelbbauchunke	yellow-bellied toad, yellowbelly toad, variegated fire-toad
Bombus distinguendus	Große Gelbhummel	great yellow bumble bee
Bombus hortorum	Gartenhummel	small garden bumble bee
Bombus humilis	Veränderliche Hummel	brown-banded carder bee
Bombus lapidarius/ Bombus lapidarius	Steinhummel	red-tailed bumble bee
Bombus lucorum	Weißschwanz-Erdhummel	white-tailed bumble bee
Bombus muscorum	Mooshummel	moss carder bee
Bombus pascuorum/ Bombus agrorum	Ackerhummel	carder bee, common carder bee
Bombus pratorum	Wiesenhummel	early bumble bee
Bombus ruderatus	Große Gartenhummel	large garden bumble bee
Bombus subterraneus	Kurzhaar-Hummel	short-haired bumble bee
Bombus sylvarum	Waldhummel	knapweed carder bee, shrill carder bee
Bombus terrestris	Erdhummel	buff-tailed bumble bee
Bombycidae	Seidenspinner, Echte Spinner	silkworm moths
Bombycilla cedrorum	Zedernseidenschwanz	cedar waxwing
Bombycilla garrulus	Seidenschwanz	bohemian waxwing
Bombycillidae	Seidenschwänze (inkl. Seidenschnäpper)	waxwings
Bombyliidae	Hummelschweber (Hummelfliegen) & Wollschweber & Trauerschweber	beeflies
Bombyx mori	Maulbeerspinner, Echter Seidenspinner	silkworm moth
Bonasa bonasia/ Tetrastes bonasia	Haselhuhn	hazel grouse
Bonasa umbellus	Kragenhuhn	ruffed grouse
Bonellia viridis (Echiura)	Igelwurm (Rüsselwurm)	
Boophilus annulatus	Nordamerikanische Rinderzecke	North American cattle tick

Boophilus decoloratus	Afrikanische Schweinezecke, Blauzecke*	blue tick
Boophilus microplus	Tropische Rinderzecke	tropical cattle tick
Boops boops	Gelbstriemen, Blöker	bogue
Boops salpa/Sarpa salpa	Goldstriemen, Goldstrieme, Ulvenfresser	saupe, salema (FAO), goldline
Boreidae	Winterhafte	snow scorpionflies
Boreogadus saida	Polardorsch	polar cod (FAO/U.K.), Arctic cod (U.S./Canada)
Boreoiulus tenuis/ Blaniulus tenuis	Schlanker Tüpfeltausendfüßer	slender snake millepede
Bos gaurus	Gaur	gaur, seladang
Bos grunniens/Bos mutus	Yak	yak
Bos javanicus	Banteng	banteng
Bos sauveli	Kouprey	kouprey, gray ox
Bos taurus/Bos primigenius	Auerochse, Ur (Hausrind)	aurochs (domestic cattle)
Boselaphus tragocamelus	Nilgauantilope	nilgai, bluebuck
Bosmina coregoni	Großer Rüsselkrebs	larger bosminid waterflea*
Bosmina longirostris	Rüsselkrebschen	lesser bosminid waterflea*
Bosminidae	Rüsselkrebse	bosminid waterfleas
Bostrichidae/Bostrychidae	Kapuzenkäfer, Kapuzinerkäfer, Bohrkäfer, Holzbohrkäfer	horned powder-post beetles, branch borers and twig borers, bostrichids (wood borers)
Botanophila gnava/ Phorbia gnava	Lattichfliege, Salatfliege	lettuce seed fly
Botaurus lentiginosus	Nordamerikanische Rohrdommel	American bittern
Botaurus stellaris	Rohrdommel	bittern
Bothidae	Butte	lefteye flounders, left-eye flounders
Bothriechis aurifer	Gelbfleck-Lanzenotter	yellow-blotched palm pit viper
Bothriechis bicolor	Guatemala-Lanzenotter	Guatemala palm pit viper
Bothriechis lateralis/ Bothrops lateralis	Grüngelbe Lanzenotter	yellow-lined palm viper, side-striped palm pit viper
Bothriechis marchi/ Bothrops marchi	Honduras-Lanzenotter	Honduran palm viper, March's palm pit viper
Bothriechis schlegelii/ Bothrops schlegelii	Schlegels Lanzenotter	Schlegel's viper, eyelash viper, eyelash palm pit viper
Bothriopsis bilineata bilineata/ Bothrops bilineata	Grüne Jararaca, Amazonas-Lanzenotter	Amazonian tree viper, Amazonian palm viper, two-striped forest pit viper
Bothriopsis taeniata	Gefleckte Lanzenotter*	speckled forest pit viper
Bothrochilus boa/ Morelia boa	Neuguinea-Zwergpython	barred python, ringed python
Bothrops alternatus	Halbmond-Lanzenotter	urutu, wutu
Bothrops asper/ Bothrops andianus asper	Terciopelo-Lanzenotter	terciopelo, fer-de-lance, barba amarilla

Bothrops atrox	Gewöhnliche Lanzenotter, Gemeine Lanzenotter	common lancehead, barba amarilla
Bothrops bilineata/ Bothriopsis bilineata	Grüne Jararaca, Amazonas-Lanzenotter	Amazonian tree viper, Amazonian palm viper, two-striped forest pit viper
Bothrops caribbaeus	St.Lucia-Lanzenotter	St.Lucia serpent, Saint Lucia lancehead
Bothrops insularis	Insel-Lanzenotter	jararaca ilhoa, Queimada Island bothrops, golden lancehead (IUCN)
Bothrops jararaca	Jararaca	jararaca
Bothrops jararacussu	Jararacussu	jararacussú
Bothrops lanceolatus	Martinique-Lanzenotter	fer-de-lance, Martinique lancehead
Bothrops lansbergii/ Porthidium lansbergi	Lansbergs Lanzenotter	Lansberg's hognose viper
Bothrops lateralis/ Bothriechis lateralis	Grüngelbe Lanzenotter	yellow-lined palm viper
Bothrops marchi/ Bothriechis marchi	Honduras-Lanzenotter	Honduran palm viper, March's palm pit viper
Bothrops nasutus/ Porthidium nasutum	Stülpnasen-Lanzenotter	horned hognose pit viper
Bothrops neuwiedi	Jararaca pintada, Neuwieds Lanzenotter	jararaca pintada, Neuwied's lancehead
Bothrops nummifer/ Porthidium nummifer	Spring-Lanzenotter	jumping viper
Bothrops schlegelii/ Bothriechis schlegelii	Schlegelsche Lanzenotter	eyelash viper, horned palm viper, eyelash palm pit viper
Bothrops spp. (see also: Lachesis/Porthidium)	Amerikanische Lanzenottern	American lance-headed vipers, lanceheads
Bothus lunatus	Pfauenaugenbutt, Pfauen-Butt	peacock flounder
Bothus podas	Weitäugiger Butt	wide-eyed flounder
Bothynoderes punctiventris/ Cleonus punctiventris	Rübenderbrüssler	beet root weevil
Botia helodes	Tiger-Prachtschmerle	tiger loach
Botia lohachata	Netz-Prachtschmerle	Pakistani loach
Botia macracanthus	Clown-Prachtschmerle	clown loach
Botia modesta	Grüne Schmerle, Blaue Prachtschmerle	orange-finned loach
Botia morletti	Aalstrich-Prachtschmerle	Hora's loach
Botia sidthimunki	Schachbrett-Prachtschmerle, Zwerg-Prachtschmerle	dwarf loach
Botia spp.	Prachtschmerlen	clown loaches
Botia striata	Zebra-Prachtschmerle, Streifen-Prachtschmerle	banded loach
Botryllus schlosseri	Stern-Seescheide, Sternascidie	star ascidian, star sea-squirt
Botula fusca	Zimtmuschel	cinnamon mussel
Bougainvillia ramosa	Bougainvillia-Polyp	Bougainvillia polyp*
Boulengerina annulata	Gebänderte Wasserkobra	banded water cobra
Boulengerina spp.	Wasserkobras	water cobras
Bourletiella hortensis	Gartenspringschwanz	garden springtail
Bovicola bovis	Rinderhaarling	cattle-biting louse

Bovicola caprae	Ziegenhaarling	goat-biting louse
Bovicola equi/ *Werneckiella equi*	Pferdehaarling	horse-biting louse
Bovicola ovis/ *Lepikentron ovis*	Sandlaus, Schaflaus	sheep-biting louse
Brachaeluridae	Blinzelhaie	blind sharks
Brachaelurus waddi	Blinzelhai	blind shark
Brachinus crepitans	Großer Bombardierkäfer	greater bombardier beetle
Brachinus explodens	Kleiner Bombardierkäfer	lesser bombardier beetle
Brachinus spp.	Bombardierkäfer	bombardier beetles
Brachiones przewalskii	Przewalski-Rennmaus	Przewalski's gerbil
Brachionichthyidae	Handfische*	handfishes (warty anglers)
Brachionycha nubeculosa	Frühlings-Rauhaareule	Rannoch sprawler
Brachionycha sphinx	Sphinxeule	sprawler, common sprawler
Brachiopoda	Brachiopoden, Armfüßer, „Lampenmuscheln"	lampshells, brachiopods
Brachyaspis curta/ *Echiopsis curta*	Bardick	bardick
Brachycaudus cardui	Große Pflaumenblattlaus, Große Pflaumenlaus	thistle aphid, greater plum aphid*
Brachycaudus helichrysi	Kleine Pflaumenblattlaus	leaf-curling plum aphid
Brachycaudus persicae	Schwarze Pfirsichblattlaus	black peach aphid
Brachycephalidae	Sattelkröten	saddleback toads
Brachycephalus ephippium	Sattelkröte	Spix's saddleback toad, golden frog, gold frog
Brachycephalus spp.	Sattelkröten	saddleback toads
Brachycera (Diptera)	Fliegen	true flies
Brachydanio albolineatus	Schillerbärbling	pearl danio
Brachydanio frankei	Leopardbärbling	leopard danio
Brachydanio nigrofasciatus	Tüpfelbärbling	spotted danio
Brachydanio rerio/ *Danio rerio*	Zebrabärbling	zebra danio (FAO), zebrafish
Brachydeuterus auritus	Großaugen-Angola-Meerbrasse	bigeye grunt
Brachygobius xanthozona	Goldringelgrundel	bumblebee fish
Brachyistius frenatus	Tangbarsch	kelp perch
Brachylophus fasciatus	Kurzkammleguan	Fiji banded iguana
Brachylophus spp.	Kammleguane	Fiji iguanas, banded iguanas
Brachylophus vitiensis	Fiji Kammleguan	Fiji crested iguana
Brachymystax lenok	Lenok	lenok
Brachynotus sexdentatus	Mittelmeer-Felsenkrabbe	Mediterranean crab, Mediterranean rock crab
Brachypelma albopilosum	Kräuselhaar-Vogelspinne*	curlyhair tarantula
Brachypelma auratum	Goldknie-Vogelspinne	Mexican flameknee tarantula
Brachypelma emilia	Rotbein-Vogelspinne	Mexican redleg tarantula
Brachypelma smithi	Rotfüßige Vogelspinne	Mexican redknee tarantula
Brachypelma vagans	Schwarzrote Vogelspinne	Mexican redrump tarantula
Brachypteraciidae	Erdracken	ground-rollers
Brachypteryx montana	Bergkurzflügel	blue shortwing
Brachytarsomys albicauda	Kurzfuß-Inselratte	Madagascan rat

Brachyteles arachnoides	Muriki	woolly spider monkey, muriqui
Brachytron pratense/ Brachytron hafniense	Kleine Mosaikjungfer	lesser hairy dragonfly, hairy hawker
Brachyura	Kurzschwanzkrebse, Echte Krabben	crabs
Brachyuromys spp.	Madagaskar-Inselratten	Madagascan rats
Brachyurophis spp.	Gürtelschlangen	girdled snakes
Braconidae	Brackwespen	braconids, braconid wasps
Bradybaena fruticum	Genabelte Strauchschnecke	bush snail
Bradybenidae	Strauchschnecken	bush snails
Bradypodion spp.	Zwergchamäleons u.a.	dwarf chameleons
Bradypodion taeniabronchum	Smiths Zwergchamäleons	Smith's dwarf chameleon
Bradypodion ventrale	Östliches Zwergchamäleon	eastern dwarf chameleon
Bradypterus baboecalus	Sumpfbuschsänger	African sedge warbler
Bradypus spp.	Dreifinger-Faultiere	three-toed sloths
Bradypus torquatus	Kragenfaultier	maned sloth, Brazilian three-toed sloth
Bradypus tridactylus	Dreifingerfaultier, Ai	pale-throated sloth
Bradypus variegatus	Braunes Faultier	brown-throated sloth
Brama brama	Brachsenmakrele	pomfret, Atlantic pomfret (FAO), Ray's bream
Bramidae	Brachsenmakrelen, Pomfrets, Goldköpfe	pomfrets
Branchiobdella parasitica	Kiemenegel, Krebsegel	fish-gill leech a.o.
Branchiobdellidae	Kiemenegel, Krebsegel	branchiobdellid leeches
Branchiostegidae	Ziegelbarsche	tilefishes
Branchiostoma lanceolatum	Lanzettfischchen	lancelet
Branchiostoma spp.	Lanzettfischchen	lancelets
Branchipus stagnalis/ Branchipus schaefferi	Echter Kiemenfuß, Sommerkiemenfuß, Sommer-Kiemenfuß	fairy shrimp
Branchiura/Argulida	Kiemenschwänze, Karpfenläuse, Fischläuse	fish lice
Branta bernicla	Ringelgans	brent goose
Branta canadensis	Kanadagans	Canada goose
Branta leucopsis	Weißwangengans	barnacle goose
Branta rufucoliis	Rothalsgans	red-breasted goose
Branta sandvicensis	Hawaiigans, Nene	Hawaiian goose, nene
Braula coeca	Bienenlaus	bee louse
Braulidae	Bienenläuse	bee lice
Bregmaceros mcclellandi	Einhorndorsch	unicorn cod
Bregmacerotidae	Einhorndorsche	codlets
Brenthis ino	Violetter Silberfalter, Mädesüß-Perlmutterfalter	lesser marbled fritillary
Brentidae/Brenthidae	Langkäfer	staright-snouted weevils
Breviceps adspersus	Gesprenkelter Kurzkopffrosch	Transvaal short-headed frog
Breviceps gibbosus	Kurzkopffrosch	rain frog, Cape rain frog (IUCN), South-African short-headed frog

Breviceps macrops	Wüsten-Regenfrosch*	desert rain frogs, Boulenger's short-headed frog
Breviceps spp.	Kurzkopffrösche	short-headed frogs
Brevicoryne brassicae	Mehlige Kohlblattlaus	cabbage aphid, mealy cabbage aphid
Brevipalpus spp.	Falsche Spinnmilben, Unechte Spinnmilben	false spider mites, false red spider mites
Breviraja caerulea/ Neoraja caerulea	Blaurochen	blue ray
Brevoortia patronus	Golf-Menhaden	Gulf menhaden (FAO), mossbunker, large-scale menhaden
Brevoortia tyrannus	Nordwestatlantischer Menhaden	Atlantic menhaden (FAO), bunker
Briareum asbestinum	Karibische Asbestkoralle*	corky sea fingers
Brintesia circe	Weißer Waldportier	great banded grayling, greater wood nymph
Brissopsis lyrifera	Leierherzigel	lyriform heart-urchin, fiddle heart-urchin
Brissus unicolor	Grauer Herzigel, Einfarbiger Herzigel	grey heart-urchin, grooved burrowing urchin
Brookesia minima	Erdchamäleon, Madagassisches Erdchamäleon	minute leaf chameleon, dwarf chameleon
Brookesia spp.	Stummelschwanz-Chamäleons	stump-tailed chameleons, leaf chameleons
Brookesia stumpfii	Stachelchamäleon	plated leaf chameleon
Brookesia superciliaris	Augenzipfel-Stummelschwanzchamäleon	brown leaf chameleon
Brookesia tuberculata	Madagassisches Zwergchamäleon	Mount d'Ambre leaf chameleon
Brosme brosme	Brosme, Lumb	cusk, torsk, tusk (FAO) (European cusk)
Brotogeris cyanoptera	Kobaltflügelsittich	cobalt-winged parakeet
Brotogeris jugularis	Tovisittich, Goldkinnsittich	orange-chinned parakeet
Brotogeris pyrrhoptera	Feuerflügelsittich	grey-cheeked parakeet, gray-cheeked parakeet
Brotogeris versicolurus	Weißflügelsittich	canary-winged parakeet
Brotuninae	Brotulas	brotulas
Bruchidae	Samenkäfer	seed beetles (pea and bean weevils), pulse beetles
Bruchus lentis	Linsenkäfer	lentil weevil
Bruchus pisorum	Erbsenkäfer, Großer Erbsenkäfer	pea weevil, pea seed weevil
Bruchus rufimanus	Ackerbohnenkäfer, Pferdebohnenkäfer	broadbean weevil, bean weevil
Bryobia graminum	Grasmilbe, Gras-Spinnmilbe	grass mite
Bryobia praetiosa	Kleemilbe, Klee-Spinnmilbe	clover mite
Bryobia ribis	Stachelbeermilbe, Stachelbeer-Spinnmilbe	gooseberry mite, gooseberry bryobia, gooseberry red spider mite
Bryobia rubrioculus	Braune Spinnmilbe	brown mite, apple and pear bryobia

Bryodema tuberculata	Gefleckte Schnarrschrecke	speckled grasshopper
Bryozoa> Ectoprocta/Polyzoa	Moostierchen, Bryozoen	moss animals, lace animals, bryozoans
Bubalornithinae	Büffelweber	buffalo weaver
Bubalus bubalis/ Bubalus arnee	Asiatischer Büffel, Wasserbüffel	Asian water buffalo, wild water buffalo, carabao
Bubalus depressicornis	Tieflandanoa	lowland anoa
Bubalus mindorensis	Tamarau, Mindoro-Büffel	tamaraw, taumarau
Bubalus quarlesi	Berganoa	mountain anoa
Bubalus spp.	Asiatische Büffel, Asiatische Wasserbüffel	Asian water buffaloes, anoas
Bubo africanus	Fleckenuhu	spotted eagle-owl
Bubo bubo	Uhu	eagle owl
Bubo lacteus	Blassuhu, Milchuhu	giant eagle-owl
Bubo virginianus	Virginia-Uhu, Amerikanischer Uhu	great horned owl
Bubo zeylonensis	Fischuhu	brown fish owl
Bubulcus ibis/Ardeola ibis	Kuhreiher	cattle egret, buff-backed heron
Bucanetes githagineus	Wüstengimpel	trumpeter finch
Buccinidae	Hornschnecken	whelks
Buccinulum corneum	Spindelhorn	horn whelk, spindle euthria whelk
Buccinum leucostoma	Gelbmund-Wellhornschnecke	yellow-mouthed whelk, yellow-lipped buccinum
Buccinum undatum	Wellhornschnecke, Gemeine Wellhornschnecke	common whelk, edible European whelk, waved whelk, buckie, common northern whelk
Buccinum zelotes	Hohe Wellhornschnecke	superior buccinum
Bucco capensis	Halsband-Faulvogel	collared puffbird
Bucconidae	Faulvögel	puffbirds
Bucculatricidae	Zwergwickler	patches, ribbed case-makers, ribbed case-bearers
Bucephala albeola	Büffelkopfente	bufflehead
Bucephala clangula	Schellente	common goldeneye
Bucephala islandica	Spatelente	Barrow's goldeneye
Buceros bicornis	Doppelhornvogel	great Indian hornbill
Buceros vigil/Rhinoplax vigil	Schildschnabel	helmeted hornbill
Bucerotidae	Nashornvögel, Hornvögel, Tokos	hornbills
Bucorvidae	Hornraben	ground-hornbills
Bucorvus cafer	Kaffernhornrabe	ground hornbill
Budorcas taxicolor	Takin, Rindergemse, Gnuziege	takin
Buenia jeffreysii	Jeffrey-Grundel	Jeffreys's goby
Bufo americanus	Amerikanische Kröte	American toad
Bufo arunco	Andenkröte	Andes toad

Bufo asper	Raukröte	Java toad
Bufo blombergi	Kolumbianische Riesenkröte	Colombian giant toad
Bufo boreas	Nordkröte	western toad
Bufo bufo	Erdkröte	European common toad
Bufo calamita	Kreuzkröte	natterjack toad, natterjack, British toad
Bufo canorus	Singkröte, Kalifornische Hochgebirgskröte	Yosemite toad
Bufo cognatus	Präriekröte	Great Plains toad
Bufo debilis	Grüne Kröte, Grüne Texas-Kröte	green toad
Bufo exsul	Schwarze Kröte	black toad
Bufo hemiophrys	Manitoba-Kröte	Canadian toad
Bufo macrotis	Großohrkröte	big-eared toad
Bufo marinus	Aga-Kröte	giant toad, marine toad, cane toad, South American Neotropical toad
Bufo mauretanicus	Berberkröte	berber toad
Bufo melanosticus	Schwarznarbenkröte	black-spined toad
Bufo periglenes	Goldkröte	golden toad
Bufo punctatus	Fleckenkröte	red-spotted toad
Bufo quercicus	Eichenkröte	oak toad
Bufo regularis	Pantherkröte	panther toad
Bufo typhonius	Südamerikanische Baumkröte	Neotropical leaf toad
Bufo valliceps	Golfkröte	Gulf Coast toad
Bufo viridis	Wechselkröte, Grüne Kröte	green toad (variegated toad)
Bufo woodhousei	Woodhouse-Kröte	Woodhouse's toad
Bufonaria margaritula/ Bufonaria nobilis/ Bursa margaritula	Edle Froschschnecke	noble frogsnail
Bufonaria bufo/Bursa bufo	Kastanien-Froschschnecke*	chestnut frogsnail
Bufonaria crumena/ Bursa crumena	Täschchen-Froschschnecke*	purse frogsnail, frilled frogsnail
Bufonaria echinata/ Bursa echinata	Stachel-Froschschnecke, Stachelige Froschschnecke	spiny frogsnail
Bufonaria elegans/ Bursa elegans	Schön-Froschschnecke	elegant frogsnail
Bufonaria rana/Bursa rana	Gemeine Froschschnecke	common frogsnail
Bufonidae	Echte Kröten	toads, true toads
Buglossidium luteum/ Solea lutea	Zwergzunge, Zwergseezunge	solenette (FAO), pygmy sole
Bugula avicularia	Vogelköpfchen*, Vogelkopf-Moostierchen*	bird's head coralline
Bugula plumosa	Feder-Moostierchen	plumose coralline
Bulbus fragilis	Zerbrechliche Mondschnecke	fragile moonsnail
Bulbus islandicus/ Amauropsis islandica	Isländische Mondschnecke, Island-Mondschnecke	Iceland moonsnail
Bulgarica cana	Graue Schließmundschnecke	grey door snail*
Bulimulidae	Baumschnecken	treesnails
Bulla ampulla	Ampullenschnecke, Ampullen-Blasenschnecke	flask bubble snail, ampulle bulla

Bulla botanica	Australische Blasenschnecke	Australian bubble snail
Bulla solida	Dickschalige Blasenschnecke	solid bubble snail
Bulla striata	Gemeine Blasenschnecke	striate bubble snail, common Atlantic bubble
Bullata bullata	Blasige Randschnecke	blistered marginsnail (blistered margin shell), bubble marginella
Bullia mauritiana	Mauritische Bullia	Mauritian bullia, Mauritius bullia
Bullia tranquebarica	Gestreifte Sandschnecke	lined bullia, Belanger's bullia
Bullina exquisita	Besondere Blasenschnecke	exquisite bubble snail
Bullina lineata	Streifen-Blasenschnecke, Streifenblase	lined bubble snail
Bulweria bulwerii	Bulwersturmvogel	Bulwer's petrel
Bunestomum spp.	Hakenwürmer u.a.	hookworms a.o.
Bungarus caeruleus	Gewöhnlicher Bungar, Indischer Krait	Indian krait
Bungarus candidus	Malayischer Krait, Blauer Krait	Malayan krait, blue krait
Bungarus fasciatus	Gelber Bungar, Bänderkrait, Bänder-Krait	banded krait
Bungarus flaviceps	Rotkopf-Krait	red-headed krait
Bungarus multicinctus	Südchinesischer Vielbindenbungar, Vielbindenkrait	many-banded krait
Bungarus spp.	Bungars, Kraits	kraits
Bunocephalidae	Bratpfannenwelse	banjo catfishes
Bunodactis verrucosa	Edelsteinrose, Warzenrose	wartlet anemone, gem anemone
Bunolagus monticularis	Buschmannhase	Bushman rabbit, Bushman hare, riverine rabbit
Bupalus piniaria	Gemeiner Kiefernspanner	pine moth, pine looper moth, bordered white beauty
Buphaginae	Madenhacker	oxpeckers
Buphagus africanus	Gelbschnabel-Madenhacker	yellow-billed oxpecker
Buphagus erythrorhynchus	Rotschnabel-Madenhacker	red-billed oxpecker
Buprestidae	Prachtkäfer	metallic wood boring beetles, metallic wood borers, splendour beetles, buprestids
Burhinidae	Triele	stone curlews, thick-knees
Burhinus oedicnemus	Triel	stone curlew
Burramys parvus	Berg-Zwerggleitbeutler	mountain pygmy possum
Bursa bubo	Knotige Froschschnecke, Riesenfroschschnecke	giant frogsnail
Bursa granularis	Körnige Froschschnecke	granulate frogsnail
Bursa grayana	Schöne Froschschnecke	elegant frogsnail
Bursa rubeta	Rotmund-Froschschnecke*	ruddy frogsnail
Bursa scrobilator	Pocken-Froschschnecke	pitted frogsnail
Bursa verrucosa	Warzige Froschschnecke	warty frogsnail

Bursidae	Froschschnecken	frog snails (frog shells)
Busarellus nigricollis	Fischbussard	black-collared hawk
Busycon candelabrum	Kandelaber-Wellhorn	splendid whelk, candelabrum whelk
Busycon carica	Warzige Wellhornschnecke*	knobbed whelk
Busycon coarctatum	Rüben-Wellhorn*	turnip whelk
Busycon contrarium	Blitzschnecke, Linksgewundene Wellhornschnecke	lightning whelk, left-handed whelk
Busycon perversum	Ungewöhnliche Wellhornschnecke*	perverse whelk
Busycon spp.	Wellhornschnecken	whelks
Busyconidae	Wellhornschnecken, Helmschnecken	whelks
Busycotypus canaliculatus/ Busycon canaliculatum	Gefurchte Wellhornschnecke*	channeled whelk
Busycotypus spiratus/ Busycon spiratum	Birnenschnecke, Birnenwellhorn	pearwhelk, pear whelk, true pear whelk
Butastur rufipennis	Heuschreckenteesa	grasshopper buzzard-eagle
Buteo albicaudatus	Weißschwanzbussard	white-tailed hawk
Buteo albonotatus	Schwarzbussard	zone-tailed hawk
Buteo brachyurus	Kurzschwanzbussard	short-tailed hawk
Buteo buteo	Mäusebussard	common buzzard
Buteo galapagoensis	Galapagosbussard	Galapagos hawk
Buteo jamaicensis	Rotschwanzbussard	red-tailed hawk
Buteo lagopus	Raufußbussard	rough-legged buzzard
Buteo lineatus	Rotschulterbussard	red-shouldered hawk
Buteo nitidus/Asturina nitida	Zweibindenbussard	gray hawk, grey hawk
Buteo platypterus	Breitflügelbussard	broad-winged hawk
Buteo regalis	Königsbussard	ferruginous hawk
Buteo rufinus	Adlerbussard	long-legged buzzard
Buteo rufofuscus	Felsenbussard	jackal buzzard
Buteo solitarius	Hawaiibussard	Hawaiian hawk, io
Buteo swainsoni	Präriebussard	Swainson's hawk
Buteogallus anthracinus	Krabbenbussard	common black hawk
Buteogallus meridionalis	Savannenbussard	savannah hawk
Buthus occitanus	Okzitanischer Skorpion, Gelber Mittelmeerskorpion, Gemeiner Mittelmeerskorpion	common yellow scorpion
Butis butis	Spitzkopfgrundel	duckbill sleeper
Butorides striatus	Mangrovenreiher	green-backed heron, green heron
Byctiscus betulae	Rebenstecher, Zigarrenwickler	hazel leaf roller weevil
Byctiscus populi	Pappelblattroller	poplar leaf roller weevil
Byrrhidae	Pillenkäfer	pill beetles
Byrsocrypta ulmi	Rüsternblasenlaus	fig gall aphid
Bythinella spp.	Quellschnecken	springsnails
Bythiospeum acicula	Kleine Brunnenschnecke	lesser springsnail*
Bythiospeum helveticum	Schweizer Brunnenschnecke	Swiss springsnail

Bythitidae	Lebendgebärende Brotulas	bythitids, viviparous brotulas
Byturidae	Himbeerkäfer, Blütenfresser	fruitworm beetles
Byturus tomentosus	Himbeerkäfer, Europäischer Himbeerkäfer	European raspberry fruitworm, raspberry beetle
Byturus unicolor	Amerikanischer Himbeerkäfer	American raspberry fruitworm

Cabassous spp.	Nacktschwanzgürteltiere	naked-tailed armadillos
Cabera boryi	Fächer-Moostierchen	palmate bryozoan*
Cabestana cutacea	Borken-Tritonshorn	Mediterranean bark triton
Cacajao calvus	Scharlachgesicht (Uakari)	white uakari, bald uakari
Cacajao melanocephalus	Schwarzkopfuakari	black-headed uakari
Cacajao rubicundus	Roter Uakari	red uakari
Cacatua alba	Weißhaubenkakadu	white cockatoo
Cacatua galerita	Gelbhaubenkakadu	sulphur-crested cockatoo
Cacatua haematuropygia	Rotsteißkakadu	red-vented cockatoo
Cacatua moluccensis	Molukkenkakadu	salmon-crested cockatoo
Cacatua pastinator	Wühlerkakadu	long-billed corella
Cacatua sanguinea	Nacktaugenkakadu	little corella
Cacatua tenuirostris	Nasenkakadu	long-billed corella
Cacatuidae	Kakadus	cockatoos (incl. corellas)
Cacophryne borbonica	Urwaldkröte	jungle toad
Cacospongia scalaris	Lederschwamm	leather sponge
Cactoblastis cactorum	Kaktusmotte	cactus moth
Cadlina laevis	Weiße Sternschnecke	white sea slug*
Cadra calidella	Rosinenmotte	raisin moth
Cadra cautella	Dattelmotte, Dörrobstmotte, Tropische Speichermotte	dried-fruit knot-horn, dried-fruit moth, almond moth
Cadra figulilella	Feigenmotte (*auch:* Rosinenmotte)	raisin moth (*also:* fig moth)
Caeciliidae/Caeciliaidae	Wurmwühlen	common caecilians
Caecobarbus geertsi	Blindbarbe	blind barb
Caenolestes spp.	Opossummäuse	common "shrew" opossums
Caenorhinus aeneovirens	Eichenknospenstecher	
Caenorhinus interpunctatus	Blattrippenstecher, Erdbeerrippenstecher	
Caenorhinus pauxillus	Obstbaumblattrippenstecher	
Caenoscelis subdeplanata	Gelbroter Schimmelkäfer	
Caesio caerulaurea	Blaugoldener Füsilier, Gelbstreifen-Füsilier	blue-and-gold fusilier (FAO), gold-banded fusilier
Caesio lunaris	Blauer Füsilier, Himmelblauer Füsilier, Schwarzfleck-Füsilier	lunar fusilier
Caesio teres	Gelbrücken-Füsilier, Gelbschwanz-Füsilier	beautiful fusilier, yellowtail fusilier, yellow and blueback fusilier (FAO)
Caesio tile	Streifenfüsilier	bartail fusilier
Caesio xanthonotus	Gelbflossen-Füsilier	yellow-fin fusilier
Caesionidae	Füsiliere	fusiliers
Caiman crocodilus	Krokodilkaiman	spectacled caiman
Caiman latirostris	Breitschnauzenkaiman	broad-nosed caiman
Caiman yacare	Paraguay-Kaiman	Paraguayan caiman
Cairina moschata	Moschusente	muscovy duck
Cairina scutulata/ Asarcornis scutulatus	Weißflügelente	white-winged wood duck
Calabaria reinhardtii	Erdpython	Calabar python, African burrowing python

Calamaria linnaei	Linnés Zwergschlange	Linné's reed snake
Calamaria spp.	Zwergschlangen	reed snakes
Calamia tridens	Grünculc	
Calamobius filum/ Calamobius gracilis	Getreidebock	grain borer
Calamoichthys calabaricus	Flösselaal	reedfish
Calamus calamus	Großaugen-Brasse	saucereye porgy
Calamus pennatula	Pluma-Brasse	pluma
Calandrella brachydactyla	Kurzzehenlerche	short-toed lark
Calandrella rufescens	Stummellerche	lesser short-toed lark
Calappa flammea	Schamkrabbe	shame-faced crab, flame box crab
Calappa granulata	Mittelmeer-Schamkrabbe, Hahnenkammkrabbe	Mediterranean shame-faced crab
Calappidae	Schamkrabben	box crabs a.o.
Calathus fuscipes	Braunfüßiger Breithalskäfer, Kreiselkäfer	
Calathus melanocephalus	Hellschildiger Breithalskäfer	
Calcarea	Kalkschwämme	calcareous sponges
Calcarius lapponicus	Spornammer	Lapland bunting
Calceola sandalina	Sandalenkoralle, Pantoffelkoralle	sandal coral
Calcinus tubularis/ Calcinus ornatus	Bunter Einsiedler	variegated hermit crab, ornate hermit crab
Calidris acuminata	Spitzschwanz-Strandläufer	sharp-tailed sandpiper
Calidris alba	Sanderling	sanderling
Calidris alpina	Alpenstrandläufer	dunlin
Calidris bairdii	Bairdstrandläufer	Baird's sandpiper
Calidris canutus	Knutt	knot
Calidris ferruginea	Sichelstrandläufer	curlew sandpiper
Calidris fuscicollis	Weißbürzel-Strandläufer	white-rumped sandpiper
Calidris maritima	Meerstrandläufer	purple sandpiper
Calidris mauri	Bergstrandläufer	Western sandpiper
Calidris melanotos	Graubrust-Strandläufer	pectoral sandpiper
Calidris minuta	Zwergstrandläufer	little stint
Calidris minutilla	Wiesenstrandläufer	least sandpiper
Calidris pusilla	Sandstrandläufer	semipalmated sandpiper
Calidris ruficollis	Rotkehl-Strandläufer	red-necked stint
Calidris subminuta	Langzehen-Strandläufer	long-toed stint
Calidris temminckii	Temminckstrandläufer	Temminck's stint
Caliprobola speciosa	Pracht-Schwebfliege	
Caliroa annulipes	Eichenblattwespe*	oak slug sawfly, oak slugworm
Caliroa cerasi	Schwarze Kirschenblattwespe	pear sawfly, pear slug sawfly, pear and cherry sawfly, pear and cherry slugworm; *larva*: pearslug
Callaeatidae/Callaeidae	Neuseeländische Lappenvögel	New Zealand wattlebirds
Callagur borneoensis	Callagur-Schildkröte	painted batagur, painted terrapin, three-striped batagur, biuku

Callanthias ruber	Papageien-Fahnenbarsch, Tiefenrötling	parrot seaperch (FAO), barberfish
Callanthiidae	Goldfahnenbarsche	goldies
Callaphididae/ Drepanosiphonidae	Zierblattläuse, Zierläuse	drepanosiphonid plant-lice
Calliactis parasitica/ Peachia parasitica/ Adamsia rondeletii	Einsiedlerrose (Schmarotzerrose)	hermit crab anemone (parasitic anemone)
Callianassa gigas	Große Sandgarnele	giant sandprawn, giant ghost shrimp
Callianassa kraussi	Gemeine Südafrikanische Sandgarnele	common South-African sandprawn
Callianassidae	Sandgarnelen	ghost shrimps, sandprawns
Callicanthus lituratus/ Naso lituratus	Kuhkopffisch, Kuhkopf-Doktorfisch, Hornloser Einhornfisch	smooth-headed unicornfish, orangespine unicornfish (FAO), green unicornfish
Callicarpa chazaliei	Federmoos*	plumed hydroid
Callicebus moloch	Grauer Springaffe	dusky titi
Callicebus personatus	Masken-Springaffe	masked titi
Callicebus spp.	Springaffen	titi monkeys
Callichthyidae	Panzerwelse, Schwielenwelse	callichthyid armored catfishes
Callichthys callichthys	Schwielen-Panzerwels	armored catfish, hassar, cascarudo (FAO)
Callidium aeneum	Erzfarbener Scheibenbock	brazen tanbark beetle
Callidium violaceum	Blauer Scheibenbock, Violetter Scheibenbock, Blauvioletter Scheibenbock, Veilchenbock	violet tanbark beetle
Callimico goeldii	Springtamarin, Callimico	Goeldi's marmoset
Callimorpha quadripunctaria/ Euplagia quadripunctaria	Russischer Bär, Spanische Fahne	Jersey tiger (moth), Russian tiger (moth)
Callimorpha dominula/ Panaxia dominula	Schönbär, Spanische Flagge, Grüner Bär, Spanische Fahne	scarlet tiger (moth)
Callinectes arcuatus	Bogen-Schwimmkrabbe	arched swimming crab
Callinectes danae	Dana-Blaukrabbe	Dana swimming crab
Callinectes sapidus	Blaukrabbe, Blaue Schwimmkrabbe	blue crab, Chesapeake Bay swimming crab
Callionymidae	Leierfische (Spinnenfische)	dragonets
Callionymus fasciatus	Balkenstreifiger Leierfisch	barred dragonet
Callionymus lyra	Gestreifter Leierfisch, Gemeiner Leierfisch, Gewöhnlicher Leierfisch, Europäischer Leierfisch	common dragonet
Callionymus maculatus	Gefleckter Leierfisch	spotted dragonet
Callionymus pusillus/ Callionymus festivus	Geschmückter Leierfisch*, Schmuckleierfisch	festive dragonet
Callionymus reticulatus	Gebänderter Leierfisch	banded dragonet, reticulated dragonet (FAO)
Calliophis spp.	Schmuckottern	Oriental coral snakes
Calliostoma annulatum	Gebänderte Kreiselschnecke	purple-ring topsnail, blue-ring topsnail, ringed topsnail (ringed top shell)
Calliostoma conulum	Rotbraune Kreiselschnecke	rusty topsnail
Calliostoma jujubinum	Gefleckte Kreiselschnecke*	jujube topsnail, mottled topsnail

Calliostoma monile	Rotgepunktete Kreiselschnecke*	monile topsnail, Australian necklace topsnail
Calliostoma occidentale	Perlkreiselschnecke*	boreal topsnail, pearly topsnail
Calliostoma zizyphinum	Bunte Kreiselschnecke	painted topsnail, European painted topsnail
Callipepla californica	Schopfwachtel	California quail
Callipepla gambelii	Helmwachtel, Gambelwachtel	Gambel's quail
Calliphlox amethystina	Amethystkolibri	amethyst woodstar
Calliphora spp.	Blaue Schmeißfliegen	bluebottles
Calliphoridae	Schmeißfliegen	blowflies
Calliptamus italicus	Schönschrecke, Italienische Schönschrecke	Italian locust
Callirhipidae	Zedernkäfer*	cedar beetles
Callisaurus draconoides	Gitterschwanzleguan	zebratail lizard, zebra-tailed lizard
Callista chione/ Meretrix chione	Braune Venusmuschel, Glatte Venusmuschel	brown callista, brown venus
Callista erycina	Rote Venusmuschel	red callista, red venus
Callistege mi	Mi-Eule	Mother Shipton
Callistus lunatus	Mondfleckläufer, Mondfleck	
Calliteara pudibunda/ Dasychira pudibunda/ Olene pudibunda/ Elkneria pudibunda	Buchenrotschwanz, Buchen-Streckfuß, Rotschwanz	pale tussock, red-tail moth
Callithrix argentata melanura	Braunes Silberäffchen	black-tailed marmoset
Callithrix argentata	Silberäffchen	silvery marmoset
Callithrix aurita	Ohrbüschel-Seidenäffchen*	buffy tufted-ear marmoset
Callithrix flaviceps	Gelbkopf-Seidenäffchen	buffy-headed marmoset
Callithrix geoffroyi	Weißgesichts-Seidenäffchen	white-faced marmoset
Callithrix humeralifer	Weißschulter-Seidenäffchen	Santarem marmoset
Callithrix jacchus	Weißbüscheläffchen	common marmoset
Callithrix penicillata	Schwarzbüscheläffchen	black-plumed marmoset
Callithrix saterei	Satere-Marmosette	Satere marmoset
Callithrix spp.	Marmosetten, Seidenäffchen	short-tusked marmosets, titis
Callocephalon fimbriatum	Helmkakadu	gang-gang cockatoo
Callochiton laevis (Polyplacophora)	Rote Käferschnecke	red chiton
Callochiton septemvalvis (Polyplacophora)	Siebenschild-Käferschnecke*	seven-plated chiton*
Callophrys rubi	Brombeerzipfelfalter	green hairstreak
Callopistes maculatus	Chile-Teju	spotted false monitor
Callopistes spp.	Unechte Warane, Südamerikanische Tejus	false monitors
Callopistria juventina	Adlerfarneule	bracken fern moth*
Callorhinus ursinus	Nördlicher Seebär	northern fur seal
Callorhynchidae	Pflugnasenchimären, Elefantenchimären	elephantfishes, plownose chimaeras
Callorhynchus capensis/ Callorhinchus capensis	Totenkopfchimäre, Elefantenchimäre	Cape elephantfish (FAO), silver trumpeter (tradename)

Callosamia promethea	Prometheusmotte*	promethea moth, spicebush silk moth
Callosciurus notatus	Plantagenhörnchen, Plantagen-Schönhörnchen	plantation squirrel
Callosciurus prevosti	Prevost-Schönhörnchen	Prevost's squirrel
Callosciurus spp.	Eigentliche Schönhörnchen	beautiful squirrels, tricolored squirrels
Calloselasma rhodostoma/ Agkistrodon rhodostoma	Malayen-Mokassinschlange	Malayan pit viper
Callosobruchus chinensis	Kundekäfer	cowpea weevil
Callyspongia vaginalis	Verzweigter Vasenschwamm	tube sponge, branching vase sponge
Calocoris fulvomaculatus	Hopfenwanze	hop capsid bug, needle-nosed hop bug, shy bug
Calocoris norvegicus	Zweipunktige Grünwanze	potato capsid bug
Calomys spp.	Vespermäuse	vesper mice
Calomyscus bailwardi	Mausartiger Zwerghamster, Mäuseartiger Zwerghamster	mouselike hamster
Calonectris diomedea	Gelbschnabel-Sturmtaucher	Cory's shearwater
Calophasia lunula		toadflax brocade moth
Caloprymnus campestris	Nacktbrustkänguru(h)	desert "rat"-kangaroo
Calopterygidae/Agrionidae	Prachtlibellen	demoiselles, broad-winged damselflies
Calopteryx splendens/ Agrion splendens	Gebänderte Prachtlibelle	banded blackwings, banded agrion
Calopteryx spp.	Prachtlibellen	demoiselles
Calopteryx virgo	Blauflügel-Prachtlibelle	bluewing, demoiselle agrion
Caloptilia azaleella	Azaleenmotte	azalea leaf miner
Caloptilia rufipennella	Ahornmotte	maple moth, maple leaf miner
Calosoma auropunctatum	Goldpunktierter Puppenräuber	goldendot caterpillar hunter*
Calosoma inquisitor	Kleiner Puppenräuber, Kleiner Kletterlaufkäfer	oakwood ground beetle
Calosoma maderae/ Calosoma auropunctatum	Goldpunkt-Puppenräuber	
Calosoma reticulatum	Genetzter Puppenräuber, Smaragdgrüner Puppenräuber	
Calosoma spp.	Puppenräuber	caterpillar hunters
Calosoma sycophanta	Großer Puppenräuber	greater caterpillar hunter
Calotermes flavicollis/ Kalotermes flavicollis	Gelbhalstermite	yellow-necked dry-wood termite
Calotes calotes	Baumechse, Gewöhnliche Schönechse	common tree lizard
Calotes liocephalus	Löwenkopf-Schönechse	lionhead agama, spineless forest lizard (IUCN)
Calotes spp.	Schönechsen	garden lizards, varied lizards, "bloodsuckers"
Calotes versicolor	Indische Schönechse	common bloodsucker, Indian variable lizard, variable agama, "chameleon"
Calpensia nobilis	Krustiges Moostierchen	crusty bryozoan
Caluromys spp.	Wollbeutelratten	woolly opossums

Calycella syringa	Kriechendes Glockenmoos*	creeping bell hydroid
Calypte anna	Annakolibri	Anna's hummingbird
Calypte helenae	Bienenelfe, Bienenkolibri	bee hummingbird
Calyptorhynchus funereus	Rußkakadu	black cockatoo
Calyptorhynchus magnificus	Rabenkakadu	red-tailed cockatoo
Calyptraea centralis	Rund-Mützenschnecke*	circular cup-and-saucer limpet, circular Chinese hat
Calyptraea chinensis	Chinesenhut, Chinesenhütchen	Chinaman's hat, Chinese hat, Chinese cup-and-saucer limpet
Calyptraeidae/Crepidulidae	Mützenschnecken, Haubenschnecken	cup-and-saucer limpets, slipper limpets, slipper shells
Camarhynchus pallidus	Spechtfink	woodpecker finch
Cambarus affinis/ Orconectes limosus	Amerikanischer Flusskrebs	American crayfish, striped crayfish, American river crayfish
Camelus bactrianus	Trampeltier, Zweihöckriges Kamel	Bactrian camel, two-humped camel
Camelus dromedarius	Dromedar, Einhöckriges Kamel	dromedary, one-humped camel
Cameraria ohridella	Rosskastanien-Miniermotte	horse chestnut leafminer*
Campanularia spp.		wine-glass hydroids
Campanulotes bidentatus compar	Kleiner Taubeneckkopf	small pigeon louse
Campephagidae	Kuckuckswürger, Stachelbürzler	cuckoo-shrikes
Campephilus imperialis	Kaiserspecht	imperial woodpecker
Campephilus principalis	Elfenbeinspecht	ivory-billed woodpecker
Campethera nubica	Nubierspecht	Nubian woodpecker
Campogramma glaycos/ Campogramma vadigo/ Campogramma lirio/ Lichia vadigo	Vadigo	vadigo
Camponotus herculeanus	Riesen-Holzameise, Riesenholzameise	giant carpenter ant
Camponotus spp.	Rossameisen, Riesenameisen	carpenter ants
Camponotus vagus	Eichen-Holzameise	oak carpenter ant*
Camptogramma bilineata	Ockergelber Blattspanner	yellow shell
Camptopus lateralis	Sichelbein (Wanze)	
Campylaea planospirum/ Chilostoma planospirum	Flache Felsenschnecke	
Campylopterus ensipennis	Weißschwanz-Degenflügel	white-tailed sabrewing
Campylorhynchus brunneicapillus	Kaktuszaunkönig	cactus wren
Canaceridae/Canacidae	Strandfliegen	beach flies
Cancellaria cancellata	Echte Gitterschnecke, Gegitterte Muskatnuss	cancellate nutmeg, lattice nutmeg
Cancellaria nodulifera	Knorrige Gitterschnecke	knobbed nutmeg
Cancellaria piscatoria	Fischernetz-Gitterschnecke	fishnet snail, fisherman's nutmeg, fishnet nutmeg
Cancellaria pulchra	Hübsche Gitterschnecke	beautiful nutmeg
Cancellaria reticulata	Ostatlantische Gitterschnecke, Gemeine Muskatnuss	common nutmeg, common East-Atlantic nutmeg

Cancellaria spirata	Spiral-Gitterschnecke	spiral nutmeg, spiraled nutmeg
Cancellariidae	Gitterschnecken	nutmeg snails, nutmegs (nutmeg shells)
Cancer antennarius	Pazifischer Taschenkrebs	Pacific rock crab
Cancer borealis	Jonahkrabbe	jonah crab
Cancer gracilis	Hübscher Taschenkrebs*	graceful rock crab, graceful crab
Cancer irroratus	Atlantischer Taschenkrebs	Atlantic rock crab
Cancer magister	Kalifornischer Taschenkrebs, Pazifischer Taschenkrebs	Dungeness crab, Californian crab, Pacific crab
Cancer pagurus	Taschenkrebs	European edible crab
Cancer productus	Bogenkrabbe	red rock crab, red crab
Cancilla praestantissima	Hohe Mitra	superior mitre, superior miter
Cancridae	Taschenkrebse	rock crabs, edible crabs
Candidula gigaxii	Helle Heideschnecke	eccentric snail
Candidula intersecta/ Helicella caperata	Gefleckte Heideschnecke	wrinkled snail
Candidula unifasciata	Quendelschnecke	
Candoia bibroni	Fidschi-Boa, Große Pazifik-Boa	Pacific boa
Candoia carinata	Pazifik-Boa	tree boa, Solomon's ground boa
Candoia spp.	Pazifik-Boas, Südseeboas	Pacific boas
Canis adustus	Streifenschakal	side-striped jackal
Canis aureus	Goldschakal	golden jackal
Canis familiaris	Haushund	domestic dog
Canis latrans	Kojote	coyote
Canis lupus	Wolf	gray wolf
Canis mesomelas	Schabrackenschakal	black-backed jackal
Canis rufus	Rotwolf	red wolf
Canis simensis	Abessinischer Fuchs, Äthiopischer Wolf	Simien jackal, Ethiopian wolf, Simien fox
Cannomys badius	Kleine Bambusratte	lesser bamboo rat
Cantharidae	Weichkäfer	soldier beetles & sailor beetles
Cantharidus opalus	Opal-Kreiselschnecke	opal jewel topsnail
Cantharidus striatus	Gefurchte Kreiselschnecke*	grooved topsnail
Cantharis fusca	Gemeiner Weichkäfer	common cantharid, common soldier beetle
Cantharis livida	Variabler Weichkäfer	variable cantharid, variable soldier beetle
Cantharis spp.	Weichkäfer	soldier beetles
Cantharus assimilis	Grauer Pokal	gray goblet whelk
Cantharus erythrostomus	Rotmund-Pokal	red-mouth goblet whelk
Cantharus melanostomus	Schwarzmund-Pokal	black-mouthed goblet whelk
Cantharus undosus	Wellenpokal	waved goblet whelk, wavy goblet whelk
Cantherinus macrocerus	Weißflecken-Feilenfisch	white-spotted filefish

Canthigaster rostrata	Karibik-Spitzkopfkugelfisch	sharpnose puffer
Canthigasteridae	Krugfische, Spitzkopfkugelfische	sharpnose pufferfishes
Capensibufo rosei	Muizenberg-Kröte	Muizenberg Cape toad, Cape mountain toad (IUCN)
Caperea marginata	Zwergglattwal	pygmy right whale
Capitellidae	Kopfringler	capitellid worms
Capitonidae	Bartvögel	New World barbets
Capnea lucida/ Heteractis lucida	Atlantische Glasperlenanemone	Atlantic beaded anemone
Capnella thyrsoidea	Blumenkohl-Weichkoralle	cauliflower soft coral
Capniinae	Capniden, Kleine Winter-Steinfliegen	small winter stoneflies
Capoeta spp.	Quermundbarben	butterfly barbs*
Capra aegagrus	Bezoarziege	wild goat
Capra caucasica	Westkaukasischer Steinbock	West-Caucasian tur
Capra cylindricornis	Ostkaukasischer Steinbock	East-Caucasian tur
Capra falconeri	Schraubenziege, Markhor	markhor
Capra hircus	Hausziege	domestic goat
Capra ibex	Alpen-Steinbock, Alpensteinbock	Alpine ibex
Capra pyrenaica	Iberiensteinbock, Spanischer Steinbock	Spanish ibex
Capra walie	Äthiopischer Steinbock, Waliasteinbock	Walia ibex
Caprella linearis	Gestreifter Gespenstkrebs (Widderkrebs)	linear skeleton shrimp
Caprellidae	Gespenstkrebse	skeleton shrimps
Capreolus capreolus	Reh, Europäisches Reh	roe deer
Capricornis spp.	Seraus	serow
Caprimulgidae	Nachtschwalben	nightjars & nighthawks
Caprimulgiformes	Nachtschwalben	nightjars & goatsuckers & oilbirds
Caprimulgus aegyptius	Pharaonenziegenmelker	Egyptian nightjar
Caprimulgus europaeus	Ziegenmelker	nightjar
Caprimulgus nubicus	Nubischer Ziegenmelker	Nubian nightjar
Caprimulgus ruficollis	Rothals-Ziegenmelker	red-necked nightjar
Caproidae	Eberfische	boarfishes
Caprolagus hispidus	Borstenkaninchen	bristly rabbit, hispid "hare", Assam rabbit
Capromyidae	Baumratten, Ferkelratten	hutias & nutria
Capromys spp.	Baumratten (Kuba-Baumratten & Jamaika-Baumratten & Bahama-Baumratten)	hutias (Cuban hutias & Jamaica hutias & Bahamian hutias)
Capros aper	Eberfisch	boarfish
Capua reticulana/ Adoxophyes orana	Apfelschalenwickler, Fruchtschalenwickler	summer fruit tortrix
Capulidae	Haubenschnecken (Mützenschnecken) (inkl. Haarschnecken)	cap limpets, capsnails (incl. hairysnails)
Capulus incurvus	Gekrümmte Haubenschnecke*	incurved capsnail

Capulus ungaricus	Ungarkappe	Hungarian capsnail, fool's cap (bonnet shell)
Carabidae	Laufkäfer	ground beetles
Carabus alpestris	Alpen-Laufkäfer	alpine ground beetle*
Carabus arvensis/ Carabus arcensis	Acker-Laufkäfer, Hügel-Laufkäfer	field ground beetle*
Carabus auratus	Gold-Laufkäfer, Goldlaufkäfer, Goldschmied, Goldhenne	golden ground beetle, gilt ground beetle
Carabus auronitens	Goldglänzender Laufkäfer	
Carabus cancellatus	Gitterlaufkäfer, Körnerwarze, Großer Feldlaufkäfer	cancellate ground beetle
Carabus clathratus	Ufer-Laufkäfer	latticed ground beetle
Carabus convexus	Konvexer Laufkäfer, Kurzgewölbter Laufkäfer	convex ground beetle
Carabus coriaceus	Leder-Laufkäfer, Lederkäfer	leatherback ground beetle*
Carabus depressus	Breiter Alpenlaufkäfer	broad Alpine ground beetle*
Carabus glabratus	Glatter Laufkäfer	smooth ground beetle*
Carabus granulatus	Gemeiner Feldlaufkäfer, Körniger Laufkäfer, Gekörnter Laufkäfer	field ground beetle
Carabus hortensis	Garten-Laufkäfer, Goldgruben-Laufkäfer	garden ground beetle
Carabus intricatus	Dunkelblauer Laufkäfer, Blauer Laufkäfer	blue ground beetle, darkblue ground beetle
Carabus irregularis	Schluchtwald-Laufkäfer, Berglaufkäfer	montane ground beetle*
Carabus linnei	Zarter Bergwaldlaufkäfer	
Carabus menetriesi	Hochmoor-Laufkäfer, Waldmoor-Laufkäfer, Moorlaufkäfer	bog ground beetle*
Carabus monilis	Feingestreifter Laufkäfer	
Carabus nemoralis	Hain-Laufkäfer	forest ground beetle
Carabus nitens	Heide-Laufkäfer	shiny ground beetle*, heathland ground beetle*
Carabus nodulosus	Schwarzer Grubenlaufkäfer	
Carabus problematicus	Kleiner Kettenlaufkäfer, Blauvioletter Wald-Laufkäfer	
Carabus scheidleri	Scheidlers Laufkäfer, Gleichgestreifter Laufkäfer	Scheidler's ground beetle
Carabus silvestris	Bergwald-Laufkäfer, Robuster Bergwaldlaufkäfer	
Carabus variolosus	Schwarzer Grubenkäfer, Grubenlaufkäfer	black ditchbeetle*
Carabus violaceus	Purpur-Laufkäfer, Goldleiste	violet ground beetle
Caracal caracal/ Felis caracal	Wüstenluchs, Karakal	caracal
Caracanthidae	Pelzgroppen, Pelzbarsche	coral crouchers, orbicular velvetfishes
Carangidae	Stachelmakrelen, Pferdemakrelen	jacks (jack mackerels) & pompanos, kingfishes, horse mackerels
Carangoides bartholomaei	Gelbschwanz-Makrele	yellow jack

Carangoides ferdau	Streifenmakrele	horse mackerel, Ferdau's trevally, blue kingfish, blue trevally (FAO)
Carangoides ruber/ *Caranx ruber*	Blaurücken-Makrele, Schwarzrücken-Stachelmakrele	bar jack
Caranx caballus	Grüne Makrele*, Grünmakrele*	green jack
Caranx crysos	Blaue Stachelmakrele, Blaumakrele*, Rauchflossenmakrele	blue runner (FAO), blue runner jack
Caranx elacate/ *Caranx sexfasciatus*	Stachelmakrele	bigeye trevally (FAO), large-mouth trevally
Caranx hippos	Caballa, Pferdemakrele, Pferde-Makrele, Pferde-Stachelmakrele	crevalle jack (FAO), Samson fish
Caranx ignobilis	Dickkopf-Makrele	giant trevally (FAO), giant kingfish
Caranx latus	Goldaugen-Makrele, Großaugen-Makrele	horse-eye jack (FAO), horse-eye trevally
Caranx lugubris	Schwarze Makrele, Dunkle Stachelmakrele	black jack (FAO), black kingfish
Caranx melampygus/ *Caranx bixanthopterus*	Blauflossen-Makrele	bluefin kingfish, bluefin trevally (FAO), blue jack, spotted trevally
Caranx rhonchus	Gelbe Stachelmakrele	false scad (FAO), yellow jack, yellow horse mackerel
Caranx ruber/ *Carangoides ruber*	Blaurücken-Makrele, Schwarzrücken-Stachelmakrele	bar jack
Caranx sem/ *Caranx heberi*	Schwarzspitzen-Stachelmakrele	blacktip trevally (FAO), blacktip kingfish
Caranx speciosus/ *Gnathanodon speciosus*	Goldene Königsmakrele	golden trevally
Caranx tille	Tilles Makrele, Südafrikanische Makrele*	tille trevally (FAO), tille kingfish
Carapidae/Fierasferidae	Eigentliche Eingeweidefische, Nadelfische	pearlfishes, carapids
Carapus acus/Fierasfer acus	Nadelfisch, Fierasfer	pearlfish
Carassius auratus	Goldfisch, Goldkarausche	goldfish (FAO), common carp
Carassius auratus gibelio	Giebel	gibel carp, Prussian carp (FAO)
Carassius carassius	Karausche	Crucian carp
Carcharhinidae	Blauhaie, Grundhaie	requiem sharks
Carcharhiniformes	Echte Haie	tiger sharks & catsharks & sand sharks & requiem sharks & hammerheads and others
Carcharhinus *amblyrhynchos/* *Carcharhinus wheeleri*	Grauer Riffhai	gray reef shark
Carcharhinus acronotus	Schwarznasenhai	blacknose shark
Carcharhinus albimarginatus	Silberspitzenhai, Riff-Weißspitzenhai	silvertip shark
Carcharhinus altimus	Großnasenhai	bignose shark
Carcharhinus brachyurus	Kupferhai, Bronzehai	copper shark (FAO), bronze whaler, narrowtooth shark

Carcharhinus brevipinna/ *Carcharhinus maculipinnis*	Großer Schwarzspitzenhai, Langnasenhai	spinner shark (FAO), long-nose grey shark
Carcharhinus falciformis/ *Carcharhinus menisorrah*	Seidenhai	silky shark
Carcharhinus galapagensis	Galapagos-Hai	Galapagos shark
Carcharhinus isodon/ *Aprionodon isodon*	Feinzähniger Hai	finetooth shark
Carcharhinus leucas	Stierhai, Bullenhai, Gemeiner Grundhai	bull shark
Carcharhinus limbatus	Kleiner Schwarzspitzenhai	blacktip shark
Carcharhinus longimanus/ *Carcharhinus maou*	Hochsee-Weißflossenhai, Weißspitzenhai, Weißspitzen-Hochseehai, Langflossen-Hai	oceanic whitetip shark (FAO), whitetip shark, whitetip oceanic shark
Carcharhinus melanopterus	Schwarzspitzen-Riffhai, Schwarzflossenhai, Schwarzflossen-Riffhai	blacktip reef shark
Carcharhinus obscurus/ *Carcharhinus macrurus*	Schwarzhai, Schwarzer Hai	dusky shark (FAO), black whaler
Carcharhinus plumbeus/ *Carcharhinus milberti*	Sandbankhai, Atlantischer Braunhai, Großrückenflossenhai	sandbar shark
Carcharhinus signatus	Atlantischer Nachthai	night shark
Carcharhinus spp.	Grundhaie, Braunhaie	requiem sharks
Carcharias ferox/ *Odontaspis ferox*	Schildzahnhai	smalltooth sand tiger (FAO), fierce shark, shovelnose shark, bumpytail ragged-tooth shark
Carcharias taurus/ *Carcharias arenarius/* *Odontaspis taurus/* *Eugomphodus taurus*	Sandhai, Echter Sandhai, Sandtiger	sand shark, sand tiger shark (FAO), sandtiger shark, gray nurse shark
Carchariidae/Odontaspididae	Sandhaie	sand sharks, sand tigers, ragged-tooth sharks
Carcharodon carcharias/ *Carcharodon rondeletii*	Weißhai, Weißer Hai, Menschenhai	white shark, great white shark (FAO)
Carcharodus alceae	Malven-Dickkopffalter	mallow skipper
Carcinus aestuarii/ *Carcinus mediterraneus*	Mittelmeer-Strandkrabbe	Mediterranean green crab
Carcinus maenas	Strandkrabbe, Nordatlantik-Strandkrabbe	green shore crab, green crab, North Atlantic shore crab
Cardiidae	Herzmuscheln	cockles (cockle shells)
Cardinalinae	Kardinäle	cardinals
Cardinalis cardinalis	Rotkardinal	common cardinal
Cardiocranius paradoxus	Fünfzehen-Zwergspringmaus	five-toed dwarf jerboa
Cardioderma cor	Herznasenfledermaus	African false vampire bat
Cardioglossa pulchra	Schöner Herzzüngler	black long-fingered frog
Cardioglossa spp.	Langfinger, Herzzüngler	long-fingered frogs
Cardisoma spp.	Riesenlandkrabben, Karibische Landkrabben	giant land crabs, great land crabs
Carditidae	Trapezmuscheln	carditas, cardita clams
Cardium costatum	Gerippte Herzmuschel	costate cockle

Cardium edule/	Essbare Herzmuschel,	common cockle,
Cerastoderma edule	Gemeine Herzmuschel	edible cockle,
		common European cockle
Carduelinae	Stieglitzvögel,	cardueline finches,
	Gimpel	goldfinches,
		crossbills
Carduelis cannabina	Bluthänfling	linnet
Carduelis carduelis	Stieglitz	goldfinch,
		European goldfinch
Carduelis chloris	Grünling	greenfinch
Carduelis cucullata/	Kapuzenzeisig	red siskin
Spinus cucullatus		
Carduelis flammea	Birkenzeisig	redpoll,
		common redpoll
Carduelis flavirostris	Berghänfling	twite
Carduelis hornemanni	Polarbirkenzeisig	Arctic redpoll,
		hoary redpoll
Carduelis pinus	Kiefernzeisig	pine siskin
Carduelis psaltria/	Mexikanerzeisig	lesser goldfinch
Spinus psaltria		
Carduelis spinus	Erlenzeisig	siskin
Carduelis tristis/Spinus tristis	Goldzeisig	American goldfinch
Caretta caretta	Unechte Karettschildkröte,	loggerhead sea turtle,
	Unechte Karette	loggerhead
Carettochelys insculpta	Papua-Schildkröte,	pig-nosed turtle,
	Papua-Weichschildkröte	pitted-shell turtle,
		New Guinea plateless turtle,
		pignosed softshell turtle
Cariamidae	Seriemas	seriemas
Carinaria cristata	Glasboot	glassy nautilus
Carinaria lamarcki	Kielfußschnecke,	Lamarck's nautilus
	Schwimmende Mütze	
Caristiidae	Caristiiden	manefishes
Carnegiella marthae	Schwarzschwingen-	black-winged hatchetfish
	Beilbauchfisch	
Carnegiella strigata	Marmorierter Beilbauchfisch	marbled hatchetfish
Carnivora	Karnivoren, Raubtiere	carnivores
Carollia perspillicata	Brillenblattnase	Seba's short-tailed bat
Carollia spp.	Kurzschwanz-Blattnasen	short-tailed leaf-nosed bats
Carphophis amoenus	Wurmnatter	worm snake
Carpodacus erythrinus	Karmingimpel	scarlet rosefinch,
		common rosefinch
Carpodacus mexicanus	Hausgimpel,	house finch,
	Mexikanischer Karmingimpel	California linnet
Carpodacus purpureus	Purpurgimpel	purple finch
Carpodacus rubicilla	Berggimpel	great rosefinch
Carpodacus synoicus	Einödgimpel	Sinai rosefinch
Carpoglyphidae	Backobstmilben	driedfruit mites
Carpoglyphus lactis	Backobstmilbe	driedfruit mite
Carpomys spp.	Luzon-Ratten	Luzon rats
Carpophilus dimidiatus	Getreidesaftkäfer	corn sap beetle
Carpophilus hemipterus	Backobstkäfer	driedfruit beetle
Carterocephalus palaemon	Gelbwürfeliger Dickkopffalter	chequered skipper (butterfly)
Carterodon sulcidens	Brasilianische Stachelratte u.a.	Eastern Brasilian spiny rat

Cartodere elongata	Gemeiner Moderkäfer	European plaster beetle*
Cartodere filiformis	Hefekäfer	yeast beetle
Carybdea alata	Seewespe u.a.	sea wasp a.o.
Carybdea marsupialis	Mittelmeer Seewespe, Mittelmeer-Würfelqualle	Mediterranean seawasp
Carychium minimum	Bauchige Zwergschnecke, Bauchige Zwerghornschnecke	short-toothed herald snail, herald thorn snail (U.S.), herald snail, sedge snail
Carychium tridentatum	Dreizahn-Zwerghornschnecke	long-toothed herald snail, dentate thorn snail (U.S.), slender herald snail (Br.)
Caryedon fuscus	Erdsamenrüssler, Tamarindenrüssler	groundnut borer, tamarind weevil
Caryedon serratus	Erdnusssamenkäfer, Westafrikanischer Erdnusssamenkäfer	groundnut bruchid, West-African groundnut borer
Caryophyllaeus laticeps	Nelkenwurm, Nelkenbandwurm	
Caryophyllia calix	Atlantische Becherkoralle	Atlantic cupcoral
Caryophyllia cornuformis	Kleine Nelkenkoralle*	lesser cupcoral, lesser horncoral
Caryophyllia inormata	Runde Nelkenkoralle	circular cupcoral
Caryophyllia smithii	Kreiselkoralle, Ovale Nelkenkoralle	Devonshire cupcoral
Caryophyllia spp.	Nelkenkorallen	cupcorals
Casarea dussumieri	Mauritius-Boa	Round Island boa
Cassida nebulosa	Nebliger Schildkäfer	beet tortoise beetle, clouded shield beetle
Cassida viridis	Grüner Schildkäfer	green tortoise beetle
Cassida vittata	Glanzstreifiger Schildkäfer	
Cassidae	Sturmhauben, Helmschnecken	helmet snails (helmet shells)
Cassidaria echinophora/ Galeoda echinophora	Stachelige Helmschnecke, Stachel-Helmschnecke	spiny helmet snail, Mediterranean spiny bonnet
Cassidinae	Schildkäfer	tortoise beetles, shield beetles
Cassidula aurisfelis	Katzenohr	cat's ear cassidula
Cassidula nucleus	Gebänderte Ohrenschnecke	nucleus cassidula
Cassiopeia andromedra	Saugschirmqualle	sucker upsidedown jellyfish, suction cup jellyfish
Cassiopeia xamachana	Mangrovenqualle	mangrove upsidedown jellyfish
Cassis cornuta	Große Sturmhaube, Gehörnte Helmschnecke	horned helmet, giant helmet
Cassis flammea	Flammen-Helmschnecke	flame helmet, princess helmet
Cassis madagascariensis	Kaiserliche Helmschnecke	cameo helmet, queen helmet, emperor helmet
Cassis nana	Zwerg-Helmschnecke	dwarf helmet
Cassis saburon	Afrikanische Helmschnecke	African helmet
Cassis sulcosa	Gefurchte Sturmhaube, Gefurchte Helmschnecke	grooved helmet
Cassis tuberosa	Königshelm	Caribbean helmet, king helmet
Castor canadensis	Kanadischer Biber	North American beaver, Canadian beaver

Castor fiber	Altwelt-Biber	Eurasian beaver, European beaver
Castor spp.	Biber	beavers
Casuariidae	Kasuare und Emu	cassowaries and emu
Casuarius bennetti	Bennettkasuar	dwarf cassowary
Casuarius casuarius	Helmkasuar	double-wattled cassowary
Catablema vesicarium	Korsettqualle*	constricted jellyfish
Catagonus wagneri	Chacopekari	Chacoan peccary, tagua
Catamblyrhynchinae	Plüschkopftangare	plush-capped finch, plushcap
Catantopidae	Knarrschrecken	catantopid grasshoppers
Catarrhina	Altweltaffen, Schmalnasenaffen	old-world monkeys (incl. apes)
Cathartes aura	Truthahnsgeier	turkey vulture
Cathartidae	Neuweltgeier	New World vultures
Catharus aurantiirostris	Goldschnabel-Musendrossel	orange-billed nightingale thrush
Catharus fuscescens	Weidendrossel, Wilsondrossel	veery
Catharus guttatus	Einsiedlerdrossel	hermit thrush
Catharus minimus	Grauwangendrossel	grey-cheeked thrush
Catharus ustulatus	Zwergdrossel	Swainson's thrush
Catherpes mexicanus	Schluchtenzaunkönig	canyon wren
Catinella arenaria	Salzbernsteinschnecke	sand-bowl amber snail, sandbowl snail, sand amber snail
Catla catla	Catla, Theila, Tambra	catla
Catlocarpio siamensis	Riesenbarbe	giant barb (giant "carp")
Catocala electa	Weidenkarmin	rosy underwing
Catocala elocata	Pappelkarmin	poplar underwing
Catocala fraxini	Blaues Ordensband	Clifden Nonpareil
Catocala fulminea	Gelbes Ordensband	yellow underwing
Catocala nupta	Rotes Ordensband	red underwing
Catocala promissa	Kleines Eichenkarmin, Kleines Eichenkarmin-Ordensband, Kleines Karminrotes Eichenhain-Ordensband	light crimson underwing
Catocala sponsa	Großes Eichenkarmin, Großes Eichenkarmin-Ordensband, Großes Karminrotes Eichenmischwald-Ordensband	dark crimson underwing
Catopidae/Leptodiridae	Nestkäfer, Erdaaskäfer	small carrion beetles
Catoprion mento	Schuppenräuber, Wimpelpiranha	wimple piranha
Catoprioninae	Schuppenräuber	
Catorama tabaci	Großer Tabakkäfer	catorama beetle
Catostomidae	Sauger	suckers
Catostomus commersoni	Sauger	white sucker (FAO), freshwater mullet
Catreus wallichii	Schopffasan, Wallichfasan	cheer pheasant

Caudiverbera caudiverbera	Helmkopf	helmeted water toad
Caudofoveata	Schildfüßer	caudofoveates
Caulolatilus princeps	Ozeanmaräne*	ocean whitefish
Caulolepis longidens/ Anoplogaster cornuta	Fangzähner	common fangtooth
Causa holosericum/ Isognomostoma holosericum	Genabelte Maskenschnecke	
Causus bilineatus	Gestreifte Nachtotter	lined night adder
Causus lichtensteini	Lichtensteins Nachtotter	Lichtenstein's night adder
Causus maculatus	Westafrikanische Nachtotter	West African night adder
Causus resimus	Grüne Krötenotter, Grüne Nachtotter	green night adder
Causus rhombeatus	Pfeilotter, Krötenotter, Nachtotter	rhombic night adder, common night adder
Causus spp.	Krötenottern, Krötenvipern	night adders
Cavernularia obesa	Walzen-Seefeder	barrel sea-pen*
Cavernulina cylindrica	Zylindrische Seefeder	cylindrical sea-pen
Cavia spp.	Meerschweinchen	cavies, guinea pigs
Caviidae	Meerschweinchen	cavies & Patagonian "hares"
Cavolinia tridentata	Dreizahn-Seeschmetterling	three-tooth cavoline
Cavolinia uncinata	Haken-Seeschmetterling	uncinate cavoline, hooked cavoline
Cebidae	Kapuzinerartige	New World monkeys
Cebrionidae		robust click beetles
Cebuella pygmaea	Zwergseidenäffchen	pygmy marmoset
Cebus albifrons	Weißstirnkapuziner	white-fronted capuchin, brown pale-fronted capuchin
Cebus apella	Apella, Faunaffe, Gehaubter Kapuziner	black-capped capuchin
Cebus capucinus	Weißschulterkapuziner, Weißschulteraffe, Kapuziner	white-throated capuchin
Cebus nigrivittatus	Brauner Kapuziner	weeper capuchin, wedge-capped capuchin
Cebus spp.	Kapuzineraffen, Kapuziner, Rollaffen	capuchins, ring-tailed monkeys
Cecidomyia baeri	Nadelnknickende Kieferngallmücke	pine needle gall midge
Cecidomyia pini	Kiefernharz-Gallmücke	pine resin midge
Cecidomyiidae	Gallmücken	gall midges, gall gnats
Cecidophyopsis ribis	Johannisbeergallmilbe	big bud gall mite
Cecilioides acicula	Blindschnecke, Blinde Turmschnecke, Nadelschnecke	blind awlsnail, blind snail, agate snail
Cecilioides aperta	Stumpfe Turmschnecke*	obtuse awlsnail
Celaenia distincta	Australische Bola-Spinne	Australian bolas spider
Celaenomys silaceus	Luzon-Spitzmausratte*	Luzon shrew-rat
Celastrina argiolus	Faulbaum-Bläuling	holly blue (butterfly)
Celastrina ladon	Frühjahrs-Bläuling	spring azure (butterfly)
Celerio galii	Labkrautschwärmer	bedstraw hawkmoth
Celes variabilis	Pferdeschrecke	horse grasshopper*

Celeus castaneus	Kastanienspecht	chestnut-coloured woodpecker
Cellana exarata	Schwarze Napfschnecke	black limpet
Cellana talcosa	Talkum-Napfschnecke	talcum limpet
Cellepora ramulosa	Geweih-Moostierchen	rameous bryozoan*, rameous false coral
Cemophora coccinea	Scharlachnatter	scarlet snake
Centracanthidae	Laxierfische, Schnauzenbrassen	picarels
Centracanthus cirrus		imperial jerret
Centrarchidae	Sonnenbarsche, Sonnenfische	sunfishes
Centrarchus macropterus	Pfauenaugenbarsch, Pfauenaugensonnenbarsch	flier
Centriscidae	Schnepfenmesserfische	shrimpfishes
Centrocercus urophasianus	Beifußhuhn	sage grouse
Centrolabrus exoletus	Kleinmündiger Lippfisch, Kleinmäuliger Lippfisch	rock cock, rock cook (FAO), small-mouthed wrasse
Centrolabrus trutta	Forellen-Lippfisch*	trout wrasse
Centrolene geckoideum	Gecko-Glasfrosch	Pacific giant glass frog
Centrolene spp.	Groß-Glasfrösche	giant glass frogs
Centrolenidae	Glasfrösche	glass frogs, centrolenids (leaf frogs)
Centrolophidae	Schwarzfische	medusafishes, barrelfishes
Centrolophus niger	Schwarzfisch	blackfish
Centronycteris maximiliani	Strubbelige Fledermaus	shaggy-haired bat
Centrophorus granulosus	Schlingerhai, Rauer Dornhai	gulper shark (FAO), rough shark
Centrophorus lusitanicus	Lusitanischer Schlingerhai	lowfin gulper shark
Centrophorus squamosus	Blattschuppiger Schlingerhai, Düsterer Dornhai	leafscale gulper shark
Centrophorus uyato	Kleiner Schlingerhai	little gulper shark
Centrophryne spinulosa	Tiefsee-Angler	deep-sea anglerfish
Centropodidae	Laufkuckucke, Spornkuckucke	coucals
Centropomidae	Nilbarsche, Snooks	snooks
Centropomus undecimalis	Olivgrüner Snook	snook, common snook (FAO)
Centropristis striata	Schwarzer Sägebarsch	black seabass (FAO), black sea bass
Centropyge argi	Blauer Zwergkaiser, Blauer Karibischer Zwergkaiserfisch	cherubfish
Centropyge bicolor	Blaugelber Zwergkaiser, Schwarzgelber Zwergkaiserfisch	bicolor angelfish
Centropyge bispinosus	Gestreifter Zwergkaiser	twospined angelfish (FAO), dusky angelfish
Centropyge flavissimus	Zitronen-Zwergkaiser, Zitronengelber Zwergkaiserfisch	lemonpeel angelfish
Centroscyllium fabricii	Fabricius Dornhai	black dogfish
Centroscymnus coelolepis	Portugieserhai, Portugiesischer Dornhai	Portuguese dogfish (FAO), Portuguese shark
Centroscymnus crepidater	Langnasen-Dornhai	longnose velvet dogfish
Centroscymnus cryptacanthus	Kurznasen-Dornhai	shortnose velvet dogfish

Centrostephanus longispinus	Mittelmeer-Diademseeigel, Langstachliger Seeigel	Mediterranean hatpin urchin, needle-spined urchin
Centrotus cornutus	Dornzikade, Graubraune Dornzikade	horned treehopper
Centruroides spp.	Rindenskorpione*	bark scorpions
Centurio senex	Greisengesicht, Greisenhaupt	wrinkle-faced bat, lattice-winged bat
Cepaea hortensis	Gartenschnirkelschnecke	white-lip gardensnail, white-lipped snail, garden snail, smaller banded snail
Cepaea nemoralis	Schwarzmündige Bänderschnecke, Hain-Schnirkelschnecke, Hainbänderschnecke	brown-lipped snail, grove snail, grovesnail, English garden snail, larger banded snail, banded wood snail
Cepaea spp.	Schnirkelschnecken	banded snails, garden snails
Cephalaspidea/Kephalaspidea	Kopfschildschnecken, Kopfschildträger	bubble snails
Cephalcia abietis	Gemeine Fichtengespinstblattwespe	web-spinning spruce sawfly, spruce webspinner
Cephalcia lariciphila	Lärchengespinstblattwespe	web-spinning larch sawfly, larch webspinner
Cephalochordata (Amphioxiformes)	Cephalochordaten, Lanzettfischchen	cephalochordates, lancelet
Cephaloidae	Cephaloiden	false longhorn beetles
Cephalopholis miniata	Juwelenbarsch	blue-spotted rockcod, coral trout, coral hind (FAO)
Cephalopholis sexmaculata	Sechsfleckenbarsch	sixblotch hind (FAO), six-barred grouper, sixspot grouper
Cephalophus adersi	Adersducker	Aders' duiker
Cephalophus callipygus	Petersducker, Harveyducker, Schönsteißducker	Peters' duiker
Cephalophus dorsalis	Schwarzrückenducker	bay duiker
Cephalophus jentinki	Jentinkducker	Jentink's duiker
Cephalophus leucogaster	Weißbauchducker, Gabunducker	white-bellied duiker
Cephalophus monticola/ Philantomba monticola	Blauducker, Blauböckchen	blue duiker
Cephalophus natalensis	Rotducker	red forest duiker
Cephalophus niger	Schwarzducker	black duiker
Cephalophus nigrifrons	Schwarzstirnducker	black-fronted duiker
Cephalophus ogilbyi	Ogilbyducker	Ogilby's duiker
Cephalophus rufilatus	Rotflankenducker, Blaurückenducker	red-flanked duiker
Cephalophus spadix	Abbottducker	Abbott's duiker
Cephalophus spp.	Schopfducker	duikers
Cephalophus sylvicultor	Gelbrückenducker, Riesenducker	yellow-backed duiker
Cephalophus weynsi	Weynsducker	Weyns' duiker
Cephalophus zebra	Zebraducker	banded duiker
Cephalopoda	Kopffüßer, Cephalopoden	cephalopods

Cephalorhynchus commersonii	Jacobita, Commerson-Delphin	Commerson's dolphin, Tonina dolphin, piebald dolphin, black and white dolphin
Cephalorhynchus eutropia	Chile-Delphin	black dolphin, black Chilean dolphin, white-bellied dolphin
Cephalorhynchus heavisidii	Haviside-Delphin, Kapdelphin	South African dolphin, Haviside's dolphin, Benguela dolphin
Cephalorhynchus hectori	Hector-Delphin, Neuseeland-Delphin	Hector's dolphin, New Zealand white-front dolphin
Cephalorhynchus spp.	Schwarz-Weiß-Delphine	southern dolphins, piebald dolphins
Cephaloscyllium spp.	Schwellhaie	swell sharks
Cephenomyia auribarbis	Hirschrachenbremse	deer nostril fly
Cephenomyia stimulator	Rehrachenbremse	roe deer nostril fly
Cephenomyia trompe	Rentierrachenbremse	reindeer nostril fly
Cephenomyia ulrichi	Elchrachenbremse	elk nostril fly (Br.), moose nostril fly (U.S.)
Cephenomyiinae	Rachenbremsen	nostril flies
Cephidae	Halmwespen	stem sawflies
Cephus pygmeus/ Cephus pygmaeus	Getreidehalmwespe	wheat stem sawfly, European wheat stem sawfly (U.S.)
Cepola macrophthalma/ Cepola rubescens	Roter Bandfisch	red bandfish
Cepolidae	Bandfische	bandfishes
Cepphus grylle	Gryllteiste	black guillemot
Ceramaster placenta	Fladenstern	placental sea star*
Cerambycidae	Bockkäfer, Böcke	longhorn beetles, long-horned beetles
Cerambyx cerdo	Heldbock, Eichenbock, Großer Eichenbock	great capricorn beetle, oak cerambyx
Cerambyx scopolii	Buchenspießbock, Kleiner Eichenbock	beech capricorn beetle, small oak capricorn beetle
Cerastes cerastes	Hornviper	horned viper, African desert horned viper
Cerastes vipera	Avicennaviper	Sahara sand viper, Avicenna's viper
Cerastoderma edule/ Cardium edule	Essbare Herzmuschel, Gemeine Herzmuschel	common cockle, edible cockle, common European cockle
Cerastoderma glaucum	Lagunen-Herzmuschel	lagoon cockle
Ceratias holboelli	Riesenangler	deep-sea angler
Ceratiidae	Tiefsee-Anglerfische	seadevils
Ceratitis capitata	Mittelmeerfruchtfliege	Mediterranean fruit fly
Ceratobatrachus guentheri	Salomonischer Zipfelfrosch	Gunther's triangle frog
Ceratodontidae	Lurchfische	Australian lungfish
Ceratogyrus darlingi	Afrikanische Horn-Vogelspinne*	African horned tarantula
Ceratophora spp.	Nashornagamen	horned agamas
Ceratophora tennentii	Nashornagame	rhinoceros agamas
Ceratophrys calcarata	Kolumbianischer Hornfrosch	Colombian horned frog

Ceratophrys cornuta	Surinam-Hornfrosch, Gehörnter Hornfrosch	Surinam horned frog
Ceratophrys ornata	Schmuck-Hornfrosch	ornate horned frog, ornate horned toad, escuerzo
Ceratophrys spp.	Südamerikanische Hornfrösche	common horned frogs, South American horned frogs, wide-mouthed toads
Ceratophyllus gallinae	Hühnerfloh, Europäischer Hühnerfloh	European chicken flea, common chicken flea, hen flea
Ceratopogonidae	Gnitzen, Bartmücken	biting midges, punkies, no-see-ums
Ceratostoma foliata	Blattförmige Dornen-Stachelschnecke	leafy hornmouth
Ceratotherium simum	Breitmaulnashorn, Breitlippennashorn, „Weißes Nashorn"	white rhinoceros, square-lipped rhinoceros, grass rhinoceros
Cerberus microlepis	Bockadam	bockadam
Cerberus rhynchops	Hundskopf-Wassertrugnatter	dog-faced water snake, bockadam
Cerceris arenaria	Sandknotenwespe	sand-tailed digger wasp
Cercidia prominens	Erdkreuzspinne	ground spider*
Cercocebus albigena	Mantel-Mangabe	gray-cheeked mangabey, white-cheeked mangabey
Cercocebus aterrimus/ Lophocebus aterrimus	Schopf-Mangabe	black-crested mangabey, black mangabey, crested mangabey
Cercocebus galeritus	Hauben-Mangabe	agile mangabey, Tana mangabey
Cercocebus spp.	Mangaben	mangabeys
Cercocebus torquatus	Halsband-Mangabe	red-crowned mangabey, red-capped mangabey, white-collared mangabey
Cercomela melanura	Schwarzschwanz	blackstart
Cercopidae	Schaumzikaden (Schaumzirpen)	froghoppers, spittlebugs
Cercopis vulnerata	Rotschwarze Schaumzikade, Blutzikade	red-and-black froghopper
Cercopithecidae	Meerkatzenverwandte	Old World monkeys
Cercopithecus aethiops	Grüne Meerkatze	grivet monkey, savanna monkey, green monkey
Cercopithecus albogularis	Weißkehlmeerkatze	Sykes monkey
Cercopithecus ascanius	Rotschwanzmeerkatze	black-cheeked white-nosed monkey, Schmidt's guenon, red-tailed monkey
Cercopithecus campbelli	Campbells Meerkatze	Campbell's monkey
Cercopithecus cephus	Blaumaulmeerkatze	moustached monkey
Cercopithecus denti	Dent-Meerkatze, Dents Meerkatze	Dent's monkey
Cercopithecus diana	Dianameerkatze	Diana monkey, Diana guenon
Cercopithecus dryas	Dryasmeerkatze	dryas monkey
Cercopithecus erythrogaster	Rotbauchmeerkatze	red-bellied guenon, Nigerian white-throat monkey, white-throated monkey

Cercopithecus erythrotis	Rotohrmeerkatze	russet-eared guenon, red-eared monkey, red-eared nose-spotted guenon
Cercopithecus hamlyni	Eulenkopfmeerkatze	owl-faced monkey
Cercopithecus l'hoesti	Vollbartmeerkatze	L'Hoest's monkey
Cercopithecus lowei	Lowes Meerkatze	Lowe's monkey
Cercopithecus mitis	Diademmeerkatze	blue monkey, diademed monkey, gentle monkey
Cercopithecus mona	Monameerkatze	mona monkey
Cercopithecus neglectus	Brazzameerkatze	De Brazza's monkey
Cercopithecus nictitans	Große Weißnasenmeerkatze	greater white-nosed guenon, putty-nosed monkey
Cercopithecus petaurista	Kleine Weißnasenmeerkatze	lesser white-nosed guenon, lesser spot-nosed monkey
Cercopithecus pogonias	Kronenmeerkatze	crowned guenon, crowned monkey
Cercopithecus preussi	Preuß-Bartmeerkatze	Preuss's monkey, Preuss's guenon
Cercopithecus pygerythrus	Vervetmeerkatze, Südafrikanische Grünmeerkatze	vervet monkey
Cercopithecus sabaeus	Gelbgrüne Meerkatze	green monkey, callithrix monkey
Cercopithecus salango	Zaire-Dianameerkatze	Zaire Diana monkey
Cercopithecus sclateri	Nigeria Blaumaulmeerkatze	Sclater's monkey
Cercopithecus solatus	Leuchtschwanz-Meerkatze	sun-tailed monkey
Cercopithecus spp.	Meerkatzen	guenons
Cercopithecus tantalus	Tantalus-Meerkatze	tantalus monkey
Cercopithecus wolfi	Wolfs Meerkatze, Wolf-Meerkatze	Wolf's monkey
Cercosaura ocellata	Augenfleck-Teju	ocellated tegu
Cercotrichas galactotes	Heckensänger	rufous scrub robin, rufous-tailed scrub robin, rufous warbler
Cerdocyon thous/Dusicyon t.	Waldfuchs	crab-eating fox, common zorro
Cerebratulus fuscus (Nemertini)	Brauner Schnurwurm	brown ribbon worm
Cerebratulus marginatus (Nemertini)	Schwarzer Schnurwurm, Schwarze Nemertine	black ribbon worm
Cereus pedunculatus	Seemaßliebchen, Seemannsliebchen, Sonnenrose	daisy anemone
Ceriagrion tenellum	Späte Adonislibelle, Scharlachlibelle	small red damselfly
Ceriantharia	Zylinderrosen	tube anemones, cerianthids
Cerianthopsis americanus	Amerikanische Zylinderrose	sand anemone, burrowing sea anemone, tube-dwelling sea anemone
Cerianthus lloydii	Nordsee-Zylinderrose	North Sea cerianthid
Cerianthus membranaceus	Mittelmeer-Zylinderrose	Mediterranean cerianthid
Cerianthus spp.	Zylinderrosen	cerianthids, cylinder anemones
Ceriodaphnia reticulata	Wabenwasserfloh	honeycomb waterflea*
Cerion lindeni	Pokal-Azurjungfer	

Cerithidea cingulata	Gürtel-Schlammschnecke	girdled hornsnail (girdled horn shell)
Cerithidea costata	Gerippte Schlammschnecke	costate hornsnail
Cerithiidae	Nadelschnecken, Hornschnecken	ceriths, cerithids, hornsnails (horn shells), needle whelks
Cerithium atratum	Dunkle Nadelschnecke	dark cerith, Florida cerith
Cerithium eburneum	Elfenbein-Nadelschnecke	ivory cerith
Cerithium litteratum	Veränderliche Nadelschnecke	stocky cerith
Cerithium lutosum	Atlantische Zwerg-Nadelschnecke	dwarf Atlantic cerith
Cerithium muscarum	Fliegenfleck-Nadelschnecke	flyspeck cerith, fly-specked cerith
Cerithium noduloum	Große Höckrige Nadelschnecke	giant knobbed cerith
Cerithium vulgatum/ Gourmya vulgata	Gemeine Nadelschnecke, Gemeine Hornschnecke, Gemeine Seenadel	European cerith, common cerith
Cernina fluctuata	Gewellte Mondschnecke	wavy moonsnail
Cernuella neglecta	Rotmündige Heideschnecke	red-mouthed banded snail
Cernuella virgata	Gebänderte Heideschnecke, Veränderliche Trockenschnecke	banded snail, striped snail
Cerophytidae	Mulmkäfer	rare click beetles
Cerorhinca monocerata	Nashornalk	rhinoceros auklet, horn-billed puffin
Certhia brachydactyla	Gartenbaumläufer	short-toed treecreeper
Certhia familiaris	Waldbaumläufer	treecreeper
Certhiini	Eigentliche Baumläufer	northern creepers
Ceruchus chrysomelinus	Rindenschröter	
Cerura bifida/ Harpyia bifida/ Harpyia hermelina/ Furcula bifida	Kleiner Gabelschwanz	sallow kitten
Cerura erminea	Hermelinspinner, Weißer Großer Gabelschwanz	
Cerura vinula	Gabelschwanz, Großer Gabelschwanz	puss moth
Cervidae	Hirsche	deer
Cervimunida johni	Gelber Scheinhummer, Südlicher Scheinhummer	yellow squat lobster
Cervus albirostris	Weißlippenhirsch	Thorold's deer
Cervus dama/Dama dama	Damhirsch	fallow deer
Cervus duvauceli	Zackenhirsch, Barasingha, Barasinghahirsch	barasingha, swamp deer
Cervus elaphus	Rothirsch, Edelhirsch	red deer, wapiti, elk (U.S.)
Cervus eldi	Leierhirsch, Thamin	thamin, brow-antlered deer, Eld's deer
Cervus mariannus	Philippinensambar	Philippine sambar
Cervus nippon	Sika-Hirsch, Sikahirsch	sika deer
Cervus schomburgki	Schomburgkhirsch	Schomburgk's deer
Cervus spp.	Edelhirsche	red deers

Cervus timorensis	Mähnenhirsch	Sunda sambar
Cervus unicolor	Indischer Sambar, Pferdehirsch	sambar
Ceryle alcyon	Gürtelfischer	belted kingfisher
Ceryle rudis	Graufischer	pied kingfisher
Cerylidae	Graufischer	cerylid kingfishers
Cerylon histeroides	Gemeiner Glattrindenkäfer	
Cerylonidae	Glattrindenkäfer	minute bark beetles
Cestidea	Venusgürtel	cestids
Cestoda	Bandwürmer, Cestoden	tapeworms, cestodes
Cestus veneris	Venusgürtel	Venus's girdle
Cetacea	Wale & Delphine	cetaceans: whales & porpoises & dolphins
Cetengraulis edentulus	Atlantische Anchoveta	Atlantic anchoveta
Cetengraulis mysticetus	Pazifische Anchoveta	Pacific anchoveta
Cetomimidae	Walköpfe	flabby whalefishes
Cetonia aurata	Rosenkäfer, Goldkäfer	rose chafer
Cetopsidae	Walwelse*	whalelike catfishes
Cetorhinidae	Riesenhaie	basking sharks
Cetorhinus maximus	Riesenhai, Reusenhai	basking shark
Ceutorhynchus assimilis	Kohlschotenrüssler	cabbage seedpod weevil
Ceutorhynchus napi	Großer Kohltriebrüssler, Großer Rapsstengelrüssler	greater cabbage curculio
Ceutorhynchus obstrictus	Kohlschotenrüssler	
Ceutorhynchus picitarsis	Schwarzer Kohltriebrüssler	rape winter stem weevil
Ceutorhynchus pleurostigma	Kohlgallenrüssler	turnip gall weevil, cabbage gall weevil
Ceutorhynchus quadridens	Gefleckter Kohltriebrüssler, Kleiner Kohltriebrüssler	cabbage seedstalk curculio
Chabertia ovina	Großmaul-Stuhlwurm*	large-mouth bowel worm
Chaca chaca	Großkopfwels	squarehead catfish (FAO), tadpole-headed catfish
Chaceon maritae	Rote Tiefseekrabbe	West African geryonid crab, red crab
Chacidae	Großkopfwelse, Großmaulwelse	squarehead catfishes, angler catfishes, frogmouth catfishes
Chaenichthyidae/ Channichthyidae	Eisfische	crocodile icefishes
Chaenocephalus aceratus	Scotia-See-Eisfisch	Scotia Sea icefish, blackfin icefish (FAO)
Chaenodraco wilsoni	Wilsons Eisfisch	Wilson icefish, spiny icefish (FAO)
Chaenopsidae	Hechtschleimfische	pikeblennies, tubeblennies, flagblennies
Chaeropus ecaudatus	Schweinsfuß	pig-footed bandicoot
Chaetocnema concinna	Mangolderdfloh, Rübenerdfloh	mangold flea beetle, mangel flea beetle, beet flea beetle
Chaetoderma nitidulum (Aplacophora)	Gemeiner Schildfuß	glistenworm a.o.
Chaetodermis penicilligera	Fransen-Feilenfisch	prickly leatherjacket, Queensland leafy leatherjacket
Chaetodipterus faber	Spatenfisch	Atlantic spadefish

Chaetodon auriga | Fähnchen-Falterfisch, Faden-Falterfisch | threadfin butterflyfish (FAO), spined butterflyfish
Chaetodon capistratus | Vieraugen-Falterfisch, Pfauenaugen-Falterfisch | foureye butterflyfish
Chaetodon collare | Halsband-Falterfisch | redtail butterflyfish
Chaetodon ephippium | Sattelfleck-Falterfisch | saddleback butterflyfish, saddle butterflyfish (FAO), black-blotched butterflyfish
Chaetodon falcula | Keilfleck-Falterfisch | sickle butterflyfish, blackwedged butterflyfish (FAO)
Chaetodon fasciatus | Rotmeer-Falterfisch, Rotmeer-Schmetterlingsfisch, Tabak-Falterfisch | Red Sea raccoon butterflyfish, diagonal butterflyfish (FAO)
Chaetodon lunula | Mond-Falterfisch | raccoon butterflyfish (FAO), red-striped butterflyfish
Chaetodon melannotus | Schwarzrücken-Falterfisch | striped butterflyfish, blackback butterflyfish (FAO)
Chaetodon meyeri | Meyers Falterfisch, Gebänderter Falterfisch | scrawled butterflyfish (FAO), maypole butterflyfish
Chaetodon ocellatus | Nördlicher Falterfisch | spotfin butterflyfish
Chaetodon oxycephalus | Falscher Gitterfalterfisch | false lined butterflyfish, spot-nape butterflyfish (FAO)
Chaetodon rafflesii | Großschuppen-Falterfisch | Raffles butterflyfish, latticed butterflyfish (FAO)
Chaetodon sedentarius | Riff-Falterfisch | reef butterflyfish
Chaetodon selene | Goldtupfen-Falterfisch | moon butterflyfish, yellowdotted butterflyfish (FAO)
Chaetodon semilarvatus | Halbmasken-Falterfisch | red-lined butterflyfish, bluecheek butterflyfish (FAO)
Chaetodon spp. | Falterfische, Schmetterlingsfische | butterflyfishes
Chaetodon striatus | Gestreifter Falterfisch, Schwarzbinden-Falterfisch | banded butterflyfish
Chaetodon trifasciatus | Rippelstreifen-Falterfisch, Rotsaum-Falterfisch, Dreistreifen-Falterfisch | melon butterflyfish
Chaetodon unimaculatus | Chinesen-Falterfisch, Gelber Tränentropf-Falterfisch | lime-spot butterflyfish, teardrop butterflyfish (FAO)
Chaetodon xanthocephalus | Gelbkopf-Falterfisch | yellowhead butterflyfish
Chaetodontidae | Falterfische, Borstenzähner | butterflyfishes
Chaetognatha | Pfeilwürmer, Borstenkiefer, Chaetognathen | arrow worms, chaetognathans
Chaetomys subspinosus | Borsten-Baumstachler | thin-spined porcupine, bristle-spined porcupine
Chaetopelma aegyptiacum | Ägyptische Vogelspinne | Egyptian tarantula
Chaetophractus spp./ *Euphractus* spp. | Borstengürteltiere | hairy armadillos, peludos
Chaetophractus villosus | Braunzottiges Borstengürteltier, Braunes Borstengürteltier, Braunborsten-Gürteltier | larger hairy armadillo
Chaetopleura apiculata (Polyplacophora) | Gemeine Ostatlantische Käferschnecke, Gerippte Westindische Käferschnecke, Bienenchiton | eastern beaded chiton, bee chiton, common eastern chiton, West Indian ribbed chiton (common American Atlantic coast chiton)
Chaetopleura papilio (Polyplacophora) | Schmetterlingschiton | butterfly chiton

Chaetopoda	Chaetopoden, Borstenwürmer (Polychaeten & Oligochaeten)	chaetopods, bristle worms (annelids with chaetae: polychetes & oligochetes)
Chaetopsylla globiceps	Fuchsfloh	fox flea
Chaetopsylla trichosa	Dachsfloh	badger flea
Chaetopteridae	Pergamentwürmer	parchment tubeworms, chaetopterid worms
Chaetopterus variopedatus	Pergamentwurm	parchment tubeworm, parchment worm
Chaetosiphon fragaefolii	Knotenhaarlaus	strawberry aphid
Chaetura pelagica	Schornsteinsegler	chimney swift
Chalarodon madagascariensis	Madagaskar-Leguan	Madagascar iguana
Chalcalburnus chalcoides	Mairenke, Schemaja	Danubian bleak, Danube bleak (FAO), shemaya
Chalceus erythrurus	Gelbflossen-Großschuppensalmler, Glanzsalmler	yellow-finned chalceus
Chalceus macrolepidatus	Rosaflossen-Großschuppensalmler	pink-tailed chalceus
Chalcides bedriagai	Spanischer Walzenskink	Bedriaga's skink
Chalcides chalcides	Erzschleiche	three-toed skink, Algerian cylindrical skink
Chalcides mauritanicus	Walzenskink	cylindrical skink
Chalcides ocellatus	Gefleckter Walzenskink	ocellated skink
Chalcides spp.	Walzenskinks	cylindrical skinks
Chalcides viridanus	Kanarenskink	Canarian cylindrical skink
Chalcidoidea	Erzwespen	parasitic wasps
Chalcolestes viridis	Weidenjungfer	
Chalcophora mariana	Großer Kiefernprachtkäfer, Marienprachtkäfer	European sculptured pine borer
Chalcopsitta atra	Schwarzlori	black lory
Chalicodoma parietinum/ Chalicoderma muraria/ Megachile parietina	Mörtelbiene	wall bee, mason bee
Chalinolobus morio	Schokoladen-Fledermaus	chocolate bat
Chalinolobus spp.	Australische Schmetterlings-Fledermaus	lobe-lipped bats, groove-lipped bats, wattled bats
Chama brassica	Kohl-Juwelendose	cabbage jewel box
Chama gryphoides	Gemeine Lappenmuschel, Juwelendose	common jewel box
Chama lazarus	Lazarusklappe, Lazarus-Schmuckkästchen, Stachelige Hufmuschel, Stachelige Gienmuschel	Lazarus jewel box
Chamaea fasciata	Chaparraltimalie	wren-tit, wrentit
Chamaeini	Chaparraltimalie	wrentit
Chamaeleo africanus	Basilisken-Chamäleon, Afrikanisches Chamäleon	African chameleon
Chamaeleo bitaeniatus	Zweistreifen-Chamäleon	two-striped chameleon, two-lined chameleon
Chamaeleo calyptratus	Jemen-Chamäleon	Yemen chameleon, cone-headed chameleon, veiled chameleon

Chamaeleo chamaeleon	Europäisches Chamäleon, Gemeines Chamäleon, Gewöhnliches Chamäleon	Mediterranean chameleon, African chameleon, common chameleon
Chamaeleo cristatus	Kammchamäleon	crested chameleon
Chamaeleo dilepis	Lappenchamäleon	flap-necked chameleon, flapneck chameleon
Chamaeleo furcifer	Gabelchamäleon	forked chameleon, forknose chameleon
Chamaeleo jacksoni	Ostafrikanisches Dreihornchamäleon	Jackson's chameleon, Jackson's three-horned chameleon
Chamaeleo lateralis	Teppichchamäleon	jewelled chameleon
Chamaeleo melleri	Mellers Chamäleon	Meller's chameleon
Chamaeleo montium	Bergchamäleon	Cameroon sailfin chameleon, sail-finned chameleon, mountain chameleon
Chamaeleo namaquensis	Wüstenchamäleon	desert chameleon, Namaqua chameleon
Chamaeleo oustaleti	Riesenchamäleon	Oustalet's giant chameleon, Oustalet's chameleon
Chamaeleo oweni	Owens Chamäleon	Owen's three-horned chameleon, Owen's chameleon
Chamaeleo pardalis	Pantherchamäleon	panther chameleon
Chamaeleo parsoni	Parsons Chamäleon	Parson's chameleon
Chamaeleo pumilis	Buntes Zwergchamäleon	variegated dwarf chameleon
Chamaeleo quadricornis	Vierhornchamäleon	four-horned chameleon
Chamaeleo tenuis	Weichhornchamäleon	slender chameleon
Chamaeleo tigris	Tigerchamäleon	tiger chameleon
Chamaeleo verrucosus	Warzenchamäleon	giant chameleon, warty chameleon
Chamaeleonidae	Chamäleons	chameleons
Chamaemyiidae/ Ochthiphilidae	Blattlausfliegen	aphid flies
Chamaesaura spp.	Schlangen-Gürtelechsen	sweepslangs
Chamelea gallina/ Venus gallina/ Chione gallina	Gemeine Venusmuschel, Strahlige Venusmuschel	striped venus, chicken venus
Chamidae	Gienmuscheln, Juwelendosen	jewel boxes, jewelboxes
Champsocephalus gunnari	Bändereisfisch	Antarctic icefish, mackerel icefish (FAO)
Champsodontidae	Großmäuler*	gapers
Chanda ranga	Indischer Glasbarsch	Indian glassfish
Chandidae/Ambassidae	Asiatische Glasbarsche	Asiatic glassfishes
Chanidae	Milchfische	milkfish
Channa micropeltes	Indonesischer Schlangenkopffisch	Indonesian snakehead
Channa spp.	Schlangenkopffische	snakeheads
Channa striatus	Gestreifter Schlangenkopffisch	striped snakehead
Channichthyidae/ Chaenichthyidae	Eisfische	crocodile icefishes
Channichthys rhinoceratus	Langschnauzen-Eisfisch	long-snouted icefish, unicorn icefish (FAO)
Channidae/Ophiocephalidae	Schlangenköpfe, Schlangenkopffische	snakeheads

Characidae	Salmler (Echte Amerikanische Salmler)	characins
Characidiinae	Grundsalmler, Bodensalmler	South American darters
Characiformes	Salmler	characins: tetras & piranhas
Charadriidae	Regenpfeifer	plovers, lapwings
Charadriiformes	Möwenvögel & Watvögel & Alken	gulls & shorebirds & auks
Charadrius alexandrinus	Seeregenpfeifer	Kentish plover
Charadrius asiaticus	Wermutregenpfeifer	Caspian plover
Charadrius dubius	Flussregenpfeifer	little ringed plover
Charadrius hiaticula	Sandregenpfeifer	ringed plover
Charadrius leschenaultii	Wüstenregenpfeifer	greater sand plover
Charadrius mongolus	Mongolenregenpfeifer	lesser sand plover
Charadrius morinellus	Mornellregenpfeifer	dotterel
Charadrius vociferus	Keilschwanz-Regenpfeifer	killdeer
Charaxes jasius	Erdbeerbaumfalter	two-tailed pasha
Charina bottae	Gummiboa, Nordamerikanische Sandboa	rubber boa
Charina spp.	Zwergboas	rubber boas
Charina trivirgata/ Lichanura trivirgata	Rosenboa, Zwergboa, Dreistreifen-Rosenboa	rosy boa
Charmosyna josefinae	Josefinenlori	Josephine's lory
Charonia lampas/ Tritonium nodiferum	Trompetenschnecke, Kinkhorn, Knotiges Tritonshorn	knobbed triton
Charonia tritonis	Pazifisches Tritonshorn, Echtes Tritonshorn	trumpet triton, Pacific triton, giant triton, triton's trumpet
Charonia variegata	Atlantisches Tritonshorn	Atlantic triton
Charopus flavipes	Schwarzer Zipfelkäfer	
Charybdis feriatus	Kreuzkrabbe	striped swimming crab
Chasiempis sandwichensis	Elepaio	elepaio
Chasmodes bosquianus	Streifen-Schleimfisch*	striped blenny
Chauliodontidae	Viperfische, Vipernfische	viperfishes
Chaunacidae	Chaunaciden	coffinfishes, sea toads
Chazara briseis	Berghexe, Steppenpförtner	the hermit
Cheilea equestris	Falsche Mützenschnecke*	false cup-and-saucer limpet
Cheilochromis euchilus	Sauglippen-Buntbarsch	euchilus
Cheilodactylidae	Morwongs	fingerfins, morwongs
Cheilodactylus bergi	Bergs Morwong	castaneta
Cheilodactylus variegatus	Prächtiger Morwong	pintadilla
Cheilodipterus lineatus lineatus	Gestreifter Kardinalfisch	tiger cardinalfish
Cheilopogon agoo/ Cypselurus agoo	Japanischer Flugfisch, Japanischer Fliegender Fisch	Japanese flyingfish
Cheilopogon cyanopterus/ Cypselurus cyanopterus	Geränderter Flugfisch	margined flyingfish
Cheilopogon exsiliens/ Cypselurus exsiliens/	Gestreiftflügel-Flugfisch	bandwing flyingfish
Cheilopogon heterurus/ Cypselurus heterurus	Vierflügel-Flugfisch (Fliegender Fisch)	Mediterranean flyingfish
Cheilopogon melanurus/ Cypselurus melanurus	Atlantischer Flugfisch, Atlantischer Fliegender Fisch	Atlantic flyingfish

Cheilosia pictipennis	Flügelfleck-Erzschwebfliege	
Cheilostomata	Lippenmünder, Lappenmünder	cheilostomates
Cheimerius nufar	Nufar-Seebrasse	soldier, santer seabream (FAO)
Cheiracanthium mordax		pale leaf spider
Cheiracanthium punctorium	Dornfinger, Ammen-Dornfinger	European sac spider
Cheirodon axelrodi/ Paracheirodon axelrodi	Roter Neon, „Falscher" Neon	cardinal tetra
Cheirogaleidae	Katzenmakis	mouse lemurs
Cheirogaleus major	Großer Katzenmaki	greater dwarf lemur
Cheirogaleus medius	Mittlerer Katzenmaki, Fettschwanzmaki	lesser dwarf lemur, fat-tailed dwarf lemur
Cheirogaleus spp.	Echte Katzenmakis	dwarf lemurs
Cheiromeles spp.	Nacktfledermäuse	naked bats, hairless bats
Chelicerata	Cheliceraten, Chelizeraten, Scherenhörnler	chelicerates
Chelidae	Schlangenhalsschildkröten	Austro-American side-necked turtles, snake-necked turtles
Chelidonichthys capensis	Kap-Knurrhahn	Cape gurnard
Chelidonichthys kumu	Blauflossen-Knurrhahn	blue-fin gurnard
Chelidoptera tenebrosa	Schwalbenfaulvogel	swallow-wing puffbird
Chelidurella acanthopygia	Waldohrwurm	forest earwig
Chelifer cancroides	Bücherskorpion	house pseudoscorpion
Chelisochidae	Schwarze Ohrwürmer	black earwigs
Chelmon rostratus	Kupfer-Pinzettfisch	copper-banded butterflyfish, copperband butterflyfish (FAO), long-nosed butterflyfish, beaked coralfish
Chelodina expansa	Riesen-Schlangenhalsschildkröte	giant snake-necked turtle, broad-shell snakeneck, broad-shelled tortoise
Chelodina longicollis	Glattrückige Schlangenhalsschildkröte, Glattrücken-Schildkröte	common snake-necked turtle, common snakeneck, long-necked tortoise
Chelodina novaeguineae	Neuguinea-Schlangenhalsschildkröte	New Guinea snake-necked turtle
Chelodina oblonga	Schmalbrust-Schlangenhalsschildkröte	oblong tortoise, narrow-breasted snake-necked turtle
Chelodina parkeri	Parkers Schlangenhalsschildkröte	Parker's side-necked turtle
Chelodina rugosa	Nordaustralische Schlangenhalsschildkröte	northern snakeneck, northern Australian snake-necked turtle, northern long-necked tortoise
Chelodina siebenrocki	Siebenrock-Schlangenhals-schildkröte, Siebenrocks Schlangenhalsschildkröte	Siebenrock's snake-necked turtle
Chelodina spp.	Australische Schlangenhalsschildkröten	Australian snake-necked turtles
Chelodina steindachneri	Westaustralische Schlangenhalsschildkröte	dinner-plate turtle, Steindachner's side-necked turtle
Chelon labrosus/ Mugil chelo/ Mugil provensalis	Dicklippige Meeräsche	thick-lipped grey mullet, thicklip grey mullet (FAO)
Chelonariidae	Schildkrötenkäfer	turtle beetles

Chelonia depressa	Australische Plattschildkröte, Australische Suppenschildkröte	flatback sea turtle, flatback turtle, Australian green turtle
Chelonia mydas	Suppenschildkröte	green turtle (rock turtle, meat turtle)
Chelonibia testudinaria	Schildkröten-Seepocke, Schildkrötenpocke, Schildseepocke	turtle barnacle
Cheloniidae	Meeresschildkröten	sea turtles
Chelonoidis carbonaria/ Testudo carbonaria/ Geochelone carbonaria	Köhlerschildkröte	red-footed tortoise, coal tortoise
Chelonoidis chilensis/ Testudo chilensis/ Geochelone chilensis	Argentinische Landschildkröte	Chilean tortoise, Chaco tortoise, Argentine tortoise (IUCN)
Chelonoidis denticulata/ Testudo denticulata/ Geochelone denticulata	Waldschildkröte	yellow-footed tortoise, Brazilian giant tortoise, American forest tortoise, South American yellow-footed tortoise (IUCN), Hercules tortoise
Chelonoidis spp.	Südamerikanische Landschildkröten	South American tortoises
Chelosania brunnea	Chamäleon-Agame	chameleon dragon
Chelura terebrans	Holz-Flohkrebs, Scherenschwanz	wood-boring amphipod
Chelus fimbriatus	Fransenschildkröte, Matamata	matamata
Chelydra serpentina	Schnappschildkröte	snapping turtle, American snapping turtle
Chelydridae	Alligatorschildkröten	snapping turtles
Chelyletidae	Raubmilben	quill mites, predatory mites
Cheramoeca leucosterna	Weißrückenschwalbe	white-backed swallow
Cherax destructor	Kleiner Australkrebs, Yabbie	yabbie
Cherax tenuimanus	Großer Australkrebs, Marron	marron
Chernetidae	Chernetiden	chernetids
Chersina angulata	Afrikanische Schnabelbrustschildkröte	South African bowsprit tortoise
Chersophilus duponti	Dupontlerche	Dupont's lark
Chersydrus granulatus/ Acrochordus granulatus	Indische Warzenschlange, Kleine Warzenschlange	Indian wart snake
Chesias rufata	Ginsterheiden-Silberstreifenspanner	broom-tip
Chettusia gregaria	Steppenkiebitz	sociable plover
Chettusia leucura	Weißschwanzkiebitz	white-tailed plover
Cheyletiella parasitivorax	Kaninchenpelzmilbe	rabbit fur mite a.o.
Cheyletiella yasguri	Hundepelzmilbe	dog fur mite
Cheyletiellidae	Pelzmilben	fur mites
Chiasmia clathrata	Gitterspanner	
Chiasmodontidae	Schwarze Schlinger	swallowers
Chicomurex superbus	Fantastische Stachelschnecke, Superbe Stachelschnecke	superb murex
Chicoreus asianus	Japanische Stachelschnecke	Asian murex

Chicoreus cornucervi	Hirschgeweih-Stachelschnecke	monodon murex
Chicoreus dilectus		lace murex
Chicoreus nobilis	Edle Stachelschnecke	noble murex
Chicoreus palmarosae	Rosenzweigschnecke, Rosenzweig-Stachelschnecke	rose-branch murex, rose-branched murex
Chicoreus pomum/ Phyllonotus pomum	Apfel-Stachelschnecke	apple murex
Chicoreus ramosus	Riesenstachelschnecke, Riesen-Murex	giant murex, ramose murex, branched murex
Chicoreus saulii	Sauls Stachelschnecke	Saul's murex
Chicoreus spectrum	Gespenst-Stachelschnecke	ghost murex, spectre murex
Chicoreus torrefactus	Sengfarbenschnecke	scorched murex
Chicoreus venustulus	Liebliche Stachelschnecke	lovely murex
Chicoreus virgineus	Jungfräuliche Stachelschnecke	virgin murex
Chilo suppressalis/ Chilo simplex	Gestreifter Reisstengelbohrer	Asiatic rice borer, striped riceborer
Chilocorus spp.	Schwarze Marienkäfer*	black ladybird beetles
Chilodontidae	Kopfsteher	headstanders
Chilodus punctatus punctatus	Punktierter Kopfsteher	spotted headstander
Chilomeniscus cinctus	Gebänderte Sandschlange	banded sand snake
Chilomeniscus spp.	Gebänderte Sandschlangen	banded sand snakes
Chilomys instans	Kolumbianische Waldmaus	Colombian forest mouse
Chilopoda	Hundertfüßer, Chilopoden	centipedes, chilopodians
Chiloscyllium griseum	Grauer Bambushai	grey carpet shark, grey bambooshark
Chiloscyllium indicum	Indischer Bambushai	Indian carpet shark, slender bambooshark (FAO)
Chiloscyllium punctatum	Gebänderter Bambushai, Brauner Lippenhai, Sattelfleckenhai	brown carpet shark, brownbanded bambooshark (FAO)
Chilostoma planospirum/ Campylaea planospirum	Flache Felsenschnecke	
Chimabacche fagella	Buchenmotte, Sängerin	beech moth*
Chimaera monstrosa	Seeratte, Seekatze, Spöke	rabbit fish (FAO), European ratfish, chimaera
Chimaeridae	Seekatzen	shortnose chimaeras, ratfishes
Chimarrogale spp.	Asiatische Wasserspitzmäuse	Asiatic water shrews
Chinchilla brevicaudata	Kurzschwanz-Chinchilla	short-tailed chinchilla
Chinchilla lanigera	Langschwanz-Chinchilla	long-tailed chinchilla
Chinchilla spp.	Chinchillas	chinchillas
Chinchillula sahamae	Chinchilla-Maus	Altiplano chinchilla mouse
Chinemys kwangtungensis	Rothalsschildkröte	Chinese red-necked pond turtle
Chinemys megalocephala	Chinesische Dickkopfschildkröte	Chinese broad-headed pond turtle
Chinemys reevesii	Chinesische Dreikielschildkröte	Reeves' turtle, Chinese three-keeled pond turtle
Chioglossa lusitanica	Goldstreifensalamander	golden-striped salamander, gold-striped salamander
Chionactis occipitalis	Schaufelnasennatter	western shovelnose snake

Chionactis palarostris	Sonora-Schaufelnasennatter	Sonoran shovelnose snake
Chionactis spp.	Schaufelnasennattern	shovelnose snakes
Chionaspis salicis	Weidenschildlaus	willow scale
Chione cancellata	Querstreifige Venusmuschel*	cross-barred venus
Chione ovata/Timoclea ovata	Ovale Venusmuschel	oval venus
Chione subimbricata	Stufige Venusmuschel	stepped venus
Chionididae	Scheidenschnäbel	sheathbills
Chionodraco rastrospinosus	Augenfleck-Eisfisch	Kathleen's icefish, ocellated icefish (FAO)
Chionoecetes opilio	Schneekrabbe, Nordische Eismeerkrabbe, Arktische Seespinne	Atlantic snow spider crab, Atlantic snow crab, queen crab
Chionoecetes spp.	Schneekrabben	snow crabs
Chiridota laevis	Seidige Seegurke*	silky sea cucumber
Chirocentridae	Wolfsheringe	wolf-herrings, wolf herrings
Chirocentrus dorab	Großer Wolfshering, Indischer Hering	wolf-herring, dorab wolf-herring (FAO)
Chirocephalus grubei	Frühjahrskiemenfuß	springtime fairy shrimp*
Chirocephalus spp.	Kiemenfußkrebse	fairy shrimps a.o.
Chiroderma spp.	Großaugenfledermaus*	big-eyed bats, white-lined bats
Chirolophis ascanii	Stachelrücken-Schleimfisch	Yarrell's blenny
Chiromantis rufescens	Rauhäutiger Baumfrosch	African foam-nest treefrog
Chiromantis spp.	Baumfrösche, Afrikanische Greiffrösche	foam-nest treefrogs
Chiromantis xerampelina	Grauer Baumfrosch	African gray treefrog
Chiromyscus chiropus	Feas Baumratte	Fea's tree rat
Chironax melanocephalus	Schwarzköpfiger Flughund	black-capped fruit bat
Chironectes minimus	Schwimmbeutler, Yapok (Hbz. Lampara de agua = Lampenratte)	water opossum, yapok
Chironemidae	Seetangfische*	kelpfishes
Chironex fleckeri	Australische Würfelqualle, Seewespe u.a.	Australian box jellyfish, box jelly, deadly sea wasp
Chironius bicarinatus	Doppelkopf*	two-headed sipo
Chironius carinatus	Sipo	sipo
Chironius exoletus	Linnés Baumnatter	Linnaeus' sipo
Chironius scurulus	Grünkopf-Baumnatter	Wagler's sipo
Chironius spp.	Sipos	sipos
Chironomidae	Zuckmücken, Schwarmmücken (Tanzmücken)	nonbiting midges and gnats
Chiropodomys spp.	Pinselschwanz-Baummäuse	pencil-tailed tree mice
Chiropotes albinasus	Weißnasensaki	white-nosed saki
Chiropotes satans	Satansaffe	red-backed saki
Chiropsalmus quadrigatus	Seewespe u.a.	sea wasp a.o.
Chiroptera	Fledermäuse, Flattertiere	bats, chiropterans
Chiropterotriton magnipes	Großfuß-Schwielensalamander	bigfoot splayfoot salamander
Chiropterotriton spp.	Schwielensalamander	splayfoot salamanders
Chiroteuthis veranyi	Anglerkalmar	
Chiruromys forbesi	Große Baumratte	greater tree mouse
Chiruromys lamia	Breitkopf-Baumratte	broad-skulled tree mouse

Chiruromys spp.	Baumratten	tree mice
Chiruromys vates	Kleine Baumratte	lesser tree mouse
Chitala chitala/ *Notopterus chitala*	Indischer Fähnchen-Messerfisch, Gebänderter Messerfisch	clown knifefish
Chiton articulatus (Polyplacophora)	Glatte Panama-Käferschnecke	smooth Panama chiton
Chiton marmoratus (Polyplacophora)	Marmor-Käferschnecke, Marmorchiton	marbled chiton
Chiton olivaceus (Polyplacophora)	Bunte Käferschnecke, Mittelmeer-Chiton	variable chiton, Mediterranean chiton
Chiton squamosus (Polyplacophora)	Schuppige Käferschnecke	squamose chiton
Chiton striatus (Polyplacophora)	Prachtchiton	striate chiton
Chiton tuberculatus (Polyplacophora)	Grüne Westindische Käferschnecke, Grüne Westindische Chiton	West Indian green chiton, common West Indian chiton
Chiton tulipa (Polyplacophora)	Tulpen-Käferschnecke	tulip chiton
Chitra chitra	Gestreifte Kurzkopf-Weichschildkröte	striped narrow-headed softshell turtle
Chitra indica	Kurzkopf-Weichschildkröte	narrow-headed softshell turtle
Chlaenius nigricornis	Schwarzfühler-Grünkäfer	
Chlaenius vestitus	Gelbrand-Grünkäfer, Gelbgeranderter Grünkäfer	
Chlamydera nuchalis	Graulaubenvogel	great grey bowerbird
Chlamydogobius eremius	Wüstengrundel	desert goby (Austr.)
Chlamydosaurus kingi	Kragenechse	frilled lizard, frill-necked lizard, Australian frilled lizard
Chlamydoselachidae	Krausenhaie, Kragenhaie	frill sharks, frilled sharks
Chlamydoselachus anguineus	Krausenhai, Kragenhai	frill shark, frilled shark (FAO)
Chlamydotis undulata	Kragentrappe	Houbara bustard
Chlamyphorus retusa/ *Burmeisteria retusa*	Nördlicher Gürtelmull	Chacoan fairy armadillo, greater fairy armadillo, Burmeister's armadillo
Chlamyphorus spp.	Gürtelmulle	pichiciegos
Chlamyphorus truncatus	Gürtelmull, Gürtelmaus, Schildwurf	pink fairy armadillo, lesser fairy armadillo, lesser pichi ciego, pichiciego
Chlamys asperrima	Stachlige Kammmuschel	prickly scallop
Chlamys australis	Südliche Kammmuschel	austral scallop
Chlamys delicatula	Dünnschalige Kammmuschel	delicate scallop
Chlamys dianae	Diana Kammmuschel	Diana's scallop
Chlamys distorta	Höckerige Kammmuschel	hunchback scallop
Chlamys islandica	Isländische Kammmuschel, Island-Kammmuschel	Iceland scallop
Chlamys luculenta	Weißstreifen-Kammmuschel	white-streaked scallop
Chlamys senatoria/ *Chlamys nobilis*	Feine Kammmuschel, Edle Kammmuschel	senate scallop, noble scallop
Chlamys spp.	Kammmuscheln, Kamm-Muscheln	scallops
Chlamys squamosa	Schuppige Kammmuschel	squamose scallop

Chlamys tigerina	Tiger-Kammmuschel	tiger scallop
Chlamys tincta	Färber-Kammmuschel	tinted scallop
Chlamys varia	Bunte Kammmuschel	variegated scallop
Chlidonias hybridus	Weißbart-Seeschwalbe	whiskered tern
Chlidonias leucopterus	Weißflügel-Seeschwalbe	white-winged black tern
Chlidonias niger	Trauerseeschwalbe	black tern
Chloebia gouldiae	Gouldamadine	Gouldian finch, Lady Gouldian finch
Chloephaga picta	Magellangans	Magellan goose
Chlopsidae/Xenocongridae	Unechte Muränen	false morays, false moray eels
Chlopsis bicolor	Falsche Muräne	bicoloured false moray
Chloroclysta truncata	Heckenkirschenspanner	common marbled carpet
Chloroclystis rectangulata	Obstgartenspanner	green pug
Chlorohydra viridissima/ Hydra viridis	Grüne Hydra, Grüner Süßwasserpolyp	green hydra
Chloroperlidae	Grüne Steinfliegen	green stoneflies
Chlorophidia	Ahnenschlangen	
Chlorophis spp.	Grünnattern	green snakes
Chlorophorus annularis/ Rhaphuma annularis	Bambusbohrer	bamboo borer, yellow bamboo longhorn beetle
Chlorophthalmidae	Grünaugen	greeneyes
Chlorophthalmus agassizi	Grünauge	green-eye, shortnose greeneye (FAO)
Chloropidae	Halmfliegen, Gelbkopffliegen	chloropids, chloropid flies (eye gnats, grass flies, eye flies)
Chlorops pumilionis	Gelbe Halmfliege, Gelbe Weizenhalmfliege	gout fly
Chloroscombrus chrysurus	Schwanzfleck-Stachelmakrele, Atlantische Goldmakrele*	Atlantic bumper
Choeroniscus spp.	Godmans Langnasen-Fledermaus	Godman's long-nosed bats
Choeronycteris mexicana	Langnasenfledermaus	Mexican long-nosed bat, hog-nosed bat
Choeropsis liberiensis/ Hexaprotodon liberiensis	Zwergflusspferd	pygmy hippopotamus
Choloepus spp.	Zweifinger-Faultiere	two-toed sloths
Chologaster cornutus	Trugkärpfling	swampfish
Chondrichthyes	Knorpelfische	cartilaginous fishes, chondrichthians
Chondrina avenacea	Haferkornschnecke	
Chondrina clienta	Feingerippte Haferkornschnecke	
Chondrinidae	Kornschnecken	
Chondrodactylus angulifer	Sandgecko	whorled sand gecko
Chondropython spp.	Baumpythons	tree pythons
Chondrosia reniformis	Nierenschwamm, Lederschwamm	chicken liver sponge, kidney sponge
Chondrostei	Altfische	primitive ray-finned bony fishes
Chondrostoma genei	Lau	South European nase, casca (*ital.*)
Chondrostoma nasus	Nase	nase
Chondrostoma phoxinus	Elritzennäsling	minnow nase
Chondrostoma toxostoma	Savetta, Strömer, Südwesteuropäischer Näsling	soiffe (*Fr.*), soffie (*Fr.*)

Chordata	Chordatiere, Chordaten, Rückgrattiere	chordates
Chordeiles acutipennis	Texas-Nachtschwalbe	lesser nighthawk
Chordeiles minor	Falkennachtschwalbe	common nighthawk
Chordeilinae	Falkennachtschwalben	nighthawks
Chordeumatidae	Samenfüßer	chordeumatid millepedes
Choreutidae	Rundstirnmotten	choreutid moths (skeletonizer moths)
Choriaster spp.	Nadelkissensterne u.a.	pincushion starfish a.o.
Chorioptes bovis	Ochsenschwanz-Räudemilbe*, Pferdefuß-Räudemilbe*	chorioptic mange mite, horse foot mange mite, oxtail mange mite, symbiotic mange mite
Choristoneura fumiferana	Nordamerikanischer Fichtentriebwickler	spruce budworm (tortrix)
Choromytilus chorus	Chormuschel	chorus mussel
Chorosoma schillingi	Grasgespenst	
Chorthippus albomarginatus	Weißrandiger Grashüpfer	lesser marsh grasshopper
Chorthippus alticola	Höhengrashüpfer	eastern Alpine grasshopper
Chorthippus apricarius	Feld-Grashüpfer	upland field grasshopper
Chorthippus biguttulus	Nachtigall-Grashüpfer	bow-winged grasshopper
Chorthippus binotatus	Zweifleckiger Grashüpfer	twospotted grasshopper
Chorthippus brunneus	Brauner Grashüpfer, Feldheuschrecke	field grasshopper, common field grasshopper
Chorthippus cialancensis	Cialancia-Grashüpfer	Piedmont grasshopper
Chorthippus dorsatus	Wiesen-Grashüpfer	meadow grasshopper
Chorthippus mollis	Verkannter Grashüpfer	lesser field grasshopper
Chorthippus montanus	Sumpf-Grashüpfer	marsh-meadow grasshopper
Chorthippus parallelus	Gemeiner Grashüpfer	common meadow grasshopper
Chorthippus pullus	Kiesbank-Grashüpfer	gravel grasshopper
Chorthippus spp.	Grashüpfer, Heuhüpfer	field grasshoppers, meadow grasshoppers
Chorthippus vagans	Steppen-Grashüpfer	dryland grasshopper (heath grasshopper)
Chriopeops goodei/ Lucania goodei	Rotschwanzkärpfling	bluefin killifish (FAO), blue-fin killy
Christinus marmoratus/ Phyllodactylus marmoratus	Marmorierter Blattfingergecko	marbled gecko, Australian marbled gecko (Australian clawed gecko)
Chromaphis juglandicola	Kleine Walnussblattlaus	small walnut aphid
Chromidotilapia guentheri	Günthers Prachtbuntbarsch	Guenther's mouthbrooder
Chromis caerulea	Grünes Schwalbenschwänzchen, Grüner Schwalbenschwanz, Grünling	blue puller, green chromis (FAO)
Chromis chromis	Mönchsfisch	damsel fish, damselfish (FAO)
Chromis cyanea	Blauer Hochseebarsch	blue chromis
Chromis dimidiata	Zweifarben-Demoiselle	bicolor damselfish, half-and-half chromis, chocolatedip chromis (FAO)
Chromogobius quadrivittatus	Vierband-Grundel*	banded goby
Chrotogale owstoni	Fleckenroller	Owston's palm civet
Chrotomys whiteheadi	Luzon Streifenratte	Luzon striped rat
Chrotopterus auritus	Peters Wollgesicht-Fledermaus	Peter's woolly false vampire bat

Chrysaora hysoscella	Kompassqualle	compass jellyfish, red-banded jellyfish
Chrysaora melanaster	Braune Nesselqualle, Pazifische Braune Nesselqualle	brown jellyfish, brown giant jellyfish
Chrysaora quinquecirrha	Seenessel	sea nettle
Chrysaora spp.	Nesselquallen	sea nettles
Chrysemys nelsoni/ Pseudemys rubriventris nelsoni	Florida-Zierschildkröte	Florida redbelly turtle
Chrysemys picta picta	Zierschildkröte	painted turtle, eastern painted turtle
Chrysemys see Pseudemys		
Chrysididae	Goldwespen	cuckoo wasps
Chrysis cuprea	Kupfer-Goldwespe	copper wasp
Chrysis ignita	Feuer-Goldwespe, Feuergoldwespe, Gemeine Goldwespe	common gold wasp, ruby-tail, ruby-tailed wasp
Chrysoblephus gibbiceps	Rote Stumpfnasenbrasse	red stumpnose
Chrysobothris affinis	Goldgruben-Eichenprachtkäfer	
Chrysobothris femorata	Flachkopf-Apfelbaum-Bohrwurm (Prachtkäfer)	flatheaded appletree borer
Chrysochloridae	Goldmulle	golden moles
Chrysochloris asiatica	Kapgoldmull	Cape golden mole
Chrysochraon brachypterus/ Euthystira brachyptera	Kleine Goldschrecke	small gold grasshopper
Chrysochraon dispar	Große Goldschrecke	large gold grasshopper
Chrysococcyx caprius	Goldkuckuck	dudric cuckoo
Chrysocolaptes lucidus	Sultanspecht	greater flame-backed woodpecker
Chrysocyon brachyurus	Mähnenwolf	maned wolf
Chrysogaster solstitialis	Gemeine Goldbauch-Schwebfliege	
Chrysolina cerealis	Regenbogen-Blattkäfer	rainbow leaf beetle
Chrysolina menthastri	Minzenblattkäfer	mint leaf beetle
Chrysolophus amherstiae	Diamantfasan	Lady Amherst's pheasant
Chrysolophus pictus	Goldfasan	golden pheasant
Chrysomela populi/ Melasoma populi	Pappelblattkäfer, Roter Pappelblattkäfer	red poplar leaf-beetle, poplar leaf beetle, poplar beetle
Chrysomela tremulae/ Melasoma tremulae	Aspenblattkäfer	aspen leaf beetle
Chrysomelidae	Blattkäfer	leaf beetles
Chrysomya albiceps	Weißkopf-Dasselfliege*	banded blow fly
Chrysomya bezziana	Schraubenwurmfliege, Altwelt-Schraubenwurmfliege	screwworm, Old World screw worm
Chrysomya megacephala	Orient-Latrinenfliege	Oriental latrine fly
Chrysomya rufifacies	Behaarte Maden-Dasselfliege*	hairy maggot blow fly
Chrysopa carnea/ Anisochrysa carnea	Gemeine Florfliege	common green lacewing
Chrysopa perla	Perlige Florfliege*	pearly green lacewing
Chrysopelea ornata	Goldschlange	ornate flying snake
Chrysopelea paradisi	Paradies-Baumschlangen	paradise tree snake
Chrysopelea spp.	Schmuck-Baumschlangen, Schmuckbaumnattern	flying snakes, Asian parrot snakes

Chrysopidae	Florfliegen (Goldaugen, Stinkfliegen, Blattlauslöwen)	green lacewings
Chrysops caecutiens	Blindfliege	blinding breeze fly
Chrysops spp.	Blindfliegen	deerflies
Chrysoteuchia culmella		garden grass veneer moth, grass moth
Chrysotoxum bicinctum	Zweiband-Wespen-Schwebfliege	
Chrysotoxum cautum	Gemeine Wespen-Schwebfliege	
Chthamalus stellatus	Sternseepocke	star barnacle
Chydorus sphaericus	Linsenfloh	
Cicada orni	Mannazikade, Eschenzikade	
Cicadellidae	Zwergzikaden	leaf hoppers
Cicadetta montana	Bergzikade	New Forest cicada
Cicadidae	Singzikaden (Singzirpen)	cicadas
Cichla ocellaris	Augenfleck-Kammbarsch	peacock bass (FAO), peacock cichlid
Cichlasoma arnoldi	Schwarzer Buntbarsch	black cichlid
Cichlasoma bimaculatum	Zweifleck-Buntbarsch	black acara
Cichlasoma biocellatum/ Cichlasoma octofasciatum	Schwarzgebänderter Buntbarsch	Jack Dempsey
Cichlasoma facetum	Chanchito	chanchito, chameleon cichlid (FAO)
Cichlasoma meeki	Feuermaul-Buntbarsch	firemouth cichlid
Cichlasoma nigrofasciatum	Zebra-Buntbarsch, Grünflossen-Buntbarsch	zebra cichlid, convict cichlid (FAO)
Cichlasoma severum	Augenfleck-Buntbarsch	banded cichlid (FAO), convict fish, deacon, severum, striped cichlid
Cichlidae	Buntbarsche	cichlids
Cicindela arenaria	Südöstlicher Sandlaufkäfer*	southeastern European tiger beetle*
Cicindela campestris	Feld-Sandlaufkäfer	green tiger beetle
Cicindela germanica	Deutscher Sandlaufkäfer	German tiger beetle
Cicindela hybrida	Dünen-Sandlaufkäfer, Brauner Sandlaufkäfer, Kupferbrauner Sandlaufkäfer	dune tiger beetle
Cicindela marginipennis	Pflasterstein-Sandlaufkäfer	cobblestone tiger beetle
Cicindela maritima	Küsten-Sandlaufkäfer	coastal tiger beetle
Cicindela silvatica	Wald-Sandlaufkäfer	wood tiger beetle
Cicindelidae	Sandlaufkäfer (Tigerkäfer)	tiger beetles
Cicinnurus regius	Königsparadiesvogel	king bird-of-paradise
Ciconia ciconia	Weißstorch	white stork
Ciconia nigra	Schwarzstorch	black stork
Ciconiidae	Störche	storks
Ciconiiformes	Schreitvögel, Stelzvögel	herons & storks & ibises & allies
Cidaridae	Lanzenseeigel	cidarids
Cidaris cidaris	Grauer Lanzenseeigel (Kaiserigel)	piper, king of the sea-eggs
Cidaroida	Lanzenseeigel	cidaroids
Ciidae/Cisidae	Schwammkäfer, Baumschwammkäfer	minute tree-fungus beetles
Ciliata	Wimpertierchen, Ciliaten	ciliates

Ciliata mustela	Fünfbärtelige Seequappe	five-bearded rockling
Ciliata septentrionalis	Nordische Seequappe	northern rockling
Cilix glauca	Silberspinner, Silberspinnerchen	Chinese character
Cimbex femoratus	Große Birkenblattwespe, Birkenknopfhorn-Blattwespe	birch sawfly
Cimbicidae	Knopfhornblattwespen	cimbicids
Cimex colombarius	Taubenwanze, Geflügelwanze	pigeon bug, fowl bug
Cimex hemipterus	Tropische Bettwanze	tropical bedbug
Cimex lectularius	Gemeine Bettwanze	bedbug, common bedbug (wall-louse)
Cimex pipistrelli	Fledermauswanze	bat bug
Cimicidae	Plattwanzen (Hauswanzen)	bedbugs
Cinara fresai	Amerikanische Wacholderrindenlaus	American juniper aphid
Cinara kochiana	Große Lärchenrindenlaus	giant larch aphid
Cinara laricis	Gefleckte Lärchenrindenlaus	spotted larch aphid
Cinara pilicornis	Braune Tannenrindenlaus	brown spruce aphid, spruce shoot aphid
Cinara pinea	Große Kiefernrindenlaus	large pine aphid
Cinclidae	Wasseramseln	dippers
Cinclocerthia ruficauda	Zitterdrossel	brown trembler
Cinclosomatinae	Flöter	quail-thrushes, whipbirds
Cinclus cinclus	Wasseramsel	dipper, white-throated dipper
Cinclus mexicanus	Grauwasseramsel	North American dipper
Cingulata/Loricata (Xenarthra)	Gürteltiere	armadillos
Cinnyricinclus leucogaster	Amethystglanzstar	violet starling
Ciona intestinalis	Schlauch-Ascidie, Durchsichtige Seeurne	sea vase
Cionella lubrica/ *Cochlicopa lubrica*	Gemeine Achatschnecke, Gemeine Glattschnecke, Glatte Achatschnecke	slippery moss snail, glossy pillar snail (U.S.)
Cionella lubricella/ *Cochlicopa lubricella*	Kleine Achatschnecke, Kleine Glattschnecke	lesser slippery moss snail, thin pillar snail (U.S.)
Cionella morseana/ *Cochlicopa morseana*	Appalachen-Achatschnecke	Appalachian pillar snail (U.S.)
Cionella nitens/ *Cochlicopa nitens*	Glänzende Achatschnecke, Glänzende Glattschnecke	robust slippery moss snail, robust pillar snail (U.S.)
Cionella repentina/ *Cochlicopa repentina*	Mittlere Achatschnecke, Mittlere Glattschnecke	intermediate moss snail, intermediate pillar snail
Cionellidae/Cochlicopidae	Achatschnecken, Glattschnecken	slippery moss snails, pillar snails
Cionus scrophulariae	Braunwurzschaber	figwort weevil
Circaetus gallicus	Schlangenadler	short-toed eagle
Circe scripta	Schrift-Venusmuschel	script venus
Circus aeruginosus	Rohrweihe	marsh harrier
Circus cyaneus	Kornweihe	hen harrier, marsh hawk, northern harrier
Circus macrourus	Steppenweihe	pallid harrier
Circus melanoleucos	Elsterweihe	pied harrier

Circus pygargus	Wiesenweihe	Montagu's harrier
Cirratulidae	Rankenwürmer, Fadenkiemer	cirratulid worms, cirratulids
Cirratulus cirratus/ Cirratulus borealis	Fadenbüschelwurm, Nordischer Rankenwurm	northern cirratule, redthreads
Cirrhigaleus asper/ Squalus asper	Rauer Dornhai	roughskin spurdog (FAO), roughskin spiny dogfish
Cirrhinus molitorella	Schlammkarpfen	mud carp
Cirrhitichthys aprinus	Gepunkteter Korallenwächter	blotched hawkfish, spotted hawkfish (FAO)
Cirrhitidae	Büschelbarsche, Korallenwächter	hawkfishes
Cirripathes anguina/ Cirrhipathes anguina	Gewundene Dörnchenkoralle, Gewundene Drahtkoralle	coiled wire coral, whip coral
Cirripathes rumphii/ Cirrhipathes rumphii	Große Dörnchenkoralle, Große Drahtkoralle	giant whip coral
Cirripathes spiralis/ Cirrhipathes spiralis	Spiralen-Dörnchenkoralle, Spiralige Drahtkoralle	spiraled whip coral
Cirripedia	Rankenfüßler, Rankenfüßer, Cirripeden, Cirripedier	barnacles, cirripedes
Cirroteuthis muelleri	Arktischer Wunderschirm	cirrate octopus
Cirrothauma murrayi	Blinder Wunderschirm	
Cis boleti	Gemeiner Schwammkäfer	
Cissa chinensis	Jagdelster	green magpie
Cisticola juncidis	Cistensänger	fan-tailed warbler
Cisticola textrix	Pinkpink-Cistensänger	tink-tink cisticola
Cisticolidae	Afrikanische Cistensänger	African warblers
Cistothorus palustris	Sumpfzaunkönig	marsh wren, long-billed marsh wren
Cistothorus platensis	Seggenzaunkönig	sedge wren, short-billed marsh wren
Cistugo spp.	Drüsenflügel-Mausohr	wing-gland bats
Citellus citellus/ Spermophilus citellus	Europäisches Ziesel, Schlichtziesel	European ground squirrel, European suslik, European souslik
Citellus spp./ *Spermophilus* spp.	Ziesel	ground squirrels, susliks, sousliks
Citharidae	Cithariden	largescale flounders, citharids
Citharinidae	Geradsalmler	citharinids
Citharus linguatula	Großschuppige Scholle	spotted flounder
Citheronia regalis	Königsmotte*, Königliche Walnussmotte*	regal moth (*larva*: hickory horned devil), regal walnut moth, royal walnut moth
Civettictis civetta	Afrika-Zibetkatze	African civet
Cladius pectinicornis	Geweih-Rosenblattwespe*	antler sawfly
Cladocera	Wasserflöhe	water fleas, cladocerans
Cladocora arbuscula	Röhrchenkoralle, Röhrenkoralle	tube coral
Cladocora caespitosa	Rasenkoralle, Rasige Röhrchenkoralle	caespitose tube coral, lawn coral
Cladocora debilis	Feine Röhrchenkoralle*	thin tube coral

Cladonema radiatum	Aquarienmeduse	aquarium medusa*
Clamator glandarius	Häherkuckuck	great spotted cuckoo
Clambidae	Punktkäfer	fringe-winged beetles, minute beetles
Clanculus pharaonius	Pharaonenklapper	strawberry top
Clangula hyemalis	Eisente	long-tailed duck
Clarias batrachus	Froschwels, Büschelwels	walking catfish
Clarias cavernicola	Höhlen-Raubwels	cave catfish
Clarias spp.	Raubwelse, Luftatmende Welse	labyrinth catfishes, air-breathing catfishes
Clariidae	Raubwelse, Kiemensackwelse	airbreathing catfishes, clarid catfishes
Clathria procera	Oranger Fingerschwamm	orange tree sponge, orange antler sponge*
Clathrina clathrus	Gelber Gitterkalkschwamm	yellow lattice sponge
Clathrina coriacea	Weißer Gitterkalkschwamm	white lattice sponge
Claudius angustatus	Großkopf-Schlammschildkröte	narrow-bridged musk turtle
Clausilia bidentata	Zweizähnige Schließmundschnecke	two-toothed door snail, common door snail
Clausilia cruciata	Scharfgerippte Schließmundschnecke	sharp-ribbed door snail*
Clausilia dubia	Gitterstreifige Schließmundschnecke	craven door snail
Clausilia parvula	Kleine Schließmundschnecke, Zierliche Schließmundschnecke	dwarf door snail
Clausilia pumila	Keulige Schließmundschnecke	clublike door snail*
Clausiliidae	Schließmundschnecken	door snails, clausiliids
Clausinella fasciata/ Venus fasciata	Gebänderte Venusmuschel	banded venus
Clava spp.	Keulenpolypen	club hydroids, club polyps
Clavelina borealis	Nordische Keulensynascidie	northern light-bulb sea-squirt
Clavelina caerulea	Blaue Seescheide	blue light-bulb sea-squirt
Clavelina lepadiformis	Gelbe Seescheide, Keulen-Synascidie, Keulensynascidie, Glaskeulen-Seescheide	light-bulb sea-squirt
Clavelina nana	Kleine Büschel-Seescheide	dwarf light-bulb sea-squirt
Clavelina spp.	Glaskeulen-Seescheiden, Keulensynascidien	light-bulb tunicates, light-bulb sea-squirts
Claviger testaceus	Rotbrauner Keulenkäfer	redbrown clavigerid
Clavigeridae	Keulenkäfer	clavigerid beetles
Cleithracara maronii	Schlüsselloch-Buntbarsch, Maroni	keyhole cichlid
Clelia clelia	Mussurana	mussurana
Clemmys caspica	Kaspische Wasserschildkröte	Caspian pond turtle
Clemmys guttata	Tropfenschildkröte	spotted turtle
Clemmys insculpta	Waldbachschildkröte	wood turtle
Clemmys marmorata	Pazifik-Wasserschildkröte	Pacific pond turtle, western pond turtle
Clemmys muhlenbergii	Amerikanische Hochmoor-Schildkröte, Mühlenberg-Schildkröte	bog turtle, Muhlenberg's turtle

Clepsis senecionana		rustic tortrix moth, rustic twist
Clepsis spectrana	Geflammter Rebenwickler	straw-coloured tortrix moth, cyclamen tortrix moth, fern tortrix
Cleptes spp.	Diebswespen	
Cleridae	Buntkäfer	checkered beetles, clerid beetles, clerids
Clerus mutillarius	Eichenbuntkäfer	oak clerid
Clerus mutillarius/ Pseudoclerops mutillarius	Eichenbuntkäfer (Großer Ameisenkäfer)	
Clethrionomys glareolus	Rötelmaus, Waldwühlmaus	bank vole
Clethrionomys occidentalis	Kalifornische Rötelmaus	red-backed mouse
Clethrionomys rufocanus	Graurötelmaus	gray-sided vole
Clethrionomys rutilus	Polarrötelmaus	northern red-backed vole, ruddy vole
Clethrionomys spp.	Rötelmäuse	red-backed mice, bank voles
Clibanarius erythropus	Felsküsten-Einsiedler	rocky shore hermit crab*
Climacteridae	Baumrutscher	Australo-Papuan treecreepers
Clinidae	Beschuppte Schleimfische	scaled blennies, klipfishes, clinids
Clinitrachus argentatus	Haubenschleimfisch	Mediterranean clinid
Clinocardium ciliatum	Isländische Herzmuschel	hairy cockle, Iceland cockle
Cliona celata	Bohrschwamm, Zellenbohrschwamm	yellow boring sponge, sulfur sponge
Cliona delitrix	Oranger Bohrschwamm	orange boring sponge
Cliona lampa	Roter Bohrschwamm	red boring sponge
Cliona spp.	Bohrschwämme	boring sponges
Cliona viridis	Grüner Bohrschwamm	green boring sponge
Clione limacina	Nackter Seeschmetterling*	naked sea butterfly
Clitellata	Clitellaten, Gürtelwürmer	clitellata
Clivina fossor	Schwarzbrauner Fingerkäfer, Spreizläufer, Gewöhnlicher Grabspornläufer	
Clivina impressifrons		slender seedcorn beetle
Cloeon dipterum	Fliegenhaft	pond olive dun
Cloeotis percivali	Afrikanische Dreispitznäsige Fledermaus, Percivals Dreispitz-Fledermaus	African trident-nosed bat
Clonophis kirtlandi	Kirtlands Wasserschlange	Kirtland's snake
Clonorchis sinensis	Asiatischer Leberegel*	Asian liver fluke
Closia lilacina	Flieder-Randschnecke	lilac marginsnail, lilac marginella
Clossiana dia/Boloria dia	Hainveilchen-Perlmutterfalter, Kleinster Perlmutterfalter	violet fritillary, Weaver's fritillary
Clossiana euphrosyne/ Boloria euphrosyne	Veilchen-Perlmutterfalter, Silberfleck-Perlmuttfalter, Früher Perlmuttfalter	pearl-bordered fritillary
Clossiana selene/ Boloria selene	Braunscheckiger Perlmutterfalter, Braunfleckiger Perlmutterfalter	small pearl-bordered fritillary

Clostera anachoreta	Schwarzfleck-Erpelschwanz, Einsiedler	
Clostera anastomosis	Weidenspinner, Rostbrauner Erpelschwanz	
Clostera curtula	Erpelschwanz	chocolate-tip
Clostera pigra	Espenspinner, Kleiner Erpelschwanz	
Clubionidae	Sackspinnen, Röhrenspinnen	sac spiders, two-clawed hunting spiders, foliage spiders, clubiones, clubionids
Clupea harengus	Hering, Atlantischer Hering	herring, Atlantic herring (FAO) (digby, mattie, slid, yawling, sea herring)
Clupea pallasi	Pazifischer Hering	Pacific herring
Clupeidae	Heringe	herrings (shads, sprats, sardines, pilchards & menhadens)
Clupeiformes	Heringsfische, Heringsverwandte	herrings & relatives
Clupeonella cultriventris	Kilka, Tyulka-Sardelle	clupeonella, kilka, Black Sea sprat (FAO), Tyulka sprat
Clupeonella cultriventris cultriventris	Schwarzmeer-Kilka	Black Sea sardelle, Sea of Azov sardelle
Clupisudis niloticus/ Heterotis niloticus	Afrikanischer Knochenzüngler	African bonytongue, heterotis (FAO)
Clymenella spp.	Bambuswürmer u.a.	bamboo worm a.o.
Clypeaster humilis	Gewöhnlicher Sanddollar	common sand dollar
Clypeaster rosaceus	Brauner Sanddollar, Brauner Schildseeigel, Kleiner Sanddollar	brown sand dollar, brown sea biscuit
Clypeaster subdepressus	Großer Sanddollar	flat sea biscuit, giant sand dollar*
Clypeasteroida	Sanddollars, Schildseeigel	sand dollars, true sand dollars
Clysia ambiguella/ Cochylis ambiguella/ Eupoecilia ambiguella	Einbindiger Traubenwickler (Heuwurm)	European grapevine moth, small brown-barrel conch, grapeberry moth
Clythiidae/Platypezidae	Pilzfliegen	flat-footed flies
Clytra quadripunctata	Sackkäfer, Vierpunkt-Sackblattkäfer	
Clytus arietis	Widderbock, Gemeiner Widderbock	wasp beetle
Cnemophilus macgregorii	Furchenparadiesvogel	sickle-crested bird-of-paradise
Cnidaria/Coelenterata	Nesseltiere, Hohltiere, Cnidarien, Coelenteraten	cnidarians, coelenterates
Cobitidae	Dorngrundeln	loaches
Cobitis aurata	Gold-Steinbeißer	golden loach
Cobitis elongata	Großer Steinbeißer	Balkan loach
Cobitis taenia	Steinbeißer, Europäischer Steinbeißer	spined loach, spotted weatherfish
Coccidae	Napfschildläuse	soft scales & wax scales & tortoise scales (scale insects)

Coccinella quinquepunctata	Fünfpunkt	five-spot ladybird, fivespot ladybird, 5-spot ladybird
Coccinella septempunctata	Siebenpunkt	seven-spot ladybird, sevenspot ladybird, 7-spot ladybird
Coccinella undecimpunctata	Elfpunkt	eleven-spot ladybird, elevenspot ladybird, 11-spot ladybird
Coccinellidae	Marienkäfer	ladybirds, ladybird beetles, lady beetles, "ladybugs"
Coccoidea	Schildläuse (inkl. Wollschildläuse)	scale insects & mealybugs
Coccothraustes coccothraustes	Kernbeißer	hawfinch
Coccothrautes vespertinus	Abendkernbeißer	evening grosbeak
Coccus hesperidum	Weiche Schildlaus, Braune Gewächshaus-Napfschildlaus*	brown soft scale
Coccyzidae	Regenkuckucke, Amerikanische Kuckucke	American cuckoos, New World cuckoos
Coccyzus americanus	Gelbschnabelkuckuck	yellow-billed cuckoo
Coccyzus erythrophthalmus	Schwarzschnabelkuckuck	black-billed cuckoo
Cochleariidae	Kahlschnabel	boatbill heron
Cochlicella acuta	Spitzschnecke*	pointed snail, pointed helicellid
Cochlicopa lubrica/ Cionella lubrica	Gemeine Achatschnecke, Gemeine Glattschnecke, Glatte Achatschnecke	slippery moss snail, glossy pillar snail (U.S.)
Cochlicopa lubricella/ Cionella lubricella	Kleine Achatschnecke, Kleine Glattschnecke	lesser slippery moss snail, thin pillar snail (U.S.)
Cochlicopa morseana/ Cionella morseana	Appalachen-Achatschnecke	Appalachian pillar snail (U.S.)
Cochlicopa nitens/ Cionella nitens	Glänzende Achatschnecke, Glänzende Glattschnecke	robust slippery moss snail, robust pillar snail (U.S.)
Cochlicopa repentina/ Cionella repentina	Mittlere Achatschnecke, Mittlere Glattschnecke	intermediate moss snail, intermediate pillar snail
Cochlicopidae/Cionellidae	Achatschnecken, Glattschnecken	slippery moss snails, pillar snails
Cochlidion limacodes/ Apoda avellana	Kleiner Asselspinner, Große Schildmotte	festoon
Cochliomyia hominivorax	Schraubenwurmfliege, Neuwelt-Schraubenwurmfliege	screwworm fly, New World screwworm fly
Cochlodina costata	Berg-Schließmundschnecke	mountain door snail
Cochlodina fimbriata	Bleiche Schließmundschnecke	pallid door snail, (pale door snail)
Cochlodina laminata	Glatte Schließmundschnecke	plaited door snail
Cochlodina orthostoma	Geradmund-Schließmundschnecke	straightmouth door snail*
Cochlostoma septemspirale	Kleine Walddeckelschnecke	
Cochoa viridis	Smaragdschnäpperdrossel	green cochoa
Cochylis ambiguella/ Eupoecilia ambiguella/ Clysia ambiguella	Einbindiger Traubenwickler (Heuwurm)	European grapevine moth, small brown-barrel conch, grapeberry moth
Codophila varia	Fleckenwanze	
Coelacanthidae	Hohlstachler	coelacanths

Coelacanthiformes	Hohlstachler	coelacanths
Coelioxys spp.	Kegelbienen	
Coelopa spp.	Tangfliegen	seaweed flies
Coelophora inaequalis	Australischer Marienkäfer	common Australian lady beetle
Coelopidae	Tangfliegen	seaweed flies
Coelorhynchus coelorhynchus/ Caelorhynchus caelorhincus	Schwarzschwanz-Grenadier, Hohlmaul-Panzerratte	hollowsnout grenadier (FAO), hollowsnout rattail, black-spotted grenadier
Coelotes atropos	Bodentrichterspinne	
Coenagrion armatum	Hauben-Azurjungfer	Norfolk damselfly (Br.)
Coenagrion hastulatum	Speer-Azurjungfer	northern damselfly, northern blue damselfly (Br.)
Coenagrion hylas	Sibirische Azurjungfer	Siberian damselfly
Coenagrion lunulatum	Mond-Azurjungfer	Irish damselfly, lunular damselfly*
Coenagrion mercuriale	Helm-Azurjungfer	southern damselfly (Br.)
Coenagrion ornatum	Vogel-Azurjungfer	ornate damselfly
Coenagrion puella	Hufeisen-Azurjungfer	common coenagrion, azure damselfly
Coenagrion pulchellum	Fledermaus-Azurjungfer	variable damselfly
Coenagrion scitulum	Gabel-Azurjungfer	dainty damselfly
Coenagrionidae	Schlanklibellen	narrow-winged damselflies
Coendou spp.	Greifstachler	prehensile-tailed porcupines, coendous
Coenobita spp.	Land-Einsiedlerkrebse	land hermit crabs
Coenobitidae	Land-Einsiedlerkrebse	land hermit crabs
Coenocalpe lapidata		slender striped rufous
Coenomyia ferruginea	Stinkfliege	stink fly*
Coenonympha arcania	Perlgrasfalter	pearly heath
Coenonympha glycerion	Rostbraunes Wiesenvögelchen, Rotbraunes Wiesenvögelchen	chestnut heath
Coenonympha hero	Braunes Wiesenvögelchen	scarce heath
Coenonympha oedippus	Wald-Wiesenvögelchen	false ringlet butterfly
Coenonympha pamphilus	Kleiner Heufalter	small heath
Coenonympha tullia/ Coenonympha typhon	Großer Heufalter, Großes Gelbes Wiesenvögelchen	large heath
Coenorhinus aequatus/ Rhynchites aequatus	Rotbrauner Fruchtstecher	apple fruit rhynchites
Coenothecalia/Helioporida	Blaukorallen	blue corals
Coereba flaveola	Zuckervogel	bananaquit
Coerebidae	Bananaquit, Zuckervogel	bananaquit
Coilia mystus	Grenadieranchovy	rat-tail anchovy, Osbeck's grenadier anchovy (FAO)
Colangia immersa	Kleine Gefleckte Nelkenkoralle	lesser speckled cupcoral
Colaphellus sophiae	Senfkäfer, Senfblattkäfer	
Colaptes auratus	Goldspecht	common flicker
Colaptes campestris	Feldspecht	campo flicker
Coleoidea/Dibranchiata	Tintenfische	coleoids
Coleonyx brevis	Texas-Krallengecko	Texas banded gecko
Coleonyx spp.	Krallengeckos, Wüstengeckos	banded geckos
Coleonyx switaki	Barfuß-Gecko	barefoot gecko
Coleonyx variegatus	Gebänderter Krallengecko	western banded gecko

Coleophora alcyonipennella	Kleesamenmotte	clover-seed moth*
Coleophora anatipennella	Kirschblattmotte*	cherry pistol casebearer moth
Coleophora fuscidenella	Erlenknospenmotte	alder bud moth*
Coleophora gryphipennella	Rosenfutteralmotte, „Rosenschabe"	rose casebearer moth*
Coleophora hemerobiella	Knospenminiermotte, Obstblattmotte, Gemeine Obstbaumschabe, Obstbaum-Sackmotte	fruit-tree budmining moth*
Coleophora laricella	Lärchenminiermotte	larch casebearer, larch leaf miner
Coleophora lutipennella	Eichenknospenmotte	oak bud moth*
Coleophora serratella	Birkenminiermotte*	birch casebearer, cigar casebearer
Coleophora spinella/ Coleophora coracipennella	Apfel-Pflaume-Miniermotte	apple and plum casebearer moth, cigar casebearer
Coleophora spissicornis	Kleesamenmotte	
Coleophoridae	Sackträgermotten, Sackmotten, Futteralmotten	casebearers, casebearer moths
Coleoptera	Käfer	beetles
Coleura afra	Afrikanische Freischwanzfledermaus	African sheath-tailed bat
Coleura seychellensis	Seychellen-Freischwanzfledermaus	Arabian sheath-tailed bat
Colias australis/ Colias alfacariensis	Südlicher Heufalter	Berger's clouded yellow
Colias croceus	Posthörnchen, Postillion, Wandergelbling	clouded yellow
Colias eurytheme	Amerikanischer Luzernenheufalter	alfalfa butterfly (alfalfa caterpillar)
Colias hyale	Goldene Acht, Gemeiner Heufalter, Gelber Heufalter, Weißklee-Gelbling	pale clouded yellow
Colias palaeno	Hochmoor-Gelbling, Moorgelbling, Zitronengelber Heufalter	moorland clouded yellow
Colias spp.	Gelblinge, Heufalter	yellows
Coliidae	Mausvögel, Buschkletterer	mousebirds, colies
Coliiformes	Mausvögel	mousebirds, colies
Coliinae	Mausvögel	typical mousebirds
Colinus virginianus	Virginiawachtel	northern bobwhite
Colisa lalia	Zwergfadenfisch	dwarf gourami
Colisa sota	Honigfadenfisch	honey gourami
Collembola	Springschwänze, Collembolen	springtails, garden fleas
Colletes spp.	Seidenbienen	plasterer bees
Colletidae	Seidenbienen	plasterer bees & yellow-faced bees, plumed bees, colletid bees
Collichthys crocea	Asiatischer Gelber Umberfisch	large yellow croaker
Collisella digitalis	Finger-Napfschnecke	fingered limpet
Colobocentrotus atratus	Schild-Seeigel	shield urchin

Colobus angolensis	Angolaguereza	Angolan colobus, Angola pied colobus
Colobus badius/ Piliocolobus badius	Roter Stummelaffe	red colobus, western red colobus
Colobus guereza/ Colobus abyssinicus	Guereza, Mantelaffe (Hbz. Scheitelaffe)	guereza, guereza colobus (eastern black-and-white colobus)
Colobus kirkii/ Piliocolobus kirkii	Sansibarstummelaffe	Kirk's colobus, Zanzibar red colobus
Colobus polykomos	Bärenstummelaffe, Südlicher Guereza (Hbz. Scheitelaffe)	king colobus, western pied colobus (western black-and-white colobus)
Colobus satanas	Schwarzer Guereza	black colobus
Colocasia coryli	Haseleule	nut-tree tussock
Coloeus monedula/ Corvus monedula	Dohle	jackdaw
Cololabis saira	Kurzschnabel-Makrelenhecht	Pacific saury (FAO), mackerel-pike
Colomys goslingi	Afrikanische Waldbachmaus	African water rat, velvet rat
Colonidae	Kolonistenkäfer	colonid beetles
Coloradia pandora	Pandoramotte	pandora moth
Colossoma bidens/ Piaractus brachypomus	Riesenpacu, Gamitana-Scheibensalmler, Mühlsteinsalmler	pirapatinga (FAO), cachama
Colostethus inguinalis	Raketenfrosch, Panama-Baumsteiger	common rocket frog
Colostethus subpunctatus	Tüpfelbaumsteiger, Tüpfel-Raketenfrosch	Bogota rocket frog
Colotois pennaria	Haarrückenspanner	feathered thorn
Colpidium colpoda	Busentierchen	
Colpoda cucullus	Kappentierchen	
Colpophyllia amaranthus	Grobe Atlantische Hirnkoralle	coarse Atlantic brain coral
Colpophyllia natans	Gerillte Atlantische Hirnkoralle	boulder brain coral, large-grooved brain coral
Coluber algirus	Algerische Zornnatter	Algerian whip snake
Coluber constrictor	Schwarznatter	racer
Coluber cypriensis	Zypern-Schlanknatter	Cyprus whip snake
Coluber elegantissimus	Pracht-Zornnatter	beautiful whip snake
Coluber flagellum/ Masticophis flagellum	Peitschennatter, Kutschenpeitschen-Natter	coachwhip
Coluber florulentus	Ägyptische Zornnatter	Egyptian whip snake
Coluber gemonensis/ Coluber laurenti	Balkan-Zornnatter	Balkan whip snake
Coluber hippocrepis	Hufeisennatter	horseshoe snake, horseshoe whip snake
Coluber jugularis	Pfeilnatter	green whip snake, large whip snake
Coluber najadum	Schlanknatter (Dahl'sche Natter/ Steignatter)	light-green whip snake, Dahl's whip snake
Coluber nummifer	Münzennatter	coin snake
Coluber ventromaculatus	Gefleckte Zornnatter	spotted whip snake
Coluber viridiflavus	Gelbgrüne Zornnatter	European whip snake, Western European whip snake, dark-green whipsnake

Colubraria tortuosa	Verdrehter Zwergtriton	twisted dwarf triton
Colubrariidae	Zwerg-Tritonschnecken	dwarf tritons, false tritons
Colubridae	Nattern	colubrine snakes, common snakes
Columba fasciata	Bandtaube, Schuppenhalstaube	band-tailed pigeon
Columba junoniae	Lorbeertaube	laurel pigeon
Columba leuconota	Schneetaube	snow pigeon
Columba livia f. domestica	Straßentaube	domestic pigeon
Columba livia	Felsentaube	feral rock dove
Columba mayeri/ Nesoenas mayeri	Rosentaube	pink pigeon
Columba oenas	Hohltaube	stock dove
Columba palumbus	Ringeltaube	woodpigeon
Columba trocaz	Silberhalstaube	long-toed pigeon
Columbarium pagoda	Pagodenschnecke, Taubenschlag-Pagodenschnecke	common pagoda snail
Columbarium spinicinctum	Stachelige Pagodenschnecke	spiny pagoda snail
Columbella haemastoma	Blutfleck-Täubchenschnecke	blood-stained dovesnail (blood-stained dove shell)
Columbella mercatoria	Gemeine Täubchenschnecke	West Indian dovesnail, common dove snail (common dove shell)
Columbella rustica	Schlichte Täubchenschnecke	rustic dovesnail (rustic dove shell)
Columbella strombiformis	Strombus-Täubchenschnecke	stromboid dovesnail (stromboid dove shell)
Columbellidae/Pyrenidae	Täubchenschnecken	dove snails (dove shells)
Columbidae	Tauben	pigeons and doves
Columbiformes	Taubenvögel	doves & pigeons and allies
Columbina passerina	Sperlingstäubchen	common ground dove
Columella aspera	Raue Windelschnecke	rough whorl snail
Columella columella	Hohe Windelschnecke	mellow column snail (U.S.), hightopped chrysalis snail*
Columella edentula	Zahnlose Windelschnecke	toothless column snail (U.S.), toothless chrysalis snail
Colus gracilis	Röhrenhorn	graceful colus, slender colus
Colydiidae	Rindenkäfer	cylindrical bark beetles
Colydium filiforme	Fadenkäfer	filiform beetle
Colymbetes fuscus	Teichschwimmer	
Comatulida	Haarsterne, Federsterne, Comatuliden	feather stars, comatulids
Comephoridae	Ölfische, Baikal-Ölfische	Baikal oilfishes
Comephorus baicalensis	Großer Ölfisch	greater Baikal oilfish
Comephorus dybowskii	Kleiner Ölfisch	lesser Baikal oilfish
Comibaena bajularia/ Comibaena pustulata	Pustelspanner, Gelbgrüner Eichenmittelwaldspanner	blotched emerald
Comma c-album/ Polygonia c-album	Weißes C	comma
Comma egea/Polygonia egea	Gelber C-Falter	southern comma

Comma l-album	Weißes L	false comma, Compton tortoiseshell
Concentricycloidea	Seegänseblümchen	sea daisies, concentricycloids, concentricycloideans
Conchifera	Schalenweichtiere	conchiferans
Conchoderma spp.	Wal-Seepocken	whale barnacles
Concholepas concholepas	Hasenohr	barnacle rocksnail (barnacle rock shell, hare's ear shell)
Conchostraca	Muschelschaler	clam shrimps
Condylactis aurantiaca	Sandgoldrose, Goldrose, Goldfarbige Seerose	golden anemone, golden sand anemone
Condylactis gigantea	Riesen-Goldrose, Karibische Goldrose	giant Caribbean anemone, pink-tipped anemone, "condy" anemone, Atlantic anemone
Condylura cristata	Sternmull	star-nosed mole
Conepatus spp.	Schweinsnasenskunks	hog-nosed skunks
Conger conger	Meeraal, Gemeiner Meeraal, Seeaal, Congeraal	conger eel, European conger (FAO)
Conger oceanicus	Amerikanischer Meeraal, Amerikanischer Conger	American conger
Conger orbignyanus	Argentinischer Meeraal	Argentine conger
Congeria cochleata	Brackwasser-Dreieckmuschel, Brackwasserdreiecksmuschel	brackish water wedge clam
Congiopodidae	Schweinsfische, Congiopodiden	horsefishes
Congridae	Meeraale	conger eels, congers
Congrogadidae	Congrogadiden	snakelets
Conicera tibialis	Gräberfliege	coffin fly
Conidae	Kegelschnecken	cone snails, cone shells, cones
Conilurus spp.	Kaninchenratten	tree rats, rabbit rats
Coniopterygidae	Staubhafte	white lacewings
Conistra vaccinii	Braune Heidelbeereule	chestnut (moth)
Connochaetes gnou	Weißschwanzgnu	black wildebeest (white-tailed gnu)
Connochaetes taurinus	Streifengnu	blue wildebeest (brindled gnu, white-bearded wildebeest)
Conocephalidae	Schwertschrecken, Kegelköpfe	meadow grasshoppers
Conocephalus conocephalus	Südliche Schwertschrecke	southern conehead
Conocephalus discolor	Langflüglige Schwertschrecke	long-winged conehead
Conocephalus dorsalis	Kurzflüglige Schwertschrecke	short-winged conehead
Conodon nobilis	Gebänderte Süßlippe (Gestreifte Süßlippe)	barred grunt
Conolophus subcristatus	Drusenkopf	Galapagos land iguana
Conopidae	Dickkopffliegen	thick-headed flies
Conopophagidae	Mückenfresser, Mückenfänger	gnateaters
Conosoma testaceum	Gemeiner Spitzleibkurzflügler	

Conraua goliath	Goliathfrosch	goliath frog
Contarinia baeri/	Nadelnknickende	pine needle gall midge
Cecidomyia baeri	Kieferngallmücke	
Contarinia coryli	Haselkätzchen-Gallmücke	hazel-catkin gall midge
Contarinia medicaginis	Luzerneblüten-Gallmücke	lucerne flower midge
Contarinia merceri	Gelbe Fuchsschwanz-Gallmücke	foxtail midge, cocksfoot midge
Contarinia nasturtii	Kohldrehherzmücke, Kohldrehherz-Gallmücke	swede midge, cabbage midge
Contarinia pirivora	Birnengallmücke	pear gall midge
Contarinia pisi	Erbsengallmücke	pea gall midge
Contarinia rubicola	Brombeer-Gallmücke	bramble gall midge*
Contarinia tritici	Gelbe Weizengallmücke	yellow wheat blossom midge, wheat blossom midge
Contia tenuis	Dornschwanzschlange, Spitzschwanzschlange	sharptail snake
Conus advertex		reference cone
Conus amadis	Amadis-Kegel	Amadis cone
Conus ammiralis	Admiral, Admiralskegel	admiral cone, ammiralis cone
Conus arenatus	Sandkegel	sand-dusted cone
Conus argillaceus	Tonerden-Kegel*	clay cone
Conus augur	Augur-Kegel	Augur cone
Conus aulicus	Röhrenförmiger Kegel	princely cone, court cone, courtly cone
Conus aurantius	Goldkegel, Goldfleckkegel	golden cone
Conus auricomus	Goldblattkegel*	gold leafed cone
Conus australis	Australkegel	austral cone
Conus barthelemyi	Barthelemy-Kegelschnecke	Bartholomew's cone
Conus bengalensis	Bengalische Kegelschnecke	Bengal cone
Conus betulinus	Birken-Kegelschnecke	birch cone (*erroneously:* beech cone)
Conus bulbus	Zwiebelkegel	onion cone
Conus bullatus	Blasen-Kegelschnecke	bubble cone
Conus californicus	Kalifornische Kegelschnecke	California cone
Conus cancellatus	Gitter-Kegelschnecke	cancellate cone
Conus capitaneus	Kapitäns-Kegelschnecke	captain cone
Conus cardinalis	Kardinal-Kegelschnecke	cardinal cone
Conus cedonulli	Unvergleichliche Kegelschnecke, Einzigartiger Kegel	matchless cone
Conus chaldeus	Holzkegel	wood cone, vermiculate cone
Conus circumcisus	Ringkegel	circumcision cone
Conus coccineus	Scharlachkegel	scarlet cone, berry cone
Conus cylindraceus	Zylinderkegel	cylindrical cone
Conus dalli	Dalls Kegelschnecke	Dall's cone
Conus daucus	Mohrrüben-Kegelschnecke*, Karotten-Kegel	carrot cone
Conus dorreensis	Papstkronen-Kegelschnecke, Papstkrone	pontifical cone

Conus ebraeus	Hebräische Kegelschnecke, Israelische Kegelschnecke	Hebrew cone
Conus eburneus	Elfenbeinkegel	ivory cone
Conus encaustus	Brandspurkegel*	burnt cone
Conus episcopus	Bischofs-Kegelschnecke	episcopal cone
Conus ermineus	Achat-Kegelschnecke	agate cone
Conus erythraeensis	Rotmeer-Kegelschnecke	Red Sea cone
Conus figulinus	Feigen-Kegelschnecke	fig cone
Conus flamingo	Flamingo-Kegelschnecke	flamingo cone
Conus flavescens	Flammen-Kegelschnecke	flame cone
Conus fulmen	Fulmen-Kegelschnecke	Fulmen's cone
Conus generalis	General-Kegelschnecke	general cone
Conus genuanus	Strumpfbandkegel	garter cone, genuanus cone
Conus geographus	Landkarten-Kegelschnecke, Landkartenkegel	geography cone, geographic cone
Conus gloriamaris	Ruhm des Meeres	glory-of-the-sea cone
Conus granulatus	Ruhm des Atlantiks Kegelschnecke	glory-of-the-Atlantic cone
Conus gubernator	Gouverneur-Kegelschnecke	governor cone
Conus ichinoseana		ichinose cone
Conus imperialis	Königliche Kegelschnecke, Kaiserkegel	imperial cone
Conus leopardus	Leoparden-Kegelschnecke	leopard cone
Conus lithoglyphus	Lithographische Kegelschnecke	lithograph cone
Conus litteratus	Tiger-Kegelschnecke	lettered cone
Conus lucidus	Spinnennetzkegel	spiderweb cone
Conus magus	Zauberkegel	magical cone
Conus maldivus	Malediven-Kegelschnecke	Maldive cone
Conus marmoreus	Marmorkegel	marble cone, marbled cone
Conus marmoreus bandanus	Gebänderter Marmorkegel	banded marble cone
Conus mediterraneus	Mittelmeerkegel	Mediterranean cone
Conus mercator	Händlerkegel	trader cone
Conus miles	Soldatenkegel	soldier cone
Conus milneedwardsi	Ruhm von Indien Kegelschnecke	glory-of-India cone
Conus mus	Mäusekegel	mouse cone
Conus mustelinus	Marder-Kegelschnecke	weasel cone
Conus nobilis	Edle Kegelschnecke	noble cone
Conus patricius	Birnenkegel	pear cone
Conus pennaceus	Federförmige Kegelschnecke	feathered cone
Conus pertusus	Pertusus-Kegel	pertusus cone
Conus praecellens	Wunderbarer Kegel	admirable cone
Conus praelatus	Prälaten-Kegelschnecke	prelate cone
Conus princeps	Prinzenkegel	prince cone, princely cone
Conus pulcher	Schmetterlingskegel	butterfly cone
Conus pulicarius	Flohstichkegel	flea-bite cone
Conus purpurascens	Purpurkegel	purple cone
Conus rattus	Rattenkegel	rat cone
Conus regius	Kronen-Kegelschnecke	crown cone

Conus retifer	Netzkegel	netted cone
Conus rutilus	Glanzkegel*	burnished cone
Conus spectrum	Spiegel-Kegelschnecke, Geisterkegel	spectral cone, spectre cone
Conus sponsalis	Hochzeitskegel	marriage cone
Conus spp.	Kegelschnecken	cone snails, cones (cone shells)
Conus spurius	Alphabet-Kegelschnecke	alphabet cone
Conus stercusmuscarum	Fliegenfleck-Kegelschnecke	fly-specked cone
Conus striatus	Gestreifte Kegelschnecke	striated cone
Conus sulcatus	Gefurchte Kegelschnecke	furrowed cone, sulcate cone
Conus taeniatus	Ringelkegel	ringed cone
Conus tessulatus	Mosaikkegel	tessellate cone
Conus textile	Netz-Kegelschnecke, Gewebte Kegelschnecke, Weberkegel	textile cone
Conus thalassiarchus	Astkegel*	bough cone
Conus trigonus	Dreieckskegel, Dreiecks-Kegelschnecke	trigonal cone
Conus tulipa	Tulpenkegel	tulip cone
Conus vexillum	Fahnenkegel*	flag cone
Conus virgo	Jungfernkegel	virgin cone
Conus vittatus	Gebänderte Kegelschnecke, Schleifen-Kegel*	ribboned cone
Conus ximenes	Unterbrochene Kegelschnecke*	interrupted cone
Conus zeylanicus	Plumpkegel*	obese cone
Conus zonatus	Zonenkegel	zoned cone
Copeina guttata	Forellenpunktsalmler	red-spotted copeina
Copella arnoldi/ Copeina arnoldi	Spritzsalmler	spraying characin, splashing tetra, splash tetra (FAO)
Copepoda	Ruderfußkrebse, Ruderfüßer	copepods
Cophopodisma pyrenaea	Pyrenäen-Gebirgsschrecke	magnificent Pyrenean grasshopper*
Cophosaurus texanus	Große Taubechse	greater earless lizard
Cophotis ceylanica	Ceylon-Taubagame	Sri Lanka prehensile-tail lizard, Ceylon deaf agama
Cophotis spp.	Taubagamen	prehensile-tailed lizards
Cophyla phyllodactyla	Pfeifender Baumfrosch	whistling treefrog
Copris lunaris	Mondhornkäfer	tumblebug, English scarab
Copsychus saularis	Dajaldrossel	magpie robin
Coptosoma scutellatum	Kugelwanze	
Coracias garrulus	Blauracke	European roller
Coraciidae	Racken	rollers
Coraciiformes	Rackenvögel	kingfishers & bee-eaters & hoopoes & rollers & hornbills
Coracina azurea	Azurraupenfänger	African blue cuckoo shrike
Coracina novaehollandiae	Schwarzgesicht-Raupenfänger	black-faced cuckoo shrike
Coracinidae	Galjoen-Fische	galjoens, galjoen fishes
Coracopsis vasa	Vasapapagei	vasa parrot

Coragyps atratus	Rabengeier	black vulture
Coralliophila galea	Helm-Korallengast	helmet coralsnail
Coralliophila meyendorffi	Lamellen-Korallengast, Lamellenschnecke	lamellose coralsnail
Coralliophila pyriformis/ Coralliophila radula	Birnenförmiger Korallengast	pear-shaped coralsnail
Coralliophila scalariformis	Treppenhaus-Korallengast	staircase coralsnail
Coralliophila violacea/ Coralliophila neritoidea	Violetter Korallengast	violet coralsnail, purple coral snail
Coralliophilidae	Latiaxis-Schnecken	coralsnails, latiaxis snails
Corallium rubrum	Edelkoralle, Rote Edelkoralle	red coral, precious red coral (jewel coral)
Corallus annulatus	Geringelte Hundskopfboa	annulated tree boa, annulated boa, boa arboricola
Corallus caninus	Grüne Hundskopfboa, Grüner Hundskopfschlinger	emerald tree boa
Corallus enydris	Gartenboa, Hundskopfboa	garden tree boa, Cook's tree boa, Amazon tree boa
Corallus spp.	Hundskopfboas	tree boas
Coranus subapterus	Raubwanze	heath assassin bug
Corbicula fluminea	Asiatische Körbchenmuschel	Asian clam, Asian basket clam, Asian corbicula
Corbula contracta	Schrumpfmuschel*	contracted corbula, contracted box clam
Corbula gibba	Korbmuschel, Körbchenmuschel	common corbula, common basket clam
Corbula luteola	Westamerikanische Korbmuschel, Gelbe Korbmuschel	common western corbula, yellow basket clam
Corbulidae	Korbmuscheln	box clams, little basket clams
Corcoracinae	Drosselhäher, Schlammnestkrähen	Australian chough, apostlebird
Corcorax melanorhamphos	Drosselkrähe	white-winged chough
Corculum cardissa	Herzmuschel, Echte Herzmuschel, Flache Herzmuschel	true heart cockle
Corcyra cephalonica	Reismotte	rice moth, raisin honey
Cordulegaster boltoni	Zweigestreifte Quelljungfer	golden-ringed dragonfly
Cordulegastridae	Quelljungfern	biddies
Cordulia aenea	Gemeine Smaragdlibelle	downy emerald
Corduliidae	Falkenlibellen	green-eyed skimmers
Cordylidae	Gürtelechsen	girdle-tailed lizards & plated lizards
Cordylobia anthropophaga	Tumbufliege	tumbu fly (mango fly)
Cordylophora caspia	Keulenpolyp	freshwater hydroid
Cordylus cataphractus	Panzer-Gürtelschweif	armadillo girdle-tailed lizard, armadillo girdled lizard (IUCN), armadillo lizard, armadillo spinytail lizard
Cordylus cordylus	Gewöhnlicher Gürtelschweif	common girdled lizard, common spinytail lizard, Lord Derby lizard

Cordylus giganteus	Riesengürtelschweif	sungazer, giant zonure, giant spinytail lizard, giant girdled lizard (IUCN),
Cordylus spp.	Gürtelschweife	girdle-tailed lizards, spinytail lizards
Coregoninae	Renken	coregonines, lake whitefishes
Coregonus albula	Kleine Maräne, Zwergmaräne	vendace
Coregonus artedii	Amerikanische Kleine Maräne	lake cisco, lake herring
Coregonus autumnalis	Arktische Maräne, Arktischer Cisco	Arctic cisco
Coregonus clupeaformis	Nordamerikanisches Felchen	whitefish, common whitefish, lake whitefish (FAO), humpback
Coregonus fera	Sandfelchen, Große Bodenrenke	broad whitefish
Coregonus hiemalis	Leman-Felchen	Lake Geneva whitefish
Coregonus lavaretus	Große Maräne, Bodenrenke, Große Schwebrenke, Wandermaräne, Lavaret	freshwater houting, powan, common whitefish (FAO)
Coregonus macrophthalmus	Gangfisch	gangfish, European whitefish (Lake Neuchâtel whitefish)
Coregonus nasus	Tschirr	broad whitefish
Coregonus oxyrhynchus/ Coregonus oxyrinchus	Kleine Schwebrenke, Schnepel, Schnäpel, Edelmaräne	houting
Coregonus peled	Peledmaräne	northern whitefish, big powan, peled (FAO)
Coregonus pidschian	Kleine Bodenrenke	humpback whitefish (FAO), humpbacked whitefish
Coregonus sardinella	Sardinenmaräne*, Sardinen-Cisco	sardine cisco
Coregonus spp.	Renken, Maränen	whitefishes, lake whitefishes
Coregonus tugun	Tugun	tugun
Coregonus wartmanni	Blaufelchen	pollan (freshwater herring)
Coreidae	Lederwanzen, Randwanzen	leaf-footed bugs, coreid bugs
Coreoidea	Lederwanzen (Randwanzen)	squashbugs
Coreus marginatus/ Mesocerus marginatus	Saumwanze, Lederwanze	squash bug
Coriarachne depressa	Wanzenspinne	
Coris angulata/Coris aygula	Spiegelfleck-Lippfisch, Orangefleck-Lippfisch	clown wrasse (FAO), clown coris, red-blotched rainbowfish
Coris gaimard gaimard	Roter Clownjunker, Roter Clown-Lippfisch	yellowtail coris (FAO), clown wrasse
Coris julis	Meerjunker, Pfauenfederfisch	rainbow wrasse
Corixidae	Ruderwanzen (Wasserzikaden)	water boatmen

Cormura brevirostris	Wagners Flügeltaschen-Fledermaus	Wagner's sac-winged bat
Cornacuspongiae	Hornschwämme, Netzfaserschwämme	horny sponges
Cornu aspersum/ Cryptomphalus aspersus/ Helix aspersa	Gefleckte Weinbergschnecke	brown garden snail, brown gardensnail, common garden snail, European brown snail
Cornularia cornucopiae	Füllhornkoralle	cornucopic coral*
Coronata	Kranzquallen, Tiefseequallen	coronate medusas
Coronella austriaca	Glattnatter, Schlingnatter, Gewöhnliche Glattnatter	smooth snake
Coronella girondica	Girondische Glattnatter, Gironde-Natter, Girondische Schlingnatter	southern smooth snake, Bordeaux snake
Corophium volutator	Schlickkrebs, Wattenkrebs, Wattkrebs	European mud scud, mud dwelling amphipod
Corucia zebrata	Salomonen-Riesenskink	giant Solomon Island skink
Corvidae	Rabenvögel	crows & magpies & jays & nutcrackers
Corvus albus	Schildrabe	pied crow
Corvus brachyrhynchos	Amerikanerkrähe	American crow
Corvus corax	Kolkrabe	common raven
Corvus corone	Aaskrähe	carrion crow
Corvus frugilegus	Saatkrähe	rook
Corvus macrorhynchos	Dschungelrabe	jungle crow
Corvus monedula/ Coloeus monedula	Dohle	jackdaw
Corvus ossifragus	Fischkrähe	fish crow
Corvus rhipidurus	Borstenrabe	fan-tailed raven
Corvus ruficollis	Wüstenrabe	brown-necked raven
Corydoras aeneus	Metall-Panzerwels	bronze corydoras (FAO), aeneus catfish
Corydoras arcuatus	Stromlinien-Panzerwels	arched corydoras, skunk corydoras (FAO)
Corydoras barbatus	Schabracken-Panzerwels	banded corydoras (FAO), giant corydoras
Corydoras hastatus	Sichelfleck-Zwergpanzerwels	dwarf catfish, dwarf corydoras (FAO)
Corydoras julii	Leopard-Panzerwels	leopard corydoras
Corydoras myersi	Schillernder Panzerwels*	iridescent corydoras
Corydoras nattereri	Blauer Panzerwels	blue corydoras
Corydoras paleatus	Punktierter Panzerwels, Marmorierter Panzerwels	peppered corydoras
Corydoras rabauti	Rostpanzerwels	rust corydoras
Corydoras undulatus	Gewellter Panzerwels	wavy catfish
Corylobium avellanae	Haselnusslaus	large hazel aphid
Corylophidae/Orthoperidae	Faulholzkäfer, Schimmelkäfer	minute fungus beetles
Corynactis californica	Erdbeeranemone	strawberry anemone
Corynactis viridis	Grüne Juwelenanemone	green jewel anemone
Coryne pusilla	Roter Kölbchenpolyp	red flowerhead polyp

Coryne spp.	Kölbchenpolypen, Kolbenpolypen	flowerhead polyps
Corynopoma riisei	Zwergdrachenflosser	swordtail characin
Coryphaena equisetis	Pompano-Goldmakrele	pompano dolphinfish
Coryphaena hippurus	Große Goldmakrele, Gemeine Goldmakrele	dolphinfish, common dolphinfish (FAO), dorado, mahi-mahi
Coryphaenidae	Goldmakrelen	dolphinfishes
Coryphaenoides rupestris	Rundkopf-Grenadier, Rundkopf-Panzerratte, Langschwanz	roundhead rattail, roundnose grenadier (FAO), rock grenadier
Coryphoblennius galerita	Marmorierter Schleimfisch	Montagu's blenny
Coryphopterus glaucofraenum	Zügel-Grundel	bridled goby
Coryphopterus nicholsi	Schwarzaugen-Grundel*	blackeye goby
Corystes cassivelaunus	Maskenkrabbe, Antennenkrebs	masked crab, helmet crab
Corystidae	Maskenkrabben	helmet crabs
Corythaeola cristata	Riesenturako	great blue turaco
Corythaixoides concolor	Graulärmvogel	go-away bird
Corythoichthys nigripectus	Harlekin Seenadel, Schwarzbrust-Seenadel	gilded pipefish
Corythucha ciliata	Platanen-Netzwanze	
Corytophanes cristatus	Glatter Helmleguan	smooth helmeted iguana
Corytophanes spp.	Helmleguane	helmeted iguanas
Coscinasterias tenuispina	Dornenseestern, Blauer Seestern	thorny sea star*
Coscinia cribraria	Weißer Grasbär	speckled footman (moth)
Coscinocera hercules	Herkulesspinner	Hercules moth
Cosmia diffinis	Weißflecken-Ulmeneule	white-spotted pinion moth
Cosmia pyralina	Violettbraune Ulmeneule, Birnbaumeule	
Cosmia trapezina	Trapezeule	dun-bar
Cossidae	Holzbohrer	carpenter moths & leopard moths
Cossus cossus	Weidenbohrer	goat moth
Cossypha natalensis	Natalrötel	red-capped robin chat
Cosymbia punctaria/ Cyclophora punctaria	Punktfleckspanner, Eichenbuschspanner	maiden's blush moth
Cotingidae	Schmuckvögel, Kotingas	cotingas, plantcutters, sharpbill
Cotinis nitida	Grüner Junikäfer*	green June beetle
Cottidae	Groppen	sculpins
Cottocomephoridae	Baikalgroppen	Lake Baikal sculpins
Cottunculus microps	Polargroppe*	polar sculpin
Cottunculus thompsoni	Blasse Groppe*	pallid sculpin
Cottus bairdi	Fleckengroppe*	mottled sculpin
Cottus bubalis/ Myoxocephalus bubalis> Taurulus bubalis	Seebulle, Seebull, Ulk (Seeskorpion)	long-spined sculpin, long-spined sea scorpion, longspined bullhead (FAO)
Cottus cognatus	Schleimgroppe*	slimy sculpin
Cottus gobio	Groppe, Koppe, West-Groppe, Kaulkopf, Mühlkoppe	Miller's thumb, bullhead (FAO)

Cottus poecilopus	Bandflossen-Koppe, Buntflossenkoppe, Buntflossen-Groppe, Sibirische Groppe	Siberian bullhead, alpine bullhead (FAO)
Cottus scorpius/ Myoxocephalus scorpius	Seeskorpion, Seeteufel	Father Lasher, shorthorn sculpin (FAO), bull rout, bull-rout, short-spined seascorpion
Coturnix chinensis	Zwergwachtel	Indian blue quail
Coturnix coturnix	Wachtel	quail
Cotylorhiza tuberculata	Knollenqualle, Spiegelei-Qualle	fried-egg jellyfish
Cotylosauria	Stammreptilien	stem reptiles, cotylosaurs
Crabro cribrarius	Silbermundwespe	slender-bodied digger wasp
Cracidae	Hokkos	guans, chachalacas, currassows
Cracticidae	Würgerkrähen	butcherbirds
Crambe crambe	Höckeriger Krustenschwamm	red encrusting sponge, bumping encrusting sponge
Crambidae	Grasmotten	grass moths, grass-veneers
Crambus perlellus	Weißer Graszünsler	yellow satin grass-veneer
Crambus pratella	Dunkelfleckiger Graszünsler	dark-inlaid grass-veneer
Crambus spp.	Grasmotten	grass moths, grass-veneers
Cranchia scabra	Warziger Gallertschirm	warty gelatin-squid*
Cranchiidae	Gallertkalmare	gelatin-squids*
Crangon crangon	Nordseegarnele, Granat, Porre	common shrimp, common European shrimp (brown shrimp)
Crangon franciscorum	Pazifik-Sandgarnele	California bay shrimp, Pacific sand shrimp
Crangon septemspinosa	Sandgarnele	sand shrimp
Crangonidae	Sandgarnelen	crangonid shrimp
Cranoglanididae	Panzerkopfwelse	armorhead catfishes
Craseonycteris thonglongyai	Hummelfledermaus, Schweinsschnauzenfledermaus, Kittis Kleinstfledermaus	Kitti's hog-nosed bat (*smallest mammal!*)
Craspedacusta sowerbyi/ Microhydra sowerbyi (Microhydra ryderi)	Süßwassermeduse, Süßwasserqualle (>Zwergpolyp)	freshwater jellyfish, Regent's Park medusa (freshwater polyp)
Crassostrea angulata/ Gryphaea angulata	Portugiesische Auster, Greifmuschel	Portuguese oyster
Crassostrea gigas	Riesenauster, Pazifische Auster	Pacific oyster, giant Pacific oyster, Japanese oyster
Crassostrea rhizophorae	Pazifische Felsenauster	Pacific cupped oyster, mangrove cupped oyster
Crassostrea virginica/ Gryphaea virginica	Amerikanische Auster	American oyster, eastern oyster, blue point oyster, American cupped oyster
Crataerina pallida	Mauerseglerlausfliege	swift parasitic fly, swallow parasitic fly
Craterolophus convolvulus	Stielqualle	stalked jellyfish
Craterolophus tethys	Schlanke Becherqualle, Tethys-Becherqualle	Tethys jelly*
Crateromys heaneyi	Panay-Borkenratte	Panay cloud rat

Crateromys paulus	Pauls Borkenratte, Ilin-Borkenratte	Ilin Island cloud rat
Crateromys schadenbergi	Buschschwanz-Borkenratte, Schadenbergs Borkenratte	bushy-tailed cloud rat, giant bushy-tailed cloud rat
Crax rubra	Tuberkelhokko	great curassow
Creadion carunculatus	Sattelvogel	saddleback
Creatophora cinerea	Lappenstar	wattled starling
Creediidae/Limnichthyidae	Sandgräberfische	sandburrowers
Cremastogaster scutellaris	Rotkopfameise	
Crenatula picta	Bunte Baummuschel	painted tree-oyster
Crenilabrus bailloni	Baillons Lippfisch	Baillon's wrasse
Crenilabrus cinereus/ Symphodus cinereus	Grauer Lippfisch	grey wrasse
Crenilabrus exoletus/ Centrolabrus exoletus	Kleinmündiger Lippfisch, Kleinmäuliger Lippfisch	rock cock, rock cook (FAO), small-mouthed wrasse
Crenilabrus mediterraneus/ Symphodus mediterraneus	Mittelmeer-Lippfisch	axillary wrasse
Crenilabrus melanocercus	Schwarzschwanz-Lippfisch	black-lip wrasse
Crenilabrus melops/ Symphodus melops	Goldmaid	corkwing wrasse
Crenilabrus ocellatus/ Symphodus ocellatus	Augenfleck-Lippfisch, Augenflecklippfisch, Augenlippfisch	spotted wrasse, ocellated wrasse
Crenilabrus pavo/ Symphodus tinca	Pfauenlippfisch, Meerschlei	peacock wrasse
Crenilabrus scina/ Coricus rostratus/ Symphodus rostratus	Schnauzen-Lippfisch, Schnauzenlippfisch	
Crenuchidae	Prachtsalmler, Segelflossensalmler	crenuchids
Creophilus maxillosus	Aas-Raubkurzflügler	hairy rove beetle
Crepidophryne epiotica	Kostarica-Kröte	Cerro Utyum toad
Crepidula aculeata	Stachelige Pantoffelschnecke	spiny slippersnail
Crepidula convexa	Konvex-Pantoffelschnecke	convex slippersnail
Crepidula fornicata	Pantoffelschnecke, Porzellanpantoffel	American slipper limpet, common Atlantic slippersnail
Crepidula gibbosa	Mittelmeer-Pantoffelschnecke	Mediterranean slippersnail, Mediterranean slipper limpet
Crepidula grandis	Große Pantoffelschnecke	great slippersnail
Crepidula plana	Flache Pantoffelschnecke	eastern white slippersnail
Crepidula unguiformis	Fingernagel-Pantoffelschnecke, Fingernagel	Mediterranean slippersnail
Crex crex	Wachtelkönig	corncrake
Cribellatae	Kräuselfadenwebspinnen	hackled band spiders
Cricetelus griseus	Chinesischer Streifenhamster	Chinese striped hamster
Cricetelus migratorius	Grauer Zwerghamster	gray hamster
Cricetomys spp.	Riesenhamsterratten	African giant pouched rats
Cricetulus spp.	Graue Zwerghamster	ratlike hamsters
Cricetus cricetus	Hamster, Feldhamster (Hbz. Hamster)	common hamster, black-bellied hamster
Cricosaura typica	Kuba-Nachtechse	Cuban night lizard
Crinia bilingua	Blökender Frosch	bleating froglet

Crinia desertcola	Sperlings-Zirpfrosch	sparrow froglet
Crinia glauerti	Rasselfrosch	rattle froglet
Crinia signifera	Ostaustralischer Zirpfrosch	common eastern froglet
Crinia spp.	Australische Zirpfrösche	Australian froglets
Crinoidea	Seelilien, Crinoiden (inkl. Haarsterne = Federsterne)	sea lilies, crinoids (incl. feather stars)
Criocephalus rusticus/ Arhopalus rusticus	Dunkelbrauner Halsgrubenbock	rusty longhorn beetle
Crioceris asparagi	Spargelhähnchen, Spargelkäfer	asparagus beetle
Crioceris duodecimpunctata	Zwölfpunkt-Spargelhähnchen, Zwölfpunkt-Spargelkäfer	spotted asparagus beetle
Crioceris spp.	Spargelhähnchen, Spargelkäfer	asparagus beetles
Criorhina floccosa	Pelz-Schwebfliege	
Crisia eburnea	Elfenbein-Moostierchen, Stacheliges Moostierchen	ivory bryozoan*
Crocallis elinguaria	Mordspanner	scalloped oak
Crocidura leucodon	Feldspitzmaus	bicoloured white-toothed shrew
Crocidura russula	Hausspitzmaus	greater white-toothed shrew
Crocidura spp.	Wimperspitzmäuse	white-toothed shrews
Crocidura suaveolens	Gartenspitzmaus	lesser white-toothed shrew
Crocodilia	Krokodile, Panzerechsen	crocodiles
Crocodilurus lacertinus	Krokodilschwanzechse	crocodile tegu, dragon lizard
Crocodylus acutus	Spitzkrokodil	American crocodile
Crocodylus cataphractus	Panzerkrokodil	African slender-snouted crocodile, long-snouted crocodile
Crocodylus intermedius	Orinoko-Krokodil	Orinoco crocodile
Crocodylus johnstoni	Australien-Krokodil	Australian freshwater crocodile
Crocodylus mindorensis	Mindoro-Krokodil	Philippine crocodile, Philippines crocodile, Mindoro crocodile
Crocodylus moreletii	Beulenkrokodil	Morelet's crocodile
Crocodylus niloticus	Nilkrokodil	Nile crocodile
Crocodylus novaeguineae	Neuguinea-Krokodil	New Guinea crocodile
Crocodylus palustris	Sumpfkrokodil	mugger crocodile, mugger, marsh crocodile, broad-snouted crocodile
Crocodylus porosus	Leistenkrokodil	saltwater crocodile, estuarine crocodile
Crocodylus rhombifer	Rautenkrokodil, Kuba-Krokodil	Cuban crocodile
Crocodylus siamensis	Siam-Krokodil	Siamese crocodile
Crocothemis erythraea	Feuerlibelle	
Crocuta crocuta	Tüpfelhyäne	spotted hyena
Croesia bergmanniana	Rosenwickler	
Croesia forskaleana	Ahornwickler	
Cromeria nilotica	Nil-Larvenfisch	naked shellear
Cromeriidae	Larvenfische	cromeriids
Crossarchus spp.	Kusimansen	cusimanses
Crossaster papposus	Stachelsonnenstern, Warziger Sonnenstern	common sun star, spiny sun star, spiny sunstar

Crossata californica	Kalifornische Froschschnecke	California frogsnail
Crossomys moncktoni	Moncktons Schwimmratte	earless water rat
Crossopterygii	Quastenflosser	lobe-finned fishes, crossopterygians
Crotalidae	Grubenottern	pit vipers
Crotalus adamanteus	Diamantklapperschlange	eastern diamondback rattlesnake
Crotalus atrox	Texas-Klapperschlange	western diamondback rattlesnake
Crotalus basiliscus	Basiliskenklapperschlange	Mexican west-coast rattlesnake
Crotalus catalinensis	Santa-Catalina-Klapperschlange	Santa Catalina rattlesnake
Crotalus cerastes	Gehörnte Klapperschlange, Seitenwinder-Klapperschlange	sidewinder
Crotalus durissus/ Crotalus terrificus	Tropische Klapperschlange, Schreckensklapperschlange, Schauerklapperschlange, Cascabel, Südamerikanische Klapperschlange	neotropical rattlesnake, cascabel
Crotalus horridus	Waldklapperschlange	timber rattlesnake
Crotalus lepidus	Felsenklapperschlange	rock rattlesnake
Crotalus mitchelli	Gefleckte Klapperschlange	speckled rattlesnake
Crotalus molossus	Schwarzschwanz-Klapperschlange	black-tailed rattlesnake, blacktail rattlesnake
Crotalus pricei	Prices Klapperschlange	twin-spotted rattlesnake
Crotalus ruber	Rote Diamantklapperschlange	red diamond rattlesnake
Crotalus scutulatus	Mojave-Klapperschlange	Mojave rattlesnake
Crotalus spp.	Echte Klapperschlangen	rattlesnakes, rattlers
Crotalus tigris	Tigerklapperschlange	tiger rattlesnake
Crotalus unicolor	Aruba-Klapperschlange	aruba rattlesnake
Crotalus viridis	Prärieklapperschlange	western rattlesnake
Crotalus willardi	Willards Klapperschlange, Kantenkopf-Klapperschlange	ridgenose rattlesnake, ridge-nosed rattlesnake
Crotaphopeltis hotamboeia	Rotlippenschlange	herald snake, cat-eyed snake
Crotaphopeltis spp.	Rotlippenschlangen	herald snakes, cat-eyed snakes, tropical water snakes
Crotaphytus collaris	Halsbandleguan, Kugelechse	collared lizard
Crotaphytus insularis	Wüsten-Halsbandleguan, Schwarzhalsbandleguan	black-collared lizard, desert collared lizard
Crotaphytus reticulatus	Gegitterter Halsbandleguan	reticulate collared lizard
Crotaphytus spp.	Halsbandleguane	collared lizards
Crotophaga sulcirostris	Riefenschnabelani	groove-billed ani
Crotophaginae	Madenkuckucke, Guira-Kuckuck	anis, Guira cuckoo
Crucibulum striatum	Gestreifte Mützenschnecke	striate cup-and-saucer limpet
Crunomys spp.	Philippinische Sumpfratten	Philippine swamp rats
Cryphalus abietis	Gekörnter Fichtenborkenkäfer	granular spruce beetle*
Cryphalus piceae	Kleiner Tannenborkenkäfer	white spruce beetle
Cryphia domestica	Flechteneule	marbled beauty
Cryptobranchidae	Riesensalamander	giant salamanders
Cryptobranchus alleganiensis	Hellbender, Amerikanischer Riesensalamander	hellbender

Cryptoceridae	Braunhelmschaben*	brown-hooded cockroaches
Cryptochiridae	Gallkrabben*	gall crabs
Cryptochiton stelleri (Polyplacophora)	Große Mantel-Käferschnecke	giant Pacific chiton, gumboot chiton (U.S.)
Cryptococcus fagisuga	Buchenwolllaus, Buchenschildlaus	beech scale, felted beech scale
Cryptodendrum adhaesivum	Noppenrand-Anemone, Noppenrand-Meerblume, Pizza-Anemone	pizza anemone, nap-edged anemone
Cryptolaemus montrouzieri	Australischer Marienkäfer	mealybug ladybird (mealybug destroyer)
Cryptolestes ferrugineus	Rotbrauner Leistenkopfplattkäfer	rusty grain beetle, rust-red grain beetle
Cryptolestes pusillus	Kleiner Leistenkopfplattkäfer	flat grain beetle
Cryptolestes turcicus	Türkischer Leistenkopfplattkäfer	Turkish grain beetle
Cryptolithodes sitchensis	Schildkrötenkrabbe*	umbrella crab, umbrella-backed crab, turtle crab
Cryptolithodes typicus	Schmetterlingskrabbe*	butterfly crab
Cryptolybia olivacea	Olivbartvogel	green barbet
Cryptomphalus aspersus/ Cornu aspersum/ Helix aspersa	Gefleckte Weinbergschnecke	brown garden snail, brown gardensnail, common garden snail, European brown snail
Cryptomys spp.	Graumulle	common mole-rats
Cryptomyzus galeopsidis	Bleichstreifige Beerenblattlaus	black currant aphid, currant blister aphid
Cryptomyzus ribis	Johannisbeerblasenlaus	red currant blister aphid, currant aphid
Cryptonatica affinis/ Natica clausa	Arktische Mondschnecke, Arktische Nabelschnecke	Arctic moonsnail
Cryptopecten pallium	Königsmantel	royal cloak scallop
Cryptopecten phrygium	Spatenmuschel	spathate scallop
Cryptophagidae	Schimmelkäfer	silken fungus beetles
Cryptoprocta ferox	Fossa, Frettkatze	fossa
Cryptops hortensis	Gartenskolopender	European garden scolopendra
Cryptorhynchus lapathi	Weidenrüssler, Erlenrüssler	poplar-and-willow borer, osier weevil, willow weevil
Cryptospira elegans	Fantastische Randschnecke	elegant marginella, elegant marginsnail (elegant margin shell)
Cryptospira strigata	Gestreifte Randschnecke	striped marginella, striped marginsnail (striped margin shell)
Cryptospira ventricosa	Breite Randschnecke	broad marginella, broad marginsnail (broad margin shell)
Cryptosula pallasiana	Flaches Krusten-Moostierchen	red crust
Cryptotis parva	Nordamerikanische Kleinohrspitzmaus	least shrew
Cryptotis spp.	Kleinohrspitzmäuse	small-eared shrews
Crypturellus cinnamomeus	Buschtinamu, Zimttao	thicket tinamou
Crypturellus soui	Brauntinamu, Brauntao	little tinamou

Crypturgus cinereus	Kleiner Kiefernborkenkäfer	
Crystallogobius linearis	Kristallgrundel	crystal goby
Ctenidae	Kammspinnen, Wanderspinnen	wandering spiders, running spiders
Cteniopus spp.	Schwefelkäfer	
Cteniza sauvagei	Falltürspinne	trapdoor spider
Ctenizidae	Falltürspinnen, Miniervogelspinnen	trapdoor spiders
Ctenobrycon spilurus	Talerfisch, Hochrückensalmler	silver tetra
Ctenocephalides canis	Hundefloh	dog flea
Ctenocephalides felis	Katzenfloh	cat flea
Ctenochaetus strigosus	Blaustreifen-Bürstenzahnseebader	blue-striped surgeon fish
Ctenodactylidae	Kammfinger, Gundis	gundis
Ctenodactylus spp.	Gundis	gundis
Ctenodiscus crispatus	Schlickstern*	mud star
Ctenoides scabra/ Lima scabra	Raue Feilenmuschel	rough lima, rough fileclam (rough file shell, Atlantic rough file shell)
Ctenolabrus rupestris/ Ctenolabrus suillus	Felsenbarsch, Klippenbarsch	goldsinny, goldsinny wrasse
Ctenoluciidae	Hechtsalmler	American pike-characids
Ctenomyidae	Kammratten	tuco-tucos, ctenomyids
Ctenomys spp.	Tukotukos	tuco-tucos
Ctenopharyngodon idella	Graskarpfen, Amurkarpfen	grass carp
Ctenophora	Ctenophoren, Kammquallen, Rippenquallen	ctenophores (sea gooseberries, sea combs, comb jellies, sea walnuts)
Ctenophora ornata	Kammschnake	
Ctenophorus maculatus/ Amphibolurus maculatus	Gefleckte Bodenagame	spotted dragon, spotted bearded dragon
Ctenophorus maculosus/ Amphibolurus maculosus	Lake-Eyre-Agame	Lake Eyre dragon
Ctenophryne geayi	Brauner Eierfrosch	brown egg frog
Ctenopoma acutirostre	Leopard-Buschfisch	leopard bushfish
Ctenopoma ansorgii/ Microctenopoma ansorgii	Prachtbuschfisch, Orange-Buschfisch	ornate climbing perch, ornate ctenopoma (FAO)
Ctenopoma fasciolatum/ Microctenopoma fasciolatum	Gebänderter Buschfisch	banded climbing perch, banded ctenopoma (FAO)
Ctenopoma spp.	Buschfische	climbing perches
Ctenosaura pectinata	Schwarzleguan	spiny-tailed iguana, pectinate ctenosaur
Ctenosaura spp.	Schwarzleguane	spiny-tailed iguanas, greater spinytail iguanas
Ctenostomata	Kammmünder	ctenostomates
Cubiceps gracilis	Langflossen-Zigarrenfisch*	longfin cigarfish
Cubomedusae/Cubozoa	Würfelquallen	cubomedusas, box jellies, sea wasps, fire medusas

Cuclotogaster heterographus/ *Gallipeurus heterographus*	Hühner-Kopflaus	chicken head louse
Cucujidae	Plattkäfer, Schmalkäfer	flat bark beetles
Cucujus cinnaberinus	Scharlachkäfer	
Cuculidae	Kuckucke	cuckoos & roadrunners & anis
Cuculiformes	Kuckucksvögel	cuckoos & turacos & allies
Cuculinae	Altwelt-Kuckucke, Eigentliche Kuckucke	Old World cuckoos
Cucullaea labiata	Hauben-Archenmuschel	hooded ark (hooded ark shell)
Cucullia absinthii	Fahler Wermut-Mönch	wormwood moth
Cucullia argentea	Silbermönch	silver shark (moth)
Cucullia artemisiae	Beifußmönch, Grauer Beifußmönch	scarce wormwood shark
Cucullia asteris	Asternmönch, Aster-Goldrutenheiden-Braunmönch	starwort moth
Cucullia umbratica	Schattenmönch, Grauer Mönch	shark moth, common shark
Cucullia verbasci	Brauner Mönch, Wollkrauteule	mullein moth
Cucullus canorus	Kuckuck	cuckoo
Cucullus saturatus	Hopfkuckuck	Oriental cuckoo
Cucumaria elongata	Sichel-Seegurke	sickled sea cucumber*
Cucumaria frondosa	Orangenfuß-Seegurke*	orange-footed sea cucumber, "pudding"
Cucumaria miniata	Rote Seegurke	red sea cucumber
Cucumaria planci	Kletterholothurie, Kletterseewalze	climbing sea cucumber*
Culcita spp.	Stachel-Kissenstern	pincushion star, pincushion starfish a.o.
Culex pipiens	Hausmücke, Gemeine Stechmücke	house mosquito, northern common house mosquito, common gnat (Br.), house gnat (Br.)
Culex quinquefasciatus	Südliche Hausstechmücke	southern house mosquito
Culicidae	Stechmücken, Moskitos	mosquitoes, gnats (Br.)
Culiseta annulata/ *Theobaldia annulata*	Ringelschnake	banded house mosquito, banded mosquito, ring-footed gnat
Cuma lacera	Gekielte Felsschnecke	keeled rocksnail, carinate rocksnail
Cumacea	Kumazeen	cumaceans
Cuniculus spp./*Agouti* spp.	Pakas	pacas
Cuon alpinus	Rothund	dhole, red dog, Asiatic wild dog
Cuora amboinensis	Amboina-Scharnierschildkröte	Malayan box turtle, Amboina box turtle, South Asian box turtle (IUCN)
Cuora flavomarginata	Gelbköpfchen	yellow-margined box turtle, snake-eating turtle
Cuora galbinifrons	Hinterindische Scharnierschildkröte, Vietnamesische Scharnierschildkröte	Indochinese box turtle

Cuora spp.	Scharnierschildkröten	box turtles
Cuora trifasciata	Dreistreifen-Scharnierschildkröte	Chinese three-striped box turtle, three-banded box turtle
Cuora yunnanensis	Yünnan-Scharnierschildkröte	Yunnan box turtle
Cupedidae	Cupediden	reticulated beetles
Cupido minimus	Zwergbläuling	small blue (butterfly)
Curculio glandium	Eichelbohrer	acorn weevil
Curculio nucum	Haselnussbohrer	nut weevil
Curculio villosus	Eichengallenrüssler	oak gall weevil
Curculionidae	Rüsselkäfer	snout beetles, weevils (true weevils)
Curimatidae	Breitlingssalmler, Barbensalmler	curimatids, toothless characins
Curimopsis nigrita	Geschwärzter Furchenbauch-Pillenkäfer	mire pill beetle
Cursorius cursor	Rennvogel	cream-coloured coursor
Cuspidaria glacialis	Gletschermuschel*	glacial dipperclam
Cuterebridae	Amerikanische Dasselfliegen	robust bot-flies
Cuvierina columnella	Cuvier-Seeschmetterling*	cigar pteropod
Cyamus boopis	Wal-Laus	whale-louse
Cyamus spp.	Wal-Läuse	whale-lice
Cyanea capillata	Gelbe Feuerqualle, Gelbe Haarqualle	lion's mane, giant jellyfish, hairy stinger, sea blubber, sea nettle, pink jellyfish
Cyanea lamarcki	Blaue Nesselqualle, Blaue Feuerqualle	blue lion's mane, cornflower jellyfish
Cyanerpes cyaneus	Türkisnaschvogel	red-legged honeycreeper
Cyaniris semiargus	Dunkelbläuling	mazarine blue (butterfly)
Cyanocitta cristata	Blauhäher	blue jay
Cyanoliseus patagonus	Felsensittich	Patagonian conure
Cyanolyca argrntigula	Silberhäher	silvery-throated jay
Cyanopica cyana	Blauelster	azure-winged magpie
Cyanopsitta spixii	Spixara	little blue macaw, Spix's macaw
Cyanoptila cyanomelana	Blauschnäpper	blue and white flycatcher
Cyanoramphus novaezelandiae	Ziegensittich	red-fronted parakeet
Cybiosarda elegans	Gefleckter Bonito	leaping bonito
Cybister laterimarginalis	Gaukler, Breitleibschwimmer	
Cybocephalidae	Schildlauskäfer	cybocephalids (sap beetles a.o.)
Cybocephalus politus	Dunkler Schildlauskäfer	
Cychrus attenuatus	Schmaler Schaufelläufer, Gestreifter Schaufelläufer	
Cychrus caraboides	Gewöhnlicher Schaufelläufer, Körniger Schaufelläufer	
Cyclagras gigas	Brasilianische Glattnatter	Brazilian smooth snake, "false water cobra"

Cyclanorbis elegans	Rückenflecken-Weichschildkröte, Gefleckte Klappen-Weichschildkröte	Nubian soft-shelled turtle, Nubian flapshell turtle
Cyclanorbis senegalensis	Senegal-Weichschildkröte	Senegal soft-shelled turtle, Senegalese flapshell turtle, Senegal flapshell turtle
Cyclanorbis spp.	Mittelafrikanische Klappen-Weichschildkröten	sub-Saharan flapshells
Cyclemys dentata	Malayische Dornschildkröte	Asian leaf turtle
Cyclemys mouhoti/ Pyxidea mouhotii	Indische Dornschildkröte	Indian leaf turtle, jagged-shell turtle, Indian keel-backed terrapin, keeled box turtle
Cyclemys spp.	Dornschildkröten	leaf turtles
Cyclichthys schoepfi	Gestreifter Igelfisch	striped burrfish
Cyclichthys spilostylus	Gelbflecken-Igelfisch	yellowspotted burrfish
Cycloderma aubryi	Rotrückige Klappen-Weichschildkröte	Aubry's soft-shelled turtle, Aubry's flapshell turtle
Cycloderma frenatum	Graurückige Klappen-Weichschildkröte, Graue Klappen-Weichschildkröte	Zambesi soft-shelled turtle, Zambezi flapshell turtle
Cycloderma spp.	Zentralafrikanische Klappen-Weichschildkröten	Central African flapshells
Cyclograpsus integer	Kuglige Strandkrabbe	globose shore crab
Cyclopes didactylus	Zwergameisenbär	silky anteater, pygmy anteater
Cyclophora annulata	Ringspanner	mocha
Cyclophora linearia	Gelber Buchenspanner	clay tripelines moth
Cyclophora pendularia/ Cyclophora orbicularia	Blasser Ringelfleck-Gürtelpuppenspanner, Pendelspanner, Weißgrauer Ringfleckspanner	dingy mocha
Cyclophoridae	Turmdeckelschnecken	cyclophorid snails
Cyclopidae	Hüpferlinge (Schwimmer)	
Cyclops spp.	Hüpferlinge	
Cyclopteridae	Lumpfische, Seehasen	lumpfishes (lumpsuckers)
Cyclopterus lumpus	Seehase (Lump, Lumpfisch)	lumpsucker (FAO), lumpfish, hen-fish, henfish, sea hen
Cyclorana platycephala	Wasserreservoirfrosch	common water-holding frog
Cyclorana spp.	Wasserreservoirfrösche	water-holding frogs
Cyclorrhapha (Muscomorpha)	Deckelschlüpfer	circular-seamed flies
Cyclosa conica	Kreisspinne, Konische Radspinne	trashline orbweaver
Cyclosa spp.	Kreisspinnen	trashline orbweavers
Cyclostomata	Rundmäuler, Kreismünder	cyclostomes
Cyclura cornuta	Nashornleguan	rhinoceros iguana
Cyclura cyclura	Andros-Wirtelschwanzleguan, Fels-Wirtelschwanzleguan	Andros ground iguana, Exuma Island ground iguana
Cyclura nubila	Kuba-Wirtelschwanzleguan	Cayman Islands ground iguana, Cuban ground iguana (IUCN)
Cyclura pinguis	Anegada-Wirtelschwanzleguan	Anegada ground iguana

Cyclura spp.	Wirtelschwanzleguane	West Indian rock iguanas, rhinoceros iguanas
Cydia conicolana	Kiefernzapfenwickler	pine cone moth
Cydia coniferana	Kiefernharzwickler	pine resin moth
Cydia fagiglandana	Buchenwickler	beech seed moth
Cydia funebrana/ Laspeyresia funebrana	Pflaumenwickler, Pflaumenmade	plum fruit moth, plum moth, red plum maggot
Cydia nigricana/ Cydia rusticella/ Laspeyresia nigricana	Erbsenwickler, Olivbrauner Erbsenwickler	pea moth
Cydia pactolana	Fichtenrindenwickler	spruce bark moth
Cydia pomonella/ Laspeyresia pomonella/ Carpocapsa pomonella	Apfelwickler (Obstmade)	apple moth (apple worm), codling moth, codlin moth
Cydia splendana	Eichelwickler, Kastanienwickler	acorn moth
Cydia strobilella	Fichtenzapfenwickler	spruce cone moth
Cydia zebeana	Lärchengallenwickler, Lärchenrindenwickler	larch gall moth*, larch bark moth*
Cydippidea	Cydippen	cydippids (comb jellies)
Cydnidae	Erdwanzen	burrower bugs, cydnid bugs
Cyema atrum	Schwarzer Tiefseeaal	deepsea arrow-eel
Cyematidae/Cyemidae	Tiefseeaale, Schwarze Tiefseeaale	bobtail snipe eels, arrow eels
Cygninae	Schwäne	swans
Cygnus atratus	Schwarzschwan	black swan
Cygnus columbianus	Zwergschwan	tundra swan
Cygnus cygnus	Singschwan	whooper swan
Cygnus olor	Höckerschwan	mute swan
Cylichna alba	Weisse Kelchschnecke*	white chalice-bubble snail
Cylichna cylindracea	Zylinderschnecke, Zylinder-Kelchschnecke*	cylindrical barrel-bubble
Cylichna occulta	Verborgene Kelchschnecke*	concealed chalice-bubble snail
Cylinder bullatus	Blasenzylinder*	bubble cone
Cylindera germanica	Deutscher Sandkäfer, Deutscher Sandläufer	
Cylindroiulus britannicus	Kleiner Gewächshaus-Tausendfüßer*	lesser glasshouse millepede
Cylindroiulus londinensis/ Cylindroiulus teutonicus	Gemeiner Tausendfüßer, Luzernentausendfüßer	black millepede, snake millepede
Cylindroiulus punctatus	Waldboden-Tausendfüßer*	woodland floor millepede
Cylindronotus laevioctostriatus/ Cylindronotus caraboides	Baumritzenkäfer	
Cylindrophis rufus	Rote Walzenschlange, Rotschwanz-Rollschlange	pipe snake
Cylindrophis spp.	Walzenschlangen, Asiatische Rollschlangen	pipe snakes
Cylindrotomidae	Moosmücken	moss craneflies*
Cymatiidae/Ranellidae	Tritons, Tritonen, Tritonschnecken	tritons, rock whelks
Cymatium aquitile		aquitile hairy triton

Cymatium caudatum	Krummhals-Triton*	bent-neck triton
Cymatium comptum	Zwerg-Triton	dwarf triton
Cymatium corrugatum	Haar-Triton	hairy triton, corrugated triton
Cymatium femorale	Kanten-Triton*	angular triton
Cymatium flaveolum	Breitband-Triton	broad-banded triton
Cymatium hepaticum	Streifen-Triton	liver-colored triton, black-striped triton
Cymatium labiosum	Lippen-Triton*	lip triton, wide-lipped triton
Cymatium martinianum		hairy triton
Cymatium muricinum	Warzen-Triton*	knobbed triton
Cymatium nicobaricum	Goldmund-Triton, Nicobaren-Triton	goldmouth triton, Nicobar hairy triton
Cymatium parthenopeum	Riesen-Triton, Neapolitanische Triton*	giant triton, Neapolitan triton
Cymatium pileare	Atlantische Haar-Triton	Atlantic hairy triton, common hairy triton
Cymatium retusum	Stumpfe Triton	blunted triton
Cymatium rubeculum	Rote Triton, Rotkehlchen-Triton	red triton, robin redbreast triton
Cymatium spp.	Tritons, Tritonen, Tritonschnecken, Tritonshörner	tritons
Cymatium tabulatum	Stufen-Triton*	shouldered triton
Cymatium tenuiliratum	Schlank-Triton*	slender triton
Cymatium testudinarium	Schildkröten-Triton	tortoise triton
Cymatium trigonum	Dreikant-Haar-Triton	trigonal hairy triton
Cymatium tripus	Dreifuß-Triton	tripod triton
Cymatium vestitum	Gewand-Triton*	garment triton
Cymatophorima diluta	Violettgrauer Eulenspinner	
Cymbiola aulica	Ohr-Walzenschnecke	princely volute
Cymbiola imperialis	Fürstliche Walzenschnecke*	imperial volute
Cymbiola magnifica	Prächtige Walzenschnecke*	magnificent volute
Cymbiola nivosa	Schneeflocken-Walzenschnecke*	snowy volute
Cymbiola nobilis	Edle Walzenschnecke	noble volute
Cymbiola rutila	Blutrote Walzenschnecke	blood-red volute
Cymbiola vespertilio	Fledermaus-Walzenschnecke	bat volute
Cymbiolacca pulchra	Hübsche Walzenschnecke	beautiful volute
Cymbium cucumis	Gurkenwalze, Gurken-Walzenschnecke	cucumber volute
Cymbium cymbium	Falsche Elefantenschnauzen-Walze*	false elephant's snout volute
Cymbium glans	Elefantenschnauzen-Walze*	elephant's snout volute
Cymbium olla	Kahnschnecke	Olla volute
Cymbium papillatum	Ohrenwalze	
Cymbium pepo	Afrikanische Neptun-Walzenschnecke	African neptune volute
Cymindis humeralis	Schulter-Nachtläufer, Rotschulteriger Nachtkäfer	
Cymindis macularis	Gefleckter Nachtläufer	
Cymindis vaporariorum	Rauchbrauner Nachtläufer	

Cynictis penicillata	Fuchsmanguste	yellow mongoose
Cynipidae	Gallwespen	gallwasps
Cynips divisa	Eichen-Glanzgallwespe (>Braune Glanzgalle)	oak bud red-gall cynipid wasp, oak bud cherry-gall cynipid wasp
Cynips longiventris	Eichenstreifgallwespe	oak leaf striped-gall cynipid wasp
Cynips quercusfolii	Gemeine Eichengallwespe	common oak gallwasp, oak leaf cherry-gall cynipid (>cherry gall)
Cynocephalus variegatus	Malaien-Gleitflieger, Temminck-Gleitflieger	Malayan flying lemur
Cynocephalus volans	Philippinen-Gleitflieger	Philippine flying lemur
Cynogale bennettii	Mampalon	otter civet
Cynoglossidae	Hundszungen	tonguefishes
Cynoglossus sinusarabici	Rotmeer-Zunge	Red Sea tongue-sole, Red Sea tonguesole
Cynolebias bellottii	Blauer Fächerfisch	pavito, Argentine pearlfish
Cynolebias nigripinnis	Schwarzflossen-Fächerfisch, Sternhimmelfisch	Argentine pearlfish
Cynomys gunnisoni	Gunnisons Präriehund	Gunnison's prairie dog
Cynomys leucurus	Weißschwanz-Präriehund	white-tailed prairie dog
Cynomys ludovicianus	Schwarzschwanz-Präriehund	black-tailed prairie dog, Plains prairie dog
Cynomys mexicanus	Mexikanischer Prairiehund	Mexican prairie dog
Cynomys parvidens	Utah-Präriehund	Utah prairie dog
Cynoponticus ferox	Guinea-Messerzahnaal	fierce conger, Guinea pike conger (FAO)
Cynops ensicauda	Schwertschwanzmolch	swordtail newt, sword-tailed newt
Cynops orientalis	Chinesischer Zwergmolch	Chinese dwarf newt, Chinese fire bellied newt
Cynops pyrrhogaster	Japanischer Feuerbauchmolch	Japanese firebelly newt, Japanese fire bellied newt
Cynopterus spp.	Kurznasen-Flughunde	short-nosed bats, dog-faced fruit bats
Cynoscion analis	Peruanischer Umberfisch	Peruvian weakfish
Cynoscion nebulosus	Gefleckter Umberfisch	spotted sea trout, spotted weakfish (FAO)
Cynoscion regalis	Königs-Corvina	grey weakfish, gray weakfish (FAO)
Cynoscion striatus	Gestreifter Umberfisch	striped weakfish
Cynthia cardui/ Vanessa cardui	Distelfalter	painted lady, thistle (butterfly)
Cyphoma gibbosa	Flamingozunge	flamingo tongue
Cyphophthalmi	Zwergweberknechte	mite harvestmen
Cyphostethus tristriatus	Buntrock (Wanze)	
Cyphotilapia frontosa	Beulenkopf-Buntbarsch, Buckelkopf-Buntbarsch, Zebra-Beulenkopf, „Frontosa"	humphead cichlid (FAO), 'frontosa'
Cypraea achatidea	Achatkauri	agate cowrie
Cypraea annulus	Ringkauri, Goldringer	ring cowrie, goldringer
Cypraea arabica	Arabische Kauri, Araberkauri	Arabian cowrie
Cypraea arabicula	Kleine Araberkauri	little Arabian cowrie

Cypraea argus	Argusaugenkauri, Augenfleck-Kaurischnecke	eyed cowrie, hundred-eyed cowrie
Cypraea asellus	Breitbandkauri*	asellus cowrie
Cypraea auriantium	Orangenkauri, Goldene Kauri	golden cowrie
Cypraea bistrinotata	Gelbbraunpunkt-Kauri	treble-spotted cowrie
Cypraea camelopardalis	Giraffenkauri	giraffe cowrie
Cypraea caputdraconis	Drachenkopfkauri	dragon-head cowrie, dragon's head cowrie
Cypraea caputserpentis	Schlangenkopfkauri	snake's head cowrie, serpent's head cowrie
Cypraea caurica	Caurica-Kauri	caurica cowrie
Cypraea cervinetta	Panama-Hirschkauri	Panamanian deer cowrie, little deer cowrie
Cypraea cervus/ Macrocypraea cervus	Atlantische Hirschkauri	Atlantic deer cowrie
Cypraea chinensis	China-Kauri	Chinese cowrie
Cypraea cicercula	Kichererbsen-Kauri*	chick-pea cowrie
Cypraea cinerea/ Luria cinerea/ Talparia cinerea	Graue Atlantik-Kauri	Atlantic grey cowrie, Atlantic gray cowrie (U.S.), ashen cowrie,
Cypraea coxeni	Coxs Kauri	Cox's cowrie
Cypraea cribraria	Siebkauri	sieve cowrie
Cypraea diluculum	Dämmerungskauri	dawn cowrie
Cypraea eburnea	Elfenbeinkauri*	pure white cowrie
Cypraea edentula/ Cypraeovula edentula	Zahnlose Kauri	toothless cowrie
Cypraea errones	Wandernde Kauri	wandering cowrie
Cypraea felina	Katzenkauri	cat cowrie, kitten cowrie
Cypraea globulus	Kugelkauri*	globose cowrie
Cypraea gracilis/ Cypraea notata/ Cypraea irescens	Zierliche Kauri	graceful cowrie
Cypraea granulata	Granula-Kauri	granulated cowrie
Cypraea guttata	Weißpunktkauri	white-spot cowrie, great spotted cowrie
Cypraea helvola	Honigkauri	honey cowrie
Cypraea hesitata/ Umbilia hesitata	Wunderkauri	umbilicate cowrie, undecided cowrie
Cypraea hirundo	Schwalbenkauri	swallow cowrie
Cypraea isabella/ Luria isabella	Isabellfarbene Kauri	isabelline cowrie, dirty-yellow cowrie
Cypraea lamarckii	Lamarcks Kauri	Lamarck's cowrie
Cypraea lurida/Luria lurida	Braune Kauri, Braune Maus	lurid cowrie
Cypraea lynx/Lyncina lynx	Lynxkauri	lynx cowrie, bobcat cowrie
Cypraea mappa	Landkartenkauri, Landkartenschnecke	map cowrie
Cypraea mauritiana	Buckelkauri, Buckelschnecke	hump-backed cowrie, humpback cowrie
Cypraea miliaris	Hirsekauri*	millet cowrie
Cypraea moneta/ Monetaria moneta	Geldkauri, Geldschnecke	money cowrie

*C*ypraea

Cypraea mus	Mauskauri, Mausschnecke	mouse cowrie
Cypraea nivosa	Schneeflockenschnecke*, Schneeflocken-Kauri*, Wolkige Kauri	cloudy cowrie
Cypraea nucleus	Nukleus-Kauri	nucleus cowrie
Cypraea obvelata		walled cowrie, Tahiti gold ringer
Cypraea ocellata	Augenfleck-Kauri	ocellated cowrie, ocellate cowrie,
Cypraea onyx	Onyxkauri	onyx cowrie
Cypraea ovum	Orangezahnkauri*	orange-toothed cowrie
Cypraea pantherina	Pantherkauri, Pantherschnecke	panther cowrie
Cypraea poraria	Porige Kauri, Porenkauri	porous cowrie
Cypraea pulchra/ Luria pulchra	Hübschkauri	lovely cowrie
Cypraea pyriformis	Birnenförmige Kauri	pear-shaped cowrie
Cypraea pyrum/ Zonaria pyrum	Birnenkauri, Birnenschnecke, Birnenporzellane	pear cowrie
Cypraea quadrimaculata	Vierpunktkauri	four-spot cowrie, four-spotted cowrie
Cypraea scurra	Skurril-Kauri*, Narrenkauri*	jester cowrie
Cypraea spadicea/ Zonaria spadicea	Kastanienkauri	chestnut cowrie
Cypraea spp.	Kauris, Kaurischnecken, Porzellanschnecken, Porzellanen	cowries (*sg* cowrie or cowry)
Cypraea spurca	Variable Kauri	European yellow cowrie
Cypraea staphylaea	Traubenkauri	grape cowrie
Cypraea stolida	Stolidkauri	stolid cowrie
Cypraea subviridis	Grünliche Kauri	green-tinted cowrie, greenish cowrie
Cypraea talpa/ Talparia talpa	Maulwurfskauri	mole cowrie
Cypraea teres	Zulaufende Kauri*	tapering cowrie
Cypraea tessellata/ Luria tessellata	Karierte Kauri	checkerboard cowrie, checkered cowrie
Cypraea tigris	Tigerkauri, Tigerschnecke	tiger cowrie
Cypraea turdus	Taubenkauri*	thrush cowrie
Cypraea ursellus	Kleine Bärenkauri	little bear cowrie
Cypraea vitellus	Pazifische Rehschnecke	Pacific deer cowrie, little-calf cowrie
Cypraea xanthodon	Gelbzahnkauri	yellow-toothed cowrie
Cypraea zebra/ Macrocypraea zebra	Zebrakauri, Gemaserte Kauri*	zebra cowrie, measled cowrie
Cypraea ziczac	Zickzackkauri	zigzag cowrie
Cypraea zonaria/ Zonaria zonaria	Zonenkauri	zoned cowrie
Cypraecassis rufa	Rote Porzellanschnecke, Rote Helmschnecke, Feuerofen	bullmouth helmet, bull's-mouth conch, red helmet

Cypraecassis testiculus	Netz-Helmschnecke*	reticulate cowrie-helmet
Cypraeidae	Kaurischnecken, Porzellanschnecken	cowries
Cyprinidae	Karpfenfische	minnows, carps
Cypriniformes	Karpfenfische, Karpfenartige	carps & characins & minnows & suckers & loaches
Cyprinodon variegatus	Edelsteinkärpfling	sheepshead minnow
Cyprinodontidae	Eierlegende Zahnkarpfen, Killifische	pupfishes
Cyprinodontiformes	Kleinkärpflinge	killifishes
Cyprinus carpio	Karpfen, Flusskarpfen	carp, common carp (FAO), European carp
Cypselurus agoo/ Cheilopogon agoo	Japanischer Flugfisch, Japanischer Fliegender Fisch	Japanese flyingfish
Cypselurus comatus	Hellflügel-Flugfisch*	clearwing flyingfish
Cypselurus cyanopterus/ Cheilopogon cyanopterus	Geränderter Flugfisch	margined flyingfish
Cypselurus exsiliens/ Cheilopogon exsiliens	Gestreiftflügel-Flugfisch	bandwing flyingfish
Cypselurus furcatus/ Cheilopogon furcatus	Gefleckter Flugfisch, Gefleckter Fliegender Fisch	spotfin flyingfish
Cypselurus melanurus/ Cheilopogon melanurus	Atlantischer Flugfisch, Atlantischer Fliegender Fisch	Atlantic flyingfish
Cyrtoclytus capra	Haarbock	
Cyrtodactylus kotschyi/ Mediodactylus kotschyi	Ägäischer Nacktfinger	Kotschy's gecko
Cyrtodactylus spp.	Nacktfingergeckos, Bogenfingergeckos	bow-fingered geckos
Cyrtonyx montezumae	Montezumawachtel	Montezuma's quail
Cyrtophora citricola	Opuntienspinne	fig-cactus spider
Cyrtopleura costata	Engelsflügel	angel's wings
Cystobranchus mammillatus	Ruttenegel, Quappenegel	burbot leech
Cystobranchus respirans	Barbenegel, Platter Fischegel	flat fish leech
Cystophora cristata	Klappenmütze, Klappmütze	hooded seal (bladder-nose)
Cytodites nudus	Fallschirm-Milbe*	air-sac mite

Daboia lebetina/ *Vipera lebetina*	Levante-Otter, Levanteotter	blunt-nosed viper, levantine viper
Daboia palaestinae/ *Vipera palaestinae*	Palästinaviper	Palestinian viper
Daboia russelli/ *Vipera russelli*	Kettenviper, Russells Viper	Russell's viper
Daboia xanthina/ *Vipera xanthina*	Bergotter	European coastal viper, Ottoman viper, Near East viper
Dacelo novaeguineae	Jägerliest, Lachender Hans	laughing kookaburra
Dacelonidae	Lieste, Jägerlieste	dacelonid kingfishers
Dacne bipustulata	Zweifleckiger Pilzkäfer	
Dacnomys millardi	Millard-Ratte	large-toothed giant rat, Millard's rat
Dactylomys spp.	Fingerratten	coro-coros
Dactylopius coccus	Cochenillelaus	cochineal insect
Dactylopsila spp.	Streifenbeutler, Streifenphalanger, Streifenkletterbeutler	striped possums
Dactylopterus volitans	Flughahn	flying gurnard
Dactyloscopidae	Sandsterngucker	sand stargazers
Dalatias licha/ *Scymnorhinus licha/* *Squalus licha*	Schokoladenhai	kitefin shark, seal shark, darkie charlie
Dalatiidae	Unechte Dornhaie, Eishaie	sleeper sharks
Dalophis boulengeri	Delphinaal	Boulenger's snake eel*
Dalophis imberbis	Armloser Delphinaal*	armless snake eel
Dalopius marginatus	Geränderter Schnellkäfer	
Dama dama/Cervus dama	Damhirsch	fallow deer
Damaliscus dorcas	Buntbock, Blessbock	bontebok, blesbok
Damaliscus hunteri	Hunters Leierantilope	Hunter's antelope, Hunter's hartebeest, hirola, sassaby
Damaliscus lunatus	Leierantilope	topi, tsessebi
Danaus plexippus	Monarchfalter	monarch, milkweed
Danio aequipinnatus/ *Brachydanio aequipinnatus*	Königsbärbling, Malabarbärbling	giant danio
Danio spp.	Danio-Bärblinge	danios
Daphnia pulex	Gemeiner Wasserfloh	common water flea
Daphnis nerii	Oleanderschwärmer	oleander hawkmoth
Daption capense	Kapsturmvogel	pintado petrel
Daptomys spp.	Südamerikanische Fischratten u.a.	fish-eating rats a.o., aquatic rats a.o.
Dardanus calidus	Großer Roter Einsiedlerkrebs, Mittelmeer-Dardanus	great red hermit crab, Mediterranean hermit crab
Dardanus deformis	Felsen-Dardanus	rock hermit crab
Dardanus lagopodes	Schwarzaugen-Dardanus	black-eyed hermit crab
Dardanus venosus	Sternaugen-Einsiedlerkrebs	stareye hermit crab
Darioconus textile/ *Conus textile*	Netz-Kegelschnecke	textile cone
Dascillidae	Moorweichkäfer, Dascilliden	soft-bodied plant beetles

Dascyllus aruanus	Preußenfisch	black-and-white damselfish, zebra humbug, white-tailed humbug, whitetail dascyllus (FAO)
Dascyllus carneus	Rückenfleck-Preußenfisch	two-striped damselfish, two-bar humbug, cloudy dascyllus
Dascyllus melanurus	Perl-Preußenfisch	blacktail humbug (FAO), blacktail dascyllus
Dascyllus spp.	Preußenfische	humbugs
Dascyllus trimaculatus	Schwarzer Preußenfisch, Samtjungferchen	domino damselfish, white-spotted damselfish, threespot dascyllus (FAO)
Dasia spp.	Baumskinke	dasias
Dasineura abietiperda	Fichtentrieb-Gallmücke	spruce-bud gall midge*
Dasineura affinis	Veilchenblattroll-Gallmücke	violet leaf midge
Dasineura alopecuri	Rote Fuchsschwanz-Gallmücke	foxtail midge
Dasineura brassicae	Kohlschoten-Gallmücke	cabbage pod midge
Dasineura laricis/ Dasineura kellneri	Lärchenknospen-Gallmücke	larch bud midge
Dasineura plicatrix	Brombeerblatt-Gallmücke	blackberry leaf midge, bramble leaf midge
Dasineura rhodophaga	Rosengallmücke	rose midge
Dasineura rosaria	Weidenrosen-Gallmücke	willowrose gall midge*
Dasineura saliciperda	Weidenholz-Gallmücke	willow-wood gall midge*
Dasineura salicis	Weidenruten-Gallmücke	willow-twig gall midge
Dasineura tetensi/ Perrisia tetensi	Johannisbeerblatt-Gallmücke	black currant leaf midge
Dasineura trifolii	Kleeblatt-Gallmücke	clover leaf midge
Dasyatidae	Stechrochen, Stachelrochen	stingrays (butterfly rays)
Dasyatis akajei	Pazifischer Stechrochen, Pazifischer Peitschenrochen	whip stingray, whip ray, red stingray (FAO)
Dasyatis americana	Amerikanischer Stechrochen	southern stingray (FAO), sting ray
Dasyatis brevicaudata	Indopazifischer Stechrochen	short-tail stingray, giant short-tail stingray of Australia
Dasyatis centroura	Rauer Stechrochen, Brucko	roughtail stingray
Dasyatis dipterurus	Diamant-Stechrochen	diamond stingray
Dasyatis pastinaca	Stechrochen, Gewöhnlicher Stechrochen	blue stingray, European stingray, common stingray (FAO)
Dasyatis sabina	Atlantischer Stechrochen	Atlantic stingray
Dasyatis sayi	Stumpfnasen-Stechrochen	bluntnose stingray
Dasyatis violacea	Violetter Stechrochen	pelagic stingray (FAO), violet stingray
Dasycercus cristicauda	Kammschwanzbeutelmaus	mulgara
Dasyceridae	Moosschimmelkäfer	minute scavenger beetles
Dasychira abietis	Tannenstreckfuß	fir-and-spruce tussock
Dasychira pudibunda/ Calliteara pudibunda/ Elkneria pudibunda	Buchenrotschwanz, Rotschwanz, Streckfuß	pale tussock, red-tail moth
Dasymutilla occidentalis	Schnelle Spinnenameise*, „Kuhtöter"	cow killer
Dasymutilla spp.	Amerikanische Spinnenameisen	velvet-ants

Dasymys incomtus	Afrikanische Sumpfratte, Wollhaarratte	shaggy swamp rat
Dasyneura see Dasineura		
Dasyornithinae	Lackvögel	bristlebirds
Dasypeltis scabra	Afrikanische Eierschlange	egg-eating snake, African egg-eating snake
Dasypoda spp.	Hosenbienen	hairy-legged bees
Dasypolia templi	Graugelbe Rauhaareule, Bärenklau-Rauhaareule	
Dasyprocta aguti	Goldrückenaguti	orange-rumped agouti
Dasyprocta punctata	Mittelamerikanisches Aguti	Central American agouti
Dasyprocta spp.	Stummelschwanzagutis	agoutis
Dasyproctidae	Agutis	agoutis
Dasypus novemcinctus	Neunbinden-Gürteltier	nine-banded armadillo
Dasypus septemcinctus	Siebenbinden-Gürteltier	seven-banded armadillo
Dasypus spp.	Weichgürteltiere	long-nosed armadillos
Dasytes plumbeus	Bleischwarzer Wollhaarkäfer	
Dasytidae	Wollhaarkäfer	dasytids, dasytid beetles (soft-winged flower beetles)
Dasyuroides byrnei	Doppelkammbeutelmaus	kowari
Dasyurus geoffroii	Chuditch	chuditch, western quoll
Dasyurus maculatus	Großer Tüpfelbeutelmarder	large spotted quoll, large spotted native "cat"
Dasyurus spp.	Tüpfelbeutelmarder	quolls, native "cats", tiger "cats"
Daubentonia madagascariensis	Fingertier	aye-aye
Davainea proglottina	Hausgeflügel-Bandwurm*	poultry tapeworm
Deania calcea	Schnabeldornhai	shovelnose shark, birdbeak dogfish (FAO)
Decapoda	Decapoden, Zehnfußkrebse	decapods
Decapoda/Decabrachia	Zehnarmige Tintenschnecken	cuttlefish & squids
Decapterus macarellus	Gelbschwanz-Stachelmakrele	mackerel scad
Decapterus maruadsi	Langflossen-Stachelmakrele	long-fin scad, Japanese scad (FAO)
Decapterus punctatus	Gefleckte Stachelmakrele, Ohrfleck-Heringsmakrele	round scad
Decapterus russelli	Indische Stachelmakrele	Indian scad
Decticus albifrons	Südlicher Warzenbeißer	white-faced bushcricket, Mediterranean wart-biter
Decticus verrucivorus	Warzenbeißer	wart-biter, wart-biter bushcricket
Deilephila elpenor	Mittlerer Weinschwärmer	elephant hawkmoth
Deilephila porcellus	Kleiner Weinschwärmer	small elephant hawkmoth
Deinagkistrodon acutus/ Agkistrodon acutus	Chinesische Nasenotter	sharp-nosed pit viper, sharp-nosed viper, hundred-pace viper, hundred pacer
Deinopidae/Dinopidae	Ogerspinnen, Kescherspinnen	ogre-faced spiders
Deinopis longipes	Ogerspinne, Kescherspinne	net-casting spider
Deirochelys reticularia	Langhals-Schmuckschildkröte	chicken turtle

Delanymys brooksi	Delanys Sumpfklettermaus	Delany's swamp mouse
Delia antiqua/ Phorbia antiqua/ Hylemyia antiqua	Zwiebelfliege	onion fly, onion maggot
Delia coarctata/ Phorbia coarctata	Brachfliege	wheat bulb fly
Delia floralis/ Hylemia floralis/ Phorbia floralis/ Chortophila floralis	Große Kohlfliege	cabbage-root fly, radish fly, turnip maggot
Delia platura/ Hylemia platura/ Phorbia platura	Bohnenfliege, Kammschienen-Wurzelfliege, Schalottenfliege	seed-corn fly, seed-corn maggot, bean-seed fly, shallot fly
Delia radicum/ Hylemia brassicae/ Chortophila brassicae/ Paregle radicum	Kleine Kohlfliege, Wurzelfliege, Radieschenfliege	cabbage fly, cabbage maggot, radish fly
Delichon urbica	Mehlschwalbe	common house martin
Delphacidae	Spornzikaden, Stirnhöckerzirpen	delphacid planthoppers
Delphinapterus leucas	Weißwal, Beluga	white whale, beluga
Delphinus delphis	Gemeiner Delphin	common dolphin, crisscross dolphin, saddleback(ed) dolphin,
Deltentosteus quadrimaculatus	Vierflecken-Grundel	fourspot goby
Deltentosteus colonianus	Zackengrundel*	toothed goby
Deltocyathus calcar	Tiefsee-Sternkoralle*	deepsea star coral
Deltote bankiana		silver-barred moth
Deltote uncula	Ried-Grasmotteneulchen	
Demansia atra	Schwärzliche Braunschlange	black whipsnake
Demansia olivacea	Oliven-Braunschlange*	olive whipsnake
Demansia psammophis	Gelbschnauzen-Braunschlange*	yellow-faced whipsnake
Demansia spp.	Australische Braunschlangen	Australian whipsnakes, Australian brown snakes, venomous whip snakes
Demetrias atricapillus	Schwarzköpfiger Scheunenkäfer, Schwarzköpfiger Halmläufer	
Demetrias monostigma	Ried-Halmläufer	
Demodex bovis	Rinder-Haarbalgmilbe	cattle follicle mite
Demodex brevis	Talgfollikelmilbe*	lesser follicle mite, sebaceous follicle mite
Demodex canis	Hunde-Haarbalgmilbe	dog follicle mite
Demodex cati	Katzen-Haarbalgmilbe	cat follicle mite
Demodex folliculorum	Haarbalgmilbe	follicle mite, human follicle mite
Demodex phylloides	Schweine-Haarbalgmilbe	hog follicle mite, pig follicle mite, pig head mange mite
Demodicidae	Haarbalgmilben	follicle mites
Dendragapus canadensis	Fichtenhuhn	spruce grouse
Dendraster excentricus	Exzentrischer Sanddollar	eccentric sand dollar
Dendrelaphis spp.	Bronzenattern	Australian tree snakes, painted bronzebacks

Dendroaspis angusticeps	Blattgrüne Mamba, Gewöhnliche Mamba	eastern green mamba, common mamba
Dendroaspis polylepis	Schwarze Mamba	black mamba
Dendroaspis spp.	Mambas	mambas
Dendroaspis viridis	Grüne Mamba	green mamba, western green mamba
Dendrobates auratus	Goldbaumsteiger	green & black poison-arrow frog, green & black poison frog
Dendrobates azureus	Blauer Pfeilgiftfrosch	blue poison-arrow frog
Dendrobates histrionicus	Punktierter Pfeilgiftfrosch, Bunter Baumsteiger	red-and-black poison-arrow frog, harlequin poison frog
Dendrobates pumilio	Erdbeerfröschchen, Kleiner Erdbeerfrosch	strawberry poison-arrrow frog, red-and-blue poison-arrrow frog, flaming poison-arrrow frog
Dendrobates spp.	Baumsteigerfrösche	poison-arrow frogs, poison frogs
Dendrobates tinctorius	Färberfrosch	dyeing poison-arrow frog
Dendrobatidae	Baumsteigerfrösche, Farbfrösche, Pfeilgiftfrösche	dendrobatids, poison dart frogs, poison arrow frogs, arrow-poison frogs
Dendrobium pertinax/ Anobium pertinax	Trotzkopf	furniture beetle, furniture borer, common furniture beetle
Dendrochirus brachypterus	Roter Zwergfeuerfisch, Gewöhnlicher Zwergfeuerfisch	dwarf lionfish
Dendrochirus spp.	Zwergfeuerfische	lionfishes
Dendrochirus zebra	Zebra-Zwergfeuerfisch	zebra lionfish
Dendrocolaptinae	Baumsteiger, Baumkletterer	woodcreepers
Dendrocopos leucotos/ Picoides leucotos	Weißrückenspecht	white-backed woodpecker
Dendrocopos major/ Picoides major	Buntspecht	great spotted woodpecker
Dendrocopos medius/ Picoides medius	Mittelspecht	middle spotted woodpecker
Dendrocopos minor/ Picoides minor	Kleinspecht	lesser spotted woodpecker
Dendrocopos syriacus	Blutspecht	Syrian woodpecker
Dendroctonus micans	Riesenbastkäfer	European spruce beetle, great spruce bark beetle
Dendrocygna viduata	Witwenpfeifgans	white-faced whistling duck
Dendrocygnidae	Pfeifgänse (Baumenten)	whistling-ducks
Dendrodoa carnea	Bluttropfen-Seescheide	blood drop sea-squirt
Dendrodoa grossularia	Stachelbeer-Seescheide	gooseberry sea-squirt
Dendrogale spp.	Bergtupaias	small smooth-tailed tree shrews
Dendrogyra cylindrus	Säulenkoralle*	pillar coral
Dendrohyrax spp.	Baumschliefer	tree hyraxes, bush hyraxes
Dendroica coronata	Kronwaldsänger	yellow-rumped warbler
Dendroica discolor/ Agreocantor discolor	Rotscheitel-Waldsänger	prairie warbler
Dendroica petechia	Goldwaldsänger	yellow warbler
Dendroica striata	Streifenwaldsänger	blackpoll warbler
Dendrolagus spp.	Baumkängurus	tree kangaroos

Dendrolimus pini	Kiefernspinner	pine lappet, pine-tree lappet, European pine moth
Dendromus spp.	Aalstrich-Klettermäuse, Afrikanische Klettermäuse	African climbing mice, tree mice
Dendronanthus indicus	Baumstelze	forest wagtail
Dendronotacea	Bäumchenschnecken, Baumschnecken	dendronotacean snails, dendronotaceans
Dendronotus frondosus	Bäumchenschnecke	
Dendrophryniscus brevipollicatus	Kurzdaumige Baumkröte	coastal tree toad
Dendrophyllia cornigera	Bäumchenkoralle, Hornige Baumkoralle	horny tree coral*
Dendrophyllia ramea	Gelbe Baumkoralle, Baumkoralle, Astkoralle	yellow tree coral*
Dendropicos fuscescens	Kardinalspecht	cardinal woodpecker
Dendropoma corrodens	Geringelte Wurmschnecke	ringed wormsnail
Dendroprionomys rousseloti	Rousselot-Baummaus	Congo tree mouse
Denisonia maculata	Ornamentschlange	ornamental snake
Denisonia spp.	Ornamentschlangen	ornamental snakes
Dentaliidae (Scaphopoda)	Zahnschnecken, Elefantenzähne	tusks, tuskshells
Dentalium aprinum (Scaphopoda)	Eberzahnschnecke	boar's tusk
Dentalium corneum (Scaphopoda)	Horn-Zahnschnecke	horned tusk, horned tuskshell
Dentalium elephantinum (Scaphopoda)	Elefantenzahnschnecke	elephant's tusk, elephant tusk, elephant's tuskshell
Dentalium entale/ Antalis entale/ Antalis entalis (Scaphopoda)	Nordeuropäische Elefantenzahnschnecke	North European elephant tusk, common elephant tusk, entale tusk
Dentalium pretiosum/ Antalis pretiosum (Scaphopoda)	Indischer Elefantenzahn, Kostbarer Elefantenzahn	Indian money tusk, wampum tuskshell
Dentalium vulgare/ Antalis tarentinum (Scaphopoda)	Gemeiner Elefantenzahn, Gemeine Zahnschnecke	common elephant's tusk, common tusk (common tuskshell)
Dentex dentex	Zahnbrassen	dentex, common dentex (FAO)
Dentex gibbosus	Rosa Zahnbrassen	pink dentex
Dentex macrophthalmus	Großaugen-Zahnbrassen	large-eyed dentex, large-eye dentex (FAO)
Dentex maroccanus	Marokko-Brassen, Marokko-Zahnbrassen	Morocco dentex
Denticeps clupeoides	Stachelhering, Süßwasserhering	denticle herring (FAO), spiny herring
Denticipitidae/Denticipidae	Stachelheringe, Süßwasserheringe	denticle herrings
Denticollis linearis	Dornhals-Schnellkäfer	
Deomys ferrugineus	Insektenfressende Waldmaus	Congo forest mouse, link rat
Deporaus betulae	Birkentrichterwickler, Birkenblattroller	birch leafroller weevil
Deporaus tristis	Ahornblattroller	maple leafroller weevil

Depressaria nervosa	Kümmelmotte, Kümmelpfeifer, Möhrenschabe	carrot and parsnip flat-body moth
Derichthyidae	Langhalsaale	longneck eels
Dermacentor albipictus	Winterzecke	winter tick, shingle tick
Dermacentor andersoni	Rocky Mountain Holzzecke*	Rocky Mountains wood tick
Dermacentor marginatus	Schafzecke	European sheep tick
Dermacentor nigrolineatus	Braune Winterzecke*	brown winter tick
Dermacentor nitens/ *Anocentor nitens*	Tropische Pferdezecke	tropical horse tick
Dermacentor occidentalis	Pazifikküsten-Zecke*	Pacific Coast tick
Dermacentor reticulatus	Marschzecke*	marsh tick
Dermacentor spp.	Holzzecken*	eastern wood ticks
Dermacentor variabilis	Amerikanische Hundezecke	American dog tick
Dermanyssidae	Vogelmilben	dermanyssid mites, poultry mites
Dermanyssus gallinae	Hühnermilbe, Rote Vogelmilbe	red mite of poultry, poultry red mite, red chicken mite, chicken mite
Dermaptera	Ohrwürmer	earwigs
Dermasterias imbricata	Lederstern*	leather star
Dermatemys mawii	Tabasco-Schildkröte	Central American river turtle
Dermatobia hominis	Menschendasselfliege	human bot fly, torsalo
Dermatophagoides pteronyssinus	Hausstaubmilbe, Europäische Hausstaubmilbe	European house dust mite
Dermatophagoides farinae	Amerikanische Hausstaubmilbe	American house dust mite
Dermestes ater	Aas-Dornspeckkäfer	black larder beetle
Dermestes haemorrhoidales	Afrikanischer Speckkäfer	African larder beetle, black larder beetle
Dermestes lardarius	Gemeiner Speckkäfer	larder beetle, common larder beetle, bacon beetle
Dermestes maculatus	Dornspeckkäfer	hide beetle, common hide beetle, leather beetle
Dermestidae	Speckkäfer & Pelzkäfer	larder beetles, skin beetles & carpet beetles
Dermochelys coriacea	Lederschildkröte	leatherback sea turtle, leatherback, leathery turtle, luth turtle
Dermogenys pusillus	Hechtköpfiger Halbschnäbler	wrestling halfbeak
Dermophis oaxacae	Lafrentz-Hautwühle	Oaxacan caecilian
Dermophis spp.	Mexikanische Hautwühlen	Mexican caecilians
Dermoptera	Riesengleiter, Pelzflatterer, Riesengleitflieger	gliding lemurs, flying lemurs, colugos, dermopterans
Deroceras agreste/ *Agriolimax agreste*	Einfarbige Ackerschnecke, Graue Ackerschnecke	field slug, grey field slug, grey slug
Deroceras heterura	Sumpfschnegel	marsh slug (U.S.)
Deroceras laeve/ *Agriolimax laevis*	Wasserschnegel, Farnschnecke	meadow slug (U.S.), marsh slug, brown slug, smooth slug

Deroceras panormitanum/ Deroceras carnanae	Mittelmeer-Ackerschnecke, Kastanienschnecke*	longneck fieldslug (U.S.), chestnut slug (Br.), Carnana's slug, Sicilian slug
Deroceras reticulatum/ Agriolimax reticulatus	Genetzte Ackerschnecke, Gartenschnecke	netted slug, gray fieldslug (U.S.), grey field slug (Br.), milky slug
Deroceras spp.	Ackerschnecken	field slugs
Derodontidae	Knopfkäfer, Derodontiden	tooth-necked fungus beetles
Deroptyus accipitrinus	Fächerpapagei	hawk-headed parrot
Desidae	Meeresspinnen*	marine spiders
Desmana moschata	Russischer Desman	Russian desman
Desmodilliscus braueri	Brauers Sandrennmaus	pouched gerbil
Desmodillus auricularis	Namaqua-Sandrennmaus, Kurzschwanz-Sandrennmaus	Cape short-eared gerbil, short-tailed gerbil, Namaqua gerbil
Desmodontidae	Echte Vampire	vampire bats, common vampire bats
Desmodus rotundus	Gemeiner Vampir, Gewöhnlicher Vampir	vampire bat, common vampire bat
Desmognathus quadramaculatus	Schwarzbäuchiger Bachsalamander	black-bellied salamander
Desmognathus aeneus	Cherokee-Bachsalamander	seepage salamander
Desmognathus auriculatus	Süd-Bachsalamander	southern dusky salamander
Desmognathus fuscus	Brauner Bachsalamander	dusky salamander
Desmognathus monticola	Robben-Bachsalamander	seal salamander
Desmognathus ochrophaeus	Alleghany-Bachsalamander	mountain dusky salamander
Desmognathus spp.	Bachsalamander	dusky salamanders
Desmognathus wrighti	Zwergbachsalamander	pygmy salamander
Deuterostomia	Neumundtiere, Neumünder, Zweitmünder	deuterostomes
Diachrysia chrysitis	Messingeule	burnished brass
Diachrysia chryson	Goldfleckeule	scarce burnished brass moth
Diacria trispinosa	Dreistachlige Diacria	three-spine cavoline
Diacrisia sannio	Rotrandbär	clouded buff moth
Diadema antillorum	Antillen-Diademseeigel	long-spined sea urchin
Diadema savignyi	Savignys Diademseeigel	Savigny's hatpin urchin
Diadema setosum	Langstacheliger Diademseeigel	hatpin urchin, longspined sea urchin
Diadematidae	Diademseeigel	hatpin urchins
Diadophis punctatus	Halsbandnatter	ringneck snake
Diadophis spp.	Halsbandnattern	ringneck snakes
Diadumene cincta	Brackwasseraktinie	orange anemone
Diadumene leucolena	Geisteraktinie*	white anemone, ghost anemone
Diadumene luciae	Strandrose, Hafenrose	orange-striped anemone
Diaea dorsata	Grüne Krabbenspinne	green crab spider
Diaemus youngi	Weißflügelvampir	white-winged vampire bat
Dialeurodes chittendeni	Rhododendron-Weißfliege*	rhododendron whitefly
Diamanus montanus	Eichhörnchen-Floh	squirrel flea

Diaphana minuta	Arktische Papierblase	Arctic paperbubble
Diaphora mendica	Hellgrauer Fleckleib-Bär	muslin moth
Diarthronomyia chrysanthemi/ Rhopalomyia chrysanthemi	Chrysanthemen-Gallmücke	chrysanthemum gall midge
Diaspididae	Austernschildläuse, Deckelschildläuse, Echte Schildläuse	armored scales
Diaspis boisduvalii	Orchideenschildlaus	orchid scale
Diaspis bromeliae	Ananasschildlaus	pineapple scale
Diatraea saccharalis	Zuckerrohrzünsler	sugar cane borer
Diazona violacea	Fußball-Seescheide	football sea-squirt
Dibamidae	Schlangenschleichen	blind lizards
Dibamus spp.	Schlangenschleichen	Asian blind lizards
Dibranchus atlanticus/ Halieutaea senticosa	Atlantischer Fledermausfisch	Atlantic batfish
Dicaeidae	Mistelfresser, Blütenpicker	flowerpeckers
Dicamptodon atterimus	Idaho-Riesen-Querzahnmolch	Idaho giant salamander
Dicamptodon copei	Copes Riesen-Querzahnmolch	Cope's giant salamander
Dicamptodon ensatus	Kalifornischer Riesen-Querzahnmolch	California giant salamander
Dicamptodon tenebrosus	Pazifischer Riesen-Querzahnmolch	Pacific giant salamander
Dicentrarchus labrax/ Roccus labrax/ Morone labrax	Wolfsbarsch, Seebarsch	bass, sea bass
Dicentrarchus punctatus	Gefleckter Wolfsbarsch	spotted bass
Diceratiidae	Diceratiden	horned anglers
Dicerca alni	Kupferfarbener Erlenprachtkäfer	
Dicerorhinus sumatrensis	Sumatranashorn	Sumatran rhinoceros, hairy rhinoceros
Diceros bicornis	Spitzmaulnashorn, Spitzlippennashorn, „Schwarzes" Nashorn	black rhinoceros, hooked-lipped rhinoceros, browse rhinoceros
Dichocoenia stellaris	Pfannkuchen-Koralle*	pancake star coral
Dichocoenia stokesi	Ananaskoralle*	pineapple coral
Dichonia aprilina/ Griposia aprilina	Aprileule, Grüne Eicheneule, Lindeneule	merveille-du-jour, green owlet moth, common merveille-du-jour,
Diclidurus spp.	Gespenstfledermaus	ghost bats, white bats
Dicologlossa cuneata	Cuneata-Seezunge, Keilzunge*	wedge sole (FAO), Senegal sole
Dicologlossa hexophthalma	Sechsaugenzunge*	six-eyed sole
Dicrocoelium dendriticum/ Dicrocoelium lanceolatum/ Fasciola lanceolatum	Kleiner Leberegel (Lanzettegel)	lancet fluke, common lancet fluke
Dicrocoelium hospes	Afrikanischer Leberegel (Afrikanischer Lanzettegel)	African lancet fluke
Dicrostonyx spp.	Halsbandlemminge	collared lemmings, varying lemmings
Dicruridae	Drongos	drongos
Dicrurus forficatus	Gabeldrongo	crested drongo
Dicrurus hottentottus	Haarbuschdrongo	hair-crested drongo

Dicrurus paradiseus	Flaggendrongo	greater racquet-tailed drongo
Dictyna civica	Mauerspinne	dictynid spider
Dictynidae	Kräuselspinnen, Eigentliche Kräuselspinnen	dictynid spiders
Dictyophara europaea	Europäischer Laternenträger	European lanternfly
Dictyoptera aurora	Scharlachroter Netzkäfer	
Dicycla oo	Nulleneule, O-Eule, Eichen-Nulleneule, Eichenhochwald-Doppelkreiseule	heart moth
Dicyrtomidae	Spinnenspringer	dicyrtomid springtails
Didelphidae	Beutelratten	opossums, American opossums
Didelphis marsupialis	Südopossum	southern opossum
Didelphis spp.	Amerikanische Opossums	large American opossums
Didelphis virginiana	Nordopossum	Virginian opossum
Didemnum molle	Grüne-Riffseescheide	green reef sea-squirt
Didinium nasutum	Nasentierchen	
Didymomyia tiliacea	Linden-Gallmücke	lime leaf gall midge
Dienerella filum	Herbarkäfer*	herbarium beetle
Dileptus anser	Gänsetierchen	
Diloba caeruleocephala	Blaukopf	figure of eight moth
Dilophus febrilis	Strahlenmücke	fever fly, blossom fly
Dinapate wrighti	Palmenbohrer	palm borer
Dinaromys bogdanovi	Bergmaus	Martino's snow vole, Balkan snow vole
Dinemellia dinemelli	Starweber	white-headed buffalo weaver
Dinoderus bifoveolatus	Wurzelholzbohrer	root borer, West African ghoon beetle
Dinoderus minutus	Bambusbohrer	bamboo borer
Dinoflagellata	Panzergeißler, Dinoflagellaten	dinoflagellates
Dinolestes lewini	Langflossenhecht*	long-finned pike
Dinomyidae	Pakaranas	pacaranas
Dinomys branickii	Pakarana	pacarana
Dinopercidae	Höhlenbarsche	cavebass
Dinornis maximus	Riesenmoa	giant moa
Dinornithiformes	Moas	moas
Dioctophyme renale	Nierenwurm	kidney worm
Diodon holacanthus	Brauner Igelfisch	long-spine porcupinefish (FAO), balloon porcupinefish, balloonfish
Diodon hystrix	Gewöhnlicher Igelfisch, Gemeiner Igelfisch, Kosmopolit-Igelfisch, Gepunkteter Igelfisch	common porcupinefish, spotted porcupinefish, spot-fin porcupinefish (FAO)
Diodontidae	Igelfische, Zweizähner	porcupinefishes (burrfishes)
Diodora aspera	Rauschalige Schwellenschnecke, Raue Schwellenschnecke	rough keyhole limpet
Diodora gibberula	Bucklige Schwellenschnecke	humped keyhole limpet
Diodora graeca/ Diodora apertura	Europäische Schwellenschnecke	common keyhole limpet

Diogenes pugilator	Sand-Einsiedler, Strandeinsiedler, Kleiner Einsiedlerkrebs	Roux's hermit crab
Diogenidae	Linkshänder-Einsiedlerkrebse	left-handed hermit crabs
Diomedea albatrus	Kurzschnabelalbatros	short-tailed albatross
Diomedea exulans	Wanderalbatros	wandering albatross
Diomedea irrorata	Galapagosalbatros	waved albatross
Diomedea melanophris	Schwarzbrauenalbatros	black-browed albatross
Diomedeidae	Albatrosse	albatrosses
Diomys crumpi	Crump-Maus	Manipur mouse, Crump's mouse
Diopsidae	Stielaugenfliegen	stalk-eyed flies
Dioryctria abietella	Fichtenzapfenzünsler, Fichtentriebzünsler	pine knothorn moth
Diphasia pinastrum	Kiefernmoos*	sea pine hydroid
Diphasia rosacea	Lilienmoos*	lily hydroid
Diphylla ecaudata	Kleiner Blutsauger, Kammzahnvampir	hairy-legged vampire bat
Diphyllobothrium latum	Fischbandwurm, Breiter Bandwurm	broad fish tapeworm, broad tapeworm
Diphyllobothrium spp.	Fischbandwürmer	fish tapeworms
Diplecogaster bimaculata	Zweiflecken-Ansauger	two-spotted clingfish
Diplectrum formosum	Sand-Zackenbarsch, Sandbarsch	sand perch
Diploastrea heliopora	Blumensternkoralle	flowery star coral*
Diplobatis ommata	Bullaugen-Zitterrochen*	ocellated electric ray (FAO), bull's eye torpedo
Diplodactylus ciliaris	Stachelschwanzgecko	spiny-tailed gecko, spinytail gecko
Diplodactylus spp.	Doppelfingergeckos	two-toed geckos, spinytail geckos
Diplodactylus stenodactylus	Kronengecko	crowned gecko, long-fingered gecko
Diplodactylus taenicauda	Goldschwanzgecko	golden-tailed gecko, goldentail gecko
Diplodactylus vittatus	Steingecko	wood gecko, stone gecko
Diplodactylus williamsi	Williams-Gecko	Williams' spinytail gecko, Williams' diplodactyl
Diplodillus spp.	Kurzschwanz-Rennmäuse	short-tailed gerbils
Diplodus annularis	Ringelbrassen	annular seabream (FAO), annular bream, annular gilthead
Diplodus cervinus	Zebrabrassen, Bänderbrassen	zebra seabream
Diplodus holbrookii	Holbrooks Brassen	spottail pinfish (FAO), spot-tail seabream
Diplodus puntazzo	Spitzbrassen	sharpsnout seabream (FAO), sheepshead bream
Diplodus sargus	Weißbrassen, Großer Geißbrassen	white bream, white seabream (FAO)
Diplodus vulgaris	Zweibinden-Brassen	two-banded bream, common two-banded seabream (FAO)
Diploglossus costatus	Hispaniola-Schleiche	common galliwasp

Diploglossus spp.	Doppelzungenschleichen, Gallwespenschleichen	galliwasps
Diplolaemus bibronii	Bibrons Eidechse	Bibron's iguana
Diplolaemus darwinii	Darwins Eidechse	Darwin's iguana
Diplolaemus spp.	Patagonische Erdleguane	Patagonia iguanas
Diplolepis eglanteriae	Glattkugel-Rosengallwespe*	rose smooth pea-gall cynipid wasp (>smooth pea gall)
Diplolepis nervosa	Dornen-Rosengallwespe*	rose spiked pea-gall cynipid wasp (>spiked pea gall)
Diplolepis rosae	Gemeine Rosengallwespe	mossy rose gall wasp, bedeguar gall wasp (>bedeguar gall/Robin's pincushion)
Diplomesodon pulchellum	Gescheckte Spitzmaus	piebald shrew, Turkestan desert shrew
Diplomys labilis	Gleit-Borstenratte*	gliding spiny rats
Diplomys spp.	Baumstachelratten	forest spiny rats
Diploria clivosa	Knotige Hirnkoralle	knobby brain coral
Diploria labyrinthiformis	Labyrinthkoralle, Gefurchte Hirnkoralle, Feine Atlantische Hirnkoralle	labyrinthine brain coral, grooved brain coral, depressed brain coral
Diploria spp.	Hirnkorallen u.a.	brain corals a.o.
Diploria strigosa	Symmetrische Hirnkoralle, Gewöhnliche Hirnkoralle	symmetrical brain coral
Diplosaurus dorsalis	Wüstenleguan	desert iguana
Diplostraca/Onychura	Doppelschaler, Krallenschwänze	clam shrimps & water fleas
Diplothrix legata/ Rattus legata	Ryukyu-Ratte	Ryukyu Island rat
Dipluridae	Trichternetz-Vogelspinnen	funnel-web spiders, funnel-web tarantulas, sheetweb building tarantulas
Dipnoi	Lungenfische	lungfishes
Dipodidae	Springmäuse	jerboas
Dipodomys ingens	Riesen-Kängururatte	giant kangaroo rat
Dipodomys spectabilis	Bannerschwanz-Kängururatte	bannertail
Dipodomys spp.	Taschenspringer, Kängururatten	kangaroo rats, California kangaroo rats
Dipoena spp.	Ameisenspinnen u.a.	ant spiders a.o.
Diporiphora bilineata	Zweistreifen-Bodenagame	two-lined dragon
Diporiphora spp.	Zweiporen-Bodenagamen	two-pored dragons
Diprion pini	Gemeine Kiefernbuschhornblattwespe	pine sawfly
Diprionidae	Buschhornblattwespen	conifer sawflies
Dipsas indica	Südamerikanische Dickkopfnatter	Indian snail-eating snake
Dipsas spp.	Dickkopfnattern, Schneckennattern	snail-eating snakes
Dipsas variegata variegata	Schneckennatter, Veränderliche Dickkopfnatter	snail-eating snake, thirst snake
Dipturus batis/Raja batis	Europäischer Glattrochen, Spiegelrochen	common skate, common European skate, grey skate, blue skate, skate (FAO)
Dipturus laevis/Raja laevis	Westatlantischer Glattrochen, Spitznasen-Rochen	barndoor skate (FAO), sharpnose skate

Dipturus linteus/*Raja lintea*	Nordatlantischer Weißrochen, Nördlicher Weißrochen	sailray (FAO), sharpnose skate
Dipturus nasutus/ *Raja nasuta*	Australoasiatischer Raurochen, Raurochen	rough skate (Australasian)
Dipus sagitta	Raufuß-Springmaus	rough-legged jerboa, northern three-toed jerboa
Dipylidium caninum	Gurkenkernbandwurm	double-pored dog tapeworm
Diretmidae	Silberköpfe	spinyfins
Dirofilaria immitis	Hundeherzwurm	heartworm, dog heartworm
Dirofilaria spp.	Herzwürmer	heartworms
Discoglossidae	Scheibenzüngler	disk-tongued toads
Discoglossus nigriventer	Israelischer Scheibenzüngler	Israel painted frog
Discoglossus pictus	Gemalter Scheibenzüngler	painted frog
Discoglossus sardus	Sardischer Scheibenzüngler	Tyrrhenian painted frog
Discosoma spp.	Scheibenanemonen	disc anemones, disc corallimorphs
Discus perspectivus	Gekielte Schüsselschnecke	keeled disc snail
Discus rotundatus	Gefleckte Schüsselschnecke	rounded snail, rotund disc snail, radiated snail
Discus ruderatus	Braune Schüsselschnecke	brown disc snail
Dispholidus typus	Boomslang, Grüne Baumschlange	boomslang
Dissostichus eleginoides	Schwarzer Seehecht, Schwarzer Zahnfisch	Patagonian toothfish
Dissostichus mawsoni	Antarktischer Zahnfisch	Antarctic toothfish
Disteira spp.	Seeschlangen	common sea snakes
Distoechurus pennatus	Federschwanzbeutler	feather-tailed possum
Distorsio anus	Gemeine Distorsio	common distorsio
Distorsio clathrata	Atlantische Distorsio	Atlantic distorsio
Ditoma crenata	Gekielter Rindenkäfer	
Ditylenchus angustus	Reisstengelälchen	rice nematode (ufra disease of rice)
Ditylenchus destructor	Kartoffelkrätzeälchen (Nematodenfäule der Kartoffel)	potato rot nematode
Ditylenchus dipsaci	Luzerneälchen, Stengelälchen, Stockälchen (Rübenkopfälchen)	lucerne stem nematode, stem-and-bulb eelworm, stem and bulb nematode, stem nematode, bulb nematode (potato tuber eelworm)
Ditylenchus radicicola/ *Subanguina radicicola*	Graswurzelälchen	grass nematode
Dobsonia spp.	Nacktrücken-Flughund	bare-backed fruit bats
Dociotaurus hispanicus/ *Ramburiella hispanica*	Spanischer Grashüpfer	Iberian cross-backed grasshopper
Dociotaurus maroccanus	Marokkanische Wanderheuschrecke	Moroccan locust
Dogania subplana/ *Trionyx subplanus*	Malayische Weichschildkröte, Malayen-Weichschildkröte	Malayan softshell turtle
Dolichoderinae	Schuppenameisen, Drüsenameisen	dolichoderine ants
Dolichonabis limbatus	Sumpfräuber	marsh damsel bug
Dolichonyx oryzivorus	Bobolink	bobolink
Dolichopodidae	Langbeinfliegen	long-headed flies

Dolichotis spp.	Maras	Patagonian cavies, Patagonian "hares", maras
Dolichovespula maculata	Gefleckte Hornisse	bold-faced hornet
Dolichovespula media	Mittlere Hornisse	medium wasp*
Dolichovespula norwegica	Norwegische Wespe	Norwegian wasp, Norway wasp
Dolichovespula saxonica	Kleine Hornisse, Sächsische Wespe	Saxon wasp*
Dolichovespula spp.	Langkopfwespen	long-headed wasps
Dolichovespula sylvestris	Waldwespe	tree wasp, wood wasp
Dologale dybowskii	Listige Manguste	African tropical savannah mongoose
Dolomedes fimbriatus	Gebänderte Listspinne, Gerandete Jagdspinne	fimbriate fishing spider*
Dolomedes plantarius	Moor-Jagdspinne	fen raft spider, great raft spider
Dolomedes spp.	Jagdspinnen	fishing spiders
Dolomedes triton	Sechspunkt-Jagdspinne	sixspotted fishing spider
Dolycoris baccarum	Beerenwanze	sloe bug, sloebug
Donacia spp.	Rohrkäfer, Schilfkäfer	reed beetles
Donacia vulgaris	Grünkupferner Rohrkäfer	shiny-green reed beetle*
Donacidae	Stumpfmuscheln, Dreiecksmuscheln, Dreieckmuschel, Sägezähnchen	wedge clams, donax clams (wedge shells)
Donax cuneatus	Wiegenmuschel	cuneate wedge clam, cuneate beanclam, cradle donax
Donax gouldii	Goulds Dreiecksmuschel	Gould's wedge clam, Gould beanclam
Donax scortum/ Hecuba scortum	Ledrige Dreiecksmuschel	leather donax
Donax serra	Südafrikanische Riesen-Dreiecksmuschel	white mussel, giant South African wedge clam
Donax trunculus	Gestutzte Dreiecksmuschel, Sägezahnmuschel	truncate donax, truncated wedge clam
Donax variabilis	Schmetterlings-Dreiecksmuschel	variable coquina, coquina clam, pompano (coquina shell, butterfly shell)
Donax vittatus	Gebänderte Dreiecksmuschel, Gebänderte Sägemuschel, Sägezähnchen	banded wedge clam
Doradidae	Dornwelse	thorny catfishes
Doras niger/ Oxydoras niger	Liniendornwels	ripsaw catfish
Doratopsis vermicularis	Anglerkalmarlarve	worm squid*
Dorcadion fuliginator	Variabler Erdbock	
Dorcatragus megalotis	Beira	beira antilope
Dorcopsis spp.	Buschkängurus	New Guinean forest wallabies
Dorcus parallelopipedus	Balkenschröter	lesser stag beetle
Doridacea/ Holohepatica	Sternschnecken, Warzenschnecken	doridacean snails, doridaceans
Dorippe lanata/ Medorippe lanata	Schirmkrabbe	demon-faced porter crab

Dorippidae	Gespenstkrabben	sumo crabs, demon-faced crabs
Doris verrucosa	Gelbe Sternschnecke, Warzige Schwammschnecke*	sponge slug, sponge seaslug
Dormitator latifrons/ Eleotris latifrons	Breitkopf-Schläfergrundel	broad-headed sleeper
Dormitator maculatus	Gefleckte Schläfergrundel	spotted sleeper, fat sleeper (FAO)
Dorosoma cepedianum	Fadenflossige Alse	gizzard shad, American gizzard shad (FAO)
Doru spp.		spine-tailed earwigs
Dorylinae	Treiberameisen, Wanderameisen	legionary ants, army ants
Dorymenia vagans (Aplacophora)	Walzen-Furchenfuß	barrel solenogaster*
Dosidicus gigas	Riesenkalmar u.a., Jumbokalmar	jumbo flying squid, jumbo squid
Dosima fascicularis	Bojen-Seepocke	buoy barnacle
Dosinia anus	Greisinnenmuschel	old-woman dosinia
Dosinia elegans	Zauberhafte Artmuschel	elegant dosinia
Dosinia exoleta	Gemeine Artmuschel, Artemismuschel	rayed dosinia, rayed artemis
Dosinia lupinus	Glatte Artmuschel, Glatte Artemis	smooth dosinia, smooth artemis
Dosinia variegata	Bunte Artmuschel, Verschiedenfarbige Artmuschel, Bunte Artemis	variegated dosinia, variegated artemis
Dracaena guianensis	Krokodilteju	caiman lizard
Draco spp.	Flugdrachen	flying dragons, flying lizards
Dracunculus medinensis	Medinawurm, Guineawurm, Drachenwurm	fiery serpent, medina worm, guinea worm
Drassodidae/Gnaphosidae	Plattbauchspinnen, Glattbauchspinnen	ground spiders
Dreissena polymorpha	Dreikantmuschel, Wandermuschel	zebra mussel, many-shaped dreissena
Dremomys spp.	Rotwangenhörnchen	red-cheeked squirrels
Drepana binaria	Eichensichler, Eichen-Sichelspinner, Zweipunkt-Sichelflügler	oak hook-tip
Drepana cultraria/ Platypteryx cultraria	Buchen-Sichelflügler	barred hook-tip
Drepana falcataria	Birkensichler, Gemeiner Sichelflügler, Weiden-Sichelspinner	pebble hook-tip
Drepane africana	Afrikanischer Sichelflosser	African sicklefish
Drepane punctata	Geperlter Sichelflosser	spotted sicklefish
Drepaneidae	Sichelflosser	sicklefishes
Drepanidae	Sichelflügler	hooktip moths
Drepanididae	Kleidervögel	Hawaiian honeycreepers
Drepanophorus crassus (Nemertini)	Band-Schnurwurm, Band-Nemertine	
Drepanotrema kermatoides		crested ramshorn
Dreyfusia nordmanniana/ Adelges nordmanniana	Tannentrieblaus, Weißtannen-Trieblaus, Nordmannstannen-Trieblaus	silver fir migratory adelges

Dreyfusia piceae/ *Adelges piceae*	Europäische Tannen-Stammlaus	silver fir adelges
Drimo elegans/ *Gnathophyllum elegans*	Goldpunktgarnele	golden-spotted shrimp*
Dromadidae	Reiherläufer	crab-plover
Dromaiidae	Emus	emus
Dromaius novaehollandiae	Emu	emu
Dromia personata/ *Dromia vulgaris*	Wollkrabbe	sponge crab, common sponge crab, sleepy sponge crab, Linnaeus's sponge crab, sleepy crab, little hairy crab
Dromia spp.	Wollkrabben	sponge crabs
Dromidia antillensis	Haarige Wollkrabbe	hairy sponge crab
Dromiidae	Wollkrebse	sponge crabs
Dromius agilis	Lebhafter Rindenläufer, Dunkelbrauner Rennkäfer	
Dromius linearis	Schmaler Rindenläufer, Gestreifter Rennkäfer	
Dromius quadrimaculatus	Vierfleck-Rindenläufer, Vierfleckiger Rennkäfer	
Drosophila funebris	Große Essigfliege	greater vinegar fly, greater fruit fly
Drosophila melanogaster	Essigfliege, Kleine Essigfliege, Taufliege	vinegar fly, fruit fly
Drosophilidae	Essigfliegen, Obstfliegen, Taufliegen	"fruit flies", pomace flies, small fruit flies, ferment flies
Drupa clathrata	Gitter-Purpurschnecke	clathrate drupe
Drupa grossularia	Orangemündige Pazifische Purpurschnecke	finger drupe
Drupa lobata	Gelappte Purpurschnecke	lobate drupe
Drupa morum	Schwarze Igelschnecke, Schwarzer Igel, Violettmündige Pazifische Purpurschnecke	purple drupe, purple Pacific drupe
Drupa ricinus	Großstachlige Pazifische Igelschnecke	prickly drupe, prickly Pacific drupe
Drupa rubusidaeus	Erdbeer-Igelschnecke, Erdbeer-Igel	rose drupe, strawberry drupe
Dryinidae	Zikadenwespen	dryinids, dryinid wasps
Drymarchon corais	Cribo	cribo
Drymarchon corais couperi	Indigonatter	indigo snake
Drymarchon spp.	Indigoschlangen, Indigonattern	indigo snakes
Drymobius margaritiferus margaritiferus	Perlnatter, Gesprenkelte Bodenschlange	speckled racer
Drymonema dalmatinum	Dalmatinische Fahnenqualle	stinging cauliflower, Dalmatian mane jelly*
Dryocoetes autographus	Zottiger Fichtenborkenkäfer	spruce bark beetle*
Dryocopus galeatus/ *Ceophloeus galeatus*	Wellenohrspecht	helmeted woodpecker
Dryocopus martius	Schwarzspecht	black woodpecker
Dryocopus pileatus	Helmspecht	pileated woodpecker
Dryomys laniger	Türkischer Baumschläfer	Turkish dormouse

Dryomys nitedula	Baumschläfer	forest dormouse
Dryomys spp.	Baumschläfer	forest dormice
Dryomyzidae	Baumfliegen	dryomyzid flies
Dryopidae	Hakenkäfer, Klauenkäfer	long-toed water beetles
Duberria lutrix	Schneckennatter	common slug-eater
Ducula aenea	Bronzefruchttaube	green imperial pigeon
Dulus dominicus (Dulidae)	Palmenschwätzer (Palmenschmätzer)	palmchat
Dumetella carolinensis	Katzenvogel	catbird
Dusicyon australis	Falklandfuchs, Falklandwolf	Falkland Island wolf
Dusicyon griseus/ Pseudalopex griseus	Graufuchs, Argentinischer Graufuchs	gray zorro, Argentine gray fox, South American gray fox
Dussumieria acuta	Regenbogen-Rundhering	rainbow sardine (FAO), round herring
Dynamena pumila/ Dynamena cavolini/ Sertularia pumila	Kleines Seemoos, Zwergmoos	sea oak, minute garland hydroid, minute hydroid
Dynastes ssp.	Nashornkäfer	unicorn beetles, hercules beetles
Dynastinae	Riesenkäfer	rhinoceros beetles & hercules beetles & others
Dypterygia scabriuscula	Knötericheule	bird's wing
Dysaphis apiifolia	Kreuzdorn-Petersilien-Blattlaus	hawthorn-parsley aphid
Dysaphis crataegi	Weißdorn-Karotten-Röhrenlaus	hawthorn-carrot aphid
Dysaphis devecta	Rosige Apfelfaltenlaus	rosy leaf-curling aphid
Dysaphis plantaginea	Mehlige Apfellaus, Mehlige Apfelblattlaus	rosy apple aphid
Dysaphis pyri	Mehlige Birnblattlaus	mealy pear aphid, pear-bedstraw aphid
Dysaphis radicola	Apfel-Ampfer-Blattrolllaus	apple-dock aphid
Dysaphis tulipae	Tulpen-Röhrenlaus	tulip bulb aphid
Dysauxea ancilla	Braunwidderchen	brown burnet
Dyschirius globosus	Handkäfer, Kugliger Klumphandkäfer	
Dyscia fagaria	Heidekraut-Fleckenspanner, Heidekraut-Punktstreifenspanner	grey scalloped bar (moth)
Dyscophus antongilli	Tomatenfrosch	tomato frog
Dysdera crocota	Asselspinne*	woodlouse spider
Dysderidae	Sechsaugenspinnen, Dunkelspinnen, Walzenspinnen	dysderids
Dysichthys coracoideus	Zweifarbiger Bratpfannenwels	guitarrita
Dysidea arenaria	Kraterschwamm	crater sponge*
Dysommidae	Pfeilzahnaale	arrowtooth eels, mustard eels
Dyspessa ulula	Zwiebelbohrer, Lauchzwiebelbohrer	
Dytiscidae	Schwimmkäfer	predaceous diving beetles, carnivorous water beetles
Dytiscus latissimus	Breitrand, Breitrandkäfer	broad diving beetle*
Dytiscus marginalis	Gelbrand, Gelbrandkäfer	great diving beetle
Dytiscus spp.	Tauchkäfer*	diving beetles, diving water beetles

Eacles imperialis	Kaisermotte*	imperial moth
Earias chlorana	Weidenkahneule	cream-bordered green pea
Ebalia cranchii	Cranch-Kugelkrabbe*	Cranch's nut crab
Ebalia intermedia	Glattschalige Kugelkrabbe	smooth nut crab
Ebalia tuberosa	Höckerige Kugelkrabbe	Pennant's nut crab
Ebalia tumefacta	Bryer-Kugelkrabbe*	Bryer's nut crab
Echelus myrus	Kurzschnauziger Schlangenaal, Stumpfnasen-Schlangenaal	blunt-nosed snake eel
Echeneidae (Echeneididae)	Schiffshalter, Saugfische	remoras (sharksuckers)
Echeneis naucrates	Schiffshalter	whitefin sharksucker, Indian remora, suckerfish, live sharksucker (FAO)
Echidna nebulosa	Sternmuräne	snowflake moray (FAO), starry moray (Austr.)
Echidnophaga gallinacea/ Sarcopsylla gallinacea	Hühnerkammfloh, Hühnerfloh	sticktight flea
Echiichthys vipera/ Trachinus vipera	Kleines Petermännchen, Zwergpetermännchen, Viperqueise	lesser weever
Echimyidae	Stachelratten, Lanzenratten	spiny rats
Echimys spp.	Kammstachelratten	arboreal spiny rats
Echinarachnius parma		common sand dollar
Echinaster sepositus	Purpurstern, Purpurseestern, Roter Seestern, Blutstern, Roter Mittelmeerseestern	red Mediterranean sea star
Echinidae	Echte Seeigel	sea urchins
Echinocardium cordatum	Kleiner Herzigel, Herzigel	common heart-urchin, sea potato, heart urchin
Echinocardium flavescens	Gelber Herzigel	yellow sea potato
Echinocardium pennatifidum	Großer Herzigel	sea potato
Echinoclathria gigantea	Wabenschwamm*	honeycomb sponge
Echinococcus granulosus	Hundebandwurm, Dreigliedriger Hundebandwurm	dwarf dog tapeworm, dog tapeworm, hydatid tapeworm
Echinococcus multilocularis	Kleiner Fuchsbandwurm	lesser fox tapeworm, alveolar hydatid tapeworm
Echinocyamus pusillus	Kleiner Schildigel, Zwergseeigel, Zwerg-Seeigel	green sea urchin, green urchin, pea urchin
Echinodermata	Stachelhäuter, Echinodermen	echinoderms
Echinodiscus auritus	Zweikerben-Sanddollar	two-slit sand dollar
Echinoidea	Seeigel, Echinoiden	sea urchins, echinoids
Echinometra lucunter	Atlantischer Bohrseeigel	Atlantic boring sea urchin*, rock-boring urchin
Echinometra mathaei	Riffdach-Seeigel	reef-flat sea urchin*
Echinometra viridis	Grüner Seeigel	green sea urchin*
Echinoprocta rufescens	Bergstachler	Upper Amazon porcupine
Echinorhinidae	Nagelhaie, Alligatorenhaie, Alligatorhaie	bramble sharks

Echinorhinus brucus	Nagelhai, Brombeerhai, Stachelhai, Alligatorhai	bramble shark
Echinostoma ilocanum	Echinostomose-Darmegel	intestinal echinostome
Echinothrix diadema	Schwarzer Diadem-Seeigel	black hatpin urchin
Echinothuridae	Lederseeigel	leather urchins*
Echinotriton andersoni	Japanischer Bergmolch	Anderson's newt
Echinus acutus	Rotgelber Seeigel	red-yellow sea urchin*
Echinus esculentus	Essbarer Seeigel	edible sea urchin, common sea urchin
Echinus melo	Melonen-Seeigel	melon urchin
Echiodon drummondi	Perlfisch*	pearlfish
Echiopsis curta/ Brachyaspis curta	Bardick	bardick snake
Echiothrix leucura	Celebes-Stachelratte	Celebes spiny rat, Celebes shrew rat
Echis carinatus	Sandrasselotter, Efa	saw-scaled viper, saw-scaled adder
Echis coloratus	Arabische Sandrasselotter	Arabic saw-scaled viper, Palestine saw-scaled viper
Echiura	Igelwürmer, Stachelschwänze, Echiuriden	spoon worms, echiuroid worms
Echymipera spp.	Stachelnasenbeutler	New Guinea spiny bandicoot
Eclectus roratus	Edelpapagei	eclectus parrot
Ecribellatae	Klebfadenwebspinnen	viscid band spiders
Ectobius lapponicus	Gemeine Waldschaben	dusky cockroach
Ectobius panzeri	Küsten-Waldschabe, Heideschabe	lesser cockroach
Ectobius silvestris	Podas Waldschabe	Poda's cockroach
Ectobius spp.	Waldschaben	ectobid cockroaches
Ectomyelois ceratoniae	Johannisbrotmotte	carob moth
Ectophylla alba	Weiße Fledermaus	white bat
Ectopistes migratoria†	Wandertaube	passenger pigeon
Ectoprocta/Polyzoa (Bryozoa)	Moostierchen, Bryozoen	moss animals, lace animals, bryozoans
Ectropis crepuscularia/ Ectropis bistortata	Lärchenspanner, Beerenkrautspanner, Heidelbeerspanner, Tannenspanner, Pflaumenspanner	larch looper, blueberry lopper, fir looper, plum looper
Ectypa glyphica/ Euclidia glyphica	Braune Tageule, Luzerneule	burnet companion
Edentata/Xenarthra	Zahnarme, Nebengelenktiere	edentates, "toothless" mammals, xenarthrans
Edwardsia elegans	Schöne Grabanemone*	elegant burrowing anemone
Edwardsia ivelli	Ivells Seeanemone*	Ivell's sea anemone
Edwardsiana rosae	Rosenzikade	rose leafhopper
Egernia depressa	Plattschwänzchen	pigmy spinytail skink
Egernia frerei		major skink
Egernia kingii	Kings Skink	King's skink
Egernia major	Stachelskink	land mullet, major skink

Egernia spp.	Stachelskinke	spinytail skinks
Egernia stokesii	Dornschwanzskink	spiny-tailed Australian skink
Egernia striata	Nachtskink	elliptical-eye skink
Egernia striolata	Baumstachelskink	tree skink
Egretta alba	Silberreiher	great white egret
Egretta garzetta	Seidenreiher	little egret
Egretta gularis	Küstenreiher	western reef heron
Eidolon helvum	Palmenflughund	straw-colored fruit bat
Eigenmannia virescens	Grüner Messerfisch	green knifefish
Eilema complana	Flechtenspinner	scarce footman
Eilema lurideola	Laubholzflechtenspinner	common footman
Eilema pygmaeola	Blassstirniges Flechtenbärchen	pygmy footman
Eilema sericea	Nördliches Flechtenbärchen	northern footman
Eilema sororcula	Orange Flechtenbärchen	orange footman
Eira barbara	Tayra	tayra
Eirenis collaris	Halsband-Zwergnatter	collared dwarf snake
Eirenis lineomaculatus	Längsgepunktete Zwergnatter	striped dwarf snake
Eirenis modestus	Kopfbinden-Zwergnatter	Asia Minor dwarf snake
Eisenia fetida/Eisenia foetida	Mistwurm, Dungwurm, Kompostwurm	brandling, manure worm
Eisenia rosea	Rosenwurm*	rosy worm
Elacatinus oceanops/ Gobiosoma oceanops	Neongrundel, Neon-Grundel, Blaue Putzgrundel	neon goby
Elachistidae	Grasminiermotten	grass miners
Elachistocleis ovalis	Ovalfrosch	oval frog, common oval frog
Elachistodon westermanni	Indische Eierschlange	Indian egg-eating snake
Elagatis bipinnulata	Regenbogen-Stachelmakrele	rainbow runner
Elanoides forficatus	Schwalbenweih	swallow-tailed kite
Elanus caeruleus	Gleitaar	black-shouldered kite
Elaphe carinata	Stinknatter	stink snake
Elaphe dione	Steppennatter, Dione-Natter	Dione's snake
Elaphe guttata	Kornnatter	corn snake
Elaphe hohenackeri	Transkaukasische Kletternatter	Transcaucasian rat snake, Caucasian snake
Elaphe longissima	Äskulapnatter	Aesculapian snake
Elaphe obsoleta	Erdnatter, Schwarze Erdnatter	black rat snake, eastern rat snake
Elaphe oxycephala	Spitzkopfnatter	red-tailed snake
Elaphe prasina	Grüne Baumnatter	green trinket snake
Elaphe quadrivirgata	Japanische Vierstreifen-Rattennatter	Japanese four-lined rat snake
Elaphe quatuorlineata	Vierstreifennatter	four-lined snake, yellow rat snake
Elaphe radiata	Strahlennatter	radiated rat snake, copperhead racer
Elaphe scalaris	Treppennatter	ladder snake
Elaphe schrenckii schrenckii	Amurnatter	Russian rat snake
Elaphe situla	Leopardnatter	leopard snake
Elaphe spp.	Kletternattern	rat snakes, ratsnakes

Elaphe subocularis	Trans-Peco-Natter	Trans-Pecos rat snake
Elaphe taeniura	Streifenkletternatter	Taiwan beauty snake
Elaphe triaspis	Grüne Kletternatter	green rat snake
Elaphe vulpina	Fuchsnatter	fox snake
Elaphodus cephalophus	Schopfhirsch	tufted deer
Elaphrus cupreus	Kupferfarbener Uferläufer, Bronzefarbener Raschkäfer	copper ground beetle*
Elaphrus riparius	Kleiner Raschkäfer, Kleiner Uferläufer	lesser wetland ground beetle*
Elaphrus spp.	Uferkäfer, Uferläufer	wetland ground beetles*
Elaphrus viridis	Grüner Raschkäfer	delta green ground beetle
Elaphurus davidianus	Davidshirsch, Milu	Père David's deer
Elapidae	Giftnattern	front-fanged snakes
Elapognathus minor	Kleine Braunschlange	little brown snake
Elaps spp.	Afrikanische Korallenschlangen	African dwarf garter snakes
Elapsoidea spp.	Afrikanische Gift-Bänder-Korallenschlange, Afrikanische Bänder-Korallenschlange	venomous garter snakes
Elapsoidea sundevallii	Afrikanische Bänder-Korallenschlange	Sundevall's garter snake, African garter snake
Elasmobranchii	Plattenkiemer (Haie & Rochen)	sharks & rays & skates
Elasmostethus interstinctus	Bunte Blattwanze	birch bug
Elasmucha ferrugata	Heidelbeerwanze	
Elasmucha fieberi	Gezähnte Brutwanze	
Elasmucha grisea	Fleckige Brutwanze, Birkenwanze, Erlenwanze	parent bug, mothering bug
Elassoma evergladei	Zwergsonnenbarsch	pygmy sunfish
Elassomatidae/Elassomidae	Zwergsonnenbarsche	pygmy sunfishes
Elateridae	Schnellkäfer (Schmiede, Schuster)	click beetles
Elatobium abietinum	Fichtenröhrenlaus, Sitkalaus	green spruce aphid
Electra pilosa	Zottige Seerinde	hairy sea-mat
Electrophoridae	Zitteraale	electric knifefishes
Electrophorus electricus	Zitteraal	electric knifefish
Eledone cirrosa/ Ozeana cirrosa	Zirrenkrake	lesser octopus, curled octopus, horned octopus (FAO)
Eledone moschata/ Ozeana moschata	Moschuskrake, Moschuspolyp	white octopus, musky octopus
Eleginops maclovinus	Patagonischer Zahnfisch	Patagonian mullet, Patagonian blenny (FAO)
Eleginus gracilis	Fernöstliche Nawaga, Fernöstliche Navaga	saffron cod (FAO), Far Eastern navaga
Eleginus navaga	Europäische Nawaga, Europäische Navaga	navaga (FAO), wachna cod, Atlantic navaga, Arctic cod
Eleotridae	Schläfergrundeln	sleepers
Eleotris fusca	Schwärzliche Schläfergrundel	dusky sleeper
Eleotris pisonis	Stachelwangen-Schläfergrundel	spinycheek sleeper

Elephantulus spp.	Elefantenspitzmäuse	long-eared elephant shrews, small elephant shrews
Elephas maximus	Asiatischer Elefant, Indischer Elefant	Asiatic elephant, Indian elephant
Eleutherodactylus caryophyllaeus	Blatt-Antillen-Pfeiffrosch	La Loma robber frog
Eleutherodactylus marmoratus/ Leptodactylus marmoratus	Marmor-Pfeiffrosch	marbled robber frog
Eleutherodactylus planirostris	Gewächshausfrosch	greenhouse frog
Eleutherodactylus augusti/ Hylactophryne augusti	Mexikanischer Klippenfrosch	common robber frog, barking frog
Eleutherodactylus coqui	Puerto-Rico-Pfeiffrosch	Puerto Rican coqui
Eleutherodactylus cruentus	Blutfleck-Antillen-Pfeiffrosch	Chiriqui robber frog
Eleutherodactylus decoratus	Schmuck-Antillen-Pfeiffrosch	adorned robber frog
Eleutherodactylus jasperi	Lebendgebärender Puerto Rico-Pfeiffrosch	Cayey robber frog, Puerto Rican live-bearing frog, golden cocqui frog
Eleutherodactylus limbatus/ Sminthillus limbatus	Kuba-Zwergfrosch	Habana robber frog
Eleutherodactylus marnockii/ Syrrhophus marnockii	Marnock-Frosch, Felsen-Zwitscherfrosch*	cliff frog, cliff chirping frog
Eleutherodactylus ricordii	Ricordis Gewächshausfrosch	Ricordi's robber frog
Eleutherodactylus spp./ *Trachyphrynus* spp.	Antillen-Pfeiffrösche, Antillenfrösche	robber frogs
Eleutheronema tetradactylum	Riesenfadenfisch	four-finger threadfin
Elgaria kingii nobilis/ Gerrhonotus kingii nobilis	Arizona-Alligatorschleiche	Arizona alligator lizard
Elgaria kingii/ Gerrhonotus kingii	Madrea-Alligatorschleiche	Madrean alligator lizard
Elgaria multicarinata/ Gerrhonotus multicarinatus	Südliche Alligatorschleiche	Southern alligator lizard
Elgaria multicarinata multicarinata/ Gerrhonotus multicarinata multicarinata	Kalifornische Alligatorschleiche	California alligator lizard, California legless lizard
Elgaria spp.	Alligatorschleichen u.a.	western alligator lizards
Eligmodonta ziczac/ Notodonta ziczac	Zickzackspinner, Kamelspinner, Uferweiden-Zahnspinner	pebble prominent
Eligmodontia typus	Hochland-Wüstenmaus	highland desert mouse
Eliomys melanurus	Südwestasiatischer Gartenschläfer	Southwest Asian garden dormouse
Eliomys quercinus	Gartenschläfer	garden dormouse
Eliurus spp.	Bilchschwänze	Madagascan rats
Elkneria pudibunda/ Calliteara pudibunda/ Dasychira pudibunda/ Olene pudibunda	Rotschwanz, Buchenrotschwanz, Buchen-Streckfuß	pale tussock, red-tail moth
Ellobiidae	Küstenschnecken	coastal snails*
Ellobium aurismidae	Midasohr, Eselohr des Midas	Midas ear cassidula (snail)
Ellobius spp.	Mull-Lemminge	mole-voles, mole-lemmings
Elmidae	Hakenkäfer	drive beetles, riffle beetles
Elminia longicauda	Türkiselminie	blue flycatcher

Elminius modestus	Austral-Seepocke	modest barnacle
Elopichthys bambusa	Scheltostscheck	sheltostshek, yellow-cheek
Elopidae	Frauenfische	ladyfishes, tenpounders
Elopiformes	Tarpunähnliche	tarpons
Elops affinis		machete
Elops saurus	Frauenfisch	ladyfish
Elseya dentata	Nord-Australische Elseya-Schildkröte	northern Australian snapping turtle, northern snapping tortoise
Elseya dentata novaeguineae/ Elseya novaeguineae	Neuguinea Elseya-Schildkröte, Neuguinea-Schnapper	New Guinea snapping turtle
Elseya latisternum	Zacken-Elseya-Schildkröte, Breitbrustschildkröte	saw-shelled snapping turtle, serrated snapping turtle, East Australian snapping turtle
Elysia crispata/ Tridachia crispata	Salatschnecke, Kräuselschnecke	lettuce slug
Elysia splendida/ Thuridilla hopei	Pracht-Samtschnecke	splendid velvet snail, splendid elysia
Elysia viridis	Grüne Samtschnecke	green velvet snail*, green elysia
Emarginula conica	Rosa Schlitzschnecke, Rosa Ausschnittsschnecke	pink slit limpet, pink emarginula
Emarginula crassa	Dickschalige Schlitzschnecke, Dickschalige Ausschnittsschnecke	thick slit limpet, thick emarginula
Emarginula fissura	Schlitzschnecke, Ausschnittsschnecke	slit limpet
Emarginula spp.	Schlitzschnecken u.a., Ausschnittsschnecken u.a.	slit limpets a.o.
Ematurga atomaria	Heidekrautspanner	common heath
Emballonura spp.	Altwelt-Freischwanzfledermaus	Old World sheath-tailed bats
Emballonuridae	Glattnasen-Freischwänze	sheath-tailed bats, sac-winged bats & ghost bats
Emberiza aureola	Weidenammer	yellow-breasted bunting
Emberiza bruniceps	Braunkopfammer	red-headed bunting
Emberiza buchanani	Steinortolan	grey-necked bunting
Emberiza caesia	Grauortolan	Cretzschmar's bunting
Emberiza chrysophrys	Gelbbrauenammer	yellow-browed bunting
Emberiza cia	Zippammer	rock bunting
Emberiza cineracea	Türkenammer	cinereous bunting
Emberiza cirlus	Zaunammer	cirl bunting
Emberiza citrinella	Goldammer	yellowhammer
Emberiza hortulana	Ortolan	ortolan bunting
Emberiza leucocephalos	Fichtenammer	pine bunting
Emberiza melanocephala	Kappenammer	black-headed bunting
Emberiza pallasi	Pallasammer	Pallas's reed bunting
Emberiza pusilla	Zwergammer	little bunting
Emberiza rustica	Waldammer	rustic bunting
Emberiza schoeniclus	Rohrammer	reed bunting
Emberiza spodocephala	Maskenammer	black-faced bunting
Emberiza striolata	Hausammer	house bunting
Emberizidae	Ammern	buntings & cardinals & tanagers & longspurs & towhees

Embioptera	Tarsenspinner, Fußspinner, Embien	embiids, webspinners, footspinners
Embiotocidae	Brandungsbarsche	surfperches
Emerita analoga	Pazifik-Brandungskrebs	Pacific sand crab
Emerita talpoida	Atlantik-Brandungskrebs	Atlantic sand crab, mole crab
Emmelia trabealis	Windeneulchen	spotted sulphur
Emmelichthyidae	Emmelichthyiden	rovers
Emoia atrocostata	Mangrovenskink	marine skink
Empetrichthyinae	Nordamerikanische Quellkärpflinge	springfishes & poolfishes
Empicoris vagabunda	Mückenwanze	
Empididae	Tanzfliegen (Rennfliegen)	dance flies
Empidonax virescens	Buchentyrann	acadian flycatcher
Empoasca decipiens	Europäische Kartoffelzikade	green leafhopper
Emus hirtus	Zottiger Raubkäfer, Behaarter Kurzflügler	
Emydidae	Sumpfschildkröten	pond terrapins (pond and river turtles), emydid turtles
Emydocephalus annulatus	Indonesische Schildkrötenkopf- Seeschlange, Eierfressende Seeschlange	ringed turtlehead sea snake, annulated sea snake
Emydocephalus ijimae	Schildkrötenköpfige Seeschlange, Schildkrötenkopf-Seeschlange	turtlehead sea snake
Emydocephalus spp.	Schildkrötenköpfige Seeschlangen, Schildkrötenkopf-Seeschlangen	turtlehead sea snakes
Emydoidea blandingii	Amerikanische Sumpfschildkröte	Blanding's turtle
Emydura australis	Australische Spitzkopfschildkröte	Australian big-headed side-necked turtle
Emydura kreffti	Krefft-Spitzkopfschildkröte	Krefft's river turtle, Krefft's tortoise
Emydura macquarrii	Breitrand-Spitzkopfschildkröte	Murray River turtle, Macquarie tortoise
Emydura signata	Ost-Spitzkopfschildkröte	Brisbane short-necked turtle, Eastern Australian short-necked tortoise
Emydura spp.	Spitzkopfschildkröten	short-necked turtles, short-necked tortoises
Emydura subglobosa	Rotbauchschildkröte	red-bellied short-necked turtle
Emys orbicularis	Europäische Sumpfschildkröte	European pond turtle, European pond terrapin
Ena montana	Große Turmschnecke	mountain bulin
Ena obscura	Kleine Turmschnecke	lesser bulin
Enallagma civile		bluet damselfly
Enallagma cyathigera	Becher-Azurjungfer	common blue damselfly, common bluet damselfly
Enarmonia formosana	Gummiwickler, Rindenwickler	cherry bark tortrix (moth)
Enchelyopus cimbrius/ Rhinonemus cimbrius	Vierbärtelige Seequappe	fourbeard rockling (FAO), four-bearded rockling
Enchytraeidae	Enchyträen	potworms, aster worms, white worms

Enchytraeus albidus	Topfwurm, Weißer Topfwurm	white potworm
Endomychidae	Pilzkäfer, Stäublingskäfer, Pilzfresser (Puffpilzkäfer)	handsome fungus beetles
Endomychus coccineus	Scharlachroter Stäublingskäfer, Stockkäfer	
Endromididae/Endromidae	Birkenspinner, Frühlingsspinner	endromid moths
Endromis versicolora	Birkenspinner	Kentish glory
Endrosis sarcitrella	Kleistermotte	white-shouldered house moth
Engina mendicaria		striped engina
Engraulidae	Sardellen	anchovies
Engraulis anchoita	Argentinische Sardelle	anchoita, Argentine anchovy
Engraulis australis	Australische Sardelle	Australian anchovy
Engraulis capensis	Südafrikanische Sardelle	Southern African anchovy
Engraulis encrasicholus	Anchovis, Sardelle, Europäische Sardelle	anchovy, European anchovy
Engraulis japonicus	Japanische Sardelle	Japanese anchovy
Engraulis mordax	Amerikanische Sardelle	North Pacific anchovy, Californian anchovy (FAO)
Engraulis ringens	Peru-Sardelle, Anchoveta	anchoveta (FAO), Peruvian anchovy
Enhydra lutris	Seeotter (Hbz. Kalan)	sea otter
Enhydrina schistosa	Südostasiatische Seeschlange	beaked sea snake
Enhydris punctata	Gesprenkelte Wassertrugnatter	spotted water snake
Enicocephalidae		unique-headed bugs, gnat bugs
Enicognathus ferrugineus	Smaragdsittich	austral conure
Enicognathus leptorhynchus	Langschnabelsittich	slender-billed conure
Enicurus leschenaulti	Weißscheitel-Scherenschwanz	white-crowned forktail
Enidae	Turmschnecken	bulins
Enneacanthus chaetodon/ Mesogonistius chaetodon	Scheibenbarsch	blackbanded sunfish (FAO), chaetodon
Enneacanthus gloriosus	Diamantbarsch	bluespotted sunfish
Ennomos alniaria	Erlenspanner	alder thorn*
Ennomos autumnaria	Zackenspanner, Herbstspanner, Herbstlaubspanner	large thorn
Ennomos fuscantaria	Eschenspanner	dusky thorn
Ennucula tenuis/ Nucula tenuis	Dünnschalige Nussmuschel, Glatte Nussmuschel	smooth nutclam
Enoplognatha ovata	Rotweiß-Spinne*	red-and-white spider
Enoplometopus antillensis	Atlantischer Riffhummer	flaming reef lobster
Enoplometopus spp.	Riffhummer	reef lobsters
Enoplosus armatus		oldwife
Enoplus meridionalis	Meernematode	
Ensatina eschscholtzi	Ensatina, Eschscholtz-Salamander	ensatina
Ensifera ensifera	Schwertschnabel	sword-billed hummingbird
Ensis directus	Atlantische Schwertmuschel, Amerikanische Schwertmuschel, Gerade Scheidenmuschel	Atlantic jackknife clam

Ensis ensis	Gemeine Schwertmuschel, Schwertförmige Messerscheide, Schwertförmige Scheidenmuschel	common razor clam, narrow jackknife clam, sword razor
Ensis siliqua	Schotenförmige Schwertmuschel, Schotenförmige Messerscheide, Taschenmesser-Muschel, Schotenmuschel	pod razor clam, giant razor clam
Entacmaea quadricolor	Knubbelanemone	four-colored anemone, bubble-tip anemone, bulb-tip anemone, maroon anemone, bulb-tentacle sea anemone
Entamoeba histolytica	Ruhramöben (*Amöbiasis*)	dysentery ameba (*amebic dysentery/amebiasis*)
Entelurus aequoreus	Große Schlangennadel	snake pipefish
Enterobius vermicularis	Madenwurm (Springwurm, Pfriemenschwanz)	pinworm (of man), seatworm
Enteropneusta	Eichelwürmer, Enteropneusten	acorn worms, enteropneusts
Entodinium caudatum	Geschwänztes Pansenwimpertierchen	
Entomobryidae	Laufspringer	entomobryid springtails
Entoprocta	Kelchwürmer, Nicktiere, Kamptozooen	kamptozoans, entoprocts
Enyalius catenatus	Wieds Leguan	Wied's fathead anole
Enyalius spp.	Großkopf-Anolis	fathead anoles
Eobania vermiculata	Divertikelschnecke	vermiculate snail, chocolate-band snail
Eohippolysmata ensirostris	Jagdgarnele*	hunter shrimp
Eolagurus luteus/ Lagurus luteus	Gelblemming	yellow steppe lemming
Eolophus roseicapillus	Rosakakadu	galah
Eonycteris spp.	Asiatische Höhlenflughunde	dawn bats
Eopsaltriidae	Südseeschnäpper, Südseesänger	Australasian robins
Eopsetta jordani	Kalifornische Scholle	petrale sole
Eos bornea	Rotlori	red lory
Eos cyanogenia	Schwarzflügellori	black-winged lory
Eos histrio	Rotblaulori	red-and-blue lory
Eos reticulata	Strichellori	blue-streaked lory
Eothenomys spp.	Père-Davids-Wühlmäuse	Père David's voles, Pratt's voles
Eozapus setchuanus	Chinesische Hüpfmaus	Chinese jumping mouse
Epacromius tergestinus	Fluss-Strandschrecke	eastern longwinged grasshopper, riverside grasshopper*
Epalzeorhynchos bicolor/ Labeo bicolor	Feuerschwanz	redtail sharkminnow (FAO), red-tailed "shark", red-tailed black "shark"
Epalzeorhynchus kallopterus	Schönflossenbarbe	flying fox
Epeolus variegatus	Filzbiene	
Ephebopus murinus		skeleton tarantula
Ephemeroptera	Eintagsfliegen	mayflies

Ephestia cautella	Tropische Speichermotte, Dattelmotte, Mandelmotte	almond moth (date moth, tropical warehouse moth)
Ephestia elutella	Speichermotte, Heumotte, Tabakmotte, Kakao-Motte, Kakaomotte	tobacco moth, cocoa moth, chocolate moth, flour moth, warehouse moth, cinereous knot-horn
Ephestia kuehniella/ Anagasta kuehniella	Mehlmotte	Mediterranean flour moth
Ephippidae	Spatenfische, Fledermausfische	batfishes, spadefishes
Ephippiger ephippiger	Rebensattelschrecke, Steppen-Sattelschrecke, Gemeine Sattelschrecke	common saddle-backed bushcricket, tizi
Ephippiger terrestris	Südalpen-Sattelschrecke	Alpine saddle-backed bushcricket
Ephippigerida taeniata	Große Streifen-Sattelschrecke	large striped bushcricket
Ephippigeridae	Sattelschrecken	saddle-backed bushcrickets
Ephippion guttiferum	Bennetts Kofferfisch*	Bennett's pufferfish
Ephistemus globulus	Kurzovaler Schimmelkäfer	
Ephoron virgo/ Polymitarcis virgo	Uferaas (Weißwurm)	virgin mayfly
Ephydatia fluviatilis	Flussschwamm, Großer Süßwasserschwamm, Klumpenschwamm	greater freshwater sponge
Ephydatia muelleri	Kleiner Süßwasserschwamm, Blasenzellenschwamm	lesser freshwater sponge
Ephydridae	Salzfliegen, Sumpffliegen	shoreflies, shore flies
Epiactis arctica	Arktische Brutanemone*	Arctic brooding anemone
Epiactis prolifera	Westpazifische Brutanemone	proliferating anemone
Epiblema tedella/ Epinotia tedella	Fichtennestwickler, Hohlnadelwickler	streaked pine bell, cone moth
Epicauta fabricii	Aschgrauer Blasenkäfer*	ashgray blister beetle
Epicauta vittata	Gestreifter Blasenkäfer*	striped blister beetle
Epicrates angulifer	Kuba-Schlankboa	Cuban tree boa, Cuban boa (IUCN), maja
Epicrates cenchria cenchria	Regenbogenboa	Brazilian rainbow boa
Epicrates inornatus	Puerto-Rico-Boa	Puerto Rican boa (IUCN), culebra grande
Epicrates spp.	Schlankboas	rainbow boas
Epicrates striatus	Haiti-Boa	Fischer's tree boa
Epicrates subflavus	Jamaika-Boa	Jamaican boa, yellow snake
Epigonus telescopus	Teleskop-Kardinalfisch	black cardinal fish, bull's-eye (FAO)
Epimenia verrucosa (Aplacophora)	Warziger Furchenfuß	warty solenogaster*
Epimyrma spp.	Lappenameisen	
Epinephelus adscensionis	Karibischer Felsenbarsch	rock hind
Epinephelus aeneus	Weißer Zackenbarsch	white grouper
Epinephelus alexandrinus	Gold-Zackenbarsch, Goldener Zackenbarsch, Spitzkopf-Zackenbarsch	golden grouper
Epinephelus analogus	Pazifischer Fleckenbarsch	spotted grouper

Epinephelus caninus	Grauer Zackenbarsch	dogtooth grouper (FAO), dog-toothed grouper
Epinephelus coioides	Estuar-Zackenbarsch	orange-spotted grouper
Epinephelus cruentatus		graysby
Epinephelus fasciatus	Baskenmützen-Zackenbarsch, Baskenmützenbarsch	blacktipped grouper, blacktip grouper (FAO), blacktip rockcod
Epinephelus flavocaeruleus	Blaugelber Zackenbarsch	blue and yellow grouper (FAO), purple rock-cod
Epinephelus guttatus	Trauerrand-Zackenbarsch	red hind
Epinephelus itajara	Riesen-Zackenbarsch (Judenfisch)	giant grouper (jewfish)
Epinephelus lanceolatus	Schwarzer Zackenbarsch, Dunkler Riesenzackenbarsch	giant grouper (FAO), giant seabass
Epinephelus malabaricus/ Epinephelus salmonoides	Malabar-Zackenbarsch, Malabar-Riffbarsch	Malabar grouper (FAO), Malabar reefcod
Epinephelus marginatus	Brauner Zackenbarsch	dusky grouper, dusky perch
Epinephelus microdon/ Epinephelus polyphekadion	Marmorierter Zackenbarsch	mottled grouper, camouflage grouper (FAO)
Epinephelus morio	Roter Zackenbarsch	red grouper
Epinephelus spp.	Zackenbarsche	groupers
Epinephelus striatus	Nassau Zackenbarsch	Nassau grouper
Epinephelus tukula	Gefleckter Riesenzackenbarsch	potato grouper (FAO), potato bass
Epinotia tedella/ Epiblema tedella	Fichtennestwickler, Hohlnadelwickler	streaked pine bell, cone moth
Epinotia tenerana	Haselnusswickler	nut bud tortrix (moth)
Epione parallelaria/ Epione vespertaria	Espen-Saumbandspanner, Weiden-Saumbandspanner, Espenfrischgehölz-Saumbandspanner, Birken-Braunhalsspanner	dark bordered beauty (moth)
Epiplatys annulatus	Ringelhechtling	rocket panchax
Epiplatys dageti monroviae	Querbandhechtling, Monrovia-Hechtling, Rotkehlhechtling	firemouth epiplatys
Episinus spp.	Seilspinnen	
Episyrphus balteatus	Hain-Schwebfliege	
Epitheca bimaculata/ Libellula bimaculata	Zweifleck, Zweifleck-Libelle	two-spotted dragonfly*
Epitoniidae	Wendeltreppen	wentletraps
Epitonium acuminatum	Spitze Wendeltreppe	pointed wentletrap
Epitonium angulatum	Kantige Wendeltreppe	angulate wentletrap
Epitonium aurita	Geohrte Wendeltreppe	eared wentletrap
Epitonium clathratulum	Weiße Wendeltreppe	white wentletrap
Epitonium clathrum	Unechte Wendeltreppe, Gemeine Wendeltreppe	false wentletrap, common European wentletrap
Epitonium imperialis	Kaiserliche Wendeltreppe	imperial wentletrap
Epitonium perplexa	Perplex-Wendeltreppe	perplexed wentletrap
Epitonium pyramidale	Pyramiden-Wendeltreppe	pyramid wentletrap
Epitonium scalare/ Scalaria pretiosa	Echte Wendeltreppe	precious wentletrap
Epitonium turtonae	Feinrippige Wendeltreppe, Turtons Wendeltreppe	finely ribbed wentletrap
Epixerus ebii	Großes Rotschenkelhörnchen	African palm squirrel

Epizoanthus

Epizoanthus arenaceus	Weichboden-Krustenanemone	gray zoanthid, gray encrusting anemone
Epizoanthus paxii	Braune Krustenanemone	brown zoanthid, brown encrusting anemone
Epizoanthus scotinus	Orangene Krustenanemone	orange zoanthid, orange encrusting anemone
Epomophorus spp.	Afrikanische Epauletten-Flughunde u.a.	epauleted fruit bats
Epomops spp.	Afrikanische Epauletten-Flughunde u.a.	epauleted bats
Eptesicus fuscus	Große Braune Fledermaus	big brown bat
Eptesicus nilssoni	Nordfledermaus, Nordische Fledermaus	northern bat
Eptesicus serotinus	Europäische Breitflügelfledermaus, Spätfliegende Fledermaus	serotine bat
Eptesicus spp.	Breitflügelfledermäuse	serotines, serotine bats (big brown bats, house bats)
Epuraea depressa	Flacher Glanzkäfer	
Equetus lanceolatus	Gebänderter Ritterfisch	jackknife fish
Equetus punctatus	Tüpfel-Ritterfisch	spotted jackknife fish
Equus africanus	Afrikanischer Wildesel	African wild ass
Equus asinus	Wildesel	donkey, burro
Equus burchelli	Burchell-Steppenzebra	Burchell's zebra
Equus caballus	Hauspferd	horse
Equus grevyi	Grevy-Zebra	Grevy's zebra
Equus hemionus	Asiatischer Halbesel, Kulan, Khur, Onager	kulan, khur, onager
Equus kiang	Kiang	kiang
Equus przewalski	Przewalski-Pferd	Przewalski's horse
Equus quagga	Steppenzebra	quagga
Equus zebra	Bergzebra	mountain zebra
Erannis defoliaria	Großer Frostspanner, Hainbuchenspanner	mottled umber
Erebia christi	Simplon-Mohrenfalter	Raetzer's ringlet (butterfly)
Erebia epiphron	Knochs Mohrenfalter	mountain ringlet (butterfly)
Erebia ligea	Milchfleck, Weißband-Mohrenfalter	arran brown (ringlet butterfly)
Erebia pandrose	Lappländischer Schwärzling, Graubrauner Mohrenfalter	dewy ringlet (butterfly)
Erebia spp.	Mohrenfalter	ringlets a.o.
Eremias argus	Mongolischer Wüstenrenner	Mongolia racerunner
Eremias arguta/ Ommateremias arguta	Steppenrenner, Steppeneidechse	stepperunner, arguta
Eremias lugubris	Trauer-Wüstenrenner*	mourning racerunner
Eremias multiocellata	Vielfleckiger Wüstenrenner	multi-ocellated racerunner
Eremias pleskei	Pleskes Wüstenrenner, Transkaukasischer Wüstenrenner	Pleske's racerunner
Eremias spp.	Wüstenrenner	racerunners
Eremias strauchi	Strauchs Wüstenrenner	Strauch's racerunner
Eremias velox	Schneller Wüstenrenner	rapid racerunner

Eremophila alpestris	Ohrenlerche	shore lark, shore horned lark
Eremophila bilopha	Saharaohrenlerche	Temminck's horned lark
Eremopterix nigriceps	Weißstirnlerche	black-crowned sparrow-lark
Eresidae	Röhrenspinnen	eresid spiders*
Eresus niger	Marienkäferspinne*	ladybird spider
Erethizon dorsatum	Urson	North American porcupine
Erethizontidae	Baumstachler	New World porcupines
Eretmochelys imbricata	Karettschildkröte, Karette, Echte Karettschildkröte	hawksbill turtle, hawksbill sea turtle
Eretmochelys imbricata bissa	Pazifische Karettschildkröte	Pacific hawksbill turtle
Eretmochelys spp.	Echte Karettschildkröten, Karetten	hawksbill turtles, hawksbill sea turtles
Ergates faber	Mulmbock	carpenter longhorn
Erignathus barbatus	Bartrobbe	bearded seal
Erigoninae	Zwergspinnen	dwarf spiders
Erinaceidae	Igel	hedgehogs & gymnures
Erinaceus albiventris/ Atelerix albiventris	Weißbauchigel	four-toed hedgehog
Erinaceus algirus/ Atelerix algirus	Algerischer Igel, Mittelmeerigel	Algerian hedgehog
Erinaceus concolor	Ostigel, Weißbrustigel	eastern hedgehog
Erinaceus europaeus	Westigel, Braunbrustigel	western hedgehog (European hedgehog)
Erinaceus frontalis/ Atelerix frontalis	Kap-Igel	South African hedgehog
Erinaceus sclateri/ Atelerix sclateri	Sclaters Igel, Somalia-Igel	Somali hedgehog
Erinnidae/ Xylophagidae	Holzfliegen	xylophagid flies
Eriocheir sinensis	Chinesische Wollhandkrabbe	Chinese mitten crab
Eriocnemis nigrivestris	Schwarzbrust-Schneehöschen	black-breasted puffleg
Eriococcidae	Woll-Schildläuse, Wollläuse	mealybugs
Eriocrania sparmannella	Birkenminiermotte	birch leaf miner*
Eriocraniidae	Trugmotten	eriocraniid moths
Eriogaster lanestris	Wollafter, Birkenwollafter	small eggar
Eriophyes erinea/ Eriophyes tristriatus	Walnusspockenmilbe	walnut blister mite
Eriophyes essigi/ Acalitus essigi	Brombeermilbe	redberry mite, blackberry mite
Eriophyes macrorhynchus	Ahorngallmilbe	maple nail-gall mite (>maple nail gall)
Eriophyes pyri	Birnenpockenmilbe	pear blister mite
Eriophyes ribis	Johannisbeergallmilbe	currant gall mite
Eriophyes sheldoni	Zitrusknospenmilbe	citrus bud mite
Eriophyes similis	Pflaumenblatt-Beutelgallmilbe, Pflaumenbeutelgallmilbe	plum pouch-gall mite
Eriophyes tiliae	Lindengallmilbe	lime nail-gall mite (>lime nail gall)
Eriophyes vitis/ Colomerus vitis	Rebenpockenmilbe	grapeleaf blister mite

Eriophyiidae	Gallmilben	gall mites, eriophyiid mites
Eriopygodes imbecilla	Braune Feuchtwieseneule, Braune Bergeule	Silurian moth
Eriosoma lanigerum	Blutlaus	woolly aphid, "American blight"
Eriosoma lanuginosum	Birnenblutlaus, Ulmenbeutelgallenlaus	elm balloon-gall aphid
Eriosoma ulmi	Ulmenblattrollenlaus	currant root aphid, elm-currant aphid, elm leaf aphid
Eriosomatidae/ Pemphigidae	Blasenläuse	aphids
Eriphia verrucosa	Gelbe Krabbe, Italienischer Taschenkrebs, Gemeine Krabbe	warty xanthid crab, Italian box crab, yellow box crab
Eristalinus sepulchralis	Schwarze Augenfleck-Schwebfliege	
Eristalis arbustorum	Kleine Bienenschwebfliege	lesser drone fly
Eristalis intricaria	Pelzige Bienenschwebfliege	
Eristalis tenax	Schlammfliege, Große Bienenschwebfliege, Mistbiene (Rattenschwanzlarve)	drone fly (rattailed maggot)
Eristicophis macmahoni	McMahon-Viper, Wüstenviper	McMahon's viper, leaf-nosed viper
Erithacus rubecula	Rotkehlchen	robin (European robin)
Ernobius mollis	Weicher Nagekäfer	pine bark anobiid
Ernoporus fagi	Kleiner Buchenborkenkäfer	
Eropeplus canus	Celebes Weichfellratte*	Celebes soft-furred rat
Erophylla spp.	Braune Blumenfledermäuse	brown flower bats
Erosaria acicularis	Atlantische Gelbkauri, Gelbe Kaurischnecke	Atlantic yellow cowrie
Erosaria caputserpentis/ *Cypraea caputserpentis*	Schlangenkopfkauri, Kleines Schlangenköpfchen	snake's head cowrie, serpent's head cowrie
Erotylidae	Südamerikanische Baumschwammkäfer, Südamerikanischer Pilzkäfer	pleasing fungus beetles
Erpeton tentaculatum	Fühlerschlange	tentacled snake, fishing snake
Erpobdella octoculata/ *Herpobdella octoculata*	Rollegel, Achtäugiger Schlundegel	eight-eyed leech
Erronea errones	Falsche Kauri	mistaken cowrie
Erronea ovum/*Ovula ovum*	Gemeine Eischnecke	egg cowrie, common egg cowrie
Erronea pulchella	Schönkauri*	pretty cowrie
Erymnochelys madagascariensis	Madagassische Schienenschildkröte	Madagascan big-headed side-necked turtle, Madagascan big-headed turtle (IUCN)
Erynnis tages	Dunkler Dickkopffalter, Grauer Dickkopf	dingy skipper
Erythrinidae	Forellensalmler, Raubsalmler	trahiras
Erythrocebus patas	Husarenaffe	Patas monkey, red guenon, red monkey

Erythrolamprus bizonus	Falsche Korallenschlange	false coral snake
Erythromma najas	Großes Granatauge	red-eyed damselfly
Erythromma viridulum	Kleines Granatauge	lesser red-eyed damselfly (*Continental*)
Eryx colubrinus	Ägyptische Sandboa	Egyptian sand boa
Eryx conicus	Rauschuppen-Sandboa	rough-scaled sand boa
Eryx jaculus	Westliche Sandboa, Sandschlange	Javelin sand boa
Eryx jayakari	Arabische Sandboa	Arabian sand boa
Eryx johnii johnii	Indische Sandboa	John's sand boa
Eryx miliaris	Wüsten-Sandboa	desert sand boa
Eryx tataricus	Große Sandboa	tartar sand boa
Eschrichtius robustus/ Eschrichtius gibbosus	Grauwal	gray whale
Esocidae	Hechte	pikes
Esomus spp.	Flugbarben	flying barbs
Esox americanus	Rotflossenhecht	redfin pickerel (FAO), redfin, grass pickerel
Esox lucius	Hecht, Flusshecht	pike, northern pike (FAO)
Esox masquinongy	Muskellunge	muskellunge
Esox niger	Kettenhecht	chain pickerel
Esox reicherti	Amur-Hecht	Amur pike
Esox vermiculatus	Maß-Hecht	grass pickerel
Estrilda astrild	Wellenastrild	common waxbill
Estrildidae	Prachtfinken	estrildine finches, waxbills
Etheostomatini	Grundbarsche, Darter	North American darters
Ethmalosa fimbriata	Bonga-Hering	bonga shad
Ethmidium maculatum	Pazifischer Menhaden	Pacific menhaden
Ethusa mascarone americana	Gepäckträgerkrabbe	stalkeye sumo crab, stalkeye porter crab
Etmopterus perryi	Perrys Schwarzer Dornhai	dwarf dogshark, dwarf lanternshark
Etmopterus princeps	Großer Schwarzer Dornhai	great lanternshark (FAO), greater lanternshark
Etmopterus pusillus	Glatter Schwarzer Dornhai	smooth lanternshark
Etmopterus spinax	Kleiner Schwarzer Dornhai, Schwarzer Hundfisch	velvet-belly, velvet belly (FAO)
Etmopterus spp.	Dornhaie u.a.	lanternsharks
Etroplus maculatus	Punktierter Buntbarsch, Punktierter Indischer Buntbarsch	orange chromide
Etroplus suratensis	Gebänderter Indischer Buntbarsch	banded chromide
Etropus crossotus		fringed flounder
Etrumerus teres	Rundhering, Gemeiner Rundhering	round herring
Etrumeus acuminatus	Kalifornischer Rundhering	California round herring
Etrumeus sadina	Atlantischer Rundhering	Atlantic round herring
Euastacus serratus	Australischer Flusskrebs	Australian crayfish

Eubalaena australis/ *Balaena glacialis australis*	Südkaper, Südlicher Glattwal	southern right whale
Eubalaena glacialis/ *Balaena glacialis*	Nordkaper	northern right whale, black right whale, Pacific right whale
Eubalaena japonica/ *Balaena glacialis japonica*	Nordpazifik-Glattwal	North Pacific right whale
Eublepharus macularius	Leopardgecko, Panthergecko	leopard gecko
Euborellia annulipes	Ringelbein-Ohrwurm*	ringlegged earwig
Eubucco bourcierii	Andenbartvogel	red-headed barbet
Eucera spp.	Langhornbienen	
Euceraphis punctipennis	Gemeine Birkenzierlaus	downy birch aphid, birch aphid
Euchoreutes naso	Riesenohr-Springmaus	long-eared jerboa
Euchorthippus declivus	Dickkopf-Grashüpfer	sharptailed grasshopper
Euchorthippus pulvinatus	Gelber Grashüpfer	straw-coloured grasshopper
Eucidaris tribuloides	Karibischer Lanzenseeigel	mine urchin, club urchin
Eucinetidae	Flachschenkelkäfer*	plate-thigh beetles
Eucinostomus melanopterus	Schwarzflossen-Mojarra, Schwarzflossen-Silberling	flagfin mojarra
Euclichthys polynemus		eucla cod
Euclidia glyphica/ *Ectypa glyphica*	Braune Tageule, Luzerneule	burnet companion
Eucnemidae	Schienenkäfer	false click beetles
Euconulidae	Kegelchen	
Eucosma conterminana	Grauer Salatsamenwickler	
Eucrate loricata	Paarzweig-Moostierchen	
Eudendrium album	Weißer Bäumchenpolyp	white stickhydroid
Eudendrium annulatum	Ringel-Bäumchenpolyp	annulate stickhydroid
Eudendrium carneum	Roter Bäumchenpolyp	red stickhydroid
Eudendrium ramosum	Bäumchenpolyp	stickhydroid, stick hydroid
Eudendrium tenue	Schlanker Bäumchenpolyp	slender stickhydroid
Euderma maculatum	Gefleckte Fledermaus	spotted bat, pinto bat
Eudia pavonia/ *Saturnia pavonia*	Kleines Nachtpfauenauge	emperor moth
Eudiscopus denticulus	Diskusfüßige Fledermaus	disk-footed bat
Eudistoma hepaticum	Seeleber*	sea liver
Eudocimus ruber	Scharlachsichler	scarlet ibis
Eudontomyzon danfordi	Karpatenneunauge (Donauneunauge)	Carpathian lamprey (FAO), Carpathian brook lamprey (Hungarian lamprey/ Danubian lampern)
Eudromia elegans	Perlsteißhuhn	elegant crested-tinamou
Eudyptes chrysocome	Felsenpinguin	rockhopper penguin
Eudyptes chrysolophus	Goldschopfpinguin	macaroni penguin
Eudyptes pachyrhynchus	Dickschnabelpinguin	Victoria penguin
Eudyptes schlegeli	Haubenpinguin	royal penguin
Eudyptula minor	Zwergpinguin	little penguin
Eugenes fulgens	Dickschnabelkolibri, Rivolikolibri	magnificent hummingbird

Euglandina rosea	Rosa Räuberschnecke	rosy predator snail
Euglenidae	Ameisenblattkäfer*	antlike leaf beetles
Euglenophyta	Augenflagellaten	euglenoids, euglenids
Eugomphodus taurus/	Sandhai,	sand shark,
Carcharias taurus/	Echter Sandhai,	sand tiger shark (FAO),
Odontaspis taurus	Sandtiger	sandtiger shark,
		gray nurse shark
Eugraphe sigma	Waldmantel-Erdeule	
Eugraphe subrosea	Rosiges Erdeulchen	rosy marsh (moth)
Eulalia viridis	Grüner Blattwurm	greenleaf worm,
		green paddle worm
Eulamellibranchia	Lamellenkiemer,	eulamellibranch bivalves
	Blattkiemer	
Eulecanium corni	Große Obstbaumschildlaus	brown scale,
		brown fruit scale,
		European fruit lecanium,
		European peach scale
Euleia heraclei	Selleriefliege	celery fly
Eulemur coronatus/	Kronenmaki	crowned lemur
Lemur coronatus/		
Petterus coronatus		
Eulemur mongoz/	Mongozmaki	mongoose lemur
Lemur mongoz/		
Petterus mongoz		
Eulithidium affine affine/	Karierte Fasanenschnecke	checkered pheasant (snail)
Tricolia affinis affinis		
Eulithis prunata	Schlehdornspanner	phoenix moth
Eumastacidae	Flügellose	monkey grasshoppers
	Tropen-Feldheuschrecken*	
Eumeces algeriensis	Berberskink	Algerian skink
Eumeces anthracinus	Kohlenskink	coal skink
Eumeces egregius	Maulwurfskink	mole skink
Eumeces fasciatus	Streifenskink	five-lined skink
Eumeces inexpectatus	Fünfbinden-Skink	southeastern five-lined skink
Eumeces laticeps	Breitkopfskink	broadheaded skink,
		broad-headed skink,
		"scorpion"
Eumeces multivirgatus	Mehrstreifen-Skink	many-lined skink
Eumeces obsoletus	Spitzkopfskink, Prärie-Skink	Great Plains skink
Eumeces schneideri	Tüpfelskink	Berber skink
Eumeces septentrionalis	Prärie-Skink	prairie skink
Eumeces skiltonianus	Westlicher Skink	western skink
Eumeces spp.	Amerikanische Skinks	eyelid skinks
Eumeces tetragrammus	Vierstreifen-Skink	four-lined skink
Eumenes coarctatus	Heide-Töpferwespe,	heath potter wasp,
	Glockenwespe	potter wasp
Eumenes spp.	Töpferwespen, Pillenwespen	potter wasps
Eumenidae	Lehmwespen & Pillenwespen	mason wasps,
		potter wasps
Eumerus strigatus	Gemeine	onion bulb fly,
	Zwiebelmondschwebfliege	small narcissus fly
Eumerus tuberculatus	Höcker-	tuberculate bulb fly,
	Zwiebelmondschwebfliege	lesser bulb fly (U.S.)
Eumetopias jubatus	Stellers Seelöwe	northern sea lion,
		Steller sea lion

Eumops spp.	Bulldogg-Fledermäuse	mastiff bats, bonneted bats
Eunectes murinus	Große Anakonda, Grüne Anakonda	anaconda, water boa
Eunectes notaeus	Süd-Anakonda, Gelbe Anakonda, Paraguay-Anakonda	yellow anaconda
Eunectes spp.	Anakondas	anacondas
Euneomys spp.	Patagonische Chinchillamäuse	Patagonian chinchilla mice
Eunice aphroditois	Wunderwurm*	wonder-worm
Eunice fucata	Atlantischer Palolo	Atlantic palolo worm
Eunice gigantea	Riesenborster	giant eunice
Eunice harassii	Kieferwurm	
Eunice schemacephala	Westindischer Palolo	West Indian palolo worm
Eunice viridis/Palola viridis	Samoa-Palolo	Pacific palolo worm, Samoan palolo worm
Eunicea calyculata	Höckerige Hornkoralle	knobby candelabrum
Eunicella cavolini	Gelbe Hornkoralle, Gelbe Gorgonie	yellow horny coral, yellow sea fan
Eunicella singularis	Weiße Hornkoralle, Weiße Gorgonie, Gestreckte Hornkoralle	white horny coral, white sea fan
Eunicella spp.	Seefächer, Hornkorallen u.a.	horny corals a.o., sea fans
Eunicella stricta	Weiße Mittelmeer-Hornkoralle	Mediterranean white horny coral
Eunicella verrucosa	Warzenkoralle, Warzige Hornkoralle, Kleiner Seefächer	warty coral, pink sea fan
Euoticus elegantulus/ Galago elegantulus	Westlicher Kielnagelgalago, Südlicher Kielnagelgalago	western needle-clawed bush baby, elegant needle-clawed galago
Euoticus pallidus	Nördlicher Kielnagelgalago	pallid needle-clawed galago
Eupetaurus cinereus	Fels-Gleithörnchen	woolly flying squirrel
Euphagus carolinus	Roststärling	rusty blackbird
Euphagus cyanocephalus	Purpurstärling	Brewer's blackbird
Euphausia superba	Südlicher Krill, Antarktischer Krill	whale krill, Antarctic krill
Euphausiacea	Leuchtkrebse & Krill	euphausiaceans (krill & allies)
Euphausiidae	Leuchtkrebse & Krill	euphausiids, krill
Euphlyctis cyanophlyctis/ Rana cyanophlyctis	Indischer Wasserfrosch	Cyan five-fingered frog
Euphlyctis spp.	Fünffingerfrösche	five-fingered frogs
Eupholidoptera chabrieri	Grüne Strauchschrecke	Chabrier's bushcricket
Euphonia laniirostris	Dickschnabelorganist	thick-billed euphonia
Euphractus pichiy/ Zaedyus pichiy	Zwerggürteltier	pichi
Euphractus sexcinctus	Weißborsten-Gürteltier, Sechsbinden-Gürteltier	six-banded armadillo
Euphractus spp./ *Chaetophractus* spp.	Borstengürteltiere	hairy armadillos, peludos
Euphractus villosus	Braunzottiges Borstengürteltier, Braunes Borstengürteltier	larger hairy armadillo
Euphydryas maturna/ Hypodryas maturna	Kleiner Maivogel	scarce fritillary
Euphyllia ancora	Hammerkoralle	hammer-tooth coral*
Euphyllia spp.	Große Bukettkorallen	tooth corals

Eupithecia abietaria	Fichtenzapfenspanner, Zapfenspanner	
Eupithecia assimilata	Johannisbeerzapfenspanner	currant pug moth
Eupithecia centaureata	Trockenrasen-Blütenspanner	lime-speck pug
Eupithecia egenaria		pauper pug
Eupithecia vulgata	Gemeiner Blütenspanner	common pug
Euplagia quadripunctaria/ Callimorpha quadripunctaria	Russischer Bär	Jersey tiger (moth), Russian tiger (moth)
Euplectella aspergillum	Gießkannenschwamm	Venus' flower basket
Euplectes jacksoni	Leierschwanzwida	Jackson's whydah
Euplectes orix	Oryxweber	red bishop
Eupleres goudotii goudotii	Kleinfanaluk	eastern fanalouc
Eupleres goudotii major	Großfanaluk	western fanalouc
Eupleres goudotii	Graubraune Madagassische Schleichkatze, Ungefleckte Fanaloka, 'Ameisenschleichkatze' (Fanaloka, Ridaridy; siehe auch: *Fossa fossa*)	falanouc (IUCN), fanalouc (*better*: grey-brown Malagasy civet; *native common names*: fanaloka, ridaridy)
Eupodotis bengalensis	Barttrappe	Bengal florican
Eupoecilia ambiguella/ Cochylis ambiguella/ Clysia ambiguella	Einbindiger Traubenwickler (Heuwurm)	European grapevine moth, small brown-barrel conch, grapeberry moth
Eupolymnia nebulosa	Polymnia, Erdbeerwurm*	strawberry worm
Euproctis chrysorrhoea	Goldafter	brown-tail moth, brown-tail
Euproctis similis/ Porthesia similis/ Sphrageidus similis	Schwan	yellow-tail moth, gold-tail
Euproctus asper	Pyrenäen-Gebirgsmolch	Pyrenean brook salamander, Pyrenean mountain newt
Euproctus montanus	Korsischer Gebirgsmolch	Corsian brook salamander, Corsican mountain newt
Euproctus platycephalus	Sardinischer Gebirgsmolch, Hechtkopf-Gebirgsmolch	Sardinian brook salamander, Sardinian mountain newt
Euproctus spp.	Europäische Gebirgsmolche	European brook salamanders, European mountain salamanders
Euprotomicrus bispinatus	Zwerghai	pygmy shark (FAO), slime shark
Eupsilia transversa	Satelliteule	satellite
Eurodryas aurinia/ Melitaea aurinia	Skabiosen-Scheckenfalter, Goldener Scheckenfalter	marsh fritillary
Eurois occulta	Große Heidelbeereule	
Eurrhypara hortulata	Nesselzünsler	small magpie (moth)
Eurycea bislineata	Nördlicher Zweistreifiger Gelbsalamander	two-lined salamander
Eurycea longicauda	Langschwänziger Gelbsalamander, Langschwanzsalamander	long-tailed salamander
Eurycea lucifuga	Höhlensalamander, Höhlen-Gelbsalamander	cave salamander
Eurycea neotenes	Texas-Gelbsalamander, Texas-Höhlensalamander	Texas salamander
Eurycea quadridigitata	Vierzehen-Gelbsalamander, Zwergsalamander	dwarf salamander
Eurycea spp.	Gelbsalamander	American brook salamanders

Euryceros prevostii	Helmvanga	helmet bird
Eurydema dominulus	Zierliche Gemüsewanze	
Eurydema oleraceum	Kohlwanze	brassica bug
Eurydema ornatum	Kohlschmuckwanze, Schwarzrückige Gemüsewanze	ornate cabbage bug
Eurygaster maura	Gras-Schildwanze	
Eurygaster testudinaria	Schildkrötenwanze	
Eurylaimidae	Breitrachen, Breitmäuler	broadbills
Eurynome aspera	Erdbeerkrabbe*	strawberry crab
Eurypharyngidae/ Eupharyngidae	Echte Pelikanaale, Großmäuler	pelican eels, gulpers
Eurypharynx pelecanoides	Pelikanaal	pelican eel
Eurypterida	Seeskorpione	sea scorpions, eurypterids
Eurypyga helias	Sonnenralle	sunbittern
Eurypygidae	Sonnenralle	sunbittern
Eurystomus orientalis	Dollarvogel	eastern broad-billed roller, dollar bird
Eurythoe complanata	Feuerwurm	fireworm
Eurythyrea austriaca	Metallgrüner Tannenprachtkäfer	
Eurytoma amygdali	Steinobstsamenwespe	
Eurytoma gibbus/ *Bruchophagus gibbus*	Kleesamenwespe	lucerne chalcid wasp
Eurytoma orchidearum	Orchideenfliege	orchidfly, orchid wasp, cattleya 'fly'
Euryzygomatomys spinosus	Suira	guiara
Euscorpius flavicaudis	Gelbschwänziger Skorpion	yellowtail scorpion
Euscorpius italicus	Italienischer Skorpion	Italian scorpion
Eusmilia fastigiata	Glatte Blumenkoralle	smooth flower coral
Eusphalerum minutum	Kleiner Blütenkurzflügler	
Eusphyra blochii		winghead shark
Euspira heros/*Lunatia heros*	Nördliche Mondschnecke	northern moonsnail, common northern moonsnail, sand collar moon snail
Euspira nana/*Neverita nana*	Kleine Mondschnecke	tiny moonsnail
Euspira pallida/ *Polinices pallidus*	Blasse Mondschnecke	pale moonsnail
Euspira poliana/*Lunatia p.*	Glänzende Mondschnecke	shiny moonsnail
Eustroma reticulata/ *Lygris reticulata*	Springkraut-Netzspanner, Weißgerippter Haarbuschspanner	netted carpet (moth)
Eutamias sibiricus/ *Tamias sibiricus*	Burunduk (Hbz.), Eurasisches Erdhörnchen, Östliches Streifenhörnchen, Sibirisches Streifenhörnchen	Siberian chipmunk
Eutamias spp.	Streifenhörnchen	Siberian & Western American chipmunks
Euthynnus affinis	Pazifische Thonine	kawakawa
Euthynnus alletteratus	Thonine, Falscher Bonito, Kleine Thonine	little tunny (FAO), little tuna, mackerel tuna (bonito)
Euthynnus lineatus	Schwarze Thonine	black skipjack (FAO)
Euthynnus pelamis/ *Katsuwonus pelamis*	Gestreifter Thun, Echter Bonito	skipjack tuna (FAO), bonito, stripe-bellied bonito

Eutrigla gurnardus	Grauer Knurrhahn	grey gurnard (FAO), gray searobin
Eutrombicula alfreddugesi/ Trombicula alfreddugesi	Amerikanische Erntemilbe*	American chigger mite, common chigger mite
Eutromula pariana/ Choreutis pariana	Apfelblattmotte	apple leaf skeletonizer
Eutropiella debauwi	Kongo-Glaswels	African glass catfish
Euura mucronata	Weidenknospen-Blattwespe	willow bud sawfly
Euxerus erythropus/ Xerus erythropus	Rotfuß-Erdhörnchen, Streifen-Borstenhörnchen	striped ground squirrel, Geoffroy's ground squirrel
Euxiphipops asfur/ Arusetta asfur/ Pomacanthus asfur	Asfur-Kaiserfisch	Arabian angelfish
Euxiphipops navarchus/ Pomacanthus navarchus	Traumkaiserfisch	blue-girdled angelfish
Euxoa nigricans	Schwarzeule	garden dart moth
Evacanthus interruptus	Hopfenzikade	hop leafhopper
Evania appendigaster	Hungerwespe	
Evaniidae	Hungerwespen	ensign wasps
Everes argiades	Kleebläuling	short-tailed blue
Evergestis forficalis	Kohlzünsler, Meerrettichzünsler	garden pebble moth
Evermannellidae	Säbelzahnfische	sabertooth fishes, sabretoothed fishes
Evermannichthys metzelaari	Rauschwanzgrundel*	roughtail goby
Evermannichthys spongicola	Schwämmchengrundel*	sponge goby
Evetria buoliana/ Rhyacionia buoliana	Kieferntriebwickler	gemmed shoot moth, pine-sprout tortrix, European pine shoot moth
Evodinus clathratus	Fleckenbock	
Evorthodus lyricus/ Mugilostoma gobio	Lyra-Grundel	lyre goby
Exilisciurus spp.	Asiatische Zwerghörnchen u.a.	Asian pygmy squirrels a.o. (Borneo/Philippines)
Exocentrus adspersus	Weißgefleckter Wimpernhornbock	
Exocoetidae	Fliegende Fische, Flugfische	flyingfishes
Exocoetus monocirrhus	Bartel-Flugfisch, Zweiflügelfisch	barbel flyingfish (FAO), two-wing flyingfish
Exocoetus obtusirostris	Ozean-Flugfisch	oceanic two-wing flyingfish
Exocoetus volitans	Fliegender Fisch, Flugfisch, Gemeiner Flugfisch, Meerschwalbe	tropical two-wing flyingfish
Exodon paradoxus	Zweitupfen-Raubsalmler	two-blotched tetra, bucktooth tetra (FAO)
Exoteleia dodecella	Kiefernknospentriebmotte	European pine bud moth*

Fabriciana adippe/ *Argynnis adippe*	Märzveilchenfalter, Bergmatten-Perlmutterfalter, Hundsveilchen-Perlmutterfalter	high brown fritillary
Facciolella oxyrhyncha	Facciola-Aal	Facciola's sorcerer
Facelina auriculata/ *Facelina drummondi*	Drummonds Fadenschnecke	Drummond's facelina
Facelina bostoniensis	Boston-Fadenschnecke	Boston facelina
Faculiferidae	Gefiedermilben	feather mites*
Fagesia lineata	Liniierte Anemone*	lined anemone
Falcaria lacertinaria	Eidechsenschwanz	scalloped hook-tip
Falcidens crossotus (Aplacophora)	Gemeiner Zangenschildfuß	glistenworm
Falcidens gutturosus (Aplacophora)	Einfacher Zangenschildfuß	Mediterranean glistenworm a.o.
Falco biarmicus	Lannerfalke	lanner falcon
Falco cenchroides	Graubartfalke	Australian kestrel
Falco cherrug	Würgfalke	saker falcon
Falco columbarius	Merlin	merlin
Falco concolor	Schieferfalke	sooty falcon
Falco eleonorae	Eleonorenfalke	Eleonora's falcon
Falco femoralis	Aplomadofalke	Aplomado falcon
Falco mexicanus	Präriefalke	prairie falcon
Falco naumanni	Rötelfalke	lesser kestrel
Falco peregrinoides	Wüstenfalke	barbary falcon
Falco peregrinus	Wanderfalke	peregrine
Falco punctatus	Mauritiusfalke	Mauritius kestrel
Falco rusticolus	Gerfalke	gyrfalcon
Falco sparverius	Buntfalke	American kestrel, sparrow hawk
Falco subbuteo	Baumfalke	hobby, northern hobby
Falco tinnunculus	Turmfalke	kestrel, common kestrel
Falco vespertinus	Rotfußfalke	red-footed falcon
Falconidae	Falken	falcons, caracaras
Falconiformes	Greifvögel	diurnal birds of prey (falcons and others)
Fannia canicularis	Kleine Stubenfliege	lesser house fly
Fannia scalaris	Latrinenfliege	latrine fly
Farancia abacura	Schlammnatter	mud snake
Farancia erythrogramma	Regenbogennatter	rainbow snake
Farfantepenaeus aztecus/ *Penaeus aztecus*	Braune Garnele	brown shrimp, northern brown shrimp
Farfantepenaeus brasiliensis/ *Penaeus brasiliensis*	Rotpunktgarnele	pinkspotted shrimp, redspotted shrimp
Farfantepenaeus brevirostris/ *Penaeus brevirostris*	Kristall-Geißelgarnele	crystal shrimp
Farfantepenaeus *californiensis/* *Penaeus californiensis*	Kalifornische Geißelgarnele	yellowleg shrimp, yellow-leg shrimp
Farfantepenaeus duorarum/ *Penaeus duorarum*	Nördliche Rosa-Garnele, Rosa Golfgarnele, Nördliche Rosa Geißelgarnele	pink shrimp, northern pink shrimp
Farfantepenaeus notialis/ *Penaeus notialis*	Südliche Rosa Geißelgarnele, Senegal-Garnele	southern pink shrimp, candied shrimp
Farlowella acus	Lanzettförmiger Schnabelwels, Gemeiner Nadelwels	twig catfish

Fasciola gigantica	Riesenleberegel	giant liver fluke
Fasciola hepatica	Großer Leberegel	sheep liver fluke
Fasciolaria lignaria	Tarentinische Spindelschnecke	Tarentine spindle snail
Fasciolaria lilium	Gebänderte Tulpenschnecke	banded tulip (banded tulip shell)
Fasciolaria tulipa	Echte Tulpenschnecke	true tulip, tulip spindle snail, (tulip spindle shell)
Fasciolariidae	Spindelschnecken, Tulpenschnecken, Pferdeschnecken, Bündelhörner	spindle snails (spindle shells) & tulip shells
Fascioloides magna	Großer Amerikanischer Leberegel*	large American liver fluke
Fasciolopsis buski	Riesendarmegel, Großer Darmegel	giant intestinal fluke
Favia favus	Knopfkoralle	honeycomb coral, pineapple coral
Favia fragum	Kleine Sternkoralle	golfball coral, star coral
Favites spp.	Wabenkorallen*	honeycomb corals
Felicola subrostratus	Katzenhaarling	cat louse
Felidae	Katzen	cats
Felis badia/Catopuma badia	Borneo-Goldkatze	Borneo golden cat, Bornean bay cat, bay cat, Bornean red cat
Felis bengalensis	Bengalkatze, Leopardkatze	leopard cat
Felis bieti	Graukatze	Chinese desert cat
Felis caracal	Wüstenluchs, Karakal	caracal
Felis catus/ Felis silvestris f. catus	Hauskatze	domestic cat
Felis chaus	Rohrkatze (Hbz. Dschungelkatze)	jungle cat
Felis colocolo/ Lynchailurus pajeros	Pampaskatze	pampas cat
Felis concolor/ Puma concolor	Puma, Silberlöwe	cougar, puma, mountain lion
Felis euptilura	Amur-Katze	Amur cat
Felis geoffroyi	Kleinfleckkatze, Salzkatze	Geoffroy's cat
Felis guigna	Chilenische Waldkatze	kodkod
Felis iriomotensis/ Mayailurus iriomotensis	Iriomoto-Katze	Iriomote cat
Felis jacobita	Andenkatze	Andean cat (mountain cat)
Felis lynx/Lynx lynx	Nordluchs (Hbz. Luchs)	lynx
Felis lynx canadensis	Kanadaluchs (Hbz. Silberluchs)	Canadian lynx, silver lynx
Felis manul	Manul	Pallas's cat
Felis margarita	Sandkatze	sand cat
Felis marmorata/ Pardofelis marmorata	Marmorkatze	marbled cat
Felis nigripes	Schwarzfußkatze	black-footed cat
Felis pardalis	Ozelot	ocelot

Felis pardinus/Lynx pardinus	Spanische Wildkatze	Spanish lynx, pardel
Felis planiceps/ Ictailurus planiceps	Flachkopfkatze	flat-headed cat
Felis rubiginosa/ Prionailurus rubiginosus	Rostkatze	rusty-spotted cat
Felis rufus/Lynx rufus	Rotluchs (Hbz. Luchskatze)	bobcat
Felis serval/ Leptailurus serval	Serval	serval
Felis silvestris	Wildkatze	wild cat
Felis temmincki/ Profelis temmincki	Asiatische Goldkatze, Temminckkatze	Asian golden cat
Felis tigrina	Ozelotkatze, Oncille	little spotted cat
Felis viverrina/ Prionailurus viverrinus	Fischkatze	fishing cat
Felis wiedii	Langschwanzkatze, Margay	margay
Felis yagouaroundi	Wieselkatze, Jaguarundi	jaguarundi
Felovia vae	Senegal-Gundi, Senegalkammfinger	felou gundi, Senegal gundi
Fennecus zerda	Fennek	fennec fox
Fenneropenaeus chinensis/ Penaeus chinensis	Hauptmannsgarnele	fleshy prawn
Fenneropenaeus indicus/ Penaeus indicus	Indische Hauptmannsgarnele	Indian prawn
Fenneropenaeus merguiensis/ Penaeus merguiensis	Bananen-Garnele	banana prawn
Fenneropenaeus penicillatus/ Penaeus penicillatus	Rotschwanzgarnele	redtail prawn
Feresa attenuata	Zwerggrindwal, Zwergschwertwal	pygmy killer whale, slender blackfish
Feroculus feroculus	Kelaarts Langkrallenspitzmaus	Kelaart's long-clawed shrew
Ferrissia spp.	Septenmützenschnecken	ancylids a.o.
Ferrissia wautieri	Flache Mützenschnecke	flat ancylid
Ferrussaciidae	Turmschnecken	awlsnails a.o.
Festilyria festiva	Festliche Leierschnecke*	festive volute, festive lyria
Feyliniidae	Afrikanische Schlangenechsen	feylinias
Ficedula albicollis	Halsbandschnäpper	collared flycatcher
Ficedula hypoleuca	Trauerschnäpper	pied flycatcher
Ficedula parva	Zwergschnäpper	red-breasted flycatcher
Ficedula semitorquata	Halbringschnäpper	semi-collared flycatcher
Ficidae	Feigenschnecken	fig snails (fig shells)
Ficus communis	Atlantik-Feigenschnecke, Gemeine Feigenschnecke	Atlantic figsnail (common fig shell)
Filibranchia	Fadenkiemer	filibranch bivalves
Fimbrios klossi	Lippennatter	bearded snake
Fissidentalium vernedei (Scaphopoda)	Vernedes Zahnschnecke	Vernede's tusk
Fissipedia	Landraubtiere	terrestrial carnivores
Fissurella aperta	Doppelkanten-Schlüssellochschnecke	double-edged keyhole limpet
Fissurella barbadensis	Barbados-Schlüssellochschnecke	Barbados keyhole limpet

Fissurella gemmata	Weiße Schlüssellochschnecke	white keyhole limpet
Fissurella nodosa	Knotige Schlüssellochschnecke	knobby keyhole limpet, knobbed keyhole limpet
Fissurella volcano	Vulkan-Schlüssellochschnecke	volcano keyhole limpet
Fissurellidae	Schlüssellochschnecken, Schlitzschnecken	keyhole limpets, slit limpets
Fistularia petimba	Flötenfisch	red cornetfish (FAO), flutemouth
Fistulariidae	Flötenmünder, Flötenmäuler, Pfeifenfische	flutemouths, cornetfishes
Flabellina affinis	Violette Fadenschnecke	Mediterranean violet aeolid
Flabellina iodinea	Kalifornische Fadenschnecke	Spanish shawl
Flabellum macandrewi	Spalt-Fächerkoralle*	splitting fan coral
Flabellum spp.	Fächerkorallen u.a.	fan corals
Flectonotus goeldii/ Fritziana goeldii	Schüsselrücken-Laubfrosch	Colonia Alpina treefrog
Flustra foliacea	Blätter-Moostierchen	broad-leaved hornwrack
Fluvicola pica	Elstertyrann	pied water tyrant
Fodiator acutus	Scharfkinn-Flugfisch	sharpchin flyingfish
Foraminiferida	Foraminiferen	foraminiferans, forams
Forcipiger flavissimus	Röhrenmaul-Pinzettfisch, Gelber Maskenpinzettfisch	longnose butterflyfish (FAO), long-beaked butterflyfish
Forcipiger longirostris	Langschnauzen-Pinzettfisch, Langmaul-Pinzettfisch	long-snouted forceps fish
Forcipiger spp.	Pinzettfische	forceps fishes
Forcipulatida	Zangensterne	forcipulatids
Fordonia leucobalia	Krebsstrugnatter	white-bellied mangrove snake, whitebelly mangrove snake
Forficula auricularia	Gemeiner Ohrwurm	common earwig, European earwig
Forficulidae	Europäische Ohrwürmer	European earwigs
Formica exsecta/ Coptoformica exsecta	Buchtenkopf-Waldameise, Kerbameise	narrow-headed ant
Formica candica	Moorameise	bog ant
Formica fusca	Grauschwarze Sklavenameise, Furchtsame Hilfswaldameise	negro ant
Formica polyctena	Kleine Rote Waldameise, Kahlrückige Rote Waldameise	small red wood ant
Formica pratensis/ Formica nigricans	Dunkle Wiesenameise, Dunkle Waldameise	black-backed meadow ant
Formica rufa	Rote Waldameise, Große Rote Waldameise	wood ant
Formica rufibarbis	Rotbärtige Sklavenameise	red-barbed ant
Formica sanguinea	Blutrote Raubameise	blood-red ant
Formica spp.	Waldameisen und Raubameisen	wood ants and predatory ants
Formicariidae	Ameisenvögel	ground antbirds
Formicidae	Ameisen	ants
Formicinae	Schuppenameisen	carpenter ants and others
Formicomus pedestris	Ameisen-Blütenmulmkäfer	
Formicoxenus nitidulus	Gastameise, Glänzende Gastameise	
Formio niger	Schwarzer Pomfret	black pomfret
Forpus conspicillatus	Brillenpapagei	spectacled parrotlet

Fossa fossa	Fanaloka, Gefleckte Madagassische Schleichkatze (siehe auch: *Eupleres goudotii*)	spotted Malagasy civet, spotted fanaloka, spotted fanalouc (see also: *Eupleres goudotii*)
Foudia madagascariensis	Madagaskarweber	Madagascan red fody
Fragum unedo	Erdbeer-Herzmuschel	strawberry cockle
Francolinus afer	Rotkehlfrankolin	red-necked spurfowl
Francolinus bicalcaratus	Doppelspornfrankolin	double-spurred francolin
Francolinus francolinus	Halsbandfrankolin	black francolin
Francolinus pondicerianus	Wachtelfrankolin	Indian grey francolin
Frankliniella intonsa	Blütenthrips, Gemeiner Blütenthrips	flower thrips
Fratercula arctica	Papageitaucher	puffin, Atlantic puffin
Fratercula cirrhata/ Lunda cirrhata	Gelbschopflund	tufted puffin
Fregata andrewsi	Weißbauchfregattvogel	Christmas Islands frigate bird, Christmas Island frigatebird
Fregata ariel	Arielfregattvogel	lesser frigate bird
Fregata magnificens	Prachtfregattvogel	magnificent frigatebird
Fregata minor	Bindenfregattvogel	great frigate bird
Fregatidae	Fregattvögel	frigatebirds
Fringilla coelebs	Buchfink	chaffinch
Fringilla montifringilla	Bergfink	brambling
Fringilla teydea	Teydefink	blue chaffinch
Fringillidae	Finken, Finkenvögel	finches and allies
Fringillinae	Edelfinken	chaffinches & brambling
Frondipora verrucosa	Höckerzweig-Moostierchen, Verwachsenes Moostierchen	warty leaf-bryozoan*
Frontinella pyramitela/ Frontinella communis	Gemeine Frontinella*	bowl and doily spider
Fulgora laternaria	Großer Laternenträger	greater lanternfly
Fulgoraria rupestris/ Fulgoraria fulminata	Asiatische Feuerwalzenschnecke	Asian flame volute
Fulgoraria hirasei	Hirases Walzenschnecke	Hirase's volute
Fulgoridae	Laternenträger, Leuchtzikaden	lanternflies, lantern flies, fulgorid planthoppers
Fulica americana	Indianerblässhuhn	American coot
Fulica atra	Blässhuhn, Bläßhuhn	coot
Fulica cristata	Kammblässhuhn	crested coot
Fulmarus glacialis	Eissturmvogel	fulmar
Funambulus spp.	Gestreifte Palmenhörnchen	Asiatic striped palm squirrels
Fundulus catenatus	Kettenkärpfling	northern studfish
Fundulus chrysotus	Goldohr, Goldauge	golden ear killifish, golden topminnow (FAO)
Fundulus grandis	Golfkärpfling	Gulf killifish
Fundulus heteroclitus	Zebrakärpfling, Blaubandkärpfling	killifish, mummichog (FAO)
Fungia fungites	Pilzkoralle	mushroom coral
Funicula quadrangularis	Seepeitsche	sea whip a.o.

Funisciurus spp.	Rotschenkelhörnchen	African striped squirrels, African side-striped squirrels, rope squirrels
Furcula bicuspis/ *Harpyia bicuspis*	Birken-Gabelschwanz	
Furcula bifida/ *Cerura bifida/* *Harpyia hermelina/* *Harpyia bifida*	Kleiner Gabelschwanz	sallow kitten
Furcula furcula/ *Harpyia furcula*	Salweiden-Gabelschwanz, Buchen-Gabelschwanz	
Furipteridae	Stummeldaumen-Fledermäuse	smoky bats, thumbless bats
Furnariidae	Töpfervögel, Ofenvögel	ovenbirds
Fusigobius longispinis	Orangeflecken-Grundel	butterfly goby
Fusinus colus	Weberspindel	distaff spindle
Fusinus longissimus	Lange Spindelschnecke	long spindle
Fusinus nicobaricus	Nikobaren-Spindelschnecke	Nicobar spindle
Fusinus syracusanus	Sizilianische Spindelschnecke	Sicilian spindle
Fusinus undatus	Gewellte Spindelschnecke	wavy spindle
Fusitriton magellanicum	Magellan-Triton	Magellanic triton
Fusitriton oregonensis	Oregon-Triton	Oregon triton
Fusus antiqua/ *Neptunea antiqua*	Gemeines Neptunshorn, Gemeine Spindelschnecke	ancient whelk, ancient neptune, common spindle snail, neptune snail, red whelk, buckie
Fusus rostratus/ *Neptunea rostratus*	Zierliche Spindelschnecke, Geschnäbeltes Spindelhorn	rostrate whelk, rostrate neptune

Gadiculus argenteus	Silberdorsch	silvery pout
Gadiculus argenteus thori	Mittelmeer-Silberdorsch	Mediterranean silvery pout
Gadidae	Dorsche & Schellfische	cod or cods (both *pl.*)
Gadiformes	Dorschfische	codfishes & haddock & hakes
Gadus macrocephalus	Pazifik-Dorsch, Pazifischer Kabeljau	Pacific cod (FAO), gray cod, grayfish
Gadus morhua	Kabeljau (*Ostsee/Jungform*: Dorsch)	cod, Atlantic cod (*young*: codling)
Gadus ogac	Grönland-Kabeljau, Grönland-Dorsch, Fjord-Dorsch	Greenland cod
Gafrarium divaricatum	Verzweigte Venusmuschel	forked venus
Gaidropsarus granti	Grant-Seequappe	Grant's rockling
Gaidropsarus mediterraneus	Mittelmeer-Seequappe	shore rockling
Gaidropsarus vulgaris	Dreibartelige Seequappe	three-bearded rockling
Galago alleni	Buschwaldgalago	Allen's bush baby, Allen's squirrel galago, black-tailed bush baby
Galago crassicaudatus/ Otolemur crassicaudatus	Riesengalago	thick-tailed bush baby, greater galago, greater bush baby
Galago demidovii/ Galagoides demidoff	Zwerggalago, Urwaldgalago	dwarf bush baby, Demidoff's bush baby, Demidoff's galago
Galago elegantulus/ Euoticus elegantulus	Westlicher Kielnagelgalago, Südlicher Kielnagelgalago	western needle-clawed bush baby, elegant needle-clawed galago
Galago gallarum	Somalia-Galago, Somaligalago	Somali galago
Galago matschiei/ Galago inustus	Östlicher Kielnagelgalago	eastern needle-clawed bush baby, spectacled galago
Galago moholi	Südafrikanischer Galago	southern lesser bush baby, South African galago
Galago senegalensis	Senegalgalago, Steppengalago	Senegal bush baby, lesser bush baby, Senegal galago
Galago spp.	Galagos	galagos, bush babies
Galagoides spp.	Zwerggalagos	dwarf bush baby, dwarf galagos
Galathea squamifera	Furchenkrebs, Schuppiger Springkrebs	Leach's squat lobster
Galathea strigosa	Blaustreifen-Springkrebs, Blaugestreifter Springkrebs, Bunter Furchenkrebs, Bunter Springkrebs	strigose squat lobster
Galatheidae	Furchenkrebse, Scheinhummer	squat lobsters
Galaxea astreata	Sternkoralle, Skalpellkoralle	scalpel coral*
Galaxea fascicularis	Kristallkoralle	crystal coral*
Galaxias truttaceus	Forellen-Hechtling*	spotted trout minnow, spotted mountain trout (FAO)
Galaxiidae	Hechtlinge	galaxiids
Galba truncatula	Leberegelschnecke, Kleine Sumpfschnecke	dwarf pond snail, dwarf mud snail
Galbula dea	Paradiesglanzvogel	paradise jacamar
Galea spp.	Wieselmeerschweinchen	yellow-toothed cavies, cuis

Galeichthys ater	Schwarzer Meereswels*	black seacatfish
Galeichthys feliceps	Katzen-Kreuzwels, Weißer Meereswels*	white seacatfish, white baggar (FAO)
Galemys pyrenaicus	Pyrenäendesman	Pyrenean desman
Galenomys garleppi	Bolivianische Blattohrmaus	Bolivian leaf-eared mouse
Galeocerdo cuvieri	Tigerhai	tiger shark
Galeodea echinophora	Stachlige Helmschnecke, Stachel-Helmschnecke	spiny helmet snail, Mediterranean spiny bonnet
Galeodea rugosa	Knotige Helmschnecke, Knotenschnecke, Knotenschelle	rugose helmet snail, Mediterranean rugose bonnet
Galeolaria truncata	Helmqualle	helmet jelly*
Galeorhinus galeus/ Galeorhinus zygopterus/ Eugaleus galeus	Australischer Hundshai, Hundshai, Biethai (Suppenflossenhai)	tope shark (FAO), tope, soupfin shark, school shark
Galerida cristata	Haubenlerche	crested lark
Galerida theklae	Theklalerche	thekla lark
Galeruca tanaceti	Rainfarnblattkäfer	tansy beetle
Galerucella nymphaeae	Erdbeerkäfer, Seerosen-Blattkäfer	waterlily leaf beetle, waterlily beetle, pond-lily leaf-beetle
Galeus melastomus/ Pristiurus melanostomus	Fleckhai, Sägeschwanz	blackmouth catshark (FAO), black-mouthed dogfish
Galeus murinus	Sägeschwanz-Katzenhai	mouse catshark
Galictis cuja	Kleingrison	little grisón
Galictis vittata	Großgrison	greater grisón
Galidia elegans	Ringelmungo, Ringelschwanzmungo	Malagasy ring-tailed mongoose
Galidictis fasciata fasciata	Bändermungo	Malagasy striped mongoose
Galidictis fasciata striata	Breitstreifenmungo	Malagasy broad-striped mongoose
Galleria mellonella	Große Wachsmotte, Bienenwolf, Rankmade (Larve)	greater wax moth, honeycomb moth, bee moth
Gallinago gallinago	Bekassine	common snipe
Gallinago media	Doppelschnepfe	great snipe
Gallinago stenura	Speißbekassine	pintail snipe
Gallinula chloropus	Teichhuhn	moorhen
Gallinula martinica/ Porphyrula martinica/ Porphyrio martinica	Zwergsultanshuhn	purple gallinule, American purple gallinule
Gallirallus sylvestris/ Sylvestrornis sylvestris	Waldralle	Lord Howe rail, Lord Howe Island woodhen
Galloperdix bicalcarata	Weißkehl-Spornhuhn	Ceylon spurfowl
Gallotia atlantica	Atlantik-Eidechse, Purpurarien-Eidechse	Atlantic lizard
Gallotia simonyi	Simonys Eidechse	Hierro giant lizard (IUCN), Simony's lizard
Gallus gallus	Bankivahuhn	red jungle-fowl
Gambelia silus	Stumpfnasen-Leopardleguan, Stumpfnasenleguan	bluntnose leopard lizard, blunt-nosed leopard lizard (IUCN)
Gambelia spp.	Leopardleguane	leopard lizards
Gambelia wislizenii	Langnasen-Leopardleguan	longnose leopard lizard
Gambusia affinis	Moskitofisch, Koboldkärpfling, Gambuse	mosquito fish, mosquitofish (FAO)

Gammarus fossarum	Gebirgs-Bachflohkrebs	mountain riverine amphipod
Gammarus insensibilis	Lagunen-Flohkrebs	lagoon sand shrimp
Gammarus locusta	Gemeiner Flohkrebs	locust amphipod, common intertidal amphipod
Gammarus oceanicus	Ozeanischer Flohkrebs*	oceanic scud
Gammarus pulex	Gemeiner Flohkrebs, Bachflohkrebs	common freshwater amphipod, common riverine amphipod, common freshwater shrimp,
Gammarus roeseli	Flussflohkrebs	lacustrine amphipod, lacustrine shrimp
Gammarus spp.	Flohkrebse u.a.	scuds a.o.
Gampsocleis glabra	Heideschrecke	heath bushcricket, Continental heath bushcricket
Gari depressa/ Psammobia depressa	Flache Sandmuschel, Große Flache Sandmuschel	large sunsetclam, flat sunsetclam
Gari fervensis	Violettgestreifte Sandmuschel	Faroe sunsetclam
Gari fucata	Gezeichnete Sandmuschel	painted sunsetclam
Gari ornata	Platte Sandmuschel	ornate sunsetclam
Garra cambodgiensis	Kambodscha-Saugbarbe	stonelapping minnow
Garrulus glandarius	Eichelhäher	jay
Gasteracantha cancriformis	Krebsspinne*	spinybacked spider
Gasteracantha spp.	Stachelspinnen	spiny orbweavers
Gasteracanthinae	Stachelspinnen	spiny-bellied spiders
Gasterochisma melampus	Schmetterlings-Thunfisch	butterfly kingfish
Gasteropelecidae	Beilbäuche, Beilbauchfische, Beilbauchsalmler	freshwater hatchetfishes, flying characins
Gasteropelecus sternicla	Silber-Beilbauchfisch, Gemeiner Silberbeilbauchfisch	common hatchetfish, silver hatchetfish, river hatchetfish (FAO)
Gasterophilidae	Magendasseln, Magenbremsen (Dasselfliegen)	botflies, bot flies (horse botflies)
Gasterophilus haemorrhoidalis	Nasendassel	nose botfly, lip botfly, rectal botfly
Gasterophilus intestinalis	Pferdemagenbremse	horse botfly, common horse botfly, common botfly
Gasterophilus nasalis	Rachendassel	throat botfly
Gasterophilus pecorum	Pflanze-zu-Tier-Dasselfliege*	plant-animal botfly
Gasterosteidae	Stichlinge	sticklebacks
Gasterosteiformes	Stichlingsartige, Stichlingverwandte	sticklebacks (and sea horses)
Gasterosteus aculeatus	Dreistachliger Stichling, Gemeiner Stichling	three-spined stickleback
Gasteruptiidae/ Gasteruptionidae	Gichtwespen, Schmalbauchwespen	gasteruptiids, gasteruptionid wasps
Gasteruption spp.	Gichtwespen	gasteruptid wasps
Gastridium geographus/ Conus geographus	Landkarten-Kegelschnecke	geography cone
Gastrochaena dubia/ Rocellaria dubia	Europäische Gastrochaena	flask shell, European flask shell
Gastrodes abietum	Fichtenwanze	spruce bug
Gastrodes grossipes	Kiefernzapfenwanze	pinecone bug

Gastrodiscoides hominis	Schweinegel, Rattenegel	pig intestinal fluke, rat fluke
Gastrodiscus aegyptiacus		equine intestinal fluke
Gastropacha quercifolia	Kupferglucke	lappet
Gastrophryne carolinensis/ Microhyla carolinensis	Carolina-Engmundkröte	Carolina narrowmouthed toad, eastern narrowmouth toad
Gastrophryne spp.	Amerikanische Engmundkröten	American narrowmouth toads
Gastropoda	Schnecken, Bauchfüßer, Gastropoden	gastropods, snails
Gastropteron rubrum	Rote Fledermausflügel-Meeresnacktschnecke	bat-wing sea-slug
Gastrotheca galeata	Bizarrer Beutelfrosch	helmeted marsupial frog
Gastrotheca marsupiata	Beutelfrosch	common marsupial frog
Gastrotheca ovifera	Riesen-Beutelfrosch	giant marsupial frogs
Gastrotheca spp.	Beutelfrösche	marsupial frogs
Gastrotricha	Bauchhaarlinge, Bauchhärlinge, Flaschentierchen, Gastrotrichen	gastrotrichs
Gaurotes virginea	Blaubock	
Gavia adamsii	Gelbschnabeltaucher	white-billed diver, yellow-billed loon
Gavia arctica	Prachttaucher	black-throated diver, Arctic loon
Gavia immer	Eistaucher	great northern diver, common loon
Gavia stellata	Sterntaucher	red-throated diver, red-throated loon
Gavialis gangeticus	Ganges-Gavial	gharial, gavial, Indian gharial
Gazella cuvieri	Edmigazelle	Edmi gazelle
Gazella dama	Damagazelle	Addra gazelle
Gazella dorcas	Dorkasgazelle	Dorcas gazelle
Gazella gazella	Echtgazelle	mountain gazelle
Gazella granti	Grantgazelle	Grant's gazelle
Gazella leptoceros	Dünengazelle	sand gazelle, rhim, slender-horned gazelle
Gazella rufifrons	Rotstirngazelle	red-fronted gazelle
Gazella rufina	Rotgazelle	red gazelle
Gazella soemmerringii	Sömmerringgazelle	Soemmerring's gazelle
Gazella spekei	Spekegazelle	Speke's gazelle
Gazella spp.	Gazellen	gazelles
Gazella subgutturosa	Kropfgazelle	goitred gazelle
Gazella thomsonii	Thomsongazelle	Thomson's gazelle
Gecarcinidae	Landkrabben	land crabs
Gecarcinus lateralis/ Gigantinus lateralis	Schwarze Landkrabbe, Rote Landkrabbe	blackback land crab, black land crab, red land crab
Gecarcinus ruricola	Antillen-Landkrabbe	purple land crab, mountain crab
Geckolepis spp.	Schuppengeckos	large-scaled geckos
Geckolepis typica	Grandidiers Schuppengecko	Grandidier's gecko
Geckonia chazaliae	Helmkopfgecko	helmethead gecko
Gehyra mutilata	Pazifikgecko	stump-toed dtella, common house gecko

Gekko gecko	Tokay, Tokee	tokay gecko, tokee
Gekko vittatus	Streifengecko	lined gecko
Gekkonidae	Geckos	geckos
Gelanga succincta	Gürtel-Tritonshorn	lesser girdled triton
Gelastocoridae	Krötenwanzen*	toad bugs, gelastocorids
Gelochelidon nilotica	Lachseeschwalbe	gull-billed tern
Gemma gemma	Amethystmuschel*	amethyst gemclam
Gempylidae	Schlangenmakrelen	snake mackerels
Genetta genetta	Kleinfleck-Ginsterkatze	small-spotted genet
Genetta spp.	Ginsterkatzen	genets
Genicanthus caudovittatus	Zebra-Kaiserfisch*	zebra angelfish
Genitoconia rosea (Aplacophora)	Rosa Stilett-Leistenfuß	rosy solenogaster*
Genyonemus lineatus	Weißer Kalifornia-Umberfisch	white croaker
Genypterus blacodes	Rosa Kingklip	pink cusk-eel, pink ling (FAO)
Genypterus capensis	Königs-Bartmännchen, Südafrikanischer Kingklip	kingklip
Genypterus chilensis	Chilenischer Kingklip	red cusk-eel
Geocapromys brownii	Jamaika-Baumratte	Jamaican hutia (Indian coney)
Geocapromys ingrahami	Bahama-Baumratte	Bahamian hutia
Geochelone carbonaria/ Testudo carbonaria/ Chelonoidis carbonaria	Köhlerschildkröte	red-footed tortoise, South American red-footed tortoise, coal tortoise
Geochelone chilensis/ Testudo chilensis/ Chelonoidis chilensis	Argentinische Landschildkröte	Argentine tortoise (IUCN), Chaco tortoise, Chilean tortoise
Geochelone denticulata	Gelbfuß-Landschildkröte	South American yellow-footed tortoise
Geochelone elegans elegans	Sternschildkröte	Indian star tortoise, starred tortoise
Geochelone elephantopus/ Geochelone nigra/ Testudo elephantopus/ Chelonoides elephantopus	Galapagos-Riesenschildkröte, Elefantenschildkröte	Galapagos giant tortoise
Geochelone elongata elongata/ Indotestudo elongata elongata	Gelbkopf-Landschildkröte	yellow tortoise, elongated tortoise (IUCN)
Geochelone emys/ Manouria emys	Braune Landschildkröte	brown tortoise, Burmese brown tortoise, Asian giant tortoise (IUCN)
Geochelone gigantea/ Testudo gigantea/ Megalochelys gigantea	Seychellen-Riesenschildkröte	Seychelles giant tortoise, Aldabran giant tortoise, Aldabra giant tortoise (IUCN)
Geochelone impressa/ Manouria impressa	Hinterindische Landschildkröte	impressed tortoise
Geochelone pardalis pardalis	Pantherschildkröte	leopard tortoise
Geochelone platynota/ Geochelone elegans platynota	Burma-Landschildkröte	Burmese star tortoise, Burmese starred tortoise
Geochelone radiata/ Testudo radiata/ Asterochelys radiata	Madagassische Strahlenschildkröte	radiated tortoise
Geochelone sulcata	Spornschildkröte	African spurred tortoise
Geochelone travancorica/ Indotestudo forsteni	Travancore-Spornschildkröte	Travancore tortoise

Geochelone yniphora/ *Testudo yniphora*/ *Asterochelys yniphora*	Madagskar-Spornschildkröte, Madagassische Schnabelbrustschildkröte	Madagascan spurred tortoise, Northern Madagascar spur tortoise, angonoka, Madagascar tortoise (IUCN)
Geoclemys hamiltonii	Strahlen-Dreikielschildkröte	black pond turtle, spotted pond turtle
Geococcyx californianus	Wegekuckuck, Erdkuckuck	roadrunner, greater roadrunner
Geococcyx velox	Rennkuckuck	lesser roadrunner
Geocolaptes olivaceus	Erdspecht	ground woodpecker
Geocoris bullatus	Großaugenwanze	large bigeyed bug
Geocoris grylloides	Grillenwanze	cricket bug, crickety bigeyed bug*
Geocrinia alba	Australischer Weißbauchfrosch	white ground froglet, white-bellied frog (IUCN)
Geocrinia vitellina	Australischer Gelbbauchfrosch	Australian ground froglet, yellow-bellied frog (IUCN)
Geodia cydonium/ *Geodia gigas*	Stinkender Ankerschwamm, Riesenschwamm	giant sponge*
Geodia gibberosa	Weißschwamm*	white sponge
Geoemyda leytensis	Philippinische Teichschildkröte	Philippine pond turtle
Geoemyda silvatica	Erd-Sumpfschildkröte	Cochin Forest cane turtle, forest turtle
Geoemyda spengleri	Zacken-Erdschildkröten	black-breasted leaf turtle
Geoffroyus geoffroyi	Rotkopfpapagei	red-cheeked parrot
Geogale aurita	Großohr-Tanrek	large-eared tenrec
Geolycosa spp.		burrowing wolf spiders
Geomalacus maculosus	Gelbgefleckte Nacktschnecke	kerry slug, spotted kerry slug, spotted Irish slug
Geometra papilionaria	Grünes Blatt, Grünling	large emerald
Geometridae	Spanner	geometer moths, geometers
Geomyidae	Taschenratten	pocket gophers
Geomys spp.	Flachland-Taschenratten	eastern pocket gophers
Geophagus brasiliensis	Perlmutter-Erdfresser, Brasilperlmutterfisch	pearl cichlid
Geophagus jurupari/ *Satanoperca jurupari*	Teufelsangel-Erdfresser, Teufelsangel	demon eartheater (FAO), eartheating devilfish
Geophagus steindachneri	Steindachners Rothauben-Erdfresser	redhump eartheater
Geophagus surinamensis	Surinamperlfisch	Surinam pearl cichlid
Geophilidae	Erdläufer	wire centipedes, garden centipedes
Geophilomorpha	Erdläufer	geophilomorphs
Geophilus electricus	Leuchtender Erdläufer	luminous centipede
Geopsittacus occidentalis	Höhlensittich	night parrot
Georychus capensis	Kap-Blessmull	Cape mole-rat
Georyssidae	Uferschlammkäfer	minute mud-loving beetles
Geosciurus inauris/ *Xerus inauris*	Südafrikanisches Erdhörnchen	South African ground squirrel, Cape ground squirrel
Geosciurus princeps/ *Xerus princeps*	Damara Erdhörnchen	Damara ground squirrel, Kaokoveld ground squirrel, mountain ground squirrel
Geositta cunicularia	Kaninchenerdhacker	common miner
Geospiza magnirostris	Großgrundfink	large ground finch

Geothlypis trichas	Weidengelbkehlchen	common yellowthroat
Geotrupes spp.	Mistkäfer, Rosskäfer	dor beetles
Geotrupes stercorarius	Mistkäfer	common dor beetle
Geotrupidae	Mistkäfer	dung beetles
Geotrygon montana	Bergtaube	ruddy quail dove
Geotrypetes spp.	Westafrikanische Erdwühlen	West African caecilians
Geranospiza caerulescens	Sperberweihe	crane hawk
Gerardia savaglia	Strauchkoralle	bushy crust coral*
Gerbillurus spp.	Namib-Rennmäuse	southern pygmy gerbils, hairy-footed gerbils
Gerbillus spp.	Eigentliche Rennmäuse	northern pygmy gerbils
Geronticus eremita	Waldrapp	bald ibis, northern bald ibis
Gerreidae/Gerridae	Mojarras	pursemouths, mojarras
Gerrhonotus coerulens/ Elgaria coerulea	Nördliche Alligatorschleiche	northern alligator lizard
Gerrhonotus imbricatus	Kiel-Krokodilschleiche	rain alligator lizard
Gerrhonotus kingii/ Elgaria kingii	Arizona-Alligatorschleiche	Arizona alligator lizard
Gerrhonotus liocephalus	Texas-Alligatorschleiche	Texas alligator lizard
Gerrhonotus multicarinatus multicarinatus/ Elgaria multicarinata m.	Kalifornische Alligatorschleiche	California legless lizard
Gerrhonotus multicarinatus/ Elgaria multicarinata	Südliche Alligatorschleiche	southern alligator lizard
Gerrhonotus spp.	Krokodilschleichen, Alligatorschleichen u.a.	eastern alligator lizards
Gerridae	Teichläufer (Wasserläufer, Schneider)	pond skaters, water striders, pond skippers
Gersemia rubiformis	See-Erdbeere*	red soft coral, sea strawberry
Geryonia proboscidalis	Rüsselqualle, Rüsselmeduse	trunked jellyfish
Geryonidae	Geryoniden (Tiefseekrabben u.a.)	geryonid crabs, deepsea crabs
Gibberichthyidae	Schnabelfische	gibberfishes
Gibberula lavalleeana	Schneeflocken-Randschnecke*	snowflake marginella
Gibberula miliaris	Hirsekorn-Kreiselschnecke	millet topsnail
Gibbium psylloides	Kugelkäfer, Buckelkäfer	smooth spider beetle (bowl beetle)
Gibbula cineraria	Friesenkopf, Aschfarbene Kreiselschnecke, Aschgraue Kreiselschnecke	grey topsnail
Gibbula magus	Knorrige Kreiselschnecke	turban topsnail
Gibbula pennanti	Pennant-Kreiselschnecke	Pennant's topsnail
Gibbula umbilicalis	Flache Kreiselschnecke	flat topsnail, purple topsnail, umbilical trochid
Gigantorana goliath/ Conraua goliath	Goliathfrosch	goliath frog
Giganturidae	Teleskopfische (Riesenschwänze)	telescopefishes
Gilletteella cooleyi	Sitkafichten-Gallenlaus	Douglas fir adelges

Ginglymostoma cirratum	Atlantischer Ammenhai, Karibischer Ammenhai	nurse shark
Ginglymostomatidae	Ammenhaie	nurse sharks
Giraffa camelopardalis	Giraffe	giraffe
Girardinus metallicus	Metallkärpfling	girardinus, metallic livebearer (FAO)
Girella nigricans	Opalauge	opaleye
Glabella harpaeformis	Harfen-Randschnecke	harplike marginella
Glabella pseudofaba	Königin-Randschnecke	queen marginella
Glacicavicola bathysciodes	Blinder Höhlenkäfer	blind cave beetle
Glareola nordmanni	Schwarzflügel-Brachschwalbe	black-winged pratincole
Glareola pratincola	Rotflügel-Brachschwalbe	collared pratincole
Glaucidium gnoma	Gnomenkauz	pygmy owl, Eurasian pygmy owl
Glaucidium passerinum	Sperlingskauz	pygmy owl
Glaucidium perlatum	Perlkauz	pearl-spotted owlet
Glaucomys sabrinus	Nord-Assapan	northern flying squirrel
Glaucomys spp.	Neuwelt-Gleithörnchen	New World flying squirrels
Glaucomys volans	Süd-Assapan	southern flying squirrel
Glauconycteris spp.	Afrikanische Schmetterlings-Fledermäuse	butterfly bats, silvered bats
Glaucopsyche alexis	Alexis-Bläuling	green underside blue
Glaucosomatidae	Perlbarsche*	pearl perches
Gliridae	Bilche und Schläfer	dormice
Glironia venusta	Buschschwanzbeutelratte	bushy-tailed opossum
Glirulus japonicus	Japanischer Schläfer	Japanese dormouse
Glis glis	Siebenschläfer	edible dormouse, edible commoner dormouse, fat dormouse, squirrel-tailed dormouse
Glischropus spp.	Dickdaumen-Fledermäuse	thick-thumbed bats
Globicephala melaena	Gewöhnlicher Grindwal, Langflossen-Grindwal	northern pilot whale, Atlantic pilot whale, blackfish, pothead, long-finned pilot whale
Globicephala macrorhynchus/ Globicephala seiboldii	Indo-Pazifischer Grindwal, Kurzflossen-Grindwal	Indo-Pacific pilot whale, short-finned pilot whale, blackfish
Globodera pallida	Weißer Kartoffelnematode	white potato cyst nematode, pale potato cyst nematode, potato root eelworm
Globodera rostochiensis	Gelber Kartoffelnematode	yellow potato cyst nematode, golden nematode, potato root eelworm
Globularia fluctuata/ Cernina fluctuata	Gewellte Mondschnecke	wavy moonsnail
Glomeridae	Saftkugler, Kugelasseln, Kugeltausendfüßer	pill millipedes, glomerid millepedes
Glomeridellidae	Zwergkugler	lesser pill millipedes
Glomeris spp.	Saftkugler	pill millepedes, European pill millepedes
Gloripallium pallium	Mantel-Kammmuschel	mantle scallop
Gloripallium sanguinolenta	Blutfleck-Kammmuschel	blood-stained scallop
Glossanodon semifasciatus/ Argentina semifasciata	Japanisches Glasauge, Goldlachs	Pacific argentine, deepsea smelt (FAO)

Glossina spp.	Tsetse-Fliegen, Tsetsefliegen	tsetse flies
Glossinae	Tsetse-Fliegen, Tsetsefliegen	tsetse flies
Glossiphonia complanata	Großer Schneckenegel	snail leech, greater snail leech
Glossiphonia heteroclita	Kleiner Schneckenegel	small snail leech
Glossiphoniidae	Knorpelegel, Plattegel	glossiphoniid leeches
Glossobalanus minutus (Enteropneusta)	Kleiner Eichelwurm	lesser acorn worm
Glossolepis incisus	Roter Regenbogenfisch	red rainbowfish
Glossophaga soricina	Spitzmaus-Langzüngler	shrew-like long-tongued bat
Glossophaga spp.	Langzungen-Fledermäuse u.a.	long-tongued bats
Glossopsitta porphyreocephala	Blauscheitellori	purple-crowned lorikeet
Glossus humanus	Ochsenherz, Menschenherz	ox heart, heart cockle
Gloydius blomhoffi/ Agkistrodon blomhoffi	Mamushi, Japanische Grubenotter	Japanese mamushi
Gloydius saxatilis/ Agkistrodon saxatilis	Amurotter	brown mamushi
Gluphisia crenata	Kleiner Pappelauen-Zahnspinner	
Glycera spp.	Glyzerinwürmer	glycerine worm
Glycymeridae/ Glycymerididae	Samtmuscheln	dog cockles, bittersweets, bittersweet clams (U.S.)
Glycymeris americana/ Glycymeris gigantea	Riesensamtmuschel	giant bittersweet, American bittersweet
Glycymeris cor	Braune Pastetenmuschel	brown bittersweet*
Glycymeris decussata	Gekreuztrippige Samtmuschel*	decussate bittersweet
Glycymeris formosa	Schöne Samtmuschel	beautiful bittersweet
Glycymeris gigantea/ Glycymeris americana	Riesensamtmuschel	giant bittersweet, American bittersweet
Glycymeris glycymeris	Gemeine Samtmuschel, Archenkammmuschel, Mandelmuschel, Meermandel, Englisches Pastetchen	dog cockle, orbicular ark (comb-shell)
Glycymeris pectinata	Kamm-Samtmuschel	comb bittersweet
Glycymeris pectiniformis	Kammmuschelartige Samtmuschel	scalloplike bittersweet
Glycymeris pilosa	Echte Samtmuschel, Violette Pastetenmuschel	hairy dog cockle, hairy bittersweet
Glycymeris spp.	Samtmuscheln	dog cockles (Br.), bittersweet clams (U.S.)
Glycymeris subobsoleta	Pazifik-Samtmuschel	Pacific Coast bittersweet, West Coast bittersweet
Glycymeris undata	Gewellte Samtmuschel, Atlantik-Samtmuschel	wavy bittersweet, Atlantic bittersweet, lined bittersweet
Glycyphagus domesticus	Polstermilbe, Hausmilbe	furniture mite
Glyphipterygidae	Rundstirnmotten Wippmotten)	glyphipterygid moths
Glyphis gangeticus	Ganges-Hai	Ganges shark
Glyphocrangonidae	Panzergarnelen	armored shrimps
Glyphodon spp.	Australische Nackenband-Giftnatter	Australian collared snakes
Glyphotes spp.	Borneo-Zwerghörnchen*	Bornean pygmy squirrels

Glyptocephalus cynoglossus	Rotzunge, Hundszunge, Zungenbutt	witch
Glyptocephalus zachirus	Amerikanische Scholle	rex sole
Gnaphosa spp.	Echte Plattbauchspinnen	hunting spiders a.o.
Gnaphosidae/Drassodidae	Plattbauchspinnen, Glattbauchspinnen	hunting spiders, ground spiders
Gnathanodon speciosus/ Caranx speciosus	Goldene Königsmakrele	golden trevally
Gnathobdelliformes	Kieferegel	jawed leeches
Gnathocerus cornutus	Vierhornkäfer	broad-horned flour beetle
Gnathodentex aureolineatus	Goldstreifenbrasse	gold-lined bream, large-eyed bream, striped large-eye bream (FAO)
Gnathonemus petersii	Tapirfisch, Elefanten-Rüsselfisch	Peter's worm-jawed mormyrid, Peter's elephant nose, elephantnose fish (FAO)
Gnathonemus tamandua/ Campylomormyrus tamandua	Ameisenbärfisch, Spitzbartfisch	worm-jawed mormyrid
Gnathophis mystax	Schwarzschwanz-Meeraal*	black-tailed conger, blacktailed conger, thinlip conger (FAO)
Gnathophyllidae	Hummelgarnelen	bumblebee shrimps
Gnathophyllum americanum	Hummelgarnele	bumblebee shrimp
Gnathophyllum elegans/ Drimo elegans	Goldpunktgarnele, Gepunktete Garnele	golden-spotted shrimp*
Gnathostoma spp.	Magenzystenwürmer	stomach cyst worms, stomach cyst nematodes (of swine)
Gnathostomata	Kiefermünder	jawed vertebrates, jaw-mouthed animals, gnathostomatans
Gnathostomulida	Kiefermäuler, Kiefermündchen, Gnathostomuliden	gnathostomulids
Gobiescociformes	Saugfischverwandte, Schildfische, Spinnenfischartige	clingfishes
Gobiesocidae	Schildfische, Schildbäuche	clingfishes
Gobiesox strumosus	Bratpfannenfisch*	skilletfish
Gobiidae	Grundeln	gobies
Gobio albipinnatus	Flachlandgründling, Weißflossiger Gründling	Russian whitefin gudgeon, white-finned gudgeon (FAO)
Gobio gobio	Gründling, Belingi-Gründling	gudgeon
Gobio uranoscopus	Steingressling	Danubian gudgeon
Gobiodon citrinus	Zitronengrundel	lemon goby
Gobioides broussoneti	Lila Aalgrundel	violet goby
Gobionellus boleosoma	Pfeilgrundel*	darter goby
Gobionellus hastatus	Spitzschwanzgrundel	sharptail goby
Gobionellus shufeldti	Süßwasser-Grundel	freshwater goby
Gobionellus smaragdus	Smaragdgrundel	emerald goby
Gobiopterus chuno	Glasgrundel	glas goby, translucent goby, clear goby
Gobiosoma evelynae	Hainasen-Grundel	sharknosed goby

Gobiosoma ginsburgi	Ginsburg-Grundel	seabord goby
Gobiosoma oceanops/ *Elacatinus oceanops*	Blaue Putzgrundel, Neon-Grundel, Neongrundel	neon goby
Gobius auratus	Goldgrundel	golden goby
Gobius bucchichi	Anemonengrundel	Bucchich's goby
Gobius cobitis	Riesengrundel, Große Meergrundel	giant goby
Gobius cruentatus	Blutlippengrundel, Rotmaulgrundel	red-mouthed goby (FAO), red-mouth goby
Gobius flavescens/ *Gobius ruthensparri/* *Gobiusculus flavescens*	Schnappküling, Schwimmgrundel, Zweistreifen-Grundel	two-spotted goby (FAO), two-spot goby
Gobius fluviatilis/ *Neogobius fluviatilis*	Flussgrundel	river goby, monkey goby (FAO)
Gobius geniporus	Schlankgrundel	slender goby
Gobius lota/ *Gobius ophiocephalus/* *Zosterisessor ophiocephalus*	Schlangenkopfgrundel	grass goby
Gobius marmoratus/ *Proterorhinus marmoratus*	Marmorierte Grundel	tubenose goby
Gobius niger	Schwarzgrundel, Schwarzküling	black goby
Gobius paganellus	Paganellgrundel, Felsgrundel, Felsengrundel	rock goby
Gobius vittatus	Streifengrundel	striped goby
Gobiusculus flavescens/ *Gobius flavescens/* *Gobius ruthensparri*	Schnappküling, Schwimmgrundel, Zweistreifen-Grundel	two-spotted goby (FAO), two-spot goby
Goheria fusca	Braune Mehlmilbe	brown fluor mite
Golunda ellioti	Elliot Buschratte, Indische Buschratte	Indian bush rat, coffee rat
Gomphidae	Flussjungfern	clubtails
Gomphocerus rufus	Rote Keulenschrecke	rufous grasshopper
Gomphocerus sibiricus/ *Aeropus sibiricus*	Sibirische Keulenschrecke	club-legged grasshopper, Sibirian grasshopper*
Gomphus flavipes	Asiatische Keiljungfer	Asian gomphus
Gomphus pulchellus	Westliche Keiljungfer	western European gomphus
Gomphus simillimus	Gelbe Keiljungfer	yellow gomphus, Mediterranean gomphus
Gomphus vulgatissimus	Gemeine Keiljungfer	club-tailed dragonfly
Gonatodes albogularis	Gelbkopfgecko	yellow-headed gecko
Gonatus fabricii	Köderkalmar	boreoatlantic armhook squid
Goneplacidae	Rhombenkrabben	goneplacid crabs, angular crabs
Goneplax rhomboides	Rhombenkrabbe	angular crab, square crab (mud runner)
Gonepteryx cleopatra	Kleopatrafalter, Kleopatra	cleopatra
Gonepteryx rhamni	Zitronenfalter	brimstone
Gongylonema pulchrum	Ösophagus-Wurm	gullet worm, zigzag worm
Goniocotes gallinae	Flaumlaus	fluff louse
Goniodes dissimilis/ *Oulocrepis dissimilis*	Braune Hühnerlaus	brown chicken louse

Goniodes gigas/ Stenocrotaphus gigas	Große Hühnerlaus	large chicken louse
Gonionemus vertens	Ansaugqualle*	clinging jellyfish, angled hydromedusa
Goniopora lobata	Grobe Porenkoralle	coarse porous coral, flowerpot coral
Goniopora planulata	Anemonenkoralle	anemone coral*
Goniopsis cruentata	Mangroven-Wurzelkrabbe	Mangrove root crab
Gonocephalus grandis	Große Winkelkopfagame	giant forest dragon
Gonocephalus spp.	Winkelkopfagamen u.a.	humphead forest dragons
Gonodactylus falcatus	Schlagender Fangschreckenkrebs*	clubbing mantis shrimp
Gonodera luperus	Veränderlicher Pflanzenkäfer	
Gonorhynchidae	Sandfische	beaked sandfishes
Gonorhynchiformes	Sandfische, Milchfischverwandte	milkfishes and relatives
Gonorynchus gonorynchus	Sandfisch	beaked salmon (FAO), sand eel
Gonostomatidae	Borstenmünder, Borstenmäuler	bristlemouths
Gonyosoma oxycephalum	Spitzkopfnatter	mangrove ratsnake
Goodeidae	Zwischenkärpflinge, Hochlandkärpflinge	
Gopherus agassizii	Wüstenschildkröte, Kalifornische Gopherschildkröte	desert tortoise, western gopher tortoise, California desert tortoise
Gopherus berlandieri	Texanische Gopherschildkröte	Texas tortoise, Texas gopher tortoise
Gopherus flavomarginatus	Mexikanische Gopherschildkröte	Mexican gopher tortoise, Bolson tortoise (IUCN)
Gopherus polyphemus	Florida-Gopherschildkröte	gopher tortoise, Florida gopher tortoise
Gordius aquaticus (Nematomorpha)	Wasserkalb, Rosshaarwurm, Gemeiner Wasserdrahtwurm	gordian worm, common horsehair worm, common hair worm
Gorgonacea/Gorgonaria	Hornkorallen, Rindenkorallen	gorgonians, gorgonian corals, horny corals
Gorgonia flabellum/ Gorgonia ventalina/ Rhipidogorgia flabellum	Venusfächer, Großer Seefächer, Große Fächerkoralle, Große Netzgorgonie	Venus sea fan, Venus' fan, common sea fan
Gorgonia spp.	Seefächer u.a., Fächerkorallen u.a.	sea fans a.o.
Gorgonocephalidae	Medusenhäupter	medusa-head stars, basket stars
Gorgonocephalus arcticus	Arktischer Medusenstern	northern basket star
Gorgonocephalus caput-medusae	Gorgonenhaupt	medusa-head star, Gorgon's head, basket star
Gorilla gorilla beringei	Berggorilla	mountain gorilla
Gorilla gorilla gorilla	Flachlandgorilla	lowland gorilla
Gorilla gorilla	Gorilla	gorilla
Gortyna borelii	Haarstrangwurzeleule	Fisher's estuarine (moth)
Gortyna flavago	Markeule	frosted orange
Gossyparia spuria	Ulmenwollschildlaus	European elm scale

Goura victoria	Victoria-Krontaube	Victoria crowned-pigeon
Gourmya gourmyi	Gourmya-Nadelschnecke	gourmya cerith
Gourmya rupestris	Felsnadelschnecke, Felsnadel	rocky-shore cerith*
Gourmya vulgata/ Cerithium vulgatum	Gemeine Nadelschnecke, Gemeine Hornschnecke, Gemeine Seenadel	European cerith
Gracillaria syringella	Fliedermotte	lilac leaf miner
Gracillariidae	Blatttütenmotten (Miniermotten)	leaf blotch miners
Gracula religiosa	Beo	southern grackle, hill mynah
Graellsia isabellae/ Graellsia isabelae	Spanisches Nachtpfauenauge	Spanish moon moth
Grallina cyanoleuca	Drosselstelze	magpie lark
Grammatidae/Grammidae	Feenbarsche	basslets
Grammatorcynus bilineatus	Doppellinien-Makrele	double-lined mackerel
Grammicolepididae	Papierschupper	tinselfishes, grammicolepidids
Grammistes sexlineatus	Goldstreifenbarsch	six-lined grouper, golden-striped grouper, sixline soapfish (FAO)
Grammistidae	Seifenfische, Streifenbarsche	soapfishes
Grammostola burzaquensis/ Grammostola argentinensis	Argentinische Busch-Vogelspinne	Argentinean rose tarantula
Grammostola spatulata	Chile-Vogelspinne	Chilean rose tarantula
Grampus griseus	Rissos Delphin, Rundkopfdelphin	Risso's dolphin, Gray grampus, white-headed grampus
Granaria frumentum	Wulstige Kornschnecke	
Grandala coelicolor	Grandala	Hodgson's grandala
Graneledone verrucosa	Warzenkrake	warty octopus
Grantia compressa	Beutel-Kalkschwamm	purse sponge
Graomys spp.	Großohrmäuse u.a.	big-eared mice a.o.
Graphiurus nanus	Zwergschläfer	pygmy dormouse
Graphiurus spp.	Pinselschwanz-Bilche	African dormice
Graphoderus zonatus	Gelbrandiger Breitflügel-Tauchkäfer	spangled water beetle
Graphosoma lineatum	Streifenwanze	
Graphosoma semipunctatum	Fleckige Streifenwanze	
Grapsidae	Felsenkrabben	shore crabs, marsh crabs & talon crabs
Grapsus albolineatus	Schnelle Felsenkrabbe	swift-footed crab
Grapsus grapsus	Felsenkrabbe	Sally Lightfoot crab, mottled shore crab
Graptemys pseudogeographica	Falsche Landkartenschildkröte, Falsche Landkarten-Höckerschildkröte	false map turtle
Graptemys pseudogeographica versa	Texanische Landkartenschildkröte	Texas map turtle
Graptemys barbouri	Barbours Höckerschildkröte	Barbour's map turtle
Graptemys flavimaculata	Gelbfleck-Landkartenschildkröte, Pascagoula-River-Höckerschildkröte	yellow-blotched map turtle (IUCN), yellow-blotched sawback

Graptemys geographica	Landkartenschildkröte, Landkarten-Höckerschildkröte	map turtle
Graptemys kohnii	Mississippi-Landkartenschildkröte, Mississippi-Höckerschildkröte	Mississippi map turtle
Graptemys nigrinoda	Schwarzhöckerschildkröte	black-knobbed map turtle
Graptemys oculifera	Pracht-Höckerschildkröte	ringed map turtle
Graptemys ouachitensis	Ouachita-Landkartenschildkröte	Ouachita map turtle
Graptemys pulchra	Schmuck-Höckerschildkröte	Alabama map turtle
Graptemys spp.	Landkartenschildkröten	map turtles
Graptemys versa/ Graptemys pseudogeographica versa	Texanische Landkartenschildkröte	Texas map turtle
Graptolithina	Graptolithen	graptolites
Grasseichthyidae	Zwerglarvenfische	grasseichthyids
Griposia aprilina/ Dichonia aprilina	Aprileule, Grüne Eicheneule, Lindeneule	Merveille-du-Jour
Gromphadorina portentosa		Madagascar hissing cockroach
Grus americana	Schneekranich, Schreikranich	whooping crane
Grus antigone	Saruskranich	sarus crane
Grus canadensis	Kanadakranich	sandhill crane
Grus grus	Kranich	common crane
Grus japonensis	Mandschurenkranich	red-crowned crane, Japanese crane, Manchurian crane
Grus leucogeranus	Nonnenkranich	great white crane, Siberian crane, Siberian white crane
Grus nigricollis	Schwarzhalskranich	black-necked crane
Grus rubicunda	Brolgakranich	brolga
Gryllacrididae	Flügellose Langfühlerschrecken, Grillenschrecken	cave crickets & camel crickets (wingless long-horned grasshoppers)
Gryllidae	Grillen	crickets, true crickets
Grylloblattidae	Grillenschaben	rock crawlers, grylloblattids
Gryllotalpa gryllotalpa	Maulwurfsgrille, Werre	mole cricket
Gryllotalpa major	Prärie-Maulwurfsgrille	prairie mole-cricket
Gryllotalpa vineae	Weinberg-Maulwurfsgrille	vineyard mole-cricket
Gryllotalpidae	Maulwurfsgrillen	mole crickets
Gryllus bimaculatus	Mittelmeer-Feldgrille	two-spotted cricket, Mediterranean field cricket
Gryllus campestris	Feldgrille	field cricket
Guaruba guarouba	Goldsittich	golden parakeet, golden conure
Guignotus pusillus	Gelbbrauner Zwergschwimmer	
Guiraca caerulea	Azurbischof	blue grosbeak, Guira cuckoo
Gulo gulo	Vielfraß	wolverine
Gyalopion canum	Mexikanische Hakennasen-Natter	western hooknose snake
Gymnacanthus tricuspis	Arktischer Hirschgeweihgroppe	Arctic staghorn sculpin

Gymnachirus melas	Nacktzunge	naked sole
Gymnammodytes semisquamatus	Nackt-Sandaal, Nacktsandaal	smooth sandeel
Gymnammodytes cicerelus/ Gymnammodytes cicerellus	Mittelmeer-Nacktsandaal	Mediterranean sand lance, Mediterranean sand-eel (FAO), Mediterranean smooth sandeel
Gymnamoebia	Nacktamöben	naked amebas
Gymnarchidae	Großnilhechte, Nilaale	abas
Gymnarchus niloticus	Großer Nilhecht, Großnilhecht	aba
Gymnobelideus leadbeateri	Hörnchen-Kletterbeutler	Leadbeater's possum
Gymnocephalus cernuus	Kaulbarsch	ruffe (FAO), pope
Gymnocephalus schraetzer	Schrätzer	striped ruffe, schraetzer (FAO), Danube ruffe
Gymnocorymbus ternetzi	Trauermantelsalmler	black tetra
Gymnocranius spp.	Imperatorfische, Großaugen-Brassen	large-eye breams
Gymnodactylus geckoides	Nacktfingergecko	naked-toed gecko
Gymnodactylus spp.	Nacktfingergeckos	naked-toed geckos
Gymnogeophagus meridionalis	La-Plata-Erdfresser	La-Plata eartheater
Gymnogeophagus gymnogenys	Dunkler Perlmutterbuntbarsch	smoothcheek eartheater
Gymnogeophagus balzanii	Paraguay-Maulbrüter, Paraguay-Erdfresser	Argentine humphead
Gymnogeophagus rhabdotus	Streifen-Erdfresser	stripefin eartheater
Gymnogyps californianus	Kalifornischer Kondor	California condor
Gymnolaemata/Stelmatopoda	Kreiswirbler	gymnolaemates, "naked throat" bryozoans
Gymnopholus lichenifer	Flechtenkäfer	lichen weevil
Gymnopleurus geoffroyi	Seidiger Pillenwälzer	
Gymnorhamphichthys hypostomus	Sandmesserfisch	cuchillo (Venezuela)
Gymnorhina tibicen	Flötenvogel	black-backed magpie
Gymnosarda unicolor/ Orcynopsis unicolor	Rüppell's Thunfisch	scaleless tuna, plain bonito, pigtooth tuna, Rüppell's bonito, dogtooth tuna (FAO)
Gymnosomata	Ruderschnecken, Nackte Flossenfüßer (Flügelschnecken)	naked pteropods
Gymnothorax favagineus	Große Netzmuräne	honeycomb moray
Gymnothorax funebris	Grüne Muräne, Westatlantische Muräne	green moray
Gymnothorax javanicus	Riesenmuräne	giant moray
Gymnothorax permistus	Kleine Netzmuräne	black-blotched moray
Gymnothorax tile	Süßwassermuräne	freshwater moray eel
Gymnothorax unicolor	Braune Muräne	brown moray
Gymnotidae	Echte Messeraale, Gestreifte Messeraale	naked-back knifefishes

Gymnotus carapo	Amerikanischer Messerfisch, Gebänderter Messerfisch	American knife fish, striped knife fish, banded knifefish (FAO)
Gymnura altavela	Schmetterlingsrochen	spiny butterfly ray
Gymnura micrura	Glatter Schmetterlingsrochen	smooth butterfly ray
Gymnura poecilura	Langschwanz-Schmetterlingsrochen	long-tailed butterfly ray
Gymnura spp.	Schmetterlingsrochen	butterfly rays
Gymnuridae	Schmetterlingsrochen, Falter-Stechrochen	butterfly rays
Gymnuromys roberti	Roberts Nacktschwanzmaus	voalavoanala
Gypaetus barbatus	Bartgeier	lammergeier
Gypohierax angolensis	Palmgeier	palm-nut vulture
Gypopsitta vulturina	Kahlkopfpapagei	vulturine parrot
Gyps coprotheres	Kapgeier	Cape vulture, Cape griffon
Gyps fulvus	Gänsegeier	griffon vulture
Gyps rueppellii	Sperbergeier	Ruppell's griffon
Gyraulus acronicus	Verbogenes Posthörnchen	contorted ramshorn*
Gyraulus albus	Weißes Posthörnchen, Weiße Tellerschnecke	white ramshorn
Gyraulus crista	Zwergposthörnchen	dwarf ramshorn, nautilus ramshorn, star gyro
Gyraulus gigantea/ Ranella gigantea	Große Krötenschnecke, Taschenschnecke, Großes Ochsenauge, Ölkrug	giant frogsnail, oil-vessel triton
Gyraulus laevis	Glattes Posthörnchen, Glänzende Tellerschnecke	shiny ramshorn, smooth ramshorn
Gyraulus parvus	Kleines Posthörnchen	lesser ramshorn, ash gyro
Gyraulus riparius	Flaches Posthörnchen	river ramshorn
Gyrineum gyrinum	Braunweißband-Triton*, Kaulquappentriton	tadpole triton
Gyrineum perca/ Biplex perca	Ahornblatt-Triton	maple leaf triton, winged triton
Gyrineum pusillum	Purpur-Tritonshorn	purple gyre triton
Gyrineum roseum	Rosige Triton	rose triton
Gyrinidae	Taumelkäfer	whirligig beetles
Gyrinocheilidae	Algenfresser, Saugschmerlen	algae eaters
Gyrinophilus palleucus	Tennessee-Höhlensalamander	Tennessee cave salamander
Gyrinophilus porphyriticus	Porphyrsalamander	spring salamander, northern spring salamander
Gyrinophilus spp.	Quellensalamander	spring salamanders
Gyrinophilus subterraneus	West Virginia-Quellensalamander	West Virginia spring salamander
Gyrinus spp.	Taumelkäfer	whirligig beetles
Gyrinus substriatus	Gemeiner Taumelkäfer	common whirligig beetle
Gyrostoma helianthus	Rote Riesenaktinie	giant red anemone*

Hadogenes spp.	Spaltenskorpione	South African rock scorpions
Hadogenes troglodytes	Südafrikanischer Spaltenskorpion	South African rock scorpion
Hadrurus arizonensis	Behaarter Arizona-Skorpion*	Arizona hairy scorpion
Haemadipsa zeylandica	Ceylonegel, Sri Lanka-Egel	Sri Lanka leech
Haemadipsidae	Landegel	terrestrial leeches, haemadipsid leeches
Haemaphysalis leachi	Gelbe Hundezecke	yellow dog tick
Haemaphysalis longicornis	Neuseeländische Rinderzecke	New Zealand cattle tick
Haematobia minuta	Afrikanische Hornfliege	African horn-fly
Haematobosca stimulans/ Haematobia stimulans/ Siphona stimulans	Kleine Stechfliege, Herbststechfliege	cattle biting fly (Texas fly)
Haematopinidae	Tierläuse	wrinkled sucking lice
Haematopinus asini macrocephalus	Pferdelaus	horse sucking louse
Haematopinus asini asini	Esellaus	donkey sucking louse
Haematopinus eurysternus	Kurzköpfige Rinderlaus, Kurznasige Rinderlaus	shortnosed cattle louse
Haematopinus suis	Schweinelaus	hog louse
Haematopodini	Austernfischer	oystercatchers
Haematopota spp./ *Chrysozona* spp.	Regenbremsen, Blinde Fliegen	clegs, stouts
Haematopota pluvialis	Regenbremse	cleg-fly, cleg
Haematopus ostralegus	Austernfischer	oystercatcher
Haementeria costata	Schildkrötenegel	turtle leech
Haementeria ghilianii	Riesenegel	giant turtle leech, giant leech
Haemodipsus ventricosus	Kaninchenlaus	rabbit louse
Haemonchus placei	Großer Rindermagennematode, Großer Rindermagenwurm	Barber's poleworm, large stomach worm (of cattle), wire worm
Haemopis marmorata	Amerikanischer Pferdeegel	American horse leech
Haemopis sanguisuga	Pferdeegel, Vielfraßegel	European horse leech
Haemulidae	Grunzer	grunts (rubberlips & grunters)
Haemulon sciurus	Blaustreifengrunzer, Blaustreifen-Grunzerfisch	bluestriped grunt
Haemulon spp.	Grunzer	grunts
Haeromys spp.	Zwergmäuse	pygmy tree mice
Hahniidae	Bodenspinnen	hahniids
Haideotriton wallacei	Blindsalamander	Georgia blind salamander
Halacaridae	Meeresmilben	sea mites
Halaelurus vincenti	Golf-Katzenhai	Gulf catshark
Halcampoides purpureus	Purpur-Anemone	purple anemone*
Halcyon smyrnensis	Braunliest	white-breasted kingfisher
Halecium halecinum	Fiederzweigpolyp	herringbone hydroid
Halecium muricatum	Stacheliger Fiederzweigpolyp*	sea hedgehog hydroid
Haliaeetus albicilla	Seeadler	white-tailed eagle
Haliaeetus leucocephalus	Weißkopfseeadler	bald eagle
Haliaeetus leucoryphus	Bindenseeadler	Palla's fish eagle
Haliaeetus vocifer	Schreiseeadler	African fish eagle

Haliaeetus vociferoides	Madagaskarseeadler	Madagascar fish eagle
Haliastur indus	Brahaminenweih	Brahminy kite
Halichoeres biocellatus	Orangegrüner Junker, Zweifleck-Junker	red-lined wrasse
Halichoeres hortulanus	Schachbrett-Junker	checkerboard wrasse
Halichoerus grypus	Kegelrobbe	gray seal
Halichondria panicea	Brotkrustenschwamm, Brotkrumenschwamm, Meerbrot	breadcrumb sponge, crumb of bread sponge
Haliclona limbata	Klumpenschwamm	
Haliclona mediterranea	Mittelmeer-Zylinderschwamm, Rosa Röhrenschwamm	Mediterranean tube sponge
Haliclona oculata	Geweihschwamm	mermaid's glove, deadman's finger, finger sponge, "eyed sponge", encrusting turret sponge
Haliclona permollis	Lila Schwamm*	purple encrusting sponge, purple sponge
Haliclona rubens	Roter Geweihschwamm*	red sponge
Haliclona spp.	Geweihschwämme, Röhrenschwämme u.a.	
Haliclona viscosa	Fleischschwamm	fleshy sponge, flesh sponge
Haliclystus octoradiatus	Achtstrahlige Becherqualle, Kleine Becherqualle	octaradiate stalked jellyfish*
Haliclystus salpinx	Trompeten-Becherqualle	trumpet stalked jellyfish
Haliclystus spp.	Trompeten-Becherquallen	trumpet stalked jellyfish
Halictus spp.	Furchenbienen, Schmalbienen	sweat bees, flower bees, halictid bees
Halieutaea indica	Indische Seefledermaus	Indian handfish
Halieutaea stellata	Stern-Seefledermaus	starry handfish
Halieutichthys aculeatus	Pfannkuchenfisch*	pancake batfish
Halimochirurgus centriscoides	Langschnäuziger Seebader	longsnout spikefish
Haliochoeres hortulanus	Augenfleck-Junker	chequerboard wrasse
Haliotis asinina	Eselsohr-Abalone	ass's ear abalone, donkey's ear abalone
Haliotis australis	Silberne Abalone, Silbernes Meerohr	silver abalone, silver paua
Haliotis corrugata	Rosa Abalone, Rosafarbenes Meerohr	pink abalone
Haliotis cracherodii	Schwarze Abalone, Schwarzes Meerohr	black abalone
Haliotis diversicolor	Vielfarbige Abalone, Vielfarbiges Meerohr	multicolored abalone
Haliotis fulgens	Grüne Abalone, Grünes Meerohr	green abalone
Haliotis gigantea	Riesenabalone, Riesenmeerohr, Riesen-Seeohr, Japanisches Riesenmeerohr	giant abalone
Haliotis lamellosa	Mittelmeer-Abalone, Mittelländisches Seeohr	Mediterranean ormer, common ormer
Haliotis midae	Südafrikanische Abalone, Südafrikanisches Meerohr	perlemoen abalone

Haliotis rufescens	Rote Abalone, Rotes Meerohr	red abalone
Haliotis scalaris	Treppen-Abalone, Treppenhaus-Meerohr	staircase abalone
Haliotis sorenseni	Weiße Abalone	white abalone
Haliotis spp.	Abalones, Seeohren, Meerohren	abalones (U.S.), ormers (Br.)
Haliotis tuberculata	Gemeine Abalone, Gemeines Meerohr, Gemeines Seeohr	common ormer, European edible abalone
Haliotis walallensis	Flache Abalone, Flaches Seeohr	flat abalone
Haliplanella luciae	Streifenanemone*	striped anemone
Haliplidae	Wassertreter, Wassertretkäfer	crawling water beetles
Haliplus ruficollis	Tropfenförmiger Wassertreter	
Haliporoides diomedeae	Chilenische Messergarnele	Chilean knife shrimp
Haliporoides sibogae	Rote Messergarnele	pink prawn, royal red prawn
Haliporoides triarthrus	Messergarnele	knife shrimp
Halisarca dujardini	Gallertschwamm	Dujardin's slime sponge
Halitholus cirratus	Ballonqualle	balloon jelly*
Halocynthia papillosa/ Tethyum papillosum	Rote Seescheide, Roter Seepfirsich*	red sea-squirt, red sea peach*
Halocynthia aurantia	Westpazifischer Seepfirsich	sea peach, Western Pacific sea peach
Halocynthia pyriformis	Seebirne*, Meerbirne* (Seepfirsich*, Meerespfirsich*)	sea pear* (a sea peach)
Halocynthia spinosa	Runzelige Seescheide	spiny sea peach*
Halocynthia spp.	Seepfirsiche*	sea peaches
Halophryne gangene/ Cottus grunniens	Gewöhnlicher Froschfisch	common toadfish
Halosauridae	Halosauriden	halosaurs
Halotydeus destructor	Rotbeinige Erdmilbe, Schwarze Sandmilbe	redlegged earth mite, black sand mite
Halteria grandinella	Springwimperling	
Hamearis lucina	Brauner Würfelfalter, Frühlingsscheckenfalter, Perlbinde, Schlüsselblumen-Würfelfalter	Duke of Burgundy
Hapalemur aureus	Goldener Bambus-Halbmaki	golden bamboo lemur
Hapalemur griseus	Grauer Halbmaki	gray gentle lemur
Hapalemur simus	Breitschnauzenhalbmaki	broad-nosed gentle lemur
Hapalemur spp.	Halbmakis	gentle lemurs
Hapalocarcinidae	Gallenkrabben	coral gall crabs
Hapalochlaena maculosa	Blaubandkrake, Blaugeringelter Krake	blue-ringed octopus
Hapalomys spp.	Asiatische Klettermäuse	Asiatic climbing rats, marmoset rats
Hapalopus incei	Trinidad Zwergvogelspinne, Tiger-Zwergvogelspinne	Trinidad olive tarantula
Haplodiplosis equestris	Sattelmücke	saddle gall midge
Haplopelma lividum	Kobaltblaue Vogelspinne*	cobalt blue tarantula
Haplopelma minax	Schwarze Thai-Vogelspinne*	Thailand black tarantula

Haplotaxidae	Brunnenwürmer	haplotaxid worms
Hardella thurjii	Diademschildkröte	Brahminy river turtle, crowned river turtle (IUCN)
Harengula clupeola	Falscher Hering, Kleinhering	false pilchard
Harengula jaguana	Geschuppter Kleinhering	scaled sardine
Harmandia loewi	Pappelblattgallmücke*	poplar leaf gall midge
Harmandia tremulae	Zitterpappel-Blattgallmücke*	aspen leaf gall midge
Harmothoe imbricata	Gemeiner Schuppenwurm	common fifteen-scaled worm
Harmothoe spp.	Schuppenwürmer	fifteen-scaled worms
Harpa articularis	Gegliederte Harfenschnecke	articulate harp
Harpa costata	Gerippte Harfenschnecke	imperial harp
Harpa doris	Rosenharfe	rosy harp, rose harp
Harpa harpa	Gemeine Harfenschnecke, Gewöhnliche Harfenschnecke	common harp
Harpa major/ Harpa ventricosa	Davidsharfe, Große Harfe	major harp, swollen harp
Harpactea hombergi	Schleichspinne*	sneak spider
Harpadon nehereus	Bombay-Ente	Bombay duck, bumalo, bummalow
Harpadontidae	Bombay-Enten (>Laternenfische)	Bombay ducks (>lanternfishes)
Harpagiferidae	Harpagiferiden	plunderfishes
Harpagoxenus sublaevis	Braune Raubknotenameise	
Harpalus affinis/ Harpalus aeneus	Haarrand-Schnellläufer, Metallfarbener Schnellläufer	
Harpalus latus	Schwarzglänzender Schnellläufer	
Harpalus rubripes	Rotbeiniger Schnellläufer	
Harpalus rufipes	Behaarter Schnellläufer	strawberry seed beetle
Harpalus spp.	Schnellläufer (Schnellläufer)	black-lustred ground beetles
Harpia harpyja	Harpyie	harpy eagle
Harpidae	Harfenschnecken	harp snails (harp shells)
Harpiocephalus harpia	Behaartflüglige Fledermaus	hairy-winged bat
Harpulina lapponica	Gestrichelte Walzenschnecke	brown-lined volute
Harpyhaliaetus coronatus	Zaunadler (Streitadler)	crowned solitary eagle
Harpyhaliaetus solitarius	Einsiedleradler	black solitary eagle
Harpyia hermelina/ Harpyia bifida/ Cerura bifida/Furcula bifida	Kleiner Gabelschwanz	sallow kitten
Harpyionycteris spp.	Spitzzahn-Flughunde	harpy fruit bats
Harpyopsis novaeguineae	Papuaadler	New Guinea harpy eagle
Harriotta raleighana	Gewöhnliche Langnasenchimäre, Gemeine Langnasenchimäre	long-nosed chimaera
Hartigiola annulipes	Buchengallmücke, Buchen-Haargallmücke*	beech pouch-gall midge, beech hairy-pouch-gall midge*
Hasemania nana	Kupfersalmler	silver-tipped tetra
Hebridae	Zwergwasserläufer, Uferläufer	velvet water bugs, sphagnum bugs, hebrids
Hebrus pusillus	Gefleckter Uferläufer	
Hebrus spp.	Zwergwasserläufer, Uferläufer	hebrids, velvet water bugs, sphagnum bugs

Hecuba scortum/ *Donax scortum*	Leder-Koffermuschel, Ledrige Dreiecksmuschel	leather donax
Hediste diversicolor/ *Nereis diversicolor*	Schillernder Seeringelwurm	estuary ragworm
Hedobia imperialis	Kaiserlicher Bohrkäfer	
Hedobia regalis	Kleiner Bohrkäfer	
Hedya nubiferana/ *Hedya dimidioalba*	Grauer Knospenwickler	marbled orchard tortrix (moth), fruit tree tortrix (green budworm, spotted apple budworm)
Hedya pruniana	Pflaumenknospenwickler	plum tortrix (moth)
Hedychrum rufilans	Rötliche Goldwespe	
Helarctos malayanus	Sonnenbär, Malaienbär	sun bear, Malayan sun bear
Helcion pellucidum	Durchsichtige Napfschnecke, Durchscheinende Häubchenschnecke	blue-rayed limpet
Helcomyzidae	Strandfliegen	seabeach flies, helcomyzid flies
Heleioporus australiacus	Riesen-Grabfrosch	giant burrowing frog
Heleioporus eyrei	Gähnender Grabfrosch, Gähnfrosch*	moaning frog
Heleioporus spp.	Grabfrösche	giant burrowing frogs
Heleomyzidae	Sumpffliegen, Scheufliegen, Dunkelfliegen	heleomyzid flies
Heleophryne rosei	Roses Gespenstfrosch	Skeleton Gorge ghost frog, Table Mountain ghost frog
Heleophryne spp.	Gespenstfrösche	ghost frogs
Heleophrynidae	Gespenstfrösche	ghost frogs
Heliacus cylindricus	Zylinder-Sonnenuhrschnecke, Atlantische Zylinder-Sonnenuhr	Atlantic cylinder sundial
Heliacus stramineus	Stroh-Sonnenuhrschnecke, Stroh-Sonnenuhr	straw sundial
Heliacus variegatus	Verschiedenfarbige Sonnenuhrschnecke	variegated sundial
Helicella itala	Gemeine Heideschnecke	common heath snail
Helicella obvia	Weiße Heideschnecke	white heath snail
Helicellinae	Heideschnecken	heath snails
Helicidae	Schnirkelschnecken (Gehäuseschnecken)	
Helicigona lapicida	Steinpicker	lapidary snail
Helicodonta obvoluta	Riemenschnecke, Eingerollte Zahnschnecke	cheese snail
Helicodonta spp.	Riemenschnecken	
Helicolenus dactylopterus	Blaumaul	bluemouth, blackbelly rosefish (FAO)
Helicomyia saliciperda	Weidenholzgallmücke	willow shot-hole midge, willow wood midge
Helicops angulatus	Scheelaugennatter	mountain keelback
Helicops carinicaudus	Kielschwanznatter	Wied's keelback
Helicopsis striata	Gestreifte Heideschnecke	striated heath snail
Helicotylenchus dihystera/ *Helicotylenchus nannus*	Spiralälchen	Cobb's spiral nematode, Steiner's spiral nematode (on cowpea/tomato)

Helicoverpa armigera/ *Heliothis armigera/* *Heliothis obsoleta*	Altweltlicher Baumwoll-Kapselwurm	scarce bordered straw moth, Old World cotton bollworm, African bollworm, corn earworm, cotton bollworm
Helicoverpa zea/ *Heliothis zea*	Maiseule, Maiseulenfalter*, Maiskolbenbohrer*, Maismotte*	corn earworm, bollworm, tomato fruitworm
Heliobolis spp.	Südafrikanische Buschechse	bushveld lizards
Heliodinidae	Sonnenmotten	heliodinid moths
Heliophobius *argenteocinereus*	Silbergrauer Erdbohrer	silvery mole-rat, sand rat
Heliophobus reticulatus	Haldenflur-Nelkeneule, Haldenflur-Netzeule, Hellgerippte Garteneule	bordered gothic (moth)
Heliopora coerulea	Blaue Koralle, Blaue Feuerkoralle	blue coral, blue fire coral
Heliopora spp.	Blaue Korallen	blue corals
Heliornithidae	Binsenhühner, Binsenrallen	sungrebes, finfoots
Heliosciurus spp.	Sonnenhörnchen	sun squirrels
Heliothis maritima warneckei	Warneckes Heidemoor-Sonneneule	shoulder-striped clover (moth)
Heliothis obsoleta/ *Heliothis armigera/* *Helicoverpa armigera*	Altweltlicher Baumwoll-Kapselwurm	scarce bordered straw moth, Old World cotton bollworm; tomato moth, tomato fruitworm, tomato grub
Heliothis virescens	Amerikanische Tabakeule, Amerikanische Tabakknospeneule	tobacco budworm
Heliothis viriplaca/ *Heliothis dipsacea*	Kardeneule, Steppenkräuterhügel-Sonneneule	marbled clover
Heliothis zea/ *Helicoverpa zea*	Maiseule	corn earworm, bollworm, tomato fruitworm
Heliothrips haemorrhoidalis	Gewächshausblasenfuß, Schwarze Fliege	glasshouse thrips, greenhouse thrips
Heliozelidae	Erzglanzmotten	shield bearers, leaf miners
Heliozoa	Sonnentierchen, Heliozoen	sun animalcules, heliozoans
Helix aperta	Zischende Weinbergschnecke, Singende Weinbergschnecke	singing snail, green gardensnail
Helix aspersa/ *Cornu aspersum/* *Cryptomphalus aspersus*	Gefleckte Weinbergschnecke	brown garden snail, brown gardensnail, common garden snail, European brown snail
Helix pomatia	Weinbergschnecke	Roman snail, escargot, escargot snail, edible snail, apple snail, grapevine snail, vineyard snail, vine snail
Helobdella stagnalis	Zweiäugiger Plattenegel, Zweiäugiger Plattegel	
Heloderma horridum	Skorpions-Krustenechse, Skorpion-Giftechse	Mexican beaded lizard

Heloderma suspectum	Gila-Krustenechse, Gila-Tier	gila monster
Helodermatidae	Krustenechsen, Giftechsen	beaded lizards
Helodidae	Sumpfkäfer	marsh beetles
Helogale spp.	Zwergmangusten, Zwergmungos	dwarf mongooses
Helogenidae/Helogeneidae	Fähnchenwelse	helogene catfishes
Helophilus pendulus	Gemeine Sonnenschwebfliege	
Helophorus flavipes	Gemeiner Furchenwasserkäfer, Buckelwasserkäfer	
Helophorus nubilus	Weizen-Furchenwasserkäfer	wheat shoot beetle, wheat mud beetle
Helops caerulens	Violetter Weidenkäfer*	violet willow beetle
Helostoma temmincki	Küssender Gurami	kissing gourami
Helostomatidae	Küssende Guramis	kissing gouramis
Hemachatus haemachatus/ Haemachatus haemachatus	Ringhalskobra	ringhals, ringneck spitting cobra
Hemaris fuciformis	Hummelschwärmer	broad-bordered bee hawkmoth
Hemaris tityus	Skabiosenschwärmer	narrow-bordered bee hawkmoth
Hemerobiidae	Taghafte	brown lacewings
Hemianax ephippiger	Schabrackenlibelle	
Hemichordata/ Branchiotremata	Kragentiere, Hemichordaten	hemichordates
Hemichromis bimaculatus/ Hemichromis guttatus	Roter Buntbarsch, Prachtbuntbarsch	jewel fish, jewelfish (FAO), red jewel fish, red cichlid
Hemicoccinae	Kugelschildläuse	gall-like coccids
Hemidactylium scutatum	Vierzehensalamander	four-toed salamander
Hemidactylus brookii	Brooks Halbzehengecko	Brook's half-toed gecko, Brook's gecko
Hemidactylus flaviviridis	Gelbgrüner Halbzehengecko	Indian wall gecko, Indian leaf-toed gecko
Hemidactylus turcicus	Europäischer Halbfinger, Halbfingergecko	Turkish gecko, Mediterranean gecko
Hemiemblemaria similis	Mimikry-Hechtschleimfisch	wrasse blenny
Hemifusus colosseus/ Pugilina colossea	Riesentreppenschnecke	colossal false fusus, giant stair shell
Hemigaleidae	Wieselhaie	weasel sharks
Hemigalus spp.	Bänderroller	banded palm civets
Hemigrammocypris lini	Venusfisch	garnet minnow
Hemigrammus bleheri	Rotkopfsalmler	firehead tetra
Hemigrammus caudovittatus	Rautenflecksalmler	Buenos Aires tetra
Hemigrammus erythrozonus	Glühlichtsalmler	glowlight tetra
Hemigrammus hyanuary	Grüner Neon, Costello-Salmler	January tetra
Hemigrammus ocellifer	Laternensalmler, Schlusslichtsalmler, Fleckensalmler	beacon fish, head-and-taillight tetra (FAO)
Hemigrammus pulcher	Karfunkelsalmler	garnet tetra (FAO), pretty tetra
Hemigrammus rhodostomus	Rotmaulsalmler	red-nosed tetra, rummy-nose tetra (FAO), red-nosed characin, rummy-nosed characin

Hemigrammus rodwayi/ *Hemigrammus armstrongi*	Goldtetra, Kirschfleckensalmler, Glanztetra, Messingsalmler	golden tetra
Hemigrammus ulreyi	Flaggensalmler, Ulreys Salmler	Ulrey's tetra
Hemigrammus unilineatus	Schwanzstrichsalmler	featherfin tetra
Hemigrapsus nudus	Violette Strandkrabbe	purple shore crab
Hemilepidotus hemilepidotus		Red Irish Lord
Hemilepistus ssp.	Wüstenasseln	desert woodlice
Hemiodontidae/Hemiodidae	Halbzähner, Schlanksalmler	
Hemiodopsis semitaenita/ *Hemiodus semitaeniatus*	Fleckstrichsalmler	halfline hemiodus
Hemipristis elongata	Schlanker Wieselhai	snaggletooth shark
Hemiprocnidae	Baumsegler	treeswifts, crested swifts
Hemiptera/Rhynchota (Heteroptera & Homoptera)	Halbflügler, Schnabelkerfe	hemipterans, bugs
Hemiramphidae/ Hemirhamphidae	Halbschnäbler, Halbschnabelhechte	halfbeaks
Hemiramphus archipelagicus	Archipel-Halbschnabelhecht	island halfbeak
Hemiramphus balao	Balao-Halbschnäbler	balao, balao halfbeak (FAO)
Hemiramphus brasiliensis	Ballyhoo, Brasilianischer Halbschnabelhecht, Brasilianischer Halbschnäbler	ballyhoo (FAO), ballyhoo halfbeak
Hemiramphus far	Gefleckter Halbschnabelhecht	blackbarred halfbeak (FAO), black-barred garfish, spotted halfbeak
Hemiramphus sajori	Japanischer Halbschnäbler	Japanese halfbeak
Hemiramphus saltator	Langflossen-Halbschnäbler	longfin halfbeak
Hemirhamphodon *pogonognathus*	Zahnleisten-Halbschnäbler	South East Asian livebearing halfbeak, bearded halfbeak
Hemiscylliidae	Bambushaie, Gebänderte Katzenhaie	bamboo sharks, longtailed carpet sharks
Hemiscyllium punctatum/ *Chiloscyllium punctatum*	Brauner Bambushai, Sattelfleckenhai	brown carpet shark, brownbanded bambooshark (FAO)
Hemiscyllium spp.	Epauletten-Haie	epaulette sharks
Hemiscyllium trispeculare	Gesprenkelter Epaulettenhai	speckled carpetshark (FAO), speckled catshark
Hemisepius typicus	Halbsepie	semi-cuttlefish*
Hemisus spp.	Ferkelfrösche	shovelnose frogs
Hemithea aestivaria	Schlehen-Grünflügelspanner	common emerald
Hemitheconyx caudicinctus	Fettschwanzgecko	fat-tailed gecko
Hemitragus hylocrius	Nilgiri-Tahr	Nilgiri tahr
Hemitragus jayakari	Arabischer Tahr	Arabian tahr
Hemitragus spp.	Tahre	tahrs
Hemitripterus americanus	Seerabe, Atlantik-Seerabe	Atlantic sea raven
Heniochus acuminatus	Wimpelfisch	pennant coralfish
Heniochus monoceros	Gehörnter Wimpelfisch	horned coachman, masked bannerfish (FAO)
Henria psalliotae	Pilzmücke*	mushroom midge

Henricia oculata	Gepunkteter Blutstern	bloody henry
Henricia sanguinolenta	Blutstern	blood star, northern henricia
Heodes alciphron/ Loweia alciphron	Sauerampferfeuchthalden-Goldfalter	purple-shot copper
Heodes tityrus/Loweia tityrus	Schwefelvögelchen	sooty copper
Heodes virgaureae/ Lycaena vigaureae/ Chrysophanus virgaureae	Dukatenfalter, Feuervogel	scarce copper
Heosemys depressa	Flache Erdschildkröte	Arakan forest turtle, Arakan terrapin
Heosemys grandis	Riesenerdschildkröte	giant Asian pond turtle, giant spined terrapin
Heosemys leytensis	Leyte-Erdschildkröte	Leyte pond turtle
Heosemys silvatica	Cochin-Erdschildkröte	Cochin forest cane turtle
Heosemys spinosa	Stachel-Erdschildkröte	spiny turtle, common spined terrapin
Heosemys spp.	Stachel-Erdschildkröten	spiny turtles, spined turtles
Hepatus epheliticus	Calico-Kastenkrabbe	calico box crab
Hepialidae	Wurzelbohrer	swift moths (swifts and ghost moths)
Hepialus lupulinus/ Korscheltellus lupulinus	Wurzelspinner, Kleiner Hopfenwurzelbohrer, Queckenwurzelspinner	common swift, garden swift moth
Hepialus hecta	Heidekrautwurzelbohrer	gold swift
Hepialus humuli	Hopfenspinner, Hopfenmotte, Hopfenwurzelbohrer	ghost swift, ghost moth
Hepsetidae	Afrikanische Hechtsalmler	African pike characins, hepsetids
Hepsetus odoe	Wasserhund, Afrikanischer Hechtsalmler	pike characin, kafue pike (FAO)
Heptranchias perlo	Siebenkiemer-Grauhai, Spitzkopf-Siebenkiemer, Perlonhai	seven-gilled shark, sharpnose sevengill shark (FAO)
Herbstia condyliata	Runzelige Seespinne	rugose spider crab
Heriades truncorum	Löcherbiene	
Herklotsichthys punctatus	Rotmeer-Hering	spotback herring
Herminia tarsicrinalis	Schattierte Bogenlinien-Spannereule*	shaded fan-foot (moth)
Herminia tarsipennalis	Federfußeule	fan-foot
Hermodice carunculata	Grüner Feuerwurm	green fireworm
Herotilapia multispinosa	Regenbogen-Buntbarsch	rainbow cichlid
Herpestes spp.	Echte Mungos	mongooses
Herpetotheres cachinnans	Lachfalke	laughing falcon
Herpobdella octoculata/ Erpobdella octoculata	Hundegel, Rollegel, Achtäugiger Schlundegel	eight-eyed leech
Herpolitha limax	Pantoffelkoralle*	slipper coral
Herpyllus ecclesiasticus	Parson-Spinne	Parson spider
Herse convolvuli/ Sphinx convolvuli/ Agrius convolvuli	Windenschwärmer	convolvulus hawkmoth, morning glory sphinx moth
Hesione pantherina	Pantherwurm	panther worm
Hespererato columbella/ Erato columbella	Pazifische Tauben-Kaurischnecke, Taubenkauri	pigeon erato, columbelle erato

Hesperia comma	Kommafalter	silver-spotted skipper
Hesperiidae	Dickkopffalter (Dickköpfe)	skippers
Hesperornithiformes	Zahntaucher	western birds
Hetaerius ferrugineus	Ameisenstutzkäfer, Rostroter Stutzkäfer	ant hister beetle
Heteractis aurora	Glasperlenanemone, Glasperlen-Anemone	beaded sea anemone, mat anemone, aurora anemone, button anemone
Heteractis crispa	Indopazifische Lederanemone	leather anemone
Heteractis magnifica	Prachtanemone	magnificent anemone, magnificent sea anemone
Heterakis gallinae	Geflügel-Pfriemenschwanz	cecal worm, caecal worm
Heterakis spp.	Pfriemenschwänze	pinworms a.o.
Heterandria formosa	Zwergkärpfling	least killifish
Heterocarpus ensifer	Bewaffnete Kantengarnele	armed nylon shrimp
Heterocarpus laevigatus	Glattschalige Kantengarnele	smooth nylon shrimp
Heterocarpus reedei	Chilenische Kantengarnele	Chilean nylon shrimp
Heterocarpus spp.	Kantengarnelen, Kanten-Tiefseegarnelen	nylon shrimps
Heterocentrotus mammillatus	Griffelseeigel, Griffel-Seeigel	slate pencil urchin
Heterocephalus glaber	Nacktmull	naked mole-rat, sand puppy
Heterocera	Motten	moths
Heteroceridae	Sägekäfer	variegated mud-loving beetles
Heterocongridae/ Heterocongrinae	Röhrenaale	garden eels
Heterodera avenae/ Heterodera major	Haferzystenälchen, Getreidezystenälchen, Hafernematode	oat cyst nematode, cereal cyst nematode, cereal root nematode
Heterodera rostochiensis/ Globodera rostochiensis	Kartoffelzystenälchen, Gelbe Kartoffelnematode	golden nematode (golden nematode disease of potato)
Heterodera glycine	Sojazystenälchen	soybean cyst nematode
Heterodera goettingiana	Erbsenzystenälchen	pea cyst nematode ("pea sickness")
Heterodera punctata	Gräserzystenälchen	grass cyst nematode
Heterodera schachtii	Rübenzystenälchen	beet eelworm, sugar-beet eelworm, beet cyst nematode
Heterodera trifolii	Kleezystenälchen	clover cyst nematode
Heterodon nasicus	Westliche Hakennatter	western hognose snake
Heterodon platyrhinos	Gewöhnliche Hakennatter	eastern hognose snake
Heterodon simus	Südliche Hakennatter	southern hognose snake
Heterodon spp.	Hakennattern, Schweinsnasen, Hakennasen-Nattern	hognose snakes
Heterodonax bimaculatus	Atlantische Falsche Dreiecksmuschel*	false-bean clam
Heterodonax pacificus	Pazifische Falsche Dreiecksmuschel*	Pacific false-bean clam
Heterodontidae	Stierkopfhaie, Hornhaie, Doggenhaie	hornsharks, bullhead sharks, Port Jackson sharks
Heterodontiformes	Stierkopfhaiartige, Hornhaiartige, Doggenhaiartige	hornsharks, bullhead sharks

Heterodontus francisci	Kalifornischer Stierkopfhai, Kalifornischer Hornhai	hornshark (FAO), California hornshark
Heterodontus spp.	Stierkopfhaie, Hornhaie	hornsharks
Heterodontus zebra	Zebra-Stierkopfhai, Zebra-Doggenhai	zebra bullhead shark
Heterogaster urticae	Brennnesselwanze	nettle groundbug
Heterogenea asella	Kleine Schildmotte	triangle (moth)
Heterohyrax spp.	Buschschliefer	gray hyraxes, yellowspotted hyraxes
Heteromastus filiformis	Kotpillenwurm	
Heterometrus spp.	Asiatische Waldskorpione	Asian forest scorpions
Heteromeyenia baleyi	Borkenschwamm, Smaragdschwamm	emerald sponge*
Heteromyidae	Taschenmäuse	pocket mice & kangaroo rats & kangeroo mice
Heteromys spp.	Wald-Stacheltaschenmäuse	forest spiny pocket mice
Heterophyes spp.	Zwergdarmegel	heterophyid flukes
Heterophyidae	Zwergdarmegel	heterophyid flukes
Heteropneustes fossilis	Sackkiemer	airsac catfish
Heteropneustidae	Sackkiemer, Kiemenschlauchwelse	airsac catfishes
Heteropoda venatoria	Bananenspinne	banana spider, large brown spider, huntsman spider
Heteropodidae (Eusparassidae, Sparassidae)	Jagdspinnen, Riesenkrabbenspinnen	giant crab spiders, huntsman spiders
Heteroptera (Hemiptera)	Wanzen	heteropterans, true bugs
Heteropterus morpheus	Spiegelfleck-Dickkopffalter, Bruchwald-Dickkopf	large chequered skipper
Heterotis niloticus/ Clupisudis niloticus	Afrikanischer Knochenzüngler	African bonytongue, heterotis (FAO)
Hexabranchus sanguineus	Spanische Tänzerin	Spanish dancer
Hexacorallia	Hexakorallen	hexacorallians, hexacorals
Hexactinellida	Glasschwämme, Hexactinelliden	glass sponges
Hexagrammidae	Grünlinge	greenlings
Hexanchidae	Grauhaie	cow sharks
Hexanchus griseus	Großer Grauhai, Sechskiemer-Grauhai	bluntnosed shark, six-gilled shark, sixgill shark, grey shark, bluntnose sixgill shark (FAO)
Hexanchus spp.	Grauhaie, Sechskiemer	sixgill sharks
Hexaplex brassica/ Phyllonotus brassica	Kohlschnecke, Kohl-Stachelschnecke	cabbage murex
Hexaplex cichoreus	Endivienschnecke	endive murex
Hexaplex erythrostomus	Rosenmundschnecke	pink-mouthed murex
Hexaplex fulvescens	Große Amerikanische Ostküstenmurex	giant eastern murex
Hexaplex radix/ Muricanthus radix	Wurzelschnecke	radish murex
Hexaplex regius	Königsschnecke*	royal murex, regal murex
Hexaplex trunculus	Purpurschnecke u.a.	trunk murex, trunculus murex

Hexaprotodon liberiensis/ *Choeropsis liberiensis*	Zwergflusspferd	pygmy hippopotamus
Hexarthrum culinaris	Grubenholzkäfer	
Hexathelidae	Röhrenvogelspinnen	Australian funnelweb spiders
Hexatrygonidae	Sechskiemenstechrochen	sixgill stingrays
Hiatella arctica	Arktischer Felsenbohrer	Arctic hiatella, red-nose clam, red nose, wrinkled rock borer, Arctic rock borer
Hiatella rugosa/ *Hiatella striata*	Gemeiner Felsenbohrer	common rock borer
Hiatellidae	Felsenbohrer	rock borers
Hiatula diphos	Diphos-Sandmuschel	diphos sunset clam, diphos sanguin
Hieraaetus fasciatus	Habichtsadler	Bonelli's eagle
Hieraaetus pennatus	Zwergadler	booted eagle
Hieremys annandalei	Tempelschildkröte	yellow-headed temple turtle
Hieremys spp.	Tempelschildkröten	temple turtles
Hilsa kelee	Kelee-Alse	kelee shad
Himantolophidae	Himantolophiden	footballfishes
Himantolophus groenlandicus	Atlantischer Tiefseeangler, Grönlandangler	Atlantic football fish
Himantopus himantopus	Stelzenläufer	black-winged stilt
Himantura chaophraya	Großer Süßwasserrochen	giant freshwater stingray, freshwater whipray (FAO)
Himantura gerrardi	Spitznasen-Stechrochen*	sharpnose stingray
Himantura imbricata	Schuppiger Peitschenrochen	scaly whipray
Himantura uarnak	Leopard-Stechrochen	honeycomb stingray (FAO), thornycomb stingray
Hinia incrassata	Dickmundreusenschnecke	thick-lipped dog whelk, angulate nassa
Hinia pygmaea	Zwergreusenschnecke	small dog whelk
Hinia reticulata	Netzreusenschnecke, Gemeine Netzreuse	netted dog whelk
Hinnites giganteus	Pazifische Riesen-Felskammmuschel	giant rock scallop
Hiodon alosoides	Goldauge	goldeye
Hiodon tergisus	Mondauge	mooneye
Hiodontidae	Zahnheringe, Mondaugen	mooneyes
Hipparchia fagi	Großer Waldportier	woodland grayling
Hipparchia semele	Rostbinde	grayling
Hipparchia statilinus/ *Neohipparchia statilinus*	Eisenfarbiger Samtfalter	tree grayling
Hippasteria phrygiana	Pferdestern, Pferde-Kissenstern	horse star, rigid cushion star
Hippeutis complanatus	Linsenförmige Tellerschnecke	flat ramshorn
Hippidae	Sandkrebse	sand crabs, mole crabs
Hippobosca equina	Pferdelausfliege	forest-fly
Hippoboscidae	Lausfliegen	louseflies (forest flies & sheep keds)
Hippocamelus spp.	Andenhirsche, Huemuls	huemuls, guemals

Hippocampus guttulatus/ *Hippocampus ramulosus*	Langschnauziges Seepferdchen	seahorse, European seahorse, long-snouted seahorse (FAO)
Hippocampus hippocampus	Kurzschnauziges Seepferdchen	short-snouted sea horse
Hippocampus kuda	Krönchen-Seepferd	golden seahorse, yellow seahorse, spotted seahorse (FAO)
Hippocampus zosterae	Zwergseepferdchen	pygmy seahorse, dwarf seahorse (FAO)
Hippodiplosia foliacea	Blattzweig-Moostierchen, Elchgeweih-Moostierchen	antler bryozoan, leafy horse-bryozoan*
Hippoglossoides dubius	Japanischer Heilbutt	Pacific false halibut, flathead flounder (FAO)
Hippoglossoides elassodon	Heilbuttscholle	flathead sole
Hippoglossoides platessoides	Raue Scholle, Raue Scharbe, Doggerscharbe	long rough dab
Hippoglossus hippoglossus	Heilbutt	Atlantic halibut
Hippoglossus stenolepis	Pazifischer Heilbutt	Pacific halibut
Hippolais caligata	Buschspötter	booted warbler
Hippolais icterina	Gelbspötter	icterine warbler
Hippolais languida	Dornspötter	Upcher's warbler
Hippolais olivetorum	Olivenspötter	olive-tree warbler
Hippolais pallida	Blassspötter	olivaceous warbler
Hippolais polyglotta	Orpheusspötter	melodious warbler
Hippolyte coerulescens	Himmelblaue Sargassumgarnele	cerulean sargassum shrimp
Hippolyte inermis/ *Hippolyte varians*	Seegrasgarnele	seaweed chameleon prawn
Hippolyte varians	Farbwechselnde Garnele, Chamäleongarnele	chameleon prawn
Hipponicidae	Hufschnecken	hoof limpets, horsehoof limpets, bonnet limpets
Hipponix antiquatus	Weiße Hufschnecke	white hoofsnail
Hipponix subrufus	Orange Hufschnecke	orange hoofsnail
Hippopotamus amphibius	Flusspferd, Großflusspferd	hippopotamus ("hippo")
Hippopus hippopus	Pferdehufmuschel	bear's paw clam, bear paw clam (IUCN), horseshoe clam, horse's hoof, strawberry clam
Hippopus porcellanus	Chinesische Hufmuschel	China clam
Hipposideridae	Rundblattnasen	Old World leaf-nosed bats, (incl. trident-nosed bats)
Hipposideros spp.	Rundblattnasen	Old World leaf-nosed bats
Hippospongia canaliculata	Wollschwamm	wool sponge, sheepswool sponge
Hippospongia communis	Pferdeschwamm	horse sponge
Hippospongia lachne	Schafwollschwamm*	sheep's wool sponge
Hippotion celerio	Großer Weinschwärmer	silver-striped hawkmoth
Hippotragus equinus	Pferdeantilope, Roan	roan antelope
Hippotragus leucophaeus	Blaubock	blue buck
Hippotragus niger	Rappenantilope	sable antelope
Hirudinea	Egel, Hirudineen	leeches, hirudineans

Hirudo medicinalis	Blutegel, Medizinischer Blutegel	medicinal leech
Hirundapus caudacutus	Stachelschwanzsegler	needle-tailed swift
Hirundichthys rondeletii/ Exonautes rondeleti/ Exocoetus rondeleti	Schwarzflügelfisch*, Rondelets Flugfisch	blackwing flyingfish (FAO), subtropical flyingfish
Hirundichthys affinis	Indopazifischer Flugfisch, Indopazifischer Fliegender Fisch	fourwing flyingfish
Hirundichthys speculiger	Spiegelflügelfisch*	mirrorwing flyingfish
Hirundinidae	Schwalben	swallows, martins
Hirundo atrocaerulea	Stahlschwalbe	blue swallow
Hirundo daurica	Rötelschwalbe	red-rumped swallow
Hirundo rustica	Rauchschwalbe	swallow, barn swallow
Hispa atra	Stachelkäfer, Dorn-Blattkäfer	spiny leaf beetle, leaf-mining beetle, wedge-shaped leaf beetle
Hispella atra	Schwarzer Stachelkäfer	
Hister cadaverinus	Aasstutzkäfer	
Hister quadrimaculatus	Viergefleckter Stutzkäfer	
Histeridae	Stutzkäfer	hister beetles, clown beetles
Histiogaster carpio	Essigmilbe, Karpfenschwanzmilbe	vinegar mite*, carptail mite*
Histiostoma feroniarum	Feuchtmilbe*	damp mite
Histiostoma spp.	Abwassermilben*	sewage mites
Histioteuthis bonellii	Segelkalmar	umbrella squid
Histioteuthis dofleini	Blumentopfkalmar*	flowervase jewel squid
Histioteuthis elongata	Länglicher Segelkalmar*	elongate jewel squid
Histiotus spp.	Braune Großohrfledermäuse	big-eared brown bats
Histrio histrio	Sargasso-Fisch	Sargassumfish
Histrionicus histrionicus	Kragenente	harlequin duck
Hodotermitidae	Erntetermiten	harvester termites
Hofmannophila pseudospretella	Samenmotte	brown house-moth, brown house moth
Holacanthus tricolor	Karibenkaiserfisch, Felsenschönheit	rock beauty
Holaspis guentheri	Günthers Stacheleidechse	sawtail lizard
Holbrookia lacerata	Gefleckter Taubleguan	spot-tailed earless lizard, spot-tail earless lizard
Holbrookia maculata	Kleiner Taubleguan	lesser earless lizard
Holbrookia propinqua	Gekielter Taubleguan	keeled earless lizard
Holbrookia spp.	Taubleguane	earless lizards
Holbrookia subcaudalis	Südlicher Taubleguan	southern earless lizard
Holocanthus ciliaris	Königin-Engelsfisch	queen angelfish
Holocentridae	Eichhörnchenfische & Soldatenfische	squirrelfishes & soldierfishes
Holocentrinae	Eichhörnchenfische	squirrelfishes
Holocentrus rubrum/ Sargocentron rubrum	Roter Husar, Roter Soldatenfisch, Rotgestreifter Soldatenfisch, Gewöhnlicher Eichhornfisch	red soldierfish
Holocentrus diadema	Diadem-Soldatenfisch	crowned soldierfish

Holocentrus rufus	Langstachel-Husar, Karibischer Eichhörnchenfisch	longspine squirrelfish
Holocentrus spp.	Soldatenfische	squirrelfishes
Holocephali	Seedrachen, Seekatzen, Chimären	chimaeras, ratfishes, rabbit fishes
Holochilus spp.	Südamerikanische Sumpfratte*	web-footed rats, marsh rats
Holothuria atra	Schwarze Seegurke	black sea cucumber
Holothuria edulis	Essbare Seegurke	edible sea cucumber
Holothuria forskali	Variable Seegurke	cotton spinner
Holothuria impatiens	Schlanke Seegurke	slender sea cucumber
Holothuria nobilis	Weißfuß-Seegurke	whitefoot sea cucumber*
Holothuria tubulosa	Röhrenholothurie, Röhrenseegurke	Mediterranean trepang
Holothuriidae	Seegurken, Seewalzen	sea cucumbers, holothurids
Holothuroidea	Seewalzen, Seegurken, Holothurien	sea cucumbers, holothurians
Homalopoma albidum	Weiße Zwerg-Turbanschnecke	white dwarf-turban
Homalopoma indutum	Zweiseitige Zwerg-Turbanschnecke	two-faced dwarf-turban
Homalopsis buccata	Boa-Trugnatter	masked water snake
Homalopteridae/Balitoridae	Plattschmerlen, Karpfenschmerlen	river loaches
Homarinus capensis	Kap-Hummer, Südafrikanischer Hummer	Cape lobster
Homarus americanus	Amerikanischer Hummer	northern lobster, American clawed lobster
Homarus gammarus	Hummer, Europäischer Hummer	common lobster, European clawed lobster, Maine lobster
Hominoidea	Menschenartige	apes, anthropoid apes
Homo heidelbergensis (*Homo erectus heidelbergensis*)	Heidelbergmensch	Heidelberg man
Homo sapiens sapiens	Mensch	people, human beings, humans
Homo sapiens neanderthalensis	Neanderthalmensch	Neanderthals, Neanderthal people, Neanderthal man
Homola barbata	Behaarte Kastenkrabbe	hairy box crab
Homola spinifrons see *Eriphia verrucosa*		
Homolidae	Kastenkrabben, Taschenkrebse u.a.	carrier crabs (box crabs a.o.)
Homolocantha scorpio	Skorpion-Stachelschnecke	scorpion murex
Homoptera	Pflanzensauger	homopterans (cicadas & aphids & scale insects)
Homopus areolatus	Areolen-Flachschildkröte, Papageischnabel-Flachschildkröte	parrot-beaked tortoise, beaked Cape tortoise
Homopus belliana	Bells Flachschildkröte	Bell's hingeback tortoise
Homopus boulengeri	Boulenger-Flachschildkröte	donner-weer tortoise, Boulenger's Cape tortoise

Homopus femoralis	Sporn-Flachschildkröte	Karroo tortoise, Karroo Cape tortoise
Homopus signatus	Gesägte Flachschildkröte	speckled Cape tortoise
Homopus spp.	Flachschildkröten	Cape tortoises
Hopkinsia rosacea	Hopkins Rose	Hopkins' rose
Hoplangia durotrix		Weymouth carpet coral
Hoplia philanthus	Großer Gras-Laubkäfer	Welsh chafer
Hoplias malabaricus	Jagdsalmler	tigerfish, tararira, haimara, trahira (FAO)
Hoplichthyidae	Hoplichthyiden	spiny flatheads, ghost flatheads
Hoplobatrachus tigerinus/ Rana tigerina	Tigerfrosch, Asiatischer Ochsenfrosch	tiger Peters frog, Indian bullfrog
Hoplocampa brevis	Birnensägewespe	pear sawfly
Hoplocampa flava	Gelbe Pflaumensägewespe	plum sawfly
Hoplocampa minuta	Schwarze Pflaumensägewespe	black plum sawfly
Hoplocampa testudinea	Apfelsägewespe	European apple sawfly
Hoplocarida/Stomatopoda	Fangschreckenkrebse, Maulfüßer	mantis shrimps
Hoplocephalus bitorquatus	Australische Blasskopf-Giftnatter	pale-headed snake
Hoplocephalus bungaroides	Australische Gelbfleck-Giftnatter	Australian yellow-spotted snake, broadhead snake, broad-headed snake (IUCN)
Hoplocephalus spp.	Australische Breitkopf-Giftnatter	Australian broad-headed snakes, pale-headed snakes
Hoplocercus spinosus	Stachelschwanzleguan	weapontail
Hoplomys gymnurus	Lanzenratte	armored rat, thick-spined rat
Hoplophryne spp.	Afrikanische Bananenfrösche	African banana frogs
Hoplophryne uluguruensis	Afrikanischer Bananenfrosch (Uluguru-Schwarzfrosch)	African banana frog
Hoplopleura acanthopus	Feldmauslaus	vole louse
Hoplopleura pacifica	Tropische Rattenlaus	tropical rat louse
Hoplopleuridae	Kleinsäugerläuse	small mammal sucking lice
Hoplopterus indicus	Rotlappenkiebitz	red-wattled plover
Hoplopterus spinosus	Spornkiebitz	spur-winged plover
Hoplostethus atlanticus	Granatbarsch, Atlantischer Sägebauch	orange roughie
Hoplostethus mediterraneus	Kaiserbarsch, Mittelmeer-Kaiserbarsch	rough-fish, roughfish, rosy soldierfish, Mediterranean slimehead (FAO)
Horaichthyidae	Glaskärpflinge	horaichthyids
Horaichthys setnai	Indischer Glaskärpfling	Malabar ricefish
Hormathia nodosa		rugose anemone
Hornera frondiculata	Farn-Moostierchen, Feinästiges Moostierchen	fern bryozoan
Hucho hucho	Huchen	Danube salmon, huchen
Huso dauricus	Sibirischer Hausen	freshwater kaluga
Huso huso	Europäischer Hausen	great sturgeon, volga sturgeon, beluga (FAO)

Hyaena

Hyaena brunnea	Schabrackenhyäne	brown hyena
Hyaena hyaena	Streifenhyäne	striped hyena
Hyaenidae	Hyänen	hyenas
Hyalina pallida	Farblose Randschnecke*	pallid marginella
Hyalina pergrandis	Rosa Randschnecke	pink marginella
Hyalinobatrachium spp.	Glasfrösche	glass frogs
Hyalonema spp.		glass rope sponges
Hyalophora cecropia/ Platysamia cecropia	Cecropiaspinner	cecropia silk moth, cecropia moth
Hyalopterus pruni	Mehlige Zwetschgenblattlaus, Pflaumenblattlaus, Hopfenblattlaus, Hopfenlaus	damson-hop aphid, hop aphid
Hyas araneus	Atlantische Seespinne, Nordische Seespinne	Atlantic lyre crab, great spider crab, toad crab
Hyas coarctatus	Arktische Seespinne	Arctic lyre crab, lesser toad crab
Hyas lyratus	Pazifische Seespinne	Pacific lyre crab
Hybocodon pendulus	Einarmqualle*	one-arm jellyfish
Hybomys spp.	Streifen-Waldmäuse	back-striped mice, hump-nosed mice
Hybomys univittatus	Peters Einstreifenmaus	Peter's striped mouse
Hydatigera taeniaeformis/ Taenia taeniaeformis	Dickhalsiger Bandwurm, Katzenbandwurm u.a.	common cat tapeworm
Hydatina amplustre/ Aplustrum amplustre	Prächtige Papierblase	royal paperbubble
Hydatina albocincta	Streifen-Papierblase	white-banded paperbubble
Hydatina physis	Meeresrose	brown-line paperbubble, green-lined paperbubble, green paperbubble
Hydatina zonata	Gebänderte Papierblase	zoned paperbubble
Hydnophora rigida	Stachelkoralle	
Hydra oligactis/ Pelmatohydra oligactis	Graue Hydra, Grauer Süßwasserpolyp	grey hydra*, brown hydra
Hydra americana	Weiße Hydra, Weißer Süßwasserpolyp	white hydra
Hydra viridis/ Chlorohydra viridissima/ Hydra viridissima	Grüne Hydra	green hydra
Hydra vulgaris	Braune Hydra, Brauner Süßwasserpolyp	brown European hydra
Hydrachnellae/Hydrachnidia	Süßwassermilben	freshwater mites
Hydractinia echinata	Stachelpolyp	snailfur, snail fur
Hydraecia micacea	Markeule, Schachtelhalmeule, Kartoffelbohrer, Rübenbohrer	rosy rustic (moth)
Hydraecia osseola	Marsch-Markeule*	marsh mallow (moth)
Hydraenidae	Langtaster-Wasserkäfer	minute moss beetles
Hydrallmania falcata	Korallenmoos	sickle hydroid, sickle coralline, coral moss
Hydrelaps darwiniensis	Darwins Seeschlange	Darwin's sea snake
Hydrelia sylvata/ Hydrellia testaceata	Erlen-Blattspanner	sylvan waved carpet (moth)

Hydrellia griseola	Graue Gerstenminierfliege	smaller rice leafminer (U.S.)
Hydrobates pelagicus	Sturmschwalbe	storm petrel
Hydrobatidae	Sturmschwalben	storm-petrels
Hydrobia spp.	Wattschnecken	mud snails
Hydrobia ulvae	Wattschnecke	laver spire snail, laver mud snail
Hydrobia ventrosa/ Hydrobia stagnalis	Hängende Wattschnecke, Bauchige Wattschnecke	spine snail, hanging mud snail*
Hydrobiidae	Wasserdeckelschnecken	mud snails
Hydrochaeridae	Wasserschweine	capybaras
Hydrochaeris hydrochaeris	Capybara	capybara, "water hog"
Hydrochara caraboides	Kleiner Kolben-Wasserkäfer, Schwarzbeiniger Stachel-Wasserkäfer, Stachelwasserkäfer, Großer Teichkäfer	lesser silver water beetle
Hydrocinidae	Schwarmsalmler	hydrocinids
Hydrocorisae/Hydrocorizae	Wasserwanzen	water bugs
Hydrocynus goliath	Riesentigerfisch	giant tigerfish
Hydrocynus lineatus	Kleiner Tigerfisch	lesser tigerfish
Hydrodynastes gigas	Brasilianische Glattnatter	South American water 'cobra', beach 'cobra', false cobra
Hydroida	Hydrozoen	hydrozoans, hydra-like animals, hydroids
Hydroides uncinata	Atlantischer Röhrenwurm	Atlantic tubeworm
Hydrolagus colliei	Gefleckte Pazifische Seeratte, Amerikanische Spöke, Amerikanische Chimäre	spotted ratfish
Hydromantes genei/ Speleomantes genei	Sardinischer Schleuderzungensalamander	Sardinian cave salamander, brown cave salamander
Hydromantes brunus	Kalkstein-Höhlensalamander	limestone salamander
Hydromantes italicus	Italienischer Schleuderzungensalamander	Italian cave salamander
Hydromantes platycephalus	Mount Lyell-Höhlensalamander	Mount Lyell salamander
Hydromantes shastae	Shasta-Höhlensalamander	Shasta salamander
Hydromantes spp.	Schleuderzungensalamander, Höhlensalamander	web-toed salamander
Hydromedusa maximiliani	Brasilianische Schlangenhalsschildkröte	Maximilian's snake-necked turtle, Maximilian's snake-headed turtle
Hydromedusa spp.	Amerikanische Schlangenhalsschildkröten, Amerikanische Schlangenhälse	American snake-necked turtles, American snake-headed turtles
Hydromedusa tectifera	Argentinische Schlangenhalsschildkröte	South American snake-necked turtle, South American snake-headed turtle
Hydrometra stagnorum	Teichläufer	water measurer, marsh treader
Hydrometridae	Teichläufer	water measurers, marsh treaders
Hydromys spp.	Schwimmratten	water rats, beaver rats
Hydrophiidae	Seeschlangen	sea snakes
Hydrophilidae	Wasserkäfer, Kolbenwasserkäfer	water scavenger beetles, herbivorous water beetles

Hydrophilus piceus/ *Hydrous piceus*	Großer Schwarzer Kolbenwasserkäfer, Großer Kolbenwasserkäfer	greater silver beetle, great black water beetle, great silver water beetle, diving water beetle
Hydrophis cyanocinctus	Blaugebänderte Ruderschlange	annulated sea snake, ringed sea snake, banded sea snake
Hydrophis fasciatus	Gebänderte Ruderschlange	banded small-headed sea snake, striped sea snake
Hydrophis ornatus	Riff-Ruderschlange	reef sea snake, spotted sea snake
Hydrophis spiralis	Ruderschlange	yellow sea snake, banded sea snake
Hydrophis spp.	Ruderschlangen	Asian sea snakes, Australian sea snakes
Hydroporus spp.	Zwergschwimmer	
Hydropotes inermis	Wasserreh, Chinesisches Wasserreh	Chinese water deer
Hydroprogne caspia/ *Sterna caspia*	Raubseeschwalbe	Caspian tern
Hydrosaurus pustulatus	Philippinische Segelechse	Philippine sail-fin lizard
Hydrosaurus spp.	Segelechsen	sail-fin lizards
Hydroscaphidae		skiff beetles
Hydrous piceus/ *Hydrochara piceus*	Großer Kolbenwasserkäfer, Großer Schwarzer Kolbenwasserkäfer	great black water beetle, great silver water beetle, greater silver beetle, diving water beetle
Hydrous spp./ *Hydrophilus* spp.	Kolbenwasserkäfer	water scavenger beetles, herbivorous water beetles
Hydrurga leptonyx	Seeleopard	leopard seal
Hyemoschus aquaticus	Wassermoschustier, Afrikanisches Hirschferkel	water chevrotain
Hygrobia spp.	Schlammschwimmer, Feuchtkäfer	screech beetle
Hygromia cinctella	Gürtelschnecke*	girdled snail
Hygromia limbata	Heckenschnecke*	hedge snail
Hyla andersoni	Anderson-Laubfrosch	pine barrens treefrog
Hyla annectans	Assam-Laubfrosch	Assam treefrog
Hyla arborea	Europäischer Laubfrosch	European treefrog, common treefrog, Central European treefrog
Hyla avivoca	Vogelzwitscher-Laubfrosch*	bird-voiced treefrog
Hyla cinerea	Grüner Laubfrosch	green treefrog
Hyla crepitans	Rasselnder Laubfrosch	rattle-voiced treefrog, emerald-eyed treefrog
Hyla crucifer/ *Pseudacris crucifer*	Wasserpfeifer	spring peeper
Hyla ebraccata	Bromelienlaubfrosch	hourglass treefrog
Hyla faber	Schmied	blacksmith treefrog
Hyla meridionalis	Mittelmeer-Laubfrosch	stripeless treefrog, Mediterranean treefrog
Hyla miliaria	Fransen-Laubfrosch	Cope's brown treefrog
Hyla regilla	Pazifik-Laubfrosch	Pacific treefrog
Hyla spp.	Laubfrösche	common treefrogs
Hyla squirella	Eichhörnchen-Laubfrosch	squirrel treefrog

Hyla versicolor	Grauer Laubfrosch	gray treefrog
Hylactophryne augusti/ *Eleutherodactylus augusti*	Mexikanischer Klippenfrosch	barking frog
Hylaeus spp.	Maskenbienen	yellow-faced bees
Hylastes cunicularius	Schwarzer Fichtenbastkäfer	
Hylastinus obscurus	Kleeborkenkäfer, Kleewurzelborkenkäfer	large broom bark beetle
Hylecoetus dermestoides	Buchenwerftkäfer, Bohrkäfer, Sägehörniger Werftkäfer	large timberworm, European sapwood timberworm*
Hyles euphorbiae	Wolfsmilchschwärmer	spurge hawkmoth
Hyles gallii	Labkrautschwärmer	bedstraw hawkmoth
Hyles lineata	Linienschwärmer	striped hawkmoth
Hylesinus crenatus	Großer Schwarzer Eschenbastkäfer	greater ash bark beetle
Hylidae	Laubfrösche	tree frogs, true tree frogs
Hylobates agilis	Ungka, Schlankgibbon	dark-handed gibbon, agile gibbon
Hylobates concolor	Schopfgibbon, Weißwangengibbon	crested gibbon
Hylobates hoolock	Hulock, Weißbrauengibbon	white-browed gibbon, Hoolock gibbon
Hylobates klossii/ *Symphalangus klossi*	Zwergsiamang, Biloh, Mentawai-Gibbon	Kloss's gibbon, dwarf siamang, Mentawai gibbon, beeloh
Hylobates lar	Lar, Weißhandgibbon	common gibbon, white-handed gibbon
Hylobates moloch	Silbergibbon	silvery gibbon, Javan gibbon
Hylobates muelleri	Grauer Gibbon, Borneo-Gibbon	gray gibbon, Müller's gibbon
Hylobates pileatus	Kappengibbon	capped gibbon, pileated gibbon
Hylobates syndactylus/ *Symphalangus syndactulus*	Siamang	siamang
Hylobius abietis	Großer Brauner Rüsselkäfer, Großer Fichtenrüssler	fir-tree weevil, pine weevil
Hylobius piceus	Großer Kiefernrüssler	larch weevil
Hylobius pinastri	Kleiner Brauner Rüsselkäfer	small fir-tree weevil
Hylochoerus meinertzhageni	Riesenwaldschwein	giant forest hog
Hylocichla mustelina	Walddrossel (Mäusedrossel)	wood thrush
Hyloicus pinastri/ *Sphinx pinastri*	Kiefernschwärmer, Kleiner Fichtenrüssler	pine hawkmoth
Hylomyscus spp.	Afrikanische Waldmäuse u.a.	African wood mice
Hylonycteris underwoodi	Underwood-Langzungenfledermaus	Underwood's long-tongued bat
Hylopetes spp.	Asiatische Gleithörnchen, Pfeilschwanz-Gleithörnchen	arrow-tailed flying squirrels
Hylophila prasinana/ *Pseudoips fagana/* *Bena prasinana/* *Bena fagana*	Kahnspinner, Kleiner Kahnspinner, Buchenwickler, Buchenkahneule, Jägerhütchen	green silverlines, scarce silverlines
Hylophilidae	Mulmkäfer	hylophilid beetles

Hylotrupes bajulus	Hausbock, Hausbockkäfer	house longhorn beetle
Hylurgopinys rufipes	Amerikanischer Ulmenborkenkäfer	American elm bark beetle
Hylurgus ligniperda	Rothaariger Kiefernbastkäfer	
Hymenocera elegans	Westliche Harlekingarnele	western harlequin shrimp
Hymenocera picta	Östliche Harlekingarnele	eastern harlequin shrimp
Hymenochirus spp.	Zwergkrallenfrösche	dwarf clawed frogs
Hymenolepis spp.	Zwerg-Darm-Bandwürmer	dwarf intestinal tapeworms
Hymenopenaeus triarthrus/ Haliporoides triarthrus	Messergarnele	knife shrimp
Hymenoptera	Hautflügler	hymenopterans
Hynobiidae	Winkelzahnmolche	Asiatic salamanders, hynobiids
Hynobius spp.	Echte Winkelzahnmolche	Asian salamanders
Hynobius stejnegeri	Bernstein-Winkelzahnmolche	amber-coloured salamander
Hyomys dammermani	Westliche Weißohr-Riesenratte	western white-eared giant rat
Hyomys goliath	Goliathratte, Östliche Weißohr-Riesenratte	eastern white-eared giant rat
Hyosciurus heinrichi	Ferkelhörnchen	Celebes long-nosed squirrel
Hyostrongylus rubidus	Roter Magenwurm	red stomach worm (of swine)
Hypebaeus flavipes	Hainbuchen-Zipfelkäfer*	Moccas beetle
Hypena obsitalis		Bloxworth snout
Hypena proboscidalis	Zünslereule, Nesselschnabeleule	common snout (moth)
Hypena rostralis	Hopfeneule	buttoned snout (moth)
Hyperacrius fertilis	Kaschmir-Wühlmaus, True-Wühlmaus	Kashmir vole, True's vole
Hyperacrius spp.	Kaschmir-/Punjab-Wühlmäuse	Kashmir voles, Punjab voles
Hyperia spp.	Quallenflohkrebse	jellyfish amphipods
Hyperoglyphe antarctica	Antarktischer Schwarzfisch	bluenose warehou, Antarctic butterfish (FAO)
Hyperoglyphe perciformis	Tonnenfisch*	barrelfish
Hyperolius concolor	Spitzkopf-Riedfrosch	variable reed frog
Hyperolius horstockii	Aronstabfrosch, Aronstab-Riedfrosch	arum reed frog
Hyperolius nasutus	Scharfnasen-Riedfrosch	longnose reed frog
Hyperolius pusillus	Wasserlilienfrosch, Wasserlilien-Riedfrosch	dwarf reed frogs
Hyperolius spp.	Riedfrösche, Afrikanische Riedfrösche	African reed frogs
Hyperoodon ampullatus	Nördlicher Entenwal, Dögling	northern bottlenosed whale, North-Atlantic bottle-nosed whale
Hyperoodon planifrons	Südlicher Entenwal	southern bottlenosed whale, southern flat-headed bottle-nosed whale
Hyperoplus lanceolatus/ Ammodytes lanceolatus	Großer Sandspierling, Großer Sandaal	greater sandeel, lance, sandlance
Hyphantria cunea	Amerikanischer Webebär	fall webworm
Hyphessobrycon bifasciatus	Gelber Salmler, Gelber von Rio	yellow tetra
Hyphessobrycon callistus	Blutsalmler	jewel tetra

Hyphessobrycon eos	Sonnensalmler	dawn tetra
Hyphessobrycon erythrostigma/ Hyphessobrycon rubrostigma	Sherrysalmler, Perez-Salmler, Kirschflecksalmler	bleeding-heart tetra
Hyphessobrycon flammeus	Roter von Rio	flame tetra
Hyphessobrycon herbertaxelrodi	Schwarzer Flaggensalmler, Schwarzer Neon	black neon
Hyphessobrycon heterorhabdus	Dreibandsalmler, „Falscher" Ulrey	flag tetra
Hyphessobrycon loretoensis	Loretosalmler	Loreto tetra
Hyphessobrycon metae	Rio-Meta-Salmler	Rio Meta tetra
Hyphessobrycon ornatus	Schmucksalmler	ornate tetra
Hyphessobrycon pulchripinnis	Schönflossensalmler, Zitronensalmler	lemon tetra
Hyphessobrycon rubrostigma/ Hyphessobrycon erythrostigma	Sherrysalmler	bleeding-heart tetra
Hyphessobrycon scholzei	Schwarzbandsalmler	black-line tetra
Hyphydrus ovatus	Kugelschwimmer	
Hypnale hypnale/ Agkistrodon hypnale	Indische Nasenotter	Indian humpnose viper, hump-nosed viper
Hypnidae	Australische Zitterrochen	coffin rays
Hypochilidae/ Hypochelidae	Vierlungenspinnen*	four-lunged spiders
Hypocolius ampelinus	Seidenwürger, Nachtschattenfresser	grey hypocolius
Hypoderma actaeon	Hirschdasselfliege	deer warble fly a.o.
Hypoderma bovis	Rinderdasselfliege (Hautdasselfliege, Große Dasselfliege)	ox warble fly a.o., northern cattle grub (U.S.)
Hypoderma diana	Rehdasselfliege	deer warble fly a.o.
Hypoderma lineatum	Kleine Dasselfliege	lesser ox warble fly, lesser ox botfly, common cattle grub (U.S.)
Hypodryas maturna/ Euphydryas maturna	Kleiner Maivogel	scarce fritillary
Hypogastruridae	Kurzspringer	hypogastrurid springtails
Hypogeomys antimena	Votsotsa	Malagasy giant rat
Hypogeophis rostratus	Cuviers Erdwühle	Frigate Island caecilian
Hypolophus sephen/ Pastinachus sephen/ Dasyatis sephen	Federschwanz-Stechrochen	cowtail stingray
Hypomesus olidus	Süßwasser-Stint	pond smelt
Hypomesus pretiosus	Kleinmäuliger Kalifornischer Seestint	surf smelt
Hypomma bituberculatum		money spider
Hyponephele lycaon/ Maniola lycaon	Kleines Ochsenauge	dusky meadow brown
Hyponomeuta malinellus	Apfelgespinstmotte, Apfelbaumgespinstmotte	apple moth, Adkin's apple ermel
Hyponomeuta padellus/ Yponomeuta padellus	Pflaumengespinstmotte	common hawthorn ermel, small ermine moth
Hyponomeutidae (Yponomeutidae)	Gespinstmotten	ermine moths
Hypopachus variolosus	Schafsfrosch	sheep frog

Hypophthalmichthys molitrix	Silberkarpfen, Gewöhnlicher Tolstolob	silver carp (FAO), tolstol
Hypophthalmichthys nobilis/ Aristichthys nobilis	Marmorkarpfen, Edler Tolstolob	bighead carp
Hypophthalmidae	Tolstoloben	lookdown catfishes, loweye catfishes
Hypopta caestrum	Spargelbohrer	
Hypoptychus dybowskii	Dybowski-Sandaal, Koreanischer Sandaal	Dybowski's sand eel, Korean sandeel (FAO)
Hyporhamphus acutus	Pazifik-Halbschnabelhecht	Pacific halfbeak
Hyporhamphus rosae	Kalifornischer Halbschnabelhecht, Kalifornischer Halbschnäbler	California halfbeak
Hyporhamphus unifasciatus	Silberstreif-Halbschnabelhecht, Silberstreif-Halbschnäbler	silverstripe halfbeak
Hypostomus punctatus	Punktierter Schilderwels	spotted hypostomus
Hypsiglena spp.	Nachtschlangen	night snakes
Hypsignathus monstrosus	Hammerkopfflughund, Hammerkopf	hammer-headed fruit bat
Hypsiprymnodon moschatus	Moschusrattenkänguru(h)	musky "rat"-kangaroo
Hypsoblennius hentzi	Gefederter Schleimfisch*	feather blenny
Hypsopygia costalis	Heuzünsler	gold fringe, clover hayworm
Hypsugo savii/ Pipistrellus savii	Alpenfledermaus	Savi's pipistrelle
Hyptiotes cavatus	Amerikanische Dreieckspinne	American triangle spider
Hyptiotes paradoxus	Europäische Dreieckspinne	European triangle spider, triangle weaver
Hysterocrates gigas	Afrikanische Riesenvogelspinne	Cameroon red tarantula
Hystricidae	Altwelt-Stachelschweine	Old World porcupines
Hystricomorpha	Stachelschweinverwandte	porcupinelike rodents
Hystrix brachyura/ Acanthion brachyura	Kurzschwanz-Stachelschwein	Malayan porcupine
Hystrix cristata	Stachelschwein, Gewöhnliches Stachelschwein	African porcupine, crested porcupine
Hystrix hodgsoni	Kammloses Himalaya- Stachelschwein*	crestless Himalayan porcupine
Hystrix spp.	Stachelschweine, Eigentliche Stachelschweine	Old World porcupines

Ia io	Ostasiatischer Frühabendsegler	great evening bat
Iago omanensis	Großaugen-Hundshai, Großaugen-Marderhai	bigeye hound shark
Ibacus novemdentatus	Glattschaliger Bärenkrebs	smooth fan lobster
Ibidorhynchidae	Ibisschnabel	ibisbill
Icelus bicornis	Arktische Panzergroppe	two-horned sculpin
Icelus spatula	Spatelgroppe*	spatulate sculpin
Icerya purchasi	Australische Wollschildlaus, Orangenschildlaus	cottony cushion scale, fluted scale
Ichneumia albicauda	Weißschwanzichneumon	white-tailed mongoose
Ichneumonidae	Schlupfwespen, Echte Schlupfwespen	ichneumon flies, ichneumons
Ichnotropis spp.	Rauschuppen-Eidechse	rough-scaled lizard
Ichthyobdellidae/ Piscicolidae	Fischegel	fish leeches
Ichthyomys spp.	Fischratten	fish-eating rats, aquatic rats
Ichthyophiidae	Fischwühlen	fish caecilians
Ichthyophis glandulosus	Basilian-Wühle	Abungabung caecilian
Ichthyophis glutinosus	Ceylonwühle	Ceylon caecilian
Ichthyophis kohtaoensis	Kohtao-Wühle	Koh Tao Island caecilian
Ichthyophis spp.	Fischwühlen	Asian caecilians
Ichthyornithiformes	Fischvögel	fish birds
Ichthyosauria	Fischsaurier	fish-reptiles, ichthyosaurs (ocean-living reptiles)
Icosteus aenigmaticus	Lappenfisch*	ragfish
Ictailurus planiceps/ Felis planiceps	Flachkopfkatze	flat-headed cat
Ictaluridae	Amerikanische Zwergwelse, Katzenwelse	North American freshwater catfishes
Ictalurus catus	Weißer Katzenwels	fork-tailed catfish
Ictalurus furcatus	Blauer Katzenwels	blue catfish
Ictalurus melas	Schwarzer Katzenwels, Schwarzer Zwergwels	black bullhead
Ictalurus natalis	Gelber Katzenwels	yellow bullhead
Ictalurus nebulosus/ Ameiurus nebulosus	Amerikanischer Zwergwels, Brauner Zwergwels, Langschwänziger Katzenwels	horned pout, American catfish, brown bullhead (FAO), "speckled catfish"
Ictalurus punctatus	Getüpfelter Gabelwels	channel catfish
Icteria virens	Gelbbrust-Waldsänger	yellow-breasted chat
Icteridae	Stärlinge	troupials, New World blackbirds, meadowlarks
Icterus galbula	Baltimoretrupial	northern oriole
Icterus pectoralis	Tropfentrupial, Schwarzbrusttrupial	spot-breasted oriole
Icterus pustulatus	Piroltrupial	streak-backed oriole
Icterus spurius	Gartentrupial	orchard oriole
Ictinaetus malayensis	Malaienadler	Indian black eagle
Ictonyx striatus	Zorilla	zorilla, striped polecat
Idaea aversata	Mausohrspanner	riband wave (moth)

Idaea contiguaria	Fetthennen-Felsflur-Kleinspanner	Weaver's wave (moth)
Idaea degeneraria	Veränderlicher Magerrasen-Kleinspanner	Portland ribbon wave (moth)
Idaea dilutaria	Punktierter Welklaub-Kleinspanner	silky wave (moth)
Idaea serpentata	Rostgelber Magerrasen-Kleinspanner	ochraceous wave (moth)
Idiacanthidae	Schwarze Drachenfische	sawtail-fishes
Idiopterus nephrelepidis	Farnblattlaus	fern aphid
Idiurus spp.	Gleitbilche	pygmy scaly-tailed flying squirrel
Idotea baltica	Baltische Meerassel, Ostsee-Meerassel	Baltic isopod, Baltic Sea "centipede"
Idotea chelipes	Klappenassel, Krallenfüßige Meerassel	clawfooted marine isopod
Idotea granulosa	Körnige Meerassel	granular marine isopod*
Idotea linearis	Stabförmige Meerassel	rod-shaped marine isopod*
Idotea metallica	Metallische Meerassel	metallic marine isopod
Iguana delicatissima	Grüner Inselleguan	delicate green Carribean Island iguana, lesser Antillean iguana (IUCN)
Iguana iguana	Grüner Leguan	green iguana, common iguana
Iguanidae	Leguane	iguanas
Ilia nucleus	Mittelmeer-Kugelkrabbe, Mittelmeer-Nusskrabbe	leucosian nut crab, Mediterranean nut crab
Ilisha africana	Westafrikanische Ilisha	West African ilisha
Ilisha elongata	Schlank-Ilisha	elongate ilisha
Illadopsis rufipennis	Grauwangen-Buschdrossling	pale-breasted thrush babbler
Illex argentinus	Argentinischer Kurzflossenkalmar	Argentine shortfin squid
Illex coindetii	Roter Kalmar, Breitschwanz-Kurzflossenkalmar	southern shortfin squid, broadtail shortfin squid
Illex illecebrosus	Nördlicher Kurzflossenkalmar	northern shortfin squid (FAO), common shortfin squid, northern squid, boreal squid
Illex oxygonius	Pfeilflossenkalmar	sharptail shortfin squid, arrow-finned squid
Ilybius fuliginosus/ Ilybius fenestratus	Schlammschwimmer	
Ilyocoris cimicoides	Schwimmwanze	saucer bug
Ilyophidae/Ilyophinae	Schlickaale	arrowtooth eels, mustard eels
Imantodes cenchoa	Riemennatter	blunt-headed tree snake
Inachis io/Nymphalis io	Tagpfauenauge	peacock moth, peacock
Inachus dorsettensis	Kurzkopfige Gespensterkrabbe	scorpion spider crab
Inachus leptochirus	Dünnbeinige Gespensterkrabbe	slender-legged spider crab
Inachus phalangium	Anemonen-Gespenstkrabbe, Mittelmeer-Gespenstkrabbe	Leach's spider crab, Mediterranean spider crab
Incurvaria capitella	Johannisbeermotte	currant bud moth
Incurvaria koerneriella	Schildkrötenmotte	tortoise moth*
Incurvaria rubiella	Himbeermotte, Himbeerschabe	raspberry bud moth

Incurvariidae	Miniersackmotten	yucca moths & fairy moth and others
Indicator indicator	Schwarzkehl-Honiganzeiger	black-throated honeyguide
Indicator xanthonotus	Gelbbürzel-Honiganzeiger	Indian honeyguide
Indicatoridae	Honiganzeiger	honeyguides
Indopacetus pacificus	Indopazifischer Schnabelwal	Indo-Pacific beaked whale
Indotestudo forsteni/ Geochelone travancorica	Travancore-Spornschildkröte	Travancore tortoise
Indotestudo elongata elongata/ Geochelone elongata elongata	Gelbkopf-Landschildkröte	yellow tortoise, elongated tortoise (IUCN)
Indri indri	Indri	indri
Inermiidae	Inermiiden	bonnetmouths
Inia geoffrensis	Amazonas-Delphin, Butu, Inia	Amazon dolphin, Amazon River dolphin, bouto, boutu
Inpaichthys kerri	Königssalmler	blue emperor
Invertebrata	Wirbellose, Wirbellose Tiere, Invertebraten, Evertebraten	invertebrates
Iomys horsfieldi	Horsefields Flughörnchen	Horsfield's flying squirrel
Iothia fulva	Gelbbraune Napfschnecke	tawny limpet
Iphiclides podalirius	Segelfalter	scarce swallowtail, kite swallowtail
Ipnopidae	Netzaugenfische	ipnopids
Ips cembrae	Großer Lärchenborkenkäfer, Großer Achtzähniger Lärchenborkenkäfer	larch bark beetle, Siberian fir bark-beetle
Ips sexdentatus	Großer Kiefernborkenkäfer	greater European pine engraver, six-toothed pine bark beetle (Br.)
Ips typographus	Buchdrucker, Großer Borkenkäfer, Achtzähniger Borkenkäfer	engraver beetle, common European engraver, spruce bark beetle (Br.)
Irania gutturalis	Weißkehlsänger	irania
Ircinia campana	Vasenschwamm	vase sponge, bell sponge
Ircinia fasciculata	Krustenlederschwamm	stinker sponge
Ircinia muscarum	Schwarzer Lederschwamm	black sponge*
Ircinia strobilina	Schwarzer Schwammball	loggerhead sponge, "cake sponge"
Ircinia variabilis	Variabler Lederschwamm	variable loggerhead sponge
Iredalina mirabilis	Goldene Walzenschnecke	golden volute
Irenidae	Blattvögel, Feenvögel	fairy-bluebirds, leafbirds
Irenomys tarsalis	Chile-Ratte	Chilean rat
Iridomyrmex humilis	Argentinische Ameise	Argentine ant
Irus irus/Venerupis irus	Irusmuschel	irus clam
Isacia conceptionis	Südostpazifische Süßlippe	Southeast Pacific grunt
Isaurus tuberculatus	Schlangenpolyp*	snake polyps
Ischnochiton albus/ Stenosemus albus (Polyplacophora)	Weiße Käferschnecke, Helle Käferschnecke, Weißer Chiton	northern white chiton, white chiton, white northern chiton

Ischnochiton contractus (Polyplacophora)	Gitterchiton	contracted chiton
Ischnochiton erythronotus (Polyplacophora)	Gesprenkelte Käferschnecke*, Gesprenkelte Chiton*	multihued chiton
Ischnodemus sabuleti	Getreidewanze, Europäische Getreidewanze, Schmalwanze	European chinch bug
Ischnura elegans	Große Pechlibelle	common ischnura, blue-tailed damselfly
Ischnura pumilio	Kleine Pechlibelle	lesser ischnura, scarce blue-tailed damselfly
Ischyropsalididae	Schneckenkanker	
Isidella elongata	Weiße Koralle	white coral*
Isis hippuris	Goldener Seefächer, Königsgliederkoralle	golden sea fan
Isistius brasiliensis	Ausstecherhai, Keksausstecherhai, Keksstecherhai	cookiecutter
Isocrinida	Zirrentragende Seelilien	sea lilies with cirri
Isodictya palmata	Palmenschwamm*	common palmate sponge
Isognomon alatus	Flache Baumauster*	flat tree-oyster
Isognomon bicolor	Zweifarben-Baumauster*	bicolor purse-oyster
Isognomon janus	Dünnschalige Baumauster	thin purse-oyster
Isognomon radiatus	Strahlenauster*, Strahlenförmige Baumauster*	radial purse-oyster
Isognomon recognitus	Westliche Baumauster*	purple purse-oyster, western tree-oyster (purse shell)
Isognomostoma isognomostoma	Maskenschnecke	
Isognomostoma holosericum/ Causa holosericum	Genabelte Maskenschnecke	
Isonidae/ Notocheiridae	Flügelährenfische	surf sprites
Isoodon auratus	Goldener Kurznasenbeutler	golden bandicoot
Isoodon spp. & *Thylacis* spp.	Kurznasenbeutler	short-nosed bandicoots
Isophya kraussi	Plumpschrecke	speckled bushcricket
Isophya pyrenea	Große Plumpschrecke	large speckled bushcricket
Isophyllia multiflora	Kleine Kaktuskoralle	lesser cactus coral
Isophyllia sinuosa	Gewellte Kaktuskoralle	sinuous cactus coral
Isopoda	Asseln	isopods (incl. pill bugs, woodlice, sowbugs)
Isothrix spp.	Toros	toros
Isotoma nivalis	Schneefloh	snow flea, snow springtail
Isotoma saltans	Gletscherfloh	glacier flea
Isotoma viridis	Grüner Springschwanz	green springtail
Isotomidae	Gleichringler (Springschwänze)	isotomid springtails
Isotomurus palustris	Moorspringer	marsh springtail
Issoria lathonia/ Argynnis lathonia	Kleiner Perlmutterfalter	Queen of Spain fritillary
Istiblennius edentulus	Doppelbinden-Kammzähner	rippled rockskipper
Istiblennius periophthalmus	Kugelkopf-Kammzähner	red-dotted blenny

Istiophoridae	Segelfische	sailfishes, billfishes (incl. spearfishes & marlins)
Istiophorus albicans	Atlantischer Fächerfisch, Atlantischer Segelfisch	Atlantic sailfish
Istiophorus platypterus	Indopazifischer Fächerfisch, Indopazifischer Segelfisch	Indo-Pacific sailfish
Isuridae	Makrelenhaie	mackerel sharks
Isurus oxyrinchus/ Isurus oxyrhynchus/ Isurus glaucus	Kurzflossen-Mako	shortfin mako (FAO), mako shark, mako
Isurus paucus	Langflossen-Mako	longfin mako
Ithaginis cruentus	Blutfasan	blood pheasant
Ixobrychus minutus	Zwergdommel	little bittern
Ixodes canisuga	Hundezecke	dog tick, British dog tick
Ixodes hexagonus	Igelzecke	hedgehog tick
Ixodes holocyclus	Australische Lähmezecke*	Australian paralysis tick
Ixodes pacificus	Kalifornischer Holzbock	California black-eyed tick, western black-legged tick
Ixodes persulcatus	Taigazecke	Taiga tick
Ixodes ricinus	Holzbock, Gemeiner Holzbock	castor bean tick, European castor bean tick, European sheep tick
Ixodes rubicundus	Südafrikanische Lähmezecke*	South African paralysis tick
Ixodes scapularis/ Ixodes dammini	Hirschzecke	deer tick, black-legged tick
Ixodidae	Schildzecken (Holzböcke)	hard ticks, hardbacked ticks
Ixodides	Zecken	ticks
Ixonotus guttatus	Fleckenbülbül	spotted greenbul
Ixoreus naevius	Halsbanddrossel	varied thrush

Jaapiella medicaginis	Luzerneblatt-Gallmücke	lucerne leaf midge
Jacamerops aurea	Breitmaul-Glanzvogel	great jacamar
Jacanidae	Blatthühnchen	jacanas, lily-trotters
Jacquinotia edwardsii	Südliche Seespinne	southern spider crab
Jaculus spp.	Wüstenspringmäuse	desert jerboas
Janthina exigua	Zwerg-Veilchenschnecke	dwarf janthina
Janthina golbosa	Kuglige Veilchenschnecke	globular janthina, elongate janthina
Janthina janthina	Veilchenschnecke, Floßschnecke	large violet snail, common purple sea snail, violet seasnail, common janthina
Janthina nitens	Mittelmeer-Veilchenschnecke	Mediterranean purple sea snail
Janthina pallida	Blasse Veilchenschnecke	pallid janthina
Janthinidae	Veilchenschnecken	violet snails
Janus compressus	Birnentriebwespe	pear stem sawfly
Janus luteipes	Weidentriebwespe	willow stem sawfly
Japalura spp.	Bergagamen	mountain lizards
Jasus edwardsii	Neuseeländische Languste	red rock lobster
Jasus frontalis	Juan-Fernandez-Languste	Australian spiny lobster, Juan Fernandez rock lobster
Jasus lalandei	Kap-Languste, Afrikanische Languste	Cape rock crawfish, Cape rock lobster
Jasus novaehollandiae	Austral-Languste	Australian rock lobster
Jasus tristani	Tristan Languste	Tristan rock lobster
Jasus verreauxi	Ostaustralische Languste	green rock lobster
Javesella pellucida/ Liburnia pellucida	Glasflügelzikade (Wiesenzirpe)	cereal leafhopper
Jenneria pustulata (Ovulidae)	Falsche Noppenkauri*, Pustel-„Kauri"	pustuled cowrie, pustulose false cowrie
Jenynsia lineata	Linienkärpfling	onesided livebearer
Jodia croceago/ Xanthia croceago	Safran-Wintereule, Safraneule, Eichenbuschwald-Safraneule	orange upperwing moth
Jordanella floridae	Floridakärpfling, Amerikanischer Flaggenkilli	flagfish, American flagfish
Jordanella pulchra	Yucatankärpfling, Yucatan-Flaggenkilli	Yucatan flagfish
Jorunna tomentosa	Graue Sternschnecke	grey sea slug*
Joturus pichardi	Bobo-Meeräsche	bobo mullet
Jujubinus exasperatus	Kegelige Kreiselschnecke	rough topsnail
Jujubinus striatus	Gestreifte Kreiselschnecke, Gefurchte Kreiselschnecke	grooved topsnail
Julidae	Schnurfüßer	juliform millepedes
Julidochromis ornatus	Gelber Schlankcichlide	ornate julie, julie ornatus, golden julie (FAO)
Julidochromis transcriptus	Schwarzweißer Schlankcichlide	black-and-white julie, masked julie (FAO)
Junceela fragilis	Zerbrechliche Seepeitsche, Helle Seepeitsche	fragile sea whip
Junceela juncea	Seepeitsche u.a., Peitschen-Gorgonie u.a.	sea whip a.o.
Junco hyemalis	Junko	dark-eyed junco
Juscelinomys candango	Brasilia Grabmaus	Brasilia burrowing mouse
Jynx torquilla	Wendehals	wryneck

Kachuga dhongoka	Dreistreifen-Dachschildkröte	three-striped roof turtle, three-stripe roofed turtle
Kachuga kachuga	Bunte Dachschildkröte, Bengalische Dachschildkröte	red-crowned roofed turtle (IUCN), painted roofed turtle, Bengal roof turtle
Kachuga spp.	Dachschildkröten	roofed turtles
Kachuga tecta	Indische Dachschildkröte	Indian tent turtle, Indian roofed turtle
Kachuga tentoria	Mittel-Indische Dachschildkröte	Mid-Indian tent turtle
Kachuga trivittata	Burma-Dachschildkröte	Burmese roofed turtle (IUCN)
Kakothrips pisivorus	Erbsenblasenfuß	pea thrips
Kalotermitidae	Trockenholztermiten	drywood termites & powderpost termites
Kaltenbachiola strobi	Fichtenzapfenschuppen-Gallmücke	spruce cone gall midge
Kannabateomys amblyonyx	Bambus-Fingerratte	rato de Taquara
Kassina maculata/ Hylambates maculata	Flecken-Rennfrosch	spotted running frog
Kassina senegalensis	Senegal-Rennfrosch, Senegal-Streifenfrosch	Senegal running frog
Kassina spp.	Rennfrösche	running frogs
Katsuwonus pelamis/ Euthynnus pelamis	Gestreifter Thun, Echter Bonito	skipjack tuna (FAO), bonito, stripe-bellied bonito
Kaupifalco monogrammicus	Sperberbussard	lizard buzzard
Kerivoula spp.	Wollfledermäuse	painted bats, woolly bats
Kermes quercus	Eichenschildlaus	oak scale
Kermes spp.	Färberschildläuse	kermes coccids, kermes scales
Kermes vermilio	Kermeslaus	Mediterranean kermes coccid
Kermidae/Kermesidae	Eichenschildläuse, Eichennapfläuse	gall-like coccids
Kerodon rupestris	Felsenmeerschweinchen, Moko	rock cavy, moco
Kinixys belliana	Glattrand-Gelenkschildkröte	Bell's hingeback tortoise
Kinixys erosa	Stachelrand-Gelenkschildkröte	serrated hinged tortoise, serrated hingeback tortoise, common hinged tortoise, eroded hingeback tortoise
Kinixys homeana	Stutz-Gelenkschildkröte	Home's hingeback tortoise
Kinixys natalensis	Natal-Gelenkschildkröte	Natal hingeback tortoise
Kinixys spp.	Gelenkschildkröten	hinge-backed tortoises, hingeback tortoises, hinged tortoises
Kinorhyncha	Hakenrüssler	kinorhynchs
Kinosternidae	Schlammschildkröten	American mud and musk turtles
Kinosternon depressum/ Sternotherus depressus	Flache Moschusschildkröte	flattened musk turtle
Kinosternon minor/ Sternotherus minor	Zwerg-Moschusschildkröte	loggerhead musk turtle
Kinosternon odoratum/ Sternotherus odoratus	Gewöhnliche Moschusschildkröte	common musk turtle, stinkpot
Kinosternon acutum	Tabasco-Klappschildkröte	Tabasco mud turtle
Kinosternon alamosae	Alamos-Klappschildkröte	Alamos mud turtle
Kinosternon angustipons	Schmalkiel-Klappschildkröte	narrow-bridged mud turtle

Kinosternon baurii baurii	Dreistreifen-Klappschildkröte	striped mud turtle
Kinosternon creaseri	Yucatan-Klappschildkröte	Creaser's mud turtle
Kinosternon flavescens flavescens	Gelbe Klappschildkröte, Gelbliche Klappschildkröte	yellow mud turtle
Kinosternon hirtipes hirtipes	Raufuß-Klappschildkröte	Mexican rough-footed mud turtle, Mexican mud turtle
Kinosternon integrum	Mexikanische Moschusschildkröte	Mexican musk turtle
Kinosternon leucostomum	Weißmaul-Klappschildkröte	white-lipped mud turtle
Kinosternon scorpioides scorpioides	Skorpions-Klappschildkröte	scorpion mud turtle
Kinosternon sonoriense	Sonora-Klappschildkröte	Sonora mud turtle
Kinosternon spp.	Klappschildkröten	mud turtles
Kinosternon subrubrum subrubrum	Pennsylvania-Klappschildkröte	mud turtle, common mud turtle, eastern mud turtle
Kissophagus hederae	Efeuborkenkäfer	ivy bark beetle
Klauberina riversiana/ Xantusia riversiana	Insel-Nachtechse	Island night lizard
Kleidocerys resedae	Birkenwanze, Birkwanze	birch bug
Knemidokoptes gallinae	Federabwurfmilbe*	depluming mite
Knemidokoptes mutans	Vogelschuppenbeinmilbe*	scalyleg mite (of fowl)
Knemidokoptes pilae	Wellensittich-Schuppenbeinmilbe*	scalyleg mite (of budgerigars)
Kneriidae	Ohrenfische, Schlankfische	kneriids
Knipowitschia longecaudata	Langschwanz-Grundel, Langschwanzgrundel	long-tail goby
Knipowitschia panizzae	Lagunen-Grundel	lagoon goby
Knulliana cincta		banded hickory borer, belted chion beetle
Kobus ellipsiprymnus	Wasserbock, Hirschantilope	waterbuck
Kobus kob	Moorantilope, Kob-Antilope	kob
Kobus leche	Litschi-Wasserbock, Litschi-Moorantilope	lechwe
Kobus megaceros	Weißnacken-Moorantilope, Nil-Lechwe	Nile lechwe
Kobus vardoni	Puku-Antilope	puku
Kogia breviceps	Zwergpottwal	pygmy sperm whale, lesser cachalot
Kogia simus	Pazifik-Pottwal, Kleinpottwal	dwarf sperm whale
Komodomys rintjanus	Komodo-Ratte	Komodo rat
Korscheltellus lupulinus/ Hepialus lupulina	Wurzelspinner, Kleiner Hopfenwurzelbohrer	common swift, garden swift moth
Korynetes coeruleus	Blauer Fellkäfer	blue ham beetle*
Korynetidae/Corynetidae (>Cleridae)	Jagdraubkäfer, Fellkäfer	ham beetles
Kraemeriidae	Lanzenfische	sand darts, sandfishes, sand gobies
Kryptophaneron alfredi	Atlantik-Laternenfisch	Atlantic flashlightfish

Kryptopterus bicirrhis	Indischer Glaswels	glass catfish (FAO), ghost catfish
Kryptopterus macrocephalus	Phantom-Glaswels	poor man's glass catfish, striped glass catfish (FAO)
Kuhliidae/Duleidae	Kuhlien, Fahnenschwänze	flagtails (aholeholes)
Kunsia spp.	Südamerikanische Riesenratten	South American giant rats
Kurtidae	Kurter	nurseryfishes
Kyarranus sphagnicolus	Sphagnumfrosch, Torfmoosfrosch	sphagnum frog
Kyphosidae	Pilotbarsche, Steuerbarsche	sea chubs
Kyphosus bigibbus	Großer Pilotbarsch	grey sea chub (FAO), grey chub, buffalo bream
Kyphosus sectatrix	Gelblinien Pilotbarsch, Bermuda-Ruderfisch	Bermuda chub, Bermuda sea chub (FAO)

*L*abeo

Labeo bicolor/ *Epalzeorhynchus bicolor*	Feuerschwanz	redtail sharkminnow (FAO), red-tailed "shark", red-tailed black "shark"
Labeo cylindricus	Rotaugen-Fransenlipper	redeye labeo
Labeo spp.	Fransenlipper	labeos
Labia minor	Kleiner Ohrwurm, Zwergohrwurm	lesser earwig, little earwig
Labidesthes siccula	Rotmaul-Ährenfisch	brook silverside
Labidura bidens	Gestreifter Ohrwurm	striped earwig
Labidura riparia	Sandohrwurm, Dünenohrwurm	tawny earwig, giant earwig
Labiduridae	Langhorn-Ohrwürmer	long-horned earwigs, striped earwigs
Labiidae	Kleine Ohrwürmer	little earwigs
Labiobarbus festivus	Signalbarbe	signal barb (FAO), sailfin shark
Labridae	Lippfische	wrasses
Labroides dimidiatus	Putzerlippfisch	cleaner wrasse
Labroides spp.	Putzerfische, Putzerlippfische	cleaner wrasses
Labrus bergylta	Gefleckter Lippfisch	ballan wrasse
Labrus bimaculatus/ *Labrus mixtus*	Kuckuckslippfisch	cuckoo wrasse
Labrus merula	Brauner Lippfisch	brown wrasse
Labrus viridis	Grüner Lippfisch, Amsellippfisch, Meerdrossel	green wrasse
Lacanobia oleracea/ *Mamestra oleracea*	Gemüseeule	tomato moth (bright-line brown-eye moth)
Laccifer lacca	Indische Lackschildlaus	Indian lac insect
Lacciferidae (Tachardiidae)	Lackschildläuse	lac insects
Lacerta perspicillata/ *Podarcis perspicillata*	Brilleneidechse	Moroccan rock lizard
Lacerta agilis	Zauneidechse	sand lizard
Lacerta bedriagae	Bedriagas Gebirgseidechse, Tyrrhenische Gebirgseidechse	Bedriaga's rock lizard
Lacerta echinata	Stachelschwanzeidechse	Cope's spinytail lizard
Lacerta graeca	Griechische Felseidechse	Greek rock lizard
Lacerta horvathi	Kroatische Gebirgseidechse	Horvath's rock lizard
Lacerta laevis	Libanon-Eidechse	Lebanon lizard
Lacerta lepida	Perleidechse	ocellated lizard, ocellated green lizard, eyed lizard, jewelled lizard
Lacerta monticola	Iberische Gebirgseidechse	Iberian rock lizard
Lacerta mosorensis	Mosor-Eidechse	Mosor rock lizard
Lacerta muralis/ *Podarcis muralis*	Mauereidechse	common wall lizard
Lacerta oxycephala	Spitzkopfeidechse, Dalmatinische Spitzkopfeidechse	sharp-snouted rock lizard
Lacerta parva	Zwerg-Zauneidechse	dwarf lizard
Lacerta praticola	Wieseneidechse	pontic lizard, meadow lizard
Lacerta princeps	Zagros-Eidechse	Zagrosian lizard

Lacerta saxicola	Felseidechse	rock lizard
Lacerta schreiberi	Iberische Smaragdeidechse, Schreibers Smaragdeidechse	Schreiber's green lizard
Lacerta strigata	Kaspische Smaragdeidechse, Streifensmaragdeidechse	Caspian green lizard, Caucasus emerald lizard
Lacerta taurica/ Podarcis taurica	Krimeidechse	Crimean wall lizard, Crimean lizard
Lacerta trilineata	Riesen-Smaragdeidechse	Balkan green lizard, Balkan emerald lizard
Lacerta viridis	Smaragdeidechse	green lizard, emerald lizard
Lacerta vivipara	Waldeidechse, Bergeidechse, Mooreidechse	viviparous lizard, European common lizard
Lacertidae/Lacertilia (Squamata)	Eidechsen	lizards
Lachesilla pedicularia	Gemeine Getreidestaublaus	cosmopolitan grain psocid
Lachesis mutus/ Lachesis muta	Buschmeister	bushmaster
Lachesis spp. (see also: *Bothrops/Porthidium*)	Lanzenottern	American lance-headed vipers
Lachnidae	Baumläuse, Rindenläuse	lachnids, lachnid plantlice
Lachnolaimus maximus	Schweinsfisch, Westatlantischer Lippfisch	hogfish (FAO), hogsnapper
Laciniaria biplicata	Zweifaltenrandige Schließmundschnecke	two-lipped door snail
Laciniaria plicata	Faltenrandige Schließmundschnecke	single-lipped door snail*
Lactarius lactarius	Lactarius	false trevally (FAO), white fish
Lactoria cornuta	Langhorn-Kofferfisch	longhorn cowfish
Lacuna crassior	Dickschalige Grübchenschnecke	thick lacuna, thick chink snail
Lacuna marmorata	Gefleckte Grübchenschnecke	chink snail
Lacuna pallidula	Flache Grübchenschnecke	pallid lacuna, pallid chink snail
Lacuna parva	Kleine Grübchenschnecke	tiny lacuna, small chink snail, least chink snail
Lacuna spp.	Grübchenschnecken	lacuna snails, chink snails
Lacuna vincta/ Lacuna divaricata	Gebänderte Grübchenschnecke	northern lacuna, banded chink snail
Lacunidae	Grübchenschnecken, Lacuniden	lacuna snails, chink snails (chink shells)
Ladoga camilla/ Limenitis camilla	Kleiner Eisvogel	white admiral
Laelaps echidnina	Stachlige Rattenmilbe	spiny rat mite
Laemanctus longipes	Östlicher Helmkopfbasilisk	eastern casquehead iguana
Laemanctus serratus	Helmkopfbasilisk	serrated casquehead iguana
Laemanctus spp.	Helmkopfbasilisken	casquehead iguana
Laemostenus terricola/ Pristonychus terricola	Kellerkäfer, Grotten-Dunkelkäfer	cellar beetle*, European cellar beetle
Laetacara curviceps	Tüpfelbuntbarsch	flag acara, flag cichlid (U.S.)

Laevicardium attenuatum	Schlanke Herzmuschel	attenuated cockle
Laevicardium crassum	Norwegische Herzmuschel	Norway cockle, Norwegian cockle
Laevicardium elatum	Pazifische Riesen-Herzmuschel	giant eggcockle, giant Pacific eggcockle
Laevicardium laevigatum	Atlantische Herzmuschel	eggcockle
Laevicardium oblongum	Lange Herzmuschel	oblong cockle
Lagenodelphis hosei	Borneo-Delphin, Kurzschnabeldelphin, Frasers Delphin	shortsnouted whitebelly dolphin, Fraser's dolphin, Sarawak dolphin, Bornean dolphin
Lagenorhynchus acutus	Weißseitendelphin	Atlantic white-sided dolphin
Lagenorhynchus albirostris	Weißschnauzendelphin	white-beaked dolphin
Lagenorhynchus australis	Süddelphin, Schwarzkinndelphin, Peales Delphin	Peale's dolphin, Peale's black-chinned dolphin, blackchin dolphin, southern dolphin
Lagenorhynchus cruciger	Kreuzdelphin, Sanduhrdelphin	hourglass dolphin, southern white-sided dolphin
Lagenorhynchus obliquidens	Weißstreifendelphin, Pazifischer Weißseiten-Delphin	Pacific white-sided dolphin
Lagenorhynchus obscurus	Dunkler Delphin	Gray's dusky dolphin
Lagenorhynchus spp.	Flaschenschnabel-Delphine	white-sided dolphins
Lagidium spp.	Hasenmäuse	mountain viscachas
Lagis koreni/ Amphictene koreni/ Pectinaria koreni	Köcherwurm	
Lagocephalus lagocephalus	Atlantischer Kofferfisch	Atlantic pufferfish, pufferfish, oceanic puffer (FAO)
Lagonosticta senegala	Senegalamarant	red-billed fire finch
Lagopus lagopus	Moorschneehuhn	willow grouse
Lagopus mutus	Alpenschneehuhn	ptarmigan
Lagorchestes spp.	Hasenkängurus	hare wallabies
Lagostomus maximus	Feld-Viscacha, Pampas-Viscacha (Hbz. Viscacha)	plains viscacha
Lagostrophus fasciatus	Bänderkänguru	banded hare wallaby, munning
Lagothrix flavicauda	Gelbschwanzwollaffe	yellow-tailed woolly monkey
Lagothrix lagotricha	Wollaffe	common woolly monkey
Lagothrix spp.	Wollaffen	woolly monkeys
Lagria hirta	Gemeiner Wollkäfer	
Lagriidae	Wollkäfer	long-jointed beetles
Lagurus curtatus/ Lemmiscus curtatus	Beifuß-Steppenlemming	sagebrush vole
Lagurus lagurus	Graulemming	steppe lemming
Lagurus luteus/ Eolagurus luteus	Gelblemming	yellow steppe lemming
Lagurus spp.	Steppenlemminge	sagebrush voles, steppe lemmings
Lalage sueurii	Weißflügellalage	white-winged triller
Lama glama	Lama	llama
Lama guanicoe	Guanako	guanaco
Lama pacos	Alpaka	alpaca

Lambis arthritica	Arthritische Spinnenschnecke	arthritic spider conch
Lambis chiragra	Chiragra-Spinnenschnecke, Großer Bootshaken	chiragra spider conch, gouty spider conch
Lambis crocata	Rote Spinnenschnecke, Orange Spinnenschnecke	orange spider conch
Lambis digitata	Langgestreckte Spinnenschnecke	finger spider conch, elongate spider conch
Lambis lambis	Krabben-Fechterschnecke	common spider conch, smooth spider conch
Lambis millepeda	Tausenfüßlerschnecke	milleped spider conch
Lambis scorpius	Skorpionschnecke	scorpion spider conch
Lambis spp.	Spinnenschnecken, Spinnen-Fechterschnecken	spider conch snails
Lambis truncata	Große Teufelskralle, Gemeine Spinnenschnecke	giant spider conch, wild-vine root
Lambis violacea	Violette Spinnenschnecke	violet spider conch
Lamellaria perspicua	Durchsichtige Blättchenschnecke	transparent lamellaria
Lamellariinae	Blättchenschnecken u.a.	lamellarias, ear snails (ear shells)
Lamellaxis micra	Kleine Turmschnecke*	tiny awlsnail
Lamellitrochus lamellosus	Blättrige Sonnenschnecke	lamellose solarelle
Lamia textor	Weberbock	
Laminosioptes cysticola	Geflügel-Zystenmilbe*	fowl cyst mite, flesh mite, subcutaneous mite
Lamna ditropis	Pazifischer Heringshai	Pacific porbeagle, salmon shark (FAO)
Lamna nasus	Heringshai	porbeagle (FAO), mackerel shark
Lamnidae	Heringshaie	mackerel sharks
Lampanyctodes hectoris	Hektor-Laternenfisch	lanternfish
Lampetra appendix/ Lethenteron appendix	Amerikanischer Bachneunauge	American brook lamprey
Lampetra fluvialis	Flussneunauge	river lamprey, lampern, European river lamprey (FAO)
Lampetra planeri	Bachneunauge	brook lamprey, European brook lamprey (FAO)
Lampides boeticus	Langschwänziger Bläuling, Großer Wander-Bläuling	long-tailed blue (butterfly)
Lampra dives	Metallfarbener Weidenprachtkäfer	
Lampridae	Gotteslachse, Glanzfische	opahs
Lampriformes	Gotteslachsverwandte, Glanzfischartige, Glanzfische	moonfishes
Lamprohiza splendidula	Gemeiner Leuchtkäfer, Kleiner Leuchtkäfer, Johanniskäfer, Johanniswürmchen	small lightning beetle
Lampropeltis calligaster	Prärie-Königsnatter	prairie kingsnake
Lampropeltis getulus	Kettennatter, Östliche Kettennatter	common kingsnake
Lampropeltis mexicana	Trans-Peco-Königsnatter	Trans Peco kingsnake
Lampropeltis pyromelana	Berg-Königsnatter	mountain kingsnake

Lampropeltis spp.	Königsnattern	kingsnakes
Lampropeltis triangulum	Dreiecksnatter, Rote Königsnatter	milk snake, eastern milk snake
Lampropeltis zonata	Korallen-Königsnatter	coral kingsnake, California mountain kingsnake
Lamprophis fuliginosus/ Boaedon fuliginosus	Braune Hausschlange	common house snake, common brown house snake
Lamprophis lineatus/ Boaedon lineatus	Afrikanische Hausnatter, Gewöhnliche Hausschlange	African house snake, African brown house snake
Lamprophis spp./ *Boaedon* spp.	Boazähner, Hausschlangen, Hausnattern	house snakes
Lamprotornis splendidus	Prachtglanzstar	splendid glossy starling
Lampyridae	Leuchtkäfer (Glühwürmchen, Johanniswürmchen)	glowworms, fireflies, lightning "bugs"
Lampyris noctiluca	Großer Leuchtkäfer, Großes Glühwürmchen, Gelbhals-Leuchtkäfer	glowworm, glow-worm, great European glow-worm beetle
Langaha alluaudi	Süd-Blattnasennatter	southern leafnose snake
Langaha nasuta	Nord-Blattnasennatter	northern leafnose snake
Langaha spp.	Blattnasennattern	leafnose snakes
Languria mozardi	Kleestengelbohrer*	clover stem borer
Languriidae	Echsenkäfer*	lizard beetles
Lanice conchilega	Muschelsammlerin, Sandröhrenwurm, Bäumchenröhrenwurm	sand mason
Laniidae	Würger	shrikes, true shrikes
Lanius collaris	Fiskalwürger	fiscal shrike
Lanius collurio	Rotrückenwürger, Neuntöter	red-backed shrike
Lanius excubitor	Raubwürger	great grey shrike, northern shrike
Lanius isabellinus	Isabellwürger	isabelline shrike
Lanius ludovicianus	Louisianawürger	loggerhead shrike
Lanius minor	Schwarzstirnwürger	lesser grey shrike
Lanius nubicus	Maskenwürger	masked shrike
Lanius schach	Schachwürger	black-headed shrike
Lanius senator	Rotkopfwürger	woodchat shrike
Lanthanotidae	Taubwaran	earless monitor
Lanthanotus borneensis	Borneo-Taubwaran	Bornean earless lizard, Borneo earless monitor
Laomedea geniculata/ Obelia geniculata	Zickzack-Glockenpolyp	knotted thread hydroid, zig-zag wine-glass hydroid
Laothoe populi	Pappelschwärmer	poplar hawkmoth
Lapemis curtus	Shaw-Seeschlange, Kurze Seeschlange, Plump-Seeschlange	Shaw's sea snake
Laricobius erichsoni	Gelbbrauner Lärchenkäfer	
Laridae	Möwen & Seeschwalben	gulls & terns
Larinae	Möwen	gulls
Larinioides sclopetarius/ Araneus sclopetarius	Brückenkreuzspinne, Brückenspinne	bridge orbweaver
Larinioides cornutus/ Araneus cornutus	Schilfradspinne	furrow orbweaver

Lariscus hosei	Vierstreifen-Erdhörnchen	four-striped ground squirrel
Lariscus insignis	Dreistreifen-Erdhörnchen	three-striped ground squirrel
Lariscus spp.	Streifen-Erdhörnchen	Malaysian striped ground squirrels
Lartetia spp.	Zwerghöhlenschnecke	
Larus argentatus	Silbermöve	herring gull
Larus atricilla	Aztekenmöve	laughing gull
Larus audouinii	Korallenmöve	Audouin's gull
Larus cachinnans	Weißkopfmöve	yellow-legged gull
Larus canus	Sturmmöve	common gull
Larus delawarensis	Ringschnabelmöve	ring-billed gull
Larus fuscus	Heringsmöve	lesser black-backed gull
Larus genei	Dünnschnabelmöve	slender-billed gull
Larus glaucoides	Polarmöve	Iceland gull
Larus hyperboreus	Eismöve	glaucous gull
Larus ichthyaetus	Fischmöve	great black-headed gull
Larus marinus	Mantelmöve	great black-backed gull
Larus melanocephalus	Schwarzkopfmöve	Mediterranean gull
Larus minutus	Zwergmöve	little gull
Larus philadelphia	Bonapartemöve	Bonaparte's gull
Larus pipixcan	Präriemöve	Franklin's gull
Larus relictus	Lönnbergmöve	relict gull
Larus ridibundus	Lachmöve	black-headed gull
Larus sabini	Schwalbenmöve	Sabine's gull
Lasaeidae	Münzmuscheln*	coin shells
Lasiocampa quercus	Eichenspinner, Quittenvogel	oak eggar
Lasiocampa trifolii/ Pachygastria trifolii	Kleespinner	grass eggar
Lasiocampidae	Glucken, Wollraupenspinner	lackeys & eggars, lappet moths (tent caterpillars)
Lasioderma serricorne	Kleiner Tabakkäfer	cigarette beetle (tobacco beetle)
Lasiodora parahybana	Parahyba-Vogelspinne, Brasilien-Riesenvogelspinne	Brazilian salmon tarantula
Lasiommata maera	Braunauge, Rispenfalter	large wall brown, wood-nymph
Lasiommata megera	Mauerfuchs	wall, wall brown
Lasiommata petropolitana	Braunscheckauge	northern wall brown
Lasionycteris noctivagans	Silberhaarige Fledermaus	silver-haired bat
Lasioptera rubi	Brombeersaummücke, Himbeergallmücke	blackberry stem gall midge, raspberry stem gall midge
Lasiorhinus spp.	Haarnasenwombats	hairy-nosed wombats, soft-furred wombats
Lasiurus borealis	Rote Fledermaus	red bat
Lasiurus cinereus	Silberfledermaus	hoary bat
Lasiurus ega	Gelbe Fledermaus	yellow bat
Lasiurus spp.	Nordamerikanische Glattnasen	hairy-tailed bats
Lasius brunneus	Braune Wegameise	brown ant
Lasius flavus	Gelbe Wiesenameise, Gelbe Wegameise	mound ant, yellow turf ant, yellow ant, yellow meadow ant

Lasius fuliginosus	Glänzendschwarze Holzameise, Schwarze Holzameise, Kartonnestameise	jet ant, shining jet black ant
Lasius niger	Schwarze Gartenameise, Schwarzgraue Wegameise, Schwarzbraune Wegameise	black ant, common black ant, garden ant
Lasius spp.	Wiesenameisen & Holzameisen & Wegameisen	field ants (black ants)
Laspeyresia funebrana/ Cydia funebrana	Pflaumenwickler	plum fruit moth, plum moth, red plum maggot
Laspeyresia nigricana/ Cydia nigricana/ Cydia rusticella	Erbsenwickler, Olivbrauner Erbsenwickler	pea moth
Laspeyresia pomonella/ Cydia pomonella/ Carpocapsa pomonella	Apfelwickler (Apfelmade/Obstmade)	apple moth, codling moth, codlin moth
Lateolabrax japonicus	Japanischer Barsch	Japanese seabass
Laternula anatina	Entenlaterne	duck lantern clam
Laternulidae	Laternenmuscheln	lantern clams
Lates calcarifer	Barramundi	barramundi (FAO), giant sea perch
Lates niloticus	Nilbarsch (Viktoriabarsch)	Nile perch (Sangara)
Lathamus discolor	Schwalbensittich	swift parrot
Latheticus oryzae	Rundköpfiger Reismehlkäfer	longheaded flour beetle
Lathridiidae	Moderkäfer	plaster beetles, minute brown scavenger beetles
Lathridius nodifer	Schwarzbrauner Rippenmoderkäfer	black-brown plaster beetle*, black-brown fungus beetle
Lathrimaeum atrocephalum	Schwarzköpfiger Rindenkurzflügler	
Lathrobium fulvipenne	Rotbrauner Uferkurzflügler	
Latiaxis mawae	Mawes Latiaxis	Mawe's latiaxis
Laticauda colubrina	Nattern-Plattschwanz	yellow-lipped sea krait, yellow-lipped sea snake, banded sea snake
Laticauda laticaudata	Gewöhnlicher Plattschwanz	common banded sea snake, black-banded sea snake
Laticauda semifasciata	Halbgebänderter Plattschwanz	Chinese sea snake (banded sea snake)
Laticauda spp.	Plattschwanz-Seeschlangen, Plattschwänze	banded sea snakes, sea kraits
Latimeria chalumnae	Komoren-Quastenflosser, Gombessa	gombessa
Latimeridae	Quastenflosser	coelacanths
Latirus angulatus	Kurzschwanz-Latirus*	short-tail latirus
Latirus carinifer	Rollenschnecke	yellow latirus, trochlear latirus
Latirus craticulatus	Rotgerippte Latirus*	red-ripped latirus
Latirus infundibulum	Braunbandschnecke	brown-line latirus
Latirus nodatus	Knotenschnecke*, Knotige Latirus*	knobbed latirus
Latreutes parvulus	Sargassumgarnele	sargassum shrimp
Latridae	Trompeterfische	trumpeters
Latridopsis ciliaris	Trompeterfisch	blue moki

Latrodectus bishopi	Rotbeinwitwe*, Rote Witwe*	red-legged widow, red widow spider
Latrodectus geometricus	Braune Witwe	brown widow spider
Latrodectus hesperus	West-Amerikanische Schwarze Witwe	western black widow spider
Latrodectus mactans hasselti	Katipo	katipo widow
Latrodectus mactans tredecimguttatus	Schwarze Witwe, Europäische Schwarze Witwe, Malmignatte	European black widow, southern black widow
Latrodectus mactans	Schwarze Witwe	European black widow (hourglass spider, shoe button spider, po-ko-moo spider)
Latrodectus spp.	Witwen	widow spiders
Latrodectus variolus	Nördliche Amerikanische Schwarze Witwe	northern black widow spider
Latrunculia magnifica	Prachtfeuerschwamm	magnificent fire sponge
Lauria cylindracea	Genabelte Puppenschnecke	chrysalis snail, common chrysalis snail
Lauxaniidae	Polierfliegen, Faulfliegen	lauxaniids
Lavia frons	Gelbflüglige Großblattnase	African yellow-winged bat
Leandria perarmata	Panzerchamäleon	armored chameleon
Leandrites cyrtorhynchus	Putzende Partnergarnele	cleaning anemone shrimp
Lebetus scorpioides	Skorpiongrundel*	diminutive goby
Lebiasinidae	Spitzsalmlerverwandte	lebiasinids (voladoras, pencil fishes)
Lebistes reticulatus/ Poecilia reticulata	Guppy, Millionenfisch	guppy
Lechriodus spp.	Kannibalenfrösche	cannibal frogs
Ledra aurita	Ohrenzikade, Ohrzikade	ear cicada*
Leggadina forresti	Forrest-Kleinmaus	Forrest's mouse
Leggadina lakedownensis	Lakeland Down-Kleinmaus	Lakeland Downs mouse
Leggadina spp.	Australische Kleinmäuse	Australian dwarf mice
Lehmannia flavus	Kellerschnecke, Bierschnegel, Gelber Schnegel, Gelbe Egelschnecke	tawny garden slug, yellow gardenslug, yellow slug (cellar slug, dairy slug, house slug)
Lehmannia marginata/ Limax marginatus	Baumschnegel, Baum-Egelschnecke	tree slug
Lehmannia valentiana	Spanische Egelschnecke, Valencianische Egelschnecke	threeband gardenslug, Iberian slug, Canadian slug
Leimacomys buettneri	Buettners Waldmaus	groove-toothed forest mouse, Buettner's forest mouse, Togo mouse
Leiocassis siamensis	Asiatischer Ringelwels	Asiatic bumblebee catfish
Leiocephalus carinatus	Rollschwanzleguan	curly-tailed lizard
Leiocephalus personatus	Maskenleguan	Haitian curlytail lizard
Leiocephalus schreibersii	Glattkopfleguan	red-sided curlytail lizard
Leiocephalus spp.	Glattkopfleguane	curly-tailed lizards, curlytail lizards, curly-tails
Leiodidae/Anisotomidae	Trüffelkäfer, Schwammkugelkäfer	round fungus beetles

Leiognathidae	Ponyfische, Schlupfmäuler	ponyfishes, slimys, slipmouths, soapies
Leiognathus klunzingeri	Klunzingers Ponyfisch	ponyfish
Leiognathus spp.	Ponyfische	ponyfishes
Leioheterodon madagascariensis	Menarana Madagaskar-Braunnatter	Madagascar Menarana snake
Leioheterodon modestus	Madagaskar-Braunnatter	Madagascar brown snake
Leiolopisma grande	Australischer Riesenskink	giant skink
Leiopathes glaberrima	Glatte Schwarzkoralle*	smooth black coral
Leiopelma hamiltoni	Hamilton-Frosch	Hamilton's frog
Leiopelma spp.	Neuseeländische Urfrösche	New Zealand frogs
Leiopelmidae/Leiopelmatidae	Neuseeländische Urfrösche	ribbed frogs, leiopelmids (tailed frogs & New Zealand frogs)
Leiopus nebulosus	Braungrauer Splintbock, Braungrauer Laubholzbock	
Leiostomus xanthurus	Punkt-Umberfisch	spot, spot croaker
Leiostyla anglica	Englische Puppenschnecke	English chrysalis snail
Leiostyla cylindracea	Gemeine Puppenschnecke, Zylindrische Puppenschnecke	common chrysalis snail
Leiothrix lutea	Sonnenvogel	Pekin robin
Leipoa ocellata	Thermometerhuhn	mallee fowl
Leirus quinquestriatus	Fünfstreifenskorpion	fivekeeled gold scorpion, African gold scorpion
Leistus ferrugineus	Gewöhnlicher Bartläufer, Rostfarbiger Bartkäfer	
Leistus rufomarginatus	Rotrandiger Bartläufer, Rotrandiger Bartkäfer	
Leistus spinibarbis	Dornbartkäfer, Blauer Bartläufer	
Lema melanopus/ *Oulema melanopus*	Rothalsiges Getreidehähnchen, Blatthähnchen	cereal leaf beetle (oat leaf beetle, barley leaf beetle)
Lema trilinea	Dreistreifen-Kartoffelkäfer	threelined potato beetle
Lemintina arenaria	Große Wurmschnecke	Mediterranean wormsnail
Lemmiscus curtatus/ *Lagurus curtatus*	Beifuß-Steppenlemming	sagebrush vole
Lemmus lemmus	Berglemming	Norway lemming
Lemmus spp.	Echte Lemminge	true lemmings
Lemniscomys spp.	Afrikanische Streifen-Grasmäuse	striped grass mice, zebra mice
Lemoniidae	Herbstspinner, Wiesenspinner	lemoniid moths
Lemur albifrons	Weißkopfmaki	white-headed lemur
Lemur catta	Katta	ring-tailed lemur
Lemur coronatus/ *Petterus coronatus*/ *Eulemur coronatus*	Kronenmaki	crowned lemur
Lemur fulvus/*Petterus fulvus*	Schwarzkopfmaki	brown lemur
Lemur macaco/ *Petterus macaco*	Mohrenmaki	black lemur
Lemur mongoz/ *Petterus mongoz*/ *Eulemur mongoz*	Mongozmaki	mongoose lemur
Lemur rubriventer/ *Petterus rubriventer*	Rotbauchmaki	red-bellied lemur
Lemuridae	Lemurenartige	lemurs

Lenomys meyeri	Sulawesi Riesenratte	trefoil-toothed giant rat, Sulawesi giant rat
Lenothrix canus	Graue Baumratte	gray tree rat
Lenoxus apicalis	Andenratte	Andean rat
Leontopithecus rosalia/ Leontideus rosalia	Löwenaffe, Goldgelbes Löwenäffchen, Goldenes Löwenäffchen	lion tamarin, golden lion tamarin
Leontopithecus chrysomelas	Goldkopf-Löwenäffchen	golden-headed lion tamarin, gold and black lion tamarin
Leontopithecus chrysopygus	Goldsteiß-Löwenäffchen	black lion tamarin, golden rump lion tamarin, golden-rumped lion tamarin
Leopoldamys spp.	Asiatische Langschwanz-Riesenratten	long-tailed giant rats
Lepadidae	Entenmuscheln	goose barnacles
Lepadogaster microcephalus/ Apletodon microcephalus> Apletodon dentatus	Kleinköpfiger Ansauger, Kleinkopf-Schildfisch	small-headed clingfish (FAO), small-headed sucker
Lepadogaster candollei	Rotfleckiger Ansauger, Rotflecken-Ansauger, Rotfleck-Schildfisch	Connemara clingfish (FAO), Connemara sucker
Lepadogaster lepadogaster	Blaufleckiger Ansauger, Blaufleck-Schildfisch	shore clingfish (FAO), Cornish clingfish
Lepas anatifera	Gemeine Entenmuschel, Glatte Entenmuschel	common goose barnacle
Lepas fascicularis	Schwebende Entenmuschel	float goose barnacle, buoy-making barnacle, short-stalked goose barnacle
Leperisinus varius/ Leperisinus fraxini/ Hylesinus fraxini	Eschenbastkäfer, Bunter Eschenbastkäfer, Kleiner Bunter Eschenbastkäfer (>Eschenrose)	ash bark beetle, lesser ash bark beetle
Lepeta caeca	Nördliche Blindkäferschnecke*	northern blind limpet
Lepetidae	Blindkäferschnecken*	blind limpets, eyeless limpets
Lepidobatrachus asper	Chaco-Pfeiffrosch	Paraguay horned frog
Lepidocephalus thermalis	Kleine Schmerle	lesser loach
Lepidochelys kempii	Kemps Bastardschildkröte, Atlantik-Bastardschildkröte, Atlantische Bastardschildkröte	Kemp's ridley (sea turtle), Atlantic ridley turtle
Lepidochelys olivacea	Pazifische Bastardschildkröte	olive ridley (sea turtle), Pacific ridley turtle
Lepidochitona cinerea (Polyplacophora)	Aschgraue Käferschnecke, Rändel-Käferschnecke	cinereous chiton
Lepidochitona raymondi (Polyplacophora)	Zwitter-Käferschnecke	Raymond's chiton
Lepidogalaxiidae	Salamanderfische*	salamanderfishes
Lepidoglyphus destructor/ Glycyphagus destructor/ Lepidoglyphus cadaverum	Pflaumenmilbe	cosmopolitan food mite, grocers' itch mite
Lepidonotus squamatus	Flacher Schuppenwurm	twelve-scaled worm
Lepidophyma flavimaculatum	Gelbgefleckte Nachtechse	yellow-spotted night lizard
Lepidopleurus asellus (Polyplacophora)	Assel-Käferschnecke	coat-of-mail chiton, pill chiton*
Lepidopleurus cajetanus (Polyplacophora)	Rippen-Käferschnecke	ribbed chiton*
Lepidopleurus cancellatus (Polyplacophora)	Kugel-Käferschnecke	Arctic cancellate chiton

Lepidopleurus intermedius (Polyplacophora)	Sand-Käferschnecke	intermediate chiton, sand chiton
Lepidopsetta bilineata	Pazifische Scholle	rock sole
Lepidoptera	Schuppenflügler (Schmetterlinge & Motten)	lepidopterans (butterflies & moths)
Lepidopus caudatus	Strumpfbandfisch	silver scabbardfish (FAO), ribbonfish, frostfish
Lepidorhombus boscii	Gefleckter Flügelbutt, Vierfleckbutt	four-spot megrim
Lepidorhombus whiffiagonis	Flügelbutt, Scheefsnut	megrim (FAO), sail-fluke, whiff
Lepidosaphes ulmi	Komma-Schildlaus, Gemeine Kommaschildlaus, Obstbaumkommaschildlaus	oystershell scale, mussel scale
Lepidosauria	Schuppenkriechtiere	lepidosaurs
Lepidosiren paradoxa	Südamerikanischer Lungenfisch	South American lungfish
Lepidosirenidae	Molchfische	South American lungfish
Lepidoteuthis grimaldi	Schuppenkalmar	scaly squid*
Lepidotrigla cavillone	Stachel-Knurrhahn	large-scale gurnard
Lepidotrigla dieuzeidei	Stachliger Knurrhahn	spiny gurnard
Lepidozona regularis (Polyplacophora)	Gewöhnliche Kalifornische Käferschnecke	regular chiton
Lepidurus apus/ Lepidurus productus	Kleiner Rückenschaler, Schuppenschwanz, Langschwänziger Flossenkrebs, Frühjahrs-Kieferfuß	
Lepikentron ovis/ Bovicola ovis	Sandlaus, Schaflaus	sheep biting louse
Lepilemur mustelinus	Großer Wieselmaki	greater weasel lemur, sportive lemur, greater sportive lemur
Lepilemur ruficaudatus	Kleiner Wieselmaki	red-tailed sportive lemur
Lepilemur spp.	Wieselmakis	weasel lemurs, sportive lemurs
Lepisma saccharina	Silberfischchen	silverfish
Lepismatidae	Fischchen	silverfish
Lepismodes inquilinus/ Thermobia domestica	Ofenfischchen	firebrat
Lepisosteidae	Knochenhechte, Kaimanfische	gars
Lepisosteiformes	Knochenhechte, Kaimanfische	gars
Lepisosteus oculatus	Gefleckter Knochenhecht	spotted gar
Lepisosteus osseus	Knochenhecht, Gemeiner Knochenhecht, Schlanker Knochenhecht	longnose gar
Lepisosteus platyrhinchus	Florida-Knochenhecht	Florida gar
Lepisosteus platystomus	Kurznasen-Knochenhecht	shortnose gar
Lepisosteus spatula/ Atractosteus spatula	Mississippi-Knochenhecht	alligator gar
Lepisosteus tristoechus	Kaimanfisch, Alligatorfisch	Cuban gar
Lepomis auritus	Rotbrust-Sonnenbarsch, Großohriger Sonnenfisch	redbreast sunfish (FAO), red-breasted sunfish

Lepomis cyanellus	Grüner Sonnenbarsch, Grasbarsch	green sunfish
Lepomis gibbosus	Gemeiner Sonnenbarsch	pumpkin-seed sunfish, pumpkinseed (FAO)
Lepomis gulosus	Warmouth-Sonnenbarsch*	warmouth
Lepomis macrochirus	Stahlblauer Sonnenbarsch	bluegill
Leporidae	Hasenartige	leporids (rabbits & hares)
Leporillus conditor	Große Australische Häschenratte	greater stick-nest rat
Leporillus spp.	Australische Häschenratten	Australian stick-nest rats
Leposternon microcephalum	Kleinköpfige Doppelschleiche	smallhead worm lizard
Leptagonus decagonus		Atlantic poacher
Leptailurus serval/ Felis serval	Serval	serval
Leptasterias littoralis	Grüner Schlank-Seestern*	green slender seastar
Leptasterias tenera	Schlank-Seestern*	slender seastar
Leptidea sinapis	Senfweißling, Tintenfleck-Weißling, Tintenfleck	wood white butterfly, wood white
Leptinidae	Pelzflohkäfer, Mausflohkäfer	leptinids (mammal-nest beetles & beaver parasites, beaver beetles, rodent beetles)
Leptinotarsa decemlineata	Kartoffelkäfer (Koloradokäfer)	Colorado potato beetle, Colorado beetle, potato beetle
Leptinus testaceus	Mäusefloh	mouse flea
Leptobrama muelleri	Strandlachs*	beachsalmon
Leptocharias smithii	Bärteliger Hundshai	barbelled houndshark
Leptochariidae	Bärtelige Hundshaie, Schlankhaie	barbelled houndsharks
Leptochilichthyidae	Leptochilichthyiden	leptochilichthyids
Leptoclinus maculatus	Schlangen-Schleimfisch*	snake blenny
Leptoconus gloriamaris/ Conus gloriamaris	Ruhm des Meeres	glory-of-the-sea cone
Leptodactylidae	Südfrösche	southern frogs, tropical frogs, leptodactylids
Leptodactylodon spp.	Eierfrösche	egg frogs
Leptodactylus albilabris	Gunthers Weißlippen-Pfeiffrosch	Gunther's white-lipped frog
Leptodactylus bufonius	Kröten-Pfeiffrosch	Vizcacheras' frog
Leptodactylus labialis	Copes Weißlippen-Pfeiffrosch, Mexikanischer Weißlippen-Pfeiffrosch	Cope's white-lipped frog
Leptodactylus melanonotus	Schwarzrückiger Pfeiffrosch	Sabinal frog
Leptodactylus ocellatus	Augenpfeifer	Criolla frog
Leptodactylus pentadactylus	Südamerikanischer Ochsenfrosch, Fünffingriger Pfeiffrosch	South American bullfrog
Leptodactylus spp.	Echte Pfeiffrösche	white-lipped frogs, nest-building frogs
Leptodeira annulata	Katzenaugennatter	banded cat-eyed snake
Leptodeira septentrionalis	Gebänderte Katzenaugennatter	cat-eyed snake
Leptodeira spp.	Katzenaugennattern	cat-eyed snakes

Leptodiridae

Leptodiridae/Catopidae	Nestkäfer, Erdaaskäfer	small carrion beetles
Leptogasteridae	Schlankfliegen	grass flies, leptogasterid flies
Leptogorgia virgulata	Seepeitsche u.a., Peitschen-Gorgonie u.a.	whip coral, sea whip a.o.
Leptomicrurus spp.	Korallenschlangen	slender coral snakes, thread coral snakes
Leptomys elegans	Langfuß-Wasserratte	long-footed hydromyine
Leptonidae	Leptoniden	coin shells
Leptonychotes weddelli	Weddell-Robbe	Weddell seal
Leptonycteris spp.	Saussure-Langnasenfledermäuse	Saussure's long-nosed bats
Leptopelis bufonides	Krötenähnlicher Waldsteigerfrosch	Savannah forest treefrog
Leptopelis cynnamomeus	Braunrückiger Waldsteigerfrosch	Angola forest treefrog
Leptopelis spp.	Waldsteiger, Waldsteigerfrösche	forest treefrogs
Leptophis ahaetulla	Schlanknatter	parrot snake
Leptophis spp.	Dünnschlangen	parrot snakes
Leptophyes albovittata	Gestreifte Zartschrecke	striped bushcricket
Leptophyes punctatissima	Punktierte Zartschrecke	speckled bushcricket
Leptophyes spp.	Zartschrecken	bushcrickets
Leptopodidae	Stachelwanzen	spiny-legged bugs, spiny shore bugs
Leptopoecile elegans	Schopfhähnchen	crested tit warbler
Leptopoecile sophiae	Purpurhähnchen	Severtzov's tit warbler
Leptopsammia pruvoti	Gelbe Nelkenkoralle	sunset star coral
Leptopsylla segnis	Hausmausfloh	European mouse flea, house mouse flea
Leptopterna dolobrata	Graswanze	meadow plant bug
Leptoptilos crumeniferus	Marabu	marabou stork
Leptopus marmoratus	Stachelwanze	spiny-legged bug, spiny shore bug
Leptoscopidae	Südliche Sandfische	southern sandfishes
Leptoseris cucullata	Wellenkoralle	sunray lettuce coral
Leptosomidae/ Leptosomatidae	Kurol	cuckoo-roller
Leptosomus discolor	Kurol	courol
Leptosynapta inhaerens	Wurmholothurie, Kletten-Holothurie, Kletten-Seegurke	common white synapta
Leptothorax spp.	Schmalbrustameisen	
Leptothorax tuberum	Zwergameise	
Leptotila verreauxi	Blauringtaube	white-fronted dove
Leptotrombidium akamushi/ Trombicula akamushi	Japanische Fleckfiebermilbe (Tsutsugamushi-Fieber-Erntemilbe)	scrub typhus chigger mite, Japanese scrub typhus chigger mite*
Leptotyphlopidae	Schlankblindschlangen	slender blind snakes
Leptotyphlops macrorhynchus	Hakenschnabel-Schlankblindschlange	hookbilled blindsnake
Leptotyphlops cairi	Ägyptische Schlankblindschlange	Cairo earthsnake

Leptotyphlops dulcis	Texas-Wurmschlange, Texas-Schlankblindschlange, New Mexico-Wurmschlange	Texas blindsnake
Leptotyphlops humilis	Mexikanische Schlankblindschlange	Mexican blindsnake
Lepus alleni	Allens Eselhase, Antilopenhase	antelope jack rabbit
Lepus americanus	Schneeschuhhase (Hbz. Schneehase)	snowshoe hare, varying hare
Lepus arcticus	Polarhase	arctic hare
Lepus brachyurus	Japanischer Hase	Japanese hare
Lepus californicus	Kalifornischer Eselhase	black-tailed jack rabbit
Lepus callotis	Weißflanken-Jackrabbit, Mexikohase	white-sided jack rabbit
Lepus capensis	Kaphase, Wüstenhase	Cape hare, brown hare
Lepus castroviejoi	Ginsterhase, Spanischer Hase	broom hare
Lepus europaeus	Europäischer Hase, Europäischer Feldhase	European hare
Lepus spp.	Echte Hasen	hares, jack rabbits
Lepus timidus	Eurasischer Schneehase (Hbz. Schneehase)	blue hare, mountain hare
Lepus townsendii	Weißschwanzeselhase, Präriehase	white-tailed jack rabbit
Lesbia victoriae	Schwarzschwanzlesbia	black-tailed trainbearer
Lestes barbatus	Südliche Binsenjungfer	southern European emerald damselfly
Lestes dryas	Glänzende Binsenjungfer	scarce emerald damselfly
Lestes macrostigma	Dunkle Binsenjungfer	dusky emerald damselfly
Lestes sponsa	Gemeine Binsenjungfer	green lestes, emerald damselfly
Lestes virens	Kleine Binsenjungfer	lesser emerald damselfly
Lestidae	Teichjungfern	spread-winged damselflies
Lestoros inca	Peru-Opossummaus	Peruvian "shrew" opossum
Lesuerigobius friesii	Fries-Grundel	Fries's goby
Lethenteron appendix/ Lampetra appendix	Amerikanischer Bachneunauge	American brook lamprey
Lethrinidae	Ruderfische, „Straßenkehrer"	emperors, emperor breams
Lethrinus kallopterus	Gelbflossen-Ruderfisch	yellow-spotted emperor
Lethrus apterus	Rebschneider, Rebenschneider	
Leucandra heath	Heaths Kalkschwamm*	Heath's sponge
Leucandra nivea	Fladen-Kalkschwamm	pancake sponge*
Leucaspius delineatus	Moderlieschen	moderlieschen, belica (FAO), sunbleak
Leucetta philippinensis	Gelber Kalkschwamm	yellow calcareous sponge
Leucetta primigenia	Rostroter Kalkschwamm	ferruginius sponge*
Leucichthys spp.	Maränen, Felchen	ciscoes
Leucilla nuttingi	Nuttings Schwamm	Nutting's sponge
Leuciscus cephalus	Döbel	chub
Leuciscus idus	Aland, Orfe	ide (FAO), orfe

Leuciscus leuciscus	Hasel	dace
Leuciscus souffia	Strömer	vairone (FAO), telestes (U.S.), souffie
Leucochlaena oditis		beautiful gothic (moth)
Leucodonta bicoloria	Weißer Zahnspinner, Weißer Birkenzahnspinner,	
Leucoma salicis	Atlasspinner, Atlas, Pappelspinner, Weidenspinner	satin moth
Leuconia aspera	Knollen-Kalkschwamm	knobby calcareous sponge*
Leucophaea maderae	Madeira-Schabe	Madeira cockroach
Leucophytia bidentata	Zweizahn-Weißschnecke*	two-toothed white snail
Leucopsar rothschildi	Balistar	Rothschild's starling, Bali mynah
Leucopternis schistacea	Schieferbussard	slate-coloured hawk
Leucoraja erinacea/ Raja erinacea/	Kleiner Igelrochen	little skate (FAO), hedgehog skate
Leucoraja ocellata/ Raja ocellata/	Winterrochen, Westatlantischer Winterrochen	winter skate
Leucorrhinia albifrons	Östliche Moosjungfer	
Leucorrhinia caudalis	Zierliche Moosjungfer	
Leucorrhinia dubia	Kleine Moosjungfer	white-faced darter, white-faced dragonfly
Leucorrhinia pectoralis	Große Moosjungfer	greater white-faced darter
Leucorrhinia rubicunda	Nördliche Moosjungfer	
Leucorrhinia spp.	Moosjungfern	white-faced darters
Leucosiidae	Kugelkrabben	purse crabs & nut crabs & pebble crabs
Leucosolenia botryoides	Traubiger Röhrenkalkschwamm	organ-pipe sponge
Leucosolenia fragilis	Zarter Sack-Kalkschwamm	fragile calcareous sponge*
Leucosticte arctoa	Rosenbauch-Schneegimpel	rosy finch
Leucothea multicornis/ Eucharis multicornis	Glas-Lappenrippenqualle	vitreous lobate comb-jelly, warty comb jelly
Leucozonia ocellata	Weißgetupfte Bandschnecke	white-spot latirus, white-spotted latirus
Leuresthes tenuis	Grunion	grunion
Leurognathus marmoratus	Marmorierter Bachsalamander	marbled salamander
Lialis spp.	Spitzkopf-Flossenfüße	sharpsnouted snake lizards
Liasis amethistimus/ Morelia amethistima	Amethystpython	amethystine python, scrub python
Liasis childreni/ Morelia childreni/ Antaresia childreni	Childrens Python	Children's python, Children's rock python
Liasis fuscus/Morelia fusca (Liasis mackloti)	Brauner Wasserpython	brown water python (white-lipped python)
Liasis mackloti/ Morelia mackloti	Macklots Python	water python, Macklot's python
Liasis olivaceus	Olivenpython*	olive python
Liasis perthensis/ Morelia perthensis/ Antaresia perthensis	Zwergpython	Perth pigmy python, pygmy python
Libellula depressa	Plattbauch	broad-bodied libellula, broad-bodied chaser

Libellula fulva	Spitzenfleck	scarce chaser dragonfly, scarce libellula
Libellula quadrimaculata	Vierflecklibelle, Vierfleck	four-spotted libellula, four-spotted chaser, four spot
Libellulidae	Segellibellen	common skimmers
Liberiictis kuhni	Liberia-Kusimanse, Liberianische Manguste	Liberian mongoose
Libidoclea granaria	Chilenische Seespinne	southern spider crab
Libinia dubia	Zweifelhafte Seespinne*	longnose spider crab, doubtful spider crab
Libinia emarginata	West-Atlantische Seespinne	portly spider crab, common spider crab
Libinia erinacea	Tang-Seespinne*	seagrass spider crab
Libythea celtis	Zürgelbaum-Schnauzenfalter, Zürgelbaumfalter	nettle-tree butterfly
Libytheidae	Schnauzenfalter	snout butterflies
Lichanura spp.	Rosenboas	rosy boas
Lichanura trivirgata	Rosenboa, Dreistreifen-Rosenboa	rosy boa, striped rosy boa
Lichanura trivirgata roseofusca	Küsten-Rosenboa	coastal rosy boa
Lichanura trivirgata gracia	Wüsten-Rosenboa	desert rosy boa
Lichenopora radiata	Korallen-Moostierchen	coralline bryozoan*
Lichia amia	Große Gabelmakrele	leerfish
Licinus hoffmannseggi	Berg-Stumpfzangenläufer, Glänzendschwarzer Bodenkäfer	
Ligia oceanica	Klippenassel, Strandassel	great sea-slater, sea slater (quay-louse)
Ligia spp.	Klippenasseln, Strandasseln	slaters, sea-slaters, rock lice
Ligidium hypnorum	Sumpfassel	moss slater
Ligiidae	Klippenasseln, Strandasseln	sea slaters, rock lice
Ligula intestinalis	Riemenwurm, Riemenbandwurm	
Liguus fasciatus	Florida-Baumschnecke	Florida treesnail
Lima inflata	Bauchige Feilenmuschel	inflated fileclam, inflated lima
Lima lima	Gewöhnliche Feilenmuschel	spiny fileclam, spiny lima (frilled file shell)
Lima scabra/ Ctenoides scabra	Raue Feilenmuschel	rough lima, rough fileclam (rough file shell, Atlantic rough file shell)
Lima vulgaris	Pazifische Feilenmuschel	Pacific fileclam
Limacidae	Egelschnecken, Schnegel	keeled slugs
Limacina inflata	Aufgeblasener Flossenfüßer*	planorbid pteropod
Limacina lesueurii	Lesueurs Flossenfüßer	Lesueur's pteropod
Limacina retroversa	Umgekehrter Flossenfüßer*	retrovert pteropod
Limacodidae (Cochlidiidae)	Asselspinner, Mottenspinner (Schildmotten)	slug caterpillars & saddleback caterpillars
Limanda aspera	Raue Kliesche, Gelbflossenzunge*	yellowfin sole
Limanda ferruginea	Gelbschwanzflunder	yellowtail flounder
Limanda herzensteini/ Pseudopleuronectes herzensteini	Japanische Flunder	Japanese dab, yellow-striped flounder, littlemouth flounder (FAO)

Limanda limanda	Kliesche (Scharbe)	dab, common dab
Limaria hians	Klaffende Feilenmuschel	gaping fileclam
Limax valentianus/ *Lehmannia valentiana*	Gewächshausschnegel, Iberische Egelschnecke	glasshouse slug, Iberian slug
Limax cinereoniger	Schwarzer Schnegel	ash-black slug
Limax flavus	Kellerschnecke, Bierschnegel, Gelber Schnegel, Gelbe Egelschnecke	yellow gardenslug, yellow slug (cellar slug, dairy slug, house slug, tawny garden slug)
Limax maculatus	Grüne Egelschnecke, Grüner Schnegel	green slug
Limax marginatus/ *Lehmannia marginata*	Baumschnegel, Baum-Egelschnecke	tree slug
Limax maximus	Große Egelschnecke, Großer Schnegel	giant gardenslug, European giant gardenslug, great grey slug, spotted garden slug
Limax tenellus	Gelbe Wald-Egelschnecke	slender slug, tender slug
Limenitis camilla/ *Ladoga camilla*	Kleiner Eisvogel	white admiral
Limenitis populi	Großer Eisvogel	poplar admiral
Limenitis reducta	Blauschwarzer Eisvogel	southern white admiral
Limicola falcinellus	Sumpfläufer	broad-billed sandpiper
Limidae	Feilenmuscheln	file clams (file shells)
Limifossor fratula (Aplacophora)	Südlicher Schlamm-Maulwurf (Schildfuß)	southern mole glistenworm
Limifossor talpoideus (Aplacophora)	Nördlicher Schlamm-Maulwurf (Schildfuß)	northern mole glistenworm, mole solenogaster
Limnebiidae	Teichkäfer, Sumpfkäfer	minute moss beetles
Limnichidae		minute marsh-loving beetles
Limnodromus griseus	Kleiner Schlammläufer	short-billed dowitcher
Limnodromus scolopaceus	Großer Schlammläufer	long-billed dowitcher
Limnodynastes dorsalis	Banjo-Frosch	banjo frog
Limnodynastes spp.	Sumpffrösche	swamp frogs
Limnodynastes tasmaniensis	Tasmanischer Sumpffrosch	spotted grass frog
Limnogale mergulus	Wasser-Borstenigel	aquatic tenrec
Limnomys sibuanus	Mindanao-Ratte u.a.	Mindanao rat (Mount Apo)
Limnonectes macrodon	Zahnfrosch	Malayan wart frog
Limnonectes spp.	Warzenfrösche	wart frogs
Limnoria lignorum	Bohrassel, Holzbohrassel	gribble
Limoniidae	Stelzmücken, Sumpfmücken	short-palped craneflies
Limoniscus violaceus	Veilchenblauer Wurzelhals-Schnellkäfer	violet click beetle
Limosa lapponica	Pfuhlschnepfe	bar-tailed godwit
Limosa limosa	Uferschnepfe	black-tailed godwit
Limothrips cerealium	Gewitterfliege, Getreideblasenfuß	corn thrips, grain thrips, black wheat thrips, cereal thrips
Limulodidae		horseshoe crab beetles
Linatella caudata	Bändertriton	ringed triton

Linatella succincta	Kleine Bändertriton	lesser girdled triton
Linckia laevigata	Blauer Seestern	blue star, blue seastar
Linckia multifora	Kometenstern	comet star*, comet seastar
Lineatriton lineola	Veracruz-Salamander	Veracruz worm salamander
Lineus geniculatus (Nemertini)	Gebänderter Schnurwurm	banded bootlace
Lineus longissimus (Nemertini)	Langer Schnurwurm, Engländischer Langwurm	giant bootlace worm, sea longworm
Lineus ruber (Nemertini)	Roter Schnurwurm	red bootlace
Lineus viridis (Nemertini)	Grüner Schnurwurm	green bootlace
Linguatula serrata	Gemeiner Zungenwurm	common tongue worm
Linognathidae	Glattläuse*	smooth sucking lice
Linognathus ovillus	Schaflaus, Schaf-Gesichtslaus	sheep face louse, sheep sucking louse
Linognathus setosus	Hundelaus	dog sucking louse
Linognathus stenopsis	Ziegenlaus	goat sucking louse
Linognathus vituli	Langköpfige Rinderlaus	longnosed cattle louse
Linophryne arborifera	Kinnbart-Laternenangler*	lanternfish a.o.
Linuche unguiculata	Fingerhutqualle*	thimble jellyfish
Linyphiidae	Deckennetzspinnen (Baldachinspinnen) & Zwergspinnen	sheet-web weavers, sheet-web spinners, line-weaving spiders, line weavers, money spiders
Liocarcinus holsatus/ Portunus holsatus	Gemeine Schwimmkrabbe, Ruderkrabbe	common swimming crab, flying swimming crab
Liocarcinus corrugatus	Runzelige Schwimmkrabbe	wrinkled swimming crab
Liocarcinus depurator	Hafenkrabbe	harbour crab, harbour swimming crab
Liocarcinus marmoreus	Marmorierte Schwimmkrabbe	marbled swimming crab
Lioconcha castrensis	Zeltlagermuschel	chocolate flamed venus, camp pitar-venus
Liocranidae	Feldspinnen	liocranid spiders
Liodidae/Anisotomidae	Schwammkugelkäfer	liodid beetles
Liolaemus spp.	Erdleguane	tree iguanas
Liomys spp.	Stacheltaschenmäuse	spiny pocket mice
Lionychus quadrillum	Gelbgefleckter Krallenkäfer, Gelbfleckiger Krallenkäfer	
Liopepis belliana	Schmetterlingsagame	butterfly lizard
Liophis frenata	Gewöhnliche Bodennatter	common ground snake
Liophis lineatus	Gestreifte Bodennatter	lined tropical snake
Liophis poecilogyrus	Buntnatter, Bunte Bodennatter	Wied's ground snake
Liophis spp.	Amazonas Bodennattern, Buntnattern	Amazon ground snakes
Liopsetta putnami	Glattflunder	smooth flounder
Lipara lucens	Zigarrenfliege, Schilfgallenfliege	reed gall fly
Liparididae	Scheibenbäuche	snailfishes
Liparis atlanticus	Westatlantischer Scheibenbauch	Atlantic sea snail
Liparis liparis	Großer Scheibenbauch	common seasnail, sea snail, striped seasnail (FAO)

Liparis montagui	Kleiner Scheibenbauch	Montagu's seasnail (FAO), Montagu's sea-snail
Lipeurus caponis	Flügellaus	wing louse, chicken wing louse
Liphistiidae	Gliederspinnen	trap-door spiders
Liponyssoides sanguineus/ Allodermanyssus sanguineus	Hausmausmilbe	house-mouse mite
Lipophrys basiliscus	Zebra-Schleimfisch	zebra blenny, basilisk blenny
Lipophrys canevae/ Blennius canevae	Gelbkehl-Schleimfisch, Gelbwangen-Schleimfisch	yellow-throat blenny, Caneva's blenny
Lipophrys dalmatinus	Zwergschleimfisch, Dalmatinischer Schleimfisch	Dalmatian blenny
Lipophrys fluviatilis/ Blennius fluviatilis/ Salaria fluviatilis	Süßwasser-Schleimfisch, Fluss-Schleimfisch	freshwater blenny (FAO), river blenny
Lipophrys nigriceps/ Blennius nigriceps	Schwarzkopf-Schleimfisch	black-headed blenny
Lipophrys pavo/ Blennius pavo/Salaria pavo	Pfauenschleimfisch	peacock blenny
Lipophrys pholis	Schan, Schleimlerche, Atlantischer Schleimfisch	shanny
Lipophrys trigloides	Grauer Schleimfisch	gurnard blenny
Lipoptena cervi	Hirschlausfliege	deer ked, deer fly
Liposcelidae	Bücherläuse	booklice
Liposcelis divinatorius	Bücherlaus	booklouse
Lipotes vexillifer	Chinesischer Flussdelphin, Yangtse-Delphin, Beiji	whitefin dolphin, Chinese River dolphin, Yangtse River dolphin, baiji, pei chi
Lirioceris lilii	Lilienhähnchen	lily beetle
Lirioceris merdigera	Zwiebelhähnchen	onion beetle*
Lirioceris spp.	Lilienhähnchen	lily beetles
Lissa chiraga	Rote Seespinne	red masked crab*
Lissemys punctata	Indische Klappen-Weichschildkröte	Indo-Gangetic flapshell, Indian flapshell turtle
Lissemys scutata	Burmesische Klappen-Weichschildkröte, Burma-Klappen-Weichschildkröte	Burmese flapshell turtle
Lissocarcinus orbicularis	Seegurken-Schwimmkrabbe	sea cucumber swimming crab
Lissodelphis borealis	Nördlicher Glattdelphin	Pacific right whale dolphin, northern right-whale dolphin
Lissodelphis peronii	Südlicher Glattdelphin	southern right-whale dolphin, Delfin liso
Listrophoridae	Haarmilben	fur mites
Listrophorus gibbus	Kaninchen-Haarmilbe, Kaninchenmilbe	rabbit fur mite a.o.
Lithobiidae	Steinläufer	garden centipedes
Lithobiomorpha	Steinläufer	lithobiomorphs
Lithobius forficatus	Brauner Steinläufer	common garden centipede
Lithodes aequispina	Gold-Königskrabbe	golden king crab
Lithodes antarctica/ Lithodes santolla	Antarktische Königskrabbe	southern king crab

Lithodes couesi	Tiefsee-Königskrabbe	scarlet king crab, deep-sea crab, deep-sea king crab
Lithodes maja	Nördliche Steinkrabbe	northern stone crab, Norway crab, devil's crab (prickly crab), stone king crab
Lithodes murrayi	Murray Steinkrabbe, Subantarktische Steinkrabbe	Murray king crab, Subantarctic stone crab
Lithodidae	Steinkrabben & Königskrabben	stone crabs & king crabs
Lithoglyphus naticoides	Flusssteinkleber	
Lithognathus mormyrus/ Pagellus mormyrus	Marmorbrassen	marmora, striped seabream (FAO)
Lithophaga aristata	Scheren-Seedattel*	scissor datemussel
Lithophaga lithophaga (see: Petrophaga lithographica)	Seedattel, Steindattel, Meeresdattel	datemussel, common date mussel
Lithophaga spp.	Seedatteln, Meeresdatteln	datemussels, date mussels
Lithophane ornitopus	Holzeule	
Lithopoma americanum	Amerikanische Stern-Turbanschnecke	American starsnail
Lithopoma caelatum	Gemeißelte Stern-Turbanschnecke*	carved starsnail
Lithopoma gibberosum	Rote Stern-Turbanschnecke	red starsnail, red turbansnail, red turban
Lithopoma tectum	Westindische Stern-Turbanschnecke	West Indian starsnail
Lithopoma tuber	Grüne Stern-Turbanschnecke	green starsnail
Lithopoma undosum	Gewellte Stern-Turbanschnecke	wavy starsnail, wavy turbansnail
Lithostege griseata	Sophienkrautspanner	grey carpet moth
Litocranius walleri	Giraffengazelle, Gerenuk	gerenuk
Litopenaeus occidentalis/ Penaeus occidentalis	Pazifische Weiße Garnele	western white shrimp, Central American white shrimp
Litopenaeus setiferus/ Penaeus setiferus	Atlantische Weiße Garnele, Nördliche Weiße Geißelgarnele	white shrimp, lake shrimp, northern white shrimp
Litopenaeus vannamei/ Penaeus vannamei	Weißbein-Garnele	whiteleg shrimp, Central American shrimp
Litopterna	Glattferser	litopternas
Litoria aurea/Hyla aurea	Goldlaubfrosch	green & golden bell frog
Litoria brevipalmata	Grünschenkelfrosch	green-thighed frog
Litoria caerulea/ Hyla caerulea/ Pelodryas caerulea	Korallenfinger	White's treefrog
Litoria castanea	Gelbfleckfrosch	cream-spotted treefrog, yellow-spotted treefrog
Litoria chloris	Rotaugenfrosch	Australian red-eyed treefrog
Litoria dentata	Blökfrosch*	bleating treefrog
Litoria spp.	Australasische Baumfrösche	Australasian treefrogs
Litoria verreauxii	Pfeifender Baumfrosch	fawn treefrog, whistling tree frog
Littorina coccinea	Rote Strandschnecke*	scarlet periwinkle

Littorina irrorata	Sumpf-Strandschnecke*, Sumpf-Uferschnecke*	marsh periwinkle
Littorina littoralis	Zwergstrandschnecke	dwarf periwinkle
Littorina littorea	Gemeine Strandschnecke, Gemeine Uferschnecke, „Hölker"	common periwinkle, common winkle, edible winkle
Littorina mariae	Flache Strandschnecke	flat periwinkle
Littorina neritoides	Zwergstrandschnecke, Blaue Strandschnecke, Gewöhnliche Strandschnecke	small periwinkle
Littorina obtusata	Stumpfe Strandschnecke, Stumpfkegelige Uferschnecke	flat periwinkle, yellow periwinkle, northern yellow periwinkle
Littorina planaxis/ Littorina striata/ Littorina keenae	Graue Strandschnecke	gray periwinkle, "eroded" periwinkle
Littorina saxatilis	Raue Strandschnecke, Spitze Strandschnecke, Dunkle Strandschnecke	rough periwinkle
Littorina scabra	Mangroven-Strandschnecke	rough periwinkle, variegated periwinkle, mangrove winkle
Littorina ziczac/ Nodilittorina ziczac	Zebra-Strandschnecke	zebra periwinkle, zebra winkle
Littorinidae	Strandschnecken	winkles, periwinkles
Livonia mammilla	Falsche Melonenwalze	false melon volute, mammal volute
Liza aurata/Mugil auratus	Gold-Meeräsche, Goldmeeräsche, Goldäsche	golden grey mullet (FAO), golden mullet
Liza haematochila	So-iny-Meeräsche	so-iny mullet
Liza ramada/Mugil capito	Dünnlippige Meeräsche	thinlip grey mullet, thin-lipped grey mullet, thinlip mullet (FAO)
Liza saliens/Mugil saliens	Springmeeräsche	leaping mullet
Lo vulpinus/Siganus vulpinus	Fuchsgesicht	foxface
Loa loa	Afrikanischer Augenwurm, Wanderfilarie	African eye worm
Lobata	Lappenrippenquallen	lobate comb-jellies, lobates
Lobesia botrana	Bekreuzter Traubenwickler, Weinmotte (Sauerwurm/ Gelbköpfiger Sauerwurm)	grape fruit moth, vine moth, European vine moth, grape-berry moth
Lobodon carcinophagus	Krabbenesser	crabeater seal
Lobophora halterata	Lappenspanner	seraphim
Lobophyllia corymbosa	Gelappte Becherkoralle, Faltenbecherkoralle	lobed cup coral, meat coral
Lobophyllia costata	Gezackte Becherkoralle	dentate flower coral, meat coral
Lobotes surinamensis	Schwarzer Dreischwanzbarsch	tripletail
Lobotidae	Dreischwanzbarsche	tripletails
Lochmaea suturalis	Heideblattkäfer	heather beetle
Locusta migratoria	Wanderheuschrecke	migratory locust
Locustella certhiola	Streifenschwirl	Pallas's grasshopper warbler
Locustella fluviatilis	Schlagschwirl	river warbler

Locustella lanceolata	Strichelschwirl	lanceolated warbler
Locustella luscinioides	Rohrschwirl	Savi's warbler
Locustella naevia	Feldschwirl	grasshopper warbler
Loddigesia mirabilis	Wundersylphe	marvellous spatuletail
Loimia medusa	Medusenwurm, Spaghetti-Wurm	medusa worm*
Loliginidae	Kalmare	inshore squids
Loligo edulis	Schwertspitzenkalmar*	swordtip squid
Loligo forbesi	Nordischer Kalmar, Forbes' Kalmar	long-finned squid, northern squid, veined squid (FAO)
Loligo opalescens	Pazifischer Opalkalmar, Opalisierender Kalmar	opalescent inshore squid, opalescent squid (American "market squid"), common Pacific squid
Loligo pealei	Nordamerikanischer Kalmar, Langflossen-Schelfkalmar	longfin inshore squid, Atlantic long-fin squid, Atlantic long-finned squid
Loligo reynaudi	Kap-Kalmar	Cape Hope squid, chokker squid
Loligo vulgaris	Gemeiner Kalmar, Roter Gemeiner Kalmar	common squid
Lolliguncula brevis	Gedrungener Kalmar, Kurz-Kalmar*	Atlantic brief squid, brief squid, brief thumbstall squid, small squid
Lomaspilis marginata		clouded border moth
Lomographa bimaculata	Zweifleckspanner	
Lomographa temerata		clouded silver
Lonchaeidae	Lanzenfliegen u.a.	lonchaeids
Lonchopteridae	Lanzenfliegen u.a.	spear-winged flies
Lonchorhina aurita	Speernase	Tomes' long-eared bat
Lonchorhina spp.	Schwertnasen	sword-nosed bats
Longitarsus ferrugineus	Minzen-Erdfloh*	mint flea beetle
Longitarsus parvulus	Schwarzer Flachserdfloh	black flea beetle, flax flea beetle, linseed flea beetle
Lopha cristagalli	Hahnenkammauster	cock's-comb oyster, cockscomb oyster, coxcomb oyster
Lophelia pertusa	Augenkoralle	eye coral*
Lophiidae	Anglerfische	monks, goosefishes
Lophiiformes	Armflosser, Anglerfische	anglerfishes
Lophiomys imhausi	Mähnenratte	maned rat, crested rat
Lophiotoma indica	Indische Turmschnecke	Indian turret
Lophius americanus	Amerikanischer Angler	American goosefish
Lophius budegassa	Schwarzbauch-Seeteufel	black-bellied angler
Lophius gastrophysus	Schwarzflossen-Angler*	blackfin goosefish
Lophius piscatorius	Atlantischer Seeteufel, Atlantischer Angler	Atlantic angler fish, angler (FAO), monkfish
Lophocebus aterrimus/ Cercocebus aterrimus	Schopf-Mangabe	black mangabey
Lophocebus albigena	Grauwangen-Mangabe	grey-cheeked mangabey

Lophogorgia ceratophyta	Orangerote Gorgonie	
Lophogorgia miniata	Karminroter Seereisig	
Lopholatilus chamaeleonticeps	Blauer Ziegelbarsch	tilefish
Lophophorus impeyanus	Rostschwanzmonal	Himalayan monal pheasant
Lophopoda/Phylactolaemata	Armwirbler, Süßwasserbryozoen	phylactolaemates, "covered throat" bryozoans, freshwater bryozoans
Lophorina superba	Kragenparadiesvogel	superb bird of paradise
Lophortyx californica	Schopfwachtel	California quail
Lophostrix cristata	Haubenkauz	crested owl
Lophotidae	Schopffische, Schopfköpfe, Einhornfische	crestfishes
Lophura edwardsi	Edwardsfasan	Edward's pheasant
Lophura imperialis	Kaiserfasan	imperial pheasant
Lophura nycthemera	Silberfasan	silver pheasant
Lophuromys spp.	Afrikanische Bürstenhaarmäuse	brush-furred mice, crest-tailed rats
Lopinga achine	Bacchantin, Gelbringfalter	woodland brown
Lorentzimys nouhuysi	Neuguinea-Hüpfmaus	New Guinea jumping mouse
Loricariidae	Harnischwelse	suckermouth armored catfishes
Loricera pilicornis	Borstenhornläufer, Schwarzer Krummhornkäfer	
Loricifera	Loriciferen, Korsetttierchen, Panzertierchen	corset bearers, loriciferans
Loriculus galgulus	Blaukrönchen	blue-crowned hanging parrot
Loriidae	Loris	lories, lorikeets
Loris tardigradus	Schlanklori	slender loris
Lorisidae	Loris	lorises & pottos & galagos
Lorius lory	Frauenlori	black-capped lory
Lota lota	Quappe, Rutte, Trüsche	burbot
Lottia gigantea	Eulen-Napfschnecke*	owl limpet
Lovenia cordiformis	Herzförmiger Herzigel	heart urchin, cordiform heart urchin
Lovenia elongata	Länglicher Herzigel	elongate heart urchin
Loxechinus albus	Chilenischer Seeigel	Chilean sea urchin
Loxia curvirostra	Fichtenkreuzschnabel	common crossbill
Loxia leucoptera	Bindenkreuzschnabel	two-barred crossbill
Loxia pytyopsittacus	Kiefernkreuzschnabel	parrot crossbill
Loxia scotica	Schottischer Kreuzschnabel	Scottish crossbill
Loxocemus bicolor	Spitzkopfpython	burrowing python, ground python, New World python
Loxodon macrorhinus	Schlitzaugen-Grundhai	sliteye shark
Loxodonta africana	Afrikanischer Elefant	African elephant
Loxorhynchus grandis	Kalifornische Schafskrabbe	Californian sheep crab
Loxosceles deserta	Wüsten-Braunspinne	desert loxosceles, desert brown spider
Loxosceles laeta	Südamerikanische Braunspinne	South American brown spider

Loxosceles reclusa	Braune Spinne	violin spider, brown recluse spider, fiddleback spider
Loxostege sticticalis/ Margaritia sticticalis	Wiesenzünsler, Rübenzünsler	beet webworm
Lucania goodei/ Chriopeops goodei	Rotschwanzkärpfling	blue-fin killifish, blue-fin killy, bluefin killifish (FAO)
Lucania parva	„Rainwater" Kärpfling	rainwater killifish
Lucanidae	Schröter, Hirschkäfer	stag beetles
Lucanus cervus	Hirschkäfer	stag beetle, European stag beetle
Lucanus elaphus	Amerikanischer Hirschkäfer	giant stag beetle
Lucernaria quadricornis	Breite Becherqualle	horned stalked jellyfish
Lucifer faxoni	Teufelsgarnele*	lucifer shrimp
Lucilia bufonivora	Krötenfliege	toad blowfly*
Lucilia sericata	Schafs-Goldfliege	sheep maggot fly, sheep blowfly, greenbottle
Lucilia spp.	Goldfliegen	greenbottles
Lucina pectinata	Dicke Mondmuschel	thick lucine
Lucinidae	Mondmuscheln	lucina clams, lucines (hatchet shells)
Lucinisca muricata	Stachlige Mondmuschel	spinose lucine
Luciocephalidae	Hechtköpfe	pikeheads
Luciocephalus pulcher	Hechtkopf	pikehead
Luciosoma spilopleura (Hbz. *L. setigerum*)	Seitenfleckbärbling	apollo shark, apollo sharkminnow (FAO) (side-spot barb)
Luidia ciliaris	Siebenarmiger Seestern, Schmalarmiger Großplattenstern, Großplattenseestern	seven-armed starfish, seven-rayed starfish
Luidia clathrata	Schlanker Seestern*	slender seastar
Lullula arborea	Heidelerche	woodlark
Lumbricidae	Regenwürmer	earthworms
Lumbricus rubellus	Roter Waldregenwurm	red earthworm, red worm
Lumbricus terrestris	Gemeiner Regenwurm, Tauwurm	earthworm, common earthworm; lob worm, dew worm, squirreltail worm, twachel (Br.)
Lumpenidae	Bandfische	snake blennies
Lumpenus lampretaeformis	Bandfisch, Spitzschwänziger Bandfisch	snake blenny
Lunatia catena/Natica catena	Halsband-Mondschnecke, Halsband-Nabelschnecke, Große Nabelschnecke	large necklace snail, large necklace moonsnail (large necklace shell)
Lunatia lewisi	Pazifische Mondschnecke	Lewis' moonsnail, western moon shell
Lunatia poliana	Glänzende Mondschnecke	shiny moonsnail
Lunatia spp.	Mondschnecken, Nabelschnecken	moonsnails, moon snails (moon shells)
Luperina nickerlii	Nickerls Graswurzeleule	sandhill rustic (moth)

Luperina testacea	Gelbliche Wieseneule, Grasstengeleule	flounced rustic (moth)
Luperus lyperus	Schwarzer Weidenblattkäfer	
Luscinia luscinia	Sprosser	thrush nightingale
Luscinia megarhynchos	Nachtigall	nightingale
Luscinia svecica/ Cyanosylvia svecia	Blaukehlchen	bluethroat
Lutjanidae	Schnapper	snappers
Lutjanus analis	Hammel-Schnapper	mutton snapper
Lutjanus apodus	Schoolmaster	schoolmaster snapper
Lutjanus argentimaculatus	Mangrovenbarsch, Mangroven-Schnapper	river snapper, mangrove red snapper (FAO)
Lutjanus argentiventris	Armadillo-Schnapper, Gelber Schnapper	armadillo snapper, yellow snapper (FAO)
Lutjanus buccanella	Schwarzflossen-Schnapper	blackfin snapper
Lutjanus campechanus	Roter Schnapper, Nördlicher Schnapper	red snapper, northern red snapper (FAO)
Lutjanus carponotatus	Braunstreifen-Schnapper	stripey, Spanish flag snapper (FAO)
Lutjanus decussatus	Gitterschnapper	checkered snapper
Lutjanus gibbus	Buckelschnapper	paddletail, humpback snapper, humpback red snapper (FAO)
Lutjanus griseus	Grauer Schnapper	gray snapper, grey snapper
Lutjanus jocu	Hundsschnapper	dog snapper
Lutjanus kasmira	Goldstreifenschnapper	blue-banded snapper, moonlighter, common bluestripe snapper (FAO)
Lutjanus mahogoni	Mahagoni-Schnapper	mahogany snapper
Lutjanus purpureus	Südlicher Schnapper	southern red snapper
Lutjanus sebae	Kaiserschnapper	emperor snapper, red emperor, emperor red snapper (FAO)
Lutjanus synagris	Rotschwanzschnapper	lane snapper
Lutjanus vivanus	Seidenschnapper	silk snapper
Lutra felina	Südamerikanischer Meerotter	marine otter (sea cat)
Lutra lutra	Europäischer Fischotter	European river otter
Lutra maculicollis	Fleckenhalsotter (Hbz. Afrikanischer Otter)	African river otter
Lutra provocax	Südamerikanischer Flussotter	southern river otter
Lutra spp.	Fischotter	river otters
Lutra sumatrana	Haarnasenotter (Hbz. Sumatra-Otter)	Sumatra river otter, hairy-nosed otter
Lutraria lutraria	Ottermuschel	common otter clam
Lutrariidae	Ottermuscheln	otter clams
Lutreolina crassicaudata	Dickschwanzbeutelratte	thick-tailed opossum
Luvaridae	Hahnenfische	louvars
Luvarus imperialis	Hahnenfisch, Dianafisch	luvar, louvar
Lybia tessalata	Boxerkrabbe	anemone carrying crab, common boxing crab
Lybiidae	Afrikanische Bartvögel	African barbets
Lybius bidentatus	Doppelzahn-Bartvogel	double-toothed barbet

Lycaeides idas	Ginster-Bläuling	Idas blue
Lycaena dispar	Großer Feuerfalter	large copper
Lycaena hippothoe/ Palaeochrysophanus hippothoe	Lilagold-Feuerfalter	purple-edged copper
Lycaena phlaeas	Kleiner Feuerfalter	small copper
Lycaenidae	Bläulinge (Feuerfalter) & Zipfelfalter	blues & hairstreaks & coppers & harvesters & metalmarks
Lycalopex vetulus/ Pseudalopex vetulus	Brasilianischer Kampffuchs (Brasilfuchs, Rio-Fuchs)	hoary fox, Brasilian fox
Lycaon pictus	Afrikanischer Wildhund	African wild dog, African hunting dog, African painted wolf
Lycengraulis grossidens	Atlantische Säbelzahn-Sardelle	river anchoita
Lycia hirtaria	Kirschenspanner	brindled beauty (moth)
Lycia zonaria	Trockenrasen-Spinnerspanner	belted beauty (moth)
Lycidae	Rotdeckenkäfer, Rotdecken-Käfer	net-winged beetles
Lycodes esmarkii	Esmarks Wolffisch	Esmark's eelpout
Lycodes vahlii	Vahls Wolffisch	Vahl's eelpout
Lycodon aulicus	Gewöhnliche Wolfszahnnatter	wolf snake
Lycodon spp.	Wolfszahnnattern	wolf snakes
Lycophidion capense	Kap-Wolfsnatter	African wolf snake
Lycophidion spp.	Wolfsnattern	African wolf snakes
Lycophotia porphyrea	Porphyr-Erdeule	true lovers knot moth
Lycoriidae	Trauermücken	dark-winged fungus gnats, root gnats
Lycosa narbonensis	Tarantel	European tarantula
Lycosa rabida	Tollwut-Tarantel*	rabid tarantula, rabid wolf spider
Lycosa tarentula	Apulische Tarantel	Apulian tarantula
Lycosidae	Wolfsspinnen, Wolfspinnen	wolf spiders, ground spiders
Lycoteuthis diadema/ Thaumatolampas diadema	Wunderlampe	deepsea luminescent squid a.o.
Lyctidae	Splintholzkäfer	powder-post beetles & shot-hole borers
Lyctocoris campestris	Geflügelte Bettwanze	debris bug, stack bug, field anthocoris
Lyctus brunneus	Brauner Splintholzkäfer	brown powderpost beetle
Lyctus linearis	Parkettkäfer	true powderpost beetle, European powderpost beetle
Lyctus planicollis	Amerikanischer Splintholzkäfer	southern lyctus beetle, European powderpost beetle
Lygaeidae	Bodenwanzen, Langwanzen (Ritterwanzen)	ground bugs, seed bugs
Lygaeus equestris	Ritterwanze	
Lygaeus kalmii	Kleine Wolfsmilchwanze*	small milkweed bug
Lygaeus pandurus	Pandur	
Lygephila craccae		scarce blackneck moth
Lygocoris pabulinus	Grüne Futterwanze	common green capsid

Lygus pratensis/ *Lygus rugulipennis*	Gemeine Wiesenwanze	European tarnished plant bug, tarnished plant bug, bishop bug
Lymantria dispar	Schwammspinner	gipsy moth
Lymantria monacha	Nonne	black arches
Lymantriidae	Wollspinner, Trägspinner, Schadspinner	tussock moths & gypsy moths & others
Lymexylon navale	Eichenwerftkäfer, Schiffswerftkäfer	ship timberworm
Lymexylonidae	Werftkäfer	ship-timber beetles
Lymnaea auricularia	Ohren-Schlammschnecke	eared pondsnail, ear pond snail
Lymnaea glabra/ *Stagnicola glabra*	Längliche Sumpfschnecke	oblong pondsnail, mud pondsnail, mud snail
Lymnaea palustris	Sumpf-Schlammschnecke	marsh pondsnail, marsh snail
Lymnaea peregra	Gewöhnliche Schlammschnecke	wandering pondsnail, common pondsnail
Lymnaea stagnalis	Spitzschlammschnecke, Spitzhorn, Spitzhorn-Schlammschnecke, Große Schlammschnecke	great pondsnail, swamp lymnaea
Lymnaea truncatula	Zwerg-Schlammschnecke	dwarf pondsnail
Lymnaeidae	Schlammschnecken	pondsnails, pond snails
Lymnocryptes minimus	Zwergschnepfe	jack snipe
Lynchailurus pajeros/ *Felis colocolo*	Pampaskatze	pampas cat
Lyncodon patagonicus	Zwerggrison	Patagonian weasel
Lyneus brachyurus	Dickbauchkrebs	
Lynx lynx/*Felis lynx*	Nordluchs (Hbz. Luchs)	lynx
Lyonetia clerkella	Schlangenminiermotte, Obstbaumminiermotte	apple leaf miner, Clerk's snowy bentwing
Lyonetiidae	Langhorn-Blattminiermotten	lyonetiid moths
Lyria delessertiana	Delesserts Leierschnecke	
Lyria lyraeformis	Leierschnecke	lyre-formed lyria
Lyria mitraeformis	Bischofmützen-Leierschnecke*	miter-shaped lyria
Lyriocephalus scutatus	Leierkopfagame	lyrehead lizard
Lyristes plebeja	Gemeine Zikade	common Southern European cicada
Lyrocardium lyratum	Leier-Herzmuschel	lyre cockle
Lyroderma lyra/ *Megaderma lyra*	Falscher Vampir	false vampire bat
Lyrodus pedicellatus	Schwarzspitzen-Schiffsbohrwurm*	blacktip shipworm
Lyropecten corallinoides	Korallen-Kammmuschel	coral scallop
Lyropecten subnodosa	Pazifische Löwenpranke	Pacific lion's paw
Lyrurus tetrix/ *Tetrao tetrix*	Birkhuhn	black grouse
Lysandra bellargus/ *Polyommatus bellargus*	Himmelblauer Bläuling	adonis blue (butterfly)
Lysandra coridon	Silbergrüner Bläuling, Steppenheidebläuling	chalkhill blue (butterfly)

Lysiosquilla maculata	Gepunkteter Fangschreckenkrebs	dotted mantis shrimp
Lysmata seticaudata	Mittelmeer-Putzergarnele	Mediterranean cleaner shrimp, Mediterranean rock shrimp, Monaco cleaner shrimp
Lysmata wurdemanni	Pfefferminzgarnele	peppermint shrimp, Carribbean cleaner shrimp, veined shrimp
Lystrophis dorbignyi	Südamerikanische Hakennatter	South American hognose snake
Lystrophis spp.	Südamerikanische Hakennattern	South American hognose snakes
Lytechinus variegatus	Weißer Westindischer Seeigel, Verschiedenfarbiger Seeigel	variegated urchin
Lytocarpia myriophyllum	Fasanenfedermoos*	pheasant-tail hydroid (pheasant's tail coralline)
Lytocarpus nuttingi	Brennmoos*	stinging hydroid
Lytocarpus philippinus	Fieder-Hydrozoe	white stinging hydroid
Lytorhynchus diadema	Gekrönte Schnauzennatter	awl-headed snake
Lytta vesicatoria	Spanische Fliege	Spanish fly, blister beetle

Mabuya heathi	Brasilianische Mabuya, Brasilianische Mabuye	Brazilian skink, Brazilian mabuya
Mabuya quinquetaeniata	Fünfstreifen-Mabuye	African five-lined skink, rainbow skink
Mabuya wrightii	Große Seychellen-Mabuye	greater Seychelles skink, Wright's mabuya
Macaca arctoides	Bärenmakak, Stumpfschwanzmakak	bear macaque, stump-tailed macaque
Macaca assamensis	Bergrhesus, Assamrhesus, Assammakak	Assam macaque
Macaca cyclopis	Formosamakak, Rundgesichtsmakak	Taiwan macaque
Macaca fascicularis/ Macaca irus	Javaneraffe, Langschwanzmakak	crab-eating macaque
Macaca fuscata	Rotgesichtsmakak	Japanese macaque
Macaca maura	Mohrenmakak	moor macaque
Macaca mulatta	Rhesusaffe, Rhesusmakak	rhesus monkey
Macaca nemestrina	Schweinsaffe	pigtail macaque
Macaca nigra	Schopfmakak, Schopfaffe	Celebes ape
Macaca ochreata	Grauarmmakak	booted macaque
Macaca radiata	Indischer Hutaffe	bonnet macaque
Macaca silenus	Bartaffe, Wanderu	liontail macaque, lion-tailed macaque
Macaca sinica	Ceylon-Hutaffe	toque macaque
Macaca spp.	Makaken	macaques
Macaca sylvanus	Berberaffe, Magot	barbary ape, barbary macaque
Macaca thibetana	Tibetmakak	Père David's stump-tailed macaque, Tibetan stump-tailed macaque
Macaca tonkeana	Tonkeanamakak	Tonkean macaque
Macaria liturata	Veilgrauer Kiefernspanner	
Machaehamphus alcinus	Fledermausaar	bat hawk
Machilidae	Felsenspringer	jumping bristletails
Macolor niger	Schwarzer Schnapper	black beauty, mottled snapper, black and white snapper (FAO)
Macoma baltica	Baltische Tellmuschel, Baltische Plattmuschel, Rote Bohne	Baltic macoma
Macoma loveni	Blasige Tellmuschel	inflated macoma
Macoma moesta	Flache Tellmuschel	flat macoma
Macracanthorhynchus hirudinaceus (Acanthocephala)	Riesenkratzer	thorny headed worm
Macristiidae	Kammfische	
Macrobdella decora	Amerikanischer Blutegel, Amerikanischer Medizinischer Blutegel	American medicinal leech
Macrobrachium carcinus	Großscheren-Süßwassergarnele*	bigclaw river shrimp
Macrobrachium rosenbergii	Rosenberg-Garnele, Rosenberg Süßwassergarnele, Hummerkrabbe (Hbz.)	Indo-Pacific freshwater prawn, giant river shrimp, giant river prawn, blue lobster (tradename)

Macrocephalon maleo	Hammerhuhn	maleo
Macrocheira kaempferi	Japanische Riesenkrabbe	giant spider crab
Macroclemys temminckii	Geierschildkröte	alligator snapping turtle
Macrocoeloma septemspinosum	Siebenspitz-Schmuckkrabbe*	thorny decorator crab
Macrocoeloma trispinosum	Dreispitz-Schmuckkrabbe*	spongy decorator crab
Macrocyclops fuscus	Riesenhüpferling	
Macrocypraea cervus/ Cypraea cervus	Atlantische Hirschkauri	Atlantic deer cowrie
Macrocypraea zebra/ Cypraea zebra	Zebrakauri, Gemaserte Kauri*	zebra cowrie, measled cowrie
Macrodactyla doreensis	Korkenzieheranemone, Fadenanemone	corkscrew anemone, long tentacle anemone (L.T.A.), red base anemone
Macrodactylus subspinosus	Rosenkäfer*	rose chafer
Macroderma gigas	Australische Gespenstfledermaus	Australian giant false vampire bat, ghost bat
Macrodon ancylodon	Südamerikanischer Königs-Umberfisch	king weakfish
Macrogalidia musschenbroeki	Celebes-Roller	Celebes palm civet
Macrogastra lineolata	Mittlere Schließmundschnecke	lined door snail
Macrogastra plicatula	Gefältelte Schließmundschnecke	plicate door snail
Macrogastra rolphii	Spindelförmige Schließmundschnecke	Rolph's door snail
Macrogastra ventricosa	Bauchige Schließmundschnecke	ventricose door snail
Macroglossum stellatarum	Taubenschwänzchen	hummingbird hawkmoth
Macroglossus spp.	Langzungen-Flughunde	long-tongued fruit bats
Macron aethiops	Dunkles Riesenhorn	dusky macron
Macronectes giganteus	Riesensturmvogel	giant petrel
Macronyssidae	Nagermilben	rodent mites
Macronyx capensis	Großspornpieper	Cape longclaw
Macrophyllum macrophyllum	Langbein-Fledermaus	long-legged bat
Macropis labiata	Schenkelbiene	
Macropodia rostrata	Langbeinige Spinnenkrabbe, Gespensterkrabbe	long-legged spider crab
Macropodia tenuirostris	Schlankhörnige Spinnenkrabbe	slender spider crab
Macropodinae	Kampffische, Paradiesfische	paradisefishes
Macropodus opercularis	Makropode, Paradiesfisch	paradise fish, paradisefish
Macropodus spp.	Paradiesfische	paradisefishes
Macroposthonia xenoplax	Ringälchen (der Weinrebe)	ring nematode (on grapevine)
Macroprotodon cucullatus	Kapuzennatter	hooded snake, false smooth snake
Macropus agilis	Flinkes Känguru, Sandwallaby	agile wallaby
Macropus antilopinus	Antilopenkänguru	antilopine wallaroo
Macropus bernardus	Schwarzes Bergkänguru, Schwarzes Wallaruh	black wallaroo, Bernard's wallaroo
Macropus dorsalis	Rückenstreifkänguru, Aalstrichwallaby	black-striped wallaby
Macropus eugenii	Derbywallaby, Tammarwallaby	tammar wallaby, dama wallaby

Macropus fuliginosus	Westliches Graues Riesenkänguru	western gray kangaroo
Macropus giganteus	Graues Riesenkänguru	eastern gray kangaroo
Macropus greyi	Greys Wallaby	toolache wallaby
Macropus irma	Westliches Strauchwallaby	western brush wallaby
Macropus parma	Parmawallaby	parma wallaby
Macropus parryi	Schönwallaby, Hübschgesichtwallaby	whiptail wallaby, pretty-face wallaby
Macropus robustus	Bergkänguru, Wallaruh	wallaroo, common wallaroo, euro
Macropus rufogriseus	Bennettkänguru	red-necked wallaby
Macropus rufus	Rotes Riesenkänguru	red kangaroo
Macroramphosidae	Schnepfenfische	snipefishes
Macroramphosus scolopax	Schnepfenfisch	snipefish
Macroscelidea	Rüsselspringer	African elephant shrews
Macroscelides proboscideus	Kurzohrrüsselspringer	short-eared elephant shrew
Macroscelididae	Rüsselspringer, Rohrrüssler	elephant shrews
Macrosiphum avenae/ Sitobion avenae	Große Getreideblattlaus	greater cereal aphid, European grain aphid, English grain aphid
Macrosiphum rosae	Rosenblattlaus, Große Rosenblattlaus, Große Rosenlaus	rose aphid, "greenfly"
Macrotarsomys spp.	Inselmäuse	Madagascan rats
Macrothylacia rubi	Brombeerspinner	fox moth
Macrotis spp.	Kaninchennasenbeutler, Ohrenbeuteldachse	bilbies, rabbit-eared bandicoots
Macrotus spp.	Großohrfledermäuse	big-eared bats
Macrouridae	Grenadierfische, Rattenschwänze	grenadiers, rattails
Macrourus berglax	Rauer Grenadier, Raukopf-Panzerratte, Nordatlantik-Grenadier	rough rattail, rough-headed grenadier, rough-head grenadier, onion-eye grenadier (FAO)
Macrozoarces americanus	Amerikanische „Aalmutter"	ocean pout
Macruromys spp.	Neuguinea-Ratten, Langschwanzratten	New Guinean rats
Macruronidae	Süd-Seehechte	southern hakes
Macruronus magellanicus	Patagonischer Grenadier	Patagonian grenadier
Macruronus novaezealandiae	Neuseeländischer Grenadier	blue grenadier
Mactra cinerea/ Mactra stultorum	Gemeine Trogmuschel, Strahlenkörbchen	white trough clam
Mactra corallina corallina	Weiße Trogmuschel	rayed trough clam
Mactra dolabriformis	Beilmuschel*, Beil-Trogmuschel*	hatchet surfclam
Mactra sachalinensis	Sachalin-Trogmuschel	hen clam
Mactrellona alata	Flügel-Trogmuschel	winged surfclam
Mactrellona exoleta	Reife Trogmuschel	mature surfclam
Mactridae	Trogmuscheln	mactras, trough clams (trough shells)
Mactrotoma californica	Kalifornische Trogmuschel	California surfclam, Californian mactra

Maculinea alcon	Lungenenzian-Ameisen-Bläuling, Kleiner Moorbläuling	alcon blue (butterfly)
Maculinea arion	Schwarzfleckiger Bläuling, Arion-Bläuling, Quendel-Ameisenbläuling	large blue (butterfly)
Maculinea telejus	Augenbläuling, Großer Moorbläuling	scarce large blue (butterfly)
Madoqua spp.	Windspielantilopen, Dikdiks	dik-diks
Madracis decactis	Knotige Koralle, Felderkoralle	ten-ray star coral
Madracis mirabilis	Zweigkoralle	yellow pencil coral
Madracis pharensis	Krustenkoralle	crusty pencil coral
Madrepora oculata	Augenkoralle, Weiße Koralle	ocular coral, white coral
Madreporaria/Scleractinia	Riffkorallen, Steinkorallen	stony corals, madreporarian corals, scleractinians
Maena maena/Spicara maena	Gefleckter Schnauzenbrassen, Laxierfisch	menola, menole, mendole, blotched picarel (FAO)
Maja squinado/ Maia squinado	Große Seespinne, Teufelskrabbe	common spider crab, thorn-back spider crab
Maja verrucosa/ Maia verrucosa/ Maja crispata	Kleine Seespinne	small spider crab, lesser spider crab
Majidae	Seespinnen	spider crabs
Makaira indica	Schwarzer Marlin	black marlin
Makaira nigricans	Blauer Marlin	blue marlin, Atlantic blue marlin (FAO)
Malacanthidae	Ziegelfische	sand tilefishes
Malacanthus brevirostris	Streifenschwanz-Ziegelbarsch	banded blanquillo, quakerfish (FAO)
Malacanthus latovittatus	Blaukopf-Ziegelbarsch	sand tilefish
Malachiidae	Zipfelkäfer, Warzenkäfer	malachiid beetles, flower beetles
Malachius bipustulatus	Rotzipfelkäfer	red-tipped flower beetle
Malaclemys terrapin	Diamantschildkröte	diamondback terrapin
Malacocephalus laevis	Weichkopf-Grenadier	softhead rattail
Malacochersus tornieri	Spaltenschildkröte	pancake tortoise, African pancake tortoise
Malacolimax tenellus	Pilzschnegel	slender slug
Malacomys spp.	Langfußratten	big-eared swamp rats, long-eared marsh rats, long-footed rats
Malaconotinae	Buschwürger	bush-shrikes
Malacoraja senta/Raja senta	Glattrochen	smooth skate
Malacosoma castrensis	Wolfsmilch-Ringelspinner, Grasheiden-Ringelspinner, Wolfsmilchspinner	ground lackey moth
Malacosoma franconica	Frankfurter Ringelspinner	Francfort lackey moth
Malacosoma neustria	Ringelspinner, Gemeiner Ringelspinner, Obsthain-Ringelspinner	lackey, European lackey moth, common lackey
Malacosteidae	Zungenkiemer, Weichstrahlenfische	loosejaws

Malacothrix typica	Südafrikanische Langohrmaus	gerbil mouse, long-eared mouse
Maladera castanea	Asiatischer Gartenkäfer*	Asiatic garden beetle
Maladera holosericea	Dunkler Seidenkäfer	dark silky beetle*
Malapteruridae	Elektrische Welse, Zitterwelse	electric catfishes
Malapterurus electricus	Elektrischer Wels, Zitterwels	electric catfish
Malayemys subtrijuga	Malaiische Sumpfschildkröte, Malaien-Sumpfschildkröte	Malayan snail-eating turtle
Malea pomum	Apfelschnecke	apple tun, Pacific grinning tun
Malea ringens	Grinsendes Fass	great grinning tun
Malimbus scutatus	Schildweber	red-vented malimbe
Malleidae	Hammermuscheln	hammer oysters
Malleus albus	Weiße Hammermuschel	white hammer-oyster
Malleus candeanus	Amerikanische Hammermuschel, Karibische Hammermuschel	Caribbean hammer-oyster, American hammer oyster, American malleus
Malleus malleus	Häufige Hammermuschel, Schwarze Hammermuschel	common hammer-oyster
Malleus spp.	Hammermuscheln	hammer oysters
Mallomys rothschildi	Rothschild-Goliathratte	giant tree rat
Mallophaga	Läuslinge, Kieferläuse (Federlinge & Haarlinge)	chewing lice, biting lice, bird lice
Mallotus villosus	Lodde	capelin
Malpolon moilensis	Moilanatter	moila snake
Malpolon monspessulans	Eidechsennatter, Europäische Eidechsennatter	Montpellier snake
Maluridae	Staffelschwänze, Australische Sänger	fairywrens, Australian wrens
Malurus cyaneus	Prachtstaffelschwanz	blue wren
Mamestra brassicae	Kohleule, Herzeule	cabbage moth
Mamestra oleracea/ Lacanobia oleracea	Gemüseeule	tomato moth (bright-line brown-eye moth)
Mamestra persicariae	Schwarze Garteneule	
Mamestra pisi	Erbseneule	
Mammalia	Säugetiere, Säuger	mammals
Mammilla melanostoma	Schwarzmündige Nabelschnecke	black-mouth moonsnail
Mandrillus leucophaeus/ Papio leucophaeus	Drill	drill
Mandrillus sphinx/ Papio sphinx	Mandrill	mandrill
Mangora acalypha	Streifenkreuzspinne	lined orbweaver
Mangora gibberosa	Höckrige Streifenkreuzspinne*	gibbose lined orbweaver
Manica rubida/ Myrmica rubida	Gefährliche Knotenameise	
Manicina areolata areolata	Rosenkoralle*	rose coral
Maniola jurtina	Großes Ochsenauge	meadow brown (butterfly)
Manis crassicaudata	Vorderindisches Schuppentier	Indian pangolin
Manis gigantea	Riesen-Schuppentier	large African pangolin
Manis javanica	Javanisches Schuppentier	Malayan pangolin

Manis pentadactyla	Chinesisches Ohren-Schuppentier	Chinese pangolin
Manis spp.	Schuppentiere	scaly anteaters, pangolins
Manis temmincki	Steppen-Schuppentier	Cape pangolin
Manis tetradactyla	Langschwanz-Schuppentier	phatagin
Manis tricuspis	Weißbauch-Schuppentier	African tree pangolin
Mannophryne trinitatis/ Colostethus trinitatis	Trinidad-Baumsteiger (Venezuela-Blattsteiger)	Trinidad poison frog
Manouria emys	Asiatische Riesenschildkröte, Braune Landschildkröte	Asian giant tortoise (IUCN), Asian brown tortoise, Burmese brown tortoise
Manouria impressa/ Geochelone impressa	Hinterindische Landschildkröte	impressed tortoise
Manta birostris	Manta, Riesenmanta, Großer Teufelsrochen	manta, giant manta (FAO), Atlantic manta, giant devil ray
Manta hamiltoni	Ostpazifisch-Mittelamerikanischer Manta	Eastern Pacific manta
Mantella aurantiaca	Goldfröschchen	golden frog
Mantis religiosa	Gottesanbeterin	European preying mantis
Mantispidae	Fanghafte	mantis flies, mantidflies
Mantodea/Mantoptera	Fangschrecken & Gottesanbeterinnen	mantids
Manucodia keraudrenii/ Phonygammus keraudrenii	Trompeterparadiesvogel	trumpet bird
Marcusenius longianalis/ Brienomyrus longianalis	Schmaler Nilhecht	slender Nile pike*, mpanda
Marcusenius macrolepidotus	Großschuppiger Nilhecht	bulldog (FAO & South Africa), mputa, big-scale Nile pike
Margaretamys spp.	Margareta-Ratte	Margareta's rats
Margarites costalis		boreal rosy margarite
Margarites groenlandicus	Grönland-Kreiselschnecke	Greenland margarite, Greenland topsnail (Greenland top shell)
Margarites helicinus	Glatte Perlkreiselschnecke	spiral margarite, smooth margarite, pearly topsnail (pearly top shell)
Margarites olivaceus	Oliven-Kreiselschnecke	olive margarite, olive topsnail (olive top shell)
Margarites pupillus	Puppen-Kreiselschnecke	puppet margarite (puppet top shell)
Margarites spp.	Perlkreiselschnecken	margarites, pearly topsnails (pearly top shells)
Margaritifera falcata	Westliche Flussperlmuschel	western pearlshell, western freshwater pearl mussel
Margaritifera margaritifera	Flussperlmuschel	freshwater pearl mussel (Scottish pearl mussel), eastern pearlshell
Margarodidae	Höhlenschildläuse	giant coccids & ground pearls
Margaroperdix madagarensis	Perlwachtel	Madagascar partridge

Marginella see also: *Prunum*

Marginella accola/ *Persicula accola*	Panama-Randschnecke*	twinned marginella, twinned marginsnail (twinned margin shell)
Marginella cingulata/ *Marginella lineata/* *Persicula cingulata*	Gürtel-Randschnecke, Bänder-Randschnecke	girdled marginella, girdled marginsnail (girdled margin shell)
Marginella cornea/ *Persicula cornea*	Blasse Randschnecke	plain marginella, pale marginella, pale marginsnail (pale margin shell)
Marginella glabella	Glatte Randschnecke	shiny marginella, smooth marginella, smooth marginsnail (smooth margin shell)
Marginella labiata	Königliche Randschnecke	royal marginella, royal marginsnail (royal margin shell)
Marginella muscaria/ *Austroginella muscaria*	Fliegenschnecke, Fliegen-Randschnecke	fly marginella, fly marginsnail (fly margin shell)
Marginella nebulosa	Nebel-Randschnecke, Wolken-Randschnecke	cloudy marginella, clouded marginella, clouded marginsnail (clouded margin shell, cloudy margin shell)
Marginella persicula/ *Persicula persicula*	Getupfte Randschnecke	spotted marginella, spotted marginsnail (spotted margin shell)
Marginella strigata/ *Cryptospira strigata*	Gestreifte Randschnecke	striped marginella, striped marginsnail (striped margin shell)
Marginellidae	Randschnecken	marginellas
Marmoronetta angustirostris	Marmelente	marbled teal
Marmosa spp.	Zwergbeutelratten	murine opossums, mouse opossums
Marmota bobak	Bobak (Hbz. Murmel)	bobac marmot
Marmota caligata	Eisgraues Murmeltier	hoary marmot
Marmota flaviventris	Gelbbäuchiges Murmeltier	yellow-bellied marmot
Marmota marmota	Alpenmurmeltier (Hbz. Murmel)	alpine marmot
Marmota monax	Waldmurmeltier	woodchuck, ground hog
Marmota spp.	Murmeltiere	marmots
Marphysa sanguinea	Blutkieferwurm	rock worm, red rock worm, red-gilled marphysa
Marpissa radiata	Schilf-Springspinne	reed slender spider*, reed spider*
Marstoniopsis steini	Kammkiemenschnecke	
Marsupenaeus japonicus/ *Penaeus japonicus*	Radgarnele	Kuruma shrimp
Marsupialia	Beuteltiere	marsupials, pouched mammals
Martes americana	Fichtenmarder (Hbz. Nordamerikanischer Zobel, Kanadischer Zobel)	American pine marten
Martes flavigula	Buntmarder (Hbz. Charsa)	yellow-throated marten

Martes foina	Steinmarder	beech marten, stone marten
Martes gwatkinsi	Indischer Charsa	Nilgiri marten
Martes martes	Baummarder, Edelmarder	European pine marten
Martes melampus	Japanischer Marder	Japanese marten
Martes pennanti	Fischermarder (Hbz. Fisher/Pekan/ Virginischer Iltis)	fisher, pekan
Martes zibellina	Zobel	sable
Marthasterias glacialis	Eisseestern, Eisstern, Warzenstern	spiny starfish
Marumba quercus/ Smerinthus quercus	Eichenschwärmer	oak hawkmoth
Masaridae	Honigwespen	shining wasps, masarid wasps
Massoutiera mzabi	Sahara-Gundi, Langhaargundi, Sahara-Kammfinger	Mzab gundi, fringe-eared gundi
Mastacembelidae	Stachelaale, Pfeilaale	spiny eels
Mastacomys fuscus	Breitzahnratte*	broad-toothed rat
Masticophis bilineatus	Sonora-Peitschenatter, Sonora-Zornnatter	Sonora whipsnake
Masticophis flagellum flagellum/ Coluber flagellum	Peitschennatter, Peitschenschlange, Rote Kutschenpeitschen-Natter	coachwhip, San Joaquin coachwhip
Masticophis lateralis	Gestreifte Peitschennatter, Streifen-Zornnatter	striped racer, striped whipsnake
Masticophis spp.	Peitschennattern	whipsnakes
Masticophis taeniatus	Wüsten-Peitschennatter, Bandnatter	desert striped whipsnake, desert whipsnake
Mastigophora	Geißeltierchen, Geißelträger, Flagellaten	flagellates, mastigophorans
Mastigoproctus giganteus	Essigskorpion, Geißelskorpion	giant vinegaroon
Mastomys spp.	Vielzitzenmäuse	multimammate rats
Mastophora bisaccata	Amerikanische Bola-Spinne, Lasso-Spinne	American bolas spider
Mastophora spp.	Bola-Spinnen, Lasso-Spinnen	bolas spiders
Mastotermes darwiniensis	Riesentermite	Darwin termite
Mastotermitidae	Riesentermiten	Darwin termites
Maticora bivirgata	Blaue Bauchdrüsenotter	blue Malaysian coral snakes
Maticora intestinalis	Rotschwanz-Bauchdrüsenotter	banded Malaysian coral snakes
Maticora spp.	Bauchdrüsenottern	long-glanded coral snakes, Malaysian coral snakes
Mauligobius maderensis	Madeira-Grundel	Madeiran goby
Maurea tigris	Tigerschnecke	tiger topsnail, tiger maurea (tiger top shell)
Mauremys caspica	Kaspische Bachschildkröte	Caspian turtle
Mauremys japonica	Japanische Bachschildkröte	Japanese turtle
Mauremys leprosa	Maurische Bachschildkröte	Maurish turtle, Mediterranean turtle
Mauremys nigricans	Dreikiel-Wasserschildkröte	Asian yellow pond turtle

Mauremys spp.	Eurasiatische Wasserschildkröten	Eurasian pond turtles
Mauritia arabica arabica	Arabische Kauri	Arabian cowrie
Mauritia eglantina		eglantine cowrie
Mauritia histrio	Geschichtskauri*	history cowrie
Mauritia mappa/ Cypraea mappa	Landkartenkauri, Landkartenschnecke	map cowrie
Mauritia mauritiana	Buckelkauri, Buckelschnecke, Großer Schlangenkopf	hump-backed cowrie, humpback cowrie
Maurolicus muelleri		pearlsides (FAO), sheppy argentine
Maxillopoda	Kieferfüßer	maxillopods
Maxomys spp.	Stachelratten	Rajah rats, spiny rats
Mayailurus iriomotensis/ Felis iriomotensis	Iriomoto-Katze	Iriomote cat
Mayermys ellermani	Ellermans Maus, Shaw-Mayer-Maus	Shaw-Mayer's mouse, one-toothed shrew-mouse
Mayetiola destructor	Hessenmücke, Hessenfliege	Hessian fly
Mazama americana	Großer Roter Spießhirsch	red brocket
Mazama chunyi	Kleiner Grauer Spießhirsch	dwarf brocket
Mazama gouazoubira	Großer Grauer Spießhirsch	brown brocket
Mazama rufina	Kleiner Roter Spießhirsch	little red brocket
Mazama spp.	Spießhirsche	brocket deer
Meandrina meandrites meandrites	Labyrinthkoralle*, Mäanderkoralle	maze coral (brainstone)
Meconema thalassinum	Eichenschrecke, Gemeine Eichenschrecke	oak bushcricket
Mecoptera	Schnabelfliegen	scorpion flies, mecopterans
Mecostethus alliaceus/ Parapleurus alliaceus	Lauchschrecke	leek grasshopper
Mecostethus grossus/ Stethophyma grossum	Sumpfschrecke	large marsh grasshopper
Mecynogea lemniscata	Basilikaspinne*	basilica spider, basilica orbweaver
Mediodactylus kotschyi/ Cyrtodactylus kotschyi	Ägäischer Nacktfinger	Kotschy's gecko
Medorippe lanata/ Dorippe lanata	Schirmkrabbe	demon-faced porter crab
Megabalanus psittacus	Riesen-Seepocke	giant Chilean barnacle
Megacardita incrassata	Dicke Trapezmuschel	thickened cardita
Megachasma pelagios	Riesenmaulhai, Großmaulhai	megamouth shark
Megachasmidae	Riesenmaulhaie, Großmaulhaie	megamouth sharks
Megachile centuncularis/ Megachile versicolor	Gemeine Blattschneiderbiene, Rosenblattschneiderbiene	common leafcutter bee, common leafcutting bee, rose leaf-cutting bee
Megachile parietina/ Chalicodoma parietina/ Chalicodoma muraria	Mörtelbiene	wall bee, mason bee
Megachile rotundata	Alfalfa-Biene, Luzernen-Blattschneiderbiene	alfalfa leafcutter bee

Megachile spp.	Blattschneiderbienen	leafcutting bees, leaf-cutter bees
Megachilidae	Blattschneiderbienen	leafcutting bees, leaf-cutter bees
Megadendromus nikolausi	Riesenbaummaus, Riesenklettermaus	giant climbing mouse
Megaderma spp.	Großblattnasen, Eigentliche Großblattnasen	Asian false vampire bats
Megadermatidae	Großblattnasen	false vampire bats, yellow-winged bats
Megadyptes antipodes	Gelbaugenpinguin	yellow-eyed penguin
Megalaima asiatica	Blauwangen-Bartvogel	blue-throated barbet
Megalaimoidae	Asiatische Bartvögel	Asian barbets
Megalaspis cordyla	Torpedo Stachelmakrele	torpedo scad
Megalochelys gigantea/ Testudo gigantea/ Geochelone gigantea	Seychellen-Riesenschildkröte	Seychelles giant tortoise, Aldabran giant tortoise, Aldabra giant tortoise (IUCN)
Megaloglossus woermanni	Afrikanischer Langzungen-Flughund	African long-tongued fruit bat
Megalomys spp.	West-Indische Riesenreisratte	West Indian giant rice rats
Megalopidae	Tarpune	tarpons
Megalops atlanticus/ Tarpon atlanticus	Atlantischer Tarpon, Atlantischer Tarpun	Atlantic tarpon
Megalops cyprinoides	Ochsenauge, Indopazifischer Tarpon, Indopazifischer Tarpun	Indopacific tarpon
Megaloptera	Schlammfliegen	megalopterans: dobsonflies / fishflies / alderflies (neuropterans)
Megalopyge opercularis		puss caterpillar
Megalurinae	Grassteiger	grass-warblers
Megamerinidae	Schenkelfliegen	megamerinid flies
Megamuntiacus vuquagensis	Riesenmuntjak	giant muntjac
Meganola strigula	Ungebändertes Eichen-Kleinbärchen	small black arches (moth)
Meganyctiphanes norvegica	Nördlicher Krill, Norwegischer Krill	Norwegian krill
Megaoryzomys curioi	Galapagos-Riesenreisratte	Galapagos giant rat, Galapagos giant rice rat
Megaphobema velvetosoma	Braune Ecuador-Samt-Vogelspinne*	Ecuadorian brownvelvet tarantula
Megapodiidae	Großfußhühner	megapodes
Megaptera novaeangliae	Buckelwal	humpback whale
Megarhynchus pitangua	Bauchschnabeltyrann	boat-billed flycatcher
Megascolides australis	Australischer Riesenregenwurm	giant Gippsland earthworm, karmai
Megastigmus aculeatus	Rosensamenwespe	rose seed wasp
Megastigmus bipunctatus	Wacholdersamenwespe	juniper seed wasp
Megastigmus brevicaudis	Vogelbeersamenwespe	rowan seed wasp
Megastigmus pinus	Silbertannen-Samenwespe	silver fir seed wasp
Megastigmus pistaciae	Pistaziensamenwespe	pistachio seed wasp
Megastigmus spermotrophus	Douglasien-Samenwespe	Douglas fir seed wasp, Douglas fir seedfly
Megastigmus spp.	Samenwespen	seed wasps
Megastigmus strobilobius	Fichtensamenwespe	spruce seed wasp

Megastigmus suspectus	Tannensamenwespe	fir seed wasp
Megathura crenulata	Riesen-Schlüssellochschnecke, Riesen-Lochschnecke	giant keyhole limpet
Megophrys montana nasuta	Zipfelkrötenfrosch	Asian spadefoot toad
Megophrys spp.	Zipfelfrösche	Asian spadefoot toads, horned-nosed frogs, horned toads
Megopis scabricornis	Körnerbock	
Mehelya capensis	Kap-Feilennatter	Cape file snake
Mehelya spp.	Feilennattern	African file snakes
Melamphaeidae	Großschuppenfische	bigscale fishes, ridgeheads
Melanargia galathea	Schachbrett, Damenbrett	marbled white
Melanchra persicariae	Schwarze Garteneule, Knötericheule, Flohkrauteule, Blumeneule	dot moth
Melanchra pisi	Erbseneule	broom moth
Melandrya caraboides	Laufkäferartiger Düsterkäfer	
Melandryidae (Serropalpidae)	Düsterkäfer, Schwarzkäfer	false darkling beetles
Melanerpes candidus	Weißspecht	white woodpecker
Melanerpes erythrocephalus	Rotkopfspecht	red-headed woodpecker
Melanerpes formicivorus	Eichelspecht	acorn woodpecker
Melanitta fusca	Samtente	velvet scoter
Melanitta nigra	Trauerente	black scoter
Melanitta perspicillata	Brillenente	surf scoter
Melanobatrachus indicus	Indischer Schwarzfrosch	Kerala Hills frog (black frog)
Melanobatrachus spp.	Schwarzfrösche	Malaysian treefrogs, black frogs
Melanocetidae	Schwarze Angler, Schwarzangler, Tiefseeteufel	devil-anglers
Melanochelys spp.	Indische Erdschildkröte	Indian terrapins
Melanochelys tricarinata	Dreikiel-Erdschildkröte	tricarinate hill turtle, three-keeled land tortoise (IUCN)
Melanochelys trijuga trijuga	Schwarzbauch-Erdschildkröte	Indian black turtle, hard-shelled terrapin
Melanochlora sultanea	Sultansmeise	sultan tit
Melanochromis auratus/ Pseudotropheus auratus	Türkisgoldbarsch	golden mbuna, auratus
Melanochromis johanni/ Pseudotropheus johanni	Kobaltorangebarsch	bluegray mbuna
Melanochromis labrosus	Wulstlippen-Buntbarsch	labrosus
Melanochromis vermivorus	Lila Mbuna	purple mbuna
Melanocorypha bimaculata	Bergkalanderlerche	bimaculated lark
Melanocorypha calandra	Kalanderlerche	calandra lark
Melanocorypha leucoptera	Weißflügellerche	white-winged lark
Melanocorypha yeltoniensis	Schwarzlerche	black lark
Melanogrammus aeglefinus	Schellfisch	haddock (chat, jumbo)
Melanogryllus desertus	Steppengrille	steppe cricket
Melanonidae	Melanoniden	melanonids, pelagic cods

Melanophila acuminata	Bleischwarzer Prachtkäfer	
Melanophryniscus stelzneri	Argentinische Rotbauchkröte, Argentinischer Stummelfußfrosch	redbelly toad
Melanoplus borealis		northern grasshopper
Melanoplus devastator		devastating grasshopper
Melanoplus frigidus	Nordische Gebirgsschrecke	northern migratory grasshopper
Melanoplus sanguinipes	Wander-Gebirgsschrecke*	migratory grasshopper, lesser migratory grasshopper
Melanostomiidae/ Melanostomiatidae	Schuppenlose Drachenfische	scaleless dragonfishes
Melanosuchus niger	Mohrenkaiman	black caiman
Melanotaenia spp.	Regenbogenfische	rainbowfishes
Melanotaeniidae	Regenbogenfische	rainbow fishes, rainbowfishes
Melanotus communis	Mais-Schnellkäfer*	corn wireworm, corn click beetle
Melanotus rufipes	Rotfüßiger Schnellkäfer	redfooted click beetle*
Melasmothrix naso	Kleine Celebes-Spitzmausratte*	Celebes lesser shrew rat
Melasoma populi/ *Chrysomela populi*	Pappelblattkäfer, Roter Pappelblattkäfer	red poplar leaf-beetle, poplar leaf beetle, poplar beetle
Meleagridinae	Truthühner	turkeys, wild turkeys
Meleagris gallopavo	Truthuhn	common turkey
Melecta luctuosa	Weißfleckige Trauerbiene	
Melecta punctata/ *Melecta armata*	Gemeine Trauerbiene	
Meles meles	Europäischer Dachs	Old World badger
Melibe leonina	Löwen-Nacktschnecke	lion nudibranch
Melicertus canaliculatus/ *Penaeus canaliculatus*	Hexengarnele	witch shrimp
Melicertus kerathurus/ *Penaeus kerathurus*	Furchengarnele	caramote prawn
Melicertus latisuculatus/ *Penaeus latisuculatus*	Königsgeißelgarnele	western king prawn
Melichthys indicus	Schwarzer Drückerfisch	blackfinned triggerfish, Indian triggerfish (FAO)
Melichthys niger/ *Balistes radula*	Dunkler Ringens, Schwarzer Weißbindendrücker	black durgon
Melichtys vidua/ *Balistes vidua*	Braune Witwe	pinktail triggerfish
Melierax metabates	Graubürzel-Singhabicht	dark chanting goshawk
Meligethes aeneus	Rapsglanzkäfer	pollen beetle
Meliphagidae	Honigesser, Honigfresser	honeyeaters & *Ephthianura*, *Ashbyia*
Melipona spp.	Stachellose Bienen*	stingless bees
Meliscaeva cinctella	Gemeine Zart-Schwebfliege	
Melitaea cinxia	Gemeiner Scheckenfalter, Wegerich-Scheckenfalter	Glanville fritillary
Melitaea cynthia/ *Euphydryas cynthia*	Veilchen-Scheckenfalter, Weißfleckenfalter	Cynthia's fritillary
Melitaea didyma	Roter Scheckenfalter	spotted fritillary
Melitaea phoebe	Flockenblumen-Scheckenfalter, Großer Scheckenfalter	knapweed fritillary

Mellinus arvensis	Kotwespe	field digger wasp
Mellisuga minima	Zwergelfe, Zwergkolibri	vervain hummingbird
Mellitidae	Schlüsselloch-Sanddollars	keyhole urchins
Mellivora capensis	Honigdachs	honey badger, rattel
Melo melo	Indische Walzenschnecke	baler snail (baler shell), Indian volute
Meloë proscarabaeus	Schwarzblauer Ölkäfer	
Meloë rugosus	Mattschwarzer Maiwurmkäfer	
Meloë spp.	Ölkäfer (Maiwurm)	oil beetles, blister beetles
Melogale moschata	Chinesischer Sonnendachs (Hbz. Pahmi)	Chinese ferret badger
Melogale spp.	Sonnendachse	ferret badgers
Meloidae	Blasenkäfer (inkl. Ölkäfer)	blister beetles (incl. oil beetles)
Meloidogyne graminis	Graswurzelgallenälchen	grass root-knot nematode
Meloidogyne hapla	Nördliches Wurzelgallenälchen	northern root-knot nematode
Meloidogyne mali	Apfelgallenälchen	apple root-knot nematode
Meloidogyne marioni/ Meloidogyne incognita	Südliches Wurzelgallenälchen	southern root-knot nematode
Meloidogyne naasi	Weizenwurzelgallenälchen	root-knot nematode (on wheat)
Meloidogyne spp.	Wurzelgallenälchen	root-knot nematodes, root-knot eelworms
Meloinae	Ölkäfer	oil beetles
Melolontha hippocastani	Waldmaikäfer	field maybeetle
Melolontha melolontha	Maikäfer, Feldmaikäfer	common cockchafer, maybug
Melomys spp.	Melomys-Ratten	mosaic-tailed rats, banana rats
Melongena corona	Gemeine Kronenschnecke, Florida-Kronenschnecke	common crown conch, Florida crown conch, American crown conch
Melongena melongena	Westindische Kronenschnecke, Westindische Krone	West Indian crown conch
Melongena patula	Pazifik-Kronenschnecke	Pacific crown conch
Melongenidae	Kronenschnecken	crown snails (crown shells)
Melophagus ovinus	Schaflaus, Schaflausfliege	sheep ked (fly)
Melopsittacus undulatus	Wellensittich	budgerigar ("budgie", parakeet)
Melospiza georgiana	Sumpfammer	swamp sparrow
Melospiza lincolnii	Lincolnammer	Lincoln's sparrow
Melospiza melodia	Singammer	song sparrow
Melursus ursinus/ Ursus ursinus	Lippenbär	sloth bear
Melyridae	Melyriden	soft-winged flower beetles
Membracidae	Buckelzirpen	treehoppers
Membranipora membranacea	Seerinde	sea-mat
Membranipora pilosa	Bedornte Seerinde	thorny sea-mat
Menacanthus stramineus/ Eomenacanthus stramineus	Körperlaus, Hühner-Körperlaus	chicken body louse
Mene maculata	Fleckenmondfisch	moonfish
Menesia bipunctata	Schwarzbock	

Menetes berdmorei	Berdmores Palmhörnchen	Berdmore's palm squirrel, multistriped palm squirrel
Menetus dilatatus	Trompetenschnecke*	trumpet ramshorn
Menidae	Mondfische*	moonfish
Menidia beryllina	Binnen-Ährenfisch	inland silverside
Menidia menidia	Mondährenfisch, Gezeiten-Ährenfisch	Atlantic silverside
Menippe mercenaria	Große Steinkrabbe	stone crab, black stone crab
Menopon gallinae	Schaftlaus, Hühnerfederling	shaft louse, chicken shaft louse
Menticirrhus americanus	Amerikanischer Umberfisch	southern kingfish, southern kingcroaker (FAO)
Menticirrhus littoralis	Golf-Umberfisch	Gulf kingcroaker
Menticirrhus saxatilis	Königs-Umberfisch	northern kingfish
Menticirrhus spp.	Stumme Umberfische	kingcroakers
Menuridae	Leierschwänze	lyrebirds
Meoma ventricosa	Roter Herzigel	sea pussy, cake urchin, red heart urchin*
Mephitis macroura	Langschwanzskunk	hooded skunk
Mephitis mephitis	Streifenskunk	striped skunk
Mercenaria campechiensis	Südliche Quahog-Muschel	southern quahog
Mercenaria mercenaria	Nördliche Quahog-Muschel	northern quahog, quahog (hard clam)
Mercuria confusa	Geschwollene Wasserdeckelschnecke*	swollen spire snail
Meretrix lusoria	Japanische Venusmuschel	Japanese hard clam
Merganetta armata	Sturzbachente	torrent duck
Mergelia truncata (Brachiopoda)	Gestutzte Stielmuschel	
Mergus albellus	Zwergsäger	smew
Mergus cucullatus	Kappensäger	hooded merganser
Mergus merganser	Gänsesäger	common merganser, goosander
Mergus serrator	Mittelsäger	red-breasted merganser
Meriones spp.	Sandrennmäuse, Sandrennratten (Wüstenmäuse)	jirds
Meriones unguiculatus	Mongolen-Rennmaus	Mongolian gerbil, clawed jird
Merlangius merlangus	Wittling, Merlan	whiting
Merlangius merlangus euxinus	Schwarzmeer-Wittling	Black Sea whiting
Merlucciidae	Seehechte	hakes, merluccid hakes
Merluccius albidus	Großäugiger Seehecht	offshore silver hake, offshore hake (FAO)
Merluccius australis	Südlicher Seehecht	southern hake
Merluccius bilinearis	Nordamerikanischer Seehecht, Silberhecht, Silber-Wittling	silver hake (FAO), silver whiting
Merluccius capensis	Flachwasser-Seehecht, Stockfisch, Kap-Hecht	Cape hake, stokvis (S. Africa), stock fish, stockfish, shallow-water Cape hake (FAO)

Merluccius gayi	Chilenischer Seehecht, Pazifischer Seehecht	Chilean hake, Peruvian hake, South Pacific hake (FAO)
Merluccius hubbsi	Patagonischer Seehecht, Argentinischer Seehecht	Argentine hake (FAO), Southwest Atlantic hake
Merluccius merluccius	Seehecht, Hechtdorsch, Europäischer Seehecht	hake, European hake (FAO), North Atlantic hake
Merluccius paradoxus	Tiefenwasser-Kapseehecht	deepwater hake, deepwater Cape hake (FAO)
Merluccius polli	Benguela Seehecht	Benguela hake
Merluccius productus	Pazifik-Seehecht, Nordpazifischer Seehecht	Pacific hake, North Pacific hake (FAO)
Merluccius senegalensis	Senegalesischer Seehecht	Senegalese hake (FAO), black hake
Merodon equestris	Gemeine Zwiebelschwebfliege	large narcissus fly, large bulb fly
Meroles spp.	Scharreidechsen	desert lizards
Meromyza saltatrix	Grüne Schenkelfliege	grass fly
Meropidae	Bienenesser, Bienenfresser, Spinte	bee-eaters
Merops apiaster	Bienenfresser	bee-eater
Merops orientalis	Smaragdspint	little green bee-eater
Merops ornatus	Regenbogenspint	Australian bee eater
Merops persicus	Blauwangenspint	blue-cheeked bee-eater
Mertensia ovum	Seenuss	sea nut, Arctic sea gooseberry
Mertensiella caucasica/ Mertensia caucasica	Kaukasus-Salamander	Caucasian salamander
Mertensiella luschani/ Salamandra luschani	Kleinasiatischer Salamander, Lykischer Salamander	Luschan's salamander, Lycian salamander
Mertensophryne micranotis	Loveridge-Schnauzenfrosch	Loveridge's snouted frog
Mesapamea secalis/ Apamea secalis	Getreidewurzeleule, Getreideeule, Getreidesaateule, Roggeneule	common rustic moth
Mesembrina meridiana	Mittagsfliege	noon-fly, noonfly
Mesembriomys spp.	Baumratten	tree rats
Mesitornis variegata	Kurzfuß-Stelzenralle	white-breasted mesite
Mesitornithidae	Stelzenrallen, Stelzrallen, Madagaskar-Rallen	mesites, monias, roatelos
Mesoacidalia aglaja/ Argynnis aglaja	Großer Permutterfalter	dark green fritillary
Mesocapromys auritus	Großohr-Baumratte	large-eared hutia
Mesocapromys nanus	Zwerg-Baumratte	dwarf hutia
Mesoclemmys gibba/ Phrynops gibbus	Buckelschildkröte	gibba turtle
Mesocricetus auratus	Syrischer Goldhamster	golden hamster
Mesocricetus newtoni	Rumänischer Hamster	Romanian hamster
Mesocricetus spp.	Mittelhamster	golden hamsters
Mesogobius batrachocephalus	Krötengrundel	knout goby
Mesogonistius chaetodon/ Enneacanthus chaetodon	Scheibenbarsch	blackbanded sunfish (FAO), chaetodon

Mesoleuca albicillata	Himbeerspanner	beautiful carpet
Mesonauta festivus/ Heros festivus	Flaggen-Buntbarsch	flag cichlid
Mesoplodon bidens	Zweizahnwal, Sowerby-Zweizahnwal	Sowerby's whale, Sowerby's North Sea beaked whale
Mesoplodon bowdoini	Bowdoins Schnabelwal	deep-crested beaked whale
Mesoplodon carlhubbsi	Hubbs Schnabelwal	arch-beaked whale, Hubbs' beaked whale
Mesoplodon densirostris	Blainville-Zweizahnwal	Blainville's dense-beaked whale
Mesoplodon europaeus	Gervais-Zweizahnwal	Gervais' beaked whale, Gulfstream beaked whale
Mesoplodon ginkgodens	Japanischer Schnabelwal, Ginkgozahn-Schnabelwal	ginkgo-toothed beaked whale
Mesoplodon grayi	Gray-Zweizahnwal	Gray's beaked whale, scamperdown beaked whale
Mesoplodon hectori	Hectors-Zweizahnwal	Hector's beaked whale, skew-beaked whale
Mesoplodon layardi	Layard-Schnabelwal	strap-toothed beaked whale, Layard's beaked whale
Mesoplodon mirus	True-Wal, True-Zweizahnwal	True's beaked whale, True's North Atlantic beaked whale
Mesoplodon spp.	Zweizahnwale	beaked whales
Mesoplodon stejnegeri	Stejneger-Schnabelwal	Bering Sea beaked whale, North Pacific beaked whale, saber-toothed whale
Mesosa curculionides	Großer Augenfleckenbock	
Mesosa nebulosa	Binden-Augenfleckenbock	
Mesovelia furcata	Hüftwasserläufer	pondweed bug, water treader
Mesoveliidae	Hüftwasserläufer	water treaders
Mespilia globulus	Kugel-Seeigel	sphere urchin
Messor spp.	Ernteameisen, „Getreideameisen"	harvester ants
Meta menardi	Höhlenkreuzspinne	cave orbweaver
Meta segmentata	Herbstspinne	autumn orbweaver*
Metachirops spp.	Vieraugenbeutelratten	"four-eyed" opossums
Metachirus nudicaudatus/ Philander nudicaudatus	Nacktschwanzbeutelratte	brown "four-eyed" opossum
Metagonimus yokogawai	Zwerg-Darmegel, Kleiner Darmegel	Yokogawa fluke
Metanephrops andamanicus	Andamanen-Kaisergranat, Andamanen-Schlankhummer	southern langoustine, Andaman lobster
Metanephrops binghami	Karibik-Kaisergranat	Caribbean lobsterette
Metanephrops challengeri	Neuseeländischer Kaisergranat	New Zealand scampi, deep water scampi
Metapenaeopsis goodei	Samtgarnele	velvet shrimp
Metapenaeus brevicornis	Gelbe Geißelgarnele	yellow prawn, yellow shrimp
Metapenaeus endeavouri	Braune Geißelgarnele	endeavour shrimp
Metapenaeus ensis	Graue Geißelgarnele	greasyback shrimp
Metapenaeus joyneri	Shiba-Geißelgarnele	shiba shrimp
Metapenaeus monoceros	Einhorn-Geißelgarnele	speckled shrimp
Metapenaeus spp.	Geißelgarnelen u.a.	metapenaeus shrimps
Metatheria/Didelphia	Beutelsäuger	pouched mammals, metatherians

Metazoa

Metazoa	Mitteltiere, Gewebetiere, „Vielzeller"	metazoans
Metidae	Herbstspinnen	autumn orbweavers*
Metoecus paradoxus	Wespenkäfer	
Metopolophium dirhodum	Bleiche Getreideblattlaus	rose-grain aphid
Metridium giganteum	Pazifische Riesenanemone	gigantic anemone, white-plumed anemone
Metridium senile	Seenelke	clonal plumose anemone, frilled anemone, plumose sea anemone, brown sea anemone, plumose anemone
Metrioptera bicolor	Zweifarbige Beißschrecke	twocoloured bushcricket
Metrioptera brachyptera	Kurzflüglige Beißschrecke	bog bushcricket
Metrioptera roeselii	Roesels Beißschrecke	Roesel's bushcricket
Metrioptera saussureana	Gebirgs-Beißschrecke	Saussure's bushcricket
Metynnis argenteus	Scheibensalmler, Silberdollar	silver dollar a.o.
Metynnis hypsauchen	Dickkopf-Scheibensalmler	silver dollar a.o.
Metynnis spp.	Amerikanische Scheibensalmler	silver dollars
Mezium affine	Kapuzenkugelkäfer	shiny spider beetle
Mezium americanum	Amerikanischer Buckelkäfer, Schwarzer Kapuzenkugelkäfer	American spider beetle, black spider beetle
Micaria spp.	Ameisenspinnen u.a.	ant spiders a.o.
Micrastur ruficollis	Sperberwaldfalke	barred forest falcon
Micrathene whitneyi/ Glaucidium whitneyi	Elfenkauz, Kaktuskauz	elf owl
Micrelaps muelleri	Muellers Erdviper	Mueller's ground-viper
Microbates cinereiventris	Graubauch-Degenschnäbler	half-collared gnatwren
Microbatrachella capensis	Microfrosch	micro frog
Microcavia spp.	Zwergmeerschweinchen	mountain cavies
Microcebus murinus	Grauer Mausmaki, Gewöhnlicher Mausmaki	lesser mouse lemur, grey mouse lemur
Microcebus spp.	Mausmakis, Zwergmakis	mouse lemurs
Microcephalophis gracilis/ Hydrophis gracilis	Schmuck-Zwergkopf-Seeschlange	graceful small-headed sea snake, graceful sea snake
Microcephalophis spp.	Zwergkopf-Seeschlangen	small-headed sea snake
Microchirus ocellatus/ Solea ocellata	Augen-Seezunge, Augenfleck-Seezunge	four-eyed sole, foureyed sole (FAO)
Microchirus theophila/ Microchirus azevia/ Solea azevia	Azevia-Seezunge, Bastardzunge	bastard sole (FAO) (Jewish sole)
Microchirus variegatus/ Solea variegata	Bastardzunge, Dickhäutige Seezunge*	thickback sole (FAO), thick-backed sole
Microciona prolifera	Roter Moosschwamm*	red bread sponge, red moss sponge, red sponge
Microcosmus sulcatus/ Microkosmus sabatieri	Essbare Seescheide, Mikrokosmos	grooved sea-squirt, sea-egg, edible sea-squirt
Microdesmidae/Cerdalidae	Wurmfische & Pfeilgrundeln	wormfishes & dartfishes
Microdesminae	Wurmfische	wormfishes
Microdipodops spp.	Kängurumäuse	kangaroo mice

Microgadus proximus	Pazifischer Tomcod	Pacific tomcod
Microgadus tomcod	Tomcod, Atlantischer Tomcod	Atlantic tomcod
Microgale spp.	Spitzmaus-Borstenigel, Langschwanz-Borstenigel	shrew-tenrecs, long-tailed tenrecs
Microgeophagus ramirezi/ Apistogramma ramirezi/ Papiliochromis ramirezi	Schmetterlings-Zwergbuntbarsch, Schmetterlingsbuntbarsch	Ramirez's dwarf cichlid, Ramirezi, Ram cichlid (FAO)
Microgeophagus spp.	Schmetterlingsbuntbarsche	butterfly cichlids
Microgobius thalassinus	Grüne Grundel, Grüngrundel	green goby
Microhierax caerulescens	Rotkehlfälkchen	collared falconet
Microhydra ryderi (Craspedacusta sowerbyi)	Zwergpolyp	
Microhydromys richardsoni	Furchenzahn-Spitzmausratte*	groove-toothed shrew rat, moss mouse
Microhylidae	Engmundkröten, Engmaulkröten	narrowmouth toads, microhylids
Microlestes minutulus	Schmaler Zwergstutzläufer	
Micromalthidae	Micromalthiden	telephone-pole beetles
Micromelo undatus	Zwerg-Melo	miniature melo
Micromesistius poutassou	Blauer Wittling, Blauwittling	blue whiting (FAO), poutassou
Micrommata rosea	Grüne Huschspinne, Grasgrüne Huschspinne	green spider*
Micromys minutus	Eurasiatische Zwergmaus	Old World harvest mouse
Micronisus gabar	Gabarhabicht	Gabar goshawk
Micronycteris spp.	Kleine Großohr-Fledermäuse	little big-eared bats
Micropalama himantopus	Bindenstrandläufer	stilt sandpiper
Micropeplus porcatus	Schwarzer Rippenkäfer	
Micropeza corrigiolata	Stelzfliege	stilt-legged fly
Micropezidae/Tylidae	Stelzfliegen	stilt-legged flies
Microphis boaja	Große Süßwassernadel	freshwater pipefish
Micropochis spp.	Pazifik-Korallenschlange	Pacific coral snakes
Micropogonias furnieri	Weißmaul-Umberfisch	white-mouth croaker
Micropogonias undulatus	Atlantischer Umberfisch	Atlantic croaker
Micropsitta pusio	Braunstirn-Spechtpapagei	buff-faced pygmy parrot
Micropteropus spp.	Zwerg-Epaulettenflughunde	dwarf epauletted bats
Micropterus dolomieui	Schwarzbarsch	smallmouth bass
Micropterus salmoides	Forellenbarsch	largemouth black bass, largemouth bass (FAO)
Micropterygidae	Urmotten	mandibulate moths
Microsaura spp.	Zwergchamäleons u.a.	pygmy chameleon
Microsciurus spp.	Neuweltliche Zwerghörnchen	Neotropical dwarf squirrels, pygmy tree squirrels
Microsorex hoyi	Amerikanische Zwergspitzmaus	pygmy shrew, American pygmy shrew
Microspalax leucodon/ Nanospalax leucodon	Westblindmaus	lesser mole-rat
Microstomus kitt	Echte Rotzunge, Limande	lemon sole
Microtus agrestis	Erdmaus	field vole, short-tailed vole
Microtus arvalis	Feldmaus	common vole

Microtus bavaricus/ *Pitymys bavaricus*	Bayerische Kurzohrmaus	Bavarian pine vole
Microtus brandti	Brandt-Steppenwühlmaus	Brandt's vole
Microtus breweri	Küstenwühlmaus	beach meadow vole, beach vole
Microtus cabrerae	Cabreramaus	Cabrera's vole
Microtus longicaudus	Langschwanz-Wühlmaus	long-tailed vole, long-tailed meadow mouse
Microtus lusitanicus	Iberien-Wühlmaus	Iberian vole, Spanish vole
Microtus miurus	Alaska-Wühlmaus	singing vole
Microtus montanus	Rocky-Mountains-Wühlmaus	Rocky-Mountains vole, mountain meadow mouse
Microtus nivalis	Schneemaus	snow vole
Microtus ochrogaster	Präriewühlmaus	prairie vole
Microtus oeconomus/ *Microtus ratticeps*	Sumpfmaus, Nordische Wühlmaus	root vole, tundra vole, northern vole
Microtus pennsylvanicus	Wiesenwühlmaus	meadow vole
Microtus pinetorum/ *Pitymys pinetorum*	Kiefernwühlmaus	American pine vole
Microtus pyrenaicus	Pyrenäen-Kleinwühlmaus	Pyrenean pine vole
Microtus rossiaemeridionalis/ *Microtus epiroticus*	Südfeldmaus, Epirus-Feldmaus	sibling vole
Microtus savii/Pitymys savii	Italienische Kleinwühlmaus, Savi-Kleinwühlmaus	Savi's pine vole
Microtus socialis/ *Microtus guentheri*	Levante-Wühlmaus, Mittelmeer-Feldmaus	Mediterranean vole, Gunther's vole
Microtus spp.	Feldmäuse & Wühlmäuse	voles, meadow mice
Microtus subterraneus/ *Pitymys subterraneus*	Kleine Wühlmaus, Kleinwühlmaus, Kurzohrmaus, Kleinäugige Wühlmaus	European pine vole
Microtus thomasi/ *Pitymys thomasi*	Balkan-Kurzohrmaus, Thomas-Kleinwühlmaus	Thomas's pine vole
Microtus xanthognathus	Gelbwangenwühlmaus	yellow-cheek vole
Micruroides euryxanthus *euryxanthus*	Arizona-Korallenschlange	Sonoran coral snake, Arizona coral snake
Micrurus annellatus	Geringelte Korallenschlange	annellated coral snake
Micrurus corallinus	Gewöhnliche Korallenschlange, „Kobra-Korallenschlange"	common coral snake, painted coral snake
Micrurus diastema	Atlantische Korallenschlange	Atlantic coral snake, variable coral snake
Micrurus distans	Breitgestreifte Korallenschlange	broad-banded coral snake, West Mexican coral snake
Micrurus frontalis	Südamerikanische Korallenschlange, Kobra-Korallenschlange	southern coral snake
Micrurus fulvius	Harlekin-Korallenschlange	harlequin coral snake, eastern coral snake
Micrurus mipartitus	Schwarzgeringelte Korallenschlange	black-ringed coral snake, redtail coral snake
Micrurus nigrocinctus	Schwarzgebänderte Korallenschlange	black-banded coral snake, Central American coral snake
Micrurus spixii	Amazonas-Korallenschlange	Amazonian coral snake, Amazon coral snake

Micrurus spp.	Amerikanische Korallenschlangen, Echte Korallenottern	American coral snakes, common coral snakes
Micryphantidae	Zwergspinnen	dwarf spiders
Mictyridae	Grenadierkrabben, Soldatenkrabben	soldier crabs, grenadier crabs
Mictyris longicarpus	Grenadierkrabbe, Soldatenkrabbe	soldier crab
Mikiola fagi	Buchenblattgallmücke	beech leaf gall midge, beech pouch-gall midge
Milacidae	Kielnacktschnecken, Kielschnegel	keeled slugs
Milax budapestensis/ Tandonia budapestensis	Boden-Kielnacktschnecke	Budapest slug
Milax gagates	Dunkle Kielnacktschnecke	keeled slug, greenhouse slug, small black slug, jet slug
Milax nigricans	Schwarze Kielnacktschnecke	black slug
Milax rusticus/ Tandonia rustica	Große Kielnacktschnecke	bulb-eating slug, root-eating slug
Milax sowerbyi/ Tandonia sowerbyi	Sowerbys Kielnacktschnecke	Sowerby's slug, keeled slug
Miliaria calandra	Grauammer	corn bunting
Milichiidae	Nistfliegen	milichiids
Millardia spp.	Millardia-Ratten, Millard-Mäuse	Asian soft-furred rats, cutch rats
Millepora alcicornis	Verzweigte Feuerkoralle	branching fire coral
Millepora complanata	Blättrige Feuerkoralle	bladed fire coral
Millepora dichotoma	Netz-Feuerkoralle	ramified fire coral, reticulated fire coral
Millepora platyphylla	Brettartige Feuerkoralle, Platten-Feuerkoralle	sheet fire coral
Millepora spp.	Feuerkorallen	fire corals, stinging corals
Millepora squarrosa	Krustige Feuerkoralle	crustal fire coral
Millepora tortuosa	Verzweigte Feuerkoralle	ramified fire coral*
Milleporina	Feuerkorallen	milleporine hydrocorals, stinging corals, fire corals
Millericrinida	Zirrenlose Seelilien	sea lilies without cirri
Millerigobius macrocephalus	Millers Grundel	Miller's goby
Miltochrista miniata	Rosenmotte	rosy footman
Milvago chimachima	Gelbkopfkarakara	yellow-headed caracara
Milvus migrans	Schwarzmilan	black kite
Milvus milvus	Rotmilan	red kite
Mimas tiliae	Lindenschwärmer	lime hawkmoth
Mimetidae	Spinnenfresser	pirate spiders, spider-hunting spiders
Mimetillus moloneyi	Moloney-Flachkopf-Fledermaus	Moloney's flat-headed bat, narrow-winged bat
Mimidae	Spottdrosseln	mockingbirds & thrashers & catbirds
Mimon spp.	Speernasen-Fledermaus	Gray's spear-nosed bats
Mimus polyglottos	Spottdrossel	northern mockingbird

Mindarus abietinus	Weißtannentrieblaus	balsam twig aphid
Mindarus obliquus	Fichtentrieblaus	spruce twig aphid
Miniopterus schreibersi	Schreibers Langflügelfledermaus	Schreiber's bat
Miniopterus spp.	Langflügel-Fledermäuse	long-winged bats, bent-winged bats, long-fingered bats
Minoa murinata	Mausspanner, Mäuschen	drab looper (moth)
Minois dryas/ Satyrus dryas	Blauäugiger Waldportier	dryad
Miopithecus ogouensis	Ogoue Zwergmeerkatze	northern talapoin
Miopithecus talapoin	Zwergmeerkatze, Angola-Zwergmeerkatze	talapoin, southern talapoin
Mirafra apiata	Grasklapperlerche	clapper lark
Mirafra javanica	Horsfieldlerche	eastern singing bush lark
Miramella alpina	Alpine Gebirgsschrecke	green mountain grasshopper, Alpine migratory grasshopper*
Mirapecten mirificus	Wundervolle Kammmuschel	miraculous scallop
Mirapinnidae	Wunderflosser	mirapinnids (incl. hairyfishes & tapetails)
Miridae	Weichwanzen, Blindwanzen	mirids, capsid bugs, plant bugs
Mirounga angustirostris	Nördlicher See-Elefant	northern elephant seal
Mirounga leonina	Südlicher See-Elefant	southern elephant seal
Mirounga spp.	Elefantenrobben	elephant seals
Mirza coquereli	Coquerels Katzenmaki	Coquerel's dwarf lemur
Misgurnus anguillicaudatus	Ostasiatischer Schlammpeitzger	Japanese weatherfish
Misgurnus fossilis	Schlammpeitzger	weatherfish
Misumena spp.	Blüten-Krabbenspinne*	flower crab spiders
Misumena vatia	Veränderliche Krabbenspinne	goldenrod crab spider
Mitella pollicipes/ Pollicipes cornucopia	Felsen-Entenmuschel	rocky shore goose barnacle
Mithrax hispidus	Korallen-Seespinne	coral clinging crab
Mithrax sculptus	Grüne Seespinne	green clinging crab
Mithrax spinosissimus	Stachelige Seespinne	channel clinging crab, spiny spider crab
Mitra cardinalis	Kardinalsmütze	cardinal mitre
Mitra ebenus/ Vexillum ebenus	Ebenholz-Mitra, Brauner Stufenturm	ivory mitre, brown miter
Mitra fraga	Erdbeer-Mitra	strawberry mitre
Mitra fusiformis f. zonata	Breitgebänderte Mitra	zoned mitre
Mitra incompta	Mosaik-Mitra*	tessellate mitre
Mitra mitra/ Mitra episcopalis	Bischofsmütze	episcopal mitre
Mitra nigra	Schwärzliche Mitra	black mitre, black miter
Mitra nodulosa	Knotige Mitra	beaded mitre
Mitra papalis	Papstkrone	papal mitre
Mitra puncticulata	Gesprenkelte Mitra	punctured mitre, dotted miter (U.S.)
Mitra retusa	Stumpfe Mitra	blunt mitre
Mitra sanguisuga	Blut-Mitra*	blood-sucking mitre

Mitra spp.	Mitra-Schnecken, Mitras	mitres a.o. (Br.), miters a.o. (U.S.)
Mitra stictica	Gepunzte Mitra	pontifical mitre
Mitra tricolor	Dreifarben-Mitra	three-color mitre
Mitra zonata/ Mitra fusiformis f. zonata	Kronenschnecke	zoned miter (zoned mitra shell)
Mitrella lunata	Mond-Täubchen*, Mond-Täubchenschnecke	lunar dovesnail, crescent mitrella
Mitrella ocellata	Weißpunkt-Täubchenschnecke*	white-spot dovesnail
Mitrella scripta/ Mitrella flaminea	Schlanke Birnenschnecke, Schrifttäubchen	music dovesnail
Mitridae	Mitraschnecken, Straubschnecken, Kronenschnecken	miters (U.S.), mitres (Br.), mitras (mitra shells)
Mitrocoma cellularia	Kreuzqualle*	cross jellyfish
Mitsukurina owstoni/ Scapanorhynchus owstoni	Koboldhai, Japanischer Nasenhai	goblin shark
Mitsukurinidae/ Scapanorhynchidae	Koboldhaie, Nasenhaie	goblin sharks
Mitu mitu	Mitu	Alagoas curassow, razor-billed curassow
Mixophyes balbus	Stotterfrosch	gray barred frog, stuttering frog (IUCN)
Mixophyes fasciolatus	Großer Balkenfrosch	great barred frog
Mixophyes iteratus	Riesen-Balkenfrosch	giant barred frog
Mnemiopsis leidyi	Meerwalnuss, Katzenauge	sea walnut, cat's eye
Mniotilta varia	Kletterwaldsänger	black-and-white warbler
Mobula diabola	Indopazifischer Teufelsrochen, Zwergteufelsrochen	devil ray
Mobula hypostoma	Westatlantischer Teufelsrochen	West Atlantic devil ray, lesser devil ray (FAO)
Mobula mobular	Kleiner Teufelsrochen, Ostatlantischer Teufelsrochen, Meeresteufel	devil fish (FAO), devilfish, devil ray
Mobulidae	Teufelsrochen, Mantas	mantas (manta rays & devil rays)
Mochocidae/Mochokidae	Fiederbartwelse	squeakers, upside-down catfishes
Modiolaria tumida	Marmor-Bohnenmuschel*	marbled crenella
Modiolula phaseolina	Bohnenförmige Miesmuschel, Bohnenmiesmuschel*	bean horse mussel
Modiolus barbatus/ Lithographa barbatus	Haarige Miesmuschel, Bartige Miesmuschel, Bartmuschel	bearded horse mussel
Modiolus modiolus	Große Miesmuschel	horse mussel, northern horsemussel
Modulidae	Mass-Schnecken	modulus snails
Modulus modulus	Atlantische Maßschnecke	buttonsnail, Atlantic modulus
Modulus tectum	Bedeckte Maßschnecke	knobby snail
Moelleria costulata	Gerippte Turbanschnecke	ribbed moelleria
Moenkhausia comma	Kommasalmler	comma tetra
Moenkhausia intermedia	Scherenschwanzsalmler	
Moenkhausia oligolepis	Schwanztupfensalmler	glass tetra

Moenkhausia pittieri	Brillantsalmler	diamond tetra, Pittier's tera
Moenkhausia sanctaefilomenae	Rotaugensalmler	redeye tetra, red-eyed moenkhausia
Moenkhausia takasei	Schwarzmantelsalmler*	black-jacket tetra
Mogurnda mogurnda	Tüpfelgrundel	northern trout gudgeon
Mogurnda striata/ Mogurnda adspersa	Lila-Tüpfelgrundel	trout gudgeon, purple-spotted gudgeon (FAO)
Moina brachiata	Tümpelwasserfloh	
Moira atropus	Schlammigel*, Schlamm-Herzigel*	mud heart urchin
Mola mola	Mondfisch	ocean sunfish
Molgula spp.	Meertrauben*	sea grapes
Molidae	Mondfische (Sonnenfische), Klumpfische	ocean sunfishes
Mollienesia latipinna/ Poecilia latipinna	Breitflossenkärpfling	sailfin molly
Mollienesia sphenops/ Poecilia sphenops	Spitzmaulkärpfling	marbled molly
Mollusca	Weichtiere, Mollusken	mollusks, molluscs
Moloch horridus	Wüstenteufel, Dornteufel, Moloch	moloch, horny devil, thorny devil
Molops piceus	Braunfüßiger Striemenkäfer	
Molorchus minor	Kleiner Wespenbock	
Molossidae	Bulldogg-Fledermäuse	free-tailed bats, mastiff bats
Molossus spp.	Freischwanz-Samt-Fledermäuse (Bulldogg-Fledermäuse)	velvety free-tailed bats
Molothrus aeneus	Rotaugen-Kuhstärling	bronzed cowbird
Molothrus ater	Braunkopf-Kuhstärling	brown-headed cowbird
Molva dipterygia macrophthalma/ Molva macrophthalma/ Molva elongata	Mittelmeer-Leng, Spanischer Leng	Mediterranean ling, Spanish ling (FAO)
Molva dipterygia dipterygia	Blauleng	blue ling
Molva molva	Leng, Lengfisch	ling (FAO), European ling
Moma alpinum	Seladoneule	scarce Merveille du jour
Momotidae	Sägeracken, Motmots	motmots
Momotus momota	Blauscheitelmotmot	blue-crowned motmot
Momphidae	Fransenmotten	momphid moths
Monacanthidae	Feilenfische	filefishes
Monacanthus chinensis	Fächerbauch	fan-bellied leatherjacket
Monacanthus hispidus	Flachkopf-Feilenfisch	planehead filefish
Monacha cantiana	Große Kartäuserschnecke	Kentish snail
Monacha cartusiana	Kartäuserschnecke	Carthusian snail, Chartreuse snail
Monacha granulata	Samtschnecke*	silky snail, Ashford's hairy snail
Monachus monachus	Mittelmeer-Mönchsrobbe	Mediterranean monk seal
Monachus schauinslandi	Laysan-Mönchsrobbe	Hawaiian monk seal
Monachus spp.	Mönchsrobben	monk seals

Monachus tropicalis	Karibische Mönchsrobbe	Caribbean monk seal
Monanchora arbuscula	Roter Krustenschwamm	red-white marbled sponge
Monarcha melanopsis	Maskenmonarch	pearly-winged monarch
Monarchidae	Monarchen	monarchs, magpie-larks
Monasa morphoeus	Weißstirntrappist	white-fronted nunbird
Monia patelliformis	Gerippte Sattelauster, Rippen-Sattelauster	ribbed saddle oyster
Monoacanthus tuckeri	Schlanker Feilenfisch	slender filefish
Monocentridae	Tannenzapfenfische	pineapple fishes, pinecone fishes
Monochamus galloprovincialis	Kiefernbock, Bäckerbock, Gefleckter Langhornbock	pine sawyer beetle
Monochamus sartor	Schneiderbock	carpenter sawyer beetle*
Monochamus sutor	Schusterbock, Einfarbiger Langhornbock	larch sawyer beetle
Monochirus hispidus	Schnurrbart-Zunge, Pelz-Seezunge	whiskered sole
Monochirus luteus	Gelbzunge	yellow sole
Monocirrhus polyacanthus	Blattfisch, Südamerikanischer Blattfisch	South American leaf-fish
Monodactylidae	Silberflossenblätter, Flossenblätter	moonies, moonfishes (fingerfishes), monos
Monodactylus argenteus	Silberflossenblatt, Mondfisch	silver mono, moonfish, silver moony (FAO), diamondfish, fingerfish, kilefish, butter-bream
Monodaeus couchi	Couchs Krabbe*	Couch's crab
Monodelphis spp.	Spitzmausbeutelratten	short-tailed opossums
Monodon monoceros	Narwal	narwhal, unicorn whale
Monodonta labio	Dicklippige Buckelschnecke	thick-lipped topsnail, labio
Monodonta lineata	Gestrichelte Buckelschnecke	lined topsnail (lined top shell, toothed top shell, thick top shell)
Monodonta turbinata	Gewürfelte Kreiselschnecke*, Würfelturban	checkered topsnail, one-toothed turban
Monognathidae	Einkieferaale	singlejaw eels
Monommatidae	Opossumkäfer	opossum beetles
Monomorium pharaonis	Pharaoameise, Pharao-Ameise	Pharaoh ant, Pharaoh's ant, little red ant
Monopeltis capensis	Kap-Keilschnauzen-Wurmschleiche	Cape wedge-snouted worm lizard
Monophadnoides geniculatus	Erdbeerblattwespe	geum sawfly (Br.), raspberry sawfly (U.S.)
Monopis rusticella	Fellmotte	skin moth, fur moth, wool moth
Monoplacophora	Einplatter, Urmützenschnecken, Monoplacophoren	monoplacophorans
Monotoma picipes	Gemeiner Detrituskäfer	

Monotomidae	Detrituskäfer	small flattened bark beetles
Monotremata (Prototheria)	Kloakentiere	monotremes (prototherians)
Monouria spp.	Hinterindische Landschildkröten	Indochinese land tortoises
Montacuta bidentata	Zweizähnige Linsenmuschel	bidentate montacutid
Montastrea annularis	Knopfkoralle	boulder star coral
Montastrea cavernosa	Große Sternkoralle	great star coral, large star coral
Monticola saxatilis	Steinrötel	mountain rock thrush
Monticola solitarius	Blaumerle	blue rock thrush
Montifringilla nivalis	Schneefink	snow finch
Montipora foliosa	Folien-Mikroporenkoralle, Folien-Koralle	leaf coral
Montipora monasteriata	Monasteria-Mikroporenkoralle	Monasteria microporous coral*
Montipora tuberculosa	Mikroporenkoralle	microporous coral
Mopalia ciliata (Polyplacophora)	Haarige Käferschnecke	hairy chiton, hairy mopalia
Mopalia lignosa (Polyplacophora)	Hölzerne Käferschnecke	woody chiton, woody mopalia
Mopalia muscosa (Polyplacophora)	Moos-Käferschnecke	mossy chiton, mossy mopalia
Mora mora	Ostatlantischer Tiefseedorsch	hakeling
Mordellidae	Stachelkäfer	tumbling flower beetles
Morelia albertisii	Weißlippenpython	D'Albertis' python
Morelia amethistima/ Liasis amethistimus	Amethystpython	amethystine python
Morelia argus/ Morelia spilota spilota	Rautenpython	diamond python
Morelia boa/ Bothrochilus boa	Neuguinea-Zwergpython	barred python, ringed python
Morelia childreni/ Liasis childreni	Gefleckter Python	Children's python, Children's rock python
Morelia mackloti/ Liasis mackloti	Macklots Python	water python, Macklot's python
Morelia perthensis/ Liasis perthensis	Zwergpython	pygmy python
Morelia spilota spilota/ Morelia argus	Rautenpython	diamond python
Morelia spilota variegata	Teppichpython	carpet python
Morelia viridis	Grüner Baumpython	green tree python
Morellia simplex	Rinderschweißfliege, Schweißfliege	cattle sweat fly, sweat fly
Morenia ocellata	Hinterindische Pfauenaugen-Sumpfschildkröte	Burmese eyed turtle
Morenia petersi	Vorderindische Pfauenaugen-Sumpfschildkröte	Indian eyed turtle
Moridae	Tiefseedorsche	deepsea cods, morid cods, moras
Morimus funereus	Trauerbock	dusky longicorn*
Moringua ferruginea	Schlank-Wurmaal	slender worm-eel
Moringua spp.	Wurmaale	worm-eels
Moringuidae	Spaghettiaale, Wurmaale	spaghetti eels
Mormo maura	Schwarzes Ordensband	old lady

Mormoopidae	Nacktrücken-Fledermäuse, Kinnblatt-Fledermäuse	moustached bats, naked-backed bats, ghost-faced bats
Mormoops spp.	Flachkinn-Fledermäuse, Blattkinn-Fledermäuse	leaf-chinned bats
Mormyridae	Nilhechte, Elefantenfische	elephantfishes
Mormyriformes	Nilhechte, Elefantenfische	mormyrids
Mormyrus kannume	Tapirrüsselfisch	elephantsnout fish
Morone americana/ Roccus americanus	Amerikanischer Streifenbarsch	white perch
Morone chrysops/ Roccus chrysops	Weißer Sägebarsch	white bass
Morone labrax/ Dicentrachus labrax/ Roccus labrax	Wolfsbarsch, Seebarsch	bass, sea bass
Morone mississippiensis	Mississippi-Barsch	yellow bass
Morone saxatilis/ Roccus saxatilis	Nordamerikanischer Streifenbarsch, Felsenbarsch	striped bass, striper, striped sea-bass (FAO)
Moronidae/Percichthyidae	Seebarsche, Wolfsbarsche, Streifenbarsche	temperate basses
Morphnus guianensis	Würgadler	Guiana crested eagle
Morum cancellatum	Gitterhelm	cancellate morum
Morum dennisoni	Dennisons Maulbeerschnecke	Dennison morum
Morum exquisitum	Besondere Maulbeerschnecke*	exquisite morum
Morum grande	Riesen-Maulbeerschnecke	giant morum
Morum lamarcki	Rosenmund-Maulbeerschnecke	rose-mouth morum
Morum oniscus	Atlantische Maulbeerschnecke	Atlantic morum
Moschus spp.	Moschustiere, Moschushirsche	musk deer
Motacilla alba	Bachstelze	pied wagtail, pied white wagtail
Motacilla flava	Schafstelze	yellow wagtail
Motacillidae	Stelzen & Pieper	wag-tails, pipits
Muelleripicus pulverulentus	Puderspecht	great slaty woodpecker
Mugil auratus/Liza aurata	Goldmeeräsche, Gold-Meeräsche	golden grey mullet
Mugil capito/Liza ramada	Dünnlippige Meeräsche	thin-lipped grey mullet, thinlip grey mullet, thinlip mullet (FAO)
Mugil capurrii	Afrikanische Meeräsche	narrow-head mullet, leaping African mullet
Mugil cephalus	Gemeine Meeräsche, Flachköpfige Meeräsche, Großkopfmeeräsche	striped gray mullet, striped mullet, common grey mullet, flat-headed grey mullet, flathead mullet (FAO)
Mugil chelo/ Mugil provensalis/ Chelon labrosus	Dicklippige Meeräsche	thick-lipped grey mullet, thick-lip grey mullet, thicklip grey mullet (FAO)
Mugil liza	Brasilianische Meeräsche	Brazilian mullet
Mugil saliens/Liza saliens	Springmeeräsche	leaping mullet

*M*ugilidae

Mugilidae	Meeräschen	mullets, grey mullets
Mugiloididae	Sandbarsche	sandsmelts, sandperches
Mullidae	Meerbarben	goatfishes
Mulloidichthys flavolineatus	Seitenfleck-Barbe, Seitenfleck-Meerbarbe	yellowstripe goatfish
Mulloidichthys vanicolensis	Gelbflossen-Barbe, Gelbflossen-Meerbarbe	yellowfin goatfish
Mullus auratus	Nördlicher Ziegenfisch	golden goatfish
Mullus barbatus	Meerbarbe, Gewöhnliche Meerbarbe, Rote Meerbarbe	red mullet (FAO), plain red mullet
Mullus surmuletus	Gestreifte Meerbarbe, Streifenbarbe	striped red mullet
Multiceps multiceps/ Taenia multiceps	Quesenbandwurm, Hundebandwurm u.a.	dog tapeworm a.o. (>gid/sturdy in sheep)
Mundia rugosa	Langarmiger Springkrebs, Tiefwasser-Springkrebs	rugose squat lobster
Mungos gambianus	Gambia-Kusimanse	Gambian mongoose
Mungos mungo	Zebramanguste	banded mongoose
Mungotictis decemlineata	Schmalstreifenmungo	Malagasy narrow-striped mongoose
Muntiacus crinifrons	Schwarzer Muntjak	black muntjac
Muntiacus reevesi	Chinesischer Muntjak, Zwergmuntjak	Chinese muntjac, Reeve's muntjac
Muntiacus spp.	Muntjaks	muntjacs, barking deer
Muraena helena	Mittelmeer-Muräne	Mediterranean moray (FAO), European moray
Muraena pardalis	Drachenmuräne	leopard moray eel
Muraenescocidae	Messerzahnaale	pike congers, pike eels
Muraenesox cinereus	Batavia-Hechtmuräne, Hechtmuräne	daggertooth pike conger (FAO), sharp-toothed eel
Muraenidae/Heteromyridae	Muränen, Muränenaale	moray eels, morays
Murex aduncospinosus	Gebogenstachelmurex*	bent-spined murex
Murex brandaris/ Bolinus brandaris	Brandhorn, Herkuleskeule	dye murex, purple dye murex
Murex brevispina	Kurzstachelmurex	brevispined murex, short-spine murex
Murex chrysosoma	Goldmund-Stachelschnecke, Goldmundmurex	goldmouth murex
Murex cornutus	Afrikanische Hornschnecke	African horned murex
Murex haustellum/ Haustellum haustellum	Schnepfenschnabel, Schnepfenschnabel-Stachelschnecke	snipe's bill murex
Murex nigrispinosus	Schwarzspitzenmurex, Schwarzspitzen-Stachelschnecke	black-spined murex
Murex palmarosae/ Chicoreus palmarosae	Rosenzweigschnecke, Rosenzweig-Stachelschnecke	rose-branch murex, rose-branched murex
Murex pecten/ Murex tenuispina	Venuskamm, Venuskamm-Stachelschnecke, Skelettspindel, Spinnenkopf	Venus comb murex, Venus comb, thorny woodcock
Murex scolopax	Waldschnepfen-Stachelschnecke*	woodcock murex
Murex spp.	Stachelschnecken	murex snails

Murex trapa	Schlicht-Stachelschnecke*	rare spined murex
Murex troscheli	Troschels Stachelschnecke	Troschel's murex
Murexia spp.	Neuguinea-Beutelmäuse	short-tailed marsupial "mice"
Murexsul octagonus	Oktagon-Stachelschnecke	octagon murex
Muricanthus nigritus	Trauerschnecke, Schwarze Stachelschnecke	black murex
Muricanthus radix/ Hexaplex radix	Wurzelschnecke	root murex
Muricea muricata	Kratzige Seepeitsche	spiny sea fan, spiny sea whip (gorgonians)
Muricidae	Stachelschnecken, Muriciden	murex snails (murex shells, rock shells)
Muriculus imberbis	Rückenstreifenmaus*	stripe-backed mouse, Wurch mouse
Muridae	Mäuseartige, Echte Mäuse, Langschwanzmäuse	mice family
Murina spp.	Insektenfressende Röhrennasen-Fledermäuse	tube-nosed insectivorous bats
Mus musculus	Hausmaus	house mouse
Mus spicilegus/ Mus hortulans	Ährenmaus	steppe mouse
Mus spp.	Echte Mäuse	mice
Mus spretus	Heckenhausmaus	Algerian mouse, Lataste's mouse
Musca autumnalis	Gesichtsfliege, Augenfliege	face fly
Musca domestica	Stubenfliege, Große Stubenfliege	house fly
Muscardinus avellanarius	Haselmaus	common dormouse, hazel dormouse
Muscicapa dauurica	Braunschnäpper	brown flycatcher
Muscicapa striata	Grauschnäpper	spotted flycatcher
Muscicapidae	Sänger, Eigentliche Fliegenschnäpper	Old World flycatchers
Muscidae	Echte Fliegen	houseflies, house flies
Muscina stabulans	Hausfliege, Stallfliege	false stable fly
Muscisaxicola albifrons	Klippentyrann	white-fronted ground tyrant
Musculium lacustre/ Sphaerium lacustre	Haubenmuschel, Häubchenmuschel, Teich-Kugelmuschel	lake orb mussel, lake fingernailclam, capped orb mussel
Musculium transversum	Längliche Haubenmuschel	oblong orb mussel
Musculus discors	Grüne Bohnenmuschel	discordant mussel, green crenella
Musculus marmoratus	Marmorierte Bohnenmuschel	spotted mussel, marbled mussel, marbled musculus
Musculus niger	Schwarze Bohnenmuschel	black mussel, black musculus, little black mussel
Musonycteris harrisoni	Bananenfledermaus	banana bat, Colima long-nosed bat
Musophaga violacea	Schildturako	violet turaco
Musophagidae	Turakos & Lärmvögel	turacos & plaintain-eaters
Mussa angulosa	Atlantische Dickstielige Doldenkoralle	spiny flower coral

Mustela

Mustela africana	Tropisches Wiesel	tropical weasel
Mustela altaica	Alpenwiesel, Altaiwiesel	mountain weasel
Mustela erminea	Hermelin, Großwiesel	ermine, stoat
Mustela eversmanni	Steppeniltis	steppe polecat
Mustela felipei	Kolumbianisches Wiesel, Kolumbienwiesel	Colombian weasel
Mustela frenata	Langschwanzwiesel	long-tailed weasel
Mustela kathiah	Gelbbauchwiesel	yellow-bellied weasel
Mustela lutreola	Nerz, Europäischer Nerz	European mink
Mustela nigripes	Schwarzfußiltis	black-footed ferret
Mustela nivalis	Mauswiesel	least weasel
Mustela nudipes	Malaienwiesel	Malayan weasel
Mustela putorius f. furo	Frettchen	domestic polecat
Mustela putorius	Europäischer Iltis, Waldiltis	European polecat
Mustela sibirica	Sibirisches Feuerwiesel, Kolonok	Siberian weasel
Mustela strigidorsa	Rückenstreifenwiesel	back-striped weasel
Mustela vison	Amerikanischer Nerz, Mink	American mink
Mustelidae	Wiesel & Dachse & Skunks & Otter	weasels & badgers & skunks & otters
Mustelus antarcticus	Australischer Glatthai	gummy shark
Mustelus asterias	Weißgefleckter Glatthai, Nördlicher Glatthai, Gefleckter Glatthai	stellate smooth-hound, starry smoothhound (FAO)
Mustelus canis	Westatlantischer Glatthai	dusky smooth-hound (FAO), Western Atlantic smoothhound
Mustelus mustelus	Grauer Glatthai, Südlicher Glatthai, Mittelmeer-Glatthai	smooth-hound (FAO), smoothhound, smooth dogfish
Mustelus punctulatus/ Mustelus mediterraneus	Schwarzgefleckter Glatthai, Schwarzpunkt-Glatthai, Mittelmeer-Glatthai	blackspotted smooth-hound
Mustelus schmitti	Schmalnasen-Glatthai, Patagonischer Glatthai	narrownose smooth-hound (FAO), Patagonian smooth-hound
Mutilla europaea	Große Spinnenameise	European velvet-ant
Mutillidae	Bienenameisen, Spinnenameisen, Ameisenwespen	velvet-ants
Mya arenaria/ Arenomya arenaria	Sandmuschel, Sandklaffmuschel, Strandauster, Große Sandklaffmuschel	sand gaper, soft-shelled clam, softshell clam, large-neck clam, steamer
Mya spp.	Klaffmuscheln	gaper clams
Mya truncata	Gestutzte Klaffmuschel, Gestutzte Sandklaffmuschel	blunt gaper clam, truncate softshell (clam)
Myadestes genibarbis	Bartklarino	rufous-throated solitaire
Myathropa florea	Gemeine Dolden-Schwebfliege	
Mycale ovulum	Polsterschwamm	padded sponge*
Mycedium elephantotus	Chinakohl-Koralle	Chinese cabbage coral*

Mycetaea hirta	Behaarter Stäublingskäfer, Pilzkäfer	hairy fungus beetle
Mycetophagidae	Baumschwammkäfer	hairy fungus beetles
Mycetophilidae	Pilzmücken	fungus gnats
Mycetophyllia danaana	Niedrighöckerige Kaktuskoralle	lowridge cactus coral
Mycetophyllia ferox	Raue Kaktuskoralle, Lamarcks Kaktuskoralle	rough cactus coral
Mycetophyllia lamarckiana	Faltenkoralle, Höckerige Kaktuskoralle	ridged cactus coral
Mycteria americana	Waldstorch	American wood ibis
Myctophidae	Laternenfische	lanternfishes
Myctophiformes	Laternenfische	lanternfishes & blackchins
Myctophum punctatum	Laternenfisch	lantern fish
Mydaus spp.	Stinkdachse	stink badgers
Myiarchus crinitus	Schnäppertyrann	great crested flycatcher
Myidae	Klaffmuscheln	gaper clams
Myiophoneus caerulens	Purpurpfeifdrossel	Himalayan whistling thrush
Myiopsitta monachus	Mönchsittich	monk parakeet
Myiornis ecaudatus	Stummelschwanz-Zwergtyrann	short-tailed pygmy tyrant
Myleus pacu	Pacu	pacu
Myleus rubripinnis	Hakenscheibensalmler	redhook myleus
Myliobatidae	Adlerrochen	eagle rays
Myliobatiformes	Stechrochenartige	stingrays
Myliobatis aquila	Europäischer Adlerrochen, Gewöhnlicher Adlerrochen	eagle ray
Myliobatis californica	Fledermausrochen	California bat ray
Myllaena intermedia	Gemeiner Kahnkurzflügler	
Mylomys dybowski/ Mylomys dybowskyi	Dybowski-Ratte	African groove-toothed rat, mill rat
Mylopharyngodon piceus	Schwarzer Karpfen	black carp
Mylossoma aureum	Goldener Mühlsteinsalmler	golden mylossoma
Mymaridae	Zwergwespen	fairy flies
Myobatrachidae	Australische Südfrösche	myobatrachid frogs
Myobatrachus gouldii	Schildkrötenfrosch	turtle frog
Myocastor coypus	Sumpfbiber, Nutria	coypu, nutria
Myomimus personatus	Dünnschwanz-Mausschläfer	Middle-Eastern mouse-tailed dormouse*, mouselike dormouse
Myomimus roachi	Europäischer Mausschläfer	European mouse-tailed dormouse
Myomimus spp.	Mausschläfer	mouse-tailed dormice, mouselike dormice
Myomorpha	Mäuseverwandte	ratlike rodents
Myomys spp.	Afrikanische Feldmäuse	African meadow rats
Myonycteris spp.	Kragenflughunde	little collared fruit bats
Myoprocta acouchy (Husson '78)	Grüner Acouchi	green acouchi
Myoprocta exilis (Husson '78)	Roter Acouchi	red acouchi
Myoprocta spp.	Acouchis	acouchis
Myopus schisticolor	Waldlemming	wood lemming
Myosciurus pumilio	Afrikanisches Zwerghörnchen	African pygmy squirrel

Myosorex spp.	Afrikanische Walsspitzmäuse	mouse shrews, forest shrews
Myospalax spp.	Blindmulle	mole-rats, zokors
Myotis bechsteini	Bechsteinfledermaus	Bechstein's bat
Myotis blythi	Kleinmausohr	lesser mouse-eared bat
Myotis brandti	Große Bartfledermaus, Brandt-Fledermaus	Brandt's bat
Myotis capaccinii	Langfußfledermaus	long-fingered bat
Myotis dasycneme	Teichfledermaus	pond bat
Myotis daubentoni	Wasserfledermaus	Daubenton's bat
Myotis emarginata	Wimperfledermaus	Geoffroy's bat
Myotis mehelyi	Mehely-Hufeisennase	Mehely's horseshoe bat
Myotis myotis	Großes Mausohr (Riesenfledermaus)	greater mouse-eared bat
Myotis mystacina	Kleine Bartfledermaus	whiskered bat
Myotis nattereri	Fransenfledermaus	Natterer's bat
Myotis spp.	Mausohr-Fledermäuse	little brown bats
Myotis vivesi	Fischende Fledermaus	fishing bat
Myoxocephalus aenaeus		grubby
Myoxocephalus bubalis/ Cottus bubalis/ Taurulus bubalis	Seebulle, Seebull, Ulk (Seeskorpion)	long-spined sculpin, long-spined sea scorpion, longspined bullhead (FAO)
Myoxocephalus octodecemspinosus	Langhorn-Seeskorpion, Amerikanischer Seeskorpion	longhorn sculpin
Myoxocephalus quadricornis/ Oncocottus quadricornis/ Triglopsis quadricornis	Vierhörniger Seeskorpion	four-horn sculpin, four-horned sculpin, fourhorn sculpin (FAO)
Myoxocephalus scorpioides	Arktischer Seeskorpion	Arctic sculpin (FAO), northern sculpin
Myoxocephalus scorpius/ Cottus scorpius	Seeskorpion, Seeteufel	Father Lasher, shorthorn sculpin, bull rout, bull-rout, short-spined seascorpion
Myra fugax	Kieselkrabbe*	pebble crab
Myriapoda	Tausendfüßler, Tausendfüßer, Doppelfüßer, Diplopoden, Myriapoden	millepedes (Br.), millipedes (U.S.) ("thousand-leggers"), diplopods, myriapodians
Myriapora truncata	Trugkoralle, Hundskoralle (Falsche Koralle)	false coral
Myrichthys acuminatus/ Muraena acuminata	Gelbgefleckter Schlangenaal	sharptail eel
Myripristinae	Soldatenfische	soldierfishes
Myripristis adustus	Schwarzbinden-Soldatenfisch	shadowfin soldierfish (FAO), blacktip soldierfish, blue squirrelfish
Myripristis jacobus	Karibischer Soldatenfisch	blackbar soldierfish
Myripristis murdjan/ Myripristis axillares	Roter Eichhörnchenfisch, Roter Soldatenfisch, Kurzstacheliger Soldatenfisch	pinecone soldierfish (FAO), crimson soldierfish, white-edge soldierfish
Myripristis spp.	Riffhörnchenfische	soldierfishes
Myrmarachne formicaria	Ameisenspringspinne, Ameisenspinne	ant spider a.o.
Myrmecia forficata	Schwarze Stachelameise, Schwarze Bulldogameise*	black bulldog ant

Myrmecobius fasciatus	Ameisenbeutler	numbat, banded anteater
Myrmecocichla arnoti	Arnotschmätzer	white-headed black-chat
Myrmecophaga tridactyla	Großer Ameisenbär	giant anteater
Myrmecophila acervorum	Ameisengrille	ant's-nest cricket
Myrmecoris gracilis	Ameisenwanze	ant capsid bug
Myrmeleonidae	Ameisenlöwen	antlions
Myrmeleotettix maculatus	Gefleckte Keulenschrecke	mottled grasshopper
Myrmica laevinodis	Kurzdornige Rote Knotenameise, Kurzdornige Rote Gartenameise	shortsting red myrmicine ant*
Myrmica rubra	Rote Knotenameise, Rotgelbe Knotenameise	red myrmicine ant, red ant
Myrmica ruginodis	Langdornige Rote Knotenameise, Langdornige Rote Gartenameise	longsting red myrmicine ant*
Myrmica scabrinodis	Trockenrasen-Knotenameise	
Myrmica spp.	Rote Knotenameisen, Gemeine Knotenameisen	red myrmicine ants
Myrmicidae	Knotenameisen, Stachelameisen	harvester ants & others
Myrocongridae	Weißaale	myrocongrids
Myrophinae	Wurmaale	worm-eels
Mysidacea	Spaltfüßer	opossum shrimps
Mysis spp.	Schwebegarnelen	opossum shrimps
Mystacina tuberculata	Neuseeland-Fledermaus	New Zealand short-tailed bat
Mysticeti	Bartenwale	whalebone whales, baleen whales
Mystromys albicaudatus	Weißschwänziger Hamster, Afrika-Hamster	white-tailed rat, white-tailed mouse
Mystus nemurus	Rotflossen-Stachelwels	Asian redtail catfish
Mystus spp.	Stachelwelse	featherbacks
Mystus vittatus/ Hypselobagrus tengara	Indischer Stachelwels, Indischer Goldstreifenwels, Indischer Streifenwels	striped catfish, striped dwarf catfish (FAO)
Mythimna albipuncta	Wegericheule, Trockenrasenhalden-Weißfleckeule, Weißfleckige Schilfgraseule	
Mythimna comma	Kommaeule	shoulder-striped wainscot
Mythimna conigera	Zapfeneule	cone wainscot*
Mythimna convecta	Australische Graseule*	common Australian wainscot, common armyworm (Austr.)
Mythimna favicolor	Salzwiesen-Weißadereule	Matthew's wainscot
Mythimna ferrago	Frischrasen-Weißfleckeule, Mittelwegericheule, Rötlichbraune Schilfgraseule	clay wainscot, clay moth
Mythimna impura	Seggeneule	sedge wainscot
Mythimna pallens	Weißadereule, Kräutereule	common wainscot
Mythimna separata	Asiatische Reiseule	Asian wainscot, northern armyworm (Austr.)
Mythimna turca	Rotgefranste Weißpunkteule, Marbeleule, Binsengraseule, Finkeneule, Waldmoorrasen-Türkeneule	double-lined wainscot, double line moth

Mythimna unipuncta	Amerikanische Reiseule	white-speck wainscot, American wainscot, American armyworm
Mytilicola intestinalis	Muscheldarmkrebs	mussel intestinal crab
Mytilidae	Miesmuscheln	mussels
Mytiloidea	Miesmuscheln	mussels
Mytilus californianus	Kalifornische Miesmuschel	California mussel (common mussel)
Mytilus chilensis	Chilenische Miesmuschel	Chilean mussel
Mytilus crassitesta	Koreanische Miesmuschel	Korean mussel
Mytilus edulis	Gemeine Miesmuschel	blue mussel, bay mussel, common mussel, common blue mussel
Mytilus galloprovincialis	Mittelmeer-Miesmuschel, Blaubartmuschel, Seemuschel	Mediterranean mussel
Mytilus planulatus	Australische Miesmuschel	Australian mussel
Mytilus platensis	Rio-de-la-Plata Miesmuschel	River Plate mussel
Mytilus smaragdinus	Grüne Miesmuschel	green mussel
Mytilus trossulus	Seltsame Miesmuschel*	foolish mussel
Myxas glutinosa	Mantelschnecke	glutinous snail
Myxicola infundibulum	Schlicksabelle, Schlickröhrenwurm	slime feather duster, mud feather duster
Myxilla incrustans/ Halichondria incrustans	Grubenschwamm	yellow lobed sponge
Myxine glutinosa	Inger, Atlantischer Inger	hagfish
Myxiniformes (Myxinida)	Ingerartige, Inger, Schleimaale	hagfishes
Myzopoda aurita	Madagassische Haftscheiben-Fledermaus	Old World sucker-footed bat
Myzus ascalonicus	Schalottenlaus, Zwiebellaus	shallot aphid
Myzus cerasi	Schwarze Kirschenlaus, Sauerkirschenlaus	black cherry aphid
Myzus persicae	Grüne Pfirsichblattlaus, Grüne Pfirsichlaus	peach-potato aphid, green peach aphid, greenfly, cabbage aphid

Nabidae	Sichelwanzen	damsel bugs
Nacerda melanura/ *Nacerdes melanura*	Scheinbock	wharf borer
Naenia typica	Buchdruckereule	gothic (moth)
Naididae	Wasserschlängler, Naiden	naidid oligochaetes, naidids
Naja atra	Chinesische Kobra	Chinese cobra
Naja haje	Uräusschlange	Egyptian cobra
Naja kaouthia	Monokelkobra	monocled cobra
Naja melanoleuca	Schwarzweiße Kobra, Schwarzweiße Hutschlange	forest cobra, black-and-white lipped cobra
Naja mossambica	Mosambik-Speikobra	Mozambique spitting cobra
Naja naja	Brillenschlange, Südasiatische Kobra	common cobra, Indian cobra
Naja nigricollis	Speikobra, Schwarzhalskobra, Afrikanische Speikobra	spitting cobra, black-necked spitting cobra, blackneck spitting cobra
Naja nivea	Kapkobra, Kap-Kobra	Cape cobra, yellow cobra
Naja oxiana	Mittelasiatische Kobra	Central Asiatic cobra, Central Asian cobra
Naja pallida	Afrikanische Kobra, Rote Speikobra	African cobra
Naja spp.	Kobras	cobras
Naja sputatrix	Indonesische Kobra, Sunda-Speikobra	Indonesian cobra
Naja sumatrana	Goldene Speikobra	golden spitting cobra
Nandayus nenday	Nandaysittich	black-hooded conure
Nandidae	Nanderbarsche, Nanderfische, Blattfische	leaffishes
Nandinia binotata	Pardelroller	African palm civet
Nannacara anomala	Glänzender Zwergbuntbarsch, Gestreifter Zwergbuntbarsch	golden-eyed dwarf cichlid, golden dwarf cichlid, goldeneye cichlid (FAO)
Nannopterum harrisi	Galapagosscharbe	flightless cormorant
Nannosciurus melanotis	Braunes Zwerghörnchen, Schwarzohr-Zwerghörnchen	black-eared squirrel
Nannostomus beckfordi/ *Nannostomus aripirangensis*	Längsbandsalmler	golden pencilfish
Nannostomus bifasciatus	Zweibinden-Ziersalmler, Zweibindensalmler	whiteside pencilfish
Nannostomus eques	Spitzmaul-Ziersalmler, Schrägsteher, Schrägsalmler	tube-mouthed pencilfish
Nannostomus espei	Schrägbinden-Ziersalmler	barred pencilfish
Nannostomus harrisoni	Goldmaul-Ziersalmler	Harrison's pencilfish
Nannostomus marginatus	Zwergziersalmler	dwarf pencilfish
Nannostomus trifasciatus	Dreibinden-Ziersalmler	three-striped pencilfish, three-lined pencilfish
Nannostomus unifasciatus	Einbinden-Ziersalmler, Einbinden-Schrägsteher	oneline pencilfish
Nanochromis parilus/ *Nanochromis nudiceps*	Kongo-Zwergbuntbarsch, Blauer Kongocichlide	Nudiceps, Congo dwarf cichlid*
Nanonycteris veldkampi	Veldkamps Zwergflughund	little flying cow
Nanospalax leucodon/ *Microspalax leucodon*	Westblindmaus	lesser mole-rat, western mole rat

Napaeozapus insignis	Waldhüpfmaus	woodland jumping mouse
Narcine brasiliensis	Karibischer Zitterrochen, Kleiner Zitterrochen	Brazilian electric ray
Narcine entemedor	Riesenzitterrochen	giant electric ray
Narcine timlei	Gefleckter Zitterrochen	spotted numbfish
Narcinidae/Narkidae	Kleine Zitterrochen	electric rays, numbfishes
Narcomedusae	Narkomedusen	narcomedusas
Narke dipterygia	Kleiner Zitterrochen	numbray
Nasalis concolor/ Simias concolor	Ringelschwanz-Stumpfnasenaffe, Pageh-Stumpfnasenaffe	pig-tailed snub-nosed monkey, pig-tailed snub-nosed langur
Nasalis larvatus	Nasenaffe	proboscis monkey
Nasidae/Nasinae	Nasenfische, Einhornfische	unicornfishes
Naso brevirostris	Kurznasen-Einhornfisch, Schärpen-Nasendoktor	short-nose unicornfish, short-snouted unicornfish, spotted unicornfish (FAO)
Naso hexacanthus	Sechsstachel-Doktorfisch	six-spine surgeonfish, black unicornfish, sleek unicornfish (FAO)
Naso lituratus/ Callicanthus lituratus	Kuhkopffisch, Kuhkopf-Doktorfisch, Hornloser Einhornfisch	smooth-headed unicornfish, orangespine unicornfish (FAO), green unicornfish
Naso tuberosus	Ramsnasen-Einhornfisch, Wulstnasendoktor	humpnose unicornfish
Naso unicornis	Langschnauzen-Einhornfisch, Blauklingen-Nasendoktor	long-snouted unicornfish, bluespine unicornfish (FAO)
Naso vlamingii	Masken-Doktorfisch, Maskendoktorfisch	zebra unicornfish, Vlaming's unicornfish, bignose unicornfish (FAO)
Nassariidae	Netzreusenschnecken	dogwhelks, dog whelks, nassa mud snails (basket shells)
Nassarius obsoletus/ Ilyanassa obsoleta	Östliche Reusenschnecke	eastern mudsnail, eastern nassa
Nassarius acutus	Spitze Reusenschnecke	sharp nassa
Nassarius arcularius	Schmuckkästchen-Reusenschnecke, Schmuckkästchen	casket nassa, cake nassa
Nassarius consensus	Streifenreusenschnecke	striate nassa
Nassarius coronatus	Kronenreusenschnecke	crowned nassa
Nassarius distortus	Halsketten-Reusenschnecke	necklace nassa
Nassarius fossatus	Gefurchte Reusenschnecke, Westliche Riesenreuse	channeled nassa, giant western nassa
Nassarius fraterculus	Japanische Reusenschnecke	Japanese nassa
Nassarius glans	Eichelreusenschnecke, Eichelreuse	glans nassa
Nassarius insculptus	Glatte Reusenschnecke	smooth western nassa
Nassarius marmoreus	Marmorierte Reusenschnecke	marbled nassa
Nassarius mutabilis	Wandelbare Reusenschnecke	mutable nassa
Nassarius papillosus	Pustel-Reusenschnecke	papillose nassa, pimpled nassa
Nassarius reticulatus	Netz-Reusenschnecke, Netz-Fischreuse	netted nassa, netted dogwhelk
Nassarius spp.	Reusenschnecken	dog whelks, nassas

Nassarius trivittatus	Dreistreifen-Reusenschnecke	threeline mudsnail, New England nassa
Nassarius vibex	Unterlaufene Reusenschnecke*	bruised nassa
Nasua nasua	Südamerikanischer Nasenbär, Gewöhnlicher Nasenbär	coatimundi, common coati
Nasua spp.	Nasenbären	coatis, coatimundis
Nasuella olivacea	Kleiner Nasenbär	mountain coati
Natalidae	Trichterohren-Fledermäuse	funnel-eared bats
Natalobatrachus bonebergi	Natalfrosch	Natal diving frog
Natalus spp.	Trichterohren-Fledermaus	funnel-eared bats
Natalus stramineus	Mexikanische Trichterohren-Fledermaus	Mexican funnel-eared bat
Natator depressus/ Chelonia depressa	Australische Plattschildkröte, Australische Suppenschildkröte	flatback sea turtle, flatback turtle, Australian green turtle
Natica acinonyx	Beeren-Mondschnecke*	African berry moonsnail
Natica alapapilionis	Schmetterlings-Mondschnecke	butterfly moonsnail
Natica arachnoidea	Spinnen-Mondschnecke	spider moonsnail
Natica canrena/ Naticarius canrena	Vielfarbige Mondschnecke*	colorful Atlantic moonsnail, colorful moonsnail
Natica castrensis	Netz-Mondschnecke	netted moonsnail
Natica catena/ Euspira catena/ Lunatia catena	Halsband-Mondschnecke, Halsband-Nabelschnecke, Große Nabelschnecke	large necklace snail, large necklace moonsnail (large necklace shell)
Natica clausa/ Cryptonatica affinis	Arktische Mondschnecke, Arktische Nabelschnecke	Arctic moonsnail
Natica fanel	Fanel-Mondschnecke*	fanel moonsnail, the fanel moon
Natica fasciata	Derbe Mondschnecke	solid moonsnail
Natica lineata	Linien-Mondschnecke*	lined moonsnail
Natica livida	Fahlgraue Mondschnecke	livid moonsnail
Natica maculata/ Natica hebraeus	Hebräische Mondschnecke, Hebräische Nabelschnecke	Hebrew moonsnail (Hebrew necklace shell)
Natica onca	China-Mondschnecke, Chinamond	China moonsnail
Natica rubromaculata	Rotgestreifte Mondschnecke	red-striped moonsnail
Natica stellata	Gestirnte Mondschnecke	starry moonsnail, stellate sand snail
Natica stercusmuscarum/ Naticarius stercusmuscarum	Tausendpunkt-Mondschnecke, Tausendpunkt-Nabelschnecke, Fliegendreck-Mondschnecke	fly-speck moonsnail, fly-specked moonsnail
Natica tigrina	Tiger-Mondschnecke	tiger moonsnail
Natica undulata/ Notocochlis undulata	Zebra-Mondschnecke	zebra moonsnail
Natica unifasciata	Einband-Mondschnecke*	single-banded moonsnail
Natica violacea	Violette Mondschnecke	violet moonsnail
Natica vitellus	Kalb-Mondschnecke*	calf moonsnail
Naticarius canrena	Vielfarbige Mondschnecke*	colorful Atlantic moonsnail, colorful moonsnail
Naticarius hebraeus/ Natica maculata	Hebräische Mondschnecke, Hebräische Nabelschnecke	Hebrew moonsnail (Hebrew moon shell, Hebrew necklace shell)
Naticidae	Mondschnecken, Nabelschnecken	moon snails, necklace snails

Natrix erythrogaster/ *Nerodia erythrogaster*	Rotbauch-Wassernatter	red-bellied water snake
Natrix maura	Vipernatter	viperine snake, viperine grass snake
Natrix megalocephala	Großkopf-Ringelnatter	bigheaded grass snake, large-headed water snake (IUCN)
Natrix natrix	Ringelnatter	grass snake
Natrix piscator/ *Xenochrophis piscator*	Fischernatter	chequered keelback, common scaled water snake, fishing snake
Natrix septemvittata	Königinschlange	queen snake
Natrix spp.	Kielrückennattern	grass snakes
Natrix tessellata	Würfelnatter	dice snake
Naucoridae	Schwimmwanzen	creeping water bugs, saucer bugs
Naucrates ductor	Lotsenfisch	pilot fish
Naultinus elegans	Grüner Baumgecko	green tree gecko
Naultinus spp.	Baumgeckos	tree geckos
Nausithoe punctata	Kronenqualle*	crown jellyfish
Nautiloidea	Nautilusverwandte	nautilus (*pl* nautili or nautiluses)
Nautilus macromphalus	Neukaledonisches Perlboot	New Caledonian nautilus, bellybutton nautilus (FAO)
Nautilus pompilius	Gemeines Perlboot, Gemeines Schiffsboot	chambered nautilus, emperor nautilus (FAO)
Nautilus scrobiculatus	Salomons-Perlboot	Salomon's nautilus
Nautilus spp.	Perlboote, Schiffsboote	chambered nautiluses, pearly nautiluses
Neacomys spp.	Südamerikanische Stachelratten*	bristly mice, spiny mice rats
Nebria brevicollis	Gewöhnlicher Dammläufer, Pechschwarzer Dammläufer	common black ground beetle
Nebria livida	Gelbrandiger Dammläufer	
Nebria picicornis	Flussdammläufer	
Nebrius ferrugineus/ *Nebrius concolor*	Spitzflossen-Ammenhai, Keulenhai	tawny nurse shark (FAO), giant sleepy shark
Necator americanum	Neuwelt-Hakenwurm	New World hookworm
Necora puber/ *Liocarcinus puber/* *Macropipus puber*	Samtkrabbe, Wollige Schwimmkrabbe	velvet swimming crab, velvet fiddler, devil crab
Necrobia ruficollis	Rothalsiger Schinkenkäfer	red-shouldered ham beetle
Necrobia rufipes	Rotbeiniger Kolbenkäfer, Rotbeiniger Schinkenkäfer, Koprakäfer	red-legged ham beetle, copra beetle
Necrophoridae	Totengräber	burying beetles
Nectarinia asiatica	Purpurnektarvogel	purple sunbird
Nectarinia johnstoni	Lobeliennektarvogel	red-tufted malachite sunbird
Nectarinia jugularis	Grünrücken-Nektarvogel	olive-backed sunbird
Nectarinia osea	Jerichonektarvogel	northern orange-tufted sunbird, Palestine sunbird
Nectarinia regia	Königsnektarvogel	regal sunbird
Nectarinia reichenowi	Goldschwingen-Nektarvogel	golden-winged sunbird
Nectarinia senegalensis	Rotbrust-Glanzköpfchen	scarlet-chested sunbird
Nectarinia superba	Prachtnektarvogel	superb sunbird

Nectarinia thomensis	Riesennektarvogel	Sao Thome giant sunbird
Nectarinia zeylonica	Ceylonnektarvogel	purple-rumped sunbird
Nectariniidae	Nektarvögel	sunbirds, spiderhunters
Nectogale elegans	Gebirgsbachspitzmaus, Tibetanische Wasserspitzmaus	Tibetan water shrew, web-footed water shrew
Nectomys spp.	Südamerikanische Wasserratten	Neotropical water rats
Nectophryne spp.	Afrikanische Baumkröten	African tree toads
Nectophrynoides minutus	Kleinst-Baumkröte	minute tree toad
Nectophrynoides spp.	Lebendgebärende Kröten	African tree toads, live-bearing toads
Nectophrynoides viviparus	Tanganjika-Baumkröte	Morogoro tree toad
Necturus alabamensis	Alabama-Furchenmolch	Alabama waterdog
Necturus beyeri	Golf-Furchenmolch	Gulf Coast waterdog
Necturus maculosus	Gefleckter Furchenmolch	mudpuppy, waterdog
Necturus punctatus	Zwergfurchenmolch	dwarf waterdog
Necydalis major	Großer Wespenbock	
Negaprion brevirostris	Zitronenhai, Amerikanischer Zitronenhai	lemon shark
Negaprion spp.	Zitronenhaie	lemon sharks
Nehalennia speciosa	Zwerglibelle	green damsel
Neides tipularius	Schnakenwanze, Schnakerich	
Nelsonia neotomodon	Nelsons Baumratte	Nelson's wood rat
Nemadactylus macropterus	Großflossen-Morwong	tarakihi
Nemanthus nitidus	Schmuckanemone	jewel anemone
Nemapalpus nearcticus		sugarfoot moth fly
Nemapogon granellus	Kornmotte	European grain moth
Nemastomatidae	Fadenkanker	nemastomatids
Nematistius pectoralis	Hahnfisch*	roosterfish
Nematocera (Diptera)	Mücken & Schnaken	mosquitoes
Nematoda	Fadenwürmer, Nematoden, Rundwürmer (*sensu strictu*)	roundworms, nematodes
Nematomenia banyulensis (Aplacophora)	Schlundkegel-Glattfuß	
Nematomenia corallophila (Aplacophora)	Tarnglattfuß	Mediterranean coral solenogaster*
Nematomenia flavens (Aplacophora)	Gelber Glattfuß	Mediterranean yellow solenogaster*
Nematomorpha	Saitenwürmer	horsehair worms, hairworms, gordian worms, threadworms, nematomorphans, nematomorphs
Nematostella vectensis	Sternchenanemone*	starlet sea anemone
Nematus ribesii/ Pteronidea ribesii	Gelbe Stachelbeerblattwespe, Stachelbeerfliege	gooseberry sawfly, common gooseberry sawfly
Nematus salicis	Weidenblattwespe	willow sawfly
Nemertesia antennina	Meerbart*	sea beard

Nemertini/ Nemertea/ Rhynchocoela	Schnurwürmer	nemertines, nemerteans, proboscis worms, rhynchocoelans, ribbon worms (broad/flat), bootlace worms (long)
Nemesiidae	Falltürklappenspinnen	tube trapdoor spiders, wishbone spiders
Nemestrinidae	Netzfliegen	tangle-veined flies
Nemichthyidae	Schnepfenaale	snipe eels
Nemichthys scolopaceus	Schnepfenaal	snipe eel
Nemipteridae	Scheinschnapper (inkl. Fadenflosser & Monokelbrassen)	threadfin breams (butterfly breams, spinecheeks, monocle breams & dwarf breams)
Nemipterus spp.	Scheinschnapper	thread-fin bream
Nemipterus virgatus	Goldener Scheinschnapper	golden thread-fin bream
Nemobius sylvestris	Waldgrille	wood cricket
Nemonychidae/ Rhinomaceridae	Schlankrüssler	pine-flower snout beetles
Nemopsis bachei	Amerikanische Brackwassermeduse	American brackish water medusa*
Nemorhaedus baileyi	Rotgoral, Rote Ziegenantilope	red goral
Nemorhaedus spp.	Gorals, Ziegenantilopen	gorals
Nemouridae	Nemouriden	winter stoneflies & spring stoneflies
Neobatrachus centralis	Trillerfrosch	trilling frog
Neobatrachus pelobatoides	Knoblauchkrötenähnlicher Frosch, Summfrosch*	humming frog
Neobatrachus pictus	Katzenfrosch	meeoing frog
Neobatrachus sutor	Schusterfrosch	shoemaker frog
Neobisium carcinoides	Moosskorpion	
Neoceratodus forsteri	Australischer Lungenfisch	Australian lungfish
Neodiprion sertifer	Rote Kiefernbuschhornblattwespe	fox-coloured sawfly, lesser pine sawfly, small pine sawfly, European pine sawfly (U.S.)
Neofelis nebulosa	Nebelparder	clouded leopard
Neofiber alleni	Florida-Wasserratte	round-tailed muskrat, Florida water rat
Neofibularia nolitangere	Brennschwamm, „Fass-mich-nicht-an-Schwamm"	do-not-touch-me sponge
Neogastropoda/Stenoglossa	Neuschnecken, Schmalzüngler	neogastropods: whelks & cone shells
Neohydromys fuscus	Kurzschwanz-Spitzmausratte*	short-tailed shrew mouse
Neomenia carinata (Aplacophora)	Kielmondling	keeled solenogaster*
Neomorphinae	Erdkuckucke	roadrunners, ground-cuckoos
Neomys anomalus	Sumpfspitzmaus, Kleine Wasserspitzmaus	Miller's water shrew
Neomys fodiens	Große Wasserspitzmaus	Old World water shrew
Neomys spp.	Wasserspitzmäuse	Old World water shrews
Neopanope sayi	Schlickkrebs	mud crab
Neopetrolisthes maculatus	Punkttupfen-Anemonenkrabbe	dotted anemone crab

Neophema chrysogaster	Goldbauchsittich	orange-bellied parakeet, orange-bellied parrot
Neophema splendida	Glanzsittich	scarlet-chested parakeet, splendid parakeet, splendid parrot
Neophoca cinerea	Australischer Seelöwe	Australian sea lion
Neophocaena phocaenoides	Indischer Schweinswal	finless porpoise
Neophron percnopterus	Schmutzgeier	Egyptian vulture
Neoplatymops mattogrossensis	Südamerikanische Flachkopf-Fledermaus	South American flat-headed bat
Neoraja caerulea/ Breviraja caerulea	Blaurochen	blue ray
Neornithes	Neuvögel	neornithes, true birds
Neoscona adianta	Heide-Radspinne, Heideradspinne	heathland orbweaver
Neoscona spp.	Gefleckte Radspinnen*	spotted orbweavers
Neoscopelidae	Laternenzüngler	blackchins
Neoseps reynoldsi	Florida-Sandskink	sand skink
Neosittini	Spiegelkleiber	sittellas
Neostethidae	Neostethiden (Zwergährenfische u.a.)	neostethids
Neotoma spp.	Buschratten	wood rats, woodrats, pack rats, trade rats
Neotomodon alstoni	Mexikanische Vulkanmaus	Mexican volcano mouse
Neotomys ebriosus	Anden-Sumpfratte	Andean swamp rat
Neotragus batesi	Batesböckchen	Bate's dwarf antelope
Neotragus moschatus	Suniböckchen	suni
Neotragus pygmaeus	Kleinstböckchen	royal antelope
Neotrigonia margaritacea	Australmuschel	Australian brooch clam
Neotrombicula autumnalis	Erntemilbe, Herbstmilbe, Herbstgrasmilbe	harvest mite, red bug, chiggers (*larvas*)
Neottiophilidae	Meisensauger	neottiophilids (nestling bird suckers)
Nepa cinerea	Wasserskorpion	water scorpion
Nephila clavipes	Goldene Seidenspinne, Gold-Seidenspinne	golden-silk spider, golden silk orbweaver
Nephila maculata	Gefleckte Seidenspinne, Riesen-Waldspinne	giant wood spider
Nephila spp.	Seidenspinnen, Gold-Seidenspinnen	golden-silk spiders, golden orb spiders, golden orbweavers
Nephilidae	Seidenspinnen	silk spiders
Nephropidae	Hummer	clawed lobsters
Nephrops norvegicus	Kaisergranat, Kaiserhummer, Kronenhummer, Schlankhummer, Tiefseehummer	Norway lobster, Norway clawed lobster, Dublin Bay lobster, Dublin Bay prawn (scampi, langoustine)
Nephropsis atlantica	Atlantischer Krallenhummer*	scarlet clawed lobster
Nephropsis rosea	Rosa Krallenhummer	rosy lobsterette
Nephrotoma spp.	Krähenschnaken	spotted craneflies
Nephrurus spp.	Keulenschwanzgeckos	knob-tailed geckos, knobtail geckos

Nephtyidae	Opalwürmer	catworms, shimmy worms
Nephtys spp.	Opalwürmer	catworms, shimmy worms
Nepidae	Skorpionwanzen, Skorpionswanzen	waterscorpions
Nepticulidae	Zwergmotten	dwarf eyecap moths, leaf miners, leaf-miners
Neptunea antiqua	Gemeines Neptunshorn, Gemeine Spindelschnecke	ancient whelk, ancient neptune snail, common spindle snail, red whelk, buckie
Neptunea brevicauda	Dickgeripptes Neptunshorn	thick-ribbed whelk
Neptunea contraria	Linksgewundenes Neptunshorn	left-handed whelk, left-handed neptune
Neptunea lyrata lyrata	Lyra-Neptunshorn	lyre whelk, common Northwest neptune, New England neptune
Neptunea lyrata decemcostata	Schrumpelhorn*, Schrumpeliges Neptunshorn	wrinkle whelk, wrinkled neptune
Neptunea tabulata	Flachrand-Neptunshorn	tabled whelk, tabled neptune
Neptunea ventricosa	Dickbauchiges Neptunshorn	fat whelk
Neptunea vinosa	Pazifisches Weinhorn*	wine whelk
Nereidae	Nereiden, Seeringelwürmer u.a.	nereids
Nereis diversicolor/ Hediste diversicolor	Schillernder Seeringelwurm	estuary ragworm
Nereis pelagica	Gemeiner Seeringelwurm, Schwimmender Seeringelwurm	pelagic clam worm
Nereis virens/Neanthes virens	Grüner Seeringelwurm	ragworm, sandworm, clam worm, king ragworm
Nerita exuvia	Schlangenhaut-Nixenschnecke, Schlangenhaut-Nixe	snake-skin nerite
Nerita fulgurans	Antillen-Nixenschnecke	Antillean nerite
Nerita peloronta	Blutender Zahn	bleeding tooth
Nerita polita	Glatte Nixenschnecke	polished nerite
Nerita pupa	Zebra-Nixenschnecke	zebra nerite
Nerita tessellata	Karierte Nixenschnecke	checkered nerite, tessellate nerite
Nerita versicolor	Vierzahn-Nixenschnecke	four-tooth nerite, four-toothed nerite
Neritidae	Kahnschnecken, Schwimmschnecken, Nixenschnecken	nerites, neritids
Neritina communis/ Theodoxus communis	Zickzack-Nerite	zigzag nerite, common nerite
Neritina virginea	Jungfrau-Nerite	virgin nerite
Nerodia clarkii	Salzmarschnatter	salt marsh snake
Nerodia erythrogaster/ Natrix erythrogaster	Rotbauch-Wassernatter	red-bellied water snake, plainbelly water snake, plain-bellied water snake
Nerodia fasciata fasciata	Gebänderte Wassernatter	banded water snake
Nerodia rhombifer	Rauten-Wassernatter	diamondback water snake

Nerodia sipedon	Siegelringnatter	northern water snake
Nerodia spp.	Nordamerikanische Wassernattern	water snakes
Nerodia taxispilota	Braune Wassernatter	brown water snake
Nerophis lumbriciformis	Krummschnauzige Schlangennadel, Große Wurm-Seenadel	worm pipefish
Nerophis ophidion	Kleine Schlangennadel	straight-nosed pipefish
Nesiarchus nasutus		Johnson's scabbardfish
Nesokia indica	Kurzschwanz-Maulwurfsratte	pest rat, short-tailed bandicoot rat
Nesolagus netscheri	Sumatra-Kaninchen	Sumatra short-eared rabbit, Sumatran rabbit
Nesomantis thomasseti	Thomassets Seychellenfrosch	Thomasset's Seychelles frog, Seychelle rock frog
Nesomys rufus	Inselratte	Madagascan rat
Nesoromys ceramicus	Ceramratte	Ceram Island rat
Nesovitrea hammonis	Streifen-Glanzschnecke	rayed glass snail
Nesticidae	Höhlenspinnen	cave cobweb spiders
Nestor noabilis	Kea	kea
Netta rufina	Kolbenente	red-crested pochard
Nettapus auritus	Afrikanische Zwergente	African pygmy goose
Nettastomatidae	Entenschnabelaale	witch eels, duckbill eels
Netuma thalassinus/ Arius thalassinus	Riesenmeereswels	giant catfish
Neuroptera/Planipennia	Echte Netzflügler, Hafte	neuropterans (dobson flies/antlions)
Neuroterus albipes/ Neuroterus laeviusculus	Weißfuß-Eichenlinsengallwespe (>Glatte Linsengalle/ Kleine Blattrandgalle)	oak leaf smooth-gall cynipid wasp, Schenck's gall wasp, smooth spangle gall wasp (>smooth spangle gall)
Neuroterus numismalis	Seidenknopfgallwespe (>Grüne Pustelgalle)	oak leaf blister-gall cynipid wasp, oakleaf silkbutton-spanglegall cynipid wasp (>silk button spangle gall)
Neuroterus quercusbaccarum	Eichenlinsengallwespe (>Große Linsengalle[1]), Weinbeerengallwespe (>Weinbeerengalle[2])	oak leaf spangle-gall cynipid wasp (>common spangle gall[1]), oak leaf currant-gall cynipid wasp (>currant gall[2])
Neuroterus tricolor/ Neuroterus fumipennis	Borstenkugel-Gallwespe (>Borstige Kugelgalle)	oak leaf cupped-gall cynipid wasp, oakleaf cupped-spanglegall cynipid wasp (>cupped spangle gall)
Neurotoma nemoralis	Steinobst-Gespinstblattwespe	apple web-spinning sawfly
Neurotoma saltuum	Gesellige Birnblattwespe	social pear sawfly
Neusticomys monticolus	Südamerikanische Fischratte u.a., Nord-Equador-Fischratte	fish-eating rat, aquatic rat
Neverita albumen	Weisser Mond	egg-white moonsnail
Neverita duplicata/ Policines duplicatus	Haifischauge, Große Mondschnecke	shark eye, sharkeye, lobed moonsnail
Neverita heliciodes	Spiralen-Mondschnecke*	spiral moonsnail
Neverita josephina	Josefinische Mondschnecke	Josephine's moonsnail
Neverita peselephanti	Elefantenfuß-Mondschnecke, Elefantenfußschnecke	elephant's-foot moonsnail
Neverita politiana	Glänzende Mondschnecke	polished moonsnail

Nezumia aequalis	Glatter Grenadier, Glatte Panzerratte	smooth rattail
Nibea albiflora	Weißblumen-Umberfisch	white flower croaker (FAO), yellow drum
Nicrophorus americanus	Amerikanischer Riesenaaskäfer	giant carrion beetle, American burying beetle (IUCN)
Niditinea fuscipunctella	Nestermotte	brown-dotted clothes moth, poultry house moth
Nilssonia formosa/ Trionyx formosus	Burma-Weichschildkröte	Burmese peacock softshell turtle, Burmese softshell turtle
Nimbochromis livingstonii	Schläfer	livingstoni
Ninox connivens/ Hieracoglaux connivens	Kläfferkauz	barking owl, barking hawk-owl
Niphargus aquilex	Brunnenkrebs, Höhlenflohkrebs	cavernous well shrimp* a.o.
Niphargus puteanus	Höhlenflohkrebs	cavernous well shrimp* a.o.
Niphargus spp.	Höhlenkrebse	well shrimps
Nipponia nippon	Nipponibis	crested ibis
Niptus hololeucus	Messingkäfer	golden spider beetle
Nitidella ocellata	Weißflecken-Täubchen	white-spotted dovesnail
Nitidula bipunctata	Rauchfleischglanzkäfer	
Nitidulidae	Glanzkäfer	sap beetles, sap-feeding beetles
Niveria quadripunctata/ Trivia quadripunctata	Vierpunkt-Kaurischnecke, Vierpunktkauri	fourspot trivia, four-spotted trivia
Niveria candidula/ Trivia candidula	Weiße Kaurischnecke, Weißkauri	white trivia
Niveria pediculus/ Trivia pedicula	Kaffeebohnen-Kaurischnecke	coffeebean trivia
Niveria suffusa/ Trivia suffusa	Rosa Kaurischnecke, Rosa Kauri	pink trivia
Niviventer fulvescens	Kastanienmaus	chestnut rat
Niviventer niviventer	Weißbauchratte	white-bellied rat
Niviventer spp.	Weißbauchratten	white-bellied rats
Noctilio spp.	Hasenmaul-Fledermäuse, Hasenmäuler	bulldog bats, fisherman bats
Noctilionidae	Hasenmaul-Fledermäuse, Hasenmäuler	bulldog bats, fisherman bats
Noctua comes	Primeleule	lesser yellow underwing (moth)
Noctua fimbriata	Gelbe Bandeule	broad-bordered yellow underwing
Noctua orbona/ Triphaena orbona	Kleine Bandeule, Heckenkräuterflur-Bandeule	lunar yellow underwing (moth)
Noctua pronuba	Hausmutter	large yellow underwing (moth)
Noctuidae	Eulen, Eulenfalter	noctuid moths
Nodipecten nodosus/ Lyropecten nodosa	Löwenpranke	lions-paw scallop, lion's paw
Nodipecten subnodosus	Pazifische Riesen-Löwenpranke	giant lions-paw
Noemacheilus barbulatus/ Barbatula barbatula/ Nemacheilus barbatulus	Bartgrundel, Bachschmerle, Schmerle	stone loach
Noemacheilus spp.	Bachschmerlen	stone loaches
Nola cucullatella	Kapuzenbärchen	short-cloaked moth
Nomada spp.	Wespenbienen, Kuckucksbienen	cuckoo bees

Nomeidae	Quallenfische, Driftfische, Galeerenfische	driftfishes
Nonagria typhae	Gemeine Schilfeule	bulrush wainscot
Nonnula rubecula	Rotkehl-Faulvogel	rusty-breasted nunlet
Nordmannia ilicis/ Satyrium ilicis	Brauner Eichen-Zipfelfalter, Eichenzipfelfalter, Stechpalmen-Zipfelfalter	ilex hairstreak
Nosodendridae/ Nosodendronidae	Saftkäfer	wounded-tree beetles
Nosopsyllus fasciatus	Rattenfloh, Europäischer Rattenfloh	northern rat flea
Notacanthidae	Stachelaale, Dornaale, Eigentliche Dornrückenaale	spiny eels
Notacanthus bonapartei	Kurzflossen-Dornrückenaal	short-finned spiny-eel
Notacanthus chemnitzii	Stumpfnasen-Dornrückenaal	snub-nosed spiny-eel
Notaden bennettii	Katholiken-Frosch	crucifix toad
Notaden spp.	Notaden	Australian shovelfoots
Notaspidea	Flankenkiemer	notaspideans
Notechis ater	Schwarze Tigerotter	black tiger snake
Notechis scutatus	Tigerotter	Australian tiger snake, mainland tiger snake
Notechis spp.	Tigerottern, Australische Tigerottern	Australian tiger snakes, black tiger snakes
Noteridae	Uferfeuchtkäfer, Noteriden	burrowing water beetles
Notharchus macrorhynchos	Weißhals-Faulvogel	white-necked puffbird
Nothocrax urumutum	Rothokko	nocturnal curassow
Nothomyrmecia macrops	Australische Ameise	Australian ant, dinosaur ant
Nothoprocta pentlandii	Andensteißhuhn, Pentland-Steißhuhn	Andean tinamou
Notidanidae/Hexanchidae	Grauhaie	gray sharks, hexanchids
Notiomys spp.	Maulwurfsmäuse*	long-clawed mice, mole mice
Notiophilus aquaticus	Dunkler Laubläufer	
Notiophilus biguttatus	Zweifleckiger Laubläufer, Zweigefleckter Eilkäfer	burnished ground beetle
Notiophilus palustris	Gewöhnlicher Laubläufer	
Notiophilus spp.	Eilkäfer	
Notiosorex crawfordi	Graue Wüstenspitzmaus	gray shrew
Notiosorex gigas	Große Wüstenspitzmaus	desert shrew
Notocelia uddmanniana	Brombeertriebwickler	
Notocheiridae/Isonidae	Flügelährenfische	surf sprites
Notochelys platynota	Plattrückenschildkröte	Malayan flatshell turtle, Malayan flat-shelled turtle
Notocochlis undulata	Zebramondschnecke	zebra moonsnail
Notodonta dromedarius	Dromedar, Dromedarspinner, Birkenspinner, Erlenzahnspinner, Erlenbirkenauen-Zahnspinner	iron prominent

Notodonta torva	Gelbbrauner Zahnspinner, Gelbbrauner Zickzackspinner, Auenpappelgestrüpp-Zahnspinner	large dark prominent
Notodonta ziczac/ Eligmodonta ziczac	Zickzackspinner, Uferweiden-Zahnspinner	pebble prominent
Notodontidae	Zahnspinner	prominents
Notomys alexis	Spinifex-Hüpfmaus	spinifex hopping mouse
Notomys aquilo	Nordaustralische Hüpfmaus	northern hopping mouse
Notomys cervinus	Rehbraune Hüpfmaus	fawn hopping mouse
Notomys fuscus	Dunkle Hüpfmaus*	dusky hopping mouse
Notomys spp.	Australische Hüpfmäuse, Australische Kängurumäuse	Australian hopping mice, jerboa mice, kangaroo mice
Notonecta glauca	Gemeiner Rückenschwimmer	common backswimmer
Notonecta ssp.	Rückenschwimmer	backswimmers (water boatmen)
Notonectidae	Rückenschwimmer	backswimmers (water boatmen)
Notophthalmus meridionalis	Schwarzgefleckter Wassermolch	black-spotted newt
Notophthalmus perstriatus	Gestreifter Wassermolch	striped newt
Notophthalmus viridescens	Grünlicher Wassermolch	eft, red-spotted newt, red eft, eastern newt
Notopteridae	Messerfische, Eigentliche Messerfische	featherfin knifefishes, Old World knifefishes, featherbacks
Notopteris macdonaldi	Langschwanz-Flughund	long-tailed fruit bats
Notopterus afer/ Papyrocranus afer	Afrikanischer Fähnchen-Messerfisch	reticulate knifefish
Notopterus chitala/ Chitala chitala	Indischer Fähnchen-Messerfisch, Gebänderter Messerfisch, Tausenddollarfisch	clown knifefish, Indian featherback
Notopterus notopterus	Bronze-Fähnchen-Messerfisch, Asiatischer Fähnchen-Messerfisch	bronze featherback
Notoryctes typhlops	Beutelmull, Beutelmaulwurf	marsupial "mole"
Notorynchus cepedianus/ Notorynchus maculatus	Siebenkiemiger Kammzahnhai	sevengill shark, broadnose sevengill shark (FAO), spotted sevengill shark
Notostraca	Rückenschaler	tadpole shrimps, shield shrimps
Notosudidae	Notosudiden	waryfishes, notosudids
Nototeredo norvagicus/ Teredo norvegica	Nordischer Schiffsbohrwurm	Norway shipworm, Norwegian shipworm
Notothenia coriiceps	Schwarze Notothenia	black rockcod, broad-headed notothenia, yellowbelly rockcod (FAO)
Notothenia gibberifrons/ Gobionotothen gibberifrons	Grüne Notothenia	bumphead notothenia, humped rockcod (FAO)
Notothenia kempi/ Lepidonotothen kempi	Augenstreifen-Notothenia	striped-eyed rockcod, striped-eye notothen (FAO)
Notothenia rossii	Marmorbarsch	marbled notothenia, marbled rockcod (FAO)
Notothenia squamifrons/ Lepidonotothen squamifrons	Graue Notothenia	scaled notothenia, grey rockcod (FAO)

Nototheniidae	Antarktische Dorsche, Antarktis-Eisfische	cod icefishes, Antarctic rockcod, notothenids
Notothenoips nudifrons/ Lepidonotothen nudifrons	Gelbflossen-Notothenia	naked-head notothenia, yellow-fin notie, gaudy notothen (FAO)
Nototodarus sloani	Wellington-Flugkalmar	Wellington flying squid
Notoungulata	Süd-Huftiere	notoungulates
Notropis albeolus	Weiß-Orfe	white shiner
Notropis atherinoides	Smaragd-Orfe	emerald shiner
Notropis boops	Großaugen-Orfe	bigeye shiner
Notropis chrysocephalus	Gestreifte Orfe	striped shiner
Notropis cornutus/ Luxilus cornutus	Gewöhnliche Orfe, Gemeine Orfe	common shiner
Notropis hypselopterus	Längsbandorfe	sailfin shiner
Notropis lutrensis	Rotflossenorfe	red shiner
Noturus gyrinus	Steinwels	tadpole madtom
Nucella lapillus/ Thais lapillus	Steinschnecke, Nordische Steinchenschnecke, Nördliche Purpurschnecke, Nordische Purpurschnecke	Atlantic dog whelk, northern dog whelk, Atlantic dogwinkle, northern dogwinkle
Nucella lima	Feilenschnecke*	file dogwinkle
Nucifraga caryocatactes	Tannenhäher	nutcracker
Nucifraga columbiana	Kiefernhäher	Clark's nutcracker
Nucras spp.	Nukras-Eidechse	sandveld lizards
Nuctenea cornuta	Gehörnte Spaltenkreuzspinne	furrow spider
Nuctenea umbratica/ Araneus umbraticus	Spaltenkreuzspinne	crevice spider*
Nucula crenulata	Feingekerbte Nussmuschel	crenulate nutclam
Nucula exigua	Pazifische Nussmuschel	iridescent nutclam, Pacific crenulate nutclam
Nucula nucleus	Gemeine Nussmuschel	common nutclam
Nucula proxima	Atlantische Nussmuschel	Atlantic nutclam
Nucula spp.	Nussmuscheln	nut clams, nutclams
Nucula tenuis/ Ennucula tenuis	Dünnschalige Nussmuschel, Glatte Nussmuschel	smooth nutclam
Nucula turgida	Glänzende Nussmuschel	shiny nutclam*
Nuculacea	Nussmuscheln	nut clams
Nuculana minuta	Geschnäbelte Nussmuschel	minute nutclam, beaked nutclam
Nuculana pernula	Nördliche Nussmuschel	northern nutclam, Müller's nutclam
Nuculanidae	Schwanenhalsmuscheln, Schwanenhals-Nussmuscheln	swan-neck nut clams (swan-neck shells), elongate nut clams
Nuculata sulcata	Streifen-Nussmuschel	furrowed nutclam
Nuculidae	Nussmuscheln	nut clams (nut shells)
Nudibranchia	Nacktkiemer, Meeresnacktschnecken, Nudibranchier	sea slugs, nudibranchs
Numenius arquata	Großer Brachvogel	curlew
Numenius borealis	Eskimobrachvogel	Eskimo curlew, prairie pigeon, doughbird
Numenius phaeopus	Regenbrachvogel	whimbrel

Numenius tenuirostris	Dünnschnabel-Brachvogel	slender-billed curlew
Numida meleagris	Helmperlhuhn	helmeted guineafowl
Numidinae	Perlhühner	guineafowls
Nuttallia nuttallii	Mahagoni-Muschel*	California mahogany-clam, mahogany clam
Nuttallina californica (Polyplacophora)	Kalifornische Höhlen-Käferschnecke*	California Nuttall chiton, troglodyte chiton
Nyctalus lasiopterus	Riesenabendsegler	greater noctule
Nyctalus leisleri	Kleinabendsegler, Kleiner Abendsegler	Leisler's bat
Nyctalus noctula	Abendsegler, Großer Abendsegler	noctule
Nyctalus spp.	Abendsegler	noctule bats
Nyctea scandiaca	Schnee-Eule	snowy owl
Nyctereutes procyonoides	Marderhund, Enok (Hbz. Seefuchs)	raccoon dog
Nycteribiidae	Fledermauslausfliegen, Fledermausfliegen	bat flies
Nycteris grandis	Große Schlitznase, Große Hohlnase	large slit-faced bat
Nycteris hispida	Rauhaar-Schlitznase	hairy slit-faced bat
Nycteris spp.	Schlitznasen, Hohlnasen	slit-faced bats, hollow-faced bats, hispid bats
Nycteris thebaica	Geoffroy-Schlitznase	Egyptian slit-faced bat
Nyctibiidae	Tagschläfer	potoos
Nycticebus coucang	Plumplori, Großer Plumplori	slow loris, cu lan
Nycticebus pygmaeus	Zwerg-Plumplori, Zwergplumplori	pygmy slow loris
Nycticeius spp.	Abenddämmerungs-Fledermäuse	evening bats, twilight bats
Nycticorax nycticorax	Nachtreiher	night heron
Nyctimene spp.	Röhrennasen-Flughunde	tube-nosed fruit bats
Nyctomys sumichrasti	Dämmerungsratte, Sumichrasti-Dämmerungsratte	vesper rat, Sumichrast's vesper rat
Nyctophilus spp.	Australische Großohrfledermaus	New Guinean big-eared bats & Australian big-eared bats
Nymphalidae	Fleckenfalter	brush-footed butterflies
Nymphalis antiopa	Trauermantel	Camberwell beauty
Nymphalis c-album/ Polygonia c-album/ Comma c-album	C-Falter, Weißes C	comma
Nymphalis egea/ Polygonia egea/ Comma egea	Gelber C-Falter, Glaskrautfalter	southern comma
Nymphalis io/Inachis io	Tagpfauenauge	peacock moth, peacock
Nymphalis polychloros/ Vanessa polychloros	Großer Fuchs	large tortoiseshell
Nymphicus hollandicus	Nymphensittich	cockatiel
Nymphula nymphaeata	Seerosenzünsler	brown China-mark moth, brown China-mark
Nysson spp.	Kuckucksgrabwespen	

Obelia bidentata	Zweizahn-Glockenpolyp	doubletoothed hydroid
Obelia dichotoma	Gegabelter Glockenpolyp	sea thread hydroid, thin-walled obelia
Obelia geniculata/ Laomedea geniculata	Zickzack-Glockenpolyp	knotted thread hydroid, zig-zag wine-glass hydroid
Obelia spp.	Glockenpolypen	thread hydroids, bushy wine-glass hydroids
Oberea bimaculata	Himbeerrutenbock	raspberry cane borer
Oberea linearis	Haselbock	hazel borer, hazelnut capricorn beetle
Oberea oculata	Rothalsiger Weidenbock, Weiden-Linienbock	
Oberea spp.	Linienböcke	
Oblada melanura	Brandbrasse, Oblada	saddled bream, saddled seabream (FAO)
Obrium brunneum	Reisigbock	
Obrium cantharinum	Rotbrauner Reisigbock	
Ocadia sinensis	Chinesische Streifenschildkröte	Chinese stripe-necked turtle, Chinese striped-neck turtle
Oceanites oceanicus	Buntfuß-Sturmschwalbe	Wilson's petrel
Oceanodroma castro	Madeirawellenläufer	Madeiran petrel
Oceanodroma leucorhoa	Wellenläufer	Leach's petrel
Ocenebra erinacea/ Ceratostoma erinaceum	Gerippte Felsschnecke, Gerippte Purpurschnecke, Großes Seekälbchen	European sting winkle, European oyster drill, rough tingle
Ocenebra corallina	Korallen Purpurschnecke, Korallen-Seekälbchen	coralline sting winkle*
Ochlodes venata	Rostfarbiger Dickkopf	large skipper
Ochotona spp.	Pfeifhasen	pikas, mouse hares, conies (coneys)
Ochotonidae	Pfeifhasen	pikas, mouse hares, conies (coneys)
Ochropacha duplaris	Zweipunkt-Eulenspinner	common lutestring
Ochrotomys nuttalli/ Peromyscus nuttalli	Goldmaus	golden mouse
Ochsenheimeriidae	Bohrmotten	fields, field moths
Ochteridae		velvety shore bugs
Octocorallia	Oktokorallen	octocorallians, octocorals
Octodon spp.	Strauchratten, Degus	degus
Octodontidae	Trugratten	octodonts
Octodontomys gliroides	Bori	chozchoz (*pl* chozchoris)
Octomys spp.	Viscacharatten	viscacha rats
Octopoda/Octobrachia	Achtarmige Tintenschnecken, Kraken	octopods, octopuses
Octopodoteuthis sicula	Achtarmkalmar	octopod squid
Octopus aegina	Sandvogelkrake*	sandbird octopus
Octopus bimaculatus	Zweifleck-Krake	two-spotted octopus
Octopus bimaculoides	Flachwasserkrake*	mud-flat octopus
Octopus californicus	Kalifornischer Krake	California octopus

Octopus cyaneus	Großer Blauer Krake	big blue octopus (FAO), common tropical octopus
Octopus defilippi	Lilliput-Langarmkrake	lilliput longarm octopus
Octopus dofleini	Pazifischer Riesenkrake	North Pacific giant octopus (FAO), giant Pacific octopus, common Pacific octopus
Octopus joubini	Zwergkrake, Joubins Krake	Atlantic pygmy octopus, dwarf octopus, Joubin's octopus
Octopus macropus	Langarmiger Krake	Atlantic white-spotted octopus, long-armed octopus, grass octopus, white-spotted octopus (FAO)
Octopus membranaceus	Schwimmhautkrake*	webfoot octopus
Octopus moschata/ Eledone moschata	Moschuskrake, Moschuspolyp	musky octopus, white octopus
Octopus vulgaris	Gemeiner Krake, Gemeiner Octopus, Polyp	common octopus, common Atlantic octopus, common European octopus
Oculina arbuscula	Feinverzweigte Augenkoralle	compact ivory bush coral
Oculina diffusa	Diffuse Augenkoralle, Elfenbein-Buschkoralle	diffuse ivory bush coral, ivory bush coral
Oculina robusta	Feste Augenkoralle	robust ivory tree coral
Oculina spp.	Augenkorallen	eyed coral, ivory bush corals
Oculina tenella	Zarte Augenkoralle	delicate ivory bush coral
Ocypode ceratophthalma	Indopazifische Reiterkrabbe	Indo-Pacific ghost crab, horn-eyed ghost crab
Ocypode gaudichaudii	Pazifische Reiterkrabbe	Pacific ghost crab
Ocypode quadrata	Atlantische Reiterkrabbe	Atlantic ghost crab
Ocypode spp.	Reiterkrabben	ghost crabs
Ocypodidae	Winkerkrabben & Reiterkrabben	fiddler crabs & ghost crabs
Ocypus olens/ Staphylinus olens	Schwarzer Moderkäfer, Schwarzer Moderkurzflügler	devil's coach-horse
Ocythoe tuberculata	Schmarotzerkrake, Höckeriger Seepolyp	tuberculate pelagic octopus
Ocyurus chrysurus	Gelbschwanzschnapper	yellowtail snapper
Odacantha melanura	Sumpf-Halsläufer, Schlanker Halskäfer	
Odezia atrata	Schwarzspanner, Kälberkropfspanner	
Odobenus rosmarus	Walross	walrus
Odocoileus bezoarticus/ Ozotoceros bezoarticus	Pampashirsch	pampas deer
Odocoileus dichotomus/ Blastocerus dichotomus	Sumpfhirsch	marsh deer
Odocoileus hemionus	Schwarzwedelhirsch, Maultierhirsch	mule deer
Odocoileus virginianus	Weißwedelhirsch, Virginiahirsch	white-tailed deer
Odondebuenia balearica	Korallengrundel	coralline goby
Odonestis pruni	Pflaumenglucke	
Odontaspididae/ Carchariidae	Sandhaie	sand sharks, sand tigers, ragged-tooth sharks

Odontaspis taurus/	Sandhai,	sand shark,
Eugomphodus taurus>	Echter Sandhai,	sand tiger shark (FAO),
Carcharias taurus	Sandtiger	sandtiger shark,
		gray nurse shark
Odontaspis ferox	Schildzahnhai	smalltooth sand tiger (FAO),
(Carcharias ferox)		fierce shark,
		bumpytail ragged-tooth shark,
		shovelnose shark
Odontoceti	Zahnwale	toothed whales &
		porpoises & dolphins
Odontodactylus scyllarus	Pfauen-Fangschreckenkrebs,	peacock mantis shrimp
	Bunter Fangschreckenkrebs	
Odontopera bidentata	Doppelzahnspanner	scalloped hazel moth
Odontophorinae	Zahnwachteln,	New World quails
	Neuwelt-Wachteln	
Odontoscelis fuliginosus	Große Steppenwanze	
Odonus niger/	Rotzahn,	redtooth triggerfish,
Xenodon niger/	Rotzahn-Drückerfisch,	red-toothed triggerfish,
Balistes erythrodon	Atlantischer Drückerfisch,	redfang
	Nördlicher Drückerfisch,	
	Gabelschwanz Drückerfisch,	
	Blauer Drücker	
Oecanthidae	Blütengrillen	tree crickets (Oecanthinae)
Oecanthus pellucens	Weinhähnchen	fragile whistling cricket,
		European tree cricket,
		Italian cricket
Oeciacus hirundinis	Schwalbenwanze	martin bug,
		swallow bug
Oecobiidae	Scheibennetzspinnen &	flatmesh weavers
	Zeltdachspinnen	
Oecobius spp.	Scheibennetzspinnen	saucerweb spiders*
Oecophoridae	Faulholzmotten & Palpenmotten	oecophorid moths
	und Verwandte	
Oecophylla spp.	Weberameisen	
Oedalechilus labeo	Graue Meeräsche	lesser grey mullet,
		boxlip mullet (FAO)
Oedaleus decorus	Kreuzschrecke	Mediterranean whitecross
		grasshopper*
Oedemagena tarandi	Rentier-Hautbremse,	reindeer warble fly
	Rentierdasselfliege	
Oedemeridae	Scheinbockkäfer,	false blister beetles,
	Scheinböcke	pollen-feeding beetles
	(Engdeckenkäfer)	
Oedipina spp.	Tropensalamander	worm salamanders
Oedipoda coerulescens	Blauflügelige Ödlandschrecke	blue-winged grasshopper
Oedipoda germanica	Rotflügelige Ödlandschrecke	red-winged grasshopper
Oedura spp.	Fettschwanzgeckos	velvet geckos
Oena capensis	Kaptäubchen	Namaqua dove
Oenanthe deserti	Wüstensteinschmätzer	desert wheatear
Oenanthe finschii	Felsensteinschmätzer	Finsch's wheatear
Oenanthe hispanica	Mittelmeer-Steinschmätzer	black-eared wheatear
Oenanthe isabellina	Isabellsteinschmätzer	isabelline wheatear
Oenanthe leucopyga	Saharasteinschmätzer	white-crowned black wheatear
Oenanthe leucura	Trauersteinschmätzer	black wheatear
Oenanthe lugens	Schwarzrücken-Steinschmätzer	mourning wheatear
Oenanthe moesta	Fahlbürzel-Steinschmätzer	red-rumped wheatear

Oenanthe monacha	Kappensteinschmätzer	hooded wheatear
Oenanthe oenanthe	Steinschmätzer	northern wheatear
Oenanthe pleschanka	Nonnensteinschmätzer	pied wheatear
Oenanthe xanthoprymna	Rostbürzel-Steinschmätzer	red-tailed wheatear
Oenomys hypoxanthus	Rotnasenratte, Afrikanische Rotnasenratte	rufous-nosed rat, rusty-nosed rat
Oenopota turricula/ Lora turricula/ Bela turricula/ Propebela turricula	Treppenschnecke, Kleine Treppenschnecke, Treppengiebelchen	small staircase snail*
Oesophagostomum columbianum	Schafs-Knötchenwurm	sheep nodular worm
Oesophagostomum radiatum	Knötchenwurm	nodular worm
Oesophagostomum spp.	Knötchenwürmer	nodular worms
Oestergrenia adriatica	Zerbrechliche Wurmseegurke	
Oestridae	Magendasseln & Rachendasseln & Nasenbremsen & Biesfliegen	warble flies
Oestrus ovis	Schafbiesfliege, Schafbremse, Nasendasselfliege	sheep nostril-fly, sheep bot fly, sheep nose bot fly
Ogcocephalidae	Seefledermäuse	seabats, batfishes
Ogcocephalus nasutus	Kurznasen-Seefledermaus, Amerikanische Kurznasen-Seefledermaus	shortsnout batfish, shortnose batfish (FAO)
Ogcocephalus parvus	Raurücken-Seefledermaus	roughback batfish
Ogcocephalus vespertilio	Langnasen-Seefledermaus	longsnout batfish
Ogmodon vitianus	Fidschiotter, Fiji-Kobra	Fiji snake, Fiji cobra
Ogyrides alphaerostris	Ästuar-Stielaugengarnele*	estuarine long-eyed shrimp, estuarine longeye shrimp
Ogyrididae	Stielaugengarnelen*	long-eyed shrimps, longeye shrimps
Oinophila v-flavum	Weinkellermotte, Weinmotte	yellow V carl
Oinophilidae	Weinmotten	oinophilid moths
Okapia johnstoni	Okapi	okapi
Oligochaeta	Wenigborster, Oligochaeten	oligochetes
Oligocladus sanguinolentus	Blutfleckplanarie	
Oligodon affinis	Kukrinatter	kukri snake
Oligodon spp.	Kukrinattern	kukri snakes
Oligoneuriella rhenana	Rheinmücke, Augustmücke	Rhine river mayfly
Oligonychus ununguis	Nadelbaum-Spinnmilbe	spruce spider mite, conifer spinning mite
Olindias phosphorica	Nachtaktive Hydromeduse	iridescent limnomedusian
Oliva annulata	Amethyst-Olive	amethyst olive
Oliva australis	Australische Olive	Australian olive
Oliva bulbiformis	Kugelförmige Olive	rounded olive, bulb-shaped olive
Oliva bulbosa	Bauchige Olive	inflated olive
Oliva carneola	Fleischfarbene Olive*	carnelian olive
Oliva incrassata	Winkel-Olive	angled olive, angulate olive, giant olive

Oliva miniacea	Goldmund-Olive, Rotmund-Olive	gold-mouthed olive, orange-mouthed olive, red-mouth olive, red-mouthed olive
Oliva mustelina	Wiesel-Olive	weasel olive
Oliva oliva	Gemeine Olive, Gemeine Olivenschnecke	common olive
Oliva porphyria	Porphyrwalze, Zelt-Olive	tent olive
Oliva reticularis	Netzolive	netted olive
Oliva reticularis olorinella	Perlolive	pearl olive
Oliva reticulata	Blut-Olive	blood olive
Oliva rubrolabiata	Rotlippige Olivenschnecke	red-lipped olive
Oliva rufula	Rötliche Bänder-Olive	reddish olive
Oliva sayana	Schrift-Olive	lettered olive
Oliva splendidula	Pracht-Olive	splendid olive
Oliva spp.	Oliven, Olivenschnecken	olives, olive snails
Oliva tessellata	Mosaik-Olive	tessellated olive
Oliva textilina	Textil-Olive	textile olive
Oliva tigrina	Tiger-Olive	tiger olive
Oliva tricolor	Dreifarben-Olive	tricolor olive, three-colored olive
Oliva vidua	Witwen-Olivenschnecke	widow olive, black olive
Olivancillaria contortuplicata	Wendelfalten-Olivenschnecke	twisted plait olive
Olivancillaria gibbosa	Buckelolive*, Rundliche Olivenschnecke	gibbose olive, swollen olive
Olivancillaria urceus	Bärenolive*	bear olive, bear ancilla
Olivella biplicata	Purpurne Zwergolive	purple dwarf olive, two-plaited dwarf olive
Olivella dama	Dama-Zwergolive	Dama dwarf olive
Olivella gracilis	Zierliche Zwergolive	graceful dwarf olive
Olivella volutella	Geschnörkelte Zwergolive*	volute-shaped olive, volute-shaped olivella
Olividae	Olivenschnecken	olives, olive snails (olive shells)
Olyridae	Spitzschwanzwelse	olyrids
Omalium rivulare	Gemeiner Kurzflügler	
Omalogyra atomus	Atomschnecke	atom snail
Ommastrephes bartrami/ Sthenoteuthis bartrami	Fliegender Kalmar, Flugkalmar	red flying squid, neon flying squid (FAO)
Ommastrephes pteropus	Orangerücken-Flugkalmar	orangeback flying squid
Ommastrephes sagittatus/ Todarodes sagittatus	Pfeilkalmar	flying squid, European flying squid (FAO), sagittate squid, sea-arrow, red squid
Ommateremias arguta/ Eremias arguta	Steppenrenner, Steppeneidechse	stepperunner, arguta
Ommatophoca rosii	Ross-Robbe	Ross seal
Omocestus haemorrhoidalis	Rotleibiger Grashüpfer	orange-tipped grasshopper
Omocestus rufipes	Buntbäuchiger Grashüpfer	woodland grasshopper
Omocestus viridulus	Bunter Grashüpfer	common green grasshopper

Omosita colon	Aas-Glanzkäfer	
Omosudidae	Hammerkieferfische	omosudids
Omosudis lowei	Hammerkieferfisch	omosudid
Onchidoris muricata	Raue Sternschnecke	rough doris
Onchocerca reticulata/ Onchocerca cervicalis	Cervicalis-Fadenwurm*	neck threadworm
Onchocerca volvulus	Knotenwurm	blinding nodular worm
Oncocottus quadricornis/ Myoxocephalus quadricornis/ Triglopsis quadricornis	Vierhörniger Seeskorpion	four-horn sculpin, four-horned sculpin, fourhorn sculpin (FAO)
Oncorhynchus gorbuscha	Buckellachs, Buckelkopflachs, Rosa Lachs	pink salmon
Oncorhynchus keta	Keta-Lachs, Ketalachs, Hundslachs, Chum-Lachs	chum salmon
Oncorhynchus kisutch	Coho-Lachs, Silberlachs	coho salmon (FAO), silver salmon
Oncorhynchus masou	Masu-Lachs	masu salmon, cherry salmon (FAO)
Oncorhynchus mykiss/ Salmo gairdneri	Regenbogenforelle	rainbow trout (steelhead: *sea-run & large lake populations*)
Oncorhynchus nerka	Blaurückenlachs, Blaurücken, Roter Lachs	sockeye salmon (FAO), sockeye (*lacustrine populations in U.S./Canada:* kokanee)
Oncorhynchus rhodurus	Amago-Lachs	amago salmon, amago
Oncorhynchus spp.	Pazifische Lachse	Pacific salmon
Oncorhynchus tschawytcha	Königslachs, Quinnat	chinook salmon (FAO), chinook, king salmon
Ondatra zibethica	Bisamratte (Hbz. Bisam)	muskrat
Oniscidea	Landasseln	oniscideans (pillbugs, woodlice, sowbugs, slaters)
Oniscus asellus	Mauerassel, Gemeine Mauerassel	common woodlouse, common sowbug, grey garden woodlouse
Onogadus argentatus	Dreibärtelige Silber-Seequappe	silvery rockling
Onthophagus spp.	Kotfresser	dung beetles
Onuphidae	Schopfwürmer	beachworms
Onychiuridae	Blindspringer	blind springtails
Onychodactylus spp.	Krallenfingermolche	clawed salamanders
Onychogalea spp.	Nagelkängurus	nail-tailed wallabies
Onychognathus tristramii	Tristramstar	Tristram's grackle
Onychogomphus forcipatus	Kleine Zangenlibelle	
Onychogomphus uncatus	Große Zangenlibelle	
Onychomys spp.	Grashüpfermäuse	grasshopper mice
Onychophora	Stummelfüßer, Onychophoren	velvet worms, onychophorans
Onychorhynchus coronatus	Kronentyrann	royal flycatcher

Onychoteuthis banksii	Hakenkalmar, Krallenkalmar	common clubhook squid, clawed squid, clawed calamary squid
Oodes helopioides	Eiförmiger Sumpfläufer, Mattschwarzer Straßenkäfer	
Oonopidae	Zwergsechsaugen(spinnen)	dwarf sixeyed spiders (minute jumping spiders)
Oonops domesticus	Kleine Hausspinne*	tiny house spider
Opatrum sabulosum	Staubkäfer, Gemeiner Staubkäfer	
Opeas pumilum	Zwerg-Turmschnecke*	dwarf awlsnail
Opeas pyrgula	Scharfkantige Turmschnecke*	sharp awlsnail
Opeatostoma pseudodon	Dorn-Bandschnecke, Walross-Schnecke	thorn latirus, thorned latirus
Operophtera brumata/ Cheimatobia brumata	Gemeiner Frostspanner, Obstbaumfrostspanner, Kleiner Frostspanner	winter moth, small winter moth
Operophtera fagata	Buchenfrostspanner	beech winter moth, northern winter moth
Ophelia denticulata	Sandbankwurm*	sand bar worm
Ophelia limacina	Schneckenwurm	snail opheliid*
Opheodesoma spp.	Alabaster-Wurmseegurke	alabaster sea cucumber
Opheodrys aestivus	Rauhäutige Grasnatter, Raue Grasnatter	rough green snake
Opheodrys spp.	Grasnattern	green snakes
Opheodrys vernalis	Glatthäutige Grasnatter	smooth green snake
Ophiactis balli	Kleiner Band-Schlangenstern*	small banded brittlestar
Ophichthyidae/Ophichthinae	Schlangenaale & Wurmaale & Garnelenaale	snake eels & worm eels
Ophidiaster ophidianus	Rotvioletter Seestern, Purpurstern	purple sea star*
Ophidiidae	Ophidiiden, Bartmännchen	cuskeels, cusk-eels
Ophidiiformes	Eingeweidefische	brotulas & cusk-eels & pearlfishes
Ophidion barbatum	Bartmännchen	barbed cusk-eel, snake blenny (FAO)
Ophidion holbrooki		bank cusk-eel
Ophidion scrippsae		basketweave cusk-eel
Ophiocara aporos	Manilagrundel	snake-headed gudgeon
Ophiocara porocephala	Schlangenkopfgrundel	northern mud gudgeon
Ophiocephalidae/Channidae	Schlangenkopffische, Schlangenköpfe	snakeheads
Ophiocoma echinata	Stacheliger Schlangenstern	spiny brittlestar
Ophiocoma scolopendrina	Riffdach-Schlangenstern	centipede brittlestar, reefflat brittlestar*
Ophiocomina nigra	Schwarzer Schlangenstern	black brittlestar, black serpent-star
Ophioderma longicauda/ Ophiura longicauda	Brauner Schlangenstern, Großer Schlangenstern	long-spined brittlestar
Ophioderma brevispina	Kurzstachel-Schlangenstern	short-spined brittlestar
Ophioderma rubicundum	Rubin-Schlangenstern	ruby brittlestar*
Ophiodon elongatus	Langer Grünling, Langer Terpug	lingcod
Ophiodromus pugettensis	Flügelsternwurm*	bat star worm

Ophiogomphus serpentinus/ Ophiogomphus cecilia	Grüne Keiljungfer	serpentine dragonfly
Ophiomorus punctatissimus	Gesprenkelter Schlangenskink	Greek legless skink, Greek snake skink
Ophiomorus spp.	Schlangenskinke	greater snake skinks
Ophiomyxa pentagona	Fleckiger Schlangenstern, Fünfeck-Schlangenstern	pentagonal brittlestar*
Ophionyssus natricis	Schlangenmilbe	snake mite
Ophiophagus hannah	Königskobra	king cobra, hamadryad
Ophiopholis aculeata	Gänseblümchen-Schlangenstern	daisy brittlestar, crevice brittlestar
Ophiopsila annulosa	Schotter-Schlangenstern*	gravel brittlestar
Ophiorachna incrassata	Olivgrüner Schlangenstern	green brittlestar, olivegreen brittlestar*
Ophiothrix angulata	Langstachel-Schlangenstern*	Atlantic long-spined brittlestar
Ophiothrix fragilis	Zerbrechlicher Schlangenstern	common brittlestar
Ophiothrix quinquemaculata	Fünfflecken-Schlangenstern	fivespot brittlestar
Ophiothrix spiculata	Stacheliger Schlangenstern	spiny brittlestar
Ophisaurus apodus	Scheltopusik	European glass lizard, armored glass lizard
Ophisaurus attenuatus	Schlanke Glasschleiche	slender glass lizard
Ophisaurus compressus	Insel-Glasschleiche, Küstenglasschleiche	island glass lizard (U.S.)
Ophisaurus gracilis	Burma-Glasschleiche	Burma glass lizard
Ophisaurus koellikeri	Marokko-Glasschleiche	Koelliker's glass lizard
Ophisaurus spp.	Glasschleichen & Panzerschleichen	glass lizards
Ophisaurus ventralis	Östliche Glasschleiche	eastern glass lizard (U.S.)
Ophisops elegans	Schlangenauge	snake-eyed lizard
Ophisops spp.	Schlangenaugen-Echse	snake-eyed lizards
Ophisurus rufus	Rötlicher Schlangenaal	rufous snake eel
Ophisurus serpens	Langmaul-Schlangenaal, Mittelmeer-Schlangenaal	long-jawed snake eel
Ophiura albida	Heller Schlangenstern	lesser sandstar
Ophiura texturata	Großer Schuppen-Schlangenstern	large sand brittlestar
Ophiuroidea	Schlangensterne, Ophiuroiden	brittle stars, serpent stars & basket stars (*Gorgonocephalus, Astrophyton*)
Ophlitaspongia pennata		velvety red sponge
Ophonus diffinis	Metallglanz-Haarschnellläufer	
Ophonus signaticornis	Sand-Haarschnellläufer	
Ophryacus undulatus/ Bothrops undulatus	Wellenband-Lanzenotter	undulated pit viper, Mexican horned pit viper
Opiliones/Phalangida	Weberknechte, Kanker	harvestmen, "daddy longlegs"
Opilo domesticus	Hausbuntkäfer	
Opisthobranchia	Hinterkiemenschnecken, Hinterkiemer	opisthobranch snails, opisthobranchs
Opisthocomidae	Schopfhuhn, Hoatzin	hoatzin
Opisthognathidae	Brunnenbauer, Kieferfische	jawfishes

Opisthograptis luteolata	Gelbspanner, Weißdornspanner	brimstone moth
Opisthonema libertate	Pazifischer Fadenhering	deep-body thread herring, Pacific thread herring (FAO)
Opisthonema oglinum	Atlantischer Fadenhering	Atlantic thread herring
Opisthoproctidae	Hochgucker	barreleyes, spookfishes
Opisthorchis felineus	Katzenleberegel, Katzenegel	cat liver fluke, Siberian liver fluke
Opisthothylax immaculatus	Grauaugen-Waldsteiger (Schaumnestbauender Waldsteiger)	gray-eyed frog
Opistoteuthidae	Scheibenschirme	flapjack octopuses
Oplegnathidae	Messerkieferfische	knifejaws
Oplophoridae	Tiefseegarnelen u.a.	deepsea shrimps
Oplosia fennica	Lindenbock	linden borer (longhorn beetle)
Oplurus sebae	Madagaskar-Leguan	Madagascar swift
Oplurus spp.	Madagaskar-Leguane	Madagascar swifts
Opomyza florum	Gelbe Grasfliege	yellow cereal fly
Opomyza germinationis	Gestreifte Grasfliege	dusky-winged cereal fly
Opomyzidae	Grasfliegen, Wiesenfliegen, Saftfliegen	opomyzid flies
Opopsitta diophthalma	Rotwangen-Zwergpapagei	double-eyed fig parrot
Opsanus tau	Austernfisch, Karibischer Krötenfisch	oyster toadfish
Orbatida	Hornmilben, Moosmilben	orbatid mites, beetle mites, moss mites
Orcaella brevirostris	Irrawadydelphin, Irawadi-Delphin	Irrawaddy River dolphin, snub-fin dolphin
Orchestia gammarellus	Küstenhüpfer, Sandhüpfer	shore-hopper, beach-flea, common shore-skipper, common scud
Orchestoidea californiana	Kalifornischer Strandfloh, Kalifornischer Sandhüpfer	California beach flea
Orcinus orca	Schwertwal, Killerwal, Orca	killer whale, orca
Orconectes limosus/ Cambarus affinis	Amerikanischer Flusskrebs, Kamberkrebs	spinycheek crayfish, American crayfish, American river crayfish, striped crayfish
Orcula dolium	Große Fässchenschnecke, Kleine Fassschnecke	
Orcula gularis	Schlanke Fässchenschnecke	
Orculidae	Fässchenschnecken	
Orcynopsis unicolor/ Gymnosarda unicolor	Rüppell's Thunfisch, Ungestreifte Pelamide	plain bonito (FAO), scaleless tuna, plain bonito, pigtooth tuna, Rüppell's bonito
Oreamnos americanus	Schneeziege	mountain goat
Oreaster reticulatus	Netz-Kissenstern	reticulate cushion star
Oreasteridae	Kissensterne	cushion stars

Orectochilus villosus	Bachtaumelkäfer, Haariger Taumelkäfer	hairy whirligig
Orectolobidae	Ammenhaie, Teppichhaie, Wobbegongs	carpet sharks, carpetsharks, wobbegongs
Orectolobiformes	Ammenhaie, Teppichhaie, Wobbegongs	carpet sharks, carpetsharks, wobbegongs
Orectolobus maculatus	Gefleckter Teppichhai, Gefleckter Wobbegong (Australischer Ammenhai)	spotted wobbegong (Australian carpet shark)
Orectolobus ornatus	Ornament-Wobbegong, Gezeichneter Wobbegong	ornate wobbegong (FAO), banded wobbegong
Oregonia gracilis	Westpazifische Dreiecks-Schmuckkrabbe*	triangular decorator crab
Oreochromis niloticus/ Tilapia nilotica	Nil-Buntbarsch	Nile tilapia (FAO), Nile mouthbreeder
Oreochromis andersonii	Dreipunkt-Buntbarsch*	three-spotted tilapia
Oreochromis macrochir	Langflossen-Buntbarsch*	longfin tilapia
Oreochromis mossambicus	Mozambik-Buntbarsch	Mozambique tilapia
Oreochromis placidus	Schwarzer Maulbrüter	black tilapia
Oreophasis derbianus	Zapfenguan, Bergguan	horned guan
Oreophryne anthonyi	Kletter-Engmundfrosch	Anthony's cross frog
Oreopsittacus arfaki	Arfaklori	whiskered lorikeet
Oreoscoptes montanus	Bergspottdrossel	sage thrasher
Oreosomatidae	Oreos	oreos
Oreotragus oreotragus	Klippspringer	klippspringer
Oreotrochilus estella	Andenkolibri	Andean hillstar
Orgyia antiqua/ Orgyia recens	Kleiner Bürstenspinner, Schlehenspinner, Bürstenbinder	vapourer moth, common vapourer, rusty tussock moth (U.S.)
Orgyia ericae	Heide-Bürstenspinner	
Orgyia recens/ Orgyia gonostigma/ Teia recens	Eckfleck-Bürstenspinner	scarce vapourer (moth)
Origma solitaria	Steinhuscher	rock warbler
Oriolidae	Pirole & Feigenpirol	orioles, cuckooshrikes
Oriolus oriolus	Pirol	golden oriole
Oriolus trailii	Blutpirol	maroon oriole
Orius insidiosus		insidious flower bug
Orius majusculus	Großer Putt	
Orius minutus	Kleiner Putt	
Orius vicinus	Himbeer-Putt	raspberry bug
Orlitia borneensis	Borneo-Flussschildkröte	Malaysian giant turtle
Orneodidae	Geistchen	many-plume moths
Ornithischia	Vogelbecken-Dinosaurier	bird-hipped dinosaurs, ornithischian reptiles
Ornithodoros coriaceus	Pajarello-Zecke*	pajarello tick
Ornithodoros moubata	Tampan-Zecke	tampan tick, eyeless tampan
Ornithodoros savignyi	Sandtampan	eyed tampan, sand tampan
Ornithodoros turicata	Rückfallfieberzecke (trop. Lederzecke)	relapsing fever tick

Ornithonyssus bacoti/ *Liponyssus bacoti*	Tropische Rattenmilbe, Nagermilbe	tropical rat mite
Ornithonyssus bursa	Tropische Geflügelmilbe	tropical fowl mite
Ornithonyssus natricis	Schlangenmilbe	snake mite
Ornithonyssus sylviarum	Europäische Hühnermilbe	northern fowl mite
Ornithorhychidae	Schnabeltiere	platypuses and relatives
Ornithorhynchus anatinus	Schnabeltier	platypus, duck-billed platypus
Ornithoteuthis antillarum	Atlantische Vogelsepie*	Atlantic bird squid
Ortalis vetula	Braunflügelguan	plain chachalaca
Orthetrum albistylum	Östlicher Blaupfeil	eastern European skimmer
Orthetrum brunneum	Südlicher Blaupfeil	southern European skimmer
Orthetrum cancellatum	Schwarzspitzen-Blaupfeil, Großer Blaupfeil	black-tailed skimmer
Orthetrum coerulescens	Gekielter Blaupfeil, Kleiner Blaupfeil	keeled skimmer
Orthetrum spp.	Blaupfeile	orthetrums
Orthezia insignis	Gewächshaus-Röhrenschildlaus	glasshouse orthezia
Orthezia urticae	Brennnessel-Röhrenschildlaus	stinging nettle orthezia
Ortheziidae	Röhrenschildläuse	ensign coccids
Orthogeomys spp.	Riesentaschenratten	taltuzas
Orthonychidae	Laufflöter	logrunners, chowchilla
Orthopristis chrysoptera	Gelbflossen-Süßlippe	pigfish
Orthoptera	Geradflügler	orthopterans
Orthosia cruda	Kleine Kätzcheneule	small quaker moth
Orthosia gothica	Bräunlichgraue Frühlingseule	Hebrew character (moth)
Orthosia incerta	Violettbraune Frühlingseule	clouded drab moth
Ortyxelos meiffrenii	Lerchenlaufhühnchen	quail plover
Orussidae	Parasitische Holzwespen	parasitic woodwasps
Orycteropus afer	Erdferkel	Aardvark, ant bear
Oryctes nasicornis	Nashornkäfer	European rhinoceros beetle
Oryctolagus cuniculus	Altweltliches Wildkaninchen, Europäisches Wildkaninchen (Hbz. Wildkanin)	Old World rabbit, domestic rabbit
Oryx dammah	Säbelantilope, Nordafrikanischer Spießbock	scimitar oryx, scimitar-horned oryx
Oryx gazella	Südafrikanischer Spießbock	gemsbock
Oryx leucoryx	Weißer Oryx, Arabischer Spießbock	Arabian oryx
Oryx spp.	Spießböcke	oryx and gemsbock
Oryzaephilus mercator	Erdnussplattkäfer	merchant grain beetle
Oryzaephilus surinamensis	Getreideplattkäfer	saw-toothed grain beetle
Oryzias latipes	Japankärpfling, Japanischer Goldhecht, Medaka	medaka
Oryziatidae/Oryziinae	Reisfische, Reiskärpflinge, Japan-Kärpflinge	medakas, ricefishes
Oryzomys palustris	Sumpf-Reisratte	marsh rice rat
Oryzomys spp.	Reisratten	rice rats
Oryzorictes spp.	Reiswühler	rice tenrecs
Osbornictis piscivora	Wasserschleichkatze	aquatic genet

Oscarella lobularis	Feigenbohrschwamm (Fleischschwamm)	lobate fig sponge*, flesh sponge
Oscinella frit	Fritfliege	frit-fly
Osmeridae	Stinte	smelts
Osmerus eperlanus	Stint (Spierling, Wanderstint)	smelt, European smelt (FAO)
Osmerus mordax	Regenbogenstint	lake smelt, Atlantic rainbow smelt (FAO)
Osmerus mordax dentex	Asiatischer Stint	Arctic smelt, Asiatic smelt, boreal smelt, Arctic rainbow smelt (FAO)
Osmia rufa	Rote Mauerbiene	red mason bee
Osmia spp.	Mauerbienen	mason bees
Osmoderma eremita	Eremit, Juchtenkäfer	hermit beetle
Osmylidae	Bachhafte	osmylid flies
Osphronemidae	Großguramis	giant gouramies
Osphronemus goramy	Gurami, Speisegurami, Riesengurami, Knurrender Gurami	gourami, giant gourami
Osteogeneiosus militaris	Soldatenwels	soldier catfish
Osteoglossidae	Knochenzüngler	bonytongues, osteoglossids
Osteoglossiformes	Knochenzüngler, Knochenzünglerartige	osteoglossiforms
Osteoglossum bicirrhosum	Knochenzüngler, Arawana	arawana
Osteoglossum spp.	Gabelbärte	arawanas
Osteolaemus tetraspis	Stumpfkrokodil	African dwarf crocodile, West African dwarf crocodile
Ostertagia ostertagi	Mittlerer Rindermagennematode, Mittlerer Rindermagenwurm	medium stomach worm (of cattle), brown stomach worm
Ostomidae	Flachkäfer, Jagdkäfer	ostomatid beetles, ostomine beetles, gnawing beetles (Trogositidae)
Ostraciidae/Ostraciontidae	Kofferfische	boxfishes (cowfishes & trunkfishes)
Ostracion quadricornis/ Acanthostracion quadricornis	Vierhorn-Kofferfisch	scrawled cowfish
Ostracion cubicus/ Ostracion tuberculatus	Gelber Kofferfisch, Kleiner Kofferfisch	blue-spotted boxfish, yellow boxfish (FAO), polka dot boxfish
Ostracion lentiginosum	Blauer Kofferfisch	blue boxfish
Ostracion meleagris	Weißpunkt-Kofferfisch, Schwarzer Kofferfisch	whitespotted boxfish
Ostracoda	Muschelkrebse, Ostracoden	ostracods (shell-covered crustaceans), seed shrimps
Ostracodermata	Schalenhäuter	ostracoderms
Ostrea chilensis	Chilenische Plattauster	Chilean flat oyster
Ostrea crestata	Hahnenkammauster	cock's comb oyster*
Ostrea denticulata	Gezähnte Auster	denticulate rock oyster

Ostrea edulis	Europäische Auster, Gemeine Auster	common oyster, flat oyster, European flat oyster
Ostrea equestris	Atlantische Kammauster*	crested oyster
Ostrea imbricata	Schindelauster*	imbricate oyster
Ostrea lurida	Kleine Pazifik-Auster, Pazifische Plattauster	native Pacific oyster, Olympia flat oyster, Olympic oyster
Ostrea lutaria	Neuseeland-Plattauster	New Zealand dredge oyster
Ostrea virginica	Amerikanische Auster	American oyster, eastern oyster
Ostreidae	Austern	oysters
Ostrinia nubilalis/ Pyrausta nubilalis	Maiszünsler, Hirsezünsler	European corn borer
Otala lactea	Milchige Schnecke	milk snail, milky snail
Otaria flavescens	Mähnenrobbe	southern sea lion, South American sea lion
Othius punctulatus	Punktierter Mulmkurzflügler	
Othniidae	Othniiden, Falsche Tigerkäfer	false tiger beetles
Otididae	Trappen	bustards
Otiorhynchus cribricollis	Australischer Apfel-Dickmaulrüssler	cribrate weevil, apple weevil
Otiorhynchus gemmatus	Hellgefleckter Dickmaulrüssler	lightspotted snout weevil
Otiorhynchus ligustici	Kleeluzernerüssler, Luzerne-Dickmaulrüssler	alfalfa snout beetle
Otiorhynchus niger	Schwarzer Fichten-Dickmaulrüssler	black spruce weevil*
Otiorhynchus ovatus	Ovaler Dickmaulrüssler, Erdbeerwurzelrüssler	strawberry root weevil
Otiorhynchus spp.	Lappenrüssler, Dickmaulrüssler	snout beetles, snout weevils
Otiorhynchus sulcatus	Gefurchter Lappenrüssler, Gewächshaus-Dickmaulrüssler	vine weevil, black vine weevil, European vine weevil
Otis tarda	Großtrappe	great bustard
Otitidae	Schmuckfliegen	picture-winged flies
Otobius megnini	Stachel-Ohrenzecke*	spinose ear tick
Otocyon megalotis	Löffelhund	bat-eared fox
Otolemur crassicaudatus/ Galago crassicaudatus	Riesengalago	greater bush baby, greater galago, thick-tailed bush baby
Otolemur argentatus	Silbergalago	silver galago
Otolemur garnetti	Kleinohrgalago	small-eared galago
Otolemur spp.	Großgalagos	greater bush babies
Otolithes ruber	Hundszahn-Umberfisch	long-tooth croaker, tiger-toothed croaker (FAO)
Otomops spp.	Großohr-Bulldogg-Fledermäuse	big-eared free-tailed bats
Otomys spp.	Afrikanische Lamellenzahnratten, Ohrenratten	African swamp rats, groove-toothed rats
Otonycteris hemprichi	Wüsten-Langohrfledermaus	desert long-eared bat
Otonyctomys hatti	Yukatan-Dämmerungsratte	Yucatan vesper rat
Ototylomys phyllotis	Großohr-Kletterratte*	big-eared climbing rat
Ottonia brunnea	Brauner Fischegel	brown fish worm

Otus asio	Ostkreischeule	eastern screech owl
Otus brucei	Streifenohreule	striated scops owl
Otus ireneae	Sokoke-Eule	Sokoke scops-owl
Otus kennicottii	Westliche Kreischeule	western screech owl
Otus pauliani	Komoren-Eule	Comoro scops-owl
Otus scops	Zwergohreule	scops owl
Oulema melanopus/ Lema melanopus	Getreidehähnchen, Blatthähnchen	cereal leaf beetle (oat leaf beetle, barley leaf beetle)
Oulema spp.	Getreidehähnchen	cereal leaf beetle
Oulocrepis dissimilis/ Goniodes dissimilis	Braune Hühnerlaus	brown chicken louse
Ourapteryx sambucaria	Holunderspanner, Nachtschwalbenschwanz	swallow-tailed moth
Ourebia ourebi	Bleichböckchen	oribi
Ovalipes trimaculatus/ Ovalipes punctatus	Dreipunkt-Schwimmkrabbe	three-spot swimming crab
Ovalipes ocellatus	Frauenkrabbe	lady crab
Ovatella myosotis	Mäuseöhrchen	mouse-ear-shelled snail, mouse-eared snail
Ovibos moschatus	Moschusochse	muskox
Ovis ammon	Riesenwildschaf, Argali	argali
Ovis aries	Hausschaf	domestic sheep
Ovis canadensis	Dickhornschaf	bighorn sheep, American bighorn, mountain sheep
Ovis dalli	Dallschaf, Dünnhornschaf	Dall's sheep, white sheep
Ovis musimon	Mufflon, Muffelwild	mouflon
Ovis nivicola	Schneeschaf	snow sheep, Siberian bighorn
Ovis orientalis	Orientalisches Wildschaf	Asiatic mouflon
Ovis vignei	Urial	urial
Ovophis spp.	Asiatische Grubenottern	Asian pit vipers
Ovula costellata	Rosamund-Eischnecke*	pink-mouthed egg cowrie, pinkmouth ovula
Ovula ovum/Erronea ovum	Gemeine Eischnecke	common egg cowrie
Ovulidae	Ovuliden	egg cowries (egg shells)
Oxidus gracilis	Amerikanischer Garten-Tausendfüßer	garden millipede (U.S.), garden millepede (Br.)
Oxybelis aeneus	Erzspitznatter	Mexican vine snake
Oxybelis fulgidus	Glanzspitznatter, Grüne Erzspitznatter	green vine snake
Oxybelis spp.	Spitznattern, Erzspitznattern	vine snakes
Oxybelis wilsoni	Gelbe Erzspitznatter	yellow vine snake
Oxycarenus modestus	Spitznase (Wanze)	nosy bug* (a seed bug)
Oxychilus alliarius	Knoblauch-Glanzschnecke	garlic snail, garlic glass-snail
Oxychilus cellarius	Keller-Glanzschnecke	cellar glass snail, cellar glass-snail, cellar snail
Oxychilus clarus	Farblose Glanzschnecke	clear glass snail

Oxychilus depressus	Flache Glanzschnecke	depressed glass snail
Oxychilus draparnaudi	Große Glanzschnecke, Draparnauds Glanzschnecke, Große Gewächshaus-Glanzschnecke	Draparnaud's glass snail, dark-bodied glass-snail
Oxychilus glaber	Glatte Glanzschnecke	smooth glass snail
Oxychilus helveticus	Schweizerische Glanzschnecke	glossy glass snail, Swiss glass-snail (U.S.)
Oxycirrhites typus	Langnasen-Falkenfisch	long-nosed hawkfish
Oxydoras niger/Doras niger	Liniendornwels	ripsaw catfish
Oxyeleotris marmorata/ Eleotris marmorata	Marmorgrundel	marble goby
Oxygastra curtisii	Gekielter Flussfalke	orange-spotted emerald dragonfly
Oxylebius pictus		painted greenling
Oxylipeurus polytrapezius		slender turkey louse, turkey wing louse
Oxyloma elegans/ Oxyloma dunkeri	Schlanke Bernsteinschnecke, Dunker's Bernsteinschnecke	Pfeiffer's ambersnail, Dunker's ambersnail
Oxyloma sarsii	Rötliche Bernsteinschnecke	slender amber snail, reddish ambersnail*
Oxymonacanthus longirostris	Orangeflecken-Feilenfisch	beaked leatherjacket, harlequin filefish (FAO)
Oxymycterus spp.	Südamerikanische Grabmäuse	burrowing mice
Oxynotidae	Schweinshaie, Meersau-Haie, Meersäue, Rauhaie	prickly dogfishes, oxynotids
Oxynotus centrina	Flecken-Meersau, Flecken-Schweinshai	angular roughshark (FAO), flatiron shark, humantin
Oxynotus paradoxus	Graue Meersau, Grauer Schweinshai	sailfin roughshark (FAO), sharp-back shark
Oxyopes lineatus	Gebänderte Luchsspinne	banded lynx spider
Oxyopidae	Luchsspinnen, Scharfaugenspinnen	lynx spiders
Oxypogon guerinii	Helmkolibri	bearded helmetcrest
Oxyporhamphus micropterus	Kleinflügelfisch*, Kleinflügel-Flugfisch	smallwing flyingfish
Oxyporus rufus	Roter Buntkurzflügler	
Oxyrhopus spp.	Mondnattern, Falsche Korallenschlangen	false coral snake
Oxyrhopus trigeminus	Mondnatter	Brazilian false coral snake
Oxyruncidae	Flammenkopf, Feuerkopf	sharpbill
Oxyruncus cristatus	Flammenkopf, Feuerkopf	sharpbill
Oxystele sinensis	Rosenmund-Kreiselschnecke	rosy-base topsnail
Oxyura jamaicensis	Schwarzkopf-Ruderente	ruddy duck
Oxyura leucocephala	Weißkopf-Ruderente	white-headed duck
Oxyuranus scutellatus/ Oxyuranus scutulatus	Taipan	taipan
Oxyuranus microlepidotus	Inland-Taipan, Schreckens-Schlange*	fierce snake
Oxyuris equi	Pfriemenschwanz des Pferdes	horse pinworm, equine pinworm
Ozotoceros bezoarticus/ Odocoileus bezoarticus	Pampashirsch	pampas deer

Pachycephalidae	Dickkopfschnäpper, Dickkopfvögel	whistlers, shrike-thrushes
Pachyclavularia violacea		brown star polyp
Pachygastria trifolii/ Lasiocampa trifolii	Kleespinner	grass eggar
Pachygnatha spp.	Eigentliche Kieferspinnen	thick-jawed spiders, thick-jawed orb weavers
Pachygrapsus crassipes	Gestreifte Felsenkrabbe	striped shore crab
Pachygrapsus gracilis	Dunkle Felsenkrabbe	dark shore crab
Pachygrapsus marinus	Pazifische Schwebkrabbe	drifter crab
Pachygrapsus marmoratus	Rennkrabbe, Felsenkrabbe	marbled shore crab, marbled rock crab
Pachymedusa dacnicolor	Gespenstfrosch	Mexican leaf frog
Pachyptila vittata	Großer Entensturmvogel	broad-billed prion
Pachyramphus aglaiae	Dickkopfbekarde	rose-throated becard
Pachyta quadrimaculata	Gelber Vierfleckbock	
Pachytrachis striolatus	Gestreifte Südschrecke	striated bushcricket
Pachytriton breviceps	Chinesischer Kurzfußmolch	Tsitou newt, Chinese newt
Pachyuromys duprasi	Dickschwanzmaus, Fettschwanz-Rennmaus	fat-tailed gerbil
Pacifastacus leniusculus	Signalkrebs	signal crayfish
Padda oryzivora	Reisvogel	Java sparrow
Paederus riparius	Bunter Uferkurzflügler	
Pagellus acarne	Achselfleck-Brassen, Achselbrasse, Spanische Meerbrasse	Spanish seabream, axillary bream, axillary seabream (FAO)
Pagellus bellottii	Rote Pandora	red pandora
Pagellus bogaraveo	Nordischer Meerbrassen, Graubarsch, Seekarpfen	red seabream, common seabream, blackspot seabream (FAO)
Pagellus erythrinus	Rotbrassen, Roter Meerbrassen	pandora, common pandora
Pagellus mormyrus/ Lithognathus mormyrus	Marmorbrassen	marmora, striped seabream (FAO)
Pagodulina pagodula principalis	Pagodenschnecke	pagoda snail
Pagophila eburnea	Elfenbeinmöwe	ivory gull
Pagothenia hansoni/ Trematomus hansoni	Gestreifte Notothenia	striped rockcod, green rockcod
Pagrus auriga	Rotband-Seebrassen, Rotgebänderter Meerbrassen	red-banded seabream
Pagrus caeruleostictus	Blaugefleckter Meerbrassen	blue-spotted seabream
Pagrus pagrus/ Sparus pagrus pagrus/ Pagrus vulgaris	Sackbrassen, Gemeiner Seebrassen	common seabream (FAO), red porgy, Couch's seabream
Paguma larvata	Larvenroller	masked palm civet
Paguridae	Meeres-Einsiedlerkrebse, Rechtshänder-Einsiedlerkrebse	right-handed hermit crabs
Pagurus anachoretus	Bunter Einsiedler, Gestreifter Felseneinsiedler	striped hermit crab*, multicoloured hermit crab*
Pagurus arcuatus	Behaarter Einsiedler	hairy hermit crab
Pagurus bernhardus/ Eupagurus bernhardus	Gemeiner Einsiedler, Bernhardskrebs, Bernhardseinsiedler, Nordsee-Einsiedlerkrebs	large hermit crab, common hermit crab, soldier crab, soldier hermit crab, Bernhard's hermit crab

Pagurus cuanensis	Wollhand-Einsiedler	woolly hermit crab
Pagurus prideaux/ Eupagurus prideaux	Anemonen-Einsiedler, Prideaux-Einsiedlerkrebs	Prideaux's hermit crab, smaller hermit crab
Palaemon adspersus/ Palaemon squilla/ Leander adspersus	Ostseegarnele	Baltic prawn
Palaemon elegans	Kleine Felsgarnele, Steingarnele	rockpool prawn
Palaemon serratus/ Leander serratus	Große Felsgarnele, Sägegarnele	common prawn
Palaemon squilla/ Leander squilla	Steingarnele, Große Felsgarnele, Schwertgarnele	cup shrimp
Palaemon xiphias/ Leander xiphias	Schwertgarnele	glass prawn
Palaemonetes varians	Farbwechselnde Schwimmgarnele	variegated shore shrimp
Palaemonetes vulgaris	Gemeine Schwimmgarnele	marsh grass shrimp, common shore shrimp
Palaemonidae	Felsengarnelen	palaemonid shrimps
Palaeochrysophanus hippothoe/ Lycaena hippothoe	Lilagold-Feuerfalter	purple-edged copper
Palea steindachneri/ Trionyx steindachneri	Kehllappen-Weichschildkröte, Nackendorn-Weichschildkröte	wattle-necked softshell turtle
Paleosuchus palpebrosus	Brauen-Glattstirnkaiman, Brauenkaiman	dwarf caiman, Cuvier's dwarf caiman
Paleosuchus trigonatus	Keilkopf-Glattstirnkaiman	Schneider's smooth-fronted caiman
Pales maculata	Gefleckte Schnake	
Palicidae	Stelzenkrabben	stilt crabs
Palicus spp.	Stelzenkrabben	stilt crabs
Palingenia longicauda	Theißblüte	long-tailed mayfly
Palinurellus gundlachi	Samtlanguste*	furry lobster
Palinuridae	Langusten	spiny lobsters
Palinurus elephas	Europäische Languste, Stachelhummer	common crawfish (Br.), European spiny lobster, langouste
Palinurus mauritanicus	Mauretanische Languste	pink spiny lobster
Palliolum tigerinum	Tigermuschel, Tiger-Kammmuschel	tiger scallop
Pallopteridae	Zitterfliegen	pallopterid flies
Palmar festiva	Wacholderprachtkäfer	
Palmatogekko rangei	Namib-Sandgecko (Wüstengecko u.a.)	Namib sand gecko, web-footed gecko
Palmeria dolei	Schopfkleidervogel	crested honeycreeper
Palola viridis/ Eunice viridis	Samoa-Palolo	Pacific palolo worm, Samoan palolo worm
Palomena prasina	Grüne Stinkwanze	green shield bug, common green shield bug
Palorus subdepressus	Kleinäugiger Reismehlkäfer	depressed flour beetle
Palpigradi	Palpigraden	microwhipscorpions, palpigrades
Palythoa grandis	Sonnen-Krustenanemone, Große Krustenanemone	giant colonial anemone, golden sea mat, button polyps, polyp rock, cinnamon polyp

Palythoa mammillosa	Warzige Krustenanemone	knobby zoanthid
Pamphiliidae (Lydidae)	Gespinstblattwespen (Kotsackblattwespen)	webspinning sawflies & leafrolling sawflies
Pamphobeteus ornatus	Kolumbianische Riesenvogelspinne	Colombian pinkbloom tarantula
Pampus argenteus	Silberne Pampel	silver pomfret
Pan paniscus	Bonobo, Zwergschimpanse	bonobo, pygmy chimpanzee
Pan troglodytes	Schimpanse	savanna chimpanzee, 'chimp'
Panagaeus bipustulatus	Kleiner Scheukäfer, Zweifleck-Kreuzläufer	
Panaque nigrolineatus	Schwarzstreifen-Harnischwels	royal panaque
Panaque suttoni	Blauaugen-Harnischwels	blue-eyed panaque
Panchlora exoleta	Grüne Bananenschabe	green banana roach*
Panchlora nivea	Kubanische Schabe	Cuban cockroach
Pandalidae	Tiefseegarnelen	pandalid shrimps
Pandalus borealis	Nördliche Tiefseegarnele, Grönland-Shrimp	northern shrimp, pink shrimp, northern pink shrimp
Pandalus montagui	Rosa Tiefseegarnele	Aesop shrimp, Aesop prawn, pink shrimp
Pandalus platyceros	Gefleckte Tiefwassergarnele	spot shrimp
Pandemis cerasana	Johannisbeerwickler	
Pandemis heparana	Obstwickler	
Pandinus imperator	Kaiserskorpion, Afrikanischer Riesenskorpion, Afrikanischer Waldskorpion	common emperor scorpion
Pandinus spp.	Afrikanische Riesenskorpione	African emperor scorpions
Pandion haliaetus	Fischadler	osprey (fish hawk)
Pandionidae	Fischadler	osprey (fish hawk)
Pandora glacialis	Gletscherbüchse*	glacial pandora
Pandoriana pandora/ *Argynnis pandora*	Kardinal, Pandorafalter	cardinal
Pandoridae	Büchsenmuscheln, Pandoramuscheln	pandoras, pandora's boxes
Pangasianodon gigas	Riesenwels	Mekong catfish
Pangasiidae	Riesenwelse, Schlankwelse	pangasid catfishes
Pangasius pangasius	Schwarzflossen-Haiwels	Pungas catfish
Pangasius sutchi	Haiwels	Sutchi catfish
Panolis flammea	Kieferneule, Forleule	pine beau, pine beauty moth
Panomya norvegica/ *Panomya arctica*	Arktische Klaffmuschel	Arctic roughmya
Panonychus ulmi/ *Metatetranychus ulmi*	Obstbaumspinnmilbe, Rote Spinne	fruittree red spider mite, European red mite
Panopea abrupta/ *Panopea generosa*	Pazifische Panopea	Pacific geoduck, geoduck (*pronounce*: "gouy-duck")
Panopea bitruncata	Westatlantische Panopea	Atlantic geoduck
Panopea glycymeris	Europäische Panopea	European panopea
Panorpa communis	Gemeine Skorpionsfliege	common scorpionfly
Panorpidae	Skorpionsfliegen	scorpionflies, common scorpionflies

Panthea coenobita	Mönch, Klosterfrau	pine arches
Panthera leo	Löwe	lion
Panthera onca	Jaguar	jaguar
Panthera pardus	Leopard	leopard
Panthera tigris	Tiger	tiger
Panthera uncia/Uncia uncia	Schneeleopard	snow leopard
Pantholops hodgsonii	Tschiru, Tibetantilope	chiru, Tibetan antelope
Pantodon buchholzi	Schmetterlingsfisch, Afrikanischer Schmetterlingsfisch	butterfly fish, butterflyfish, freshwater butterflyfish (FAO)
Pantodontidae	Schmetterlingsfische	butterflyfishes
Panulirus argus	Amerikanische Languste, Karibische Languste	West Indies spiny lobster, Caribbean spiny lobster, Caribbean spiny crawfish
Panulirus cygnus	Australische Languste	Australian spiny lobster
Panulirus delagoae	Natal-Languste	Natal spiny lobster, Natal deepsea lobster
Panulirus gilchristi	Gilchrists Languste	South Coast spiny lobster, South Coast rock lobster
Panulirus guttatus	Fleckenlanguste	spotted spiny lobster
Panulirus homarus	Transkei-Languste	scalloped spiny crawfish
Panulirus interruptus	Kalifornische Felsenlanguste, Kalifornische Languste	California spiny lobster, California rock lobster
Panulirus japonicus	Japanische Languste	Japanese lobster
Panulirus laevicauda	Glattschwanzlanguste	smoothtail spiny lobster
Panulirus ornatus	Ornatlanguste	ornate spiny crawfish
Panulirus penicillatus	Rifflanguste	reef spiny crawfish, reef rock lobster
Panulirus polyphagus	Schlicklanguste	mud spiny crawfish
Panulirus regius	Grüne Languste, Königslanguste	royal spiny crawfish
Panulirus versicolor	Schmucklanguste, Bunte Languste	painted crawfish, painted spiny crawfish, painted rock lobster
Panurgus calcaratus	Zottelbiene, Trugbiene	
Panuridae/Paradoxornithidae	Papageischnäbel, Papageimeisen	parrotbills
Panurus biarmicus	Bartmeise	bearded reedling, bearded tit
Papagomys armandvillei	Flores-Riesenratte	Flores giant rat
Papasula abbotti/Sula abbotti	Abbott-Tölpel	Abbott's booby
Paphia alapapilionis	Schmetterlingsflügel	butterfly venus, butterfly-wing venus
Paphia rhomboides/ Venerupis rhomboides/ Tapes rhomboides	Einfache Teppichmuschel	banded carpetclam, banded venus (banded carpet shell)
Papilio antimachus	Afrikanischer Riesenschwalbenschwanz	African giant swallowtail
Papilio machaon	Schwalbenschwanz	swallowtail
Papiliochromis ramirezi/ Microgeophagus ramirezi/ Apistogramma ramirezi	Schmetterlingsbuntbarsch, Schmetterlings-Zwergbuntbarsch	Ramirez's dwarf cichlid, Ramirezi, Ram cichlid (FAO)
Papilionidae	Ritter (Edelfalter & Schwalbenschwänze)	swallowtails & apollos

Papio anubis	Grüner Pavian	olive baboon
Papio cyanocephalus	Webbipavian	yellow baboon
Papio hamadryas	Mantelpavian	hamadryas baboon, sacred baboon
Papio leucophaeus/ Mandrillus leucophaeus	Drill	drill
Papio papio	Guineapavian	Guinean baboon, western baboon
Papio sphinx/ Mandrillus sphinx	Mandrill	mandrill
Papio ursinus	Bärenpavian	chacma baboon
Pappogeomys spp.	Gelbe und Zimtfarbene Taschenratten	yellow and cinnamon pocket gophers
Papuina pulcherrima	Grüne Baumschnecke	green tree snail
Papyrocranus afer/ Notopterus afer	Afrikanischer Fähnchen-Messerfisch	reticulate knifefish
Parablennius gattorugine	Gestreifter Schleimfisch	tompot blenny
Parablennius marmoreus	Algen-Kammzähner	seaweed blenny
Parablennius pilicornis	Variabler Schleimfisch	variable blenny
Parablennius rouxi	Längsband-Schleimfisch, Längsstreifen-Schleimfisch	striped blenny
Parablennius sanguinolentus	Blutstriemen-Schleimfisch	blood-striped blenny
Parablennius tentacularis	Gehörnter Schleimfisch	horned blenny
Parabrotulidae	Unechte Gebärfische	false brotulas
Parabuteo unicinctus	Wüstenbussard	Harris' hawk
Paracentrotus lividus	Stein-Seeigel, Steinseeigel	purple sea urchin, stony sea urchin, black urchin
Paracerianthus lloydi	„Weichkoralle"	
Paracheirodon innesi/ Hyphessobrycon innesi	Neonfisch	neon tetra
Paracinema tricolor	Dreifarbschrecke	tricolor grasshopper
Paracirrhites arcatus	Büschelbarsch	arc-eyed hawkfish
Paracirrhites forsteri	Schlanker Korallenklimmer	freckled hawkfish, Forster's hawkfish, blackside hawkfish (FAO)
Paracolax derivalis		clay fan-foot (moth)
Paraconger macrops	Schwarzfleck-Meeraal*	black-spotted conger
Paracyathus pulchellus	Papillen-Nelkenkoralle*	papillose cupcoral
Paracynictis selousi	Trugmanguste	gray meerkat, Selous' mongoose
Paradiarsia sobrina	Moosheiden-Erdeule	cousin German (moth)
Paradiplosis abietis	Tannennadel-Gallmücke	fir-needle gall midge
Paradipus ctenodactylus	Kammzehen-Springmaus	comb-toed jerboa
Paradisaea decora	Lavendelparadiesvogel	Goldie's bird-of-paradise
Paradisaea raggiana	Raggiparadiesvogel	Raggiana bird-of-paradise
Paradisaea rudolphi	Blauparadiesvogel	blue bird-of-paradise
Paradisaeidae	Paradiesvögel	birds-of-paradise
Paradoxornis webbianus	Braunkopf-Papageimeise	vinous-throated parrotbill
Paradoxurus hermaphroditus	Fleckenmusang, Gefleckter Palmenroller (Hbz. Palmenroller)	common palm civet
Paradoxurus spp.	Palmenroller	palm civets, musangs, toddy cats

Paragaleus pectoralis	Atlantischer Wieselhai	Atlantic weasel shark
Paragonimus westermani	Lungenegel, Ostasiatischer Lungenegel	human lung fluke
Paragrapsus laevis	Glattschaliger Tiefsee-Einsiedler	smooth shore crab
Parahalomitra robusta	Hutkoralle	basket coral
Parahydromys asper	Gebirgs-Wasserratte	mountain water rat, coarse-haired hydromyine, waterside rat
Paralabrax clathratus	Tangbarsch*	kelp bass
Paralabrax humeralis	Peruanischer Felsbarsch	Peruvian rock bass
Paralepididae	Barrakudinas	barracudinas
Paralepis spp.	Barrakudinas	barracudinas
Paraleptomys rufilatus	Rotflanken-Wasserratte*	red-sided hydromyine, northern hydromyine
Paraleptomys wilhelmina	Kurzhaar-Wasserratte*	short-haired hydromyine
Paralichthys dentatus	Sommerflunder	summer flounder
Paralichthys oblongus	Vierfleckflunder	four-spot flounder
Paralichthys olivaceus	Olivgrüne Flunder	olive flounder, bastard halibut (FAO)
Paralipsa gularis/ *Aphomia gularis*	Samenzünsler	stored nut moth, brush-winged honey, Japanese grain moth
Paralister stercorarius	Miststutzkäfer	
Paralithodes camtschaticus	Königskrabbe (Kronenkrebs, Kamschatkakrebs), Alaska-Königskrabbe, Kamschatka-Krabbe	king crab, red king crab, Alaskan king crab, Alaskan king stone crab (Japanese crab, Kamchatka crab, Russian crab)
Paralithodes platypus	Blaue Königskrabbe	blue king crab
Paralomis granulosa	Falsche Südliche Königskrabbe	false southern king crab
Paralomis multispina	Stachelige Königskrabbe	spiny king crab
Paralonchurus peruanus	Coco-Umberfisch	Peruvian drum
Paramecium bursaria	Grünes Pantoffeltierchen	green slipper animalcule
Paramecium caudatum	Pantoffeltierchen	slipper animalcule
Paramecium spp.	Pantoffeltierchen	slipper animalcules
Paramesotriton spp.	Warzenmolche, Chinesische Warzenmolche	warty newts
Parametaria macrostoma	Kegel-Täubchenschnecke	conelike dovesnail
Paramphistomum cervi	Gemeiner Pansenegel	common stomach fluke, rumen fluke
Paramphistomum spp.	Pansenegel	stomach flukes, rumen flukes
Paramuricea chamaeleon	Violette Hornkoralle	violet horny coral
Paramuricea clavata	Farbwechselnde Hornkoralle, Farbwechselnde Gorgonie	chameleon sea fan, variable gorgonian*
Paranaja spp.	Westafrikanische Erdkobra	burrowing cobras
Paranoplocephala mamillana	Zwerg-Pferdebandwurm	dwarf equine tapeworm
Paranotothenia magellanica	Blaue Notothenia	Magellanic notothenia, Magellanic rockcod, Maori cod (FAO)
Paranthrene tabaniformis	Kleiner Pappelglasflügler, Bremsenglasflügler	lesser poplar hornet clearwing, lesser hornet moth
Paranyctimene raptor	Kleiner Röhrennasen-Flughund	lesser tube-nosed fruit bat
Parapaguridae	Tiefenwasser-Einsiedlerkrebse	deepwater hermit crabs

Parapenaeus longirostris	Rosa Tiefen-Garnele*	deepwater rose shrimp
Parapenaeus politus	Rosa Garnele	rose shrimp
Parapercidae/Pinguipedidae	Krokodilfische	sandperches
Parapercis colias	Neuseeland-Flussbarsch	New Zealand blue cod
Parapercis cylindrica	Spitzkopf-Sandbarsch, Binden-Krokodilfisch	banded weever, grubfish, cylindrical sandperch (FAO)
Parapercis polyophthalma	Vielaugen-Sandbarsch, Vielaugen-Krokodilfisch	black-tailed weever, blacktailed sandperch, speckled sandperch (FAO)
Parapercis xanthozona	Lippenfleck-Sandbarsch, Gelbfleck-Krokodilfisch	spottyfin weever, whitebar weever, yellowbar sandperch (FAO)
Paraphoxinus alepidotus	Adria-Elritze	Adriatic minnow
Paraphysa manicata	Kleine Rot-Vogelspinne*	dwarf rose tarantula
Parapleurus alliaceus/ Mecostethus alliaceus	Lauchschrecke	leek grasshopper
Parapriacanthus ransonneti	Glasfisch	slender sweeper
Parapristipoma octolineatum	Afrikanische Streifensüßlippe	African striped grunt
Pararge aegeria	Waldbrettspiel, Laubfalter	speckled wood
Parascaris equorum	Großer Pferdenematode, Großer Pferdespulwurm	large equine roundworm, horse roundworm
Parascorpaena aurita	Roter Drachenkopf	red stingfish
Parascorpididae	Parascorpididen, Jutjawfische	jutjaw fish
Parascorpis typus		jutjaw
Parascotia fuliginaria	Pilzeule	
Parascyllidae	Kragen-Teppichhaie	collared carpet sharks
Parasitidae	Käfermilben	beetle mites
Parasitus coleoptratorum	Käfermilbe	beetle mite
Parasitus fucorum	Hummelmilbe	bumblebee mite
Parastacidae	Südliche Flusskrebse	southern crawfish
Paratettix meridionalis	Mittelmeer-Dornschrecke	Mediterranean groundhopper
Paratylenchus curvitatus	Speerälchen	pin nematode, South African pin nematode
Paraxerus spp.	Afrikanische Buschhörnchen	African bush squirrels
Paraxerus vexillarius	Usambara-Buschhörnchen	Usambara squirrel
Parazen pacificus	Parazen	parazen
Parazoanthus axinellae	Gelbe Krustenanemone	yellow commensal zoanthid, yellow encrusting sea anemone
Parazoanthus parasiticus	Parasitische Krustenanemone	parasitic colonial anemone, sponge zoanthid
Parazoanthus swiftii	Gelbe Karibik-Krustenanemone	golden zoanthid, yellow Caribbean colonial anemone
Pardalotinae	Panthervögel	pardalotes
Pardofelis marmorata/ Felis marmorata	Marmorkatze	marbled cat
Pardosa spp.	Schmalbeinige Wolfspinnen*	thin-legged wolf spiders
Pareas spp.	Asiatische Schneckennattern	slug snakes
Parerythropodium coralloides	Falsche Edelkoralle, Trugkoralle	encrusting leather coral
Pareulype berberata	Berberitzenspanner	barberry carpet moth
Parexocoetus mento	Afrikanischer Segelflugfisch*	African sailfin flyingfish

Parhippolyte uveae	Zuckerrohr-Garnele	sugarcane shrimp
Paridae	Meisen	titmice, chickadees (U.S.)
Parnassius apollo	Apollofalter	apollo
Parnassius mnemosyne	Schwarzer Apollo	clouded apollo, black apollo
Paroaria coronata	Graukardinal	red-crested cardinal
Paromola cuvieri	Großer Rückenfüßler	paromola
Parotomys brantsii	Brants Pfeifratte	Brant's whistling rat
Parotomys spp.	Pfeifratten	Karroo rats, whistling rats
Parribacus antarcticus	Antarktischer Bärenkrebs	sculptured mitten lobster, sculptured slipper lobster
Parribacus spp.	Gedrungene Bärenkrebse	mitten lobsters, slipper lobsters
Parthenope angulifrons/ Lambrus angulifrons	Langarmkrabbe	long-armed crab
Parthenope pourtalesii	Pourtales Langarmkrabbe	Pourtales' long-armed crab
Parthenope serrata	Sägezahnkrabbe*	saw-toothed crab
Parthenopidae	Langarmkrabben, Ellbogenkrabben*	elbow crabs
Parula americana	Meisenwaldsänger	Northern parula
Parulidae	Waldsänger	wood warblers, New World warblers
Parupeneus barberinus	Strich-Punkt-Barbe, Strich-Punkt-Meerbarbe	dash-dot goatfish, dash-and-dot goatfish (FAO)
Parupeneus cyclostomus	Zitronenbarbe, Gelber Ziegenfisch	yellowsaddle goatfish, yellow-tailed goatfish, goldsaddle goatfish (FAO)
Parupeneus trifasciatus	Dreistreifen-Meerbarbe	threestripe goatfish
Paruroctonus mesaensis	Kalifornischer Sandskorpion	giant sand scorpion
Parus ater	Tannenmeise	coal tit
Parus atricapillus	Schwarzkopfmeise	black-capped chickadee
Parus bicolor	Indianermeise	tufted titmouse
Parus caeruleus	Blaumeise	blue tit
Parus carolinensis	Carolina-Meise	Carolina chickadee
Parus cinctus	Lapplandmeise	Siberian tit
Parus cristatus	Haubenmeise	crested tit
Parus cyanus	Lasurmeise	azure tit
Parus gambeli	Gambelmeise	mountain chickadee
Parus hudsonicus	Hudsonmeise	boreal chickadee
Parus inornatus	Schlichtmeise	plain titmouse
Parus lugubris	Trauermeise	sombre tit
Parus major	Kohlmeise	great tit
Parus montanus	Weidenmeise	willow tit
Parus niger	Schwarzmeise	southern black tit
Parus palustris	Sumpfmeise	marsh tit
Parus rufescens	Rotrückenmeise	chestnut-backed chickadee
Parus sclateri	Grauflankenmeise	Mexican chickadee, grey-sided chickadee
Parus spilonotus	Königsmeise	Chinese yellow tit

Parvicardium exiguum	Dreieckige Herzmuschel, Kleine Herzmuschel	little cockle
Parvimolge townsendi	Dunns Pygmäensalamander	Dunn's pygmy salamander
Pasiphaeidae	Glassgarnelen*	glass shrimps
Passalidae	Zuckerkäfer	bess beetles, "bessbugs", peg beetles
Passer brachydactyla	Fahlsperling	pale rock sparrow
Passer domesticus	Haussperling	house sparrow
Passer hispaniolensis	Weidensperling	Spanish sparrow
Passer moabiticus	Moabsperling	Dead Sea sparrow
Passer montanus	Feldsperling	tree sparrow
Passer petronia/ Petronia petronia	Steinsperling	rock sparrow
Passer simplex	Wüstensperling	desert sparrow
Passer xanthocollis	Gelbkehlsperling	yellow-throated sparrow
Passerella iliaca	Fuchsammer	fox sparrow
Passeridae/Passerinae	Sperlinge	sparrows, rock-sparrows
Passeriformes	Sperlingsvögel	passerines, passeriforms (perching birds)
Passerina amoena	Lazulifink	lazuli bunting
Passerina ciris	Papstfink	painted bunting
Passerina cyanea	Indigofink	indigo bunting
Pastinachus sephen/ Dasyatis sephen/ Hypolophus sephen	Federschwanz-Stechrochen	cowtail stingray
Pataecidae	Indianerfische	
Patagona gigas	Riesenkolibri	giant hummingbird
Patagonotothen brevicauda	Kurzschwanz-Notothenia	Patagonian rockcod
Patella barbara	Barbara Napfschnecke	Barbara limpet
Patella caerulea	Blaue Napfschnecke	rayed Mediterranean limpet
Patella crenata	Zacken-Napfschnecke	crenate limpet
Patella depressa	Schwarzfuß-Napfschnecke	black-footed limpet
Patella ferruginea	Rippen-Napfschnecke*, Rostrote Napfschnecke*	ribbed Mediterranean limpet
Patella granatina	Schmirgelschnecke*	sandpaper limpet
Patella granularis	Knotige Napfschnecke	granular limpet
Patella laticostata/ Patella neglecta	Riesen-Napfschnecke	giant limpet, neglected limpet
Patella longicosta	Spitzrippige Napfschnecke	spiked limpet, long-ribbed limpet, star limpet
Patella mamillaris	Warzen-Napfschnecke	mamillary limpet
Patella mexicana/ Patella gigantea/ Patella maxima	Mexikanische Riesen-Napfschnecke	giant Mexican limpet
Patella miniata/ Patella pulchra	Zinnober-Napfschnecke	cinnabar limpet
Patella oculus	Augen-Napfschnecke*	eye limpet, South African eye limpet
Patella rustica/ Patella lusitanica	Lusitanische Napfschnecke, Portugiesische Napfschnecke	Lusitanian limpet
Patella scutellaris	Schild-Napfschnecke	scutellate limpet

Patella tabularis	Abgeplattete Napfschnecke*	tabular limpet
Patella ulyssiponensis/ Patella aspera	Chinahut, Chinesische Napfschnecke	China limpet, painted limpet, European China limpet
Patella vulgata	Gemeine Napfschnecke, Gewöhnliche Napfschnecke	common limpet, common European limpet
Patellidae	Napfschnecken	limpets, true limpets
Patelloida saccharina	Süße Napfschnecke	Pacific sugar limpet
Patinopecten caurinus	Pazifische Riesen-Kammmuschel	giant Pacific scallop
Patiria miniata	Netzstern, Seefledermaus	bat star
Patrobus atrorufus	Schwarzbrauner Grubenhalskäfer, Schwarzbrauner Grubenhalsläufer	
Pauropoda	Wenigfüßer, Pauropoden	pauropods
Pavo cristatus	Pfau	common peafowl (peacock/ peahen), Indian peafowl
Pavo muticus	Ährenträgerpfau	green peafowl
Pawsonia saxicola	Pawson's Seegürkchen	sea gherkin
Payraudeautia intricata	Graue Mondschnecke	European gray moonsnail
Peachia hastata	Speer-Anemone	spear anemone
Peachia parasitica	Einsiedlerrose, Schmarotzerrose	parasitic anemone a.o., hermit crab anemone
Peachia quninquecapitata	Zwölfarmige Schmarotzerrose	twelve-tentacle parasitic anemone
Pechipogo strigilata		common fan-foot (moth)
Peckoltia spp.	Zwergschilderwelse	tiger clown pleco
Pecten imbricata	Kleine Knopf-Kammmuschel*	little knobbly scallop
Pecten jacobaeus	Jakobs-Pilgermuschel, Jakobsmuschel	St.James's scallop, great scallop
Pecten maximus	Große Pilgermuschel, Große Jakobsmuschel	great scallop, common scallop, coquille St. Jacques
Pecten meridionalis	Australische Jakobsmuschel	Australian scallop
Pecten novaezealandiae	Neuseeländische Jakobsmuschel	New Zealand scallop
Pecten sulcicostatus	Südatlantische Kammmuschel	South Atlantic scallop
Pecten ziczac	Zickzack-Kammmuschel	zigzag scallop
Pectinaria auricoma	Schillernder Goldwurm	golden trumpet worm
Pectinaria gouldii	Trompetenwurm	ice cream cone worm, trumpet worm
Pectinariidae	Kammwürmer, Kammborstenwürmer	comb worms, trumpet worms, ice-cream-cone worms, pectinariids
Pectinator spekei	Buschschwanz-Gundi	Speke's pectinator, bushy-tailed gundi
Pectinia lactuca	Knüllpapierkoralle	carnation coral
Pectinidae	Kammmuscheln, Pilgermuscheln	scallops
Pedetes capensis	Springhase	springhare, springhaas, jumping hare
Pediacus depressus	Rostbrauner Plattkäfer	

Pediculidae	Menschenläuse	human lice
Pediculus capitis (*P. humanus capitis*)	Kopflaus	head louse
Pediculus humanus (*P. humanus humanus*/ *P. humanus corporis*)	Kleiderlaus	body louse, clothes louse, "cootie", "seam squirrel"
Pedilidae	Pediliden	false antloving flower beetles
Pedinus femoralis	Kleiner Stinkkäfer	
Pedionomidae	Trappenlaufhühnchen, Steppenläufer	plains-wanderer
Pedionomus torquatus	Trappenlaufhühnchen	plains-wanderer
Pedioplanes spp.	Sandechsen	sand lizards
Pedipalpi (Uropygi & Amblypygi)	Geißelskorpione & Geißelspinnen	whipscorpions & tailless whipscorpions (incl. vinegaroons)
Pedostibes hosii	Boulengers Asien-Baumkröte	Boulenger's Asian tree toad
Pedostibes tuberculosus	Warzige Asien-Baumkröte, Malabar Baumkröte	warty Asian tree toad, Malabar tree toad
Pegasidae	Flügelrossfische, Flügelrösschen, Seeschmetterlinge	seamoths
Pegasiformes (Pegasidae)	Seemotten	seamoths
Pegea bicaudata	Riesensalpe	giant salp
Pegohylemyia gnava/ *Botanophila gnava*	Lattichfliege, Salatfliege	lettuce seed fly
Pegomya betae	Rübenfliege, Runkelfliege	beet fly, beet miner, beet leafminer
Pegusa lascaris/ *Solea lascaris*	Sandzunge, Warzen-Seezunge	sand sole
Pelagia colorata	Violette Streifenqualle*	purplestriped jellyfish
Pelagia noctiluca	Leuchtqualle	phosphorescent jellyfish, purple jellyfish, pink jellyfish, night-light jellyfish
Pelagia spp.	Lilastreifen-Quallen	purplestriped jellyfishes
Pelamis platurus	Plättchen-Seeschlange, Gelbbauch-Seeschlange	yellow-bellied sea snake, pelagic sea snake
Pelea capreolus	Rehantilope, Rehböckchen	rhebok, gray rhebok
Pelecanidae	Pelikane	pelicans
Pelecaniformes	Ruderfüßer, Ruderfüßler	totipalmate swimmers: pelicans & allies
Pelecanoididae	Lummensturmvögel, Tauchersturmvögel	diving petrels
Pelecanus conspicillatus	Brillenpelikan	Australian pelican
Pelecanus crispus	Krauskopfpelikan	Dalmatian pelican
Pelecanus erythrorhynchos	Nashornpelikan	American white pelican
Pelecanus occidentalis	Braunpelikan	brown pelican
Pelecanus onocrotalus	Rosapelikan	white pelican
Pelecanus rufescens	Rötelpelikan	pink-backed pelican
Pelecus cultratus	Ziege, Sichling (Sichelfisch)	ziege (FAO), chekhon, sabre carp, sichel

Pellona ditchela	Indische Pellona	Indian pellona
Pelmatohydra oligactis/ Hydra oligactis	Graue Hydra, Grauer Süßwasserpolyp	grey hydra*, brown hydra
Pelobates cultripes	Messerfuß	Western European spadefoot, Iberian spadefoot
Pelobates fuscus	Knoblauchkröte	common spadefoot (garlic toad)
Pelobates syriacus	Syrische Schaufelkröte	Eastern European spadefoot, Syrian spadefoot
Pelobatidae	Krötenfrösche	spadefoot toads
Pelochelys bibroni	Asiatische Riesen-Weichschildkröte	Asian giant softshell turtle
Pelodiscus sinensis/ Trionyx sinensis	Chinesische Weichschildkröte	Chinese softshell turtle
Pelodryas caerulea/ Litoria caerulea	Korallenfinger	White's treefrog
Pelodytes caucasicus	Kaukasischer Schlammtaucher	Caucasian parsley frog
Pelodytes punctatus	Westlicher Schlammtaucher	parsley frog, common parsley frog, mud-diver, spotted mud frog
Pelodytes spp.	Schlammtaucher	parsley frogs, mud frogs, mud divers
Pelomedusa spp.	Pelomedusen	side-necked turtles
Pelomedusa subrufa	Starrbrust-Pelomeduse	helmeted turtle, African helmeted turtle, marsh turtle
Pelomedusidae	Pelomedusen-Schildkröten	Afro-American side-necked turtles, Afro-American sidenecks
Pelomys fallax	Afrikanische Bachratte	creek rat, groove-toothed creek rat (groove-toothed mouse)
Pelomys spp.	Afrikanische Bachratten	groove-toothed creek rats
Pelophryne brevipes	Kurzbeinige Philippinenkröte	short-legged toadlet
Pelophryne spp.	Philippinenkröten	flathead toads, Philippine toads
Pelosia muscerda	Gepunktetes Flechtenbärchen	dotted footman (moth)
Pelosia obtusa	Schilf-Flechtenbärchen	small dotted footman (moth)
Peltocephalus dumerilianus	Dumerils Schienenschildkröte	big-headed Amazon River turtle
Peltodoris atromaculata	Leopardenschnecke, Fleck-Sternschnecke	leopard nudibranch
Peltoperlidae	Schabenartige Steinfliegen	roachlike stoneflies, forestflies
Peltophryne lemur	Tiefland-Karibikkröte	lowland Caribbean toad, Puerto Rican crested toad (IUCN)
Pelurga comitata	Gänsefußspanner	dark spinach
Pelusios adansonii	Weißbrust-Pelomeduse	Adanson's mud turtle
Pelusios bechuanicus	Okavango-Pelomeduse	Okavango mud turtle
Pelusios carinatus	Gekielte Pelomeduse, Kongo-Pelomeduse	African keeled mud turtle
Pelusios castaneus	Braune Pelomeduse	West African mud turtle
Pelusios castanoides	Ostafrikanische Gelbbauch-Pelomeduse	East African yellow-bellied mud turtle
Pelusios gabonensis	Rückenstreifen-Pelomeduse	African forest turtle, stripe-backed sidenecked turtle, Gabon turtle

Pelusios nanus	Kleine Pelomeduse	African dwarf mud turtle
Pelusios niger	Schwarze Pelomeduse	West African black forest turtle, black sidenecked turtle, African black terrapin
Pelusios rhodesianus	Vielfarbige Pelomeduse*	variable mud turtle
Pelusios seychellensis	Seychellen Pelomeduse	Seychelles mud turtle
Pelusios sinatus	Gezähnelte Pelomeduse, Gezackte Pelomeduse	serrated turtle, East African serrated mud turtle
Pelusios subniger	Dunkle Pelomeduse	East African black mud turtle
Pelusios upemba	Upemba-Pelomeduse	Upemba mud turtle
Pelvicachromis pulcher	Königscichlide, Kribensis	kribensis
Pempheridae	Gleiter	sweepers
Pemphigidae (Eriosomatidae)	Blasenläuse	aphids
Pemphigus betae	Zuckerrübenwurzellaus	sugarbeet root aphid
Pemphigus bursarius	Salatwurzellaus	lettuce root aphid, poplar-lettuce aphid, lettuce purse-gall aphid
Pemphigus filaginis/ Pemphigus populinigrae	Pappelblattrippen-Gallenlaus	poplar-cudweed aphid
Pemphigus spirothecae	Spiralgallenlaus	poplar spiral-gall aphid
Penaeidae	Geißelgarnelen	penaeid shrimps
Penaeopsis serrata	Rosagesprenkelte Garnele	pinkspeckled shrimp
Penaeus aztecus/ Farfantepenaeus aztecus	Braune Garnele	brown shrimp, northern brown shrimp
Penaeus brasiliensis/ Farfantepenaeus brasiliensis	Rotpunktgarnele	pinkspotted shrimp, redspotted shrimp
Penaeus brevirostris/ Farfantepenaeus brevirostris	Kristall-Geißelgarnele	crystal shrimp
Penaeus californiensis/ Farfantepenaeus californiensis	Kalifornische Geißelgarnele	yellowleg shrimp, yellow-leg shrimp
Penaeus canaliculatus/ Melicertus canaliculatus	Hexengarnele	witch shrimp (striped prawn)
Penaeus chinensis/ Fenneropenaeus chinensis	Hauptmannsgarnele	fleshy prawn
Penaeus duorarum/ Farfantepenaeus duorarum	Nördliche Rosa-Garnele, Rosa Golfgarnele, Nördliche Rosa Geißelgarnele	pink shrimp, northern pink shrimp
Penaeus esculentus	Braune Tigergarnele	brown tiger prawn
Penaeus indicus/ Fenneropenaeus indicus	Indische Hauptmannsgarnele	Indian prawn
Penaeus indicus	Weiße Garnele	white prawn
Penaeus japonicus/ Marsupenaeus japonicus	Radgarnele	Kuruma shrimp
Penaeus japonicus	Bambusgarnele	bamboo prawn
Penaeus kerathurus/ Melicertus kerathurus	Furchengarnele	caramote prawn
Penaeus latisuculatus/ Melicertus latisuculatus	Königsgeißelgarnele	western king prawn
Penaeus merguiensis/ Fenneropenaeus merguiensis	Bananen-Garnele	banana prawn
Penaeus monodon	Bärengarnele, Bärenschiffskielgarnele, Schiffskielgarnele	giant tiger prawn

Penaeus notialis/ *Farfantepenaeus notialis*	Südliche Rosa Geißelgarnele, Senegal-Garnele	southern pink shrimp, candied shrimp
Penaeus occidentalis/ *Litopenaeus occidentalis*	Pazifische Weiße Garnele	western white shrimp, Central American white shrimp
Penaeus penicillatus/ *Fenneropenaeus penicillatus*	Rotschwanzgarnele	redtail prawn
Penaeus semisulcatus	Grüne Tigergarnele	green tiger prawn, zebra prawn
Penaeus setiferus/ *Litopenaeus setiferus*	Atlantische Weiße Garnele, Nördliche Weiße Geißelgarnele	white shrimp, lake shrimp, northern white shrimp
Penaeus spp.	Geißelgarnelen	white shrimps
Penaeus vannamei/ *Litopenaeus vannamei*	Weißbein-Garnele	whiteleg shrimp, Central American shrimp
Penelope albipennis	Weißschwingenguan	white-winged guan
Penelope purpurascens	Rostbauchguan	crested guan
Pennahia argentata	Silberne Pennahia	silver croaker
Pennaria tiarella	Federmoos*	feathered hydroid
Pennatula phosphorea	Leuchtende Seefeder	luminescent sea-pen
Pennatula rubra	Dunkelrote Seefeder	dark-red sea-pen
Pennatularia	Seefedern	sea pens, pennatulaceans
Pennisetia hylaeiformis	Himbeerglasflügler	raspberry clearwing moth*, raspberry cane borer*
Pentacerotidae	Panzerköpfe	armourheads, armorheads
Pentalagus furnessi	Riu-Kiu-Kaninchen	Ryukyu rabbit, Amami rabbit
Pentanemus quinquarius	Königsfadenfisch	royal threadfin
Pentapora fascialis	Elchgeweih-Moostierchen, Band-Moostierchen	banded bryozoan
Pentapora foliacea	Meerrose	rose 'coral', rose-coral bryozoan, sea rose
Pentastomida	Zungenwürmer, Linguatuliden, Pentastomitiden	tongue worms, linguatulids, pentastomids
Pentatoma rufipes	Rotbeinige Baumwanze	forest bug
Pentatomidae	Baumwanzen & Schildwanzen & Stinkwanzen	shield bugs & stink bugs
Pentatrichopus fragaefolii/ *Chaetosiphon fragaefolii*	Erdbeerknotenhaarlaus, Knotenhaarlaus, Erdbeerblattlaus	strawberry aphid
Penthaleus major	Wintergetreidemilbe*	winter grain mite
Penthetor lucasii	Schwarzer Flughund*	dusky fruit bat
Peponocephala electra	Melonenkopf	many-toothed blackfish, melon-headed whale, small melon-headed blackfish
Peprilus alepidotus	Erntefisch (Amerikanischer Butterfisch)	harvestfish (FAO), North American harvestfish
Peprilus simillimus	Pazifischer Pompano, Pazifischer Butterfisch	Pacific pompano
Peprilus triacanthus	Amerikanischer Butterfisch, Atlantik-Butterfisch	butterfish, Atlantic butterfish, American butterfish (FAO)
Peracarida	Ranzenkrebse	peracarids
Peradorcas concinna	Zwergsteinkänguru, Zwergfelskänguru, Nabarlek	pygmy rock wallaby, nabarlek

Perameles gunnii	Tasmanischer Beuteldachs	Tasmanian long-nosed bandicoot
Perameles spp.	Langnasenbeutler	long-nosed bandicoots
Peramelidae	Nasenbeutler, Bandikuts	bandicoots
Perca flavescens	Gelbbarsch, Amerikanischer Flussbarsch	yellow perch (FAO), American yellow perch
Perca fluviatilis	Flussbarsch	perch, European perch (FAO), redfin perch
Percichthyidae/Moronidae	Seebarsche, Wolfsbarsche, Streifenbarsche	temperate perches
Percidae	Barsche, Echte Barsche	perches
Perciformes	Barschfische, Barschartige	perches & perchlike fishes
Percina caprodes caprodes	Logbarsch, Schweinsbarsch (Manitou-Springbarsch)	logperch
Percophidae/Percophididae/ Bembropsidae	Krokodilfische	duckbills, flatheads
Percophis brasilianus	Brasilianischer Plattkopf	Brazilian flathead
Percopsidae	Barschlachse	trout-perches
Percopsiformes	Barschlachsverwandte	trout-perches & freshwater relatives
Percopsis omiscomaycus	Östlicher Barschlachs	trout-perch
Percopsis transmontana	Sandroller, Amerikanischer Barschlachs, Gewöhnlicher Barschlachs	sand roller
Perdix perdix	Rebhuhn	grey partridge
Perforatella bidentata	Zweizähnige Laubschnecke	
Perforatella incarnata	Rötliche Laubschnecke	
Peribatodes rhomboidaria	Weidenspanner*	willow beauty
Peribatodes secundaria	Weißlicher Kiefernspanner	white pine beauty*
Periclimenes brevicarpalis	'Salami'-Garnele*	pepperoni shrimp
Periclimenes imperator	Imperator-Garnele, Verwandlungs-Partnergarnele	emperor shrimp
Periclimenes spp.	Partnergarnelen	cleaner shrimps, grass shrimps, coral shrimps, anemone shrimps
Pericrocotus ethologus	Langschwanz-Mennigvogel	long-tailed minivet
Peridea anceps	Eichenzahnspinner	great prominent
Periglypta puerpera	Jugendliche Venusmuschel*	youthful venus
Periglypta reticulata	Gebogene Venusmuschel	reticulated venus
Perileptus areolatus	Schlanker Sand-Ahlenläufer, Schwarzbrauner Flinkläufer	
Perinereis cultrifera	Seeringelwurm	marine ragworm a.o.
Peringia ulvae	Gemeine Wattschnecke	common mudflat snail*
Periophthalmidae	Schlammspringer	mudskippers, mudhoppers, climbing-fish
Periophthalmus spp.	Schlammspringer	mudskippers (FAO), mudhoppers, climbing-fish
Periplaneta americana	Amerikanische Großschabe, Amerikanische Schabe	American cockroach

Periplaneta australasiae	Australische Schabe, Südliche Großschabe	Australian cockroach
Periplaneta brunnea	Braune Schabe	brown cockroach
Periplaneta fuliginosa	Rußbraune Schabe	smoky brown cockroach, smokybrown cockroach
Perisoreus infaustus	Unglückshäher	Siberian jay
Peristediidae	Panzerknurrhähne, Panzerhähne	armed gurnards, armored gurnards, armoured searobins
Peristedion cataphractum	Panzerknurrhahn, Panzerhahn	armed gurnard, armored gurnard, African armoured searobin (FAO)
Peristernia nassatula		fine-net peristernia
Perizoma sagittata	Wiesenrauten-Blattspanner	marsh carpet (moth)
Perkinsiella saccharicida	Zuckerrohrzikade	sugarcane delphacid
Perlidae	Gewöhnliche Steinfliegen	common stoneflies
Perlodidae	Perlodiden	perlodid stoneflies
Perna canaliculus	Neuseeland-Miesmuschel, Große Streifen-Miesmuschel*	New Zealand mussel, channel mussel
Perna perna	Westatlantische Miesmuschel, Braune Miesmuschel	Mexilhao mussel, South American rock mussel, "brown mussel"
Perna viridis	Grüne Miesmuschel	green mussel
Pernis apivorus	Wespenbussard	western honey buzzard
Perodicticus potto	Potto	potto, potto gibbon
Perognathus spp.	Taschenmäuse, Eigentliche Taschenmäuse	pocket mice
Peromyscus eremicus	Kaktusmaus	cactus mouse
Peromyscus gossypinus	Baumwollmaus	cotton mouse
Peromyscus leucopus	Weißfußmaus	white-footed mouse
Peromyscus manicilatus	Hirschmaus	deer mouse
Peromyscus spp.	Weißfußmäuse	white-footed mice, deer mice
Peropteryx spp.	Hündchen-Fledermaus*	doglike bats
Peroryctes broadbenti	Riesennasenbeutler	giant bandicoot
Peroryctes spp.	Neuguinea-Nasenbeutler	New Guinean bandicoots
Perothopidae	Amerikanische Buchenkäfer	beech-tree beetles
Perotrochus amabilis	Wunderbare Schlitzbandschnecke	lovely slitsnail
Perotrochus hirasei	Millionärsschnecke	Hirase's slitsnail
Persicula catenata	Prinzessin-Randschnecke	princess marginella, princess marginsnail (princess margin shell)
Persicula cingulata/ Persicula marginata	Gürtel-Randschnecke	girdled marginella, girdled marginsnail (girdled margin shell, belted margin shell)
Persicula cornea/ Marginella cornea	Blasse Randschnecke	pale marginella, pale marginsnail (pale margin shell), plain marginella
Persicula persicula/ Marginella persicula	Getupfte Randschnecke	spotted marginella, spotted marginsnail
Perissodactyla	Unpaarhufer	odd-toed ungulates, perissodactyls

Petasina edentula/	Zahnlose Haarschnecke	edentate marginsnail,
Trichia edentula		toothless hairysnail
Petasina unidentata/	Einzähnige Haarschnecke	unidentate marginsnail,
Trichia unidentata		single-toothed hairysnail
Petaurillus spp.	Zwerggleithörnchen u.a.	pygmy flying squirrel
Petaurista elegans	Geflecktes Riesengleithörnchen	spotted giant flying squirrel
Petaurista petaurista	Taguan	red giant flying squirrel
Petaurista spp.	Riesengleithörnchen	giant flying squirrels
Petaurus australis	Großer Gleithörnchenbeutler	fluffy glider
Petaurus breviceps	Kurzkopf-Gleithörnchenbeutler	sugar glider
Petaurus norfolcensis	Mittlerer Gleithörnchenbeutler	squirrel glider
Petaurus spp.	Gleithörnchenbeutler	lesser gliding possums
Petinomys spp.	Zwerggleithörnchen u.a.	dwarf flying squirrels
Petitella georgiae	Falscher Rotmaulsalmler	false rummy-nose tetra
Petricola lapicida	Atlantische Bohrmuschel	boring petricola
Petricola pholadiformis/	Amerikanische Bohrmuschel	American piddock,
Petricolaria pholadiformis		false angelwing (U.S.)
Petricolidae	Engelsflügel	piddocks
Petrobia latens	Braune Weizenmilbe*	brown wheat mite
Petrochirus diogenes	Amerikanischer Riesen-Einsiedler	giant hermit crab
Petrodromus tetradactylus	Vierzehen-Rüsselratte	four-toed elephant shrew, forest elephant shrew
Petrogale spp.	Felskänguru	rock wallabies
Petromus typicus	Felsenratte	dassie rat
Petromyscus spp.	Felsenmäuse, Zwerg-Felsenmäuse	rock mice, pygmy rock mice
Petromyzon marinus	Meerneunauge	sea lamprey
Petromyzontida	Neunaugen, Neunaugenartige	lampreys
Petronia petronia/	Steinsperling	rock sparrow
Passer petronia		
Petronia xanthocollis	Gelbkehlsperling	yellow-throated sparrow
Petropedetes cameronensis	Kamerunschwimmer, Kamerun-Wasserfrosch	Cameroon water frog
Petropedetes natator	Schwimmer, Sierra Leone-Wasserfrosch	Sierra Leone water frog
Petrophora chlorosata	Adlerfarnspanner	bracken-fern moth*
Petrosaurus mearnsi	Gebänderter Felsenleguan	banded rock lizard
Petrosia ficiformis	Feigenhornschwamm	fig sponge a.o.*
Petrus rupestris	Gelbrote Meerbrasse	red steenbras
Pezophaps solitaria[†]	Einsiedler	Rodgriguez solitaire
Pezoporus occidentalis/	Höhlensittich	night parrot,
Geopsittacus occidentalis		Australian parrot
Pezoporus wallicus	Erdsittich	ground parrot
Phacellophora camtschatica	Dotterqualle*	eggyolk jellyfish
Phacocoerus aethiopicus	Warzenschwein	wart hog
Phaenomys ferrugineus	Rio-Reisratte	Rio de Janeiro rice rat
Phaenops cyanea	Blauer Kiefernprachtkäfer	
Phaeognathus hubrichti	Hubrichts Schleichensalamander	Red Hills salamander
Phaethon aetherus	Rotschnabel-Tropikvogel	red-billed tropic bird
Phaethon lepterus	Weißschwanz-Tropikvogel	white-tailed tropic bird

Phaethontidae	Tropikvögel	tropicbirds
Phaethornithinae	Einsiedler-Kolibris	hermits
Phainopepla nitens	Trauerseidenschnäpper	phainopepla
Phalacridae	Glattkäfer	shining flower beetles
Phalacrocoracidae	Kormorane, Scharben	cormorants, shags
Phalacrocorax aristotelis	Krähenscharbe	shag
Phalacrocorax atriceps	Blauaugenscharbe	blue-eyed cormorant
Phalacrocorax auritus	Ohrenscharbe	double-crested cormorant
Phalacrocorax bougainvillei	Guanoscharbe	Guanay cormorant
Phalacrocorax capensis	Kapscharbe	Cape cormorant
Phalacrocorax carbo	Kormoran	great cormorant
Phalacrocorax gaimardi	Buntscharbe	red-legged cormorant
Phalacrocorax pygmeus	Zwergscharbe	pygmy cormorant
Phalacrocorax varius	Elsterscharbe	pied cormorant
Phalacrus coruscus	Getreide-Glattkäfer	cereal flower beetle*
Phalacrus politus	Gleichfuß-Glattkäfer	smut beetle
Phalaenoptilus nuttallii	Poorwill, Winternachtschwalbe	poorwill, common poorwill
Phalanger spp.	Kuskuse	cuscuses
Phalangiidae	Echte Weberknechte, Schneider	daddy-long-legs
Phalangium opilio	Gemeiner Weberknecht	common harvestman
Phalaropus fulicarius	Thorshühnchen	grey phalarope
Phalaropus lobatus	Odinshühnchen	red-necked phalarope
Phalaropus tricolor	Wilsonwassertreter	Wilson's phalarope
Phalax pellucidus	Durchsichtige Messerscheide, Durchscheinende Messerscheide, Kleine Schwertmuschel	translucent piddock*
Phalera bucephala	Mondvogel, Mondfleck	buff-tip moth
Phalium areola	Karierte Helmschnecke*	chequered bonnet, checkered bonnet
Phalium fimbria	Gefranste Helmschnecke	fringed bonnet
Phalium flammiferum	Streifen-Helmschnecke*, Flammen-Helmschnecke*	striped bonnet, flammed bonnet
Phalium glaucum	Graue Helmschnecke	grey bonnet
Phalium granulatum	Schottenhaube	Scotch bonnet
Phalium strigatum	Gestreifte Helmschnecke	striped bonnet
Phallichthys amates	Guatemalakärpfling, Honduraskärpfling	merry widow
Phallichthys pittieri	Netzkärpfling, Panamakärpfling	iridescent widow, orange-dorsal live bearer, merry widow livebearer (FAO)
Phalloceros caudimaculatus	Schwanzfleckkärpfling	dusky millions fish
Phalloptychus januarius	Januarkärpfling, Kiesfisch	barred millionsfish
Phallostethidae	Kehlphallusfische, Priapiumfische, Phallostethiden (Zwergährenfische u.a.)	priapium fishes
Phallusia fumigata	Rauch-Warzenseescheide	fumigate sea-squirt*, smoke tube tunicate

Phallusia

Phallusia mammillata	Warzen-Ascidie, Weiße Warzenseescheide, Knorpelseescheide	simple sea-squirt, white-and-warty sea-squirt*
Phallusia nigra	Schwarze Seescheide	black sea-squirt
Phaner furcifer	Gabelstreifiger Katzenmaki	fork-marked mouse lemur
Phaneroptera falcata	Gemeine Sichelschrecke	sickle-bearing bushcricket
Phaneropteridae	Sichelschrecken	bush katydids & round-headed katydids
Pharaxonota kirschi	Mexikanischer Getreidekäfer	Mexican grain beetle
Pharomachrus mocinno	Quetzal	resplendent quetzal
Pharus legumen	Taschenmessermuschel	
Pharyngobdelliformes	Schlundegel	pharyngobdelliform leeches
Phascogale spp.	Pinselschwanzbeutler	brush-tailed marsupial "mice", tuans
Phascolarctos cinereus	Koalabär, Beutelbär	koala
Phascolosorex spp.	Streifenbeutelmäuse	black-stripe marsupial "mouse"
Phasianella australis	Australische Fasanenschnecke	painted lady
Phasianella spp.	Fasanenschnecken	pheasant snails (pheasant shells)
Phasianella variegata	Bunte Fasanenschnecke	variegated pheasant
Phasianellidae	Fasanenschnecken	pheasant snails (pheasant shells)
Phasianidae	Fasanenartige	pheasants & grouse & turkeys & partridges
Phasianinae	Fasane	pheasants & partridges
Phasianus colchicus	Fasan	pheasant, common pheasant, ring-necked pheasant
Phasmatidae/Phasmida	Gespenstheuschrecken & Stabheuschrecken	walking sticks, stick-insects
Phausis splendidula	Kleines Johanniswürmchen	lesser glow-worm
Pheidole megacephala	Großkopfameise*	big-headed ant
Pheidole pallidula	Bleiche Soldatenameise	
Phelsuma spp.	Taggeckos	day geckos
Phenacoccus aceris	Ahornschmierlaus	apple mealybug, maple mealybug
Phenacogrammus interruptus	Blauer Kongosalmler	Congo tetra
Phenacolimax annularis	Alpen-Glasschnecke	Alpine glass snail
Phenacolimax major	Große Glasschnecke	greater pellucid glass snail
Phenacomys longicaudus	Tannenmaus	red tree vole
Phenacomys spp.	Heidewühlmäuse, Nordamerikanische Tannenmäuse	heather voles, tree voles
Phengodidae	Phengodiden	glowworm beetles
Pheosia gnoma	Birkenporzellanspinner	lesser swallow prominent
Pheosia tremula	Pappelporzellanspinner, Pappelzahnspinner	swallow prominent
Pherbellia knutsoni	Schneckentöter*	snail-killing fly
Pheucticus ludovicianus	Rosenbrust-Kernknacker	rose-breasted grosbeak
Phigalia pilosaria/ Apocheima pilosaria	Schlehenfrostspinner	pale brindled beauty
Phigys solitarius	Einsiedlerlori	collared lory
Philaenus spumarius	Wiesenschaumzikade	common froghopper
Philander nudicaudatus/ Metachirus nudicaudatus	Nacktschwanzbeutelratte	brown "four-eyed" opossum

Philanthus triangulum	Bienenwolf	bee-killer wasp, bee-killer
Philautus spp.	Blasennestfrösche*	bubble-nest frogs
Philepittidae	Lappenpittas	asities (asitys)
Philetairus socius	Siedelweber	sociable weaver
Philine alba	Weiße Seemandel, Weiße Blasenschnecke	white paperbubble
Philine aperta	Seemandel, Offene Seemandel, Offene Blasenschnecke	paperbubble, European paperbubble, open-shelled paperbubble*
Philine lima	Feilen-Seemandel	file paperbubble
Philine quadrata	Viereckige Seemandel	quadrate paperbubble
Philine sinuata	Gebuchtete Seemandel	sinuate paperbubble
Philinidae	Mandelschnecken, Seemandeln	paperbubbles a.o.
Philippia mediterranea	Mittelmeer-Sonnenuhrschnecke	Mediterranean sundial
Philippia radiata	Strahlende Sonnenuhrschnecke, Strahlende Sonnenuhr	radial sundial
Philodromidae	Laufspinnen	running crab spiders, philodromids, philodromid spiders
Philodromus aureolus	Goldfarbige Laufspinne	
Philomachus pugnax	Kampfläufer	ruff
Philorhizus melanocephalus	Schwarzkopf-Rindenläufer	
Philorhizus sigma	Sumpf-Rindenläufer	
Philoria frosti	Baw-Baw	baw-baw frog
Philosamia cynthia/ Samia cynthia/ Platysamia cynthia	Götterbaumspinner, Ailanthusspinner	Cynthia silkmoth, Ailanthus silkworm
Philoscia muscorum	Streifenassel*	common striped woodlouse
Philudoria potatoria	Grasglucke	drinker
Phlebobranchia	Glattkiemen-Seescheiden	phlebobranchs (smooth-gill sea-squirts)
Phlebotomidae	Sandmücken	sandflies
Phlebotomus spp.	Sandmücken	sandflies
Phlegra fasciata	Gebänderte Springspinne	
Phloeomys spp.	Riesenborkenratten	slender-tailed cloud rats
Phloeonomus pusillus	Kleiner Rindenkurzflügler	
Phlogophora meticulosa	Achateule	angle shades
Phoca caspica	Kaspi-Ringelrobbe	Caspian seal
Phoca fasciata	Bandrobbe	ribbon seal
Phoca groenlandica	Sattelrobbe	harp seal
Phoca hispida	Eismeer-Ringelrobbe	ringed seal
Phoca largha	Largha-Seehund	spotted seal, largha seal
Phoca sibirica	Baikal-Ringelrobbe	Baikal seal
Phoca vitulina	Seehund	harbor seal, common seal
Phocaena dioptrica	Brillenschweinswal	spectacled porpoise, Marposa de anteojos
Phocarctos hookeri	Auckland-Seelöwe	Hooker's sea lion, New Zealand sea lion, Auckland sea lion

Phocidae	Hundsrobben	true seals, earless seals, hair seals
Phocoena phocoena	Schweinswal, Braunfisch, Kleiner Tümmler, Kleintümmler	harbor porpoise, common harbor porpoise
Phocoena sinus	Golftümmler	Gulf of California porpoise, Chochito, Vaquita gulf porpoise
Phocoena spinipinnis	Schwarzer Tümmler, Burmeisters Tümmler	black porpoise, Burmeister's porpoise, Marposa espinosa
Phocoenoides dalli	Dall-Hafenschweinswal	Dall porpoise, Dall's white-flanked porpoise
Phodilus badius	Maskeneule	bay owl
Phodilus prigoginei	Prigogine-Eule	Congo bay-owl
Phodopus spp.	Kurzschwänzige Zwerghamster	small desert hamsters, dwarf hamsters
Phodopus sungorus	Dshungarischer Zwerghamster	striped hairy-footed hamster
Phodopus sungorus sungorus	Russischer Zwerghamster	Russian striped hairy-footed hamster
Phoebetria fusca	Dunkler Rußalbatros, Dunkelalbatros	sooty albatross
Phoebetria palpebrata	Rußalbatros	light-mantled sooty albatross
Phoenicoparrus andinus	Gelbfußflamingo	Andean flamingo
Phoenicoparrus jamesi	Kurzschnabelflamingo	Puna flamingo
Phoenicoperidae	Flamingos	flamingos (*also:* flamingoes)
Phoenicopteriformes	Flamingos	flamingos and allies
Phoenicopterus chilensis	Chileflamingo	Chilean flamingo
Phoenicopterus minor	Zwergflamingo	lesser flamingo
Phoenicopterus ruber	Rosaflamingo	greater flamingo
Phoeniculidae	Baumhopfe	wood hoopoes
Phoeniculus purpureus	Baumhopf	green wood hoopoe
Phoenicurus erythrogaster	Riesenrotschwanz	Güldenstädt's redstart
Phoenicurus moussieri	Diademrotschwanz	Moussier's redstart
Phoenicurus ochruros	Hausrotschwanz	black redstart
Phoenicurus phoenicurus	Gartenrotschwanz	redstart
Pholadidae	Bohrmuscheln, Echte Bohrmuscheln	piddocks
Pholas dactylus	Gemeine Bohrmuschel, Große Bohrmuschel, Dattelmuschel	common piddock
Pholcus phalangioides	Zitterspinne	long-bodied cellar spider, longbodied cellar spider
Pholicidae	Zitterspinnen	cellar spiders
Pholidae/Pholididae	Butterfische	gunnels
Pholidichthys leucotaenia	Aalgrundel	convict blenny
Pholidoptera aptera	Alpen-Strauchschrecke	Alpine dark bushcricket
Pholidoptera fallax	Südliche Strauchschrecke	Fischer's bushcricket
Pholidoptera griseoaptera	Gewöhnliche Strauchschrecke	dark bushcricket
Pholidoptera littoralis	Küsten-Strauchschrecke	littoral bushcricket
Pholidota	Schuppentiere	pangolins, scaly anteaters
Pholis gunnellus	Butterfisch, Messerfisch	butterfish, gunnel

Phorbia coarctata/ *Delia coarctata*	Brachfliege	wheat bulb fly
Phoridae	Buckelfliegen (Rennfliegen)	humpbacked flies
Phormia regina	Glanzfliege	black blow fly
Phormictopus cancerides	Haiti-Vogelspinne	Haitian brown tarantula
Phormictopus cubensis	Kuba-Riesenvogelspinne	Cuban giant tarantula
Phorodon humuli	Hopfenblattlaus, Hopfenlaus	hop aphid, damson-hop aphid
Phoronidea	Hufeisenwürmer, Phoroniden	phoronids
Phos senticosus		phos whelk
Phosphaenus hemipterus	Kurzflügel-Leuchtkäfer, Kurzflügeliger Leuchtkäfer	short-wing lightning beetle
Photedes captiuncula	Grashaldeneule	least minor (moth)
Photedes extrema		concolorous moth
Photedes minima	Schmieleneule	
Photichthyidae/ Phosichthyidae	Leuchtfische	lightfishes
Photinus pennsylvanicus	Pennsylvanisches Glühwürmchen	Pennsylvania firefly
Photinus pyralis	Pyralis-Glühwürmchen	pyralis firefly
Photoblepharon palpebratus	Kleiner Laternenfisch	flashlight fish
Phoxinus eos/ *Chrosomus erythrogaster*	Rötling	northern redbelly dace
Phoxinus oreas	Gebirgs-Rötling	mountain redbelly dace
Phoxinus percnurus	Sumpf-Elritze	swamp minnow
Phoxinus phoxinus	Elritze (Eurasische Elritze)	minnow (Eurasian minnow)
Phractolaemidae	Afrikanische Schlammfische	snake mudheads
Phractolaemus ansorgei	Afrikanischer Schlammfisch	snake mudhead
Phragmataecia castaneae	Rohrbohrer	reed leopard moth
Phragmatobia fuliginosa	Zimtbär, Rostbär	ruby tiger (moth)
Phrenacosaurus heterodermus	Anden-Eidechse	Andean lizard
Phrixotrichus cala	Rote Chile-Vogelspinne*	Chilean rose tarantula
Phrixotrichus spatulata	Gemeine Chile-Vogelspinne*	Chilean common tarantula
Phryganea grandis	Große Köcherfliege	great red sedge
Phryneta leprosa		African brown longhorn beetle
Phrynobatrachus spp.	Silberfrösche	silver frogs
Phrynocephalus guttatus	Gefleckte Krötenkopfagame	spotted toad-headed lizard, spotted toadhead agama
Phrynocephalus helioscopus	Sonnengucker	sunwatcher toadhead agama
Phrynocephalus mystaceus	Bärtige Krötenkopfagame, Ohrenkrötenkopf	secret toadhead agama
Phrynocephalus ornatus	Schmuck-Krötenkopfagame	striped toadhead agama
Phrynocephalus spp.	Krötenkopf-Agamen	toadhead agamas, toadhead lizards, toad-headed lizards
Phrynocephalus theobaldi	Himalaja-Krötenkopfagame	Himalaya toadhead agama
Phrynodactylus spp.	Blattfingergeckos	leaf-toed geckos
Phrynohyas spp.	Goldaugen-Baumlaubfrösche*	golden-eyed treefrogs
Phrynohyas venulosa	Giftlaubfrosch	veined treefrog
Phrynomantis bifasciatus/ *Phrynomerus bifasciatus*	Südafrikanischer Wendehalsfrosch	South African snake-neck frog

Phrynophiurida	Krötenschlangensterne	phrynophiurids
Phrynops dahli	Dahls Krötenkopf-Schildkröte	Dahl's toad-headed turtle
Phrynops geoffroanus	Dunkle Krötenkopf-Schildkröte	Geoffroy's side-necked turtle
Phrynops gibbus/ Mesoclemmys gibba	Buckelschildkröte	gibba turtle
Phrynops hilarii	Froschkopf-Schildkröte	frog-headed turtle, spotted-bellied side-necked turtle
Phrynops nasutus/ Batrachemys nasuta	Gewöhnliche Froschkopf-Schildkröte, Gewöhnliche Krötenkopf-Schildkröte	common South American toad-headed turtle, toad-headed side-necked turtle
Phrynops rufipes	Rote Krötenkopf-Schildkröte	red-footed sideneck turtle, red toad-headed turtle
Phrynops spp.	Krötenkopf-Schildkröten	toad-headed turtles
Phrynops tuberculatus	Warzen-Krötenkopf-Schildkröte, Raue Krötenkopf-Schildkröte	tuberculate toad-headed turtle
Phrynops williamsi	Williams Krötenkopf-Schildkröte	William's side-necked turtle
Phrynops zuliae	Zulia Krötenkopf-Schildkröte	Zulia toad-headed turtle
Phrynorhombus norvegicus	Norwegischer Zwergbutt	Norwegian topknot
Phrynorhombus regius	Südlicher Zwergbutt	Eckström's topknot
Phrynosoma cornutum	Texas-Krötenechse	Texas horned lizard
Phrynosoma coronatum	Kronen-Krötenechse	Coast horned lizard
Phrynosoma douglasi	Kurzhorn-Krötenechse	short-horned lizard
Phrynosoma m'calli	Flachschwanz-Krötenechse	flat-tailed horned lizard
Phrynosoma modestum	Rundschwänzige Krötenechse	roundtail horned lizard, round-tailed horned lizard
Phrynosoma platyrhinos	Wüsten-Krötenechse	desert horned lizard
Phrynosoma solare	Pracht-Krötenechse	regal horned lizard
Phrynosoma spp.	Krötenechsen	horned lizards, "horny toads"
Phrynosoma xanti	Sonora-Krötenechse	Sonoran horned lizard
Phryxe vulgaris	Gammaeulen-Raupenfliege	silver Y moth parasite fly
Phthiraptera (Mallophaga & Anoplura)	Tierläuse, Lauskerfe, Läuslinge	phthirapterans
Phthirus pubis/ Pthirus pubis	Filzlaus, Schamlaus	pubic louse, crab louse, "crab"
Phthorimaea operculella	Kartoffelmotte	potato moth, potato tuber moth (potato tuberworm)
Phycis blennoides/ Urophycis blennoides	Gabeldorsch, Meertrüsche, Großer Gabeldorsch	greater forkbeard
Phycis chesteri	Langflossen-Gabeldorsch	longfin hake
Phycis phycis	Mittelmeer-Gabeldorsch, Südliche Meerschleie, Kleiner Gabeldorsch	forkbeard (FAO), lesser hake
Phycis spp.	Ostatlantische Gabeldorsche	Eastern Atlantic hakes
Phycodorus eques	Blatt-Fetzenfisch*	leafy seadragon
Phylactolaemata/Lophopoda	Süßwasserbryozoen, Armwirbler	phylactolaemates, "covered throat" bryozoans, freshwater bryozoans
Phyllangia americana	Versteckte Nelkenkoralle	hidden cupcoral
Phyllaphis fagi	Wollige Buchenlaus	beech aphid, woolly beech aphid

Phyllobates lugubris	Pracht-Giftfrosch	lovely poison-dart frog, lovely poison frog
Phyllobates spp.	Blattsteigerfrösche	golden poison frogs, poison-arrow frog
Phyllobates terribilis	Gelber Blattsteiger, Schrecklicher Giftfrosch	golden poison frog
Phyllobates vittatus	Orangeschwarzer Giftfrosch	orange-and-black poison-dart frog, Golfodulcean poison frog
Phyllobius argentatus	Goldgrüner Blattnager	silver-green leaf weevil
Phyllobius maculicornis	Grüner Blattnager	green leaf weevil
Phyllobius oblongus	Zweifarbiger Schmalbauchrüssler	brown leaf weevil
Phyllobius pomaceus	Brennnesselblattrüssler	nettle leaf weevil
Phyllobius pyri	Birnen-Grünrüssler	pear weevil
Phyllobius urticae	Nessel-Grünrüssler	nettle weevil, stinging nettle weevil
Phyllobrotica quadrimaculata	Helmkraut-Blattkäfer	
Phyllocnistidae	Saftschlürfermotten	eyecap moths
Phyllocnistis suffusella	Pappelblattminiermotte	poplar leafminer a.o.
Phyllocnistis unipunctella	Einpunkt-Pappelblattminiermotte	onespot poplar leafminer*
Phylloda foliacea/ Tellina foliacea	Goldzunge	foliated tellin, leafy tellin
Phyllodactylus europaeus	Europäischer Blattfingergecko	European leaf-toed gecko
Phyllodactylus marmoratus/ Christinus marmoratus	Marmorierter Blattfingergecko	marbled gecko, Australian marbled gecko (Australian clawed gecko)
Phyllodactylus spp.	Blattfingergeckos	leaf-toed geckos
Phyllodactylus xanti	Kalifornischer Blattfingergecko	leaf-toed gecko, American leaf-toed gecko
Phyllodecta laticollis	Kleiner Pappelblattkäfer*	small poplar leaf beetle
Phyllodecta viminalis	Veränderlicher Weidenblattkäfer	variable willow leaf beetle
Phyllodecta vulgatissima	Blauer Weidenblattkäfer	blue willow leaf beetle
Phylloderma stenops	Peters Spießblattnase	Peter's spear-nosed bat
Phyllodesma ilicifolia	Weidenglucke, Heidelbeer-Glucke	small lappet (moth)
Phyllodesma tremulifolia	Eichenglucke, Kleine Eichenglucke	oak lappet (moth)*
Phyllodoce spp.	Paddelwürmer, Ruderwürmer	paddleworms
Phyllodocidae	Paddelwürmer, Ruderwürmer	paddleworms
Phyllodromica maculata	Gefleckte Kleinschabe	
Phyllodytes spp.	Herzzüngler	heart-tongue frogs
Phyllomedusa hypochondrialis	Affenfrosch, Makifrosch, Greiffrosch	orange-legged leaf frog
Phyllomedusa bicolor	Riesen-Greiffrosch	giant monkey frog
Phyllomedusa lemur	Lemuren-Greiffrosch	lemur leaf frog
Phyllomedusa sauvagii	Buntbauch-Greiffrosch	painted-belly monkey frog
Phyllomedusa spp.	Greiffrösche, Makifrösche, Affenfrösche	leaf frogs
Phyllomedusa tomopterna	Makifrosch	barred leaf frog, tiger-striped leaf frog
Phyllonorycter coryli	Haselnussblatt-Blasenmotte*	nut leaf blister moth

Phyllonotus brassica/ *Hexaplex brassica*	Kohlschnecke, Kohl-Stachelschnecke	cabbage murex
Phyllonotus pomum	Apfel-Stachelschnecke	apple murex
Phyllopertha horticola	Gartenlaubkäfer	garden chafer
Phyllopoda/Branchiopoda	Kiemenfüßer, Blattfußkrebse	phyllopods, branchiopods
Phyllops spp.	Sichelflügel-Fledermaus*	falcate-winged bats
Phyllopteryx spp.	Fetzenfische	weedy seadragons
Phyllorhynchus spp.	Blattnasennattern	leaf-nosed snakes
Phylloscopus bonelli	Berglaubsänger	Bonelli's warbler
Phylloscopus borealis	Wanderlaubsänger	Arctic warbler
Phylloscopus collybita	Zilpzalp, Weidenlaubsänger	chiffchaff
Phylloscopus fuscatus	Dunkellaubsänger	dusky warbler
Phylloscopus humei	Tienschan-Laubsänger	Hume's yellow-browed warbler
Phylloscopus inornatus	Gelbbrauen-Laubsänger	yellow-browed warbler
Phylloscopus proregulus	Goldhähnchen-Laubsänger	Pallas's warbler
Phylloscopus schwarzi	Bartlaubsänger	Radde's warbler
Phylloscopus sibilatrix	Waldlaubsänger	wood warbler
Phylloscopus trochiloides	Grünlaubsänger	greenish warbler
Phylloscopus trochilus	Fitis	willow warbler
Phyllostomidae	Blattnasen-Fledermäuse, Neuwelt-Blattnasen, Neuwelt-Lanzennasen	American leaf-nosed bats, New World spear-nosed bats, New World leaf-nosed bats
Phyllostomus discolor	Kleiner Lanzennasen-Fruchtvampir	pale spear-nosed bat, lesser spear-nosed bat
Phyllostomus spp.	Lanzennasen	spear-nosed bats
Phyllotis spp.	Blattohrmäuse	leaf-eared mice, pericotes
Phyllotreta aerea	Kleiner Schwarzer Erdfloh*	small black flea beetle
Phyllotreta atra	Schwarzer Kohlerdfloh	black turnip flea beetle
Phyllotreta cruciferae	Grünglänzender Kohlerdfloh	
Phyllotreta diademata	Kronen-Erdfloh*	crown flea beetle
Phyllotreta nemorum	Gelbstreifiger Kohlerdfloh	large striped flea beetle
Phyllotreta nigripes	Blauseidiger Kohlerdfloh	blueish turnip flea beetle
Phyllotreta spp.	Kohlerdflöhe	flea beetles, turnip flea beetles
Phyllotreta undulata	Geschweiftstreifiger Kohlerdfloh, Gewelltstreifiger Kohlerdfloh	small striped flea beetle
Phyllotreta vittula	Getreideerdfloh	barley flea beetle
Phylloxera coccinea	Eichenzwerglaus u.a.	oak leaf phylloxera a.o.
Phylloxera glabra	Eichenzwerglaus u.a.	oak leaf phylloxera a.o.
Phylloxeridae	Zwergläuse, Zwergblattläuse	phylloxerans (aphids)
Phyllurus spp.	Blattschwanzgeckos	leaf-tailed geckos, leaftail geckos
Phymatidae	Fangwanzen, Gespensterwanzen	ambush bugs
Phymatodes testaceus	Variabler Schönbock	tanbark borer (longhorn beetle)
Physa acuta/*Physella acuta*	Blasenschnecke	pointed bladder snail
Physa fontinalis	Quellblasenschnecke, Quell-Blasenschnecke, Quellenblasenschnecke	bladder snail, common bladder snail

Physailia pellucida	Afrikanischer Glaswels	African glass catfish
Physalaemus nattereri	Augenkröte	false-eyed frog
Physalia physalis	Portugiesische Galeere, Blasenqualle, Seeblase	Portuguese man-of-war, Portuguese man-o-war, blue-bottle
Physaloptera spp.	Magenwürmer u.a.	stomach worms
Physella acuta/Physa acuta	Spitze Blasenschnecke	pointed bladder snail, European physa
Physeter macrocephalus (catodon)	Großer Pottwal, Cachalot, Kaschelot	sperm whale, great sperm whale, cachalot
Physiculus bacchus/ Pseudophycis bacchus	Neuseeland-Eisfisch	red cod, New Zealand red cod, red codling (FAO)
Physidae	Blasenschnecken	bladder snails, tadpole snails
Physignathus concincinus	Conchinchina-Wasserdrache	water dragon
Physignathus lesueurii	Gewöhnlicher Wasserdrache	eastern water dragon
Physignathus spp.	Wasserdrachen	Asian water dragons
Physocyclus globosus	Gedrungene Kellerspinne*	short-bodied cellar spider
Physokermes piceae	Fichtenquirlschildlaus	spruce bud scale
Phytodecta decemnotata	Zitterpappelblattkäfer	aspen leaf beetle
Phytodecta vitellinae	Messingfarbener Weidenblattkäfer	brassy willow leaf beetle
Phytoecia coerulescens	Dichtpunktierter Walzenhalsbock	
Phytoecia nigripes	Schwarzfüßiger Walzenhalsbock	
Phytomyza atricornis	Erbsenminierfliege	chrysanthemum leafminer (fly)
Phytomyza flavicornis	Brennnesselstengel-Minierfliege	nettle stem miner (fly)*
Phytomyza ilicis	Stechpalmenblatt-Minierfliege	holly leafminer (fly)
Phytomyza nigra	Getreideblatt-Minierfliege	cereal leafminer (fly)
Phytomyza ramosa	Kardenfliege*	teasel fly
Phytomyza rufipes	Kohlblattminierfliege*	cabbage leafminer (fly)
Phytonemus pallidus/ Tarsonemus pallidus	Alpenveilchenmilbe, Begonienmilbe	cyclamen mite, begonia mite
Phytonomus variabilis/ Hypera postica	Luzerneblattnager	lucerne weevil
Phytotoma rara	Rotschwanz-Pflanzenmäher	Chilean plantcutter
Phytotomidae	Pflanzenmäher	plantcutters
Piaractus brachypomus/ Colossoma bidens	Riesenpacu, Gamitana-Scheibensalmler, Mühlsteinsalmler	pirapatinga (FAO), cachama
Pica pica	Elster	black-billed magpie
Picathartes gymnocephalus	Gelbkopf-Felshüpfer, Gelbkopf-Stelzenkrähe	white-necked rockfowl
Picathartes oreas	Buntkopf-Felshüpfer	gray-necked rockfowl, gray-necked bald-crow, gray-necked picathartes
Picathartidae	Felshüpfer	bald crows
Picidae	Spechte	woodpeckers & wrynecks & piculets
Piciformes	Spechtvögel, Spechtartige	woodpeckers & barbets & toucans and allies
Picoides borealis	Kokardenspecht	red-cockaded woodpecker
Picoides leucotos/ Dendrocopos leucotos	Weißrückenspecht	white-backed woodpecker

Picoides major/ *Dendrocopos major*	Buntspecht	great spotted woodpecker
Picoides medius/ *Dendrocopos medius*	Mittelspecht	middle spotted woodpecker
Picoides minor/ *Dendrocopos minor*	Kleinspecht	lesser spotted woodpecker
Picoides scalaris	Texasspecht	ladder-backed woodpecker
Picoides tridactylus	Dreizehenspecht	three-toed woodpecker
Picoides villosus	Haarspecht	hairy woodpecker
Picromerus bidens	Zweispitzwanze	
Pictodentalium formosum (Scaphopoda)	Hübsche Zahnschnecke	beautiful tusk
Picumnus minutissimus	Däumlingsspecht	Guianan piculet
Picus canus	Grauspecht	grey-headed woodpecker
Picus viridis	Grünspecht	green woodpecker
Pidonia lurida	Schnürhalsbock	
Pieridae	Weißlinge	whites & sulphurs & orange-tips
Pieris brassicae	Großer Kohlweißling	large white
Pieris napi	Rapsweißling, Grünader-Weißling	green-veined white
Pieris rapae/Artogeia rapae	Rübenweißling, Kleiner Kohlweißling	small white, cabbage butterfly, imported cabbageworm (U.S.)
Piesma quadrata	Rübenwanze	beet leaf bug
Piesmatidae	Meldenwanzen, Rübenwanzen	beet bugs, ash-gray leaf bugs, piesmatids
Piezodorus lituratus	Ginster-Baumwanze	
Piliocolobus badius	Westafrikanischer Rotstummelaffe	western red colobus
Piliocolobus kirkii	Sansibar-Rotstummelaffe	Zanzibar red colobus, Kirk's colobus
Piliocolobus oustaleti	Uganda-Rotstummelaffe	Central African red colobus
Piliocolobus pennanti	Pennant-Rotstummelaffe	Pennant's red colobus
Piliocolobus preussi	Preuss-Rotstummelaffe, Kamerun-Rotstummelaffe	Preuss's red colobus
Piliocolobus rufomitratus	Ostafrikanischer Rotstummelaffe	eastern red colobus
Piliocolobus spp.	Rotstummelaffen	red colobuses
Pilosa (Xenarthra)	Faultiere	sloths
Pilumnus hirtellus	Borstenkrabbe	hairy crab, bristly xanthid crab
Pimelodidae	Antennenwelse	long-whiskered catfishes, pimelodid catfishes
Pinctada imbricata	Atlantische Perlmuschel	Atlantic pearl-oyster
Pinctada martensi	Japanische Perlmuschel	Japanese pearl-oyster, Marten's pearl oyster
Pinctada maxima	Große Perlmuschel	gold-lip pearl-oyster, golden-lip pearl oyster, black silver pearl oyster, mother-of-pearl shell, pearl button oyster
Pinctada penguin	Schwarzflügel-Perlmuschel	blackwing pearl-oyster
Pinctada radiata	Gestreifte Perlmuschel	rayed pearl-oyster

Pineus cembrae	Zirbelkieferwolllaus	Swiss stone pine adelges, Arolla pine woolly aphid
Pineus pineoides	Weißwollige Fichtenstammlaus	small spruce adelges, small spruce woolly aphid
Pineus pini	Europäische Kiefernwolllaus	Scots pine adelges, European pine woolly aphid
Pineus similis	Fichtengalllaus	spruce gall adelges
Pineus strombi	Stroben-Rindenlaus	Weymouth pine adelges
Pinguipedidae/ Parapercidae	Krokodilfische	sandperches
Pinguipes spp.	Brasilianische Sandbarsche	Brazilian sandperches
Pinicola enucleator	Hakengimpel	pine grosbeak
Pinna carnea	Bernstein-Steckmuschel, Fleischfarbene Steckmuschel	amber penshell
Pinna fragilis/Atrina fragilis	Zerbrechliche Steckmuschel	fragile penshell, fan mussel (fragile fanshell)
Pinna nobilis	Steckmuschel, Edle Steckmuschel, Große Steckmuschel	rough penshell, noble penshell
Pinna rudis	Stachelige Steckmuschel, Durchsichtige Steckmuschel	rude penshell, rough penshell
Pinna squamosa	Gemeine Steckmuschel, Schinkenmuschel	common penshell
Pinnidae	Steckmuscheln	pen shells, fan mussels
Pinnipedia	Flossenfüßer, Robben	marine carnivores (seals & sealions & walruses)
Pinnotheres pinnotheres	Steckmuschel-Erbsenkrabbe, Steckmuschelwächter	Pinna pea crab
Pinnotheres pisum	Erbsenkrabbe, Muschelwächter, Muschelwächterkrabbe	pea crab, Linnaeus's pea crab
Pinnotheridae	Erbsenkrabben, Muschelwächter	pea crabs, commensal crabs
Pionites melanocephalus	Grünzügelpapagei	black-headed caique
Pionus maximiliani	Maximilianpapagei	scaly-headed parrot
Pionus menstruus	Schwarzohrpapagei	blue-headed parrot
Piophila casei	Käsefliege	cheese-skipper, cheese maggot
Piophilidae	Käsefliegen	cheese-skippers
Pipa parva	Zwergwabenkröte	Sabana Surinam toad
Pipa pipa	Wabenkröte	Surinam toad
Pipidae	Zungenlose, Krallenfrösche	tongueless frogs (clawed frogs & Surinam toads)
Pipilo chlorurus/ Chlorura chlorura	Grünschwanz-Grundammer	green-tailed towhee
Pipilo erythrophthalmus	Rötelgrundammer	rufous-sided towhee
Pipistrellus kuhlii	Weißrandfledermaus	Kuhl's pipistrelle
Pipistrellus nanus	Bananenzwergfledermaus	banana bat
Pipistrellus nathusii	Rauhautfledermaus	Nathusius' pipistrelle
Pipistrellus pipistrellus	Zwergfledermaus	common pipistrelle
Pipistrellus savii	Alpenfledermaus	Savi's pipistrelle
Pipistrellus spp.	Zwergfledermäuse	pipistrelles
Pipreola arcuata	Bindenschmuckvogel	barred fruiteater

Pipridae	Pipras, Schnurrvögel, Manakins	manakins
Pipunculidae	Augenfliegen, Kugelkopffliegen	big-headed flies
Piranga ludoviciana	Kieferntangare, Louisianatangare	western tanager
Piranga olivacea	Scharlachtangare	scarlet tanager
Piranga rubra	Sommertangare, Feuertangare	summer tanager
Pirata piraticus	Piratenspinne	pirate spider
Pisa armata	Spitzkopfmaskenkrabbe	Gibb's spider crab
Pisa tetraodon	Vierhorn-Spitzkopfkrabbe	four-horned spider crab
Pisania maculosa/ Pisania striata	Geflecktes Klipphorn	spotted pisania
Pisaster giganteus	Riesenstern	giant seastar
Pisater ochraceus	Ockerstern	ochre seastar, ochre star
Pisaura mirabilis	Listspinne, Raubspinne	fantastic fishing spider*
Pisauridae	Raubspinnen, Jagdspinnen	nursery-web spiders, fisher spiders, fishing spiders
Pisces	Fische	fishes
Piscicola geometra	Gemeiner Fischegel	common fish leech, great tailed leech
Piscicolidae	Fischegel	fish leeches
Pisidia longicornis/ Porcellana longicornis	Langhorn-Porzellankrebs, Schwarzer Porzellankrebs, Schwarzes Porzellankrebschen	long-clawed porcelain crab, common porcelain crab, minute porcelain crab
Pisidium amnicum	Große Erbsenmuschel	giant pea mussel, river pea mussel, large pea shell, greater European peaclam (U.S.)
Pisidium casertanum	Gemeine Erbsenmuschel	common pea mussel, caserta pea mussel, ubiquitous peaclam
Pisidium conventus	See-Erbsenmuschel	Alpine peaclam
Pisidium henslowanum	Faltenerbsenmuschel, Kleine Falten-Erbsenmuschel	Henslow pea mussel, Henslow peaclam
Pisidium hibernicum	Glatte Erbsenmuschel	smooth pea mussel
Pisidium lilljeborgii	Kreisrunde Erbsenmuschel	Lilljeborg pea mussel, Lilljeborg peaclam
Pisidium milium	Eckige Erbsenmuschel	quadrangular pea mussel, quadrangular pillclam, rosy pea mussel
Pisidium nitidum	Glänzende Erbsenmuschel	shining pea mussel, shiny peaclam
Pisidium obtusale	Stumpfe Erbsenmuschel	obtuse pea mussel
Pisidium pseudosphaerium	Kugelige Erbsenmuschel	pseudospherical pea mussel*
Pisidium pulchellum	Schöne Erbsenmuschel	beautiful pea mussel
Pisidium spp.	Erbsenmuscheln	pea mussels, peaclams (U.S.), (pea shells)
Pisidium subtruncatum	Schiefe Erbsenmuschel	shortended pea mussel, shortended peaclam
Pisidium supinum	Dreieckige Erbsenmuschel	triangular pea mussel, humpbacked peaclam

Pisidium tenuilineatum	Kleinste Erbsenmuschel	miniscule pea mussel*
Pisodonophis semicinctus	Gürtel-Eingeweideaal	saddled snake eel
Pisodonophis spp.	Eingeweideaale	snake eels
Pissodes castaneus	Kiefernkulturrüssler	small banded pine weevil
Pissodes piceae	Weißtannenrüssler	white fir weevil
Pissodes pini	Gestreifter Kiefernrüssler	banded pine weevil
Pissodes validirostris	Kiefernzapfenrüssler	pinecone weevil
Pitangus sulphuratus	Bentevi	great kiskadee
Pitar dione	Kamm-Venusmuschel	royal comb venus, royal pitar
Pitar fulminatus	Blitzschlagmuschel*	lightning pitar
Pitar morrhuanus	Falsche Quahog-Muschel	false quahog
Pithecheir melanurus	Malaiische Baumratte	monkey-footed rat, Malay tree rat
Pithecia aequatorialis	Äquatorial-Saki	equatorial saki
Pithecia albicans	Schwarzrücken-Mönchsaffe	white saki
Pithecia irrorata	Kahlgesicht-Mönchsaffe	bald-faced saki
Pithecia monachus	Rotbärtiger Mönchsaffe, Zottelschweifaffe	monk saki
Pithecia pithecia	Weißkopfsaki, Blasskopfsaki	white-faced saki
Pithecia spp.	Sakis	sakis
Pithecophaga jefferyi	Affenadler	great Philippine eagle, monkey-eating eagle
Pitta angolensis	Angolapitta	African pitta
Pitta granatina	Granatpitta	garnet pitta
Pitta gurneyi	Goldkehlpitta	Gurney's pitta, jewel thrush
Pitta iris	Regenbogenpitta	rainbow pitta
Pittidae	Pittas	pittas
Pituophis catenifer	Gophernatter	gopher snake
Pituophis catenifer sayi	Bullennatter	bullsnake
Pituophis melanoleucus	Kiefernnatter	pine snake
Pituophis spp.	Kiefernnattern	pine snakes, gopher snakes
Pitymys bavaricus/ Microtus bavaricus	Bayerische Kurzohrmaus	Bavarian pine vole
Pitymys duodecimcostatus	Mittelmeer-Kleinwühlmaus	Mediterranean pine vole
Pitymys lusitanicus	Lusitanien-Kleinwühlmaus	Lusitanian pine vole
Pitymys multiplex	Fatio-Kleinwühlmaus	alpine pine vole, Fatio's pine vole
Pitymys spp./*Microtus* spp.	Feldmäuse	voles, meadow mice
Pitymys tatricus	Tatra-Kleinwühlmaus	Tatra pine vole
Pityogenes bidentatus	Hakenzähniger Kiefernborkenkäfer	bidentated bark beetle
Pityogenes chalcographus	Kupferstecher, Sechszähniger Fichtenborkenkäfer	six-dentated bark beetle
Pityohyphantes costatus	Hängemattenspinne*	hammock spider
Pityokteines curvidens/ Ips curvidens	Krummzähniger Tannenborkenkäfer	curvidentated fir bark beetle
Pityriasinae	Warzenkopf	Bornean bristlehead
Placentalia/Eutheria	Plazentatiere	placentals, eutherians

Placopecten magellanicus	Atlantische Tiefsee-Kammmuschel, Atlantischer Tiefseescallop	Atlantic deep-sea scallop, sea scallop
Placophora	Käferschnecken	placophorans (incl. chitons)
Placuna ephippium	Große Pazifische Sattelauster	greater Pacific saddle oyster
Placuna placenta	Fenstermuschel, Fensterscheibenmuschel, Glockenmuschel	windowpane oyster (window shell, jingle shell)
Placuna sella	Gemeine Sattelauster	common saddle oyster
Plagiocardium pseudolima	Riesen-Herzmuschel	giant cockle
Plagiodera versicolor	Blauer Weidenblattkäfer	imported willow leaf beetle (U.S.)
Plagiodontia spp.	Zagutis, Hispaniola-Baumratten	Hispaniolan hutias
Plagiolepis spp.	Zwergameisen	pygmy ants
Plagionotus arcuatus	Eichenwidderbock, Eichenzierbock	yellow-bowed longhorn beetle
Plagionotus detritus	Hornissenbock	
Plagiotremus rhinorhynchus	Blaustreifen-Säbelzahnschleimfisch	bluestriped fangblenny
Plagiotremus tapeinosoma	Piano-Säbelzahnschleimfisch	scale-eating blenny
Plagodis dolabraria	Hobelspanner	scorched wing
Plagusia depressa	Flachkrabbe*	flattened crab
Planaxidae	Haufenschnecken	clusterwinks, grooved snails
Planaxis sulcatus	Gerippte Haufenschnecke	ribbed clusterwink, furrowed planaxis
Planes minutus	Kolumbuskrabbe	gulfweed crab, Gulf weed crab, Columbus crab
Planigale spp.	Flachkopfbeutelmäuse	planigales, flat-skulled marsupial "mice"
Planococcus citri	Zitronenschmierlaus, Gewächshausschmierlaus	citrus mealybug
Planorbarius corneus	Posthornschnecke	horn-colored ram's horn, great ramshorn (trumpet shell)
Planorbidae	Tellerschnecken	ramshorn snails
Planorbis carinatus	Gekielte Tellerschnecke	keeled ramshorn
Planorbis planorbis	Gemeine Tellerschnecke, Flache Tellerschnecke	ramshorn, ram's horn, common ramshorn, margined ramshorn (margined trumpet shell)
Platacanthomys lasiurus	Südindischer Stachelbilch	spiny dormouse
Platalea leucorodia	Löffler	white spoonbill
Platambus maculatus	Gefleckter Schnellschwimmer	
Platanista gangetica	Ganges-Delphin, Schnabeldelphin, Susu	Ganges River dolphin, su-su
Platanista indi	Indus-Delphin, Indusdelphin	Indus River dolphin, bhulan
Platax orbicularis	Fledermausfisch, Meeressegelflosser, Einfacher Segelflosser	batfish
Platax pinnatus	Rotrand-Fledermausfisch, Rotsaum-Segelflosser	red-margined batfish

Platemys macrocephala/ Acanthochelys macrocephala	Großkopf-Pantanal-Sumpfschildkröte	big-headed Pantanal swamp turtle, large-headed Pantanal swamp turtle
Platemys pallidipectoris/ Acanthochelys pallidipectoris	Sporen-Plattschildkröte	Chaco side-necked turtle
Platemys platycephala	Rotkopf-Plattschildkröte	twist-neck turtle, twist-necked turtle
Platemys radiolata	Strahlen-Plattschildkröte, Brasilianische Plattschildkröte	Brazilian radiolated swamp turtle, Brazilian radiated swamp turtle
Platemys radiolata spixii	Stachelhals-Plattschildkröte	black spine-necked swamp turtle, black spiny-necked swamp turtle
Platemys spp.	Plattschildkröte	flat-shelled turtles
Plathelminthes	Plattwürmer, Plathelminthen	flatworms, platyhelminths (Platyhelminthes)
Platichthys flesus	Flunder, Gemeine Flunder (Sandbutt)	flounder (FAO), European flounder
Platichthys stellatus	Sternflunder, Pazifische Sternflunder	starry flounder
Platyarthrus hoffmannseggi	Ameisenassel	ant woodlouse
Platycephalidae	Flachköpfe, Plattköpfe	flatheads, flathead gurnards
Platycephalus indicus	Sandflachkopf	Indo-Pacific flathead
Platycerus caraboides/ Systenocernus caraboides	Rehschröter	
Platycerus elegans	Pennantsittich	crimson rosella
Platycheirus manicatus		potato aphid hover fly
Platychirograpsus spectabilis	Säbelkrabbe	saber crab
Platycleis affinis	Südliche Beißschrecke	tuberous bushcricket, southern bushcricket
Platycleis albopunctata/ Platycleis denticulata	Westliche Beißschrecke	western bushcricket
Platycleis grisea	Graue Beißschrecke	grey bushcricket
Platycleis modesta	Veränderte Beißschrecke	variable bushcricket
Platycleis montana	Steppen-Beißschrecke	steppe bushcricket
Platycleis sepium/ Sepiana sepium	Zaunschrecke	sepia bushcricket
Platycleis stricta	Südöstliche Beißschrecke	Italian bushcricket
Platycleis tesselata	Braunfleckige Beißschrecke	brown-spotted bushcricket
Platycnemidae	Federlibellen	platycnemid damselflies
Platycnemis latipes	Weiße Federlibelle	white damselfly*
Platycnemis pennipes	Federlibelle	white-legged damselfly
Platyctenidea	Platte Rippenquallen	platyctenids
Platydoris argo	Rotbraune Ledernacktschnecke	redbrown nudibranch
Platygyra spp.	Hirnkorallen u.a., Neptunskorallen	brain corals a.o.
Platymantis spp.	Runzelfrösche	Fijian ground frogs
Platymops setiger	Afrikanische Flachkopf-Fledermaus	flat-headed free-tailed bat
Platynereis dumerilii	Dumerils Ringelwurm	Dumeril's clam worm, comb-toothed nereid
Platynus assimilis	Schwarzer Enghalsläufer, Schwarzer Putzläufer	
Platynus dorsalis	Bunter Enghalsläufer, Buntfarbener Putzläufer	

Platyparea poeciloptera	Spargelfliege, Große Spargelfliege	asparagus fly (asparagus maggot)
Platypezidae	Plattfüßer (Sohlenfliegen, Tummelfliegen)	flat-footed flies
Platypodidae	Kernkäfer, Kernholzkäfer	pinhole borers
Platyproctidae	Eigentliche Glattkopffische	tubeshoulders
Platypsyllus castoris	Biberkäfer, Biberlaus	beaver beetle
Platypus cylindrus	Eichenkernkäfer	oak pinhole borer
Platyrhinidae	Dornrückengitarrenfische	platyrhinids
Platyrhinoidis triseriata	Dornrücken-Geigenrochen, Dornrückengitarrenfisch	thornback guitarfish
Platyrhinus resinosus	Schwammbreitrüssler	
Platyrinchus mystaceus	Weißkehl-Breitschnabel	white-throated spadebill
Platyrrhina	Neuweltaffen, Breitnasenaffen	new-world monkeys (South American monkeys & marmosets)
Platysamia cecropia/ Hyalophora cecropia	Cecropiaspinner	cecropia silk moth, cecropia moth
Platysaurus capensis	Kap-Plattgürtelechse	Cape red-tailed flat lizard
Platysaurus spp.	Plattgürtelechsen	flat lizards
Platysteiridae	Schnäpperwürger, Kleinschnäpper	wattle-eyes, puffback-flycatchers
Platysternidae	Großkopfschildkröten	big-headed turtles
Platysternon megacephalum	Großkopfschildkröte	big-headed turtle
Platytroctidae/Searsiidae	Leuchtheringe	tubeshoulders
Plea leachi/Plea minutissima	Zwergrückenschwimmer, Wasserzwerg (Wanze)	pygmy backswimmer (pleid water bug)
Plebejus argus	Geißkleebläuling, Silberfleckbläuling	silver-studded blue (butterfly)
Plebicula amanda/ Polyommatus amanda/ Agrodiaetus amanda	Prächtiger Bläuling	Amanda's blue (butterfly)
Plecoglossus altivelis	Ayu	ayu, ayu sweetfish
Plecoptera	Steinfliegen, Uferfliegen	stoneflies
Plecotus auritus	Braunes Langohr, Braune Langohrfledermaus, Großohr	brown long-eared bat, common long-eared bat
Plecotus austriacus	Graues Langohr, Graue Langohrfledermaus	grey long-eared bat
Plecotus spp.	Langohren-Fledermäuse	lump-nosed bats, long-eared bats, lappet-eared bats
Plectorhinchus chaetodonoides	Gepunktete Süßlippe, Harlekin-Süßlippe	harlequin sweetlip
Plectorhinchus gaterinus	Gemeine Süßlippe	black-spotted rubberlip
Plectorhinchus macrolepis	Dicklippiger Grunzer	biglip grunt
Plectorhinchus mediterraneus	Westmediterrane Süßlippe	rubberlip grunt
Plectorhinchus obscurus	Riesen-Süßlippe	giant sweetlip
Plectorhinchus orientalis	Orientalische Süßlippe, Orient-Süßlippe	oriental sweetlip
Plectorhynchidae/Gaterinidae	Süßlippen, Weichlipper	sweetlips
Plectrophenax nivalis	Schneeammer	snow bunting

Plectropomus leopardus	Leopardenbarsch	leopard coral trout, leopard grouper, leopard coralgrouper (FAO)
Plectropomus melanoleucus	Gelber Zackenbarsch	coral trout
Plectropomus truncatus	Pantherbarsch	squaretail grouper
Plectropterus gambensis	Sporngans	spur-winged goose
Plegadis falcinellus	Sichler	glossy ibis
Pleidae	Zwergrückenschwimmer	pygmy backswimmers, pleid water bugs, lesser water-boatmen
Plemeliella abietina	Fichtensamen-Gallmücke	spruce-seed gall midge
Pleoticus muelleri	Argentinische Rotgarnele	Argentine red shrimp
Pleoticus robustus	Königs-Rotgarnele	royal red shrimp
Plerogyra sinuosa	Blasenkoralle u.a.	bubble coral a.o.
Plesiobatis daviesi	Tiefen-Stechrochen, Tiefsee-Stechrochen	deepwater stingray
Plesiocoris rugicollis	Grüne Apfelwanze, Nordische Apfelwanze	apple capsid bug
Plesionika edwardsii	Kleine Mittelmeergarnele	soldier striped shrimp
Plesionika martia	Goldgarnele	golden shrimp
Plesionika narval	Einhorngarnele, Narval	unicorn striped shrimp, narval
Plesiopenaeus edwardsianus	Rote Riesengarnele, Atlantische Rote Riesengarnele	scarlet gamba prawn, scarlet shrimp
Plesiopidae	Rundköpfe, Mirakelbarsche	longfins, roundheads
Plethodon cinereus	Rotrücken-Waldsalamander	redback salamander, red-backed salamander
Plethodon dorsalis	Zickzack-Salamander	zigzag salamander
Plethodon glutinosus	Silber-Waldsalamander	slimy salamander, northern slimy salamander
Plethodon jordani	Jordans Waldsalamander, Appalachen-Waldsalamander	Jordan's salamander, red-cheeked salamander, Appalachian woodland salamander
Plethodon spp.	Waldsalamander	woodland salamanders
Plethodon vehiculum	Westlicher Rotrückensalamander	western red-backed salamander
Plethodontidae	Lungenlose Salamander	lungless salamanders
Pleurobrachia bachei	Pazifische Seestachelbeere	Pacific sea gooseberry, cat's eye
Pleurobrachia pileus/ Pleurobrachia rhodophis	Atlantische Seestachelbeere, Meeresstachelbeere	Atlantic sea gooseberry
Pleurobrachia spp.	Seestachelbeeren, Meeresstachelbeeren	sea gooseberries, cat's eyes
Pleurobranchus areolatus	Atlantische Flankenkiemenschnecke*	Atlantic sidegill "slug"
Pleurodeles poireti	Poiretischer Rippenmolch	Poiret's newt
Pleurodeles waltl	Spanischer Rippenmolch	sharp-ribbed salamander
Pleurodema bibroni	Vieraugenkröte, Bibronkröte	four-eyed frog
Pleurodema spp.	Vieraugenkröten	four-eyed frogs
Pleurogrammus monopterygius	Terpug, Atka-Makrele	Atka mackerel
Pleurogrammus spp.	Terpuge, Atka-Makrelen	Atka mackerels

Pleuroncodes monodon	Roter Scheinhummer	red squat lobster
Pleuroncodes planipes	Roter Hochsee-Scheinhummer*	pelagic red crab
Pleuroncodes spp.	Scheinhummer	red crabs, squat lobsters, lobster-krill
Pleuronectes quadrituberculatus	Alaska-Scholle	Alaska plaice
Pleuronectes platessus	Scholle, Goldbutt	plaice, European plaice (FAO)
Pleuronectidae	Schollen	righteye flounders
Pleuronectiformes	Plattfische	flatfishes
Pleuroploca aurantiaca	Goldene Pferdeschnecke	golden horseconch
Pleuroploca gigantea	Riesen-Pferdeschnecke	horse conch
Pleuroploca persica	Persische Bandschnecke	Persian horseconch
Pleuroploca trapezium	Trapez-Bandschnecke, Fuchskopf-Bandschnecke	trapeze horseconch, trapezium horse conch
Pleurotomariidae	Schlitzbandschnecken, Schlitzkegelschnecken, Schlitzkegel	slitsnails (slit shells)
Plexaura flava	Gelbe Seerute, Gelbe Federgorgonie	yellow sea rod
Plexaura flexuosa	Seerute, Biegsame Seerute	flexible sea rod, bent sea rod
Plexaura homomalla	Schwarze Seerute	black sea rod
Plica plica	Baumläuferleguan	tree runner
Plica spp.	Stelzenläuferleguane	racerunners
Plica umbra	Stelzenläuferleguan	harlequin racerunner
Plocamia karykina	Glattroter Schwamm*	smooth red sponge
Ploceinae	Eigentliche Weber, Eigentliche Weberfinken, Webervögel	weavers
Plocepasser mahali	Augenbrauenmahali	white-browed sparrow weaver
Ploceus bojeri	Palmenweber	golden palm weaver
Ploceus cucullatus	Textorweber	village weaver
Ploceus philippinus	Bayaweber	Baya weaver
Plodia interpunctella	Dörrobstmotte	Indian meal moth
Plotosidae	Korallenwelse	eeltail catfishes, eel catfishes
Plotosus lineatus	Gestreifter Korallenwels	striped catfish, striped eel catfish (FAO)
Plumularia setacea	Kleiner Strauch-Hydroid	little seabristle
Plusiinae	Goldeulen	gems, plusiine moths
Plutella maculipennis/ Plutella xylostella	Kohlmotte, Kohlschabe, Schleiermotte	cabbage moth, diamondback moth
Pluvialis apricaria	Goldregenpfeifer	golden plover
Pluvialis dominica	Amerikanischer Goldregenpfeifer	American golden plover
Pluvialis fulva	Pazifischer Goldregenpfeifer	Pacific golden plover
Pluvialis squatarola	Kiebitzregenpfeifer	grey plover
Pocilopora verrucosa	Warzige Buschkoralle	warty bushcoral
Pocillopora damicornis	Himbeerkoralle, Keulenkoralle	raspberry coral*
Pocillopora eydouxi	Blumenkohlkoralle	cauliflower coral*

Podarcis bocagei	Bocages Mauereidechse	Bocage's wall lizard
Podarcis dugesii/ *Lacerta dugesii*	Madeira-Mauereidechse	Madeira wall lizard
Podarcis erhardi/ *Lacerta erhardi*	Ägäische Mauereidechse, Kykladen-Eidechse	Erhard's wall lizard
Podarcis filfolensis/ *Lacerta filfolensis*	Malta-Eidechse	Maltese wall lizard, Filfola wall lizard
Podarcis hispanica/ *Lacerta hispanica*	Spanische Mauereidechse, Iberische Mauereidechse	Iberian wall lizard
Podarcis lilfordi/ *Lacerta lilfordi*	Balearen-Eidechse	Lilford's wall lizard
Podarcis melisellensis/ *Lacerta melisellensis*	Karstläufer	Dalmatian wall lizard
Podarcis milensis/ *Lacerta milensis*	Milos-Eidechse	Milos wall lizard
Podarcis muralis/ *Lacerta muralis*	Mauereidechse	common wall lizard
Podarcis peloponnesciaca/ *Lacerta peloponnesciaca*	Peloponnes-Eidechse	Peloponnese wall lizard
Podarcis perspicillata/ *Lacerta perspicillata*	Brilleneidechse	Moroccan rock lizard, Menorca wall lizard
Pociliopora verrucosa	Warzige Buschkoralle	warty bushcoral
Pocillopora damicornis	Himbeerkoralle, Keulenkoralle	raspberry coral*
Pocillopora eydouxi	Blumenkohlkoralle	cauliflower coral*
Podalonia spp.	Kurzstielsandwespen	
Podarcis bocagei	Bocages Mauereidechse	Bocage's wall lizard
Podarcis filfolensis/ *Lacerta filfolensis*	Malta-Eidechse	Maltese wall lizard, Filfola wall lizard
Podarcis pityusensis/ *Lacerta pityusensis*	Pityusen-Eidechse	Ibiza wall lizard
Podargidae	Eulenschwalme	frogmouths
Podargus strigoides	Eulenschwalm	tawny frogmouth
Podiceps auritus	Ohrentaucher	Slavonian grebe
Podiceps cristatus	Haubentaucher	great crested grebe
Podiceps dominicus	Schwarzkopftaucher	least grebe
Podiceps griseigena	Rothalstaucher	red-necked grebe
Podiceps nigricollis	Schwarzhalstaucher	black-necked grebe
Podiceps ruficollis	Zwergtaucher	little grebe
Podicipedidae	Lappentaucher	grebes
Podicipediformes	Lappentaucher	grebes
Podilymbus podiceps	Bindentaucher	pied-billed grebe
Podisma pedestris	Gewöhnliche Gebirgsschrecke	brown mountain grasshopper
Podoces panderi	Saxaulhäher	Pander's ground jay
Podocnemis erythrocephala	Amazonas-Rotkopf-Schienenschildkröte	red-headed Amazon side-necked turtle, red-headed river turtle (IUCN)
Podocnemis expansa	Arrauschildkröte, Arrau-Schienenschildkröte	Arrau River turtle, giant South American river turtle
Podocnemis lewyana	Rio Magdalena-Schienenschildkröte	Rio Magdalena River turtle, Magdalena River turtle (IUCN)
Podocnemis sextuberculata	Höcker-Schienenschildkröte	six-tubercled Amazon River turtle, six-tubercled river turtle (IUCN)
Podocnemis spp.	Schienenschildkröten	South American river turtles, podocnemids

Podocnemis unifilis	Terekay-Schildkröte	yellow-headed sideneck, yellow-spotted sideneck turtle, yellow-spotted Amazon River turtle, yellow-spotted river turtle (IUCN)
Podocnemis vogli	Savannen-Schienenschildkröte	savanna side-necked turtle, Orinoco turtle
Podocoryna carnea		smoothspined snailfur
Podomys floridanus	Florida-Maus	Florida mouse
Podops inuncta	Hakenwanze	European turtle-bug
Podoxymys roraimae	Mount Roraima Maus	Mount Roraima mouse
Podura aquatica	Schwarzer Wasserspringer	common black freshwater springtail*
Poecilia formosa	Amazonas-Kärpfling	Amazon molly
Poecilia heterandria/ Limia heterandria	Orangekärpfling	dwarf molly
Poecilia latipinna	Breitflossenkärpfling	sailfin molly
Poecilia melanogaster/ Limia melanogaster	Dreifarbiger Jamaikakärpfling	black-bellied molly, blue limia, blackbelly limia (FAO)
Poecilia nigrofasciata/ Limia nigrofasciata	Schwarzbandkärpfling	black-barred molly, blackbarred limia (FAO)
Poecilia ornata/Limia ornata	Schmuckkärpfling	ornate molly, ornate limia
Poecilia parae	Parakärpfling	twospot livebearer
Poecilia reticulata (Lebistes reticulatus)	Guppy, Millionenfisch	guppy
Poecilia sphenops/ Mollienesia sphenops	Spitzmaulkärpfling	marbled molly, molly (FAO)
Poecilia velifera	Riesen-Segelflossenkärpfling*	giant sailfin molly, sail-fin molly (FAO)
Poecilia vivipara	Augenfleckkärpfling	onespot livebearer
Poecilictis libyca	Streifenwiesel, Libysches Streifenwiesel	North African striped weasel, Libyan striped weasel
Poeciliidae	Lebendgebärende Zahnkarpfen	poeciliids
Poecilimon elegans	Kleine Buntschrecke	lesser bushcricket
Poecilimon ornatus	Südliche Buntschrecke	ornate bushcricket
Poeciliopsis gracilis	Seitenfleckkärpfling	porthole livebearer
Poeciliopsis occidentalis		gila topminnow
Poeciliopsis viriosa	Brauner Kärpfling	brown molly*, brown livebearer*
Poecilistes pleurospilus		portholefish
Poecilocampa populi	Kleine Pappelglucke	December moth
Poecilogale albinucha	Weißnackenwiesel	African striped weasel
Poecilonota variolosa	Espenprachtkäfer	
Poecilotheria fasciata	Sri Lanka Ornament-Vogelspinne	Sri Lankan ornamental tarantula
Poecilotheria ornata	Ornament-Baumvogelspinne	fringed ornamental tarantula
Poecilotheria regalis	Tiger-Vogelspinne, Indische Ornament-Vogelspinne	Indian ornamental tarantula
Poelagus marjorita	Buschkaninchen	Central African rabbit
Poephila guttata	Zebrafink	spotted-sided finch, zebra finch
Pogona barbatus/ Amphibolurus barbatus	Bartagame	bearded dragon
Pogonias cromis	Trommelfisch	black drum
Pogoniulus bilineatus	Goldbürzel-Bartvogel	yellow-rumped tinkerbird

Pogonocherus fasciculatus	Gemeiner Wimperbock, Kiefernzweigbock	conifer-wood longhorn beetle*
Pogonocherus hispidulus	Doppeldorniger Wimperbock	apple-wood longhorn beetle
Pogonocherus hispidus	Dorniger Wimperbock	
Pogonomelomys spp.	Rümmlers Mosaikschwanzratte	Rümmler's mosaic-tailed rats
Pogonomyrmex spp.	Ernteameisen	harvester ants
Pogonomys spp.	Greifschwanzratten*	prehensile-tailed rats
Pogonoperca punctata	Tüpfelseifenfisch	speckled soapfish, clown grouper
Pogonophora	Bartwürmer, Bartträger	beard worms, beard bearers, pogonophorans
Pogonus chalceus	Erzfarbener Bartträger	
Pogonus luridipennis	Heller Salzstellenläufer, Gelbbrauner Salzuferläufer	
Poiana richardsoni	Afrika-Linsang, Pojana	African linsang, oyan
Polemaetus bellicosus/ Hieraaetus bellicosus	Kampfadler	martial eagle
Polemon spp.	Schlangenfresser	snake-eaters
Polia bombycina		pale shining brown moth
Polia hepatica	Beerstrauch-Blättereule	
Polihierax semitorquatus	Halsband-Zwergfalke	African pygmy falcon
Polinices aurantius	Gold-Mondschnecke, Orangefarbene Mondschnecke	golden moonsnail
Polinices bifasciatus	Zweiband-Mondschnecke	two-banded moonsnail
Polinices catenus/ Lunatia catena	Gefleckte Halsband-Mondschnecke, Gefleckte Halsbandnabelschnecke	spotted necklace snail (spotted necklace shell), necklace moonsnail (European necklace shell)
Polinices grossularius	Senegalesische Mondschnecke	Senegal moonsnail
Polinices polianus/ Euspira poliana	Polis Halsbandnabelschnecke, Alders Halsbandnabelschnecke	Poli's necklace snail, Alder's necklace snail (Alder's necklace shell)
Polioptila caerulea	Blaumückenfänger	blue-grey gnatcatcher, blue-gray gnatcatcher
Polioptila melanura	Schwarzschwanz-Mückenfänger	black-tailed gnatcatcher
Polioptilidae	Mückenfänger	gnatcatchers, gnatwrens
Polistes gallicus	Feldwespe	paper wasp
Polistinae	Feldwespen	polistine wasps
Pollachius pollachius	Pollack, Heller Seelachs, Steinköhler	pollack (green pollack, pollack lythe)
Pollachius virens	Köhler, Seelachs, Blaufisch	saithe (FAO), pollock, Atlantic pollock, coley, coalfish
Pollenia rudis	Pollenia-Schmeißfliege	cluster fly
Pollichthys mauli	Sternauge*	stareye lightfish
Pollicipes polymerus	Blättrige Entenmuschel	leaf barnacle
Polybiinae	Schwimmkrabben u.a.	swimming crabs a.o.
Polybius henslowi	Henslow Schwimmkrabbe	Henslow's swimming crab, sardine swimming crab
Polyboroides typus	Höhlenweihe	African harrier hawk

Polyborus plancus	Karakara, Schopfkarakara	common caracara, crested caracara
Polycarpa aurata	Gold-Seescheide	golden sea-squirt
Polycarpa spongiabilis	Riesen-Seescheide	giant sea-squirt, giant tunicate
Polycera atra	Orangestachel-Hörnchenschnecke*	orange-spike polycera
Polycera quadrilineata	Gestreifte Hörnchenschnecke	four-striped polycera
Polyceridae	Hörnchenschnecken, Kopflappen-Sternschnecken	polyceras
Polychaeta	Vielborster, Borstenwürmer, Polychaeten	bristle worms, polychaetes, polychetes, polychete worms
Polychrosis botrana	Bekreuzter Traubenwickler	
Polychrus marmoratus	Marmorierter Buntleguan	many-colored bush anole
Polychrysia moneta	Münzen-Eule	golden plusia
Polycitor crystallinus	Pilz-Seescheide	fungal sea-squirt*, crystalline community sea-squirt
Polyclinidae	Krusten-Seescheiden	sea biscuits
Polyclinum aurantium/ Aplidium ficus	Meerfeige (Goldenes Meerbiskuit)	golden sea biscuit
Polyctenidae	Fledermauswanzen	bat bugs
Polydactylus virginicus	Atlantischer Fadenflosser	barbu
Polydesmidae	Bandfüßer	flat millepedes, flat-backed millepedes
Polydesmus angustus	Tüpfeldoppelfüßer	flat millepede, flat-backed millepede
Polydesmus complanatus	Abgeplatteter Bandfüßer	flat-backed millipede a.o.
Polydora ciliata	Gewöhnlicher Polydora-Wurm	common polydora worm
Polydora ligni	Holzbohrender Polydora-Wurm	polydora mud worm
Polyergus rufescens	Amazonenameise	robber ant, Amazon ant
Polygonia c-album/ Comma c-album/ Nymphalis c-album	C-Falter, Weißes C	comma
Polygonia egea/ Comma egea/ Nymphalis egea	Gelber C-Falter, Glaskrautfalter	southern comma
Polygraphus poligraphus	Doppeläugiger Fichtenbastkäfer	small spruce bark beetle
Polymastia mammillaris	Papillenschwamm	papillate sponge*, teat sponge
Polymastia robusta	Warzenschwamm*	nipple sponge
Polymixiidae	Barbudos, Bartfische	beardfishes
Polymixis xanthomista	Blaugraue Steineule	black-banded moth
Polymyces fragilis	Zwölfwurzelkoralle*	twelve-root cupcoral
Polynemidae	Federflosser, Fadenflosser	threadfins
Polynemus quadrifilis/ Polydactylus quadrifilis	Fingerfisch, Kapitänsfisch	threadfin, five-rayed threadfin, giant African threadfin (FAO)
Polynoidae	Schuppenwürmer u.a.	polynoids, scaleworms a.o.
Polyodon spathula	Löffelstör	paddlefish, Mississippi paddlefish (FAO)

Polyodontidae	Löffelstöre, Vielzähner, Schaufelrüssler	spoonbills, paddlefishes
Polyommatus amanda/ Agrodiaetus amanda/ Plebicula amanda	Prächtiger Bläuling	Amanda's blue (butterfly)
Polyommatus bellargus/ Lysandra bellargus	Himmelblauer Bläuling	adonis blue (butterfly)
Polyommatus coridon/ Meleageria coridon	Silbergrüner Bläuling	chalk-hill blue (butterfly)
Polyommatus damon/ Agrodiaetus damon	Großer Esparsetten-Bläuling, Grünblauer Bläuling	damon blue (butterfly)
Polyommatus daphnis/ Meleageria daphnis	Zahnflügel-Bläuling	Meleager's blue (butterfly)
Polyommatus icarus	Hauhechelbläuling, Wiesenbläuling, Gemeiner Bläuling	common blue (butterfly)
Polyommatus semiargus/ Cyaniris semiargus	Violetter Waldbläuling	mazarine blue (butterfly)
Polyorchis penicillatus	Pinselqualle*	penicillate jellyfish
Polyphagidae	Sandschaben*	sand cockroaches
Polyphagotarsonemus latus/ Tarsonemus latus/ Hemitarsonemus latus	Gelbe Teemilbe, Breitmilbe*	broad mite
Polyphemus pediculus	Kleiner Raubwasserfloh, Großäugiger Wasserfloh	big-eyed water flea*
Polyphylla fullo	Walker, Türkischer Maikäfer	fuller
Polyphyllia talpina	Zungenkoralle	lingual coral*
Polyplacidae	Nagetierläuse*	rodent sucking lice
Polyplacophora/ Placophora/ Loricata	Käferschnecken, Chitone	chitons, coat-of-mail shells
Polyplectana kefersteini	Braune Wurmseegurke	
Polyploca ridens	Moosgrüner Eulenspinner	frosted green (moth)
Polyprion americanus	Wrackbarsch, Atlantischer Wrackbarsch	wreckfish (FAO), stone bass
Polyprion oxygeneios	Hapuku-Wrackbarsch	Hapuku wreckfish
Polyprionidae	Wrackbarsche	wreckfishes
Polypteridae	Flösselhechte	bichirs
Polypteriformes	Flösselhechtverwandte	bichirs & reedfishes and allies
Polypterus bichir	Nil-Flösselhecht	Nile bichir
Polypterus ornatipennis	Schönflössler, Schmuckflossen-Flösselhecht	ornate bichir
Polypterus spp.	Flösselhechte	bichirs
Polysarcus denticauda/ Orphania denticauda	Wanstschrecke	large swordtailed bushcricket
Polysticta stelleri	Scheckente	Steller's eider
Polyxenus lagurus	Pinselfüßer	brushfoot millepede*
Polyzoniidae	Saugfüßer	polyzoniid millepedes
Pomacanthidae/ Pomacanthinae	Kaiserfische	angelfishes
Pomacanthus annularis	Ring-Kaiserfisch, Ringelkaiserfisch	bluering angelfish (FAO), blue-ringed angelfish
Pomacanthus arcuatus	Grauer Kaiserfisch	gray angelfish

Pomacanthus asfur/ *Euxiphipops asfur/* *Arusetta asfur*	Asfur-Kaiserfisch	Arabian angelfish
Pomacanthus chrysurus	Goldschwanz-Kaiserfisch	goldtail angelfish
Pomacanthus imperator	Großer Kaiserfisch, Imperator-Kaiserfisch, Nikobaren-Kaiserfisch	emperor angelfish (FAO), imperial angelfish
Pomacanthus maculosus	Arabischer Kaiserfisch, Sichelkaiserfisch	yellow-blotch angelfish, Red Sea angelfish, blue moon angelfish, bride of the sea, yellowbar angelfish (FAO)
Pomacanthus navarchus/ *Euxiphipops navarchus*	Traumkaiserfisch	blue-girdled angelfish
Pomacanthus paru	Franzosen-Kaiserfisch, Schwarzer Engelsfisch	French angelfish
Pomacanthus rhomboides/ *P. striatus*	Trapez-Kaiserfisch	old woman angelfish
Pomacanthus xanthometopon	Diademkaiserfisch, Gelbmaskenkaiserfisch, Blaukopfkaiserfisch	yellow-faced angelfish, blue-face angelfish, yellowface angelfish (FAO)
Pomacentridae	Korallenbarsche & Riffbarsche & Jungfernfische	damselfishes
Pomacentrus caerulens	Azurblaue Demoiselle	caerulean damsel (FAO), blue damselfish
Pomacentrus fuscus/ *Stegastes fuscus*	Blauer Teufel	Brazilian damsel (FAO), blue devil
Pomacentrus leucostictus/ *Stegastes leucostictus*	Beaugregory	beaugregory (yellow-belly)
Pomacentrus pavo	Gelbbauch	blue damsel, sapphire damsel (FAO)
Pomacentrus spp.	Demoisellen	damselfishes, demoiselles
Pomacentrus stigma	Einfleck-Jungferchen	blackspot damsel (FAO), analspot demoiselle
Pomacentrus sulfureus	Zitronengelbe Demoiselle	sulphur damsel (FAO), lemon demoiselle
Pomacentrus trilineatus	Dreilinien-Demoiselle	threeline damsel (FAO), three-lined demoiselle
Pomacentrus tripunctatus	Dreifleck-Demoiselle	threespot damsel (FAO)
Pomacentrus vaiuli	Prinzessjungferchen	princess damselfish, ocellate damselfish (FAO)
Pomacentrus variabilis	Schokoladen-Demoiselle*	cocoa damsel (FAO)
Pomadasyidae/Haemulidae	Grunzer, Grunzerfische, Schweinsfische	grunts
Pomadasys argenteus/ *Pomadasys hasta*	Silberne Süßlippe, Silbergrunzer	silver grunter, spotted javelin fish, grunter bream, lined silver grunt, silver grunt (FAO)
Pomadasys commersonnii	Brackwasser-Süßlippe, Gefleckte Süßlippe*	spotted grunter
Pomadasys incisus	Bastard-Süßlippe	bastard grunt
Pomadasys jubelini	Sompat-Süßlippe	Atlantic spotted grunt, sompat grunt (FAO)
Pomadasys striatum	Gestreifte Süßlippe	striped grunter
Pomadasys stridens	Streifen-Süßlippe	lined piggy

Pomatias elegans	Schöne Landdeckelschnecke	red-mouthed snail, round-mouthed snail
Pomatias spp.	Landdeckelschnecken	
Pomatiasidae	Landdeckelschnecken	
Pomatoceros lamarcki	Lamarck-Kielwurm	Lamarck's keelworm
Pomatoceros triqueter	Dreikantröhrenwurm, Dreikantwurm, Kielwurm	keelworm
Pomatomidae	Blaufische	bluefishes
Pomatomus saltator	Blaubarsch, Blaufisch	blue fish, bluefish (FAO), tailor, elf, elft
Pomatoschistus marmoratus	Marmorierte Grundel u.a.	marbled goby (FAO)
Pomatoschistus microps	Strandküling, Strandgrundel	common goby
Pomatoschistus minutus/ Gobius minutus	Sandküling, Sandgrundel	sand goby
Pomatoschistus norvegicus	Norwegische Grundel	Norway goby (FAO), Norwegian goby
Pomatoschistus pictus	Fleckenküling, Fleckengrundel	painted goby
Pomatostegus stellatus	Stern-Röhrenwurm*	star tubeworm
Pomatostomidae	Säbler, Weißbrauensäbler, Australische Scheitelsäbler	Australo-Papuan babblers
Pomolobus aestivalis	Blaurücken-Flusshering	blueback herring (alewife)
Pomolobus mediocris/ Alosa mediocris		hickory shad (alewife)
Pomolobus pseudoharengus/ Alosa pseudoharengus	Graurücken-Flusshering	grayback herring (alewife)
Pomolobus spp.	Nordamerikanische Flussheringe	alewifes
Pomoxis annularis	Weißer Crappie	white crappie
Pomoxis nigromaculatus/ Centrarchus hexacanthus	Schwarzer Crappie, Silberbarsch	black crappie
Pompilidae (Psammocharidae)	Wegwespen, Spinnentöter	pompilids, spider-hunting wasps, spider wasps
Pompilus plumbeus	Bleigraue Wegwespe	
Pompilus viaticus/ Anoplius viaticus/ Anoplius fuscus	Frühlings-Wegwespe	black-banded spider wasp
Ponentina subvirescens	Grüne Haarschnecke	green hairysnail, green snail
Ponera spp.	Stachelameisen, Urameisen	ponerine ants, primitive ants
Ponerinae	Stachelameisen, Urameisen	ponerine ants, primitive ants (bulldog ants)
Pongidae	Menschenaffen	great apes, pongids
Pongo pygmaeus	Orang-Utan	orangutan
Pontania proxima	Bruchweiden-Erbsengallen-Blattwespe*	willow bean-gall sawfly
Pontia callidice	Alpenweißling	peak white
Pontia daplidice	Reseda-Weißling, Resedafalter	bath white

Pontinus kuhlii	Kuhls Drachenkopf	Kuhl's scorpionfish
Pontobdella muricata	Rochenegel	ray leech*
Pontonia custos	Wächtergarnele (Muschelfreund)	
Pontophilus norvegicus	Norwegische Garnele	Norwegian shrimp
Pontoporia blainvillei	Franciscana, La Plata-Delphin	Franciscana dolphin, La Plata dolphin
Pooecetes gramineus	Abendammer	vesper sparrow
Popillia japonica	Japanischer Käfer	Japanese beetle
Porania pulvillus	Karminroter Kissenstern	red cushion star
Porcellana longicornis/ Pisidia longicornis	Langhorn-Porzellankrebs, Schwarzer Porzellankrebs, Schwarzes Porzellankrebschen	long-clawed porcelain crab, common porcelain crab, minute porcelain crab
Porcellana platycheles	Graues Porzellankrebschen, Grauer Porzellankrebs	broad-clawed porcelain crab, gray porcelain crab
Porcellanidae	Porzellankrebse	porcelain crabs
Porcellanopagurus spp.	Porzellan-Einsiedlerkrebse	porcelain hermit crabs
Porcellio scaber	Kellerassel, Raue Kellerassel	common rough woodlouse, garden woodlouse, slater, scabby sow bug
Porichthys notatus	Nördlicher Bootsmannsfisch	plainfin midshipman
Porichthys plectrodon	Atlantischer Bootsmannsfisch	Atlantic midshipman
Porites astreoides	Knotige Porenkoralle	mustard hill coral, knobby porous coral, yellow porous coral
Porites branneri	Blaue Krustenanemone	blue crust coral
Porites cylindrica	Zylindrische Porenkoralle	cylindrical finger coral, cylindrical porous coral
Porites porites	Fingerförmige Porenkoralle, Fingerkoralle	clubtip finger coral, clubbed finger coral, thick finger coral
Porites rus	Bergkoralle	mountain cupcoral*
Poroderma africanum/ Squalus africanus	Gestreifter Katzenhai	striped catshark
Porphyrio mantelli	Takahe	takahe
Porphyrio porphyrio	Purpurhuhn	purple gallinule
Porpita pacifica	Pazifische Floßqualle*	raft hydroid
Porpita porpita	Blauknopf*	blue button
Porpita umbella	Tellerqualle	umbel button jelly* (a raft hydroid)
Porthidium barbouri/ Bothrops barbouri	Barbours Lanzenotter	Barbour's pit viper
Porthidium godmani/ Bothrops godmani	Godmans Lanzenotter	Godman's pit viper
Porthidium lansbergi	Lansbergs Lanzenotter	Lansberg's hognose viper
Porthidium melanurum	Gehörnte Lanzenotter	black-tailed horned pit viper, blacktail montane viper
Porthidium nasutum/ Bothrops nasutus	Stülpnasen-Lanzenotter	horned hognose pit viper, rainforest hognose viper
Porthidium ophryomegas	Westliche Stülpnasen-Lanzenotter	western hog-nosed viper, slender hognose viper
Portumnus latipes	Pennants Schwimmkrabbe	Pennant's swimming crab
Portunidae	Schwimmkrabben u.a.	swimming crabs a.o.
Portunus anceps	Zierliche Schwimmkrabbe	delicate swimming crab

Portunus depressifrons	Flachbrauen-Schwimmkrabbe	flatface swimming crab, flat-browed crab
Portunus gibbesii	Schillernde Schwimmkrabbe	iridescent swimming crab
Portunus holsatus/ Liocarcinus holsatus/ Macropinus holsatus	Gemeine Schwimmkrabbe, Ruderkrabbe	common swimming crab
Portunus pelagicus	Blaukrabbe, Blaue Schwimmkrabbe, Große Pazifische Schwimmkrabbe	blue swimming crab, sand crab, pelagic swimming crab
Portunus sanguinolentus	Pazifische Rotpunkt-Schwimmkrabbe	blood-spotted swimming crab
Portunus sayi	Sargassum-Krabbe	sargassum swimming crab, sargassum crab
Portunus spinimanus	Flecken-Schwimmkrabbe	blotched swimming crab
Portunus trituberculatus	Gazami-Schwimmkrabbe	Gazami crab
Porzana carolina	Carolinasumpfhuhn	sora
Porzana parva	Kleines Sumpfhuhn	little crake
Porzana porzana	Tüpfelsumpfhuhn	spotted crake
Porzana pusilla	Zwergsumpfhuhn	Baillon's crake
Potamidae	Süßwasserkrabben	freshwater crabs
Potamididae	Teleskopschnecken*	telescope snails, mud whelks
Potamobius torrentium/ Austropotamobius torrentium	Steinkrebs	stone crayfish, torrent crayfish
Potamochoerus porcus	Buschschwein, Flussschwein, Pinselohrschwein	African bush pig, red river hog
Potamogalidae	Otterspitzmäuse	otter shrews
Potamon fluviatile	Gemeine Flusskrabbe, Gemeine Süßwasserkrabbe	Italian freshwater crab
Potamon ibericum	Iberische Süßwasserkrabbe	Bieberstein's freshwater crab
Potamon potamios	Ägäische Süßwasserkrabbe	Aegean freshwater crab
Potamopyrgus antipodarum/ Potamopyrgus jenkinsi	Neuseeländische Deckelschnecke, Jenkins Deckelschnecke	New Zealand spiresnail, Jenkins's spire snail, Jenkins' spiresnail (Jenkins' spire shell)
Potamotrygonidae	Südamerikanische Süßwasserstechrochen	river stingrays
Poterion neptuni	Neptunsbecher	Neptune's cup sponge
Potorous platyops	Breitkopf-Kaninchenkänguru	broad-faced potoroo
Potorous spp.	Kaninchenkängurus	potoroos
Potorous tridactylus	Langschnauzen-Kaninchenkänguru	long-nosed potoroo
Potos flavus	Wickelbär	kinkajou
Praomys spp.	Große Afrikanische Waldmäuse	African soft-furred rats
Prasinohaema virens	Lauchgrüner Skink	green-blood skink
Pratylenchus pratensis	Wiesenälchen	De Man's meadow nematode (brown rot of tobacco)
Pratylenchus spp.	Wurzelfäule-Älchen	root-lesion nematodes
Praunus flexuosus	Große Schwebegarnele	bent mysid shrimp
Premnas biaculeatus/ Amphiprion biaculeatus	Samtkorallenfisch, Samtanemonenfisch	spine-cheek anemonefish

Presbytis comata/ *Presbytis aygula*	Mützenlangur	grizzled leaf monkey, Javan leaf monkey, Sunda leaf monkey, Sunda Islands leaf monkey
Presbytis cristata	Haubenlangur	silvered leaf monkey
Presbytis entellus	Hanumanlangur, Hulman	hanuman langur
Presbytis francoisi	Tonkinlangur	Tonkin leaf monkey, François' leaf monkey
Presbytis frontata	Weißstirnlangur	white-fronted leaf monkey
Presbytis geei	Goldlangur	golden leaf monkey
Presbytis johnii	Nilgirilangur	Nilgiri langur
Presbytis melalophos	Roter Langur	banded leaf monkey
Presbytis obscura	Brillenlangur	dusky leaf monkey, spectacled leaf monkey
Presbytis phayrei	Phayres Langur	Phayre's leaf monkey
Presbytis pileata	Schopflangur	capped leaf monkey
Presbytis potenziani	Mentawailangur	Mentawai leaf monkey
Presbytis rubicunda	Maronenlangur	maroon leaf monkey
Presbytis spp.	Languren	leaf monkeys, langurs
Presbytis vetulus/ *Presbytis senex*	Weißbartlangur	purple-faced leaf monkey
Priacanthidae	Großaugenbarsche, Großaugen, Bullaugen, Catalufas	bigeyes (catalufas)
Priacanthus hamrur	Großaugenbarsch	lunar-tailed bullseye, crescent-tail bigeye, moontail bullseye (FAO)
Priacanthus macracanthus	Australischer Großaugenbarsch, Bullaugenbarsch	bull's eye perch, red bulleye, red bigeye (FAO)
Priapulida	Priapswürmer, Priapuliden	priapulans
Priapulus caudatus	Geschwänzter Keulenwurm	tailed priapulid worm
Primates/Primata	Primaten, Herrentiere	primates
Primovula carnea/ *Pseudosimnia carnea*	Rotes Vogelei, Rote Eischnecke	red dwarf ovula
Prinia gracilis	Streifenprinie	graceful warbler
Prinia maculosa	Fleckenprinie	karroo prinia
Priodontes maximus	Riesengürteltier	giant armadillo
Prionace glauca	Großer Blauhai, Menschenhai	blue shark
Prionailurus rubiginosus/ *Felis rubiginosus*	Rostkatze	rusty-spotted cat
Prionailurus viverrinus/ *Felis viverrinus*	Fischkatze	fishing cat
Prioniturus discurus	Spatelschwanzpapagei	blue-crowned racket-tailed parrot
Prionobrama filigera	Glasrotflosser	glass bloodfin
Prionodon linsang	Bänderlinsang	banded linsang
Prionodon pardicolor	Fleckenlinsang	spotted linsang
Prionomys batesi	Dollmans Baummaus	Dollman's tree mouse
Prionopinae	Brillenwürger	helmet-shrikes
Prionops plumata	Brillenwürger	long-crested helmet shrike
Prionotus tribulus	Großkopf-Knurrhahn	bighead searobin

Prionus coriarius	Sägebock, Gerberbock	prionus longhorn beetle (greater British longhorn)
Prionus imbricornis		tilehorned prionus
Prionychus ater	Mattschwarzer Pflanzenkäfer	
Pristella maxillaris/ Pristella riddlei	Sternflecksalmler, Stieglitzsalmler	X-ray tetra (FAO), X-ray fish, pristella
Pristidae	Sägerochen, Sägefische	sawfishes
Pristiformes	Sägerochen, Sägefische	sawfishes
Pristiophoridae	Sägehaie	sawsharks, saw sharks
Pristiophorus cirratus/ Squalus anisodon	Langnasen-Sägehai	longnose sawshark
Pristiphora abbreviata	Schwarze Birnenblattwespe	black pearleaf sawfly
Pristiphora abietina	Kleine Fichtenblattwespe	gregarious spruce sawfly
Pristiphora erichsonii	Große Lärchenblattwespe	large larch sawfly
Pristiphora laricis	Kleine Schwarze Lärchenblattwespe	small larch sawfly
Pristiphora pallipes	Schwarze Stachelbeerblattwespe	small gooseberry sawfly
Pristis microdon (Pristis perotteti/ Pristis zephyreus)	Langzahn-Sägerochen, Langzahn-Sägefisch	largetooth sawfish
Pristis pectinata	Gemeiner Sägefisch, Westlicher Sägefisch, Schmalzahn-Sägerochen	smalltooth sawfish (FAO), greater sawfish
Pristis pristis	Ostatlantischer Sägerochen, Sägefisch	sawfish, common sawfish (FAO)
Pristis zijsron	Langkamm-Sägefisch	longcomb sawfish
Pristonychus terricola/ Laemostenus terricola	Grotten-Dunkelkäfer, Kellerkäfer	cellar beetle*, European cellar beetle
Pristopsis leichhardti	Süßwasser-Sägefisch	freshwater sawfish
Proasellus cavaticus	Höhlenassel	cave isopod*
Proboscidea	Rüsseltiere	elephants & relatives
Probosciger aterrimus	Arakakadu	palm cockatoo
Probstmayria vivipara	Kleiner Pferdepfriemenschwanz	small pinworm (equine)
Procambarus clarkii	Louisiana-Sumpfkrebs, Louisiana-Flusskrebs, Roter Sumpfkrebs	Louisiana red crayfish, red swamp crayfish, Louisiana swamp crayfish, red crayfish
Procambarus gracilis	Prärie-Flusskrebs	prairie crayfish
Procapra gutturosa	Mongoleigazelle	zeren, Mongolian gazelle
Procapra picticaudata	Tibetgazelle	goa, Tibetan gazelle
Procapra przewalskii	Przewalski-Gazelle	Przewalski's gazelle
Procavia capensis	Klippschliefer	hyrax, rock dassie
Procellariidae	Sturmvögel	petrels, shearwaters
Procellariiformes	Röhrennasen	tubenoses, tube-nosed swimmers: albatrosses & shearwaters & petrels
Proceraea fasciata	Rot-Weiß-Blau-Wurm*	red-white-and-blue worm
Processidae	Nachtgarnelen*	night shrimps
Prochaetoderma raduliferum (Aplacophora)	Einfacher Doppelschildfuß	Mediterranean glistenworm a.o.

Prochilodus insignis/ *Semaprochilodus theraponura>* *Semaprochilodus insignis*	Nachtsalmler	night characid, kissing prochilodus
Procolobus kirkii	Sansibar-Rotstummelaffe	Zanzibar red colobus
Procolobus rufomitratus rufomitratus	Tana River-Rotstummelaffe	Tana River red colobus
Procolobus rufomitratus/ *Piliocolobus rufomitratus*	Ostafrikanischer Rotstummelaffe	eastern red colobus
Procolobus verus	Schopfstummelaffe, Grüner Stummelaffe	olive colobus
Procris statices/ *Adscita statices*	Grünwidderchen, Grasnelken-Widderchen	forester, common forester
Proctotrupidae (Serphidae)	Zehrwespen	proctotrupids
Procyon lotor	Nördlicher Waschbär	common raccoon
Procyon spp.	Waschbären	raccoons
Procyonidae	Waschbärenartige	raccoons & relatives
Prodoxidae	Yuccamotten	yucca moths
Proechimys spp.	Igelratten	spiny rats, casiraguas
Profelis temmincki/ *Felis temmincki*	Asiatische Goldkatze, Temminckkatze	Asian golden cat
Progne subis	Purpurschwalbe	purple martin
Prognichthys gibbifrons	Stupsnasen-Flugfisch*	bluntnose flyingfish
Promephales promelas		fathead minnow
Promeropidae/Promeropinae	Proteavögel	sugarbirds
Prometheomys schaposchnikowi	Prometheus-Maus, Prometheusmaus	long-clawed mole-vole
Promicrops lanceolatus/ *Epinephelus lanceolatus*	Schwarzer Zackenbarsch	giant grouper (FAO), giant seabass
Promops spp.	Kuppelgaumen-Bulldogg-Fledermäuse	domed-palate mastiff bats
Pronolagus spp.	Rotkaninchen	red rabbits, red rock rabbits
Prophalangopsidae	Prophalangopsiden	hump-winged crickets
Prophysaon andersoni	Genetzte Nacktschnecke	reticulated slug, British garden slug
Propilidium exiguum		cap limpet, curled limpet
Propithecus diadema	Diademsifaka	diadem sifaka, diademed sifaka
Propithecus spp.	Sifakas	sifakas
Propithecus tattersalli	Tattersall Sifaka, Goldkamm-Sifaka	golden-crowned sifaka, Tattersall's sifaka
Propithecus verreauxi	Larvensifaka	Verreaux's sifaka
Propustularia surinamensis	Surinam-Kauri	Suriname cowrie
Propylea quatuordecimpunctata	Vierzehnpunktiger Marienkäfer, Geballter Marienkäfer	fourteen-spot ladybird
Prosciurillus spp.	Celebes-Zwerghörnchen	Celebes dwarf squirrels
Proscylliidae	Kammzahn-Katzenhaie	finback catsharks
Proscyllium habereri	Kammzahn-Katzenhai	graceful catshark
Proserpinus proserpina	Nachtkerzenschwärmer	willowherb hawkmoth, Curzon's sphinx moth
Prosimii	Halbaffen	prosimians, lower primates

Prosopium cylindraceum	Rundfelchen*	round whitefish (FAO), menominee
Prosopium spp.	Rundfelchen*	round whitefishes
Prosorhochmus claparedi (Nemertini)	Lebendgebärender Schnurwurm	livebearing nemertean
Prostephanus truncatus	Großer Kornbohrer	larger grain borer
Prosthecereus spp.	Bandplanarien	
Prosthogonimus spp.	Eileiteregel*	oviduct flukes (of birds/poultry)
Prostoma graecense	Süßwasser-Schnurwurm	freshwater nemertean
Prostomidae	Prostomiden	jugular-horned beetles
Proteidae	Olme	mudpuppies & olms
Proteinus brachypterus	Gemeiner Plumpkurzflügler	
Proteles cristatus	Erdwolf	aardwolf
Proteopecten glaber	Glatte Kammmuschel	bald scallop, smooth scallop
Proterorhinus marmoratus/ Gobius marmoratus	Marmorierte Grundel	tubenose goby
Proteus anguinus	Grottenolm	European olm (blind salamander)
Protobranchiata	Kammkiemer, Fiederkiemer	protobranch bivalves
Protonotaria citrea	Zitronenwaldsänger	prothonotary warbler
Protopteridae	Afrikanische Lungenfische	African lungfishes
Protopterus aethiopicus	Ostafrikanischer Lungenfisch	East African lungfish
Protopterus annectens	Westafrikanischer Lungenfisch	West African lungfish
Protopterus spp.	Afrikanische Lungenfische	African lungfishes
Protoreaster nodosus	Knotiger Walzenstern	wartstar*
Protostomia	Urmundtiere, Urmünder, Erstmünder	protostomes
Protoxerus spp.	Ölpalmenhörnchen	oil palm squirrels, bar-tailed giant squirrels
Protoxerus stangeri	Afrikanisches Riesenhörnchen	African giant squirrel
Protozoa	Protozoen, Urtierchen, Urtiere, „Einzeller"	protozoans, "first animals"
Protula tubularia	Glatter Kalkröhrenwurm, Glattwandiger Kalkröhrenwurm	smooth tubeworm
Protulopsis intestinum/ Protula intestinum	Roter Kalkröhrenwurm, Blutroter Röhrenwurm	bloody tubeworm*, red calcareous tubeworm
Protura	Beintastler	proturans
Prunella collaris	Alpenbraunelle	alpine accentor
Prunella modularis	Heckenbraunelle	dunnock
Prunellidae	Braunellen	accentors, dunnock
Prunum cincta/ Marginella cincta	Ringelschnecke*, Ringel-Randschnecke*	encircled marginella
Prunum guttatum/ Marginella guttata	Weißflecken-Randschnecke*	white-spot marginella, white-spotted marginella
Prunum labiata/ Marginella labiata	Königliche Randschnecke*	royal marginella
Prunum prunum/ Marginella prunum	Pflaumenschnecke*, Pflaumen-Randschnecke*	plum marginella
Psalidopodidae	Scherenfußgarnelen*	scissorfoot shrimps
Psalidoprocne obscura	Scherenschwanzschwalbe	fantee saw-wing

Psalmopoeus cambridgei		Trinidad chevron tarantula
Psalmopoeus irminia	Venezuela Ornament-Vogelspinne	suntiger tarantula
Psaltriparus minimus	Buschmeise	common bushtit
Psammechinus microtuberculatus	Kletter-Seeigel, Kletterseeigel	
Psammechinus miliaris	Strandseeigel, Strand-Seeigel, Olivgrüner Strandigel	shore sea urchin, shore urchin, purple-tipped sea urchin
Psammobates geometrica	Geometrische Landschildkröte	geometric tortoise
Psammobates oculifera	Stachelrand-Landschildkröte	serrated star tortoise, toothed Cape tortoise
Psammobates spp.	Südafrikanische Landschildkröten	South African land tortoises
Psammobates tentoria tentoria	Höcker-Landschildkröte, Höckerschildkröte	African tent tortoise
Psammobatis spinosissima> Bathyraja spinosissima	Weißrochen, Pazifischer Weißrochen, Stachliger Tiefenrochen	white skate (FAO), spiny deep-water skate
Psammobia depressa/ Gari gari	Flache Sandmuschel, Große Flache Sandmuschel	flat sunsetclam, large sunsetclam
Psammobiidae	Sandmuscheln	sunsetclams
Psammodromus algirus	Algerischer Sandläufer	Algerian psammodromus
Psammodromus hispanicus	Spanischer Sandläufer	Spanish psammodromus
Psammodromus mucronata	Sinai Sandläufer	Sinai psammodromus
Psammodromus spp.	Sandläufer	sand lizards
Psammomys obesus	Schwere Sandrennmaus	fat sand rat
Psammophilus dorsalis	Indische Sandagame	Indian sand agama
Psammophilus spp.	Sandagamen	sand agamas
Psammophis schokari	Schlanke Sandrennnatter	Forskal's sand snake
Psammophis sibilans	Gestreifte Sandrennnatter	striped sand snake
Psammophis spp.	Sandrennnattern	sand snakes
Psarocolius decumanus	Krähenstirnvogel	crested oropendola
Pselaphidae	Palpenkäfer, Zwergkäfer	short-winged mold beetles
Pselaphognatha/Penicillata (Polyxenidae)	Pinselfüßer	pselaphognaths
Psenopsis anomala	Asiatischer Schwarzfisch	Pacific rudderfish
Psephenidae	Psepheniden, Wassermünzenkäfer	water-penny beetles
Psephotus chrysopterygius	Goldschultersittich	golden-shouldered parakeet, golden-winged parrot
Psephotus dissimilis	Hauben-Goldschultersittich	hooded parakeet, hooded parrot
Psephurus gladius	Chinesischer Schwertstör, Schwertstör	Chinese paddlefish
Psetta maxima/ Scophthalmus maximus	Steinbutt	turbot
Psettodes belcheri	Gefleckter Ebarme	spot-tail spiny turbot
Psettodes bennetti	Westafrikanischer Ebarme	spiny turbot
Psettodes erumei	Indopazifischer Ebarme	adalah, Indian halibut, Indian spiny turbot (FAO)
Psettodidae	Ebarmen, Hartstrahlenflunder	adalahs, psettodids, spiny flatfishes

Pseudacris crucifer/ *Hyla crucifer*	Wasserpfeifer, Quellpfeifer*, Quellen-Chorfrosch	spring peeper
Pseudacris spp.	Chorfrösche	chorus frogs
Pseudacris triseriata	Gestreifter Chorfrosch	striped chorus frog, western chorus frog
Pseudaletia unipuncta		armyworm
Pseudalopex gymnocercus	Pampasfuchs, Azarafuchs (Hbz. Provincia-Fuchs, Pampasfuchs)	Pampas fox
Pseudalopex spp.	Festland-Kampfüchse	South American foxes
Pseudalopex vetulus/ *Lycalopex vetulus*	Brasilianischer Kampfuchs (Brasilfuchs/Rio-Fuchs)	Brasilian fox, hoary fox
Pseudaspis cana	Maulwurfsnatter	mole snake, molslang
Pseudechis australis	Mulga-Schlange	Australian mulga snake
Pseudechis porphyriacus	Rotbauch-Schwarzotter	red-bellied black snake
Pseudechis spp.	Schwarzottern	Australian black snakes
Pseudemydura umbrina	Falsche Spitzkopfschildkröte	western swamp turtle
Pseudemys alabamensis	Alabama-Rotbauch- Schmuckschildkröte	Alabama red-bellied turtle
Pseudemys concinna *concinna*	Fluss-Hieroglyphen- Schmuckschildkröte	river cooter
Pseudemys concinna *hieroglyphica*	Hieroglyphen- Schmuckschildkröte	slider, hieroglyphic slider
Pseudemys floridana *floridana/* *Chrysemys floridana* *floridana*	Florida-Schmuckschildkröte	cooter, common cooter, coastal plain turtle, Florida cooter
Pseudemys rubriventris *nelsoni/* *Chrysemys nelsoni*	Florida-Zierschildkröte	Florida redbelly turtle, Florida red-bellied turtle
Pseudemys rubriventris *rubriventris*	Rotbauch-Schmuckschildkröte	red-bellied turtle, American red-bellied turtle
Pseudemys scripta scripta/ *Trachemys scripta scripta/* *Chrysemys scripta scripta*	Buchstaben-Schmuckschildkröte	slider, pond slider, yellow-bellied turtle
Pseudemys scripta elegans/ *Trachemys scripta elegans*	Rotwangen-Schmuckschildkröte	red-eared turtle, red-eared slider
Pseudemys terrapen/ *Trachemys terrapen*	Antillen-Schmuckschildkröte, Jamaika-Schmuckschildkröte	Jamaican slider
Pseudidae	Harlekinfrösche	harlequin frogs, pseudids
Pseudis minuta	Kleiner Schwimmfrosch*, Kleiner Harlekinfrosch	lesser swimming frog
Pseudis paradoxa	Schwimmfrosch*, Paradox-Harlekinfrosch, Großer Harlekinfrosch	paradox frog, swimming frog
Pseudoanodonta complanata	Abgeplattete Teichmuschel, Schmale Teichmuschel	compressed river mussel
Pseudobalistes fuscus	Blaustreifen-Drückerfisch, Tüpfel-Drückerfisch	rippled triggerfish, yellow-spotted triggerfish (FAO)
Pseudobranchus striatus	Zwergarmmolch, Gestreifter Zwergarmmolch	dwarf siren, mud siren
Pseudobufo subasper	Falsche Kröte	false toad
Pseudocaranx dentex	Neuseeland-Stachelmakrele	guelly jack, white kingfish, New Zealand trevally, white trevally (FAO)

Pseudocarcharias kamoharrai	Krokodilhai, Falscher Sandtiger	crocodile shark
Pseudocarcinus gigas	Australische Riesenkrabbe	Tasmanian giant crab
Pseudocerastes persicus	Westasiatische Hornviper	Persian false horned viper
Pseudochaenichthys georgianus	South-Georgia Eisfisch	South Georgia icefish
Pseudocheirus spp.	Ringelschwanz-Kletterbeutler	ring-tailed possums
Pseudochelidon sirintarae	Weißaugen-Trugschwalbe, Sirintaraschwalbe	white-eyed river martin
Pseudochermes fraxini	Eschenwolllaus	ash scale
Pseudochromidae	Zwergzackenbarsche	dottybacks
Pseudococcidae	Wollläuse, Schmierläuse	mealybugs
Pseudococcus citri	Zitronenschmierlaus, Citrus-Schmierlaus	common mealybug, citrus scale
Pseudococcus fragilis/ Pseudococcus calceolariae		citrophilus mealybug
Pseudococcus longispinus	Langdornige Schmierlaus, Mehlige Gewächshausschildlaus	longtailed mealybug
Pseudococcus maritimus	Süd-Schmierlaus	grape mealybug
Pseudococcus obscurus	Kalifornische Schmierlaus	Californian mealybug
Pseudocolochirus violaceus	Bunte Seegurke, Violette Seegurke	multicolored sea cucumber*, violet sea cucumber
Pseudocordylus microlepidotus	Falscher Kap-Gürtelschweif	leathery crag lizard
Pseudocordylus spp.	Falsche Gürtelschweife	crag lizards
Pseudocorynopoma doriae	Drachenflosser, Kehlkropfsalmler	dragon-finned characin, dragonfin tetra (FAO)
Pseudocrenilabrus multicolor	Vielfarbiger Maulbrüter, Kleiner Maulbrüter	dwarf Egyptian mouthbrooder
Pseudodoras niger/ Doras niger/ Oxydoras niger	Liniendornwels	ripsaw catfish (FAO), black-shielded catfish
Pseudoeurycea spp.	Mexiko-Salamander, Mexikosalamander	false brook salamanders
Pseudofusulus varians	Gedrungene Schließmundschnecke	
Pseudogekko smaragdinus/ Gekko smaragdinus	Smaragdgecko	Polillo false gecko
Pseudogobio rivularis/ Abbottina rivularis	Unechter Gründling	Chinese false gudgeon
Pseudogrammidae	Unechte Zwergbarsche	podges
Pseudohaje goldii	Waldkobra	Gold's tree cobra
Pseudohaje spp.	Waldkobras	tree cobras
Pseudohydromys spp.	Falsche Neuguinea-Wasserratte	New Guinean false water rats, New Guinean shrew-mice
Pseudoips fagana/ Bena prasinana/ Bena fagana/ Hylophila prasinana	Buchen-Kahnspinner, Kleiner Kahnspinner, Buchenwickler, Buchenkahneule, Jägerhütchen	green silverlines, scarce silverlines
Pseudois spp.	Blauschafe	blue sheep, bharal
Pseudolynchia canariensis	Taubenlausfliege	pigeon fly
Pseudomogoistes squamiger	Schuppige Heuschrecke	scaly cricket
Pseudomugilidae	Blauaugen	blue eyes, blueeyes

Pseudomys praeconis	Shark Bay Pseudomys-Maus	Shark Bay mouse, shaggy mouse, shaggy-haired mouse
Pseudomys spp.	Pseudomys-Mäuse	Australian native mice
Pseudonaja nuchalis	Westliche Braunschlange	western brown snake
Pseudonaja spp.	Australische Scheinkobras	venomous brown snakes
Pseudonaja textilis	Gewöhnliche Braunschlange	eastern brown snake
Pseudonereis variegata	Miesmuschel-Seeringelwurm	mussel worm
Pseudonestor xanthophrys	Papageischnabelgimpel	Maui parrotbill
Pseudonovibos spiralis	Südostasiatische Drehhornantilope	linh-duong, Southeast Asian screwhorn antilope*
Pseudophryne australis	Rotspitzen-Scheinkröte*, Australische Scheinkröte	red-crowned toadlet
Pseudophryne bibronii	Braune Scheinkröte*, Bibrons Scheinkröte	brown toadlet, Bibron's toadlet (IUCN)
Pseudophryne spp.	Australische Scheinkröten	crowned toadlets, Australian toad
Pseudophycis bacchus/ Physiculus bacchus	Neuseeland-Eisfisch	red codling (FAO), red cod, New Zealand red cod
Pseudopleuronectes americanus	Winterflunder, Amerikanische Winterflunder	winter flounder
Pseudopterogorgia spp.	Seefedern, Federgorgonien	sea plumes, feather gorgonian
Pseudorasbora parva	Blaubandbärbling	false harlequin
Pseudorca crassidens	Kleiner Schwertwal, Unechter Schwertwal, Kleiner Mörder	false killer whale, false pilot whale
Pseudorhombus cinnamoneus	Zimtflunder	cinnamon flounder
Pseudoryzomys spp.	Falsche Reisratten u.a., Ratos-do-Mato	ratos-do-mato
Pseudosciaena polyactis	Kleine Gelbcorvina	lesser yellow croaker
Pseudoscorpiones/Chelonethi	Afterskorpione, Pseudoskorpione	pseudoscorpions, false scorpions
Pseudosimnia adriatica/ Aperiovula adriatica/ Ovula adriatica	Weißes Vogelei, Weiße Eischnecke	Adriatic ovula
Pseudoterpna pruinata	Grüner Geißkleespanner	grass emerald
Pseudotheraphosa apophysis	Goliath-Vogelspinne	goliath pinkfoot tarantula
Pseudotolithus brachygnathus	Kurzkiefer-Umberfisch	law croaker
Pseudotolithus elongatus	Bobo-Umberfisch	bobo croaker
Pseudotolithus senegalensis	Senegal-Umberfisch	cassava croaker
Pseudotolithus typus	Langkopf-Umberfisch	longneck croaker
Pseudotriakidae	Falsche Katzenhaie, Falsche Marderhaie, Unechte Marderhaie	false catsharks
Pseudotriakis microdon	Falscher Katzenhai, Falscher Marderhai, Atlantischer Falscher Marderhai	false catshark (FAO), Atlantic false catshark
Pseudotriton montanus	Schlammsalamander	mud salamander
Pseudotriton ruber	Roter Wiesensalamander, Rotsalamander	red salamander
Pseudotropheus auratus/ Melanochromis auratus	Türkisgoldbarsch	golden mbuna (FAO), auratus
Pseudupeneus maculatus	Gefleckte Meerbarbe	spotted goatfish
Pseudupeneus prayensis	Westafrikanische Meerbarbe	West African goatfish

Pseustes poecilonotus	Zischnatter	puffing snake
Pseustes spp.	Zischnattern	puffing snakes
Psila rosae	Möhrenfliege, Rüeblifliege	carrot rust fly, carrot fly
Psilidae	Nacktfliegen	rust flies
Psilochorus simoni	Weinkellerspinne*	wine cellar spider
Psilopa petrolei/ Helaeomyia petrolei	Petroleumfliege	petroleum fly
Psilorhynchidae	Spindelschmerlen	mountain minnows
Psithyrus rupestris	Bergland-Schmarotzerhummel	hill cuckoo bee
Psithyrus spp.	Schmarotzerhummeln	cuckoo bees
Psittacidae	Papageien	parrots
Psittaciformes	Papageien	parrots & parakeets
Psittacula alexandri	Bartsittich, Rosenbrustsittich	moustached parakeet
Psittacula cyanocephala	Pflaumenkopfsittich	plum-headed parakeet
Psittacula echo	Mauritiussittich	Mauritius parakeet
Psittacula eupatria	Alexandersittich	Alexandrine parakeet
Psittacula krameri	Halsbandsittich	rose-ringed parakeet, ring-necked parakeet
Psittacula roseata	Rosenkopfsittich	blossom-headed parakeet
Psittacus erithacus	Graupapagei	grey parrot
Psocoptera/Copeognatha	Bücherläuse	booklice, psocids
Psophia crepitans	Grauflügel-Trompetervogel	common trumpeter
Psophiidae	Trompetervögel	trumpeters
Psophus stridulus	Rotflügelige Schnarrschrecke, Schnarrheuschrecke	rattle grasshopper
Psoroptidae	Räudemilben	mange mites, scab mites
Psyche casta	Gemeiner Sackträger	common bagworm
Psychidae	Sackträger, Sackspinner	bagworms, bagworm moths
Psychoda grisescens	Abortfliege	sewage fly
Psychodidae	Schmetterlingsmücken	mothflies, owl midges, sewage farm flies, waltzing midges (incl. sandflies)
Psychrolutidae	Fettgroppen	fatheads, flathead sculpins, tadpole sculpins
Psylla mali	Apfelsauger, Apfelblattsauger	apple sucker, apple psyllid
Psylla pyricula	Birnenblattsauger, Gelber Birnenblattsauger	pear sucker, pear psyllid
Psyllidae	Blattflöhe, Blattsauger	jumping plantlice, psyllids
Psylliodes affinis	Kartoffelerdfloh	potato flea-beetle
Psylliodes attenuata	Hopfenerdfloh, Hanferdfloh	hop flea beetle, European hop flea beetle
Psylliodes chrysocephala	Rapserdfloh	cabbage stem flea beetle, rape flea beetle
Psylloidea	Blattflöhe, Blattsauger	jumping plant lice, psyllids
Psyllophryne didactyla	Minikröte	Izecksohn's toad

Ptenopus garrulus	Pfeifgecko	barking gecko, talkative gecko, whistling gecko
Pteraster militaris	Klappenstern	winged seastar
Ptereleotrinae		dartfishes, firefishes
Ptereleotris evides	Schwarzschwanzgrundel	scissor-tailed goby, blackfin gudgeon, blackfin dartfish (FAO)
Pteria colymbus	Atlantische Flügelmuschel	Atlantic wing-oyster
Pteria hirundo	Vogelmuschel, Europäische Vogelmuschel, Flügelmuschel, Europäische Flügelmuschel	European wing-oyster (European wing shell)
Pteria longisquamosa	Schuppen-Flügelmuschel	scaly wing-oyster
Pteria penguin	Pinguin-Flügelmuschel	penguin wing-oyster
Pteria tortirostris	Gedrehte Flügelmuschel	twisted wing-oyster
Pteria vitrea	Gläserne Flügelmuschel	glassy wing-oyster
Pteridophora alberti	Wimpelträger	King of Saxony bird-of-paradise
Pteriidae	Vogelmuscheln, Perlmuscheln	wing oysters & pearl oysters
Pterinochilus murinus	Mombasa-Vogelspinne*	Mombasa golden starburst tarantula
Pternohyla dentata	Hochland-Grablaubfrosch	upland burrowing frog
Pternohyla fodiens	Tiefland-Grablaubfrosch	lowland burrowing frog
Pternohyla spp.	Grablaubfrösche	burrowing frogs
Pterobranchia	Flügelkiemer	pterobranchs
Pterocaesio chrysozona	Goldbandfüsilier	goldband fusilier
Pterocles alchata	Spießflughuhn	pin-tailed sandgrouse
Pterocles coronatus	Kronenflughuhn	crowned sandgrouse
Pterocles lichtensteinii	Wellenflughuhn	Lichtenstein's sandgrouse
Pterocles orientalis	Sandflughuhn	black-bellied sandgrouse
Pterocles senegallus	Tropfenflughuhn	spotted sandgrouse
Pteroclidae/Pteroclididae	Flughühner	sandgrouse
Pterocnemia pennata	Darwinstrauß, Kleiner Nandu	lesser rhea
Pteroctopus tetracirrhus	Vierhornkrake*	fourhorn octopus
Pterodoras granulosus		granulated catfish
Pterodoras niger see *Pseudodoras niger*		ripsaw catfish
Pterodroma cahow	Bermudasturmvogel	cahow
Pterodroma feae	Kapverdensturmvogel	gon-gon
Pterodroma madeira	Madeirasturmvogel	freira
Pteroglossus bitorquatus	Rotnackenarassari	red-necked araçari
Pterogorgia citrina	Gelbe Bänderhornkoralle, Flache Gorgonie, Seeblatt-Gorgonie	yellow sea whip, yellow ribbon
Pterogymnus lanarius	Panga-Meerbrasse	panga
Pteroidae/Pteroinae	Feuerfische, Korallenfische	lionfishes & turkeyfishes
Pteroides griseum	Graue Seefeder	gray sea-pen
Pterois antennata	Antennen-Rotfeuerfisch, Antennenfeuerfisch	spotfin lionfish, broadbarred firefish (FAO)
Pterois radiata	Strahlenfeuerfisch	radial firefish (FAO), longhorn firefish, clearfin turkeyfish

Pterois spp.	Rotfeuerfische	firefishes, turkeyfishes
Pterois volitans	Eigentlicher Rotfeuerfisch	red firefish, lionfish, devil firefish, fireworkfish, red lionfish (FAO)
Pteromylaeus bovinus/ Myliobatis bovina	Kuhrochen, Afrikanischer Adlerrochen	bull ray
Pteromys spp.	Altwelt-Gleithörnchen	Old World flying squirrels
Pteromys volans	Eurasisches Gleithörnchen	Eurasian flying squirrel
Pteromyscus pulverulentus	Ruß-Gleithörnchen	smoky flying squirrel
Pteronarcidae	Riesensteinfliegen	giant stoneflies, salmonflies
Pteronemobius heydenii	Sumpfgrille	marsh cricket
Pteronidea ribesii/ Nematus ribesii	Gelbe Stachelbeerblattwespe, Stachelbeerfliege	common gooseberry sawfly, gooseberry sawfly
Pteronotus davyi	Kleine Nacktrücken-Fledermaus	lesser naked-backed bat
Pteronotus parnelii	Schnurrbärtige Fledermaus	mustache bat, mustached bats
Pteronotus spp.	Nacktrücken-Fledermäuse	naked-backed bats
Pteronura brasiliensis	Riesenotter	giant otter
Pterophoridae	Federgeistchen, Federmotten	plume moths
Pterophorus monodactylus	Federmotte, Windling-Geistchen	common plume, sweet-potato plume moth
Pterophorus pentadactylus	Federgeistchen, Fünffedriges Geistchen, Weißes Geistchen, Schneeweiße Windenmotte	large white plume moth
Pterophyllum altum	Hoher Segelflosser	deep angelfish
Pterophyllum scalare	Segelflosser, Skalar	freshwater angelfish (FAO), longfin angel fish, black angelfish, scalare
Pteropoda	Flossenfüßer, Flügelschnecken	pteropods, seabutterflies
Pteropodidae	Flughunde, Flederhunde	Old World fruit bats
Pteropurpura trialata	Dreiflügelschnecke	three-winged murex
Pteropus giganteus	Indischer Riesenflughund	Indian flying fox
Pteropus hypomelanus	Insel-Flugfuchs	small flying fox, Island flying fox
Pteropus spp.	Eigentliche Flughunde, Flugfüchse	flying foxes
Pteropus vampyrus	Kalong, Fliegender Hund, Riesen-Flugfuchs	kalong, flying dog, large flying fox
Pterosauria	Flugsaurier	pterosaurs (extinct flying reptiles, winged reptiles)
Pteroscion peli	Boe-Umberfisch	boe drum
Pterostichus cupreus	Regen-Grabläufer*	rain beetle
Pterostichus madidus	Erdbeer-Grabläufer*	strawberry ground beetle
Pterostichus niger	Schwarzer Grabläufer	black cloaker
Pterostichus spp.	Grabläufer, Grabkäfer	cloakers

Pterostoma palpina	Palpenspinner, Schnauzenspinner	pale prominent
Pterothrissidae	Großflossengrätenfische	
Pterycombus brama	Silberbrachsen, Silberbrassen	silver pomfret, silver bramid, Atlantic fanfish (FAO)
Pterygota	Fluginsekten, Geflügelte Insekten	winged insects, pterygote insects
Pterynotus miyokoae	Miyoko-Stachelschnecke	Miyoko murex
Pterynotus pellucidus	Durchscheinende Stachelschnecke	pellucid murex
Pterynotus phyllopterus	Blattförmige Stachelschnecke	leafy-winged murex
Pterynotus pinnatus	Fiedrige Stachelschnecke	pinnate murex
Pthirus pubis/Phthirus pubis	Filzlaus, Schamlaus	crab louse, "crab", pubic louse
Ptilichthys goodei		quillfish
Ptiliidae	Federflügler & Haarflügler (Zwergkäfer)	feather-winged beetles
Ptilocercus lowii	Federschwanz	pen-tailed tree shrew, feather-tailed tree shrew
Ptilodactylidae	Ptilodactyliden	toed-winged beetles
Ptilodon capucina	Kamelspinner	coxcomb prominent
Ptilonorhynchidae	Laubenvögel	bowerbirds
Ptilonorhynchus violaceus	Seidenlaubenvogel	satin bowerbird
Ptilopachus petrosus	Felsenhenne	stone partridge
Ptilophora plumigera	Frostspinner, Haarschuppenspinner	plumed prominent
Ptilornis magnificus	Prachtparadiesvogel	magnificent riflebird
Ptilorrhoa castanonota	Buntflöter	mid-mountain rail babbler
Ptilostomus afer	Piapia	piapiac
Ptinidae	Diebkäfer, Diebskäfer	spider beetles
Ptinus claviceps	Brauner Diebkäfer	brown spider beetle
Ptinus fur	Kräuterdieb	white-marbled spider beetle, white-marked spider beetle
Ptinus tectus	Australischer Diebkäfer	Australian spider beetle
Ptinus villiger	Behaarter Diebkäfer	hairy spider beetle
Ptosima flavoguttata	Variabler Prachtkäfer	
Ptyas korros	Gelbbäuchige Rattenschlange	yellow-bellied rat snake
Ptyas mucosus	Dhaman, Indische Rattenschlange	dhaman, common rat snake, Oriental rat snake
Ptyas spp.	Asiatische Rattenschlangen	Oriental rat snakes
Ptychadena mascareniensis	Nilfrosch	Mascarene grassland frog
Ptychadena spp.	Graslandfrösche	grassland frogs
Ptychocheilus lucius	Colorado-Squawfisch	Colorado squawfish
Ptychozoon spp.	Faltengeckos	flying geckos
Ptyodactylus hasselquistii	Gelber Fächerfußgecko	fan-toed gecko, yellow fan-fingered gecko
Ptyodactylus spp.	Fächerfußgeckos	fan-toed geckos, fan-fingered geckos
Ptyonoprogne fuligula	Steinschwalbe	rock martin
Ptyonoprogne rupestris	Felsenschwalbe	crag martin
Pudu spp.	Pudus	pudus

Puffinus assimilis	Kleiner Sturmtaucher	little shearwater
Puffinus gravis	Großer Sturmtaucher	great shearwater
Puffinus griseus	Dunkler Sturmtaucher	sooty shearwater
Puffinus puffinus	Schwarzschnabel-Sturmtaucher	Manx shearwater
Puffinus tenuirostris	Kurzschwanz-Sturmtaucher	short-tailed shearwater
Puffinus yelkouan	Mittelmeer-Sturmtaucher	Yelkouan shearwater
Pugilina cochlidium	Gewundene Treppenschnecke	spiral melongena, winding stair shell
Pugilina colossea/ Hemifusus colosseus	Riesen-Treppenschnecke	colossal false fusus, giant stair shell
Pugilina morio	Riesen-Kronenschnecke	giant hairy melongena, giant melongena
Pulex irritans	Menschenfloh	human flea
Pulicidae	Menschenflöhe	common fleas, pulicid fleas
Pulmonata	Lungenschnecken, Pulmonaten	pulmonate snails (freshwater & land snails and slugs)
Pulsatrix perspicillata	Brillenkauz	spectacled owl
Punctum pygmaeum	Punktschnecke	pygmy snail, dwarf snail
Puncturella noachina	Noahs Schlüssellochschnecke	diluvian puncturella, Noah's keyhole limpet
Pungitius pungitius	Neunstachliger Stichling, Zwergstichling	nine-spined stickleback
Punomys lemminus	Punamaus	Puna mouse
Puntazzo puntazzo/ Diplodus puntazzo	Spitzbrassen	sharp-snouted seabream, sheepshead bream, sharpsnout seabream (FAO)
Puntius arulius	Prachtglanzbarbe, Dreibandbarbe	longfin barb, arulius barb (FAO)
Puntius binotatus	Zweipunktbarbe	spotted barb
Puntius conchonius (Barbus conchonius)	Prachtbarbe	rosy barb
Puntius everetti	Clownbarbe, Everetts Barbe	clown barb
Puntius fasciatus/ Barbus fasciatus> Puntius eugrammus	Glühkohlenbarbe, Bandbarbe	striped barb (FAO), zebra barb, African banded barb
Puntius lateristriga (Barbus lateristriga)	Schwarzbandbarbe, Seitenstrichbarbe	T-barb, spanner barb (FAO)
Puntius nigrofasciatus (Barbus nigrofasciatus)	Purpurkopfbarbe	black ruby barb
Puntius oligolepis (Barbus oligolepis)	Eilandbarbe	checkered barb (FAO), checkerboard, island barb
Puntius pentazona	Fünfbandbarbe	fiveband barb
Puntius partipentazona (Barbus tetrazona partipentazona)	Teilgürtelbarbe	banded barb
Puntius sachsii	Goldflossenbarbe	goldfinned barb (FAO), golden barb
Puntius schwanenfeldi/ Barbodes schwanenfeldii	Schwanenfelds Barbe, Brassenbarbe	tinfoil barb, goldfoil barb (FAO)
Puntius spp.	Zierbarben	barbs
Puntius terio/ Barbus terio	Goldfleckbarbe	onespot barb

Puntius tetrazona (*Barbus tetrazona*)	Sumatrabarbe, Viergürtelbarbe	Sumatra barb (FAO), tiger barb
Puntius titteya (*Barbus titteya*)	Bitterlingsbarbe	cherry barb
Puntius viviparus/ Barbus viviparus	Schneiderbarbe	bowstripe barb
Pupa solidula/ Solidula solidula	Derbe Puppenschnecke	solid pupa
Pupa sulcata/Solidula sulcata	Gefurchte Puppenschnecke	sulcate pupa
Puperita pupa	Zebra-Nixenschnecke	zebra nerite
Pupilla bigranata	Zweizähnige Puppenschnecke	bidentate moss snail
Pupilla muscorum	Moos-Puppenschnecke	moss snail, moss chrysalis snail, widespread column snail (U.S.)
Pupilla spp.	Puppenschnecken	moss snails, column snails (U.S.)
Pupilla sterrii	Gestreifte Puppenschnecke	striped moss snail
Pupillacea spp.	Tönnchenschnecken	
Pupillidae	Puppenschnecken	moss snails (column snails, snaggletooths, vertigos)
Purpura nodosa	Kuglige Purpurschnecke	nodose purpura
Purpura patula	Weitmäulige Purpurschnecke	wide-mouthed purpura
Purpura planospira	Flachspindel-Purpurschnecke*	eye of Judas purpura
Purpuricenus kaehleri	Blutbock, Kahler Purpurbock	
Pusia tricolor	Buntschnecke	
Pusula californiana	Kalifornische Kaurischnecke	California trivia
Pycnodonta folium/ Lopha folium/ Ostrea folium	Blattauster	leaf oyster
Pycnodonta frons/ Lopha frons/ Ostrea frons/ Dendrostrea frons	Klammerauster	coon oyster
Pycnogaster inermis		unarmed bushcricket
Pycnogonida/Pantopoda	Asselspinnen	pycnogonids, pantopods, sea spiders
Pycnonotidae	Haarvögel, Bülbüls	bulbuls, greenbuls
Pycnonotus barbatus	Graubülbül	garden bulbul
Pycnonotus jocosus	Rotohrbülbül	red-whiskered bulbul
Pycnonotus leucogenys	Weißohrbülbül	white-cheeked bulbul
Pycnonotus striatus	Streifenbülbül	striated green bulbul
Pycnonotus xanthopygos	Gelbsteißbülbül	yellow-vented bulbul
Pycnopodia helianthoides	Sonnenblumenstern	sunflower star
Pycnoscelis surinamensis	Gewächshausschabe, Surinamschabe	Surinam cockroach
Pyemotes herfsi	Mottenmilbe	moth mite*
Pyemotes tritici/ Pyemotes ventricosus	Heumilbe, Kugelbauchmilbe	hay itch mite, grain itch mite, grain mite, straw itch mite
Pygathrix avunculus	Tonkinstumpfnase	Tonkin snub-nosed monkey

Pygathrix brelichi	Weißmantelstumpfnase	Brelich's snub-nosed monkey, Guizhou snub-nosed monkey
Pygathrix nemaeus	Rotschenkel-Kleideraffe	red-shanked douc langur
Pygathrix nigripes	Schwarzschenkel-Kleideraffe	black-shanked douc langur
Pygathrix roxellanae/ Rhinopithecus roxellanae	Goldstumpfnase	Chinese snub-nosed monkey, golden monkey, snow monkey, golden snub-nosed langur, Sichuan snub-nosed langur
Pygeretmus spp.	Fettschwanz-Springmäuse	fat-tailed jerboas
Pygoderma bilabiatum	Ipanema-Fledermaus	Ipanema bat
Pygoplites diacanthus	Pfauenkaiserfisch, Herzogen-Kaiserfisch	royal angelfish (FAO), blue-banded angelfish, regal angelfish
Pygopodidae	Flossenfüße`	snake lizards, flap-footed lizards, scaly-footed lizards
Pygopus nigriceps	Westlicher Flossenfuß	western scalyfoot
Pygoscelis adeliae	Adeliepinguin	Adelie penguin
Pygoscelis antarctica	Kehlstreifpinguin	bearded penguin
Pygoscelis papua	Eselpinguin	Gentoo penguin
Pygospio elegans	Pygospio-Wurm	pygospio worm
Pygospio filiformis	Fadenförmiger Pygospio-Wurm (ein Sandröhrenwurm)	filiform pygospio
Pylodictis olivaris	Flachkopf-Katzenwels	flathead catfish
Pyralidae	Zünsler	pyralid moths, pyralids (snout moths, grass moths & others)
Pyralis farinalis	Mehlzünsler	meal moth
Pyramidella dolabrata	Beilpyramide	giant Atlantic pyrum
Pyramidula rupestris	Pyramidenschnecke	rock snail
Pyramidulidae	Pyramidenschnecken	rock snails
Pyrene flava	Gelbes Täubchen	yellow dovesnail
Pyrene ocellata	Augenfleck-Täubchen	lightning dovesnail
Pyrene phiippinarum	Philippinisches Täubchen	Philippine dovesnail
Pyrene punctata	Teleskop-Täubchen	punctate dovesnail, telescoped dovesnail
Pyrene scripta	Schrifttäubchen, Getupftes Täubchen	dotted dovesnail, music dovesnail
Pyrenidae/Columbellidae	Taubenschnecken, Täubchenschnecken	dove snails (dove shells)
Pyrgus alveus	Halbwürfelfleckfalter	large grizzled skipper
Pyrgus armoricanus	Oberthürs Würfelfleckfalter	Oberthür's grizzled skipper
Pyrgus malvae	Malven-Würfelfleckfalter, Kleiner Malvendickkopf, Kleiner Würfeldickkopffalter	grizzled skipper
Pyrgus serratulae	Schwarzbrauner Würfelfalter	olive skipper
Pyrocephalus rubinus	Purpurtyrann	Vermilion flycatcher
Pyrochroa coccinea	Scharlachroter Feuerkäfer (Feuerfliege)	scarlet fire beetle, cardinal beetle
Pyrochroidae	Feuerkäfer (Feuerfliegen, Kardinäle)	fire beetles, fire-colored beetles, cardinal beetles
Pyroglyphidae	Hausstaubmilben	house dust mites

Pyronia tithonus/ *Maniola tithonus*	Braungerändertes Ochsenauge, Rostbraunes Ochsenauge	gatekeeper, hedge brown
Pyrosoma atlanticum	Atlantische Feuerwalze	Atlantic pyrosome, Atlantic fire salp
Pyrosomida	Feuerwalzen	pyrosomes
Pyrrhalta viburni	Schneeballkäfer	viburnum beetle
Pyrrhidium sanguineum	Rothaarbock	scarlet-coated longhorn beetle
Pyrrhocorax graculus	Alpendohle	Alpine chough
Pyrrhocorax pyrrhocorax	Alpenkrähe	red-billed chough
Pyrrhocoridae	Feuerwanzen	red bugs & stainers, pyrrhocorid bugs, pyrrhocores
Pyrrhocoris apterus	Gemeine Feuerwanze	firebug
Pyrrhosoma nymphula	Frühe Adonislibelle	large red damselfly
Pyrrhula pyrrhula	Gimpel	bullfinch, Eurasian bullfinch
Pyrrhura frontalis	Braunohrsittich	maroon-bellied conure
Pyrrhura molinae	Molinasittich	green-cheeked conure
Pyrrhura perlata	Blausteißsittich	pearly conure
Pythia scarabaeus	Gemeine Ohrenschnecke	common pythia
Python anchietae	Angola-Python	Angolan python
Python curtus	Buntpython	short-tailed python, blood python
Python molurus	Tigerpython	Burmese python, Indian python (IUCN)
Python oenpelliensis/ *Morelia oenpelliensis*	Oenpelli-Python	Oenpelli python
Python regius	Königspython	ball python, royal python
Python reticulatus	Netzpython	reticulated python (diamond python, Java rock python)
Python sebae	Felsenpython	African python, water python, African rock python
Python timorensis	Timor-Python	Timor python
Pytilia melba	Buntastrild	green-winged pytilia
Pyxicephalus adspersus	Südafrikanischer Ochsenfrosch	Tschudi's African bullfrog
Pyxidea mouhotii/ *Cyclemys mouhoti*	Dreikiel-Scharnierschildkröte, Indische Dornschildkröte	Indian keel-backed terrapin, keeled box turtle, Indian leaf turtle, jagged-shell turtle
Pyxis arachnoides *arachnoides*	Spinnenschildkröte, Madagassische Spinnenschildkröte	spider tortoise, Madagascan spider tortoise

Quadraspidiotus ostreaeformis	Austernförmige Schildlaus, Zitronenfarbene Austernschildlaus	European fruit scale
Quadraspidiotus perniciosus	San-José-Schildlaus, Kalifornische Schildlaus	San Jose scale
Quelea quelea	Blutschnabelweber	red-billed quelea
Quercusia quercus/ Thecla quercus	Blauer Eichenzipfelfalter	purple hairstreak
Quiscalus major	Bootschwanzgrackel	boat-tailed grackle
Quiscalus mexicanus	Dohlengrackel	great-tailed grackle
Quiscalus quiscala	Purpurgrackel	common grackle

Rachycentron canadum	Königsbarsch	cobia (prodigal son)
Radianthus ritteri/ *Heteractis magnifica*	Rosa Riesenanemone	purple-base anemone
Radiicephalidae	Strahlenköpfe	inkfishes
Radiicephalus elongatus	Langschwanz*	tapertail
Radiolaria	Radiolarien, Strahlentierchen	radiolarians
Radix auricularia	Ohrschlammschnecke, Ohrförmige Schlammschnecke	big-eared radix
Radix ovata	Eiförmige Schlammschnecke	ovate pondsnail (pond mud-shell)
Radix peregra	Gemeine Schlammschnecke, Wandernde Schlammschnecke	wandering pondsnail
Rafetus euphraticus/ *Trionyx euphraticus*	Euphrat-Weichschildkröte	Euphrates softshell turtle
Rafetus swinhoei/ *Trionyx swinhoei*	Shanghai-Weichschildkröte	Swinhoe's softshell turtle
Raja alba/ *Rostroraja alba*	Südlicher Weißrochen, Ostatlantischer Weißrochen, Spitzrochen, Bandrochen	white skate (FAO), bottle-nosed skate
Raja asterias	Mittelmeer-Sternrochen	starry ray
Raja batis/Dipturus batis	Europäischer Glattrochen, Spiegelrochen	common skate, grey skate, common European skate, blue skate, skate (FAO)
Raja binoculata	Großer Pazifikrochen	big skate
Raja brachyura	Blonde, Kurzschwanz-Rochen	blonde ray (FAO), blond ray
Raja circularis	Sandrochen	sandy ray
Raja clavata	Nagelrochen, Keulenrochen	thornback skate, thornback ray (FAO), roker
Raja eglanteria	Glasnasenrochen*	clearnose skate
Raja erinacea/ *Leucoraja erinacea*	Kleiner Igelrochen	little skate (FAO), hedgehog skate
Raja fullonica	Chagrinrochen, Walkerrochen	shagreen ray
Raja fyllae/Rajella fyllae	Fyllas Rochen	round skate
Raja hyperborea/ *Amblyraja hyperborea*	Arktischer Rochen	Arctic skate
Raja inornata	Kalifornischer Rochen	California ray
Raja laevis/Dipturus laevis	Westatlantischer Glattrochen, Spitznasen-Rochen	barndoor skate (FAO), sharpnose skate
Raja lintea/Dipturus linteus	Nordatlantischer Weißrochen, Nördlicher Weißrochen	sailray (FAO), sharpnose skate
Raja microocellata	Kleinäugiger Rochen, Hellfleckiger Rochen	small-eyed ray (FAO), painted ray
Raja miraletus	Braunrochen, Vieraugenrochen, Augenfleckrochen	brown ray
Raja montagui	Fleckrochen, Fleckenrochen, Gefleckter Rochen	spotted ray
Raja naevus	Kuckucksrochen	cuckoo ray (FAO), butterfly skate

Raja

Raja nasuta/ *Dipturus nasutus*	Raurochen, Australo-Asiatischer Raurochen	rough skate (Australasian)
Raja ocellata/ *Leucoraja ocellata*	Winterrochen, Westatlantischer Winterrochen	winter skate
Raja oxyrhynchus/ *Raja oxyrinchus/* *Dipturus oxyrinchus*	Ostatlantischer Spitzrochen	longnosed skate (FAO), long-nosed skate
Raja radiata/ *Amblyraja radiata*	Sternrochen, Nordatlantischer Sternrochen, Atlantischer Sternrochen	starry ray, thorny skate (FAO)
Raja radula	Mittelmeer-Raurochen	Mediterranean rough ray, rough ray (FAO)
Raja senta/Malacoraja senta	Glattrochen	smooth skate
Raja stellulata	Pazifischer Rochen	thorny skate, starry skate (FAO)
Raja straeleni	Brotlaibrochen*	biscuit skate
Raja texana	Texasrochen	roundel skate
Raja undulata	Ostatlantische Marmorrochen, Scheckenrochen, Bänderrochen	undulate ray (FAO), painted ray
Raja wallacei	Gelbfleckenrochen	yellowspotted skate
Rajella fyllae/Raja fyllae	Fyllas Rochen	round skate
Rajidae	Rochen, Echte Rochen	rays & skates
Rajiformes	Rochenartige	skates & guitarfishes
Rajoidei	Echte Rochen	skates
Rallidae	Rallen	rails, gallinules, coots
Rallus aquaticus	Wasserralle	water rail
Ramphastidae	Tukane	toucans
Ramphastos sulfuratus	Fischertukan	keel-billed toucan
Ramphastos toco	Riesentukan	toco toucan
Ramphodon dohrnii/ *Glaucis dohrnii*	Bronzeschwanz-Eremit	hook-billed hermit
Ramphodon naevius	Sägeschnabel-Eremit	saw-billed hermit
Ramphotyphlops braminus	Blumentopfschlange	braminy blind snake
Ramphotyphlops spp.		common blind snakes
Rana areolata	Krebsfrosch	crawfish frog
Rana arvalis	Moorfrosch	moor frog
Rana aurora	Rotschenkelfrosch*	red-legged frog
Rana blythi/ *Limnonectes blythi*	Uferfrosch	Blyth's wart frog
Rana cancrivora	Philippinenfrosch	Philippine frog
Rana capito	Maulwurffrosch*	gopher frog
Rana catesbeiana	Ochsenfrosch	bullfrog, American bullfrog
Rana clamitans	Schreifrosch	green frog, common spring frog
Rana dalmatina	Springfrosch	agile frog, spring frog
Rana erythraea	Rotohrfrosch	red-eared frog
Rana esculenta (*Rana* kl. *esculenta*)	„Wasserfrosch", Teichfrosch	European edible frog
Rana graeca	Griechischer Frosch	stream frog, Greek frog

Rana grylio	Schweinsfrosch	pig frog
Rana holsti	Dolchfrosch	Okinawa Shima frog
Rana iberica	Spanischer Frosch	Iberian frog, Spanish frog
Rana latastei	Italienischer Springfrosch	Italian agile frog
Rana lessonae	Kleiner Teichfrosch, Kleiner Grünfrosch, Kleiner Wasserfrosch, Tümpelfrosch	pool frog (little waterfrog)
Rana limnocharis/ Limnonectes limnocharis	Südostasiatischer Warzenfrosch, Südostasiatischer Reisfrosch	Boie's wart frog
Rana macrocnemis	Kleinasiatischer Braunfrosch	Brusa frog
Rana maculata/ Rana macroglossa	Gefleckter Hochlandfrosch	highland frog
Rana onca	Relikt-Leopardenfrosch*	relict leopard frog
Rana palustris	Sumpffrosch	pickerel frog
Rana perezi/ Rana ridibunda perezi	Iberischer Wasserfrosch	Coruna frog
Rana pipiens	Leopardfrosch	leopard frog, northern leopard frog, grass frog
Rana pretiosa	Fleckenfrosch	spotted frog
Rana ridibunda	Seefrosch	marsh frog, lake frog
Rana sylvatica	Waldfrosch	wood frog
Rana temporaria	Grasfrosch	common frog, grass frog
Rana vertebralis	Eisfrosch	ice frog, water frog (IUCN)
Rana virgatipes	Zimmermannsfrosch*	carpenter frog
Ranatra linearis	Stabwanze, Wassernadel	water stick insect
Ranella olearium/ Ranella gigantea	Wandernde Tritonschnecke, Großes Argushorn, Krötenschnecke	wandering triton
Ranellidae/Cymatiidae	Tritonen	tritons, rock whelks
Rangifer tarandus	Rentier, Ren	caribou (North America), reindeer (Europe)
Raniceps raninus	Froschquappe	tadpole fish (FAO), tadpole cod
Ranidae	Echte Frösche	true frogs
Ranina ranina	Froschkrabbe	spanner crab
Raninidae	Froschkrabben	frog crabs
Ranodon sibiricus	Sibirischer Froschzahnmolch	Siberian salamander, Siberian land salamander
Ranzania laevis (Ranzania truncata/ Ranzania typus)	Schlanker Mondfisch, Stutzmondfisch, Schwimmender Kopf, Langer Mondfisch	sunfish, truncated sunfish, slender sunfish (FAO)
Rapa rapa	Rettichschnecke	bubble turnip, papery rapa
Rapana bezoar	Bezoarschnecke	bezoar rapa whelk
Rapana venosa	Geäderte Rapana, Rotmund-Wellhorn*	veined rapa whelk, Thomas's rapa whelk

Raphicerus campestris	Steinböckchen	steenbok
Raphicerus melanotis	Greisbock	Cape grysbok
Raphicerus sharpei	Sharpe-Greisbock	Sharpe's grysbok
Raphidae	Drontevögel	dodos & solitaires
Raphidiidae	Kamelhalsfliegen	snakeflies
Raphidioptera	Kamelhalsfliegen	snakeflies
Raphus cucullaria[†]	Dronte	dodo
Raptiformica sanguinea	Blutrote Raubameise	
Rasbora borapetensis	Rotschwanzbärbling	redtailed rasbora, blackline rasbora (FAO)
Rasbora dorsiocellata dorsiocellata	Augenfleckbärbling	eyespot rasbora
Rasbora elegans	Schmuckbärbling	twospot rasbora
Rasbora hengeli	Hengels Bärbling	slender harlequin fish, glowlight rasbora (FAO)
Rasbora heteromorpha	Keilfleckbärbling, Keilfleck-Bärbling	harlequin fish, harlequin rasbora (FAO)
Rasbora kalochroma	Schönflossenbärbling	clown rasbora
Rasbora lateristriata	Seitenstrichbärbling	elegant rasbora, yellow rasbora (FAO)
Rasbora maculata	Zwergbärbling	dwarf rasbora (FAO), spotted rasbora
Rasbora pauciperforata	Rotstreifenbärbling	red-striped rasbora, red-line rasbora
Rasbora rasbora	Gangesbärbling	Ganges rasbora, Gangetic scissortail rasbora (FAO)
Rasbora trilineata	Glasbärbling, Glasrasbora	scissors-tail rasbora, three-lined rasbora (FAO)
Rasbora vaterifloris	Perlmuttbärbling	pearly rasbora
Rastrelliger brachysoma	Kurze Zwergmakrele	short mackerel
Rastrelliger kanagurta	Indische Makrele, Indische Zwergmakrele	Indian mackerel
Rastrineobola argentea	Victoria-See-Karpfen	silver cyprinid
Ratitae	Flachbrustvögel, Ratiten	ratite birds (flightless birds)
Rattus argentiventer	Reisfeldratte	rice field rat
Rattus exulans	Polynesische Ratte, Kleine Pazifikratte, Kleine Burmaratte	Polynesian rat
Rattus norvegicus	Wanderratte	brown rat, common brown rat, Norway rat, common rat
Rattus rattus	Hausratte, Alexandriner Hausratte, Dachratte	black rat, roof rat, house rat, ship rat
Rattus spp.	Ratten	rats
Ratufa affinis	Königsriesenhörnchen	cream-colored giant squirrel
Ratufa spp.	Riesenhörnchen	oriental giant squirrels
Recurvirostra avosetta	Säbelschnäbler	avocet
Recurvirostridae	Stelzenläufer (inkl. Säbelschnäbler)	avocets, stilts
Redunca arundinum	Großer Riedbock	reedbuck
Redunca fulvorufula	Bergriedbock	mountain reedbuck
Redunca redunca	Riedbock, Isabellantilope	Bohar reedbuck

Redunca spp.	Riedböcke	reedbucks
Reduviidae	Raubwanzen (Schreitwanzen)	assassin bugs, conenose bugs, ambush bugs & thread-legged bugs
Reduvius personatus	Kotwanze, Staubwanze, Maskierter Strolch	masked hunter bug (fly bug)
Reduvius senilis		tan assassin bug
Reesa vespulae	Amerikanischer Wespenkäfer	
Regalecidae	Riemenfische, Bandfische	oarfishes
Regalecus glesne	Riemenfisch, Bandfisch	oarfish
Regina alleni	Allens Königinnatter	striped crayfish snake
Regina rigida	Gestreifte Wassernatter	glossy crayfish snake
Regina septemvittata	Königinnennatter, Königinnatter, Königinschlange	queen snake
Regina spp.	Königinnennattern	crayfish snakes, queen snakes
Regulidae	Goldhähnchen	kinglets
Regulus calendula	Rubingoldhähnchen	ruby-crowned kinglet
Regulus ignicapillus	Sommergoldhähnchen	firecrest
Regulus regulus	Wintergoldhähnchen	goldcrest
Regulus satrapa	Satrap	golden-crowned kinglet
Reinhardtius hippoglossoides	Schwarzer Heilbutt, Grönland-Heilbutt	Greenland halibut (FAO), Greenland turbot, black halibut
Reithrodon physodes	Kaninchenratte*	coney rat
Reithrodontomys spp.	Amerikanische Erntemäuse	American harvest mice
Remiz pendulinus	Beutelmeise	penduline-tit
Remizidae	Beutelmeisen	penduline-tits
Remora remora	Küstensauger, Remora	common remora (FAO), shark sucker
Reniera cinerea	Röhrenschwamm	redbeard sponge
Renilla spp.	Seestiefmütterchen	sea pansies
Reptilia	Kriechtiere, Reptilien	reptiles
Resseliella crataegi/ Thomasiniana crataegi	Weißdornzweig-Gallmücke	hawthorn stem midge
Resseliella oculiperda/ Thomasiniana oculiperda	Okuliergallmücke, Okuliermade	red bud borer midge (fly)
Resseliella theobaldi/ Thomasiniana theobaldi	Himbeerruten-Gallmücke	raspberry cane midge
Retepora spp./ *Reteporella* spp./ *Sertella* spp.	Neptunschleier, Netzkorallen, Neptunsmanschetten	lace 'corals', fan bryozoans
Reticulitermes flavipes	Gelbfußtermite	eastern subterranean termite
Reticulitermes lucifugus	Erdholztermite, Mittelmeertermite	common European white ant, common European termite
Retinia resinella	Kiefernharzgallenwickler	
Retropinnidae	Neuseelandlachse	New Zealand smelts
Retusa obtusa	Kopfschildschnecke	Arctic barrel-bubble, pearl bubble
Retusa sulcata	Gefurchte Kopfschildschnecke	sulcate barrel-bubble
Retusa truncatula	Abgestutzte Kopfschildschnecke	truncate barrel-bubble

Rexea solandri	Königs-Schlangenmakrele	gemfish
Rhabdomys pumilio	Afrikanische Striemen-Grasmaus	four-striped grass mouse, striped mouse
Rhabdophaga clausilia	Weidenblattroll-Gallmücke	willow leaf-folding midge
Rhabdophaga rosaria	Weidenrosen-Gallmücke	willow rosegall midge, willow rosette-gall midge, terminal rosette-gall midge
Rhabdophaga salicis	Weidenruten-Gallmücke	willow stem gall midge
Rhabdophis spp.	Asiatische Kielrückenschlangen*	Asian keelbacks
Rhabdornithidae	Trugbaumläufer	Philippine creepers
Rhabdosargus globiceps	Weiße Stumpfnase	white stumpnose
Rhacocleis germanica	Zierliche Strauchschrecke	Mediterranean bushcricket
Rhacocleis neglecta	Adria-Strauchschrecke	Adriatic bushcricket
Rhacodactylus australis	Australischer Riesengecko	Australian giant gecko
Rhacophoridae	Ruderfrösche, Flugfrösche	flying frogs, Old World tree frogs, rhacophorids
Rhacophorus spp.	Flugfrösche, Ruderfrösche	flying frogs, Old World tree frogs, rhacophors
Rhadinaea flavilata	Amerikanische Kiefernwaldnatter	pine woods snake
Rhadinoceraea micans	Irisblattwespe	iris sawfly
Rhagio scolopacea	Schnepfenfliege	snipe fly
Rhagionidae	Schnepfenfliegen	snipe flies
Rhagium bifasciatum	Gelbbindiger Zangenbock	
Rhagium inquisitor	Schrotbock	
Rhagium mordax	Kleiner Laubholzzangenbock	oak longhorn beetle
Rhagium spp.	Zangenböcke	
Rhagium sycophanta	Eichenzangenbock	
Rhagoletis alternata	Hagebuttenfliege	rose-hip fly
Rhagoletis cerasi	Kirschfruchtfliege, Kirschfliege	European cherry fruit fly
Rhagoletis cingulata	Amerikanische Kirschfruchtfliege	cherry fruit fly (U.S.), North American cherry fruit fly (cherry maggot)
Rhagoletis fausta	Schwarzkirschfruchtfliege	black cherry fruit fly
Rhagomys rufescens	Brasilianische Baummaus*	Brazilian arboreal mouse
Rhagonycha fulva	Schwarzzipfeliger Weichkäfer	black-tipped soldier beetle
Rhamnusium bicolor	Weidenbock	
Rhamphichthyidae	Amerikanische Messerfische	sand knifefishes
Rhamphichthys rostratus	Langschnabel-Messerfisch	bandfish
Rhamphiophis spp.	Schnabelnattern	beaked snakes
Rhamphocoris clotbey	Knackerlerche	thick-billed lark
Rhampholeon platyceps	Flachkopf-Blattchamäleon	Malawi stumptail chameleon
Rhampholeon spectrum	Blattchamäleon	leaf chameleon, Cameroon stumptail chameleon
Rhampholeon spp.	Stumpfschwanz-Chamäleons	stumptail chameleons
Rhaphidophlus cervicornis	Roter Fingerschwamm	red antler sponge
Rhaphigaster nebulosa	8Gartenwanze	
Rhaphuma annularis	Gelber Bambusbockkäfer*	yellow bamboo longhorn beetle

Rhea americana	Nandu	greater rhea
Rheidae	Nandus	rheas
Rheiformes	Nandus	rheas
Rheithrosciurus macrotis	Borneo-Hörnchen	groove-toothed squirrel, great-eared squirrel
Rheobatrachus silus	Magenbrüterfrosch	gastric brooding frog, platypus frog
Rheodytes leukops	Weißgesichtsschildkröte, Fitzroy-Schildkröte	Fitzroy's turtle, Fitzroy turtle
Rheomys spp.	Zentralamerikanische Wassermäuse	Central American water mice
Rheumaptera cervinalis	Berberitzenspanner	barberry looper
Rheumaptera hasta	Schwarzweißer Birken-Blattspanner	argent and sable moth
Rheumaptera undulata	Wellenspanner	undulate looper*
Rhina ancylostoma	Bogenmund-Geigenrochen, Bogenmund-Gitarrenrochen	bowmouth guitarfish
Rhinatrematidae	Nasenwühlen	beaked caecilians
Rhincodon typus	Walhai	whale shark
Rhinecanthus aculeatus	Picassofisch, Gemeiner Picassodrückerfisch	Picasso fish, humuhumu, blackbar triggerfish (FAO)
Rhinecanthus assasi	Arabischer Picassofisch	Arabian picassofish
Rhinecanthus rectangulus	Diamant-Picassofisch, Harlekin-Drückerfisch, Keil-Picassodrücker	wedge-tail triggerfish (FAO), wedge picassofish
Rhinecanthus verrucosus	Bauchbinden-Picassofisch, Schwarzbauch-Picassodrücker, „Keilfleckdrücker"	blackbelly triggerfish
Rhineura floridana	Florida-Doppelschleiche, Florida-Breitnasenwühle	Florida worm lizard
Rhingia campestris	Feld-Schnabelschwebfliege	
Rhinidae	Haiflossen-Geigenrochen	sharkfin guitarfishes
Rhinobatidae	Geigenrochen (Gitarrenfische, Gitarrenrochen)	guitarfishes
Rhinobatoidei	Geigenrochen (Gitarrenfische, Gitarrenrochen)	guitarfishes
Rhinobatos albomaculatus	Weißflecken-Geigenrochen	white-spotted guitarfish
Rhinobatos percellens	Südlicher Geigenrochen	fiddlerfish (FAO), chola guitarfish
Rhinobatos planiceps	Peruanischer Geigenrochen	Peruvian guitarfish, Pacific guitarfish (FAO)
Rhinobatos rhinobatos	Europäischer Geigenrochen, Gemeiner Geigenrochen, Hairochen	Mediterranean guitarfish, common guitarfish (FAO), pesce violino
Rhinoceros sondaicus	Javanashorn	Javan rhinoceros
Rhinoceros unicornis	Indisches Panzernashorn	greater Indian rhinoceros, great Indian rhinoceros
Rhinocerotidae	Nashörner, Rhinozerosse	rhinoceroses, rhinos
Rhinocheilus lecontei	Nasennatter	longnosed snake
Rhinochimaeridae	Rüsselchimären, Langnasenchimären	longnose chimaeras, spookfishes
Rhinoclavis asper	Raue Nadelschnecke	rough cerith
Rhinoclemmys annulata	Braune Erdschildkröte	brown wood turtle

Rhinoclemmys areolata	Furchen-Erdschildkröte	furrowed wood turtle
Rhinoclemmys diademata	Venezuela-Erdschildkröte	Venezuelan wood turtle
Rhinoclemmys funerea	Schwarze Erdschildkröte, Bauchstreifen-Erdschildkröte	black wood turtle
Rhinoclemmys melanosterna	Kolumbianische Erdschildkröte	Colombian wood turtle
Rhinoclemmys nasuta	Nasen-Erdschildkröte, Ecuador-Erdschildkröte	large-nosed wood turtle
Rhinoclemmys pulcherrima pulcherrima	Pracht-Erdschildkröte	painted wood turtle, Mexican reed turtle, Mexican wood turtle
Rhinoclemmys punctularia	Südamerikanische Erdschildkröte, Guayana-Erdschildkröte	spot-legged turtle, spotted-legged turtle
Rhinoclemmys rubida	Mexikanische Flecken-Erdschildkröte, Rückenflecken-Erdschildkröte	Mexican spotted wood turtle
Rhinoclemmys spp.	Sumpf-Erdschildkröten, Amerikanische Erdschildkröten	neotropical wood turtles, American terrapins
Rhinocodontidae	Walhaie	whale sharks
Rhinocoris iracundus	Zornige Raubwanze, Rote Mordwanze	
Rhinocryptidae	Bürzelstelzer, Buschschlüpfer	tapaculos
Rhinoderma darwinii	Darwin-Frosch, Darwin-Nasenfrosch	Darwin's frog
Rhinoderma spp.	Nasenfrösche	mouth-brooding frogs
Rhinodermatidae	Nasenfrösche	mouth-brooding frogs, mouth-breeding frogs
Rhinoestrus purpureus	Pferdebiesfliege, Nasendasselfliege, Nasen-Rachen-Dasselfliege	
Rhinolophidae	Hufeisennasen	horseshoe bats
Rhinolophopsylla unipectinata	Hufeisennasenfloh	
Rhinolophus blasii	Blasius-Hufeisennase	Blasius' horseshoe bat
Rhinolophus euryale	Mittelmeerhufeisennase, Mittelmeer-Hufeisennase	Mediterranean horseshoe bat
Rhinolophus ferrumequinum	Großhufeisennase, Große Hufeisennase	greater horseshoe bat
Rhinolophus hipposideros	Kleinhufeisennase, Kleine Hufeisennase	lesser horseshoe bat
Rhinolophus mehelyi	Mehely-Hufeisennase	Mehely's horseshoe bat
Rhinolophus spp.	Hufeisennasen, Eigentliche Hufeisennasen	horseshoe bats
Rhinomuraena spp.	Nasenmuränen	ribbon eels
Rhinonemus cimbrius/ Enchelyopus cimbrius	Vierbärtelige Seequappe	four-bearded rockling, fourbeard rockling (FAO)
Rhinonicteris aurantius	Goldene Hufeisennase	golden horseshoe bat
Rhinophis blythis	Blyths Schildschwanzschlange	Blyth's Landau shieldtail
Rhinophrynidae	Nasenkröten	burrowing toads
Rhinophrynus dorsalis	Nasenkröte	Darwin's frog, Mexican burrowing toad
Rhinopithecus roxellanae/ Pygathrix roxellanae	Goldstumpfnase	golden snub-nosed langur, Sichuan snub-nosed langur, Chinese snub-nosed monkey, golden monkey, snow monkey

Rhinopithecus spp.	Stumpfnasenaffen	snub-nosed langurs
Rhinoplocephalus bicolor	Müllers Schlange	Müller's snake
Rhinopoma spp.	Klappnasen, Mausschwanz-Fledermaus	mouse-tailed bats, long-tailed bats
Rhinopomastidae	Sichelhopfe	scimitarbills
Rhinoptera bonasus	Kuhnasenrochen	cownose ray
Rhinoptera javanica	Java-Kuhnasenrochen	flapnose ray, Javanese cownose ray (FAO)
Rhinoptera marginata	Lusitanischer Kuhnasenrochen	Lusitanian cownose ray
Rhinopteridae	Kuhnasenrochen	cownose rays
Rhinotermitidae	Nasentermiten	rhinotermites, subterranean termites
Rhinotyphlops spp.	Schnabel-Blindschlangen	beaked snakes
Rhipicephalus appendiculatus	Braune Ohrenzecke*	brown ear tick
Rhipicephalus evertsi	Rotbeinige Zecke*	red-legged tick
Rhipicephalus sanguineus	Braune Hundezecke	brown dog tick, kennel tick
Rhipiceridae	Rhipiceriden, Zikadenräuber	cicada parasite beetles
Rhipidius quadriceps	Schaben-Fächerkäfer	
Rhipidogorgia flabellum/ Gorgonia flabellum/ Gorgonia ventalina	Venusfächer	Venus sea fan, Venus' fan, common sea fan
Rhipidomys spp.	Südamerikanische Klettermäuse*	South American climbing mice
Rhipidurini	Fächerschwanzschnäpper	fantails
Rhipiphoridae	Fächerkäfer	wedge-shaped beetles
Rhithropanopeus harrisii	Zwergkrabbe	dwarf crab, dwarf xanthid crab
Rhizocaulus verticillatus	Pferdeschwanz*	horsetail hydroid
Rhizocephala	Wurzelkrebse	rhizocephalans (parasitic "barnacles")
Rhizomyidae	Wurzelratten	bamboo rats, East African mole-rats
Rhizomys spp.	Bambusratten (Hbz. Chinesische Ratten)	bamboo rats
Rhizopertha dominica	Getreidekapuziner	lesser grain borer
Rhizophagidae	Rindenkäfer, Rindenglanzkäfer	root-eating beetles
Rhizopoda	Wechseltierchen, Wurzeltierchen, Rhizopoden, Amöben	amoebas, amebas
Rhizoprionodon acutus	Spitznasen-Grundhai	milk shark (FAO), milkshark
Rhizoprionodon longuris	Ostpazifischer Spitznasen-Grundhai	Pacific sharpnose shark
Rhizoprionodon spp.	Spitznasen-Grundhaie	milk sharks, milksharks
Rhizoprionodon terraenovae	Atlantischer Spitznasen-Grundhai, Nordwest-Atlantischer Spitznasen-Grundhai, Atlantischer Spitzmaulhai	Atlantic sharpnose shark
Rhizosmilia maculata		speckled cupcoral
Rhizostoma octopus (syn. *R. pulmo*)	Blaue Lungenqualle	blue cabbage bleb

Rhizostoma pulmo (syn. *R. octopus*)	Gelbe Lungenqualle	football jellyfish, yellow Mediterranean cabbage bleb
Rhizostoma spp.	Lungenquallen, Blumenkohlquallen	cabbage bleb, marigold, blubber, football jellyfish
Rhizostomeae	Wurzelmundquallen	rhizostome medusas
Rhizothera longirostris	Langschnabelwachtel	long-billed wood partridge
Rhizotragus solstitialis/ Amphimallon solstitialis	Junikäfer	summer chafer
Rhodactis inchoata	Kleines Elefantenohr	Tonga blue mushroom anemone, blue Tonga mushroom anemone
Rhodactis indosinensis	Indochinesisches Elefantenohr	hairy mushroom anemone
Rhodactis mussoides	Gewelltes Elefantenohr	elephant ear (metallic mushroom anemone, leaf mushroom anemone)
Rhodactis rhodostoma	Rotmund-Elefantenohr	red-mouth mushroom anemone
Rhodeus amarus/ Rhodeus sericeus	Bitterling	bitterling
Rhodogaster viridis	Grüne Blattwespe	green sawfly
Rhodopechys sanguinea	Rotflügelgimpel	crimson-winged finch
Rhodospiza obsoleta	Weißflügelgimpel	desert finch
Rhodostethia rosea	Rosenmöwe	Ross's gull
Rhogeessa spp.	Kleine Gelbe Fledermäuse	Rogeessa bats, little yellow bats
Rhombomys opimus	Große Rennmaus	great gerbil
Rhombosolea plebeia	Sandflunder	sand flounder
Rhopalidae	Glasflügelwanzen	scentless plant bugs, rhopalid bugs
Rhopalocera	Tagfalter	butterflies & skippers
Rhopalomenia aglaopheniae (Aplacophora)	Schmarotzerschlauch	"parasitic" solenogaster*
Rhopalonema velatum	Schleiermeduse	veiled medusa*
Rhopalosiphoninus latysiphon	Kellerlaus, Kartoffelkellerlaus, Breitröhrige Kartoffelknollenlaus	bulb and potato aphid
Rhopalosiphoninus tulipaellus	Mietenlaus, Tulpenblattlaus	tulip aphid, iris aphid
Rhopalosiphum insertum	Apfelgraslaus	apple-grass aphid, oat-apple aphid
Rhopalosiphum maidis	Maisblattlaus	corn-leaf aphid, corn aphid, cereal leaf aphid
Rhopalosiphum nymphaeae	Sumpfpflanzenlaus	waterlily aphid
Rhopalosiphum padi	Traubenkirschenlaus, Haferblattlaus, Haferlaus	bird-cherry aphid, oat aphid, wheat aphid
Rhopobota myrtilana	Heidelbeerwickler	blueberry leaf tier (moth)
Rhopobota naevana	Stechpalmenwickler	holly tortrix, holly leaf tier, marbled single-dot bell moth, black-headed fireworm
Rhyacichthyidae	Schmerlengrundeln	loach gobies
Rhyacionia buoliana/ Evetria buoliana	Kieferntriebwickler	gemmed shoot moth, pine-sprout tortrix, European pine shoot moth
Rhyacotriton olympicus	Olymp-Querzahnmolch	Olympic salamander

Rhynchaenus fagi	Buchenspringrüssler	beech flea weevil, beech leaf mining weevil, beech leaf miner
Rhynchaenus pallicornis	Apfelspringrüssler	apple flea weevil
Rhynchaenus quercus	Eichenspringrüssler	oak flea weevil, oak leaf mining weevil
Rhynchaenus rufipes	Amerikanischer Weidenspringrüssler	willow flea weevil
Rhynchaenus salicis	Weidenspringrüssler	European willow flea weevil
Rhynchites aequatus	Apfelfruchtstecher*	apple fruit rhynchites weevil
Rhynchites auratus	Kirschfruchtstecher	cherry fruit rhynchites weevil
Rhynchites bacchus	Purpurroter Apfelfruchtstecher	
Rhynchites caeruleus	Triebstecher, Obstbaumzweigabstecher	apple twig cutter
Rhynchites cupreus	Fruchtstecher, Kupferroter Pflaumenstecher	plum borer
Rhynchites germanicus/ Coenorrhinus germanicus	Erdbeerstengelstecher	strawberry rhynchites, strawberry weevil
Rhynchobatus djiddensis	Großer Gitarrenrochen, Riesengeigenrochen, Schulterfleck-Geigenrochen, Ulavi	whitespotted wedgefish, spotted guitarfish, shovelnose, sand shark, giant guitarfish (FAO)
Rhynchobdelliformes	Rüsselegel	proboscis leeches
Rhynchocalamus melanocephalus	Schwarzkopf-Zwergnatter	black-headed dwarf snake
Rhynchocephalia (Sphenodon)	Brückenechsen	rhynchocephalians
Rhynchocinetes rugulosus	Scharnierschnabelgarnele*	hinged-beak prawn
Rhynchocyon chrysopygus	Goldenes Rüsselhündchen	golden-rumped elephant shrew
Rhynchocyon spp.	Rüsselhündchen	checkered elephant shrews, forest elephant shrews
Rhynchoelaps spp.		desert banded snakes
Rhynchogale melleri	Langschnauzenmanguste	Meller's mongoose
Rhyncholestes raphanurus	Chile-Opossummaus	Chilean "shrew" opossum
Rhynchomeles prattorum	Ceramnasenbeutler	Ceram Isand long-nosed bandicoot
Rhynchomys spp.	Luzon-Nasenratten	Luzon shrewlike rats
Rhynchonycteris naso	Nasenfledermaus	proboscis bat, sharp-nosed bat
Rhynchopsitta pachyrhyncha	Kiefernsittich, Arasittich	thick-billed parrot
Rhynchopsitta terrisi	Maronenstirnsittich	maroon-fronted parrot
Rhynochetidae	Kagu	kagu
Rhynochetos jubatus	Kagu, Rallenkranich	kagu
Rhyparia purpurata	Stachelbeerbär, Purpurbär	gooseberry tiger (moth)
Rhysodidae	Runzelkäfer	wrinkled bark beetles
Rhyzopertha dominica	Getreidekapuziner	lesser grain borer
Ricinulei	Kapuzenspinnen	ricinuleids
Ricordea florida	Falsche Floridakoralle*	Florida false coral
Riodinidae	Scheckenfalter, Würfelfalter	metalmarks
Riparia paludicola	Braunkehl-Uferschwalbe	brown-throated sand martin
Riparia riparia	Uferschwalbe	sand martin

Rissa tridactyla	Dreizehenmöve	kittiwake
Rissoa costata	Rippen-Rissoschnecke	costate risso snail
Rissoa parva	Kleine Rissoschnecke	tiny risso snail
Rivulogammarus pulex/ Gammarus pulex	Bachflohkrebs, Gemeiner Flohkrebs	freshwater shrimp
Rivulus harti	Riesenbachling	Hart's rivulus (FAO), leaping guabine
Rivulus spp.	Bachlinge	rivuluses
Roccus chrysops/ Morone chrysops	Weißer Sägebarsch	white bass
Roccus labrax/ Dicentrarchus labrax/ Morone labrax	Wolfsbarsch, Seebarsch	bass, sea bass, European seabass (FAO)
Roccus saxatilis/ Morone saxatilis	Nordamerikanischer Streifenbarsch, Felsenbarsch	striped bass, striper, striped seabass (FAO)
Roccus spp.	Streifenbarsche	striped basses
Rodentia	Nagetiere	rodents, gnawing mammals (except rabbits)
Roeboides spp.	Wachssalmler	glass headstanders
Rollandia micropterum	Titicataucher	short-winged grebe
Rollandia rolland	Rollandtaucher	white-tufted grebe
Rollulus roulroul	Straußwachtel	crested wood partridge
Romanichthyini	Spindelbarsche	Romanian spindleperch
Romanichthys valsanicola	Groppenbarsch	Romanian bullhead-perch
Romerolagus diazi	Vulkankaninchen	volcano rabbit
Rondeletiidae	Papillenfische, Rotmäulige Walkopffische	redmouth whalefishes
Rondeletiola minor/ Sepietta minor	Kleine Sepiette, Linsen-Sepiette	lentil bobtail squid
Rosalia alpina	Alpenbock	rosalia longicorn
Rosalia funebris		banded alder borer
Rossia macrosoma	Große Rossie	Ross' cuttlefish, Ross' cuttle, large-head bob-tailed squid, stout bobtail squid (FAO)
Rossia pacifica	Pazifische Rossie	Pacific bob-tailed squid, North Pacific bobtail squid (FAO)
Rossia tenera/ Semirossia tenera	Atlantische Rossie	lesser bobtail squid, Atlantic bob-tailed squid, lesser shining bobtail squid (FAO)
Rostratulidae	Goldschnepfen	painted snipe
Rostrhamus sociabilis	Schneckenweih	Everglades kite
Rostroraja alba/ Raja alba	Südlicher Weißrochen, Ostatlantischer Weißrochen, Spitzrochen, Bandrochen	white skate (FAO), bottle-nosed skate
Rotatoria	Rädertiere, Rotatorien	rotifers
Rousettus aegyptiacus	Nil-Flughund	Egyptian rousette
Rousettus spp.	Höhlenflughunde	Rousette fruit bats
Ruditapes philippinarum/ Tapes philippinarum/ Venerupis philippinarum/ Tapes japonica	Japanische Teichmuschel	Japanese littleneck, short-necked clam, Japanese clam, Manila clam

Rugosa	Runzelkorallen	rugose corals
Rumina decollata	Stumpfschnecke	decollate snail
Ruminantia	Wiederkäuer, Retroperistaltiker, Ruminantier	ruminants, "cud chewers"
Rupestes forresti	Chinesisches Langnasenhörnchen	Chinese long-nosed squirrel
Rupicapra rupicapra	Gämse, Gamswild (Gemse)	chamois (*pronounce*: sha-mee or sham-wa)
Ruthenica filograna	Zierliche Schließmundschnecke	
Rutilus pigus pigus	Donau-Rotauge	Danubian roach
Rutilus pigus virgo	Frauennerfling	virgin Danubian roach*
Rutilus rutilus	Plötze, Rotauge	roach (FAO), Balkan roach
Ruvettus pretiosus	Ölfisch	oilfish
Rynchopidae	Scherenschnäbel	skimmers

Sabella crassicornis	Gebänderter Fächerwurm*	banded feather-duster worm
Sabella pavonina	Pfauenfederwurm, Pfauenwurm	peacock worm, peacock feather-duster worm
Sabella spallanzanii/ Spirographis spallanzanii	Schraubensabelle	fanworm, feather-duster worm a.o.
Sabellaria alveolata	Trichterwurm	honeycomb worm
Sabellaria spinulosa	Pümpwurm, „Sandkoralle"	spiny feather-duster worm
Sabellaria vulgaris	Gemeiner Trichterwurm, Amerikanischer Trichterwurm*	sand-builder worm, common feather-duster worm*
Sabellastarte magnifica	Großer Federwurm	giant feather-duster worm
Sabellidae	Federwürmer u.a., Fächerwürmer u.a.	sabellids, sabellid fanworms, sabellid feather-duster worms a.o.
Sabinea septemcarinata	Siebenliniengarnele	sevenline shrimp
Sabra harpagula	Linden-Sichelspinner	scarce hook-tip (moth)
Sacalia bealei	Chinesische Pfauenaugenschildkröte	ocellate pond turtle, Beal's eyed turtle (IUCN)
Sacalia quadriocellata	Vierfleck-Pfauenaugenschildkröte	four-eyed turtle
Sacchiphantes abietis/ Adelges abietis	Gelbe Fichten-Großgallenlaus	yellow spruce pineapple-gall adelges
Sacchiphantes viridis/ Adelges viridis	Grüne Fichten-Großgallenlaus	green spruce pineapple-gall adelges
Saccoglossus cambrensis (Enteropneusta)	Roter Eichelwurm	red acorn worm
Saccopharyngidae	Sackmäuler, Schlinger	swallowers
Saccopteryx spp.	Taschenflügel-Fledermäuse u.a.	white-lined bats
Saccostomus spp.	Kurzschwanz-Hamsterratten	African pouched rats
Saccostrea cucullata	Australische Auster, Sydney Felsenauster	Sydney cupped oyster
Sacoglossa/Saccoglossa	Schlundsackschnecken, Sackschnecken, Schlauchschnecken	sacoglossans
Saga pedo	Große Sägeschrecke	predatory bushcricket
Sagartia elegans	Rosenanemone*	rosy anemone
Sagartia rhododactylos	Tangrose	kelp-rose anemone*
Sagartia troglodytes/ Actinothoe troglodytes	Kolibrirose	cave-burrowing anemone*
Sagartiogeton undulatus	Witwenrose	widow-rose anemone*
Sagitta setosa (Chaetognatha)	Küsten-Pfeilwurm	coastal arrow worm
Sagittariidae	Sekretäre	secretary birds
Sagittarius serpentarius	Sekretär	secretary bird
Saguinus bicolor	Mantelaffe, Manteläffchen	bare-faced tamarin, pied tamarin
Saguinus fuscicollis	Braunrückentamarin	saddle-back tamarin
Saguinus imperator	Kaiserschnurrbarttamarin	emperor tamarin
Saguinus inustus	Marmorgesichttamarin	mottle-faced tamarin
Saguinus labiatus	Rotbauchtamarin	white-lipped tamarin
Saguinus leucopus	Weißfußaffe	white-footed tamarin
Saguinus midas	Rothandtamarin	red-handed tamarin
Saguinus mystax	Schnurrbarttamarin	moustached tamarin
Saguinus nigricollis	Schwarzrückentamarin	black and red tamarin

Saguinus oedipus	Lisztaffe, Lisztäffchen	cotton-top tamarin
Saiga tatarica	Saiga, Saigaantilope	saiga
Saimiri oerstedii	Rotrücken-Totenkopfaffe	red-backed squirrel monkey, Central American squirrel monkey
Saimiri sciureus	Totenkopfaffe	common squirrel monkey
Saimiri spp.	Totenkopfaffen	squirrel monkeys
Saissetia coffeae	Halbkuglige Napfschildlaus	hemispherical scale
Saissetia oleae	Schwarze Ölbaumschildlaus	black scale
Salacia alba	See-Milzkraut*	sea spleenwort
Salamandra atra	Alpensalamander	Alpine salamander, European Alpine salamander
Salamandra lanzai	Großer Alpensalamander	large Alpine salamander
Salamandra salamandra	Feuersalamander	fire salamander, European fire salamander
Salamandridae	Echte Salamander & Molche	salamanders & newts
Salamandrina terdigitata	Brillensalamander	spectacled salamander
Salangidae	Nudelfische	noodlefishes, icefishes
Salanoia concolor	Schlichtmungo	Malagasy brown-tailed mongoose, salano
Salaria fluviatilis/ Lipophrys fluviatilis/ Blennius fluviatilis	Süßwasser-Schleimfisch, Fluss-Schleimfisch	freshwater blenny (FAO), river blenny
Salaria pavo/ Lipophrys pavo/ Blennius pavo	Pfauenschleimfisch	peacock blenny
Saldidae	Uferwanzen, Springwanzen	shore bugs, saldids
Saldula saltatoria	Gemeiner Hüpferling	
Salmacina dysteri	Korallenwurm	coral worm
Salmandrella keyserlingii	Dybowski-Salamander	Dybowski's salamander
Salmo clarki	Purpurforelle	cutthroat trout
Salmo gairdneri/ Oncorhynchus mykiss	Regenbogenforelle	rainbow trout
Salmo salar	Atlantischer Lachs, Salm (*Junglachse im Meer:* Blanklachs)	Atlantic salmon (FAO) (*lake pop. in U.S./Canada:* ouananiche, lake Atlantic salmon, landlocked salmon, Sebago salmon)
Salmo spp.	Atlantische Lachse & Forellen	Atlantic trouts & Atlantic salmons
Salmo trutta trutta	Meerforelle, Lachsforelle	sea trout
Salmo trutta fario	Bachforelle, Steinforelle	brown trout (river trout, brook trout)
Salmo trutta lacustris	Seeforelle	lake trout
Salmo trutta	Forelle	trout
Salmonidae	Lachse	salmonids
Salmothymus obtusirostris	Adria-Lachs	Adriatic salmon
Salomonelaps par	Solomon-Inseln-Kupferkopfnatter	Solomon Island brown snake
Salpa democratica	Gemeine Salpe	common salpid, whale blobs

Salpa fusiformis	Kleine Salpe	lesser salp
Salpa zonaria	Gegürtelte Salpe	girdled salp
Salpida	Eigentliche Salpen	salps
Salpinctes obsoletus	Felsenzaunkönig	rock wren
Salpingidae	Engtaillien-Rindenkäfer (Scheinrüssler)	narrow-waisted bark beetles
Salpingotulus michaelis	Baluchistan-Zwergspringmaus	Baluchistan pygmy jerboa
Salpingotus spp.	Koslows Zwergspringmäuse	three-toed dwarf jerboas
Salticidae	Springspinnen, Hüpfspinnen	jumping spiders
Salticus scenicus	Zebra-Springspinne, Zebraspinne	zebra jumper
Salvadora spp.	Pflasterzahnnattern	patchnose snakes, patch-nosed snakes
Salvelinus alpinus	Seesaibling, Wandersaibling	char, charr (FAO), Arctic char, Arctic charr
Salvelinus fontinalis	Bachsaibling	brook trout (FAO), brook char, brook charr
Salvelinus namaycush	Seesaibling, Stutzersaibling, Amerikanischer Seesaibling	American lake trout, Great Lake trout, lake trout (FAO)
Salvelinus spp.	Saiblinge	chars, charrs
Salvelinus svetovidovi	Langflossen-Saibling*	long-finned char
Sandalops melancholicus	Sandalenauge	sandal-eye squid
Sanguinolaria cruenta	Blutrote Sandmuschel	blood-stained sanguin
Sanguinolaria sanguinolenta	Atlantische Sandmuschel	Atlantic sanguin
Sanopus splendidus	Pracht-Krötenfisch	coral toadfish
Sanzinia madagascariensis	Madagaskar-Hundskopfboa	Madagascar tree boa
Saperda carcharias	Großer Pappelbock	large poplar borer, large willow borer, poplar longhorn, large poplar longhorn beetle
Saperda populnea	Kleiner Pappelbock, Espenbock	small poplar borer, lesser poplar borer, small poplar longhorn beetle
Saperda scalaris	Leiterbock	scalar longhorn beetle
Sapphirina spp.	Saphirkrebse	sapphirines
Sapphirinidae	Saphirkrebse	sapphirines
Sappho sparganura	Schleppensylphe	red-tailed comet
Saprinus semistriatus	Gemeiner Stutzkäfer	common hister beetle (Central European)
Sapyga clavicornis	Gemeine Keulenwespe	common European sapygid wasp
Sapygidae	Schmarotzerwespen	sapygid wasps, parasitic wasps
Sarcophaga carnaria	Graue Fleischfliege	fleshfly, flesh-fly
Sarcophagidae	Aasfliegen, Fleischfliegen	fleshflies
Sarcophilus harrisii	Beutelteufel	Tasmanian devil
Sarcophyton trocheliophorum	Elefantenohrkoralle*	elephant ear coral, yellow-green soft-coral (leather coral)
Sarcops calvus	Kahlkopfatzel	bald starling
Sarcopterygii/ Choanichthyes	Fleischflosser, Muskelflosser	fleshy-finned fishes, sarcopterygians, sarcopts

Sarcoptes scabiei	Krätzmilbe	scabies mite, itch mite
Sarcoptidae	Krätzemilben (Räudemilben)	itch mites, scabies mites, scab mites
Sarcorhamphus papa	Königsgeier	king vulture
Sarda australis	Australischer Bonito	Australian bonito
Sarda chilensis	Chilenische Pelamide	Pacific bonito, Eastern Pacific bonito (FAO)
Sarda orientalis	Westpazifische Pelamide	striped bonito
Sarda sarda	Pelamide	Atlantic bonito
Sardina pilchardus	Sardine, Pilchard	European sardine, sardine(if *small*), pilchard (if *large*), European pilchard (FAO)
Sardina spp.	Sardinen	sardines
Sardinella aurita	Ohrensardine, Große Sardine, Sardinelle	gilt sardine, Spanish sardine, round sardinella (FAO)
Sardinella fimbriata	Asiatische Kleinsardine	fringescale sardinella
Sardinella longiceps	Großkopfsardine	Indian oil sardine (FAO), oil sardine
Sardinella madarensis	Madeira-Sardinelle	Madeiran sardinella (FAO), short-body sardinella, short-bodied sardine
Sardinella zunasi (*Harengula zunasi*)	Pazifischer Kleinhering	Japanese sardinella
Sardinops caeruleus	Kalifornische Sardine, Pazifische Sardine	Californian pilchard
Sardinops melanostricta	Japanische Sardine	Japanese pilchard
Sardinops neopilchardus	Australische Sardine	Australian pilchard (FAO), picton herring
Sardinops ocellatus	Südafrikanische Sardine	South African pilchard
Sardinops sagax	Südamerikanische Sardine	Chilean pilchard, Pacific sardine, Peruvian sardine, South American pilchard (FAO)
Sargocentron rubrum/ Holocentrus ruber	Roter Soldatenfisch, Roter Husar, Rotgestreifter Soldatenfisch, Gewöhnlicher Eichhornfisch	red soldierfish, redcoat (FAO)
Sarkidiornis melanotos	Glanzente	comb duck
Saron marmoratus	Gemeine Marmorgarnele	common marble shrimp
Sarotherodon aureum/ Chromis aureus/ Tilapia aurea	Goldtilapia	golden tilapia
Sarpa salpa/Boops salpa	Goldstriemen, Ulvenfresser	saupe, salema (FAO), goldline
Sarsia tubulosa	Klappermoos*	clapper hydroid, clapper hydromedusa
Sasia abnormis	Maleienmausspecht	rufous piculet
Sassia subdistorta	Verbogene Tritonschnecke*	distorted rock triton
Saturnia pavonia/ Eudia pavonia	Kleines Nachtpfauenauge	emperor moth
Saturnia pyri	Großes Nachtpfauenauge, Wiener Nachtpfauenauge	giant peacock moth

Saturniidae	Augenspinner (Nachtpfauenaugen/ Pfauenspinner)	giant silkmoths, silkworm moths, emperor moths
Satyridae	Augenfalter	browns, satyrs
Satyrium acaciae	Akazienzipfelfalter	sloe hairstreak
Satyrium ilicis/ Nordmannia ilicis	Brauner Eichen-Zipfelfalter, Eichenzipfelfalter, Stechpalmen-Zipfelfalter	ilex hairstreak
Satyrium pruni/ Strymonium pruni	Pflaumenzipfelfalter	black hairstreak
Satyrium spini/ Strymonium spini	Schlehenzipfelfalter	blue-spot hairstreak
Satyrium w-album/ Strymon w-album/ Strymonidia w-album	Weißes W, Ulmen-Zipfelfalter, Ulmenzipfelfalter	white-letter hairstreak
Saurida gracilis	Marmor-Eidechsenfisch	graceful lizardfish
Saurida tumbil	Großer Eidechsenfisch	greater lizardfish
Saurida undosquamis	Großschuppen-Eidechsenfisch, Gefleckter Eidechsenfisch	brushtooth lizardfish (FAO), large-scale lizardfish
Saurischia	Echsenbecken-Dinosaurier	lizard-hipped dinosaurs, reptile-like dinosaurs, saurischian reptiles
Saurodactylus mauritanicus	Echsenfingergecko	Morocco lizard-fingered gecko
Sauromalus spp.	Chuckwallas	chuckwallas
Saxicola dacotiae	Kanarenschmätzer	Canary Island chat
Saxicola rubetra	Braunkehlchen	whinchat
Saxicola torquata	Schwarzkehlchen	common stonechat
Saxicolini	Schmätzer & Erdsänger & Rotschwänze etc.	chats
Saxidomus ssp.	Buttermuscheln*	butter clams
Sayornis phoebe	Phoebe	eastern phoebe
Scabricola fissurata	Netz-Mitra	reticulate mitre, reticulate miter
Scaeurgus unicirrhus	Einhornkrake*	unicorn octopus
Scaeva pyrastri	Weiße Dickkopf-Schwebfliege	cabbage aphid hover fly
Scalidae	Wendeltreppen	wentletraps
Scalopus aquaticus	Ostamerikanischer Maulwurf	eastern American mole
Scalpellum scalpellum	Samtige Entenmuschel	velvet goose barnacle
Scandentia	Spitzhörnchen	tree shrews
Scapandus spp.	Westamerikanische Maulwürfe	western American moles
Scapanorhynchidae/ Mitsukurinidae	Koboldhaie, Nasenhaie	goblin sharks
Scapanorhynchus owstoni/ Mitsukurina owstoni	Koboldhai, Japanischer Nasenhai	goblin shark
Scaphander lignarius/ Bulla lignaria	Taucherschnecke, Holzboot	woody canoe-bubblesnail
Scaphander punctostriatus	Große Taucherschnecke, Großes Holzboot	giant canoe-bubblesnail
Scaphandridae	Bootschnecken	canoe bubblesnails
Scaphella junonia	Juno-Walzenschnecke	junonia, Juno's volute
Scaphididae	Kahnkäfer	shining fungus beetles
Scaphiodontophis annulatus	Guatemala Vielzahnnatter	Guatemala neckband snake
Scaphiopus bombifrons	Flachland-Schaufelfuß	Plains spadefoot

Scaphiopus hammondii	Westlicher Schaufelfuß	western spadefoot, Hammond's spadefoot toad
Scaphiopus holbrookii	Östlicher Schaufelfuß	eastern spadefoot, Holbrook's spadefoot
Scaphiopus spp.	Schaufelfüße	spadefoot toads
Scaphirhynchus platorhynchus/ Acipenser platorynchus	Schaufelstör, Schaufelnasenstör	shovelnose sturgeon
Scaphirhynchus albus	Blasser Schaufelnasenstör	pallid sturgeon
Scaphirhynchus spp.	Amerikanische Schaufelnasenstöre	American sturgeons
Scapholebris mucronata	Kahnfahrer	
Scapteromys tumidus	Südamerikanische Wasserratte	water rat
Scarabaeidae	Blatthornkäfer	scarab beetles, lamellicorn beetles (dung beetles & chafers)
Scarabaeus sacer	Heiliger Pillendreher	sacred scarab beetle, Egyptian scarab
Scardafella inca	Inkatäubchen	Inca dove
Scardinius erythrophthalmus	Rotfeder	rudd
Scardinius graecus	Griechische Rotfeder	Greek rudd
Scaridae/Callyodontidae	Papageienfische, Papageifische	parrotfishes
Scarus ghobban	Blauband-Papageifisch	blue-barred orange parrotfish, bluebarred parrotfish (FAO)
Scarus gibbus	Rotbauch-Papageifisch	blunt-headed parrotfish, heavybeak parrotfish (FAO)
Scarus sordidus/ Chlorurus sordidus	Grünflossen-Papageifisch, Kugelkopf-Papageifisch	green-finned parrotfish, garned red parrotfish, bullethead parrotfish, daisy parrotfish (FAO)
Scarus tricolor	Masken-Papageifisch, Schwalbenschwanz-Papageifisch	tricolor parrotfish, tricolour parrotfish (FAO)
Scathophaga stercoraria	Gelbe Dungfliege	yellow dungfly
Scathophagidae	Dungfliegen, Kotfliegen	dungflies
Scatophagidae	Argusfische	scats, scatties
Scatophagus argus argus	Argusfisch, Gemeiner Argusfisch, Grüner Argusfisch	scat, argusfish, argus fish, spotted cat (FAO)
Scatophagus tetracanthus	Afrikanischer Argusfisch	African argusfish, scatty (FAO)
Scatopsidae	Dungmücken	minute black scavenger flies
Sceliphron caementarium	Spinnen-Mörtelgrabwespe	black-and-yellow mud dauber, black-and-yellow mud wasp
Sceliphron spp.	Mörtelgrabwespen	mud daubers
Sceloporus cyanogenys	Blauer Stachelleguan	blue spiny lizard
Sceloporus graciosus	Schöner Stachelleguan	sagebrush lizard
Sceloporus grammicus	Mesquite-Stachelschuppenleguan	mesquite lizard
Sceloporus jarrovi	Yarrow-Stachelleguan	Yarrow's spiny lizard
Sceloporus magister	Wüsten-Stachelleguan	desert spiny lizard
Sceloporus occidentalis	Westlicher Zaunleguan	western fence lizard
Sceloporus olivaceus	Texas-Stachelleguan	Texas spiny lizard
Sceloporus orcutti	Granit-Stachelleguan	granite spiny lizard
Sceloporus poinsetti	Felsspalten-Stachelleguan	crevice spiny lizard

Sceloporus scalaris	Zaunleguan	bunch grass lizard
Sceloporus spp.	Stachelleguane	spiny lizards, swifts
Sceloporus undulatus	Östlicher Zaunleguan	eastern fence lizard
Sceloporus variabilis	Rotbauchleguan	rose-bellied lizard
Sceloporus virgatus	Gestreifter Plateauleguan	striped plateau lizard
Scelotes spp.	Scelotes-Skinke	burrowing skinks
Scenopinidae	Fensterfliegen	window flies
Scenopinus spp.	Fensterfliegen	window flies
Schedophilus medusophagus	Cornwall Schwarzfisch*	Cornish blackfish
Schilbe mystus	Silberwels	grasscutter catfish
Schilbeidae	Eigentliche Glaswelse	schilbeid catfishes
Schistocerca gregaria	Afrikanische Wüstenschrecke, Afrikanische Wanderheuschrecke	desert locust
Schistometopum spp.	Guinea-Wühlen	Guinea caecilians
Schistometopum thomense	Buntwühle	Island caecilian
Schistomysis spiritus	Schlanke Schwebegarnele	ghost shrimp
Schistosoma spp.	Pärchenegel (Blutegel)	blood flukes
Schizaphis graminum	Getreideblattlaus	greenbug
Schizobrachiella sanguinea	Rotbraunes Krusten-Moostierchen	sanguine crustal bryozoan*, blood-red encrusting bryozoan
Schizomida/Schizopeltidia	Zwerggeißelskorpione	microwhipscorpions, schizomids
Schizoneurinae/ Eriosomatinae	Blutläuse, Blasenläuse	woolly aphids, gall aphids
Schizophyllum sabulosum	Sandschnurfüßer	sand millepede
Schizothaerus nuttalli	Sommermuschel*	summer clam, gaper
Schizotus pectinicornis	Orangefarbener Feuerkäfer	orange-coloured fire beetle*
Schoinobates volans	Riesengleitbeutler	greater gliding possum
Schrankia taenialis	Breitflügel-Motteneule	white-line snout (moth)
Sciaena antarctica	Antarktischer Umberfisch	mulloway
Sciaena aquila/ Argyrosomus regius	Adlerfisch	meagre
Sciaena cirrosa/ Johnius cirrhosus/ Umbrina cirrosa	Bartumber, Schattenfisch, Umberfisch, Umber	shi drum (FAO), corb (U.S./U.K.), sea crow (U.S.), gurbell (U.S.), croaker
Sciaena hololepidota/ Argyrosomus hololepidotus	Afrikanischer Adlerfisch	kob, southern meagre (FAO)
Sciaena umbra/ Johnius umbra	Meerrabe, Seerabe, Rabenfisch	brown meagre
Sciaenidae	Umberfische, Adlerfische, Trommler	kobs, drums (croakers)
Sciaenops ocellatus	Augenfleck-Umberfisch	red drum
Scincella himalayana	Himalaya-Schlankskink	Himalaya ground skink
Scincella lateralis	Amerikanischer Zwergskink	ground skink
Scincella laterimaculata	Seitenfleck-Schlankskink	spotted ground skink
Scincella melanosticta	Schwarzer Schlankskink	black ground skink
Scincidae	Skinke, Glattechsen	skinks
Scincopus fasciatus	Nachtskink	Peters' banded skink

Scincus scincus	Apothekerskink	sandfish
Scincus spp.	Sandskinke	sand skinks
Sciomyzidae	Netzfliegen, Hornfliegen	marsh flies
Scissurella cingulata	Gürtel-Riss-Schnecke	belt scissurelle
Scissurellidae	Riss-Schnecken	scissurelles
Sciuridae	Hörnchen	squirrels & chipmunks & marmots & prairie dogs
Sciurillus pusillus	Neotropisches Zwerghörnchen	Neotropical pygmy squirrel
Sciuromorpha	Hörnchenverwandte	squirrel-like rodents
Sciurotamias spp.	Chinesische Rothörnchen	rock squirrels
Sciurus anomalus	Kleinasiatisches Hörnchen	Persian squirrel (Asian Minor)
Sciurus carolinensis	Grauhörnchen	gray squirrel (U.S.), grey squirrel (Br.)
Sciurus granatensis	Granathörnchen	tropical red squirrel
Sciurus griseus	Westliches Grauhörnchen	western gray squirrel
Sciurus niger	Östliches Fuchshörnchen	eastern fox squirrel
Sciurus spp.	Eichhörnchen	tree squirrels
Sciurus vulgaris	Europäisches Eichhörnchen, Eurasisches Eichhörnchen (Hbz. Feh)	European red squirrel, Eurasian red squirrel
Scleropages formosus	Asiatischer Gabelbart	Asian arowana, Asian bonytongue (FAO)
Scleropages leichardti	Gepunkteter Barramundi	spotted barramundi, spotted bonytongue (FAO)
Scobicia declivis	Kalifornischer Bleikabelbohrer	short-circuit beetle
Scolanthus callimorphus	Wurmanemone*	worm anemone
Scolecomorphidae	Grabwühlen	tropical caecilians
Scoliidae	Dolchwespen, Mordwespen	scolid wasps
Scoliopterix libatrix	Zimteule, Zackeneule, Krebssuppe	herald (noctuid moth)
Scolomys melanops	Ecuador-Stachelratte*	Ecuador spiny mouse
Scolopacidae	Schnepfen, Schnepfenvögel (inkl. Wasserläufer)	sandpipers & snipes & woodcocks
Scolopax minor	Kanadaschnepfe	American woodcock
Scolopax rusticola	Waldschnepfe	woodcock, European woodcock
Scolopendra cingulata	Südeuropäischer Gürtelskolopender	southern girdled scolopendra*
Scolopendra gigas	Tropische Riesenskolopender u.a.	tropical giant scolopendra a.o.
Scolopendra morsitans	Bissiger Skolopender	biting scolopendra*
Scolopendromorpha	Riesenläufer, Skolopender	scolopendromorphs
Scoloplacidae	Stachelzwergwelse	spiny dwarf catfishes
Scoloplos armiger	Kiemenringelwurm	
Scolopsidae	Großaugen	spinecheeks
Scolopsis xenochrous	Blaufleckenbrasse	olive-spotted monocle bream
Scolymia cubensis	Artischocken-Koralle	artichoke coral
Scolymia lacera	Atlantische Pilzkoralle*	Atlantic mushroom coral

Scolytidae

Scolytidae (Ipidae)	Borkenkäfer	bark beetles & engraver beetles & ambrosia beetles, timber beetles
Scolytus intricatus	Eichensplintkäfer	oak bark beetle
Scolytus mali	Großer Obstbaumsplintkäfer	large fruit bark beetle, larger shothole borer
Scolytus multistriatus	Kleiner Ulmensplintkäfer	small elm bark beetle, small European elm bark beetle
Scolytus rugulosus	Kleiner Obstbaumsplintkäfer	fruit bark beetle, fruit bark borer, shothole borer
Scolytus scolytus	Großer Ulmensplintkäfer	large elm bark beetle, elm bark beetle, larger European bark beetle
Scolytus spp.	Splintkäfer	bark beetles
Scomber australasicus	Indopazifische Makrele	spotted chub mackerel, blue mackerel (FAO)
Scomber japonicus/ Scomber colias	Kolios, Thunmakrele, Mittelmeermakrele	chub mackerel (FAO), Spanish mackerel, Pacific mackerel
Scomber scombrus	Makrele, Europäische Makrele, Atlantische Makrele	Atlantic mackerel (FAO), common mackerel
Scomberesocidae	Makrelenhechte	sauries
Scomberesox saurus	Makrelenhecht, Atlantischer Makrelenhecht	saury, saury pike, Atlantic saury (FAO), skipper
Scomberoides commersonianus	Talang	talang queenfish
Scomberoides lysan	Lederrücken*	doublespotted queenfish (FAO), leatherback
Scomberoides tol	Spitzschuppenmakrele	needlescaled queenfish
Scomberomoridae (Cybiidae)	Pelamiden, Königsmakrelen	Spanish mackerels
Scomberomorus brasiliensis	Serra-Makrele	Serra Spanish mackerel
Scomberomorus cavalla	Königsmakrele	king mackerel (FAO), kingfish
Scomberomorus commerson	Indische Königsmakrele	Commerson's mackerel, barred mackerel, narrow-barred Spanish mackerel (FAO)
Scomberomorus concolor	Monterey-Makrele	Monterey Spanish mackerel
Scomberomorus guttatus	Indopazifische Spanische Makrele	Indo-Pacific Spanish mackerel, Indo-Pacific king mackerel (FAO)
Scomberomorus lineolatus	Gestreifte Spanische Makrele	streaked Spanish mackerel, streaked seerfish (FAO)
Scomberomorus maculatus	Gefleckte Königsmakrele, Spanische Makrele	Atlantic Spanish mackerel, Spanish mackerel (FAO)
Scomberomorus munroi	Australische Fleckenmakrele	Australian spotted mackerel
Scomberomorus niphonius	Japanische Makrele	Japanese Spanish mackerel
Scomberomorus regalis	Cero, Falsche Königsmakrele	cero
Scomberomorus semifasciatus	Tigermakrele	broadbarred king mackerel
Scomberomorus sierra	Ostpazifische Königsmakrele	Pacific sierra

Scomberomorus sinensis	Chinesische Königsmakrele	Chinese seerfish
Scomberomorus tritor	Ostatlantische Königsmakrele, Westafrikanische Königsmakrele	West African Spanish mackerel
Scombridae	Makrelen (& Thunfische)	mackerels (incl. tunas & bonitos)
Scombrolabracidae	Schwarzmakrelen	black mackerel, scombrolabracid
Scombrolabrax heterolepis	Schwarzmakrele*	black mackerel
Scombropidae	Gnomfische	gnomefishes
Scopelarchidae	Perlaugen	pearleyes
Scophthalmidae	Steinbuttverwandte	turbot fishes
Scophthalmus aquosus	Sandbutt	windowpane (FAO), windowpane flounder
Scophthalmus maximus/ Psetta maxima	Steinbutt	turbot
Scophthalmus rhombus	Glattbutt	brill
Scopidae	Hammerkopf	hamerkop (hammerhead)
Scopula nigropunctata		subangled wave (moth)
Scopula ornata	Schmuckspanner	ornate wave (moth)
Scopula rubiginata		tawny wave (moth)
Scopus umbretta	Hammerkopf	hammerkop
Scorbiculariidae	Pfeffermuscheln	furrow clams (furrow shells)
Scorbipalpa ocellatella	Rübenmotte	turnip moth*, sugarbeet moth*
Scorpaena notata/ Scorpaena ustulata	Kleiner Roter Drachenkopf	lesser red scorpionfish, little scorpionfish, small red scorpionfish (FAO)
Scorpaena porcus	Brauner Drachenkopf	brown scorpionfish black scorpionfish (FAO)
Scorpaena scrofa	Großer Roter Drachenkopf, Roter Drachenkopf, Große Meersau, Europäische Meersau	bigscale scorpionfish, red scorpionfish
Scorpaenidae	Drachenköpfe	scorpionfishes (rockfishes)
Scorpaeniformes	Drachenkopfartige, Drachenkopffischverwandte, Panzerwangen	sculpins & sea robins
Scorpaenodes guamensis	Dunkler Drachenkopf	Guam scorpionfish
Scorpaenodes xyris	Regenbogen-Drachenkopf	rainbow scorpionfish
Scorpaenopsis cirrhosa/ Dendroscorpaena cirrhosa	Verzierter Drachenkopf, Schmuck-Drachenkopf	hairy stingfish, raggy scorpionfish, weedy stingfish (FAO)
Scorpaenopsis diabolus	Falscher Drachenkopf, Teufels-Drachenkopf	false stonefish (FAO), devil stonefish
Scorpaenopsis gibbosa/ Scorpaenopsis barbata	Buckliger Drachenkopf, Buckel-Drachenkopf	humpbacked scorpionfish, humpback scorpionfish (FAO)
Scorpididae	Steinbrachsen	stonebreams
Scorpio maurus	Maurischer Skorpion	Maurish scorpion
Scorpiones	Skorpione	scorpions
Scotinomys spp.	Braunmäuse*	brown mice
Scotocerca inquieta	Wüstenprinie	scrub warbler
Scotomanes spp.	Harlekin-Fledermäuse	harlequin bats
Scotophaeus blackwalli	Amerikanische Hausspinne	American house spider

Scotophilus spp.	Gelbfledermäuse	house bats, yellow bats
Scotopteryx bipunctaria	Zweipunkt-Wellenstriemenspanner	chalk carpet (moth)
Scotopteryx chenopodiata	Platterbsenspanner	shaded broad-bar (moth)
Scraptiidae	Seidenkäfer	scraptiid beetles
Scrobicularia plana	Große Pfeffermuschel, Flache Pfeffermuschel	peppery furrow clam (peppery furrow shell)
Scrupocellaria scabra	Schild-Moostierchen*	shielded bryozoan
Scutelleridae	Schildwanzen	shield-backed bugs
Scutigera coleoptrata	Spinnenassel, Spinnenläufer	house centipede
Scutigerella immaculata (Symphyla)	Gewächshaus-Zwergfüßer, Zwergskolopender	glasshouse symphylid, garden symphylan, garden centipede
Scutigeridae	Spinnenasseln	centipedes
Scutigeromorpha/ Notostigmophora	Spinnenasseln, Spinnenläufer	scutigeromorphs
Scutisorex spp.	Panzerspitzmäuse	armored shrews, hero shrews
Scutopus megaradulatus (Aplacophora)	Korkenzieher-Schildfuß*, Großzungen-Schildfuß*	corkscrew glistenworm, corkscrew solenogaster
Scutopus ventrolineatus (Aplacophora)	Echter Schildfuß	bellylined glistenworm*
Scutus antipodes/ Scutus anatinus	Schildschnecke, Elefantenschnecke	elephant slug, duckbill shell, shield shell
Scydmaenidae	Ameisenkäfer	antlike stone beetles
Scyliorhinidae	Katzenhaie	catsharks, cat sharks
Scyliorhinus canicula/ Scyllium canicula	Kleingefleckter Katzenhai, Kleiner Katzenhai	lesser spotted dogfish, smallspotted dogfish, rough hound, smallspotted catshark (FAO)
Scyliorhinus stellaris	Großgefleckter Katzenhai	large spotted dogfish, nurse hound, nursehound (FAO), bull huss
Scylla serrata	Gezähnte Mangroven-Schwimmkrabbe	serrated mud swimming crab, serrated mangrove swimming crab, mud crab
Scyllaea pelagica	Sargassum-Nacktschnecke	sargassum nudibranch
Scyllaridae	Bärenkrebse	slipper lobsters, shovel-nosed lobsters
Scyllarides aequinoctialis	Karibischer Bärenkrebs, „Spanischer" Bärenkrebs	"Spanish" lobster, "Spanish" slipper lobster
Scyllarides astori	Kalifornischer Bärenkrebs	Californian slipper lobster
Scyllarides brasiliensis	Brasilianischer Bärenkrebs	Brasilian slipper lobster
Scyllarides latus	Großer Mittelmeer-Bärenkrebs, Großer Bärenkrebs	Mediterranean slipper lobster
Scyllarides nodifer	Gekielter Bärenkrebs	ridged slipper lobster
Scyllarus americanus	Amerikaischer Bärenkrebs	American slipper lobster
Scyllarus arctus	Kleiner Bärenkrebs, Grillenkrebs	small European locust lobster, small European slipper lobster, lesser slipper lobster
Scymnodon ringens	Messerzahn-Tiefen-Dornhai	knifetooth dogfish

Scymnorhinus licha/ Dalatias licha/ Squalus licha	Schokoladenhai	darkie charlie, kitefin shark (FAO), seal shark
Scypha ciliata/ Sycon ciliatum	Meergürkchen, Wimper-Kalkschwamm, Wimpern-Kalkschwamm	little vase sponge
Scypha compressa	Meertäschchen*	purse sponge
Scyphozoa	Schirmquallen, Scheibenquallen, Scyphozoen (echte Quallen)	cup animals, scyphozoans
Scytalina cerdale		graveldiver
Scytalinidae	Scytaliniden	graveldivers
Scythridiidae	Ziermotten	owlets, owlet moths
Scytodes thoracica	Speispinne, Leimschleuderspinne	spitting spider
Scytodidae	Speispinnen, Leimschleuderspinnen	spitting spiders
Searsiidae/Platytroctidae	Leuchtheringe	tubeshoulders
Sebastes alutus	Pazifischer Rotbarsch, Pazifik-Goldbarsch	Pacific ocean perch
Sebastes capensis	Kap-Drachenkopf	Jacopever, Cape redfish, false jacopero (FAO)
Sebastes entomelas	Witwen-Drachenkopf	widow rockfish
Sebastes flavidus	Gelbschwanz-Drachenkopf	yellowtail rockfish
Sebastes marinus	Rotbarsch, Goldbarsch (Großer Rotbarsch)	redfish, red-fish, Norway haddock, rosefish, ocean perch (FAO)
Sebastes mentella	Tiefenbarsch, Schnabelbarsch	deepwater redfish (FAO), beaked redfish
Sebastes viviparus	Norwegischer Rotbarsch	Norway haddock, lesser redfish
Securiflustra securifrons	Schmalblättriges Moostierchen	narrow-leaved hornwrack
Segestria senoculata	Kellerspinne	snake's back spider, cellar spider*
Segestriidae	Walzenspinnen	segestriids, tube-web spiders
Segmentina nitida	Glänzende Tellerschnecke	shining ramshorn snail, shiny ram's horn
Sehirus bicolor	Schwarzweiße Erdwanze	pied shieldbug
Seiurus aurocapillus	Pieperwaldsänger	ovenbird
Seiurus noveboracensis	Drosselwaldsänger	northern waterthrush
Sekeetamys calurus	Bilch-Rennmaus, Buschschwanz-Rennmaus	bushy-tailed jird
Selar boops	Ochsenaugen-Makrele	oxeye scad
Selar crumenophthalmus	Großäugiger Selar	big-eye scad
Selaroides leptolepis	Glattschwanzmakrele, Goldband-Selar	smooth-tailed trevally, yellow-stripe trevally, yellowstripe scad (FAO)
Selasphorus rufus	Zimtkolibri	rufous hummingbird
Selatosomus aeneus	Glanzschnellkäfer	shiny click beetle*
Selenarctos thibetanus/ Ursus thibetanus	Kragenbär	Asiatic black bear, Himalayan bear, Tibetan bear

Selene dorsalis	Afrikanischer Pferdekopf	African lookdown
Selene selene	Bodengucker	lookdown
Selene setapinnis/ *Vomer setapinnis*	Atlantischer Pferdekopf, Atlantischer „Mondfisch"	Atlantic moonfish
Selenia dentaria	Dreistreifiger Mondfleckspanner	early thorn
Selenia lunularia/S. lunaria	Mondspanner	lunar thorn
Selenia tetralunaria	Mondfleckspanner	purple thorn
Selenidera culik	Pfefferfresser	Guianan toucanet
Selenidera reinwardtii	Reinwardtarassari	golden-collarded toucanet
Selenocosmia javanensis	Java-Vogelspinne	Javanese whistling spider
Selenopidae	Selenopiden	selenopid crab spiders
Seleucidis melanoleuca	Fadenparadieshopf	twelve-wired bird-of-paradise
Selevinia betpakdalensis	Salzkrautbilch	desert dormouse
Semaeostomea	Fahnenquallen, Fahnenmundquallen	semeostome medusas, semaeostome medusae
Semanotus undatus	Nadelholz-Wellenbock	European pine longhorn beetle
Semaprochilodus taeniurus	Nachtsalmler, Schwanzstreifensalmler	colored prochilodus
Semele purpurascens	Purpur-Semele	purplish semele
Semiaphis dauci	Möhrenblattlaus, Mehlige Möhrenlaus	carrot aphid
Semibalanus balanoides	Gemeine Seepocke	northern rock barnacle, acorn barnacle
Semicassis granulatum undulatum	Mittelmeer-Helmschnecke	Mediterranean bonnet
Semicassis granulatum granulatum	Granula-Helmschnecke	scotch bonnet
Semilimax semilimax	Weitmündige Glasschnecke	wide-mouth glass snail
Seminatrix pygaea	Amerikanische Sumpfnatter	black swamp snake
Semiothisa carbonaria	Berg-Gitterspanner	netted mountain moth
Semiothisa clathrata	Gitterspanner	latticed heath (moth)
Semirossia equalis	Riesen-Rossie	greater bobtail squid
Semirossia tenera	Atlantische Rossie	lesser bobtail squid, Atlantic bob-tailed squid, lesser shining bobtail squid (FAO)
Semnornis ramphastinus	Tukanbartvogel	toucan barbet
Semudobia betulae	Birkensamen-Gallmücke	birch-seed gall midge
Sepia elegans	Kleine Sepie, Schlammsepie	elegant cuttlefish
Sepia officinalis	Gemeiner Tintenfisch, Gemeine Tintenschnecke, Gemeine Sepie	common cuttlefish
Sepia orbignyana	Dornsepie, Dornsepia	pink cuttlefish (FAO), Orbigny's cuttlefish
Sepia pharaonis	Pharaosepie, Pharao-Tintenfisch	pharaoh cuttlefish
Sepiana sepium/ *Platycleis sepium*	Zaunschrecke	sepia bushcricket
Sepiella ornata	Rotflecksepie*, Rotfleck-Tintenfisch	ornate cuttlefish
Sepietta minor/ *Rondeletiola minor*	Kleine Sepiette, Linsen-Sepiette	lentil bobtail squid

Sepietta oweniana	Große Sepiette	greater cuttlefish, common bobtail squid (FAO)
Sepioidea (or Sepiida)	Tintenschnecken, Eigentliche Tintenschnecken	cuttlefish & sepiolas
Sepiola atlantica	Atlantische Sepiole	Atlantic cuttlefish, little cuttlefish, Atlantic bobtail squid (FAO)
Sepiola rondeleti	Mittelmeer-Sepiole, Zwerg-Sepia, Zwergtintenfisch, Kleine Sprutte	Mediterranean dwarf cuttlefish, lesser cuttlefish, dwarf bobtail squid (FAO)
Sepioteuthis lessoniana	Großflossen-Sepiakalmar, Großflossen-Riffkalmar	bigfin reef squid
Sepioteuthis sepioidea	Echter Sepiakalmar, Karibischer Riffkalmar	Caribbean reef squid, Atlantic oval squid
Sepsidae	Schwingfliegen	black scavenger flies (spiny-legged flies)
Septibranchia	Siebkiemer, Verwachsenkiemer	septibranch bivalves, septibranchs
Seriatopora caliendrum	Nadelkoralle	
Seriatopora hystrix	Stachelige Nadelkoralle, Stachelige Buschkoralle, Dornige Reihenkoralle, Christusdorn-Koralle	needle coral, spiny row coral
Serica brunnea	Rotbrauner Laubkäfer	brown chafer (beetle)
Sericomyia silentis	Große Torf-Schwebfliege	
Sericornithini		scrubwrens
Sericus brunneus	Brauner Schnellkäfer	brown click beetle*
Serinus albogularis	Weißkehlgirlitz	white-throated seedeater
Serinus atrogularis	Angolagirlitz	yellow-rumped seedeater
Serinus canaria	Kanarengirlitz, Wilder Kanarienvogel	canary, common canary, island canary
Serinus citrinella	Zitronengirlitz	citril finch
Serinus flaviventris	Gelbbauchgirlitz	yellow canary
Serinus mozambicus	Mozambikgirlitz	yellow-fronted canary
Serinus pusillus	Rotstirngirlitz	red-fronted serin
Serinus scotops	Waldgirlitz	forest canary
Serinus serinus	Girlitz	serin
Serinus syriacus	Zederngirlitz	Syrian serin
Seriola dumerili	Bernsteinmakrele, Gabelschwanzmakrele	amberjack, greater amberjack (FAO), greater yellowtail
Seriola fasciata	Kleine Gabelschwanzmakrele	lesser amberjack
Seriola lalandei	Australische Gelbschwanzmakrele, Riesen-Gelbschwanzmakrele	giant yellowtail, yellowtail kingfish, yellowtail amberjack (FAO)
Seriola quinqueradiata	Japanische Bernsteinmakrele, Japanische Seriola	Japanese amberjack
Seriola rivoliana	Kleine Bernsteinmakrele, Augenstreifen-Königsfisch	Almaco jack (FAO), longfin yellowtail
Seriola zonata	Gestreifte Bernsteinmakrele*	banded rudderfish
Serpentes/Ophidia (Squamata)	Schlangen	snakes, serpents, ophidians
Serpophaga cinerea	Sturzbachtachuri	torrent tyrannulet

Serpula vermicularis	Bunter Kalkröhrenwurm, Kleiner Kalkröhrenwurm, Deckel-Kalkröhrenwurm	red tubeworm
Serpulidae	Röhrenwürmer	serpulids, serpulid tubeworms, serpulid worms
Serpulorbis arenarius	Riesenwurmschnecke	giant wormsnail
Serpulorbis decussatus	Gekreuzte Wurmschnecke*	decussate wormsnail
Serpulorbis imbricata	Schuppenwurmschnecke	scaly wormsnail (scaly wormshell)
Serranidae	Zackenbarsche, Sägebarsche	seabasses & rockcods (groupers)
Serranus cabrilla	Sägebarsch, Längsgestreifter Schriftbarsch, Ziegenbarsch	comber
Serranus hepatus	Brauner Sägebarsch, Zwergbarsch	brown comber
Serranus phoebe	Masken-Sägebarsch	tattler
Serranus scriba	Schriftbarsch	painted comber (FAO), banded sea-perch
Serranus subligarius	Gürtel-Sandfisch	belted sandfish
Serranus tigrinus	Zwergtigerbarsch	harlequin bass
Serrasalmidae/Serrasalminae	Sägesalmler	pacus & silver dollars & piranhas
Serrasalmus brandtii	Pirambeba	pirambeba, white piranha (FAO)
Serrasalmus nattereri	Natterers Sägesalmler, Roter Piranha	convex-headed piranha, Natterer's piranha, red piranha (FAO), red-bellied piranha
Serrasalmus niger	Schwarzer Piranha	black piranha
Serrasalmus piraya	Piraya, Piranha, Piranha, Karibenfisch	piranha
Serrasalmus rhombeus	Gefleckter Sägesalmler	white piranha, spotted piranha, redeye piranha (FAO)
Serrasalmus spilopleura	Schwarzband-Sägesalmler	dark-banded piranha
Serrasalmus spp.	Sägesalmler, Pirayas, Pirahas, Piranhas	piranhas
Serripes groenlandicus	Grönländische Herzmuschel	Greenland cockle
Serrivomer beani	Beans Sägezahnaal*	Bean's sawtoothed eel
Serrivomer parabeani	Fadenaal*	thread eel
Serrivomeridae	Sägezahnaale	sawtooth eels
Serropalpidae (Melandryidae)	Düsterkäfer, Schwarzkäfer	false darkling beetles
Sertella beaniana	Neptunschleier, Netzkoralle	net 'coral'
Sertularella polyzonias	Zahnmoos*	great tooth hydroid
Sertularella rugosa	Schneckenmoos*	snail trefoil hydroid
Sertularia argentea	Buschschwanzmoos*	squirrel's tail hydroid, sea fir
Sertularia cupressina	Seemoos, Zypressenmoos	sea cypress, whiteweed hydroid
Sertularia pumila/ Dynamena pumila/ Dynamena cavolini	Kleines Seemoos, Zwergmoos	minute hydroid, garland hydroid

Serviformica fusca/ Formica fusca	Schwarzgraue Hilfsameise	silky ant
Sesia apiformis/ Aegeria apiformis	Hornissenschwärmer, Bienenglasflügler	poplar hornet clearwing, hornet moth
Sesiidae/Aegeriidae	Glasflügler (Glasschwärmer)	clearwing moths, clear-winged moths
Setaria equina	Pferdebauchhöhlenfilarie	equine abdominal worm
Setonix brachyurus	Kurzschwanzkänguru, Quokka	quokka
Setophaga ruticilla	Schnäpperwaldsänger	American redstart
Shinisaurus crocodilurus	Krokodilschwanz-Höckerechse	Chinese xenosaur, Chinese crocodile lizard
Sialia currucoides	Berghüttensänger	mountain bluebird
Sialia mexicana	Blaukehl-Hüttensänger	western bluebird
Sialia sialis	Rotkehl-Hüttensänger	eastern bluebird
Sialidae	Wasserflorfliegen (Schlammfliegen)	alder flies
Sicariidae/Loxoscelidae	Braunspinnen, Speispinnen	violin spiders, recluse spiders, sixeyed sicariid spiders
Sicista betulina	Waldbirkenmaus	northern birch mouse
Sicista spp.	Birkenmäuse, Streifen-Hüpfmäuse	birch mice
Sicista subtilis	Steppenbirkenmaus	southern birch mouse
Sicyonia brevirostris	Braune Felsen-Kantengarnele	brown rock shrimp
Sicyonia dorsalis	Kleine Felsen-Kantengarnele	lesser rock shrimp
Sicyonia typica	Königliche Felsen-Kantengarnele	kinglet rock shrimp
Sicyoniidae	Felsen-Kantengarnelen	rock shrimps
Siderastrea radians	Vieltrichterkoralle	lesser starlet coral
Siderea grisea	Weiße Muräne	geometric moray
Siderea picta/ Gymnothorax pictus	Pfeffermuräne	peppered moray (FAO), painted moray
Siebenrockiella crassicollis	Schwarze Dickkopf-Schildkröte	black marsh turtle, Siamese temple turtle
Sigalionidae	Sigalioniden	sigalionids, scaleworms a.o.
Siganidae/Teuthidae	Kaninchenfische	rabbitfishes
Siganus argenteus	Marmorierter Kaninchenfisch	roman-nose spinefoot, streamlined spinefoot (FAO)
Siganus chrysopilos	Gepunkteter Kaninchenfisch	gold-spotted spinefoot
Siganus corallinus	Korallen-Kaninchenfisch	coral rabbitfish
Siganus luridus	Brauner Kaninchenfisch	dusky rabbitfish
Siganus rivulatus	Rotmeer-Kaninchenfisch	marbled rabbitfish, marbled spinefoot (FAO)
Siganus spp.	Kaninchenfische	rabbitfishes
Siganus stellatus	Tüpfel-Kaninchenfisch	brown-spotted spinefoot
Siganus sutor	Tarnfarben-Kaninchenfisch	shoemaker spinefoot
Siganus vulpinus/ Lo vulpinus	Fuchsgesicht	foxface
Sigmoceros lichtensteini/ Alcelaphus lichtensteini	Lichtensteins Hartebeest	Lichtenstein's hartebeest
Sigmodon spp.	Baumwollratten	cotton rats
Silicospongiae	Kieselschwämme (Demospongien)	siliceous sponges, demosponges

Siliqua costata	Atlantische Messermuschel	Atlantic razor clam
Siliqua patula	Pazifische Messermuschel	Pacific razor clam
Siliqua radiata	Sonnenstrahl-Messermuschel, Sonnenstrahl-Scheidenmuschel	sunset razor clam, sunset siliqua
Siliqua squama	Raue Atlantische Messermuschel	rough razor clam, squamate razor clam
Siliquaria anguina	Schotenschnecke, Schlangenschnecke, Aalwurmschnecke	pod snail (pod shell, wormshell)
Siliquaria squamata/ Tenagodus squamatus	Schlitz-Wurmschnecke	slit wormsnail
Siliquariidae	Schlangenschnecken	pod snails (pod shells, wormshells)
Sillaginidae	Weißlinge	sillagos (whitlings, smelt-whitlings)
Silphidae	Aaskäfer	carrion beetles, burying beetles
Silurana tropicalis	Tropen-Krallenfrosch	tropical clawed frog
Siluridae	Welse, Echte Welse	sheatfishes
Siluriformes	Welse, Welsartige	catfishes
Silurus glanis	Wels, Waller, Schaiden	European catfish, wels, sheatfish, wels catfish (FAO)
Simenchelyidae/ Simenchelyinae	Stumpfnasenaale, Parasitische Aale, Schmarotzeraale	snubnose parasitic eels
Simenchelys parasiticus	Stumpfnase, Stumpfnasenaal, Parasitischer Aal	snubnosed eel (FAO), snubnose parasitic eel
Simias concolor/ Nasalis concolor	Ringelschwanz-Stumpfnasenaffe, Pageh-Stumpfnasenaffe	pig-tailed snub-nosed monkey, pig-tailed snub-nosed langur
Simnia acicularia/ Cymbovula acicularis	West-Indische Gorgonien-Porzellanschnecke	West Indian simnia
Simnia patula	Offene Spelze	open simnia*
Simnia spelta	Gorgonien-Porzellanschnecke	gorgonian simnia*
Simoselaps spp.	Australische Korallenottern	Australian coral snakes
Simuliidae	Kriebelmücken	black flies, blackflies, buffalo gnats
Simulium spp.	Kriebelmücken	black flies, blackflies
Singa hamata	Glanzspinne, Glanzkreuzspinne	shiny orbweaver*
Singa spp.	Gestreifte Radnetzspinnen*	striped orbweavers
Sinodendron cylindricum	Kopfhornschröter, Baumschröter	rhinoceros beetle, "small European rhinoceros beetle"
Sinum perspectivum	Weiße Babyschnecke*	baby's ear moonsnail, white baby ear
Siona lineata	Schwarznervmotte*	black-veined moth
Siphonaptera/ Aphaniptera/Suctoria	Flöhe	fleas
Siphonaria pectinata	Gestreifte Unechte Napfschnecke	striped false limpet
Siphonochalina siphonella	Röhrenschwamm u.a.	tube sponge

Siphonophanes grubii/ *Chirocephalus grubei*	Frühjahrskiemenfuß, Frühjahrs-Kiemenfuß, Gemeiner Kiemenfuß	
Siphonophora	Staatsquallen	siphonophorans
Siphonops annulatus	Ringelwühle	ringed caecilian
Sipunculida	Spritzwürmer, Sternwürmer, Sipunculiden	peanut worms, sipunculoids, sipunculans
Siratus alabaster	Alabasterschnecke	alabaster murex, abyssal murex
Siratus laciniatus	Zipfelschnecke	laciniate murex
Siratus motacilla	Bachstelzen-Schnecke	wagtail murex
Siren intermedia	Kleiner Armmolch	lesser siren
Siren lacertina	Großer Armmolch	greater siren
Sirenia	Seekühe	sirenians: sea cows & manatees & dugongs
Sirenidae	Armmolche	sirens
Sirex cyaneus	Blaue Kiefernholzwespe	blue horntail
Sirex juvencus	Kiefernholzwespe, Gemeine Holzwespe	polished horntail
Siricidae	Holzwespen	woodwasps, horntails
Siro duricorius	Zwergweberknecht	mite harvestmen
Sisoridae	Saugwelse, Turkestanische Zwergwelse	sisorid catfishes
Sistrurus catenatus	Massasauga, Kettenklapperschlange	massasauga
Sistrurus miliarius	Eigentliche Zwergklapperschlange	pigmy rattlesnake
Sistrurus ravus	Mexikanische Zwergklapperschlange	Mexican pigmy rattlesnake
Sisyphus schaefferi	Matter Pillenwälzer	
Sisyridae	Schwammfliegen	brown lacewings
Sitobion avenae/ *Macrosiphum avenae*	Große Getreideblattlaus	greater cereal aphid, European grain aphid, English grain aphid
Sitodiplosis mosellana	Orangerote Weizengallmücke, Rote Weizengallmücke	orange wheat blossom midge
Sitona lineata	Gestreifter Blattrandkäfer	pea leaf weevil
Sitophilus granarius	Kornkäfer, Schwarzer Kornwurm	grain weevil
Sitophilus oryzae	Reiskäfer	rice weevil
Sitophilus zeamais	Maiskäfer	corn weevil
Sitotroga cerealella	Getreidemotte, Französische Kornmotte (Weißer Kornwurm)	Angoumois grain moth
Sitta canadensis	Kanadakleiber	red-breasted nuthatch
Sitta carolinensis	Carolinakleiber	white-breasted nuthatch
Sitta europaea	Kleiber	Eurasian nuthatch
Sitta krueperi	Türkenkleiber	Krüper's nuthatch
Sitta ledanti	Kabylenkleiber	Algerian nuthatch
Sitta neumayer	Felsenkleiber	rock nuthatch
Sitta pygmaea	Zwergkleiber	pygmy nuthatch
Sitta tephronota	Klippenkleiber	great rock nuthatch
Sitta whiteheadi	Korsenkleiber	Corsican nuthatch

Sittidae	Kleiber, Spechtmeisen	nuthatches
Smaragdia viridis	Smaragdschnecke	emerald nerite
Smaragdinella calyculata/ Smaragdinella viridis	Smaragd-Kugelschnecke	emerald bubble snail, emerald bubble
Smerinthus ocellata	Abendpfauenauge	eyed hawkmoth
Smicromyrme rufipes/ Mutilla rufipes	Rotbeinige Spinnenameise	small velvet ant
Sminthopsis spp.	Schmalfußbeutelmäuse	dunnarts, narrow-footed marsupial "mice"
Sminthuridae	Kugelspringer	globular springtails
Sminthurus viridis	Luzernefloh	lucerne flea
Smittina cervicornis/ Porella cervicornis	Hirschgeweih-Moostierchen	staghorn bryozoan, deer-antler crust*
Solariella obscura	Seltsame Solariella*	obscure solarelle
Solariella varicosa	Schwielen-Solariella*	varicose solarelle
Solariidae/Architectonicidae	Sonnenschnecken	sundials, sundial snails (sun shells, sundial shells)
Solaster endeca	Violetter Sonnenstern	smooth sun star, purple sun star, purple sunstar
Solaster papposus	Seesonne, Sonnenstern, Europäischer Quastenstern	common sun star, spiny sun star
Solasteridae	Sonnensterne	sun stars
Solea azevia/ Microchirus azevia/ Microchirus theophila	Azevia-Seezunge	bastard sole (FAO), Jewish sole
Solea elongata	Arabische Seezunge	elongate sole (FAO), Arabic sole
Solea kleinii/ Synaptura kleinii	Schwarzrand-Seezunge	Klein's sole
Solea lascaris/ Pegusa lascaris	Sandzunge, Warzen-Seezunge	sand sole
Solea lutea/ Buglossidium luteum	Zwergzunge, Zwergseezunge	pygmy sole, solenette
Solea ocellata/ Microchirus ocellatus	Augen-Seezunge, Augenfleck-Seezunge	four-eyed sole, foureyed sole (FAO)
Solea variegata/ Microchirus variegatus	Bastardzunge	thickback sole (FAO), thick-backed sole
Solea vulgaris/Solea solea	Seezunge, Gemeine Seezunge	common sole (Dover sole)
Solecurtidae	Kurze Messermuscheln, Kurze Scheidenmuscheln	short razor clams
Solecurtus strigillatus	Striegelmuschel, Gemeine Striegelmuschel	scraper clam, rosy razor
Solegnathus spinosissimus	Fetzenfeilenfisch, Schmuck-Feilenfisch	spiny pipehorse, spiny seadragon
Soleidae	Seezungen	soles
Solemya borealis	Nördliche Schotenmuschel	boreal awningclam
Solemya grandis	Grosse Schotenmuschel	grand awningclam
Solemya togata	Toga-Schotenmuschel	toga awningclam
Solemya velum	Atlantische Schotenmuschel	Atlantic awningclam
Solen marginatus	Gefurchte Scheidenmuschel	grooved razor clam
Solen vagina	Große Scheidenmuschel	European razor clam
Solen viridis	Grüne Scheidenmuschel	green jackknife clam

Solenastrea bournoni	Glatte Sternkoralle*	smooth star coral
Solenastrea hyades	Knotige Sternkoralle*	knobby star coral
Solenidae	Scheidenmuscheln	razor clams (razor shells)
Solenocera membranacea	Membran-Geißelgarnele	mud shrimp
Solenoceridae		solenocerid shrimps
Solenoconchae/ Scaphopoda (Conchifera)	Kahnfüßer, Grabfüßer, Scaphopoden	scaphopods, scaphopodians (tooth shells, tusk shells, spade-footed mollusks)
Solenodon cubanus	Kuba-Schlitzrüssler	Cuban solenodon
Solenodon paradoxus	Haiti-Schlitzrüssler	Haitian solenodon
Solenodontidae	Schlitzrüssler	solenodons
Solenogastres/ Neomeniomorpha (Aculifera)	Furchenfüßer	solenogasters
Solenopotes capillatus	Borstige Rinderlaus	tubercle-bearing louse, blue cattle louse, small blue cattle louse
Solenopsis fugax/ Diplorhoptrum fugax	Innennest-Raubameise, „Diebsameise"	European fire ant
Solenopsis geminata	Amerikanische Feuerameise	American fire ant
Solenopsis molesta	Amerikanische Diebsameise	thief ant
Solenopsis spp.	Diebsameisen	fire ants
Solenopsis xyloni	Südliche Feuerameise	southern fire ant
Solenosteira pallida	Heller Pokal	pale goblet
Solenostomidae	Röhrenmünder, Geisterpfeifenfische	ghost pipefishes
Solenostomus spp.	Röhrenmünder	ghost pipefishes
Solifugae/Solpugida	Walzenspinnen	sun spiders, false spiders, windscorpions, solifuges, solpugids
Solisorex spp.	Langkrallenspitzmäuse	long-clawed shrews
Solomys ponceleti	Poncelets Riesenratte	Poncelet's giant rat
Solomys spp.	Nacktschwanzratten* (Solomon-Inseln)	naked-tailed rats
Somateria fischeri	Plüschkopfente	spectacled eider
Somateria mollissima	Eiderente	common eider
Somateria spectabilis	Prachteiderente	king eider
Somatochlora alpestris	Alpen-Smaragdlibelle	Alpine emerald
Somatochlora artica	Arktische Smaragdlibelle	northern emerald
Somatochlora flavomaculata	Gefleckte Smaragdlibelle	yellow-spotted emerald
Somatochlora metallica	Glänzende Smaragdlibelle	brillant emerald
Somniosus microcephalus	Grönlandhai, Großer Grönlandhai, Eishai	Greenland shark (FAO), ground shark
Somniosus rostratus	Kleiner Eishai, Lemargo	little sleeper shark
Sonora semiannulata	Bodennatter	ground snake
Sooglossus spp.	Seychellenfrösche	Seychelles frogs
Sorex alpina	Alpenspitzmaus	Alpine shrew
Sorex araneus	Waldspitzmaus	common shrew, Eurasian common shrew

Sorex bendirii	Pazifische Wasserspitzmaus	Pacific water shrew
Sorex cinereus	Aschgraue Spitzmaus	masked shrew
Sorex coronatus/ Sorex gemellus	Schabrackenspitzmaus	Millet's shrew
Sorex granarius	Kastilienspitzmaus	Spanish shrew
Sorex minutissimus	Knirpsspitzmaus	least shrew
Sorex minutus	Zwergspitzmaus	Eurasian pygmy shrew
Sorex palustris	Moorspitzmaus, Nördliche Wasserspitzmaus	northern water shrew
Sorex samniticus	Apenninenspitzmaus	Apennine shrew
Sorex spp.	Rotzahnspitzmäuse, Waldspitzmäuse	red-toothed shrews, long-tailed shrews
Soricidae	Spitzmäuse	shrews
Soriculus spp.	Großklauenspitzmäuse	Asiatic shrews
Sorubim lima	Spatelwels	cucharon, shovel-nosed catfish, duckbill catfish (FAO)
Sotalia fluviatilis	Amazonas-Sotalia	Tucuxi, Amazon River dolphin, Tookashee dolphin
Sotalia guianensis	Guayana-Delphin, Karibischer Küstendelphin	Guayana river dolphin, Guiana coastal dolphin, white dolphin
Sousa chinensis	Chinesischer Weißer Delphin, Indopazifischer Buckeldelphin	Chinese humpback dolphin, Indo-Pacific hump-backed dolphin, white dolphin
Sousa teuszi	Kamerunfluss-Delphin, Kamerun-Buckeldelphin, Atlantischer Buckeldelphin	humpback dolphin, Atlantic hump-backed dolphin, West African hump-backed dolphin
Spadella cephaloptera (Chaetognatha)	Brauner Pfeilwurm	brown arrow worm
Spalacidae	Blindmäuse	blind mole-rats
Spalacopus cyanus	Cururo	coruro
Spalax graecus	Bukowinische Blindmaus	Bukowinian blind mole-rat, greater mole rat
Spalax leucodon	Kleine Blindmaus	lesser mole rat
Spalax microphthalmus	Große Blindmaus	greater mole rat
Spalax polonicus	Podolische Blindmaus	Podolian blind mole-rat
Spalax spp.	Blindmäuse	blind mole-rats
Spalerosophis diadema	Diademnatter	diadem snake
Sparassidae	Riesenkrabbenspinnen	giant crab spiders
Sparganothis pilleriana	Springwurmwickler, Laubwurm	
Sparidae	Brassen, Meerbrassen	seabreams, porgies
Sparisoma cretense	Seepapagei, Papageifisch	parrotfish (Mediterranean/Atlantic)
Sparisoma viride	Rautenpapageifisch, Signal-Papageifisch	stoplight parrotfish
Sparus auratus	Goldbrassen	gilthead, gilthead seabream (FAO)
Sparus pagrus/ Pagrus pagrus	Sackbrassen	Couch's seabream, common seabream (FAO), red porgy
Spatangoida	Herzigel, Herzseeigel	heart urchins

Spatangus purpureus	Violetter Herzigel, Violetter Herzseeigel, Purpur-Herzigel, Purpurigel	purple heart urchin
Spathodus erythrodon	Blaupunkt-Grundelbuntbarsch	
Speleogobius trigloides	Grottengrundel	grotto goby
Speleomantes flavus	Monte-Albo-Höhlensalamander	Stefani's salamander
Speleomantes genei/ Hydromantes genei	Sardinischer Schleuderzungensalamander, Brauner Höhlensalamander	Sardinian cave salamander, brown cave salamander
Speleomantes imperialis	Prächtiger Höhlensalamander	imperial salamander
Speleomantes italicus	Italienischer Höhlensalamander	Apennines salamander
Speleomantes supramontis	Supramonte-Höhlensalamander	Nuoro salamander
Speothos venaticus	Waldhund	bush dog
Spermodea lamellata	Bienenkörbchen	plaited snail
Spermophilopsis leptodactylus	Zieselmaus	long-clawed ground squirrel
Spermophilus tridecemlineatus	Dreizehnstreifenziesel	thirteen-lined ground squirrel
Spermophilus citellus/ Citellus citellus	Europäisches Ziesel, Schlichtziesel	European ground squirrel, European suslik, European souslik
Spermophilus fulvus	Sandziesel (Hbz. Petschanik)	large-toothed souslik
Spermophilus lateralis	Goldmantelziesel, Goldgestreiftes Backenhörnchen	golden-mantled ground squirrel
Spermophilus spp./ Citellus spp.	Ziesel	ground squirrels, susliks, sousliks
Spermophilus suslicus	Perlziesel (Hbz. Perlsuslik)	spotted suslik
Spermophilus undulatus	Parry-Ziesel, Langschwanzziesel (Hbz. Suslik)	Arctic ground squirrel
Spermophora meridionalis	Gedrungene Kellerspinne*	short-bodied cellar spider
Sphaerechinus granularis	Violetter Seeigel, Dunkelvioletter Seeigel	purple sea urchin, violet urchin
Sphaerias blanfordi	Gebirgs-Flughund	mountain fruit bat
Sphaerichthys osphromenoides	Schokoladengurami	chocolate gourami
Sphaeridae	Kugelkäfer	minute bog beetles
Sphaeriidae	Kugelmuscheln	orb mussels (orb shells, sphere shells)
Sphaeritidae	Falsche Clownkäfer	false clown beetles
Sphaerium corneum	Gemeine Kugelmuschel, Hornfarbene Kugelmuschel, Linsenmuschel	horny orb mussel, European fingernailclam
Sphaerium lacustre/ Musculium lacustre	Haubenmuschel, Häubchenmuschel, Teich-Kugelmuschel	lake orb mussel, lake fingernailclam
Sphaerium rivicola	Große Kugelmuschel, Ufer-Kreismuschel, Fluss-Kugelmuschel	nut orb mussel (nut orb shell), river orb mussel, nut fingernailclam
Sphaerium solidum	Dickschalige Kugelmuschel	thickshelled fingernailclam
Sphaerium spp.	Kugelmuscheln	orb mussels, fingernailclams

Sphaeroceridae	Dungfliegen, Kleindungfliegen	small dungflies, lesser dungflies
Sphaerodactylus argus	Augenfleck-Kugelfingergecko	ocellated gecko
Sphaerodactylus cinereus	Aschgrauer Kugelfingergecko	ashy gecko, gray gecko
Sphaerodactylus notatus	Florida Riffgecko	reef gecko
Sphaerodactylus spp.	Kugelfingergeckos	least geckos
Sphaeroma hookeri	Meeres-Rollassel	marine pillbug a.o.
Sphaeroma rugicaudata	Rauschwänzige Rollassel	roughtail marine pillbug
Sphaeronassa mutabilis/ Nassarius mutabilis	Wandelbare Reusenschnecke	mutable nassa
Sphaerophoria scripta	Gemeine Stiftschwebfliege	
Sphaerosoma pilosum	Kugeliger Stäublingskäfer	
Sphaerotheriidae	Riesenkugler	giant pill millepedes
Sphaerularia bombi	Hummelnematode	bumblebee nematode
Sphecidae/Sphegidae	Grabwespen	digger wasps, hunting wasps
Sphecinae	Sandwespen	sand wasps, thread-waisted wasps
Spheciospongia vesparia	Tölpelkopfschwamm	loggerhead sponge
Sphecius speciosus	Zikadentöter*	cicada killer
Sphecotheres viridis	Feigenpirol	figbird
Spheniscidae	Pinguine	penguins
Sphenisciformes	Pinguine	penguins
Spheniscus demersus	Brillenpinguin	jackass penguin
Spheniscus humboldti	Humboldtpinguin	Humboldt penguin
Spheniscus magellanicus	Magellanpinguin	Magellanic penguin
Spheniscus mendiculus	Galapagospinguin	Galapagos penguin
Sphenodon punctatus	Brückenechse	tuatara
Sphenomorphus spp.	Waldskinke	forest skinks
Sphenops sepsoides	Keilschleiche	wedge-snouted skink
Sphex rufocinctus	Heuschrecken-Sandwespe	
Sphindidae	Staubpilzkäfer	dry-fungus beetles
Sphingidae	Schwärmer	hawkmoths, sphinx moths
Sphingonotus caerulans	Blauflügelige Sandschrecke, Blauflügelschrecke, Sandschrecke	blue-winged locust
Sphinx ligustri	Ligusterschwärmer	privet hawkmoth
Sphinx pinastri/ Hyloicus pinastri	Kiefernschwärmer, Kleiner Fichtenrüssler	pine hawkmoth
Sphoeroides maculatus/ Sphaeroides maculatus	Kugelfisch, Nördlicher Kugelfisch	northern puffer (FAO), swellfish
Sphoeroides nephelus	Südlicher Kugelfisch	southern puffer
Sphoeroides pachygaster/ Sphaeroides pachygaster/ Sphoeroides cutaneus/ Sphaeroides cutaneus	Gelbflossenaufbläser	blunthead puffer (FAO), smooth pufferfish
Sphoeroides testudineus	Schildkröten-Kugelfisch	checkered puffer
Sphyradium doliolum	Kleine Fässchenschnecke	
Sphyraena afra	Guinea-Barrakuda	Guinean barracuda
Sphyraena argentea	Kalifornischer Barrakuda, Pazifik-Barrakuda	Pacific barracuda (FAO), California barracuda

Sphyraena barracuda	Atlantischer Barrakuda, Großer Barrakuda	great barracuda
Sphyraena chrysotaenia/ Sphyraena flavicauda/ Sphyraena obtusata	Gelbflossen-Barrakuda	yellowtail barracuda, obtuse barracuda, blunt-jaw barracuda, yellowstripe barracuda (FAO)
Sphyraena jello	Indischer Barrakuda, Indomalayischer Barrakuda	banded barracuda, pickhandle barracuda (FAO), Indian barracuda
Sphyraena sphyraena	Pfeilhecht, Mittelmeer-Barrakuda, Quergestreifter Pfeilhecht	Mediterranean barracuda, European barracuda (FAO)
Sphyraena spp.	Barrakudas, Pfeilhechte	barracudas
Sphyraena viridensis	Gelbmaul-Barrakuda*	yellowmouth barracuda
Sphyraenidae	Barrakudas, Pfeilhechte	barracudas
Sphyrapicus ruber	Feuerkopf-Saftlecker	red-breasted sapsucker
Sphyrna lewini	Bogenstirn-Hammerhai, Indopazifischer Hammerhai	scalloped hammerhead shark
Sphyrna mokarran/ Sphyrna tudes	Großer Hammerhai, Kleinäugiger Hammerhai	great hammerhead (FAO), giant hammerhead shark
Sphyrna tiburo	Schaufelnasen-Hammerhai	bonnet shark, bonnethead (FAO), shovel head
Sphyrna zygaena/ Zygaena malleus	Gemeiner Hammerhai, Glatter Hammerhai	common hammerhead shark, smooth hammerhead
Sphyrnidae	Hammerhaie	hammerhead sharks
Spialia sertorius	Roter Würfelfalter, Wiesenknopf-Würfeldickkopffalter, Roter Würfel-Dickkopf	red underwing skipper
Spicara alta	Großaugen-Pikarel	bigeye picarel
Spicara axillaris		windtoy
Spicara flexuosa	Garizzo	garizzo
Spicara maena/Maena maena	Gefleckter Schnauzenbrassen, Laxierfisch	menola, menole, mendole, blotched picarel (FAO)
Spicara smaris	Pikarel	picarel, zerro
Spicara spp.	Pikarels, Schnauzenbrassen	picarels
Spilogale spp.	Fleckenskunks (Hbz. Lyraskunk)	spotted skunks
Spilonota ocellana	Roter Knospenwickler	eyespotted bud moth
Spilopsyllus cuniculi	Kaninchenfloh	rabbit flea, European rabbit flea
Spilornis cheela	Schlangenweihe	crested serpent eagle
Spilosoma lubricipeda/ Spilosoma menthastri	Weiße Tigermotte, Punktierter Fleckleib-Bär, Minzenbär	white ermine moth
Spilosoma lutea	Gelbe Tigermotte, Holunderbär	buff ermine moth
Spilosoma urticae	Nesselbär, Weißer Fleckleib-Bär	water ermine moth
Spilotes pullatus pullatus	Hühneresser, Hühnerfresser	tropical chicken snake

Spinachia spinachia	Seestichling, Meerstichling	fifteen-spined stickleback
Spinturnicidae	Fledermausmilben	bat mites*
Spio quadricornis	Vierhorn-Haarwurm*	four-horned spio
Spio seticornis	Borstenfühleriger Haarwurm	seticorn spio*
Spio setosa	Sandröhren-Haarwurm	sand chimney worm
Spiochaetopterus costarum	Glasröhrenwurm*	glassy tubeworm
Spionidae	Haarwürmer	spionid worms, spios, spionids
Spirastrella cunctatrix	Oranger Strahlenschwamm	orange vented sponge, orange ray sponge
Spirastrella purpurea	Orangeroter Bohrschwamm	orangered boring sponge*, purple boring sponge
Spirinchus starksi	Nacht-Stint	night smelt
Spirinchus theleichthys	Sacramento-Stint	Sacramento smelt
Spirobranchus giganteus	Tannenbäumchen-Röhrenwurm, Weihnachtsbaum-Röhrenwurm	spiral-gilled tubeworm, Christmas tree worm
Spirobranchus tetraceros	Vierzahn-Röhrenwurm	four-tooth tubeworm
Spirocerca lupi/ Spirocerca sanguinolenta	Hunde-Spirocercose-Wurm, Ösophagus-Rollschwanz	esophageal worm (of dog)
Spirographis spallanzanii/ Sabella spallanzanii	Schraubensabelle	fanworm, feather-duster worm a.o.
Spirolaxis centrifuga	Zentrifugal-Sonnenschnecke*	exquisite false-dial
Spirorbidae	Posthörnchenwürmer	spiral tubeworms
Spirorbis spp.	Posthörnchenwürmer	spiral tubeworms
Spirula spirula	Posthörnchen	ram's horn squid, common spirula
Spisula elliptica	Kleine Trogmuschel	elliptical surfclam (elliptical trough shell)
Spisula solida	Ovale Trogmuschel, Dickschalige Trogmuschel, Dickwandige Trogmuschel	thick surfclam (thick trough shell)
Spisula solidissima/ Hemimactra gigantea	Atlantische Riesentrogmuschel	Atlantic surfclam, solid surfclam, bar clam
Spisula spp.	Trogmuscheln	surfclams a.o. (trough shells)
Spisula subtruncata	Gedrungene Trogmuschel, Dreieckige Trogmuschel	cut surfclam (cut trough shell)
Spiza americana	Dickcissel	dickcissel
Spizaetus cirrhatus	Haubenadler	crested hawk eagle
Spizaetus coronatus	Kronenadler	crowned eagle
Spizaetus isidori	Isidoradler	black-and-chestnut eagle
Spizaetus occipitalis	Schopfadler	long-crested eagle
Spizaetus ornatus	Prachtadler	ornate hawk eagle
Spizastur melanoleucus	Elsteradler	black-and-white hawk eagle
Spizella arborea	Baumammer	American tree sparrow
Spizella passerina	Schwirrammer	chipping sparrow
Spizixos canifrons	Finkenbülbül	crested finchbill
Spodoptera litura	Ägyptische Baumwollraupe	cluster caterpillar
Spondylidae	Stachelaustern	thorny oysters
Spondyliosoma cantharus	Streifenbrassen, Brandbrassen	black bream, old wife, black seabream (FAO)
Spondylis buprestoides	Waldbock	European cylinder longhorn beetle

Spondylis spp.	Waldböcke	cylinder longhorn beetles, cylinder longhorns
Spondylus americanus	Atlantik-Stachelauster, Atlantische Stachelauster, Amerikanische Stachelauster	Atlantic thorny oyster, American thorny oyster
Spondylus anacanthus	Nackte Stachelauster	nude thorny oyster
Spondylus barbatus	Bärtige Stachelauster	bearded thorny oyster
Spondylus butleri	Butlers Stachelauster	Butler's thorny oyster
Spondylus gaederopus	Stachelauster, Eselshuf	European thorny oyster
Spondylus ictericus	Fingerige Stachelauster*, Gefingerte Stachelauster*	digitate thorny oyster
Spondylus linguaefelis	Katzenzungen-Auster	cat's-tongue oyster
Spondylus princeps	Pazifik-Stachelauster, Pazifische Stachelauster, Panama-Stachelauster	Pacific thorny oyster
Spondylus regius	Königliche Stachelauster	royal thorny oyster
Spondylus squamosus	Schuppige Stachelauster	scaly thorny oyster
Spondylus tenellus	Scharlach-Stachelauster	scarlet thorny oyster
Spondylus wrighteanus	Wrights Stachelauster	Wright's thorny oyster
Spongia agaricina	Elefantenohr	elephant's ear sponge
Spongia irregularis	Gelbschwamm	yellow sponge, hardhead sponge
Spongia officinalis	Dalmatiner Schwamm, Badeschwamm, Tafelschwamm	bath sponge
Spongia officinalis lamella	Elefantenohrschwamm	elephant's ear sponge
Spongia officinalis mollissima	Feiner Levantinerschwamm	fine Levant sponge, Turkey cup sponge
Spongia zimocca	Zimokkaschwamm	brown Turkey cup sponge
Spongilla fragilis	Bruchschwamm, Sternlochschwamm	freshwater sponge
Spongilla lacustris	Teichschwamm, Stinkschwamm, (Geweihschwamm)	pond sponge, freshwater sponge
Sporophila americana	Wechselpfäffchen	wing-barred seedeater
Sporozoa	Sporentierchen, Sporozoen	spore-former, sporozoans
Spratelloides delicatulus	Delikat-Rundhering*	delicate round herring
Spratelloides gracilis	Silberstreifen-Rundhering	silver-stripe round herring
Sprattus sprattus	Sprotte (Sprott, Brisling, Breitling)	sprat (brisling), European sprat (FAO)
Spreo bicolor	Zweifarben-Glanzstar	African pied starling
Spreo superbus/ Lamprotornis superbus/ Lamprospeo suberbus	Dreifarben-Glanzstar	superb starling
Spulerina simploniella	Eichenrindenminiermotte	
Squalidae	Dornhaie	dogfishes, dogfish sharks
Squaliformes	Stachelhaie	bramble sharks & dogfish sharks & allies
Squaliolus laticaudatus	Eindorn-Zwerghai	spined pygmy shark, spined dwarf shark
Squalus acanthias (*Acanthias vulgaris*)	Dornhai, Gemeiner Dornhai	common spiny dogfish, spotted spiny dogfish, picked dogfish, spurdog, piked dogfish (FAO)

Squalus asper> *Cirrhigaleus asper*	Rauer Dornhai	roughskin spurdog (FAO), roughskin spiny dogfish

Squalus blainvillei/
  Squalus fernandinus/
  Acanthias blainvillei
	Blainvilles Dornhai,
	Blainville-Dornhai
		longnose spurdog (FAO),
		Blainville's dogfish

Squalus megalops	Großaugen-Dornhai	shortnose spurdog
Squamata	Schuppenkriechtiere *sensu stricto* (Eidechsen & Schlangen)	squamata (incl. lizards & amphisbaenians & snakes)
Squatina aculeata	Dornen-Meerengel	sawback angelshark (FAO), thorny monkfish
Squatina californica	Pazifik-Meerengel, Pazifischer Engelhai	Pacific angelshark
Squatina dumeril	Sandteufel	sand devil
Squatina nebulosa	Nebelhai	clouded angelshark
Squatina oculata	Gefleckter Meerengel	smoothback angelshark (FAO), spotted monkfish

Squatina squatina/
  Rhina squatina/
  Squatina angelus
	Gemeiner Meerengel,
	Engelhai
		angelshark (FAO),
		angel shark,
		monkfish

Squatinidae	Meerengel, Engelhaie	angelsharks (sand devils)
Squatiniformes	Engelhaie, Engelhaiartige	monkfishes, angel sharks
Squilla empusa	Amerikanischer Heuschreckenkrebs, Gewöhnlicher Heuschreckenkrebs	common mantis shrimp
Squilla mantis	Großer Heuschreckenkrebs, Fangschreckenkrebs, Gemeiner Heuschreckenkrebs	giant mantis shrimp, spearing mantis shrimp
Srilankamys ohiensis	Ceylon-Ratte	Ceylonese rat
Stagnicola corvus	Große Sumpfschnecke	giant pondsnail
Stagnicola fuscus	Braune Sumpfschnecke	brown pondsnail
Stagnicola glabra/ *Lymnaea glabra*	Längliche Sumpfschnecke	oblong pondsnail, mud pondsnail, mud snail
Stagnicola palustris/ *Galba palustris*	Gemeine Sumpfschnecke	common pondsnail, common European pondsnail, marsh snail
Stagnicola turricula	Schlanke Sumpfschnecke	slender pondsnail
Staphylinidae	Kurzflügler, Raubkäfer	rove beetles
Staphylinus caesareus	Bunter Kurzflügler	
Stauromedusae	Becherquallen, Stielquallen	stauromedusas
Stauromedusidae	Becherquallen	stauromedusids
Staurophora celsia	Malachiteule	malachite noctuid*
Staurophora mertensi	Weißstern-Qualle*	whitecross jellyfish
Stauropus fagi	Buchenspinner	lobster moth
Staurotypus salvinii	Salvin-Kreuzbrustschildkröte	Chiapas giant musk turtle, giant musk turtle (IUCN), Chiapas cross-breasted turtle
Staurotypus spp.	Kreuzbrustschildkröten	musk turtles (IUCN), cross-breasted turtles
Staurotypus triporcatus	Große Kreuzbrustschildkröte	Mexican giant musk turtle, Mexican musk turtle (IUCN), Mexican cross-breasted turtle

Steatocranus casuarius	Buckelkopf-Buntbarsch	buffalohead cichlid
Steatoda bipunctata	Fettspinne	rabbit hutch spider, two-spot spider*
Steatoda paykullianus	Falsche Witwe	false widow spider
Steatomys spp.	Fettmäuse	fat mice
Steatornithidae	Fettschwalme	oilbirds
Stegobium paniceum	Brotkäfer, Brotbohrer	drugstore beetle, drug-store weevil (biscuit beetle, bread beetle)
Stegomyia aegypti/ Aedes aegypti	Gelbfiebermücke	yellow fever mosquito
Stegostoma fasciatum	Zebrahai, Leopardenhai	zebra shark (FAO), leopard shark
Stegostomatidae	Zebra-Ammenhaie	zebra sharks
Steindachneria argentea	Leuchtender Gabeldorsch	luminous hake
Stelis aterrima/ Stelis punctulatissima	Düsterbiene	
Stellaria solaris	Sonnenstern-Lastträgerschnecke	sunburst carriersnail
Stellula calliope	Sternelfe	Calliope hummingbird
Stenella attenuata	Schlankdelphin, Fleckendelphin	bridled dolphin, pantropical spotted dolphin, white-spotted dolphin
Stenella clymene	Zügeldelphin	Atlantic spinner dolphin, clymene dolphin, helmet dolphin
Stenella coeruleoalba	Blau-Weißer Delphin, Streifendelphin	striped dolphin, blue-white dolphin, euphrosyne dolphin
Stenella longirostris	Schmalschnabeldelphin, Spinnerdelphin	pantropical spinner dolphin, spinning dolphin, long-snouted dolphin
Stenella plagiodon	Atlantischer Fleckendelphin	Atlantic spotted dolphin, Gulf Stream spotted dolphin
Steneotarsonemus laticeps	Narzissenzwiebelmilbe	bulb scale mite
Steneotarsonemus pallidus ssp. *fragariae*	Erdbeermilbe	strawberry mite
Steneotarsonemus spirifex/ Tarsonemus spirifex	Hafermilbe	oat spiral mite
Stenepteryx hirundinis	Schwalbenlausfliege	swallow parasitic fly a.o.
Steno bredanensis	Rauzahndelphin	rough-toothed dolphin
Stenobothrus lineatus	Liniierter Grashüpfer, Panzers Grashüpfer	stripe-winged grasshopper, lined grasshopper
Stenobothrus nigromaculatus	Schwarzfleckiger Grashüpfer	black-spotted grasshopper
Stenobothrus stigmaticus	Kleiner Heidegrashüpfer, Ramburs Grashüpfer	lesser mottled grasshopper
Stenocephalemys spp.	Äthiopische Schmalkopfratten*	Ethiopian narrow-headed rats, Ethiopian meadow rats
Stenocephalidae	Wolfsmilchwanzen	spurge bugs
Stenocercus spp.	Kurzschwanzleguane, Wirtelschuppenleguane*	whorltail iguanas
Stenocionops furcatus	Atlantische Schmuckkrabbe	furcate spider crab, Atlantic decorator crab
Stenocorus meridianus	Variabler Stubbenbock	
Stenocyathus vermiformis	Wurmkoralle*	worm coral
Stenoderma rufum	Roter Flughund	red fruit bat

Stenodus leucichthys	Weißlachs	inconnu (FAO), nelma
Stenolaemata/ Stenostomata	Engmünder	stenostomates, stenolaemates, "narrow throat" bryozoans
Stenolechia gemmella	Eichentriebmotte	
Stenolophus lecontei	Scheibenhals-Schnellläufer	seed-corn beetle
Stenoplax floridana (Polyplacophora)	Schlanke Käferschnecke	Florida slender chiton
Stenopus hispidus	Gebänderte Scherengarnele	banded coral shrimp, banded cleaner shrimp
Stenopus scutellatus	Gold-Scherengarnele	golden coral shrimp, golden cleaner shrimp
Stenopus spinosus	Stachelige Scherengarnele	spiny coral shrimp, spiny cleaner shrimp
Stenorhynchus lanceolatus	Atlantische Spinnenkrabbe	Atlantic arrow crab
Stenorhynchus seticornis	Karibische Spinnenkrabbe, Dreieckskrabbe	yellowline arrow crab, Caribbean arrow crab
Stenosemus albus/ Ischnochiton albus (Polyplacophora)	Weiße Käferschnecke, Weißer Chiton, Helle Käferschnecke	northern white chiton, white chiton, white northern chiton
Stenotomus chrysops	Nordamerikanische Brasse	scup
Stentor coerulens	Blaues Trompetentierchen	blue trumpet animalcule
Stentor polymorphus	Grünes Trompetentierchen	green trumpet animalcule
Stentor spp.	Trompetentierchen	trumpet animalcules
Stephanoberycidae	Dornfische	pricklefishes
Stephanocoenia michelinii	Kleinäugige Sternkoralle	blushing star coral
Stephanolepis cirrhifer	Segel-Feilenfisch	thread-sail filefish
Stercorariidae	Raubmöwen	skuas and jaegers
Stercorarius longicaudus	Falkenraubmöve	long-tailed skua
Stercorarius parasiticus	Schmarotzerraubmöve	Arctic skua
Stercorarius pomarinus	Spatelraubmöve	pomarine skua
Stercorarius skua	Skua	great skua
Stereochilus marginatus	Streifensalamander	many-lined salamander
Stereolepis gigas	Pazifik-Riesenbarsch (Kalifornischer Judenfisch)	giant sea bass
Sterna albifrons	Zwergseeschwalbe	little tern
Sterna anaethetus	Zügelseeschwalbe	bridled tern
Sterna bengalensis	Rüppellseeschwalbe	lesser crested tern
Sterna bergii	Eilseeschwalbe	crested tern
Sterna caspia/ Hydroprogne caspia	Raubseeschwalbe	Caspian tern
Sterna dougallii	Rosenseeschwalbe	roseate tern
Sterna elegans	Schmuckseeschwalbe	elegant tern
Sterna forsteri	Forsterseeschwalbe	Forster's tern
Sterna fuscata	Rußseeschwalbe	sooty tern
Sterna hirundo	Fluss-Seeschwalbe	common tern
Sterna maxima	Königsseeschwalbe	royal tern
Sterna paradisaea	Küstenseeschwalbe	Arctic tern
Sterna sandvicensis	Brandseeschwalbe	sandwich tern
Sterninae	Seeschwalben	terns
Sternoptychidae	Tiefsee-Beilfische, Silberbeilfische	hatchetfishes
Sternoptyx diaphana	Durchsichtiger Beilbauch	transparent hatchetfish

Sternostoma tracheacolum	Kanarienvogel-Lungenmilbe	canary lung mite
Sternotherus carinatus	Dach-Moschusschildkröte	razorback musk turtle, keel-backed musk turtle, razor-backed musk turtle
Sternotherus depressus/ Kinosternon depressum	Flache Moschusschildkröte	flattened musk turtle
Sternotherus minor/ Kinosternon minor	Kleine Moschusschildkröte, Zwerg-Moschusschildkröte	loggerhead musk turtle
Sternotherus odoratus/ Kinosternon odoratum	Gewöhnliche Moschusschildkröte	common musk turtle, stinkpot
Sternotherus spp.	Moschusschildkröten	musk turtles
Stethophyma grossum/ Mecostethus grossus	Sumpfschrecke	large marsh grasshopper
Sthenelais boa	Sand-Schuppenwurm	boa worm, boa-shaped scaleworm
Sthenorytis pernobilis	Edle Wendeltreppe	noble wentletrap
Sthenoteuthis bartrami/ Ommastrephes bartrami	Fliegender Kalmar	red flying squid
Stichaeidae	Stachelrücken	pricklebacks
Stichodactyla gigantea/ Stichodactyla mertensi/ Stoichactis giganteum	Riesenanemone, Mertens Anemone, Nessellose Riesenanemone	giant carpet anemone, giant anemone
Stichodactyla haddoni	Haddons Anemone, Teppichanemone	Haddon's anemone, saddle anemone, carpet anemone
Stichodactyla helianthus	Sonnenanemone	sun anemone, Atlantic carpet anemone
Stichopathes leutkeni	Spiralige Drahtkoralle	spiraling wire coral, black wire coral
Stichopus horrens	Schreckliche Seegurke	horrific sea cucumber*
Stichopus japonicus	Japanische Seegurke	Japanese sea cucumber
Stichopus regalis	Königsseegurke, Königsholothurie	royal sea cucumber
Stictia carolina/ Bembix carolina		horse guard, horseguard
Stictonetta naevosa	Affengans	freckled duck
Stigmella anomalella	Rosenminiermotte	rose leaf-miner
Stigmella aurella	Waldbeeren-Zwergminiermotte*	bramble leaf-miner
Stigmellidae (Nepticulidae)	Zwergmotten	leaf miners, leaf-miners
Stilbus testaceus	Schwarzbrauner Glattkäfer	
Stilicus rufipes	Schwarzbrauner Dünnhalskurzflügler	
Stipiturini	Borstenschwänze	emuwrens
Stizostedion canadense	Kanadischer Zander	sauger
Stizostedion lucioperca/ Sander lucioperca	Zander Hechtbarsch, Schill, Sandbarsch	pike-perch, zander (FAO)
Stizostedion marina	Meerzander	sea zander
Stizostedion vitreum	Glasaugenbarsch	walleye
Stochomys spp.	Zentralafrikanische Langhaarratten	Central African target rats
Stolidobranchia	Faltkiemen-Seescheiden	stolidobranchs, stolidobranch sea-squirts
Stoloteuthis leucoptera	Schmetterlingstintenfisch	butterfly bobtail squid

Stoloteuthis spp.	Schmetterlingstintenfische	butterfly squids, butterfly bobtail squids
Stomatellidae	Weitmundschnecken*	false ear snails, widemouth snails (widemouth shells)
Stomatopoda	Maulfüßer, Fangschreckenkrebse	mantis shrimps
Stomias boa	Boa-Drachenfisch*	boa dragonfish
Stomiidae/Stomiatidae	Schuppen-Drachenfische	scaly dragonfishes, barbeled dragonfishes
Stomiiformes	„Tiefseefische"	deep-sea hatchetfishes & relatives
Stomolophus meleagris	Kanonenkugel-Qualle	cannonball jellyfish
Stomoxys calcitrans	Stechfliege, Wadenstecher	stable fly, dog fly, "biting housefly"
Stomphia coccinea	Schwimmanemone	swimming anemone
Storeria occipitomaculata	Rotbauch-Braunschlange	red-bellied snake
Storeria spp.	Braunschlangen	brown snakes
Stratiomyidae	Waffenfliegen	soldier flies
Stratiomys chameleon	Chamäleonfliege	soldier fly, soldier-fly
Streblidae	Fledermausfliegen u.a.	bat flies a.o.
Strepera graculina	Dickschnabel-Würgerkrähe	pied currawong
Strepsiptera	Fächerflügler, Kolbenflügler	twisted-winged insects, twisted-winged parasites, stylopids
Streptopelia decaocto	Türkentaube	collared dove
Streptopelia orientalis	Orientturteltaube	rufous turtle dove
Streptopelia senegalensis	Palmtaube	palm dove
Streptopelia turtur	Turteltaube	turtle dove
Striarca lactea	Milchweiße Archenmuschel	milky-white ark
Strigidae	Eulen	owls
Strigiformes	Eulen	owls
Strigops habroptilus	Kakapo, Eulenpapagei	kakapo, owl-parrot
Strix aluco	Waldkauz	tawny owl
Strix butleri	Fahlkauz	Hume's tawny owl
Strix nebulosa	Bartkauz	great grey owl, great gray owl
Strix occidentalis	Fleckenkauz	spotted owl
Strix uralensis	Habichtskauz	Ural owl
Strix varia	Streifenkauz	barred owl
Strix woodfordii	Afrikanischer Waldkauz	African wood owl
Stromateidae	Erntefische, Butterfische	ruffs, butterfishes, harvestfishes
Stromateoidei	Erntefische	harvestfishes
Stromateus fiatola	Butterfisch, Deckfisch, Pampelfisch, Gemeine Pampel	butterfish, blue butterfish (FAO)
Strombidae	Fechterschnecken, Flügelschnecken	conchs, true conchs (conch shells)
Strombus alatus	Florida-Fechterschnecke, Florida-Flügelschnecke	Florida fighting conch

Strombus aurisdianae	Dianas Fechterschnecke, Dianas Flügelschnecke, Dianas Ohr-Fechterschnecke	Diana's conch, Diana conch
Strombus bulla	Blasen-Fechterschnecke, Blasen-Flügelschnecke	bubble conch
Strombus canarium	Hunds-Fechterschnecke, Hunds-Flügelschnecke	yellow conch, dog conch
Strombus costatus	Milchige Fechterschnecke, Milchige Flügelschnecke	milk conch
Strombus decorus	Mauritius Fechterschnecke, Mauritius Flügelschnecke	Mauritius conch, Mauritian conch
Strombus dentatus	Samar-Fechterschnecke, Samar-Flügelschnecke	Samar conch, toothed conch
Strombus epidromis	Schwanen-Fechterschnecke, Schwanen-Flügelschnecke	swan conch
Strombus galeatus	Ostpazifische Riesen-Fechterschnecke, Ostpazifische Riesen-Flügelschnecke	galeate conch, giant East Pacific conch
Strombus gallus	Hahnen-Fechterschnecke, Hahnen-Flügelschnecke	roostertail conch, rooster-tail conch
Strombus gibberulus	Buckel-Fechterschnecke, Buckel-Flügelschnecke	hump-back conch, humped conch
Strombus gigas	Riesen-Fechterschnecke, Riesen-Flügelschnecke	queen conch, pink conch
Strombus goliath	Goliath-Fechterschnecke, Goliath-Flügelschnecke	goliath conch
Strombus gracilior	Panama-Fechterschnecke, Ostpazifische Flügelschnecke	Eastern Pacific fighting conch, Panama fighting conch
Strombus granulatus	Körnige Fechterschnecke, Körnige Flügelschnecke	granulated conch
Strombus latissimus	Breite Fechterschnecke, Breite Flügelschnecke	heavy frog conch, wide-mouthed conch, broad Pacific conch, widest Pacific conch
Strombus latus/ Strombus bubonius	Beulen-Fechterschnecke*, Beulen-Flügelschnecke*	bubonian conch
Strombus lentiginosus	Silber-Fechterschnecke, Silber-Flügelschnecke	silver conch
Strombus listeri	Listers Fechterschnecke, Listers Flügelschnecke	Lister's conch
Strombus luhuanus	Erdbeer-Fechterschnecke, Erdbeer-Flügelschnecke	strawberry conch, blood-mouth conch
Strombus marginatus	Weißgeränderte Band-Fechterschnecke*, Geränderte Flügelschnecke*	marginate conch
Strombus minimus	Miniatur-Fechterschnecke, Miniatur-Flügelschnecke	minute conch
Strombus mutabilis	Veränderliche Fechterschnecke, Veränderliche Flügelschnecke	mutable conch, changeable conch
Strombus peruvianus	Peruanische Fechterschnecke, Peruanische Flügelschnecke	Peruvian conch
Strombus pipus	Schmetterlings-Fechterschnecke, Schmetterlings-Flügelschnecke	butterfly conch
Strombus pugilis	Westindische Fechterschnecke, Westindische Flügelschnecke	fighting conch, West Indian fighting conch
Strombus raninus	Habichtsflügelschnecke	hawkwing conch

Strombus sinuatus	Lavendelfarbene Fechterschnecke, Lavendelfarbene Flügelschnecke	lavender conch, laciniate conch, sinuous conch
Strombus taurus	Stier-Fechterschnecke, Stier-Flügelschnecke	bull conch
Strombus thersites	Thersites-Fechterschnecke, Thersites-Flügelschnecke	Thersites conch
Strombus tricornis	Dreihorn-Fechterschnecke, Dreihorn-Flügelschnecke	three-knobbed conch
Strombus urceus	Kleiner Bär, Kleine Bären-Fechterschnecke, Kleine Bären-Flügelschnecke	little bear conch, pitcher stromb
Strombus variabilis	Veränderliche Fechterschnecke, Veränderliche Flügelschnecke	variable conch
Strongylocentrotus droebachiensis	Purpurseeigel	green sea urchin, northern urchin, northern sea urchin
Strongylocentrotus franciscanus	Pazifischer Roter Seeigel	red sea urchin
Strongylocentrotus purpuratus	Pazifik-Purpurseeigel	purple sea urchin
Strongylognathus testaceus	Parasit-Säbelameise	
Strongyloides papillosus	Darm-Zwergfadenwurm*	intestinal threadworm (cattle, sheep, goat a.o.)
Strongyloides ransomi	Zwergfadenwurm u.a.	intestinal threadworm a.o.
Strongyloides spp.	Fadenwürmer	threadworms
Strongylura crocodila/ Tylosurus crocodilus	Krokodilhecht, Riesen-Hornhecht, Riesen-Krokodilhecht	hound needlefish (FAO), houndfish, crocodile longtom (Austr.), crocodile needlefish
Strongylura incisa	Riff-Hornhecht	reef needlefish
Strongylura marina	Atlantischer Hornhecht, Tropischer Hornhecht	Atlantic needlefish (FAO), silver gar
Strongylura notata	Rotflossen-Hornhecht	redfin needlefish
Strongylura timucu	Timucu	timucu
Strongylus gigas/ Dioctophyma renale	Nieren-Riesen-Palisadenwurm	renal blood-red nematode
Strongylus vulgaris	Roter Palisadenwurm*	bloodworm
Strophochilus oblongus	Südamerikanische Riesenschnecke	giant South American snail
Struthio camelus	Strauß	ostrich
Struthiolaria papulosa	Großer Straußenfuß	large ostrich foot
Struthiolariidae	Straußenschnecken	ostrich foot snails
Struthionidae	Strauße	ostrich
Struthioniformes	Straußenvögel, Strauße, Laufvögel	ostriches
Strymon melinus	Grauer Zipfelfalter	gray hairstreak
Strymonidia pruni	Pflaumenzipfelfalter	black hairstreak
Strymonidia w-album/ Strymon w-album/ Satyrium w-album	Weißes W, Ulmen-Zipfelfalter	white-letter hairstreak
Studeriotes longiramosa	Weihnachtsbaumkoralle*	Christmas tree coral
Sturnella magna	Lerchenstärling	eastern meadowlark
Sturnella neglecta	Wiesenstärling	western meadowlark
Sturnidae	Stare	starlings & allies

Sturnini	Stare, Mainas	starlings, mynas
Sturnira spp.	Gelbschulter-Blattnasen	yellow-shouldered bats, American epauletted bats
Sturnus pagodarum	Pagodenstar	black-headed starling
Sturnus roseus	Rosenstar	rose-coloured starling
Sturnus unicolor	Einfarbstar	spotless starling
Sturnus vulgaris	Star	starling, European starling
Styela canopus	Einsiedler-Seescheide	hermit sea-squirt
Styela clava	Warzige Seescheide	warty sea-squirt*
Styela partita/ *Styela rustica*	Raue Seescheide	rough sea-squirt
Styela plicata	Falten-Ascidie, Walzen-Ascidie	barrel-squirt*
Stylaria lacustris	Gezüngelte Naide	
Stylaster californicus	Kalifornische Filigrankoralle	Californian hydrocoral
Stylasteridae	Hydrokorallen, Stylasteriden	stylasterine hydrocorals, lace corals
Stylatula elongata	Schlanke Seefeder	slender sea-pen
Stylephoridae	Fadenschwänze, Fadenträger, Stielaugenfische	tube-eyes, thread-tails
Stylephorus chordatus	Standartenfisch	tube-eye (FAO), thread-tail
Stylobates aenus	Mantelanemone*	cloak anemone
Stylochus pilidium	Braune Fleckplanarie	brown oyster leech*
Stylocidaris affinis	Roter Lanzenseeigel	red cidarid
Styloctenium wallacei	Harlekin-Flughund, Weißstreifen-Flughund	striped-faced fruit bat
Stylodipus telum	Dickschwanz-Springmaus	thick-tailed three-toed jerboa
Stylommatophora (Pulmonata)	Landlungenschnecken	land snails
Stylophora pistillata	Griffelkoralle	pistillate coral*
Stylopidae (Strepsiptera)	Stylopiden	parasitic beetles
Subcoccinella *vigintiquatuorpunctata*	Luzernenmarienkäfer	twentyfour-spot ladybird, 24-spot ladybird
Subergorgia apressa	Fächergorgonie u.a.	sea fan a.o.
Suberites domuncula	Häuschenschwamm, Korkschwamm, Einsiedler-Korkschwamm	sulphur sponge, sulfur sponge
Suberites ficus/ *Ficulina ficus*	Feigenschwamm, Korkschwamm	fig sponge
Subulina octona	Kleinste Turmschnecke	miniature awlsnail
Succinea oblonga	Kleine Bernsteinschnecke	small amber snail
Succinea putris	Gemeine Bernsteinschnecke	rotten amber snail, large amber snail, European ambersnail (U.S.)
Succineidae	Bernsteinschnecken	amber snails, ambersnails
Sufflamen bursa/ *Hemibalistes bursa*	Weißlinien-Drückerfisch	boomerang triggerfish (FAO), drab triggerfish
Suillia tuberiperda	Trüffelfliege	truffles fly*
Suillia univittata	Knoblauchfliege	garlic fly*

Sula bassana/ *Morus bassanus*	Basstölpel	northern gannet
Sula dactylatra	Maskentölpel	blue-faced booby
Sula leucogaster	Weißbauchtölpel	brown booby
Sula sula	Rotfußtölpel	red-footed booby
Sula variegata	Guanotölpel	Peruvian booby
Sulidae	Tölpel	boobies, gannets
Suncus etruscus	Etruskerspitzmaus	Etruscan shrew, Savi's pygmy shrew, pygmy white-toothed shrew
Suncus murinus	Moschusspitzmaus, Asiatische Moschusspitzmaus	musk shrew, Indian musk shrew
Suncus spp.	Moschusspitzmäuse, Dickschwanzspitzmäuse	musk shrews, pygmy shrews, dwarf shrews
Sundasciurus spp.	Sundahörnchen	Sunda tree squirrels
Sundasciurus tenuis	Schlankhörnchen	slender squirrel
Supella supellectilium/ *Supella longipalpa*	Braunbandschabe, Möbelschabe	brown-banded cockroach, brownbanded cockroach, furniture cockroach
Suricata suricatta	Erdmännchen	suricate, slender-tailed meerkat
Surnia ulula	Sperbereule	hawk owl
Sus barbatus	Bartschwein	bearded pig
Sus salvanius	Zwergwildschwein	pygmy hog
Sus scrofa	Wildschwein	wild boar, pig
Sus verrucosus	Pustelschwein	Javan pig
Susania testudinaria	Schildkrötenschnecke	turtle snail*
Suta suta		curl snake
Swiftia exserta	Orangerote Seefeder*	orange sea fan, soft red sea fan
Sycon ciliatum/ *Scypha ciliatum*	Meergürkchen, Wimpern-Kalkschwamm, Wimper-Kalkschwamm	little vase sponge
Sycon coronatum	Kronen-Kalkschwamm	crown vase sponge*
Sycon raphanus	Rettich-Kalkschwamm, Borstiger Kalkschwamm	radish sponge*, bristly vase sponge*
Syconycteris spp.	Blütenbesuchende Fledermäuse	blossom bats
Sylicapra grimmia	Kronenducker	common duiker, gray duiker
Sylilagus idahoensis/ *Brachylagus idahoensis*	Zwergkaninchen	pygmy rabbit
Sylvia atricapilla	Mönchsgrasmücke	blackcap
Sylvia borin	Gartengrasmücke	garden warbler
Sylvia cantillans	Weißbart-Grasmücke	subalpine warbler
Sylvia communis	Dorngrasmücke	whitethroat
Sylvia conspicillata	Brillengrasmücke	spectacled warbler
Sylvia curruca	Klappergrasmücke	lesser whitethroat
Sylvia deserticola	Atlasgrasmücke	Tristram's warbler
Sylvia hortensis	Orpheusgrasmücke	orphean warbler
Sylvia leucomelaena	Akaziengrasmücke	Arabian warbler
Sylvia melanocephala	Samtkopf-Grasmücke	Sardinian warbler
Sylvia melanothorax	Schuppengrasmücke	Cyprus warbler

Sylvia mystacea	Tamariskengrasmücke	Ménétries's warbler
Sylvia nana	Wüstengrasmücke	desert warbler
Sylvia nisoria	Sperbergrasmücke	barred warbler
Sylvia rueppelli	Maskengrasmücke	Rueppell's warbler
Sylvia sarda	Sardengrasmücke	Marmora's warbler
Sylvia undata	Provencegrasmücke	Dartford warbler
Sylvicola fenestralis	Gemeine Fenstermücke, Fensterfliege, Weißer Drahtwurm	common window gnat, common window midge
Sylviidae	Grasmücken (Zweigsänger)	Old World warblers
Sylvilagus aquaticus	Wasserkaninchen	swamp rabbit
Sylvilagus audubonii	Audubon-Kaninchen	desert rabbit
Sylvilagus bachmani	Strauchkaninchen	brush rabbit
Sylvilagus brasiliensis	Brasilien-Waldkaninchen	forest rabbit
Sylvilagus floridanus	Florida-Waldkaninchen	eastern cottontail
Sylvilagus palustris	Sumpfkaninchen	marsh rabbit
Sylvilagus spp.	Baumwollschwanz-Kaninchen	cottontails
Sylvilagus transitionalis	Neu-England-Kaninchen	New England cottontail
Sylvisorex megalura	Kletterspitzmaus	climbing shrew
Sympecma annulata/ Sympecma paedisca	Sibirische Winterlibelle	Siberian winter damselfly
Sympecma fusca	Winterlibelle, Gemeine Winterlibelle	winter damselfly
Sympetrum danae	Schwarze Heidelibelle	black sympetrum
Sympetrum depressiusculum	Sumpf-Heidelibelle	eastern European sympetrum
Sympetrum flaveolum	Gelbflüglige Heidelibelle, Gefleckte Heidelibelle	yellow-winged sympetrum, yellow-winged darter
Sympetrum fonscolombei	Frühe Heidelibelle	red-veined sympetrum
Sympetrum meridionale	Südliche Heidelibelle	meridional sympetrum*, southern European sympetrum*
Sympetrum pedemontanum	Gebänderte Heidelibelle	banded sympetrum*
Sympetrum sanguineum	Blutrote Heidelibelle	ruddy sympetrum
Sympetrum spp.	Heidelibellen	sympetrums (darters)
Sympetrum striolatum	Große Heidelibelle	common sympetrum, common darter
Sympetrum vulgatum	Gemeine Heidelibelle	vagrant sympetrum
Symphalangus klossi/ Hylobates klossi	Zwergsiamang, Mentawai-Gibbon, Biloh	Kloss's gibbon, dwarf siamang, beeloh
Symphalangus syndactulus/ Hylobates syndactylus	Siamang	siamang
Symphodus bailloni	Zweifleck-Lippfisch	vielle
Symphodus cinereus	Grauer Lippfisch	grey wrasse
Symphodus mediteraneus	Achselfleck-Lippfisch, Mittelmeerlippfisch	axillary wrasse
Symphodus melops	Goldmaid	corkwing wrasse
Symphodus ocellatus/ Crenilabrus ocellatus	Augenfleck-Lippfisch, Augenflecklippfisch, Augenlippfisch	ocellated wrasse, spotted wrasse
Symphodus roissali	Fünffleckiger Lippfisch	five-spot wrasse
Symphodus rostratus	Langschnauzen-Lippfisch, Schnauzenlippfisch	long-snouted wrasse, cannadelle
Symphodus tinca	Pfauen-Lippfisch	painted wrasse

Symphurus nigrescens	Schwarzfleckzunge*	spotted tongue-sole
Symphyla	Zwergfüßer	symphylans
Symphysodon aequifasciata aequifasciata	Grüner Diskus	green discus
Symphysodon aequifasciata axelrodi	Gelbbrauner Diskus, Gewöhnlicher Diskus	brown discus
Symphysodon aequifasciata haraldi	Blauer Diskus	blue discus
Symphysodon discus discus	Echter Diskus	discus
Symphysodon spp.	Diskus-Buntbarsche	discus fishes
Symplectoteuthis luminosa/ Eucleoteuthis luminosa	Leuchtender Flugkalmar, Streifenkalmar*	luminous flying squid, striped squid
Symplectoteuthis oualaniensis	Violetter Flugkalmar	purpleback flying squid
Synageles venator	Ameisenspinne	ant spider* a.o.
Synagrops japonicus	Japanischer Kardinalfisch	Japanese splitfin
Synanceia verrucosa	Riff-Steinfisch	reef stonefish, poison toadfish, stonefish (FAO)
Synanceidae/Synanceinae	Steinfische, Teufelsfische	stonefishes
Synanthedon myopaeformis	Apfelbaumglasflügler	red-belted clearwing moth
Synanthedon spheciformis	Erlenglasflügler	alder clearwing moth, alder borer
Synanthedon tipuliformis/ Aegeria tipuliformis	Johannisbeerglasflügler	currant clearwing moth, currant borer
Synaphobranchidae	Grubenaale	cutthroat eels
Synapta maculata	Gefleckte Wurmseegurke	spotted barrel gurkhin*
Synaptomys spp.	Lemmingmäuse	bog lemmings
Synaxiidae	Samtlangusten*	furry lobsters
Synbranchidae	Kiemenschlitzaale, Sumpfaale	swamp-eels
Syncerus caffer	Afrikanischer Büffel, Kaffernbüffel	African buffalo
Synchiropus picturatus	Paletten-Leierfisch, LSD-Leierfisch	picturesque dragonet
Synchiropus splendidus	Mandarinfisch, Mandarin-Leierfisch	mandarinfish
Synchiropus spp.	Mandarinfische	mandarinfishes
Syngamus spp.	Luftröhrenwürmer	gapeworms
Syngamus trachea	Roter Luftröhrenwurm, Rotwurm, Gabelwurm	gapeworm (of poultry)
Syngnathidae	Seenadeln & Seepferdchen	pipefishes & seahorses
Syngnathiformes (Syngnathoidei)	Seenadelverwandte, Seepferdchenverwandte, Büschelkiemenartige	sea horses & pipefishes and allies
Syngnathus abaster	Kurzschnabel-Seenadel	short-snouted pipefish, shore pipefish
Syngnathus acus	Große Seenadel, Mittelmeer-Seenadel	great pipefish, greater pipefish (FAO)
Syngnathus dactyliophorus	Zebraseenadel	zebra pipefish
Syngnathus rostellatus	Kleine Seenadel	lesser pipefish, Nilsson's pipefish (FAO)

Syngnathus typhle	Schmalschnäuzige Seenadel, Grasnadel	deep-snouted pipefish, broad-snouted pipefish, deep-nosed pipefish, broadnosed pipefish (FAO)
Synodontidae	Eidechsenfische	lizardfishes
Synodontis angelicus	Engelwels, Perlhuhnwels	polka-dot African catfish, angel squeaker (FAO)
Synodontis flavitaeniatus	Gelbbraungebänderter Fiederbartwels	orangestriped squeaker
Synodontis nigriventris	Rückenschwimmender Kongowels	blackbellied upside-down catfish
Synodontis petricola	Kuckuckswels	spotted squeaker
Synodus foetens	Küsten-Eidechsenfisch*	inshore lizardfish
Synodus intermedius	Sandtaucher* (Eidechsenfisch)	sand diver
Synodus lucioceps	Kalifornischer Eidechsenfisch	California lizardfish
Synodus saurus	Atlantischer Eidechsenfisch	Atlantic lizardfish
Synodus spp.	Eidechsenfische	lizardfishes
Synodus synodus	Roter Eidechsenfisch	red lizardfish
Synodus variegatus	Vielfarbiger Eidechsenfisch	variegated lizardfish
Syntheosciurus spp.	Neotropische Berghörnchen	Neotropical montane squirrels
Syntomaspis druparum	Kernobstsamenwespe	
Syntomidae	Widderbären, Fleckenschwärmerchen	wasp moths
Syntomis phegea	Weißfleckenwidderchen, Weißfleckwidderchen, Ringelwidderchen	yellow-belted burnet
Syntomus truncatellus	Schwarzer Rennkäfer, Gewöhnlicher Zwergstreuläufer	
Syringophilidae	Federspulenmilben	quill mites
Syrinx aruanus	Australische Rifftrompete	false trumpet snail, false trumpet, Australian trumpet
Syrinx proboscidiferus	Ritterhelm	elefant's trumpet snail
Syritta pipiens	Gemeine Mist-Schwebfliege	
Syromastes rhombeus	Rautenwanze, Rhombenwanze	
Syrphidae	Schwebfliegen	hoverflies, hover flies, syrphid flies, flower flies
Syrphus ribesii	Gemeine Garten-Schwebfliege	currant hover fly
Syrrhaptes paradoxus	Steppenflughuhn	Palla's sandgrouse
Syrrhophus marnockii/ Eleutherodactylus marnockii	Marnock-Frosch, Felsen-Zwitscherfrosch*	cliff frog, cliff chirping frog
Syrrhophus spp.	Zirpfrösche	chirping frogs

Tabanidae	Bremsen	horseflies & deerflies, gadflies
Tabanus bovinus	Rinderbremse	large horsefly
Tabanus spp.	Pferdebremsen & Rinderbremsen	horseflies
Tachagra senegala	Senegaltaschagra	black-crowned tchagra
Tachinidae	Raupenfliegen (Schmarotzerfliegen)	tachinids, parasitic flies
Tachopteryx thoreyi		grayback dragonfly
Tachuris rubrigastra	Vielfarbentachuri	many-coloured rush tyrant
Tachybaptus ruficollis	Zwergtaucher	little grebe
Tachycines asynamorus	Gewächshausheuschrecke, Gewächshausschrecke	greenhouse camel cricket, glasshouse camel-cricket (Br.), greenhouse stone cricket (U.S.)
Tachycineta bicolor	Sumpfschwalbe	tree swallow
Tachycineta thalassina	Veilchenschwalbe	violet-green swallow
Tachyeres pteneres	Magellan-Dampfschiffente	flightless steamer duck
Tachyglossidae	Schnabeligel, Ameisenigel	echidnas, spiny anteaters
Tachyglossus aculeatus	Kurzschnabeligel	echidna, short-nosed echidna, spiny anteater
Tachynemis seychellensis	Seychellen Baumfrosch, Seychellen-Waldsteiger	Seychelle Islands tree frog
Tachyoryctes spp.	Afrikanische Maulwurfsratten	African mole-rats, African root-rats
Tachypodoiulus niger	Schwarzer Tausendfüßer	white-legged black millepede, black millepede
Tachyporus chrysomelinus	Zweifarbiger Schnellkurzflügler	
Tachyrhynchus erosus	Verwitterte Turmschnecke*	eroded turretsnail
Tachys quadrisignatus/ Elaphropus quadrisignatus	Viergefleckter Ahlenkäfer, Vierfleckiger Zwergahlenläufer	
Tachysuridae/Ariidae	Maulbrüterwelse, Meereswelse, Kreuzwelse	sea catfishes
Tachysurus spp.	Langstirn-Maulbrüter	barbels, sea barbels
Tachyta nana	Rinden-Ahlenkäfer	
Tadarida spp.	Faltlippen-Fledermäuse	free-tailed bats
Tadarida teniotis	Eurasische Bulldogg-Fledermaus	Eurasian free-tailed bat
Tadorna ferruginea	Rostgans	ruddy shelduck
Tadorna tadorna	Brandgans	common shelduck
Taelia felina	Seedahlie, Dickhörnige Seerose	dahlia anemone
Taenia taeniaeformis/ Hydatigera taeniaeformis	Dickhalsiger Bandwurm, Katzenbandwurm u.a.	common cat tapeworm
Taenia hydatigena/ Taenia marginata	Geränderter Bandwurm, Geränderter Hundebandwurm	margined dog tapeworm
Taenia ovis	Schafsfinnen-Bandwurm	sheep-and-goat tapeworm
Taenia pisiformis/ Taenia serrata	Gesägter Bandwurm, Gesägter Hundebandwurm	serrated dog tapeworm
Taenia saginata	Rinderbandwurm, Rinder-Menschenbandwurm	beef tapeworm
Taenia solium	Schweinebandwurm, Schweinefinnen-Bandwurm (Bewaffneter Menschenbandwurm)	pork tapeworm

Taeniarhynchus saginatatus	Rinderfinnen-Bandwurm (Unbewaffneter Menschenbandwurm)	
Taenioconger longissimus	Gartenaal*	garden eel
Taenionotus triacanthus	Gespensterfisch, Schaukelfisch	leaf scorpionfish
Taeniophoridae/ Eutaeniophorinae	Bandträger	tapetails, ribbonbearers
Taeniopterygidae	Weiden-Steinfliegen*	willowflies
Taeniothrips laricivorus	Lärchenblasenfuß	larch thrips
Taeniura grabata	Runder Stechrochen, Kugelrochen	round stingray
Taeniura lymma	Blauflecken-Stechrochen, Blaufleck-Stachelrochen, Blaufleck-Stechrochen, Blaupunktrochen	ribbontail stingray, blue-spotted stingray, blue-spotted lagoonray, fantail, bluespotted ribbontail ray (FAO)
Taeniura meyeni/ Torpedo melanospila	Schwarzpunkt-Stechrochen, Gefleckter Stechrochen	round ribbontail ray, blotched fantail ray (FAO)
Takydromus sexlineatus	Sechsstreifige Langschwanzeidechse	Asian grass lizard
Takydromus spp.	Langschwanzeidechsen	grass lizards
Talaeporiidae	Sackmotten	bagworms, bagworm moths
Talipidae	Maulwürfe	moles & shrew moles & desmans
Talitridae	Strandflöhe, Strandhüpfer	sandhoppers a.o.
Talitrus saltator	Strandfloh, Strandhüpfer	sandhopper, greater sandhopper
Talorchestia megalophthalma	Großaugen-Strandfloh, Großaugen-Strandhüpfer	big-eyed sandhopper, big-eyed beach flea
Talpa caeca	Blindmaulwurf	blind mole
Talpa europaea	Europäischer Maulwurf	European mole
Talpa romana	Römischer Maulwurf	Roman mole
Talpa spp.	Altweltmaulwürfe	Old World moles
Talparia cinerea/ Cypraea cinerea/ Luria cinerea	Graue Atlantik-Kauri	Atlantic grey cowrie, Atlantic gray cowrie (U.S.), ashen cowrie
Tamandua spp.	Tamanduas	lesser anteaters, tamanduas
Tamarisca tamarisca	Seetamariske	sea tamarisk
Tamias sibiricus/ Eutamias sibiricus	Eurasisches Erdhörnchen, Burunduk, Östliches Streifenhörnchen, Sibirisches Streifenhörnchen (Hbz. Burunduk)	Siberian chipmunk
Tamias striatus	Chipmunk, Streifenbackenhörnchen (Hbz. Chipmunk)	eastern American chipmunk
Tamiasciurus douglasii	Chikaree	Douglas squirrel
Tamiasciurus hudsonicus	Rothörnchen (Hbz. Feh)	eastern red squirrel, red squirrel
Tamiasciurus spp.	Nordamerikanische Rothörnchen, Chikarees	American red squirrels, chickarees, chickaris
Tamiops spp.	Asiatische Zwerg-Streifenhörnchen	Asiatic striped squirrels

Tanaidacea (Anisopoda)	Scherenasseln	tanaidaceans, tanaids
Tanaoceridae	Langhorn-Wüstenheuschrecken	desert long-horned grasshoppers
Tandanus tandanus	Australischer Süßwasserwels*	Australian freshwater catfish
Tandonia budapestensis/ Milax budapestensis	Boden-Kielnacktschnecke	Budapest slug
Tandonia rustica/ Milax rusticus	Große Kielnacktschnecke	bulb-eating slug, root-eating slug
Tandonia sowerbyi/ Milax sowerbyi	Sowerbys Kielnacktschnecke	Sowerby's slug
Tanea euzona/Natica euzona	Zonierte Mondschnecke	zoned moonsnail
Tanea lineata/Natica lineata	Gestreifte Mondschnecke	lined moonsnail
Tanea undulata/ Natica undulata	Zebra-Mondschnecke	zebra moonsnail (wavy moonsnail)
Tanea zelandica/ Natica zelandica	Neuseeländische Mondschnecke	New Zealand moonsnail
Tanichthys albonubes	Kardinalfisch, „Arbeiterneon"	white cloud mountain minnow (FAO), white cloud,
Tantilla spp.	Schwarzkopfnattern	blackhead snakes, black-headed snakes
Tanypezidae	Zartfliegen	tanypezids
Tapes aureus/ Venerupis aurea/ Paphia aureus	Goldene Teppichmuschel	golden carpetclam, golden venus (golden carpet shell)
Tapes decussatus/ Venerupis decussata	Gekreuzte Teppichmuschel	chequered carpetclam, chequered venus (chequered carpet shell)
Tapes litterata	Buchstaben-Teppichmuschel*	lettered carpetclam, lettered venus (lettered carpet shell)
Tapes rhomboides/ Venerupis rhomboides/ Paphia rhomboides	Einfache Teppichmuschel, Gebänderte Teppichmuschel, Essbare Venusmuschel	banded carpetclam, banded venus (banded carpet shell)
Taphozous melanopogon	Tempelfledermaus	black-bearded tomb bat
Taphozous spp.	Grabflatterer, Grabfledermäuse	tomb bats
Taphrorychus bicolor	Buchenborkenkäfer	beech bark beetle
Tapinauchenius plumipes	Trinidad Mahagoni-Vogelspinne	Trinidad mahogany tarantula
Tapinoma erraticum	Gemeine Drüsenameise	erratic ant
Tapinoma sessile	Haus-Drüsenameise	odorous house ant
Tapirus bairdii	Baird-Tapir, Mittelamerikanischer Tapir	Baird's tapir, Central American tapir
Tapirus indicus	Asiatischer Tapir	Asiatic tapir, Malayan tapir
Tapirus pinchaque	Bergtapir	mountain tapir
Tapirus spp.	Tapire	tapirs
Tapirus terrestris	Flachlandtapir	Brazilian tapir
Taractes asper	Rauer Pomfret	rough pomfret
Taractichthys longipinnis	Langflossenbarsch*	long-finned bream
Tardigrada	Bärentierchen, Bärtierchen, Tardigraden (*sg* Tardigrad *m*)	water bears, tardigrades
Tarentola mauritanica	Mauergecko, Gewöhnlicher Mauergecko	common wall gecko, Moorish gecko
Tarentola spp.	Mauergeckos	wall geckos

Taricha granulosa	Westamerikanischer Rauhautmolch	rough-skinned newt
Taricha rivularis	Rotbauchmolch	redbelly newt
Taricha spp.	Westamerikanische Wassermolche	Pacific newts
Taricha torosa	Kalifornischer Molch	California newt
Tarpon atlanticus/ Megalops atlanticus	Atlantischer Tarpon, Atlantischer Tarpun	Atlantic tarpon
Tarsipes spenserae	Honigbeutler	honey possum
Tarsius bancanus	Sunda-Koboldmaki	western tarsier, Horsfield's tarsier
Tarsius pumilus	Zwerg-Koboldmaki	pygmy tarsier
Tarsius spectrum	Celebeskoboldmaki	spectral tarsier
Tarsius spp.	Koboldmakis	tarsiers
Tarsius syrichta	Philippinen-Koboldmaki	Philippine tarsier
Tarsomys apoensis	Mount Apo-Ratte	Mount Apo rat
Tarsonemus pallidus/ Phytonemus pallidus	Zyklamenmilbe, Alpenveilchenmilbe, Begonienmilbe	cyclamen mite, begonia mite
Tarsonemus myceliophagus	Champignon-Weichhautmilbe, Pilzmyzelmilbe	mushroom mite
Tarsonemus pallidus ssp. *fragariae*	Erdbeermilbe	strawberry mite
Tasmacetus shepherdi	Shepherd-Wal, Tasmanischer Schnabelwal	Shepherd's beaked whale, Tasman whale
Tateomys rhinogradoides	Tates Spitzmausratte	Tate's rat
Tateomys spp.	Sulawesi Spitzmausratten	Sulawesi shrew rats
Tatera spp.	Nacktsohlen-Rennmäuse	large naked-soled gerbils
Taterillus spp.	Kleine Nacktsohlen-Rennmäuse	small naked-soled gerbils
Taudactylus acutirostris	Spitzmaul-Sturzbachfrosch	sharpsnout torrent frog, sharp-nosed torrent frog (IUCN)
Taudactylus spp.	Sturzbachfrösche	torrent frogs
Tauraco persa	Guineaturako	green turaco
Taurotragus derbianus	Derbyantilope, Riesen-Elenantilope	Derby eland, giant eland
Taurotragus euryceros	Bongo	bongo
Taurotragus oryx	Elenantilope	common eland
Taurulus bubalis/ Myoxocephalus bubalis/ Cottus bubalis	Seebulle, Seebull, Ulk (Seeskorpion)	long-spined sculpin, long-spined sea scorpion, longspined bullhead (FAO)
Tautoga onitis	Tautog, Austernfisch	tautog
Tautogolabrus adspersus	Amerikanischer Lippfisch	cunner
Taxidea taxus	Silberdachs	American badger
Taxomyia taxi	Eibengallmücke	yew gall midge
Tayassu pecari	Weißbartpekari, Bisamschwein	white-lipped peccary
Tayassu tajacu	Halsbandpekari	collared peccary
Tectarius coronatus	Gekrönte Strandschnecke	beaded prickly winkle, crowned prickly winkle
Tectarius pagodus	Pagoden-Strandschnecke	pagoda prickly winkle
Tectura testudinalis/ Acmaea testudinalis/ Collisella tessulata	Schildkrötenschnecke, Klippkleber	plant limpet, tortoiseshell limpet, common tortoiseshell limpet

Tectura virginea/ *Acmaea virginea*	Weiße Schildkrötenschnecke	white tortoiseshell limpet
Tectus conus	Conus-Kreiselschnecke	cone-shaped topsnail, cone-shaped top
Tectus pyramis	Pyramiden-Kreiselschnecke	pyramid topsnail, pyramid top
Tedania ignis	Feuerschwamm	fire sponge
Tegenaria gigantea/ *Tegenaria atrica*	Große Hauswinkelspinne	giant European house spider, giant house spider, larger house spider, cobweb spider
Tegenaria agrestis	Hobospinne*	hobo spider (U.S.), yard spider (Br.)
Tegenaria domestica	Hausspinne, Gemeine Hauswinkelspinne	common European house spider, lesser house spider, barn funnel weaver (U.S.)
Tegenaria duellica	Große Amerikanische Hauswinkelspinne	giant house spider
Tegenaria parietina	Kardinal-Winkelspinne	cardinal spider
Tegenaria spp.	Hauswinkelspinnen	European house spiders
Tegula excavata	Grünflächige Turbanschnecke	green-base tegula
Tegula fasciata	Seidige Turbanschnecke	silky tegula
Tegula funebralis	Schwarze Pazifische Turbanschnecke	black tegula, black topsnail
Tegula pulligo	Düstere Turbanschnecke*	dusky tegula
Teia recens/ *Orgyia recens*/ *Orgyia gonostigma*	Eckfleck-Bürstenspinner	scarce vapourer (moth)
Teiidae	Schienenechsen	whiptails & racerunners
Telegeusidae	Langlippenkäfer	long-lipped beetles
Teleostei	Moderne Knochenfische	teleosts, teleost fishes
Telescopium telescopium	Teleskopschnecke	telescope snail, telescope creeper
Telescopus dhara	Israelische Katzennatter	Israelian cat snake
Telescopus fallax	Europäische Katzennatter	cat snake, European cat snake
Tellina agilis	Nordatlantische Zwerg-Tellmuschel	northern dwarf-tellin, dwarf tellin
Tellina crassa/ *Arcopagia crassa*	Stumpfe Tellmuschel	blunt tellin
Tellina fabula/ *Angulus fabula*/ *Fabulina fabula*	Gerippte Tellmuschel	bean-like tellin, semi-striated tellin
Tellina foliacea/ *Phylloda foliacea*	Goldzunge	foliated tellin, leafy tellin
Tellina laevigata	Glatte Tellmuschel	smooth tellin
Tellina lineata	Rosenmuschel*, Rosen-Tellmuschel*	rose-petal tellin, rose tellin, rose petal
Tellina linguaefelis	Katzenzungen-Tellmuschel, Katzenzunge	cat's-tongue tellin
Tellina listeri	Listers Tellmuschel	speckled tellin
Tellina nuculoides/ *Tellina salmonea*	Pazifische Lachs-Tellmuschel	salmon tellin
Tellina pulcherrima	Zauberhafte Tellmuschel*	beautiful tellin

Tellina radiata	Strahlige Tellmuschel	sunrise tellin (rising sun)
Tellina scobinata	Raspel-Tellmuschel	rasp tellin
Tellina solidula	Hartschalige Tellmuschel	hardshell tellin
Tellina tenuis/Angulus tenuis	Platte Tellmuschel, Zarte Tellmuschel, Plattmuschel	thin tellin, plain tellin, petal tellin
Tellina virgata/ Tellina pulchella	Gestreifte Tellmuschel (Jungfräuliche Tellmuschel)	virgate tellin, striped tellin
Tellinidae	Tellmuscheln, Plattmuscheln	tellins, sunset clams (sunset shells)
Telmatherina ladigesi	Celebes-Sonnenstrahlfisch	Celebes rainbowfish
Telmatobius culeus	Titicaca-Pfeiffrosch	Lake Titicaca frog
Telmatobius marmoratus	Marmorierter Anden-Pfeiffrosch	marbled water frog
Telmatophilus typhae	Schwarzer Schimmelkäfer	
Temnostoma vespiforme	Wespen-Moderholz-Schwebfliege	
Tenagodus obtusus	Wurmschnecke, Schlangenschnecke	obtuse wormsnail
Tenagodus squamatus	Schlitz-Wurmschnecke, Schuppige Schlangenschnecke	slit wormsnail
Tenebrio molitor	Mehlkäfer, Gelber Mehlkäfer	yellow mealworm beetle
Tenebrio spp.	Mehlkäfer	mealworm beetle
Tenebrionidae	Schwarzkäfer, Dunkelkäfer	darkling beetles, flour beetles, mealworm beetles
Tenebroides mauritanicus	Schwarzer Getreidenager	cadelle, cadelle beetle
Tenrec ecaudatus	Großer Tanrek	tenrec
Tenrecidae	Borstenigel, Tanreks	tenrecs
Tentaculata	Kranzfühler, Fühlerkranztiere, Tentaculaten	tentaculates (bryozoans, phoronids, brachiopods)
Tentaculifera	Tentaculiferen, Tentakeltragende Rippenquallen (Ctenophora)	tentaculiferans, "tentaculates"
Tenthredinidae	Blattwespen	common sawflies
Tephritidae/Trypetidae	Bohrfliegen, Fruchtfliegen	fruit flies, fruitflies
Teraponidae/ Theraponidae	Tigerfische, Sägebarsche	thornfishes, grunters, tiger-perches
Terathopius ecaudatus	Gaukler	bateleur
Terebdellidae	Schopfwürmer	terebdellid worms
Terebellum terebellum	Kleine Bohrerschnecke	little auger
Terebra areolata	Fliegendreck-Schraubenschnecke	flyspotted auger
Terebra crenulata	Gekerbte Schraube	crenulated auger
Terebra dimidiata	Orange-Schraubenschnecke	dimidiate auger, divided auger
Terebra dislocata	Ostatlantische Schraubenschnecke*	eastern auger, Atlantic auger, common American auger

Terebra duplicata	Doppelte Schraubenschnecke	duplicate auger
Terebra guttata	Weißfleck-Schraubenschnecke*	white-spotted auger (spotted auger)
Terebra maculata	Gefleckte Schraubenschnecke	marlinspike auger, giant marlin spike (spotted auger, big auger)
Terebra monilis	Geperlte Schraubenschnecke	necklace auger
Terebra ornata	Ornat-Schraubenschnecke*	ornate auger
Terebra pertusa	Löchrige Schraubenschnecke*	perforated auger
Terebra pretiosa	Kostbare Schraubenschnecke	precious auger
Terebra robusta	Dickschalige Schraubenschnecke*	robust auger
Terebra spectabilis	Hübsche Schraubenschnecke	graceful auger
Terebra strigata	Zebra-Schraubenschnecke	zebra auger
Terebra subulata	Pfriemenschnecke	subulate auger, chocolate spotted auger
Terebra taurina	Flammenschraubenschnecke, Flammenschraube	flame auger
Terebra triseriata	Dreigestreifte Schraubenschnecke	triseriate auger
Terebra vinosa	Flieder-Schraubenschnecke*	lilac auger
Terebralia palustris	Schlammkriecher	mud creeper
Terebridae	Schraubenschnecken, Pfriemenschnecken	auger snails, augers (auger shells)
Teredilia	Kleine Holzwürmer	lesser wood-boring beetles
Teredinidae	Pfahlwürmer, Schiffsbohrwürmer	shipworms
Teredo megotara	Schwebende Schiffsbohrmuschel, Schiffsbohrwurm	drifting shipworm
Teredo navalis	Schiffsbohrmuschel, Pfahlwurm	naval shipworm, common shipworm, great shipworm
Teredo norvegica/ Nototeredo norvagicus	Nordische Schiffsbohrmuschel, Nordischer Schiffsbohrwurm	Norway shipworm, Norwegian shipworm
Termitidae	Nasentermiten u.a.	nasutiform termites & soldierless termites
Terpsiphone paradisi	Fahlbauch-Paradiesschnäpper	Asiatic paradise flycatcher
Terpsiphone viridis	Graubrust-Paradiesschnäpper	African paradise flycatcher
Terrapene carolina	Carolina-Dosenschildkröte	Eastern box turtle, common box turtle (IUCN)
Terrapene coahuila	Coahuila-Dosenschildkröte	Coahuilan box turtle, aquatic box turtle (IUCN)
Terrapene nelsoni	Nelsons Dosenschildkröte	Nelson's box turtle, spotted box turtle
Terrapene ornata luteola	Wüsten-Schmuck-Dosenschildkröte	desert box turtle
Terrapene ornata ornata	Schmuck-Dosenschildkröte	western box turtle, ornate box turtle (IUCN)
Terrapene spp.	Dosenschildkröte	box turtles
Tersina viridis	Schwalbentangare	swallow tanager
Tersininae	Schwalbentangare	swallow-tanager
Teskeyostrea weberi	Fadenauster*	threaded oyster
Testacea	Thekamöben, Beschalte Amöben	testate amebas

Testacella haliotidea	Rucksackschnecke	earshell slug, shelled slug
Testacellidae	Rucksackschnecken	shelled slugs, worm-eating slugs
Testudines/Chelonia	Schildkröten	turtles
Testudinidae	Landschildkröten	tortoises, true tortoises
Testudo carbonaria/ Geochelone carbonaria/ Chelonoidis carbonaria	Köhlerschildkröte	red-footed tortoise, coal tortoise
Testudo chilensis/ Geochelone chilensis/ Chelonoidis chilensis	Argentinische Landschildkröte	Argentine tortoise (IUCN), Chilean tortoise, Chaco tortoise
Testudo denticulata/ Geochelone denticulata/ Chelonoidis denticulata	Waldschildkröte	yellow-footed tortoise, Brazilian giant tortoise, American forest tortoise, Hercules tortoise, South American yellow-footed tortoise (IUCN)
Testudo elephantopus/ Geochelone elephantopus/ Chelonoides elephantopus	Galapagos-Riesenschildkröte, Elefantenschildkröte	Galapagos giant tortoise
Testudo gigantea/ Geochelone gigantea/ Megalochelys gigantea	Seychellen-Riesenschildkröte	Seychelles giant tortoise, Aldabran giant tortoise, Aldabra giant tortoise (IUCN)
Testudo graeca	Maurische Landschildkröte	spur-thighed tortoise (IUCN), Mediterranean spur-thighed tortoise, common tortoise, Greek tortoise
Testudo hermanni	Griechische Landschildkröte	Hermann's tortoise, Greek tortoise
Testudo horsfieldii/ Agrionemis horsfieldi	Steppenschildkröte	Horsfield's tortoise, four-toed tortoise, Central Asian tortoise (IUCN)
Testudo kleinmanni	Ägytische Landschildkröte	Egyptian tortoise
Testudo marginata	Breitrandschildkröte	margined tortoise, marginated tortoise (IUCN)
Testudo spp.	Eigentliche Landschildkröten	land tortoises
Tethea ocularis	Augen-Eulenspinner, Achtzigeule, Pappelhain-Eulenspinner, Schwarzgebänderter Wollrückenspinner	figure of eighty
Tethea or	Wollrückenspinner, Pappel-Eulenspinner	poplar lutestring
Tetheella fluctuosa	Birken-Eulenspinner	
Tethya aurantia/ Tethya lyncurium	Meerorange, Apfelsinenschwamm, Kugelschwamm	orange puffbal sponge, golfball sponge
Tetracerus quadricornis	Vierhornantilope	four-horned antelope, chousingha
Tetracheledone spinicirrus	Stachelhornkrake*, Augenbrauenkrake	spiny-horn octopus, eyebrow octopus
Tetradactylus spp.	Geißelschildechsen	seps
Tetragnatha spp.	Eigentliche Streckerspinnen	long-jawed orb weavers
Tetragnathidae	Streckerspinnen & Kieferspinnen	thick-jawed spiders, large-jawed orb weavers, four-jawed spiders, elongated orb weavers, horizontal orb-web weavers

Tetragonopterinae	Tetras	tetras
Tetragonopterus argenteus	Gesäumter Schillersalmler	big-eyed characin
Tetragonuridae	Eckschwänze, Quadratschwänze	squaretails
Tetragonurus atlanticus	Atlantischer Eckschwanz	Atlantic squaretail
Tetragonurus cuvieri	Cuvier-Eckschwanz	Cuvier's squaretail
Tetramorium caespitum	Rasenameise	turf ant, pavement ant, black pavement ant
Tetranychidae	Spinnmilben	spider mites, red mites
Tetranychus cinnabarinus	Karminspinnmilbe	carmine spider mite
Tetranychus urticae	Gemeine Spinnmilbe, „Rote Spinne", Bohnenspinnmilbe	twospotted spider mite
Tetranychus viennensis	Weißdornspinnmilbe	hawthorn spider mite
Tetrao mlokosiewiczi	Kaukasusbirkhuhn	Caucasian black grouse
Tetrao tetrix/Lyrurus tetrix	Birkhuhn	black grouse
Tetrao urogallus	Auerhuhn	capercaillie
Tetraodon cutcutia	Gemeiner Kugelfisch	common pufferfish
Tetraodon fluviatilis	Grüner Kugelfisch	green pufferfish
Tetraodon spp.	Kugelfische	pufferfishes
Tetraodontidae	Kugelfische	puffers
Tetraodontiformes/ Plectognathi	Kugelfischverwandte	plectognath fishes
Tetraogallus caspius	Kaspikönigshuhn	Caspian snowcock
Tetraogallus caucasicus	Kaukasuskönigshuhn	Caucasian snowcock
Tetraogallus himalayensis	Himalayakönigshuhn	Himalayan snowcock
Tetraoninae	Raufußhühner	grouse
Tetrapoda	Landwirbeltiere, Tetrapoden	terrestrial vertebrates, tetrapods
Tetrapturus albidus	Weißer Speerfisch, Weißer Marlin	white marlin
Tetrapturus angustirostris	Kurzschnauzen-Marlin, Kurzschnäuziger Speerfisch	shortbill spearfish
Tetrapturus audax	Gestreifter Marlin, Gestreifter Speerfisch	striped marlin
Tetrapturus belone	Mittelmeer-Speerfisch, Langschwänziger Speerfisch	Mediterranean spearfish (FAO), Mediterranean marlin
Tetrapturus georgii	Rundschuppen-Speerfisch	round-scaled marlin, round-scale spearfish, roundscale spearfish (FAO)
Tetrapturus pfluegeri	Langschnäuziger Speerfisch	longbill spearfish (FAO), long-billed spearfish
Tetrarogidae	Segelfische	waspfishes
Tetrastemma melanocephalum (Nemertini)	Schwarzkopf-Schnurwurm, Schwarzkopf-Nemertine	
Tetrastes bonasia/ Bonasa bonasia	Haselhuhn	hazel grouse
Tetrax tetrax	Zwergtrappe	little bustard
Tetrigidae	Dornschrecken	pygmy grasshoppers, grouse locusts
Tetrix bipunctata	Zweipunkt-Dornschrecke	twospotted groundhopper
Tetrix ceperoi	Westliche Dornschrecke	Cepero's groundhopper

Tetrix depressa	Eingedrückte Dornschrecke	depressed groundhopper, flattened groundhopper
Tetrix subulata/ Acrydium subulatum	Säbeldornschrecke	slender groundhopper
Tetrix tenuicornis/ Tetrix natans	Langfühler-Dornschrecke	longhorned groundhopper
Tetrix undulata/Tetrix vittata	Gemeine Dornschrecke, Säbeldornschrecke	common groundhopper
Tetropium castaneum	Gemeiner Fichtensplintbock	chestnut longhorn beetle
Tetrops praeusta	Pflaumenböckchen	little longhorn beetle
Tetrosomus gibbosus	Pyramiden-Kofferfisch, Buckelkofferfisch	trunkfish
Tettigometridae	Käferzikaden	tettigometrids
Tettigonia cantans	Zwitscherschrecke	twitching green bushcricket
Tettigonia caudata	Östliches Heupferd	eastern green bushcricket
Tettigonia viridissima	Großes Heupferd, Grünes Heupferd	great green bushcricket
Tettigoniidae	Singschrecken, Laubheuschrecken	long-horned grasshoppers (and katydids/bushcrickets)
Tettigonioidea	Laubheuschrecken	bushcrickets
Teuthoidea (Teuthida)	Kalmare	squids
Thainycteris aureocollaris	Siam-Goldhals-Fledermaus, Thai-Goldkehlchen	Siam golden-throated bat, Thai goldnecklet
Thais haemastoma	Rotmund-Leistenschnecke, Rotmundige Steinschnecke	redmouthed rocksnail (red-mouthed rock shell)
Thais hippocastanum	Kastanien-Leistenschnecke	chestnut rocksnail (chestnut rock shell)
Thais lapillus/ Nucella lapillus	Steinschnecke, Nordische Steinchenschnecke, Nördliche Purpurschnecke, Nordische Purpurschnecke	Atlantic dog whelk, northern dog whelk, Atlantic dogwinkle, northern dogwinkle
Thais rugosa	Raue Felsschnecke	rough rocksnail (rough rock shell)
Thais tuberosa	Buckel-Purpurschnecke	humped rocksnail (humped rock shell)
Thalassema mellita (Echiura)		keyhole urchin spoon worm
Thalassema neptuni (Echiura)	Seesandwurm	Gaertner's spoon worm
Thalassina anomala	Großer Maulwurfskrebs	tropical mole crab
Thalassoma bifasciatum	Blaukopfjunker	bluehead
Thalassoma lunare	Mondsicheljunker	moon wrasse (FAO), green wrasse, lyretail wrasse
Thalassoma pavo	Meerpfau	Turkish wrasse, peacock wrasse, sea-peacock, rainbow wrasse, ornate wrasse (FAO)
Thalassornis leuconotos	Weißrücken-Pfeifgans	white-backed duck
Thaleichthys pacificus	Kerzenstint, Kerzenfisch, Eulachon	candlefish, eulachon (FAO)
Thalera fimbrialis		Sussex emerald (moth)
Thaliacea	Salpen, Thaliaceen	salps, thaliceans
Thallomys paedulcus	Akazienmaus	acacia rat

Thalpophila matura	Rötliche Wurzeleule	straw underwing
Thamnomys spp.	Westafrikanische Gebüschratten*, Breitfuß-Gebüschratten*	thicket rats, broad-footed thicket rats
Thamnophilidae		typical antbirds
Thamnophis eques	Braune Strumpfbandnatter	Mexican garter snake
Thamnophis gigas	Große Strumpfbandnatter, Große Bändernatter	giant garter snake
Thamnophis radix	Flachland-Strumpfbandnatter	plains garter snake
Thamnophis sauritus	Bändernatter	eastern ribbon snake
Thamnophis sirtalis	Gemeine Strumpfbandnatter	common garter snake
Thamnophis spp.	Strumpfbandnattern, Bändernattern	American garter snakes, "water snakes"
Thanasimus formicarius	Ameisen-Buntkäfer, Ameisenbuntkäfer	ant beetle
Thatcheria mirabilis	Wunderkreisel, Wunderbare Japanische Turmschnecke	Japanese wonder snail (miraculous Thatcher shell)
Thaumastocoridae	Königspalmenwanzen*	royal palm bugs, thaumastocorids
Thaumatomyia notata	Kleine Halmfliege	small cluster-fly
Thaumetopoea pinivora	Kiefern-Prozessionsspinner, Fichten-Prozessionsspinner	pine processionary moth, spruce processionary moth
Thaumetopoea pityocampa	Pinien-Prozessionsspinner	pine processionary moth
Thaumetopoea processionea	Eichen-Prozessionsspinner	oak processionary moth
Thaumetopoeidae	Prozessionsspinner	processionary moths
Thayeria boehlkei	Schrägsalmler, Schrägschwimmer	blackline penguinfish
Thayeria obliqua	Pinguinsalmler	penguinfish, honey stick tetra
Thea vigintiduopunctata	Zweiundzwanzigpunkt-Marienkäfer	twentytwo-spot ladybird beetle
Theba pisana	Mittelmeer-Sandschnecke	sandhill snail, white gardensnail, Mediterranean sand snail, Mediterranean white snail
Thecadactylus rapicauda	Rübenschwanzgecko	wood slave, turniptail gecko
Thecla betulae	Birkenzipfelfalter, Nierenfleck-Zipfelfalter	brown hairstreak
Thecla quercus/ Quercusia quercus	Eichenzipfelfalter	purple hairstreak
Thecodiplosis brachyntera	Nadelkürzende Kieferngallmücke, Kiefernnadelscheiden-Gallmücke	
Thecodontia	Urwurzelzähner	thecodonts, dinosaur ancestors
Thecosomata	Seeschmetterlinge, Beschalte Flossenfüßer (Flügelschnecken)	sea butterflies, shelled pteropods
Thelaxidae	Maskenläuse, Maskenblattläuse	thelaxid plantlice
Thelazia lacrymalis	Augenwurm, Augenfadenwurm	eyeworm of horses
Thelenota ananas	Ananas-Seewalze	pineapple sea cucumber, "prickly redfish"

Thelenota anax	Riesen-Seewalze	giant sea cucumber
Thelotornis kirtlandii	Graue Baumnatter, Vogelschlange	bird snake
Thenus orientalis	Breitkopf-Bärenkrebs	Moreton Bay flathead lobster, Moreton Bay "bug"
Theodoxus corona	Kronen-Kahnschnecke	crown nerite
Theodoxus danubialis	Donau-Kahnschnecke, Sarmatische Schwimmschnecke	Danube nerite, Danube freshwater nerite
Theodoxus fluviatilis	Gemeine Kahnschnecke, Fluss-Schwimmschnecke	common freshwater nerite, river nerite, the nerite
Theodoxus transversalis	Gebänderte Kahnschnecke, Binden-Schwimmschnecke	striped freshwater nerite
Thera firmata	Roter Kiefernspanner*	red pine carpet (moth)
Thera juniperata	Wacholderspanner	juniper carpet (moth)
Thera obeliscata	Grauer Kiefernspanner*	grey pine carpet (moth)
Theragra chalcogramma	Alaska-Pollack, Pazifischer Pollack, Mintai	pollock, pollack, Alaska pollock, walleye pollock, Alaska pollack (FAO)
Theraphosa blondi	Riesen-Vogelspinne, Goliath-Vogelspinne	goliath birdeater tarantula
Theraphosidae/Aviculariidae	Taranteln, Echte Vogelspinnen, Eigentliche Vogelspinnen, Buschspinnen	tarantulas & bird spiders
Therapon jarbua	Dreistreifen-Tigerfisch	jarbua therapon
Theraponidae/Teraponidae	Tigerfische, Sägebarsche	thornfishes, grunters, tiger-perches
Therevidae	Stilettfliegen, Luchsfliegen	stiletto flies
Theridiidae	Kugelspinnen, Haubennetzspinnen	comb-footed spiders, cobweb and widow spiders, scaffolding-web spiders
Theridion spp.	Kugelspinnen	comb-footed spider
Theridiosoma radiosa	Zwergradnetzspinne	ray spider
Theridiosomatidae	Zwergradnetzspinnen	ray spiders
Thermobia domestica/ Lepismodes inquilinus	Ofenfischchen	firebrat
Thermyzon tessulatum	Entenegel	duck leech
Theropithecus gelada	Dschelada, Blutbrustpavian	gelada
Thia scutellata	Herzkrabbe	thumbnail crab
Thinocoridae	Höhenläufer	seedsnipes
Tholera decimalis	Große Raseneule	
Thomasiniana crataegi/ Resseliella crataegi	Weißdornzweiggallmücke	hawthorn stem midge
Thomasiniana oculiperda/ Resseliella oculiperda	Okuliergallmücke, Okuliermade	red bud borer midge (larva)
Thomasiniana theobaldi/ Resseliella theobaldi	Himbeerrutengallmücke	raspberry cane midge
Thomasomys spp.	Thomas-Paramomäuse	Thomas's paramo mice
Thomisidae	Krabbenspinnen	crab spiders
Thomomys homomys	Südliche Taschenratte	southern pocket gopher
Thomomys mazama	Westliche Taschenratte	western pocket gopher

Thomomys monticola	Gebirgs-Taschenratte	mountain pocket gopher
Thomomys spp.	Taschenratten, Nordamerikanische Taschenratten	pocket gophers
Thomomys talpoides	Nördliche Taschenratte	northern pocket gopher
Thoopterus nigrescens	Kurznasen-Flughund, Schwarzflügel-Flughund	short-nosed fruit bat, swift fruit bat
Thoracocharax securis	Platin-Beilbauchfisch	giant hatchetfish (FAO), greater hatchetfish, greater silver hatchfish
Thoracocharax stellatus	Diskus-Beilbauchfisch	spotfin hatchetfish (FAO), silver hatchetfish
Thorictidae	Ameisenglattkäfer	
Thorius spp.	Mexikanische Pygmäensalamander	pygmy salamanders
Thorogobius ephippiatus/ Gobius forsteri	Leopardgrundel	leopard-spotted goby
Thorogobius macrolepis	Großschuppige Grundel	large-scaled goby
Thracia devexa	Geneigte Thracia	sloping thracia
Thracia myopsis	Arktische Thracia	Arctic thracia
Thracia septentrionalis	Nördliche Thracia	northern thracia
Thraciidae	Thracia-Muscheln	thraciids (lantern shells)
Thraupidae	Tangaren	tanagers, Neotropical honeycreepers, seedeaters, flower-piercers
Threskiornis aethiopica	Heiliger Ibis	sacred ibis
Threskiornithidae	Ibisvögel	ibises, spoonbills
Thrips angusticeps	Ackerblasenfuß	field thrips, cabbage thrips, flax thrips
Thrips fuscipennis	Rosenblasenfuß	rose thrips
Thrips palmi	Palmenblasenfuß	palm thrips
Thrips spp.	Blasenfüße, Thripse	thrips
Thrips tabaci	Tabakblasenfuß, Zwiebelblasenfuß	onion thrips, potato thrips
Throscidae	Hüpfkäfer	false metallic wood-boring beetles, throscid beetles
Thryomanes bewickii	Buschzaunkönig	Bewick's wren
Thryonomys spp.	Rohrratten	cane rats
Thuiaria thuja	Flaschenbürstenmoos*	bottlebrush hydroid
Thunnus alalunga	Weißer Thun, Germon	albacore (FAO), "white" tuna, long-fin tunny, long-finned tuna, Pacific albacore
Thunnus albacares	Gelbflossen-Thunfisch	yellowfin tuna (FAO), yellow-finned tuna, yellow-fin tunny
Thunnus atlanticus	Schwarzflossen-Thunfisch, Schwarzflossenthun	blackfin tuna
Thunnus maccoyii	Südlicher Blauflossenthun	southern bluefin tuna
Thunnus obesus	Großaugen-Thunfisch	big-eyed tuna, bigeye tuna (FAO)

Thunnus thynnus	Thunfisch,	tunny,
	Großer Thunfisch,	blue-fin tuna,
	Roter Thunfisch,	blue-finned tuna,
	Roter Thun	northern bluefin tuna (FAO)
Thunnus tonggol	Langschwanz-Thun	longtail tuna
Thuridilla hopei/	Pracht-Samtschnecke	splendid velvet snail,
Elysia splendida		splendid elysia
Thyasira flexuosa	Gewellte Sichelmuschel*	wavy hatchetclam
		(wavy hatchet-shell)
Thyasira gouldi	Nord-Sichelmuschel*	northern hatchet-shell
Thyatira batis	Roseneule,	peach blossom (moth)
	Rosen-Eulenspinner	
Thyatiridae/Cymatophoridae	Eulenspinner	thyatirid moths
	(Wollrückenspinner)	
Thylacinus cynocephalus	Beutelwolf,	thylacine,
	Tasmanischer Tiger	Tasmanian wolf,
		Tasmanian tiger
Thylacis spp./*Isoodon* spp.	Kurznasenbeutler	short-nosed bandicoots
Thylogale billardierii	Rotbauchfilander	red-bellied pademelon
Thylogale brunii	Neuguineafilander	dusky pademelon
Thylogale stigmatica	Rotbeinfilander	red-legged pademelon
Thylogale thetis	Rothalsfilander	red-necked pademelon
Thymallidae	Äschen	graylings
Thymallus arcticus	Arktische Äsche	Arctic grayling
Thymallus thymallus	Äsche,	grayling
	Europäische Äsche	
Thymelicus acteon	Mattscheckiger	Lulworth skipper (butterfly)
	Braun-Dickkopffalter	
Thymelicus lineolus	Schwarzkolbiger	Essex skipper (butterfly)
	Braun-Dickkopffalter	
Thymelicus sylvestris/	Braunkolbiger	small skipper (butterfly)
Thymelicus flavus	Braun-Dickkopffalter,	
	Ockergelber Dickkopffalter	
Thyone fusus	Sand-Seegurke,	spindle thyone
	Spindel-Thyone	
Thyreocoridae	Schwarzwanzen	negro bugs
Thyreocoris scarabaeoides	Schwarzwanze	negro bug
Thyreonotus bidens	Zweizahnschrecke*	two-toothed bushcricket
Thyreophagus entomophagus	Museumsmilbe	museum mite, flour mite
Thyrididae	Fensterschwärmer	window-winged moths
Thyroptera spp.	Haftscheiben-Fledermäuse	disk-winged bats,
		New World sucker-footed bats
Thyropteridae	Amerikanische	disk-winged bats,
	Haftscheiben-Fledermäuse	New World sucker-footed bats
Thyrsites atun	Atun	snoek (FAO),
		sea pike,
		barracouta
Thyrsitops lepidopoides	Weißer Atun	white snake mackerel
Thysania agrippina	Agrippinaeule	giant owl moth
Thysanoptera	Fransenflügler,	thrips
	Blasenfüße,	
	Thripse	
Thysanosoma actinioides	Fransen-Bandwurm*	fringed tapeworm
Thysanoteuthis rhombus	Vierecks-Tintenfisch*	diamondback squid
Thysanozoon brochii	Zottenplanarie	skirt dancer

Thysanura	Borstenschwänze, Thysanuren (Felsenspringer & Fischchen)	bristletails, thysanurans
Tiaris canora	Kubafink	melodious grassquit
Tiaris olivacea	Goldbraue	yellow-faced grassquit
Tibia curta	Indische Spindelschnecke	Indian tibia
Tibia delicatula	Dünnschalige Spindelschnecke	delicate tibia
Tibia fusus	Spindelschnecke	spindle tibia, shinbone tibia
Tibia insulae-chorab	Arabische Spindelschnecke	Arabian tibia
Tibia martini	Martins Spindelschnecke	Martin's tibia
Tibia powisi	Powis Spindelschnecke	Powis's tibia
Tibicina haematodes/ Tibicen haematodes	Blutrote Zikade, Lauer, Weinzwirner	vineyard cicada
Tichodroma muraria	Mauerläufer	wallcreeper
Tichodrominae	Mauerläufer	wallcreeper
Tilapia galilaea/ Sarotherodon galilaeus	Prachtmaulbrüter	mango tilapia (FAO), St. Peter's fish
Tilapia mossambica/ Oreochromis mossambicus	Mosambikbuntbarsch	Mozambique tilapia (FAO), Mozambique mouthbreeder
Tilapia lepidura/ Oreochromis lepidura	Perlmutterbuntbarsch	
Tilapia mariae	Marienbarsch	spotted tilapia
Tilapia nilotica/ Oreochromis niloticus	Nil-Buntbarsch	Nile tilapia (FAO), Nile mouthbreeder
Tilapia sparrmanii	Sparrmans Buntbarsch	banded tilapia (FAO), peacock cichlid, Sparrman's cichlid, banded bream
Tilapia zillii	Rotbauch-Buntbarsch, Zilles Buntbarsch	redbelly tilapia (FAO), Zilli's cichlid
Tiliqua casuarinae	Buschland-Skink	scrub skink
Tiliqua gerrardii	Rosazungen-Skink	pinktongue skink, pink-tongued skink
Tiliqua spp.	Blauzungen-Skinks	bluetongue skinks, blue-tongued skinks
Timaliidae	Timalien	babblers, *Rhabdornis*
Timandra griseata/ Timandra amata	Schrägstreifiger Ampferspanner	blood-vein (moth)
Timarcha tenebricosa	Labkrautblattkäfer	bloody-nosed beetle
Timoclea ovata/ Chione ovata/ Venus ovata	Ovale Venusmuschel	oval venus
Tinamidae	Steißhühner	tinamous
Tinamiformes	Steißhühner	tinamous
Tinamus major	Großtinamu, Großtao	great tinamou
Tinca tinca	Schlei, Schleie, Schleihe	tench
Tinea pellionella	Pelzmotte	case-bearing clothes moth, case-making clothes moth
Tineidae	Echte Motten	clothes moths
Tineola bisselliella	Kleidermotte	common clothes moth, destroyer clothes moth

Tingidae	Netzwanzen (Blasenwanzen, Buckelwanzen oder Gitterwanzen)	lace bugs
Tingis cardui	Distelnetzwanze	spear thistle lace bug
Tiphia femorata	Gemeine Rollwespe, Rotschenkelige Rollwespe	common tiphiid wasp
Tiphiidae	Rollwespen	tiphiid wasps
Tipula maxima	Riesenschnake	giant cranefly
Tipula oleracea	Kohlschnake	cabbage cranefly, brown daddy-long-legs (Br.)
Tipula paludosa	Wiesenschnake	meadow cranefly, grey daddy-long-legs (Br.)
Tipulidae	Schnaken, Stelzmücken, Erdschnaken, Pferdeschnaken (Bachmücken, Langbeinmücken, Riesenschnaken)	crane flies, crane-flies, daddy-long-legs (Br.)
Tischeria ekebladella	Eichenminiermotte	red-feather carl, trumpet leaf miner
Tischeriidae	Schopfstirnmotten	apple leaf miners & others
Tityra semifasciata	Maskentitya	masked titya
Tmetothylacus tenellus	Goldpieper	golden pipit
Tockus flavirostris	Gelbschnabeltoko	yellow-billed hornbill
Todarodes sagittatus/ Ommastrephes sagittatus	Pfeilkalmar	flying squid, European flying squid, sagittate squid, sea-arrow, red squid
Todarodes pacificus	Japanischer Flugkalmar	Japanese flying squid
Todidae	Todies, Todis	todies
Todus multicolor	Vielfarbentodi	Cuban tody
Tokudaia osimensis	Riu-Kiu-Stachelratte	Ryukyu spiny rat
Tolypeutes spp.	Kugelgürteltiere	three-banded armadillos
Tolypeutes tricinctus	Kugelgürteltier, Brasilianisches Dreibinden-Gürteltier	Brazilian three-banded armadillo
Tomicus piniperda/ Blastophagus piniperda	Großer Waldgärtner, Gefurchter Waldgärtner, Kiefernmarkkäfer, Markkäfer	pine beetle, larger pith borer, large pine shoot beetle
Torpedinidae	Große Zitterrochen, Echte Zitterrochen	electric rays
Torpedo californica	Kalifornischer Zitterrrochen, Pazifik-Zitterrochen	Pacific electric ray
Torpedo marmorata	Marmorierter Zitterrochen, Marmor-Zitterrochen, Marmelzitterrochen	marbled crampfish, marbled torpedo ray, marbled electric ray (FAO)
Torpedo melanospila/ Taeniura melanospilos> Taeniura meyeni	Gefleckter Stechrochen, Schwarzpunkt-Stechrochen	round ribbontail ray, black-spotted ray, speckled stingray, blotched fantail ray (FAO)
Torpedo nobiliana	Schwarzer Zitterrochen	Atlantic electric ray, Atlantic torpedo, electric ray (FAO)

Torpedo torpedo	Augenfleck-Zitterrochen, Gefleckter Zitterrochen	torpedo ray, common torpedo (FAO), eyed electric ray
Tortricidae	Wickler	tortricids, leaf rollers, leaf tyers
Tortrix viridana	Grüner Eichenwickler	pea-green oak curl, green oak tortrix, oak leafroller, green oak roller, oak tortrix
Toxicocalamus spp.	Gift-Inselschlangen*	venomous island snakes, elongate snakes
Toxocara canis	Hundespulwurm	canine ascarid
Toxocara cati	Katzenspulwurm	feline ascarid
Toxocara spp.	Spulwürmer u.a.	arrowhead worms
Toxotus cursor	Schulterbock	
Trabutina mannipara	Manna-Schildlaus	tamarisk manna scale
Trachelus tabidus		black grain stem sawfly
Trachemys dorbignyi/ Pseudemys dorbignyi	Brasilianische Schmuckschildkröte	Brazilian slider, black-bellied slider
Trachemys decorata	Hispaniola-Schmuckschildkröte	Hispaniolan slider
Trachemys decussata	Nord-Antillen-Schmuckschildkröte	West Indian slider, North Antillean slider
Trachemys scripta elegans/ Pseudemys scripta elegans	Rotwangen-Schmuckschildkröte	red-eared turtle, red-eared slider
Trachemys scripta scripta/ Pseudemys scripta scripta/ Chrysemys scripta scripta	Buchstaben-Schmuckschildkröte	slider, common slider, pond slider, yellow-bellied turtle
Trachemys stejnegeri	Puerto Rico-Schmuckschildkröte	Puerto Rican slider, Central Antillean slider (IUCN)
Trachemys terrapen/ Pseudemys terrapen	Antillen-Schmuckschildkröte, Jamaika-Schmuckschildkröte	Jamaican slider
Trachichthyidae	Sägebäuche	slimeheads, roughies
Trachinidae	Drachenfische, Petermännchen	weeverfishes
Trachinotus blochii	Silber-Gabelmakrele	silver pompano
Trachinotus carolinus	Florida-Gabelmakrele, Gemeiner Pompano	Florida pompano (FAO), common pompano
Trachinotus falcatus	Permit	permit
Trachinotus glaucus	Weiße Gabelmakrele	glaucous pompano
Trachinotus goodei	Palometa	palometa
Trachinotus ovatus	Gabelmakrele, Bläuel, Pompano	pompano, derbio (FAO)
Trachinus araneus	Geflecktes Petermännchen, Spinnenqueise	spotted weever
Trachinus draco	Petermännchen, Großes Petermännchen	great weever (FAO), greater weever
Trachinus radiatus	Gestreiftes Petermännchen, Strahlenpetermännchen	streaked weever
Trachinus vipera/ Echiichthys vipera	Kleines Petermännchen, Viperqueise, Zwergpetermännchen	lesser weever
Trachipteridae	Sensenfische, Bandfische	ribbonfishes

Trachipterus arcticus	Spanfisch, Nordischer Bandfisch, Nordischer Spanfisch	deal-fish, dealfish
Trachipterus trachypterus	Bandfisch, Spanfisch	ribbon fish (FAO), ribbonfish, Mediterranean dealfish
Trachurus declivis	Grünrücken-Bastardmakrele	greenback horse mackerel
Trachurus japonicus	Japanische Bastardmakrele	Japanese jack mackerel
Trachurus lathami	Raue Bastardmakrele	rough scad
Trachurus mediterraneus	Mittelmeer-Stöcker, Mittelmeer-Bastardmakrele	Mediterranean horse mackerel
Trachurus murphyi	Chilenische Bastardmakrele	Chilean jack mackerel, Inca scad (FAO)
Trachurus picturatus	Blaue Bastardmakrele, Blauer Stöcker	offshore jack mackerel, blue jack mackerel (FAO)
Trachurus symmetricus	Pazifische Bastardmakrele, Pazifik-Stöcker	Pacific jack mackerel
Trachurus trachurus	Stöcker, Schildmakrele, Bastardmakrele	Atlantic horse mackerel (FAO), scad, maasbanker
Trachusa byssina/ Trachusa serratulae	Große Harzbiene, Bastardbiene	
Trachyboa boulengeri	Rauschuppenboa	rough-scaled boa
Trachydosaurus rugosus/ Trachysaurus rugosus	Stutzechse, Tannenzapfenechse	stump-tailed skink
Trachypenaeus similis/ Trachysalambria curvirostris	Raurücken-Geißelgarnele*	roughback shrimp
Trachypenaeus constrictus	Rauhals-Geißelgarnele	roughneck shrimp
Trachypithecus francoisi/ Presbytis francoisi	Tonkinlangur	Tonkin leaf monkey, François' leaf monkey
Trachypterus see Trachipterus		
Trachyrhynchus trachyrhynchus	Langschnauzen-Seeratte	roughsnout rattail, long-snouted grenadier, roughsnout grenadier (FAO)
Trachys minutus	Gemeiner Zwergprachtkäfer	
Trachysphaeridae	Stäbchenkugler	trachysphaerid millepedes
Tragelaphus angasi	Nyala	nyala
Tragelaphus buxtoni	Bergnyala, Mittelkudu	mountain nyala
Tragelaphus imberbis	Kleiner Kudu	lesser kudu
Tragelaphus scriptus	Buschbock	bushbuck
Tragelaphus spekei	Sitatunga, Sumpfantilope, Wasserkudu	sitatunga
Tragelaphus strepsiceros	Großer Kudu	graeter kudu
Tragopan blythii	Graubauchtragopan	Blyth's tragopan
Tragopan melanocephalus	Scharzkopf-Tragopan	western tragopan
Tragosoma depsarium	Zottenbock	
Tragulidae	Hirschferkel	chevrotains, mouse deer
Tragulus javanicus	Kleinkantschil	lesser Malay chevrotain
Tragulus meminna	Fleckenkantschil, Indien-Kantschil	Indian spotted chevrotain
Tragulus napu	Großkantschil	greater Malay chevrotain

Tragulus spp.	Kantschile	Asiatic chevrotains, mouse deer
Trapelus agilis/Agama agilis	Schlankagame	brilliant agama
Trapelus mutabilis/ Agama mutabilis	Wüstenagame	desert agama
Trapelus sanguinolentus	Steppenagame	steppe agama
Trapezia spp.	Korallenkrabben	coral crabs
Trapeziidae	Korallenkrabben	coral crabs
Trapezium oblongum	Lange Riffmuschel	oblong trapezium
Trechus quadristriatus	Gewöhnlicher Flinkläufer, Feld-Flinkläufer	
Trechus spp.	Flinkläufer, Flinkkäfer	
Tremarctos ornatus	Brillenbär	spectacled bear
Trematoda	Trematoden, Egel, Saugwürmer	flukes, trematodes
Trematomus hansoni	Gestreifte Notothenie	striped rockcod (FAO), green rockcod
Trematomus spp.	Hochantarktische Nototheniiden	Antarctic cod, Antarctic rockcod
Tremoctopus violaceus	Löcherkrake	common blanket octopus, common umbrella octopus
Treron calva	Rotnasen-Grüntaube	African green pigeon
Tresus capax	Dicke Klaffmuschel	fat gaper clam
Tresus nuttallii	Pazifische Klaffmuschel	Pacific gaper clam, "horseneck clam"
Triacanthidae	Dreistachler	triplespines
Triacanthodidae	Stachelfische*, Dreistachler	spikefishes
Triactis producta	Krabbenanemone	crab anemone*
Triaenodon obesus	Weißspitzen-Riffhai, Weißspitzen-Hundshai, Weißspitzen-Marderhai	blunthead shark, whitetip reef shark (FAO)
Triaenophorus nodulosus	Hechtbandwurm	pike tapeworm*
Triaenops spp.	Dreiblatt-Fledermäuse u.a.	triple nose-leaf bats a.o.
Triakidae	Glatthaie, Hundshaie	gullysharks, houndsharks, hound sharks
Triakis megalopterus	Südafrikanischer Hundshai, Gefleckter Marderhai	spotted gullyshark
Triakis semifasciatus	Leopardenhai, Kalifornischer Leopardhai	leopard shark
Trialeurodes vaporariorum	„Weiße Fliege"	greenhouse whitefly
Tribolium audax	Amerikanischer Schwarzer Reismehlkäfer	black flour beetle, American black flour beetle
Tribolium castaneum	Rotbrauner Reismehlkäfer	red flour beetle
Tribolium confusum	Amerikanischer Reismehlkäfer	confused flour beetle
Tribolium destructor	Großer Reismehlkäfer	greater flour beetle (Scandinavian flour beetle)
Tribolium madens	Schwarzbrauner Reismehlkäfer, Schwarzer Reismehlkäfer	European black flour beetle
Tribolium spp.	Reismehlkäfer	flour beetles
Tribolonotus spp.	Helmskinke	helmet skinks, casqueheads

Trichia edentula	Zahnlose Haarschnecke	toothless hairysnail
Trichia hispida/Helix hispida	Behaarte Laubschnecke, Haarschnecke, Gemeine Haarschnecke	hairy snail, hairysnail, bristly snail
Trichia plebeia/ Trichia sericea/ Trichia liberta	Seidenhaarschnecke	velvet hairysnail
Trichia spp.	Haarschnecken	hairysnails
Trichia striolata/ Trichia rufescens	Gestreifte Haarschnecke, Gestreifte Krautschnecke	strawberry snail, reddish snail, ruddy snail
Trichia unidentata/ Petasina unidentata	Einzähnige Haarschnecke	unidentate marginsnail (unidentate margin shell), single-toothed hairysnail
Trichia villosa	Zottige Haarschnecke	villous hairysnail
Trichiinae	Pinselkäfer	trichiine beetles
Trichinella spiralis	Trichine	trichina worm
Trichiosoma tibiale	Keulhornblattwespe	hawthorn sawfly
Trichiura crataegi	Weißdornspinner	pale eggar
Trichiuridae	Haarschwänze, Rinkfische	frostfishes & scabbardfishes & cutlassfishes & hairtails
Trichiurus lepturus	Degenfisch, Haarschwanz	Atlantic cutlassfish, large-eyed hairtail, largehead hairtail (FAO)
Trichius fasciatus	Pinselkäfer	bee chafer
Trichobatrachus robustus	Haarfrosch	hairy frog
Trichobatrachus spp.	Haarfrösche	hairy frogs
Trichocera spp.	Wintermücken, Winterschnaken	winter gnats, winter crane flies
Trichoceridae (Petauristidae)	Wintermücken, Winterschnaken	winter gnats, winter crane flies
Trichodectes canis	Hundehaarling	dog biting louse
Trichodes apiarius	Immenkäfer, Bienenwolf, Gemeiner Bienenkäfer	bee beetle, bee wolf
Trichodina pediculus	Polypenlaus	
Trichodontidae	Sandfische	sandfishes
Trichogaster leeri	Mosaikfadenfisch	pearl gourami
Trichogaster microlepis	Mondscheinfadenfisch, Mondschein-Gurami	moonlight gourami (FAO), moonbeam gourami
Trichogaster pectoralis	Schaufelfadenfisch	snake-skinned gourami, snakeskin gourami (FAO)
Trichogaster trichogaster sumatranus	Blauer Fadenfisch	blue gourami
Trichogaster trichopterus trichopterus	Punktierter Fadenfisch	three-spot gourami
Trichogasterinae	Guramis	gouramis
Trichoglossus haematodus	Allfarblori	rainbow lory
Trichomycteridae	Parasitenwelse, Schmarotzerwelse, Schmerlenwelse	pencil catfishes, parasitic catfishes
Trichomys apereoides	Punaré	punaré
Trichoniscus pusillus	Zwergassel*	common pygmy woodlouse
Trichonotidae	Sandtaucher	sand divers, sanddivers

Trichophaga tapetzella	Tapetenmotte	tapestry moth
Trichopsis pumilis	Zwerg-Gurami	pygmy gourami
Trichopsis vittata	Knurrender Gurami	talking gourami (FAO), croaking gourami
Trichoptera	Köcherfliegen, Haarflügler	caddis flies
Trichopteryx polycommata	Brauner Gebüsch-Lappenspanner	barred toothed stripe moth
Trichostrongylus axei	Kleiner Rindermagennematode, Kleiner Rindermagenwurm	small stomach worm (of cattle and horses), stomach hair worm
Trichosurus spp.	Kusus	brush-tailed possums
Trichosurus vulpecula	Fuchskusu (Hbz. Austr./Neuseel. Opossum)	brush-tailed possum
Trichotropidae	Haarschnecken	hairy snails, hairysnails (hairy shells)
Trichotropis bicarinata	Zweikiel-Haarschnecke	two-keel hairysnail, two-keeled hairysnail
Trichotropis borealis	Nordmeer-Haarschnecke	boreal hairysnail
Trichotropis cancellata	Gitter-Haarschnecke*	cancellate hairysnail
Trichuris spp.	Peitschenwürmer	whipworms
Trichys fasciculata	Pinselstachler, Malaiischer Pinselstachler	long-tailed porcupine
Tricolia pullas	Kleine Doppelfußschnecke, Fasanküken	lesser pheasant snail*
Tricoliidae	Doppelfußschnecken	pheasant snails a.o. (see Phasianellidae)
Tridacna crocea	Krokus-Riesenmuschel	crocus giant clam
Tridacna gigas	Große Riesenmuschel, Mördermuschel	giant clam, killer clam
Tridacna maxima	Längliche Riesenmuschel, Kleine Riesenmuschel	elongated clam, elongate clam, small giant clam (IUCN)
Tridacna squamosa	Schuppen-Riesenmuschel, Schuppige Riesenmuschel	fluted giant clam, giant fluted clam, fluted clam, scaly clam (IUCN)
Tridactylidae	Dreifingerschrecken	pygmy mole crickets
Trigla lucerna/ *Chelidonichthys lucerna*	Roter Knurrhahn (Seeschwalbenfisch)	tub gurnard (FAO), sapphirine gurnard
Trigla lyra	Pfeifen-Knurrhahn, Leierknurrhahn, Pfeifenfisch	piper, piper gurnard (FAO)
Triglidae	Knurrhähne	searobins, gurnards
Trigloporus lastoviza	Gestreifter Knurrhahn	streaked gurnard
Triglops murrayi	Schnurrbart-Seeskorpion*	moustache sculpin
Triglopsis quadricornis/ *Myoxocephalus quadricornis*/ *Oncocottus quadricornis*	Vierhörniger Seeskorpion	four-horn sculpin, four-horned sculpin, fourhorn sculpin (FAO)
Trigona spp.		sweat-bees
Trigonaspis megaptera	Eichenblatt-Nierengallwespe, Eichen-Nierengallblattwespe	oak leaf kidney-gall cynipid wasp
Trigoniidae		brooch clams
Trigoniulus lumbricinus	Rostiger Tausendfüßler*	rusty millipede
Trigonogenius globulus	Chilenischer Diebkäfer	globular spider beetle

Trigonostoma pellucida	Dreiecks-Gitterschnecke	triangular nutmeg
Trigonostoma pulchra/ Cancellaria pulchra	Hübsche Gitterschnecke	beautiful nutmeg
Trigonostoma rugosum	Runzelige Gitterschnecke	rugose nutmeg
Trilobita	Trilobiten, Dreilapper	trilobites
Trilobus gracilis	Süßwassernematode u.a.	freshwater nematode a.o.
Trilophosauria	Dreijochzahnechsen	trilophosaurs, trilophosaurians (Triassic archosauromorphs)
Trimeresurus albolabris	Weißlippen-Bambusotter	white-lipped tree viper
Trimeresurus flavoviridis	Habu-Schlange, Habu	Okinawa habu, yellow-spotted pit viper
Trimeresurus gramineus	Indische Bambusotter, Grüne Bambusotter	Indian green tree viper
Trimeresurus kanburiensis	Kanburi-Lanzenotter	Kanburian pit viper, Kanburi pit viper, tiger pit viper
Trimeresurus mucrosquamatus	Glattschuppige Lanzenotter	Chinese habu
Trimeresurus popeorum	Popes Lanzenotter	Pope's tree viper
Trimeresurus purpureomaculatus	Mangroven-Lanzenotter	mangrove viper
Trimeresurus spp.	Asiatische Lanzenottern, Bambusottern	Asian lance-headed vipers, Asian pit vipers
Trimeresurus stejnegeri	Chinesische Bambusotter	Chinese green tree viper
Trimeresurus trigonocephalus	Ceylon-Lanzenotter, Sri Lanka-Lanzenotter	Sri Lankan pit viper, Ceylon pit viper
Trimeresurus wagleri/ Tropidolaemus wagleri	Waglers Lanzenotter	Wagler's pit viper, Wagler's palm viper
Trimorphodon biscutatus	Leiernatter, Lyraschlange	lyre snake
Trimusculus carinatus		carinate gadinia
Tringa erythropus	Dunkler Wasserläufer	spotted redshank
Tringa flavipes	Kleiner Gelbschenkel	lesser yellowlegs
Tringa glareola	Bruchwasserläufer	wood sandpiper
Tringa guttifer	Kurzfuß-Wasserläufer	Nordmann's greenshank
Tringa hypoleucos/ Actitis hypoleucos	Flussuferläufer	common sandpiper
Tringa melanoleuca	Großer Gelbschenkel	greater yellowlegs
Tringa nebularia	Grünschenkel	common greenshank
Tringa ochropus	Waldwasserläufer	green sandpiper
Tringa solitaria	Einsamer Wasserläufer	solitary sandpiper
Tringa stagnatilis	Teichwasserläufer	marsh sandpiper
Tringa totanus	Rotschenkel	redshank
Tringinae	Wasserläufer u.a.	sandpipers, curlews, phalaropes
Triodon macropterus/ Triodon bursarius	Dreizähner, Bauchsackkugelfisch	threetooth puffer (FAO), three-toothed puffer
Triodontidae	Dreizähner, Bauchsackkugelfische	threetooth puffer, three-toothed puffers
Trionychidae	Echte Weichschildkröten	softshell turtles, softshells
Trionyx ater	Schwarze Weichschildkröte	black softshell turtle, Cuatro Cienegas softshell turtle

Trionyx cartilagineus/ *Amyda cartilaginea*	Knorpel-Weichschildkröte, Asiatische Weichschildkröte	black-rayed softshell turtle, Asiatic softshell turtle
Trionyx euphraticus/ *Rafetus euphraticus*	Euphrat-Weichschildkröte	Euphrates softshell turtle
Trionyx ferox/*Apalone ferox*	Wilde Dreiklaue, Florida-Weichschildkröte	Florida softshell turtle
Trionyx formosus/ *Nilssonia formosa*	Burma-Weichschildkröte	Burmese peacock softshell turtle, Burmese softshell turtle
Trionyx gangeticus/ *Aspideretes gangeticus*	Ganges-Weichschildkröte	Ganges softshell turtle, Indian softshell turtle
Trionyx hurum/ *Aspideretes gangeticus*	Pfauenaugen-Weichschildkröte	Indian peacock softshell turtle, peacock softshell turtle
Trionyx leithii	Nagpur-Weichschildkröte, Vorderindische Weichschildkröte	Nagpur softshell turtle
Trionyx muticus/ *Apalone mutica*	Glattrand-Weichschildkröte, Glattrandige Weichschildkröte	smooth softshell turtle
Trionyx nigricans/ *Aspideretes nigricans*	Schwarze Weichschildkröte, Dunkle Weichschildkröte, Tempel-Weichschildkröte	black softshell turtle, sacred black softshell turtle, Bostami turtle, dark softshell turtle
Trionyx sinensis/ *Pelodiscus sinensis*	Chinesische Weichschildkröte	Chinese softshell turtle
Trionyx spiniferus/ *Apalone spinifera*	Dornrand-Weichschildkröte	spiny softshell turtles
Trionyx spp.	Dreiklauen-Weichschildkröten	softshells, softshell turtles
Trionyx steindachneri/ *Palea steindachneri*	Kehllappen-Weichschildkröte, Nackendorn-Weichschildkröte	wattle-necked softshell turtle
Trionyx subplanus/ *Dogania subplana*	Malayische Weichschildkröte, Malayen-Weichschildkröte	Malayan softshell turtle
Trionyx triunguis	Afrikanische Weichschildkröte, Afrikanischer Dreiklauer	African softshell turtle, Nile softshell turtle
Triops cancriformis	Sommer-Kieferfuß, Großer Rückenschaler, Kiemenfuß	tadpole shrimp, freshwater tadpole shrimp, rice apus
Triopsidae	Kieferfüße, Flossenkrebse	tadpole shrimps, triopsids
Trioza apicalis	Möhrenblattsauger	carrot sucker (psyllid)
Trioza remota	Eichenblattsauger	oak leaf sucker (psyllid)
Triphoridae	Verkehrtschnecken, Linksnadelschnecken	left-handed hornsnails
Triphosa dubitata	Höhlenspanner, Kellerspanner	cave moth*, cellar moth*
Triplax russica	Schwarzfühleriger Pilzkäfer	
Tripneustes ventricosus	See-Ei	sea-egg*
Triportheus angulatus	Armbrustsalmler, Punktierter Kropfsalmler	giant hatchetfish
Triportheus elongatus	Gestreckter Kropfsalmler	elongate hatchetfish
Triprion spatulatus	Panzerkopf-Laubfrosch, Schaufel-Panzerkopf	shovel-headed treefrog
Triprion spp.	Panzerkopf-Laubfrösche	shovel-headed treefrogs
Tripterygiidae	Dreiflossen-Schleimfische, Dreiflosser	threefin blennies, triplefin blennies
Tripterygion tripteronotum	Schnabelblennide, Spitzkopfschleimfisch	triple-fin blenny
Trisateles emortualis		olive crescent moth

Trisetacus pini	Kiefernzweig-Gallmilbe*	pine twig-knot mite, pine gall mite, pine twig gall mite
Trisidos tortuosa	Propeller-Archenmuschel	propeller ark
Trisopterus esmarki	Stintdorsch	Norway pout
Trisopterus luscus	Französischer Dorsch, Franzosendorsch	bib, pout, pouting (FAO)
Trisopterus minutus	Zwergdorsch	poor-cod, poor cod (FAO)
Tritoma bipustulata	Rotfleckiger Pilzkäfer	
Triturus alpestris	Bergmolch, Alpenmolch	alpine newt
Triturus boscai	Spanischer Wassermolch	Bosca's newt
Triturus carniflex	Alpenkammmolch, Italienischer Kammmolch	alpine crested newt, Italian warty newt
Triturus cristatus	Kammmolch	warty newt, crested newt, European crested newt
Triturus helveticus	Fadenmolch (Leistenmolch)	palmate newt
Triturus italicus	Italienischer Wassermolch	Italian newt
Triturus karelini	Balkankammmolch, Persischer Kammmolch	Balkan crested newt
Triturus marmoratus	Marmormolch	marbled newt
Triturus montandoni	Karpatenmolch	Montandon's newt, Carpathian newt
Triturus vittatus	Bandmolch	banded newt
Triturus vulgaris	Teichmolch, Streifenmolch	smooth newt
Trivia quadripunctata/ Niveria quadripunctata	Vierpunkt-Kaurischnecke	fourspot trivia, four-spotted trivia
Trivia aperta/Triviella aperta	Klaffende Kaurischnecke	gaping trivia
Trivia arctica/ Trivia europaea/ Trivia monacha	Nördliche Kaurischnecke, Arktische Kaurischnecke, Gerippte Kaurischnecke	northern cowrie, spotted cowrie, European cowrie, bean cowrie, Arctic cowrie
Trivia candidula/ Niveria candidula	Weiße Kaurischnecke, Weißkauri	white trivia
Trivia nix/Niveria nix	Schnee-Kaurischnecke, Schneekauri	snowy trivia
Trivia pedicula/ Niveria pediculus	Kaffeebohnen-Kaurischnecke	coffeebean trivia
Trivia pullicina/Trivia pulex	Tränenkauri	teardrop trivia*
Trivia radians/ Pusula radians	Radians-Kaurischnecke*	radians trivia
Trivia suffusa/ Niveria suffusa	Rosa Kaurischnecke, Rosa Kauri	pink trivia
Triviidae	Scheidewegschnecken, Kaurischnecken	false cowries, bean cowries, sea buttons
Trochia cingulata	Gürtel-Felsschnecke	corded rocksnail
Trochidae	Kreiselschnecken	topsnails (top shells)
Trochilidae	Eremiten & Kolibris	hermits & hummingbirds
Trochiliformes	Kolibris	hummingbirds
Trochilinae	Kolibris	hummingbirds

Trochoidea elegans	Kegelige Heideschnecke	
Trochospongilla horrida	Rauer Süßwasserschwamm	rough freshwater sponge
Trochus maculatus	Gefleckte Kreiselschnecke	mottled topsnail
Trochus niloticus	Riesen-Kreiselschnecke	Nile topsnail, Nile trochus, commercial trochus (pearly top shell)
Trogium pulsatorium/ Atropos pulsatoria	Totenuhr (Bücherlaus, Staublaus)	larger pale booklouse, larger pale trogiid, lesser death watch beetle (booklouse, dust-louse)
Troglodytes aedon	Hauszaunkönig	house wren
Troglodytes troglodytes	Zaunkönig	wren, winter wren
Troglodytinae	Zaunkönige	wrens
Trogoderma glabrum	Lagerhauskäfer	warehouse beetle
Trogoderma granarium	Khapra-Käfer, Khaprakäfer	khapra beetle
Trogoderma inclusum	Speicherkäfer	larger cabinet beetle
Trogoderma parabile	Amerikanischer Speicherkäfer	American cabinet beetle (grain dermestid)
Trogonidae	Trogons	trogons
Trogonini	Neuwelt-Trogons	New World trogons
Trogonophidae	Spitzschwanz-Doppelschleichen, Kurzkopf-Doppelschleichen	shorthead worm lizards
Trogopterus xanthipes	Goldfuß-Flughörnchen*	complex-toothed flying squirrel
Trogosoma depsarium	Zottenbock	
Trogossitidae	Flachkäfer	bark-gnawing beetles, trogositid beetles, trogossitids
Trogulidae	Brettkanker	trogulid harvestmen, trogulids
Troilus luridus	Spitzbauchwanze	
Trombicula akamushi/ Leptotrombidium akamushi	Japanische Fleckfiebermilbe (Tsutsugamushi-Fieber-Erntemilbe)	scrub typhus chigger mite, Japanese scrub typhus chigger mite*
Trombicula alfreddugesi/ Eutrombicula alfreddugesi	Amerikanische Erntemilbe*	American chigger mite, common chigger mite
Trombicula deliensis/ Leptotrombidium deliensis	Südostasiatische Fleckfiebermilbe (Südostasiatische Tsutsugamushi-Erntemilbe)	Southeast-Asian scrub typhus chigger mite*
Trombiculidae	Laufmilben, Erntemilben, Herbstmilben	chigger mites, harvest mites
Trombidiidae	Samtmilben	velvet mites, trombidiid mites
Trombidium holosericeum	Samtmilbe, Sammetmilbe	velvet mite
Tropheus duboisi	Weißpunkt-Buntbarsch	duboisi
Tropheus moorii	Brabantbuntbarsch, Brabant-Buntbarsch	blunthead cichlid (FAO), firecracker moorii, rainbow moorii
Tropidechis carinatus	Rauschuppige Ostaustralische Giftnatter, Rauschuppenotter	rough-scaled snake
Tropideres albirostris	Eichen-Breitrüssler	
Tropidolaemus wagleri/ Trimeresurus wagleri	Waglers Lanzenotter	Wagler's pit viper, Wagler's palm viper
Tropidophis canus	Bahama-Zwergboa	Bahamas wood snake
Tropidophis melanurus	Kubanische Zwergboa	Cuban Island ground boa

Tropidophis spp.	Zwergboas, Erdboas	pygmy ground boas
Tropidurus spp.	Kielschwänze	lava lizards
Tropinota hirta	Zottiger Rosenkäfer, Zottiger Blütenkäfer	
Truncatella subcylindrica	Glatte Stutzschnecke	looping snail
Truncatellina cylindrica	Zylinderwindelschnecke	cylindrical whorl snail
Trunculariopsis trunculus/ Hexaplex trunculus	Purpurschnecke u.a.	trunk murex, trunculus murex
Trychomycteridae	Schmerlenwelse	
Tryngites subruficollis	Grasläufer	buff-breasted sandpiper
Tryphomys adustus	Mearns Luzon-Ratte	Mearns' Luzon rat
Trypocopris vernalis	Frühlingsmistkäfer	springtime dung beetle*
Trypoxylon figulus	Holzbohrwespe	black borer wasp, black borer
Tubastraea coccinea	Orange Warzenkoralle*	orange cupcoral
Tuberculoides annulatus		oak leaf aphid
Tuberolachnus salignus	Große Weidenrindenlaus	large willow aphid, giant willow aphid
Tubicellepora magnicostata	Orangenschalen-Moostierchen*	orange peel bryozoan
Tubifex costatus	Blutwurm*	bloodworm
Tubifex spp.	Schlammröhrenwürmer	sludge-worms
Tubifex tubifex	Gemeiner Schlammröhrenwurm, Gemeiner Bachröhrenwurm, Bachwurm	river worm, sludge-worm
Tubificidae	Schlammröhrenwürmer	sludge-worms
Tubipora musica	Orgelkoralle	organ-pipe coral, pipe-organ coral
Tubipora spp.	Orgelkorallen	organ-pipe corals, pipe-organ corals
Tubulanus annulatus (Nemertini)	Ringel-Schnurwurm	
Tubulanus annulatus	Ringelnemertine	football jersey worm
Tubularia crocea	Rosamundiger Röhrenpolyp*	pinkmouth tubularian, pink-mouth hydroid
Tubularia indivisa	Hochstämmiger Röhrenpolyp	tall tubularian, tall hydroid
Tubularia larynx	Röhrenpolyp	ringed tubularian, ringed tubularia
Tubularia spp.	Köpfchenpolypen	tubularian hydroids
Tubulipora flabellaris	Röhrenfächer-Moostierchen	
Tugali cicatricosa	Narbige Kerbschnecke	scarred notched limpet
Tunga penetrans	Sandfloh	chigoe flea, jigger flea, sand flea
Tunicata	Manteltiere, Tunicaten (Urochordaten)	tunicates
Tupaia glis	Rostrotes Spitzhörnchen	tree shrew
Tupaia spp.	Tupaias, Spitzhörnchen	tree shrews
Tupaiiidae	Spitzhörnchen	tree shrews
Tupinambis rufescens	Roter Grossteju	red tegu
Tupinambis teguixin	Bänderteju, Salompenter	common tegu, black tegu, banded tegu

Turbellaria	Turbellarien, Strudelwürmer	turbellarians, free-living flatworms
Turbinaria mesenterina	Gewundene Salatkoralle, Folienkoralle, Kraterkoralle	twisted lettuce coral*, crater coral*
Turbinaria reniformis	Kartoffelchipkoralle	potatochip coral
Turbinella laevigata	Brasilianische Vasenschnecke	Brazilian chank
Turbinella pyrum	Heilige Schnecke	sacred chank, Indian chank
Turbinidae	Turbanschnecken	turban snails, turbans
Turbo argyrostomus	Silbermund-Turbanschnecke	silver-mouthed turban, silvermouth turban
Turbo canaliculatus	Gefurchte Turbanschnecke	channeled turban
Turbo castanea	Kastanien-Turbanschnecke	chestnut turban
Turbo chrysostomus	Goldmund-Turbanschnecke	gold-mouthed turban
Turbo cornutus	Gehörnte Turbanschnecke	horned turban
Turbo marmoratus	Grüne Turbanschnecke, Grüner Turban	green turban, green snail, great green turban, giant green turban
Turbo petholatus	Gobelin-Turbanschnecke, Gobelinturban, Katzenaugenschnecke	tapestry turban (cat's-eye shell)
Turbo sarmaticus	Südafrikanische Turbanschnecke, Südafrikanischer Turban	South African turban, giant pearlwinkle
Turbonilla crenata	Rotes Wendeltürmchen	red pyramidsnail
Turbonilla jeffreysii	Wendeltürmchen	staircase pyramid
Turdidae	Drosseln	solitaires, thrushes
Turdoides caudatus	Langschwanzdrossling	common babbler
Turdoides fulvus	Akaziendrossling	fulvous babbler
Turdoides squamiceps	Graudrossling	Arabian babbler
Turdus falcklandii	Magellandrossel	austral thrush
Turdus grayi	Gilbdrossel	clay-coloured thrush
Turdus iliacus	Rotdrossel	redwing
Turdus merula	Amsel	blackbird
Turdus migratorius	Wanderdrossel	American robin
Turdus naumanni	Naumanndrossel	dusky thrush
Turdus obscurus	Weißbrauendrossel	eye-browed thrush
Turdus philomelos	Singdrossel	song thrush
Turdus pilaris	Wacholderdrossel	fieldfare
Turdus plumbeus	Rotfußdrossel	red-legged thrush
Turdus poliocephalus	Südseedrossel	island thrush
Turdus ruficollis	Bechsteindrossel	black-throated thrush
Turdus torquatus	Ringdrossel	ring ouzel
Turdus viscivorus	Misteldrossel	mistle thrush
Turnicidae	Kampfwachteln, Laufhühnchen	button quails, hemipodes
Turnix sylvatica	Laufhühnchen	little button-quail, Andalusian hemipode

Turris babylonia	Babylon-Turmschnecke	Babylon turrid
Turritella communis	Gemeine Turmschnecke	common turretsnail, European turretsnail (common screw shell, common tower shell, great screw shell)
Turritella terebra	Bohrer-Schraubenschnecke	auger turretsnail (auger screw shell, tower screw shell)
Turritellidae	Turmschnecken	turret snails (tower shells, auger shells, screw shells)
Turritellopsis stimpsoni	Nadel-Turmschnecke	needle turretsnail
Tursiops truncatus	Großer Tümmler, Großtümmler, Flaschennasendelphin	bottlenosed dolphin, common bottle-nosed dolphin
Tutufa bubo	Riesenfroschschnecke	giant frogsnail
Tutufa rubeta	Rotmund-Froschschnecke	red-mouthed frogsnail, ruddy frogsnail
Tylidae/Micropezidae	Stelzfliegen	stilt-legged flies
Tylodina perversa	Verkehrte Schirmschnecke, Goldschwammschnecke	
Tylomys spp.	Mittelamerikanische Baummäuse u.a.	Central American climbing rats
Tylonycteris spp.	Bambus-Fledermäuse	club-footed bats, bamboo bats (Southeast Asian flat-headed bats)
Tylosurus crocodilus/ Strongylura crocodila	Krokodilhecht, Riesen-Hornhecht, Riesen-Krokodilhecht	hound needlefish (FAO), houndfish, crocodile longtom (Austr.), crocodile needlefish
Tylosurus acus	Agujon, Nadel-Hornhecht	agujon needlefish
Tympanoctomys barrerae	Rote Viscacharatte	red viscacha rat
Tympanuchus cupido	Präriehuhn	prairie chicken
Typhaea stercorea	Behaarter Baumschwammkäfer	hairy fungus beetle
Typhlocyba rosae/ Edwardsiana rosae	Rosenzikade	rose leafhopper
Typhlomolge rathbuni	Rathbunscher Brunnenmolch, Texas-Blindsalamander	Texas blind salamander
Typhlomolge spp.	Brunnenmolche	American blind salamanders
Typhlomys cinereus	Chinesische Zwergschlafmaus, Chinesischer Zwergbilch	Chinese pygmy dormouse
Typhlonectes spp.	Schwimmwühlen	aquatic caecilians
Typhlonectidae	Schwimmwühlen	aquatic caecilians
Typhlopidae	Blindschlangen	blind snakes
Typhlops vermicularis	Blödauge, Wurmschlange	European blind snake, Greek blind snake, worm snake
Typhlosaurus spp.	Afrikanische Blindkinke	blind worms, blind legless skinks
Typhlotriton spelaeus	Grottensalamander	grotto salamander
Typhoeus typhoeus	Stierkäfer	minotaur beetle
Tyrannidae	Tyrannen	tyrant flycatchers
Tyrannus forficata	Scherentyrann	scissor-tailed flycatcher
Tyrannus tyrannus	Königstyrann	eastern kingbird

Tyrannus verticalis	Arkansastyrann	western kingbird, Arkansas kingbird
Tyria jacobaeae	Karminbär, Jakobskreuzkrautbär, Jakobskraut-Bär, Blut-Bär	cinnabar moth
Tyrophagus casei/ Tyrolichus casei	Käsemilbe	cheese mite
Tyrophagus longior	Silomilbe*	grainstack mite, cucumber mite (seed mite, French 'fly')
Tyrophagus putrescentiae	Kopramilbe	mold mite (U.S.), mould mite (Br.)
Tyrophagus similis	Grasflurmilbe*	grassland mite
Tyta luctuosa	Windeneule	four-spotted moth
Tyto alba	Schleiereule	barn owl
Tytonidae	Schleiereulen	barn owls & grass owls

Uaru amphiacanthoides	Keilfleck-Buntbarsch	uaru, chocolate cichlid, triangle cichlid (FAO)
Uca anullipes	Lagunen-Winkerkrabbe	lagoon fiddler crab
Uca beebei	Smaragdschild-Winkerkrabbe	emerald fiddler crab
Uca crassipes	Mangroven-Winkerkrabbe	mangrove fiddler crab
Uca crenulata	Mexikanische Winkerkrabbe, Kalifornische Winkerkrabbe	Mexican fiddler, California fiddler
Uca minax	Rotgelenks-Winkerkrabbe, Brackwasser-Winkerkrabbe	redjointed fiddler, brackish-water fiddler
Uca musica	Sing-Winkerkrabbe	singing fiddler crab
Uca pugilator	Atlantische Sand-Winkerkrabbe	Atlantic sand fiddler
Uca pugnax	Schlick-Winkerkrabbe	Atlantic marsh fiddler, mud fiddler
Uca rapax	Watt-Winkerkrabbe*	mudflat fiddler
Uca saltitanta	Jitterbug-Winkerkrabbe	jitterbug fiddler
Uca speciosa	Langfinger-Winkerkrabbe	longfinger fiddler
Uca spp.	Winkerkrabben	fiddler crabs
Uca stenodactyla	Langbeinige Winkerkrabbe	long-legged fiddler crab
Uca stylifera	Stiel-Winkerkrabbe	stalked fiddler crab
Uca tangeri	Tanger-Winkerkrabbe	Moroccan fiddler crab
Uca vocator	Haarige Winkerkrabbe*	hairback fiddler
Uleiota planata	Plattkäfer, Langhorn-Plattkäfer, Quetschkäfer	
Uloboridae	Kräuselradnetzspinnen	hackled orbweavers
Uloborus glomosus	Flaumbeinige Spinne*	feather-legged spider, featherlegged orbweaver
Ultrocalamus spp.		short-fanged snake
Uma inornata	Coachella-Valley-Fransenzehenleguan	Coachella Valley fringe-toed lizard
Uma notata	Arizona-Fransenzehenleguan	Colorado Desert fringe-toed lizard
Uma scoparia	Mojave-Fransenzehenleguan	Mojave fringe-toed lizard
Umbonium vestiarium	Knopfschnecke	common button topsnail
Umbra krameri	Europäischer Hundsfisch, Ungarischer Hundsfisch	European mudminnow
Umbra limi	Amerikanischer Hundsfisch	American mudminnow
Umbra pygmaea	Ost-Amerikanischer Hundsfisch, Zwerghundsfisch	eastern mudminnow
Umbraculidae	Schirmschnecken	umbrella snails, umbrella "slugs" (umbrella shells)
Umbraculum mediterraneum	Mittelmeer-Schirmschnecke, Warzige Schirmschnecke	Mediterranean umbrella snail
Umbraculum umbraculum	Atlantik-Schirmschnecke	Atlantic umbrella snail, Atlantic umbrella "slug"
Umbridae	Hundsfische	mudminnows
Umbrina canosai	Argentina-Schattenfisch	Argentine croaker
Umbrina cirrosa/ Sciaena cirrosa	Bartumber, Schattenfisch, Umberfisch, Umber	shi drum (FAO), corb (U.S./U.K.), sea crow (U.S.), gurbell (U.S.), croaker
Uncia uncia/Panthera uncia	Schneeleopard	snow leopard

Uncinaria stenocephala	Hundehakenwurm, Fuchshakenwurm	dog hookworm
Ungaliophis continentalis	Isthmus-Zwergboa	Isthmian dwarf boa
Ungaliophis panamensis	Panama-Zwergboa	Panamanian dwarf boa
Ungaliophis spp.	Mittelamerikanische Zwergboas	Central American dwarf boas, banana boas
Unio crassus	Bachmuschel, Kleine Flussmuschel, Gemeine Flussmuschel	common river mussel, common Central European river mussel
Unio pictorum/ Pollicepes pictorum	Malermuschel	painter's mussel
Unio spp.	Flussmuscheln	freshwater mussels
Unio tumidus	Große Flussmuschel	swollen river mussel
Unionicola spp.	Muschelmilben	mussel mites
Unionidae	Flussmuscheln	freshwater mussels
Upeneus asymmetricus	Gold-Meerbarbe	golden-striped goatfish
Uperoleia spp.	Australische Krötchen*, Australische Kleinkröten*	Australian toadlets
Upogebia affinis	Flachbrauen-Maulwurfkrebs*	coastal mudprawn, coastal mud shrimp, flat-browed mud shrimp
Upogebia litoralis	Flachwasser-Maulwurfskrebs	littoral mudprawn*
Upogebia pusilla	Mittelmeer-Maulwurfskrebs	Mediterranean mudprawn, Mediterranean mud lobster
Upogebiidae	Maulwurfskrebse	mudprawns, mud shrimps
Upupa epops	Wiedehopf	hoopoe
Upupidae	Wiedehopfe	hoopoes
Uraeginthus cyanocephala	Blaukopfastrild	blue-capped cordon-bleu
Uraeotyphlus spp.	Indische Schwanzwühlen	Indian caecilians
Uranomys ruddi	Weißbauch-Bürstenfellratte*	white-bellied brush-furred rat
Uranoscopidae	Himmelsgucker, Sterngucker	stargazers
Uranoscopus scaber	Himmelsgucker, Sterngucker, Meerpfaff, Sternseher	star gazer, stargazer (FAO)
Urechis caupo (Echiura)		innkeeper worm
Uria aalge	Trottellumme	guillemot, common murre
Uria lomvia	Dickschnabellumme	Brünnich's guillemot
Urocerus gigas	Riesenholzwespe	giant wood wasp, giant horntail, greater horntail
Uroctea spp.	Zeltdachspinnen	star-web spiders
Urocyon cinereoargenteus	Nordamerikanischer Festland-Graufuchs (Hbz. Grisfuchs)	mainland gray fox, North American gray fox
Urocyon littoralis	Nordamerikanischer Insel-Graufuchs	island gray fox
Urocyon spp.	Nordamerikanische Graufüchse	gray foxes
Urodela/Caudata	Schwanzlurche	urodeles (salamanders & newts)
Uroderma spp.	Gelbohr-Fledermäuse	tent-building bats
Urogale everetti	Everetts Spitzhörnchen	Philippine tree shrew

Urogymnus africanus/ *Urogymnus asperrimus*	Gelbstachel-Igelrochen, Großer Dornen-Stechrochen	yellow-spine thorny ray, porcupine ray (FAO)
Urogymnus spp.	Igelrochen	thorny rays
Urolophidae	Rund-Stechrochen	round stingrays, stingarees (Austr.)
Urolophus concentricus		spot-to-spot round ray, spot-on-spot round ray (FAO)
Urolophus cruciatus	Gekreuzter Stechrochen	crossback stingaree
Urolophus halleri	Hallers Rund-Stechrochen	Haller's round ray (FAO), Haller's round stingray
Urolophus jamaicensis	Gelber Stechrochen*, Jamaika-Stechrochen, Jamaika-Rund-Stechrochen	yellow stingray
Urolophus maculatus	Gefleckter Rund-Stechrochen	spotted round ray
Uromastyx acanthinurus	Veränderlicher Dornschwanz, Afrikanischer Dornschwanz	African spiny-tailed lizard, Dabbs mastigure
Uromastyx aegypticus	Ägyptischer Dornschwanz	Egyptian spiny-tailed lizard, Egyptian mastigure, dabb lizard
Uromastyx hardwickii	Indischer Dornschwanz	Indian spiny-tailed lizard, Indian mastigure
Uromastyx ocellatus	Augenfleck-Dornschwanz	eyed dabb lizard, ocellated mastigure
Uromastyx spp.	Dornschwänze	spiny-tailed lizards
Uromenus brevicollis	Kurzschild-Sattelschrecke*	short-backed bushcricket
Uromenus elegans	Schöne Sattelschrecke*	elegant bushcricket
Uromenus rugosicollis	Kantige Sattelschrecke	rough-backed bushcricket
Uromys spp.	Mosaikschwanz-Riesenratten	giant naked-tailed rats
Uropeltidae	Schildschwänze	shield-tailed snakes, shieldtail earth snakes, shieldtails, roughtails
Uropeltis spp.	Indische Wühlschlangen	Indian earth snakes
Urophycis blennioides/ *Phycis blennioides*	Großer Gabeldorsch	greater forkbeard
Urophycis brasiliensis	Brasilianischer Gabeldorsch	Brazilian codling
Urophycis chuss	Roter Gabeldorsch	red hake (FAO), squirrel hake
Urophycis regia	Gefleckter Gabeldorsch	spotted hake
Urophycis spp.	Westatlantische Gabeldorsche	West Atlantic hakes
Urophycis tenuis	Weißer Gabeldorsch	white hake
Uroplatus fimbriatus	Flachschwanzgecko	common flat-tail gecko
Uropodidae	Schildkrötenmilben	tortoise mites
Uropygi	Geißelskorpione	whipscorpions
Urosalpinx cinerea	Amerikanischer Austernbohrer	American sting winkle, American oyster drill, Atlantic oyster drill
Urosaurus graciosus	Amerikanischer Halbwüsten-Leguan	brush lizard, long-tailed brush lizard
Urosaurus ornatus	Baumleguan	tree lizard, common tree lizard
Ursidae	Bären	bears
Ursus americanus	Schwarzbär	American black bear
Ursus arctos	Braunbär	brown bear, grizzly bear
Ursus malayanus/ *Helarctos malayanus*	Sonnenbär, Malaienbär	sun bear, Malayan sun bear

*U*rsus

Ursus maritimus	Eisbär	polar bear
Ursus thibetanus/ *Selenarctos thibetanus*	Kragenbär	Asiatic black bear, Himalayan bear, Tibetan bear
Ursus ursinus/ *Melursus ursinus*	Lippenbär	sloth bear
Urticina columbiana	Sandrosen-Anemone	sand-rose anemone
Urticina coriacea	Lederanemone	leathery anemone, stubby rose anemone
Urticina crassicornis	Dickhörnige Seerose	mottled anemone, rose anemone, northern red anemone, crassicorn anemone, christmas anemone, painted urticina
Urticina felina/Tealia felina	Seedahlie	feline dahlia anemone, feline sea dahlia
Urticina piscivora	Fischfressende Seedahlie	piscivorous sea dahlia, fish-eating urticina
Uta stansburiana	Seitenfleckenleguan	side-blotched lizard

Valencia hispanica	Valenciakärpfling	Valencia toothcarp
Vallonia costata	Gerippte Grasschnecke	ribbed grass snail
Vallonia declivis	Große Grasschnecke	greater grass snail
Vallonia excentrica	Schiefe Grasschnecke	eccentric grass snail
Vallonia pulchella	Glatte Grasschnecke	smooth grass snail, beautiful grass snail
Vallonia spp.	Grasschnecken	grass snails
Valloniidae	Grasschnecken	grass snails
Valvata cristata	Flache Federkiemenschnecke	flat valve snail
Valvata macrostoma	Sumpf-Federkiemenschnecke	bog valve snail
Valvata naticina	Fluss-Federkiemenschnecke	livebearing European stream valve snail
Valvata piscinalis piscinalis	Teich-Federkiemenschnecke, Gemeine Federkiemenschnecke	valve snail, European stream valve snail, common valve snail, European stream valvata
Valvata pulchella	Niedergedrückte Federkiemenschnecke	large-mouthed valve snail
Valvatidae	Federkiemenschnecken	valve snails
Vampirolepis nana/ Hymenolepis nana	Zwergbandwurm	dwarf tapeworm
Vampirolepis fraterna	Ratten-Zwergbandwurm	rat tapeworm, dwarf rat tapeworm
Vampyressa spp.	Gelbohr-Fledermäuse	yellow-eared bats
Vampyrodes major	Große Streifengesicht-Fledermaus	great stripe-faced bat
Vampyromorpha	Tiefseevampire, Vampirtintenschnecken	vampire squids
Vampyrops spp.	Weißstreifen-Fledermäuse	white-lined bats
Vampyroteuthis infernalis	Vampirkalmar	vampire squid
Vampyrum spectrum	Große Spießblattnase	Linnaeus's false vampire bat, spectral vampire
Vandeleuria spp.	Langschwänzige Indische Baummäuse	long-tailed climbing mice
Vanellus vanellus	Kiebitz	lapwing
Vanessa atalanta	Admiral	red admiral
Vanessa cardui/ Cynthia cardui	Distelfalter	painted lady, thistle (butterfly)
Vanessa polychloros/ Nymphalis polychloros	Großer Fuchs	large tortoiseshell
Vangidae	Vangawürger, Blauwürger	vanga shrikes, helmet-shrikes
Vanneaugobius canariensis	Kanarische Grundel	Canary goby
Varanidae	Warane	monitor lizards, monitors
Varanus acanthurus	Stachelschwanzwaran	spiny-tailed pygmy monitor, ridge-tailed monitor, ridgetail monitor
Varanus bengalensis	Bengalenwaran, Bengalwaran	Bengal monitor, Indian monitor, common monitor
Varanus brevicauda	Kurzschwanzwaran	short-tailed monitor, short-tail monitor, short-tailed pygmy monitor
Varanus caudolineatus	Streifenschwanz-Waran, Schwanzstrichwaran	streak-tailed monitor, stripetail monitor

Varanus dumerilii	Dumerils Waran	Dumeril's monitor
Varanus eremius	Wüsten-Zwergwaran, Höhlenwaran	desert pygmy monitor, spinifex monitor
Varanus exanthematicus	Steppenwaran	Cape monitor, rock monitor, Bosc's monitor, African savannah monitor
Varanus flavescens	Gelbwaran	yellow monitor
Varanus giganteus	Großwaran, Perenty	perenty, perentie
Varanus gilleni	Gillens Waran, Gillens Zwergwaran	pygmy mulga monitor
Varanus gouldii	Goulds Waran	Gould's monitor, sand monitor, sand goanna, bungarra
Varanus griseus	Wüstenwaran	desert monitor (agra monitor)
Varanus indicus	Pazifik-Waran, Indischer Mangrovenwaran	Pacific monitor, mangrove monitor, Indian mangrove monitor
Varanus komodoensis	Komodo-Waran, Komodowaran	Komodo dragon, ora
Varanus mertensi	Mertens Waran	Mertens' water monitor
Varanus nebulosus	Nebelwaran	clouded monitor
Varanus niloticus	Nilwaran	Nile monitor
Varanus prasinus	Smaragdwaran	emerald monitor
Varanus rudicollis	Raunackenwaran	rough-necked monitor, tree lizard, harlequin monitor
Varanus salvadorii	Salvadors Waran, Papua-Waran	Salvador's monitor, Papuan monitor, crocodile monitor, tree "crocodile"
Varanus salvator	Bindenwaran	common Asiatic monitor, water monitor, Malayan monitor, common water monitor
Varanus semiremex	Rostkopfwaran, Australischer Mangrovenwaran	Australian mangrove monitor, rusty monitor
Varanus storri	Zwergwaran	dwarf monitor
Varanus timorensis	Timor-Waran	Timor tree monitor, spotted tree monitor
Varanus tristis	Trauerwaran (Gesprenkelter Waran & Schwarzkopfwaran)	mournful tree monitor, arid monitor (freckled monitor & black-headed monitor)
Varanus varius	Buntwaran	lace monitor, common tree monitor
Varecia variegata	Vari	variegated lemur, ruffed lemur
Variola albimarginata	Mondsichelbarsch	white-margined lunartail rockcod, crescent-tailed grouper, white-edged lyretail (FAO)
Variola louti	Gelbsaum Juwelenbarsch	lunartail rockcod, moontail rockcod, lyretail grouper, yellow-edged lyretail (FAO)
Varroa jacobsoni	Varroamilbe, Bienenbrutmilbe	varroa mite

Vasum capitellum	Karibische Stachelvase	spined Caribbean vase (snail)
Vasum cassiforme	Helm-Vasenschnecke	helmet vase (snail)
Vasum muricatum	Karibische Vasenschnecke	Caribbean vase (snail)
Vasum tubiferum	Kaiservase, Kaiserliche Vasenschnecke	imperial vase (snail)
Vasum turbinellum	Pazifische Vasenschnecke	Pacific top vase (snail)
Velella velella	Segelqualle	by-the-wind sailor, little sail
Velia spp.	Bachläufer, Stoßwasserläufer	water crickets, broad-shouldered water striders
Veliferidae	Segelträger	velifers
Veliidae	Bachläufer, Stoßwasserläufer	broad-shouldered water striders, ripple bugs, small water striders, water crickets
Velleius dilatatus	Hornissenkurzflügler, Hornissenkäfer, Sägehorn-Kurzflügler	
Velutina lanigera	Woll-Samtschnecke*	woolly velutina, woolly lamellaria
Velutina plicatilis	Schiefe Samtschnecke*	oblique velutina, oblique lamellaria
Velutina undata	Gewellte Samtschnecke	wavy velutina, wavy lamellaria
Velutina velutina	Glatte Samtschnecke*	smooth velutina, smooth lamellaria, velvet snail (velvet shell)
Velutinidae (incl. Lamellariinae)	Blättchenschnecken	vetulinas, lamellarias
Veneridae	Venusmuscheln	venus clams (venus shells)
Venerupis philippinarum/ Ruditapes philippinarum/ Tapes philippinarum/ Tapes japonica	Japanische Teichmuschel	Japanese littleneck, short-necked clam, Japanese clam, Manilla clam
Venerupis geographica/ Venerupis rhomboides/ Tapes rhomboides/ Paphia rhomboides	Einfache Teppichmuschel, Gebänderte Teppichmuschel, Essbare Venusmuschel	banded carpetclam, banded venus (banded carpet shell)
Venerupis pullastra/ Venerupis saxatilis/ Venerupis perforans/ Venerupis senegalensis/ Tapes pullastra	Gemeine Teppichmuschel, Kleine Teppichmuschel, Getupfte Teppichmuschel	pullet carpetclam, pullet venus (pullet carpet shell)
Venerupis aurea/ Tapes aurea/ Paphia aurea	Goldene Teppichmuschel	golden carpetclam, golden venus (golden carpet shell)
Venerupis decussata/ Tapes decussata	Venusmuschel, Große Teppichmuschel, Kreuzgemusterte Teppichmuschel	cross-cut carpetclam, cross-cut venus (cross-cut carpet shell)
Venerupis texturatus	Gewebte Teppichmuschel	textured carpetclam, textured venus
Venus fasciata/ Clausinella fasciata	Gebänderte Venusmuschel	banded venus
Venus gallina/ Chione gallina/ Chamelea gallina	Gemeine Venusmuschel, Strahlige Venusmuschel	striped venus, chicken venus

Venus ovata/ *Timoclea ovata/* *Chione ovata*	Ovale Venusmuschel	oval venus
Venus verrucosa	Warzige Venusmuschel	warty venus
Veretillum cynomarium	Gelbe Seefeder	yellow sea-pen
Vermetidae	Wurmschnecken	worm snails (worm shells)
Vermetus triquetra/ *Bivonia triquetra*	Dreikant-Wurmschnecke, Dreieckswurmschnecke	
Vermicella annulata	Bandy-Bandy	bandy-bandy
Vermicella spp.	Bandy-Bandys	bandy-bandys
Vermicularia spirata	Westindische Wurmschnecke	West Indian wormsnail, common wormsnail
Vermilingua (Xenarthra)	Ameisenbären	anteaters
Vermivora bachmanii	Gelbstirn-Waldsänger	Bachman's warbler
Vernaya spp.	Vernays Baummaus	Vernay's climbing mice
Verongia aerophoba	Goldschwamm, Farbwechselnder Zylinderschwamm	aureate sponge*
Verruca stroemia	Meerwarze	sea wart*
Vertiginidae	Windelschnecken	vertigos: whorl snails & chrysalis snails
Vertigo alpestris	Alpen-Windelschnecke	Alpine whorl snail, mountain whorl snail, tundra vertigo (U.S.)
Vertigo angustior	Schmale Windelschnecke	narrow whorl snail, narrow-mouthed whorl snail
Vertigo antivertigo	Sumpf-Windelschnecke	marsh whorl snail
Vertigo genesii	Rundmündige Windelschnecke, Blanke Windelschnecke	round-mouthed whorl snail
Vertigo geyerii	Vierzähnige Windelschnecke	fourtoothed whorl snail*
Vertigo modesta	Arktische Windelschnecke	Arctic whorl snail, cross vertigo (U.S.)
Vertigo moulinsiana	Bauchige Windelschnecke	Desmoulins' whorl snail, ventricose whorl snail
Vertigo pusilla	Mauer-Windelschnecke, Linksgewundene Windelschnecke	wall whorl snail, wry-necked whorl snail
Vertigo pygmaea	Gemeine Windelschnecke, Zwergwindelschnecke	dwarf whorl snail, common whorl snail, crested vertigo (U.S.)
Vertigo substriata	Gebänderte Windelschnecke, Gestreifte Windelschnecke	striated whorl snail
Vespa crabro	Hornisse	hornet, brown hornet, European hornet
Vesperidae/Vespidae	Soziale Faltenwespen	vespid wasps, social wasps (hornets & yellowjackets & potter wasps & paper wasps)
Vespertilio murinus	Zweifarbfledermaus	parti-colored bat
Vespertilio spp.	Zweifarbige Fledermäuse, Zweifarbfledermäuse	frosted bats, particolored bats
Vespertilionidae	Glattnasen-Fledermäuse	vespertilionid bats
Vespula maculata/ *Dolichovespula maculata*	Fleckwespe*	bold-faced hornet
Vespula austriaca	Österreichische Wespe	Austrian cuckoo wasp

Vespula germanica	Deutsche Wespe	German wasp
Vespula rufa	Rote Wespe	red wasp
Vespula spp.	Kurzkopfwespen	short-headed wasps*
Vespula vulgaris	Gemeine Wespe	common wasp
Vexillum citrinum	Königliche Mitra	regal mitre, regal miter
Vexillum coccineum	Schmuck-Mitra	ornate mitre, ornate miter
Vexillum histrio	Harlekin-Mitra	harlequin mitre, harlequin miter
Vexillum pulchellum	Hübsche Mitra	beautiful mitre, beautiful miter
Vexillum sanguisugum	Blutsauger-Mitra	bloodsucking mitre, bloodsucking miter, bloodsucker miter
Vexillum spp.	Mitras, Mitraschnecken u.a.	mitres (Br.), miters (U.S.)
Vexillum vittatum	Schwarzgebänderte Mitra, Schwarzgebänderte Bischofsmütze	black-striped mitre, black-striped miter
Vexillum vulpecula	Fuchs-Mitra, Kleiner Fuchs	little fox mitre, little fox miter
Victorella pavida	Zitternde Seerinde*	trembling sea-mat
Vicugna vicugna	Vikunja	vicuña
Vidua macroura	Dominikanerwitwe	pin-tailed whydah
Vidua paradisaea/ Steganura paradisea	Schmalschwanz-Paradieswitwe	paradise whydah
Viduidae	Witwenvögel	whydahs
Vimba vimba	Zährte	East European bream, zährte, Baltic vimba (FAO)
Vini australis	Blaukäppchen	blue-crowned lory
Vini kuhlii	Rubinlori	Kuhl's lory
Vini peruviana	Saphirlori	Tahitian lory
Vini ultramina	Ultramarinlori	ultramarine lory
Vipera monticola/ Trimeresurus monticola	Berg-Lanzenotter	mountain viper, Chinese mountain viper
Vipera ammodytes	Sandotter	sand viper, nose-horned viper
Vipera aspis	Aspisviper, Juraviper	asp viper, aspic viper
Vipera barani	Schwarzviper, Türkische Viper	black viper, Turkish viper
Vipera berus	Kreuzotter	adder, common viper, common European viper
Vipera bornmuelleri/ Vipera bornmulleri	Libanesische Bergotter	Lebanon viper
Vipera kaznakovi	Kaukasusotter	Caucasus viper, Caucasian viper (IUCN), red viper, Kaznakow's viper
Vipera latastei	Stülpnasenotter	snub-nosed viper, Lataste's viper
Vipera lebetina/ Daboia lebetina/ Macrovipera lebetina	Levante-Otter, Levanteotter	blunt-nosed viper, bluntnose viper, Levantine viper

Vipera mauritanica	Atlasotter	Sahara rock viper, Atlas Mountain viper
Vipera monticola	Atlas-Zwergotter	mountain viper
Vipera palaestinae/ Daboia palaestinae	Palästinaviper	Palestinian viper, Palestine viper
Vipera raddei	Armenische Bergotter, Radde-Viper	Radde's rock viper, Armenian viper, Russian viper
Vipera russelli/ Daboia russelli	Kettenviper	Russell's viper
Vipera seoanei	Pyrenäenotter, Iberische Viper	Iberian adder, Pyrenean viper, Portugese viper
Vipera superciliaris/ Atheris superciliaris	Afrikanische Tieflandviper	African lowland viper, lowland swamp viper
Vipera ursinii	Wiesenotter (Karstotter), Spitzkopfotter	meadow viper, Orsini's viper
Vipera xanthina/ Daboia xanthina	Bergotter	coastal viper, European coastal viper, Ottoman viper, Near East viper
Viperidae	Vipern, Ottern	vipers
Vireo flavifrons	Gelbkehlvireo	yellow-throated vireo
Vireo gilvus	Sängervireo	warbling vireo
Vireo griseus	Weißaugenvireo	white-eyed vireo
Vireo olivaceus	Rotaugenvireo	red-eyed vireo
Vireo solitarius	Graukopfvireo	solitary vireo
Vireo vicinior	Grauvireo	grey vireo
Vireolanius pulchellus	Smaragdvireo	green shrike vireo
Vireonidae	Vireos, Laubwürger	vireos, greenlets, peppershrikes, shrike-vireos
Virginia spp.	Erdschlangen	earth snakes
Virgularia mirabilis	Zauberhafte Seefeder	sea rush, wonderful sea-pen*
Viteus vitifoliae/ Viteus vitifolii	Reblaus	vine louse, grape phylloxera
Vitrea contracta	Milchige Kristallschnecke*	milky crystal snail, (contracted glass snail)
Vitrea crystallina	Gemeine Kristallschnecke	crystal snail
Vitrea spp.	Kristallschnecken	crystal snails
Vitrina pellucida	Durchsichtige Glasschnecke, Kugelige Glasschnecke	pellucid glass snail, western glass-snail
Vitrinidae	Glasschnecken	glass snails
Vitula edmandsae	Nordamerikanische Trockenobstmotte	dried-fruit moth
Vitularia salebrosa	Schuppiges Meerkalb	rugged seacalf
Viverra spp.	Echte Zibetkatzen	oriental civets
Viverricula indica	Kleine Zibetkatze	lesser oriental civet, rasse
Viviparidae	Flussdeckelschnecken, Sumpfdeckelschnecken	river snails
Viviparus acerosus	Donau-Flussdeckelschnecke	Danube river snail
Viviparus contectus	Spitze Sumpfdeckelschnecke	pointed river snail, Lister's river snail

Viviparus intertextus	Rundliche Sumpfdeckelschnecke	rotund mysterysnail
Viviparus spp.	Sumpfdeckelschnecken	river snails
Viviparus subpurpureus	Olivfarbene Sumpfdeckelschnecke	olive mysterysnail
Viviparus viviparus	Stumpfe Sumpfdeckelschnecke, Sumpfdeckelschnecke	river snail, common river snail
Volema myristica	Schwere Kronenschnecke	nutmeg melongena, heavy crown shell
Volema paradisiaca	Birnen-Kronenschnecke, Birnenkrone	pear melongena
Volucella bombylans	Pelzige Hummel-Schwebfliege	
Volucella inanis	Gelbe Hummel-Schwebfliege	
Volucella pellucens	Gemeine Hummel-Schwebfliege	
Voluta ebraea	Hebräische Walzenschnecke	Hebrew volute
Voluta musica	Notenwalze, Notenschnecke, Musikwalze	music volute
Volutidae	Walzenschnecken, Faltenschnecken, Rollschnecken	volutes, bailer snails (bailer shells)
Volutopsius norwegicus	Norwegische Wellhornschnecke	Norway whelk
Volva volva	Weberschiffchen	shuttle volva, shuttlecock volva (shuttle shell), shuttlecock, egg spindle
Vombatus ursinus	Nacktnasenwombat	common wombat, coarse-haired wombat
Vomer setapinnis/ Selene setapinnis	Atlantischer Pferdekopf, Atlantischer „Mondfisch"	Atlantic moonfish
Vormela peregusna	Tigeriltis (Hbz. Perwitzki)	marbled polecat
Vulpes bengalensis	Bengalfuchs	Bengal fox
Vulpes cana	Afghanfuchs, Canafuchs	Blanford's fox
Vulpes chama	Kapfuchs, Kamafuchs, Silberrückenfuchs	Cape fox
Vulpes corsac	Steppenfuchs	corsac fox, red fox
Vulpes ferrilata	Tibetfuchs	Tibetan sand fox
Vulpes macrotis	Kitfuchs, Großohrkitfuchs (Hbz. Kitfuchs)	kit fox
Vulpes pallida	Blassfuchs (Hbz. Libyscher Steppenfuchs)	pale fox
Vulpes rueppelli	Sandfuchs, Rüppellfuchs	sand fox
Vulpes velox	Swiftfuchs (Hbz. Kitfuchs/Swiftfuchs)	swift fox
Vulpes vulpes	Rotfuchs	red fox
Vultur gryphus	Andenkondor	Andean condor

*W*allabia

Wallabia bicolor	Sumpfwallaby	swamp wallaby
Wallagonia attu	Jagdwels	mulley
Walterinnesia aegyptia	Schwarze Wüstenkobra, Ägyptische Wüstenkobra	black desert cobra, desert black snake
Walterinnesia spp.	Wüstenkobras	desert cobras, black desert cobras
Watasenia scintillans	Leuchtkäferkalmar	lightningbug squid*
Werneckiella equi/ Bovicola equi	Pferdehaarling	horse biting louse
Wiedomys pyrrhorhinos	Rotnasenmaus	red-nosed mouse
Wuchereria bancrofti	Haarwurm, Bancroft-Filarie	Bancroftian filariae
Wyulda squamicaudata	Schuppenschwanzkusu	scaly-tailed possum

Xanthia aurago	Goldeule	golden sallow (moth)
Xanthia citrago	Streifengelbeule	striped sallow (moth)
Xanthia icteritia	Gemeine Gelbeule	sallow, the sallow (moth)
Xanthia togata	Weidengelbeule	pink-barred sallow (moth)
Xanthichthys ringens	Sargassum-Drückerfisch, Atlantischer Wangenlinien-Drückerfisch	sargassum triggerfish
Xanthidae	Rundkrabben	xanthid crabs, mud crabs
Xantho hydrophilus	Gefurchte Steinkrabbe	furrowed crab, Montagu's crab, furrowed xanthid crab
Xantho pilipes	Rissos Steinkrabbe	Risso's crab, lesser furrowed crab
Xantho poressa	Graue Steinkrabbe	grey xanthid crab
Xanthocephalus xanthocephalus	Brillenstärling	yellow-headed blackbird
Xanthogaleruca luteola	Ulmenblattkäfer	elm leaf beetle*
Xanthogramma pedissequum	Späte Gelbrand-Schwebfliege	
Xanthorhoe fluctuata	Gemeiner Blattspanner	garden carpet (moth)
Xantusia riversiana/ Klauberina riversiana	Insel-Nachtechse	Island night lizard
Xantusia henshawi	Granit-Nachtechse	granite night lizard
Xantusia vigilis	Yucca-Nachtechse	desert night lizard
Xantusiidae	Nachtechsen	night lizards
Xenarthra/Vermilingua	Ameisenbären	ant eaters
Xenesthis immanis	Kleinere Kolumbianische Riesenvogelspinne	Colombian lesserblack tarantula
Xenesthis monstrosa	Größere Kolumbianische Riesenvogelspinne	Colombian giantblack tarantula
Xenia spp.	Straußenfeder-Weichkorallen	pulse corals, waving hand polyps
Xenicidae/Acanthisittidae	Maorischlüpfer, Neuseelandpittas	New Zealand wrens
Xenoboa cropanii	Cropan-Boa	Cropan's boa
Xenocara dolichoptera/ Ancistrus dolichoptera	Blauer Antennenwels	blue-chin xenocara, blue-chin ancistrus
Xenochrophis piscator/ Natrix piscator	Indische Fischernatter	chequered keelback, fishing snake, common scaled water snake
Xenocongridae/Chlopsidae	Falsche Muränenaale	false morays
Xenodermichthys socialis	Atlantischer Gymnast*	Atlantic gymnast
Xenodermus javanicus	Javanische Höckernatter, Java-Höckernatter	xenodermine snake
Xenodon rhabdocephalus	Falsche Lanzenotter	toad-eater snake, false fer-de-lance
Xenomys nelsoni	Magdalena-Ratte	Magdalena rat
Xenomystus nigri	Afrikanischer Messerfisch, Schwarzer Messerfisch	African knifefish (FAO), false featherback
Xenopeltidae	Erdschlangen	sunbeam snakes
Xenopeltis unicolor	Regenbogenschlange, Regenbogen-Erdschlange, Irisierende Erdschlange	sunbeam snake
Xenophalium pyrum pyrum	Birnen-Helmschnecke	pear bonnet

Xenophora conchyliophora	Atlantische Lastträgerschnecke	Atlantic carriersnail
Xenophora corrugata	Rauschalige Lastträgerschnecke	rough carriersnail
Xenophora crispa	Gekräuselte Lastträgerschnecke	curly carriersnail, Mediterranean carriersnail
Xenophora pallidula	Blasse Lastträgerschnecke	pallid carriersnail
Xenophoridae	Lastträgerschnecken	carrier snails (carrier shells)
Xenopsylla cheopis	Pestfloh, Tropischer Rattenfloh	oriental rat flea
Xenopsylla vexabilis	Australischer Rattenfloh	Australian rat flea
Xenopus gilli	Kap-Krallenfrosch	Cape platanna (IUCN), Cape clawed toad
Xenopus laevis	Glatter Krallenfrosch	African clawed frog
Xenopus spp.	Krallenfrösche	clawed frogs
Xenosauridae	Höckerechsen	knob-scaled lizards, xenosaurs
Xenosaurus spp.	Eigentliche Höckerechsen	New World xenosaurs, American knob-scaled lizards
Xenuromys barbatus	Falsche Baumratte	white-tailed New Guinean rat
Xenus cinereus	Terekwasserläufer	Terek sandpiper
Xeris spectrum	Schwarze Kiefernholzwespe, Schwarze Fichtenholzwespe, Tannenholzwespe	spectrum wood wasp
Xerolentha obvia	Weiße Heideschnecke	
Xeromys myoides	Falsche Wasserratte	false water rat
Xerus rutilus	Borstenhörnchen	unstriped ground squirrel
Xerus spp.	Ostafrikanische Erdhörnchen, Ostafrikanische Borstenhörnchen	East African ground squirrels
Xestia c-nigrum	Schwarzes C, C-Eule	setaceous Hebrew character (moth)
Xestia rhomboidea	Rhombuseule*	square-spotted clay (moth)
Xestia speciosa	Grünlichgraue Erdeule, Heidelbeer-Moorheiden-Erdeule	
Xestobium rufovillosum	Totenuhr (Gescheckter Nagekäfer)	death-watch beetle
Xestospongia exigua	Geschnürter Schwamm*, Schnurschwamm*	string sponge
Xestospongia muta	Gigantischer Tonnenschwamm	gigantic barrel sponge, giant barrel sponge
Xestospongia testudinaria	Badewannenschwamm, Großer Vasenschwamm	turtleshell bath sponge, large barrel sponge
Xiphias gladius	Schwertfisch	swordfish
Xiphiidae	Schwertfische	swordfishes, billfishes
Xiphopenaeus kroyeri	Glatthorn-Garnele, Kroyers Geißelgarnele	seabob, Atlantic sea bob
Xiphophorus couchianus	Nordplaty	Monterrey platyfish
Xiphophorus helleri	Schwertträger	green swordtail
Xiphophorus maculatus	Platy, Mondplaty, Spiegelkärpfling	southern platyfish
Xiphophorus montezumae	Montezuma-Schwertträger	Montezuma swordtail
Xiphophorus pygmaeus	Zwergschwertträger	pygmy swordtail

Xiphophorus variatus	Papageienkärpfling, Veränderlicher Spiegelkärpfling	variegated platy, variatus swordtail, variable platyfish (FAO)
Xiphosura	Pfeilschwanzkrebse	horseshoe crabs
Xiphydria spp.	Schwertwespen	wood wasps
Xiphydriidae	Schwertwespen	wood wasps
Xolmis irupero	Witwenmonjita	white monjita
Xyelidae		xyelid sawflies
Xyleborus dispar	Ungleicher Holzbohrer	shot-hole borer, European shot-hole borer, broad-leaved pinhole borer
Xyleborus dryographus	Gekörnter Eichenholzbohrer, Buchennutzholz-Borkenkäfer, Buchennutzholz-Ambrosiakäfer, Laubnutzholz-Borkenkäfer	broad-leaf wood ambrosia beetle a.o.
Xyleborus monographus	Eichenholzbohrer, Gehöckerter Eichenholzbohrer (Kleiner Schwarzer Wurm)	oak wood ambrosia beetle a.o.
Xyleborus saxeseni	Saxesens Holzbohrer, Kleiner Holzbohrer	lesser shot-hole borer, fruit-tree pinhole borer
Xylena exsoleta	Graue Moderholzeule, Gemeines Moderholz	sword-grass moth
Xylocampa areola/ Dichonia areola	Holzkappeneule, Heckenkirschenwald-Streifeneule	
Xylocopa spp.	Holzbienen	large carpenter bees
Xylodrepa quadrimaculata/ Dendroxena quadrimaculata	Vierpunkt-Aaskäfer	four-spotted burying beetle
Xylophaga atlantica	Westatlantische Holzbohrmuschel*	Atlantic woodeater, Atlantic wood piddock
Xylophaga dorsalis	Holzbohrmuschel	wood piddock
Xylophagidae	Holzfliegen	xylophagid flies, xylophagids
Xyloterus domesticus	Buchen-Nutzholzborkenkäfer, Laub-Nutzholzborkenkäfer	broad-leaf wood ambrosia beetle a.o.
Xyloterus lineatus	Linierter Nutzholzborkenkäfer, Nadelholz-Ambrosiakäfer	lineate bark beetle, striped ambrosia beetle, conifer ambrosia beetle
Xyloterus signatus	Laubnutzholzambrosiakäfer, Eichen-Ambrosiakäfer	oak wood ambrosia beetle a.o.
Xylotrechus rusticus	Dunkler Holzklafterbock	rustic borer (longhorn beetle)
Xyrichthys novacula	Schermesserfisch	cleaver wrasse

Yersinella raymondi	Kleine Strauchschrecke	Raymond's bushcricket
Yllenus arenarius	Große Sandspringspinne	
Yponomeuta padellus/ *Hyponomeuta padellus*	Apfelbaumgespinstmotte, Pflaumengespinstmotte, Zwetschgengespinstmotte, Traubenkirschengespinstmotte	common hawthorn ermel, small ermine moth, small ermine

Zabrotes subfasciatus	Brasilbohnenkäfer	Mexican bean weevil
Zabrus tenebroides	Getreidelaufkäfer	corn ground beetle
Zaedyus pichiy/ Euphractus pichiy	Zwerggürteltier	pichi
Zaglossus bruijni	Langschnabeligel	long-nosed echidna, long-beaked echidna, New Guinea long-nosed spiny anteater
Zalophus californianus	Kalifornischer Seelöwe	Californian sea lion
Zanclidae	Halfterfische, Maskenfische, Maskenwimpelfische	Moorish idols
Zanclus cornutus/ Zanclus canescens/ Chaetodon canescens	Halfterfisch, Maskenfisch, Maskenwimpelfisch	Moorish idol
Zanclus flavescens	Gelber Segeldoktor	yellow tang
Zaocys carinatus	Gekielte Rattennatter, Gekielte Rattenschlange	keeled ratsnake, black-and-yellow keeled ratsnake
Zaocys spp.	Chinesische Rattennattern, Rattenschlangen	keeled ratsnakes, Chinese ratsnakes
Zapodidae	Hüpfmäuse, Eigentliche Hüpfmäuse	birch mice & jumping mice
Zaprora silenus		prowfish
Zapus hudsonius	Wiesenhüpfmaus	meadow jumping mouse
Zapus princeps	Westliche Hüpfmaus	western jumping mouse
Zapus spp.	Hüpfmäuse, Feldhüpfmäuse	jumping mice
Zapus trinotatus	Pazifik-Hüpfmaus	Pacific jumping mouse
Zebrasoma veliferum	Segelflossen-Doktorfisch	Pacific sailfin tang, sail-finned surgeonfish, sailfin tang (FAO)
Zebrasoma xanthurum	Gelbschwanz-Segeldoktorfisch	yellowtail surgeonfish, yellowtail sailfin tang, yellowtail tang (FAO)
Zebrina detrita	Weiße Turmschnecke, Große Turmschnecke, Märzenschnecke	
Zebrus zebrus	Zebra-Grundel	zebra goby
Zeidae	Petersfische	dories
Zeiformes	Petersfischartige, Petersfische & Eberfische	dories (John Dory) & others
Zeiraphera diniana	Grauer Lärchenwickler	larch tortrix, grey larch tortrix, larch bud moth
Zelotomys spp.	Breitköpfige Mäuse, Breitkopfmäuse	broad-headed mice, broad-headed stink mice
Zenaida asiatica	Weißflügeltaube	white-winged dove
Zenaida galapagoensis	Galapagostaube	Galapagos dove
Zenaida macroura	Carolina-Taube	mourning dove
Zeniontidae	Zeniontiden	zeniontids
Zenkerella insignis	Dornschwanzbilch	flightless scaly-tailed squirrel, Cameroon scaly-tail
Zenobiella subrufescens/ Perforatella subrufescens	Braunschnecke*	dusky snail, brown snail
Zenopsis conchifera	Amerikanischer Petersfisch	sail-finned dory, American John Dory
Zerynthia polyxena	Osterluzeifalter	southern festoon

Zeugophora subspinosa	Zweifarbiger Blattminierkäfer	blue-and-brown leafmining beetle*
Zeugopterus punctatus	Müllers Zwergbutt, Haarbutt	topknot
Zeus faber	Heringskönig, Petersfisch	Dory, John Dory
Zeuzera pyrina	Blausieb, Kastanienbohrer	leopard moth
Zicrona caerulea	Bläuling, Blaugrüne Baumwanze	blue shield bug, blue bug
Zingel asper	Rhône-Streber	asper, Rhône-streber, Rhone streber (FAO)
Zingel streber	Streber	streber
Zingel zingel	Zingel	zingel
Ziphiidae	Schnabelwale, Spitzschnauzenwale	beaked whales
Ziphius cavirostris	Cuvier-Schnabelwal, Gänseschnabelwal	Cuvier's goosebeaked whale
Zirfaea crispata	Krause Bohrmuschel, Raue Bohrmuschel	great piddock, oval piddock
Zoantharia	Krustenanemonen	zoanthids
Zoanthus pulchellus	Matten-Krustenanemone	mat anemone
Zoarces viviparus	Aalmutter	eel pout, viviparous blenny (FAO)
Zoarcidae	Gebärfische, Aalmütter	eelpouts
Zodariidae/ Zodarionidae	Ameisenjäger	zodariids
Zonaria annettae	Annettes Kaurischnecke	Annette's cowrie
Zonaria pyrum/ Cypraea pyrum	Birnenkauri, Birnenporzellane	pear cowrie
Zonaria spadicea/ Cypraea spadicea	Kastanienkauri	chestnut cowrie
Zonitidae	Glanzschnecken	glass snails
Zonitoides arboreus	Kleine Gewächshaus-Glanzschnecke, Gewächshaus-Dolchschnecke	quick gloss (snail), quick glass snail, bark snail
Zonitoides excavatus	Britische Dolchschnecke, Hohlschnecke*, Hohl-Glanzschnecke	hollowed gloss (snail), hollowed glass snail
Zonitoides nitidus	Moder-Glanzschnecke, Glänzende Dolchschnecke	black gloss (snail), shiny glass snail
Zonotrichia albicollis	Weißkehlammer	white-throated sparrow
Zonotrichia capensis	Morgenammer	rufous-collared sparrow
Zonotrichia leucophrys	Dachsammer	white-crowned sparrow
Zoothera dauma	Erddrossel	White's thrush
Zoothera sibirica	Schieferdrossel	Siberian thrush
Zophobas spp.		"superworms" (*larvae*)
Zosterisessor ophiocephalus/ Gobius ophiocephalus/ Gobius lota	Schlangenkopfgrundel	grass goby
Zosteropidae	Brillenvögel	white-eyes
Zosterops albogularis	Norfolk-Brillenvogel	Norfolk Island white-eye, white-breasted white-eye, white-chested white-eye

Zu cristatus	Gezackter Bandfisch*	scalloped ribbonfish
Zygaena angelicae	Schneckenklee-Widderchen, Ungeringtes Kronwicken-Blutströpfchen, Ungeringtes Kronwicken-Widderchen	
Zygaena carniolica	Esparsetten-Widderchen	
Zygaena ephialtes	Veränderliches Blutströpfchen, Veränderliches Widderchen, Veränderliches Rotwidderchen, Wickenwidderchen	variable burnet
Zygaena exulans	Hochalpen-Rotwidderchen, Schottisches Bluttröpfchen	Scotch burnet
Zygaena fausta	Glücks-Widderchen, Bergkronwicken-Widderchen	
Zygaena filipendulae/ Anthrocera filipendulae	Gemeines Blutströpfchen, Steinbrechwidderchen, Gewöhnliches Sechsfleck-Blutströpfchen, Erdeichel-Widderchen	six-spot burnet
Zygaena lonicerae	Klee-Blutströpfchen, Klee-Widderchen	narrow-bordered five-spot burnet
Zygaena loti	Beilfleck-Blutströpfchen, Beilfleck-Widderchen, Beilfleck-Rotwidderchen	slender burnet
Zygaena minos	Bibernell-Widderchen, Bibernell-Rotwidderchen	pimpernel burnet*
Zygaena osterodensis	Platterbsen-Widderchen, Platterbsen-Rotwidderchen	
Zygaena purpuralis	Thymian-Blutströpfchen, Thymian-Rotwidderchen	transparent burnet
Zygaena transalpina	Hufeisenklee-Widderchen, Hufeisenklee-Rotwidderchen, Hufeisenklee-Blutströpfchen	horseshoe vetch burnet*
Zygaena trifolii	Sumpfhornklee-Widderchen, Kleewidderchen, Feuchtwiesen-Blutströpfchen	five-spot burnet
Zygaenidae (Anthroceridae)	Blutströpfchen (Widderchen)	burnets & foresters, smoky moths
Zygaspis quadrifrons	Kalahari-Wurmschleiche	Kalahari round-snouted worm lizard
Zygiella spp.	Sektorspinnen	sector spiders
Zygiella x-notata	Sektorspinne	sector spider
Zygodontomys spp.	Rohrmäuse	cane mice
Zygogeomys trichopus	Tuza	tuza
Zygoptera	Kleinlibellen	damselflies
Zyzomys spp.	Dickschwanzratten	thick-tailed rats, rock rats

Achatschnecke> Große Achatschnecke, Gemeine Riesenschnecke, Afrikanische Riesenschnecke	*Achatina fulica*	giant African snail, giant African land snail
Achatschnecken, Glattschnecken	Cionellidae/Cochlicopidae	slippery moss snails, pillar snails
Achselfleck-Brassen, Achselbrasse, Spanischer Meerbrassen	*Pagellus acarne*	Spanish bream, Spanish seabream, axillary bream, axillary seabream (FAO)
Achselfleck-Lippfisch, Mittelmeerlippfisch	*Symphodus mediterraneus*	axillary wrasse
Achtarmige Tintenschnecken, Kraken	Octopoda/Octobrachia	octopods, octopuses
Achtarmkalmar	*Octopodoteuthis sicula*	octopod squid
Achtstrahlige Becherqualle, Kleine Becherqualle	*Haliclystus octoradiatus*	octaradiate stalked jellyfish*
Achtzigeule, Augen-Eulenspinner, Pappelhain-Eulenspinner	*Tethea ocularis*	figure of eighty
Ackerblasenfuß	*Thrips angusticeps*	field thrips, cabbage thrips, flax thrips
Ackerbohnenkäfer, Pferdebohnenkäfer	*Bruchus rufimanus*	broadbean weevil, bean weevil
Ackerhummel	*Bombus pascuorum/ Bombus agrorum*	carder bee, common carder bee
Ackerschnecken	Agriolimacidae	field slugs
Ackerschnecken	*Deroceras* spp./*Agriolimax* spp.	field slugs
Acouchis	*Myoprocta* spp.	acouchis
Actinotriche Milben	Actinotrichida	actinotrichid mites
Adeliepinguin	*Pygoscelis adeliae*	Adelie penguin
Adersducker	*Cephalophus adersi*	Aders' duiker
Adlerbussard	*Buteo rufinus*	long-legged buzzard
Adlerfarnspanner	*Petrophora chlorosata*	bracken-fern moth*
Adlerfisch, Seeadler (Adlerlachs)	*Argyrosomus regius/ Sciaena aquila*	meagre
Adlerrochen	Myliobatidae	eagle rays
Admiral	*Vanessa atalanta*	red admiral
Admiralskegel, Admiral	*Conus ammiralis*	admiral cone, Ammiralis cone
Adria-Elritze	*Paraphoxinus alepidotus*	Adriatic minnow
Adria-Lachs	*Salmothymus obtusirostris*	Adriatic salmon
Adria-Stör, Adriatischer Stör	*Acipenser naccarii*	Adriatic sturgeon
Adria-Strauchschrecke	*Rhacocleis neglecta*	Adriatic bushcricket
Affenadler	*Pithecophaga jefferyi*	great Philippine eagle, monkey-eating eagle
Affenfrosch, Greiffrosch	*Phyllomedusa hypochondrialis*	orange-legged leaf frog
Affengans	*Stictonetta naevosa*	freckled duck
Afghanfuchs, Canafuchs	*Vulpes cana*	Blanford's fox
Afrika-Linsang, Pojana	*Poiana richardsoni*	African linsang, oyan
Afrika-Zibetkatze	*Civettictis civetta*	African civet

Afrikanische Bachratte	*Pelomys fallax*	creek rat, groove-toothed creek rat (groove-toothed mouse)
Afrikanische Bachratten	*Pelomys* spp.	groove-toothed creek rats
Afrikanische Bananenfrösche	*Hoplophryne* spp.	African banana frogs
Afrikanische Bänder-Korallenschlange	*Elapsoidea sundevallii*	Sundevall's garter snake, African garter snake
Afrikanische Baumkröten	*Nectophryne* spp.	African tree toads
Afrikanische Blindskinke	*Typhlosaurus* spp.	blind worms, blind legless skinks
Afrikanische Bürstenhaarmäuse	*Lophuromys* spp.	brush-furred mice
Afrikanische Buschhörnchen	*Paraxerus* spp.	African bush squirrels
Afrikanische Dreispitznäsige Fledermaus, Percivals Dreispitz-Fledermaus	*Cloeotis percivali*	African trident-nosed bat
Afrikanische Eierschlange	*Dasypeltis scabra*	egg-eating snake, African egg-eating snake
Afrikanische Epauletten-Flughunde u.a.	*Epomops* spp. & *Epomophorus* spp.	epauleted bats
Afrikanische Erdhörnchen u.a., Afrikanische Borstenhörnchen	*Xerus* spp.	African ground squirrels
Afrikanische Fächermuschel*	*Argopecten flabellum*	African fan scallop
Afrikanische Fadenmakrele	*Alectis crinitus*	African pompano
Afrikanische Feldmäuse	*Myomys* spp.	African meadow rats
Afrikanische Flachkopffledermaus	*Platymops setiger*	flat-headed free-tailed bat
Afrikanische Freischwanzfledermaus	*Coleura afra*	African sheath-tailed bat
Afrikanische Hausnatter, Gewöhnliche Hausschlange	*Boaedon lineatus/ Lamprophis lineatus*	African house snake, African brown house snake
Afrikanische Hechtsalmler	Hepsetidae	African pike characins, hepsetids
Afrikanische Helmschnecke	*Cassis saburon*	African helmet
Afrikanische Honigbiene	*Apis mellifera adansoni*	African honey bee, Africanized honey bee
Afrikanische Horn-Vogelspinne*	*Ceratogyrus darlingi*	African horned tarantula
Afrikanische Hornfliege	*Haematobia minuta*	African horn-fly
Afrikanische Hornschnecke	*Murex cornutus*	African horned murex
Afrikanische Kobra	*Naja pallida*	African cobra
Afrikanische Korallenschlangen	*Elaps* spp.	African dwarf garter snakes
Afrikanische Lamellenzahnratten, Ohrenratten	*Otomys* spp.	African swamp rats, groove-toothed rats
Afrikanische Languste, Kap-Languste	*Jasus lalandei*	Cape rock crawfish, Cape rock lobster
Afrikanische Lungenfische	*Protopterus* spp.	African lungfishes
Afrikanische Lungenfische	Protopteridae	African lungfishes
Afrikanische Maulwurfsratten	*Tachyoryctes* spp.	African mole-rats, African root-rats

Afrikanische Meeräsche	*Mugil capurrii*	narrow-head mullet, leaping African mullet
Afrikanische Neptun-Walzenschnecke	*Cymbium pepo*	African neptune volute
Afrikanische Riesenschnecken	Achatinidae	giant African snails
Afrikanische Riesenskorpione	*Pandinus* spp.	African emperor scorpions
Afrikanische Riesenvogelspinne	*Hysterocrates gigas*	Cameroon red tarantula
Afrikanische Schlammfische	Phractolaemidae	snake mudheads
Afrikanische Schlangenechsen	Feyliniidae	feylinias
Afrikanische Schmetterlings-Fledermäuse	*Glauconycteris* spp.	butterfly bats, silvered bats
Afrikanische Schnabelbrustschildkröte	*Chersina angulata*	South African bowsprit tortoise
Afrikanische Schweinezecke, Blauzecke*	*Boophilus decoloratus*	blue tick
Afrikanische Streifen-Grasmäuse	*Lemniscomys* spp.	striped grass mice, zebra mice
Afrikanische Streifensüßlippe	*Parapristipoma octolineatum*	African striped grunt
Afrikanische Striemen-Grasmaus	*Rhabdomys pumilio*	four-striped grass mouse, striped mouse
Afrikanische Sumpfratte, Wollhaarratte	*Dasymys incomtus*	shaggy swamp rat
Afrikanische Tieflandviper, Afrikanische Tiefland-Buschviper	*Vipera superciliaris/ Atheris superciliaris*	lowland swamp viper, African lowland viper
Afrikanische Waldbachmaus	*Colomys goslingi*	African water rat, velvet rat
Afrikanische Waldmäuse u.a.	*Hylomyscus* spp.	African wood mice
Afrikanische Waldspitzmäuse	*Myosorex* spp.	mouse shrews, forest shrews
Afrikanische Wanderheuschrecke, AfrikanischeWüstenschrecke	*Schistocerca gregaria*	desert locust
Afrikanische Weichschildkröte, Afrikanischer Dreiklauer	*Trionyx triunguis*	African softshell turtle, Nile softshell turtle
Afrikanische Zwergente	*Nettapus auritus*	African pygmy goose
Afrikanischer Adlerfisch	*Argyrosomus hololepidotus*	kob, southern meagre (FAO)
Afrikanischer Argusfisch	*Scatophagus tetracanthus*	African argusfish
Afrikanischer Augenwurm, Wanderfilarie	*Loa loa*	African eye worm
Afrikanischer Büffel, Kaffernbüffel	*Syncerus caffer*	African buffalo
Afrikanischer Dornschwanz, Veränderlicher Dornschwanz	*Uromastyx acanthinurus*	African spiny-tailed lizard, Dabbs mastigure
Afrikanischer Elefant	*Loxodonta africana*	African elephant
Afrikanischer Fähnchen-Messerfisch	*Notopterus afer/ Papyrocranus afer*	reticulate knifefish
Afrikanischer Glaswels	*Physailia pellucida*	African glass catfish
Afrikanischer Knochenzüngler	*Heterotis niloticus/ Clupisudis niloticus*	African bonytongue, heterotis (FAO)
Afrikanischer Langzungen-Flughund	*Megaloglossus woermanni*	African long-tongued fruit bat

Afrikanischer Leberegel (Afrikanischer Lanzettegel)	*Dicrocoelium hospes*	African lancet fluke
Afrikanischer Messerfisch, Schwarzer Messerfisch	*Xenomystus nigri*	African knifefish (FAO), false featherback
Afrikanischer Pferdekopf	*Selene dorsalis*	African lookdown
Afrikanischer Riesenschwalbenschwanz	*Papilio antimachus*	African giant swallowtail
Afrikanischer Riesenskorpion, Afrikanischer Waldskorpion, Kaiserskorpion	*Pandinus imperator*	common emperor scorpion
Afrikanischer Schlammfisch	*Phractolaemus ansorgei*	snake mudhead
Afrikanischer Schmetterlings-Buntbarsch, Thomas-Prachtbuntbarsch	*Anomalochromis thomasi*	dwarf jewel fish
Afrikanischer Segelflugfisch*	*Parexocoetus mento*	African sailfin flyingfish
Afrikanischer Sichelflosser	*Drepane africana*	African sicklefish
Afrikanischer Speckkäfer	*Dermestes haemorrhoidales*	African larder beetle, black larder beetle
Afrikanischer Umberfisch	*Atractoscion aequidens*	African weakfish
Afrikanischer Waldkauz	*Strix woodfordii*	African wood owl
Afrikanischer Wildesel	*Equus africanus*	African wild ass
Afrikanischer Wildhund	*Lycaon pictus*	African wild dog, African hunting dog, African painted wolf
Afrikanisches Riesenhörnchen	*Protoxerus stangeri*	African giant squirrel
Afrikanisches Zwerghörnchen	*Myosciurus pumilio*	African pygmy squirrel
Afterskorpione, Pseudoskorpione	Pseudoscorpiones/Chelonethi	pseudoscorpions, false scorpions
Aga-Kröte	*Bufo marinus*	giant toad, marine toad, cane toad, South American Neotropical toad
Ägäische Mauereidechse, Kykladen-Eidechse	*Podarcis erhardii*	Erhard's wall lizard
Ägäische Süßwasserkrabbe	*Potamon potamios*	Aegean freshwater crab
Ägäischer Nacktfinger	*Cyrtodactylus kotschyi*	Kotschy's gecko
Agamen	Agamidae	agamas, chisel-teeth lizards
Agrippinaeule	*Thysania agrippina*	giant owl moth
Agujon, Nadel-Hornhecht	*Tylosurus acus*	agujon needlefish
Agutis	Dasyproctidae	agoutis
Agutis> Stummelschwanzagutis	*Dasyprocta* spp.	agoutis
Ägyptische Baumwollraupe	*Spodoptera litura*	cluster caterpillar
Ägyptische Heuschrecke, Ägyptische Wanderheuschrecke	*Anacridium aegypticum*	Egyptian grasshopper
Ägyptische Honigbiene	*Apis mellifera fasciata*	Egyptian honey bee
Ägyptische Landschildkröte	*Testudo kleinmanni*	Egyptian tortoise
Ägyptische Sandboa	*Eryx colubrinus*	Egyptian sand boa
Ägyptische Schlankblindschlange	*Leptotyphlops cairi*	Cairo earthsnake
Ägyptische Stachelmaus	*Acomys cahirinus*	Egyptian spiny mouse

Ägyptische Vogelspinne	*Chaetopelma aegyptiacum*	Egyptian tarantula
Ägyptische Zornnatter	*Coluber florulentus*	Egyptian whip snake
Ägyptischer Dornschwanz	*Uromastyx aegypticus*	Egyptian spiny-tailed lizard, Egyptian mastigure, dabb lizard
Ahlenläufer	*Bembidion* spp.	brassy ground beetles
Ahnenschlangen	Chlorophidia	
Ahornblatt-Gallmilbe	*Aceria macrochelus*	maple leaf solitary-gall mite
Ahornblatt-Triton	*Biplex perca/ Gyrineum perca*	winged triton, maple leaf triton
Ahornblattroller	*Deporaus tristis*	maple leafroller weevil
Ahorneule, Rosskastanieneule	*Acronicta aceris*	sycamore moth
Ahorngallmilbe	*Eriophyes macrorhynchus*	maple nail-gall mite (>maple nail gall)
Ahornmotte	*Caloptilia rufipennella*	maple moth, maple leaf miner
Ahornschmierlaus	*Phenacoccus aceris*	apple mealybug, maple mealybug
Ährenfisch, Sand-Ährenfisch	*Atherina presbyter*	sandsmelt, sand smelt (FAO)
Ährenfische	Atherinidae	silversides
Ährenmaus	*Mus spicilegus*	steppe mouse
Ährenträgerpfau	*Pavo muticus*	green peafowl
Ährenwickler	*Cnephasia longana*	omnivorous leaf tier (moth)
Ailanthusspinner, Götterbaumspinner	*Philosamia cynthia/ Samia cynthia/ Platysamia cynthia*	Cynthia silkmoth, ailanthus silkworm
Akaziendrossling	*Turdoides fulvus*	fulvous babbler
Akaziengrasmücke	*Sylvia leucomelaena*	Arabian warbler
Akazienmaus	*Thallomys paedulcus*	acacia rat
Akiami-Garnele	*Acetes japonicus*	Akiami paste shrimp
Alabama-Furchenmolch	*Necturus alabamensis*	Alabama waterdog
Alabama-Rotbauch-Schmuckschildkröte	*Pseudemys alabamensis*	Alabama red-bellied turtle
Alabaster-Wurmseegurke	*Opheodesoma* spp.	alabaster sea cucumber
Alabasterschnecke	*Siratus alabaster*	alabaster murex, abyssal murex
Alamos-Klappschildkröte	*Kinosternon alamosae*	Alamos mud turtle
Aland, Orfe	*Leuciscus idus*	ide (FAO), orfe
Alaska-Königskrabbe, Königskrabbe (Kronenkrebs, Kamtschatkakrebs), Kamtschatka-Krabbe	*Paralithodes camtschaticus*	king crab, red king crab, Alaskan king crab, Alaskan king stone crab (Japanese crab, Kamchatka crab, Russian crab)
Alaska-Pollack, Pazifischer Pollack, Mintai	*Theragra chalcogramma*	pollock, pollack, Alaska pollack (FAO), walleye pollock
Alaska-Scholle	*Pleuronectes quadrituberculatus*	Alaska plaice
Alaska-Wühlmaus	*Microtus miurus*	singing vole
Alders Halsbandnabelschnecke	*Policines polianus*	Alder's necklace snail (Alder's necklace shell)
Alexandersittich	*Psittacula eupatria*	Alexandrine parakeet
Alexandria-Fadenmakrele	*Alectis alexandrinus*	Alexandria pompano, African threadfish (FAO)
Alexandriner Hausratte, Dachratte, Hausratte	*Rattus rattus*	black rat, roof rat, house rat

Alexis-Bläuling	*Glaucopsyche alexis*	green underside blue
Alfalfa-Biene, Luzernen-Blattschneiderbiene	*Megachile rotundata*	alfalfa leafcutter bee
Alfonsino, Kaiserbarsch, Nordischer Schleimkopf	*Beryx decadactylus*	beryx, alfonsino, red bream
Algen-Kammzähner	*Parablennius marmoreus*	seaweed blenny
Algenfresser, Saugschmerlen	Gyrinocheilidae	algae eaters
Algerische Zornnatter	*Coluber algirus*	Algerian whip snake
Algerischer Igel, Mittelmeerigel	*Erinaceus algirus/ Atelerix algirus*	Algerian hedgehog
Algerischer Sandläufer	*Psammodromus algirus*	Algerian psammodromus
Alleghany-Bachsalamander	*Desmognathus ochrophaeus*	mountain dusky salamander
Allens Eselhase, Antilopenhase	*Lepus alleni*	antelope jack rabbit
Allens Königinnatter	*Regina alleni*	striped crayfish snake
Allfarblori	*Trichoglossus haematodus*	rainbow lory
Alligatorenhaie, Alligatorhaie, Nagelhaie	Echinorhinidae	bramble sharks
Alligatorfisch, Kaimanfisch	*Lepisosteus tristoechus*	Cuban gar
Alligatorsalamander	*Aneides lugubris*	arboreal salamander
Alligatorschildkröten	Chelydridae	snapping turtles
Alligatorschleichen, Krokodilschleichen	*Elgaria* spp. & *Gerrhonotus* spp.	alligator lizards
Alpaka	*Lama pacos*	alpaca
Alpen-Bombardierkäfer (Großer Bombardierkäfer)	*Aptinus bombarda*	Alpine bombardier beetle
Alpen-Glasschnecke	*Phenacolimax annularis*	Alpine glass snail
Alpen-Mosaikjungfer	*Aeshna coerulea*	blue aeshna, azure hawker
Alpen-Smaragdlibelle	*Somatochlora alpestris*	Alpine emerald
Alpen-Steinbock, Alpensteinbock	*Capra ibex*	alpine ibex
Alpen-Strauchschrecke	*Pholidoptera aptera*	Alpine dark bushcricket
Alpen-Windelschnecke	*Vertigo alpestris*	Alpine whorl snail, mountain whorl snail, tundra vertigo (U.S.)
Alpenbraunelle	*Prunella collaris*	alpine accentor
Alpenbock	*Rosalia alpina*	rosalia longicorn
Alpendohle	*Pyrrhocorax graculus*	Alpine chough
Alpenfledermaus	*Hypsugo savii/Pipistrellus savii*	Savi's pipistrelle
Alpenkammmolch, Italienischer Kammmolch	*Triturus carniflex*	Alpine crested newt, Italian warty newt
Alpenkrähe	*Pyrrhocorax pyrrhocorax*	red-billed chough
Alpenmolch, Bergmolch	*Triturus alpestris*	Alpine newt
Alpenmurmeltier (Hbz. Murmel)	*Marmota marmota*	Alpine marmot
Alpensalamander	*Salamandra atra*	Alpine salamander, European Alpine salamander
Alpenschneehuhn	*Lagopus mutus*	ptarmigan
Alpensegler	*Apus melba*	Alpine swift
Alpenspitzmaus	*Sorex alpina*	Alpine shrew

Alpenstrandläufer	*Calidris alpina*	dunlin
Alpenveilchenmilbe, Zyklamenmilbe, Begonienmilbe	*Phytonemus pallidus/ Tarsonemus pallidus*	cyclamen mite, begonia mite
Alpenwaldmaus	*Apodemus alpicola*	alpine wood mouse
Alpenweißling	*Pontia callidice*	peak white
Alpenwiesel	*Mustela altaica*	mountain weasel
Alphabet-Kegelschnecke	*Conus spurius*	alphabet cone
Alse, Gewöhnliche Alse, Maifisch	*Alosa alosa*	Allis shad
Altfische	Chondrostei	primitive ray-finned bony fishes
Altschnecken, Schildkiemer	Archaeogastropoda/Diotocardia	limpets & allies, archeogastropods
Altwelt-Biber	*Castor fiber*	Eurasian beaver
Altwelt-Freischwanzfledermaus	*Emballonura* spp.	Old World sheath-tailed bats
Altwelt-Gleithörnchen	*Pteromys* spp.	Old World flying squirrels
Altwelt-Hakenwurm, Grubenwurm	*Ancylostoma duodenale*	Old World hookworm
Altwelt-Stachelschweine	Hystricidae	Old World porcupines
Altweltaffen, Schmalnasenaffen	Catarrhina	Old World monkeys (incl. apes)
Altweltlicher Baumwoll-Kapselwurm	*Helicoverpa armigera/ Heliothis armigera/ Heliothis obsoleta*	scarce bordered straw moth, Old World bollworm, African bollworm, corn earworm, cotton bollworm; tomato moth, tomato fruitworm, tomato grub
Altweltliches Wildkaninchen, Europäisches Wildkaninchen (Hbz. Wildkanin)	*Oryctolagus cuniculus*	Old World rabbit, domestic rabbit
Altweltmaulwürfe	*Talpa* spp.	Old World moles
Amadis-Kegel	*Conus amadis*	Amadis cone
Amago-Lachs	*Oncorhynchus rhodurus*	amago salmon, amago
Amazonas-Bodennattern, Buntnattern	*Liophis* spp.	Amazon ground snakes
Amazonas-Delphin, Butu, Inia	*Inia geoffrensis*	Amazon dolphin, Amazon River dolphin, bouto, boutu
Amazonas-Kärpfling	*Poecilia formosa*	Amazon molly
Amazonas-Korallenschlange	*Micrurus spixii*	Amazonian coral snake
Amazonas-Lanzenotter, Grüne Jararaca	*Bothriopsis bilineata/ Bothrops bilineata*	Amazonian tree viper, Amazonian palm viper, two-striped forest pit viper
Amazonas-Rotkopf-Schienenschildkröte	*Podocnemis erythrocephala*	red-headed Amazon side-necked turtle, red-headed river turtle (IUCN)
Amazonas-Sotalia	*Sotalia fluviatilis*	Tucuxi, Amazon River dolphin, Tookashee dolphin
Amazonenameise	*Polyergus rufescens*	robber ant, Amazon ant
Amboina-Scharnierschildkröte	*Cuora amboinensis*	Malayan box turtle, Amboina box turtle, South Asian box turtle (IUCN)
Ameisen	Formicidae	ants
Ameisen-Buntkäfer, Ameisenbuntkäfer	*Thanasimus formicarius*	ant beetle

Deutsch	Wissenschaftlich	Englisch
Ameisenassel	*Platyarthrus hoffmannseggi*	ant woodlouse
Ameisenbär> Großer Ameisenbär	*Myrmecophaga tridactyla*	giant anteater
Ameisenbär> Tamanduas	*Tamandua* spp. (*T. tetradactyla* & *T. mexicana*)	lesser anteaters, tamanduas
Ameisenbär> Zwergameisenbär	*Cyclopes didactylus*	silky anteater
Ameisenbären	Vermilingua (Xenarthra)	anteaters
Ameisenbärfisch, Spitzbartfisch	*Gnathonemus tamandua/ Campylomormyrus tamandua*	worm-jawed mormyrid
Ameisenbeutler	*Myrmecobius fasciatus*	numbat, banded anteater
Ameisengrille	*Myrmecophila acervorum*	ant's-nest cricket
Ameisenigel, Schnabeligel	Tachyglossidae	echidnas, spiny anteaters
Ameisenjäger	Zodariidae/Zodarionidae	zodariids
Ameisenkäfer	Scydmaenidae	antlike stone beetles
Ameisenlöwen	Myrmeleonidae	antlions
Ameisenspinnen	Castianeirinae (Corinnidae)	ant spiders
Ameisenspinnen	*Dipoena* spp., *Synageles* spp., *Micaria* spp., *Myrmecium* spp., *Castianeira* spp. et al.	ant spiders
Ameisenspringspinne, Ameisenspinne	*Myrmarachne formicaria*	ant spider a.o.
Ameisenstutzkäfer, Rostroter Stutzkäfer	*Hetaerius ferrugineus*	ant hister beetle
Ameisenwanze	*Myrmecoris gracilis*	ant capsid bug
Ameisenwespchen	Bethylidae	bethylid wasps
Ameive	*Ameiva ameiva*	jungle runner
Ameiven	*Ameiva* spp.	jungle runners
Amerikanerkrähe	*Corvus brachyrhynchos*	American crow
Amerikanische „Aalmutter"	*Macrozoarces americanus*	ocean pout
Amerikanische Alse, Amerikanischer Maifisch	*Alosa sapidissima*	American shad
Amerikanische Auster	*Crassostrea virginica/ Gryphaea virginica*	American oyster, eastern oyster, blue point oyster, American cupped oyster
Amerikanische Bergmeeräsche	*Agonostomus monticola*	mountain mullet
Amerikanische Bohrmuschel	*Petricola pholadiformis/ Petricolaria pholadiformis*	American piddock, false angelwing (U.S.)
Amerikanische Bola-Spinne, Lasso-Spinne	*Mastophora bisaccata*	American bolas spider
Amerikanische Brackwassermeduse	*Nemopsis bachei*	American brackish water medusa*
Amerikanische Buchenkäfer	Perothopidae	beech-tree beetles
Amerikanische Buntzecke, Amerikanische Sternzecke*	*Amblyomma americanum*	lone star tick
Amerikanische Dasselfliegen	Cuterebridae	robust bot-flies
Amerikanische Diebsameise	*Solenopsis molesta*	thief ant
Amerikanische Dreieckspinne	*Hyptiotes cavatus*	American triangle spider
Amerikanische Erntemäuse	*Reithrodontomys* spp.	American harvest mice
Amerikanische Erntemilbe*	*Trombicula alfreddugesi/ Eutrombicula alfreddugesi*	American chigger mite, common chigger mite

Amerikanische Feuerameise	*Solenopsis geminata*	American fire ant
Amerikanische Golfküsten-Buntzecke	*Amblyomma maculatum*	Gulf Coast tick
Amerikanische Großschabe, Amerikanische Schabe	*Periplaneta americana*	American cockroach
Amerikanische Haftscheiben-Fledermäuse	Thyropteridae	disk-winged bats, New World sucker-footed bats
Amerikanische Hammermuschel, Karibische Hammermuschel	*Malleus candeanus*	Caribbean hammer-oyster, American hammer oyster, American malleus
Amerikanische Hausspinne	*Scotophaeus blackwalli*	American house spider
Amerikanische Hausspinne, Gewächshausspinne	*Achaearanea tepidariorum*	house spider, American common house spider, common American house spider, domestic spider
Amerikanische Hausstaubmilbe	*Dermatophagoides farinae*	American house dust mite
Amerikanische Hochmoor-Schildkröte, Mühlenberg-Schildkröte	*Clemmys muhlenbergii*	bog turtle, Muhlenberg's turtle
Amerikanische Hundezecke	*Dermacentor variabilis*	American dog tick
Amerikanische Kiefernwaldnatter	*Rhadinaea flavilata*	pine woods snake
Amerikanische Kirschfruchtfliege	*Rhagoletis cingulata*	cherry fruit fly (U.S.), North American cherry fruit fly (cherry maggot)
Amerikanische Kleine Maräne	*Coregonus artedii*	lake cisco, lake herring
Amerikanische Korallenschlangen, Echte Korallenottern	*Micrurus* spp.	American coral snakes, common coral snakes
Amerikanische Kröte	*Bufo americanus*	American toad
Amerikanische Languste, Karibische Languste	*Panulirus argus*	West Indies spiny lobster, Caribbean spiny lobster, Caribbean spiny crawfish
Amerikanische Lanzenottern	*Bothrops* spp. (see also: *Lachesis/Porthidium*)	American lance-headed vipers, lanceheads
Amerikanische Messerfische	Rhamphichthyidae	sand knifefishes
Amerikanische Querzahnmolche	*Ambystoma* spp.	American mole salamanders
Amerikanische Radwanze, Amerikanische Sägekamm-Raubwanze	*Arilus cristatus*	wheel bug
Amerikanische Sardelle	*Engraulis mordax*	North Pacific anchovy, Californian anchovy (FAO)
Amerikanische Schabe, Amerikanische Großschabe	*Periplaneta americana*	American cockroach
Amerikanische Schaufelnasenstöre	*Scaphirhynchus* spp.	American sturgeons
Amerikanische Scheibensalmler	*Metynnis* spp.	silver dollars
Amerikanische Scheunenspinne	*Araneus cavaticus*	barn orbweaver, barn spider
Amerikanische Schlangenhalsschildkröten	*Hydromedusa* spp.	American snake-necked turtles, American snake-headed turtles
Amerikanische Schlankblindschlangen	Anomalepidae	dawn blind snakes
Amerikanische Scholle	*Glyptocephalus zachirus*	rex sole

Amerikanische Sechspunktspinne*	*Araniella displicata*	sixspotted orbweaver, six-spotted orb weaver
Amerikanische Seezungen	Achiridae	American soles
Amerikanische Skinks	*Eumeces* spp.	eyelid skinks
Amerikanische Sonnenuhrschnecke	*Architectonica nobilis*	American sundial (snail), common American sundial, common sundial
Amerikanische Spinnenameisen	*Dasymutilla* spp.	velvet-ants
Amerikanische Stern-Turbanschnecke	*Lithopoma americanum*	American starsnail
Amerikanische Sternzecke*, Amerikanische Buntzecke	*Amblyomma americanum*	lone star tick
Amerikanische Sumpfnatter	*Seminatrix pygaea*	black swamp snake
Amerikanische Sumpfschildkröte	*Emydoidea blandingii*	Blanding's turtle
Amerikanische Tabakeule, Amerikanische Tabakknospeneule	*Heliothis virescens*	tobacco budworm
Amerikanische Wacholderrindenlaus	*Cinara fresai*	American juniper aphid
Amerikanische Wespenspinne	*Argiope trifasciata*	banded garden spider, American banded garden spider, banded argiope, whitebacked garden spider
Amerikanische Zwergmäuse	*Baiomys* spp.	pygmy mice
Amerikanische Zwergspitzmaus	*Microsorex hoyi*	pygmy shrew, American pygmy shrew
Amerikanische Zwergwelse, Katzenwelse	Ictaluridae	North American freshwater catfishes
Amerikanische Zylinderrose	*Cerianthopsis americanus*	sand anemone, burrowing sea anemone, tube-dwelling sea anemone
Amerikanischer Aal	*Anguilla rostrata*	American eel
Amerikanischer Angler	*Lophius americanus*	American goosefish
Amerikanischer Austernbohrer	*Urosalpinx cinerea*	American sting winkle, American oyster drill, Atlantic oyster drill
Amerikanischer Bärenkrebs	*Scyllarus americanus*	American slipper lobster
Amerikanischer Bison	*Bison bison*	American bison, buffalo
Amerikanischer Blutegel, Amerikanischer Medizinischer Blutegel	*Macrobdella decora*	American medicinal leech
Amerikanischer Buckelkäfer, Schwarzer Kapuzenkugelkäfer	*Mezium americanum*	American spider beetle, black spider beetle
Amerikanischer Butterfisch, Atlantik-Butterfisch	*Peprilus triacanthus*	American butterfish (FAO), Atlantic butterfish
Amerikanischer Flaggenkilli, Floridakärpfling	*Jordanella floridae*	American flagfish
Amerikanischer Flussbarsch, Gelbbarsch	*Perca flavescens*	yellow perch (FAO), American yellow perch
Amerikanischer Flusskrebs, Kamberkrebs	*Orconectes limosus/ Cambarus affinis*	spinycheek crayfish, American crayfish, American river crayfish, striped crayfish
Amerikanischer Garten-Tausendfüßer	*Oxidus gracilis*	garden millipede (U.S.), garden millepede (Br.)

Amerikanischer Goldregenpfeifer	*Pluvialis dominica*	American golden plover
Amerikanischer Halbwüsten-Leguan	*Urosaurus graciosus*	brush lizard, long-tailed brush lizard
Amerikanischer Heuschreckenkrebs, Gewöhnlicher Heuschreckenkrebs	*Squilla empusa*	common mantis shrimp
Amerikanischer Himbeerkäfer	*Byturus unicolor*	American raspberry fruitworm
Amerikanischer Hirschkäfer	*Lucanus elaphus*	giant stag beetle
Amerikanischer Hummer	*Homarus americanus*	northern lobster, American clawed lobster
Amerikanischer Hundsfisch	*Umbra limi*	American mudminnow
Amerikanischer Lippfisch	*Tautogolabrus adspersus*	cunner
Amerikanischer Luzernenheufalter	*Colias eurytheme*	alfalfa butterfly (alfalfa caterpillar)
Amerikanischer Maifisch, Amerikanische Alse	*Alosa sapidissima*	American shad
Amerikanischer Meeraal, Amerikanischer Conger	*Conger oceanicus*	American conger
Amerikanischer Messerfisch, Gebänderter Messerfisch	*Gymnotus carapo*	American knife fish, striped knife fish, banded knifefish (FAO)
Amerikanischer Obstbaumwickler	*Archips argyrospilus*	fruittree leafroller
Amerikanischer Pelikansfuß	*Aporrhias occidentalis*	American pelican's foot
Amerikanischer Petersfisch	*Zenopsis conchifera*	sail-finned dory, American John Dory
Amerikanischer Pfeilzahnheilbutt	*Atheresthes stomias*	arrow-tooth flounder
Amerikanischer Pferdeegel	*Haemopis marmorata*	American horse leech
Amerikanischer Reismehlkäfer	*Tribolium confusum*	confused flour beetle
Amerikanischer Riesen-Einsiedler	*Petrochirus diogenes*	giant hermit crab
Amerikanischer Riesenaaskäfer	*Nicrophorus americanus*	giant carrion beetle, American burying beetle (IUCN)
Amerikanischer Riesensalamander, Hellbender	*Cryptobranchus alleganiensis*	hellbender
Amerikanischer Sandaal	*Ammodytes americanus*	American sandlance
Amerikanischer Sandpierwurm	*Arenicola cristata*	American lugworm
Amerikanischer Schlangenhalsvogel	*Anhinga anhinga*	American darter, anhinga
Amerikanischer Schwarzer Reismehlkäfer	*Tribolium audax*	black flour beetle, American black flour beetle
Amerikanischer Seeskorpion	*Myoxocephalus octodecemspinosus*	longhorn sculpin
Amerikanischer Speicherkäfer	*Trogoderma parabile*	American cabinet beetle (grain dermestid)
Amerikanischer Splintholzkäfer	*Lyctus planicollis*	southern lyctus beetle, European powderpost beetle
Amerikanischer Stechrochen	*Dasyatis americana*	southern stingray (FAO)
Amerikanischer Streifenbarsch	*Morone americana/ Roccus americanus*	white perch

Amerikanischer Traubenerdfloh	*Altica chalybea*	grape flea beetle
Amerikanischer Uhu, Virginia-Uhu	*Bubo virginianus*	great horned owl
Amerikanischer Ulmenborkenkäfer	*Hylurgopinys rufipes*	American elm bark beetle
Amerikanischer Umberfisch	*Menticirrhus americanus*	southern kingfish, southern kingcroaker (FAO)
Amerikanischer Webebär	*Hyphantria cunea*	fall webworm
Amerikanischer Weidenspringrüssler	*Rhynchaenus rufipes*	willow flea weevil
Amerikanischer Zwergskink	*Scincella lateralis*	ground skink
Amerikanischer Zwergwels, Brauner Zwergwels, Langschwänziger Katzenwels	*Ictalurus nebulosus/ Ameiurus nebulosus*	horned pout, American catfish, brown bullhead (FAO), "speckled catfish"
Amerikanisches Salinenkrebschen	*Artemia gracilis*	American brine shrimp
Amerikanisches Tagpfauenauge	*Automeris io*	io moth
Amethyst-Olive	*Oliva annulata*	amethyst olive
Amethystglanzstar	*Cinnyricinclus leucogaster*	violet starling
Amethystkolibri	*Calliphlox amethystina*	amethyst woodstar
Amethystmuschel*	*Gemma gemma*	amethyst gemclam
Amethystpython	*Liasis amethistimus/ Morelia amethistima*	amethystine python
Ammen-Dornfinger, Dornfinger	*Cheiracanthium punctorium*	European sac spider
Ammenhaiartige, Teppichhaie	Orectolobiformes	carpet sharks, carpetsharks
Ammenhaie	Ginglymostomatidae	nurse sharks
Ammoniten	Ammonoidea	ammonites
Amöben, Wechseltierchen, Wurzeltierchen, Rhizopoden	Amoebozoa	amebas, amoebas
Ampfereule	*Acronicta rumicis*	knot grass
Amphibien, Lurche	Amphibia	amphibians
Amphizoiden	Amphizoidae	trout-stream beetles
Ampullenschnecke, Ampullen-Blasenschnecke	*Bulla ampulla*	flask bubble
Amsel	*Turdus merula*	blackbird
Amurhecht	*Esox reicherti*	Amur pike
Amurkatze	*Felis euptilura*	Amur cat
Amurkarpfen, Graskarpfen	*Ctenopharyngodon idella*	grass carp
Amurnatter	*Elaphe schrenckii schrenckii*	Russian rat snake
Amurotter	*Gloydius saxatilis/ Agkistrodon saxatilis*	brown mamushi
Anakondas	*Eunectes* spp.	anacondas
Ananas-Seewalze	*Thelenota ananas*	prickly red 'fish'
Ananaskoralle*	*Dichocoenia stokesi*	pineapple coral
Ananasschildlaus	*Diaspis bromeliae*	pineapple scale
Anchoveta, Peru-Sardelle	*Engraulis ringens*	anchoveta (FAO), Peruvian anchovy

Anchovis, Europäische Sardelle, Sardelle	*Engraulis encrasicholus*	anchovy, European anchovy
Andamanen-Kaisergranat, Andamanen-Schlankhummer	*Metanephrops andamanicus*	southern langoustine, Andaman lobster
Anden-Eidechse	*Phrenacosaurus heterodermus*	Andean lizard
Anden-Sumpfratte	*Neotomys ebriosus*	Andean swamp rat
Andenbartvogel	*Eubucco bourcierii*	red-headed barbet
Andenhirsche, Huemuls	*Hippocamelus* spp.	huemuls, guemals
Andenkatze	*Felis jacobita*	Andean cat (mountain cat)
Andenkolibri	*Oreotrochilus estella*	Andean hillstar
Andenkondor	*Vultur gryphus*	Andean condor
Andenkröte	*Bufo arunco*	Andes toad
Andenmaus	*Andinomys edax*	Andean mouse
Andenratte	*Lenoxus apicalis*	Andean rat
Andensteißhuhn, Pentland-Steißhuhn	*Nothoprocta pentlandii*	Andean tinamou
Anderson-Laubfrosch	*Hyla andersoni*	pine barrens treefrog
Andros-Wirtelschwanzleguan, Fels-Wirtelschwanzleguan	*Cyclura cyclura*	Andros ground iguana, rock iguana
Anegada-Wirtelschwanzleguan	*Cyclura pinguis*	Anegada ground iguana
Anemonen-Einsiedler, Prideaux-Einsiedlerkrebs	*Pagurus prideaux/ Eupagurus prideaux*	Prideaux's hermit crab, smaller hermit crab
Anemonen-Gespenstkrabbe, Mittelmeer-Gespenstkrabbe	*Inachus phalangium*	Leach's spider crab, Mediterranean spider crab
Anemonenfische	Amphiprioninae	clownfishes
Anemonenfische, Clownfische	*Amphiprion* spp.	anemonefishes, clownfishes
Anemonenkoralle	*Goniopora planulata*	anemone coral
Aneuriden	Aneuridae	barkbugs
Anglerfische	Lophiidae	monks, goosefishes
Anglerfische, Armflosser	Lophiiformes	anglerfishes
Anglerkalmar	*Chiroteuthis veranyi*	
Anglerkalmarlarve	*Doratopsis vermicularis*	worm squid*
Angola-Python	*Python anchietae*	Angolan python
Angola-Zwergmeerkatze	*Miopithecus talapoin*	talapoin, southern talapoin
Angolagirlitz	*Serinus atrogularis*	yellow-rumped seedeater
Angolaguereza	*Colobus angolensis*	Angolan colobus, Angola pied colobus
Angolapitta	*Pitta angolensis*	African pitta
Anmutige Turmschnecke*	*Allopeas gracile*	graceful awlsnail
Anmutiger Chiton	*Tonicia chilensis*	Chilean chiton
Annakolibri	*Calypte anna*	Anna hummingbird, Anna's hummingbird
Annam-Schildkröte	*Annamemys annamensis*	Annam leaf turtle, Vietnamese leaf turtle
Annettes Kaurischnecke	*Zonaria annettae*	Annette's cowrie
Ansaugqualle*	*Gonionemus vertens*	clinging jellyfish, angled hydromedusa
Antarktis-Drachenfische	Bathydraconidae	Antarctic dragonfishes
Antarktische Dorsche, Antarktis-Eisfische	Nototheniidae	cod icefishes, Antarctic rockcod, notothenids

Antarktische Königskrabbe	*Lithodes antarctica/ Lithodes santolla*	southern king crab
Antarktischer Bärenkrebs	*Parribacus antarcticus*	sculptured mitten lobster, sculptured slipper lobster
Antarktischer Krill, Südlicher Krill	*Euphausia superba*	whale krill, Antarctic krill
Antarktischer Schwarzfisch	*Hyperoglyphe antarctica*	bluenose warehou, Antarctic butterfish (FAO)
Antarktischer Umberfisch	*Sciaena antarctica*	mulloway
Antarktischer Zahnfisch	*Dissostichus mawsoni*	Antarctic toothfish
Antennen-Rotfeuerfisch, Antennenfeuerfisch	*Pterois antennata*	spotfin lionfish, broadbarred firefish (FAO)
Antennenwelse	Pimelodidae	long-whiskered catfishes, pimelodid catfishes
Antennenwelse	*Ancistrus* spp.	bristle-noses
Antillen-Diademseeigel	*Diadema antillorum*	long-spined sea urchin
Antillen-Landkrabbe	*Gecarcinus ruricola*	purple land crab, mountain crab
Antillen-Nixenschnecke	*Nerita fulgurans*	Antillean nerite
Antillen-Pfeiffrösche, Antillenfrösche	*Eleutherodactylus* spp./ *Trachyphrynus* spp.	robber frogs
Antillen-Schmuckschildkröte, Jamaika-Schmuckschildkröte	*Trachemys terrapen/ Pseudemys terrapen*	Jamaican slider
Antillen-Vogelspinne	*Acanthoscurria antillensis*	Antilles tarantula
Antilopen-Erdhörnchen	*Ammospermophilus* spp.	antelope ground squirrels, antelope squirrels
Antilopenhase	*Lepus alleni*	antelope jack rabbit
Antilopenkänguru	*Macropus antilopinus*	antilopine wallaroo
Apenninenspitzmaus	*Sorex samniticus*	Apennine shrew
Apfel-Ampfer-Blattrolllaus	*Dysaphis radicola*	apple-dock aphid
Apfel-Stachelschnecke	*Chicoreus pomum/ Phyllonotus pomum*	apple murex
Apfelbaumgespinstmotte, Pflaumengespinstmotte, Zwetschgengespinstmotte, Traubenkirschen-Gespinstmotte	*Yponomeuta padellus/ Hyponomeuta padellus*	common hawthorn ermel, small ermine moth, small ermine
Apfelbaumglasflügler	*Synanthedon myopaeformis*	red-belted clearwing
Apfelblattmotte	*Eutromula pariana/ Choreutis pariana*	apple leaf skeletonizer
Apfelblattsauger, Apfelsauger	*Psylla mali*	apple sucker, apple psyllid
Apfelblütenstecher	*Anthonomus pomorum*	apple blossom weevil
Apfelfruchtstecher*	*Rhynchites aequatus*	apple fruit rhynchites weevil
Apfelgallenälchen	*Meloidogyne mali*	apple root-knot nematode
Apfelgespinstmotte, Apfelbaumgespinstmotte	*Hyponomeuta malinellus*	apple moth, Adkin's apple ermel
Apfelgraslaus	*Rhopalosiphum insertum*	apple-grass aphid, oat-apple aphid
Apfelmarkschabe, Apfeltriebmotte	*Blastodacna atra/ Spuleria atra*	pith moth, apple pith moth
Apfelmotte	*Argyresthia conjugella*	apple fruit moth
Apfelrostmilbe	*Aculus schlechtendali*	apple rust mite, apple leaf and bud mite
Apfelsägewespe	*Hoplocampa testudinea*	European apple sawfly

Apfelsauger, Apfelblattsauger	*Psylla mali*	apple sucker, apple psyllid
Apfelschalenwickler, Fruchtschalenwickler	*Capua reticulana/ Adoxophyes orana*	summer fruit tortrix
Apfelschnecke	*Malea pomum*	apple tun, Pacific grinning tun
Apfelschnecke	*Ampullarius scalaris*	scaled bubble snail*
Apfelsinenschwamm, Meerorange	*Tethya aurantia/ Tethya lyncurium*	orange puffbal sponge, golfball sponge
Apfelspringrüssler	*Rhynchaenus pallicornis*	apple flea weevil
Apfeltriebmotte, Apfelmarkschabe	*Blastodacna atra/Spuleria atra*	pith moth, apple pith moth
Apfelwickler (Apfelmade/Obstmade)	*Cydia pomonella/ Laspeyresia pomonella/ Carpocapsa pomonella*	apple moth (apple worm), codling moth, codlin moth
Aploactiniden	Aploactinidae	velvetfishes
Aplomadofalke	*Falco femoralis*	Aplomado falcon
Apollofalter	*Parnassius apollo*	apollo
Apothekerskink	*Scincus scincus*	sandfish
Appalachen-Achatschnecke	*Cionella morseana/ Cochlicopa morseana*	Appalachian pillar snail (U.S.)
Appendicularien, Geschwänzte Schwimm-Manteltiere	Appendicularia/ Larvacea	appendicularians
Aprikoseneule	*Acronicta tridens*	dark dagger
Aprileule, Grüne Eicheneule, Lindeneule	*Griposia aprilina/ Dichonia aprilina*	Merveille-du-Jour
Apulische Tarantel	*Lycosa tarentula*	Apulian tarantula
Aquarienmeduse	*Cladonema radiatum*	aquarium medusa*
Äquatorial-Saki	*Pithecia aequatorialis*	equatorial saki
Arabische Erdotter	*Atractaspis microlepidota*	northern mole viper, Arabian mole viper
Arabische Kauri, Araberkauri	*Cypraea arabica/ Mauritia arabica arabica*	Arabian cowrie
Arabische Sandboa	*Eryx jayakari*	Arabian sand boa
Arabische Sandrasselotter	*Echis coloratus*	Arabic saw-scaled viper, Palestine saw-scaled viper
Arabische Seezunge	*Solea elongata*	elongate sole, Arabic sole
Arabische Spindelschnecke	*Tibia insulae-chorab*	Arabian tibia
Arabischer Kaiserfisch, Sichelkaiserfisch	*Pomacanthus maculosus*	yellow-blotch angelfish, Red Sea angelfish, blue moon angelfish, bride of the sea, yellowbar angelfish (FAO)
Arabischer Picassofisch	*Rhinecanthus assasi*	Arabian picassofish
Arabischer Spießbock, Weißer Oryx	*Oryx leucoryx*	Arabian oryx
Arabischer Tahr	*Hemitragus jayakari*	Arabian tahr
Arafura-Warzenschlange	*Acrochordus arafurae*	Arafura file snake
Arakakadu	*Probosciger aterrimus*	palm cockatoo
Arakanga	*Ara macao*	scarlet macaw
Arapaima, Piracurú	*Arapaima gigas*	pirarucu
Ararauna	*Ara ararauna*	brown & yellow macaw, blue-and-gold macaw
Arasittich, Kiefernsittich	*Rhynchopsitta pachyrhyncha*	thick-billed parrot

Deutsch	Wissenschaftlich	Englisch
Arawana, Knochenzüngler	*Osteoglossum bicirrhosum*	arawana
Arche Noah, Archenmuschel	*Arca noae*	Noah's ark (Noah's ark shell)
Archenkammmuschel, Gemeine Samtmuschel, Mandelmuschel, Meermandel, Englisches Pastetchen	*Glycymeris glycymeris*	dog cockle, orbicular ark (comb-shell)
Archenmuschel, Arche Noah	*Arca noae*	Noah's ark (Noah's ark shell)
Archenmuscheln	Arcidae	arks (ark shells)
Archipel-Halbschnabelhecht	*Hemiramphus archipelagicus*	island halfbeak
Areolen-Flachschildkröte, Papageischnabel-Flachschildkröte	*Homopus areolatus*	parrot-beaked tortoise, beaked Cape tortoise
Arfaklori	*Oreopsittacus arfaki*	whiskered lorikeet
Argentina-Schattenfisch	*Umbrina canosai*	Argentine croaker
Argentinische Ameise	*Iridomyrmex humilis*	Argentine ant
Argentinische Busch-Vogelspinne	*Grammostola burzaquensis/ Grammostola argentinensis*	Argentinean rose tarantula
Argentinische Landschildkröte	*Testudo chilensis/ Geochelone chilensis/ Chelonoidis chilensis*	Argentine tortoise (IUCN), Chilean tortoise, Chaco tortoise
Argentinische Rotgarnele	*Pleoticus muelleri*	Argentine red shrimp
Argentinische Sardelle	*Engraulis anchoita*	anchoita, Argentine anchovy
Argentinische Schlangenhalsschildkröte	*Hydromedusa tectifera*	South American snake-necked turtle, South American snake-headed turtle
Argentinische Stilett-Garnele	*Artemesia longinaris*	Argentine stiletto shrimp
Argentinischer Graufuchs	*Pseudalopex griseus*	Argentine gray fox, gray zorro
Argentinischer Kurzflossenkalmar	*Illex argentinus*	Argentine shortfin squid
Argentinischer Meeraal	*Conger orbignyanus*	Argentine conger
Argentinischer Zackenbarsch	*Acanthistius brasilianus*	Argentine seabass
Argusaugenkauri, Augenfleck-Kaurischnecke	*Cypraea argus*	eyed cowrie, hundred-eyed cowrie
Argusfasan	*Argusianus argus*	great argus pheasant
Argusfisch, Gemeiner Argusfisch, Grüner Argusfisch	*Scatophagus argus argus*	scat, argusfish, argus fish, spotted cat (FAO)
Argusfische	Scatophagidae	scats, scatties
Arielfregattvogel	*Fregata ariel*	lesser frigate bird
Arion-Bläuling, Schwarzfleckiger Bläuling, Quendel-Ameisenbläuling	*Maculinea arion*	large blue (butterfly)
Arizona-Alligatorschleiche	*Gerrhonotus kingi/ Elgaria kingi*	Arizona alligator lizard
Arizona-Fransenzehenleguan	*Uma notata*	fringe-toed lizard
Arizona-Korallenschlange	*Micruroides euryxanthus euryxanthus*	Sonoran coral snake, Arizona coral snake
Arizonanatter	*Arizona elegans*	glossy snake
Arkansastyrann	*Tyrannus verticalis*	western kingbird, Arkansas kingbird
Arktische Astarte	*Astarte arctica*	Arctic astarte
Arktische Brutanemone*	*Epiactis arctica*	Arctic brooding anemone
Arktische Garnele*	*Argis dentata*	Arctic argid

Arktische Kaurischnecke, Gerippte Kaurischnecke, Nördliche Kaurischnecke	*Trivia arctica/ Trivia europaea/ Trivia monacha*	northern cowrie, spotted cowrie, European cowrie, bean cowrie, Arctic cowrie
Arktische Klaffmuschel	*Panomya norvegica/ Panomya arctica*	Arctic roughmya
Arktische Löcher-Käferschnecke	*Amicula vestita*	concealed Arctic chiton
Arktische Maräne, Arktischer Cisco	*Coregonus autumnalis*	Arctic cisco
Arktische Mondschnecke, Arktische Nabelschnecke	*Cryptonatica affinis/ Natica clausa*	Arctic moonsnail
Arktische Panzergroppe	*Icelus bicornis*	two-horned sculpin
Arktische Papierblase	*Diaphana minuta*	Arctic paperbubble
Arktische Seespinne	*Hyas coarctatus*	Arctic lyre crab, lesser toad crab
Arktische Smaragdlibelle	*Somatochlora artica*	northern emerald
Arktische Thracia	*Thracia myopsis*	Arctic thracia
Arktische Windelschnecke	*Vertigo modesta*	Arctic whorl snail, cross vertigo (U.S.)
Arktischer Cisco, Arktische Maräne	*Coregonus autumnalis*	Arctic cisco
Arktischer Dorsch, Arktisdorsch (Polardorsch, Grönland-Dorsch)	*Arctogadus glacialis*	Arctic cod (FAO/U.K.), polar cod (U.S./Canada)
Arktischer Felsenbohrer	*Hiatella arctica*	Arctic hiatella, red-nose clam, red nose, wrinkled rock borer, Arctic rock borer
Arktischer Hirschgeweihgroppe	*Gymnacanthus tricuspis*	Arctic staghorn sculpin
Arktischer Medusenstern	*Gorgonocephalus arcticus*	northern basket star
Arktischer Rochen	*Raja hyperborea/ Amblyraja hyperborea*	Arctic skate
Arktischer Seeskorpion	*Myoxocephalus scorpioides*	Arctic sculpin (FAO), northern sculpin
Arktischer Tiefenkrake	*Bathypolypus arcticus*	spoonarm octopus, offshore octopus, Arctic deepsea octopus
Arktischer Wunderschirm	*Cirroteuthis muelleri*	cirrate octopus
Armadillo-Schnapper, Gelber Schnapper	*Lutjanus argentiventris*	armadillo snapper, yellow snapper (FAO)
Armbrustsalmler, Punktierter Kropfsalmler	*Triportheus angulatus*	giant hatchetfish
Armenische Bergotter, Radde-Viper	*Vipera raddei*	Radde's rock viper, Armenian viper, Russian viper
Armflosser, Anglerfische	Lophiiformes	anglerfishes
Armloser Delphinaal*	*Dalophis imberbis*	armless snake eel
Armmolche	Sirenidae	sirens
Armwirbler, Süßwasserbryozoen	Phylactolaemata/Lophopoda	phylactolaemates, "covered throat" bryozoans, freshwater bryozoans
Arnolds Rotaugensalmler, Afrikanischer Großschuppensalmler	*Arnoldichthys spilopterus*	red-eyed characin
Arnotschmätzer	*Myrmecocichla arnoti*	white-headed black-chat

Arrauschildkröte, Arrau-Schienenschildkröte	*Podocnemis expansa*	Arrau River turtle, giant South American river turtle
Artemismuschel, Gemeine Artmuschel	*Dosinia exoleta*	rayed dosinia, rayed artemis
Arthritische Spinnenschnecke	*Lambis arthritica*	arthritic spider conch
Arthropoden, Gliederfüßer	Arthropoda	arthropods
Artischocken-Koralle	*Scolymia cubensis*	artichoke coral
Aruba-Klapperschlange	*Crotalus unicolor*	aruba rattlesnake
Äsche, Europäische Äsche	*Thymallus thymallus*	grayling
Äschen	Thymallidae	graylings
Aschelminthen, Nemathelminthen, Schlauchwürmer, Rundwürmer *sensu lato* (Pseudocölomaten)	Aschelminthes/ Nemathelminthes	aschelminths, nemathelminths, pseudocoelomates
Aschgraue Käferschnecke, Rändel-Käferschnecke	*Lepidochitona cinerea*	cinereous chiton
Aschgraue Spitzmaus	*Sorex cinereus*	masked shrew
Aschgrauer Blasenkäfer*	*Epicauta fabricii*	ashgray blister beetle
Aschgrauer Kugelfingergecko	*Sphaerodactylus cinereus*	ashy gecko, gray gecko
Asfur-Kaiserfisch	*Pomacanthus asfur/ Euxiphipops asfur/ Arusetta asfur*	Arabian angelfish
Asiatische Büffel, Asiatische Wasserbüffel	*Bubalus* spp.	Asian water buffaloes, anoas
Asiatische Feuerwalzenschnecke	*Fulgoraria rupestris/ Fulgoraria fulminata*	Asian flame volute
Asiatische Glasbarsche	Chandidae/Ambassidae	Asiatic glassfishes
Asiatische Goldkatze, Temminckkatze	*Felis temmincki/ Profelis temmincki*	Asian golden cat
Asiatische Höhlenflughunde	*Eonycteris* spp.	dawn bats
Asiatische Keiljungfer	*Gomphus flavipes*	Asian gomphus
Asiatische Kleinsardine	*Sardinella fimbriata*	fringe-scale sardinella
Asiatische Klettermäuse	*Hapalomys* spp.	Asiatic climbing rats, marmoset rats
Asiatische Körbchenmuschel	*Corbicula fluminea*	Asian clam, Asian basket clam, Asian corbicula
Asiatische Kurzschwanzspitzmaus	*Blarinella quadraticauda*	Asiatic short-tailed shrew
Asiatische Langschwanz-Riesenratten	*Leopoldamys* spp.	long-tailed giant rats
Asiatische Lanzenottern, Bambusottern	*Trimeresurus* spp.	Asian lance-headed vipers, Asian pit vipers
Asiatische Rattenschlangen	*Ptyas* spp.	Oriental rat snakes
Asiatische Riesen-Weichschildkröte	*Pelochelys bibroni*	Asian giant softshell turtle
Asiatische Riesensalamander	*Andrias* spp.	Asiatic giant salamanders
Asiatische Riesenschildkröte, Braune Landschildkröte	*Manouria emys*	Asian giant tortoise (IUCN), Asian brown tortoise, Burmese brown tortoise
Asiatische Schabe	*Blatella asahinai*	Asian cockroach
Asiatische Schneckennattern	*Pareas* spp.	slug snakes

Asiatische Waldskorpione	*Heterometrus* spp.	Asian forest scorpions
Asiatische Wasserspitzmäuse	*Chimarrogale* spp.	Asiatic water shrews
Asiatische Weichschildkröte, Knorpel-Weichschildkröte	*Amyda cartilaginea/ Trionyx cartilagineus*	Asiatic softshell turtle, black-rayed softshell turtle
Asiatische Zwerg-Streifenhörnchen	*Tamiops* spp.	Asiatic striped squirrels
Asiatischer Büffel, Wasserbüffel	*Bubalus bubalis/ Bubalus arnee*	Asian water buffalo, wild water buffalo, carabao
Asiatischer Elefant, Indischer Elefant	*Elephas maximus*	Asiatic elephant, Indian elephant
Asiatischer Gabelbart	*Scleropages formosus*	Asian arowana, Asian bonytongue (FAO)
Asiatischer Gartenkäfer*	*Maladera castanea*	Asiatic garden beetle
Asiatischer Gelber Umberfisch	*Collichthys crocea*	large yellow croaker
Asiatischer Halbesel, Kulan, Khur, Onager	*Equus hemionus*	kulan, khur, onager
Asiatischer Leberegel*	*Clonorchis sinensis*	Asian liver fluke
Asiatischer Ochsenfrosch	*Rana tigrina/Rana tigerina*	tiger frog, Indian bullfrog
Asiatischer Ringelwels	*Leiocassis siamensis*	Asiatic bumblebee catfish
Asiatischer Schwarzfisch	*Psenopsis anomala*	Pacific rudderfish
Asiatischer Stint	*Osmerus mordax dentex*	Arctic rainbow smelt (FAO), Asiatic smelt, Arctic smelt boreal smelt
Asiatischer Tapir	*Tapirus indicus*	Asiatic tapir, Malayan tapir
Äskulapnatter	*Elaphe longissima*	Aesculapian snake
Aspenblattkäfer	*Chrysomela tremulae/ Melasoma tremulae*	aspen leaf beetle
Aspisviper, Juraviper	*Vipera aspis*	asp viper, aspic viper
Assapan> Nord-Assapan	*Glaucomys sabrinus*	northern flying squirrel
Assapan> Süd-Assapan	*Glaucomys volans*	southern flying squirrel
Assel-Käferschnecke	*Lepidopleurus asellus*	coat-of-mail chiton, pill chiton*
Asseln	Isopoda	isopods (incl. pill bugs, woodlice, sowbugs)
Asselspinne*	*Dysdera crocota*	woodlouse spider
Asselspinnen	Pycnogonida/Pantopoda	pycnogonids, pantopods, sea spiders
Asselspinner, Mottenspinner (Schildmotten)	Limacodidae (Cochlidiidae)	slug caterpillars & saddleback caterpillars
Astartiden	Astartidae	astartes
Asternmönch, Aster-Goldrutenheiden-Braunmönch	*Cucullia asteris*	starwort moth
Astkegel*	*Conus thalassiarchus*	bough cone
Astloch-Gabelmücke*	*Anopheles plumbens*	tree-hole mosquito
Ästuar-Stielaugengarnele*	*Ogyrides alphaerostris*	estuarine long-eyed shrimp, estuarine longeye shrimp

Äthiopische Schmalkopfratten*	*Stenocephalemys* spp.	Ethiopian narrow-headed rats, Ethiopian meadow rats
Äthiopischer Steinbock, Waliasteinbock	*Capra walie*	Walia ibex
Äthiopischer Wolf, Abessinischer Fuchs	*Canis simensis*	Simien jackal, Ethiopian wolf, Simien fox
Atka-Makrelen, Terpuge	*Pleurogrammus* spp.	Atka mackerels
Atlantik-Bastardschildkröte, Atlantische Bastardschildkröte, Kemps Bastardschildkröte	*Lepidochelys kempii*	Kemp's ridley (sea turtle), Atlantic ridley turtle
Atlantik-Brandungskrebs	*Emerita talpoida*	Atlantic sand crab, mole crab
Atlantik-Feigenschnecke, Gemeine Feigenschnecke	*Ficus communis*	Atlantic figsnail (common fig shell)
Atlantik-Laternenfisch	*Kryptophaneron alfredi*	Atlantic flashlightfish
Atlantik-Schirmschnecke	*Umbraculum umbraculum*	Atlantic umbrella snail, Atlantic umbrella "slug"
Atlantik-Stachelauster, Atlantische Stachelauster, Amerikanische Stachelauster	*Spondylus americanus*	Atlantic thorny oyster, American thorny oyster
Atlantische Anchoveta	*Cetengraulis edentulus*	Atlantic anchoveta
Atlantische Becherkoralle	*Caryophyllia calix*	
Atlantische Bergschrecke	*Antaxius pedestris*	Pyrenean bushcricket
Atlantische Bohrmuschel	*Petricola lapicida*	boring petricola
Atlantische Dickstielige Doldenkoralle	*Mussa angulosa*	spiny flower coral
Atlantische Distorsio	*Distorsio clathrata*	Atlantic distorsio
Atlantische Erdbeer-Herzmuschel*	*Americardia media*	Atlantic strawberry-cockle
Atlantische Falsche Dreiecksmuschel*	*Heterodonax bimaculatus*	false-bean clam
Atlantische Feuerwalze	*Pyrosoma atlanticum*	Atlantic pyrosome
Atlantische Flankenkiemenschnecke*	*Pleurobranchus areolatus*	Atlantic sidegill "slug"
Atlantische Flügelmuschel	*Pteria colymbus*	Atlantic wing-oyster
Atlantische Gelbkauri, Gelbe Kaurischnecke	*Erosaria acicularis*	Atlantic yellow cowrie
Atlantische Glasperlenanemone	*Capnea lucida/Heteractis lucida*	Atlantic beaded anemone
Atlantische Haar-Triton	*Cymatium pileare*	Atlantic hairy triton, common hairy triton
Atlantische Herzmuschel, Gemeine Herzmuschel	*Laevicardium laevigatum*	eggcockle
Atlantische Hirschkauri	*Macrocypraea cervus/ Cypraea cervus*	Atlantic deer cowrie
Atlantische Kammauster*	*Ostrea equestris*	crested oyster
Atlantische Korallenschlange	*Micrurus diastema*	Atlantic coral snake
Atlantische Lachse, Forellen	*Salmo* spp.	Atlantic trouts
Atlantische Lastträgerschnecke	*Xenophora conchyliophora*	Atlantic carriersnail
Atlantische Maß-Schnecke	*Modulus modulus*	buttonsnail, Atlantic modulus
Atlantische Maulbeerschnecke	*Morum oniscus*	Atlantic morum

Atlantische Messermuschel	*Siliqua costata*	Atlantic razor clam
Atlantische Nussmuschel	*Nucula proxima*	Atlantic nutclam
Atlantische Papiermuschel	*Amygdalum papyrium*	Atlantic papermussel
Atlantische Perlmuschel	*Pinctada imbricata*	Atlantic pearl-oyster
Atlantische Pilzkoralle*	*Scolymia lacera*	Atlantic mushroom coral
Atlantische Reiterkrabbe	*Ocypode quadrata*	Atlantic ghost crab
Atlantische Riesentrogmuschel	*Spisula solidissima/ Hemimactra gigantea*	Atlantic surfclam, solid surfclam, bar clam
Atlantische Rossie	*Rossia tenera/ Semirossia tenera*	lesser bobtail squid, Atlantic bob-tailed squid, lesser shining bobtail squid (FAO)
Atlantische Rote Riesengarnele, Rote Riesengarnele	*Plesiopenaeus edwardsianus*	scarlet gamba prawn, scarlet shrimp
Atlantische Säbelzahn-Sardelle	*Lycengraulis grossidens*	river anchoita
Atlantische Sand-Winkerkrabbe	*Uca pugilator*	Atlantic sand fiddler
Atlantische Sandmuschel	*Sanguinolaria sanguinolenta*	Atlantic sanguin
Atlantische Schmuckkrabbe	*Stenocionops furcatus*	furcate spider crab, Atlantic decorator crab
Atlantische Schotenmuschel	*Solemya velum*	Atlantic awningclam
Atlantische Schwarze Koralle	*Antipathes atlantica*	Atlantic black coral
Atlantische Schwertmuschel	*Ensis directus*	Atlantic jackknife clam
Atlantische Seespinne, Nordische Seespinne	*Hyas araneus*	Atlantic lyre crab, great spider crab, toad crab
Atlantische Seestachelbeere	*Pleurobrachia pileus/ Pleurobrachia rhodopis*	Atlantic sea gooseberry
Atlantische Sepiole	*Sepiola atlantica*	Atlantic cuttlefish, little cuttlefish, Atlantic bobtail squid (FAO)
Atlantische Spinnenkrabbe	*Stenorhynchus lanceolatus*	Atlantic arrow crab
Atlantische Tauben-Fassschnecke	*Tonna pennata/ Tonna maculosa*	Atlantic partridge tun
Atlantische Tiefsee-Kammmuschel, Atlantischer Tiefseescallop	*Placopecten magellanicus*	Atlantic deep-sea scallop, sea scallop
Atlantische Weiße Garnele, Nördliche Weiße Geißelgarnele	*Penaeus setiferus/ Litopenaeus setiferus*	white shrimp, lake shrimp, northern white shrimp
Atlantische Zwerg-Nadelschnecke	*Cerithium lutosum*	dwarf Atlantic cerith
Atlantischer Ammenhai, Karibischer Ammenhai	*Ginglymostoma cirratum*	nurse shark
Atlantischer Angler, Atlantischer Seeteufel	*Lophius piscatorius*	Atlantic angler fish, angler (FAO), monkfish
Atlantischer Barrakuda, Großer Barrakuda	*Sphyraena barracuda*	great barracuda
Atlantischer Bohrseeigel	*Echinometra lucunter*	Atlantic boring sea urchin*
Atlantischer Bootsmannsfisch	*Porichthys plectrodon*	Atlantic midshipman
Atlantischer Braunhai, Sandbankhai, Großrückenflossenhai	*Carcharhinus plumbeus/ Carcharhinus milberti*	sandbar shark

Atlantischer Buckeldelphin, Kamerunfluss-Delphin, Kamerun-Buckeldelphin	*Sousa teuszi*	humpback dolphin, Atlantic hump-backed dolphin, West African hump-backed dolphin
Atlantischer Drückerfisch, Nördlicher Drückerfisch, Gabelschwanz Drückerfisch, Blauer Drücker, Rotzahn, Rotzahn-Drückerfisch	*Odonus niger/ Xenodon niger/ Balistes erythrodon/ Odonus erythrodon*	redtooth triggerfish (FAO), red-toothed triggerfish, redfang
Atlantischer Eidechsenfisch	*Synodus saurus*	Atlantic lizardfish
Atlantischer Fächerfisch, Atlantischer Segelfisch	*Istiophorus albicans*	Atlantic sailfish
Atlantischer Fadenflosser	*Polydactylus virginicus*	barbu
Atlantischer Fadenhering	*Opisthonema oglinum*	Atlantic thread herring
Atlantischer Fleckendelphin	*Stenella plagiodon*	Atlantic spotted dolphin, Gulf Stream spotted dolphin
Atlantischer Fledermausfisch	*Dibranchus atlanticus/ Halieutaea senticosa*	Atlantic batfish
Atlantischer Flugfisch, Atlantischer Fliegender Fisch	*Cypselurus melanurus/ Cheilopogon melanurus*	Atlantic flyingfish
Atlantischer Gymnast*	*Xenodermichthys socialis*	Atlantic gymnast
Atlantischer Hornhecht, Tropischer Hornhecht	*Strongylura marina*	Atlantic needlefish (FAO), silver gar
Atlantischer Kofferfisch	*Lagocephalus lagocephalus*	Atlantic pufferfish, pufferfish, oceanic puffer (FAO)
Atlantischer Krallenhummer*	*Nephropsis atlantica*	scarlet clawed lobster
Atlantischer Lachs, Salm	*Salmo salar*	Atlantic salmon
Atlantischer Nachthai	*Carcharhinus signatus*	night shark
Atlantischer Palolo	*Eunice fucata*	Atlantic palolo worm
Atlantischer Pferdekopf, Atlantischer „Mondfisch"	*Selene setapinnis/ Vomer setapinnis*	Atlantic moonfish
Atlantischer Riesenkalmar	*Architeuthis dux*	Atlantic giant squid
Atlantischer Riffhummer	*Enoplometopus antillensis*	flaming reef lobster
Atlantischer Röhrenwurm	*Hydroides uncinata*	Atlantic tubeworm
Atlantischer Rundhering	*Etrumeus sadina*	Atlantic round herring
Atlantischer Sägebauch, Granatbarsch	*Hoplostethus atlanticus*	orange roughie
Atlantischer Seeteufel, Atlantischer Angler	*Lophius piscatorius*	Atlantic angler fish, angler (FAO), monkfish
Atlantischer Segelfisch, Atlantischer Fächerfisch	*Istiophorus albicans*	Atlantic sailfish
Atlantischer Spitznasen-Grundhai, Nordwest-Atlantischer Spitznasen-Grundhai, Atlantischer Spitzmaulhai	*Rhizoprionodon terraenovae*	Atlantic sharpnose shark
Atlantischer Stechrochen	*Dasyatis sabina*	Atlantic stingray
Atlantischer Stör	*Acipenser oxyrhynchus*	Atlantic sturgeon
Atlantischer Tarpon, Atlantischer Tarpun	*Tarpon atlanticus/ Megalops atlanticus*	Atlantic tarpon
Atlantischer Taschenkrebs	*Cancer irroratus*	Atlantic rock crab
Atlantischer Tiefseeangler, Grönlandangler	*Himantolophus groenlandicus*	Atlantic football fish
Atlantischer Tiefseehummer	*Acanthacaris caeca*	Atlantic deepsea lobster

Atlantischer Umberfisch	*Micropogonias undulatus*	Atlantic croaker
Atlantischer Violetter Seeigel	*Arbacia punctulata*	Atlantic purple urchin
Atlantischer Wieselhai	*Paragaleus pectoralis*	Atlantic weasel shark
Atlantisches Tritonshorn	*Charonia variegata*	Atlantic triton
Atlasagame	*Agama bibroni*	Bibron's agama
Atlasgrasmücke	*Sylvia deserticola*	Tristram's warbler
Atlasotter	*Vipera mauritanica*	Sahara rock viper, Atlas Mountain viper
Atlasspinner	*Attacus atlas*	atlas moth
Atlasspinner, Atlas, Pappelspinner, Weidenspinner	*Leucoma salicis*	satin moth
Atomschnecke	*Omalogyra atomus*	atom snail
Atroposviper, Bergpuffotter, Südafrikanische Bergotter	*Bitis atropos*	berg adder
Atun	*Thyrsites atun*	snoek (FAO), sea pike, barracouta
Auckland-Seelöwe	*Phocarctos hookeri*	Hooker's sea lion, New Zealand sea lion, Auckland sea lion
Aucklandente	*Anas aucklandica*	New Zealand black duck
Audubon-Kaninchen	*Sylvilagus audubonii*	desert rabbit
Auerhuhn	*Tetrao urogallus*	capercaillie
Auerochse, Ur (Hausrind)	*Bos taurus/* *Bos primigenius*	aurochs (domestic cattle)
Aufgeblasener Flossenfüßer*	*Limacina inflata*	planorbid pteropod
Augen-Eulenspinner, Achtzigeule, Pappelhain-Eulenspinner	*Tethea ocularis*	figure of eighty
Augen-Napfschnecke*	*Patella oculus*	eye limpet, South African eye limpet
Augen-Seezunge, Augenfleck-Seezunge	*Microchirus ocellatus/* *Solea ocellata*	four-eyed sole, foureyed sole (FAO)
Augenbläuling, Großer Moorbläuling	*Maculinea telejus*	scarce large blue (butterfly)
Augenbrauenkrake, Stachelhornkrake*	*Tetracheledone spinicirrus*	spiny-horn octopus, eyebrow octopus
Augenbrauenmahali	*Plocepasser mahali*	white-browed sparrow weaver
Augenfadenwurm, Augenwurm	*Thelazia lacrymalis*	eyeworm of horses
Augenfalter	Satyridae	browns, satyrs
Augenflagellaten	Euglenophyta	euglenoids, euglenids
Augenfleck-Anglerfisch, Karibischer Augenflecken-Angler	*Antennarius multiocellatus*	longlure frogfish
Augenfleck-Buntbarsch	*Cichlasoma severum*	banded cichlid (FAO), convict fish, deacon, severum, striped cichlid
Augenfleck-Dornschwanz	*Uromastyx ocellatus*	eyed dabb lizard, ocellated mastigure

Augenfleck-Eisfisch	*Chionodraco rastrospinosus*	Kathleen's icefish, ocellated icefish (FAO)
Augenfleck-Junker	*Haliochoeres hortulanus*	chequerboard wrasse
Augenfleck-Kammbarsch	*Cichla ocellaris*	peacock bass
Augenfleck-Kauri	*Cypraea ocellata*	ocellated cowrie, ocellate cowrie
Augenfleck-Kugelfingergecko	*Sphaerodactylus argus*	ocellated gecko
Augenfleck-Lippfisch, Augenflecklippfisch, Augenlippfisch	*Crenilabrus ocellatus/ Symphodus ocellatus*	spotted wrasse, ocellated wrasse
Augenfleck-Seewalze	*Bohadschia argus*	eyespot holothurian
Augenfleck-Seezunge, Augen-Seezunge	*Microchirus ocellatus/ Solea ocellata*	four-eyed sole, foureyed sole (FAO)
Augenfleck-Täubchen	*Pyrene ocellata*	lightning dovesnail
Augenfleck-Teju	*Cercosaura ocellata*	ocellated tegu
Augenfleck-Umberfisch	*Sciaenops ocellatus*	red drum
Augenfleck-Zitterrochen, Gefleckter Zitterrochen	*Torpedo torpedo*	torpedo ray, common torpedo (FAO), eyed electric ray
Augenfleckbärbling	*Rasbora dorsiocellata dorsiocellata*	eye-spot rasbora
Augenfleckkärpfling	*Poecilia vivipara*	one-spot livebearer
Augenfleckrochen, Braunrochen, Vieraugenrochen	*Raja miraletus*	brown ray
Augenfliege, Gesichtsfliege	*Musca autumnalis*	face fly
Augenfliegen, Kugelkopffliegen	Pipunculidae	big-headed flies
Augenkoralle	*Lophelia pertusa*	eye coral*
Augenkoralle, Weiße Koralle	*Madrepora oculata*	ocular coral, white coral
Augenkorallen	*Oculina* spp.	eyed corals, ivory bush corals
Augenkröte	*Physalaemus nattereri*	false-eyed frog
Augenmarienkäfer	*Anatis ocellata*	eyed ladybird, pine ladybird beetle
Augenpfeifer	*Leptodactylus ocellatus*	Criolla frog
Augenschuppenlippfisch, Sattelfleck-Lippfisch	*Acantholabrus palloni*	scale-rayed wrasse
Augenspinner (Nachtpfauenaugen/ Pfauenspinner)	Saturniidae	giant silkmoths, silkworm moths, emperor moths
Augenstreifen-Notothenia	*Notothenia kempi*	striped-eyed rockcod
Augenwurm, Augenfadenwurm	*Thelazia lacrymalis*	eyeworm of horses
Augenzipfel-Stummelschwanzchamäleon	*Brookesia superciliaris*	brown leaf chameleon
Augur-Kegel	*Conus augur*	Augur cone
Augustmücke, Rheinmücke	*Oligoneuriella rhenana*	Rhine-river mayfly
Aurorafalter	*Anthocharis cardamines*	orange-tip
Ausrufungszeichen, Ausrufezeichen, Gemeine Graseule, Braungraue Graseule	*Agrotis exclamationis*	heart and dart moth

Ausschnittsschnecken

Ausschnittsschnecken u.a., Schlitzschnecken u.a.	*Emarginula* spp.	slit limpets a.o.
Ausstecherhai, Keksausstecherhai, Keksstecherhai	*Isistius brasiliensis*	cookiecutter
Auster> Europäische Auster, Gemeine Auster	*Ostrea edulis*	common oyster, flat oyster, European flat oyster
Austern	Ostreidae	oysters
Austernfisch, Karibischer Krötenfisch	*Opsanus tau*	oyster toadfish
Austernfischer	*Haematopus ostralegus*	oystercatcher
Austernförmige Schildlaus, Zitronenfarbene Austernschildlaus	*Quadraspidiotus ostreaeformis*	European fruit scale
Austernschildläuse, Deckelschildläuse, Echte Schildläuse	Diaspididae	armored scales
Austral-Languste	*Jasus novaehollandiae*	Australian rock lobster
Austral-Seepocke	*Elminius modestus*	modest barnacle
Australasische Baumfrösche	*Litoria* spp.	Australasian treefrogs
Australien-Krokodil	*Crocodylus johnstoni*	Australian freshwater crocodile
Australische Ameise	*Nothomyrmecia macrops*	Australian ant, dinosaur ant
Australische Auster, Sydney Felsenauster	*Saccostrea cuccullata*	Sydney cupped oyster
Australische Bergagame	*Amphibolurus diemensis*	mountain dragon
Australische Bodenagamen	*Amphibolurus* spp.	bearded dragons, Australian dragon lizards
Australische Bola-Spinne	*Celaenia distincta*	Australian bolas spider
Australische Braunschlangen	*Demansia* spp.	Australian whip snakes, Australian brown snakes, venomous whip snakes
Australische Breitkopf-Giftnatter	*Hoplocephalus* spp.	Australian broad-headed snakes
Australische Fasanenschnecke	*Phasianella australis*	painted lady
Australische Fleckenmakrele	*Scomberomorus munroi*	Australian spotted mackerel
Australische Gelbfleck-Giftnatter	*Hoplocephalus bungaroides*	Australian yellow-spotted snake, broad-headed snake (IUCN)
Australische Gelbschwanzmakrele, Riesen-Gelbschwanzmakrele	*Seriola lalandei*	giant yellowtail, yellowtail kingfish, yellowtail amberjack
Australische Gespenstfledermaus	*Macroderma gigas*	Australian giant false vampire bat, ghost bat
Australische Großohrfledermaus	*Nyctophilus* spp.	New Guinean big-eared bats & Australian big-eared bats
Australische Häschenratten	*Leporillus* spp.	Australian stick-nest rats
Australische Hüpfmäuse, Australische Kängurumäuse	*Notomys* spp.	Australian hopping mice, jerboa mice
Australische Jakobsmuschel	*Pecten meridionalis*	Australian scallop
Australische Kängurumäuse, Australische Hüpfmäuse	*Notomys* spp.	Australian hopping mice, jerboa mice
Australische Kleinmäuse	*Leggadina* spp.	Australian dwarf mice
Australische Korallenottern	*Simoselaps* spp.	Australian coral snakes

Australische Kupferkopfnatter	*Austrelaps superba*	Australian copperhead
Australische Lähmezecke*	*Ixodes holocyclus*	Australian paralysis tick
Australische Languste	*Panulirus cygnus*	Australian spiny lobster
Australische Miesmuschel	*Mytilus planulatus*	Australian mussel
Australische Nackenband-Giftnatter	*Glyphodon* spp.	Australian collared snakes
Australische Olive	*Oliva australis*	Australian olive
Australische Plattschildkröte, Australische Suppenschildkröte	*Chelonia depressa/ Natator depressus*	flatback sea turtle, flatback turtle, Australian green turtle
Australische Riesenkrabbe	*Pseudocarcinus gigas*	Tasmanian giant crab
Australische Rifftrompete	*Syrinx aruanus*	false trumpet snail, false trumpet, Australian trumpet
Australische Sardelle	*Engraulis australis*	Australian anchovy
Australische Sardine	*Sardinops neopilchardus*	Australian pilchard (FAO), picton herring
Australische Schabe, Südliche Großschabe	*Periplaneta australasiae*	Australian cockroach
Australische Scheinkobras	*Pseudonaja* spp.	venomous brown snakes
Australische Scheinkröten	*Pseudophryne* spp.	crowned toadlets, Australian toads
Australische Schlangenhalsschildkröten	*Chelodina* spp.	Australian snake-necked turtles
Australische Schmetterlings-Fledermaus	*Chalinolobus* spp.	lobe-lipped bats, groove-lipped bats, wattled bats
Australische Spitzkopfschildkröte	*Emydura australis*	Australian big-headed side-necked turtle
Australische Südfrösche	Myobatrachidae	myobatrachid frogs
Australische Trichternetz-Vogelspinne	*Atrax robustus*	Australian funnelweb spider, Sydney funnelweb spider
Australische Wollschildlaus, Orangenschildlaus	*Icerya purchasi*	cottony cushion scale, fluted scale
Australische Würfelqualle, Seewespe u.a.	*Chironex fleckeri*	Australian box jellyfish, box jelly, deadly sea wasp
Australische Zirpfrösche	*Crinia* spp.	Australian froglets
Australische Zitterrochen	Hypnidae	coffin rays
Australischer Aal	*Anguilla australis*	Australian eel, shortfin eel (FAO)
Australischer Apfel-Dickmaulrüssler	*Otiorhynchus cribricollis*	cribrate weevil, apple weevil
Australischer Blutsauger (Bodenagame)	*Amphibolurus muricatus*	Australian "bloodsucker", jacky lizard (jacky)
Australischer Bonito	*Sarda australis*	Australian bonito
Australischer Diebkäfer	*Ptinus tectus*	Australian spider beetle
Australischer Flusskrebs	*Euastacus serratus*	Australian crayfish
Australischer Glatthai	*Mustelus antarcticus*	gummy shark
Australischer Großaugenbarsch, Bullaugenbarsch	*Priacanthus macracanthus*	bull's eye perch, red bulleye, red bigeye (FAO)
Australischer Hundshai, Hundshai, Biethai (Suppenflossenhai)	*Galeorhinus galeus/ Galeorhinus zygopterus/ Eugaleus galeus*	tope shark (FAO), tope, soupfin shark, school shark

Australischer Lachs	*Arripis trutta*	Australian salmon (ruff)
Australischer Lungenfisch	*Neoceratodus forsteri*	Australian lungfish
Australischer Mangrovenwaran, Rostkopfwaran	*Varanus semiremex*	Australian mangrove monitor, rusty monitor
Australischer Marienkäfer	*Coelophora inaequalis*	common Australian lady beetle
Australischer Riesenregenwurm	*Megascolides australis*	giant Gippsland earthworm, karmai
Australischer Riesenskink	*Leiolopisma grande*	giant skink
Australischer Seelöwe	*Neophoca cinerea*	Australian sea lion
Australischer Süßwasserwels*	*Tandanus tandanus*	Australian freshwater catfish
Australkegel	*Conus australis*	austral cone
Australmuschel	*Neotrigonia margaritacea*	Australian brooch clam
Australoasiatischer Raurochen	*Raja nasuta*	rough skate (Australasian)
Australspornpieper	*Anthus novaeseelandiae*	Richard's pipit
Avicennaviper	*Cerastes vipera*	Sahara sand viper, Avicenna's viper
Axishirsch, Chital	*Axis axis/Cervus axis*	spotted deer, axis deer, chital
Axolotl	*Ambystoma mexicanum*	axolotl
Ayu	*Plecoglossus altivelis*	ayu, ayu sweetfish
Azaleenmotte	*Caloptilia azaleella*	azalea leaf miner
Azevia-Seezunge	*Microchirus azevia/ Microchirus theophila/ Solea azevia*	bastard sole
Aztekenmöve	*Larus atricilla*	laughing gull
Azurbischof	*Guiraca caerulea*	blue grosbeak, Guira cuckoo
Azurblaue Demoiselle	*Pomacentrus caerulens*	caerulean damsel (FAO), blue damselfish
Azurraupenfänger	*Coracina azurea*	African blue cuckoo shrike

Deutsch	Wissenschaftlich	Englisch
Babirusa	*Babyrousa babyrussa*	babirusa
Babylon-Latiaxis-Schnecke	*Babelomurex babelis*	babylon latiaxis
Babylon-Turmschnecke	*Turris babylonia*	babylon turrid
Babylonische Türme	*Babylonia* spp.	babylon snails
Bachflohkrebs, Gemeiner Flohkrebs	*Rivulogammarus pulex/ Gammarus pulex*	freshwater shrimp
Bachforelle, Steinforelle	*Salmo trutta fario*	brown trout (river trout, brook trout)
Bachhafte	Osmylidae	osmylid flies
Bachkröte	*Ansonia* spp.	stream toads
Bachläufer, Stoßwasserläufer	Veliidae	water crickets, broad-shouldered water striders, ripple bugs, small water striders
Bachläufer, Stoßwasserläufer	*Velia* spp.	water crickets, broad-shouldered water striders
Bachlinge	*Rivulus* spp.	rivuluses
Bachmuschel, Kleine Flussmuschel, Gemeine Flussmuschel	*Unio crassus*	common river mussel, common Central European river mussel
Bachneunauge	*Lampetra planeri*	brook lamprey, European brook lamprey (FAO)
Bachsaibling	*Salvelinus fontinalis*	brook trout (FAO), brook char, brook charr
Bachsalamander	*Desmognathus* spp.	dusky salamanders
Bachschmerle, Bartgrundel	*Noemacheilus barbatulus/ Barbatula barbatula*	stone loach
Bachschmerlen	*Noemacheilus* spp.	stone loaches
Bachstelze	*Motacilla alba*	pied wagtail, pied white wagtail
Bachstelzen-Schnecke	*Siratus motacilla*	wagtail murex
Bachtaumelkäfer, Haariger Taumelkäfer	*Orectochilus villosus*	hairy whirligig
Bachwurm, Gemeiner Schlammröhrenwurm, Gemeiner Bachröhrenwurm	*Tubifex tubifex*	river worm
Bäckerbock, Kiefernbock, Gefleckter Langhornbock	*Monochamus galloprovincialis*	pine sawyer beetle
Bäckerschabe, Küchenschabe	*Blatta orientalis*	Oriental cockroach, common cockroach
Backobstkäfer	*Carpophilus hemipterus*	driedfruit beetle
Backobstmilbe	*Carpoglyphus lactis*	driedfruit mite
Backobstmilben	Carpoglyphidae	driedfruit mites
Badewannenschwamm, Großer Vasenschwamm	*Xestospongia testudinaria*	turtleshell bath sponge, barrel sponge
Bahama-Baumratte	*Geocapromys ingrahami*	Bahamian hutia
Bahama-Rennnatter	*Alsophis vudii*	West Indian racer, Bahamas racer
Bahama-Zwergboa	*Tropidophis canus*	Bahamas wood snake
Baikal-Ringelrobbe	*Phoca sibirica*	Baikal seal
Baikalgroppen	Cottocomephoridae	Lake Baikal sculpins
Baillons Lippfisch	*Crenilabrus bailloni*	Baillon's wrasse
Baird-Strandläufer	*Calidris bairdii*	Baird's sandpiper

Baird-Tapir, Mittelamerikanischer Tapir	*Tapirus bairdii*	Baird's tapir, Central American tapir
Baird-Wal	*Berardius bairdii*	Pacific giant bottlenosed whale, Baird's beaked whale, four-toothed whale
Balao-Halbschnäbler	*Hemiramphus balao*	balao, balao halfbeak
Balearen-Eidechse	*Podarcis lilfordi*	Lilford's wall lizard
Balearen-Muräne (Azoren-Muräne)	*Ariosoma balearica*	Balearic conger
Balearenkröte	*Alytes muletensis*	Mallorcan midwife toad
Balistar	*Leucopsar rothschildi*	Rothschild's starling, Bali mynah
Balkan-Kurzohrmaus, Thomas-Kleinwühlmaus	*Microtus thomasi/ Pitymys thomasi*	Thomas's pine vole
Balkan-Zornnatter	*Coluber gemonensis/ Coluber laurenti*	Balkan whip snake
Balkankammmolch, Persischer Kammmolch	*Triturus karelini*	Balkan crested newt
Balkenschröter	*Dorcus parallelopipedus*	lesser stag beetle
Balkenstreifiger Leierfisch	*Callionymus fasciatus*	barred dragonet
Ballonqualle	*Halitholus cirratus*	balloon jelly*
Ballyhoo, Brasilianischer Halbschnabelhecht, Brasilianischer Halbschnäbler	*Hemiramphus brasiliensis*	ballyhoo (FAO), ballyhoo halfbeak
Baltimoretrupial	*Icterus galbula*	northern oriole
Baltische Meerassel, Ostsee-Meerassel	*Idotea baltica*	Baltic isopod, Baltic Sea "centipede"
Baltische Tellmuschel, Baltische Plattmuschel, Rote Bohne	*Macoma baltica*	Baltic macoma
Baltischer Stör, Gemeiner Stör	*Acipenser sturio*	common sturgeon (IUCN), Atlantic sturgeon, sturgeon (FAO), European sturgeon
Baluchistan-Zwergspringmaus	*Salpingotulus michaelis*	Baluchistan pygmy jerboa
Bambus-Fingerratte	*Kannabateomys amblyonyx*	rato de Taquara
Bambus-Fledermäuse	*Tylonycteris* spp.	club-footed bats, bamboo bats (Southeast Asian flat-headed bats)
Bambusbär, Riesenpanda, Großer Panda	*Ailuropoda melanoleuca*	giant panda
Bambusbohrer u.a.	*Chlorophorus annularis/ Rhaphuma annularis*	yellow bamboo longhorn beetle, bamboo borer
Bambusbohrer u.a.	*Dinoderus minutus*	bamboo borer a.o.
Bambusgarnele	*Penaeus japonicus*	bamboo prawn
Bambushaie, Gebänderte Katzenhaie	Hemiscylliidae	bamboo sharks, longtailed carpet sharks
Bambusottern, Asiatische Lanzenottern	*Trimeresurus* spp.	Asian lance-headed vipers, Asian pit vipers
Bambusratte> Kleine Bambusratte	*Cannomys badius*	lesser bamboo rat
Bambusratten (Hbz. Chinesische Ratten)	*Rhizomys* spp.	bamboo rats
Bambuswürmer u.a.	*Clymenella* spp. & *Axiothella* spp.	bamboo worms a.o.
Bananaquit, Zuckervogel	Coerebidae	bananaquit

Bananen-Garnele	*Fenneropenaeus merguiensis/ Penaeus merguiensis*	banana prawn
Bananenfledermaus	*Musonycteris harrisoni*	banana bat, Colima long-nosed bat
Bananenfrösche	*Afrixalus* spp.	banana frogs
Bananenspinne	*Heteropoda venatoria*	banana spider, large brown spider, huntsman spider
Bananenzwergfledermaus	*Pipistrellus nanus*	dwarf banana bat
Band-Schnurwurm, Band-Nemertine	*Drepanophorus crassus* (Nemertini)	
Band-Seehase, Großer Brauner Seehase	*Aplysia fasciata*	banded sea hare
Bandbarbe, Glühkohlenbarbe	*Barbus fasciatus/ Puntius eugrammus*	striped barb, zebra barb, African banded barb
Bändereisfisch	*Champsocephalus gunnari*	Antarctic icefish, mackerel icefish (FAO)
Bänderkänguru	*Lagostrophus fasciatus*	banded hare wallaby, munning
Bänderkrait, Bänder-Krait, Gelber Bungar	*Bungarus fasciatus*	banded krait
Bänderlinsang	*Prionodon linsang*	banded linsang
Bändermungo	*Galidictis fasciata fasciata*	Malagasy striped mongoose
Bändernatter	*Thamnophis sauritus*	ribbon snake
Bänderroller	*Hemigalus* spp.	banded palm civets
Bänderteju	*Tupinambis teguixin*	common tegu, banded tegu
Bändertriton	*Linatella caudata*	ringed triton
Bandfisch, Spitzschwänziger Bandfisch	*Lumpenus lampretaeformis*	snake blenny
Bandfisch, Spanfisch	*Trachipterus trachipterus*	ribbonfish, ribbob fish (FAO), Mediterranean dealfish
Bandfische, Sensenfische	Trachipteridae	ribbonfishes
Bandfische u.a.	Lumpenidae	snake blennies
Bandfische u.a.	Cepolidae	bandfishes
Bandflossen-Koppe, Buntflossenkoppe, Buntflossen-Groppe, Sibirische Groppe	*Cottus poecilopus*	Siberian bullhead, alpine bullhead (FAO)
Bandfüßer	Polydesmidae	flat millepedes, flat-backed millepedes
Bandikutratten	*Bandicota* spp.	bandicoot rats
Bandikuts, Nasenbeutler	Peramelidae	bandicoots
Bandmolch	*Triturus vittatus*	banded newt
Bandplanarien	*Prosthecereus* spp.	
Bandrobbe	*Phoca fasciata*	ribbon seal
Bandtaube, Schuppenhalstaube	*Columba fasciata*	band-tailed pigeon
Bandträger	Taeniophoridae/ Eutaeniophorinae	tapetails, ribbonbearers
Bandwürmer, Cestoden	Cestoda	tapeworms, cestodes
Bandy-Bandys	*Vermicella* spp.	bandy-bandys
Banjofrosch	*Limnodynastes dorsalis*	banjo frog
Banjowelse, Bratpfannenwelse	Aspredinidae	banjo catfishes
Bankivahuhn	*Gallus gallus*	red jungle-fowl

Bannerschwanz-Kängururatte	*Dipodomys spectabilis*	bannertail
Banteng	*Bos javanicus*	banteng
Barbados-Schlüssellochschnecke	*Fissurella barbadensis*	Barbados keyhole limpet
Barbara Napfschnecke	*Patella barbara*	Barbara limpet
Barbe, Flussbarbe	*Barbus barbus*	barbel
Barbenegel, Platter Fischegel	*Cystobranchus respirans*	flat fish leech
Barbensalmler, Breitlingssalmler	Curimatidae	curimatids, toothless characins
Barbours Höckerschildkröte	*Graptemys barbouri*	Barbour's map turtle
Barbours Lanzenotter	*Porthidium barbouri/ Bothrops barbouri*	Barbour's pit viper
Barbudos, Bartfische	Polymixiidae	beardfishes
Bären	Ursidae	bears
Bärengarnele, Bärenschiffskielgarnele	*Penaeus monodon*	giant tiger prawn
Bärenkrebse	Scyllaridae	slipper lobsters, shovel-nosed lobsters
Bärenmakak, Stumpfschwanzmakak	*Macaca arctoides*	bear macaque, stump-tailed macaque
Bärenolive*	*Olivancillaria urceus*	bear olive, bear ancilla
Bärenpavian	*Papio ursinus*	chacma baboon
Bärenspinner (Bären)	Arctiidae	tigermoths & footman moths & ermine moths (*caterpillars*: woolly bears)
Bärenstummelaffe, Südlicher Guereza (Hbz. Scheitelaffe)	*Colobus polykomos*	king colobus, western pied colobus (western black-and-white colobus)
Bärentierchen, Bärtierchen, Tardigraden (sg Tardigrad m)	Tardigrada	water bears, tardigrades
Barfuß-Gecko	*Coleonyx switaki*	barefoot gecko
Barrakudas, Pfeilhechte	*Sphyraena* spp.	barracudas
Barrakudas, Pfeilhechte	Sphyraenidae	barracudas
Barrakudinas	Paralepididae	barracudinas
Barrakudinas	*Paralepis* spp.	barracudinas
Barramundi	*Lates calcarifer*	barramundi (FAO), giant sea perch
Barriere-Riff-Anemonenfisch	*Amphiprion akindynos*	Barrier Reef anemonefish
Barsche, Echte Barsche	Percidae	perches
Barschfische, Barschartige	Perciformes	perches & perchlike fishes
Barschlachs> Gewöhnlicher Barschlachs, Amerikanischer Barschlachs, Sandroller	*Percopsis transmontana*	sand roller
Barschlachse	Percopsidae	trout-perches
Barschlachsverwandte	Percopsiformes	trout-perches & freshwater relatives
Bartaffe, Wanderu	*Macaca silenus*	liontail macaque, lion-tailed macaque
Bartagame	*Pogona barbatus/ Amphibolurus barbatus*	bearded dragon

Bartalk	*Aethia pygmaea*	whiskered auklet
Bartel-Flugfisch, Zweiflügelfisch	*Exocoetus monocirrhus*	barbel flyingfish (FAO), two-wing flyingfish
Bärtelige Hundshaie, Schlankhaie	Leptochariidae	barbelled houndsharks
Bärteliger Hundshai	*Leptocharias smithii*	barbelled houndshark
Bartellose Welse	Ageneiosidae	bottlenose catfishes, barbelless catfishes
Bartenwale	Mysticeti	whalebone whales, baleen whales
Bartfische, Barbudos	Polymixiidae	beardfishes
Bartgeier	*Gypaetus barbatus*	lammergeier
Bartgrundel, Bachschmerle, Schmerle	*Noemacheilus barbulatus/ Barbatula barbatula/ Nemacheilus barbatulus*	stone loach
Barthelemy-Kegelschnecke	*Conus barthelemyi*	Bartholomew's cone
Bärtige Archenmuschel	*Barbatia barbata*	bearded ark (bearded ark shell)
Bärtige Krötenkopfagame, Ohrenkrötenkopf	*Phrynocephalus mystaceus*	secret toadhead agama
Bärtige Stachelauster	*Spondylus barbatus*	bearded thorny oyster
Bartkauz	*Strix nebulosa*	great grey owl, great gray owl
Bartklarino	*Myadestes genibarbis*	rufous-throated solitaire
Bartlaubsänger	*Phylloscopus schwarzi*	Radde's warbler
Bartmännchen	*Ophidion barbatum*	barbed cusk-eel, snake blenny (FAO)
Bartmännchen, Ophidiiden	Ophidiidae	cuskeels, cusk-eels
Bartmeise	*Panurus biarmicus*	bearded reedling, bearded tit
Bartmuschel, Bartige Miesmuschel	*Modiolus barbatus/ Lithographa barbatus*	bearded horse mussel
Bartrobbe	*Erignathus barbatus*	bearded seal
Bartschwein	*Sus barbatus*	bearded pig
Bartsittich, Rosenbrustsittich	*Psittacula alexandri*	moustached parakeet
Barttrappe	*Eupodotis bengalensis*	Bengal florican
Bartumber, Schattenfisch, Umberfisch	*Umbrina cirrosa/ Sciaena cirrosa*	shi drum (FAO), corb (U.S./U.K.), sea crow (U.S.), gurbell (U.S.), croaker
Bartvögel	Capitonidae	New World barbets
Bartwürmer, Bartträger	Pogonophora	beard worms, beard bearers, pogonophorans
Basilian-Wühle	*Ichthyophis glandulosus*	Abungabung caecilian
Basilikaspinne*	*Mecynogea lemniscata*	basilica spider, basilica orbweaver
Basilisken	*Basiliscus* spp.	basilisks
Basilisken-Chamäleon, Afrikanisches Chamäleon	*Chamaeleo africanus*	African chameleon
Basiliskenklapperschlange	*Crotalus basiliscus*	Mexican West-coast rattlesnake
Baskenmützen-Zackenbarsch, Baskenmützenbarsch	*Epinephelus fasciatus*	blacktip grouper (FAO), blacktip rockcod
Basstölpel	*Sula bassana/Morus bassanus*	northern gannet
Bastard-Süßlippe	*Pomadasys incisus*	bastard grunt
Bastardmakrele, Stöcker, Schildmakrele	*Trachurus trachurus*	Atlantic horse mackerel (FAO), scad, maasbanker

Bastardzunge	*Solea variegata/ Microchirus variegatus*	thickback sole (FAO), thick-backed sole
Batagur, Batagur-Schildkröte	*Batagur baska*	batagur, river terrapin
Batavia-Hechtmuräne, Hechtmuräne	*Muraenesox cinereus*	dagger-tooth pike-conger, sharp-toothed eel, pike conger (FAO)
Batesböckchen	*Neotragus batesi*	Bate's dwarf antelope
Bathymasteriden	Bathymasteridae	ronquils
Bauchbinden-Picassofisch, Schwarzbauch-Picassodrücker, „Keilfleckdrücker"	*Rhinecanthus verrucosus*	blackbelly triggerfish
Bauchdrüsenottern	*Maticora* spp.	long-glanded coral snakes, Malaysian coral snakes
Bauchfüßer, Schnecken, Gastropoden	Gastropoda	snails, gastropods
Bauchhaarlinge, Bauchhärlinge, Flaschentierchen, Gastrotrichen	Gastrotricha	gastrotrichs
Bauchige Feilenmuschel	*Lima inflata*	inflated fileclam, inflated lima
Bauchige Olive	*Oliva bulbosa*	inflated olive
Bauchige Schließmundschnecke	*Macrogastra ventricosa*	ventricose door snail
Bauchige Schnauzenschnecke, Runde Langfühlerschnecke	*Bithynia leachi*	globose bithynia, Leach's bithynia
Bauchige Windelschnecke	*Vertigo moulinsiana*	Desmoulins' whorl snail, ventricose whorl snail
Bauchige Zwergschnecke, Bauchige Zwerghornschnecke	*Carychium minimum*	short-toothed herald snail, herald thorn snail (U.S.), herald snail, sedge snail
Bauchschnabeltyrann	*Megarhynchus pitangua*	boat-billed flycatcher
Bauchstreifen-Erdschildkröte, Schwarze Erdschildkröte	*Rhinoclemmys funerea*	black wood turtle
Baumammer	*Spizella arborea*	American tree sparrow
Bäumchen-Anemone	*Actinodendron arboreum*	tree anemone
Bäumchenkoralle, Hornige Baumkoralle	*Dendrophyllia cornigera*	horny tree coral*
Bäumchenpolyp	*Eudendrium ramosum*	stickhydroid, stick hydroid
Bäumchenschnecke	*Dendronotus frondosus*	
Bäumchenschnecken, Baumschnecken	Dendronotacea	dendronotacean snails, dendronotaceans
Baumfalke	*Falco subbuteo*	hobby, northern hobby
Baumfledermaus	*Ardops nichollsi*	tree bat
Baumfliegen	Dryomyzidae	dryomyzid flies
Baumfrösche, Afrikanische Greiffrösche	*Chiromantis* spp.	foam-nest treefrogs
Baumgeckos	*Naultinus* spp.	tree geckos
Baumhopf	*Phoeniculus purpureus*	green wood hoopoe
Baumhopfe	Phoeniculidae	wood hoopoes
Baumkängurus	*Dendrolagus* spp.	tree kangaroos
Baumkoralle, Astkoralle, Gelbe Koralle	*Dendrophyllia ramea*	tree coral*
Baumkorallen	*Acropora* spp.	horn corals

Baumläuferleguan	*Plica plica*	tree runner
Baumläuse, Rindenläuse	Lachnidae	lachnids, lachnid plantlice
Baumleguan	*Urosaurus ornatus*	tree lizard, common tree lizard
Baummarder, Edelmarder	*Martes martes*	European pine marten
Baummäuse u.a.	*Chiruromys* spp.	tree mice
Baumpieper	*Anthus trivialis*	tree pipit
Baumpythons	*Chondropython* spp.	tree pythons
Baumratten u.a.	*Mesembriomys* spp.	tree rats
Baumratten u.a., Ferkelratten	Capromyidae	hutias & nutrias
Baumrutscher	Climacteridae	Australo-Papuan treecreepers
Baumsalamander, Klettersalamander	*Aneides* spp.	arboreal salamanders, climbing salamanders
Baumschläfer	*Dryomys* spp.	forest dormice
Baumschlegel, Baum-Egelschnecke	*Limax marginatus*/ *Lehmannia marginata*	tree slug
Baumschleichen	*Abronia* spp.	tree anguids, arboreal alligator lizards
Baumschliefer	*Dendrohyrax* spp.	tree hyraxes, bush hyraxes
Baumschnecken	Bulimulidae	treesnails
Baumschnegel, Baum-Egelschnecke	*Lehmannia marginata*/ *Limax marginatus*	tree slug
Baumschnüffler	*Ahaetulla prasina*	long-nosed tree snake, long-nosed whipsnake
Baumschröter, Kopfhornschröter	*Sinodendron cylindricum*	rhinoceros beetle, "small European rhinoceros beetle"
Baumschwammkäfer	Mycetophagidae	hairy fungus beetles
Baumsegler	Hemiprocnidae	treeswifts, crested swifts
Baumskinke	*Dasia* spp.	dasias
Baumstachelratten	*Diplomys* spp.	forest spiny rats
Baumstachelskink	*Egernia striolata*	tree skink
Baumstachler	Erethizontidae	New World porcupines
Baumsteiger, Baumkletterer	Dendrocolaptinae	woodcreepers
Baumsteigerfrösche	*Dendrobates* spp.	poison-arrow frogs, poison frogs
Baumsteigerfrösche, Farbfrösche, Pfeilgiftfrösche	Dendrobatidae	dendrobatids, poison dart frogs, poison arrow frogs, arrow-poison frogs
Baumstelze	*Dendronanthus indicus*	forest wagtail
Baumwanzen & Schildwanzen & Stinkwanzen	Pentatomidae	shield bugs & stink bugs
Baumweißling	*Aporia crataegi*	black-veined white
Baumwoll-Kapselwurm> Altweltlicher Baumwoll-Kapselwurm	*Helicoverpa armigera*/ *Heliothis armigera*/ *Heliothis obsoleta*	scarce bordered straw moth, Old World bollworm, African bollworm, tomato moth, corn earworm, cotton bollworm; tomato fruitworm, tomato grub
Baumwollblattlaus	*Aphis gossypii*	cotton aphid, melon aphid
Baumwollknospenstecher	*Anthonomus grandis*	cotton boll weevil
Baumwollmaus	*Peromyscus gossypinus*	cotton mouse
Baumwollratten	*Sigmodon* spp.	cotton rats
Baumwollschwanz-Kaninchen	*Sylvilagus* spp.	cottontails
Baw-Baw-Frosch	*Philoria frosti*	baw-baw frog

Bayaweber	*Ploceus philippinus*	Baya weaver
Bayerische Kurzohrmaus	*Pitymys bavaricus/ Microtus bavaricus*	Bavarian pine vole
Beans Sägezahnaal*	*Serrivomer beani*	Bean's sawtoothed eel
Beaugregory	*Pomacentrus leucostictus*	beaugregory (yellow-belly)
Becher-Azurjungfer	*Enallagma cyathigera*	common blue damselfly, common bluet damselfly
Becherquallen, Stielquallen	Stauromedusidae	stauromedusids
Becherquallen, Stielquallen	Stauromedusae	stauromedusas
Bechsteindrossel	*Turdus ruficollis*	black-throated thrush
Bechsteinfledermaus	*Myotis bechsteini*	Bechstein's bat
Bedeckte Maßschnecke	*Modulus tectum*	knobby snail
Bedornte Seerinde	*Membranipora pilosa*	thorny sea-mat
Beeren-Mondschnecke*	*Natica acinonyx*	African berry moonsnail
Beerenwanze	*Dolycoris baccarum*	sloe bug, sloebug
Begonienmilbe, Zyklamenmilbe, Alpenveilchenmilbe	*Phytonemus pallidus/ Tarsonemus pallidus*	cyclamen mite, begonia mite
Behaarte Kastenkrabbe	*Homola barbata*	hairy box crab
Behaarte Laubschnecke, Haarschnecke, Gemeine Haarschnecke	*Trichia hispida/ Helix hispida*	hairysnail, bristly snail
Behaarter Arizona-Skorpion*	*Hadrurus arizonensis*	Arizona hairy scorpion
Behaarter Baumschwammkäfer	*Typhaea stercorea*	hairy fungus beetle
Behaarter Diebkäfer	*Ptinus villiger*	hairy spider beetle
Behaarter Einsiedler	*Pagurus arcuatus*	hairy hermit crab
Behaarter Schnellläufer	*Harpalus rufipes*	strawberry seed beetle
Behaarter Stäublingskäfer, Pilzkäfer	*Mycetaea hirta*	hairy fungus beetle
Behaartflügelige Fledermaus	*Harpiocephalus harpia*	hairy-winged bat
Beifuß-Steppenlemming	*Lemmiscus curtatus/ Lagurus curtatus*	sagebrush vole
Beifußhuhn	*Centrocercus urophasianus*	sage grouse
Beifußmönch, Grauer Beifußmönch	*Cucullia artemisiae*	scarce wormwood shark
Beilbäuche, Beilbauchfische, Beilbauchsalmler	Gasteropelecidae	freshwater hatchetfishes, flying characins
Beilfleck-Blutströpfchen, Beilfleck-Widderchen	*Zygaena loti*	slender burnet
Beilmuschel*, Beil-Trogmuschel*	*Mactra dolabriformis*	hatchet surfclam
Beilpyramide	*Pyramidella dolabrata*	giant Atlantic pyrum
Beintastler	Protura	proturans
Beira	*Dorcatragus megalotis*	beira antilope
Bekassine	*Gallinago gallinago*	common snipe
Bekreuzter Traubenwickler, Weinmotte (Sauerwurm/ Gelbköpfiger Sauerwurm)	*Lobesia botrana*	grape fruit moth, vine moth, European vine moth, grape-berry moth
Belemniten	Belemnitida	belemnites
Bells Flachschildkröte	*Homopus belliana*	Bell's hingeback tortoise

Bengalfuchs	*Vulpes bengalensis*	Bengal fox
Bengalwaran, Bengalenwaran	*Varanus bengalensis*	Bengal monitor, Indian monitor, common monitor
Bengalische Dachschildkröte, Bunte Dachschildkröte	*Kachuga kachuga*	red-crowned roofed turtle (IUCN), painted roofed turtle, Bengal roof turtle
Bengalische Kegelschnecke	*Conus bengalensis*	Bengal cone
Bengalkatze, Leopardkatze	*Felis bengalensis*	leopard cat
Benguela Seehecht	*Merluccius polli*	Benguela hake
Bennettkänguru	*Macropus rufogriseus*	red-necked wallaby
Bennettkasuar	*Casuarius bennetti*	dwarf cassowary
Bennetts Kofferfisch*	*Ephippion guttiferum*	Bennett's pufferfish
Bentevi	*Pitangus sulphuratus*	great kiskadee
Beo	*Gracula religiosa*	southern grackle, hill mynah
Berberaffe, Magot	*Macaca sylvanus*	barbary ape, barbary macaque
Berberitzenspanner	*Pareulype berberata*	barberry carpet moth
Berberitzenspanner	*Rheumaptera cervinalis*	barberry looper
Berberskink	*Eumeces algeriensis*	Algerian skink
Berdmores Palmhörnchen	*Menetes berdmorei*	Berdmore's palm squirrel, multistriped palm squirrel
Berg-Königsnatter	*Lampropeltis pyromelana*	mountain kingsnake
Berg-Lanzenotter	*Vipera monticola/ Trimeresurus monticola*	mountain viper, Chinese mountain viper
Berg-Schließmundschnecke	*Cochlodina costata*	mountain door snail
Berg-Zwerggleitbeutler	*Burramys parvus*	mountain pygmy possum
Bergagame, Felsenagame	*Agama atra*	southern rock agama, South African rock agama
Bergagamen	*Japalura* spp.	mountain lizards
Berganoa	*Bubalus quarlesi*	mountain anoa
Bergbiber, Biberhörnchen, Stummelschwanzhörnchen	*Aplodontia rufa*	sewellel, mountain beaver
Bergchamäleon	*Chamaeleo montium*	Cameroon sailfin chameleon, sail-finned chameleon, mountain chameleon
Bergeidechse, Waldeidechse, Mooreidechse	*Lacerta vivipara*	viviparous lizard, European common lizard
Bergente	*Aythya marila*	scaup, greater scaup
Bergente> Kleine Bergente	*Aythya affinis*	lesser scaup
Bergfink	*Fringilla montifringilla*	brambling
Berggimpel	*Carpodacus rubicilla*	great rosefinch
Bergguan, Zapfenguan	*Oreophasis derbianus*	horned guan
Berghänfling	*Carduelis flavirostris*	twite
Berghüttensänger	*Sialia currucoides*	mountain bluebird
Bergkalanderlerche	*Melanocorypha bimaculata*	bimaculated lark
Bergkänguru, Wallaru(h)	*Macropus robustus*	wallaroo, euro, common wallaroo
Bergkoralle	*Porites rus*	
Bergkurzflügel	*Brachypteryx montana*	blue shortwing

Bergland-Schmarotzerhummel (der Steinhummel)	*Psithyrus rupestris*	hill cuckoo bee
Berglaubsänger	*Phylloscopus bonelli*	Bonelli's warbler
Berglemming	*Lemmus lemmus*	Norway lemming
Bergmaus	*Dinaromys bogdanovi*	Martino's snow vole, Balkan snow vole
Bergmolch, Alpenmolch	*Triturus alpestris*	alpine newt
Bergnyala, Mittelkudu	*Tragelaphus buxtoni*	mountain nyala
Bergotter	*Daboia xanthina/ Vipera xanthina*	European coastal viper, Ottoman viper, Near East viper
Bergpaka	*Agouti taczanowskii*	mountain paca
Bergpieper	*Anthus spinoletta*	rock pipet, water pipit
Bergpuffotter, Südafrikanische Bergotter, Atroposviper	*Bitis atropos*	berg adder
Bergrhesus, Assamrhesus, Assammakak	*Macaca assamensis*	Assam macaque
Bergriedbock	*Redunca fulvorufula*	mountain reedbuck
Bergs Morwong	*Cheilodactylus bergi*	castaneta
Bergspottdrossel	*Oreoscoptes montanus*	sage thrasher
Bergstachler	*Echinoprocta rufescens*	Upper Amazon porcupine
Bergstrandläufer	*Calidris mauri*	western sandpiper
Bergtapir	*Tapirus pinchaque*	mountain tapir
Bergtaube	*Geotrygon montana*	ruddy quail dove
Bergtupaias	*Dendrogale* spp.	small smooth-tailed tree shrews
Bergzebra	*Equus zebra*	mountain zebra
Bergzikade	*Cicadetta montana*	New Forest cicada
Bermuda-Ruderfisch, Gelblinien Pilotbarsch	*Kyphosus sectatrix*	Bermuda chub
Bermudasturmvogel	*Pterodroma cahow*	cahow
Bernhardskrebs, Bernhardseinsiedler, Nordsee-Einsiedlerkrebs, Gemeiner Einsiedler	*Pagurus bernhardus/ Eupagurus bernhardus*	large hermit crab, common hermit crab, soldier crab, soldier hermit crab, Bernhard's hermit crab
Bernstein-Steckmuschel, Fleischfarbene Steckmuschel	*Pinna carnea*	amber penshell
Bernsteinmakrele, Gabelschwanzmakrele	*Seriola dumerili*	amberjack, greater amberjack, greater yellowtail
Bernsteinschnecke> Gemeine Bernsteinschnecke	*Succinea putris*	rotten amber snail, large amber snail, European ambersnail (U.S.)
Bernsteinschnecken	Succineidae	amber snails, ambersnails
Beschuppte Schleimfische	Clinidae	scaled blennies, klipfishes, clinids
Besondere Maulbeerschnecke*	*Morum exquisitum*	exquisite morum
Besrasperber	*Accipiter virgatus*	Besra sparrowhawk

Bettwanze> Gemeine Bettwanze	*Cimex lectularius*	bedbug, common bedbug (wall-louse)
Beulen-Fechterschnecke*, Beulen-Flügelschnecke*	*Strombus latus/* *Strombus bubonius*	bubonian conch
Beulenkopf-Buntbarsch, Buckelkopf-Buntbarsch, Zebra-Beulenkopf, „Frontosa"	*Cyphotilapia frontosa*	humphead cichlid (FAO), 'frontosa'
Beulenkrokodil	*Crocodylus moreletii*	Morelet's crocodile
Beutel-Kalkschwamm	*Grantia compressa*	purse sponge
Beutelbär, Koalabär	*Phascolarctos cinereus*	koala
Beutelfrösche	*Gastrotheca* spp.	marsupial frogs
Beutelmaulwurf, Beutelmull	*Notoryctes typhlops*	marsupial "mole"
Beutelmeise	*Remiz pendulinus*	penduline tit
Beutelmeisen	Remizidae	penduline-tits
Beutelmull, Beutelmaulwurf	*Notoryctes typhlops*	marsupial "mole"
Beutelratten	Didelphidae	opossums, American opossums
Beutelsäuger	Metatheria/Didelphia	pouched mammals, metatherians
Beutelteufel	*Sarcophilus harrisii*	Tasmanian devil
Beuteltiere	Marsupialia	marsupials, pouched mammals
Beutelwolf, Tasmanischer Tiger	*Thylacinus cynocephalus*	thylacine, Tasmanian wolf, Tasmanian tiger
Bewaffnete Kantengarnele	*Heterocarpus ensifer*	armed nylon shrimp
Bezahnte Achatschnecke	*Azeca goodalli/* *Azeca menkeana*	three-toothed moss snail, three-toothed snail, glossy trident shell
Bezoarschnecke	*Rapana bezoar*	bezoar rapa whelk
Bezoarziege	*Capra aegagrus*	wild goat
Biber	*Castor* spp.	beavers
Biber> Altwelt-Biber	*Castor fiber*	Eurasian beaver
Biber> Kanadischer Biber	*Castor canadensis*	North American beaver
Biberhörnchen, Bergbiber, Stummelschwanzhörnchen	*Aplodontia rufa*	sewellel, mountain beaver
Biberkäfer, Biberlaus	*Platypsyllus castoris*	beaver beetle
Bibronkröte, Vieraugenkröte	*Pleurodema bibroni*	four-eyed frog
Bibrons Eidechse	*Diplolaemus bibronii*	Bibron's iguana
Bienen	Apidae	hive bees (bumble bees & honey bees & orchid bees)
Bienenameisen, Spinnenameisen, Ameisenwespen	Mutillidae	velvet-ants
Bienenbrutmilbe, Varroamilbe	*Varroa jacobsoni*	varroa mite
Bienenchiton, Gemeine Ostatlantische Käferschnecke	*Chaetopleura apiculata*	eastern beaded chiton, common eastern chiton, bee chiton (common American Atlantic coast chiton)
Bienenfresser, Bienenesser, Spinte	Meropidae	bee-eaters

Bienenfresser	*Merops apiaster*	bee-eater
Bienenglasflügler, Hornissenschwärmer	*Aegeria apiformis/ Sesia apiformis*	poplar hornet clearwing, hornet moth
Bienenkäfer> Immenkäfer, Gemeiner Bienenkäfer, Bienenwolf	*Trichodes apiarius*	bee beetle, bee wolf
Bienenkörbchen	*Spermodea lamellata*	plaited snail
Bienenlaus	*Braula coeca*	bee louse
Bienenläuse	Braulidae	bee lice
Bienenmilbe, Tracheenmilbe	*Acarapis woodi*	bee mite, honey bee mite
Bienenwolf, Große Wachsmotte, Rankmade (Larve)	*Galleria mellonella*	greater wax moth, honeycomb moth, bee moth
Bienenwolf, Immenkäfer, Gemeiner Bienenkäfer	*Trichodes apiarius*	bee beetle, bee wolf
Bienenwolf u.a.	*Philanthus triangulum*	bee-killer wasp, bee-killer
Bierfassschnecke	*Tonna cerevisina*	beerbarrel tun
Bierschnegel, Kellerschnecke, Gelber Schnegel, Gelbe Egelschnecke	*Lehmannia flavus*	tawny garden slug, yellow gardenslug, yellow slug (cellar slug, dairy slug, house slug)
Biethai, Australischer Hundshai, Hundshai (Suppenflossenhai)	*Galeorhinus galeus/ Galeorhinus zygopterus/ Eugaleus galeus*	tope shark (FAO), tope, soupfin shark, school shark
Bilch-Rennmaus, Buschschwanz-Rennmaus	*Sekeetamys calurus*	bushy-tailed jird
Bilche & Schläfer	Gliridae	dormice
Bilchschwänze	*Eliurus* spp.	Madagascan rats
Binden-Krokodilfisch	*Parapercis cylindrica*	banded weever, grubfish, cylindrical sandperch (FAO)
Binden-Schwimmschnecke, Gebänderte Kahnschnecke	*Theodoxus transversalis*	striped freshwater nerite
Bindenfregattvogel	*Fregata minor*	great frigate bird
Bindengrundel	*Benthophiloides brauneri*	
Bindenkreuzschnabel	*Loxia leucoptera*	two-barred crossbill
Bindenschmuckvogel	*Pipreola arcuata*	barred fruiteater
Bindenseeadler	*Haliaeetus leucoryphus*	Palla's fish eagle
Bindenstrandläufer	*Micropalama himantopus*	stilt sandpiper
Bindentaucher	*Podilymbus podiceps*	pied-billed grebe
Bindenwaran	*Varanus salvator*	common Asiatic monitor, water monitor, Malayan monitor
Binnen-Ährenfisch	*Menidia beryllina*	inland silverside
Binsengraseule, Finkeneule, Waldmoorrasen-Türkeneule, Rotgefranste Weißpunkteule, Marbeleule	*Mythimna turca*	double-lined wainscot, double line moth
Binsenhühner, Binsenrallen	Heliornithidae	sungrebes, finfoots
Binturong	*Arctictis binturong*	binturong
Birken-Hexenbesenmilbe	*Acalitus rudis*	witches' broom mite (of birch)
Birken-Kegelschnecke	*Conus betulinus*	birch cone (*erroneously called:* beech cone)
Birkenblattroller, Birkentrichterwickler	*Deporaus betulae*	birch leafroller weevil

Birkenmäuse, Streifen-Hüpfmäuse	*Sicista* spp.	birch mice
Birkenminiermotte	*Eriocrania sparmannella*	birch leaf miner*
Birkenporzellanspinner	*Pheosia gnoma*	lesser swallow prominent
Birkensamen-Gallmücke	*Semudobia betulae*	birch-seed gall midge
Birkensichler, Gemeiner Sichelflügler	*Drepana falcataria*	pebble hook-tip
Birkenspanner	*Biston betularia*	peppered moth
Birkenspinner	*Endromis versicolora*	Kentish glory
Birkenspinner, Frühlingsspinner	Endromididae/ Endromidae	endromid moths
Birkentrichterwickler, Birkenblattroller	*Deporaus betulae*	birch leafroller weevil
Birkenwanze, Birkwanze	*Kleidocerys resedae*	birch bug
Birkenwanze u.a., Erlenwanze, Fleckige Brutwanze	*Elasmucha grisea*	parent bug, mothering bug
Birkenzeisig	*Carduelis flammea*	redpoll, common redpoll
Birkenzierlaus> Gemeine Birkenzierlaus	*Euceraphis punctipennis*	downy birch aphid, birch aphid
Birkenzipfelfalter, Nierenfleck-Zipfelfalter	*Thecla betulae*	brown hairstreak
Birkhuhn	*Lyrurus tetrix/ Tetrao tetrix*	black grouse
Birnen-Grünrüssler	*Phyllobius pyri*	pear weevil
Birnen-Helmschnecke	*Xenophalium pyrum pyrum*	pear bonnet
Birnen-Kronenschnecke, Birnenkrone	*Volema paradisiaca*	pear melongena
Birnenblattsauger, Gelber Birnenblattsauger	*Psylla pyricula*	pear sucker, pear psyllid
Birnenblutlaus, Ulmenbeutelgallenlaus	*Eriosoma lanuginosum*	elm balloon-gall aphid
Birnenförmige Kauri	*Cypraea pyriformis*	pear-shaped cowrie
Birnenförmiger Korallengast	*Coralliophila pyriformis/ Coralliophila radula*	pear-shaped coralsnail
Birnengallmücke	*Contarinia pirivora*	pear gall midge
Birnenkauri, Birnenschnecke, Birnenporzellane	*Cypraea pyrum/ Zonaria pyrum*	pear cowrie
Birnenkegel	*Conus patricius*	pear cone
Birnenknospenstecher	*Anthonomus piri*	apple bud weevil
Birnenpockenmilbe	*Eriophyes pyri*	pear blister mite
Birnenprachtkäfer	*Agrilus sinuatus*	rusty peartree borer*
Birnensägewespe	*Hoplocampa brevis*	pear sawfly
Birnenschnecke, Birnenwellhorn	*Busycotypus spiratus/ Busycon spiratum*	pearwhelk, pear whelk, true pear whelk
Birnentriebwespe	*Janus compressus*	pear stem sawfly
Bisamratte (Hbz. Bisam)	*Ondatra zibethica*	muskrat
Bisamschwein, Weißbartpekari	*Tayassu pecari*	white-lipped peccary
Bischofsmützen- Leierschnecke*	*Lyria mitraeformis*	miter-shaped lyria
Bischofs-Kegelschnecke	*Conus episcopus*	episcopal cone

Bischofsmütze	*Mitra mitra/* *Mitra episcopalis*	episcopal mitre
Bissiger Skolopender	*Scolopendra morsitans*	viciously biting scolopendra*
Bitterling	*Rhodeus amarus/* *Rhodeus sericeus*	bitterling
Bitterlingsbarbe	*Puntius titteya/Barbus titteya*	cherry barb
Blainville-Zweizahnwal	*Mesoplodon densirostris*	Blainville's dense-beaked whale
Blainvilles Dornhai, Blainville-Dornhai	*Squalus blainvillei/* *Squalus fernandinus/* *Acanthias blainvillei*	longnose spurdog (FAO), Blainville's dogfish
Blasen-Fechterschnecke, Blasen-Flügelschnecke	*Strombus bulla*	bubble conch
Blasen-Kegelschnecke	*Conus bullatus*	bubble cone
Blasenfüße, Fransenflügler, Thripse	Thysanoptera	thrips
Blasenkäfer (incl. Ölkäfer)	Meloidae	blister beetles (incl. oil beetles)
Blasenkoralle u.a.	*Plerogyra sinuosa*	bubble coral a.o.
Blasenläuse	Eriosomatidae/Pemphigidae	aphids
Blasenläuse, Blutläuse	Eriosomatinae/Schizoneurinae	woolly aphids, gall aphids
Blasenqualle, Seeblase, Portugiesische Galeere	*Physalia physalis*	Portuguese man-of-war, Portuguese man-o-war, blue-bottle
Blasenschnecke	*Physa acuta/Physella acuta*	pointed bladder snail
Blasenschnecke> Gemeine Blasenschnecke	*Bulla striata*	striate bubble, common Atlantic bubble
Blasenschnecken	Physidae	bladder snails, tadpole snails
Blasenschnecken, Kugelschnecken	Ampullariidae	bubble snails (bubble shells)
Blasenzylinder*	*Cylinder bullatus*	bubble cone
Bläßgans, Blässgans	*Anser albifrons*	white-fronted goose
Bläßhuhn, Blässhuhn	*Fulica atra*	coot
Blasige Randschnecke	*Bullata bullata*	blistered marginsnail (blistered margin shell), bubble marginella
Blasige Tellmuschel	*Macoma loveni*	inflated macoma
Blasius-Hufeisennase	*Rhinolophus blasii*	Blasius' horseshoe bat
Blasse Blattspinne*	*Cheiracanthium mordax*	pale leaf spider
Blasse Fledermäuse	*Antrozous* spp.	pallid bats, desert bats
Blasse Glasrose	*Aiptasia pallida*	pale anemone
Blasse Groppe*	*Cottunculus thompsoni*	pallid sculpin
Blasse Lastträgerschnecke	*Xenophora pallidula*	pallid carriersnail
Blasse Mondschnecke	*Euspira pallida/* *Polinices pallidus*	pale moonsnail
Blasse Randschnecke	*Persicula cornea/* *Marginella cornea*	pale marginella, plain marginella, pale marginsnail (pale margin shell)
Blasse Rennechse	*Cnemidophorus exsanguis*	Chihuahuan spotted whiptail
Blasse Veilchenschnecke	*Janthina pallida*	pallid janthina
Blasser Ringelfleck-Gürtelpuppenspanner, Pendelspanner, Weißgrauer Ringfleckspanner	*Cyclophora pendularia/* *Cyclophora orbicularia*	dingy mocha
Blasser Schaufelnasenstör	*Scaphirhynchus albus*	pallid sturgeon

Blassfuchs (Hbz. Libyscher Steppenfuchs)	*Vulpes pallida*	pale fox
Blässgans, Bläßgans	*Anser albifrons*	white-fronted goose
Blässhuhn, Bläßhuhn	*Fulica atra*	coot
Blassrote Tiefseegarnele, Blaurote Garnele	*Aristeus antennatus*	blue-and-red shrimp
Blassspötter	*Hippolais pallida*	olivaceous warbler
Blassstirniges Flechtenbärchen	*Eilema pygmaeola*	pygmy footman
Blassuhu, Milchuhu	*Bubo lacteus*	giant eagle-owl
Blatt-Fetzenfisch*	*Phycodorus eques*	leafy seadragon
Blattauster	*Pycnodonta folium/ Lopha folium/ Ostrea folium*	leaf oyster
Blattchamäleon	*Rhampholeon spectrum*	leaf chameleon, Cameroon stumptail chameleon
Blättchenschnecken	Velutinidae (incl. Lamellariinae)	vetulinas, lamellarias
Blättchenschnecken u.a.	Lamellariinae	lamellarias, ear snails (ear shells)
Blätter-Moostierchen	*Flustra foliacea*	broad-leaved hornwrack
Blattfingergecko> Europäischer Blattfinger	*Phyllodactylus europaeus*	European leaf-toed gecko
Blattfingergeckos	*Phrynodactylus* spp.	leaf-toed geckos
Blattfisch, Südamerikanischer Blattfisch	*Monocirrhus polyacanthus*	South American leaf-fish
Blattflöhe, Blattsauger	Psyllidae/Psylloidea	jumping plantlice, psyllids
Blattförmige Dornen-Stachelschnecke	*Ceratostoma foliata*	leafy hornmouth
Blattförmige Stachelschnecke	*Pterynotus phyllopterus*	leafy-winged murex
Blattfußkrebse, Kiemenfüßer	Phyllopoda/Branchiopoda	phyllopods, branchiopods
Blattgrüne Buschviper	*Atheris squamiger*	green bush viper, bush viper, leaf viper
Blattgrüne Mamba, Gewöhnliche Mamba	*Dendroaspis angusticeps*	eastern green mamba
Blatthähnchen, Getreidehähnchen	*Lema melanopus/ Oulema melanopus*	cereal leaf beetle (oat leaf beetle, barley leaf beetle)
Blatthornkäfer	Scarabaeidae	scarab beetles, lamellicorn beetles (dung beetles & chafers)
Blatthühnchen	Jacanidae	jacanas, lily-trotters
Blattkäfer	Chrysomelidae	leaf beetles
Blattkiemer, Lamellenkiemer	Eulamellibranchia	eulamellibranch bivalves
Blattläuse	Aphidina	aphids
Blattläuse	Aphidoidea	aphids & greenflies etc.
Blattlausfliegen	Chamaemyiidae/Ochthiphilidae	aphid flies
Blattlausschlupfwespen	Aphidiidae	aphid parasites
Blattnasen-Fledermäuse, Neuwelt-Blattnasen, Neuwelt-Lanzennasen	Phyllostomidae	American leaf-nosed bats, New World spear-nosed bats, New World leaf-nosed bats
Blattnasennattern	*Langaha* spp. & *Phyllorhynchus* spp.	leafnose snakes, leaf-nosed snakes

Blattnematoden	*Aphelenchoides* spp.	leaf nematodes
Blattohrmäuse	*Phyllotis* spp.	leaf-eared mice, pericotes
Blättrige Archenmuschel*	*Barbatia foliata*	leafy ark (leafy ark shell)
Blättrige Entenmuschel	*Pollicipes polymerus*	leaf barnacle
Blättrige Feuerkoralle	*Millepora complanata*	bladed fire coral
Blättrige Sonnenschnecke	*Lamellitrochus lamellosus*	lamellose solarelle
Blattroller	Attelabinae	leaf-rolling weevils
Blattschneiderameisen	*Atta* spp.	leafcutting ants
Blattschneiderbienen	*Megachile* spp.	leafcutting bees, leaf-cutter bees
Blattschneiderbienen	Megachilidae	leafcutting bees, leaf-cutter bees
Blattschupper	*Anoplogaster cornuta*	fangtooth
Blattschupper	Anoplogasteridae	fangtooths
Blattschuppiger Schlingerhai, Düsterer Dornhai	*Centrophorus squamosus*	leafscale gulper shark
Blattschwanzgeckos	*Phyllurus* spp.	leaf-tailed geckos, leaftail geckos
Blattspanner> Gemeiner Blattspanner	*Xanthorhoe fluctuata*	garden carpet
Blattsteigerfrösche, Giftfrösche	*Phyllobates* spp.	golden poison frogs, poison-arrow frog
Blatttütenmotten (Miniermotten)	Gracillariidae	leaf blotch miners
Blattvögel, Feenvögel	Irenidae	fairy-bluebirds, leafbirds
Blattwespen	Tenthredinidae	common sawflies
Blattzweig-Moostierchen, Elchgeweih-Moostierchen	*Hippodiplosia foliacea*	leafy horse-bryozoan*
Blau-Weißer Delphin, Streifendelphin	*Stenella coeruleoalba*	striped dolphin, blue-white dolphin, Euphrosyne dolphin
Blauaugen	Pseudomugilidae	blue eyes, blueeyes
Blauaugen-Harnischwels	*Panaque suttoni*	blue-eyed panaque
Blauaugenscharbe	*Phalacrocorax atriceps*	blue-eyed cormorant
Blauäugige Rossie	*Allorossia glaucopsis*	blue-eyed bob-tailed squid
Blauäugiger Waldportier	*Minois dryas/Satyrus dryas*	dryad
Blaubandbärbling	*Pseudorasbora parva*	false harlequin
Blaubandkrake, Blaugeringelter Krake	*Hapalochlaena maculosa*	blue-ringed octopus
Blaubarsch	*Badis badis*	chameleonfish, badis (FAO)
Blaubarsch, Blaufisch	*Pomatomus saltator*	blue fish, bluefish (FAO), tailor, elf, elft
Blaubarsche	Badidae	chameleonfishes
Blaubartmuschel, Seemuschel, Mittelmeer-Miesmuschel	*Mytilus galloprovincialis*	Mediterranean mussel
Blaubock	*Hippotragus leucophaeus*	blue buck
Blauböckchen, Blauducker	*Cephalophus monticola/ Philantomba monticola*	blue duiker
Blaue Argentinische Vogelspinne, Weißnacken-Vogelspinne	*Aphonopelma saltator*	Argentine blue tarantula*
Blaue Bastardmakrele, Blauer Stöcker	*Trachurus picturatus*	offshore jack mackerel, blue jack mackerel (FAO)

Blaue Feuerqualle, Blaue Nesselqualle	*Cyanea lamarcki*	blue lion's mane, cornflower jellyfish
Blaue Kiefernholzwespe	*Sirex cyaneus*	blue horntail
Blaue Königskrabbe	*Paralithodes platypus*	blue king crab
Blaue Koralle, Blaue Feuerkoralle	*Heliopora coerulea*	blue coral, blue fire coral
Blaue Korallen	*Heliopora* spp.	blue corals
Blaue Krustenanemone	*Porites branneri*	blue crust coral
Blaue Lungenqualle	*Rhizostoma octopus* (syn. *R. pulmo*)	blue cabbage bleb
Blaue Napfschnecke	*Patella caerulea*	rayed Mediterranean limpet
Blaue Nesselqualle, Blaue Feuerqualle	*Cyanea lamarcki*	blue lion's mane, cornflower jellyfish
Blaue Notothenia	*Paranotothenia magellanica*	Magellanic rockcod, Maori cod (FAO)
Blaue Putzgrundel, Neon-Grundel	*Gobiosoma oceanops/ Elacatinus oceanops*	neon goby
Blaue Schmeißfliegen	*Calliphora* spp.	bluebottles
Blaue Schwimmkrabbe, Blaukrabbe	*Callinectes sapidus*	blue crab, Chesapeake Bay swimming crab
Blaue Seescheide	*Clavelina caerulea*	blue light-bulb sea-squirt
Blaue Stachelmakrele, Blaumakrele*, Rauchflossenmakrele	*Caranx crysos*	blue runner, blue runner jack
Bläuel, Pompano, Gabelmakrele	*Trachinotus ovatus*	pompano, derbio
Blauelster	*Cyanopica cyana*	azure-winged magpie
Blauer Antennenwels	*Xenocara dolichoptera/ Ancistrus dolichoptera*	blue-chin xenocara, blue-chin ancistrus
Blauer Diskus	*Symphysodon aequifasciata haraldi*	blue discus
Blauer Doktorfisch	*Acanthurus coerulens*	blue tang surgeonfish
Blauer Eichenzipfelfalter	*Quercusia quercus/ Thecla quercus*	purple hairstreak
Blauer Erlenblattkäfer	*Agelastica alni*	alder leaf-beetle
Blauer Fächerfisch	*Cynolebias bellottii*	pavito, Argentine pearlfish
Blauer Fadenfisch	*Trichogaster trichogaster sumatranus*	blue gourami
Blauer Fellkäfer	*Korynetes coeruleus*	blue ham beetle*
Blauer Füsilier, Himmelblauer Füsilier, Schwarzfleck-Füsilier	*Caesio lunaris*	lunar fusilier
Blauer Hochseebarsch	*Chromis cyanea*	blue chromis
Blauer Katfisch, Wasserkatze	*Anarhichas latifrons/ Anarhichas denticulatus*	jelly cat, jelly wolffish, northern wolffish (FAO)
Blauer Katzenwels	*Ictalurus furcatus*	blue catfish
Blauer Kofferfisch	*Ostracion lentiginosum*	blue boxfish
Blauer Kongocichlide, Kongo-Zwergbuntbarsch	*Nanochromis parilus/ Nanochromis nudiceps*	nudiceps, Congo dwarf cichlid*
Blauer Kongosalmler	*Phenacogrammus interruptus*	Congo tetra
Blauer Marlin	*Makaira nigricans*	blue marlin, Atlantic blue marlin (FAO)
Blauer Panzerwels	*Corydoras nattereri*	blue corydoras

Blauer Pfeilgiftfrosch	*Dendrobates azureus*	blue poison-arrow frog, blue poison frog
Blauer Prachtkärpfling	*Aphyosemion sjoestedti*	blue gularis (FAO), red aphyosemion, golden pheasant
Blauer Scheibenbock, Violetter Scheibenbock, Blauvioletter Scheibenbock, Veilchenbock	*Callidium violaceum*	violet tanbark beetle
Blauer Seestern	*Linckia laevigata*	blue star, blue seastar
Blauer Stachelleguan	*Sceloporus cyanogenys*	blue spiny lizard
Blauer Teufel	*Pomacentrus fuscus/ Stegastes fuscus*	Brazilian damsel (FAO), blue devil
Blauer Weidenblattkäfer	*Plagiodera versicolor*	imported willow leaf beetle (U.S.)
Blauer Weidenblattkäfer	*Phyllodecta vulgatissima*	blue willow leaf beetle
Blauer Wittling, Blauwittling	*Micromesistius poutassou*	blue whiting (FAO), poutassou
Blauer Ziegelbarsch	*Lopholatilus chamaeleonticeps*	tilefish
Blauer Zwergkaiser, Blauer Karibischer Zwergkaiserfisch	*Centropyge argi*	cherubfish
Blaues Ordensband	*Catocala fraxini*	Clifden Nonpareil
Blaues Trompetentierchen	*Stentor coerulens*	blue trumpet animalcule
Blaufelchen	*Coregonus wartmanni*	pollan (freshwater herring)
Blaufische	Pomatomidae	bluefishes
Blaufleck-Querzahnmolch	*Ambystoma laterale*	blue-spotted salamander
Blauflecken-Stechrochen, Blaufleck-Stachelrochen, Blaufleck-Stechrochen, Blaupunktrochen	*Taeniura lymma*	ribbontail stingray, fantail, blue-spotted stingray, blue-spotted lagoonray, bluespotted ribbontail ray (FAO)
Blaufleckenbrasse	*Scolopsis xenochrous*	olive-spotted monocle bream
Blaufleckiger Ansauger	*Lepadogaster lepadogaster*	shore clingfish (FAO), Cornish clingfish
Blauflossen-Knurrhahn	*Chelidonichthys kumu*	blue-fin gurnard
Blauflossen-Makrele	*Caranx melampygus/ Caranx bixanthopterus*	bluefin kingfish, bluefin trevally (FAO), blue jack, spotted trevally
Blauflügel-Prachtlibelle	*Calopteryx virgo*	bluewing, demoiselle agrion
Blauflügelente	*Anas discors*	blue-winged teal
Blauflügelige Ödlandschrecke	*Oedipoda coerulescens*	blue-winged grasshopper
Blauflügelige Sandschrecke, Blauflügelschrecke	*Sphingonotus caerulans*	blue-winged locust
Blaugebänderte Ruderschlange	*Hydrophis cyanocinctus*	annulated sea snake, ringed sea snake, banded sea snake
Blaugefleckter Meerbrassen	*Pagrus caeruleostictus*	blue-spotted seabream
Blaugelber Zackenbarsch	*Epinephelus flavocaerulens*	blue-and-yellow grouper (FAO), purple rock-cod
Blaugelber Zwergkaiser, Schwarzgelber Zwergkaiserfisch	*Centropyge bicolor*	bicolor angelfish

Blaugeringelter Krake, Blaubandkrake	*Hapalochlaena maculosa*	blue-ringed octopus
Blaugoldener Füsilier, Gelbstreifen-Füsilier	*Caesio caerulaureus*	blue-and-gold fusilier (FAO), gold-banded fusilier
Blaugraue Steineule	*Polymixis xanthomista*	black-banded moth
Blaugrüne Mosaikjungfer	*Aeshna cyanea*	blue-green darner, southern aeshna, southern hawker
Blauhäher	*Cyanocitta cristata*	blue jay
Blauhai, Menschenhai	*Prionace glauca*	blue shark
Blauhaie, Grundhaie	Carcharhinidae	requiem sharks
Blauhecht	*Antimora rostrata*	blue antimora
Blaukäppchen	*Vini australis*	blue-crowned lory
Blaukehl-Hüttensänger	*Sialia mexicana*	western bluebird
Blaukehlagame	*Agama atricollis/ Stellio atricollis*	blue-throated agama
Blaukehlchen	*Luscinia svecica/ Cyanosylvia svecia*	bluethroat
Blaukopf	*Diloba caeruleocephala*	figure of eight moth
Blaukopf-Ziegelbarsch	*Malacanthus latovittatus*	sand tilefish
Blaukopfastrild	*Uraeginthus cyanocephala*	blue-capped cordon-bleu
Blaukopfjunker	*Thalassoma bifasciatum*	bluehead
Blaukorallen	Coenothecalia/Helioporida	blue corals
Blaukrabbe, Blaue Schwimmkrabbe	*Callinectes sapidus*	blue crab, Chesapeake Bay swimming crab
Blaukrönchen	*Loriculus galgulus*	blue-crowned hanging parrot
Blauleng	*Molva dipterygia dipterygia*	blue ling
Bläuling, Blaugrüne Baumwanze	*Zicrona caerulea*	blue shield bug, blue bug
Bläulinge (Feuerfalter) & Zipfelfalter	Lycaenidae	blues & hairstreaks & coppers & harvesters & metalmarks
Blaumaskenamazone	*Amazona versicolor*	St. Lucia parrot, St. Lucia amazon
Blaumaul	*Helicolenus dactylopterus*	bluemouth, blackberry rosemouth
Blaumaul-Agamen	*Aphaniotis* spp.	earless agamas
Blaumaulmeerkatze	*Cercopithecus cephus*	moustached monkey
Blaumeise	*Parus caeruleus*	blue tit
Blaumerle	*Monticola solitarius*	blue rock thrush
Blaumückenfänger	*Polioptila caerulea*	blue-grey gnatcatcher, blue-gray gnatcatcher
Blauparadiesvogel	*Paradisaea rudolphi*	blue bird-of-paradise
Blaupfeile	*Orthetrum* spp.	orthetrums
Blaupunkt-Buntbarsch, Blaupunktbarsch	*Aequidens latifrons/ Aequidens pulcher*	blue acara
Blaupunkt-Grundelbuntbarsch	*Spathodus erythrodon*	
Blaupunktrochen, Blauflecken-Stechrochen, Blaufleck-Stachelrochen, Blaufleck-Stechrochen	*Taeniura lymma*	ribbontail stingray, fantail, blue-spotted stingray, blue-spotted lagoonray, bluespotted ribbontail ray (FAO)
Blauracke	*Coracias garrulus*	European roller
Blauringtaube	*Leptotila verreauxi*	white-fronted dove
Blaurochen	*Neoraja caerulea/ Breviraja caerulea*	blue ray

Blaurote Garnele, Blassrote Tiefseegarnele	*Aristeus antennatus*	blue-and-red shrimp
Blaurücken-Flusshering	*Pomolobus aestivalis*	blueback herring (alewife)
Blaurücken-Makrele, Schwarzrücken-Stachelmakrele	*Caranx ruber/ Carangoides ruber*	bar jack
Blaurückenhering, Kanadische Alse	*Alosa aestivalis*	blueback herring
Blaurückenlachs, Blaurücken, Roter Lachs	*Oncorhynchus nerka*	sockeye salmon (FAO), sockeye
Blauschafe	*Pseudois* spp.	blue sheep, bharal
Blauscheitellori	*Glossopsitta porphyreocephala*	purple-crowned lorikeet
Blauscheitelmotmot	*Momotus momota*	blue-crowned motmot
Blauschnäpper	*Cyanoptila cyanomelana*	blue-and-white flycatcher
Blauschwarzer Eisvogel	*Limenitis reducta*	southern white admiral
Blauseidiger Kohlerdfloh	*Phyllotreta nigripes*	blueish turnip flea beetle
Blausieb, Kastanienbohrer	*Zeuzera pyrina*	leopard moth
Blausteißsittich	*Pyrrhura perlata*	pearly conure
Blaustirnamazone, Rotbugamazone	*Amazona aestiva*	blue-fronted amazon
Blaustreifen-Bürstenzahnseebader	*Ctenochaetus strigosus*	blue-striped surgeon fish
Blaustreifen-Doktorfisch, Streifen-Doktorfisch	*Acanthurus lineatus*	clown surgeonfish, lined surgeonfish (FAO)
Blaustreifen-Drückerfisch, Tüpfeldrückerfisch	*Pseudobalistes fuscus*	rippled triggerfish, yellow-spotted triggerfish (FAO)
Blaustreifen-Säbelzahnschleimfisch	*Plagiotremus rhinorhynchus*	bluestriped fangblenny
Blaustreifengrunzer, Blaustreifen-Grunzerfisch	*Haemulon sciurus*	bluestriped grunt
Blauwal	*Balaenoptera musculus*	blue whale, sulphur bottom whale
Blauwangen-Bartvogel	*Megalaima asiatica*	blue-throated barbet
Blauwangenspint	*Merops persicus*	blue-cheeked bee-eater
Blauwittling, Blauer Wittling	*Micromesistius poutassou*	blue whiting, poutassou
Blauzecke*, Afrikanische Schweinezecke	*Boophilus decoloratus*	blue tick
Blauzügelarassari	*Aulacorhynchus sulcatus*	groove-billed toucanet
Blauzungen-Skink	*Tiliqua* spp.	bluetongue skinks, blue-tongued skinks
Blei, Brachsen, Brassen, Brasse	*Abramis brama*	common bream, freshwater bream, carp bream (FAO)
Bleichböckchen	*Ourebia ourebi*	oribi
Bleiche Getreideblattlaus	*Metopolophium dirhodum*	rose-grain aphid
Bleiche Schließmundschnecke	*Cochlodina fimbriata*	pallid door snail, plaited door snail
Bleichstreifige Beerenblattlaus	*Cryptomyzus galeopsidis*	black currant aphid, currant blister aphid
Blessbock, Buntbock	*Damaliscus dorcas*	bontebok, blesbok

Blicke, Güster, Pliete (Halbbrachsen)	*Blicca bjoerkna*	silver bream, white bream (FAO)
Blindbarbe	*Caecobarbus geertsi*	blind barb
Blinder Höhlenkäfer	*Glacicavicola bathysciodes*	blind cave beetle
Blinder Wunderschirm	*Cirrothauma murrayi*	
Blindfische	Amblyopsidae	cavefishes
Blindfliege	*Chrysops caecutiens*	blinding breeze fly
Blindfliegen	*Chrysops* spp.	deerflies
Blindhöhlenfisch	*Anoptichthys jordani/ Astyanax fasciatus mexicanus*	blind cavefish
Blindkäferschnecken*	Lepetidae	blind limpets, eyeless limpets
Blindmaulwurf	*Talpa caeca*	blind mole
Blindmäuse	Spalacidae	blind mole-rats
Blindmäuse	*Spalax* spp.	blind mole-rats
Blindmulle	*Myospalax* spp.	mole-rats, zokors
Blindsalamander	*Haideotriton wallacei*	Georgia blind salamander
Blindschlangen	*Ramphotyphlops* spp.	common blind snakes
Blindschleiche	*Anguis fragilis*	European slow worm (blindworm), slow worm
Blindschnecke, Blinde Turmschnecke, Nadelschnecke	*Cecilioides acicula*	blind awlsnail, blind snail, agate snail
Blindspringer	Onychiuridae	blind springtails
Blinzelhai	*Brachaelurus waddi*	blind shark
Blinzelhaie	Brachaeluridae	blind sharks
Blitzaugenkalmar*	*Abralia veranyi*	eye-flash squid
Blitzschlagmuschel*	*Pitar fulminatus*	lightning pitar
Blitzschnecke, Linksgewundene Wellhornschnecke	*Busycon contrarium*	lightning whelk, left-handed whelk
Blökender Frosch	*Crinia bilingua*	bleating froglet
Blödauge, Wurmschlange	*Typhlops vermicularis*	European blind snake, Greek blind snake, worm snake
Blöker, Gelbstriemen	*Boops boops*	bogue
Blonde, Kurzschwanz-Rochen	*Raja brachyura*	blonde ray (FAO), blond ray
Blumenfliegen	Anthomyiidae	anthomyids, flower flies
Blumenkohlkoralle	*Pocillopora eydouxi*	cauliflower coral*
Blumenkohlquallen, Lungenquallen	*Rhizostoma* spp.	cabbage bleb, marigold, blubber, football jellyfish
Blumennasen-Fledermaus	*Anthops ornatus*	flower-faced bat
Blumensternkoralle	*Diploastrea heliopora*	flowery star coral*
Blumentiere, Blumenpolypen, Anthozoen	Anthozoa	flower animals, anthozoans
Blumentopfschlange	*Ramphotyphlops braminus*	braminy blind snake

Blumenwanzen	Anthocoridae	minute pirate bugs
Blut-Mitra*	*Mitra sanguisuga*	blood-sucking mitre
Blut-Olive	*Oliva reticulata*	blood olive
Blutarche*, Blutrote Archenmuschel	*Anadara ovalis*	blood ark (blood ark shell)
Blutbrustpavian, Dschelada	*Theropithecus gelada*	gelada
Blutegel> Medizinischer Blutegel	*Hirudo medicinalis*	medicinal leech
Blutegel> Amerikanischer Blutegel, Amerikanischer Medizinischer Blutegel	*Macrobdella decora*	American medicinal leech
Blutegel, Pärchenegel	*Schistosoma* spp.	blood flukes
Blutegel-Tellerschnecke*, Bilharziose-Tellerschnecke*	*Biomphalaria glabrata*	bloodfluke planorb
Blüten-Krabbenspinne*	*Misumena* spp.	flower crab spiders
Blütenbesuchende Fledermäuse	*Syconycteris* spp.	blossom bats
Blutender Zahn	*Nerita peloronta*	bleeding tooth
Blütenfresser, Himbeerkäfer	Byturidae	fruitworm beetles
Blütengrillen	Oecanthidae	tree crickets (Oecanthinae)
Blütenmulmkäfer	Anthicidae	antlike flower beetles
Blütenthrips, Gemeiner Blütenthrips	*Frankliniella intonsa*	flower thrips
Blutfasan	*Ithaginis cruentus*	blood pheasant
Blutfleck-Kammmuschel	*Gloripallium sanguinolenta*	blood-stained scallop
Blutfleck-Täubchenschnecke	*Columbella haemastoma*	blood-stained dovesnail (blood-stained dove shell)
Blutfleckplanarie	*Oligocladus sanguinolentus*	
Bluthänfling	*Carduelis cannabina*	linnet
Blutkieferwurm	*Marphysa sanguinea*	rock worm, red rock worm, red-gilled marphysa
Blutlaus	*Eriosoma lanigerum*	woolly aphid, "American blight"
Blutläuse, Blasenläuse	Schizoneurinae/Eriosomatinae	woolly aphids, gall aphids
Blutlippengrundel	*Gobius cruentatus*	red-mouthed goby (FAO), red-mouth goby
Blutpirol	*Oriolus trailii*	maroon oriole
Blutrote Archenmuschel, Blutarche*	*Anadara ovalis*	blood ark (blood ark shell)
Blutrote Heidelibelle	*Sympetrum sanguineum*	ruddy sympetrum
Blutrote Raubameise	*Formica sanguinea*	blood-red ant
Blutrote Sandmuschel	*Sanguinolaria cruenta*	blood-stained sanguin
Blutrote Walzenschnecke	*Cymbiola rutila*	blood-red volute
Blutrote Zikade, Lauer, Weinzwirner	*Tibicina haematodes/ Tibicen haematodes*	vineyard cicada
Blutroter Röhrenwurm, Roter Kalkröhrenwurm	*Protulopsis intestinum/ Protula intestinum*	bloody tubeworm*
Blutroter Schnellkäfer	*Ampedus sanguineus*	bloodred click beetle*
Blutsalmler	*Hyphessobrycon callistus*	jewel tetra
Blutsauger-Mitra	*Vexillum sanguisugum*	bloodsucking mitre, bloodsucking miter, bloodsucker miter
Blutschnabelweber	*Quelea quelea*	red-billed quelea

Blutspecht	*Dendrocopos syriacus*	Syrian woodpecker
Blutstern	*Henricia sanguinolenta*	blood star, northern henricia
Blutstern> Purpurstern, Purpurseestern, Roter Seestern, Roter Mittelmeerseestern	*Echinaster sepositus*	red Mediterranean sea star
Blutstriemen-Schleimfisch	*Parablennius sanguinolentus*	blood-striped blenny
Blutströpfchen (Widderchen)	*Zygaenidae (Anthroceridae)*	burnets & foresters, smoky moths
Blutströpfchen> Gemeines Blutströpfchen, Steinbrechwidderchen, Gewöhnliches Sechsfleck-Blutströpfchen, Erdeichel-Widderchen	*Zygaena filipendulae/ Anthrocera filipendulae*	six-spot burnet
Bluttropfen-Seescheide	*Dendrodoa carnea*	blood drop sea-squirt
Blutwurm*	*Tubifex costatus*	bloodworm
Blutzikade, Rotschwarze Schaumzikade	*Cercopis vulnerata*	red-and-black froghopper
Blyths Schildschwanzschlange	*Rhinophis blythis*	Blyth's Landau shieldtail
Boa-Trugnatter	*Homalopsis buccata*	masked water snake
Boazähner, Hausschlangen, Hausnattern	*Lamprophis* spp./*Boaedon* spp.	house snakes
Bobak (Hbz. Murmel)	*Marmota bobak*	bobac marmot
Bobo-Meeräsche	*Joturus pichardi*	bobo mullet
Bobo-Umberfisch	*Pseudotolithus elongatus*	bobo croaker
Bobolink	*Dolichonyx oryzivorus*	bobolink
Bocages Mauereidechse	*Podarcis bocagei*	Bocage's wall lizard
Bockadam	*Cerberus microlepis*	bockadam
Bockkäfer, Böcke	Cerambycidae	longhorn beetles, long-horned beetles
Boden-Kielnacktschnecke	*Milax budapestensis/ Tandonia budapestensis*	Budapest slug
Bodengucker	*Selene selene*	lookdown
Bodennatter	*Sonora semiannulata*	ground snake
Bodennatter> Gewöhnliche Bodennatter	*Liophis frenata*	common ground snake
Bodenspinnen	Hahniidae	hahniids
Bodentrichterspinne	*Coelotes atropos*	
Bodenwanzen, Langwanzen (Ritterwanzen)	Lygaeidae	ground bugs, seed bugs
Boe-Umberfisch	*Pteroscion peli*	boe drum
Bogen-Schwimmkrabbe	*Callinectes arcuatus*	arched swimming crab
Bogenfingergeckos, Nacktfingergeckos	*Cyrtodactylus* spp.	bow-fingered geckos
Bogenkrabbe	*Cancer productus*	red rock crab, red crab
Bogenmund-Geigenrochen, Bogenmund-Gitarrenrochen	*Rhina ancylostoma*	bowmouth guitarfish
Bogenstirn-Hammerhai, Indopazifischer Hammerhai	*Sphyrna lewini*	scalloped hammerhead shark
Bohnen-Papiermuschel	*Amygdalum phaseolinum*	kidney-bean horse mussel, kidney-bean papermussel

Bohnenfliege, Kammschienen-Wurzelfliege, Schalottenfliege	*Delia platura/ Hylemia platura/ Phorbia platura*	seed-corn fly, seed-corn maggot, bean-seed fly, shallot fly
Bohnenförmige Miesmuschel, Bohnenmiesmuschel*	*Modiolula phaseolina*	bean horse mussel
Bohrassel, Holzbohrassel	*Limnoria lignorum*	gribble
Bohrer-Schraubenschnecke	*Turritella terebra*	auger turretsnail (auger screw shell, tower screw shell)
Bohrfliegen, Fruchtfliegen	Tephritidae/Trypetidae	fruit flies
Bohrkäfer, Buchenwerftkäfer, Sägehörniger Werftkäfer	*Hylecoetus dermestoides*	large timberworm, European sapwood timberworm*
Bohrmotten	Ochsenheimeriidae	fields, field moths
Bohrmuscheln, Echte Bohrmuscheln	Pholadidae	piddocks
Bohrschwamm, Zellenbohrschwamm	*Cliona celata*	yellow boring sponge, sulfur sponge
Bohrschwämme	*Cliona* spp.	boring sponges
Bojen-Seepocke	*Dosima fascicularis*	buoy barnacle
Bola-Spinnen, Lasso-Spinnen	*Mastophora* spp.	bolas spiders
Bolivianische Blattohrmaus	*Galenomys garleppi*	Bolivian leaf-eared mouse
Bolyerschlange	*Bolyeria multicarinata*	Round Island burrowing boa
Bombardierkäfer	*Brachinus* spp.	bombardier beetles
Bombay-Ente	*Harpadon nehereus*	Bombay duck, bumalo, bummalow
Bombay-Enten	Harpadontidae	bombay ducks
Bonapartemöve	*Larus philadelphia*	Bonaparte's gull
Bonga-Hering	*Ethmalosa fimbriata*	bonga shad
Bongo	*Taurotragus euryceros*	bongo
Bonito> Echter Bonito, Gestreifter Thun	*Euthynnus pelamis/ Katsuwonus pelamis*	skipjack tuna
Bonito> Falscher Bonito, Thonine, Kleine Thonine	*Euthynnus alletteratus/ Gymnosarda alletterata*	little tuna, little tunny (FAO), mackerel tuna, bonito
Bonito> Unechter Bonito, Melvera-Fregattmakrele	*Auxis rochei*	bullet tuna (FAO), bullet mackerel
Bonobo, Zwergschimpanse	*Pan paniscus*	bonobo, pygmy chimpanzee
Boomslang, Grüne Baumschlange	*Dispholidus typus*	boomslang
Bootschnecken	Scaphandridae	canoe bubblesnails
Bootschwanzgrackel	*Quiscalus major*	boat-tailed grackle
Bori	*Octodontomys gliroides*	chozchoz (*pl* chozchoris)
Borken-Tritonshorn	*Cabestana cutacea*	Mediterranean bark triton
Borkenkäfer	Scolytidae (Ipidae)	bark beetles, engraver beetles & ambrosia beetles, timber beetles

Borstenwürmer

Borkenratte> Buschschwanz-Borkenratte, Schadenbergs Borkenratte	*Crateromys schadenbergi*	bushy-tailed cloud rat, giant bushy-tailed cloud rat
Borkenratte> Pauls Borkenratte, Ilin-Borkenratte	*Crateromys paulus*	Ilin Island cloud rat
Borkenschwamm, Smaragdschwamm	*Heteromeyenia baleyi*	emerald sponge*
Borneo-Delphin, Kurzschnabeldelphin, Frasers Delphin	*Lagenodelphis hosei*	shortsnouted whitebelly dolphin, Fraser's dolphin, Sarawak dolphin, Bornean dolphin
Borneo-Dornauge	*Acanthophthalmus shelfordi/* *Pangio shelfordi*	Borneo loach
Borneo-Flussschildkröte	*Orlitia borneensis*	Malaysian giant turtle
Borneo-Gibbon, Grauer Gibbon	*Hylobates muelleri*	gray gibbon, Müller's gibbon
Borneo-Goldkatze	*Felis badia/* *Catopuma badia*	Borneo golden cat, Bornean bay cat, bay cat, Bornean red cat
Borneo-Hörnchen	*Rheithrosciurus macrotis*	groove-toothed squirrel
Borneo-Taubwaran	*Lanthanotus borneensis*	Bornean earless lizard, Borneo earless monitor
Borneo-Zwerghörnchen*	*Glyphotes* spp.	Bornean pygmy squirrels
Borsten-Baumstachler	*Chaetomys subspinosus*	thin-spined porcupine, bristle-spined porcupine
Borstenegel	Acanthobdelliformes	bristly leeches
Borstenfühleriger Haarwurm	*Spio seticornis*	seticorn spio*
Borstengürteltiere	*Chaetophractus* spp./ *Euphractus* spp.	hairy armadillos, peludos
Borstenhörnchen	*Xerus rutilus*	unstriped ground squirrel
Borstenigel, Tanreks	Tenrecidae	tenrecs
Borstenkaninchen	*Caprolagus hispidus*	bristly rabbit, hispid "hare", Assam rabbit
Borstenkiefer, Pfeilwürmer, Chaetognathen	Chaetognatha	arrow worms, chaetognathans
Borstenkrabbe	*Pilumnus hirtellus*	hairy crab, bristly xanthid crab
Borstenkugel-Gallwespe (>Borstige Kugelgalle)	*Neuroterus tricolor/* *Neuroterus fumipennis*	oak leaf cupped-gall cynipid wasp, oakleaf cupped-spanglegall cynipid wasp (>cupped spangle gall)
Borstenmünder, Borstenmäuler	Gonostomatidae	bristlemouths
Borstenrabe	*Corvus rhipidurus*	fan-tailed raven
Borstenschwänze	Stipiturini	emuwrens
Borstenschwänze, Thysanuren (Felsenspringer & Fischchen)	Thysanura	bristletails, thysanurans
Borstenwürmer, Chaetopoden (Polychaeten & Oligochaeten)	Chaetopoda	chaetopods, bristle worms (annelids with chaetae: polychetes & oligochetes)
Borstenwürmer, Vielborster, Polychaeten	Polychaeta	bristle worms, polychaetes, polychetes, polychete worms

Borstige Rinderlaus	*Solenopotes capillatus*	tubercle-bearing louse, blue cattle louse, small blue cattle louse
Boston-Fadenschnecke	*Facelina bostoniensis*	Boston facelina
Bougainvillia-Polyp	*Bougainvillia ramosa*	Bougainvillia polyp*
Boulenger-Flachschildkröte	*Homopus boulengeri*	donner-weer tortoise, Boulenger's Cape tortoise
Boulengers Asien-Baumkröte	*Pedostibes hosii*	Boulenger's Asian tree toad
Bowdoins Schnabelwal	*Mesoplodon bowdoini*	deep-crested beaked whale
Boxerkrabbe	*Lybia tessalata*	anemone carrying crab, common boxing crab
Boyers Ährenfisch, Kleiner Ährenfisch	*Atherina boyeri*	big-scale sand smelt (FAO), big-scaled sandsmelt, Boyer's sand smelt
Brabantbuntbarsch, Brabant-Buntbarsch	*Tropheus moorii*	blunthead cichlid (FAO), firecracker moorii, rainbow moorii
Brachfliege	*Phorbia coarctata/ Delia coarctata*	wheat bulb fly
Brachiopoden, Armfüßer, „Lampenmuscheln"	Brachiopoda	lampshells, brachiopods
Brachpieper	*Anthus campestris*	tawny pipit
Brachsen, Blei, Brassen, Brasse	*Abramis brama*	common bream, freshwater bream, carp bream (FAO)
Brachsenmakrele	*Brama brama*	pomfret, Atlantic pomfret (FAO), Ray's bream
Brachsenmakrelen, Pomfrets, Goldköpfe	Bramidae	pomfrets
Brachvogel> Großer Brachvogel	*Numenius arquata*	curlew
Brackwasser-Dreiecksmuschel	*Congeria cochleata*	brackish water wedge clam
Brackwasser-Süßlippe, Gefleckte Süßlippe*	*Pomadasys commersonnii*	spotted grunter
Brackwasser-Winkerkrabbe, Rotgelenks-Winkerkrabbe	*Uca minax*	redjointed fiddler, brackish-water fiddler
Brackwasseraktinie	*Diadumene cincta*	orange anemone
Brackwespen	Braconidae	braconids, braconid wasps
Brahaminenweih	*Haliastur indus*	Brahminy kite
Brandbrasse, Oblada	*Oblada melanura*	saddled bream, saddled seabream (FAO)
Brandgans	*Tadorna tadorna*	common shelduck
Brandhorn, Herkuleskeule	*Murex brandaris/ Bolinus brandaris*	dye murex, purple dye murex
Brandmaus, Feldmaus	*Apodemus agrarius*	Old World field mouse, striped field mouse
Brandseeschwalbe	*Sterna sandvicensis*	Sandwich tern
Brandspurkegel*	*Conus encaustus*	burnt cone
Brandt-Fledermaus, Große Bartfledermaus	*Myotis brandti*	Brandt's bat
Brandt-Steppenwühlmaus	*Microtus brandti*	Brandt's vole
Brandungsbarsche	Embiotocidae	surfperches

Brants Pfeifratte	*Parotomys brantsii*	Brant's whistling rat
Brasilbohnenkäfer	*Zabrotes subfasciatus*	Mexican bean weevil
Brasilia Grabmaus	*Juscelinomys candango*	Brasilia burrowing mouse
Brasilianische Baummaus*	*Rhagomys rufescens*	Brazilian arboreal mouse
Brasilianische Glattnatter	*Hydrodynastes gigas/ Cyclagras gigas*	South American water 'cobra', beach 'cobra', false cobra, Brazilian smooth snake, "false water cobra"
Brasilianische Mabuya, Brasilianische Mabuye	*Mabuya heathi*	Brazilian skink, Brazilian mabuya
Brasilianische Meeräsche	*Mugil liza*	Brazilian mullet
Brasilianische Plattschildkröte, Strahlen-Plattschildkröte	*Platemys radiolata*	Brazilian radiolated swamp turtle, Brazilian radiated swamp turtle
Brasilianische Sandbarsche	*Pinguipes* spp.	Brazilian sandperches
Brasilianische Schabe	*Blaberus giganteus*	Brazilian cockroach
Brasilianische Schlangenhalsschildkröte	*Hydromedusa maximiliani*	Maximilian's snake-necked turtle, Maximilian's snake-headed turtle
Brasilianische Schmuckschildkröte	*Trachemys dorbignyi/ Pseudemys dorbignyi*	Brazilian slider, black-bellied slider
Brasilianische Stachelratte u.a.	*Carterodon sulcidens*	eastern Brasilian spiny rat
Brasilianische Vasenschnecke	*Turbinella laevigata*	Brazilian chank
Brasilianischer Bärenkrebs	*Scyllarides brasiliensis*	Brasilian slipper lobster
Brasilianischer Gabeldorsch	*Urophycis brasiliensis*	Brazilian codling
Brasilianischer Kampfuchs (Brasilfuchs/Rio-Fuchs)	*Pseudalopex vetulus/ Lycalopex vetulus*	Brasilian fox, hoary fox
Brasilianischer Plattkopf	*Percophis brasilianus*	Brazilian flathead
Brasilianischer Wühlhamster	*Blarinomys breviceps*	Brazilian shrew-mouse
Brasilien-Riesenvogelspinne, Parahyba-Vogelspinne	*Lasiodora parahybana*	Brazilian salmon tarantula
Brasilien-Waldkaninchen	*Sylvilagus brasiliensis*	forest rabbit
Brassen, Brasse, Brachsen, Blei	*Abramis brama*	common bream, freshwater bream, carp bream (FAO)
Brassen, Meerbrassen	Sparidae	seabreams, porgies
Bratpfannenfisch*	*Gobiesox strumosus*	skilletfish
Bratpfannenwelse	Bunocephalidae	banjo catfishes
Brauen-Glattstirnkaiman, Brauenkaiman	*Paleosuchus palpebrosus*	dwarf caiman, Cuvier's dwarf caiman
Brauers Sandrennmaus	*Desmodilliscus braueri*	pouched gerbil
Braunanolis, Kuba-Anolis	*Anolis sagrei*	brown anole, Cuban anole
Braunauge	*Lasiommata maera*	large wall brown, wood-nymph
Braunbandschabe, Möbelschabe	*Supella supellectilium/ Supella longipalpa*	brown-banded cockroach, brownbanded cockroach, furniture cockroach
Braunbandschnecke	*Latirus infundibulum*	brown-line latirus
Braunbär	*Ursus arctos*	brown bear (grizzly bear)
Braunbauch-Dickichtvogel	*Atrichornis clamosus*	noisy scrub-bird
Braune Baumschlange	*Boiga irregularis*	brown tree snake

Braune Blumenfledermäuse

Braune Blumenfledermäuse	*Erophylla* spp.	brown flower bats
Braune Ecuador-Samt-Vogelspinne*	*Megaphobema velvetosoma*	Ecuadorian brownvelvet tarantula
Braune Erdschildkröte	*Rhinoclemmys annulata*	brown wood turtle
Braune Erdwanze	*Aethus flavicornis*	brown groundbug*
Braune Feenlämpchenspinne	*Agroeca brunnea*	
Braune Felsen-Kantengarnele	*Sicyonia brevirostris*	brown rock shrimp
Braune Feuchtwieseneule, Braune Bergeule	*Eriopygodes imbecilla*	Silurian moth
Braune Fleckplanarie	*Stylochus pilidium*	brown oyster leech*
Braune Garnele	*Penaeus aztecus/ Farfantepenaeus aztecus*	brown shrimp, northern brown shrimp
Braune Geißelgarnele	*Metapenaeus endeavouri*	endeavour shrimp
Braune Gewächshaus-Napfschildlaus*, Braune Napfschildlaus*	*Coccus hesperidum*	brown soft scale
Braune Großohrfledermäuse	*Histiotus* spp.	big-eared brown bats
Braune Hausschlange	*Boaedon fuliginosus/ Lamprophis fuliginosus*	common house snake, common brown house snake
Braune Heidelbeereule	*Conistra vaccinii*	chestnut
Braune Hühnerlaus	*Oulocrepis dissimilis/ Goniodes dissimilis*	brown chicken louse
Braune Hundezecke	*Rhipicephalus sanguineus*	brown dog tick, kennel tick
Braune Hüpfmaus> Dunkle Hüpfmaus	*Notomys fuscus*	dusky hopping mouse
Braune Hydra, Brauner Süßwasserpolyp	*Hydra vulgaris*	brown European hydra
Braune Kauri, Braune „Maus"	*Cypraea lurida/ Luria lurida*	lurid cowrie
Braune Krustenanemone	*Epizoanthus paxii*	brown zoanthid, brown encrusting anemone
Braune Landschildkröte	*Geochelone emys/ Manouria emys*	brown tortoise, Burmese brown tortoise, Asian giant tortoise (IUCN)
Braune Mehlmilbe	*Goheria fusca*	brown fluor mite
Braune Miesmuschel, Westatlantische Miesmuschel	*Perna perna*	Mexilhao mussel, South American rock mussel, "brown mussel"
Braune Mosaikjungfer	*Aeshna grandis*	brown aeshna, brown hawker, great dragonfly
Braune Muräne	*Gymnothorax unicolor*	brown moray
Braune Nadelschnecke	*Acicula fusca/Acme fusca*	brown point snail (point shell)
Braune Nesselqualle, Pazifische Braune Nesselqualle	*Chrysaora melanogaster*	brown jellyfish, brown giant jellyfish
Braune Ohrenzecke*	*Rhipicephalus appendiculatus*	brown ear tick
Braune Pastetenmuschel	*Glycymeris cor*	brown bittersweet*
Braune Pelomeduse	*Pelusios castaneus*	West African mud turtle
Braune Schabe	*Periplaneta brunnea*	brown cockroach
Braune Schüsselschnecke	*Discus ruderatus*	brown disc snail
Braune Spinne	*Loxosceles reclusa*	violin spider, brown recluse spider, fiddleback spider

Braune Spinnmilbe	*Bryobia rubrioculus*	brown mite, apple and pear bryobia
Braune Strandschrecke	*Aiolopus strepens*	southern longwinged grasshopper
Braune Sumpfschnecke	*Stagnicola fuscus*	brown pondsnail
Braune Tageule, Luzerneule	*Ectypa glyphica/ Euclidia glyphica*	burnet companion
Braune Tannenrindenlaus	*Cinara pilicornis*	brown spruce aphid, spruce shoot aphid
Braune Tigergarnele	*Penaeus esculentus*	brown tiger prawn
Braune Venusmuschel, Glatte Venusmuschel	*Callista chione/ Meretrix chione*	brown callista, brown venus
Braune Wassernatter	*Nerodia taxispilota*	brown water snake
Braune Wegameise	*Lasius brunneus*	brown ant
Braune Wegschnecke	*Arion subfuscus*	dusky slug, dusky arion
Braune Weizenmilbe*	*Petrobia latens*	brown wheat mite
Braune Winterzecke*	*Dermacentor nigrolineatus*	brown winter tick
Braune Witwe	*Latrodectus geometricus*	brown widow spider
Braune Witwe	*Melichtys vidua/ Balistes vidua*	pinktail triggerfish
Braunellen	Prunellidae	accentors, dunnock
Brauner Bachsalamander	*Desmognathus fuscus*	dusky salamander
Brauner Bambushai, Sattelfleckenhai	*Hemiscyllium punctatum/ Chiloscyllium punctatum*	brown carpet shark, brownbanded bambooshark (FAO)
Brauner Bär	*Arctia caja*	garden tiger (moth)
Brauner Brüllaffe	*Alouatta fusca*	brown howler
Brauner Diebkäfer	*Ptinus claviceps*	brown spider beetle
Brauner Drachenkopf	*Scorpaena porcus*	brown scorpionfish (black scorpionfish)
Brauner Eichen-Zipfelfalter	*Satyrium ilicis/ Nordmannia ilicis*	ilex hairstreak
Brauner Fischegel	*Ottonia brunnea*	brown fish worm
Brauner Gebüsch-Lappenspanner	*Trichopteryx polycommata*	barred toothed stripe moth
Brauner Grashüpfer, Feldheuschrecke	*Chorthippus brunneus*	field grasshopper, common field grasshopper
Brauner Igelfisch	*Diodon holacanthus*	long-spine porcupinefish (FAO), long-spined porcupinefish, balloon porcupinefish, balloonfish
Brauner Kaninchenfisch	*Siganus luridus*	dusky rabbitfish
Brauner Kapuziner	*Cebus nigrivittatus*	weeper capuchin, wedge-capped capuchin
Brauner Kärpfling	*Poeciliopsis viviosa*	brown molly*, brown livebearer*
Brauner Katzenhai	*Apristurus brunneus*	brown catshark
Brauner Lippenhai, Sattelfleckenhai, Gebänderter Bambushai	*Chiloscyllium punctatum*	brown carpet shark, brownbanded bambooshark (FAO)
Brauner Lippfisch	*Labrus merula*	brown wrasse
Brauner Mönch, Wollkrauteule	*Cucullia verbasci*	mullein moth
Brauner Pfeilwurm	*Spadella cephaloptera* (Chaetognatha)	brown arrow worm
Brauner Röhrenschwamm	*Agelas conifera*	Pan Pipe sponge

Brauner Rübenaaskäfer	*Blithophaga opaca*	beet carrion beetle
Brauner Sägebarsch, Zwergbarsch	*Serranus hepatus*	brown comber
Brauner Sanddollar, Brauner Schildseeigel, Kleiner Sanddollar	*Clypeaster rosaceus*	brown sand dollar, brown sea biscuit
Brauner Schlangenstern, Großer Schlangenstern	*Ophioderma longicauda/ Ophiura longicauda*	long-spined brittlestar
Brauner Schnellkäfer	*Sericus brunneus*	brown click beetle*
Brauner Schnurwurm	*Cerebratulus fuscus* (Nemertini)	brown ribbon worm
Brauner Splintholzkäfer	*Lyctus brunneus*	brown powderpost beetle
Brauner Steinläufer	*Lithobius forficatus*	common garden centipede
Brauner Süßwasserpolyp, Braune Hydra	*Hydra vulgaris*	brown European hydra
Brauner Uferschnegel	*Agriolimax caruanae/ Limax brunneus/ Deroceras panormitanum*	brown slug
Brauner Waldvogel	*Aphantopus hyperantus*	ringlet
Brauner Wasserpython	*Liasis fuscus/Morelia fusca* (*Liasis mackloti*)	brown water python (white-lipped python)
Brauner Würfelfalter	*Hamearis lucina*	Duke of Burgundy
Brauner Zackenbarsch	*Epinephelus marginatus*	dusky grouper, dusky perch
Brauner Zwergwels, Langschwänziger Katzenwels	*Ameiurus nebulosus/ Ictalurus nebulosus*	brown bullhead (FAO), "speckled catfish", horned pout, American catfish
Braunes Faultier	*Bradypus variegatus*	brown-throated sloth
Braunes Knallkrebschen, Brauner Pistolenkrebs	*Alpheus armatus*	brown snapping prawn, brown pistol shrimp
Braunes Langohr, Braune Langohrfledermaus, Großohr	*Plecotus auritus*	brown long-eared bat, common long-eared bat
Braunes Papierboot	*Argonauta hians*	winged argonaut, brown paper nautilus
Braunes Silberäffchen	*Callithrix argentata melanura*	black-tailed marmoset
Braunes Zwerghörnchen, Schwarzohr-Zwerghörnchen	*Nannosciurus melanotis*	black-eared squirrel
Braunfleckige Beißschrecke	*Platycleis tesselata*	brown-spotted bushcricket
Braunflügelguan	*Ortalis vetula*	plain chachalaca
Braungebänderte Ohrenqualle	*Aurelia limbata*	brown banded moon jelly
Braungerändertes Ochsenauge	*Pyronia tithonus*	gatekeeper, hedge brown
Braunhaie, Grundhaie	*Carcharhinus* spp.	requiem sharks
Braunkehl-Uferschwalbe	*Riparia paludicola*	brown-throated sand martin
Braunkehlchen	*Saxicola rubetra*	whinchat
Braunkolbiger Braun-Dickkopffalter, Ockergelber Dickkopffalter	*Thymelicus sylvestris/ Thymelicus flavus*	small skipper
Braunkopf-Klammeraffe	*Ateles fusciceps*	brown-headed spider monkey
Braunkopf-Kuhstärling	*Molothrus ater*	brown-headed cowbird
Braunkopf-Papageimeise	*Paradoxornis webbianus*	vinous-throated parrotbill
Braunkopfammer	*Emberiza bruniceps*	red-headed bunting
Bräunlichgraue Frühlingseule	*Orthosia gothica*	Hebrew character (moth)

Bräunlichrote Samtmilbe	*Allothrombium fulginosum*	red velvet mite
Braunliest	*Halcyon smyrnensis*	white-breasted kingfisher
Braunmaina, Dschungelmaina	*Acridotheres fuscus*	Indian jungle mynah
Braunmäuse*	*Scotinomys* spp.	brown mice
Braunohrsittich	*Pyrrhura frontalis*	maroon-bellied conure
Braunpelikan	*Pelecanus occidentalis*	brown pelican
Braunrandiger Frostspanner	*Agriopis marginaria*	dotted border moth
Braunrochen, Vieraugenrochen, Augenfleckrochen	*Raja miraletus*	brown ray
Braunrückentamarin	*Saguinus fuscicollis*	saddle-back tamarin
Braunrückiger Waldsteigerfrosch	*Leptopelis cynnamomeus*	Angola forest treefrog
Braunscheckauge	*Lassiommata petropolitana*	northern wall brown
Braunscheckiger Perlmutterfalter, Braunfleckiger Perlmutterfalter	*Clossiana selene/ Boloria selene*	small pearl-bordered fritillary
Braunschlange> Gewöhnliche Braunschlange	*Pseudonaja textilis*	eastern brown snake
Braunschlangen	*Storeria* spp.	brown snakes
Braunschnäpper	*Muscicapa dauurica*	brown flycatcher
Braunschnecke*	*Zenobiella subrufescens/ Perforatella subrufescens*	dusky snail, brown snail
Braunspinnen	Sicariidae/ Loxoscelidae	violin spiders, recluse spiders, sixeyed sicariid spiders
Braunstirn-Spechtpapagei	*Micropsitta pusio*	buff-faced pygmy parrot
Braunstreifen-Schnapper	*Lutjanus carponotatus*	stripey
Brauntinamu, Brauntao	*Crypturellus soui*	little tinamou
Braunwangensittich	*Aratinga pertinax*	brown-throated conure
Braunweißband-Triton*, Kaulquappentriton	*Gyrineum gyrinum*	tadpole triton
Braunwidderchen	*Dysauxea ancilla*	brown burnet
Braunwurzschaber	*Cionus scrophulariae*	figwort weevil
Braunzottiges Borstengürteltier, Braunes Borstengürteltier, Braunborsten-Gürteltier	*Chaetophractus villosus/ Euphractus villosus*	larger hairy armadillo
Brautente	*Aix sponsa*	wood duck
Brazzameerkatze	*Cercopithecus neglectus*	De Brazza's monkey
Breitband-Triton	*Cymatium flaveolum*	broad-banded triton
Breite Becherqualle	*Lucernaria quadricornis*	horned stalked jellyfish
Breite Fechterschnecke, Breite Flügelschnecke	*Strombus latissimus*	heavy frog conch, wide-mouthed conch, broad Pacific conch, widest Pacific conch
Breite Randschnecke	*Cryptospira ventricosa*	broad marginella, broad marginsnail (broad margin shell)
Breitflossenkärpfling	*Mollienesia latipinna/ Poecilia latipinna*	sailfin molly
Breitflügel-Motteneule	*Schrankia taenialis*	white-line snout (moth)
Breitflügelbussard	*Buteo platypterus*	broad-winged hawk

Breitflügelfledermäuse	*Eptesicus* spp.	serotines, serotine bats (big brown bats, house bats)
Breitfußbeutelmäuse	*Antechinus* spp.	broad-footed marsupial "mice"
Breitfüßler, Breitrüssler, Maulkäfer	Anthribidae	fungus weevils
Breitgebänderte Mitra	*Mitra fusiformis f. zonata*	zoned mitre
Breitgestreifte Korallenschlange	*Micrurus distans*	broad-banded coral snake
Breitkopf-Bärenkrebs	*Thenus orientalis*	Moreton Bay flathead lobster, Moreton Bay "bug"
Breitkopf-Baumratte	*Chiruromys lamia*	broad-skulled tree mouse
Breitkopf-Kaninchenkänguru	*Potorous platyops*	broad-faced potoroo
Breitkopf-Schläfergrundel	*Dormitator latifrons/ Eleotris latifrons*	broad-headed sleeper
Breitköpfige Mäuse, Breitkopfmäuse	*Zelotomys* spp.	broad-headed mice, broad-headed stink mice
Breitkopfskink	*Eumeces laticeps*	broadheaded skink, broad-headed skink, "scorpion"
Breitlingssalmler, Barbensalmler	Curimatidae	curimatids, toothless characins
Breitmaul-Glanzvogel	*Jacamerops aurea*	great jacamar
Breitmaulnashorn, Breitlippennashorn, „Weißes" Nashorn	*Ceratotherium simum*	white rhinoceros, square-lipped rhinoceros, grass rhinoceros
Breitrachen, Breitmäuler	Eurylaimidae	broadbills
Breitrand, Breitrandkäfer	*Dytiscus latissimus*	broad diving beetle*
Breitrand-Spitzkopfschildkröte	*Emydura macquarrii*	Murray River turtle, Macquarie tortoise
Breitrandschildkröte	*Testudo marginata*	margined tortoise, marginated tortoise
Breitschnauzenhalbmaki	*Hapalemur simus*	broad-nosed gentle lemur
Breitschnauzenkaiman	*Caiman latirostris*	broad-nosed caiman
Breitschulterbock	*Akimerus schaefferi/ Acimerus schaefferi*	broad-shouldered longhorn beetle
Breitstreifenmungo	*Galidictis fasciata striata*	Malagasy broad-striped mongoose
Breitstreifensardelle	*Anchoa hepsetus*	striped anchovy, broad-striped anchovy
Breitwarzige Fadenschnecke	*Aeolidia papillosa*	maned nudibranch, plumed sea slug, grey sea slug
Breitzahnratte*	*Mastacomys fuscus*	broad-toothed rat
Bremsen	Tabanidae	horseflies & deerflies, gadflies
Bremsenglasflügler, Kleiner Pappelglasflügler	*Paranthrene tabaniformis*	lesser poplar hornet clearwing, lesser hornet moth
Brennmoos*	*Lytocarpus nuttingi*	stinging hydroid
Brennnesselblattrüssler	*Phyllobius pomaceus*	nettle leaf weevil
Brennnesseleulen, Nesseleulen	*Abrostola* spp.	dark spectacles
Brennnesselröhrenschildlaus	*Orthezia urticae*	stinging nettle orthezia
Brennnesselstengel-Minierfliege	*Phytomyza flavicornis*	nettle stem miner (fly)*
Brennnesselwanze	*Heterogaster urticae*	nettle groundbug

Brennschwamm, „Fass-mich-nicht-an-Schwamm"	*Neofibularia nolitangere*	do-not-touch-me sponge
Brettartige Feuerkoralle, Platten-Feuerkoralle	*Millepora platyphylla*	sheet fire coral
Brettkanker	Trogulidae	trogulid harvestmen, trogulids
Brillantsalmler	*Moenkhausia pittieri*	diamond tetra (FAO), Pittier's tetra
Brillenbär	*Tremarctos ornatus*	spectacled bear
Brillenblattnase	*Carollia perspillicata*	Seba's short-tailed bat
Brilleneidechse	*Podarcis perspicillata/ Lacerta perspicillata*	Moroccan rock lizard, Menorca wall lizard
Brillenente	*Melanitta perspicillata*	surf scoter
Brillengrasmücke	*Sylvia conspicillata*	spectacled warbler
Brillenkauz	*Pulsatrix perspicillata*	spectacled owl
Brillenlangur	*Presbytis obscura*	dusky leaf monkey, spectacled leaf monkey
Brillenpapagei	*Forpus conspicillatus*	spectacled parrotlet
Brillenpelikan	*Pelecanus conspicillatus*	Australian pelican
Brillenpinguin	*Spheniscus demersus*	jackass penguin
Brillensalamander	*Salamandrina terdigitata*	spectacled salamander
Brillenschlange, Südasiatische Kobra	*Naja naja*	common cobra, Indian cobra
Brillenschweinswal	*Phocaena dioptrica*	spectacled porpoise, Marposa de anteojos
Brillenstärling	*Xanthocephalus xanthocephalus*	yellow-headed blackbird
Brillenvögel	Zosteropidae	white-eyes
Brillenwürger	*Prionops plumata*	long-crested helmet-shrike
Brillenwürger	Prionopinae	helmet-shrikes
Britische Dolchschnecke, Hohlschnecke*, Hohl-Glanzschnecke	*Zonitoides excavatus*	hollowed gloss (snail), hollowed glass snail
Brolgakranich	*Grus rubicunda*	brolga
Brombeerblattgallmücke	*Dasineura plicatrix*	blackberry leaf midge, bramble leaf midge
Brombeerhai, Stachelhai, Alligatorhai, Nagelhai	*Echinorhinus brucus*	bramble shark
Brombeermilbe	*Eriophyes essigi/ Acalitus essigi*	redberry mite, blackberry mite
Brombeersaummücke, Himbeergallmücke	*Lasioptera rubi*	blackberry stem gall midge, raspberry stem gall midge
Brombeerspinner	*Macrothylacia rubi*	fox moth
Brombeerzipfelfalter	*Callophrys rubi*	green hairstreak
Bromelienlaubfrosch	*Hyla ebraccata*	hourglass treefrog
Bronze Fähnchen-Messerfisch, Asiatischer Fähnchen-Messerfisch	*Notopterus notopterus*	bronze featherback
Bronzefruchttaube	*Ducula aenea*	green imperial pigeon
Bronzehai, Kupferhai	*Carcharhinus brachyurus*	copper shark (FAO), bronze whaler, narrowtooth shark
Bronzenattern	*Dendrelaphis* spp.	Australian tree snakes, painted bronzebacks

Bronzeschwanz-Eremit	*Ramphodon dohrnii/ Glaucis dohrnii*	hook-billed hermit
Brooks Halbzehengecko	*Hemidactylus brookii*	Brook's half-toed gecko
Brosme, Lumb	*Brosme brosme*	cusk, torsk (European cusk), tusk (FAO)
Brotkäfer, Brotbohrer	*Stegobium paniceum*	drugstore beetle, drug store weevil (biscuit beetle, bread beetle)
Brotkrustenschwamm, Brotkrumenschwamm, Meerbrot	*Halichondria panicea*	breadcrumb sponge, crumb of bread sponge
Brotlaibrochen*	*Raja straeleni*	biscuit skate
Brotulas	Brotuninae	brotulas
Bruchschwamm, Sternlochschwamm	*Spongilla fragilis*	freshwater sponge
Bruchwasserläufer	*Tringa glareola*	wood sandpiper
Bruchweiden-Erbsengallenblattwespe*	*Pontania proxima*	willow bean-gall sawfly
Brückenechse	*Sphenodon punctatus*	tuatara
Brückenechsen	Rhynchocephalia (*Sphenodon*)	rhynchocephalians
Brückenkreuzspinne, Brückenspinne	*Larinioides sclopetarius/ Araneus sclopetarius*	bridge orbweaver
Brucko, Rauer Stechrochen	*Dasyatis centroura*	roughtail stingray
Brüllaffen	*Alouatta* spp.	howler monkeys
Brummspinne*	*Anyphaena accentuata*	buzzing spider
Brunnenbauer, Kieferfische	Opisthognathidae	jawfishes
Brunnenkrebse, Höhlenflohkrebse	*Niphargus* spp.	well shrimps
Brunnenkrebs, Höhlenflohkrebs	*Niphargus aquilex*	cavernous well shrimp*
Brunnenmolche	*Typhlomolge* spp.	American blind salamanders
Brunnenwürmer	Haplotaxidae	haplotaxid worms
Brydes Wal	*Balaenoptera edeni*	Bryde's whale
Bryer-Kugelkrabbe*	*Ebalia tumefacta*	Bryer's nut crab
Buchdrucker, Großer Borkenkäfer, Achtzähniger Borkenkäfer	*Ips typographus*	engraver beetle, common European engraver, spruce bark beetle (Br.)
Buchdruckereule	*Naenia typica*	gothic
Buchen-Kahnspinner, Kleiner Kahnspinner, Buchenwickler, Buchenkahneule, Jägerhütchen	*Pseudoips fagana/ Bena prasinana/ Bena fagana/ Hylophila prasinana*	green silverlines, scarce silverlines
Buchen-Nutzholzborkenkäfer, Buchen-Nutzholzambrosiakäfer, Laubnutzholzborkenkäfer	*Xyloterus domesticus*	broad-leaf wood ambrosia beetle a.o.
Buchen-Nutzholzborkenkäfer, Buchen-Nutzholzambrosiakäfer, Laubnutzholzborkenkäfer> Gekörnter Eichenholzbohrer	*Xyleborus dryographus*	broad-leaf wood ambrosia beetle a.o.
Buchen-Sichelflügler	*Drepana cultraria/ Platypteryx cultraria*	barred hook-tip

Buchenblattgallmücke	*Mikiola fagi*	beech leaf gall midge, beech pouch-gall midge
Buchenborkenkäfer	*Taphrorychus bicolor*	beech bark beetle
Buchenfrostspanner	*Operophtera fagata*	beech winter moth, northern winter moth
Buchengallmücke, Buchen-Haargallmücke*	*Hartigiola annulipes*	beech pouch-gall midge, beech hairy-pouch-gall midge*
Buchenmotte, Sängerin	*Chimabacche fagella*	beech moth*
Buchenprachtkäfer	*Agrilus viridis*	beech borer, flat-headed wood borer
Buchenrotschwanz, Buchen-Streckfuß, Rotschwanz	*Calliteara pudibunda/ Dasychira pudibunda/ Olene pudibunda/ Elkneria pudibunda*	pale tussock, red-tail moth
Buchenspießbock, Kleiner Eichenbock	*Cerambyx scopolii*	beech capricorn beetle, small oak capricorn beetle
Buchenspinner	*Stauropus fagi*	lobster moth
Buchenspringrüssler	*Rhynchaenus fagi*	beech flea weevil, beech leaf mining weevil, beech leaf miner
Buchentyrann	*Empidonax virescens*	acadian flycatcher
Buchenwerftkäfer, Bohrkäfer, Sägehörniger Werftkäfer	*Hylecoetus dermestoides*	large timberworm, European sapwood timberworm*
Buchenwickler	*Cydia fagiglandana*	beech seed moth
Buchenwickler, Buchenkahneule, Kahnspinner, Kleiner Kahnspinner, Jägerhütchen	*Bena prasinana/ Bena fagana/ Hylophila prasinana/ Pseudoips fagana*	green silverlines, scarce silverlines
Buchenwolllaus, Buchenschildlaus	*Cryptococcus fagisuga*	beech scale, felted beech scale
Bücherlaus	*Liposcelis divinatorius*	booklouse
Bücherläuse	Liposcelidae	booklice
Bücherläuse	Psocoptera/Copeognatha	booklice, psocids
Bücherskorpion	*Chelifer cancroides*	house pseudoscorpion
Buchfink	*Fringilla coelebs*	chaffinch
Büchsenmuscheln, Pandoramuscheln	Pandoridae	pandoras, pandora's boxes
Buchstaben-Schmuckschildkröte	*Trachemys scripta scripta/ Pseudemys scripta scripta/ Chrysemys scripta scripta*	slider, common slider, pond slider, yellow-bellied turtle
Buchstaben-Teppichmuschel*	*Tapes litterata*	lettered carpetclam, lettered venus (lettered carpet shell)
Buchtenkopf-Waldameise, Kerbameise	*Formica exsecta/ Coptoformica exsecta*	narrow-headed ant
Buckel-Fechterschnecke, Buckel-Flügelschnecke	*Strombus gibberulus*	hump-back conch, humped conch
Buckel-Purpurschnecke	*Thais tuberosa*	humped rocksnail (humped rock shell)
Buckelfliegen (Rennfliegen)	Phoridae	humpbacked flies
Buckelkäfer, Kugelkäfer	*Gibbium psylloides*	smooth spider beetle (bowl beetle)
Buckelkauri, Buckelschnecke, Großer Schlangenkopf	*Mauritia mauritiana/ Cypraea mauritiana*	hump-backed cowrie, humpback cowrie

Buckelkofferfisch, Pyramiden-Kofferfisch	*Tetrosomus gibbosus*	trunkfish
Buckelkopf-Buntbarsch	*Steatocranus casuarius*	buffalohead cichlid
Buckelkopf-Buntbarsch> Zebra-Beulenkopf, „Frontosa", Beulenkopf-Buntbarsch	*Cyphotilapia frontosa*	humphead cichlid (FAO), 'frontosa'
Buckelkopf-Papageifisch, Rotbauch-Papageifisch	*Bolbometopon muricatum*	green humphead parrotfish (FAO), humphead parrotfish
Buckellachs, Buckelkopflachs, Rosa Lachs	*Oncorhynchus gorbuscha*	pink salmon
Buckelolive*, Rundliche Olivenschnecke	*Olivancillaria gibbosa*	gibbose olive, swollen olive
Buckelschildkröte	*Phrynops gibbus/ Mesoclemmys gibba*	gibba turtle
Buckelschnapper	*Lutjanus gibbus*	paddletail, humpback snapper
Buckelwal	*Megaptera novaeangliae*	humpback whale
Buckelzirpen	Membracidae	treehoppers
Bucklige Schwellenschnecke	*Diodora gibberula*	humped keyhole limpet
Buckliger Drachenkopf, Buckel-Drachenkopf	*Scorpaenopsis gibbosa/ Scorpaenopsis barbata*	humpbacked scorpionfish
Buettners Waldmaus	*Leimacomys buettneri*	groove-toothed forest mouse, Buettner's forest mouse, Togo mouse
Büffelkopfente	*Bucephala albeola*	bufflehead
Büffelweber	Bubalornithinae	buffalo weaver
Bukowinische Blindmaus	*Spalax graecus*	Bukowinian blind mole-rat, greater mole rat
Bullaugen-Zitterrochen	*Diplobatis ommata*	ocellated electric ray (FAO), bull's eye torpedo
Bulldogg-Fledermäuse	*Eumops* spp. & *Otomops* spp. u.a.	mastiff bats, bonneted bats
Bulldogg-Fledermäuse	Molossidae	mastiff bats, free-tailed bats
Bullenhai, Stierhai, Gemeiner Grundhai	*Carcharhinus leucas*	bull shark
Bulwersturmvogel	*Bulweria bulwerii*	Bulwer's petrel
Bungar> Gewöhnlicher Bungar, Indischer Krait	*Bungarus caeruleus*	Indian krait
Bungars	*Bungarus* spp.	kraits, Indian kraits
Buntastrild	*Pytilia melba*	green-winged pytilia
Buntbarsche	Cichlidae	cichlids
Buntbauch-Greiffrosch	*Phyllomedusa sauvagii*	painted-belly monkey frog
Buntbäuchiger Grashüpfer	*Omocestus rufipes*	woodland grasshopper
Buntbock, Blessbock	*Damaliscus dorcas*	bontebok, blesbok
Bunte Artmuschel, Verschiedenfarbige Artmuschel, Bunte Artemis	*Dosinia variegata*	variegated dosinia, variegated artemis
Bunte Baummuschel	*Crenatula picta*	painted tree-oyster
Bunte Blattwanze	*Elasmostethus interstinctus*	birch bug

Bunte Bodennatter, Buntnatter	*Liophis poecilogyrus*	Wied's ground snake
Bunte Dachschildkröte, Bengalische Dachschildkröte	*Kachuga kachuga*	red-crowned roofed turtle (IUCN), painted roofed turtle, Bengal roof turtle
Bunte Fasanenschnecke	*Phasianella variegata*	variegated pheasant
Bunte Käferschnecke, Mittelmeer-Chiton	*Chiton olivaceus*	variable chiton, Mediterranean chiton
Bunte Kammmuschel	*Chlamys varia*	variegated scallop
Bunte Kammmuschel, Kleine Pilgermuschel	*Aequipecten opercularis*	queen scallop
Bunte Kreiselschnecke	*Calliostoma zizyphinum*	painted topsnail, European painted topsnail
Bunte Seegurke, Violette Seegurke	*Pseudocolochirus violaceus*	multicolored sea cucumber*, violet sea cucumber
Bunter Einsiedler u.a.	*Pagurus anachoretus*	multicoloured hermit crab*
Bunter Einsiedler u.a.	*Calcinus tubularis/ Calcinus ornatus*	variegated hermit crab, ornate hermit crab
Bunter Furchenkrebs, Bunter Springkrebs, Blaugestreifter Springkrebs	*Galathea strigosa*	strigose squat lobster
Bunter Grashüpfer	*Omocestus viridulus*	common green grasshopper
Bunter Kalkröhrenwurm, Kleiner Kalkröhrenwurm, Deckel-Kalkröhrenwurm	*Serpula vermicularis*	red tubeworm
Bunter Prachtkärpfling, Fahnenhechtling, Kap Lopez	*Aphyosemion australe*	lyretail, Cape Lopez lyretail, lyretail panchax (FAO)
Buntes Zwergchamäleon	*Chamaeleo pumilis*	variegated dwarf chameleon
Buntfalke	*Falco sparverius*	American kestrel, sparrow hawk
Buntflossen-Groppe, Bandflossen-Koppe, Buntflossenkoppe, Sibirische Groppe	*Cottus poecilopus*	Siberian bullhead, alpine bullhead (FAO)
Buntflöter	*Ptilorrhoa castanonota*	mid-mountain rail babbler
Buntfuß-Sturmschwalbe	*Oceanites oceanicus*	Wilson's petrel
Buntkäfer	Cleridae	checkered beetles, clerid beetles, clerids
Buntkopf-Felshüpfer	*Picathartes oreas*	gray-necked rockfowl, grey-necked bald-crow, gray-necked picathartes
Buntmarder (Hbz. Charsa)	*Martes flavigula*	yellow-throated marten
Buntnatter, Bunte Bodennatter	*Liophis poecilogyrus*	Wied's ground snake
Buntpython	*Python curtus*	short-tailed python, blood python
Buntscharbe	*Phalacrocorax gaimardi*	red-legged cormorant
Buntschnecke	*Pusia tricolor*	
Buntspecht	*Dendrocopos major/ Picoides major*	great spotted woodpecker
Buntwaran	*Varanus varius*	lace monitor, common tree monitor
Buntwühle	*Schistometopum thomense*	Island caecilian
Buntzecke	*Amblyomma hebraeum*	bont tick

Burchell-Steppenzebra	*Equus burchelli*	Burchell's zebra
Burma-Dachschildkröte	*Kachuga trivittata*	Burmese roofed turtle (IUCN)
Burma-Landschildkröte	*Geochelone platynota/* *Geochelone elegans platynota*	Burmese star tortoise, Burmese starred tortoise
Burma-Weichschildkröte	*Nilssonia formosa/* *Trionyx formosus*	Burmese peacock softshell turtle, Burmese softshell turtle
Burmesische Klappen-Weichschildkröte, Burma-Klappen- Weichschildkröte	*Lissemys scutata*	Burmese flapshell turtle
Bürstenbinder, Kleiner Bürstenspinner, Schlehenspinner	*Orgyia antiqua/* *Orgyia recens*	vapourer moth, common vapourer, rusty tussock moth
Bürstenhornblattwespen	Argidae	stout-bodied sawflies, argid sawflies
Bürstenkängurus	*Bettongia* spp.	short-nosed "rat"-kangaroos
Bürstenrattenkänguru	*Bettongia penicillata*	woylie
Burunduk, Eurasisches Erdhörnchen, Östliches Streifenhörnchen, Sibirisches Streifenhörnchen (Hbz. Burunduk)	*Eutamias sibiricus/* *Tamias sibiricus*	Siberian chipmunk
Bürzelstelzer, Buschschlüpfer	Rhinocryptidae	tapaculos
Busch-Felsratten	*Aethomys* spp.	bush rats, rock rats
Buschbock	*Tragelaphus scriptus*	bushbuck
Büschelbarsch	*Paracirrhites arcatus*	arc-eyed hawkfish
Büschelbarsche, Korallenwächter	Cirrhitidae	hawkfishes
Büschelbrauenotter	*Bitis cornuta*	hornsman adder, many-horned adder
Büschelkiemenartige, Seenadelverwandte, Seepferdchenverwandte	Syngnathiformes	sea horses & pipefishes and allies
Büschelohriger Katzenmaki, Kleiner Katzenmaki	*Allocebus trichotis*	hairy-eared dwarf lemur
Büschelwels, Froschwels	*Clarias batrachus*	walking catfish
Buschfische	*Ctenopoma* spp.	climbing perches
Buschgrille	*Arachnocephalus vestitus*	Mediterranean wingless shrub cricket
Buschhäher	*Aphelocoma coerulescens*	scrub jay
Buschhornblattwespen	Diprionidae	conifer sawflies
Buschhuhn	*Alectura lathami*	brush turkey
Buschkängurus	*Dorcopsis* spp.	New Guinean forest wallabies
Buschkaninchen	*Poelagus marjorita*	Central African rabbit
Buschmannhase	*Bunolagus monticularis*	Bushman rabbit, Bushman hare, riverine rabbit
Buschmeise	*Psaltriparus minimus*	common bushtit
Buschmeister	*Lachesis mutus/* *Lachesis muta*	bushmaster
Buschratten	*Neotoma* spp.	wood rats, woodrats, pack rats, trade rats

Buschrohrsänger	*Acrocephalus dumetorum*	Blyth's reed warbler
Buschschliefer	*Heterohyrax* spp.	gray hyraxes, yellowspotted hyraxes
Buschschlüpfer, Bürzelstelzer	Rhinocryptidae	tapaculos
Buschschwanz-Beutelratte	*Glironia venusta*	bushy-tailed opossum
Buschschwanz-Gundi	*Pectinator spekei*	Speke's pectinator, bushy-tailed gundi
Buschschwanz-Rennmaus, Bilch-Rennmaus	*Sekeetamys calurus*	bushy-tailed jird
Buschschwanzmoos*	*Sertularia argentea*	squirrel's tail hydroid, sea fir
Buschschwein, Flussschwein, Pinselohrschwein	*Potamochoerus porcus*	African bush pig, red river hog
Buschspötter	*Hippolais caligata*	booted warbler
Buschtinamu, Zimttao	*Crypturellus cinnamomeus*	thicket tinamou
Buschvipern	*Atheris* spp.	African bush vipers
Buschwaldgalago	*Galago alleni*	Allen's bush baby, Allen's squirrel galago, black-tailed bush baby
Buschwürger	Malaconotinae	bush-shrikes
Buschzaunkönig	*Thryomanes bewickii*	Bewick's wren
Busentierchen	*Colpidium colpoda*	
Butlers Stachelauster	*Spondylus butleri*	Butler's thorny oyster
Butte	Bothidae	lefteye flounders, left-eye flounders
Butterfisch, Deckfisch, Pampelfisch, Gemeine Pampel	*Stromateus fiatola*	butterfish, blue butterfish (FAO)
Butterfisch, Messerfisch	*Pholis gunnellus*	butterfish, gunnel
Butterfische	Pholidae/Pholididae	gunnels
Buttermuscheln*	*Saxidomus* spp.	butter clams
Butu, Amazonas-Delphin, Inia	*Inia geoffrensis*	Amazon dolphin, Amazon River dolphin, bouto, boutu

C-Eule, Schwarzes C	*Xestia c-nigrum*	setaceous Hebrew character
C-Falter, Weißes C	*Nymphalis c-album/ Polygonia c-album/ Comma c-album*	comma
Caballa, Pferdemakrele, Pferde-Makrele, Pferde-Stachelmakrele	*Caranx hippos*	crevalle jack (FAO), Samson fish
Cabreramaus	*Microtus cabrerae*	Cabrera's vole
Cachalot, Kaschelot, Großer Pottwal	*Physeter macrocephalus (catodon)*	sperm whale, great sperm whale, cachalot
Calabar-Bärenmaki	*Arctocebus calabarensis*	Calabar angwantibo, golden potto
Calico-Kastenkrabbe	*Hepatus epheliticus*	Calico box crab
Calico-Pilgermuschel	*Argopecten gibbus*	Calico scallop
Callagur-Schildkröte	*Callagur borneoensis*	painted batagur, painted terrapin, three-striped batagur, biuku
Campbells Meerkatze	*Cercopithecus campbelli*	Campbell's monkey
Capniden, Kleine Winter-Steinfliegen	Capniinae	small winter stoneflies
Capybara	*Hydrochaeris hydrochaeris*	capybara, "water hog"
Caristiiden	Caristiidae	manefishes
Carolina-Dosenschildkröte	*Terrapene carolina*	eastern box turtle, common box turtle (IUCN)
Carolina-Kleiber	*Sitta carolinensis*	white-breasted nuthatch
Carolina-Meise	*Parus carolinensis*	Carolina chickadee
Carolina-Sumpfhuhn	*Porzana carolina*	sora
Carolina-Taube	*Zenaida macroura*	mourning dove
Catla, Theila, Tambra	*Catla catla*	catla
Caurica-Kauri	*Cypraea caurica*	caurica cowrie
Cayenne-Zecke	*Amblyomma cajennense*	Cayenne tick
Cecropiaspinner	*Platysamia cecropia/ Hyalophora cecropia*	cecropia silk moth, cecropia moth
Celebes Weichfellratte*	*Eropeplus canus*	Celebes soft-furred rat
Celebes-Roller	*Macrogalidia musschenbroeki*	Celebes palm civet
Celebes-Sonnenstrahlfisch	*Telmatherina ladigesi*	Celebes rainbowfish
Celebes-Stachelratte	*Echiothrix leucura*	Celebes spiny rat, Celebes shrew rat
Celebes-Zwerghörnchen	*Prosciurillus* spp.	Celebes dwarf squirrels
Celebeskoboldmaki	*Tarsius spectrum*	spectral tarsier
Cephalochordaten, Lanzettfischchen	Cephalochordata (Amphioxiformes)	cephalochordates, lancelet
Ceramnasenbeutler	*Rhynchomeles prattorum*	Ceram Isand long-nosed bandicoot
Ceramratte	*Nesoromys ceramicus*	Ceram Island rat
Cero, Falsche Königsmakrele	*Scomberomorus regalis*	cero
Cervicalis-Fadenwurm*	*Onchocerca reticulata/ Onchocerca cervicalis*	neck threadworm
Ceylon-Egel, Sri Lanka-Egel	*Haemadipsa zeylandica*	Sri Lanka leech
Ceylon-Hechtling, Grüner Streifenhechtling	*Aplocheilus dayi*	Ceylon killifish

Ceylon-Hutaffe	*Macaca sinica*	toque macaque
Ceylon-Lanzenotter, Sri Lanka-Lanzenotter	*Trimeresurus trigonocephalus*	Sri Lankan pit viper, Ceylon pit viper
Ceylon-Makropode	*Belontia signata*	Ceylonese combtail
Ceylon-Nektarvogel	*Nectarinia zeylonica*	purple-rumped sunbird
Ceylon-Ratte	*Srilankamys ohiensis*	Ceylonese rat, Sri Lanka rat
Ceylon-Taubagame	*Cophotis ceylanica*	Sri Lanka prehensile-tail lizard
Ceylon-Wühle	*Ichthyophis glutinosus*	Ceylon caecilian
Chacopekari	*Catagonus wagneri*	Chacoan peccary, tagua
Chacunda	*Anodontostoma chacunda*	chacunda gizzard shad
Chagrinrochen, Walkerrochen	*Raja fullonica*	shagreen ray
Chamäleon> Europäisches Chamäleon, Gemeines Chamäleon, Gewöhnliches Chamäleon	*Chamaeleo chamaeleon*	Mediterranean chameleon, African chameleon, common chameleon
Chamäleon-Agame	*Chelosania brunnea*	chameleon dragon
Chamäleonfliege	*Stratiomys chameleon*	soldier fly, soldier-fly
Chamäleongarnele, Farbwechselnde Garnele	*Hippolyte varians*	chameleon prawn
Chamäleons	Chamaeleonidae	chameleons
Chamignon-Weichhautmilbe, Pilzmyzelmilbe	*Tarsonemus myceliophagus*	mushroom mite
Chanchito	*Cichlasoma facetum*	chanchito
Chaparraltimalie	*Chamaea fasciata*	wren-tit
Chaunaciden	Chaunacidae	coffinfishes, sea toads
Chelizeraten, Scherenhörnler	Chelicerata	chelicerates
Chernetiden	Chernetidae	chernetids
Cherokee-Bachsalamander	*Desmognathus aeneus*	seepage salamander
Chihuahua-Querzahnmolch	*Ambystoma rosaceum*	Chihuahuan mole salamanders
Chikaree	*Tamiasciurus douglasii*	Douglas squirrel
Childrens Python	*Antaresia childreni/ Liasis childreni/ Morelia childreni*	Children's python, Children's rock python
Chile-Delphin	*Cephalorhynchus eutropia*	black dolphin, black Chilean dolphin, white-bellied dolphin
Chile-Opossummaus	*Rhyncholestes raphanurus*	Chilean "shrew" opossum
Chile-Ratte	*Irenomys tarsalis*	Chilean rat
Chile-Teju	*Callopistes maculatus*	spotted false monitor
Chile-Vogelspinne	*Grammostola spatulata*	Chilean rose tarantula
Chileflamingo	*Phoenicopterus chilensis*	Chilean flamingo
Chilenische Bastardmakrele	*Trachurus murphyi*	Chilean jack mackerel
Chilenische Kantengarnele	*Heterocarpus reedei*	Chilean nylon shrimp
Chilenische Messergarnele	*Haliporoides diomedeae*	Chilean knife shrimp
Chilenische Miesmuschel	*Mytilus chilensis*	Chilean mussel
Chilenische Pelamide	*Sarda chilensis*	Pacific bonito, Eastern Pacific bonito (FAO)
Chilenische Plattauster	*Ostrea chilensis*	Chilean flat oyster
Chilenische Seespinne	*Libidoclea granaria*	southern spider crab
Chilenische Waldkatze	*Felis guigna*	kodkod

Chilenischer Diebkäfer	*Trigonogenius globulus*	globular spider beetle
Chilenischer Kingklip	*Genypterus chilensis*	red cusk-eel
Chilenischer Seehecht, Pazifischer Seehecht	*Merluccius gayi*	Chilean hake, Peruvian hake, South Pacific hake (FAO)
Chilenischer Seeigel	*Loxechinus albus*	Chilean sea urchin
China-Alligator	*Alligator sinensis*	Chinese alligator
China-Kauri	*Cypraea chinensis*	Chinese cowrie
China-Mondschnecke, Chinamond	*Natica onca*	China moonsnail
Chinahut, Chinesische Napfschnecke	*Patella ulyssiponensis/ Patella aspera*	China limpet, European China limpet, painted limpet
Chinakohl-Koralle	*Mycedium elephantotus*	Chinese cabbage coral*
Chinchilla-Maus	*Chinchillula sahamae*	Altiplano chinchilla mouse
Chinchilla> Kurzschwanz-Chinchilla	*Chinchilla brevicaudata*	short-tailed chinchilla
Chinchilla> Langschwanz-Chinchilla	*Chinchilla lanigera*	long-tailed chinchilla
Chinchillaratten	*Abrocoma* spp.	chinchilla rats, chinchillones
Chinchillas	*Chinchilla* spp.	chinchillas
Chinesen-Falterfisch, Gelber Tränentropf-Falterfisch	*Chaetodon unimaculatus*	lime-spot butterflyfish, teardrop butterflyfish (FAO)
Chinesenhut, Chinesenhütchen	*Calyptraea chinensis*	Chinaman's hat, Chinese hat, Chinese cup-and-saucer limpet,
Chinesische Bambusotter	*Trimeresurus stejnegeri*	Chinese green tree viper
Chinesische Dickkopfschildkröte	*Chinemys megalocephala*	Chinese broad-headed pond turtle
Chinesische Dreikielschildkröte	*Chinemys reevesii*	Reeves' turtle, Chinese three-keeled pond turtle
Chinesische Hufmuschel	*Hippopus porcellanus*	China clam
Chinesische Hüpfmaus	*Eozapus setchuanus*	Chinese jumping mouse
Chinesische Kobra	*Naja atra*	Chinese cobra
Chinesische Königsmakrele	*Scomberomorus sinensis*	Chinese seerfish
Chinesische Nasenotter	*Agkistrodon acutus/ Deinagkistrodon acutus*	sharp-nosed pit viper, sharp-nosed viper, hundred-pace viper, hundred pacer
Chinesische Pfauenaugenschildkröte	*Sacalia bealei*	ocellate pond turtle, Beal's eyed turtle (IUCN)
Chinesische Ratten (Hbz.), Bambusratten	*Rhizomys* spp.	bamboo rats
Chinesische Rattennattern	*Zaocys* spp.	keeled ratsnakes, Chinese ratsnakes
Chinesische Rotbauchunke	*Bombina orientalis*	Oriental fire-bellied toad
Chinesische Rothörnchen	*Sciurotamias* spp.	rock squirrels
Chinesische Streifenschildkröte	*Ocadia sinensis*	Chinese stripe-necked turtle, Chinese striped-neck turtle
Chinesische Warzenmolche	*Paramesotriton* spp.	warty newts
Chinesische Weichschildkröte	*Pelodiscus sinensis/ Trionyx sinensis*	Chinese softshell turtle
Chinesische Wollhandkrabbe	*Eriocheir sinensis*	Chinese mitten crab

Chinesische Zwergschlafmaus, Chinesischer Zwergbilch	*Typhlomys cinereus*	Chinese pygmy dormouse
Chinesischer Flussdelphin, Yangtse-Delphin, Beiji	*Lipotes vexillifer*	whitefin dolphin, Chinese River dolphin, Yangtse River dolphin, baiji, pei chi
Chinesischer Kurzfußmolch	*Pachytriton breviceps*	Tsitou newt, Chinese newt
Chinesischer Muntjak, Zwergmuntjak	*Muntiacus reevesi*	Chinese muntjac, Reeve's muntjac
Chinesischer Riesensalamander	*Andrias davidianus*	Chinese giant salamander
Chinesischer Schwertstör, Schwertstör	*Psephurus gladius*	Chinese paddlefish
Chinesischer Sonnendachs (Hbz. Pahmi)	*Melogale moschata*	Chinese ferret badger
Chinesischer Stör	*Acipenser sinensis*	Chinese sturgeon
Chinesischer Streifenhamster	*Cricetelus griseus*	Chinese striped hamster
Chinesischer Weißer Delphin, Indopazifischer Buckeldelphin	*Sousa chinensis*	Chinese humpback dolphin, Indo-Pacific hump-backed dolphin, white dolphin
Chinesischer Zwergmolch	*Cynops orientalis*	Chinese fire bellied newt
Chinesisches Ohren-Schuppentier	*Manis pentadactyla*	Chinese pangolin
Chipmunk, Streifenbackenhörnchen (Hbz. Chipmunk)	*Tamias striatus*	Eastern American chipmunk
Chiragra-Spinnenschnecke, Großer Bootshaken	*Lambis chiragra*	chiragra spider conch, gouty spider conch
Chordatiere, Rückgrattiere, Chordaten	Chordata	chordates
Chorfrösche	*Pseudacris* spp.	chorus frogs
Chormuschel	*Choromytilus chorus*	chorus mussel
Christusdorn-Koralle, Stachelige Nadelkoralle, Stachelige Buschkoralle, Dornige Reihenkoralle	*Seriatopora hystrix*	needle coral, spiny row coral
Chrysanthemengallmücke	*Diarthronomyia chrysanthemi/ Rhopalomyia chrysanthemi*	chrysanthemum gall midge
Chuckwallas	*Sauromalus* spp.	chuckwallas
Chuditch	*Dasyurus geoffroii*	chuditch, western quoll
Chukarhuhn	*Alectoris chukar*	chukar
Chukarsteinhuhn	*Alectoris chukar*	chukar partridge
Chum-Lachs, Keta-Lachs, Ketalachs, Hundslachs	*Oncorhynchus keta*	chum salmon
Cialancia-Grashüpfer	*Chorthippus cialancensis*	Piedmont grasshopper
Cistensänger	*Cisticola juncidis*	fan-tailed warbler
Cithariden	Citharidae	largescale flounders, citharids
Clown-Prachtschmerle	*Botia macracanthus*	clown loach
Clownbarbe, Everetts Barbe	*Puntius everetti*	clown barb

Clownfische, Anemonenfische	*Amphiprion* spp.	anemonefishes, clownfishes
Coachella-Valley-Fransenzehenleguan	*Uma inornata*	Coachella Valley fringe-toed lizard
Coahuila-Dosenschildkröte	*Terrapene coahuila*	Coahuilan box turtle, aquatic box turtle (IUCN)
Cochenillelaus	*Dactylopius coccus*	cochineal insect
Cochin-Erdschildkröte	*Heosemys silvatica*	Cochin forest cane turtle
Coco-Umberfisch	*Paralonchurus peruanus*	Peruvian drum
Coho-Lachs, Silberlachs	*Oncorhynchus kisutch*	coho salmon (FAO), silver salmon
Colorado-Squawfisch	*Ptychocheilus lucius*	Colorado squawfish
Commerson-Delphin, Jacobita	*Cephalorhynchus commersonii*	Commerson's dolphin, Tonina dolphin, piebald dolphin, black-and-white dolphin
Conchinchina-Wasserdrache	*Physignathus concincinus*	water dragon
Congrogadiden	Congrogadidae	snakelets
Conus-Kreiselschnecke	*Tectus conus*	cone-shaped topsnail, cone-shaped top
Copes Riesen-Querzahnmolch	*Dicamptodon copei*	Cope's giant salamander
Copes Weißlippen-Pfeiffrosch	*Leptodactylus labialis*	Cope's white-lipped frog
Coquerels Katzenmaki	*Mirza coquereli*	Coquerel's dwarf lemur
Couchs Krabbe*	*Monodaeus couchi*	Couch's crab
Coxs Kauri	*Cypraea coxeni*	Cox's cowrie
Cranch-Kugelkrabbe*	*Ebalia cranchii*	Cranch's nut crab
Cranchs Seespinne	*Achaeus cranchii*	Cranch's spider crab
Cribo	*Drymarchon corais*	cribo
Cropan-Boa	*Xenoboa cropanii*	Cropan's boa
Crump-Maus	*Diomys crumpi*	Manipur mouse, Crump's mouse
Ctenophoren, Kammquallen, Rippenquallen	Ctenophora	ctenophores (sea gooseberries, sea combs, comb jellies, sea walnuts)
Cuba-Riesenvogelspinne	*Phormictopus cubensis*	Cuban giant tarantula
Cuneata-Seezunge, Keilzunge*	*Dicologlossa cuneata*	wedge sole (FAO), Senegal sole
Cupediden	Cupedidae	reticulated beetles
Cururo	*Spalacopus cyanus*	coruro
Cuvier-Schnabelwal, Gänseschnabelwal	*Ziphius cavirostris*	Cuvier's goosebeaked whale
Cuvier-Seeschmetterling*	*Cuvierina columnella*	cigar pteropod
Cuviers Erdwühle	*Hypogeophis rostratus*	Frigate Island caecilian
Cydippen	Cydippidea	cydippids (comb jellies)

Dach-Moschusschildkröte	*Sternotherus carinatus*	razorback musk turtle, keel-backed musk turtle, razor-backed musk turtle
Dachratte, Hausratte, Alexandriner Hausratte	*Rattus rattus*	black rat, roof rat, house rat
Dachs, Europäischer Dachs	*Meles meles*	Old World badger
Dachsammer	*Zonotrichia leucophrys*	white-crowned sparrow
Dachschildkröten	*Kachuga* spp.	roofed turtles
Dachsfloh	*Chaetopsylla trichosa*	badger flea
Dahls Krötenkopf-Schildkröte	*Phrynops dahli*	Dahl's toad-headed turtle
Dajaldrossel	*Copsychus saularis*	magpie robin
Dall-Hafenschweinswal	*Phocoenoides dalli*	Dall porpoise, Dall's white-flanked porpoise
Dalls Kegelschnecke	*Conus dalli*	Dall's cone
Dallschaf, Dünnhornschaf	*Ovis dalli*	Dall's sheep, white sheep
Dalmatiner Schwamm, Badeschwamm, Tafelschwamm	*Spongia officinalis*	bath sponge
Dalmatinische Spitzkopfeidechse	*Lacerta oxycephala*	sharp-snouted rock lizard
Dalmatinische Fahnenqualle	*Drymonema dalmatinum*	Dalmatian mane jelly*
Dalmatinischer Schleimfisch, Zwergschleimfisch	*Lipophrys dalmatinus*	Dalmatian blenny
Dama-Zwergolive	*Olivella dama*	Dama dwarf olive
Damagazelle	*Gazella dama*	Addra gazelle
Damara Erdhörnchen	*Geosciurus princeps*	Damara ground squirrel
Damenbrett, Schachbrett	*Melanargia galathea*	marbled white
Damhirsch	*Dama dama/ Cervus dama*	fallow deer
Dämmerungskauri	*Cypraea diluculum*	dawn cowrie
Dämmerungsratte, Sumichrasti-Dämmerungsratte	*Nyctomys sumichrasti*	vesper rat, Sumichrast's vesper rat
Dammläufer> Gewöhnlicher Dammläufer, Pechschwarzer Dammläufer	*Nebria brevicollis*	common black ground beetle
Damons Walzenschnecke	*Amoria damonii*	Damon's volute
Dana-Blaukrabbe	*Callinectes danae*	Dana swimming crab
Danio-Bärblinge	*Danio* spp.	danios
Darlingtons Beutelfrosch, Darlingtonfrosch	*Assa darlingtoni*	pouched frog
Darm-Zwergfadenwurm*	*Strongyloides papillosus*	intestinal threadworm (cattle, sheep, goat a.o.)
Darwin-Frosch, Darwin-Nasenfrosch	*Rhinoderma darwini*	Darwin's frog
Darwin-Strauß, Kleiner Nandu	*Pterocnemia pennata*	lesser rhea
Darwins Eidechse	*Diplolaemus darwinii*	Darwin's iguana
Dattelmotte, Dörrobstmotte, Tropische Speichermotte	*Cadra cautella*	dried-fruit knot-horn, dried-fruit moth, almond moth

Dattelmuschel, Gemeine Bohrmuschel, Große Bohrmuschel	*Pholas dactylus*	common piddock
Däumlingsspecht	*Picumnus minutissimus*	Guianan piculet
Davidsharfe, Große Harfe	*Harpa major/ Harpa ventricosa*	major harp, swollen harp
Davidshirsch, Milu	*Elaphurus davidianus*	Père David's deer
Decapoden, Zehnfußkrebse	Decapoda	decapods
Deckelschlüpfer	Cyclorrhapha (Muscomorpha)	circular-seamed flies
Deckennetzspinnen (Baldachinspinnen) & Zwergspinnen	Linyphiidae	sheet-web weavers, sheet-web spinners, line-weaving spiders, line weavers, money spiders
Degenfisch, Haarschwanz	*Trichiurus lepturus*	Atlantic cutlassfish, large-eyed hairtail, largehead hairtail (FAO)
Degus, Strauchratten	*Octodon* spp.	degus
Delanys Sumpfklettermaus	*Delanymys brooksi*	Delany's swamp mouse
Delesserts Leierschnecke	*Lyria delessertiana*	
Delikat-Rundhering*	*Spratelloides delicatulus*	delicate round herring
Delphin> Gemeiner Delphin	*Delphinus delphis*	common dolphin, saddleback(ed) dolphin, crisscross dolphin
Delphinaal	*Dalophis boulengeri*	Boulenger's snake eel*
Delphinschnecke	*Angaria delphinus*	dolphin snail (dolphin shell)
Delphinschnecken	Angariidae	dolphin snails (dolphin shells)
Demoisellen	*Pomacentrus* spp.	damselfishes, demoiselles
Dennisons Maulbeerschnecke	*Morum dennisoni*	Dennison morum
Dent-Meerkatze, Dents Meerkatze	*Cercopithecus denti*	Dent's monkey
Derbe Mondschnecke	*Natica fasciata*	solid moonsnail
Derbe Puppenschnecke	*Pupa solidula/ Solidula solidula*	solid pupa
Derbyantilope, Riesen-Elenantilope	*Taurotragus derbianus*	Derby eland, giant eland
Derbywallaby, Tammarwallaby	*Macropus eugenii*	tammar wallaby, dama wallaby
Desman> Russischer Desman	*Desmana moschata*	Russian desman
Detrituskäfer	Monotomidae	small flattened bark beetles
Deutsche Schabe	*Blattella germanica*	German cockroach (croton-bug, shiner, steamfly)
Deutsche Wespe	*Vespula germanica*	German wasp
Deutscher Sandlaufkäfer	*Cicindela germanica*	German tiger beetle
Dhaman	*Ptyas mucosus*	dhaman, common rat snake, Oriental rat snake
Diadem-Soldatenfisch	*Holocentrus diadema*	crowned soldierfish
Diademkaiserfisch, Gelbmaskenkaiserfisch, Blaukopfkaiserfisch	*Pomacanthus xanthometopon*	yellow-faced angelfish, blue-face angelfish, yellowface angelfish (FAO)
Diademmeerkatze	*Cercopithecus mitis*	blue monkey, diademed monkey, gentle monkey

Diademnatter	*Spalerosophis diadema*	diadem snake
Diademrotschwanz	*Phoenicurus moussieri*	Moussier's redstart
Diademschildkröte	*Hardella thurjii*	Brahminy river turtle, crowned river turtle (IUCN)
Diademseeigel	Diadematidae	hatpin urchins
Diademsifaka	*Propithecus diadema*	diadem sifaka, diademed sifaka
Diamant-Picassofisch, Harlekin-Drückerfisch, Keil-Picassodrücker	*Rhinecanthus rectangulus*	wedge-tail triggerfish (FAO), wedge picassofish
Diamant-Stechrochen	*Dasyatis dipterurus*	diamond stingray
Diamantbarsch	*Enneacanthus gloriosus*	bluespotted sunfish
Diamantfasan	*Chrysolophus amherstiae*	Lady Amherst's pheasant
Diamantklapperschlange	*Crotalus adamanteus*	eastern diamondback rattlesnake
Diamantschildkröte	*Malaclemys terrapin*	diamondback terrapin
Diana Kammmuschel	*Chlamys dianae*	Diana's scallop
Dianafisch, Hahnenfisch	*Luvarus imperialis*	luvar, louvar
Dianameerkatze	*Cercopithecus diana*	Diana monkey, Diana guenon
Dianas Fechterschnecke, Dianas Flügelschnecke, Dianas Ohr-Fechterschnecke	*Strombus aurisdianae*	Diana's conch, Diana conch
Dianas Lippfisch	*Bodianus diana*	Diana's wrasse
Dibatag, Lamagazelle, Stelzengazelle	*Ammodorcas clarkei*	dibatag
Diceratiden	Diceratiidae	horned anglers
Dickbauchige Täubchenschnecke*	*Anachis obesa*	fat dovesnail
Dickbauchiges Neptunshorn	*Neptunea ventricosa*	fat whelk
Dickbauchkrebs	*Lyneus brachyurus*	
Dickcissel	*Spiza americana*	dickcissel
Dickdaumen-Fledermäuse	*Glischropus* spp.	thick-thumbed bats
Dicke Klaffmuschel	*Tresus capax*	fat gaper clam
Dicke Mondmuschel	*Lucina pectinata*	thick lucine
Dicke Trapezmuschel	*Megacardita incrassata*	thickened cardita
Dickgeripptes Neptunshorn	*Neptunea brevicauda*	thick-ribbed whelk
Dickhalsiger Bandwurm, Katzenbandwurm u.a.	*Hydatigera taeniaeformis/ Taenia taeniaeformis*	common cat tapeworm
Dickhörnige Seerose	*Urticina crassicornis*	mottled anemone, rose anemone, northern red anemone, crassicorn anemone, christmas anemone, painted urticina
Dickhornschaf	*Ovis canadensis*	bighorn sheep, American bighorn, mountain sheep
Dickichtschlüpfer, Dickichtvögel	Atrichornithidae	scrub-birds
Dickkopf-Grashüpfer	*Euchorthippus declivus*	sharptailed grasshopper
Dickkopf-Makrele	*Caranx ignobilis*	giant trevally (FAO), giant kingfish
Dickkopf-Scheibensalmler	*Metynnis hypsauchen*	silver dollar a.o.
Dickkopfanolis	*Anolis cybotes*	large-headed anole

Dickkopfbekarde	*Pachyramphus aglaiae*	rose-throated becard
Dickkopffalter (Dickköpfe)	Hesperiidae	skippers
Dickkopffliegen	Conopidae	thick-headed flies
Dickkopfnattern, Schneckennattern	*Dipsas* spp.	snail-eating snakes
Dickkopfschnäpper, Dickkopfvögel	Pachycephalidae	whistlers, shrike-thrushes
Dicklippige Buckelschnecke	*Monodonta labio*	thick-lipped topsnail, labio
Dicklippige Meeräsche	*Chelon labrosus/ Mugil chelo/ Mugil provensalis*	thick-lipped grey mullet, thicklip grey mullet (FAO)
Dicklippiger Grunzer	*Plectorhinchus macrolepis*	biglip grunt
Dickmaulrüssler, Lappenrüssler	*Otiorhynchus* spp.	snout beetles, snout weevils
Dickmundreusenschnecke	*Hinia incrassata*	thick-lipped dog whelk, angulate nassa
Dickschalige Grübchenschnecke	*Lacuna crassior*	thick lacuna, thick chink snail
Dickschalige Schraubenschnecke*	*Terebra robusta*	robust auger
Dickschalige Schlitzschnecke, Dickschalige Ausschnittsschnecke	*Emarginula crassa*	thick slit limpet, thick emarginula
Dickschalige Blasenschnecke	*Bulla solida*	solid bubble
Dickschalige Kugelmuschel	*Sphaerium solidum*	thickshelled fingernailclam
Dickschalige Pilgermuschel	*Argopecten solidulus*	solid scallop
Dickschnabel-Rohrsänger	*Acrocephalus aedon*	thick-billed warbler
Dickschnabel-Würgerkrähe	*Strepera graculina*	pied currawong
Dickschnabellumme	*Uria lomvia*	Brünnich's guillemot
Dickschnabelkolibri, Rivolikolibri	*Eugenes fulgens*	magnificent hummingbird
Dickschnabelorganist	*Euphonia laniirostris*	thick-billed euphonia
Dickschnabelpinguin	*Eudyptes pachyrhynchus*	Victoria penguin
Dickschwanz-Springmaus	*Stylodipus telum*	thick-tailed three-toed jerboa
Dickschwanzbeutelratte	*Lutreolina crassicaudata*	thick-tailed opossum
Dickschwanzmaus, Fettschwanz-Rennmaus	*Pachyuromys duprasi*	fat-tailed gerbil
Dickschwanzratten	*Zyzomys* spp.	thick-tailed rats, rock rats
Dickschwanzskorpion	*Androctonus australis*	fattailed scorpion, fat-tailed scorpion, African fat-tailed scorpion
Diebkäfer, Diebskäfer	Ptinidae	spider beetles
Diebsameisen	*Solenopsis* spp.	fire ants
Diffuse Augenkoralle	*Oculina diffusa*	diffuse ivory bush coral
Dikdiks, Windspielantilopen	*Madoqua* spp.	dik-diks
Dione-Natter, Steppennatter	*Elaphe dione*	Dione's snake
Diphos-Sandmuschel	*Hiatula diphos*	diphos sunset clam, diphos sanguin
Diskus-Beilbauchfisch	*Thoracocharax stellatus*	spotfin hatchetfish (FAO), silver hatchetfish
Diskus-Buntbarsche	*Symphysodon* spp.	discus fishes
Diskus> Echter Diskus	*Symphysodon discus discus*	discus

Diskusfüßige Fledermaus	*Eudiscopus denticulus*	disk-footed bat
Distelfalter	*Cynthia cardui/Vanessa cardui*	painted lady, thistle (butterfly)
Distelnetzwanze	*Tingis cardui*	spear thistle lace bug
Distinkte Wegschnecke*	*Arion distinctus*	darkface slug, darkface arion
Distorsio> Gemeine Distorsio	*Distorsio anus*	common distorsio
Divertikelschnecke	*Eobania vermiculata*	vermiculate snail, chocolate-band snail
Döbel	*Leuciscus cephalus*	chub
Doggenhaiartige	Heterodontiformes	horn sharks
Doggerscharbe, Raue Scholle, Raue Scharbe	*Hippoglossoides platessoides*	long rough dab
Dohle	*Coloeus monedula/ Corvus monedula*	jackdaw
Dohlengrackel	*Quiscalus mexicanus*	great-tailed grackle
Dohlenkrebs	*Austropotamobius pallipes*	white-clawed crayfish, freshwater white-clawed crayfish, river crayfish
Doktorfische & Einhornfische	Acanthuridae	surgeonfishes & unicornfishes
Doktorfische	Acanthurinae	surgeonfishes
Dolchfrosch	*Rana holsti*	Okinawa Shima frog
Dolchwespen, Mordwespen	Scoliidae	scolid wasps
Dollarvogel	*Eurystomus orientalis*	eastern broad-billed roller, dollar bird
Dollmans Baummaus	*Prionomys batesi*	Dollman's tree mouse
Dominikanerwitwe	*Vidua macroura*	pin-tailed whydah
Donau-Brachse, Zobel	*Abramis sapa*	Danube bream, Danubian bream, white-eye bream (FAO)
Donau-Flussdeckelschnecke	*Viviparus acerosus*	Danube river snail
Donau-Hering, Pontische Alse	*Alosa pontica*	Pontic shad (FAO), Black Sea shad
Donau-Kahnschnecke, Sarmatische Schwimmschnecke	*Theodoxus danubialis*	Danube nerite, Danube freshwater nerite
Donau-Rotauge	*Rutilus pigus pigus*	Danubian roach
Doppeläugiger Fichtenbastkäfer	*Polygraphus poligraphus*	small spruce bark beetle
Doppelbinden-Kammzähner	*Istiblennius edentulus*	rockskipper
Doppeldorniger Wimperbock	*Pogonocherus hispidulus*	apple-wood longhorn beetle
Doppelfingergeckos	*Diplodactylus* spp.	two-toed geckos, spinytail geckos
Doppelfüßer, Diplopoden, Tausendfüßler, Myriapoden	Myriapoda	millipedes, diplopods, myriapodians
Doppelfußschnecken	Tricoliidae	pheasant snails a.o. (see Phasianellidae)
Doppelhornvogel	*Buceros bicornis*	great Indian hornbill
Doppelkammbeutelmaus	*Dasyuroides byrnei*	kowari
Doppelkanten-Schlüssellochschnecke	*Fissurella aperta*	double-edged keyhole limpet
Doppelkopf	*Chironius bicarinatus*	two-headed sipo

Doppellinien-Makrele	*Grammatorcynus bilineatus*	double-lined mackerel
Doppelschaler, Krallenschwänze	Diplostraca/Onychura	clam shrimps & water fleas
Doppelschleichen, Wurmschleichen	Amphisbaenia	worm lizards, amphisb(a)enids, amphisbenians
Doppelschnepfe	*Gallinago media*	great snipe
Doppelspornfrankolin	*Francolinus bicalcaratus*	double-spurred francolin
Doppelte Schraubenschnecke	*Terebra duplicata*	duplicate auger
Doppelzahn-Bartvogel	*Lybius bidentatus*	double-toothed barbet
Doppelzahnspanner	*Odontopera bidentata*	scalloped hazel moth
Dorkasgazelle	*Gazella dorcas*	Dorcas gazelle
Dorn-Bandschnecke, Walross-Schnecke	*Opeatostoma pseudodon*	thorn latirus, thorned latirus
Dorn-Blattkäfer, Stachelkäfer	*Hispa atra*	spiny leaf beetle, leaf-mining beetle, wedge-shaped leaf beetle
Dornaale, Stachelaale, Eigentliche Dornrückenaale	Notacanthidae	spiny eels
Dornauge	*Acanthophthalmus kuhli/ Pangio kuhli*	coolie loach
Dörnchenkorallen, Dornkorallen, Schwarze Edelkorallen	Antipatharia	thorny corals, black corals, antipatharians
Dornen-Meerengel	*Squatina aculeata*	sawback angelshark (FAO), thorny monkfish
Dornen-Rosengallwespe*	*Diplolepis nervosa*	rose spiked pea-gall cynipid wasp (>spiked pea gall)
Dornenkronenseestern	*Acanthaster planci*	crown-of-thorns starfish
Dornenstern, Dornenseestern, Blauer Seestern	*Coscinasterias tenuispina*	thorny sea star*
Dornfinger, Ammen-Dornfinger	*Cheiracanthium punctorium*	European sac spider
Dornfische	Stephanoberycidae	pricklefishes
Dorngrasmücke	*Sylvia communis*	whitethroat
Dorngrundeln	Cobitidae	loaches
Dornhai, Gemeiner Dornhai	*Squalus acanthias/ Acanthias vulgaris*	common spiny dogfish, spotted spiny dogfish, piked dogfish (FAO), spurdog
Dornhaie	Squalidae	dogfishes, dogfish sharks
Dornige Herzmuschel	*Acanthocardia echinata*	prickly cockle, thorny cockle
Dornkorallen, Dörnchenkorallen, Schwarze Edelkorallen	Antipatharia	thorny corals, black corals, antipatharians
Dornrand-Weichschildkröte	*Apalone spinifera/ Trionyx spiniferus*	spiny softshell turtle
Dornrücken-Geigenrochen, Dornrückengitarrenfisch	*Platyrhinoidis triseriata*	thornback guitarfish
Dornrückengitarrenfische	Platyrhinidae	platyrhinids
Dornschildkröten	*Cyclemys* spp.	leaf turtles
Dornschrecke> Gemeine Dornschrecke, Säbeldornschrecke	*Tetrix undulata/ Tetrix vittata*	common groundhopper

Dornschrecken	Tetrigidae	pygmy grasshoppers, grouse locusts
Dornschwanzbilch	Zenkerella insignis	flightless scaly-tailed squirrel, Cameroon scaly-tail
Dornschwanzhörnchen	Anomaluridae	scaly-tailed squirrels, anomalures
Dornschwanzhörnchen	Anomalurus spp.	scaly-tailed flying squirrels, anomalures
Dornschwanzschlange, Spitzschwanzschlange	Contia tenuis	sharptail snake
Dornsepie, Dornsepia	Sepia orbignyana	pink cuttlefish, Orbigny's cuttlefish
Dornspeckkäfer	Dermestes maculatus	hide beetle, common hide beetle, leather beetle
Dornspötter	Hippolais languida	Upcher's warbler
Dornwels	Acanthodoras spinosissimus	talking catfish (FAO), spiny catfish, croaking catfish
Dornwelse	Doradidae	thorny catfishes
Dornzikade, Graubraune Dornzikade	Centrotus cornutus	horned treehopper
Dörrobstmotte, Indische Dörrobstmotte	Plodia interpunctella	Indian meal moth
Dörrobstmotte, Tropische Speichermotte, Dattelmotte	Cadra cautella	dried-fruit knot-horn, dried-fruit moth, almond moth
Dorsch (*Ostsee/Jungform*), *eigentlich:* Kabeljau	Gadus morhua	cod, Atlantic cod (*young:* codling)
Dorsche & Schellfische	Gadidae	cods
Dorschfische	Gadiformes	codfishes & haddock & hakes
Dosenschildkröte	Terrapene spp.	box turtles
Dotterqualle*	Phacellophora camtschatica	eggyolk jellyfish
Douglasien-Samenwespe	Megastigmus spermotrophus	Douglas fir seed wasp, Douglas fir seedfly
Drachenfische, Petermännchen	Trachinidae	weeverfishes
Drachenflosser, Kehlkropfsalmler	Pseudocorynopoma doriae	dragon-finned characin, dragonfin tetra (FAO)
Drachenkopfartige, Drachenkopffischverwandte, Panzerwangen	Scorpaeniformes	sculpins & sea robins
Drachenköpfe	Scorpaenidae	scorpionfishes (rockfishes)
Drachenkopfkauri	Cypraea caputdraconis	dragon-head cowrie, dragon's head cowrie
Drachenmuräne	Muraena pardalis	leopard moray eel
Drachenwurm, Guineawurm, Medinawurm	Dracunculus medinensis	fiery serpent, medina worm, guinea worm
Draparnauds Glanzschnecke, Große Glanzschnecke, Große Gewächshaus-Glanzschnecke	Oxychilus draparnaudi	Draparnaud's glass snail, dark-bodied glass-snail
Drechselschnecke, Europäische Drechselschnecke	Acteon tornatilis	European acteon

Drechselschnecken	Acteonidae	barrel snails, baby bubbles (barrel shells, small bubble shells)
Drehhornantilope> Südostasiatische Drehhornantilope	*Pseudonovibos spiralis*	linh-duong
Dreiband-Anemonenfisch	*Amphiprion tricinctus*	three-band anemonefish
Dreibandbarbe, Prachtglanzbarbe	*Puntius arulius*	longfin barb
Dreibandsalmler, „Falscher" Ulrey	*Hyphessobrycon heterorhabdus*	flag tetra
Dreibartelige Seequappe	*Gaidropsarus vulgaris*	three-bearded rockling
Dreibärtelige Silber-Seequappe	*Onogadus argentatus*	silvery rockling
Dreibinden-Anemonenfisch, Goldflößchen, Gelbschwanzringelfisch, Sebaes Anemonenfisch	*Amphiprion sebae*	Sebae's anemonefish
Dreibinden-Ziersalmler	*Nannostomus trifasciatus*	three-striped pencilfish
Dreiblatt-Fledermäuse u.a.	*Triaenops* spp.	triple nose-leaf bats a.o.
Dreieckige Erbsenmuschel	*Pisidium supinum*	triangular pea mussel, humpbacked peaclam
Dreieckige Herzmuschel, Kleine Herzmuschel	*Parvicardium exiguum*	little cockle
Dreiecks-Gitterschnecke	*Trigonostoma pellucida*	triangular nutmeg
Dreieckskegel, Dreiecks-Kegelschnecke	*Conus trigonus*	trigonal cone
Dreieckskrabbe, Karibische Spinnenkrabbe	*Stenorhynchus seticornis*	yellowline arrow crab, Caribbean arrow crab
Dreiecksmuscheln, Dreieckmuschel, Stumpfmuscheln, Sägezähnchen	Donacidae	wedge clams, donax clams (wedge shells)
Dreiecksnatter, Rote Königsnatter	*Lampropeltis triangulum*	milk snake, eastern milk snake
Dreieckswurmschnecke, Dreikant-Wurmschnecke	*Vermetus triquetra/ Bivonia triquetra*	
Dreifarben-Glanzstar	*Spreo superbus/ Lamprotornis superbus/ Lamprospeo suberbus*	superb starling
Dreifarben-Mitra	*Mitra tricolor*	three-color mitre
Dreifarben-Olive	*Oliva tricolor*	tricolor olive, three-colored olive
Dreifarbiger Jamaikakärpfling	*Poecilia melanogaster/ Limia melanogaster*	black-bellied molly, blue limia, blackbelly limia (FAO)
Dreifarbschrecke	*Paracinema tricolor*	tricolor grasshopper
Dreifingerfaultier, Ai	*Bradypus tridactylus*	pale-throated sloth
Dreifingerfaultiere	*Bradypus* spp.	three-toed sloths
Dreifingerschrecken	Tridactylidae	pygmy mole crickets
Dreifleck-Demoiselle	*Pomacentrus tripunctatus*	trispot damsel*
Dreifleckiger Engelfisch	*Apolemichthys trimaculatus*	threespot angelfish
Dreiflossen-Schleimfische, Dreiflosser	Tripterygiidae	threefin blennies, triplefin blennies
Dreiflügelschnecke	*Pteropurpura trialata*	three-winged murex
Dreifuß-Triton	*Cymatium tripus*	tripod triton

Dreigestreifte Schraubenschnecke	*Terebra triseriata*	triseriate auger
Dreigliedriger Hundebandwurm, Hundebandwurm	*Echinococcus granulosus*	dwarf dog tapeworm, dog tapeworm, hydatid tapeworm
Dreihorn-Fechterschnecke, Dreihorn-Flügelschnecke	*Strombus tricornis*	three-knobbed conch
Dreijochzahnechsen	Trilophosauria	trilophosaurs, trilophosaurians (Triassic archosauromorphs)
Dreikant-Haar-Triton	*Cymatium trigonum*	trigonal hairy triton
Dreikant-Wurmschnecke, Dreieckswurmschnecke	*Vermetus triquetra/ Bivonia triquetra*	trigonal wormsnail
Dreikantmuschel, Wandermuschel	*Dreissena polymorpha*	zebra mussel, many-shaped dreissena
Dreikantröhrenwurm, Dreikantwurm, Kielwurm	*Pomatoceros triqueter*	keelworm
Dreikiel-Erdschildkröte	*Melanochelys tricarinata*	tricarinate hill turtle, three-keeled land tortoise (IUCN)
Dreikiel-Scharnierschildkröte, Indische Dornschildkröte	*Pyxidea mouhotii/ Cyclemys mouhoti*	Indian keel-backed terrapin, keeled box turtle, Indian leaf turtle, jagged-shell turtle
Dreikiel-Wasserschildkröte	*Mauremys nigricans*	Asian yellow pond turtle
Dreiklauen-Weichschildkröten	*Trionyx* spp.	softshells, softshell turtles
Dreilinien-Demoiselle	*Pomacentrus trilineatus*	threeline damsel (FAO), three-lined demoiselle
Dreipunkt-Buntbarsch*	*Oreochromis andersonii*	three-spotted tilapia
Dreischwanzbarsche	Lobotidae	tripletails
Dreispitz-Schmuckkrabbe*	*Macrocoeloma trispinosum*	spongy decorator crab
Dreistacheliger Seifenfisch	*Rypticus saponaceus*	greater soapfish
Dreistachler	Triacanthidae	triplespines
Dreistachler	Triacanthodidae	spikefishes
Dreistachlige Diacria	*Diacria trispinosa*	three-spine cavoline
Dreistachliger Stichling, Gemeiner Stichling	*Gasterosteus aculeatus*	three-spined stickleback
Dreistreifen-Anemonenfisch, Clarks Anemonenfisch	*Amphiprion clarki*	Clark's anemonefish, goldbelly, yellowtail clownfish (FAO)
Dreistreifen-Dachschildkröte	*Kachuga dhongoka*	three-striped roof turtle, three-stripe roofed turtle
Dreistreifen-Erdhörnchen	*Lariscus insignis*	three-striped ground squirrel
Dreistreifen-Kartoffelkäfer	*Lema trilinea*	threelined potato beetle
Dreistreifen-Klappschildkröte	*Kinosternon baurii baurii*	striped mud turtle
Dreistreifen-Meerbarbe	*Parupeneus trifasciatus*	threestripe goatfish
Dreistreifen-Reusenschnecke	*Nassarius trivittatus*	threeline mudsnail, New England nassa
Dreistreifen-Rosenboa	*Lichanura trivirgata*	striped rosy boa
Dreistreifen-Scharnierschildkröte	*Cuora trifasciata*	Chinese three-striped box turtle, three-banded box turtle
Dreistreifen-Tigerfisch	*Therapon jarbua*	jarbua therapon
Dreistreifen-Zwergbuntbarsch	*Apistogramma trifasciata*	blue apistogramma
Dreistreifiger Mondfleckspanner	*Selenia dentaria*	early thorn
Dreizack-Blattnasen	*Asellia* spp.	trident leaf-nosed bats

Dreizahn-Seeschmetterling	*Cavolinia tridentata*	three-tooth cavoline
Dreizahn-Zwerghornschnecke	*Carychium tridentatum*	long-toothed herald snail, dentate thorn snail (U.S.), slender herald snail
Dreizähner, Bauchsackkugelfische	Triodontidae	threetooth puffer, three-toothed puffers
Dreizähner, Bauchsackkugelfisch	*Triodon bursarius/ Triodon macropterus*	threetooth puffer (FAO), three-toothed puffer
Dreizehen-Aalmolch	*Amphiuma tridactylum*	three-toed amphiuma
Dreizehenmöve	*Rissa tridactyla*	kittiwake
Dreizehenspecht	*Picoides tridactylus*	three-toed woodpecker
Dreizehnstreifenziesel	*Spermophilus tridecemlineatus*	thirteen-lined ground squirrel
Drescher, Fuchshai	*Alopias vulpinus*	thresher shark, thintail thresher (FAO), thintail thresher shark, fox shark
Drescherhaie, Fuchshaie	Alopiidae	thresher sharks
Driftfische, Quallenfische, Galeerenfische	Nomeidae	driftfishes
Drill	*Papio leucophaeus/ Mandrillus leucophaeus*	drill
Dromedar, Dromedarspinner, Birkenspinner, Erlenzahnspinner, Erlenbirkenauen-Zahnspinner	*Notodonta dromedarius*	iron prominent
Dromedar, Einhöckriges Kamel	*Camelus dromedarius*	dromedary, one-humped camel
Drongos	Dicruridae	drongos
Dronte	*Raphus cucullatus*	dodo
Drontevögel	Raphidae	dodos & solitaires
Drosselhäher, Schlammnestkrähen	Corcoracinae	Australian chough, apostlebird
Drosselkrähe	*Corcorax melanorhamphos*	white-winged chough
Drosseln	Turdidae	solitaires, thrushes
Drosselrohrsänger	*Acrocephalus arundinaceus*	great reed warbler
Drosselstelze	*Grallina cyanoleuca*	magpie-lark
Drosselstelzen	Grallinidae	magpie-larks
Drosseluferläufer	*Actitis macularia*	spotted sandpiper
Drosselwaldsänger	*Seiurus noveboracensis*	northern waterthrush
Drückerfische	Balistidae	triggerfishes
Drummonds Fadenschnecke	*Facelina auriculata/ Facelina drummondi*	Drummond's facelina
Drüsenflügel-Mausohr	*Cistugo* spp.	wing-gland bats
Dryasmeerkatze	*Cercopithecus dryas*	dryas monkey
Dschelada, Blutbrustpavian	*Theropithecus gelada*	gelada
Dschungelkatze (Hbz.), Rohrkatze	*Felis chaus*	jungle cat
Dschungelmaina, Braunmaina	*Acridotheres fuscus*	Indian jungle mynah
Dschungelrabe	*Corvus macrorhynchos*	jungle crow

Dshungarischer Zwerghamster	*Phodopus sungorus*	striped hairy-footed hamster
Dubois' Seeschlange	*Aipysurus duboisii*	Dubois's sea snake
Dukatenfalter, Feuervogel	*Heodes virgaureae/ Lycaena vigaureae/ Chrysophanus virgaureae*	scarce copper
Dumerils Boa	*Acrantophis dumerili/ Boa dumerili*	Dumeril's boa
Dumerils Ringelwurm	*Platynereis dumerilii*	Dumeril's clam worm, comb-toothed nereid
Dumerils Schienenschildkröte	*Peltocephalus dumerilianus*	big-headed Amazon River turtle
Dünen-Sandlaufkäfer, Brauner Sandlaufkäfer, Kupferbrauner Sandlaufkäfer	*Cicindela hybrida*	dune tiger beetle
Dünengazelle	*Gazella leptoceros*	sand gazelle, slender-horned gazelle, rhim
Dünenohrwurm, Sandohrwurm	*Labidura riparia*	tawny earwig, giant earwig
Dungfliegen> Kleine Dungfliegen	Sphaeroceridae	small dungflies, lesser dungflies
Dungfliegen, Kotfliegen	Scathophagidae	dungflies
Dungmücken	Scatopsidae	minute black scavenger flies
Dungwurm, Mistwurm, Kompostwurm	*Eisenia fetida/ Eisenia foetida*	brandling, manure worm
Dunkelalbatros, Dunkler Rußalbatros	*Phoebetria fusca*	sooty albatross
Dunkelblauer Laufkäfer, Blauer Laufkäfer	*Carabus intricatus*	blue ground beetle, darkblue ground beetle
Dunkelbläuling	*Cyaniris semiargus*	mazarine blue
Dunkelbrauner Halsgrubenbock	*Criocephalus rusticus/ Arhopalus rusticus*	rusty longhorn beetle
Dunkelente	*Anas rubripes*	black duck, North American black duck
Dunkelfleckiger Graszünsler	*Crambus pratella*	dark-inlaid grass-veneer
Dunkelgrauer Flachserdfloh	*Aphthona euphorbiae*	large flax flea beetle
Dunkellaubsänger	*Phylloscopus fuscatus*	dusky warbler
Dunkelrote Seefeder	*Pennatula rubra*	dark-red sea-pen
Dunkelroter Ara, Grünflügelara	*Ara chloroptera*	green-winged
Dunkelspinnen, Walzenspinnen, Sechsaugenspinnen	Dysderidae	dysderids
Dunker's Bernsteinschnecke, Schlanke Bernsteinschnecke	*Oxyloma elegans/ Oxyloma dunkeri*	Pfeiffer's ambersnail, Dunker's ambersnail
Dunkle Binsenjungfer	*Lestes macrostigma*	dusky emerald damselfly
Dunkle Felsenkrabbe	*Pachygrapsus gracilis*	dark shore crab
Dunkle Hüpfmaus*	*Notomys fuscus*	dusky hopping mouse
Dunkle Kielnacktschnecke	*Milax gagates*	keeled slug, greenhouse slug, small black slug, jet slug
Dunkle Krötenkopf-Schildkröte	*Phrynops geoffroanus*	Geoffroy's side-necked turtle
Dunkle Nadelschnecke	*Cerithium atratum*	dark cerith, Florida cerith

Dunkle Pelomeduse	*Pelusios subniger*	East African black mud turtle
Dunkle Wiesenameise, Dunkle Waldameise	*Formica pratensis/ Formica nigricans*	black-backed meadow ant
Dunkler Perlmutterbuntbarsch	*Gymnogeophagus gymnogenys*	smoothcheek eartheater
Dunkler Delphin	*Lagenorhynchus obscurus*	Gray's dusky dolphin
Dunkler Dickkopffalter, Grauer Dickkopf	*Erynnis tages*	dingy skipper
Dunkler Drachenkopf	*Scorpaenodes guamensis*	Guam scorpionfish
Dunkler Holzklafterbock	*Xylotrechus rusticus*	rustic borer (longhorn beetle)
Dunkler Kanalkäfer	*Amara apricaria*	dusky sun beetle*
Dunkler Pelzkäfer	*Attagenus megatoma*	black carpet beetle
Dunkler Riesenzackenbarsch	*Epinephelus lanceolatus*	giant grouper
Dunkler Ringens, Schwarzer Weißbindendrücker	*Melichthys niger/ Balistes radula*	black durgon
Dunkler Rußalbatros, Dunkelalbatros	*Phoebetria fusca*	sooty albatross
Dunkler Seidenkäfer	*Maladera holosericea*	dark silky beetle*
Dunkler Sturmtaucher	*Puffinus griseus*	sooty shearwater
Dunkler Zierbock	*Anaglyptus mysticus*	grey-coated longhorn beetle
Dunkler Wasserläufer	*Tringa erythropus*	spotted redshank
Dunkles Riesenhorn	*Macron aethiops*	dusky macron
Dünnbeinige Gespensterkrabbe	*Inachus leptochirus*	slender-legged spider crab
Dünnhornschaf, Dallschaf	*Ovis dalli*	Dall's sheep, white sheep
Dünnlippige Meeräsche	*Mugil capito/Liza ramada*	thinlip mullet (FAO), thinlip grey mullet
Dunns Pygmäensalamander	*Parvimolge townsendi*	Dunn's pygmy salamander
Dünnschalige Spindelschnecke	*Tibia delicatula*	delicate tibia
Dünnschalige Baumauster	*Isognomon janus*	thin purse-oyster
Dünnschalige Kammmuschel	*Chlamys delicatula*	delicate scallop
Dünnschalige Nussmuschel, Glatte Nussmuschel	*Ennucula tenuis/Nucula tenuis*	smooth nutclam
Dünnschlangen	*Leptophis* spp.	parrot snakes
Dünnschnabel-Brachvogel	*Numenius tenuirostris*	slender-billed curlew
Dünnschnabelmöve	*Larus genei*	slender-billed gull
Dünnschwanz-Mausschläfer	*Myomimus personatus*	mouselike dormouse
Dupontlerche	*Chersophilus duponti*	Dupont's lark
Durchscheinende Stachelschnecke	*Pterynotus pellucidus*	pellucid murex
Durchsichtige Blättchenschnecke	*Lamellaria perspicua*	transparent lamellaria
Durchsichtige Glasschnecke, Kugelige Glasschnecke	*Vitrina pellucida*	pellucid glass snail, western glass-snail
Durchsichtige Messerscheide, Durchscheinende Messerscheide, Kleine Schwertmuschel	*Phalax pellucidus*	translucent piddock*
Durchsichtige Napfschnecke, Durchscheinende Häubchenschnecke	*Helcion pellucidum*	blue-rayed limpet
Durchsichtiger Beilbauch	*Sternoptyx diaphana*	transparent hatchetfish

Düsterbock	*Asemum striatum*	pine longhorn beetle
Düstere Turbanschnecke*	*Tegula pulligo*	dusky tegula
Düsterer Dornhai, Blattschuppiger Schlingerhai	*Centrophorus squamosus*	leafscale gulper shark
Düsterer Humusschnellkäfer	*Agriotes obscurus*	dusky click beetle
Düsterer Trauerschweber*	*Anthrax anthrax*	dusky beefly*
Düsterkäfer, Schwarzkäfer	Melandryidae (Serropalpidae)	false darkling beetles
Dybowski-Ratte, Dybowsky-Ratte	*Mylomys dybowski/ Mylomys dybowskyi*	African groove-toothed rat, mill rat
Dybowski-Sandaal, Koreanischer Sandaal	*Hypoptychus dybowskii*	Dybowski's sand eel, Korean sandeel (FAO)

Ebarmen, Hartstrahlenflunder	Psettodidae	adalahs, psettodids, spiny flatfishes
Ebenholz-Mitra	*Mitra ebenus/Vexillum ebenus*	ivory mitre
Eberfisch	*Capros aper*	boarfish
Eberfische	Caproidae	boarfishes
Eberzahnschnecke	*Dentalium aprinum* (Scaphopoda)	boar's tusk
Echinostomose-Darmegel	*Echinostoma ilocanum*	intestinal echinostome
Echte Achatschnecke	*Achatina achatina*	common African snail
Echte Fliegen	Muscidae	houseflies, house flies
Echte Frösche	Ranidae	true frogs
Echte Gitterschnecke, Gegitterte Muskatnuss	*Cancellaria cancellata*	cancellate nutmeg, lattice nutmeg
Echte Karettschildkröten, Karetten	*Eretmochelys* spp.	hawksbill turtles, hawksbill sea turtles
Echte Katzenmakis	*Cheirogaleus* spp.	dwarf lemurs
Echte Kröten	Bufonidae	toads, true toads
Echte Motten	Tineidae	clothes moths
Echte Netzflügler, Hafte	Neuroptera/Planipennia	neuropterans (dobson flies/antlions)
Echte Pilzzungensalamander	*Bolitoglossa* spp.	tropical lungless salamanders
Echte Plattbauchspinnen	*Gnaphosa* spp.	hunting spiders a.o.
Echte Rotzunge, Limande	*Microstomus kitt*	lemon sole
Echte Salamander & Molche	Salamandridae	salamanders & newts
Echte Samtmuschel, Violette Pastetenmuschel	*Glycymeris pilosa*	hairy dog cockle, hairy bittersweet
Echte Tulpenschnecke	*Fasciolaria tulipa*	true tulip, tulip spindle snail (tulip spindle shell)
Echte Weberknechte, Schneider	Phalangiidae	daddy-long-legs
Echte Weichschildkröten	Trionychidae	softshell turtles, softshells
Echte Wendeltreppe	*Epitonium scalare/ Scalaria pretiosa*	precious wentletrap
Echter Diskus	*Symphysodon discus discus*	discus
Echter Kiemenfuß, Sommerkiemenfuß, Sommer-Kiemenfuß	*Branchipus stagnalis/ Branchipus schaefferi*	fairy shrimp
Echter Schildfuß	*Scutopus ventrolineatus* (Aplacophora)	bellylined glistenworm*
Echter Sepiakalmar, Karibischer Riffkalmar	*Sepioteuthis sepioidea*	Caribbean reef squid, Atlantic oval squid
Echtgazelle	*Gazella gazella*	mountain gazelle
Eckige Erbsenmuschel	*Pisidium milium*	quadrangular pea mussel, quadrangular pillclam, rosy pea mussel
Eckschwänze, Quadratschwänze	Tetragonuridae	squaretails
Eckschwanzsperber	*Accipiter striatus*	sharp-shinned hawk
Ecuador-Stachelratte*	*Scolomys melanops*	Ecuador spiny mouse
Edelfinken	Fringillinae	chaffinches, brambling

Edelhirsch, Rothirsch	*Cervus elaphus*	red deer, wapiti, elk (U.S.)
Edelhirsche	*Cervus* spp.	red deers
Edelkoralle, Rote Edelkoralle	*Corallium rubrum*	red coral, precious red coral
Edelkrebs	*Astacus astacus*	noble crayfish
Edellibellen, Teufelsnadeln	Aeshnidae	darners (U.S.), large dragonflies
Edelmaräne, Kleine Schwebrenke, Schnepel, Schnäpel	*Coregonus oxyrhynchus/ Coregonus oxyrinchus*	houting
Edelpapagei	*Eclectus roratus*	eclectus parrot
Edelsteinkärpfling	*Cyprinodon variegatus*	sheepshead minnow
Edelsteinrose, Warzenrose	*Bunodactis verrucosa*	wartlet anemone, gem anemone
Edle Froschschnecke	*Bufonaria margaritula/ Bufonaria nobilis/ Bursa margaritula/ Bursa nobilis*	noble frogsnail
Edle Kammmuschel, Feine Kammmuschel	*Chlamys senatoria/ Chlamys nobilis*	senate scallop, noble scallop
Edle Kegelschnecke	*Conus nobilis*	noble cone
Edle Stachelschnecke	*Chicoreus nobilis*	noble murex
Edle Walzenschnecke	*Aulica nobilis/ Cymbiola nobilis*	noble volute
Edle Wendeltreppe	*Sthenorytis pernobilis*	noble wentletrap
Edler Tolstolob, Marmorkarpfen	*Hypophthalmichthys nobilis*	bighead carp
Edmigazelle	*Gazella cuvieri*	Edmi gazelle
Edwardsfasan	*Lophura edwardsi*	Edward's pheasant
Efeuborkenkäfer	*Kissophagus hederae*	ivy bark beetle
Egel> Hirudineen	Hirudinea	leeches, hirudineans
Egel> Saugwürmer, Trematoden	Trematoda	flukes, trematodes
Egelschnecken, Schnegel	Limacidae	keeled slugs
Eichblatt-Radspinne, Eichenblatt-Radnetzspinne	*Araneus ceropegius/ Aculepeira ceropegia*	oakleaf orbweaver*
Eichelbohrer	*Curculio glandium*	acorn weevil
Eichelhäher	*Garrulus glandarius*	jay
Eichelreusenschnecke, Eichelreuse	*Nassarius glans*	glans nassa
Eichelspecht	*Melanerpes formicivorus*	acorn woodpecker
Eichelwickler, Kastanienwickler	*Cydia splendana*	acorn moth
Eichelwürmer, Enteropneusten	Enteropneusta	acorn worms, enteropneusts
Eichen-Glanzgallwespe (>Braune Glanzgalle)	*Cynips divisa*	oak bud red-gall cynipid wasp, oak bud cherry-gall cynipid wasp
Eichen-Nierengallblattwespe, Eichenblatt-Nierengallwespe	*Trigonaspis megaptera*	oak leaf kidney-gall cynipid wasp

Eichen-Nulleneule, Nulleneule, O-Eule, Eichenhochwald-Doppelkreiseule	*Dicycla oo*	heart moth
Eichen-Sichelspinner, Eichensichler, Eichen-Sichelflügler, Zweipunkt-Sichelflügler	*Drepana binaria*	oak hook-tip
Eichenblatt-Austerngallwespe	*Andricus ostreus*	oak leaf oyster-gall wasp
Eichenblatt-Nierengallwespe, Eichen-Nierengallblattwespe	*Trigonaspis megaptera*	oak leaf kidney-gall cynipid wasp
Eichenblatt-Radnetzspinne, Eichblatt-Radspinne	*Araneus ceropegius/ Aculepeira ceropegia*	oakleaf orbweaver*
Eichenblattroller, Roter Eichenkugelrüssler, Tönnchenwickler	*Attelabus nitens*	oak leaf roller, red oak roller
Eichenblattsauger	*Trioza remota*	oak leaf sucker (psyllid)
Eichenbock, Großer Eichenbock, Heldbock	*Cerambyx cerdo*	great capricorn beetle, oak cerambyx
Eichenbuntkäfer	*Clerus mutillarius*	oak clerid
Eichenbuschspanner, Punktfleckspanner	*Cosymbia punctaria/ Cyclophora punctaria*	maiden's blush moth
Eichenerdfloh	*Altica quercetorum*	oak flea beetle
Eichengallenrüssler	*Curculio villosus*	oak gall weevil
Eichengallwespe> Gemeine Eichengallwespe	*Cynips quercusfolii*	common oak gallwasp, oak leaf cherry-gall cynipid (>cherry gall)
Eichengallwespen	*Andricus* spp.	oak gall wasps, oak gall cynipids
Eichenholzbohrer, Gehöckerter Eichenholzbohrer (Kleiner Schwarzer Wurm)	*Xyleborus monographus*	oak wood ambrosia beetle
Eichenkernkäfer	*Platypus cylindrus*	oak pinhole borer
Eichenknopperngallwespe	*Andricus quercuscalicis*	acorn cup gall wasp (>knopper gall)
Eichenknospenmotte	*Coleophora lutipennella*	oak bud moth*
Eichenkragengallwespe (>Kragengalle)	*Andricus curvator*	oak bud collared-gall wasp
Eichenlinsengallwespe (>Große Linsengalle[1]), Weinbeerengallwespe (>Weinbeerengalle[2])	*Neuroterus quercusbaccarum*	oak leaf spangle-gall cynipid wasp (>common spangle gall[1]), oak leaf currant-gall cynipid wasp (>currant gall[2])
Eichenminiermotte	*Tischeria ekebladella*	red-feather carl, trumpet leaf miner
Eichenprozessionsspinner	*Thaumetopoea processionea*	oak processionary moth
Eichenrosengallwespe	*Andricus fecundator/ Andricus foecundatrix*	artichoke gall wasp, larch cone gall cynipid, hop gall wasp (>artichoke gall)
Eichenschildlaus	*Kermes quercus*	oak scale
Eichenschildläuse, Eichennapfläuse	Kermidae/Kermesidae	gall-like coccids
Eichenschrecke, Gemeine Eichenschrecke	*Meconema thalassinum*	oak bushcricket
Eichenschwammgallwespe, Schwammgallwespe	*Biorrhiza pallida/ Biorhiza pallida*	oak-apple gall wasp (>oak apple)

Eichenschwärmer	*Marumba quercus/ Smerinthus quercus*	oak hawkmoth
Eichensichler, Eichen-Sichelspinner, Eichen-Sichelflügler, Zweipunkt-Sichelflügler	*Drepana binaria*	oak hook-tip
Eichenspinner, Quittenvogel	*Lasiocampa quercus*	oak eggar
Eichensplintkäfer	*Scolytus intricatus*	oak bark beetle
Eichenspringrüssler	*Rhynchaenus quercus*	oak flea weevil, oak leaf mining weevil
Eichenstreifgallwespe	*Cynips longiventris*	oak leaf striped-gall cynipid wasp
Eichenwerftkäfer, Schiffswerftkäfer	*Lymexylon navale*	ship timberworm
Eichenwidderbock, Eichenzierbock	*Plagionotus arcuatus*	yellow-bowed longhorn beetle
Eichenwurzelknoten-Gallwespe	*Andricus testaceipes*	oak red barnacle-gall wasp
Eichenzahnspinner	*Peridea anceps*	great prominent
Eichenzierbock, Eichenwidderbock	*Plagionotus arcuatus*	yellow-bowed longhorn beetle
Eichenzipfelfalter> Blauer Eichenzipfelfalter	*Thecla quercus/ Quercusia quercus*	purple hairstreak
Eichenzipfelfalter> Brauner Eichenzipfelfalter, Stechpalmen-Zipfelfalter	*Satyrium ilicis/ Nordmannia ilicis*	ilex hairstreak
Eichenzwerglaus u.a.	*Phylloxera coccinea & Phylloxera glabra*	oak leaf phylloxera a.o.
Eichhörnchen-Floh	*Diamanus montanus*	squirrel flea
Eichhörnchen-Laubfrosch	*Hyla squirella*	squirrel treefrog
Eichhörnchenfische	Holocentrinae	squirrelfishes
Eichhörnchenfische & Soldatenfische	Holocentridae	squirrelfishes & soldierfishes
Eichhornfisch> Gewöhnlicher Eichhornfisch, Roter Husar, Roter Soldatenfisch, Rotgestreifter Soldatenfisch	*Holocentrus rubrum*	red soldierfish
Eidechsen	Lacertilia (Squamata)/ Lacertidae	lizards
Eidechsen & Schlangen> Schuppenkriechtiere	Squamata	squamata (incl. lizards & amphisbaenians & snakes)
Eidechsenfische	*Synodus* spp.	lizardfishes
Eidechsenfische	Synodontidae	lizardfishes
Eidechsennatter, Europäische Eidechsennatter	*Malpolon monspessulans*	Montpellier snake
Eidechsenschwanz	*Falcaria lacertinaria*	scalloped hook-tip
Eiderente	*Somateria mollissima*	common eider
Eierlegende Zahnkarpfen, Killifische	Cyprinodontidae	pupfishes
Eiförmige Schlammschnecke	*Radix ovata*	ovate pondsnail (pond mud-shell)
Eigentliche Baumläufer	Certhiini	northern creepers
Eigentliche Eingeweidefische, Nadelfische	Carapidae/ Fierasferidae	pearlfishes, carapids

Eigentliche Flughunde, Flugfüchse	*Pteropus* spp.	flying foxes
Eigentliche Glattkopffische	Platyproctidae	tubeshoulders
Eigentliche Höckerechsen	*Xenosaurus* spp.	New World xenosaurs, American knob-scaled lizards
Eigentliche Kieferspinnen	*Pachygnatha* spp.	thick-jawed spiders, thick-jawed orb weavers
Eigentliche Kreuzspinnen	Araneinae	typical orbweavers
Eigentliche Landschildkröten	*Testudo* spp.	land tortoises
Eigentliche Salpen	Salpida	salps
Eigentliche Störe	Acipenseridae	sturgeons
Eigentliche Streckerspinnen	*Tetragnatha* spp.	long-jawed orb weavers
Eigentliche Weber, Eigentliche Weberfinken, Webervögel	Ploceinae	weavers
Eigentlicher Rotfeuerfisch	*Pterois volitans*	red firefish, red lionfish (FAO), devil firefish, fireworkfish
Eilandbarbe	*Puntius oligolepis/ Barbus oligolepis*	checkered barb (FAO), checkerboard, island barb
Eileiteregel*	*Prosthogonimus* spp.	oviduct flukes (of birds/poultry)
Eilseeschwalbe	*Sterna bergii*	crested tern
Einarmqualle*	*Hybocodon pendulus*	one-arm jellyfish
Einband-Mondschnecke*	*Natica unifasciata*	single-banded moonsnail
Einbinden-Ziersalmler, Einbinden-Schrägsteher	*Nannostomus unifasciatus*	oneline pencilfish
Einbindiger Traubenwickler (Heuwurm)	*Cochylis ambiguella/ Eupoecilia ambiguella/ Clysia ambiguella*	European grapevine moth, small brown-barrel conch, grapeberry moth
Eindorn-Zwerghai	*Squaliolus laticaudatus*	spined pygmy shark (FAO), spined dwarf shark
Einfache Teppichmuschel, Gebänderte Teppichmuschel, Essbare Venusmuschel	*Tapes rhomboides/ Venerupis rhomboides/ Paphia rhomboides/ Venerupis geographica*	banded carpetclam, banded venus (banded carpet shell)
Einfacher Doppelschildfuß	*Prochaetoderma raduliferum* (Aplacophora)	Mediterranean glistenworm a.o.
Einfacher Zangenschildfuß	*Falcidens gutturosus* (Aplacophora)	Mediterranean glistenworm a.o.
Einfarbige Ackerschnecke, Graue Ackerschnecke	*Deroceras agreste/ Agriolimax agreste*	field slug, grey field slug, grey slug
Einfarbige Rennechse, Schlichter-Rennechse	*Cnemidophorus inornatus*	little striped whiptail
Einfarbiger Herzigel, Grauer Herzigel	*Brissus unicolor*	grey heart-urchin
Einfarbstar	*Sturnus unicolor*	spotless starling
Einfleck-Anglerfisch*	*Antennarius radiosus*	big-eyed frogfish, singlespot frogfish
Einfleck-Jungferchen	*Pomacentrus stigma*	blackspot damsel, analspot demoiselle
Eingedrückte Dornschrecke	*Tetrix depressa*	depressed groundhopper, flattened groundhopper
Eingekrümmte Haubenschnecke*	*Capulus incurvus*	incurved capsnail

Eingeweideaale	*Pisodonophis* spp.	snake eels
Eingeweidefische	Ophidiiformes	brotulas & cusk-eels & pearlfishes
Einhorn-Geißelgarnele	*Metapenaeus monoceros*	speckled shrimp
Einhorn-Lederjacke	*Alutera monoceros*	unicorn leatherjacket
Einhorndorsch	*Bregmaceros mcclellandi*	unicorn cod
Einhorndorsche	Bregmacerotidae	codlets
Einhorngarnele, Narval	*Plesionika narval*	unicorn striped shrimp, narval
Einhornschnecke	*Acanthina monodon*	unicorn, one-toothed thais
Einhornschnecken	*Acanthina* spp.	unicorns, unicorn snails
Einkieferaale	Monognathidae	singlejaw eels
Einödgimpel	*Carpodacus synoicus*	Sinai rosefinch
Einplatter, Urmützenschnecken, Monoplacophoren	Monoplacophora	monoplacophorans
Einpunkt-Pappelblattminiermotte	*Phyllocnistis unipunctella*	poplar leaf miner
Einsamer Wasserläufer	*Tringa solitaria*	solitary sandpiper
Einsiedler-Seescheide	*Styela canopus*	hermit sea-squirt
Einsiedlerdrossel	*Catharus guttatus/ Hylocichla guttata*	hermit thrush
Einsiedlerlori	*Phigys solitarius*	collared lory
Einsiedlerrose (Schmarotzerrose)	*Calliactis parasitica/ Peachia parasitica/ Adamsia rondeletii*	hermit crab anemone (parasitic anemone a.o.)
Eintagsfliegen	Ephemeroptera	mayflies
Einzähnige Haarschnecke	*Petasina unidentata/ Trichia unidentata*	unidentate marginsnail (unidentate margin shell), single-toothed hairysnail
Einzehen-Aalmolch	*Amphiuma pholeter*	one-toed amphiuma
Eisbär	*Ursus maritimus*	polar bear
Eischnecke> Gemeine Eischnecke	*Erronea ovum*	egg cowrie
Eisenfarbiger Samtfalter	*Hipparchia statilinus/ Neohipparchia statilinus*	tree grayling
Eisente	*Clangula hyemalis*	long-tailed duck
Eisfische	Channichthyidae/ Chaenichthyidae	crocodile icefishes
Eisfuchs, Polarfuchs (Hbz. Weißfuchs/Blaufuchs)	*Alopex lagopus*	arctic fox
Eishaie, Unechte Dornhaie	Dalatiidae	sleeper sharks
Eismeer-Ringelrobbe	*Phoca hispida*	ringed seal
Eismöve	*Larus hyperboreus*	glaucous gull
Eisseestern, Eisstern, Warzenstern	*Marthasterias glacialis*	spiny starfish
Eissturmvogel	*Fulmarus glacialis*	fulmar
Eistaucher	*Gavia immer*	great northern diver, common loon (U.S.)
Eisvogel	*Alcedo atthis*	kingfisher
Eisvögel & Fischer & Lieste	Alcedinidae	kingfishers
Elch	*Alces alces*	elk, moose (U.S.)
Elchgeweih-Moostierchen, Band-Moostierchen	*Pentapora fascialis*	banded bryozoan

Elchgeweihkoralle, Elchhornkoralle	*Acropora palmata*	elkhorn coral
Elchrachenbremse	*Cephenomyia ulrichi*	elk nostril fly (Br.), moose nostril fly (U.S.)
Elefant> Afrikanischer Elefant	*Loxodonta africana*	African elephant
Elefant> Asiatischer Elefant, Indischer Elefant	*Elephas maximus*	Asiatic elephant, Indian elephant
Elefanten-Rüsselfisch, Tapirfisch	*Gnathonemus petersii*	Peter's worm-jawed mormyrid, elephantnose fish (FAO)
Elefantenchimäre, Totenkopfchimäre	*Callorhynchus capensis/ Callorhinchus capensis*	elephantfish, silver trumpeter (tradename)
Elefantenchimären, Pflugnasenchimären	Callorhynchidae	elephantfishes, plownose chimaeras
Elefantenfische, Nilhechte	Mormyridae	elephantfishes
Elefantenfuß-Mondschnecke, Elefantenfußschnecke	*Neverita peselephanti*	elephant's-foot moonsnail
Elefantenohr	*Spongia agaricina*	elephant's ear sponge*
Elefantenohrkoralle*	*Sarcophyton trocheliophorum*	elephant ear coral, yellow-green soft-coral (leather coral)
Elefantenohrschwamm u.a.	*Agelas clathrodes*	orange wall sponge
Elefantenohrschwamm	*Spongia officinalis lamella*	elephant's ear sponge
Elefantenrobben	*Mirounga* spp.	elephant seals
Elefantenschildkröte, Galapagos-Riesenschildkröte	*Testudo elephantopus/ Geochelone elephantopus/ Chelonoides elephantopus*	Galapagos giant tortoise
Elefantenschnauzen-Walze*	*Cymbium glans*	elephant's snout volute
Elefantenschnecke, Schildschnecke	*Scutus antipodes/ Scutus anatinus*	elephant slug, duckbill shell, shield shell
Elefantenspitzmäuse	*Elephantulus* spp.	long-eared elephant shrews, small elephant shrews
Elefantenvögel, Madagaskarstrauße	Aepyornithiformes	elephant birds
Elefantenzähne, Zahnschnecken	Dentaliidae (Scaphopoda)	tusks, tuskshells
Elefantenzahnschnecke	*Dentalium elephantinum* (Scaphopoda)	elephant's tusk, elephant's tuskshell
Elegante Seescheide	*Aplidium elegans*	elegant sea-squirt
Elektrische Welse, Zitterwelse	Malapteruridae	electric catfishes
Elektrischer Wels, Zitterwels	*Malapterurus electricus*	electric catfish
Elenantilope	*Taurotragus oryx*	common eland
Eleonorenfalke	*Falco eleonorae*	Eleonora's falcon
Elepaio	*Chasiempis sandwichensis*	elepaio
Elfenbein-Buschkoralle, Feste Augenkoralle	*Oculina robusta*	robust ivory tree coral, ivory bush coral
Elfenbein-Moostierchen, Stacheliges Moostierchen	*Crisia eburnea*	ivory bryozoan*
Elfenbein-Nadelschnecke	*Cerithium eburneum*	ivory cerith
Elfenbein-Seepocke	*Balanus eburneus*	ivory barnacle
Elfenbeinkauri*	*Cypraea eburnea*	pure white cowrie
Elfenbeinkegel	*Conus eburneus*	ivory cone

Elfenbeinmöwe	*Pagophila eburnea*	ivory gull
Elfenbeinsittich	*Aratinga canicularis/* *Eupsittula canicularis*	orange-fronted parakeet, orange-fronted conure
Elfenbeinspecht	*Campephilus principalis*	ivory-billed woodpecker
Elfenkauz, Kaktuskauz	*Micrathene whitneyi/* *Glaucidium whitneyi*	elf owl
Elfpunkt	*Coccinella undecimpunctata*	eleven-spot ladybird, elevenspot ladybird, 11-spot ladybird
Elliots Tupaia	*Anathana ellioti*	Indian tree shrew
Elliots Walzenschnecke	*Amoria ellioti*	Elliot's volute
Elliptische Astarte	*Astarte elliptica*	elliptical astarte
Eloisenschnecke	*Acteon eloisae*	Eloise's acteon
Elritze (Eurasische Elritze)	*Phoxinus phoxinus*	minnow (Eurasian minnow)
Elritzennäsling	*Chondrostoma phoxinus*	minnow nase
Elster	*Pica pica*	black-billed magpie
Elsteradler	*Spizastur melanoleucus*	black & white hawk eagle
Elsterscharbe	*Phalacrocorax varius*	pied cormorant
Elstertyrann	*Fluvicola pica*	pied water tyrant
Elsterweihe	*Circus melanoleucos*	pied harrier
Emmelichthyiden	Emmelichthyidae	rovers
Emu	*Dromaius novaehollandiae*	emu
Emus	Dromaiidae	emus
Enchyträen	Enchytraeidae	potworms, aster worms, white worms
Endivienschnecke	*Hexaplex cichoreus*	endive murex
Engelhai, Gemeiner Meerengel	*Squatina squatina/* *Rhina squatina/* *Squatina angelus*	angelshark (FAO), angel shark, monkfish
Engelhaie, Engelhaiartige	Squatiniformes	monkfishes, angel sharks
Engelhaie, Meerengel	Squatinidae	angelsharks (sand devils)
Engelsflügel	*Cyrtopleura costata*	angel's wings
Engelsflügel	Petricolidae	piddocks
Engelwels, Perlhuhnwels	*Synodontis angelicus*	polka-dot African catfish, angel squeaker (FAO)
Engländischer Langwurm, Langer Schnurwurm	*Lineus longissimus* (Nemertini)	giant bootlace worm, sea longworm
Englische Puppenschnecke	*Leiostyla anglica*	English chrysalis snail
Engmaulsalmler, Kopfsteher	Anostomidae	anostomids
Engmünder	Stenolaemata/Stenostomata	stenostomates, stenolaemates, "narrow throat" bryozoans
Engmundkröten, Engmaulkröten	Microhylidae	narrowmouth toads, microhylids
Engmundkröten> Amerikanische Engmundkröten	*Gastrophryne* spp.	American narrowmouth toads
Engtaillien-Rindenkäfer (Scheinrüssler)	Salpingidae	narrow-waisted bark beetles
Ensatina, Eschscholtz-Salamander	*Ensatina eschscholtzi*	ensatina
Entenartige> Enten & Gänse & Schwäne	Anatidae	ducks & geese & swans
Entenegel	*Thermyzon tessulatum*	duck leech
Entenlaterne	*Laternula anatina*	duck lantern clam

„Entenmuschel", Flache Teichmuschel	*Anodonta anatina*	duck mussel
Entenmuscheln	Lepadidae	goose barnacles
Entenschnabelaale	Nettastomatidae	witch eels, duckbill eels
Entenschnabelrochen, Gefleckter Adlerrochen	*Aetobatus narinari*	spotted eagle ray
Entenvögel, Gänsevögel	Anseriformes	screamers & waterfowl (ducks & geese & swans)
Entenwal> Nördlicher Entenwal, Dögling	*Hyperoodon ampullatus*	northern bottlenosed whale, North-Atlantic bottle-nosed whale
Entenwal> Südlicher Entenwal	*Hyperoodon planifrons*	southern bottlenosed whale, southern flat-headed bottle-nosed whale
Epauletten-Haie	*Hemiscyllium* spp.	epaulette sharks
Erbsenblasenfuß	*Kakothrips pisivorus*	pea thrips
Erbseneule	*Melanchra pisi*	broom moth
Erbsengallmücke	*Contarinia pisi*	pea gall midge
Erbsenkäfer, Großer Erbsenkäfer	*Bruchus pisorum*	pea weevil, pea seed weevil
Erbsenkrabbe, Muschelwächter, Muschelwächterkrabbe	*Pinnotheres pisum*	pea crab, Linnaeus's pea crab
Erbsenkrabben, Muschelwächter	Pinnotheridae	pea crabs, commensal crabs
Erbsenminierfliege	*Phytomyza atricornis*	chrysanthemum leaf miner (fly)
Erbsenmuschel> Gemeine Erbsenmuschel	*Pisidium casertanum*	common pea mussel, caserta pea mussel, ubiquitous peaclam
Erbsenmuscheln	*Pisidium* spp.	pea mussels, peaclams (U.S.)
Erbsenwickler, Olivbrauner Erbsenwickler	*Laspeyresia nigricana/ Cydia nigricana/ Cydia rusticella*	pea moth
Erbsenzystenälchen	*Heterodera goettingiana*	pea cyst nematode ("pea sickness")
Erd-Sumpfschildkröte	*Geoemyda silvatica*	Cochin Forest cane turtle, forest turtle
Erdaaskäfer, Nestkäfer	Leptodiridae/ Catopidae	small carrion beetles
Erdbeer-Fechterschnecke, Erdbeer-Flügelschnecke	*Strombus luhuanus*	strawberry conch, blood-mouth conch
Erdbeer-Grabläufer*	*Pterostichus madidus*	strawberry ground beetle
Erdbeer-Herzmuschel	*Fragum unedo*	strawberry cockle
Erdbeer-Igelschnecke, Erdbeer-Igel	*Drupa rubusidaeus*	rose drupe, strawberry drupe
Erdbeer-Mitra	*Mitra fraga*	strawberry mitre
Erdbeer-Mottenschildlaus, Geißblatt-Mottenschildlaus	*Aleyrodes fragariae/ Aleyrodes lonicerae*	strawberry whitefly, honeysuckle whitefly
Erdbeerälchen	*Aphelenchoides fragariae*	strawberry leaf nematode, spring crimp nematode
Erdbeeranemone	*Corynactis californica*	strawberry anemone
Erdbeerbaumfalter	*Charaxes jasius*	two-tailed pasha
Erdbeerblattwespe	*Monophadnoides geniculatus*	geum sawfly (Br.), raspberry sawfly (U.S.)

Erdbeerblütenstecher, Himbeerblütenstecher, Beerenstecher	Anthonomus rubi	berry blossom weevil, strawberry blossom weevil
Erdbeerfröschchen, Kleiner Erdbeerfrosch	Dendrobates pumilio	strawberry poison-arrrow frog, red-and-blue poison-arrrow frog, flaming poison-arrrow frog
Erdbeerkäfer, Seerosen-Blattkäfer	Galerucella nymphaeae	waterlily beetle, pond-lily leaf-beetle
Erdbeerknotenhaarlaus, Knotenhaarlaus, Erdbeerblattlaus	Pentatrichopus fragaefolii/ Chaetosiphon fragaefolii	strawberry aphid
Erdbeerköpfchen, Fischers Unzertrennlicher	Agapornis fischeri	Fischer's lovebird
Erdbeerkrabbe*	Eurynome aspera	strawberry crab
Erdbeermilbe	Steneotarsonemus pallidus ssp. fragariae/ Tarsonemus pallidus ssp. fragariae	strawberry mite
Erdbeerrose	Actinia fragacea	strawberry anemone
Erdbeerspinne	Araneus alsiae	strawberry spider
Erdbeerstengelstecher	Rhynchites germanicus/ Coenorrhinus germanicus	strawberry rhynchites, strawberry weevil
Erdbeerzikade	Aphrodes bicinctus	strawberry leafhopper
Erdchamäleon> Madagassisches Erdchamäleon, Madagassisches Zwergchamäleon	Brookesia minima	minute leaf chameleon, dwarf chameleon
Erddrossel	Zoothera dauma	White's thrush
Erdeule, Graswurzeleule Saateule, Wintersaateule	Agrotis segetum	turnip moth (common cutworm)
Erdferkel	Orycteropus afer	aardvark, ant bear
Erdhase, Erdhäschen, Kleine Fünfzehenspringmaus, Zwergpferdespringer	Alactagulus pumilio	lesser five-toed jerboa, little earth hare
Erdholztermite, Mittelmeertermite	Reticulitermes lucifugus	common European white ant, common European termite
Erdhörnchen> Streifen-Erdhörnchen	Lariscus spp.	Malaysian striped ground squirrels
Erdhummel	Bombus terrestris	buff-tailed bumble bee
Erdkreuzspinne	Cercidia prominens	ground spider*
Erdkröte	Bufo bufo	European common toad
Erdkuckuck, Wegekuckuck	Geococcyx californianus	roadrunner, greater roadrunner
Erdkuckucke	Neomorphinae	roadrunners, ground-cuckoos
Erdläufer	Geophilidae	wire centipedes, garden centipedes
Erdleguane	Liolaemus spp.	tree iguanas
Erdmännchen	Suricata suricatta	suricate, slender-tailed meerkat
Erdmaus	Microtus agrestis	field vole
Erdnatter, Schwarze Erdnatter	Elaphe obsoleta	black rat snake, eastern rat snake
Erdnussplattkäfer	Oryzaephilus mercator	merchant grain beetle

Erdnusssamenkäfer, Westafrikanischer Erdnusssamenkäfer	*Caryedon serratus*	groundnut bruchid, Westafrican groundnut borer
Erdotter> Gewöhnliche Erdotter	*Atractaspis irregularis*	variable mole viper
Erdottern, Erdvipern	*Atractaspis* spp.	mole vipers
Erdpython	*Calabaria reinhardtii*	Calabar python
Erdracken	Brachypteraciidae	ground-rollers
Erdsamenrüssler, Tamarindenrüssler	*Caryedon fuscus*	groundnut borer, tamarind weevil
Erdschlangen	*Virginia* spp.	earth snakes
Erdschlangen	Xenopeltidae	sunbeam snakes
Erdsittich	*Pezoporus wallicus*	ground parrot
Erdspecht	*Geocolaptes olivaceus*	ground woodpecker
Erdwanzen	Cydnidae	burrower bugs, cydnid bugs
Erdwolf	*Proteles cristatus*	aardwolf
Erdwolfspinne*	*Geolycosa* spp.	burrowing wolf spiders
Eremit, Juchtenkäfer	*Osmoderma eremita*	hermit beetle
Erlen-Blattspanner	*Hydrelia sylvata/ Hydrellia testaceata*	sylvan waved carpet (moth)
Erleneule	*Acronicta alni*	alder moth
Erlenglasflügler	*Synanthedon spheciformis*	alder borer, alder clearwing moth
Erlenknospenmotte	*Coleophora fuscidenella*	alder bud moth*
Erlenrüssler, Weidenrüssler	*Cryptorhynchus lapathi*	poplar-and-willow borer, osier weevil, willow weevil
Erlenschaumzikade	*Aphrophora alni*	alder froghopper
Erlenspanner	*Ennomos alniaria*	alder thorn*
Erlenzeisig	*Carduelis spinus*	siskin
Ernteameisen u.a., "Getreideameisen"	*Pogonomyrmex* spp. & *Messor* spp.	harvester ants a.o.
Erntefisch (Amerikanischer Butterfisch)	*Peprilus paru/ Peprilus alepidotus*	harvestfish (FAO), North American harvestfish
Erntefische	Stromateoidei	harvestfishes
Erntefische, Butterfische	Stromateidae	ruffs, butterfishes, harvestfishes
Erntemilbe, Herbstmilbe, Herbstgrasmilbe	*Neotrombicula autumnalis*	harvest mite, red 'bug', chiggers
Erntemilben, Laufmilben, Herbstmilben	Trombiculidae	chigger mites, harvest mites
Erntetermiten	Hodotermitidae	harvester termites
Erpelschwanz	*Clostera curtula*	chocolate-tip
Erstmünder, Urmundtiere, Urmünder	Protostomia	protostomes
Erz-Kanalkäfer, Erzfarbener Kanalkäfer	*Amara aenea*	brazen sun beetle
Erzfarbener Scheibenbock	*Callidium aeneum*	brazen tanbark beetle
Erzglanzmotten	Heliozelidae	shield bearers, leaf miners
Erznektarvogel	*Anthreptes platurus*	pygmy sunbird
Erzsalamander	*Aneides aeneus*	green salamander

German	Scientific	English
Erzschleiche	*Chalcides chalcides*	three-toed skink, Algerian cylindrical skink
Erzspitznatter, Spitznatter	*Oxybelis aeneus*	Mexican vine snake
Erzspitznattern, Spitznattern	*Oxybelis* spp.	vine snakes
Erzwespen	Chalcidoidea	parasitic wasps
Eschenbastkäfer	*Leperisinus varius*	ash bark beetle
Eschenspanner	*Ennomos fuscantaria*	dusky thorn
Eschenwolllaus	*Pseudochermes fraxini*	ash scale
Esel> Wildesel	*Equus asinus*	donkey, burro
Esellaus	*Haematopinus asini asini*	donkey sucking louse
Eselohr des Midas, Midasohr	*Ellobium aurismidae*	Midas ear cassidula (snail)
Eselpinguin	*Pygoscelis papua*	Gentoo penguin
Eselshuf, Stachelauster	*Spondylus gaederopus*	European thorny oyster
Eselsohr-Abalone	*Haliotis asinina*	ass's ear abalone, donkey's ear abalone
Eskimobrachvogel	*Numenius borealis*	Eskimo curlew, prairie pigeon, doughbird
Esmarks Wolffisch	*Lycodes esmarkii*	Esmark's eelpout
Espen-Saumbandspanner, Weiden-Saumbandspanner, Espenfrischgehölz-Saumbandspanner, Birken-Braunhalsspanner	*Epione parallelaria*/ *Epione vespertaria*	dark bordered beauty (moth)
Espen-Schillerfalter, Kleiner Schillerfalter	*Apatura ilia*	lesser purple emperor
Espenblattnest-Blattlaus	*Asiphum tremulae*/ *Pachypappa tremulae*	spruce root aphid
Espenbock, Kleiner Pappelbock	*Saperda populnea*	small poplar borer, lesser poplar borer, small poplar longhorn beetle
Essbare Herzmuschel, Gemeine Herzmuschel	*Cerastoderma edule*/ *Cardium edule*	common cockle, common European cockle, edible cockle
Essbare Seegurke	*Holothuria edulis*	edible sea cucumber
Essbare Seescheide, Mikrokosmos	*Microcosmus sulcatus*/ *Microkosmus sabatieri*	grooved sea-squirt, sea-egg
Essbarer Seeigel	*Echinus esculentus*	edible sea urchin, common sea urchin
Essigälchen	*Anguillula aceti*/ *Turbatrix aceti*	vinegar eel, vinegar nematode
Essigfliege, Kleine Essigfliege, Taufliege	*Drosophila melanogaster*	vinegar fly, fruit fly
Essigfliegen, Obstfliegen, Taufliegen	Drosophilidae	"fruit flies", pomace flies, small fruit flies, ferment flies
Essigmilbe, Karpfenschwanzmilbe	*Histiogaster carpio*	vinegar mite*, carptail mite*
Essigskorpion, Geißelskorpion	*Mastigoproctus giganteus*	giant vinegaroon
Etruskerspitzmaus	*Suncus etruscus*	Etruscan shrew, Savi's pygmy shrew, pygmy white-toothed shrew

Eulen	Strigiformes/Strigidae	owls
Eulen, Eulenfalter	Noctuidae	noctuid moths
Eulen-Napfschnecke*	*Lottia gigantea*	owl limpet
Eulenkopfmeerkatze	*Cercopithecus hamlyni*	owl-faced monkey
Eulenpapagei, Kakapo	*Strigops habroptilus*	kakapo, owl-parrot
Eulenschwalme	Podargidae	frogmouths
Eulenspinner (Wollrückenspinner)	Thyatiridae/Cymatophoridae	thyatirid moths
Euphrat-Weichschildkröte	*Rafetus euphraticus/ Trionyx euphraticus*	Euphrates softshell turtle
Eurasiatische Wasserschildkröten	*Mauremys* spp.	Eurasian pond turtles
Eurasiatische Zwergmaus	*Micromys minutus*	Old World harvest mouse
Eurasische Bulldogg-Fledermaus	*Tadarida teniotis*	Eurasian free-tailed bat
Eurasischer Schneehase (Hbz. Schneehase)	*Lepus timidus*	blue hare, mountain hare
Eurasisches Eichhörnchen, Europäisches Eichhörnchen (Hbz. Feh)	*Sciurus vulgaris*	European red squirrel, Eurasian red squirrel
Eurasisches Erdhörnchen, Burunduk, Östliches Streifenhörnchen, Sibirisches Streifenhörnchen (Hbz. Burunduk)	*Tamias sibiricus*	Siberian chipmunk
Eurasisches Gleithörnchen	*Pteromys volans*	Eurasian flying squirrel
Europäische Auster, Gemeine Auster	*Ostrea edulis*	common oyster, flat oyster, European flat oyster
Europäische Breitflügelfledermaus, Spätfliegende Fledermaus	*Eptesicus serotinus*	serotine bat
Europäische Dreieckspinne	*Hyptiotes paradoxus*	European triangle spider, triangle weaver
Europäische Eidechsennatter	*Malpolon monspessulans*	Montpellier snake
Europäische Gastrochaena	*Gastrochaena dubia/ Rocellaria dubia*	flask shell, European flask shell
Europäische Gebirgsmolche	*Euproctus* spp.	European brook salamanders, European mountain salamanders
Europäische Hühnermilbe	*Ornithonyssus sylviarum*	northern fowl mite
Europäische Kartoffelzikade	*Empoasca decipiens*	green leafhopper
Europäische Katzennatter	*Telescopus fallax*	cat snake, European cat snake
Europäische Kiefernwolllaus	*Pineus pini*	Scots pine adelges, European pine woolly aphid
Europäische Languste, Stachelhummer	*Palinurus elephas*	crawfish, common crawfish (Br.), European spiny lobster, spiny lobster, langouste
Europäische Nawaga, Europäische Navaga	*Eleginus navaga*	navaga (FAO), wachna cod, Atlantic navaga, Arctic cod
Europäische Ohrwürmer	Forficulidae	European earwigs
Europäische Panopea	*Panopea glycymeris*	European panopea

Europäische Sattelauster, Sattelmuschel, Zwiebelmuschel, Zwiebelschale	*Anomia ephippium*	European saddle oyster (European jingle shell)
Europäische Schwellenschnecke	*Diodora graeca/ Diodora apertura*	common keyhole limpet
Europäische Stachel-Käferschnecke	*Acanthochitona crinita/ Acanthochitona fascicularis/ Chiton fascicularis*	bristly mail chiton (bristly mail shell)
Europäische Sumpfschildkröte	*Emys orbicularis*	European pond turtle, European pond terrapin
Europäische Tannen-Stammlaus	*Dreyfusia piceae/Adelges piceae*	silver fir adelges
Europäischer Adlerrochen, Gewöhnlicher Adlerrochen	*Myliobatis aquila*	eagle ray
Europäischer Blattfinger	*Phyllodactylus europaeus*	European leaf-toed gecko
Europäischer Fischotter	*Lutra lutra*	European river otter
Europäischer Flossenloser Aal*	*Apterichthus caecus*	European finless eel
Europäischer Flussaal	*Anguilla anguilla*	eel, European eel (FAO), river eel
Europäischer Fransenfinger, Gewöhnlicher Fransenfinger	*Acanthodactylus erythrurus*	spiny-footed lizard
Europäischer Geigenrochen, Gemeiner Geigenrochen, Hairochen	*Rhinobatos rhinobatos*	Mediterranean guitarfish, common guitarfish (FAO), pesce violino
Europäischer Glattrochen, Spiegelrochen	*Raja batis/ Dipturus batis*	common skate, skate (FAO), common European skate, grey skate, blue skate
Europäischer Halbfinger, Halbfingergecko	*Hemidactylus turcicus*	Turkish gecko, Mediterranean gecko
Europäischer Hase, Europäischer Feldhase	*Lepus europaeus*	European hare
Europäischer Hausen	*Huso huso*	great sturgeon, volga sturgeon, beluga (FAO)
Europäischer Hundsfisch, Ungarischer Hundsfisch	*Umbra krameri*	European mudminnow
Europäischer Iltis, Waldiltis	*Mustela putorius*	European polecat
Europäischer Laternenträger	*Dictyophara europaea*	European lanternfly
Europäischer Laubfrosch	*Hyla arborea*	European treefrog, common treefrog, Central European treefrog
Europäischer Quastenstern, Sonnenstern, Seesonne	*Solaster papposus*	common sun star, spiny sun star
Europäisches Chamäleon, Gemeines Chamäleon, Gewöhnliches Chamäleon	*Chamaeleo chamaeleon*	Mediterranean chameleon, African chameleon, common chameleon
Europäisches Eichhörnchen, Eurasisches Eichhörnchen (Hbz. Feh)	*Sciurus vulgaris*	European red squirrel, Eurasian red squirrel
Europäisches Ziesel, Schlichtziesel	*Citellus citellus/ Spermophilus citellus*	European ground squirrel, European suslik, European souslik
Everetts Spitzhörnchen	*Urogale everetti*	Philippine tree shrew
Exzentrischer Sanddollar	*Dendraster excentricus*	eccentric sand dollar

Fabricius-Dornhai	*Centroscyllium fabricii*	black dogfish
Facciola-Aal	*Facciolella oxyrhyncha*	Facciola's sorcerer
Fächer-Moostierchen	*Cabera boryi*	palmate bryozoan*
Fächerbauch	*Monacanthus chinensis*	fan-bellied leatherjacket
Fächerflügler, Kolbenflügler	Strepsiptera	twisted-winged insects, twisted-winged parasites, stylopids
Fächerfußgeckos	*Ptyodactylus* spp.	fan-toed geckos, fan-fingered geckos
Fächerkäfer	Rhipiphoridae	wedge-shaped beetles
Fächerkorallen u.a., Seefächer u.a.	*Gorgonia* spp. & *Flabellum* spp. et al.	sea fans, fan corals a.o.
Fächerpapagei	*Deroptyus accipitrinus*	hawk-headed parrot
Fächerröhrenwurm, Doppelköpfiger Palmfächerwurm	*Bispira voluticornis*	twin-fanworm
Fächerschwanzschnäpper	Rhipidurini	fantails
Faden-Falterfisch, Fähnchen-Falterfisch	*Chaetodon auriga*	threadfin butterflyfish (FAO), spined butterflyfish
Fadenaal*	*Serrivomer parabeani*	thread eel
Fadenauster*	*Teskeyostrea weberi*	threaded oyster
Fadenbüschelwurm, Nordischer Rankenwurm	*Cirratulus cirratus/ Cirratulus borealis*	northern cirratule, redthreads
Fadenflosser, Federflosser	Polynemidae	threadfins
Fadenflossige Alse	*Dorosoma cepedianum*	gizzard shad, American gizzard shad (FAO)
Fadenförmiger Pygospio-Wurm	*Pygospio filiformis*	filiform pygospio
Fadenkäfer	*Colydium filiforme*	filiform beetle
Fadenkanker	Nemastomatidae	nemastomatids
Fadenkiemer (Muscheln)	Filibranchia	filibranch bivalves
Fadenkiemer, Rankenwürmer	Cirratulidae	cirratulid worms, cirratulids
Fadenmakrele	*Alectis ciliaris*	African pompano (FAO) threadfin mirrorfish, pennant trevally
Fadenmakrelen	*Alectis* spp.	pompanos
Fadenmolch (Leistenmolch)	*Triturus helveticus*	palmate newt
Fadenparadieshopf	*Seleucidis melanoleuca*	twelve-wired bird-of-paradise
Fadenprachtkärpfling	*Aphyosemion filamentosum*	plumed lyretail
Fadenqualle*	*Apolemia uvaria*	string jelly
Fadenschnecken	Aeolidiacea/ Eolidiacea	aeolidacean snails, aeolidaceans
Fadenschwänze, Fadenträger, Stielaugenfische	Stylephoridae	tube-eyes, thread-tails
Fadenschwanzrochen	Anacanthobatidae	smooth skates, leg skates
Fadensegelfische	Aulopiformes (Aulopodidae)	aulopiform fishes (aulopodids)
Fadensegelfische	*Aulopus* spp.	flagfins, aulopus
Fadenstern, Fadenförmiger Schlangenstern, Fadenarm-Schlangenstern	*Amphiura filiformis*	filiform burrowing brittlestar

Fadensterne, Fadenförmige Schlangensterne	*Amphiura* spp.	filiform burrowing brittlestars
Fadenwürmer	*Strongyloides* spp.	threadworms
Fadenwürmer, Nematoden, Rundwürmer (*sensu strictu*)	Nematoda	roundworms, nematodes
Fahlbauch-Paradiesschnäpper	*Terpsiphone paradisi*	Asiatic paradise flycatcher
Fahlbürzel-Steinschmätzer	*Oenanthe moesta*	red-rumped wheatear
Fahler Wermut-Mönch	*Cucullia absinthii*	wormwood moth
Fahlgraue Mondschnecke	*Natica livida*	livid moonsnail
Fahlkauz	*Strix butleri*	Hume's tawny owl
Fahlsegler	*Apus pallidus*	pallid swift
Fahlsperling	*Passer brachydactyla*	pale rock sparrow
Fähnchen-Falterfisch, Faden-Falterfisch	*Chaetodon auriga*	threadfin butterflyfish (FAO), spined butterflyfish
Fähnchenwelse	Helogenidae/Helogeneidae	helogene catfishes
Fahnenbarsche, Rötlinge	Anthiidae	fairy basslets
Fahnenhechtling, Kap Lopez, Bunter Prachtkärpfling	*Aphyosemion australe*	lyretail, lyretail panchax (FAO), Cape Lopez lyretail
Fahnenkegel*	*Conus vexillum*	flag cone
Fahnenquallen, Fahnenmundquallen	Semaeostomea	semeostome medusas, semaeostome medusae
Fahnenschwänze, Kuhlien	Kuhliidae/Duleidae	flagtails (aholeholes)
Falken	Falconidae	falcons, caracaras
Falkenlibellen	Corduliidae	green-eyed skimmers
Falkennachtschwalben	Chordeilinae	nighthawks
Falkennachtschwalbe	*Chordeiles minor*	common nighthawk
Falkenraubmöve	*Stercorarius longicaudus*	long-tailed skua
Falklandfuchs, Falklandwolf	*Dusicyon australis*	Falkland Island wolf
Fallschirm-Milbe*	*Cytodites nudus*	air-sac mite
Falltürklappenspinnen	Nemesiidae	tube trapdoor spiders, wishbone spiders
Falltürspinne	*Cteniza sauvagei*	trapdoor spider
Falltürspinnen, Miniervogelspinnen	Ctenizidae	trapdoor spiders
Falsche Baumratte	*Xenuromys barbatus*	white-tailed New Guinean rat
Falsche Clownkäfer	Sphaeritidae	false clown beetles
Falsche Dornwelse	Auchenipteridae	driftwood catfishes
Falsche Edelkoralle, Trugkoralle	*Parerythropodium coralloides*	encrusting leather coral
Falsche Elefantenschnauzen-Walze*	*Cymbium cymbium*	false elephant's snout volute
Falsche Floridakoralle*	*Ricordea florida*	Florida false coral
Falsche Gürtelschweife	*Pseudocordylus* spp.	crag lizards
Falsche Katzenhaie, Falsche Marderhaie, Unechte Marderhaie	Pseudotriakidae	false catsharks
Falsche Kauri	*Erronea errones*	mistaken cowrie
Falsche Korallenschlangen	*Erythrolamprus* spp.	false coral snakes

Falsche Kröte	*Pseudobufo subasper*	false toad
Falsche Landkartenschildkröte, Falsche Landkarten-Höckerschildkröte	*Graptemys pseudogeographica*	false map turtle
Falsche Lanzenotter	*Xenodon rhabdocephalus*	toad-eater snake, false fer-de-lance
Falsche Melonenwalze	*Livonia mammilla*	false melon volute, mammal volute
Falsche Muräne	*Chlopsis bicolor*	false moray, bicoloured false moray (FAO)
Falsche Muränenaale	Xenocongridae/Chlopsidae	false morays
Falsche Mützenschnecke*	*Cheilea equestris*	false cup-and-saucer limpet
Falsche Neuguinea-Wasserratte	*Pseudohydromys* spp.	New Guinean false water rats, New Guinean shrew-mice
Falsche Noppenkauri*, Pustel-„Kauri"	*Jenneria pustulata* (Ovulidae)	pustuled cowrie, pustulose false cowrie
Falsche Quahog-Muschel	*Pitar morrhuanus*	false quahog
Falsche Reisratten u.a., Ratos-do-Mato	*Pseudoryzomys* spp.	ratos-do-mato
Falsche Siedleragame	*Agama paragama*	false agama
Falsche Spinnmilben, Unechte Spinnmilben	*Brevipalpus* spp.	false spider mites, false red spider mites
Falsche Spitzkopfschildkröte	*Pseudemydura umbrina*	western swamp turtle
Falsche Südliche Königskrabbe	*Paralomis granulosa*	false southern king crab
Falsche Wasserratte	*Xeromys myoides*	false water rat
Falsche Witwe	*Steatoda paykullianus*	false widow spider
Falscher Drachenkopf, Teufels-Drachenkopf	*Scorpaenopsis diabolus*	false stonefish (FAO), devil stonefish
Falscher Gitterfalterfisch	*Chaetodon oxycephalus*	false lined butterflyfish, spot-nape butterflyfish (FAO)
Falscher Hering, Kleinhering	*Harengula clupeola*	false pilchard
Falscher Kap-Gürtelschweif	*Pseudocordylus microlepidotus*	leathery crag lizard
Falscher Katzenhai, Falscher Marderhai, Atlantischer Falscher Marderhai	*Pseudotriakis microdon*	false catshark (FAO), false cat shark, Atlantic false catshark
Falscher Kohlerdfloh	*Altica oleracea*	false turnip flea beetle
Falscher Putzerfisch, Putzernachahmer	*Aspidontus taeniatus*	mimic blenny, false cleanerfish (FAO)
Falscher Sandtiger, Krokodilhai	*Pseudocarcharias kamoharrai*	crocodile shark
Falscher Vampir	*Lyroderma lyra/ Megaderma lyra*	false vampire bat
Faltbauchfische, Silberbeile	*Argyropelecus* spp.	hatchetfishes
Falten-Ascidie, Walzen-Ascidie	*Styela plicata*	barrel-squirt*
Faltenerbsenmuschel, Kleine Falten-Erbsenmuschel	*Pisidium henslowanum*	Henslow pea mussel, Henslow peaclam
Faltengeckos	*Ptychozoon* spp.	flying geckos
Faltenkoralle, Höckerige Kaktuskoralle	*Mycetophyllia lamarckiana*	ridged cactus coral
Faltenrandige Schließmundschnecke	*Laciniaria plicata*	single-lipped door snail*

Faltenschnecken, Rollschnecken, Walzenschnecken	Volutidae	volutes
Falter-Stechrochen, Schmetterlingsrochen	Gymnuridae	butterfly rays
Falterfische, Borstenzähner	Chaetodontidae	butterflyfishes
Falterfische, Schmetterlingsfische	*Chaetodon* spp.	butterflyfishes
Faltkiemen-Seescheiden	Stolidobranchia	stolidobranchs, stolidobranch sea-squirts
Faltlippen-Fledermäuse	*Tadarida* spp.	free-tailed bats
Faltttürspinnen	Antrodiaetidae	folding trapdoor spiders
Fanaloka, Gefleckte Madagassische Schleichkatze (*siehe auch*: *Eupleres goudotii*)	*Fossa fossa*	spotted Malagasy civet, spotted fanaloka, spotted fanalouc (see also: *Eupleres goudotii*)
Fanaloka> Graubraune Madagassische Schleichkatze, Ungefleckte Fanaloka, 'Ameisenschleichkatze' (Fanaloka, Ridaridy; siehe auch: *Fossa fossa*)	*Eupleres goudotii*	falanouc (IUCN), fanalouc (*better*: grey-brown Malagasy civet; *native common names*: fanaloka, ridaridy)
Fanaloka> Großfanaloka	*Eupleres goudotii major*	western fanalouc
Fanaloka> Kleinfanaloka	*Eupleres goudotii goudotii*	eastern fanalouc
Fanghafte	Mantispidae	mantis flies, mantidflies
Fangschrecken & Gottesanbeterinnen	Mantodea/Mantoptera	mantids
Fangschreckenkrebs, Gemeiner Heuschreckenkrebs, Großer Heuschreckenkrebs	*Squilla mantis*	giant mantis shrimp, spearing mantis shrimp
Fangschreckenkrebse, Maulfüßer	Hoplocarida/Stomatopoda	mantis shrimps
Fangwanzen, Gespensterwanzen	Phymatidae	ambush bugs
Fangzähner	*Caulolepis longidens/ Anoplogaster cornuta*	common fangtooth
Fantastische Randschnecke	*Cryptospira elegans*	elegant marginella, elegant marginsnail (elegant margin shell)
Fantastische Stachelschnecke, Superbus-Stachelschnecke	*Chicomurex superbus*	superb murex
Färber-Kammmuschel	*Chlamys tincta*	tinted scallop
Färberfrosch	*Dendrobates tinctorius*	dyeing poison-arrow frog, dyeing poison frog
Färberscharteneule	*Acosmetia caliginosa*	reddish buff (moth)
Färberschildläuse	*Kermes* spp.	kermes coccids, kermes scales
Farblose Glanzschnecke	*Oxychilus clarus*	clear glass snail
Farblose Randschnecke*	*Hyalina pallida*	pallid marginella
Farbwechselnde Garnele, Chamäleongarnele	*Hippolyte varians*	chameleon prawn
Farbwechselnde Hornkoralle, Farbwechselnde Gorgonie	*Paramuricea clavata*	variable gorgonian*
Farbwechselnde Schwimmgarnele	*Palaemonetes varians*	variegated shore shrimp

Farnblattlaus	*Idiopterus nephrelepidis*	fern aphid
Farnmoos*	*Abietinaria filicula*	fern hydroid
Farnschnecke, Wasserschnegel	*Deroceras laeve/ Agriolimax laevis*	meadow slug (U.S.), marsh slug, brown slug, smooth slug
Fasan	*Phasianus colchicus*	pheasant, common pheasant, ring-necked pheasant
Fasane	Phasianinae	pheasants and partridges
Fasanenartige	Phasianidae	pheasants & grouse & turkeys & partridges
Fasanenfedermoos*	*Lytocarpia myriophyllum*	pheasant-tail hydroid
Fasanenschnecken	*Phasianella* spp.	pheasant snails (pheasant shells)
Fasanenschnecken	Phasianellidae	pheasant snails (pheasant shells)
Fasanküken, Kleine Doppelfußschnecke	*Tricolia pullas*	lesser pheasant snail*
Fässchenschnecken	Orculidae	
Fassschnecken	Tonnidae	tun snails (tun shells, cask shells)
Fatio-Kleinwühlmaus	*Pitymys multiplex*	alpine pine vole
Faulbaum-Bläuling	*Celastrina argiolus*	holly blue
Faulfliegen, Polierfliegen	Lauxaniidae	lauxaniids
Faulholzkäfer, Schimmelkäfer	Corylophidae/Orthoperidae	minute fungus beetles
Faulholzmotten & Palpenmotten & Verwandte	Oecophoridae	oecophorid moths
Faultiere	Pilosa (Xenarthra)	sloths
Faultiere> Dreifingerfaultiere	*Bradypus* spp.	three-toed sloths
Faultiere> Zweifinger-Faultiere	*Choloepus* spp.	two-toed sloths
Faulvögel	Bucconidae	puffbirds
Feas Baumratte	*Chiromyscus chiropus*	Fea's tree rat
Feaviper, Fea-Viper	*Azemiops feae*	Fea's viper
Fechterschnecken, Flügelschnecken	Strombidae	conchs, true conchs (conch shells)
Feder-Moostierchen	*Bugula plumosa*	plumose coralline
Federabwurfsmilbe*	*Knemidokoptes gallinae*	depluming mite
Federflosser, Fadenflosser	Polynemidae	threadfins
Federflügler & Haarflügler (Zwergkäfer)	Ptiliidae	feather-winged beetles
Federförmige Kegelschnecke	*Conus pennaceus*	feathered cone
Federfußeule	*Herminia tarsipennalis*	fan-foot
Federgeistchen, Federmotten	Pterophoridae	plume moths
Federgeistchen, Fünffedriges Geistchen, Weißes Geistchen, Schneeweiße Windenmotte	*Pterophorus pentadactylus*	large white plume moth
Federkiemenschnecken	Valvatidae	valve snails
Federkiemenschnecke> Gemeine Federkiemenschnecke, Teich-Federkiemenschnecke	*Valvata piscinalis piscinalis*	valve snail, European stream valve snail, common valve snail, European stream valvata

Federlibelle	*Platycnemis pennipes*	white-legged damselfly
Federlibellen	Platycnemidae	platycnemid damselflies
Federlinge & Haarlinge	Mallophaga	biting lice, chewing lice
Federmilben	Analgidae	feather mites
Federmoos*	*Pennaria tiarella*	feathered hydroid
Federmotte, Windling-Geistchen	*Pterophorus monodactylus*	common plume, sweet-potato plume moth
Federmotten, Federgeistchen, Geistchen	Alucitidae/ Pterophoridae	plumed moths, many-plume moths, alucitids
Federpolyp, Nesselpolyp, Nesselfarn	*Aglaophenia tubulifera*	plume hydroid, plumed hydroid
Federschwanz	*Ptilocercus lowii*	pen-tailed tree shrew, feather-tailed tree shrew
Federschwanz-Stechrochen	*Pastinachus sephen/ Dasyatis sephen/ Hypolophus sephen*	cowtail stingray
Federschwanzbeutler	*Distoechurus pennatus*	feather-tailed possum
Federspulenmilben	Syringophilidae	quill mites
Federsterne, Haarsterne, Comatuliden	Comatulida	feather stars, comatulids
Federwürmer u.a., Fächerwürmer u.a.	Sabellidae	sabellids, sabellid fanworms, sabellid feather-duster worms a.o.
Feen-Kaiserbuntbarsch	*Aulonocara jacobfreibergi*	fairy cichlid (FAO)
Feenbarsche	Grammatidae/Grammidae	basslets
Feh (Hbz.)> Europäisches Eichhörnchen, Eurasisches Eichhörnchen	*Sciurus vulgaris*	European red squirrel, Eurasian red squirrel
Feh (Hbz.)> Rothörnchen	*Tamiasciurus hudsonicus*	eastern red squirrel, red squirrel
Feigen-Kegelschnecke	*Conus figulinus*	fig cone
Feigenbohrschwamm (Fleischschwamm)	*Oscarella lobularis*	lobate fig sponge*
Feigengallwespe	*Blastophaga psenes*	caprifig wasp
Feigenhornschwamm	*Petrosia ficiformis*	bread sponge, fig sponge
Feigenmilbe	*Aceria ficus/Aceria fici*	fig mite
Feigenmotte (*auch*: Rosinenmotte)	*Cadra figulilella*	raisin moth (*also*: fig moth)
Feigenpirol	*Sphecotheres viridis*	figbird
Feigenschnecke> Gemeine Feigenschnecke, Atlantik-Feigenschnecke	*Ficus communis*	Atlantic figsnail (common fig shell)
Feigenschnecken	Ficidae	fig snails (fig shells)
Feigenschwamm, Korkschwamm	*Suberites ficus/ Ficulina ficus*	fig sponge
Feigenwespen	Agaonidae	fig wasps
Feilen-Seemandel	*Philine lima*	file paperbubble
Feilenfische	Monacanthidae	filefishes
Feilenmuschel> Gewöhnliche Feilenmuschel	*Lima lima*	spiny fileclam, spiny lima (frilled file shell)
Feilenmuscheln	Limidae	file clams (file shells)

Feilennattern	*Mehelya* spp.	African file snakes
Feilenschnecke*	*Nucella lima*	file dogwinkle
Feine Kammmuschel, Edle Kammmuschel	*Chlamys senatoria/ Chlamys nobilis*	senate scallop, noble scallop
Feine Röhrchenkoralle*	*Cladocora debilis*	thin tube coral
Feiner Levantinerschwamm	*Spongia officinalis mollissima*	fine Levant sponge, Turkey cup sponge
Feinfliegen	Asteiidae	asteiids
Feingekerbte Nussmuschel	*Nucula crenulata*	crenulate nutclam
Feingerippte Haferkornschnecke	*Chondrina clienta*	
Feinrippige Wendeltreppe, Turtons Wendeltreppe	*Epitonium turtonae*	finely ribbed wentletrap
Feinverzweigte Augenkoralle	*Oculina arbuscula*	compact ivory bush coral
Feinzähniger Hai	*Carcharhinus isodon/ Aprionodon isodon*	finetooth shark
Feld-Grashüpfer	*Chorthippus apricarius*	upland field grasshopper
Feld-Sandlaufkäfer	*Cicindela campestris*	green tiger beetle
Feld-Viscacha, Pampas-Viscacha (Hbz. Viscacha)	*Lagostomus maximus*	plains viscacha
Felderkoralle, Knotige Koralle	*Madracis decactis*	ten-ray star coral
Feldgrille	*Gryllus campestris*	field cricket
Feldhamster, Hamster (Hbz. Hamster)	*Cricetus cricetus*	common hamster, black-bellied hamster
Feldhase> Europäischer Feldhase	*Lepus europaeus*	European hare
Feldheuschrecke, Brauner Grashüpfer	*Chorthippus brunneus*	field grasshopper, common field grasshopper
Feldheuschrecken (Grashüpfer/Heuhüpfer)	Acrididae/Acridoidea	grasshoppers (short-horned grasshoppers)
Feldhüpfmäuse	*Zapus* spp.	jumping mice
Feldlerche	*Alauda arvensis*	skylark
Feldmaus	*Microtus arvalis*	common vole
Feldmaus> Brandmaus	*Apodemus agrarius*	Old World field mouse, striped field mouse
Feldmäuse, Wühlmäuse	*Pitymys* spp./*Microtus* spp.	voles, meadow mice
Feldmauslaus	*Hoplopleura acanthopus*	vole louse
Feldrohrsänger	*Acrocephalus agricola*	paddyfield warbler
Feldschwirl	*Locustella naevia*	grasshopper warbler
Feldspecht	*Colaptes campestris*	campo flicker
Feldsperling	*Passer montanus*	tree sparrow
Feldspinnen	Liocranidae	liocranid spiders
Feldspitzmaus	*Crocidura leucodon*	bicoloured white-toothed shrew
Feldwespe	*Polistes gallicus*	paper wasp
Feldwespen	Polistinae	polistine wasps
Fellkäfer, Jagdraubkäfer	Korynetidae/Corynetidae (>Cleridae)	ham beetles
Fellmotte	*Monopis rusticella*	skin moth, fur moth, wool moth
Fels-Gleithörnchen	*Eupetaurus cinereus*	woolly flying squirrel
Fels-Wirtelschwanzleguan, Andros-Wirtelschwanzleguan	*Cyclura cyclura*	Andros ground iguana, rock iguana

Felseidechse	*Lacerta saxicola*	rock lizard
Felsen-Dardanus	*Dardanus deformis*	rock hermit crab
Felsen-Entenmuschel	*Mitella pollicipes/ Pollicipes cornucopia*	rocky shore goose barnacle
Felsen-Kantengarnelen	Sicyoniidae	rock shrimps
Felsenagame, Bergagame	*Agama atra*	southern rock agama, South African rock agama
Felsenbarsch, Klippenbarsch	*Ctenolabrus rupestris*	goldsinny, goldsinny wrasse
Felsenbarsch, Nordamerikanischer Streifenbarsch	*Morone saxatilis/ Roccus saxatilis*	striped bass, striper
Felsenbohrer	Hiatellidae	rock borers
Felsenbohrer> Gemeiner Felsenbohrer	*Hiatella rugosa/ Hiatella striata*	common rock borer
Felsenbussard	*Buteo rufofuscus*	jackal buzzard
Felsengarnelen	Palaemonidae	palaemonid shrimps
Felsengrundel, Felsgrundel, Paganellgrundel	*Gobius paganellus*	rock goby
Felsenhenne	*Ptilopachus petrosus*	stone partridge
Felsenhuhn	*Alectoris barbara*	barbary partridge
Felsenklapperschlange	*Crotalus lepidus*	rock rattlesnake
Felsenkleiber	*Sitta neumayer*	rock nuthatch
Felsenkrabbe	*Grapsus grapsus*	Sally Lightfoot crab, mottled shore crab
Felsenkrabben	Grapsidae	shore crabs, marsh crabs & talon crabs
Felsenmaus	*Apodemus mystacinus*	rock mouse
Felsenmäuse, Zwerg-Felsenmäuse	*Petromyscus* spp.	rock mice, pygmy rock mice
Felsenmeerschweinchen, Moko	*Kerodon rupestris*	rock cavy, moco
Felsenpinguin	*Eudyptes chrysocome*	rockhopper penguin
Felsenpython	*Python sebae*	African python, water python, African rock python
Felsenratte	*Petromus typicus*	dassie rat
Felsenschwalbe	*Ptyonoprogne rupestris*	crag martin
Felsensittich	*Cyanoliseus patagonus*	Patagonian conure
Felsenspringer	Archaeognatha/ Machilidae	jumping bristletails
Felsensteinschmätzer	*Oenanthe finschii*	Finsch's wheatear
Felsentaube	*Columba livia*	feral rock dove
Felsenzaunkönig	*Salpinctes obsoletus*	rock wren
Felshüpfer	Picathartidae	bald crows
Felskänguru	*Petrogale* spp.	rock wallabies
Felsküsten-Einsiedler	*Clibanarius erythropus*	rocky-shore hermit crab*
Felsnadelschnecke, Felsnadel	*Gourmya rupestris*	rocky-shore cerith*
Felsrose	*Actinothoe sphyrodeta*	cliff anemone*
Felsspalten-Stachelleguan	*Sceloporus poinsetti*	crevice spiny lizard
Fennek, Wüstenfuchs	*Fennecus zerda/Vulpes zerda*	fennec fox

Fensterfliege, Gemeine Fenstermücke, Weißer Drahtwurm	*Sylvicola fenestralis*	common window gnat, common window midge
Fensterfliegen	Scenopinidae	window flies
Fensterfliegen	*Scenopinus* spp.	window flies
Fenstermuschel, Fensterscheibenmuschel, Glockenmuschel	*Placuna placenta*	windowpane oyster (window shell, jingle shell)
Fensterschwärmer	Thyrididae	window-winged moths
Ferkelfrösche	*Hemisus* spp.	shovelnose frogs
Ferkelhörnchen	*Hyosciurus heinrichi*	Celebes long-nosed squirrel
Fernöstliche Nawaga, Fernöstliche Navaga	*Eleginus gracilis*	saffron cod (FAO), Far Eastern navaga
Feste Augenkoralle, Elfenbein-Buschkoralle	*Oculina robusta*	robust ivory tree coral
Festland-Kampfüchse	*Pseudalopex* spp.	South American foxes
Festliche Leierschnecke*	*Festilyria festiva*	festive volute, festive lyria
Fettgroppen	Psychrolutidae	fatheads, flathead sculpins, tadpole sculpins
Fetthennen-Felsflur-Kleinspanner	*Idaea contiguaria*	Weaver's wave
Fettmäuse	*Steatomys* spp.	fat mice
Fettschwalme	Steatornithidae	oilbirds
Fettschwanz-Rennmaus, Dickschwanzmaus	*Pachyuromys duprasi*	fat-tailed gerbil
Fettschwanz-Springmäuse	*Pygeretmus* spp.	fat-tailed jerboas
Fettschwanzgecko	*Hemitheconyx caudicinctus*	fat-tailed gecko
Fettschwanzgeckos	*Oedura* spp.	velvet geckos
Fettspinne	*Steatoda bipunctata*	rabbit hutch spider, two-spot spider*
Fettzünsler	*Aglossa pinguinalis*	large stable tabby, tabby moth, grease moth
Fetzenfeilenfisch, Schmuck-Feilenfisch	*Solegnathus spinosissimus*	spiny pipehorse (FAO), spiny seadragon
Fetzenfische	*Phyllopteryx* spp.	weedy seadragons
Feuchtkäfer, Schlammschwimmer	*Hygrobia* spp.	screech beetle
Feuchtmilbe*	*Histiostoma feroniarum*	damp mite
Feuchtwiesen-Blutströpfchen, Sumpfhornklee-Widderchen	*Zygaena trifolii*	five-spot burnet
Feuchtwiesen-Staubeule, Wiesen-Staubeule	*Athetis pallustris*	marsh moth
Feuer-Goldwespe, Feuergoldwespe, Gemeine Goldwespe	*Chrysis ignita*	common gold wasp, ruby-tail, ruby-tailed wasp
Feuerfische, Korallenfische	Pteroidae/Pteroinae	lionfishes & turkeyfishes
Feuerflügelsittich	*Brotogeris pyrrhoptera*	grey-cheeked parakeet, gray-cheeked parakeet
Feuerkäfer (Feuerfliegen, Kardinäle)	Pyrochroidae	fire beetles, fire-colored beetles, cardinal beetles
Feuerkopf-Saftlecker	*Sphyrapicus ruber*	red-breasted sapsucker

Feuerkorallen	*Millepora* spp.	fire corals, stinging corals
Feuerkorallen	Milleporina	fire corals, milleporine hydrocorals, stinging corals
Feuerkröten, Unken	*Bombina* spp.	fire-bellied toads, firebelly toads
Feuermaulbuntbarsch	*Cichlasoma meeki*	firemouth cichlid
Feuersalamander	*Salamandra salamandra*	fire salamander, European fire salamander
Feuerschwamm	*Tedania ignis*	fire sponge
Feuerschwanz	*Epalzeorhynchus bicolor/ Labeo bicolor*	redtail sharkminnow, red-tailed "shark", red-tailed black "shark"
Feuertangare, Sommertangare	*Piranga rubra*	summer tanager
Feuerwalzen	Pyrosomida	pyrosomes
Feuerwanzen	Pyrrhocoridae	red bugs & stainers, pyrrhocorid bugs, pyrrhocores
Feuerwürmer	Amphinomidae	firewoms, amphinomids
Feuerwürmer u.a.	*Amphinome* spp.	firewoms a.o.
Feuriger Perlmuttfalter, Märzveilchenfalter, Bergmatten-Perlmutterfalter, Hundsveilchen-Perlmutterfalter	*Argynnis adippe/ Fabriciana adippe*	high brown fritillary
Fichtenammer	*Emberiza leucocephalos*	pine bunting
Fichtenhuhn	*Dendragapus canadensis*	spruce grouse
Fichtenkreuzschnabel	*Loxia curvirostra*	common crossbill
Fichtenmarder (Hbz. Nordamerikanischer Zobel, Kanadischer Zobel)	*Martes americana*	American pine marten
Fichtennestwickler, Hohlnadelwickler	*Epiblema tedella/ Epinotia tedella*	streaked pine bell, cone moth
Fichtenprozessionsspinner, Kiefernprozessionsspinner	*Thaumetopoea pinivora*	pine processionary moth
Fichtenquirlschildlaus	*Physokermes piceae*	spruce bud scale
Fichtenrindenwickler	*Cydia pactolana*	spruce bark moth
Fichtenröhrenlaus, Sitkalaus	*Elatobium abietinum*	green spruce aphid
Fichtensamen-Gallmücke	*Plemeliella abietina*	spruce-seed gall midge
Fichtensamenwespe	*Megastigmus strobilobius*	spruce seed wasp
Fichtensplintbock> Gemeiner Fichtensplintbock	*Tetropium castaneum*	chestnut longhorn beetle
Fichtentrieblaus	*Mindarus obliquus*	spruce twig aphid
Fichtenwanze	*Gastrodes abietum*	spruce bug
Fichtenzapfenschuppen-Gallmücke	*Kaltenbachiola strobi*	spruce cone gall midge
Fichtenzapfenwickler	*Cydia strobilella*	spruce cone moth
Fichtenzapfenzünsler, Fichtentriebzünsler	*Dioryctria abietella*	pine knothorn moth
Fidschi-Boa, Große Pazifik-Boa	*Candoia bibroni*	Pacific boa
Fidschi-Kobra, Fijiotter	*Ogmodon vitianus*	Fiji snake, Fiji cobra

Fiebermücken, Malariamücken, Gabelmücken	*Anopheles* spp.	malaria mosquitoes
Fieder-Hydrozoe	*Lytocarpus philippinus*	white stinging hydroid
Fiederbartwelse	Mochocidae/ Mochokidae	squeakers, upside-down catfishes
Fiederkiemer, Kammkiemer	Protobranchiata	protobranch bivalves
Fiederzweigpolyp	*Halecium halecinum*	herringbone hydroid
Fiedrige Stachelschnecke	*Pterynotus pinnatus*	pinnate murex
Fiji Kammleguan	*Brachylophus vitiensis*	Fiji crested iguana
Filzlaus, Schamlaus	*Phthirus pubis/ Pthirus pubis*	pubic louse, crab louse, "crab"
Filzwurm, Gemeine Seemaus, Wollige Seemaus, Seeraupe	*Aphrodita aculeata*	European sea mouse
Filzwürmer, Seemäuse	*Aphrodita* spp.	sea mice
Finger-Napfschnecke	*Collisella digitalis*	fingered limpet
Fingerfisch, Kapitänsfisch	*Polynemus quadrifilis*	threadfin, five-rayed threadfin, giant African threadfin (FAO)
Fingerförmige Porenkoralle, Fingerkoralle	*Porites porites*	clubtip finger coral, clubbed finger coral
Fingerhutqualle*	*Linuche unguiculata*	thimble jellyfish
Fingerhutqualle> Rosa Fingerhutqualle	*Aglantha digitalis*	pink helmet
Fingerige Stachelauster*, Gefingerte Stachelauster*	*Spondylus ictericus*	digitate thorny oyster
Fingerkoralle, Fingerförmige Porenkoralle	*Porites porites*	clubtip finger coral, clubbed finger coral
Fingernagel- Pantoffelschnecke, Fingernagel	*Crepidula unguiformis*	Mediterranean slippersnail
Fingerottern, Kapottern (Hbz. Afrikanische Otter)	*Aonyx* spp.	clawless otters
Fingerratten	*Dactylomys* spp.	coro-coros
Fingertier	*Daubentonia madagascariensis*	aye-aye
Finken, Finkenvögel	Fringillidae	finches & allies
Finkenbülbül	*Spizixos canifrons*	crested finchbill
Finnwal	*Balaenoptera physalus*	fin whale, common rorqual
Finsterspinnen	Amaurobiidae	white-eyed spiders
Finte	*Alosa fallax/Alosa finta*	twaite shad (FAO), twaite, finta shad
Fischadler	*Pandion haliaetus*	osprey
Fischadler	Pandionidae	osprey (fish hawk)
Fischassel (Peterassel)	*Aega psora*	fish louse
Fischbandwurm, Breiter Bandwurm	*Diphyllobothrium latum*	broad fish tapeworm, broad tapeworm
Fischbandwürmer	*Diphyllobothrium* spp.	fish tapeworms
Fischbussard	*Busarellus nigricollis*	black-collared hawk
Fischchen	Lepismatidae	silverfish
Fische	Pisces	fishes
Fischegel	Ichthyobdellidae/Piscicolidae	fish leeches

Fischende Fledermaus	*Myotis vivesi*	fishing bat
Fischermarder (Hbz. Fischer, Pekan, Virginischer Iltis)	*Martes pennanti*	fisher, pekan
Fischernatter, Indische Fischernatter	*Xenochrophis piscator/ Natrix piscator*	chequered keelback, common scaled water snake, fishing snake
Fischernetz-Gitterschnecke	*Cancellaria piscatoria*	fishnet snail, fisherman's nutmeg, fishnet nutmeg
Fischers Unzertrennlicher, Erdbeerköpfchen	*Agapornis fischeri*	Fischer's lovebird
Fischertukan	*Ramphastos sulfuratus*	keel-billed toucan
Fischfressende Seedahlie	*Urticina piscivora*	piscivorous sea dahlia, fish-eating urticina
Fischkatze	*Felis viverrina/ Prionailurus viverrinus*	fishing cat
Fischkrähe	*Corvus ossifragus*	fish crow
Fischkratzer	*Acanthocephalus lucii*	thorny-headed worm of fish
Fischläuse	*Argulus* spp.	fish lice
Fischläuse, Kiemenschwänze, Karpfenläuse	Branchiura/ Argulida (Argulidae)	fish lice
Fischmöve	*Larus ichthyaetus*	great black-headed gull
Fischotter	*Lutra* spp.	river otters
Fischotter> Europäischer Fischotter	*Lutra lutra*	European river otter
Fischratten	*Ichthyomys* spp.	fish-eating rats, aquatic rats
Fischuhu	*Bubo zeylonensis*	brown fish owl
Fischvögel	*Ichthyornithiformes*	fish birds
Fischwühlen	*Ichthyophiidae*	fish caecilians
Fiskalwürger	*Lanius collaris*	fiscal shrike
Fitis	*Phylloscopus trochilus*	willow warbler
Fitzroy-Schildkröte, Weißgesichtsschildkröte	*Rheodytes leukops*	Fitzroy's turtle, Fitzroy turtle
Flachbrauen-Maulwurfkrebs*	*Upogebia affinis*	coastal mud shrimp, coastal mudprawn, flat-browed mud shrimp
Flachbrauen-Schwimmkrabbe	*Portunus depressifrons*	flatface swimming crab, flat-browed crab
Flachbrustvögel, Ratiten	Ratitae	ratite birds (flightless birds)
Flache Abalone, Flaches Seeohr	*Haliotis walallensis*	flat abalone
Flache Baumauster*	*Isognomon alatus*	flat tree-oyster
Flache Erdschildkröte	*Heosemys depressa*	Arakan forest turtle, Arakan terrapin
Flache Federkiemenschnecke	*Valvata cristata*	flat valve snail
Flache Felsenschnecke	*Chilostoma planospirum/ Campylaea planospirum*	
Flache Glanzschnecke	*Oxychilus depressus*	depressed glass snail
Flache Grübchenschnecke	*Lacuna pallidula*	pallid lacuna, pallid chink snail
Flache Kreiselschnecke	*Gibbula umbilicalis*	flat topsnail, purple topsnail, umbilical trochid

Flache Moschusschildkröte	*Sternotherus depressus/ Kinosternon depressum*	flattened musk turtle
Flache Mützenschnecke	*Ferrissia wautieri*	flat ancylid
Flache Pantoffelschnecke	*Crepidula plana*	eastern white slippersnail
Flache Sandmuschel, Große Flache Sandmuschel	*Gari depressa/ Psammobia depressa*	large sunsetclam, flat sunsetclam
Flache Strandschnecke	*Littorina mariae*	flat periwinkle
Flache Teichmuschel, „Entenmuschel"	*Anodonta anatina*	duck mussel
Flache Tellmuschel	*Macoma moesta*	flat macoma
Flacher Schuppenwurm	*Lepidonotus squamatus*	twelve-scaled worm
Flaches Krusten-Moostierchen	*Cryptosula pallasiana*	red crust
Flaches Posthörnchen	*Gyraulus riparius*	river ramshorn
Flachkäfer	Trogossitidae	bark-gnawing beetles, trogossitid beetles, trogosssitids
Flachkinn-Fledermäuse, Blattkinn-Fledermäuse	*Mormoops* spp.	leaf-chinned bats
Flachkopf- Apfelbaum-Bohrwurm (Prachtkäfer)	*Chrysobothris femorata*	flatheaded appletree borer
Flachkopf-Blattchamäleon	*Rhampholeon platyceps*	Malawi stumptail chameleon
Flachkopf-Feilenfisch	*Monacanthus hispidus*	planehead filefish
Flachkopf-Katzenwels	*Pylodictis olivaris*	flathead catfish
Flachkopfbeutelmäuse	*Planigale* spp.	planigales, flat-skulled marsupial "mice"
Flachköpfe, Plattköpfe	Platycephalidae	flatheads, flathead gurnards
Flachkopfkatze	*Ictailurus planiceps/ Felis planiceps*	flat-headed cat
Flachkrabbe*	*Plagusia depressa*	flattened crab
Flachkrebse, Flohkrebse	Amphipoda	beach hoppers, sand hoppers, scuds and relatives
Flachland-Schaufelfuß	*Scaphiopus bombifrons*	Plains spadefoot
Flachland-Taschenratten	*Geomys* spp.	eastern pocket gophers
Flachlandgründling, Weißflossiger Gründling	*Gobio albipinnatus*	Russian whitefin gudgeon, white-finned gudgeon (FAO)
Flachlandtapir	*Tapirus terrestris*	Brazilian tapir
Flachrand-Neptunshorn	*Neptunea tabulata*	tabled whelk, tabled neptune
Flachschildkröten	*Homopus* spp.	Cape tortoises
Flachschwanz-Krötenechse	*Phrynosoma m'calli*	flat-tailed horned lizard
Flachspindel- Purpurschnecke*	*Purpura planospira*	eye of Judas purpura
Flachwasser-Maulwurfskrebs	*Upogebia litoralis*	littoral mudprawn
Flachwasser-Seehecht, Stockfisch, Kap-Hecht	*Merluccius capensis*	Cape hake, stock fish, stockfish, stokvis (S. Africa), shallow-water Cape hake (FAO)
Flachwasserkrake*	*Octopus bimaculoides*	mud-flat octopus
Fladen-Kalkschwamm	*Leucandra nivea*	pancake sponge*
Fladenstern	*Ceramaster placenta*	placental sea star*

Flaggen-Buntbarsch	*Mesonauta festivus*	flag cichlid
Flaggendrongo	*Dicrurus paradiseus*	greater racquet-tailed drongo
Flaggenmuschel, Flaggen-Steckmuschel	*Atrina vexillum*	flag penshell
Flaggensalmler, Ulreys Salmler	*Hemigrammus ulreyi*	Ulrey's tetra
Flamingo-Kegelschnecke	*Conus flamingo*	flamingo cone
Flamingos	Phoenicoperidae	flamingos (*also:* flamingoes)
Flamingos	Phoenicopteriformes	flamingos and allies
Flamingozunge	*Cyphoma gibbosa*	flamingo tongue
Flammen-Helmschnecke*, Streifen-Helmschnecke	*Phalium flammiferum*	flammed bonnet, striped bonnet
Flammen-Helmschnecke	*Cassis flammea*	flame helmet, princess helmet
Flammen-Kegelschnecke	*Conus flavescens*	flame cone
Flammenfisch	*Apogon maculatus*	flamefish
Flammenkopf	*Oxyruncus cristatus*	sharpbill
Flammenkopf, Feuerkopf	Oxyruncidae	sharpbill
Flammenschraubenschnecke, Flammenschraube	*Terebra taurina*	flame auger
Flankenkiemer	Notaspidea	notaspideans
Flaschenbürstenmoos*	*Thuiaria thuja*	bottlebrush hydroid
Flaschenschnabel-Delphine	*Lagenorhynchus* spp.	white-sided dolphins
Flaschentierchen, Bauchhaarlinge, Bauchhärlinge, Gastrotrichen	Gastrotricha	gastrotrichs
Flaumbeinige Spinne*	*Uloborus glomosus*	feather-legged spider, featherlegged orbweaver
Flaumlaus	*Goniocotes gallinae*	fluff louse
Flechteneule	*Cryphia domestica*	marbled beauty
Flechtenkäfer	*Gymnopholus lichenifer*	lichen weevil
Flechtenspinner	*Eilema complana*	scarce footman
Fleck-Sternschnecke	*Peltodoris atromaculata*	
Flecken-Meersau, Flecken-Schweinshai	*Oxynotus centrina*	angular roughshark (FAO), flatiron shark, humantin
Flecken-Rennfrosch	*Kassina maculata*/ *Hylambates maculata*	spotted running frog
Flecken-Schweinshai, Flecken-Meersau	*Oxynotus centrina*	angular roughshark (FAO), flatiron shark, humantin
Flecken-Schwimmkrabbe	*Portunus spinimanus*	blotched swimming crab
Fleckenbülbül	*Ixonotus guttatus*	spotted greenbul
Fleckendelphin, Schlankdelphin	*Stenella attenuata*	bridled dolphin, pantropical spotted dolphin, white-spotted dolphin
Fleckenfalter	Nymphalidae	brush-footed butterflies
Fleckenfrosch	*Rana pretiosa*	spotted frog
Fleckengroppe*	*Cottus bairdi*	mottled sculpin
Fleckenhalsotter (Hbz. Afrikanischer Otter)	*Lutra maculicollis*	African river otter
Fleckenkantschil, Indien-Kantschil	*Tragulus meminna*	Indian spotted chevrotain

Fleckenkauz	*Strix occidentalis*	spotted owl
Fleckenkröte	*Bufo punctatus*	red-spotted toad
Fleckenküling, Fleckengrundel	*Pomatoschistus pictus*	painted goby
Fleckenlanguste	*Panulirus guttatus*	spotted spiny lobster
Fleckenlinsang	*Prionodon pardicolor*	spotted linsang
Fleckenmondfisch	*Mene maculata*	moonfish
Fleckenmusang, Gefleckter Palmenroller (Hbz. Palmenroller)	*Paradoxurus hermaphroditus*	common palm civet
Fleckenprinie	*Prinia maculosa*	karroo prinia
Fleckenquerzahnmolch	*Ambystoma maculatum*	spotted salamander
Fleckenroller	*Chrotogale owstoni*	Owston's palm civet
Fleckenskunks (Hbz. Lyraskunk)	*Spilogale* spp.	spotted skunks
Fleckenuhu	*Bubo africanus*	spotted eagle-owl
Fleckhai, Sägeschwanz	*Galeus melastomus/ Pristiurus melanostomus*	blackmouth catshark (FAO), black-mouthed dogfish
Fleckige Brutwanze, Birkenwanze, Erlenwanze	*Elasmucha grisea*	parent bug, mothering bug
Fleckiger Schlangenstern, Fünfeck-Schlangenstern	*Ophiomyxa pentagona*	pentagonal brittlestar*
Fleckrochen, Fleckenrochen, Gefleckter Rochen	*Raja montagui*	spotted ray
Fleckscheren-Schmuckkrabbe	*Microphrys bicornuta*	speck-claw decorator crab
Fleckstrichsalmler	*Hemiodus semitaeniatus/ Hemiodopsis semitaenita*	halfline hemiodus
Flederhunde, Flughunde	Pteropodidae	Old World fruit bats
Fledermaus-Azurjungfer	*Coenagrion pulchellum*	variable damselfly
Fledermaus-Lausfliegen, Fledermausfliegen	Nycteribiidae	bat flies
Fledermaus-Walzenschnecke	*Cymbiola vespertilio*	bat volute
Fledermausaar	*Machaehamphus alcinus*	bat hawk
Fledermäuse, Flattertiere	Chiroptera	bats, chiropterans
Fledermausfisch, Meeressegelflosser, Einfacher Segelflosser	*Platax orbicularis*	batfish
Fledermausfliegen	Streblidae	bat flies
Fledermausmilben	Spinturnicidae	bat mites*
Fledermausrochen	*Myliobatis californica*	California bat ray
Fledermauswanze	*Cimex pipistrelli*	bat bug
Fledermauswanzen	Polyctenidae	bat bugs
Fleischfarbene Olive*	*Oliva carneola*	carnelian olive
Fleischfliegen, Aasfliegen	Sarcophagidae	fleshflies
Fleischflosser, Muskelflosser	Sarcopterygii/ Choanichthyes	fleshy-finned fishes, sarcopterygians, sarcopts
Fleischschwamm	*Haliclona viscosa*	fleshy sponge
Flieder-Randschnecke	*Closia lilacina*	lilac marginsnail, lilac marginella
Flieder-Schraubenschnecke*	*Terebra vinosa*	lilac auger
Fliedermotte	*Gracillaria syringella*	lilac leaf miner

Fliege> Große Stubenfliege	*Musca domestica*	housefly, house fly
Fliege> Kleine Stubenfliege	*Fannia canicularis*	lesser housefly
Fliegen	Brachycera (Diptera)	true flies
Fliegen> Echte Fliegen	Muscidae	houseflies, house flies
Fliegende Fische, Flugfische	Exocoetidae	flyingfishes
Fliegender Fisch, Flugfisch, Gemeiner Flugfisch, Meerschwalbe	*Exocoetus volitans*	tropical two-wing flyingfish
Fliegender Fisch> Vierflügel-Flugfisch	*Cheilopogon heterurus/ Cypselurus heterurus*	Mediterranean flyingfish
Fliegender Hund, Riesen-Flugfuchs, Kalong	*Pteropus vampyrus*	kalong, flying dog, large flying fox
Fliegender Kalmar	*Ommastrephes bartrami/ Sthenoteuthis bartrami*	red flying squid, neon flying squid (FAO)
Fliegendreck-Schraubenschnecke	*Terebra areolata*	flyspotted auger
Fliegenfleck-Kegelschnecke	*Conus stercusmuscarum*	flyspeck cone, fly-specked cone
Fliegenfleck-Nadelschnecke	*Cerithium muscarum*	flyspeck cerith, fly-specked cerith
Fliegenhaft	*Cloeon dipterum*	pond olive dun
Fliegenschnäpper> Eigentliche Fliegenschnäpper, Sänger	Muscicapidae	Old World flycatchers
Fliegenschnecke	*Austroginella muscaria*	fly marginella
Flinkes Känguru, Sandwallaby	*Macropus agilis*	agile wallaby
Flinkläufer, Flinkkäfer	*Trechus* spp.	
Floh> Menschenfloh	*Pulex irritans*	human flea
Flöhe	Siphonaptera/ Aphaniptera/Suctoria	fleas
Flohkrebs> Gemeiner Flohkrebs	*Gammarus locusta*	locust amphipod, common freshwater shrimp
Flohkrebse, Flachkrebse	Amphipoda	beach hoppers, sand hoppers, scuds and relatives
Flohstichkegel	*Conus pulicarius*	flea-bite cone
Flores-Riesenratte	*Papagomys armandvillei*	Flores giant rat
Florfliege> Gemeine Florfliege	*Chrysopa carnea/ Anisochrysa carnea*	common green lacewing
Florfliegen (Goldaugen, Stinkfliegen, Blattlauslöwen)	Chrysopidae	green lacewings
Florida Riffgecko	*Sphaerodactylus notatus*	reef gecko
Florida-Baumschnecke	*Liguus fasciatus*	Florida treesnail
Florida-Doppelschleiche, Florida-Breitnasenwühle	*Rhineura floridana*	Florida worm lizard
Florida-Fechterschnecke, Florida-Flügelschnecke	*Strombus alatus*	Florida fighting conch
Florida-Gabelmakrele, Gemeiner Pompano	*Trachinotus carolinus*	Florida pompano (FAO), common pompano
Florida-Gopherschildkröte	*Gopherus polyphemus*	gopher tortoise, Florida gopher tortoise
Florida-Knochenhecht	*Lepisosteus platyrhinchus*	Florida gar

Florida-Maus	*Podomys floridanus*	Florida mouse
Florida-Sandskink	*Neoseps reynoldsi*	sand skink
Florida-Schmuckschildkröte	*Pseudemys floridana floridana/ Chrysemys floridana floridana*	cooter, common cooter, coastal plain turtle, Florida cooter
Florida-Waldkaninchen	*Sylvilagus floridanus*	eastern cottontail
Florida-Wasserratte	*Neofiber alleni*	round-tailed muskrat, Florida water rat
Florida-Weichschildkröte, Wilde Dreiklaue	*Apalone ferox/Trionyx ferox*	Florida softshell turtle
Florida-Zierschildkröte	*Pseudemys rubriventris nelsoni/ Chrysemys nelsoni*	Florida redbelly turtle, Florida red-bellied turtle
Floridakärpfling, Amerikanischer Flaggenkilli	*Jordanella floridae*	American flagfish
Floßschnecke, Veilchenschnecke	*Janthina janthina*	large violet snail, common purple sea snail, violet seasnail, common janthina
Flösselaal	*Calamoichthys calabaricus*	reedfish
Flösselhechte	Polypteridae	bichirs
Flösselhechte	*Polypterus* spp.	bichirs
Flösselhechtverwandte	Polypteriformes	bichirs & reedfishes and allies
Flossenfüße	Pygopodidae	snake lizards, flap-footed lizards, scaly-footed lizards
Flossenfüßer, Flügelschnecken	Pteropoda	pteropods, seabutterflies
Flossenfüßer, Robben	Pinnipedia	marine carnivores (seals, sealions, walruses)
Flossenkrebse	Triopsidae	tadpole shrimps, triopsids
Flötenfisch	*Fistularia petimba*	red cornetfish (FAO), flutemouth
Flötenmünder, Flötenmäuler, Pfeifenfische	Fistulariidae	flutemouths, cornetfishes
Flötenvogel	*Gymnorhina tibicen*	black-backed magpie
Flöter	Cinclosomatinae	quail-thrushes, whipbirds
Flugbarben	*Esomus* spp.	flying barbs
Flugdrachen	*Draco* spp.	flying dragons, flying lizards
Flügel-Trogmuschel	*Mactrellona alata*	winged surfclam
Flügelährenfische	Notocheiridae/Isonidae	surf sprites
Flügelbutt, Scheefsnut	*Lepidorhombus whiffiagonis*	megrim (FAO), sail-fluke, whiff
Flügelkiemer	Pterobranchia	pterobranchs
Flügellaus	*Lipeurus caponis*	wing louse, chicken wing louse
Flügelmuschel, Europäische Flügelmuschel, Vogelmuschel, Europäische Vogelmuschel	*Pteria hirundo*	European wing-oyster (European wing shell)
Flügelrossfische, Flügelrösschen, Seeschmetterlinge	Pegasidae	seamoths

Flügelschnecken, Flossenfüßer	Pteropoda	pteropods, seabutterflies
Flügelsternwurm*	*Ophiodromus pugettensis*	bat star worm
Flugfisch, Gemeiner Flugfisch, Fliegender Fisch, Meerschwalbe	*Exocoetus volitans*	tropical two-wing flyingfish
Flugfrösche	*Rhacophorus* spp.	flying frogs
Flughahn	*Dactylopterus volitans*	flying gurnard
Flughähne, Flatterfische	Dactylopteridae	helmet gurnards, flying gurnards
Flughühner	Pteroclidae/Pteroclididae	sandgrouse
Flughunde, Flederhunde	Pteropodidae	Old World fruit bats
Fluginsekten, Geflügelte Insekten	Pterygota	winged insects, pterygote insects
Flugsaurier	Pterosauria	pterosaurs (extinct flying reptiles/winged reptiles)
Flunder, Gemeine Flunder (Sandbutt)	*Platichthys flesus*	flounder (FAO), European flounder
Fluss-Federkiemenschnecke	*Valvata naticina*	livebearing European stream valve snail
Fluss-Hieroglyphen-Schmuckschildkröte	*Pseudemys concinna concinna*	river cooter
Fluss-Kugelmuschel, Große Kugelmuschel, Ufer-Kreismuschel	*Sphaerium rivicola*	nut orb mussel (nut orb shell), nut fingernailclam
Fluss-Schleimfisch, Süßwasser-Schleimfisch	*Salaria fluviatilis*	freshwater blenny
Fluss-Schwimmschnecke, Gemeine Kahnschnecke	*Theodoxus fluviatilis*	common freshwater nerite, river nerite, the nerite
Flussbarbe, Barbe, Gewöhnliche Barbe	*Barbus barbus*	barbel
Flussbarsch	*Perca fluviatilis*	perch, European perch (FAO), redfin perch
Flussdeckelschnecken, Sumpfdeckelschnecken	Viviparidae	river snails
Flussflohkrebs	*Gammarus roeseli*	lacustrine amphipod, lacustrine shrimp
Flussgrundel	*Gobius fluviatilis*	river goby, monkey goby (FAO)
Flusshecht, Hecht	*Esox lucius*	pike, northern pike (FAO)
Flussjungfern	Gomphidae	clubtails
Flusskarpfen, Karpfen	*Cyprinus carpio*	carp, common carp (FAO), European carp
Flusskrebse	Astacidae	crayfishes, river crayfishes
Flussmuschel> Kleine Flussmuschel, Gemeine Flussmuschel, Bachmuschel	*Unio crassus*	common river mussel, common Central European river mussel
Flussmuscheln	*Unio* spp.	freshwater mussels
Flussmuscheln	Unionidae	freshwater mussels

Flussnapfschnecke, Gemeine Flussnapfschnecke	Ancylus fluviatilis	river limpet, common river limpet
Flussnapfschnecken	Ancylidae	river limpets, freshwater limpets
Flussneunauge	Lampetra fluvialis	river lamprey, lampern, European river lamprey (FAO)
Flussperlmuschel	Margaritifera margaritifera	freshwater pearl mussel, eastern pearlshell
Flusspferd, Großflusspferd	Hippopotamus amphibius	hippopotamus ("hippo")
Flussregenpfeifer	Charadrius dubius	little ringed plover
Flussschwamm, Großer Süßwasserschwamm, Klumpenschwamm	Ephydatia fluviatilis	greater freshwater sponge
Flussschwein, Buschschwein, Pinselohrschwein	Potamochoerus porcus	African bush pig, red river hog
Flussseeschwalbe	Sterna hirundo	common tern
Flusssteinkleber	Lithoglyphus naticoides	
Flussuferläufer	Actitis hypoleucos/ Tringa hypoleucos	common sandpiper
Flusswelse	Akysidae	stream catfishes
Folien-Mikroporenkoralle, Folien-Koralle	Montipora foliosa	leaf coral
Foraminiferen	Foraminiferida	foraminiferans, forams
Forbes' Kalmar, Nordischer Kalmar	Loligo forbesii	long-finned squid, northern squid
Forelle	Salmo trutta	trout
Forellen & Atlantische Lachse	Salmo spp.	Atlantic trouts
Forellen-Hechtling	Galaxias truttaceus	spotted trout minnow, spotted mountain trout (FAO)
Forellen-Lippfisch*	Centrolabrus trutta	trout wrasse
Forellenbarsch	Micropterus salmoides	largemouth black bass, largemouth bass (FAO)
Forellenpunktsalmler	Copeina guttata	red-spotted copeina
Forellensalmler, Raubsalmler	Erythrinidae	trahiras
Forleule, Kieferneule	Panolis flammea	pine beau, pine beauty moth
Formosamakak, Rundgesichtsmakak	Macaca cyclopis	Taiwan macaque
Forrest-Kleinmaus	Leggadina forresti	Forrest's mouse
Forsterseeschwalbe	Sterna forsteri	Forster's tern
Fossa, Frettkatze	Cryptoprocta ferox	fossa
Franciscana, La Plata-Delphin	Pontoporia blainvillei	Franciscana dolphin, La Plata dolphin
Fransen-Bandwurm*	Thysanosoma actinioides	fringed tapeworm
Fransen-Feilenfisch	Chaetodermis penicilligera	prickly leatherjacket (FAO), Queensland leafy leatherjacket
Fransenfinger	Acanthodactylus spp.	fringe-toed lacertids, fringe-fingered lizards
Fransenfinger> Europäischer Fransenfinger, Gewöhnlicher Fransenfinger	Acanthodactylus erythrurus	spiny-footed lizard, fringe-fingered lizard
Fransenfledermaus	Myotis nattereri	Natterer's bat

Fransenflügler, Blasenfüße, Thripse	Thysanoptera	thrips
Fransenlipper	*Labeo* spp.	labeos
Fransenmotten	Momphidae	momphid moths
Fransenschildkröte, Matamata	*Chelus fimbriatus*	matamata
Franzosen-Kaiserfisch, Schwarzer Engelsfisch	*Pomacanthus paru*	French angelfish
Französische Kornmotte, Getreidemotte (Weißer Kornwurm)	*Sitotroga cerealella*	Angoumois grain moth
Französischer Dorsch, Franzosendorsch	*Trisopterus luscus*	bib, pout, pouting (FAO)
Frauenfisch	*Elops saurus*	ladyfish
Frauenfische	Elopidae	ladyfishes, tenpounders
Frauenkrabbe	*Ovalipes ocellatus*	lady crab
Frauenlori	*Lorius lory*	black-capped lory
Frauennerfling	*Rutilus pigus virgo*	Danubian roach
Fregattmakrele, Fregattenmakrele, Unechter Bonito	*Auxis thazard*	frigate tuna (FAO), frigate mackerel
Fregattvögel	Fregatidae	frigatebirds
Freischwanz-Samt-Fledermäuse (Bulldogg-Fledermäuse)	*Molossus* spp.	velvety free-tailed bats
Frettchen	*Mustela putorius f. furo*	domestic polecat
Frettkatze, Fossa	*Cryptoprocta ferox*	fossa
Fries-Grundel	*Lesuerigobius friesii*	Fries's goby
Friesenkopf, Aschfarbene Kreiselschnecke, Aschgraue Kreiselschnecke	*Gibbula cineraria*	grey topsnail
Fritfliege	*Oscinella frit*	frit-fly
Froschfische	Batrachoidiformes	toadfishes
Froschfische, Krötenfische	Batrachoididae	toadfishes
Froschkopf-Schildkröten	*Batrachemys* spp.	toad-headed turtles
Froschkopf-Schildkröte	*Phrynops hilarii*	frog-headed turtle, spotted-bellied side-necked turtle
Froschkrabbe	*Ranina ranina*	spanner crab
Froschkrabben	Raninidae	frog crabs
Froschlurche (Frösche & Kröten)	Anura/Salientia	frogs & toads, anurans
Froschquappe	*Raniceps raninus*	tadpole fish (FAO), tadpole cod
Froschschnecke> Gemeine Froschschnecke	*Bufonaria rana/Bursa rana*	common frogsnail
Froschschnecken	Bursidae	frog snails (frog shells)
Froschwels, Büschelwels	*Clarias batrachus*	walking catfish
Frostfisch*	*Benthodesmus simonyi*	frostfish

Frostspanner>
  Gemeiner Frostspanner,
  Obstbaumfrostspanner,
  Kleiner Frostspanner
	Operophtera brumata/
	Cheimatobia brumata
		winter moth,
		small winter moth

Frostspanner>
  Großer Frostspanner,
  Hainbuchenspanner
	Erannis defoliaria	mottled umber (moth)

Frostspinner, Haarschuppenspinner	*Ptilophora plumigera*	plumed prominent
Fruchtschalenwickler, Apfelschalenwickler	*Adoxophyes orana/* *Capua reticulana*	summer fruit tortrix
Fruchtstecher, Kupferroter Pflaumenstecher	*Rhynchites cupreus*	plum borer
Frühe Adonislibelle	*Pyrrhosoma nymphula*	large red damselfly
Frühe Heidelibelle	*Sympetrum fonscolombei*	red-veined sympetrum
Frühjahrs-Bläuling	*Celastrina ladon*	spring azure

Frühjahrs-Kieferfuß,
  Kleiner Rückenschaler,
  Schuppenschwanz,
  Langschwänziger
  Flossenkrebs
	Lepidurus apus/
	Lepidurus productus	

Frühjahrskiemenfuß,
  Frühjahrs-Kiemenfuß,
  Gemeiner Kiemenfuß
	Chirocephalus grubei/
	Siphonophanes grubii
		springtime fairy shrimp*

Frühlings-Rauhaareule	*Brachionycha nubeculosa*	Rannoch sprawler
Frühlings-Wegwespe	*Pompilus viaticus/* *Anoplius viaticus/* *Anoplius fuscus*	black-banded spider wasp
Fuchs-Mitra, Kleiner Fuchs	*Vexillum vulpecula*	little fox mitre, little fox miter
Fuchsammer	*Passerella iliaca*	fox sparrow
Fuchsfloh	*Chaetopsylla globiceps*	fox flea
Fuchsgesicht	*Lo vulpinus/* *Siganus vulpinus*	foxface
Fuchshai, Drescher	*Alopias vulpinus*	thresher shark, thintail thresher (FAO), thintail thresher shark, fox shark
Fuchshaie, Drescherhaie	Alopiidae	thresher sharks
Fuchshakenwurm, Hundehakenwurm	*Uncinaria stenocephala*	dog hookworm

Fuchskusu
  (Hbz. Austr./
  Neuseel. Opossum)
	Trichosurus vulpecula	brush-tailed possum

Fuchsmanguste	*Cynictis penicillata*	yellow mongoose
Fuchsnatter	*Elaphe vulpina*	fox snake
Fühlerfisch, Streifen-Anglerfisch	*Antennarius striatus/* *Antennarius scaber*	striated frogfish (FAO), splitlure frogfish
Fühlerfische, Krötenfische	Antennariidae	anglers, frogfishes
Fühlerschlange	*Erpeton tentaculatum*	tentacled snake, fishing snake
Füllhornkoralle	*Cornularia cornucopiae*	cornucopic coral*
Fulmen-Kegelschnecke	*Conus fulmen*	Fulmen's cone
Fünfbärtelige Seequappe	*Ciliata mustela*	five-bearded rockling
Fünfbinden-Skink	*Eumeces inexpectatus*	southeastern five-lined skink

Fünfeck-Schlangenstern, Fleckiger Schlangenstern	*Ophiomyxa pentagona*	pentagonal brittlestar*
Fünfeckstern, Kleiner Scheibenstern, Kleiner Buckelstern	*Asterina gibbosa*	cushion star, cushion starlet
Fünfflecken-Schlangenstern	*Ophiothrix quinquemaculata*	fivespot brittlestar
Fünffleckiger Lippfisch	*Symphodus roissali*	five-spot wrasse
Fünfgürtelbarbe	*Barbus pentazona pentazona*	five-banded barb, fiveband barb (FAO), tiger barb
Fünfpunkt	*Coccinella quinquepunctata*	five-spot ladybird, fivespot ladybird, 5-spot ladybird
Fünfstreifen-Mabuye	*Mabuya quinquetaeniata*	African five-lined skink, rainbow skink
Fünfstreifenskorpion	*Leirus quinquestriatus*	fivekeeled gold scorpion, African gold scorpion
Fünfzehen-Zwergspringmaus	*Cardiocranius paradoxus*	five-toed dwarf jerboa
Furchen-Erdschildkröte	*Rhinoclemmys areolata*	furrowed wood turtle
Furchenbienen, Schmalbienen	*Halictus* spp.	sweat bees, flower bees, halictid bees
Furchenfüßer	Solenogastres/ Neomeniomorpha (Aculifera)	solenogasters
Furchengarnele	*Melicertus kerathurus/ Penaeus kerathurus*	caramote prawn
Furchenkrebs, Schuppiger Springkrebs	*Galathea squamifera*	Leach's squat lobster
Furchenkrebse, Scheinhummer	Galatheidae	squat lobsters
Furchenparadiesvogel	*Cnemophilus macgregorii*	sickle-crested bird-of-paradise
Furchenschwimmer	*Acilius* spp.	pond beetles
Furchenwale	Balaenopteridae	rorquals
Furchenzahn-Spitzmausratte*	*Microhydromys richardsoni*	groove-toothed shrew rat, moss mouse
Fürstliche Walzenschnecke*	*Cymbiola imperialis*	imperial volute
Fußball-Seescheide	*Diazona violacea*	football sea-squirt
Füsiliere	Caesionidae	fusiliers
Futteralmotten, Sackträgermotten, Sackmotten	Coleophoridae	casebearers, casebearer moths
Fyllas Rochen	*Raja fyllae/ Rajella fyllae*	round skate

Gabarhabicht	*Micronisus gabar*	Gabar goshawk
Gabel-Azurjungfer	*Coenagrion scitulum*	dainty damselfly
Gabelbärte	*Osteoglossum* spp.	arawanas
Gabelbock, Gabelhornantilope	*Antilocapra americana*	pronghorn
Gabelchamäleon	*Chamaeleo furcifer*	forked chameleon, forknose chameleon
Gabeldorsch, Großer Gabeldorsch, Meertrüsche	*Phycis blennoides/ Urophycis blennoides*	greater forkbeard
Gabeldrongo	*Dicrurus forficatus*	crested drongo
Gabelhornantilope, Gabelbock	*Antilocapra americana*	pronghorn
Gabelmakrele, Bläuel, Pompano	*Trachinotus ovatus*	pompano, derbio (FAO)
Gabelmücken, Malariamücken, Fiebermücken	*Anopheles* spp.	malaria mosquitoes
Gabelschwanz> Kleiner Gabelschwanz	*Cerura bifida/ Harpyia bifida/ Harpyia hermelina/ Furcula bifida*	sallow kitten
Gabelschwanz> Großer Gabelschwanz	*Cerura vinula*	puss moth
Gabelschwanzmakrele, Bernsteinmakrele	*Seriola dumerili*	amberjack, greater amberjack (FAO), greater yellowtail
Gabelstreifiger Katzenmaki	*Phaner furcifer*	fork-marked mouse lemur
Gabelwurm, Roter Luftröhrenwurm, Rotwurm	*Syngamus trachea*	gapeworm (of poultry)
Gabunviper	*Bitis gabonica*	Gaboon viper
Gähnfrosch*, Gähnender Grabfrosch	*Heleioporus eyrei*	moaning frog
Galagos	*Galago* spp.	galagos, bush babies
Galapagos-Hai	*Carcharhinus galapagensis*	Galapagos shark
Galapagos-Riesenreisratte	*Megaoryzomys curioi*	Galapagos giant rat, Galapagos giant rice rat
Galapagos-Riesenschildkröte, Elefantenschildkröte	*Testudo elephantopus/ Geochelone elephantopus/ Chelonoides elephantopus*	Galapagos giant tortoise
Galapagosalbatros	*Diomedea irrorata*	waved albatross
Galapagosbussard	*Buteo galapagoensis*	Galapagos hawk
Galapagospinguin	*Spheniscus mendiculus*	Galapagos penguin
Galapagosscharbe	*Nannopterum harrisi*	flightless cormorant
Galapagostaube	*Zenaida galapagoensis*	Galapagos dove
Galizier, Sumpfkrebs	*Astacus leptodactylus*	long-clawed crayfish
Galjoen-Fische	Coracinidae	galjoens, galjoen fishes
Gallenkrabben	Hapalocarcinidae	coral gall crabs
Gallert-Moostierchen	*Alcyonidium gelatinosum*	gelatinous bryozoan
Gallertkalmare	Cranchiidae	gelatin-squids*
Gallertschwamm	*Halisarca dujardini*	Dujardin's slime sponge
Gallkrabben*	Cryptochiridae	gall crabs
Gallmilben	Eriophyiidae	gall mites, eriophyiid mites

Gallmücken	Cecidomyiidae	gall midges, gall gnats
Gallwespen	Cynipidae	gallwasps
Gambelmeise	*Parus gambeli*	mountain chickadee
Gambelwachtel, Helmwachtel	*Callipepla gambelii*	Gambel's quail
Gambia-Kusimanse	*Mungos gambianus*	Gambian mongoose
Gammaeule	*Autographa gamma*	silver Y
Gammaeulen-Raupenfliege	*Phryxe vulgaris*	silver Y moth parasite fly
Gämse, Gamswild (Gemse)	*Rupicapra rupicapra*	chamois
Ganges-Delphin, Susu, Schnabeldelphin	*Platanista gangetica*	Ganges River dolphin, su-su
Ganges-Gavial	*Gavialis gangeticus*	gharial, gavial, Indian gharial
Ganges-Hai	*Glyphis gangeticus*	Ganges shark
Ganges-Weichschildkröte	*Aspideretes gangeticus/ Trionyx gangeticus*	Indian softshell turtle, Ganges softshell turtle
Gangesbärbling	*Rasbora rasbora*	Ganges rasbora
Gangfisch	*Coregonus macrophthalmus*	gangfish, European whitefish (FAO), Lake Neuchâtel whitefish
Gänseblümchen-Schlangenstern	*Ophiopholis aculeata*	daisy brittlestar, crevice brittlestar
Gänsefußspanner	*Pelurga comitata*	dark spinach
Gänsefußstern	*Anseropoda placenta*	goose-foot starfish, goosefoot starfish
Gänsegeier	*Gyps fulvus*	griffon vulture
Gänsesäger	*Mergus merganser*	common merganser, goosander
Gänseschnabelwal, Cuvier-Schnabelwal	*Ziphius cavirostris*	Cuvier's goosebeaked whale
Gänsetierchen	*Dileptus anser*	
Gänsevögel, Entenvögel	Anseriformes	waterfowl (ducks & geese & swans)
Garizzo	*Spicara flexuosa*	garizzo
Garten-Laufkäfer, Goldgruben-Laufkäfer	*Carabus hortensis*	garden ground beetle
Gartenaal*	*Taenioconger longissimus*	garden eel
Gartenbaumläufer	*Certhia brachydactyla*	short-toed treecreeper
Gartengrasmücke	*Sylvia borin*	garden warbler
Gartenhaarmücke	*Bibio hortulanus*	garden march fly
Gartenhummel	*Bombus hortorum*	small garden bumble bee
Gartenkreuzspinne, Gemeine Kreuzspinne	*Araneus diadematus*	cross orbweaver, European garden spider, cross spider
Gartenlaubkäfer	*Phyllopertha horticola*	garden chafer
Gartenrotschwanz	*Phoenicurus phoenicurus*	redstart
Gartenschläfer	*Eliomys quercinus*	garden dormouse
Gartenschnecke> Garten-Wegschnecke, Gartenwegschnecke	*Arion hortensis*	garden slug, garden arion, common garden slug, black field slug
Gartenschnecke> Genetzte Ackerschnecke	*Deroceras reticulatum/ Agriolimax reticulatus*	netted slug, milky slug, gray fieldslug (U.S.), grey field slug (Br.)

Gartenschnirkelschnecke	*Cepaea hortensis*	white-lip gardensnail, white-lipped snail, garden snail, smaller banded snail
Gartenskolopender	*Cryptops hortensis*	European garden scolopendra
Gartenspitzmaus	*Crocidura suaveolens*	lesser white-toothed shrew
Gartenspringschwanz	*Bourletiella hortensis*	garden springtail
Gartentrupial	*Icterus spurius*	orchard oriole
Gaukler	*Terathopius ecaudatus*	bateleur
Gaur	*Bos gaurus*	gaur, seladang
Gazami-Schwimmkrabbe	*Portunus trituberculatus*	Gazami crab
Gazellen	*Gazella* spp.	gazelles
Geäderte Rapana, Rotmund-Wellhorn*	*Rapana venosa*	veined rapa whelk, Thomas's rapa whelk
Gebänderte Dreiecksmuschel, Gebänderte Sägemuschel, Sägezähnchen	*Donax vittatus*	banded wedge clam
Gebänderte Grübchenschnecke	*Lacuna vincta*/ *Lacuna divaricata*	northern lacuna, banded chink snail
Gebänderte Heidelibelle	*Sympetrum pedemontanum*	banded sympetrum*
Gebänderte Heideschnecke, Veränderliche Trockenschnecke	*Cernuella virgata*	banded snail, striped snail
Gebänderte Kahnschnecke, Binden-Schwimmschnecke	*Theodoxus transversalis*	striped freshwater nerite
Gebänderte Katzenaugennatter	*Leptodeira septentrionalis*	cat-eyed snake
Gebänderte Kegelschnecke, Schleifen-Kegel*	*Conus vittatus*	ribboned cone
Gebänderte Kreiselschnecke	*Calliostoma annulatum*	purple-ring topsnail, blue-ring topsnail, ringed top snail (ringed top shell)
Gebänderte Listspinne, Gerandete Jagdspinne	*Dolomedes fimbriatus*	fimbriate fishing spider*
Gebänderte Luchsspinne	*Oxyopes lineatus*	banded lynx spider
Gebänderte Papierblase	*Hydatina zonata*	zoned paperbubble
Gebänderte Prachtlibelle	*Calopteryx splendens*/ *Agrion splendens*	banded blackwings, banded agrion
Gebänderte Ruderschlange	*Hydrophis fasciatus*	banded small-headed sea snake, striped sea snake
Gebänderte Sandbiene*	*Andrena gravida*	banded mining bee
Gebänderte Sandschlange	*Chilomeniscus cinctus*	banded sand snake
Gebänderte Sandschlangen	*Chilomeniscus* spp.	banded sand snakes
Gebänderte Scherengarnele	*Stenopus hispidus*	banded coral shrimp
Gebänderte Springspinne	*Phlegra fasciata*	
Gebänderte Süßlippe (Gestreifte Süßlippe)	*Conodon nobilis*	barred grunt, barred grunter
Gebänderte Teppichmuschel, Einfache Teppichmuschel, Essbare Venusmuschel	*Venerupis geographica*/ *Venerupis rhomboides*	banded carpetclam, banded venus (banded carpet shell)
Gebänderte Tulpenschnecke	*Fasciolaria lilium*	banded tulip (banded tulip shell)
Gebänderte Venusmuschel	*Clausinella fasciata*/ *Venus fasciata*	banded venus

Gebänderte Wassernatter	*Nerodia fasciata fasciata*	banded water snake
Gebänderte Windelschnecke, Gestreifte Windelschnecke	*Vertigo substriata*	striated whorl snail
Gebänderte Zylinderrose	*Arachnanthus oligopodus*	banded tube anemone*
Gebänderter Bambushai, Brauner Lippenhai, Sattelfleckenhai	*Chiloscyllium punctatum*	brown carpet shark, brownbanded bambooshark (FAO)
Gebänderter Buschfisch	*Ctenopoma fasciolatum*	banded climbing fish, banded climbing perch, banded ctenopoma (FAO)
Gebänderter Fächerwurm*	*Sabella crassicornis*	banded feather-duster worm
Gebänderter Felsenleguan	*Petrosaurus mearnsi*	banded rock lizard
Gebänderter Indischer Buntbarsch	*Etroplus suratensis*	banded chromide
Gebänderter Krallengecko	*Coleonyx variegatus*	western banded gecko
Gebänderter Leierfisch	*Callionymus reticulatus*	banded dragonet, reticulated dragonet (FAO)
Gebänderter Marmorkegel	*Conus marmoreus bandanus*	banded marble cone
Gebänderter Ritterfisch	*Equetus lanceolatus*	jackknife fish
Gebänderter Schnurwurm	*Lineus geniculatus* (Nemertini)	banded bootlace
Gebärfische, Aalmütter	Zoarcidae	eelpouts
Gebirgs-Beißschrecke	*Metrioptera saussureana*	Saussure's bushcricket
Gebirgs-Flughund	*Sphaerias blanfordi*	mountain fruit bat
Gebirgs-Rötling	*Phoxinus oreas*	mountain redbelly dace
Gebirgs-Taschenratte	*Thomomys monticola*	mountain pocket gopher
Gebirgs-Wasserratte	*Parahydromys asper*	mountain water rat, coarse-haired hydromyine
Gebirgsbachspitzmaus, Tibetanische Wasserspitzmaus	*Nectogale elegans*	Tibetan water shrew, web-footed water shrew
Gebirgsmolche> Europäische Gebirgsmolche	*Euproctus* spp.	European brook salamanders, European mountain salamanders
Gebirgsschrecke> Gewöhnliche G.	*Podisma pedestris*	brown mountain grasshopper
Gebirgsseemolche	*Batrachuperus* spp.	mountain salamanders
Gebirgswühlmäuse	*Alticola* spp.	high mountain voles
Gebogene Venusmuschel	*Periglypta reticulata*	reticulated venus
Gebogenstachelmurex*	*Murex aduncospinosus*	bent-spined murex
Gebrannte Mandel	*Barbatia amygdalumtostum*	burnt-almond ark
Gebuchtete Seemandel	*Philine sinuata*	sinuate paperbubble
Geburtshelferkröte (Glockenfrosch, Fresser, Steinkröte)	*Alytes obstetricans*	midwife toad
Gebüsch-Ohrwurm	*Apterygida albipennis*	short-winged earwig, apterous earwig
Gecko-Glasfrosch	*Centrolene geckoideum*	Pacific giant glass frog
Geckos	Gekkonidae	geckos
Gedrehte Flügelmuschel	*Pteria tortirostris*	twisted wing-oyster
Gedrehte Scheinolive	*Agaronia contortuplicata*	twisted ancilla
Gedrungene Bärenkrebse	*Parribacus* spp.	mitten lobsters, slipper lobsters
Gedrungene Kellerspinne*	*Physocyclus globosus/ Spermophora meridionalis*	short-bodied cellar spider
Gedrungene Schließmundschnecke	*Pseudofusulus varians*	

Gedrungene Trogmuschel, Dreieckige Trogmuschel	*Spisula subtruncata*	cut surfclam (cut trough shell)
Gedrungener Kalmar, Kurz-Kalmar*	*Lolliguncula brevis*	Atlantic brief squid, brief squid, brief thumbstall squid, small squid
Gefältelte Schließmundschnecke	*Macrogastra plicatula*	plicate door snail
Gefederter Schleimfisch*	*Hypsoblennius hentzi*	feather blenny
Gefiedermilben	Faculiferidae	feather mites*
Gefiederte Anemone	*Actinodendron plumosum*	hell's fire sea anemone, pinnate anemone
Geflammter Rebenwickler	*Clepsis spectrana*	straw-coloured tortrix moth, cyclamen tortrix moth, fern tortrix
Gefleckte Baum-Lanzenotter	*Bothriopsis taeniata*	speckled forest pit viper
Gefleckte Bodenagame	*Amphibolurus maculatus*	spotted bearded dragon
Gefleckte Fassschnecke	*Tonna dolium*	spotted tun
Gefleckte Fledermaus	*Euderma maculatum*	spotted bat, pinto bat
Gefleckte Gewächshauslaus	*Aulacorthum circumflexum*	mottled arum aphid
Gefleckte Grübchenschnecke	*Lacuna marmorata*	chink snail
Gefleckte Halsband-Mondschnecke, Gefleckte Halsbandnabelschnecke	*Polinices catenus/ Lunatia catena*	spotted necklace snail (spotted necklace shell), necklace moonsnail (European necklace shell)
Gefleckte Heidelibelle, Gelbflüglige Heidelibelle	*Sympetrum flaveolum*	yellow-winged sympetrum, yellow-winged darter
Gefleckte Heideschnecke	*Candidula intersecta/ Helicella caperata*	wrinkled snail
Gefleckte Hornisse	*Dolichovespula maculata*	bold-faced hornet
Gefleckte Kartoffellaus	*Aulacorthum solani*	glasshouse aphid, potato aphid
Gefleckte Keulenschrecke	*Myrmeleotettix maculatus*	mottled grasshopper
Gefleckte Klapperschlange	*Crotalus mitchelli*	speckled rattlesnake
Gefleckte Königsmakrele, Spanische Makrele	*Scomberomorus maculatus*	Atlantic Spanish mackerel, Spanish mackerel (FAO)
Gefleckte Kreiselschnecke	*Trochus maculatus*	mottled topsnail
Gefleckte Krötenkopfagame, Gefleckter Krötenkopf	*Phrynocephalus guttatus*	spotted toad-headed lizard, spotted toadhead agama
Gefleckte Lärchenrindenlaus	*Cinara laricis*	spotted larch aphid
Gefleckte Pazifische Seeratte, Amerikanische Spöke, Amerikanische Chimäre	*Hydrolagus colliei*	spotted ratfish
Gefleckte Radspinnen*	*Neoscona* spp.	spotted orbweavers
Gefleckte Scheinolive	*Agaronia nebulosa*	blotchy ancilla
Gefleckte Schläfergrundel	*Dormitator maculatus*	spotted sleeper, fat sleeper (FAO)
Gefleckte Schnarrschrecke	*Bryodema tuberculata*	speckled grasshopper
Gefleckte Schnirkelschnecke, Baumschnecke	*Arianta arbustorum*	orchard snail, copse snail
Gefleckte Schraubenschnecke	*Terebra maculata*	marlinspike auger, giant marlin spike (spotted auger, big auger)
Gefleckte Schüsselschnecke	*Discus rotundatus*	rounded snail, rotund disc snail, radiated snail
Gefleckte Seidenspinne, Riesen-Waldspinne	*Nephila maculata*	giant wood spider

Gefleckte Smaragdlibelle	*Somatochlora flavomaculata*	yellow-spotted emerald
Gefleckte Stachelmakrele, Ohrfleck-Heringsmakrele	*Decapterus punctatus*	round scad
Gefleckte Tiefwassergarnele	*Pandalus platyceros*	spot shrimp
Gefleckte Walzenschnecke, Nordostatlantische Fleckenwalze*	*Ampulla priamus*	spotted flask
Gefleckte Weinbergschnecke	*Helix aspersa/ Cornu aspersum/ Cryptomphalus aspersus*	brown garden snail, brown gardensnail, common garden snail, European brown snail
Gefleckte Wurmschleiche	*Amphisbaena fuliginosa*	speckled worm lizard
Gefleckte Wurmseegurke	*Synapta maculata*	spotted synapta, spotted barrel gherkin*
Gefleckte Zornnatter	*Coluber ventromaculatus*	spotted whip snake
Gefleckter Adlerrochen, Entenschnabelrochen	*Aetobatus narinari*	spotted eagle ray
Gefleckter Anglerfisch	*Antennarius nummifer*	coin-bearing frogfish
Gefleckter Blattwurm	*Anaitides maculata*	spotted leafworm*
Gefleckter Bonito	*Cybiosarda elegans*	leaping bonito
Gefleckter Drückerfisch	*Balistes punctatus*	spotted triggerfish
Gefleckter Ebarme	*Psettodes belcheri*	spot-tail spiny turbot
Gefleckter Flügelbutt, Vierfleckbutt	*Lepidorhombus boscii*	four-spot megrim
Gefleckter Flugfisch, Gefleckter Fliegender Fisch	*Cypselurus furcatus/ Cheilopogon furcatus*	spotfin flyingfish
Gefleckter Furchenmolch	*Necturus maculosus*	mudpuppy, waterdog
Gefleckter Gabeldorsch	*Urophycis regia*	spotted hake
Gefleckter Halbschnabelhecht	*Hemiramphus far*	black-barred garfish, blackbarred halfbeak (FAO)
Gefleckter Knochenhecht	*Lepisosteus oculatus*	spotted gar
Gefleckter Kohltriebrüssler, Kleiner Kohltriebrüssler	*Ceutorhynchus quadridens*	cabbage seedstalk curculio
Gefleckter Kreuzwels	*Arius maculatus*	spotted catfish
Gefleckter Leierfisch	*Callionymus maculatus*	spotted dragonet
Gefleckter Lippfisch	*Labrus bergylta*	ballan wrasse
Gefleckter Marderhai, Südafrikanischer Hundshai	*Triakis megalopterus*	spotted gullyshark
Gefleckter Meerengel	*Squatina oculata*	smoothback angelshark (FAO), spotted monkfish
Gefleckter Python	*Antaresia maculosa*	spotted python
Gefleckter Riesenzackenbarsch	*Epinephelus tukula*	potato grouper (FAO), potato bass
Gefleckter Rund-Stechrochen	*Urolophus maculatus*	spotted round ray
Gefleckter Sägesalmler	*Serrasalmus rhombeus*	white piranha, spotted piranha, redeye piranha (FAO)
Gefleckter Schnauzenbrassen, Laxierfisch	*Maena maena/ Spicara maena*	menola, menole, mendole, blotched picarel (FAO)
Gefleckter Seehase	*Aplysia depilans*	spotted seahare
Gefleckter Seewolf, Gefleckter Katfisch	*Anarhichas minor*	spotted wolffish (FAO), spotted sea-cat, spotted catfish, spotted cat

Gefleckter Stechrochen, Schwarzpunkt-Stechrochen	*Torpedo melanospila/ Taeniura melanospilos/ Taeniura meyeni*	speckled stingray, black-spotted ray, blotched fantail ray (FAO)
Gefleckter Taubleguan	*Holbrookia lacerata*	spot-tailed earless lizard, spot-tail earless lizard
Gefleckter Teppichhai, Gefleckter Wobbegong (Australischer Ammenhai)	*Orectolobus maculatus*	spotted wobbegong (Australian carpet shark)
Gefleckter Umberfisch	*Cynoscion nebulosus*	spotted sea trout, spotted weakfish (FAO)
Gefleckter Walzenskink	*Chalcides ocellatus*	ocellated skink
Gefleckter Wolfsbarsch	*Dicentrarchus punctatus*	spotted bass
Gefleckter Zitterrochen	*Narcine timlei*	spotted numbfish
Gefleckter Zitterrochen> Augenfleck-Zitterrochen	*Torpedo torpedo*	torpedo ray, common torpedo (FAO), eyed electric ray
Geflecktes Klipphorn	*Pisania maculosa/ Pisania striata*	spotted pisania
Geflecktes Petermännchen, Spinnenqueise	*Trachinus araneus*	spotted weever
Geflecktes Riesengleithörnchen	*Petaurista elegans*	spotted giant flying squirrel
Geflecktflügliger Flughund, Fleckenflügel-Flughund	*Balionycteris maculata*	spotted-winged fruit bat
Geflügel-Pfriemenschwanz	*Heterakis gallinae*	cecal worm, caecal worm
Geflügel-Zystenmilbe*	*Laminosioptes cysticola*	fowl cyst mite, flesh mite, subcutaneous mite
Geflügelte Bettwanze	*Lyctocoris campestris*	debris bug, stack bug, field anthocoris
Geflügelwanze, Taubenwanze	*Cimex colombarius*	pigeon bug, fowl bug
Geflügelzecken	*Argas* spp.	fowl ticks
Gefranste Helmschnecke	*Phalium fimbria*	fringed bonnet
Gefurchte Fassschnecke	*Tonna sulcosa*	banded tun
Gefurchte Fassschnecke	*Tonna cepa*	canaliculated tun
Gefurchte Hirnkoralle, Labyrinthkoralle, Feine Atlantische Hirnkoralle	*Diploria labyrinthiformis*	labyrinthine brain coral, grooved brain coral
Gefurchte Kegelschnecke	*Conus sulcatus*	furrowed cone, sulcate cone
Gefurchte Kopfschildschnecke	*Retusa sulcata*	sulcate barrel-bubble
Gefurchte Kreiselschnecke*	*Cantharidus striatus*	grooved topsnail
Gefurchte Pupa	*Pupa sulcata/Solidula sulcata*	sulcate pupa
Gefurchte Reusenschnecke, Westliche Riesenreuse	*Nassarius fossatus*	channeled nassa, giant western nassa
Gefurchte Scheidenmuschel	*Solen marginatus*	grooved razor clam
Gefurchte Steinkrabbe	*Xantho hydrophilus*	furrowed crab, furrowed xanthid crab, Montagu's crab
Gefurchte Sturmhaube, Gefurchte Helmschnecke	*Cassis sulcosa*	grooved helmet
Gefurchte Turbanschnecke	*Turbo canaliculatus*	channeled turban

Gefurchte Wellhornschnecke*	Busycotypus canaliculatus/ Busycon canaliculatum	channeled whelk
Gefurchter Lappenrüssler, Gewächshaus-Dickmaulrüssler	Otiorhynchus sulcatus	vine weevil, black vine weevil, European vine weevil
Gegabelter Glockenpolyp	Obelia dichotoma	sea thread hydroid
Gegitterter Halsbandleguan	Crotaphytus reticulatus	reticulate collared lizard
Gegliederte Harfenschnecke	Harpa articularis	articulate harp
Gegürtelte Salpe	Salpa zonaria	girdled salp
Gehaubter Kapuziner, Apella, Faunaffe	Cebus apella	black-capped capuchin
Gehörnte Klapperschlange, Seitenwinder-Klapperschlange	Crotalus cerastes	sidewinder
Gehörnte Kreuzspinne	Araneus angulatus	horned orbweaver*
Gehörnte Lanzenotter	Porthidium melanurum	black-tailed horned pit viper, blacktail montane viper
Gehörnte Puffotter	Bitis caudalis	horned puff adder, Cape horned viper
Gehörnte Spaltenkreuzspinne	Nuctenea cornuta	furrow spider
Gehörnte Turbanschnecke	Turbo cornutus	horned turban
Gehörnter Schleimfisch	Parablennius tentacularis	horned blenny
Gehörnter Wimpelfisch	Heniochus monocerus	horned coachman, masked bannerfish (FAO)
Geierperlhuhn	Acryllium vulturinum	vulturine guineafowl
Geierschildkröte	Macroclemys temminckii	alligator snapping turtle
Geigenrochen (Gitarrenfische, Gitarrenrochen)	Rhinobatoidei/ Rhinobatidae	guitarfishes
Geigenrochen> Europäischer Geigenrochen, Gemeiner Geigenrochen, Hairochen	Rhinobatos rhinobatos	Mediterranean guitarfish, common guitarfish (FAO), pesce violino
Geißblattfedermotte	Alucita hexadactyla	many-plumed moth, twenty plumed moth
Geißelgarnelen	Penaeidae	penaeid shrimps
Geißelgarnelen u.a.	Penaeus spp.	white shrimps
Geißelgarnelen u.a.	Metapenaeus spp.	metapenaeus shrimps
Geißelschildechsen	Tetradactylus spp.	seps
Geißelskorpion> Riesen-Geißelskorpion, Essigskorpion	Mastigoproctus giganteus	giant vinegaroon
Geißelskorpione	Uropygi	whipscorpions
Geißelskorpione & Geißelspinnen	Pedipalpi (Uropygi & Amblypygi)	whipscorpions & tailless whipscorpions (incl. vinegaroons)
Geißelspinnen	Amblypygi	tailless whipscorpions
Geißeltierchen, Geißelträger, Flagellaten	Mastigophora	flagellates, mastigophorans
Geißkleebläuling, Silberfleckbläuling	Plebejus argus	silver-studded blue (butterfly)

Geistchen, Federmotten	Alucitidae/Orneodidae	plumed moths, many-plume moths, alucitids
Geisteraktinie*	Diadumene leucolena	white anemone, ghost anemone
Geisterkegel, Spiegel-Kegelschnecke	Conus spectrum	spectral cone, spectre cone
Gekerbte Schraube	Terebra crenulata	crenulated auger
Gekielte Felsschnecke	Cuma lacera	keeled rocksnail, carinate rocksnail (keeled rock shell)
Gekielte Pelomeduse, Kongo-Pelomeduse	Pelusios carinatus	African keeled mud turtle
Gekielte Rattennatter	Zaocys carinatus	keeled ratsnake, black-and-yellow keeled ratsnake
Gekielte Schüsselschnecke	Discus perspectivus	keeled disc snail
Gekielte Tellerschnecke	Planorbis carinatus	keeled ramshorn
Gekielter Bärenkrebs	Scyllarides nodifer	ridged slipper lobster
Gekielter Blaupfeil, Kleiner Blaupfeil	Orthetrum coerulescens	keeled skimmer
Gekielter Flussfalke	Oxygastra curtisii	orange-spotted emerald dragonfly
Gekielter Seestern	Asteropsis carinifera	keeled seastar*
Gekielter Taubleguan	Holbrookia propinqua	keeled earless lizard
Gekörnter Eichenholzbohrer	Xyleborus dryographus	oakwood ambrosia beetle*
Gekräuselte Lastträgerschnecke	Xenophora crispa	curly carriersnail, Mediterranean carriersnail
Gekreuzte Teppichmuschel	Tapes decussatus/ Venerupis decussata	chequered carpetclam, chequered venus (chequered carpet shell)
Gekreuzte Wurmschnecke*	Serpulorbis decussatus	decussate wormsnail
Gekreuzter Stechrochen	Urolophus cruciatus	crossback stingaree
Gekrönte Schnauzennatter	Lytorhynchus diadema	awl-headed snake
Gekrönte Strandschnecke	Tectarius coronatus	beaded prickly winkle, crowned prickly winkle
Gelappte Purpurschnecke	Drupa lobata	lobate drupe
Gelbaugenpinguin	Megadyptes antipodes	yellow-eyed penguin
Gelbbarsch, Amerikanischer Flussbarsch	Perca flavescens	yellow perch (FAO), American yellow perch
Gelbbauch, Gelbbauch-Demoiselle	Pomacentrus pavo	blue damsel, sapphire damsel (FAO)
Gelbbauch-Seeschlange, Plättchen-Seeschlange	Pelamis platurus	yellow-bellied sea snake, pelagic sea snake
Gelbbauchgirlitz	Serinus flaviventris	yellow canary
Gelbbäuchige Rattenschlange, Indomalaysische Rattenschlange	Ptyas korros	yellow-bellied rat snake
Gelbbäuchiges Murmeltier	Marmota flaviventris	yellow-bellied marmot
Gelbbauchunke	Bombina variegata	yellow-bellied toad, yellowbelly toad, variegated fire-toad
Gelbbauchwiesel	Mustela kathiah	yellow-bellied weasel
Gelbbeiniger Kanalkäfer, Geselliger Kanalkäfer	Amara familiaris	social sun beetle

Gelbbrauen-Laubsänger	*Phylloscopus inornatus*	yellow-browed warbler
Gelbbrauenammer	*Emberiza chrysophrys*	yellow-browed bunting
Gelbbraune Napfschnecke	*Iothia fulva*	tawny limpet
Gelbbrauner Diskus, Gewöhnlicher Diskus	*Symphysodon aequifasciata axelrodi*	brown discus
Gelbbrauner Zahnspinner, Gelbbrauner Zickzackspinner, Auenpappelgestrüpp-Zahnspinner	*Notodonta torva*	large dark prominent
Gelbbraungebänderter Fiederbartwels	*Synodontis flavitaeniatus*	orangestriped squeaker
Gelbbraunpunkt-Kauri	*Cypraea bistrinotata*	treble-spotted cowrie
Gelbbrust-Waldsänger	*Icteria virens*	yellow-breasted chat
Gelbbürzel-Honiganzeiger	*Indicator xanthonotus*	Indian honeyguide
Gelbe Bänderhornkoralle, Flache Gorgonie, Seeblatt-Gorgonie	*Pterogorgia citrina*	yellow sea whip, yellow ribbon
Gelbe Bandeule	*Noctua fimbriata*	broad-bordered yellow underwing
Gelbe Baumkoralle, Baumkoralle, Astkoralle	*Dendrophyllia ramea*	yellow tree coral*
Gelbe Dungfliege	*Scathophaga stercoraria*	yellow dungfly
Gelbe Erzspitznatter	*Oxybelis wilsoni*	yellow vine snake
Gelbe Federgorgonie, Gelbe Seerute	*Plexaura flava*	yellow sea rod
Gelbe Feuerqualle, Gelbe Haarqualle	*Cyanea capillata*	lion's mane, giant jellyfish, hairy stinger, sea blubber, sea nettle, pink jellyfish
Gelbe Fichten-Großgallenlaus	*Sacchiphantes abietis/ Adelges abietis*	yellow spruce pineapple-gall adelges
Gelbe Fledermaus	*Lasiurus ega*	yellow bat
Gelbe Fuchsschwanz-Gallmücke	*Contarinia merceri*	foxtail midge, cocksfoot midge
Gelbe Geißelgarnele	*Metapenaeus brevicornis*	yellow prawn, yellow shrimp
Gelbe Grasfliege	*Opomyza florum*	yellow cereal fly
Gelbe Haarqualle, Gelbe Feuerqualle	*Cyanea capillata*	lion's mane, giant jellyfish, hairy stinger, sea blubber, sea nettle, pink jellyfish
Gelbe Halmfliege, Gelbe Weizenhalmfliege	*Chlorops pumilionis*	gout fly
Gelbe Hornkoralle, Gelbe Gorgonie	*Eunicella cavolini*	yellow horny coral
Gelbe Hundezecke	*Haemaphysalis leachi*	yellow dog tick
Gelbe Karibik-Krustenanemone	*Parazoanthus swiftii*	golden zoanthid, yellow Caribbean colonial anemone
Gelbe Keiljungfer	*Gomphus simillimus*	yellow gomphus, Mediterranean gomphus
Gelbe Klappschildkröte, Gelbliche Klappschildkröte	*Kinosternon flavescens flavescens*	yellow mud turtle

Gelbe Krabbe, Italienischer Taschenkrebs, Gemeine Krabbe	*Eriphia verrucosa*	warty xanthid crab, Italian box crab, yellow box crab
Gelbe Krustenanemone	*Parazoanthus axinellae*	yellow commensal zoanthid, yellow encrusting sea anemone
Gelbe Lungenqualle	*Rhizostoma pulmo* (syn. *R. octopus*)	football jellyfish, yellow Mediterranean cabbage bleb
Gelbe Nelkenkoralle	*Leptopsammia pruvoti*	sunset star coral
Gelbe Pflaumensägewespe	*Hoplocampa flava*	plum sawfly
Gelbe Rosenbürsthornwespe	*Arge ochropus*	rose sawfly
Gelbe Seefeder	*Veretillum cynomarium*	yellow sea-pen
Gelbe Seerute, Gelbe Federgorgonie	*Plexaura flava*	yellow sea rod
Gelbe Seescheide, Keulen-Synascidie, Keulensynascidie, Glaskeulen-Seescheide	*Clavelina lepadiformis*	light-bulb sea-squirt
Gelbe Stachelbeerblattwespe, Stachelbeerfliege	*Pteronidea ribesii*/ *Nematus ribesii*	common gooseberry sawfly, gooseberry sawfly
Gelbe Stachelmakrele	*Caranx rhonchus*	false scad (FAO), yellow jack, yellow horse mackerel
Gelbe Sternschnecke, Warzige Schwammschnecke*	*Doris verrucosa*	sponge slug, sponge seaslug
Gelbe Teemilbe, Breitmilbe*	*Polyphagotarsonemus latus*/ *Tarsonemus latus*/ *Hemitarsonemus latus*	broad mite
Gelbe Tigermotte, Holunderbär	*Spilosoma lutea*	buff ermine moth
Gelbe und Zimtfarbene Taschenratten	*Pappogeomys* spp.	yellow and cinnamon pocket gophers
Gelbe Wald-Egelschnecke	*Limax tenellus*	slender slug, tender slug
Gelbe Wegameise, Gelbe Wiesenameise	*Lasius flavus*	mound ant, yellow turf ant, yellow meadow ant, yellow ant
Gelbe Weizengallmücke	*Contarinia tritici*	yellow wheat blossom midge, wheat blossom midge
Gelbe Wiesenameise, Gelbe Wegameise	*Lasius flavus*	mound ant, yellow turf ant, yellow meadow ant, yellow ant
Gelber Bambusbockkäfer*	*Rhaphuma annularis*	yellow bamboo longhorn beetle
Gelber Buchenspanner	*Cyclophora linearia*	clay tripelines moth
Gelber Bungar, Bänderkrait, Bänder-Krait	*Bungarus fasciatus*	banded krait
Gelber C-Falter, Glaskrautfalter	*Polygonia egea*/*Comma egea*/ *Nymphalis egea*	southern comma
Gelber Doktorfisch	*Acanthurus olivaceus*	olive surgeonfish, orangespot surgeonfish (FAO)
Gelber Einsiedler, Glattscheriger Einsiedler	*Anapagurus laevis*	yellow hermit crab
Gelber Fächerfußgecko	*Ptyodactylus hasselquistii*	fan-toed gecko, yellow fan-fingered gecko
Gelber Gitterkalkschwamm	*Clathrina clathrus*	yellow lattice sponge
Gelber Glattfuß	*Nematomenia flavens* (Aplacophora)	Mediterranean yellow solenogaster*
Gelber Grashüpfer	*Euchorthippus pulvinatus*	straw-coloured grasshopper
Gelber Herzigel	*Echinocardium flavescens*	yellow sea potato

Gelber Kalkschwamm	*Leucetta philippinensis*	yellow calcareous sponge
Gelber Kartoffelnematode	*Globodera rostochiensis*	yellow potato cyst nematode, golden nematode, potato root eelworm
Gelber Katzenwels	*Ictalurus natalis*	yellow bullhead
Gelber Kofferfisch, Kleiner Kofferfisch	*Ostracion cubicus/ Ostracion tuberculatus*	blue-spotted boxfish, yellow boxfish (FAO), polka dot boxfish
Gelber Krötenfisch	*Antennarius commersoni/ Antennarius moluccensis*	Commerson's frogfish
Gelber Mittelmeerskorpion, Okzitanischer Skorpion, Gemeiner Mittelmeerskorpion	*Buthus occitanus*	common yellow scorpion
Gelber Salmler, Gelber von Rio	*Hyphessobrycon bifasciatus*	yellow tetra
Gelber Scheinhummer, Südlicher Scheinhummer	*Cervimunida johni*	yellow squat lobster
Gelber Schlankcichlide	*Julidochromis ornatus*	ornate julie, golden julie (FAO)
Gelber Schnapper, Armadillo-Schnapper	*Lutjanus argentiventris*	armadillo snapper, yellow snapper (FAO)
Gelber Segeldoktor	*Zanclus flavescens*	yellow tang
Gelber Stechrochen*, Jamaika-Stechrochen, Jamaika-Rund-Stechrochen	*Urolophus jamaicensis*	yellow stingray
Gelber Zackenbarsch	*Plectropomus melanoleucus*	coral trout
Gelber Ziegenfisch	*Parupeneus cyclostomus*	yellowsaddle goatfish, goldsaddle goatfish (FAO)
Gelbes Ordensband	*Catocala fulminea*	yellow underwing
Gelbes Täubchen	*Pyrene flava*	yellow dovesnail
Gelbfiebermücke	*Aedes aegypti/ Stegomyia aegypti*	yellow fever mosquito
Gelbfleck-Krokodilfisch, Lippenfleck-Sandbarsch	*Parapercis xanthozona*	spottyfin weever, whitebar weever, yellowbar sandperch (FAO)
Gelbfleck- Landkartenschildkröte, Pascagoula-River- Höckerschildkröte	*Graptemys flavimaculata*	yellow-blotched map turtle (IUCN), yellow-blotched sawback
Gelbfleck-Lanzenotter	*Bothriechis aurifer*	yellow-blotched palm pit viper
Gelbflecken-Igelfisch	*Cyclichthys spilostylus*	yellowspotted burrfish
Gelbfleckenrochen	*Raja wallacei*	yellowspotted skate
Gelbfledermäuse	*Scotophilus* spp.	house bats, yellow bats
Gelbflossen-Barrakuda	*Sphyraena chrysotaenia/ Sphyraena flavicauda/ Sphyraena obtusata*	yellowtail barracuda, obtuse barracuda, blunt-jaw barracuda, yellowstripe barracuda (FAO)
Gelbflossen-Füsilier	*Caesio xanthonotus*	yellow-fin fusilier
Gelbflossen-Makrele	*Atule mate/Caranx mate*	yellowfin scad
Gelbflossen-Nothothenia	*Nothotheniops nudifrons*	naked-head nothothenia, yellow-fin notie, gaudy nothothen (FAO)
Gelbflossen-Ruderfisch	*Lethrinus kallopterus*	yellow-spotted emperor
Gelbflossen-Süßlippe	*Orthopristis chrysoptera*	pigfish
Gelbflossen-Thunfisch	*Thunnus albacares*	yellowfin tuna (FAO), yellow-finned tuna, yellow-fin tunny

Gelbflossenaufbläser	*Sphoeroides pachygaster/ Sphaeroides pachygaster/ Sphoeroides cutaneus/ Sphaeroides cutaneus*	blunthead puffer (FAO), smooth pufferfish
Gelbflüglige Großblattnase	*Lavia frons*	African yellow-winged bat
Gelbflüglige Heidelibelle, Gefleckte Heidelibelle	*Sympetrum flaveolum*	yellow-winged sympetrum, yellow-winged darter
Gelbfühleriger Spanner	*Apocheima hispidaria*	small brindled beauty
Gelbfuß-Landschildkröte	*Geochelone denticulata*	South American yellow-footed tortoise
Gelbfußflamingo	*Phoenicoparrus andinus*	Andean flamingo
Gelbfußtermite	*Reticulitermes flavipes*	eastern subterranean termite
Gelbgefleckte Nachtechse	*Lepidophyma flavimaculatum*	yellow-spotted night lizard
Gelbgefleckte Nacktschnecke	*Geomalacus maculosus*	kerry slug, spotted kerry slug, spotted Irish slug
Gelbgefleckter Schlangenaal	*Myrichthys acuminatus/ Muraena acuminata*	sharptail eel
Gelbgrüne Meerkatze	*Cercopithecus sabaeus*	green monkey, callithrix monkey
Gelbgrüne Zornnatter	*Coluber viridiflavus*	European whip snake, Western European whip snake, dark-green whipsnake
Gelbgrüner Halbzehengecko	*Hemidactylus flaviviridis*	Indian wall gecko, Indian leaf-toed gecko
Gelbhalsmaus	*Apodemus flavicollis*	yellow-necked mouse
Gelbhalstermite	*Calotermes flavicollis/ Kalotermes flavicollis*	yellow-necked dry-wood termite
Gelbhaubenkakadu	*Cacatua galerita*	sulphur-crested cockatoo
Gelbkehl-Schleimfisch, Gelbwangen-Schleimfisch	*Lipophrys canevae/ Blennius canevae*	yellow-throat blenny, Caneva's blenny
Gelbkehlsperling	*Petronia xanthocollis/ Passer xanthocollis*	yellow-throated sparrow
Gelbkehlvireo	*Vireo flavifrons*	yellow-throated vireo
Gelbkopf-Falterfisch	*Chaetodon xanthocephalus*	yellowhead butterflyfish
Gelbkopf-Felshüpfer, Gelbkopf-Stelzenkrähe	*Picathartes gymnocephalus*	white-necked rockfowl
Gelbkopf-Landschildkröte	*Geochelone elongata elongata/ Indotestudo elongata elongata*	yellow tortoise, elongated tortoise (IUCN)
Gelbkopf-Seidenäffchen	*Callithrix flaviceps*	buffy-headed marmoset
Gelbkopf-Stelzenkrähe, Gelbkopf-Felshüpfer	*Picathartes gymnocephalus*	white-necked rockfowl
Gelbköpfchen	*Cuora flavomarginata*	yellow-margined box turtle, snake-eating turtle
Gelbkopfgecko	*Gonatodes albogularis*	yellow-headed gecko
Gelbkopfkarakara	*Milvago chimachima*	yellow-headed caracara
Gelblemming	*Eolagurus luteus/Lagurus luteus*	yellow steppe lemming
Gelbliche Wieseneule, Grasstengeleule	*Luperina testacea*	flounced rustic moth
Gelblinge, Heufalter	*Colias* spp.	yellows
Gelblinien Pilotbarsch, Bermuda-Ruderfisch	*Kyphosus sectatrix*	Bermuda chub, Bermuda sea chub (FAO)
Gelbmaul-Barrakuda*	*Sphyraena viridensis*	yellowmouth barracuda
Gelbmund-Wellhornschnecke	*Buccinum leucostoma*	yellow-mouthed whelk, yellow-lipped buccinum

Gelbohr-Fledermäuse	*Vampyressa* spp.	yellow-eared bats
Gelbrandiger Breitflügel-Tauchkäfer	*Graphoderus zonatus*	spangled water beetle
Gelbrandkäfer, Gelbrand	*Dytiscus marginalis*	great diving beetle
Gelbrücken-Füsilier, Gelbschwanz-Füsilier	*Caesio teres*	beautiful fusilier, yellowtail fusilier, yellow and blueback fusilier (FAO)
Gelbrückenducker, Riesenducker	*Cephalophus sylvicultor*	yellow-backed duiker
Gelbsalamander	*Eurycea* spp.	American brook salamanders
Gelbsaum-Juwelenbarsch	*Variola louti*	lunartail rockcod, moontail rockcod, lyretail grouper, yellow-edged lyretail (FAO)
Gelbscheitelamazone	*Amazona ochrocephala*	yellow-crowned parrot, yellow-headed parrot, yellow-crowned amazon
Gelbschnabel-Madenhacker	*Buphagus africanus*	yellow-billed oxpecker
Gelbschnabel-Sturmtaucher	*Calonectris diomedea*	Cory's shearwater
Gelbschnabelkuckuck	*Coccyzus americanus*	yellow-billed cuckoo
Gelbschnabeltaucher	*Gavia adamsii*	white-billed diver, yellow-billed loon
Gelbschnabeltoko	*Tockus flavirostris*	yellow-billed hornbill
Gelbschopflund	*Fratercula cirrhata/ Lunda cirrhata*	tufted puffin
Gelbschulter-Blattnasen	*Sturnira* spp.	yellow-shouldered bats, American epauleted bats
Gelbschwamm	*Spongia irregularis*	yellow sponge, hardhead sponge
Gelbschwanz-Drachenkopf	*Sebastes flavidus*	yellowtail rockfish
Gelbschwanz-Makrele	*Carangoides bartholomaei*	yellow jack
Gelbschwanz-Segeldoktorfisch	*Zebrasoma xanthurum*	yellowtail surgeonfish, yellowtail sailfin tang, yellowtail tang (FAO)
Gelbschwanz-Stachelmakrele	*Decapterus macarellus*	mackerel scad
Gelbschwanzflunder	*Limanda ferruginea*	yellowtail flounder
Gelbschwänziger Skorpion	*Euscorpius flavicaudis*	yellowtail scorpion
Gelbschwanzschnapper	*Ocyurus chrysurus*	yellowtail snapper
Gelbschwanzwollaffe	*Lagothrix flavicauda*	yellow-tailed woolly monkey
Gelbspanner, Weißdornspanner	*Opisthograptis luteolata*	brimstone moth
Gelbspötter	*Hippolais icterina*	icterine warbler
Gelbstachel-Igelrochen, Großer Dornen-Stechrochen	*Urogymnus africanus/ Urogymnus asperrimus*	yellow-spine thorny ray, porcupine ray (FAO)
Gelbsteißbülbül	*Pycnonotus xanthopygos*	yellow-vented bulbul
Gelbstirn-Waldsänger	*Vermivora bachmanii*	Bachman's warbler
Gelbstreifige Wegschnecke	*Arion fasciatus*	orange-banded slug, orange-banded arion, banded slug
Gelbstreifiger Kohlerdfloh	*Phyllotreta nemorum*	large striped flea beetle
Gelbstriemen, Blöker	*Boops boops*	bogue
Gelbwangenwühlmaus	*Microtus xanthognathus*	yellow-cheek vole
Gelbwaran	*Varanus flavescens*	yellow monitor
Gelbwürfeliger Dickkopffalter	*Carterocephalus palaemon*	chequered skipper
Gelbzahnkauri	*Cypraea xanthodon*	yellow-toothed cowrie

Geldkauri, Geldschnecke	*Cypraea moneta/ Monetaria moneta*	money cowrie
Geldspinne*	*Hypomma bituberculatum*	money spider
Gelenkschildkröten	*Kinixys* spp.	hinge-backed tortoises, hingeback tortoises, hinged tortoises
Gelippte Tellerschnecke	*Anisus spirorbis*	
Gemalter Scheibenzüngler	*Discoglossus pictus*	painted frog
Gemeine Abalone, Gemeines Meerohr, Gemeines Seeohr	*Haliotis tuberculata*	common ormer, European edible abalone
Gemeine Achatschnecke, Gemeine Glattschnecke, Glatte Achatschnecke	*Cochlicopa lubrica/ Cionella lubrica*	slippery moss snail, glossy pillar snail (U.S.)
Gemeine Artmuschel, Artemismuschel	*Dosinia exoleta*	rayed dosinia, rayed artemis
Gemeine Atlantische Pfeffermuschel	*Abra aequalis*	common Atlantic abra (common Atlantic furrow-shell)
Gemeine Bernsteinschnecke	*Succinea putris*	rotten amber snail, large amber snail, European ambersnail (U.S.)
Gemeine Bettwanze	*Cimex lectularius*	bedbug, common bedbug (wall-louse)
Gemeine Binsenjungfer	*Lestes sponsa*	green lestes, emerald damselfly
Gemeine Birkenzierlaus	*Euceraphis punctipennis*	downy birch aphid, birch aphid
Gemeine Blasenschnecke	*Bulla striata*	striate bubble, common Atlantic bubble
Gemeine Blattschneiderbiene, Rosenblattschneiderbiene	*Megachile centuncularis/ Megachile versicolor*	common leafcutter bee, common leafcutting bee, rose leaf-cutting bee
Gemeine Bohrmuschel, Große Bohrmuschel, Dattelmuschel	*Pholas dactylus*	common piddock
Gemeine Chile-Vogelspinne*	*Phrixotrichus spatulata*	Chilean common tarantula
Gemeine Dornschrecke, Säbeldornschrecke	*Tetrix undulata/Tetrix vittata*	common groundhopper
Gemeine Drüsenameise	*Tapinoma erraticum*	erratic ant
Gemeine Eichengallwespe	*Cynips quercusfolii*	common oak gallwasp, oak leaf cherry-gall cynipid (>cherry gall)
Gemeine Eischnecke	*Ovula ovum/Erronea ovum*	egg cowrie, common egg cowrie
Gemeine Entenmuschel, Glatte Entenmuschel	*Lepas anatifera*	common goose barnacle
Gemeine Erbsenmuschel	*Pisidium casertanum*	common pea mussel, caserta pea mussel, ubiquitous peaclam
Gemeine Federkiemenschnecke, Teich-Federkiemenschnecke	*Valvata piscinalis piscinalis*	valve snail, European stream valve snail, common valve snail, European stream valvata
Gemeine Feigenschnecke, Atlantik-Feigenschnecke	*Ficus communis*	Atlantic figsnail (common fig shell)
Gemeine Fenstermücke, Fensterfliege (Weißer Drahtwurm)	*Sylvicola fenestralis*	common window gnat, common window midge

Gemeine Feuerwanze	*Pyrrhocoris apterus*	firebug
Gemeine Fichtengespinstblattwespe	*Cephalcia abietis*	web-spinning spruce sawfly, spruce webspinner
Gemeine Fiebermücke	*Anopheles atroparvus*	common malarial anopheles mosquito, common anopheles mosquito
Gemeine Florfliege	*Chrysopa carnea/ Anisochrysa carnea*	common green lacewing
Gemeine Flusskrabbe, Gemeine Süßwasserkrabbe	*Potamon fluviatile*	Italian freshwater crab
Gemeine Frontinella*	*Frontinella pyramitela/ Frontinella communis*	bowl and doily spider
Gemeine Froschschnecke	*Bufonaria rana/Bursa rana*	common frogsnail
Gemeine Garten-Schwebfliege	*Syrphus ribesii*	currant hover fly
Gemeine Gelbeule	*Xanthia icteritia*	sallow, the sallow (moth)
Gemeine Getreidestaublaus	*Lachesilla pedicularia*	cosmopolitan grain psocid
Gemeine Harfenschnecke, Gewöhnliche Harfenschnecke	*Harpa harpa*	common harp
Gemeine Heidelibelle	*Sympetrum vulgatum*	vagrant sympetrum
Gemeine Heideschnecke	*Helicella itala*	common heath snail
Gemeine Herzmuschel, Essbare Herzmuschel	*Cardium edule/ Cerastoderma edule*	common cockle, common European cockle, edible cockle
Gemeine Kahnschnecke, Fluss-Schwimmschnecke	*Theodoxus fluviatilis*	common freshwater nerite, river nerite, the nerite
Gemeine Keiljungfer	*Gomphus vulgatissimus*	club-tailed dragonfly
Gemeine Keulenwespe	*Sapyga clavicornis*	common European sapygid wasp
Gemeine Kiefernbuschhornblattwespe	*Diprion pini*	pine sawfly
Gemeine Kreuzspinne, Gartenkreuzspinne	*Araneus diadematus*	cross orbweaver, European garden spider, cross spider
Gemeine Kristallschnecke	*Vitrea crystallina*	crystal snail
Gemeine Kronenschnecke, Florida-Kronenschnecke	*Melongena corona*	common crown conch, Florida crown conch, American crown conch
Gemeine Kugelmuschel, Hornfarbene Kugelmuschel, Linsenmuschel	*Sphaerium corneum*	horny orb mussel, European fingernailclam
Gemeine Lappenmuschel, Juwelendose	*Chama gryphoides*	common jewel box
Gemeine Marmorgarnele	*Saron marmoratus*	common marble shrimp
Gemeine Meeräsche, Flachköpfige Meeräsche, Großkopfmeeräsche	*Mugil cephalus*	striped gray mullet, striped mullet, flathead mullet (FAO), common grey mullet, flat-headed grey mullet
Gemeine Miesmuschel	*Mytilus edulis*	blue mussel, bay mussel, common mussel, common blue mussel
Gemeine Nadelschnecke, Gemeine Hornschnecke, Gemeine Seenadel	*Cerithium vulgatum/ Gourmya vulgata*	European cerith, common cerith
Gemeine Napfschnecke, Gewöhnliche Napfschnecke	*Patella vulgata*	common limpet, common European limpet

Gemeine Nussmuschel	*Nucula nucleus*	common nutclam
Gemeine Olive, Gemeine Olivenschnecke	*Oliva oliva*	common olive
Gemeine Ostatlantische Käferschnecke, Bienenchiton	*Chaetopleura apiculata*	eastern beaded chiton, common eastern chiton, bee chiton (common American Atlantic coast chiton)
Gemeine Pelzbiene	*Anthophora acervorum*	common Central European flower bee
Gemeine Puppenschnecke, Zylindrische Puppenschnecke	*Leiostyla cylindracea*	common chrysalis snail
Gemeine Rollassel, Kugelassel	*Armadillidium vulgare*	common woodlouse, common pillbug, sow bug
Gemeine Rollwespe, Rotschenkelige Rollwespe	*Tiphia femorata*	common tiphiid wasp
Gemeine Rosengallwespe	*Diplolepis rosae*	mossy rose gall wasp, bedeguar gall wasp (>bedeguar gall/Robin's pincushion)
Gemeine Rundkrabbe	*Atelecyclus rotundatus*	circular crab
Gemeine Salpe	*Salpa democratica*	common salpid, whale blobs
Gemeine Samtmuschel, Archenkammmuschel, Mandelmuschel, Meermandel, Englisches Pastetchen	*Glycymeris glycymeris*	dog cockle, orbicular ark (comb-shell)
Gemeine Sattelauster	*Placuna sella*	common saddle oyster
Gemeine Schilfeule	*Nonagria typhae*	bulrush wainscot
Gemeine Schlammschnecke, Gewöhnliche Schlammschnecke, Wandernde Schlammschnecke	*Radix peregra/ Lymnaea peregra*	wandering pondsnail, common pondsnail
Gemeine Schließmundschnecke	*Balea biplicata*	common door snail, Thames door snail (Br.)
Gemeine Schnauzenschnecke, Große Langfühlerschnecke	*Bithynia tentaculata*	common bithynia, faucet snail
Gemeine Schwertmuschel, Schwertförmige Messerscheide, Schwertförmige Scheidenmuschel	*Ensis ensis*	common razor clam, narrow jackknife clam, sword razor
Gemeine Schwimmgarnele	*Palaemonetes vulgaris*	marsh grass shrimp, common shore shrimp
Gemeine Schwimmkrabbe, Ruderkrabbe	*Portunus holsatus/ Liocarcinus holsatus/ Macropinus holsatus*	common swimming crab, flying swimming crab
Gemeine Seemaus, Filzwurm, Wollige Seemaus, Seeraupe	*Aphrodita aculeata*	European sea mouse
Gemeine Seepocke	*Semibalanus balanoides/ Balanus balanoides*	northern rock barnacle, acorn barnacle, common rock barnacle
Gemeine Seescheide, Gemeine Ascidie	*Ascidiella scabra*	common sea-squirt

Gemeine Sichelschrecke	*Phaneroptera falcata*	sickle-bearing bushcricket
Gemeine Skorpionsfliege	*Panorpa communis*	common scorpionfly
Gemeine Smaragdlibelle	*Cordulia aenea*	downy emerald
Gemeine Spinnmilbe, „Rote Spinne", Bohnenspinnmilbe	*Tetranychus urticae*	twospotted spider mite
Gemeine Stachel-Käferschnecke	*Acanthochitona communis*	velvety mail chiton (velvety mail shell)
Gemeine Steckmuschel, Schinkenmuschel	*Pinna squamosa*	common penshell
Gemeine Steinlaus (Pschyrembel)	*Petrophaga lithographica* (see: *Lithophaga lithophaga*)	common stonelouse
Gemeine Strandschnecke, Gemeine Uferschnecke, „Hölker"	*Littorina littorea*	common periwinkle, common winkle, edible winkle
Gemeine Sumpfschnecke	*Stagnicola palustris/ Galba palustris*	common pondsnail, common European pondsnail, marsh snail
Gemeine Süßlippe	*Plectorhinchus gaterinus*	black-spotted rubberlip
Gemeine Süßwasserkrabbe, Gemeine Flusskrabbe	*Potamon fluviatile*	Italian freshwater crab
Gemeine Täubchenschnecke	*Columbella mercatoria*	West Indian dovesnail, common dove snail (common dove shell)
Gemeine Teichmuschel	*Anodonta anodonta*	common pond mussel*
Gemeine Tellerschnecke, Flache Tellerschnecke	*Planorbis planorbis*	ramshorn, common ramshorn, margined ramshorn, ram's horn (margined trumpet shell)
Gemeine Teppichmuschel, Kleine Teppichmuschel, Getupfte Teppichmuschel	*Venerupis pullastra/ Venerupis saxatilis/ Venerupis perforans/ Venerupis senegalensis/ Tapes pullastra*	pullet carpetclam, pullet venus (pullet carpet shell)
Gemeine Trogmuschel, Strahlenkörbchen	*Mactra cinerea/ Mactra stultorum*	white trough clam
Gemeine Turmschnecke	*Turritella communis*	common turretsnail, European turretsnail (common screw shell, common tower shell, great screw shell)
Gemeine Venusmuschel, Strahlige Venusmuschel	*Venus gallina/ Chione gallina/ Chamelea gallina*	striped venus, chicken venus
Gemeine Vogelspinne	*Avicularia avicularia*	pinktoe tarantula
Gemeine Waldschaben	*Ectobius lapponicus*	dusky cockroach
Gemeine Wattschnecke	*Peringia ulvae*	common mudflat snail*
Gemeine Wespe	*Vespula vulgaris*	common wasp
Gemeine Wiesenwanze	*Lygus pratensis/ Lygus rugulipennis*	European tarnished plant bug, tarnished plant bug, bishop bug
Gemeine Windelschnecke, Zwergwindelschnecke	*Vertigo pygmaea*	dwarf whorl snail, common whorl snail, crested vertigo (U.S.)
Gemeine Zahnschnecke, Gemeiner Elefantenzahn	*Dentalium vulgare/ Antalis tarentinum* (Scaphopoda)	common tusk, common tuskshell
Gemeine Zikade	*Lyristes plebeja*	common southern European cicada

Gemeine Zwiebelmondschwebfliege	*Eumerus strigatus*	onion bulb fly, small narcissus fly
Gemeine Zwiebelschwebfliege	*Merodon equestris*	large bulb fly, large narcissus fly, narcissus bulb fly (U.S.)
Gemeiner Bienenkäfer, Bienenwolf, Immenkäfer	*Trichodes apiarius*	bee beetle, bee wolf
Gemeiner Blattspanner	*Xanthorhoe fluctuata*	garden carpet
Gemeiner Blütenmulmkäfer	*Anthicus floralis*	narrownecked grain beetle, narrow-necked harvest beetle
Gemeiner Blütenspanner	*Eupithecia vulgata*	common pug
Gemeiner Elefantenzahn, Gemeine Zahnschnecke	*Dentalium vulgare/ Antalis tarentinum* (Scaphopoda)	common elephant's tusk, common tusk (common tuskshell)
Gemeiner Feldlaufkäfer, Körniger Laufkäfer, Gekörnter Laufkäfer	*Carabus granulatus*	field ground beetle
Gemeiner Felsenbohrer	*Hiatella rugosa/ Hiatella striata*	common rock borer
Gemeiner Fichtensplintbock	*Tetropium castaneum*	chestnut longhorn beetle
Gemeiner Fischegel	*Piscicola geometra*	common fish leech, great tailed leech
Gemeiner Flohkrebs, Bachflohkrebs	*Gammarus pulex/ Rivulogammarus pulex*	freshwater shrimp
Gemeiner Frostspanner, Obstbaumfrostspanner, Kleiner Frostspanner	*Operophtera brumata/ Cheimatobia brumata*	winter moth, small winter moth
Gemeiner Furchenschwimmer	*Acilius sulcatus*	pond beetle, common pond beetle
Gemeiner Grashüpfer	*Chorthippus parallelus*	common meadow grasshopper
Gemeiner Hammerhai, Glatter Hammerhai	*Sphyrna zygaena/ Zygaena malleus*	common hammerhead shark, smooth hammerhead (FAO)
Gemeiner Hechtling	*Aplocheilus panchax*	blue panchax
Gemeiner Kalmar, Roter Gemeiner Kalmar	*Loligo vulgaris*	common squid
Gemeiner Kanalkäfer, Gewöhnlicher Kanalkäfer	*Amara plebeja*	common sun beetle
Gemeiner Kiefernspanner	*Bupalus piniaria*	pine moth, pine looper moth, bordered white beauty
Gemeiner Krake, Gemeiner Octopus, Polyp	*Octopus vulgaris*	common octopus, common Atlantic octopus, common European octopus
Gemeiner Kugelfisch	*Tetraodon cutcutia*	common pufferfish
Gemeiner Leuchtkäfer, Kleiner Leuchtkäfer, Johanniskäfer, Johanniswürmchen, Glühwürmchen	*Lamprohiza splendidula*	small lightning beetle
Gemeiner Meerengel, Engelhai	*Squatina squatina/ Rhina squatina/ Squatina angelus*	angelshark (FAO), angel shark, monkfish
Gemeiner Nagekäfer, Gewöhnlicher Nagekäfer, Holzwurm	*Anobium punctatum*	furniture beetle, woodworm (*larva*)

Gemeiner Octopus, Gemeiner Krake, Polyp	*Octopus vulgaris*	common octopus, common Atlantic octopus, common European octopus
Gemeiner Ohrwurm	*Forficula auricularia*	common earwig, European earwig
Gemeiner Pansenegel	*Paramphistomum cervi*	common stomach fluke, rumen fluke
Gemeiner Regenwurm, Tauwurm	*Lumbricus terrestris*	common earthworm, earthworm; lob worm, dew worm, squirreltail worm, twachel (Br.)
Gemeiner Rückenschwimmer	*Notonecta glauca*	common backswimmer
Gemeiner Sägefisch, Westlicher Sägefisch, Schmalzahn-Sägerochen	*Pristis pectinata*	smalltooth sawfish (FAO), greater sawfish
Gemeiner Samenstecher	*Apion vorax*	bean flower weevil
Gemeiner Scheckenfalter	*Mellicta athalia*	heath fritillary
Gemeiner Schildfuß	*Chaetoderma nitidulum* (Aplacophora)	glistenworm a.o.
Gemeiner Schlammröhrenwurm, Gemeiner Bachröhrenwurm, Bachwurm	*Tubifex tubifex*	river worm
Gemeiner Schuppenwurm	*Harmothoe imbricata*	common fifteen-scaled worm
Gemeiner Seeringelwurm	*Nereis pelagica*	pelagic clam worm
Gemeiner Seestern	*Asterias rubens*	common starfish, common European seastar
Gemeiner Sonnenbarsch	*Lepomis gibbosus*	pumpkin-seed sunfish, pumpkinseed (FAO)
Gemeiner Speckkäfer	*Dermestes lardarius*	larder beetle, common larder beetle, bacon beetle
Gemeiner Stutzkäfer	*Saprinus semistriatus*	common hister beetle
Gemeiner Taumelkäfer	*Gyrinus substriatus*	common whirligig beetle
Gemeiner Tausendfüßer, Luzernentausendfüßer	*Cylindroiulus londinensis/ Cylindroiulus teutonicus*	black millepede, snake millepede
Gemeiner Tintenfisch, Gemeine Tintenschnecke, Gemeine Sepie	*Sepia officinalis*	common cuttlefish
Gemeiner Trichterwurm, Amerikanischer Trichterwurm*	*Sabellaria vulgaris*	sand-builder worm, common feather-duster worm*
Gemeiner Tüpfeltausendfüßer, Gefleckter Doppelfüßer, Getüpfelter Tausendfuß	*Blaniulus guttulatus*	spotted snake millepede, snake millepede
Gemeiner Wasserfloh	*Daphnia pulex*	common water flea
Gemeiner Weberknecht	*Phalangium opilio*	common harvestman
Gemeiner Weichkäfer	*Cantharis fusca*	common cantharid, common soldier beetle
Gemeiner Wimperbock, Kiefernzweigbock	*Pogonocherus fasciculatus*	conifer-wood longhorn beetle*
Gemeiner Zangenschildfuß	*Falcidens crossotus* (Aplacophora)	glistenworm
Gemeiner Zungenwurm	*Linguatula serrata*	common tongue worm

Gemeines Moderholz, Graue Moderholzeule	*Xylena exsoleta*	sword-grass moth
Gemeines Neptunshorn, Gemeine Spindelschnecke	*Fusus antiqua/ Neptunea antiqua*	ancient whelk, ancient neptune, common spindle snail, neptune snail, red whelk, buckie
Gemeines Perlboot, Gemeines Schiffsboot	*Nautilus pompilius*	chambered nautilus, emperor nautilus (FAO)
Gemeißelte Stern-Turbanschnecke*	*Lithopoma caelatum*	carved starsnail
Gemse, Gamswild (*jetzt*: Gämse)	*Rupicapra rupicapra*	chamois (*pronounce*: sha-mee or sham-wa)
Gemüseeule	*Lacanobia oleracea/ Mamestra oleracea*	tomato moth (bright-line brown-eye moth)
Genabelte Maskenschnecke	*Causa holosericum/ Isognomostoma holosericum*	
Genabelte Puppenschnecke	*Lauria cylindracea*	chrysalis snail
Genabelte Strauchschnecke	*Bradybaena fruticum*	bush snail
Genarbte Archenmuschel, Westpazifische Archenmuschel	*Anadara granosa*	granular ark (granular ark shell)
Geneigte Thracia	*Thracia devexa*	sloping thracia
General-Kegelschnecke	*Conus generalis*	general cone
Genés Höhlensalamander, Sardinischer Schleuderzungensalamander	*Hydromantes genei/ Speleomantes genei*	Sardinian cave salamander, brown cave salamander
Genetzte Ackerschnecke, Gartenschnecke	*Deroceras reticulatum/ Agriolimax reticulatus*	netted slug, gray fieldslug (U.S.), grey field slug (Br.), milky slug
Genetzte Nacktschnecke	*Prophysaon andersoni*	reticulated slug, British garden slug
Genetzter Querzahnmolch	*Ambystoma cingulatum*	reticulated salamander, flatwood salamander
Geoffroy-Klammeraffe	*Ateles geoffroyi*	black-headed spider monkey
Geoffroy-Schlitznase	*Nycteris thebaica*	Egyptian slit-faced bat
Geoffroys Langnasen-Fledermaus	*Anoura* spp.	Geoffroy's long-nosed bats
Geohrte Wendeltreppe	*Epitonium aurita*	eared wentletrap
Geometrische Landschildkröte	*Psammobates geometrica*	geometric tortoise
Georgia-Lachs	*Arripis georgiana*	ruff
Gepäckträgerkrabbe	*Ethusa mascarone americana*	stalkeye sumo crab, stalkeye porter crab
Gepanzerter Kammstern	*Astropecten armatus*	armored seastar, armored comb star
Gepard	*Acinonyx jubatus*	cheetah
Geperlte Schraubenschnecke	*Terebra monilis*	necklace auger
Geperlter Sichelflosser	*Drepane punctata*	spotted sicklefish
Gepfriemter Zwergkalmar	*Alloteuthis subulata*	little squid
Gepunktete Süßlippe, Harlekin-Süßlippe	*Plectorhinchus chaetodonoides*	harlequin sweetlip
Gepunkteter Barramundi	*Scleropages leichardti*	spotted barramundi
Gepunkteter Blutstern	*Henricia oculata*	bloody henry

Gepunkteter Fangschreckenkrebs	*Lysiosquilla maculata*	dotted mantis shrimp
Gepunkteter Kaninchenfisch	*Siganus chrysopilos*	gold-spotted spinefoot
Gepunkteter Korallenwächter	*Cirrhitichthys aprinus*	blotched hawkfish, spotted hawkfish (FAO)
Gepunktetes Flechtenbärchen	*Pelosia muscerda*	dotted footman (moth)
Gepunzte Mitra	*Mitra stictica*	pontifical mitre
Geradflügler	Orthoptera	orthopterans
Geradmund-Schließmundschnecke	*Cochlodina orthostoma*	straightmouth door snail*
Geradsalmler	Citharinidae	citharinids
Geränderte Scheinolive*	*Amalda marginata*	margin ancilla
Geränderter Bandwurm, Geränderter Hundebandwurm	*Taenia hydatigena/ Taenia marginata*	margined dog tapeworm
Geränderter Flugfisch	*Cypselurus cyanopterus/ Cheilopogon cyanopterus*	margined flyingfish
Gerandete Jagdspinne, Gebänderte Listspinne	*Dolomedes fimbriatus*	fimbriate fishing spider*
Gerberbock, Sägebock	*Prionus coriarius*	prionus longhorn beetle (greater British longhorn)
Gerenuk, Giraffengazelle	*Litocranius walleri*	gerenuk
Gerfalke	*Falco rusticolus*	gyrfalcon
Gerillte Atlantische Hirnkoralle	*Colpophyllia natans*	boulder brain coral
Geringelte Hundskopfboa	*Corallus annulatus*	annulated tree boa, boa arboricola
Geringelte Korallenschlange	*Micrurus annellatus*	annellated coral snake
Geringelte Wurmschnecke	*Dendropoma corrodens*	ringed wormsnail
Gerippte Fassschnecke	*Tonna allium*	costate tun
Gerippte Felsschnecke, Gerippte Purpurschnecke, Großes Seekälbchen	*Ocenebra erinacea/ Ceratostoma erinaceum*	European sting winkle, European oyster drill, rough tingle
Gerippte Grasschnecke	*Vallonia costata*	ribbed grass snail
Gerippte Harfenschnecke	*Harpa costata*	imperial harp
Gerippte Haufenschnecke	*Planaxis sulcatus*	ribbed clusterwink, furrowed planaxis
Gerippte Herzmuschel	*Cardium costatum*	costate cockle
Gerippte Hydromeduse	*Aequorea aequorea*	many-ribbed hydromedusa
Gerippte Sattelauster, Rippen-Sattelauster	*Monia patelliformis*	ribbed saddle oyster
Gerippte Schlammschnecke	*Cerithidea costata*	costate hornsnail
Gerippte Tellmuschel	*Angulus fabula/ Fabulina fabula/ Tellina fabula*	bean-like tellin, semi-striated tellin
Gerippte Turbanschnecke	*Moelleria costulata*	ribbed moelleria
Gerippte Westindische Käferschnecke	*Chaetopleura apiculata*	West Indian ribbed chiton
Geronimo-Ringelschleiche	*Anniella geronimensis*	Baja California legless lizard
Gervais-Zweizahnwal	*Mesoplodon europaeus*	Gervais' beaked whale, Gulfstream beaked whale
Geryoniden (Tiefseekrabben u.a.)	Geryonidae	geryonid crabs, deepsea crabs
Gesägte Flachschildkröte	*Homopus signatus*	speckled Cape tortoise

Gesägter Bandwurm, Gesägter Hundebandwurm	*Taenia pisiformis/ Taenia serrata*	serrated dog tapeworm
Gesäumter Schillersalmler	*Tetragonopterus argenteus*	big-eyed characin
Gescheckte Spitzmaus	*Diplomesodon pulchellum*	piebald shrew, Turkestan desert shrew
Geschichtskauri*	*Mauritia histrio*	history cowrie
Geschmückter Leierfisch*, Schmuckleierfisch	*Callionymus pusillus/ Callionymus festivus*	festive dragonet
Geschnäbelte Kreiselwespe	*Bembix rostrata/ Epibembix rostrata*	rostrate bembix wasp*
Geschnäbelte Nussmuschel	*Nuculana minuta*	minute nutclam, beaked nutclam
Geschnäbeltes Spindelhorn, Zierliche Spindelschnecke	*Fusus rostratus/ Neptunea rostratus*	rostrate whelk, rostrate netune
Geschnörkelte Zwergolive*	*Olivella volutella*	volute-shaped olive, volute-shaped olivella
Geschnürter Schwamm*, Schnurschwamm*	*Xestospongia exigua*	string sponge
Geschuppter Kleinhering	*Harengula jaguana*	scaled sardine
Geschwänzter Keulenwurm	*Priapulus caudatus*	tailed priapulid worm
Geschwänztes Pansenwimpertierchen	*Entodinium caudatum*	
Geschwärzter Furchenbauch-Pillenkäfer	*Curimopsis nigrita*	mire pill beetle
Geschweiftstreifiger Kohlerdfloh, Gewelltstreifiger Kohlerdfloh	*Phyllotreta undulata*	small striped flea beetle
Gesellige Birnblattwespe	*Neurotoma saltuum*	social pear sawfly
Gesichtsfliege, Augenfliege	*Musca autumnalis*	face fly
Gespenst-Stachelschnecke	*Chicoreus spectrum*	ghost murex, spectre murex
Gespensterfisch, Schaukelfisch	*Taenionotus triacanthus*	leaf scorpionfish
Gespensterwanzen, Fangwanzen	Phymatidae	ambush bugs
Gespenstfledermaus	*Diclidurus* spp.	ghost bats, white bats
Gespenstfrosch	*Pachymedusa dacnicolor*	Mexican leaf frog
Gespenstfrösche	*Heleophryne* spp.	ghost frogs
Gespenstfrösche	Heleophrynidae	ghost frogs
Gespensheuschrecken & Stabheuschrecken	Phasmida/Phasmatidae	walking sticks, stick-insects
Gespenstkrabben	Dorippidae	sumo crabs, demon-faced crabs
Gespenstkrebse	Caprellidae	skeleton shrimps
Gespinstblattwespen (Kotsackblattwespen)	Pamphiliidae (Lydidae)	webspinning sawflies & leafrolling sawflies
Gespinstmotten	Hyponomeutidae (Yponomeutidae)	ermine moths
Gesprenkelte Käferschnecke*, Gesprenkelte Chiton*	*Ischnochiton erythronotus*	multihued chiton
Gesprenkelte Mitra	*Mitra puncticulata*	punctured mitre, dotted miter (U.S.)
Gesprenkelte Wassertrugnatter	*Enhydris punctata*	spotted water snake

Gesprenkelter Epaulettenhai	*Hemiscyllium trispeculare*	speckled carpetshark (FAO), speckled catshark
Gesprenkelter Kurzkopffrosch	*Breviceps adspersus*	Transvaal short-headed frog
Gesprenkelter Schlangenskink	*Ophiomorus punctatissimus*	Greek legless skink, Greek snake skink
Gestelzter Seehase*	*Aplysia juliana*	walking sea hare
Gestirnte Mondschnecke	*Natica stellata*	starry moonsnail, stellate sand snail
Gestreifte Bernsteinmakrele*	*Seriola zonata*	banded rudderfish
Gestreifte Bodennatter	*Liophis lineatus*	lined tropical snake
Gestreifte Drechselschnecke	*Acteon virgatus*	striped acteon
Gestreifte Felsenkrabbe	*Pachygrapsus crassipes*	striped shore crab
Gestreifte Grasfliege	*Opomyza germinationis*	dusky-winged cereal fly
Gestreifte Guatemala-Vogelspinne	*Aphonopelma seemanni*	Costa Rican zebra tarantula
Gestreifte Haarschnecke, Gestreifte Krautschnecke	*Trichia striolata/ Trichia rufescens*	strawberry snail, reddish snail, ruddy snail
Gestreifte Heideschnecke	*Helicopsis striata*	striated heath snail
Gestreifte Helmschnecke	*Phalium strigatum*	striped bonnet
Gestreifte Hochland-Rennechse	*Cnemidophorus velox*	plateau striped whiptail
Gestreifte Hörnchenschnecke	*Polycera quadrilineata*	four-striped polycera
Gestreifte Kegelschnecke	*Conus striatus*	striated cone
Gestreifte Kreiselschnecke, Gefurchte Kreiselschnecke	*Jujubinus striatus*	grooved topsnail
Gestreifte Kurzkopf-Weichschildkröte	*Chitra chitra*	striped narrow-headed softshell turtle
Gestreifte Meerbarbe, Streifenbarbe	*Mullus surmuletus*	striped red mullet
Gestreifte Mondschnecke	*Tanea lineata*	lined moonsnail
Gestreifte Mützenschnecke	*Crucibulum striatum*	striate cup-and-saucer limpet
Gestreifte Nachtotter	*Causus bilineatus*	lined night adder
Gestreifte Nadelschnecke	*Acicula lineata*	striped point snail
Gestreifte Notothenia	*Pagothenia hansoni*	striped rockcod (FAO), green rockcod
Gestreifte Orfe	*Notropis chrysocephalus*	striped shiner
Gestreifte Palmenhörnchen	*Funambulus* spp.	Asiatic striped palm squirrels
Gestreifte Peitschennatter	*Masticophis lateralis*	striped whipsnake, striped racer
Gestreifte Perlmuschel	*Pinctada radiata*	rayed pearl-oyster
Gestreifte Puppenschnecke	*Pupilla sterrii*	striped moss snail
Gestreifte Radnetzspinnen*	*Singa* spp.	striped orbweavers
Gestreifte Randschnecke	*Cryptospira strigata*	striped marginella, striped marginsnail (striped margin shell)
Gestreifte Sandrennnatter	*Psammophis sibilans*	striped sand snake
Gestreifte Sandschnecke	*Bullia tranquebarica*	lined bullia, Belanger's bullia
Gestreifte Spanische Makrele	*Scomberomorus lineolatus*	streaked Spanish mackerel, streaked seerfish (FAO)
Gestreifte Südschrecke	*Pachytrachis striolatus*	striated bushcricket
Gestreifte Süßlippe	*Pomadasys striatum*	striped grunter

Gestreifte Tellmuschel (Jungfräuliche Tellmuschel)	*Tellina virgata/ Tellina pulchella*	virgate tellin, striped tellin
Gestreifte Unechte Napfschnecke	*Siphonaria pectinata*	striped false limpet
Gestreifte Wassernatter	*Regina rigida*	glossy crayfish snake
Gestreifte Windelschnecke, Gebänderte Windelschnecke	*Vertigo substriata*	striated whorl snail
Gestreifte Zartschrecke	*Leptophyes albovittata*	striped bushcricket
Gestreifter Anglerfisch, Krötenfisch	*Antennarius hispidus*	anglerfish, fishing frog, shaggy angler (FAO)
Gestreifter Blasenkäfer*	*Epicauta vittata*	striped blister beetle
Gestreifter Blattrandkäfer	*Sitona lineata*	pea leaf weevil
Gestreifter Blindskink	*Acantophiops lineatus*	woodbush legless skink
Gestreifter Drückerfisch, Grüner Drückerfisch	*Balistapus undulatus*	undulate triggerfish, orange-lined triggerfish (FAO)
Gestreifter Falterfisch, Schwarzbinden-Falterfisch	*Chaetodon striatus*	banded butterflyfish
Gestreifter Felseneinsiedler	*Pagurus anachoretus*	striped hermit crab*
Gestreifter Gespenstkrebs (Widderkrebs)	*Caprella linearis*	linear skeleton shrimp
Gestreifter Igelfisch	*Cyclichthys schoepfi*	striped burrfish
Gestreifter Kardinalfisch	*Cheilodipterus lineatus lineatus*	tiger cardinalfish
Gestreifter Katzenhai	*Poroderma africanum/ Squalus africanus*	striped catshark
Gestreifter Kiefernrüssler	*Pissodes pini*	banded pine weevil
Gestreifter Knurrhahn	*Trigloporus lastoviza*	streaked gurnard
Gestreifter Korallenwels	*Plotosus lineatus*	striped catfish, striped eel-catfish
Gestreifter Leierfisch, Gemeiner Leierfisch, Gewöhnlicher Leierfisch, Europäischer Leierfisch	*Callionymus lyra*	common dragonet
Gestreifter Marlin, Gestreifter Speerfisch	*Tetrapturus audax*	striped marlin
Gestreifter Ohrwurm	*Labidura bidens*	striped earwig
Gestreifter Plateauleguan	*Sceloporus virgatus*	striped plateau lizard
Gestreifter Reisstengelbohrer	*Chilo suppressalis/ Chilo simplex*	Asiatic rice borer, striped riceborer
Gestreifter Schlangenkopffisch	*Channa striatus*	striped snakehead
Gestreifter Schleimfisch	*Parablennius gattorugine*	tompot blenny
Gestreifter Thun, Echter Bonito	*Katsuwonus pelamis/ Euthynnus pelamis*	skipjack tuna (FAO), bonito, stripe-bellied bonito
Gestreifter Umberfisch	*Cynoscion striatus*	striped weakfish
Gestreifter Wassermolch	*Notophthalmus perstriatus*	striped newt
Gestreifter Zwergkaiser	*Centropyge bispinosus*	two-spined angelfish (FAO), dusky angelfish
Gestreiftes Petermännchen, Strahlenpetermännchen	*Trachinus radiatus*	streaked weever
Gestreiftflügel-Flugfisch	*Cypselurus exsiliens/ Cheilopogon exsiliens*	bandwing flyingfish

Gestrichelte Buckelschnecke	*Monodonta lineata*	lined topsnail (lined top shell, toothed top shell, thick top shell)
Gestrichelte Walzenschnecke	*Harpulina lapponica*	brown-lined volute
Gestutzte Dreiecksmuschel, Sägezahnmuschel	*Donax trunculus*	truncate donax, truncated wedge clam
Gestutzte Klaffmuschel, Gestutzte Sandklaffmuschel	*Mya truncata*	blunt gaper clam, truncate softshell (clam)
Gestutzte Stielmuschel	*Mergelia truncata* (Brachiopoda)	
Getreide-Glattkäfer	*Phalacrus coruscus*	cereal flower beetle*
Getreideblasenfuß, Gewitterfliege	*Limothrips cerealium*	corn thrips, grain thrips, black wheat thrips, cereal thrips
Getreideblattlaus	*Schizaphis graminum*	greenbug
Getreideblattminierfliege	*Phytomyza nigra*	cereal leaf miner (fly)
Getreidebock	*Calamobius filum/ Calamobius gracilis*	grain borer
Getreideerdfloh	*Phyllotreta vittula*	barley flea beetle
Getreideeule, Getreidesaateule, Roggeneule (Getreidewurzeleule)	*Apamea secalis/ Mesapamea secalis*	common rustic
Getreidehähnchen	*Oulema* spp.	cereal leaf beetle
Getreidehähnchen, Blatthähnchen	*Lema melanopus/ Oulema melanopus*	cereal leaf beetle (oat leaf beetle, barley leaf beetle)
Getreidehalmwespe	*Cephus pygmeus/ Cephus pygmaeus*	wheat stem sawfly, European wheat stem sawfly (U.S.)
Getreidekapuziner	*Rhizopertha dominica/ Rhyzopertha dominica*	lesser grain borer
Getreidelaufkäfer	*Zabrus tenebroides*	corn ground beetle
Getreidemotte, Französische Kornmotte (Weißer Kornwurm)	*Sitotroga cerealella*	Angoumois grain moth
Getreideplattkäfer	*Oryzaephilus surinamensis*	saw-toothed grain beetle
Getreiderostmilbe	*Abacarus hystrix*	grain rust mite, cereal rust mite
Getreidesaftkäfer	*Carpophilus dimidiatus*	corn sap beetle
Getreidespitzwanze, Spitzling	*Aelia acuminata*	bishop's mitre, bishop's mitre bug
Getreidestaublaus> Gemeine Getreidestaublaus	*Lachesilla pedicularia*	cosmopolitan grain psocid
Getreidewanze> Amerikanische Getreidewanze	*Blissus leucopterus*	chinch bug, American chinch bug
Getreidewanze> Europäische Getreidewanze, Schmalwanze	*Ischnodemus sabuleti*	European chinch bug
Getreidewickler	*Cnephasia pumicana*	cereal leaf tier (moth)
Getreidewurzeleule, Getreideeule, Getreidesaateule, Roggeneule	*Mesapamea secalis/ Apamea secalis*	common rustic moth
Getreidewurzeleule, Wurzelfresser	*Apamea monoglypha*	dark arches

Getreidezystenälchen, Haferzystenälchen, Hafernematode	*Heterodera avenae/ Heterodera major*	oat cyst nematode, cereal cyst nematode, cereal root nematode
Getüpfelter Gabelwels	*Ictalurus punctatus*	channel catfish
Getupfte Randschnecke	*Persicula persicula*	spotted marginella, spotted marginsnail (spotted margin shell)
Getupfter Seehase, Kleingepunkteter Seehase	*Aplysia rosea/ Aplysia punctata*	small rosy sea hare*, common sea hare
Gewächshaus-Blasenfuß, Schwarze Fliege	*Heliothrips haemorrhoidalis*	glasshouse thrips, greenhouse thrips
Gewächshaus-Dickmaulrüssler, Gefurchter Lappenrüssler	*Otiorhynchus sulcatus*	vine weevil, black vine weevil, European vine weevil
Gewächshaus-Dolchschnecke, Kleine Gewächshaus-Glanzschnecke	*Zonitoides arboreus*	quick gloss (snail), quick glass snail, bark snail
Gewächshaus-Heuschrecke, Gewächshausschrecke	*Tachycines asynamorus*	greenhouse camel cricket, glasshouse camel-cricket (Br.), greenhouse stone cricket (U.S.)
Gewächshaus-Röhrenschildlaus	*Orthezia insignis*	glasshouse orthezia
Gewächshaus-Schmierlaus, Zitronenschmierlaus	*Planococcus citri*	citrus mealybug
Gewächshaus-Zwergfüßer, Zwergskolopender	*Scutigerella immaculata* (Symphyla)	glasshouse symphylid, garden symphylan, garden centipede
Gewächshausfrosch	*Eleutherodactylus planirostris*	greenhouse frog
Gewächshausschabe, Surinamschabe	*Pycnoscelis surinamensis*	Surinam cockroach
Gewächshausschnegel, Iberische Egelschnecke	*Limax valentianus/ Lehmannia valentiana*	glasshouse slug, Iberian slug
Gewächshausspinne, Amerikanische Hausspinne	*Achaearanea tepidariorum*	house spider, American common house spider, common American house spider, domestic spider
Gewand-Triton*	*Cymatium vestitum*	garment triton
Gewebte Teppichmuschel	*Venerupis texturatus*	textured carpetclam, textured venus
Geweih-Moostierchen	*Cellepora ramulosa*	rameous bryozoan
Geweih-Rosenblattwespe*	*Cladius pectinicornis*	antler sawfly
Geweihschwamm	*Haliclona oculata*	mermaid's glove, deadman's finger, finger sponge, "eyed sponge"
Geweihschwämme u.a.	*Haliclona* spp.	tube sponges
Geweihschwämme u.a.	*Axinella* spp.	antler sponges
Gewellte Kaktuskoralle	*Isophyllia sinuosa*	sinuous cactus coral
Gewellte Mondschnecke	*Globularia fluctuata/ Cernina fluctuata*	wavy moonsnail
Gewellte Samtmuschel, Atlantik-Samtmuschel	*Glycymeris undata*	wavy bittersweet, Atlantic bittersweet, lined bittersweet
Gewellte Samtschnecke	*Velutina undata*	wavy velutina, wavy lamellaria
Gewellte Sichelmuschel*	*Thyasira flexuosa*	wavy hatchetclam (wavy hatchet-shell)

Gewellte Spindelschnecke	*Fusinus undatus*	wavy spindle
Gewellte Stern-Turbanschnecke	*Lithopoma undosum*	wavy starsnail, wavy turbansnail
Gewellter Panzerwels	*Corydoras undulatus*	wavy catfish
Gewelltes Elefantenohr	*Rhodactis mussoides*	elephant ear (metallic mushroom anemone, leaf mushroom anemone)
Gewitterfliege, Getreideblasenfuß	*Limothrips cerealium*	corn thrips, grain thrips, black wheat thrips, cereal thrips
Gewöhnliche Feilenmuschel	*Lima lima*	spiny fileclam, spiny lima (frilled file shell)
Gewöhnliche Froschkopf-Schildkröte, Gewöhnliche Krötenkopf-Schildkröte	*Phrynops nasutus/ Batrachemys nasuta*	common South American toad-headed turtle, toad-headed side-necked turtle
Gewöhnliche Gebirgsschrecke	*Podisma pedestris*	brown mountain grasshopper
Gewöhnliche Hakennatter	*Heterodon platyrhinos*	eastern hognose snake
Gewöhnliche Hirnkoralle, Symmetrische Hirnkoralle	*Diploria strigosa*	symmetrical brain coral
Gewöhnliche Kalifornische Käferschnecke	*Lepidozona regularis*	regular chiton
Gewöhnliche Langnasenchimäre, Gemeine Langnasenchimäre	*Harriotta raleighana*	long-nosed chimaera
Gewöhnliche Moschusschildkröte	*Sternotherus odoratus/ Kinosternon odoratum*	common musk turtle, stinkpot
Gewöhnliche Orfe, Gemeine Orfe	*Notropis cornutus*	common shiner
Gewöhnliche Sattelmuschel	*Anomia simplex*	common saddle oyster (common jingle shell)
Gewöhnliche Schlammschnecke, Gemeine Schlammschnecke, Wandernde Schlammschnecke	*Lymnaea peregra/ Radix peregra*	wandering pondsnail, common pondsnail
Gewöhnliche Steinfliegen	Perlidae	common stoneflies
Gewöhnliche Strauchschrecke	*Pholidoptera griseoaptera*	dark bushcricket
Gewöhnlicher Dammläufer, Pechschwarzer Dammläufer	*Nebria brevicollis*	common black ground beetle
Gewöhnlicher Fransenfinger, Europäischer Fransenfinger	*Acanthodactylus erythrurus*	spiny-footed lizard
Gewöhnlicher Froschfisch	*Halophryne gangene/ Cottus grunniens*	common toadfish
Gewöhnlicher Gürtelschweif	*Cordylus cordylus*	common girdled lizard, Lord Derby lizard
Gewöhnlicher Igelfisch, Gemeiner Igelfisch, Kosmopolit-Igelfisch, Gepunkteter Igelfisch	*Diodon hystrix*	common porcupinefish, spotted porcupinefish, spot-fin porcupinefish (FAO)
Gewöhnlicher Polydora-Wurm	*Polydora ciliata*	common polydora worm
Gewöhnlicher Sanddollar	*Clypeaster humilis*	common sand dollar
Gewöhnlicher Wasserdrache	*Physignatus lesueurii*	eastern water dragon

*G*ewundene Dörnchenkoralle

Gewundene Dörnchenkoralle, Gewundene Drahtkoralle	*Cirripathes anguina/ Cirrhipathes anguina*	whip coral a.o.
Gewundene Salatkoralle, Folienkoralle, Kraterkoralle	*Turbinaria mesenterina*	folded lettuce coral, scroll coral
Gewundene Treppenschnecke	*Pugilina cochlidium*	spiral melongena, winding stair shell
Gewürfelte Kreiselschnecke*, Würfelturban	*Monodonta turbinata*	checkered topsnail, one-toothed turban
Gezackter Bandfisch*	*Zu cristatus*	scalloped ribbonfish
Gezähnelte Pelomeduse, Gezackte Pelomeduse	*Pelusios sinatus*	serrated turtle, East African serrated mud turtle
Gezähnte Auster	*Ostrea denticulata*	denticulate rock oyster
Gezähnte Mangroven-Schwimmkrabbe	*Scylla serrata*	serrated mud swimming crab, serrated mangrove swimming crab, mud crab
Gezeichnete Sandmuschel	*Gari fucata*	painted sunsetclam
Gezeichneter Feilenfisch	*Aluterus scriptus*	long-tailed filefish, scrawled filefish (FAO), scribbled filefish
Gezeiten-Ährenfisch, Mondährenfisch	*Menidia menidia*	Atlantic silverside
Gezüngelte Naide	*Stylaria lacustris*	
Gichtwespen	*Gasteruption* spp.	gasteruptid wasps
Gichtwespen, Schmalbauchwespen	Gasteruptiidae/ Gasteruptionidae	gasteruptiids, gasteruptionid wasps
Giebel	*Carassius auratus gibelio*	gibel carp, Prussian carp (FAO)
Gienmuscheln, Juwelendosen	Chamidae	jewel boxes, jewelboxes
Gießkannenschwamm	*Euplectella aspergillum*	Venus' flower basket
Giftechsen, Krustenechsen	Helodermatidae	beaded lizards
Giftfrösche, Blattsteigerfrösche	*Phyllobates* spp.	golden poison frogs, poison-arrow frog
Giftlaubfrosch	*Phrynohyas venulosa*	veined treefrog
Giftnattern	Elapidae	front-fanged snakes
Gigantischer Tonnenschwamm	*Xestospongia muta*	gigantic barrel sponge
Gila-Krustenechse, Gila-Tier	*Heloderma suspectum*	gila monster
Gilbdrossel	*Turdus grayi*	clay-coloured thrush
Gilchrists Languste	*Panulirus gilchristi*	South Coast spiny lobster
Gillens Waran, Gillens Zwergwaran	*Varanus gilleni*	pygmy mulga monitor
Gimp	*Blepharidopterus angulatus*	black-kneed capsid
Gimpel	*Pyrrhula pyrrhula*	bullfinch, Eurasian bullfinch
Ginsburg-Grundel	*Gobiosoma ginsburgi*	seabord goby
Ginster-Bläuling	*Lycaeides idas*	Idas blue
Ginsterheiden- Silberstreifenspanner	*Chesias rufata*	broom-tip
Ginsterkatzen	*Genetta* spp.	genets
Gipskraut-Kapseleule	*Hadena irregularis*	viper's bugloss
Giraffe	*Giraffa camelopardalis*	giraffe
Giraffengazelle, Gerenuk	*Litocranius walleri*	gerenuk
Giraffenkauri	*Cypraea camelopardalis*	giraffe cowrie

Girlitz	*Serinus serinus*	serin
Girondische Glattnatter, Gironde-Natter, Girondische Schlingnatter	*Coronella girondica*	southern smooth snake, Bordeaux snake
Gitter-Haarschnecke*	*Trichotropis cancellata*	cancellate hairysnail
Gitter-Kegelschnecke	*Conus cancellatus*	cancellate cone
Gitter-Purpurschnecke	*Drupa clathrata*	clathrate drupe
Gitterchiton	*Ischnochiton contractus*	contracted chiton
Gitterhelm	*Morum cancellatum*	cancellate morum
Gitterlaufkäfer, Körnerwarze, Großer Feldlaufkäfer	*Carabus cancellatus*	cancellate ground beetle
Gitterschnapper	*Lutjanus decussatus*	checkered snapper
Gitterschnecken	Cancellariidae	nutmeg snails, nutmegs (nutmeg shells)
Gitterschwanzleguan	*Callisaurus draconoides*	zebratail lizard, zebra-tailed lizard
Gitterspanner	*Semiothisa clathrata*	latticed heath
Gitterstreifige Schließmundschnecke	*Clausilia dubia*	craven door snail
Gladiatorkrabbe*	*Acanthocarpus alexandri*	gladiator box crab
Glänzende Achatschnecke, Glänzende Glattschnecke	*Cionella nitens/ Cochlicopa nitens*	robust slippery moss snail, robust pillar snail (U.S.)
Glänzende Binsenjungfer	*Lestes dryas*	scarce emerald damselfly
Glänzende Dolchschnecke, Moder-Glanzschnecke	*Zonitoides nitidus*	black gloss (snail), shiny glass snail
Glänzende Erbsenmuschel	*Pisidium nitidum*	shining pea mussel, shiny peaclam
Glänzende Mondschnecke	*Lunatia poliana/Euspira poliana*	shiny moonsnail (Poli's necklace shell)
Glänzende Mondschnecke> Pazifische Glänzende Mondschnecke	*Neverita politiana*	polished moonsnail (U.S.)
Glänzende Nussmuschel	*Nucula turgida*	shiny nutclam*
Glänzende Pfeffermuschel	*Abra nitida*	shiny abra (glossy furrow-shell)
Glänzende Smaragdlibelle	*Somatochlora metallica*	brillant emerald
Glänzende Tellerschnecke	*Segmentina nitida*	shining ramshorn snail, shiny ram's horn
Glänzender Getreideschimmelkäfer	*Alphitobius diaperinus*	lesser mealworm
Glänzender Knurrhahn, Rauer Knurrhahn	*Aspitrigla obscura*	longfin gurnard (FAO), shining gurnard
Glänzender Zwergbuntbarsch, Gestreifter Zwergbuntbarsch	*Nannacara anomala*	golden-eyed dwarf cichlid, goldeneye cichlid (FAO)
Glänzendschwarze Holzameise, Schwarze Holzameise, Kartonnestameise	*Lasius fuliginosus*	jet ant, shining jet black ant
Glanzente	*Sarkidiornis melanotos*	comb duck
Glanzfliege	*Phormia regina*	black blow fly
Glanzkäfer	Nitidulidae	sap beetles, sap-feeding beetles
Glanzkegel*	*Conus rutilus*	burnished cone
Glanzkreuzspinne, Glanzspinne	*Singa hamata*	shiny orbweaver*
Glanzschnecken	Zonitidae	glass snails

Glanzschnellkäfer

Glanzschnellkäfer	*Selatosomus aeneus*	shiny click beetle*
Glanzsittich	*Neophema splendida*	scarlet-chested parakeet, splendid parakeet, splendid parrot
Glanzspinne, Glanzkreuzspinne	*Singa hamata*	shiny orbweaver*
Glanzspitznatter, Grüne Erzspitznatter	*Oxybelis fulgidus*	green vine snake
Glanztetra, Goldtetra, Messingsalmler, Kirschfleckensalmler	*Hemigrammus rodwayi*	golden tetra
Glanzvögel	Galbulidae	jacamars
Glas-Lappenqualle	*Bolinopsis infundibulum*	common northern comb jelly
Glasauge> Großes Glasauge	*Argentina silus*	greater argentine
Glasauge> Kleines Glasauge	*Argentina sphyraena*	argentine (FAO), lesser argentine
Glasaugenbarsch	*Stizostedion vitreum*	walleye
Glasbärbling, Glasrasbora	*Rasbora trilineata*	scissors-tail rasbora
Glasbarsche	Ambassidae	glassies
Glasboot	*Carinaria cristata*	glassy nautilus
Gläserne Flügelmuschel	*Pteria vitrea*	glassy wing-oyster
Glasfisch	*Parapriacanthus ransonneti*	slender sweeper
Glasflügelwanzen	Rhopalidae	scentless plant bugs, rhopalid bugs
Glasflügelzikade (Wiesenzirpe)	*Javesella pellucida/ Liburnia pellucida*	cereal leafhopper
Glasflügler (Glasschwärmer)	Sesiidae/ Aegeriidae	clearwing moths, clear-winged moths
Glasfrösche	Centrolenidae	leaf frogs, glass frogs, centrolenids
Glasfrösche	*Hyalinobatrachium* spp.	glass frogs
Glasgarnelen*	Pasiphaeidae	glass shrimps
Glasgrundel u.a., Klargrundel*	*Gobiopterus chuno*	glas goby, translucent goby, clear goby
Glasgrundel, Glasküling	*Aphia minuta*	transparent goby
Glasgrundeln	Aphyonidae	aphyonids
Glaskärpflinge	Horaichthyidae	horaichthyids
Glaskeulen-Seescheiden, Keulensynascidien	*Clavelina* spp.	light-bulb tunicates, light-bulb sea-squirts
Glaskrautfalter, Gelber C-Falter	*Polygonia egea/ Comma egea/ Nymphalis egea*	southern comma
Glasküling, Glasgrundel	*Aphia minuta*	transparent goby
Glasnasenrochen*	*Raja eglanteria*	clearnose skate
Glasperlenanemone, Glasperlen-Anemone	*Heteractis aurora*	beaded sea anemone, mat anemone, aurora anemone, button anemone
Glasröhrenwurm*	*Spiochaetopterus costarum*	glassy tubeworm
Glasrose, Siebanemone	*Aiptasia mutabilis*	trumpet anemone, glassrose anemone
Glasrotflosser	*Prionobrama filigera*	glass bloodfin
Glasschleichen & Panzerschleichen	*Ophisaurus* spp.	glass lizards

Glasschnecken	Vitrinidae	glass snails
Glasschwämme, Hexactinelliden	Hexactinellida	glass sponges
Glaswelse> Eigentliche Glaswelse	Schilbeidae	schilbeid catfishes
Glattbauchspinnen, Plattbauchspinnen	Drassodidae/Gnaphosidae	ground spiders
Glattbutt	*Scophthalmus rhombus*	brill
Glattdelphin> Nördlicher Glattdelphin	*Lissodelphis borealis*	Pacific right whale dolphin, northern right-whale dolphin
Glattdelphin> Südlicher Glattdelphin	*Lissodelphis peronii*	southern right-whale dolphin, Delfin liso
Glattdick	*Acipenser nudiventris*	ship sturgeon
Glatte Artmuschel, Glatte Artemis	*Dosinia lupinus*	smooth dosinia, smooth artemis
Glatte Blumenkoralle	*Eusmilia fastigiata*	smooth flower coral
Glatte Entenmuschel, Gemeine Entenmuschel	*Lepas anatifera*	common goose barnacle
Glatte Ententrogmuschel	*Anatina anatina*	smooth duckclam
Glatte Erbsenmuschel	*Pisidium hibernicum*	smooth pea mussel
Glatte Glanzschnecke	*Oxychilus glaber*	smooth glass snail
Glatte Grasschnecke	*Vallonia pulchella*	smooth grass snail, beautiful grass snail
Glatte Kammmuschel	*Proteopecten glaber*	bald scallop, smooth scallop
Glatte Nixenschnecke	*Nerita polita*	polished nerite
Glatte Panama-Käferschnecke	*Chiton articulatus*	smooth Panama chiton
Glatte Perlkreiselschnecke	*Margarites helicinus*	spiral margarite, smooth margarite, pearly topsnail (pearly top shell)
Glatte Randschnecke	*Marginella glabella*	smooth marginella, smooth marginsnail (smooth margin shell)
Glatte Reusenschnecke	*Nassarius insculptus*	smooth western nassa
Glatte Samtschnecke*	*Velutina velutina*	smooth velutina, smooth lamellaria, velvet snail (velvet shell)
Glatte Schließmundschnecke	*Cochlodina laminata*	plaited door snail
Glatte Schwarzkoralle*	*Leiopathes glaberrima*	smooth black coral
Glatte Sternkoralle*	*Solenastrea bournoni*	smooth star coral
Glatte Stutzschnecke	*Truncatella subcylindrica*	looping snail
Glatte Tellmuschel	*Tellina laevigata*	smooth tellin
Glatte Venusmuschel, Braune Venusmuschel	*Callista chione/ Meretrix chione*	brown callista, brown venus
Glattechsen, Skinke	Scincidae	skinks
Glatter Grenadier, Glatte Panzerratte	*Nezumia aequalis*	smooth rattail
Glatter Hammerhai, Gemeiner Hammerhai	*Sphyrna zygaena/ Zygaena malleus*	common hammerhead shark, smooth hammerhead (FAO)
Glatter Helmleguan	*Corytophanes cristatus*	smooth helmeted iguana
Glatter Kalkröhrenwurm, Glattwandiger Kalkröhrenwurm	*Protula tubularia*	smooth tubeworm
Glatter Krallenfrosch	*Xenopus laevis*	African clawed frog

Glatter Schmetterlingsrochen	*Gymnura micrura*	smooth butterfly ray
Glatter Schwarzer Dornhai	*Etmopterus pusillus*	smooth lanternshark
Glattes Posthörnchen, Glänzende Tellerschnecke	*Gyraulus laevis*	shiny ramshorn, smooth ramshorn
Glattferser	Litopterna	litopternas
Glattflunder	*Liopsetta putnami*	smooth flounder
Glatthaie, Hundshaie	Triakidae	gullysharks, houndsharks, hound sharks
Glatthäutige Grasnatter	*Opheodrys vernalis*	smooth green snake
Glatthorn-Garnele, Kroyers Geißelgarnele	*Xiphopenaeus kroyeri*	seabob, Atlantic sea bob
Glattkäfer	Phalacridae	shining flower beetles
Glattkiemen-Seescheiden	Phlebobranchia	phlebobranchs (smooth-gill sea-squirts)
Glattkopf, Bairds Glattkopffisch	*Alepocephalus bairdii*	Baird's smooth-head
Glattkopffische, Schwarzköpfe	Alepocephalidae	slickheads
Glattkopfleguan	*Leiocephalus schreibersii*	red-sided curlytail lizard
Glattkopfleguane	*Leiocephalus* spp.	curly-tailed lizards, curlytail lizards, curly-tails
Glattkugel-Rosengallwespe*	*Diplolepis eglanteriae*	rose smooth pea-gall cynipid wasp (>smooth pea gall)
Glattläuse*	Linognathidae	smooth sucking lice
Glattmaul-Kreuzwels	*Arius heudeloti*	smoothmouth sea catfish
Glattnasen-Fledermäuse	Vespertilionidae	vespertilionid bats
Glattnasen-Freischwänze	Emballonuridae	sheath-tailed bats, sac-winged bats & ghost bats
Glattnatter, Schlingnatter	*Coronella austriaca*	smooth snake
Glattrand-Gelenkschildkröte	*Kinixys belliana*	Bell's hingeback tortoise
Glattrand-Weichschildkröte, Glattrandige Weichschildkröte	*Trionyx muticus/Apalone mutica*	smooth softshell turtle
Glattrindenkäfer	Cerylonidae	minute bark beetles
Glattrochen	*Raja senta*	smooth skate
Glattroter Schwamm*	*Plocamia karykina*	smooth red sponge
Glattrückige Schlangenhalsschildkröte, Glattrücken-Schildkröte	*Chelodina longicollis*	common snake-necked turtle, common snakeneck, long-necked tortoise
Glattschalige Kantengarnele	*Heterocarpus laevigatus*	smooth nylon shrimp
Glattschalige Kugelkrabbe	*Ebalia intermedia*	smooth nut crab
Glattschalige Sonnenuhrschnecke	*Architectonia laevigata*	smooth sundial (snail)
Glattschaliger Bärenkrebs	*Ibacus novemdentatus*	smooth fan lobster
Glattschaliger Tiefsee-Einsiedler	*Paragrapsus laevis*	smooth shore crab
Glattscheriger Einsiedler, Gelber Einsiedler	*Anapagurus laevis*	yellow hermit crab
Glattschuppige Lanzenotter	*Trimeresurus mucrosquamatus*	Chinese habu
Glattschwanzlanguste	*Panulirus laevicauda*	smoothtail spiny lobster

Glattschwanzmakrele, Goldband-Selar	*Selaroides leptolepis*	smooth-tailed trevally, yellow-stripe trevally, yellowstripe scad (FAO)
Glattwale	Balaenidae	right whales
Gleichfuß-Glattkäfer	*Phalacrus politus*	smut beetle
Gleichringler (Springschwänze)	Isotomidae	isotomid springtails
Gleit-Borstenratte*	*Diplomys labilis*	gliding spiny rats
Gleitaar	*Elanus caeruleus*	black-shouldered kite
Gleitbilche	*Idiurus* spp.	pygmy scaly-tailed flying squirrel
Gleiter	Pempheridae	sweepers
Gleithörnchen> Asiatische Gleithörnchen, Pfeilschwanz-Gleithörnchen	*Hylopetes* spp.	arrow-tailed flying squirrels
Gleithörnchenbeutler	*Petaurus* spp.	lesser gliding possums
Gletscherbüchse*	*Pandora glacialis*	glacial pandora
Gletscherfloh	*Isotoma saltans*	glacier flea
Gliederfüßer, Arthropoden	Arthropoda	arthropods
Gliederspinnen	Liphistiidae	trap-door spiders
Gliedertiere, Articulaten	Articulata	articulates, articulated animals
Glockenmuschel, Fenstermuschel, Fensterscheibenmuschel	*Placuna placenta*	windowpane oyster (window shell, jingle shell)
Glockenpolypen	*Obelia* spp.	thread hydroids, bushy wine-glass hydroids
Glockenwespe, Heide-Töpferwespe	*Eumenes coarctatus*	heath potter wasp, potter wasp
Glotzauge	*Boleophthalmus boddarti*	mud-hopper, Boddart's goggle-eyed goby (FAO)
Glucken, Wollraupenspinner	Lasiocampidae	lackeys & eggars, lappet moths (tent caterpillars)
Gluckente	*Anas formosa*	Baikal teal
Glühkohlenbarbe, Bandbarbe	*Barbus fasciatus/ Puntius eugrammus*	striped barb (FAO), zebra barb, African banded barb
Glühkohlenfisch, Glühkohlen-Anemonenfisch	*Amphiprion ephippium*	saddle anemonefish (FAO), red saddleback anemonefish
Glühlichtsalmler	*Hemigrammus erythrozonus*	glowlight tetra
Glühwürmchen, Johanniswürmchen> Leuchtkäfer	Lampyridae	glowworms, fireflies, lightning "bugs"
Gnitzen, Bartmücken	Ceratopogonidae	biting midges, punkies, no-see-ums
Gnomenkauz	*Glaucidium gnoma*	pygmy owl, Eurasian pygmy owl
Gnomfische	Scombropidae	gnomefishes
Gobelin-Turbanschnecke, Gobelinturban, Katzenaugenschnecke	*Turbo petholatus*	tapestry turban (cat's-eye shell)
Godmans Langnasen-Fledermaus	*Choeroniscus* spp.	Godman's long-nosed bats
Godmans Lanzenotter	*Porthidium godmani/ Bothrops godmani*	Godman's pit viper
Gofferfrosch	*Rana capito*	gopher frog

Gold-Bärenmaki	*Arctocebus aureus*	golden angwantibo
Gold-Königskrabbe	*Lithodes aequispina*	golden king crab
Gold-Laufkäfer, Goldlaufkäfer, Goldschmied, Goldhenne	*Carabus auratus*	golden ground beetle, gilt ground beetle
Gold-Meeräsche, Goldmeeräsche, Goldäsche	*Liza aurata/ Mugil auratus*	golden grey mullet, golden mullet
Gold-Meerbarbe	*Upeneus asymmetricus*	golden-striped goatfish
Gold-Mondschnecke, Orangefarbene Mondschnecke	*Polinices aurantius*	golden moonsnail
Gold-Napfschnecke, Goldnapf	*Nacella deaurata*	golden limpet
Gold-Scherengarnele	*Stenopus scutellatus*	golden coral shrimp
Gold-Seescheide	*Polycarpa aurata*	golden sea-squirt
Gold-Steinbeißer	*Cobitis aurata*	golden loach
Gold-Wespenspinne	*Argiope aurantia*	black-and-yellow argiope, black-and-yellow garden spider, yellow garden spider, writing spider
Gold-Zackenbarsch, Goldener Zackenbarsch, Spitzkopf-Zackenbarsch	*Epinephelus alexandrinus*	golden grouper
Goldafter	*Euproctis chrysorrhoea*	brown-tail moth, brown-tail
Goldammer	*Emberiza citrinella*	yellowhammer
Goldäsche, Gold-Meeräsche, Goldmeeräsche	*Liza aurata/ Mugil auratus*	golden grey mullet (FAO), golden mullet
Goldauge	*Hiodon alosoides*	goldeye
Goldaugen-Baumfrösche*	*Phrynohyas* spp.	golden-eyed treefrogs
Goldaugen-Makrele, Großaugen-Makrele	*Caranx latus*	horse-eye jack (FAO), horse-eye trevally
Goldbandfüsilier	*Pterocaesio chrysozona*	goldband fusilier
Goldbarsch, Rotbarsch (Großer Rotbarsch)	*Sebastes marinus*	redfish, red-fish, Norway haddock, rosefish, ocean perch (FAO)
Goldbauchsittich	*Neophema chrysogaster*	orange-bellied parakeet, orange-bellied parrot
Goldbaumsteiger	*Dendrobates auratus*	green & black poison-arrow frog, green & black poison frog
Goldblattkegel*	*Conus auricomus*	gold leafed cone
Goldbrassen	*Sparus auratus*	gilthead, gilthead seabream (FAO)
Goldbraue	*Tiaris olivacea*	yellow-faced grassquit
Goldbürzel-Bartvogel	*Pogoniulus bilineatus*	yellow-rumped tinkerbird
Goldbutt, Scholle	*Pleuronectes platessa*	plaice, European plaice (FAO)
Goldene Acht, Gemeiner Heufalter, Gelber Heufalter, Weißklee-Gelbling	*Colias hyale*	pale clouded yellow
Goldene Hufeisennase	*Rhinonicteris aurantius*	golden horseshoe bat
Goldene Kauri, Orangenkauri	*Cypraea auriantium*	golden cowrie
Goldene Königsmakrele	*Gnathanodon speciosus/ Caranx speciosus*	golden trevally

Goldene Pferdeschnecke	*Pleuroploca aurantiaca*	golden horseconch
Goldene Scheinolive*	*Ancilla glabrata*	golden ancilla
Goldene Seidenspinne	*Nephila clavipes*	golden-silk spider, golden-silk orbweaver
Goldene Stachelmakrele	*Alepes djedaba*	shrimp scad (FAO), golden scad
Goldene Teppichmuschel	*Venerupis aurea/ Tapes aurea/ Paphia aurea*	golden carpetclam, golden venus (golden carpet shell)
Goldene Walzenschnecke	*Iredalina mirabilis*	golden volute
Goldener Bambus-Halbmaki	*Hapalemur aureus*	golden bamboo lemur
Goldener Kurznasenbeutler	*Isoodon auratus*	golden bandicoot
Goldener Scheinschnapper	*Nemipterus virgatus*	golden thread-fin bream
Goldener Seefächer, Königsgliederkoralle	*Isis hippuris*	golden sea fan
Goldenes Rüsselhündchen	*Rhynchocyon chrysopygus*	golden-rumped elephant shrew
Goldeule	*Xanthia aurago*	golden sallow (moth)
Goldeulen	Plusiinae	gems, plusiine moths
Goldfahnenbarsche	Callanthiidae	goldies
Goldfarbige Laufspinne	*Philodromus aureolus*	
Goldfasan	*Chrysolophus pictus*	golden pheasant
Goldfisch, Goldkarausche	*Carassius auratus*	goldfish (FAO), common carp
Goldfleckbarbe	*Puntius terio/Barbus terio*	onespot barb
Goldfleckeule	*Diachrysia chryson*	scarce burnished brass moth
Goldfliegen	*Lucilia* spp.	greenbottles
Goldflossenbarbe	*Puntius sachsii*	goldfinned barb (FAO), golden barb
Goldfröschchen	*Mantella aurantiaca*	golden frog
Goldfuß-Flughörnchen*	*Trogopterus xanthipes*	complex-toothed flying squirrel
Goldgarnele	*Plesionika martia*	golden shrimp
Goldgelbe Messingbarbe	*Barbus schuberti*	golden barb
Goldgelbes Löwenäffchen, Goldenes Löwenäffchen	*Leontopithecus rosalia/ Leontideus rosalia*	lion tamarin, golden lion tamarin
Goldgestreiftes Backenhörnchen, Goldmantelziesel	*Spermophilus lateralis*	golden-mantled ground squirrel
Goldgrundel	*Gobius auratus*	golden goby
Goldgrüner Blattnager	*Phyllobius argentatus*	silver-green leaf weevil
Goldhähnchen	Regulidae	kinglets
Goldhähnchen-Laubsänger	*Phylloscopus proregulus*	Pallas's warbler
Goldhamster> Syrischer Goldhamster	*Mesocricetus auratus*	golden hamster
Goldhaubensittich	*Aratinga auricapilla*	golden-capped parakeet, golden-capped conure
Goldkäfer, Rosenkäfer	*Cetonia aurata*	rose chafer
Goldkamm-Sifaka, Tattersall Sifaka	*Propithecus tattersalli*	golden-crowned sifaka, Tattersall's sifaka
Goldkarausche, Goldfisch	*Carassius auratus*	goldfish (FAO), common carp
Goldkegel, Goldfleckkegel	*Conus aurantius*	golden cone
Goldkehlpitta	*Pitta gurneyi*	Gurney's pitta, jewel thrush
Goldkinnsittich, Tovisittich	*Brotogeris jugularis*	orange-chinned parakeet
Goldknie-Vogelspinne	*Brachypelma auratum*	Mexican flameknee tarantula

Goldkopf-Löwenäffchen	*Leontopithecus chrysomelas*	golden-headed lion tamarin, gold and black lion tamarin
Goldköpfchen	*Auriparus flaviceps*	verdin
Goldkröte	*Bufo periglenes*	golden toad
Goldkuckuck	*Chrysococcyx caprius*	dudric cuckoo
Goldlachs, Japanisches Glasauge	*Glossanodon semifasciatus/ Argentina semifasciata*	Pacific argentine, deepsea smelt (FAO)
Goldlachse	Argentinidae	argentines, herring smelts
Goldlangur	*Presbytis geei*	golden leaf monkey
Goldlaubfrosch	*Litoria aurea/Hyla aurea*	green & golden bell frog
Goldmaid	*Crenilabrus melops/ Symphodus melops*	corkwing wrasse
Goldmakrele> Große Goldmakrele, Gemeine Goldmakrele	*Coryphaena hippurus*	dolphinfish, common dolphinfish (FAO), dorado, mahi-mahi
Goldmakrelen	Coryphaenidae	dolphinfishes
Goldmantelziesel, Goldgestreiftes Backenhörnchen	*Spermophilus lateralis*	golden-mantled ground squirrel
Goldmaul-Ziersalmler	*Nannostomus harrisoni*	Harrison's pencilfish
Goldmaus	*Ochrotomys nuttalli/ Peromyscus nuttalli*	golden mouse
Goldmeeräsche, Goldäsche	*Mugil auratus/ Liza aurata*	golden grey mullet
Goldmulle	Chrysochloridae	golden moles
Goldmund-Olive, Rotmund-Olive	*Oliva miniacea*	gold-mouthed olive, orange-mouthed olive, red-mouth olive, red-mouthed olive
Goldmund-Stachelschnecke, Goldmundmurex	*Murex chrysosoma*	goldmouth murex
Goldmund-Triton, Nicobaren-Triton	*Cymatium nicobaricum*	goldmouth triton, Nicobar hairy triton
Goldmund-Turbanschnecke	*Turbo chrysostomus*	gold-mouthed turban
Goldohr, Goldauge	*Fundulus chrysotus*	golden ear killifish, golden topminnow (FAO)
Goldpieper	*Tmetothylacus tenellus*	golden pipit
Goldpunktgarnele	*Drimo elegans/ Gnathophyllum elegans*	golden-spotted shrimp*
Goldpunktierter Puppenräuber	*Calosoma auropunctatum*	goldendot caterpillar hunter*
Goldregenpfeifer	*Pluvialis apricaria*	golden plover
Goldregenpfeifer> Pazifischer Goldregenpfeifer	*Pluvialis fulva*	Pacific golden plover
Goldringelgrundel	*Brachygobius xanthozona*	bumblebee fish
Goldringer, Ringkauri	*Cypraea annulus*	ring cowrie, goldringer
Goldrose, Sandgoldrose, Goldfarbige Seerose	*Condylactis aurantiaca*	golden anemone, golden sand anemone
Goldrückenaguti	*Dasyprocta aguti*	orange-rumped agouti
Goldsaum-Buntbarsch	*Aequidens rivulatus*	green terror
Goldschakal	*Canis aureus*	golden jackal
Goldschlange	*Chrysopelea ornata*	ornate flying snake

Goldschmied, Goldhenne, Gold-Laufkäfer, Goldlaufkäfer	*Carabus auratus*	golden ground beetle, gilt ground beetle
Goldschnabel-Musendrossel	*Catharus aurantiirostris*	orange-billed nightingale thrush
Goldschnepfen	Rostratulidae	painted snipe
Goldschopfpinguin	*Eudyptes chrysolophus*	macaroni penguin
Goldschultersittich	*Psephotus chrysopterygius*	golden-shouldered parakeet, golden-winged parrot
Goldschwamm, Farbwechselnder Zylinderschwamm	*Verongia aerophoba*	aureate sponge*
Goldschwammschnecke, Verkehrte Schirmschnecke	*Tylodina perversa*	
Goldschwanz-Kaiserfisch	*Pomacanthus chrysurus*	goldtail angelfish
Goldschwanzgecko	*Diplodactylus taenicauda*	golden-tailed gecko, goldentail gecko
Goldschwingen-Nektarvogel	*Nectarinia reichenowi*	golden-winged sunbird
Goldsittich	*Aratinga guarouba/ Guaruba guarouba*	golden parakeet, golden conure
Goldspecht	*Colaptes auratus*	common flicker
Goldsteiß-Löwenäffchen	*Leontopithecus chrysopygus*	black lion tamarin, golden rump lion tamarin, golden-rumped lion tamarin
Goldstirn-Klammeraffe	*Ateles belzebuth*	long-haired spider monkey
Goldstreifen-Scheinolive*	*Ancilla cingulata*	honey-banded ancilla
Goldstreifenbarsch	*Grammistes sexlineatus*	six-lined grouper, golden-striped grouper, sixline soapfish (FAO)
Goldstreifenbrasse	*Gnathodentex aureolineatus*	gold-lined bream, large-eyed bream, striped large-eye bream (FAO)
Goldstreifensalamander	*Chioglossa lusitanica*	golden-striped salamander, gold-striped salamander
Goldstreifenschnapper	*Lutjanus kasmira*	common bluestripe snapper (FAO), moonlighter
Goldstriemen, Ulvenfresser	*Sarpa salpa/ Boops salpa*	saupe, salema (FAO), goldline
Goldstumpfnasenaffe, Goldstumpfnase	*Pygathrix roxellanae/ Rhinopithecus roxellanae*	Chinese snub-nosed monkey, golden monkey, snow monkey, golden snub-nosed langur, Sichuan snub-nosed langur
Goldtetra, Kirschfleckensalmler, Glanztetra, Messingsalmler	*Hemigrammus rodwayi/ Hemigrammus armstrongi*	golden tetra
Goldtilapia	*Sarotherodon aureum/ Chromis aureus/Tilapia aurea*	golden tilapia
Goldtupfen-Falterfisch	*Chaetodon selene*	moon butterflyfish
Goldwaldsänger	*Dendroica petechia*	yellow warbler
Goldwespe> Gemeine Goldwespe, Feuer-Goldwespe, Feuergoldwespe	*Chrysis ignita*	common gold wasp, ruby-tail, ruby-tailed wasp
Goldwespen	Chrysididae	cuckoo wasps
Goldzeisig	*Carduelis tristis/Spinus tristis*	American goldfinch

Goldzunge	*Phylloda foliacea/ Tellina foliacea*	foliated tellin, leafy tellin
Golf-Furchenmolch	*Necturus beyeri*	Gulf Coast waterdog
Golf-Katzenhai	*Halaelurus vincenti*	Gulf catshark
Golf-Menhaden	*Brevoortia patronus*	Gulf menhaden (FAO), large-scale menhaden, mossbunker
Golf-Umberfisch	*Menticirrhus littoralis*	Gulf kingcroaker
Golfkärpfling	*Fundulus grandis*	Gulf killifish
Golfkröte	*Bufo valliceps*	Gulf Coast toad
Goliath-Fechterschnecke, Goliath-Flügelschnecke	*Strombus goliath*	goliath conch
Goliath-Vogelspinne, Riesen-Vogelspinne	*Theraphosa blondi*	goliath birdeater tarantula
Goliathfrosch	*Gigantorana goliath/ Conraua goliath*	goliath frog
Goliathratte, Östliche Weißohr-Riesenratte	*Hyomys goliath*	eastern white-eared giant rat
Gorals, Ziegenantilopen	*Nemorhaedus* spp.	gorals
Gorgonenhaupt	*Gorgonocephalus caput-medusae*	medusa-head star, Gorgon's head, basket star
Gorgonien-Porzellanschnecke	*Simnia spelta*	gorgonian simnia*
Gorilla	*Gorilla gorilla*	gorilla
Gorilla> Berggorilla	*Gorilla gorilla beringei*	mountain gorilla
Gorilla> Flachlandgorilla	*Gorilla gorilla gorilla*	lowland gorilla
Götterbaumspinner, Ailanthusspinner	*Philosamia cynthia/ Samia cynthia/ Platysamia cynthia*	Cynthia silkmoth, ailanthus silkworm
Gottesanbeterin	*Mantis religiosa*	European preying mantis
Gotteslachse, Glanzfische	Lampridae	opahs
Gotteslachsverwandte, Glanzfischartige, Glanzfische	Lampriformes	moonfishes
Gouldamadine	*Chloebia gouldiae*	Gouldian finch, Lady Gouldian finch
Gouldnektarvogel	*Aethopyga gouldiae*	Mrs. Gould's sunbird
Goulds Dreiecksmuschel	*Donax gouldii*	Gould's wedge clam, Gould beanclam
Goulds Waran	*Varanus gouldii*	Gould's monitor, sand monitor, sand goanna, bungarra
Gourmya-Nadelschnecke	*Gourmya gourmyi*	gourmya cerith
Gouverneur-Kegelschnecke	*Conus gubernator*	governor cone
Grabender Schlangenstern*	*Amphiodia occidentalis*	burrowing brittlestar
Gräberfliege	*Conicera tibialis*	coffin fly
Grabflatterer, Grabfledermäuse	*Taphozous* spp.	tomb bats
Grabfrösche	*Heleioporus* spp.	giant burrowing frogs
Grabfüßer, Kahnfüßer, Scaphopoden	Solenoconchae/Scaphopoda (Conchifera)	scaphopods, scaphopodians (tooth shells, tusk shells, spade-footed mollusks)
Grablaubfrösche	*Pternohyla* spp.	burrowing frogs

Grabläufer, Grabkäfer	*Pterostichus* spp.	cloakers
Grabwespen	Sphecidae/Sphegidae	digger wasps, hunting wasps
Graeffes Seewalze	*Bohadschia graeffei*	Graeffe's holothurian
Granat, Nordseegarnele, Porre	*Crangon crangon*	brown shrimp, common shrimp, common European shrimp
Granatbarsch, Atlantischer Sägebauch	*Hoplostethus atlanticus*	orange roughie
Granathörnchen	*Sciurus granatensis*	tropical red squirrel
Granatpitta	*Pitta granatina*	garnet pitta
Grandala	*Grandala coelicolor*	Hodgson's grandala
Grandidiers Schuppengecko	*Geckolepis typica*	Grandidier's gecko
Granit-Nachtechse	*Xantusia henshawi*	granite night lizard
Granit-Stachelleguan	*Sceloporus orcutti*	granite spiny lizard
Grant-Seequappe	*Gaidropsarus granti*	Grant's rockling
Grantgazelle	*Gazella granti*	Grant's gazelle
Grants Waldratte	*Batomys granti*	Grant's Luzon forest rat
Granula-Helmschnecke	*Semicassis granulatum granulatum*	scotch bonnet
Granula-Kauri	*Cypraea granulata*	granulated cowrie
Grasanolis	*Anolis auratus*	grass anole
Gräserzystenälchen	*Heterodera punctata*	grass cyst nematode
Graseule> Gemeine Graseule, Ausrufungszeichen, Ausrufezeichen	*Agrotis exclamationis*	heart and dart moth
Grasfliegen, Wiesenfliegen, Saftfliegen	Opomyzidae	opomyzid flies
Grasflurmilbe*	*Tyrophagus similis*	grassland mite
Grasfrosch	*Rana temporaria*	common frog, grass frog
Grasglucke	*Philudoria potatoria*	drinker
Grasgrüne Huschspinne, Grüne Huschspinne	*Micrommata rosea*	green spider*
Grashaldeneule	*Photedes captiuncula*	least minor (moth)
Grashüpfer, Heuhüpfer	*Chorthippus* spp.	field grasshoppers, meadow grasshoppers
Grashüpfermäuse	*Onychomys* spp.	grasshopper mice
Grasläufer	*Tryngites subruficollis*	buff-breasted sandpiper
Graskarpfen, Amurkarpfen	*Ctenopharyngodon idella*	grass carp
Grasklapperlerche	*Mirafra apiata*	clapper lark
Grasmilbe, Gras-Spinnmilbe	*Bryobia graminum*	grass mite
Grasminiermotten	Elachistidae	grass miners
Grasmotten	Crambidae	grass moths, grass-veneers
Grasmotten	*Crambus* spp.	grass moths, grass-veneers
Grasmücken (Zweigsänger)	Sylviidae	Old World warblers
Grasnattern	*Opheodrys* spp.	green snakes
Grasnelken-Widderchen, Grünwidderchen	*Procris statices/ Adscita statices*	forester, common forester
Grasschlüpfer	Amytornithinae	grasswrens

Grasschnecken	*Vallonia* spp.	grass snails
Grasschnecken	Valloniidae	grass snails
Grassteiger	Megalurinae	grass-warblers
Grasstengeleule, Gelbliche Wieseneule	*Luperina testacea*	flounced rustic moth
Graswanze	*Leptopterna dolobrata*	meadow plant bug
Graswurzelälchen	*Ditylenchus radicicola/ Subanguina radicicola*	grass nematode
Graswurzeleule, Trockenrasen-Graswurzeleule	*Apamea lithoxylea*	common light arches
Graswurzelgallenälchen	*Meloidogyne graminis*	grass root-knot nematode
Grätenfisch	*Albula vulpes*	bonefish
Grätenfische	Albulidae	bonefishes
Grauammer	*Miliaria calandra*	corn bunting
Grauarmmakak	*Macaca ochreata*	booted macaque
Graubarsch, Seekarpfen, Nordischer Meerbrassen	*Pagellus bogaraveo*	red seabream, common seabream, blackspot seabream (FAO)
Graubartfalke	*Falco cenchroides*	Australian kestrel
Graubauch-Degenschnäbler	*Microbates cinereiventris*	half-collared gnatwren
Graubauchtragopan	*Tragopan blythii*	Blyth's tragopan
Graubraune Madagassische Schleichkatze, Ungefleckte Fanaloka, 'Ameisenschleichkatze' (Fanaloka, Ridaridy; siehe auch: *Fossa fossa*)	*Eupleres goudotii*	falanouc (IUCN), fanalouc (*better*: grey-brown Malagasy civet; *native common names*: fanaloka, ridaridy)
Graubrust-Paradiesschnäpper	*Terpsiphone viridis*	African paradise flycatcher
Graubrust-Strandläufer	*Calidris melanotos*	pectoral sandpiper
Graubülbül	*Pycnonotus barbatus*	garden bulbul
Graubürzel-Singhabicht	*Melierax metabates*	dark chanting goshawk
Graudrossling	*Turdoides squamiceps*	Arabian babbler
Graue Ackerschnecke, Einfarbige Ackerschnecke	*Deroceras agreste/ Agriolimax agreste*	field slug, grey field slug, grey slug
Graue Atlantik-Kauri	*Cypraea cinerea/ Luria cinerea/ Talparia cinerea*	Atlantic grey cowrie, Atlantic gray cowrie (U.S.), ashen cowrie
Graue Baumnatter, Vogelschlange	*Thelotornis kirtlandii*	bird snake
Graue Baumratte	*Lenothrix canus*	gray tree rat
Graue Beißschrecke	*Platycleis grisea*	grey bushcricket
Graue Fleischfliege	*Sarcophaga carnaria*	fleshfly, flesh-fly
Graue Geißelgarnele	*Metapenaeus ensis*	greasyback shrimp
Graue Gerstenminierfliege	*Hydrellia griseola*	smaller rice leafminer (U.S.)
Graue Helmschnecke	*Phalium glaucum*	grey bonnet
Graue Hydra, Grauer Süßwasserpolyp	*Pelmatohydra oligactis/ Hydra oligactis*	grey hydra*, brown hydra
Graue Langohrfledermaus, Graues Langohr	*Plecotus austriacus*	grey long-eared bat
Graue Meeräsche	*Oedalechilus labeo*	lesser grey mullet, boxlip mullet (FAO)

Graue Meersau, Grauer Schweinshai	*Oxynotus paradoxus*	sailfin roughshark (FAO), sharp-back shark
Graue Moderholzeule, Gemeines Moderholz	*Xylena exsoleta*	sword-grass moth
Graue Mondschnecke	*Payraudeautia intricata*	European grey moonsnail
Graue Notothenia	*Notothenia squamifrons*	scaled notothenia, grey rockcod
Graue Scheinolive	*Agaronia hiatula*	olive-gray ancilla
Graue Schließmundschnecke	*Bulgarica cana*	grey door snail*
Graue Seefeder	*Pteroides griseum*	gray sea-pen
Graue Steinkrabbe	*Xantho poressa*	grey xanthid crab
Graue Sternschnecke	*Jorunna tomentosa*	grey sea slug*
Graue Strandschnecke	*Littorina planaxis/ Littorina striata/ Littorina keenae*	grey periwinkle, "eroded" periwinkle
Graue Wegschnecke	*Arion circumscriptus*	white-soled slug, grey garden slug, brown-banded arion
Graue Zwerghamster	*Cricetulus* spp.	ratlike hamsters
Grauer Bambushai	*Chiloscyllium griseum*	grey carpet shark, grey bambooshark (FAO)
Grauer Gibbon, Borneo-Gibbon	*Hylobates muelleri*	grey gibbon, Müller's gibbon
Grauer Glatthai, Südlicher Glatthai, Mittelmeer-Glatthai	*Mustelus mustelus*	smooth-hound (FAO), smoothhound
Grauer Halbmaki	*Hapalemur griseus*	grey gentle lemur
Grauer Herzigel, Einfarbiger Herzigel	*Brissus unicolor*	grey heart-urchin
Grauer Kaiserfisch	*Pomacanthus arcuatus*	grey angelfish
Grauer Kiefernspanner*	*Thera obeliscata*	grey pine carpet moth
Grauer Knospenwickler	*Hedya nubiferana/ Hedya dimidioalba*	marbled orchard tortrix (moth), fruit tree tortrix (green budworm, spotted apple budworm)
Grauer Knurrhahn	*Eutrigla gurnardus*	grey gurnard, gray searobin (U.S.)
Grauer Lanzenseeigel (Kaiserigel)	*Cidaris cidaris*	piper, king of the sea-eggs
Grauer Lärchenwickler	*Zeiraphera diniana*	larch tortrix, grey larch tortrix, larch bud moth
Grauer Laubfrosch	*Hyla versicolor*	gray treefrog
Grauer Lippfisch	*Crenilabrus cinereus/ Symphodus cinereus*	grey wrasse
Grauer Pokal	*Cantharus assimilis*	grey goblet whelk
Grauer Riffhai	*Carcharhinus amblyrhynchos/ Carcharhinus wheeleri*	grey reef shark
Grauer Schleimfisch	*Lipophrys trigloides*	gurnard blenny
Grauer Schnapper	*Lutjanus griseus*	grey snapper (FAO), gray snapper
Grauer Schweinshai, Graue Meersau	*Oxynotus paradoxus*	sailfin roughshark (FAO), sharp-back shark
Grauer Springaffe	*Callicebus moloch*	dusky titi
Grauer Süßwasserpolyp, Graue Hydra	*Hydra oligactis/ Pelmatohydra oligactis*	grey hydra*, brown hydra

Grauer Wollrückenspanner, Schlehenfrostspinner	*Apocheima pilosaria/ Phigalia pilosaria*	pale brindled beauty
Grauer Zackenbarsch	*Epinephelus caninus*	dogtooth grouper (FAO), dog-toothed grouper
Grauer Zipfelfalter	*Strymon melinus*	grey hairstreak
Grauer Zwerghamster	*Cricetelus migratorius*	grey hamster
Graues Langohr, Graue Langohrfledermaus	*Plecotus austriacus*	grey long-eared bat
Graues Porzellankrebschen, Grauer Porzellankrebs	*Porcellana platycheles*	broad-clawed porcelain crab, grey porcelain crab
Graues Riesenkänguru	*Macropus giganteus*	eastern grey kangaroo
Graufischer	Cerylidae	cerylid kingfishers
Graufischer	*Ceryle rudis*	pied kingfisher
Grauflankenmeise	*Parus sclateri*	Mexican chickadee, grey-sided chickadee
Grauflügel-Trompetervogel	*Psophia crepitans*	common trumpeter
Graufuchs, Argentinischer Graufuchs	*Dusicyon griseus/ Pseudalopex griseus*	gray zorro, Argentine gray fox, South American gray fox
Graugans	*Anser anser*	graylag goose, grey lag goose
Grauhaie	Notidanidae/ Hexanchidae	gray sharks, hexanchids, cow sharks
Grauhaie, Sechskiemer	*Hexanchus* spp.	sixgill sharks
Grauhörnchen	*Sciurus carolinensis*	gray squirrel (U.S.)
Graukardinal	*Paroaria coronata*	red-crested cardinal
Graukatze	*Felis bieti*	Chinese desert cat
Graukopfvireo	*Vireo solitarius*	solitary vireo
Graulärmvogel	*Corythaixoides concolor*	go-away bird
Graulaubenvogel	*Chlamydera nuchalis*	great grey bowerbird
Graulemming	*Lagurus lagurus*	steppe lemming
Graumulle	*Cryptomys* spp.	common mole-rats
Grauortolan	*Emberiza caesia*	Cretzschmar's bunting
Graupapagei	*Psittacus erithacus*	grey parrot
Graureiher	*Ardea cinerea*	grey heron
Graurötelmaus	*Clethrionomys rufocanus*	grey-sided vole
Graurücken-Flusshering	*Pomolobus pseudoharengus/ Alosa pseudoharengus*	grayback herring (alewife)
Graurückige Klappen-Weichschildkröte, Graue Klappen-Weichschildkröte	*Cycloderma frenatum*	Zambesi soft-shelled turtle, Zambezi flapshell turtle
Grauschnäpper	*Muscicapa striata*	spotted flycatcher
Grauschwarze Sklavenameise, Furchtsame Hilfswaldameise	*Formica fusca*	negro ant
Grauspecht	*Picus canus*	grey-headed woodpecker
Grauvireo	*Vireo vicinior*	grey vireo
Grauwal	*Eschrichtius robustus/ Eschrichtius gibbosus*	gray whale
Grauwangen-Buschdrossling	*Illadopsis rufipennis*	pale-breasted thrush babbler
Grauwangen-Mangabe	*Lophocebus albigena*	grey-cheeked mangabey
Grauwangendrossel	*Catharus minimus*	grey-cheeked thrush

Grauwasseramsel	*Cinclus mexicanus*	North American dipper
Gray-Zweizahnwal	*Mesoplodon grayi*	Gray's beaked whale, scamperdown beaked whale
Greiffrosch, Affenfrosch	*Phyllomedusa hypochondrialis*	orange-legged leaf frog
Greiffrösche, Affenfrösche, Makifrösche	*Phyllomedusa* spp.	leaf frogs
Greifschwanzratten*	*Pogonomys* spp.	prehensile-tailed rats
Greifstachler	*Coendou* spp.	prehensile-tailed porcupines, coendous
Greifvögel	Falconiformes	diurnal birds of prey (falcons and others)
Greisbock	*Raphicerus melanotis*	Cape grysbok
Greisengesicht, Greisenhaupt	*Centurio senex*	wrinkle-faced bat, lattice-winged bat
Greisinnenmuschel	*Dosinia anus*	old-woman dosinia
Grenadier	*Acanthisitta chloris*	rifleman
Grenadieranchovy	*Coilia mystus*	rat-tail anchovy, Osbeck's grenadier anchovy (FAO)
Grenadierfische, Rattenschwänze	Macrouridae	grenadiers, rattails
Grenadierkrabbe, Soldatenkrabbe	*Mictyris longicarpus*	soldier crab
Grenadierkrabben, Soldatenkrabben	Mictyridae	soldier crabs, grenadier crabs
Grevy-Zebra	*Equus grevyi*	Grevy's zebra
Greys Wallaby	*Macropus greyi*	toolache wallaby
Griechische Felseidechse	*Lacerta graeca*	Greek rock lizard
Griechische Honigbiene	*Apis mellifera cecropia*	Greek honey bee
Griechische Landschildkröte	*Testudo hermanni*	Hermann's tortoise, Greek tortoise
Griechische Rotfeder	*Scardinius graecus*	Greek rudd
Griechischer Apoll	*Archon apollinus*	false apollo
Griechischer Frosch	*Rana graeca*	stream frog, Greek frog
Griffelkoralle	*Stylophora pistillata*	pistillate coral*
Griffelseeigel, Griffel-Seeigel	*Heterocentrotus mammillatus*	slate pencil urchin
Grillen	Gryllidae	crickets, true crickets
Grillenfrösche (Heuschreckenfrösche)	*Acris* spp.	cricket frogs
Grillenkrebs, Kleiner Bärenkrebs	*Scyllarus arctus*	small European locust lobster, small European slipper lobster, lesser slipper lobster
Grillenschaben	Grylloblattidae	rock crawlers, grylloblattids
Grillenwanze	*Geocoris grylloides*	cricket bug
Grindwal> Gewöhnlicher Grindwal, Langflossen-Grindwal	*Globicephala melaena*	northern pilot whale, Atlantic pilot whale, blackfish, pothead, long-finned pilot whale
Grindwal> Indo-Pazifischer Grindwal, Kurzflossen-Grindwal	*Globicephala seiboldii/ Globicephala macrorhynchus*	Indo-Pacific pilot whale, short-finned pilot whale, blackfish
Grinsendes Fass	*Malea ringens*	great grinning tun

Grisfuchs (Hbz.), Nordamerikanischer Festland-Graufuchs	*Urocyon cinereoargenteus*	mainland gray fox, North American gray fox
Grobe Porenkoralle	*Goniopora lobata*	coarse porous coral, flowerpot coral
Grönland-Heilbutt, Schwarzer Heilbutt	*Reinhardtius hippoglossoides*	Greenland halibut (FAO), Greenland turbot, black halibut
Grönland-Kabeljau, Grönland-Dorsch, Fjord-Dorsch	*Gadus ogac*	Greenland cod
Grönland-Kreiselschnecke	*Margarites groenlandicus*	Greenland margarite, Greenland topsnail (Greenland top shell)
Grönland-Shrimp, Nördliche Tiefseegarnele	*Pandalus borealis*	northern shrimp, pink shrimp, northern pink shrimp
Grönlandangler, Atlantischer Tiefseeangler	*Himantolophus groenlandicus*	Atlantic football fish
Grönlandhai, Großer Grönlandhai, Eishai	*Somniosus microcephalus*	Greenland shark (FAO), ground shark
Grönländische Herzmuschel	*Serripes groenlandicus*	Greenland cockle
Grönlandrochen	*Bathyraja spinicauda*	spiny-tail skate, spinetail ray (FAO)
Grönlandwal	*Balaena mysticetus*	bowhead whale, Greenland right whale, Arctic right whale
Groppe, Koppe, West-Groppe, Kaulkopf, Mühlkoppe	*Cottus gobio*	Miller's thumb, bullhead (FAO)
Groppen	Cottidae	sculpins
Groppenbarsch	*Romanichthys valsanicola*	Romanian bullhead-perch
Groß-Glasfrösche	*Centrolene* spp.	giant glass frogs
Großaugen	Scolopsidae	spinecheeks
Großaugen-Angola-Meerbrasse	*Brachydeuterus auritus*	bigeye grunt
Großaugen-Brasse	*Calamus calamus*	saucereye porgy
Großaugen-Dornhai	*Squalus megalops*	shortnose spurdog
Großaugen-Fuchshai, Großäugiger Fuchshai	*Alopias superciliosus*	bigeye thresher
Großaugen-Hundshai, Großaugen-Marderhai	*Iago omanensis*	bigeye hound shark
Großaugen-Orfe	*Notropis boops*	bigeye shiner
Großaugen-Pikarel	*Spicara alta*	bigeye picarel
Großaugen-Seequappe*	*Antonogadus macrophthalmus*	big-eyed rockling
Großaugen-Strandfloh, Großaugen-Strandhüpfer	*Talorchestia megalophthalma*	big-eyed sandhopper, big-eyed beach flea
Großaugen-Thunfisch	*Thunnus obesus*	big-eyed tuna, bigeye tuna (FAO)
Großaugen-Zahnbrassen	*Dentex macrophthalmus*	large-eyed dentex, large-eye dentex (FAO)
Großaugenbarsch	*Priacanthus hamrur*	lunar-tailed bullseye, crescent-tail bigeye, moontail bullseye (FAO)

Großaugenbarsche, Großaugen, Bullaugen, Catalufas	Priacanthidae	bigeyes (catalufas)
Großaugenfledermaus*	*Chiroderma* spp.	big-eyed bats, white-lined bats
Großaugensalmler (Echte Afrikanische Salmler)	Alestidae/Alestiinae	African tetras
Großaugenwanze	*Geocoris bullatus*	large bigeyed bug
Großäugiger Seehecht	*Merluccius albidus*	offshore silver hake
Großäugiger Selar	*Selar crumenophthalmus*	big-eye scad
Großblattnasen	Megadermatidae	false vampire bats, yellow-winged bats
Großblattnasen, Eigentliche Großblattnasen	*Megaderma* spp.	Asian false vampire bats
Große Achatschnecke, Gemeine Riesenschnecke, Afrikanische Riesenschnecke	*Achatina fulica*	giant African snail, giant African land snail
Große Afrikanische Waldmäuse	*Praomys* spp.	African soft-furred rats
Große Amerikanische Hauswinkelspinne	*Tegenaria duellica*	giant house spider
Große Amerikanische Ostküstenmurex	*Hexaplex fulvescens*	giant eastern murex
Große Anakonda, Grüne Anakonda	*Eunectes murinus*	anaconda, water boa
Große Australische Häschenratte	*Leporillus conditor*	greater stick-nest rat
Große Bandikutratte	*Bandicota indica*	greater bandicoot rat
Große Bartfledermaus, Brandt-Fledermaus	*Myotis brandti*	Brandt's bat
Große Baumratte	*Chiruromys forbesi*	greater tree mouse
Große Birkenblattwespe, Birkenknopfhorn-Blattwespe	*Cimbex femoratus*	birch sawfly
Große Bodenrenke, Sandfelchen	*Coregonus fera*	broad whitefish
Große Braune Fledermaus	*Eptesicus fuscus*	big brown bat
Große Bukettkorallen	*Euphyllia* spp.	tooth corals
Große Dörnchenkoralle, Große Drahtkoralle	*Cirripathes rumphii/ Cirrhipathes rumphii*	giant whip coral
Große Egelschnecke, Großer Schnegel	*Limax maximus*	giant gardenslug, European giant gardenslug, great grey slug, spotted garden slug
Große Erbsenmuschel	*Pisidium amnicum*	giant pea mussel, river pea mussel, large pea shell, greater European peaclam (U.S.)
Große Essigfliege	*Drosophila funebris*	greater vinegar fly, greater fruit fly
Große Fässchenschnecke, Kleine Fassschnecke	*Orcula dolium*	
Große Fassschnecke	*Tonna galea*	giant tun
Große Felsgarnele, Sägegarnele	*Palaemon serratus/ Leander serratus*	common prawn
Große Flussmuschel	*Unio tumidus*	swollen river mussel
Große Gabelmakrele	*Lichia amia*	leerfish

Große Gartenhummel	*Bombus ruderatus*	large garden bumble bee
Große Gelbhummel	*Bombus distinguendus*	great yellow bumble bee
Große Getreideblattlaus	*Sitobion avenae/* *Macrosiphum avenae*	greater cereal aphid, European grain aphid, English grain aphid
Große Glanzschnecke, Draparnauds Glanzschnecke, Große Gewächshaus-Glanzschnecke	*Oxychilus draparnaudi*	Draparnaud's glass snail, dark-bodied glass-snail
Große Glasschnecke	*Phenacolimax major*	greater pellucid glass snail
Große Goldmakrele, Gemeine Goldmakrele	*Coryphaena hippurus*	dolphinfish, common dolphinfish (FAO), dorado, mahi-mahi
Große Goldschrecke	*Chrysochraon dispar*	large gold grasshopper
Große Grasschnecke	*Vallonia declivis*	greater grass snail
Große Hauswinkelspinne	*Tegenaria gigantea/* *Tegenaria atrica*	giant European house spider, giant house spider, larger house spider, cobweb spider
Große Heidelibelle	*Sympetrum striolatum*	common sympetrum, common darter
Große Herzmuschel, Stachelige Herzmuschel	*Acanthocardia aculeata*	spiny cockle
Große Höckerschrecke, Großer Band-Grashüpfer	*Arcyptera fusca*	large banded grasshopper
Große Höckrige Nadelschnecke	*Cerithium noduloum*	giant knobbed cerith
Große Hühnerlaus	*Goniodes gigas/* *Stenocrotaphus gigas*	large chicken louse
Große Jakobsmuschel, Große Pilgermuschel	*Pecten maximus*	great scallop, common scallop, coquille St. Jacques
Große Kartäuserschnecke	*Monacha cantiana*	Kentish snail
Große Kieferngespinstblattwespe	*Acantholyda posticalis*	pine web-spinning sawfly
Große Kiefernrindenlaus	*Cinara pinea*	large pine aphid
Große Kielnacktschnecke	*Milax rusticus/* *Tandonia rustica*	bulb-eating slug, root-eating slug
Große Köcherfliege	*Phryganea grandis*	great red sedge
Große Kohlfliege	*Delia floralis/* *Hylemia floralis/* *Phorbia floralis/* *Chortophila floralis*	cabbage-root fly, radish fly, turnip maggot
Große Königslibelle	*Anax imperator*	emperor dragonfly
Große Kreuzbrustschildkröte	*Staurotypus triporcatus*	Mexican giant musk turtle, Mexican musk turtle (IUCN), Mexican cross-breasted turtle
Große Krötenschnecke, Taschenschnecke, Großes Ochsenauge, Ölkrug	*Gyraulus gigantea/* *Ranella gigantea*	giant frogsnail, oil-vessel triton
Große Krustenanemone, Sonnen-Krustenanemone	*Palythoa grandis*	giant colonial anemone, golden sea mat, button polyps, polyp rock, cinnamon polyp

Große Kugelmuschel, Ufer-Kreismuschel, Fluss-Kugelmuschel	*Sphaerium rivicola*	nut orb mussel (nut orb shell), nut fingernailclam
Große Kugelschnecke	*Ampullarius gigas*	giant bubble snail (giant bubble shell)
Große Lärchenblattwespe	*Pristiphora erichsonii*	large larch sawfly
Große Lärchenrindenlaus	*Cinara kochiana*	giant larch aphid
Große Mantel-Käferschnecke	*Cryptochiton stelleri*	giant Pacific chiton, gumboot chiton (U.S.)
Große Maräne, Große Schwebrenke, Wandermaräne, Lavaret, Bodenrenke	*Coregonus lavaretus*	freshwater houting, powan, common whitefish (FAO)
Große Miesmuschel	*Modiolus modiolus*	horse mussel, northern horsemussel
Große Moosjungfer	*Leucorrhinia pectoralis*	greater white-faced darter
Große Nabelschnecke, Halsband-Mondschnecke, Halsband-Nabelschnecke	*Lunatia catena/ Natica catena*	large necklace snail, large necklace moonsnail (large necklace shell)
Große Netzmuräne	*Gymnothorax favagineus*	honeycomb moray
Große Obstbaumschildlaus	*Eulecanium corni*	brown scale, brown fruit scale, European fruit lecanium, European peach scale
Große Pantoffelschnecke	*Crepidula grandis*	great slippersnail
Große Pazifik-Boa, Fidschi-Boa	*Candoia bibroni*	Pacific boa
Große Pazifische Sattelauster	*Placuna ephippium*	greater Pacific saddle oyster
Große Pechlibelle	*Ischnura elegans*	common ischnura, blue-tailed damselfly
Große Perlmuschel	*Pinctada maxima*	gold-lip pearl-oyster, golden-lip pearl oyster, black silver pearl oyster, mother-of-pearl shell, pearl button oyster
Große Pfeffermuschel, Flache Pfeffermuschel	*Scrobicularia plana*	peppery furrow clam (peppery furrow shell)
Große Pflaumenblattlaus, Große Pflaumenlaus	*Brachycaudus cardui*	thistle aphid, greater plum aphid*
Große Pilgermuschel, Große Jakobsmuschel	*Pecten maximus*	great scallop, common scallop, coquille St. Jacques
Große Plumpschrecke	*Isophya pyrenea*	large speckled bushcricket
Große Rennmaus	*Rhombomys opimus*	great gerbil
Große Riesenmuschel, Mördermuschel	*Tridacna gigas*	giant clam, killer clam
Große Rossie	*Rossia macrosoma*	Ross' cuttlefish, Ross' cuttle, large-head bob-tailed squid, stout bobtail squid (FAO)
Große Rote Wegschnecke	*Arion rufus*	large red slug, greater red slug, chocolate arion
Große Sägeschrecke	*Saga pedo*	predatory bush cricket
Große Sandboa	*Eryx tataricus*	tartar sand boa
Große Sandspringspinne	*Yllenus arenarius*	
Große Scheidenmuschel	*Solen vagina*	European razor clam

Große Schildmotte, Kleiner Asselspinner	*Apoda limacodes/ Cochlidion limacodes/ Apoda avellana*	festoon
Große Schlangennadel	*Entelurus aequoreus*	snake pipefish
Große Schlitznase, Große Hohlnase	*Nycteris grandis*	large slit-faced bat
Große Schotenmuschel	*Solemya grandis*	grand awningclam
Große Schwarze Gleithörnchen	*Aeromys* spp.	large black flying squirrels
Große Schwebegarnele	*Praunus flexuosus*	bent mysid shrimp
Große Seenadel, Mittelmeer-Seenadel	*Syngnathus acus*	great pipefish, greater pipefish
Große Seeperlmuschel	*Pinctada margaritifera/ Pteria margaritifera/ Meleangrina margaritifera*	Pacific pearl-oyster, black-lipped pearl oyster, black-lip pearl oyster
Große Seepocke	*Balanus balanus*	rough barnacle
Große Seespinne, Teufelskrabbe	*Maja squinado/ Maia squinado*	common spider crab, thorn-back spider crab
Große Sepiette	*Sepietta oweniana*	greater cuttlefish, common bobtail squid (FAO)
Große Seychellen-Mabuye	*Mabuya wrightii*	greater Seychelles skink, Wright's mabuya
Große Spießblattnase	*Vampyrum spectrum*	Linnaeus's false vampire bat, spectral vampire
Große Spinnenameise	*Mutilla europaea*	European velvet-ant
Große Steingarnele, Große Felsgarnele, Schwertgarnele	*Palaemon squilla/ Leander squilla*	cup shrimp
Große Steinkrabbe	*Menippe mercenaria*	stone crab, black stone crab
Große Sternkoralle	*Montastrea cavernosa*	great star coral, large star coral
Große Streifen-Sattelschrecke	*Ephippigerida taeniata*	large striped bushcricket
Große Streifengesicht-Fledermaus	*Vampyrodes major*	great stripe-faced bat
Große Sturmhaube, Gehörnte Helmschnecke	*Cassis cornuta*	horned helmet, giant helmet
Große Sumpfschnecke	*Stagnicola corvus*	giant pondsnail
Große Süßwassernadel	*Microphis boaja*	freshwater pipefish
Große Taubechse	*Cophosaurus texanus*	greater earless lizard
Große Teichmuschel, Schwanenmuschel	*Anodonta cygnea*	swan mussel
Große Teppichmuschel, Venusmuschel, Kreuzgemusterte Teppichmuschel	*Venerupis decussata/ Tapes decussata*	cross-cut carpetclam, cross-cut venus (cross-cut carpet shell)
Große Teufelskralle, Gemeine Spinnenschnecke	*Lambis truncata*	giant spider conch, wild-vine root
Große Turmschnecke	*Ena montana*	mountain bulin
Große Wachsmotte, Bienenwolf, Rankmade (Larve)	*Galleria mellonella*	greater wax moth, honeycomb moth, bee moth
Große Weidenrindenlaus	*Tuberolachnus salignus*	large willow aphid, giant willow aphid
Große Weißnasenmeerkatze	*Cercopithecus nictitans*	greater white-nosed guenon, putty-nosed monkey
Große Winkelkopfagame	*Gonocephalus grandis*	giant forest dragon

Große Wurmschnecke	*Lemintina arenaria*	Mediterranean wormsnail
Große Zitterrochen, Echte Zitterrochen	Torpedinidae	electric rays
Großer Abendsegler	*Nyctalus noctula*	noctule
Großer Ährenfisch	*Atherina hepsetus*	sand-smelt, silverside, Mediterranean sand smelt (FAO)
Großer Alpensalamander	*Salamandra lanzai*	large Alpine salamander
Großer Amerikanischer Leberegel*	*Fascioloides magna*	large American liver fluke
Großer Armmolch	*Siren lacertina*	greater siren
Großer Australkrebs, Marron	*Cherax tenuimanus*	marron
Großer Band-Grashüpfer, Große Höckerschrecke	*Arcyptera fusca*	large banded grasshopper
Großer Bärenkrebs, Großer Mittelmeer-Bärenkrebs	*Scyllarides latus*	Mediterranean slipper lobster
Großer Blauer Krake	*Octopus cyaneus*	big blue octopus
Großer Blauer Erdfloh	*Altica lythri*	large blue flea beetle
Großer Bombardierkäfer	*Brachinus crepitans*	greater bombardier beetle
Großer Bootshaken, Chiragra-Spinnenschnecke	*Lambis chiragra*	chiragra spider conch, gouty spider conch
Großer Brachvogel	*Numenius arquata*	curlew
Großer Brauner Rüsselkäfer, Großer Fichtenrüssler	*Hylobius abietis*	fir-tree weevil, pine weevil
Großer Breitkäfer	*Abax parallelepipedus*	parallel-sided ground beetle
Großer Darmegel, Riesendarmegel	*Fasciolopsis buski*	giant intestinal fluke
Großer Eichenkarmin	*Astiotes sponsa*	dark crimson underwing, scarlet underwing moth
Großer Eidechsenfisch	*Saurida tumbil*	greater lizardfish
Großer Eisvogel	*Limenitis populi*	poplar admiral
Großer Entensturmvogel	*Pachyptila vittata*	broad-billed prion
Großer Federwurm	*Sabellastarte magnifica*	giant feather-duster worm
Großer Feuerfalter	*Lycaena dispar*	large copper
Großer Fleckenkiwi	*Apteryx haastii*	great spotted kiwi
Großer Frostspanner, Hainbuchenspanner	*Erannis defoliaria*	mottled umber
Großer Fuchs	*Nymphalis polychloros/ Vanessa polychloros*	large tortoiseshell
Großer Gabeldorsch	*Urophycis blennioides/ Phycis blennioides*	greater forkbeard
Großer Gelbschenkel	*Tringa melanoleuca*	greater yellowlegs
Großer Gitarrenrochen, Riesengeigenrochen, Schulterfleck-Geigenrochen, Ulavi	*Rhynchobatus djiddensis*	whitespotted wedgefish, spotted guitarfish, shovelnose, sand shark, giant guitarfish (FAO)
Großer Gleithörnchenbeutler	*Petaurus australis*	fluffy glider
Großer Gras-Laubkäfer	*Hoplia philanthus*	Welsh chafer
Großer Grauhai, Sechskiemer-Grauhai	*Hexanchus griseus*	bluntnosed shark, six-gilled shark, sixgill shark, grey shark, bluntnose sixgill shark (FAO)
Großer Hammerhai	*Sphyrna mokarran*	great hammerhead shark
Großer Herzigel	*Echinocardium pennatifidum*	sea potato

Großer Heufalter, Großes Gelbes Wiesenvögelchen	*Coenonympha tullia/ Coenonympha typhon*	large heath
Großer Heuschreckenkrebs, Fangschreckenkrebs, Gemeiner Heuschreckenkrebs	*Squilla mantis*	giant mantis shrimp, spearing mantis shrimp
Großer Kaiserfisch, Imperator-Kaiserfisch, Nikobaren-Kaiserfisch	*Pomacanthus imperator*	emperor angelfish (FAO), imperial angelfish
Großer Katzenmaki	*Cheirogaleus major*	greater dwarf lemur
Großer Kiefernborkenkäfer	*Ips sexdentatus*	greater European pine engraver, six-toothed pine bark beetle (Br.)
Großer Kiefernprachtkäfer, Marienprachtkäfer	*Chalcophora mariana*	European sculptured pine borer
Großer Kiefernrüssler	*Hylobius piceus*	larch weevil
Großer Kohltriebrüssler, Großer Rapsstengelrüssler	*Ceutorhynchus napi*	greater cabbage curculio
Großer Kohlweißling	*Pieris brassicae*	large white
Großer Kolbenwasserkäfer, Großer Schwarzer Kolbenwasserkäfer	*Hydrous piceus/ Hydrochara piceus*	great black water beetle, great silver water beetle, greater silver water beetle, diving water beetle
Großer Kornbohrer	*Prostephanus truncatus*	larger grain borer
Großer Lärchenborkenkäfer, Großer Achtzähniger Lärchenborkenkäfer	*Ips cembrae*	larch bark beetle, Siberian fir bark-beetle
Großer Laternenträger	*Fulgora laternaria*	greater lanternfly
Großer Leberegel	*Fasciola hepatica*	sheep liver fluke
Großer Leuchtkäfer, Großes Glühwürmchen	*Lampyris noctiluca*	glowworm, glow-worm, great European glow-worm beetle
Großer Maulwurfskrebs	*Thalassina anomala*	tropical mole crab
Großer Mittelmeer-Bärenkrebs, Großer Bärenkrebs	*Scyllarides latus*	Mediterranean slipper lobster
Großer Nilhecht, Großnilhecht	*Gymnarchus niloticus*	aba
Großer Obstbaumsplintkäfer	*Scolytus mali*	large fruit bark beetle, larger shothole borer
Großer Obstbaumwickler	*Archips podana*	large fruittree tortrix moth
Großer Ölfisch	*Comephorus baicalensis*	greater Baikal oilfish
Großer Pappelbock	*Saperda carcharias*	large poplar borer, large willow borer, poplar longhorn, large poplar longhorn beetle
Großer Pazifikrochen	*Raja binoculata*	big skate
Großer Permutterfalter	*Mesoacidalia aglaja/ Argynnis aglaja*	dark green fritillary
Großer Pferdespulwurm, Großer Pferdenematode	*Parascaris equorum*	large equine roundworm, horse roundworm
Großer Pilotbarsch	*Kyphosus bigibbus*	grey sea chub (FAO), grey chub, buffalo bream
Großer Puppenräuber	*Calosoma sycophanta*	greater caterpillar hunter
Großer Purpur-Zipfelfalter	*Atlides halesus*	great purple hairstreak
Großer Rapsstengelrüssler, Großer Kohltriebrüssler	*Ceutorhynchus napi*	greater cabbage curculio

Großer Reismehlkäfer	Tribolium destructor	greater flour beetle (Scandinavian flour beetle)
Großer Riedbock	Redunca arundinum	reedbuck
Großer Rindermagennematode, Großer Rindermagenwurm	Haemonchus placei	Barber's poleworm, large stomach worm (of cattle), wire worm
Großer Roter Drachenkopf, Roter Drachenkopf, Große Meersau, Europäische Meersau	Scorpaena scrofa	bigscale scorpionfish, red scorpionfish
Großer Roter Einsiedlerkrebs, Mittelmeer-Dardanus	Dardanus calidus	great red hermit crab, Mediterranean hermit crab
Großer Rückenfüßler	Paromola cuvieri	paromola
Großer Rüsselkrebs	Bosmina coregoni	larger bosminid waterflea*
Großer Sandaal, Großer Sandspierling, Großer Tobiasfisch	Hyperoplus lanceolatus/ Ammodytes lanceolatus	greater sandeel, lance, sandlance, great sandeel (FAO)
Großer Sanddollar	Clypeaster subdepressus	giant sand dollar*
Großer Sandspierling, Großer Sandaal, Großer Tobiasfisch	Hyperoplus lanceolatus/ Ammodytes lanceolatus	greater sandeel, lance, sandlance, great sandeel (FAO)
Großer Scheckenfalter	Melitaea phoebe	knapweed fritillary
Großer Scheibenbauch	Liparis liparis	sea snail, common seasnail
Großer Schillerfalter	Apatura iris	purple emperor
Großer Schlammläufer	Limnodromus scolopaceus	long-billed dowitcher
Großer Schneckenegel	Glossiphonia complanata	snail leech, greater snail leech
Großer Schuppen-Schlangenstern	Ophiura texturata	large sand brittlestar
Großer Schuppeneinsiedler	Aniculus maximus	large hermit crab
Großer Schwarzer Kolbenwasserkäfer, Großer Kolbenwasserkäfer	Hydrophilus piceus/ Hydrous piceus	greater silver beetle, great black water beetle, great silver water beetle, diving water beetle
Großer Schwarzer Dornhai	Etmopterus princeps	great lanternshark (FAO), greater lanternshark
Großer Schwarzspitzenhai, Langnasenhai	Carcharhinus brevipinna/ Carcharhinus maculipinnis	spinner shark (FAO), long-nose grey shark
Großer Soldatenfisch	Adioryx spinifer	scarlet-fin soldierfish, spiny squirrelfish sabre squirrelfish (FAO)
Großer Sonnenröschen-Bläuling	Aricia artaxerxes	northern brown argus
Großer Steinbeißer	Cobitis elongata	Balkan loach
Großer Straußenfuß	Struthiolaria papulosa	large ostrich foot
Großer Sturmtaucher	Puffinus gravis	great shearwater
Großer Süßwasserrochen	Himantura chaophraya	giant freshwater stingray, freshwater whipray (FAO)
Großer Tabakkäfer	Catorama tabaci	catorama beetle
Großer Teufelsrochen, Riesenmanta, Manta	Manta birostris	manta, Atlantic manta, giant devil ray
Großer Totenkäfer	Blaps mortisaga	giant churchyard beetle, giant cellar beetle
Großer Tümmler, Großtümmler, Flaschennasendelphin	Tursiops truncatus	bottlenosed dolphin, common bottle-nosed dolphin

Großer Tüpfelbeutelmarder	*Dasyurus maculatus*	large spotted quoll, large spotted native "cat"
Großer Ulmensplintkäfer	*Scolytus scolytus*	large elm bark beetle, elm bark beetle, larger European bark beetle
Großer Waldgärtner, Gefurchter Waldgärtner, Kiefernmarkkäfer, Markkäfer	*Blastophagus piniperda/ Tomicus piniperda*	pine beetle, larger pith borer, large pine shoot beetle
Großer Waldportier	*Hipparchia fagi*	woodland grayling
Großer Weinschwärmer	*Hippotion celerio*	silver-striped hawkmoth
Großer Wolfshering, Indischer Hering	*Chirocentrus dorab*	wolf-herring, dorab wolf-herring (FAO)
Größere Kolumbianische Riesenvogelspinne	*Xenesthis monstrosa*	Colombian giantblack tarantula
Großes Eichenkarmin, Großes Eichenkarmin-Ordensband, Großes Karminrotes Eichenmischwald-Ordensband	*Catocala sponsa*	dark crimson underwing
Großes Elefantenohr	*Amplexidiscus fenestrafer*	giant elephant's ear*
Großes Glasauge, Goldlachs	*Argentina silus*	greater argentine
Großes Granatauge	*Erythromma najas*	red-eyed damselfly
Großes Heupferd, Grünes Heupferd	*Tettigonia viridissima*	great green bushcricket
Großes Jungfernkind, Jungfernsohn	*Archiearis parthenias*	orange underwing moth
Großes Mausohr (Riesenfledermaus)	*Myotis myotis*	greater mouse-eared bat
Großes Nachtpfauenauge	*Saturnia pyri*	giant peacock moth
Großes Rotschenkelhörnchen	*Epixerus ebii*	African palm squirrel
Großfleckiger Seehase	*Aplysia dactylomela*	large-spotted sea hare
Großflossen-Morwong	*Nemadactylus macropterus*	tarakihi
Großflossengrätenfische	Pterothrissidae	
Großfuß-Schwielensalamander	*Chiropterotriton magnipes*	bigfoot splayfoot salamander
Großfußhühner	Megapodiidae	megapodes
Großgalagos	*Otolemur* spp.	greater bush babies
Großgefleckter Katzenhai	*Scyliorhinus stellaris*	large spotted dogfish, nurse hound, bull huss, nursehound (FAO)
Großgrison	*Galictis vittata*	greater grisón
Großgrundfink	*Geospiza magnirostris*	large ground finch
Großguramis	Osphronemidae	giant gouramies
Großhufeisennase, Große Hufeisennase	*Rhinolophus ferrumequinum*	greater horseshoe bat
Großkantschil	*Tragulus napu*	greater Malay chevrotain
Großklauenspitzmäuse	*Soriculus* spp.	Asiatic shrews
Großkopf	*Acronicta megacephala*	poplar grey
Großkopf-Anolis	*Enyalius* spp.	fathead anoles
Großkopf-Knurrhahn	*Prionotus tribulus*	bighead searobin
Großkopf-Plattschildkröte, Großkopf-Pantanal-Sumpfschildkröte	*Platemys macrocephala/ Acanthochelys macrocephala*	Pantanal swamp turtle, big-headed Pantanal swamp turtle, large-headed Pantanal swamp turtle

Großkopf-Ringelnatter	*Natrix megalocephala*	bigheaded grass snake, large-headed water snake
Großkopf-Schlammschildkröte	*Claudius angustatus*	narrow-bridged musk turtle
Großkopfsardine	*Sardinella longiceps*	Indian oil sardine (FAO), oil sardine
Großkopfschildkröten	Platysternidae	big-headed turtles
Großkopfschildkröte	*Platysternon megacephalum*	big-headed turtle
Großkopfwels	*Chaca chaca*	squarehead catfish (FAO), tadpole-headed catfish
Großkopfwelse, Großmaulwelse	Chacidae	squarehead catfishes, angler catfishes, frogmouth catfishes
Großlibellen	Anisoptera	dragonflies; hawkers (Europe)
Großmaul-Stuhlwurm*	*Chabertia ovina*	large-mouth bowel worm
Großmäuler*	Champsodontidae	gapers
Großmaulhaie, Riesenmaulhaie	Megachasmidae	megamouth sharks
Großnasenhai	*Carcharhinus altimus*	bignose shark
Großnilhechte, Nilaale	Gymnarchidae	abas
Großohr-Bulldogg-Fledermäuse	*Otomops* spp.	big-eared free-tailed bats
Großohr-Kletterratte*	*Ototylomys phyllotis*	big-eared climbing rat
Großohr-Tanrek	*Geogale aurita*	large-eared tenrec
Großohrfledermäuse	*Macrotus* spp.	big-eared bats
Großohrkitfuchs, Kitfuchs (Hbz.)	*Vulpes macrotis*	kit fox
Großohrmäuse u.a.	*Graomys* spp.	big-eared mice a.o.
Großscheren-Knallkrebschen, Großscheren-Pistolenkrebs	*Alpheus macrocheles*	snapping prawn, snapping shrimp, big-claw snapping prawn*
Großscheren-Süßwassergarnele*	*Macrobrachium carcinus*	bigclaw river shrimp
Großschuppen-Eidechsenfisch, Gefleckter Eidechsenfisch	*Saurida undosquamis*	brushtooth lizardfish (FAO), large-scale lizardfish
Großschuppen-Falterfisch	*Chaetodon rafflesii*	Raffles butterflyfish, latticed butterflyfish (FAO)
Großschuppenfische	Melamphaeidae	bigscale fishes, ridgeheads
Großschuppige Grundel	*Thorogobius macrolepis*	large-scaled goby
Großschuppige Scholle	*Citharus linguatula*	spotted flounder
Großschuppiger Nilhecht	*Marcusenius macrolepidotus*	bulldog (FAO & South Africa), mputa, big-scale Nile pike
Großspornpieper	*Macronyx capensis*	Cape longclaw
Großstachlige Pazifische Igelschnecke	*Drupa ricinus*	prickly drupe, prickly Pacific drupe
Großtao, Großtinamu	*Tinamus major*	great tinamou
Großtrappe	*Otis tarda*	great bustard
Großwaran, Perenty	*Varanus giganteus*	perenty, perentie
Großwels	*Bagarius bagarius*	goonch
Grotten-Dunkelkäfer, Kellerkäfer	*Pristonychus terricola/ Laemostenus terricola*	cellar beetle*, European cellar beetle

Grottengrundel	*Speleogobius trigloides*	grotto goby
Grottenolm	*Proteus anguinus*	European olm (blind salamander)
Grottensalamander	*Typhlotriton spelaeus*	grotto salamander
Grübchenschnecken	*Lacuna* spp.	lacuna snails, chink snails
Grübchenschnecken, Lacuniden	Lacunidae	lacuna snails, chink snails (chink shells)
Grubenaale	Synaphobranchidae	cutthroat eels
Grubenottern	Crotalidae	pit vipers
Grubenschwamm	*Myxilla incrustans/ Halichondria incrustans*	yellow lobed sponge
Grubenwurm, Altwelt-Hakenwurm	*Ancylostoma duodenale*	Old World hookworm
Grünauge	*Chlorophthalmus agassizi*	green-eye, shortnose greeneye (FAO)
Grünaugen	Chlorophthalmidae	greeneyes
Grundbarsche, Darter	Etheostomatini	North American darters
Grundeln	Gobiidae	gobies
Grundhaie, Braunhaie	*Carcharhinus* spp.	requiem sharks
Gründling, Belingi-Gründling	*Gobio gobio*	gudgeon
Grundsalmler, Bodensalmler	Characidiinae	South American darters
Grüne Abalone, Grünes Meerohr	*Haliotis fulgens*	green abalone
Grüne Apfellaus	*Aphis pomi*	apple aphid, green apple aphid
Grüne Apfelwanze, Nordische Apfelwanze	*Plesiocoris rugicollis*	apple capsid bug
Grüne Baumnatter	*Elaphe prasina*	green trinket snake
Grüne Baumschnecke	*Papuina pulcherrima*	green tree snail
Grüne Blattwespe	*Rhodogaster viridis*	green sawfly
Grüne Bohnenmuschel	*Musculus discors*	discordant mussel, green crenella
Grüne Buschviper	*Atheris chloroechis*	plain green bush viper
Grüne Egelschnecke, Grüner Schnegel	*Limax maculatus*	green slug
Grüne Erbsenblattlaus, Grüne Erbsenlaus	*Acyrthosiphon pisum*	pea aphid
Grüne Erzspitznatter, Glanzspitznatter	*Oxybelis fulgidus*	green vine snake
Grüne Fichten-Großgallenlaus	*Sacchiphantes viridis/ Adelges viridis*	green spruce pineapple-gall adelges
Grüne Futterwanze	*Lygocoris pabulinus*	common green capsid
Grüne Grundel, Grüngrundel	*Microgobius thalassinus*	green goby
Grüne Haarschnecke	*Ponentina subvirescens*	green hairysnail, green snail
Grüne Heidelbeereule	*Anaplectoides prasina*	green arches
Grüne Hundskopfboa, Grüner Hundskopfschlinger	*Corallus caninus*	emerald tree boa
Grüne Huschspinne, Grasgrüne Huschspinne	*Micrommata rosea*	green spider*
Grüne Hydra, Grüner Süßwasserpolyp	*Chlorohydra viridissima/ Hydra viridis/Hydra viridissima*	green hydra

Grüne Jararaca, Amazonas-Lanzenotter	*Bothriopsis bilineata bilineata*/ *Bothrops bilineata*	Amazonian tree viper, Amazonian palm viper, two-striped forest pit viper
Grüne Juwelenanemone	*Corynactis viridis*	green jewel anemone
Grüne Keiljungfer	*Ophiogomphus serpentinus*/ *Ophiogomphus cecilia*	serpentine dragonfly
Grüne Kletternatter	*Elaphe triaspis*	green rat snake
Grüne Königslibelle	*Anax junius*	green darner, green emperor dragonfly
Grüne Krabbenspinne	*Diaea dorsata*	green crab spider
Grüne Kröte, Wechselkröte	*Bufo viridis*	green toad, European green toad (variegated toad)
Grüne Kröte, Grüne Texas-Kröte	*Bufo debilis*	green toad
Grüne Krötenotter, Grüne Nachtotter	*Causus resimus*	green night adder
Grüne Languste, Königslanguste	*Panulirus regius*	royal spiny crawfish
Grüne Makrele*, Grünmakrele*	*Caranx caballus*	green jack
Grüne Mamba	*Dendroaspis viridis*	western green mamba
Grüne Meerkatze	*Cercopithecus aethiops*	grivet monkey, savanna monkey, green monkey
Grüne Miesmuschel	*Perna viridis*/ *Mytilus smaragdinus*	green mussel
Grüne Mosaikjungfer	*Aeshna viridis*	green hawker
Grüne Muräne, Westatlantische Muräne	*Gymnothorax funebris*	green moray
Grüne Nachtotter, Grüne Krötenotter	*Causus resimus*	green night adder
Grüne Nothenia	*Notothenia gibberifrons*	bumphead notothenia, humped rockcod (FAO)
Grüne Pfirsichblattlaus, Grüne Pfirsichlaus	*Myzus persicae*	peach-potato aphid, green peach aphid, greenfly, cabbage aphid
Grüne Riesenanemone*	*Anthopleura xanthogrammica*	giant green anemone, great green anemone
Grüne Samtschnecke	*Elysia viridis*	green velvet snail*, green elysia
Grüne Scheidenmuschel	*Solen viridis*	green jackknife clam
Grüne Schenkelfliege	*Meromyza saltatrix*	grass fly
Grüne Schmerle, Blaue Prachtschmerle	*Botia modesta*	orange-finned loach
Grüne Seespinne	*Mithrax sculptus*	green clinging crab
Grüne Steinfliegen	Chloroperlidae	green stoneflies
Grüne Stern-Turbanschnecke	*Lithopoma tuber*	green starsnail
Grüne Stinkwanze	*Palomena prasina*	green shield bug, common green shield bug
Grüne Strandschrecke	*Aiolopus thalassinus*/ *Epacromia thalassina*	longwinged grasshopper
Grüne Strauchschrecke	*Eupholidoptera chabrieri*	Chabrier's bushcricket
Grüne Tigergarnele	*Penaeus semisulcatus*	green tiger prawn, zebra prawn

Grüne Turbanschnecke, Grüner Turban	*Turbo marmoratus*	green turban, great green turban, giant green turban, green snail
Grüne Westindische Käferschnecke, Grüne Westindische Chiton	*Chiton tuberculatus*	West Indian green chiton, common West Indian chiton
Grüne-Riffseescheide	*Didemnum molle*	green reef sea-squirt
Grüner Acouchi	*Myoprocta acouchy* (Husson '78)	green acouchi
Grüner Baumgecko	*Naultinus elegans*	green tree gecko
Grüner Baumpython	*Morelia viridis*	green tree python
Grüner Blattnager	*Phyllobius maculicornis*	green leaf weevil
Grüner Blattwurm	*Eulalia viridis*	greenleaf worm, green paddle worm
Grüner Bohrschwamm	*Cliona viridis*	green boring sponge
Grüner Diskus	*Symphysodon aequifasciata aequifasciata*	green discus
Grüner Eichenwickler	*Tortrix viridana*	pea-green oak curl, green oak tortrix, oak leafroller, green oak roller, oak tortrix
Grüner Feuerwurm	*Hermodice carunculata*	green fireworm
Grüner Geißkleespanner	*Pseudoterpna pruinata*	grass emerald
Grüner Inselleguan	*Iguana delicatissima*	delicate green Carribean Island iguana, Lesser Antillean iguana (IUCN)
Grüner Junikäfer*	*Cotinis nitida*	green June beetle
Grüner Kugelfisch	*Tetraodon fluviatilis*	green pufferfish
Grüner Laubfrosch	*Hyla cinerea*	green treefrog
Grüner Leguan	*Iguana iguana*	green iguana, common iguana
Grüner Lippfisch, Amsellippfisch, Meerdrossel	*Labrus viridis*	green wrasse
Grüner Messerfisch	*Eigenmannia virescens*	green knifefish
Grüner Neon, Costello-Salmler	*Hemigrammus hyanuary*	January tetra
Grüner Pavian	*Papio anubis*	olive baboon
Grüner Raschkäfer	*Elaphrus viridis*	delta green ground beetle
Grüner Salmler	*Alestes chaperi*	Chaper's characin
Grüner Schildkäfer	*Cassida viridis*	green tortoise beetle
Grüner Schlank-Seestern*	*Leptasterias littoralis*	green slender seastar
Grüner Schnegel, Grüne Egelschnecke	*Limax maculatus*	green slug
Grüner Schnurwurm	*Lineus viridis* (Nemertini)	green bootlace
Grüner Seeigel	*Echinometra viridis*	green sea urchin*
Grüner Seeringelwurm	*Nereis virens/ Neanthes virens*	ragworm, sandworm, clam worm, king ragworm
Grüner Sonnenbarsch, Grasbarsch	*Lepomis cyanellus*	green sunfish
Grüner Springschwanz	*Isotoma viridis*	green springtail
Grüner Stör	*Acipenser medirostris*	green sturgeon
Grüner Streifenhechtling, Ceylonhechtling	*Aplocheilus dayi*	Ceylon killifish (FAO), Day's killifish

Grüner Stummelaffe, Schopfstummelaffe	*Procolobus verus*	olive colobus
Grüner Süßwasserpolyp, Grüne Hydra	*Chlorohydra viridissima/ Hydra viridis/ Hydra viridissima*	green hydra
Grünes Blatt	*Geometra papilionaria*	large emerald
Grünes Pantoffeltierchen	*Paramecium bursaria*	green slipper animalcule
Grünes Schwalbenschwänzchen, Grüner Schwalbenschwanz, Grünling	*Chromis caerulens*	blue puller, green chromis (FAO)
Grünes Trompetentierchen	*Stentor polymorphus*	green trumpet animalcule
Grünflächige Turbanschnecke	*Tegula excavata*	green-base tegula
Grünflossen-Papageifisch	*Scarus sordidus*	green-finned parrotfish, garned red parrotfish, daisy parrotfish (FAO)
Grünflügelara, Dunkelroter Ara	*Ara chloroptera*	green-winged macaw
Grüngelbe Lanzenotter	*Bothriechis lateralis/ Bothrops lateralis*	yellow-lined palm viper, side-striped palm pit viper
Grüngrundel, Grüne Grundel	*Microgobius thalassinus*	green goby
Grunion	*Leuresthes tenuis*	grunion
Grünkopf-Baumnatter	*Chironius scurulus*	Wagler's sipo
Grünkugelgallwespe (>Grüne Kugelgalle)	*Andricus inflator*	oak bud globular-gall wasp
Grünkupferner Rohrkäfer	*Donacia vulgaris*	shiny-green reed beetle*
Grünlaubsänger	*Phylloscopus trochiloides*	greenish warbler
Grünliche Kauri	*Cypraea subviridis*	green-tinted cowrie, greenish cowrie
Grünlicher Wassermolch	*Notophthalmus viridescens*	eft, red eft, red-spotted newt, eastern newt
Grünling	*Carduelis chloris*	greenfinch
Grünlinge	Hexagrammidae	greenlings
Grünnattern	*Chlorophis* spp.	green snakes
Grünrücken-Bastardmakrele	*Trachurus declivis*	green-back horse mackerel
Grünrücken-Nektarvogel	*Nectarinia jugularis*	olive-backed sunbird
Grünschenkel	*Tringa nebularia*	common greenshank
Grünschwanz-Grundammer	*Pipilo chlorurus/ Chlorura chlorura*	green-tailed towhee
Grünsittich	*Aratinga holochlora/ Psittacara holochlora*	green parakeet
Grünspecht	*Picus viridis*	green woodpecker
Grünwangenamazone	*Amazona viridigenalis*	red-crowned parrot (Mexican red-head amazon)
Grünwidderchen, Grasnelken-Widderchen	*Procris statices/ Adscita statices*	forester, common forester
Grunzer	*Haemulon* spp.	grunts
Grunzer, Grunzerfische, Schweinsfische	Pomadasyidae/ Haemulidae	grunts (rubberlips & grunters)
Grünzügelpapagei	*Pionites melanocephalus*	black-headed caique
Gryllteiste	*Cepphus grylle*	black guillemot
Guadalupe Seebär	*Arctocephalus townsendi*	Guadalupe fur seal
Guam-Kardinalfisch	*Apogon guamensis/ Apogon nubilus*	pearl cardinalfish

Guanako	*Lama guanicoe*	guanaco
Guanoscharbe	*Phalacrocorax bougainvillei*	Guanay cormorant
Guanotölpel	*Sula variegata*	Peruvian booby
Guatemala-Brüllaffe	*Alouatta villosa*	Guatemalan howler
Guatemala-Lanzenotter	*Bothriechis bicolor*	Guatemala palm pit viper
Guatemalakärpfling, Honduraskärpfling	*Phallichthys amates*	merry widow
Guayana-Delphin, Karibischer Küstendelphin	*Sotalia guianensis*	Guayana river dolphin, Guiana coastal dolphin, white dolphin
Guayana-Erdschildkröte, Südamerikanische Erdschildkröte	*Rhinoclemmys punctularia*	spot-legged turtle, spotted-legged turtle
Guayaquilsittich	*Aratinga erythrogenys*	red-masked conure, cherry head conure
Guereza, Mantelaffe (Hbz. Scheitelaffe)	*Colobus guereza/ Colobus abyssinicus*	guereza, guereza colobus (eastern black-and-white colobus)
Guinea-Barrakuda	*Sphyraena afra*	Guinean barracuda
Guinea-Messerzahnaal	*Cynoponticus ferox*	fierce conger, Guinean pike conger (FAO)
Guinea-Wühlen	*Schistometopum* spp.	Guinea caecilians
Guineapavian	*Papio papio*	Guinean baboon, western baboon
Guineaturako	*Tauraco persa*	green turaco
Guineawurm, Medinawurm, Drachenwurm	*Dracunculus medinensis*	fiery serpent, medina worm, guinea worm
Gummiboa, Nordamerikanische Sandboa	*Charina bottae*	rubber boa
Gummiwickler, Rindenwickler	*Enarmonia formosana*	cherry bark tortrix (moth)
Gundis	*Ctenodactylus* spp.	gundis
Gunnisons Präriehund	*Cynomys gunnisoni*	Gunnison's prairie dog
Gunthers Weißlippen-Pfeiffrosch	*Leptodactylus albilabris*	Gunther's white-lipped frog
Günthers Prachtbuntbarsch	*Chromidotilapia guentheri*	Guenther's mouthbrooder
Günthers Stacheleidechse	*Holaspis guentheri*	sawtail lizard
Guppy, Millionenfisch	*Lebistes reticulatus/ Poecilia reticulata*	guppy
Gurami, Speisegurami, Riesengurami, Knurrender Gurami	*Osphronemus goramy*	gourami, giant gourami
Guramis, Labyrinthfische	Belontiidae/Polyacanthidae (Trichogasterinae)	gouramis, gouramies
Gurkenkernbandwurm	*Dipylidium caninum*	double-pored dog tapeworm
Gurkenqualle, Melonenqualle	*Beroe cucumis*	melon jellyfish, melon comb jelly
Gurkenquallen, Melonenquallen	*Beroe* spp.	beroids, melon jellyfish
Gurkenwalze, Gurken-Walzenschnecke	*Cymbium cucumis*	cucumber volute
Gürtel-Eingeweideaal	*Pisodonophis semicinctus*	saddled snake eel
Gürtel-Felsschnecke	*Trochia cingulata*	corded rocksnail

Gürtel-Randschnecke	*Persicula cingulata/ Persicula marginata*	girdled marginella, girdled marginsnail (girdled margin shell, belted margin shell)
Gürtel-Riss-Schnecke	*Scissurella cingulata*	belt scissurelle
Gürtel-Rosenblattwespe, Gebänderte Rosenblattwespe	*Allantus cinctus*	banded rose sawfly
Gürtel-Sandfisch	*Serranus subligarius*	belted sandfish
Gürtel-Schlammschnecke	*Cerithidea cingulata*	girdled hornsnail (girdled horn shell)
Gürtel-Tritonshorn	*Gelanga succincta*	lesser girdled triton
Gürtelechsen	Cordylidae	girdle-tailed lizards & plated lizards
Gürtelfischer	*Ceryle alcyon*	belted kingfisher
Gürtelmull, Gürtelmaus, Schildwurf	*Chlamyphorus truncatus*	pink fairy armadillo, lesser fairy armadillo, lesser pichi ciego, pichiciego
Gürtelmull> Nördlicher Gürtelmull	*Chlamyphorus retusa/ Burmeisteria retusa*	Chacoan fairy armadillo, greater fairy armadillo, Burmeister's armadillo
Gürtelmulle	*Chlamyphorus* spp.	pichiciegos
Gürtelrose, Ringelrose	*Actinia cari*	green sea anemone
Gürtelschlangen	*Brachyurophis* spp.	girdled snakes
Gürtelschnecke*	*Hygromia cinctella*	girdled snail
Gürtelschweif> Gewöhnlicher Gürtelschweif	*Cordylus cordylus*	common girdled lizard, common spinytail lizard, Lord Derby lizard
Gürtelschweife	*Cordylus* spp.	girdle-tailed lizards, spinytail lizards
Gürtelskolopender> Südeuropäischer Gürtelskolopender	*Scolopendra cingulata*	southern girdled scolopendra*
Gürteltiere	Cingulata/Loricata (Xenarthra)	armadillos
Gürtelwürmer, Clitellaten	Clitellata	clitellata
Güster, Blicke, Pliete (Halbbrachsen)	*Blicca bjoerkna*	silver bream, white bream (FAO)

Haarbalgmilbe	*Demodex folliculorum*	follicle mite, human follicle mite
Haarbalgmilben	Demodicidae	follicle mites
Haarbuschdrongo	*Dicrurus hottentottus*	hair-crested drongo
Haarbutt, Müllers Zwergbutt	*Zeugopterus punctatus*	topknot
Haarfrosch	*Trichobatrachus robustus*	hairy frog
Haarfrösche	*Trichobatrachus* spp.	hairy frogs
Haarige Käferschnecke	*Mopalia ciliata*	hairy chiton, hairy mopalia
Haarige Miesmuschel, Bartige Miesmuschel, Bartmuschel	*Modiolus barbatus/ Lithographa barbatus*	bearded horse mussel
Haarige Winkerkrabbe*	*Uca vocator*	hairback fiddler
Haarige Wollkrabbe	*Dromidia antillensis*	hairy sponge crab
Haarmilben	Listrophoridae	fur mites
Haarmücken	Bibionidae	march flies, St.Mark's flies
Haarnasenotter (Hbz. Sumatra-Otter)	*Lutra sumatrana*	Sumatra river otter, hairy-nosed otter
Haarnasenwombats	*Lasiorhinus* spp.	hairy-nosed wombats, soft-furred wombats
Haarrückenspanner	*Colotois pennaria*	feathered thorn
Haarschnecke, Gemeine Haarschnecke, Behaarte Laubschnecke	*Trichia hispida/ Helix hispida*	hairysnail, bristly snail
Haarschnecken	*Trichia* spp.	hairysnails
Haarschnecken	Trichotropidae	hairy snails, hairysnails (hairy shells)
Haarschuppenspinner, Frostspinner	*Ptilophora plumigera*	plumed prominent
Haarschwanz, Degenfisch	*Trichiurus lepturus*	Atlantic cutlassfish, large-eyed hairtail, large-head hairtail, largehead hairtail (FAO)
Haarschwänze, Rinkfische	Trichiuridae	frostfishes, scabbardfishes, cutlassfishes, hairtails
Haarspecht	*Picoides villosus*	hairy woodpecker
Haarsterne, Federsterne, Comatuliden	Comatulida	feather stars, comatulids
Haartriton, Haar-Triton	*Cymatium corrugatum*	hairy triton, corrugated triton
Haarvögel, Bülbüls	Pycnonotidae	bulbuls, greenbuls
Haarwurm, Bancroft-Filarie	*Wuchereria bancrofti*	Bancroftian filariae
Haarwürmer	Spionidae	spionid worms, spios, spionids
Habicht	*Accipiter gentilis*	goshawk, northern goshawk
Habichtartige	Accipitrinae	hawks & eagles & accipiters & kites
Habichtsadler	*Hieraaetus fasciatus*	Bonelli's eagle

Habichtsflügelschnecke	*Strombus raninus*	hawkwing conch
Habichtskauz	*Strix uralensis*	Ural owl
Habu-Schlange, Habu	*Trimeresurus flavoviridis*	Okinawa habu, yellow-spotted pit viper
Haddons Anemone, Teppichanemone	*Stichodactyla haddoni*	Haddon's anemone, saddle anemone, carpet anemone
Hafenkrabbe	*Liocarcinus depurator*	harbour crab, harbour swimming crab
Haferblattlaus, Haferlaus, Traubenkirschenlaus	*Rhopalosiphum padi*	bird-cherry aphid, oat aphid, wheat aphid
Haferkornschnecke	*Chondrina avenacea*	
Hafermilbe	*Steneotarsonemus spirifex/ Tarsonemus spirifex*	oat spiral mite
Haferzystenälchen, Getreidezystenälchen, Hafernematode	*Heterodera avenae/ Heterodera major*	oat cyst nematode, cereal cyst nematode, cereal root nematode
Hafte, Echte Netzflügler	Neuroptera/Planipennia	neuropterans (dobson flies/antlions)
Haftscheiben-Fledermäuse	*Thyroptera* spp.	disk-winged bats, New World sucker-footed bats
Hagebuttenfliege	*Rhagoletis alternata*	rose-hip fly
Häherkuckuck	*Clamator glandarius*	great spotted cuckoo
Häherlinge	Garrulacinae	laughingthrushes
Hahnen-Fechterschnecke, Hahnen-Flügelschnecke	*Strombus gallus*	roostertail conch, rooster-tail conch
Hahnenfisch, Dianafisch	*Luvarus imperialis*	luvar, louvar
Hahnenfische	Luvaridae	louvars
Hahnenkammauster	*Lopha cristagalli*	cock's-comb oyster, cockscomb oyster, coxcomb oyster
Hahnenkammauster	*Ostrea crestata*	cock's comb oyster* a.o.
Hahnenkammkrabbe, Mittelmeer-Schamkrabbe	*Calappa granulata*	Mediterranean shame-faced crab
Hahnfisch*	*Nematistius pectoralis*	roosterfish
Haibarbe	*Balantiocheilus melanopterus*	bala shark, silver shark, tricolor sharkminnow (FAO)
Haie> Echte Haie	Carcharhiniformes	tiger sharks/catsharks & sand sharks & requiem sharks & hammerheads and others
Haifischauge, Große Mondschnecke	*Neverita duplicata/ Polinices duplicatus*	shark eye, sharkeye moonsnail, lobed moonsnail
Haiflossen-Geigenrochen	Rhinidae	guitarfishes
Hain-Laufkäfer	*Carabus nemoralis*	forest ground beetle
Hain-Schnirkelschnecke, Schwarzmündige Bänderschnecke, Hainbänderschnecke	*Cepaea nemoralis*	brown-lipped snail, grove snail, grovesnail, English garden snail, larger banded snail, banded wood snail
Hainasen-Grundel	*Gobiosoma evelynae*	sharknosed goby
Hainbuchen-Zipfelkäfer*	*Hypebaeus flavipes*	Moccas beetle

Hainbuchenspanner, Großer Frostspanner	*Erannis defoliaria*	mottled umber
Hainveilchen-Perlmutterfalter, Kleinster Perlmutterfalter	*Clossiana dia/ Boloria dia*	violet fritillary, Weaver's fritillary
Haiti-Boa	*Epicrates striatus*	Fischer's tree boa
Haiti-Riesenanolis	*Anolis ricordi*	Haitian green anole
Haiti-Schlitzrüssler	*Solenodon paradoxus*	Haitian solenodon
Haiti-Vogelspinne	*Phormictopus cancerides*	Haitian brown tarantula
Haiwels	*Pangasius sutchi*	Sutchi catfish
Haken-Seeschmetterling	*Cavolinia uncinata*	uncinate cavoline, hooked cavoline
Hakengimpel	*Pinicola enucleator*	pine grosbeak
Hakenkäfer	Elmidae	drive beetles, riffle beetles
Hakenkäfer, Klauenkäfer	Dryopidae	long-toed water beetles
Hakenkalmar, Krallenkalmar	*Onychoteuthis banksii*	common clubhook squid, clawed squid, clawed calamary squid
Hakennatter> Gewöhnliche Hakennatter, Östliche Hakennatter	*Heterodon platyrhinos*	eastern hognose snake
Hakennattern, Hakennasen-Nattern	*Heterodon* spp.	hognose snakes
Hakenrüssler	Kinorhyncha	kinorhynchs
Hakenscheibensalmler	*Myleus rubripinnis*	redhook myleus
Hakenschnabel-Schlankblindschlange	*Leptotyphlops macrorhynchus*	hookbilled blindsnake
Hakenwanze	*Podops inuncta*	European turtle-bug
Hakenwürmer u.a.	*Bunestomum* spp.	hookworms a.o.
Hakenzähniger Kiefernborkenkäfer	*Pityogenes bidentatus*	bidentated bark beetle
Halbaffen	Prosimii	prosimians, lower primates
Halbentblößte Steckmuschel*	*Atrina seminuda*	half-naked penshell
Halbfingergecko, Europäischer Halbfinger	*Hemidactylus turcicus*	Turkish gecko, Mediterranean gecko
Halbflügler, Schnabelkerfe	Hemiptera/Rhynchota (Heteroptera & Homoptera)	hemipterans, bugs
Halbgebänderte Plattschwanz-Seeschlange	*Laticauda semifasciata*	Chinese sea snake
Halbkuglige Napfschildlaus	*Saissetia coffeae*	hemispherical scale
Halbmakis	*Hapalemur* spp.	gentle lemurs
Halbmasken-Falterfisch	*Chaetodon semilarvatus*	red-lined butterflyfish, bluecheek butterflyfish (FAO)
Halbmond-Lanzenotter	*Bothrops alternatus*	urutu, wutu
Halbmotten	Acrolepiidae	smudges
Halbringschnäpper	*Ficedula semitorquata*	semi-collared flycatcher
Halbschnäbler, Halbschnabelhechte	Hemiramphidae/ Hemirhamphidae	halfbeaks
Halbsepie	*Hemisepius typicus*	semi-cuttlefish*
Halbzähner, Schlanksalmler	Hemiodontidae/Hemiodidae	hemiodontids
Haldenflur-Nelkeneule, Haldenflur-Netzeule, Hellgerippte Garteneule	*Heliophobus reticulatus*	bordered gothic (moth)

Halfterfisch, Maskenfisch, Maskenwimpelfisch	Zanclus cornutus/ Zanclus canescens/ Chaetodon canescens	moorish idol
Halfterfische, Maskenfische, Maskenwimpelfische	Zanclidae	moorish idols
Hallers Rund-Stechrochen	Urolophus halleri	Haller's round ray (FAO), Haller's round stingray
Halmfliegen, Gelbkopffliegen	Chloropidae	chloropids, chloropid flies (eye gnats, grass flies, eye flies)
Halmwespen	Cephidae	stem sawflies
Halosauriden	Halosauridae	halosaurs
Halsband-Anemonenfisch, Weißbinden-Glühkohlen-Anemonenfisch, Roter Clownfisch, Unechter Glühkohlenfisch	Amphiprion frenatus	red clownfish, tomato anemonefish, tomato clownfish (FAO)
Halsband-Falterfisch	Chaetodon collare	red-tailed butterflyfish
Halsband-Faulvogel	Bucco capensis	collared puffbird
Halsband-Mangabe	Cercocebus torquatus	red-crowned mangabey, red-capped mangabey, white-collared mangabey
Halsband-Mondschnecke, Halsband-Nabelschnecke, Große Nabelschnecke	Natica catena/ Euspira catena/ Lunatia catena	large necklace snail, large necklace moonsnail (large necklace shell)
Halsband-Zwergfalke	Polihierax semitorquatus	African pygmy falcon
Halsband-Zwergnatter	Eirenis collaris	collared dwarf snake
Halsbanddrossel	Ixoreus naevius	varied thrush
Halsbandfrankolin	Francolinus francolinus	black francolin
Halsbandleguan, Kugelechse	Crotaphytus collaris	collared lizard
Halsbandleguane	Crotaphytus spp.	collared lizards
Halsbandlemminge	Dicrostonyx spp.	collared lemmings, varying lemmings
Halsbandmaina	Acridotheres albocinctus	white-collared mynah
Halsbandnatter	Diadophis punctatus	ringneck snake
Halsbandnattern	Diadophis spp.	ringneck snakes
Halsbandpekari	Tayassu tajacu	collared peccary
Halsbandschnäpper	Ficedula albicollis	collared flycatcher
Halsbandsittich	Psittacula krameri	rose-ringed parakeet, ring-necked parakeet
Halsketten-Reusenschnecke	Nassarius distortus	necklace nassa
Halysschlange, Halys-Grubenotter, Halysotter	Agkistrodon halys/ Gloydius halys	Asiatic pit viper, Pallas' viper, Halys' viper, mamushi
Hamilton-Frosch	Leiopelma hamiltoni	Hamilton's frog
Hammel-Schnapper	Lutjanus analis	mutton snapper
Hammerhai> Gemeiner Hammerhai, Glatter Hammerhai	Sphyrna zygaena/ Zygaena malleus	common hammerhead shark, smooth hammerhead (FAO)
Hammerhaie	Sphyrnidae	hammerhead sharks
Hammerhuhn	Macrocephalon maleo	maleo
Hammerkieferfisch	Omosudis lowei	omosudid
Hammerkieferfische	Omosudidae	omosudids

Hammerkopf	Scopidae	hamerkop (hammerhead)
Hammerkopf	*Scopus umbretta*	hamerkop (hammerhead)
Hammerkopfflughund, Hammerkopf	*Hypsignathus monstrosus*	hammer-headed fruit bat
Hammerkoralle	*Euphyllia ancora*	hammer-tooth coral*
Hammermuscheln	Malleidae	hammer oysters
Hammermuscheln	*Malleus* spp.	hammer oysters
Hamster, Feldhamster (Hbz. Hamster)	*Cricetus cricetus*	common hamster, black-bellied hamster
Hamsterratte> Kleine Hamsterratte	*Beamys hindei*	long-tailed pouched rat, lesser pouched rat, lesser hamster-rat
Handfische*	Brachionichthyidae	handfishes (warty anglers)
Händlerkegel	*Conus mercator*	trader cone
Handwühlen	*Bipes* spp.	two-legged worm lizards
Hanferdfloh, Hopfenerdfloh	*Psylliodes attenuata*	hop flea beetle, European hop flea beetle
Hängemattenspinne*	*Pityohyphantes costatus*	hammock spider
Hängende Wattschnecke, Bauchige Wattschnecke	*Hydrobia ventrosa/ Hydrobia stagnalis*	spine snail, hanging mud snail*
Hanumanlangur, Hulman	*Presbytis entellus*	hanuman langur
Hapuku-Wrackbarsch	*Polyprion oxygeneios*	Hapuku wreckfish
Hardun	*Agama stellio/Stellio stellio*	roughtail rock agama, hardun
Harfen-Randschnecke	*Glabella harpaeformis*	harplike marginella
Harfenschnecken	Harpidae	harp snails (harp shells)
Harlekin Seenadel, Schwarzbrust-Seenadel	*Corythoichthys nigripectus*	gilded pipefish
Harlekin, Stachelbeerspanner	*Abraxas grossulariata*	magpie moth, currant moth
Harlekin-Fledermäuse	*Scotomanes* spp.	harlequin bats
Harlekin-Flughund, Weißstreifen-Flughund	*Styloctenium wallacei*	striped-faced fruit bat
Harlekin-Korallenschlange	*Micrurus fulvius*	harlequin coral snake
Harlekin-Mitra	*Vexillum histrio*	harlequin mitre, harlequin miter
Harlekin-Süßlippe, Gepunktete Süßlippe	*Plectorhinchus chaetodonoides*	harlequin sweetlip
Harlekinfrösche	Pseudidae	harlequin frogs, pseudids
Harnischwelse	Loricariidae	suckermouth armored catfishes
Harpagiferiden	Harpagiferidae	plunderfishes
Harpyie	*Harpia harpyja*	harpy eagle
Hartkopf-Ährenfisch	*Atherinomorus lacunosus*	hardyhead silverside
Hartschalige Tellmuschel	*Tellina solidula*	hardshell tellin
Harzbiene, Kleine Harzbiene	*Anthidium strigatum/ Anthidiellum strigatum*	small anthid bee
Hase> Europäischer Hase, Europäischer Feldhase	*Lepus europaeus*	European hare
Hasel	*Leuciscus leuciscus*	dace
Haselblattroller, Dickkopfrüssler	*Apoderus coryli*	hazel weevil

Haselbock	*Oberea linearis*	hazel borer, hazelnut capricorn beetle
Haseleule	*Colocasia coryli*	nut-tree tussock
Haselhuhn	*Bonasa bonasia/ Tetrastes bonasia*	hazel grouse
Haselmaus	*Muscaridinus avellanarius*	common dormouse, hazel mouse
Haselnussbohrer	*Curculio nucum*	nut weevil
Haselnusslaus	*Corylobium avellanae*	large hazel aphid
Haselnusswickler	*Epinotia tenerana*	nut bud tortrix (moth)
Hasen> Echte Hasen	*Lepus* spp.	hares, jack rabbits
Hasenartige	Leporidae	leporids (rabbits and hares)
Hasenkängurus	*Lagorchestes* spp.	hare wallabies
Hasenmaul-Fledermäuse, Hasenmäuler	Noctilionidae	bulldog bats, fisherman bats
Hasenmaul-Fledermäuse, Hasenmäuler	*Noctilio* spp.	bulldog bats, fisherman bats
Hasenmäuse	*Lagidium* spp.	mountain viscachas
Hasenohr	*Concholepas concholepas*	barnacle rocksnail (barnacle rock shell, hare's ear shell)
Hauben-Archenmuschel*	*Cucullaea labiata*	hooded ark (hooded ark shell)
Hauben-Azurjungfer	*Coenagrion armatum*	Norfolk damselfly (Br.)
Hauben-Goldschultersittich	*Psephotus dissimilis*	hooded parakeet, hooded parrot
Hauben-Mangabe	*Cercocebus galeritus*	agile mangabey, Tana mangabey
Haubenadler	*Spizaetus cirrhatus*	crested hawk eagle
Haubengarnele	*Athanas nitescens*	
Haubenkauz	*Lophostrix cristata*	crested owl
Haubenlangur	*Presbytis cristata*	silvered leaf monkey
Haubenlerche	*Galerida cristata*	crested lark
Haubenmaina	*Acridotheres cristatellus*	crested mynah, Chinese jungle mynah
Haubenmeise	*Parus cristatus*	crested tit
Haubenmuschel, Häubchenmuschel, Teich-Kugelmuschel	*Sphaerium lacustre/ Musculium lacustre*	lake orb mussel, lake fingernailclam
Haubennetzspinnen, Kugelspinnen	Theridiidae	comb-footed spiders, cobweb and widow spiders, scaffolding-web spiders
Haubenpinguin	*Eudyptes schlegeli*	royal penguin
Haubenschleimfisch	*Clinitrachus argentatus*	Mediterranean clinid
Haubenschnecken (Mützenschnecken) (inkl. Haarschnecken)	Capulidae	cap limpets, capsnails (incl. hairysnails)
Haubentaucher	*Podiceps cristatus*	great crested grebe
Haufenschnecken	Planaxidae	clusterwinks, grooved snails
Häufige Hammermuschel, Schwarze Hammermuschel	*Malleus malleus*	common hammer-oyster
Hauhechelbläuling, Wiesenbläuling	*Polyommatus icarus*	common blue
Hauptfeldwebelfisch, Fünfstreifen-Zebrafisch, Sergeant Major	*Abudefduf saxatlis*	sergeant major

Hauptmannsgarnele	*Fenneropenaeus chinensis/ Penaeus chinensis*	fleshy prawn
Haus-Drüsenameise	*Tapinoma sessile*	odorous house ant
Hausammer	*Emberiza striolata*	house bunting
Hausbock, Hausbockkäfer	*Hylotrupes bajulus*	house longhorn beetle
Häuschenschwamm, Korkschwamm, Einsiedler-Korkschwamm	*Suberites domuncula*	sulphur sponge, sulfur sponge
Hausfliege, Stallfliege	*Muscina stabulans*	false stable fly
Hausgeflügel-Bandwurm*	*Davainea proglottina*	poultry tapeworm
Hausgimpel, Mexikanischer Karmingimpel	*Carpodacus mexicanus*	house finch, California linnet
Hausgrille, Heimchen	*Acheta domestica*	house cricket (domestic cricket, domestic grey cricket)
Haushund	*Canis familiaris*	domestic dog
Hauskatze	*Felis catus/ Felis silvestris f. catus*	domestic cat
Hausmaus	*Mus musculus*	house mouse
Hausmausfloh	*Leptopsylla segnis*	European mouse flea, house mouse flea
Hausmausmilbe	*Liponyssoides sanguineus/ Allodermanyssus sanguineus*	house-mouse mite
Hausmilbe, Polstermilbe	*Glycyphagus domesticus*	furniture mite
Hausmücke, Gemeine Stechmücke	*Culex pipiens*	house mosquito, northern common house mosquito, common gnat (Br.), house gnat (Br.)
Hausmutter	*Noctua pronuba*	large yellow underwing
Hauspferd	*Equus caballus*	horse
Hausratte, Alexandriner Hausratte, Dachratte	*Rattus rattus*	black rat, roof rat, house rat
Hausrotschwanz	*Phoenicurus ochruros*	black redstart
Hausschaf	*Ovis aries*	domestic sheep
Hausschlange> Afrikanische Hausnatter, Gewöhnliche Hausschlange	*Boaedon lineatus/ Lamprophis lineatus*	African house snake, African brown house snake
Hausschlangen, Hausnattern, Boazähner	*Boaedon* spp./*Lamprophis* spp.	house snakes
Haussegler	*Apus affinis*	little swift
Haussperling	*Passer domesticus*	house sparrow
Hausspinne, Gemeine Hauswinkelspinne	*Tegenaria domestica*	common European house spider, lesser house spider, barn funnel weaver (U.S.)
Hausspitzmaus	*Crocidura russula*	greater white-toothed shrew
Hausstaubmilbe, Europäische Hausstaubmilbe	*Dermatophagoides pteronyssinus*	European house dust mite
Hausstaubmilben	Pyroglyphidae	house dust mites
Hauswinkelspinne> Gemeine Hauswinkelspinne, Hausspinne	*Tegenaria domestica*	common European house spider, lesser house spider, barn funnel weaver (U.S.)
Hauswinkelspinnen	*Tegenaria* spp.	European house spiders
Hauszaunkönig	*Troglodytes aedon*	house wren

Hausziege	*Capra hircus*	domestic goat
Hautflügler	Hymenoptera	hymenopterans
Haviside-Delphin, Kapdelphin	*Cephalorhynchus heavisidii*	South African dolphin, Haviside's dolphin, Benguela dolphin
Hawaiianische Baumschnecken	*Achatinella* spp.	Hawaiian tree snails
Hawaiianische Baumschnecken	Achatinellidae	Hawaiian tree snails
Hawaiibussard	*Buteo solitarius*	Hawaiian hawk, io
Hawaiigans, Nene	*Branta sandvicensis*	Hawaiian goose, nene
Heaths Kalkschwamm*	*Leucandra heath*	Heath's sponge
Hebräische Kegelschnecke, Israelische Kegelschnecke	*Conus ebraeus*	Hebrew cone
Hebräische Mondschnecke, Hebräische Nabelschnecke	*Naticarius hebraeus/ Natica maculata*	Hebrew moonsnail (Hebrew moon shell, Hebrew necklace shell)
Hebräische Walzenschnecke	*Voluta ebraea*	Hebrew volute
Hecht, Flusshecht	*Esox lucius*	pike, northern pike
Hechtbandwurm	*Triaenophorus nodulosus*	pike tapeworm*
Hechte	Esocidae	pikes
Hechtkärpfling	*Belonesox belizanus*	pike top minnow, pike killifish (FAO)
Hechtkopf	*Luciocephalus pulcher*	pikehead
Hechtköpfe	Luciocephalidae	pikeheads
Hechtköpfiger Halbschnäbler	*Dermogenys pusillus*	wrestling halfbeak
Hechtlinge u.a.	Aplocheilidae	rivulines
Hechtlinge u.a.	Galaxiidae	galaxiids
Hechtsalmler	Ctenoluciidae	American pike-characids
Hechtschleimfische	Chaenopsidae	pikeblennies, tubeblennies, flagblennies
Heckenbraunelle	*Prunella modularis*	dunnock
Heckenhausmaus	*Mus spretus*	Algerian mouse
Heckenkirschenspanner	*Chloroclysta truncata*	common marbled carpet
Heckensänger	*Agrobates galactotes/ Cercotrichas galactotes*	rufous scrub robin, rufous-tailed scrub robin, rufous warbler
Heckenschnecke*	*Hygromia limbata*	hedge snail
Hector-Delphin, Neuseeland-Delphin	*Cephalorhynchus hectori*	Hector's dolphin, New Zealand white-front dolphin
Hectors-Zweizahnwal	*Mesoplodon hectori*	Hector's beaked whale, skew-beaked whale
Hefekäfer	*Cartodere filiformis*	yeast beetle
Heide-Radspinne, Heideradspinne	*Araneus adiantus/ Neoscona adianta*	heathland orbweaver
Heide-Töpferwespe, Glockenwespe	*Eumenes coarctatus*	heath potter wasp, potter wasp
Heideblattkäfer	*Lochmaea suturalis*	heather beetle
Heidekraut-Fleckenspanner, Heidekraut-Punktstreifenspanner	*Dyscia fagaria*	grey scalloped bar (moth)
Heidekrauterdfloh	*Altica ericeti*	heather flea beetle
Heidekrauteule	*Anarta myrtilii*	beautiful yellow underwing

Heidekrautspanner	*Ematurga atomaria*	common heath
Heidekrautwurzelbohrer	*Hepialus hecta*	gold swift
Heidelbeerwickler	*Rhopobota myrtilana*	blueberry leaf tier (moth)
Heidelbergmensch	*Homo heidelbergensis* (*Homo erectus heidelbergensis*)	Heidelberg man
Heidelerche	*Lullula arborea*	woodlark
Heidelibellen	*Sympetrum* spp.	sympetrums (darters)
Heideschabe, Küsten-Waldschabe	*Ectobius panzeri*	lesser cockroach
Heideschnecken	Helicellinae	heath snails
Heideschrecke	*Gampsocleis glabra*	heath bushcricket, Continental heath bushcricket
Heidewühlmäuse, Nordamerikanische Tannenmäuse	*Phenacomys* spp.	heather voles, tree voles
Heilbutt	*Hippoglossus hippoglossus*	Atlantic halibut
Heilbuttscholle	*Hippoglossoides elassodon*	flathead sole
Heilige Schnecke	*Turbinella pyrum*	sacred chank, Indian chank
Heiliger Ibis	*Threskiornis aethiopica*	sacred ibis
Heiliger Pillendreher	*Scarabaeus sacer*	sacred scarab beetle, Egyptian scarab
Heimchen, Hausgrille	*Acheta domestica*	house cricket (domestic cricket, domestic gray cricket)
Hektor-Laternenfisch	*Lampanyctodes hectoris*	lanternfish
Heldbock, Eichenbock, Großer Eichenbock	*Cerambyx cerdo*	great capricorn beetle, oak cerambyx
Hellbender, Amerikanischer Riesensalamander	*Cryptobranchus alleganiensis*	hellbender
Helle Heideschnecke	*Candidula gigaxii*	eccentric snail
Heller Pokal	*Solenosteira pallida*	pale goblet
Heller Schlangenstern	*Ophiura albida*	lesser sandstar
Heller Schnurwurm, Helle Nemertine	*Amphiporus lactifloreus* (Nemertini)	
Heller Tiefenkatzenhai	*Apristurus aphyodes*	pale catshark
Hellfleckiger Rochen, Kleinäugiger Rochen	*Raja microocellata*	small-eyed ray (FAO), painted ray
Hellflügel-Flugfisch*	*Cypselurus comatus*	clearwing flyingfish
Hellgefleckter Dickmaulrüssler	*Otiorhynchus gemmatus*	lightspotted snout weevil
Hellgrauer Fleckleib-Bär	*Diaphora mendica*	muslin moth
Helm-Azurjungfer	*Coenagrion mercuriale*	southern damselfly (Br.)
Helm-Korallengast	*Coralliophila galea*	helmet coralsnail
Helm-Vasenschnecke	*Vasum cassiforme*	helmet vase
Helmbasilisk	*Basiliscus basiliscus*	common basilisk
Helmkakadu	*Callocephalon fimbriatum*	gang-gang cockatoo
Helmkasuar	*Casuarius casuarius*	double-wattled cassowary
Helmkolibri	*Oxypogon guerinii*	bearded helmetcrest
Helmkopf	*Caudiverbera caudiverbera*	helmeted water toad
Helmkopfbasilisk	*Laemanctus serratus*	serrated casquehead iguana
Helmkopfbasilisken	*Laemanctus* spp.	casquehead iguana

Helmkopfgecko	*Geckonia chazaliae*	helmethead gecko
Helmleguane	*Corytophanes* spp.	helmeted iguanas
Helmperlhuhn	*Numida meleagris*	helmeted guineafowl
Helmqualle	*Galeolaria truncata*	helmet jelly*
Helmschnecken, Wellhornschnecken	Busyconidae	whelks
Helmskinke	*Tribolonotus* spp.	helmet skinks, casqueheads
Helmspecht	*Dryocopus pileatus*	pileated woodpecker
Helmvanga	*Euryceros prevostii*	helmet bird
Helmwachtel, Gambelwachtel	*Callipepla gambelii*	Gambel's quail
Hengels Bärbling	*Rasbora hengeli*	slender harlequin fish
Henslow Schwimmkrabbe	*Polybius henslowi*	Henslow's swimming crab, sardine swimming crab
Herbarkäfer*	*Dienerella filum*	herbarium beetle
Herbst-Mosaikjungfer	*Aeshna mixta*	scarce aeshna, migrant hawker
Herbstgrasmilbe, Herbstmilbe, Erntemilbe	*Neotrombicula autumnalis*	harvest mite, red bug, chiggers
Herbstlaubspanner, Zackenspanner, Herbstspanner	*Ennomos autumnaria*	large thorn
Herbstmilbe, Herbstgrasmilbe, Erntemilbe	*Neotrombicula autumnalis*	harvest mite, red 'bug', (*larvas:* chiggers)
Herbstmilben, Laufmilben, Erntemilben	Trombiculidae	chigger mites, harvest mites
Herbstspinne	*Meta segmentata*	autumn orbweaver*
Herbstspinnen	Metidae	autumn orbweavers*
Herbstspinner, Wiesenspinner	Lemoniidae	lemoniid moths
Herbststechfliege, Kleine Stechfliege	*Haematobosca stimulans/ Haematobia stimulans/ Siphona stimulans*	cattle biting fly (Texas fly)
Hering, Atlantischer Hering	*Clupea harengus*	herring, Atlantic herring (FAO) (digby, mattie, slid, yawling, sea herring)
Heringe	Clupeidae	herrings (shads, sprats, sardines, pilchards & menhadens)
Heringsfische, Heringsverwandte	Clupeiformes	herrings & relatives
Heringshai	*Lamna nasus*	porbeagle (FAO), mackerel shark
Heringshaie	Lamnidae	mackerel sharks
Heringskönig, Petersfisch	*Zeus faber*	Dory, John Dory
Heringsmöve	*Larus fuscus*	lesser black-backed gull
Herkulesspinner	*Coscinocera hercules*	Hercules moth
Hermelin	*Mustela erminea*	ermine, stoat
Hermelinspinner, Weißer Großer Gabelschwanz	*Cerura erminea*	
Herrentiere, Primaten	Primata	primates

Herzeule, Kohleule	*Mamestra brassicae*	cabbage moth
Herzförmiger Herzigel	*Lovenia cordiformis*	heart urchin, cordiform heart urchin
Herzigel, Herzseeigel	Spatangoida	heart urchins
Herzigel, Kleiner Herzigel	*Echinocardium cordatum*	common heart-urchin, sea potato, heart urchin
Herzkrabbe	*Thia scutellata*	thumbnail crab
Herzmuschel, Echte Herzmuschel, Flache Herzmuschel	*Corculum cardissa*	true heart cockle
Herzmuschel> Gemeine Herzmuschel, Essbare Herzmuschel	*Cardium edule/ Cerastoderma edule*	common cockle, common European cockle, edible cockle
Herzmuscheln	Cardiidae	cockles (cockle shells)
Herznasenfledermaus	*Cardioderma cor*	African false vampire bat
Herzwürmer	*Dirofilaria* spp.	heartworms
Herzzüngler	*Phyllodytes* spp.	heart-tongue frogs
Herzzüngler> Langfinger	*Cardioglossa* spp.	long-fingered frogs
Hessenmücke, Hessenfliege	*Mayetiola destructor*	Hessian fly
Heufalter, Gelblinge	*Colias* spp.	yellows
Heufalter> Gemeiner Heufalter, Gelber Heufalter, Weißklee-Gelbling, Goldene Acht	*Colias hyale*	pale clouded yellow
Heuhüpfer, Grashüpfer	*Chorthippus* spp.	field grasshoppers, meadow grasshoppers
Heumilbe, Kugelbauchmilbe	*Pyemotes tritici/ Pyemotes ventricosus*	hay itch mite, grain itch mite, grain mite, straw itch mite
Heupferd> Großes Heupferd, Grünes Heupferd	*Tettigonia viridissima*	great green bushcricket
Heuschreckenkrebs> Amerikanischer Heuschreckenkrebs, Gewöhnlicher Heuschreckenkrebs	*Squilla empusa*	common mantis shrimp
Heuschreckenteesa	*Butastur rufipennis*	grasshopper buzzard-eagle
Heuzünsler	*Hypsopygia costalis*	gold fringe, clover hayworm
Hexakorallen	Hexacorallia	hexacorallians, hexacorals
Hexengarnele	*Melicertus canaliculatus/ Penaeus canaliculatus*	witch shrimp
Hieroglyphen-Schmuckschildkröte	*Pseudemys concinna hieroglyphica*	slider, hieroglyphic slider
Himalaja-Krötenkopfagame	*Phrynocephalus theobaldi*	Himalaya toadhead agama
Himalaya-Grubenotter	*Agkistrodon himalayanus/ Gloydius himalayanus*	Himalaya pit viper
Himalaya-Schlankskink	*Scincella himalayana*	Himalaya ground skink
Himalayakönigshuhn	*Tetraogallus himalayensis*	Himalayan snowcock
Himantolophiden	Himantolophidae	footballfishes
Himbeer-Glasflügler	*Bembecia hylaeiformis*	raspberry clearwing
Himbeerblütenstecher, Erdbeerblütenstecher, Beerenstecher	*Anthonomus rubi*	berry blossom weevil, strawberry blossom weevil

Himbeergallmücke, Brombeersaummücke	Lasioptera rubi	blackberry stem gall midge, raspberry stem gall midge
Himbeerglasflügler	Pennisetia hylaeiformis	raspberry clearwing moth*, raspberry cane borer*
Himbeerkäfer, Blütenfresser	Byturidae	fruitworm beetles
Himbeerkäfer, Europäischer Himbeerkäfer	Byturus tomentosus	European raspberry fruitworm, raspberry beetle
Himbeerkoralle, Keulenkoralle	Pocillopora damicornis	raspberry coral*
Himbeermotte, Himbeerschabe	Incurvaria rubiella	raspberry bud moth
Himbeerrutenbock	Oberea bimaculata	raspberry cane borer
Himbeerrutengallmücke	Thomasiniana theobaldi/ Resseliella theobaldi	raspberry cane midge
Himbeerspanner	Mesoleuca albicillata	beautiful carpet
Himmelblaue Sargassumgarnele	Hippolyte coerulescens	cerulean sargassum shrimp
Himmelblauer Bläuling	Lysandra bellargus	adonis blue (butterfly)
Himmelsgucker, Meerpfaff, Sternseher, Sterngucker	Uranoscopus scaber	star gazer, stargazer (FAO)
Himmelsgucker, Sterngucker	Uranoscopidae	stargazers
Himmelssylphe	Aglaiocercus kingi	long-tailed sylph
Hindutrappe, Indische Trappe	Ardeotis nigriceps	great Indian bustard
Hinterindische Landschildkröte	Manouria impressa/ Geochelone impressa	impressed tortoise
Hinterindische Landschildkröten	Monouria spp.	Indochinese land tortoises
Hinterindische Pfauenaugen- Sumpfschildkröte	Morenia ocellata	Burmese eyed turtle
Hinterindische Scharnierschildkröte, Vietnamesische Scharnierschildkröte	Cuora galbinifrons	Indochinese box turtle
Hinterkiemenschnecken, Hinterkiemer	Opisthobranchia	opisthobranch snails, opisthobranchs
Hirases Walzenschnecke	Fulgoraria hirasei	Hirase's volute
Hirnkoralle> Gewöhnliche Hirnkoralle, Symmetrische Hirnkoralle	Diploria strigosa	symmetrical brain coral
Hirnkorallen u.a., Neptunskorallen	Diploria spp. & Platygyra spp.	brain corals a.o.
Hirschantilope, Wasserbock	Kobus ellipsiprymnus	waterbuck
Hirschdasselfliege	Hypoderma actaeon	deer warble fly a.o.
Hirsche	Cervidae	deer, cervids
Hirsche> Edelhirsche	Cervus spp.	red deers
Hirschferkel	Tragulidae	chevrotains, mouse deer
Hirschgeweih-Moostierchen	Smittina cervicornis/ Porella cervicornis	staghorn bryozoan, deer-antler crust*
Hirschgeweih- Stachelschnecke	Chicoreus cornucervi	monodon murex

Hirschgeweihkoralle	*Acropora cervicornis*	staghorn coral
Hirschkäfer	*Lucanus cervus*	stag beetle, European stag beetle
Hirschkäfer, Schröter	Lucanidae	stag beetles
Hirschlausfliege	*Lipoptena cervi*	deer ked, deer fly
Hirschmaus	*Peromyscus manicilatus*	deer mouse
Hirschrachenbremse	*Cephenomyia auribarbis*	deer nostril fly
Hirschschleimfisch	*Blennius zvonimiri/ Parablennius zvonimiri*	Black Sea blenny
Hirschzecke	*Ixodes scapularis/ Ixodes dammini*	deer tick, black-legged tick
Hirschziegenantilope	*Antilope cervicapra*	blackbuck
Hirsekauri*	*Cypraea miliaris*	millet cowrie
Hirsekorn-Kreiselschnecke	*Gibberula miliaris*	millet topsnail
Hirtenmaina	*Acridotheres tristis*	common myna, common mynah
Hispaniola-Schmuckschildkröte	*Trachemys decorata*	Hispaniolan slider
Hoatzin, Schopfhuhn	Opisthocomidae	hoatzin
Hobelspanner	*Plagodis dolabraria*	scorched wing
Hobospinne*	*Tegenaria agrestis*	hobo spider (U.S.), yard spider (Br.)
Hochantarktische Nototheniiden	*Trematomus* spp.	Antarctic cod, Antarctic rockcod
Hochgucker	Opisthoproctidae	barreleyes, spookfishes
Hochland-Grablaubfrosch	*Pternohyla dentata*	upland burrowing frog
Hochland-Querzahnmolch	*Ambystoma velascoi*	plateau tiger salamander
Hochland-Wüstenmaus	*Eligmodontia typus*	highland desert mouse
Hochmoor-Gelbling, Moorgelbling, Zitronengelber Heufalter	*Colias palaeno*	moorland clouded yellow
Hochmoor-Mosaikjungfer	*Aeshna subarctica*	subarctic peat-moor hawker*
Hochsee-Weißflossenhai, Weißspitzenhai, Weißspitzen-Hochseehai, Langflossen-Hai	*Carcharhinus longimanus/ Carcharhinus maou*	oceanic whitetip shark (FAO), whitetip shark, whitetip oceanic shark
Hochstämmiger Röhrenpolyp	*Tubularia indivisa*	tall tubularian, tall tubularia
Hochzeitskegel	*Conus sponsalis*	marriage cone
Höcker-Landschildkröte, Höckerschildkröte	*Psammobates tentoria tentoria*	African tent tortoise
Höcker-Schienenschildkröte	*Podocnemis sextuberculata*	six-tubercled Amazon River turtle, six-tubercled river turtle (IUCN)
Höcker-Zwiebelmondschwebfliege	*Eumerus tuberculatus*	tuberculate bulb fly
Höckerechsen	Xenosauridae	knob-scaled lizards, xenosaurs
Höckerechsen> Eigentliche Höckerechsen	*Xenosaurus* spp.	New World xenosaurs, American knob-scaled lizards
Höckerige Hornkoralle	*Eunicea calyculata*	knobby candelabrum
Höckerige Kammmuschel	*Chlamys distorta*	hunchback scallop
Höckerige Kugelkrabbe	*Ebalia tuberosa*	Pennant's nut crab
Höckeriger Krustenschwamm	*Crambe crambe*	red encrusting sponge

Höckeriger Seepolyp, Schmarotzerkrake	Ocythoe tuberculata	tuberculate pelagic octopus
Höckerschildkröte, Höcker-Landschildkröte	Psammobates tentoria tentoria	African tent tortoise
Höckerschwan	Cygnus olor	mute swan
Höckerzweig-Moostierchen, Verwachsenes Moostierchen	Frondipora verrucosa	warty leaf-bryozoan*
Höckrige Streifenkreuzspinne*	Mangora gibberosa	gibbose lined orbweaver
Hohe Mitra	Cancilla praestantissima	superior mitre, superior miter
Hohe Wellhornschnecke	Buccinum zelotes	superior buccinum
Hohe Windelschnecke	Columella columella	mellow column snail (U.S.), hightopped chrysalis snail*
Höhengrashüpfer	Chorthippus alticola	eastern Alpine grasshopper
Höhenläufer	Thinocoridae	seedsnipes
Hoher Segelflosser	Pterophyllum altum	deep angelfish
Hohl-Glanzschnecke, Britische Dolchschnecke, Hohlschnecke*	Zonitoides excavatus	hollowed gloss (snail), hollowed glass snail
Höhlen-Gelbsalamander, Höhlensalamander	Eurycea lucifuga	cave salamander
Höhlen-Raubwels	Clarias cavernicola	cave catfish
Höhlenanolis	Anolis lucius	cave anole
Höhlenassel	Proasellus cavaticus	cave isopod*
Höhlenbarsche	Dinopercidae	cavebass
Höhlenflohkrebs u.a., Brunnenkrebs	Niphargus aquilex	cavernous well shrimp*
Höhlenflohkrebse, Höhlenkrebse	Niphargus spp.	well shrimps
Höhlenflughunde	Rousettus spp.	Rousette fruit bats
Höhlenfrosch	Arthroleptis troglodytes	cave frog
Höhlenkreuzspinne	Meta menardi	cave orbweaver
Höhlensalamander, Höhlen-Gelbsalamander	Eurycea lucifuga	cave salamander
Höhlensalamander, Schleuderzungensalamander	Hydromantes spp.	web-toed salamander
Höhlenschildläuse	Margarodidae	giant coccids & ground pearls
Höhlenschwalme, Zwergschwalme	Aegothelidae	owlet-nightjars
Höhlensittich	Pezoporus occidentalis/ Geopsittacus occidentalis	night parrot, Australian parrot
Höhlenspanner, Kellerspanner	Triphosa dubitata	cave moth*, cellar moth*
Höhlenspinnen	Nesticidae	cave cobweb spiders
Höhlenwaran, Wüsten-Zwergwaran	Varanus eremius	desert pygmy monitor, spinifex monitor
Höhlenweihe	Polyboroides typus	African harrier hawk
Hohlmaul-Panzerratte, Schwarzschwanz-Grenadier	Coelorhynchus coelorhynchus/ Caelorhynchus caelorhincus	hollowsnout grenadier (FAO), black-spotted grenadier
Hohlnadelwickler, Fichtennestwickler	Epiblema tedella/ Epinotia tedella	streaked pine bell, cone moth
Hohlstachler	Coelacanthiformes/ Coelacanthidae	coelacanths
Hohltaube	Columba oenas	stock dove

Hohltiere, Nesseltiere, Cnidarien, Coelenteraten	Cnidaria/ Coelenterata	cnidarians, coelenterates
Hokkos	Cracidae	guans, chachalacas, currassows
Holbrooks Brassen	*Diplodus holbrooki*	spottail pinfish (FAO), spot-tail seabream
Holunderbär, Gelbe Tigermotte	*Spilosoma lutea*	buff ermine moth
Holunderlaus, Schwarze Holunderblattlaus	*Aphis sambuci*	elder aphid
Holunderspanner, Nachtschwalbenschwanz	*Ourapteryx sambucaria*	swallow-tailed moth
Holzbienen	*Xylocopa* spp.	large carpenter bees
Holzbock, Gemeiner Holzbock	*Ixodes ricinus*	castor bean tick, European castor bean tick, European sheep tick
Holzbohrassel, Bohrassel	*Limnoria lignorum*	gribble
Holzbohrender Polydora-Wurm	*Polydora ligni*	polydora mud worm
Holzbohrer	Cossidae	carpenter moths & leopard moths
Holzbohrmuschel	*Xylophaga dorsalis*	wood piddock
Holzbohrwespe	*Trypoxylon figulus*	black borer wasp, black borer
Holzboot, Taucherschnecke	*Scaphander lignarius/ Bulla lignaria*	woody canoe-bubblesnail
Hölzerne Käferschnecke	*Mopalia lignosa*	woody chiton, woody mopalia
Holzfliegen	Erinnidae/Xylophagidae	xylophagid flies, xylophagids
Holzflohkrebs, Scherenschwanz	*Chelura terebrans*	wood-boring amphipod
Holzkegel	*Conus chaldeus*	wood cone
Holzschaben	Blattellidae	wood cockroaches
Holzwespe> Gemeine Holzwespe, Kiefernholzwespe	*Sirex juvencus*	polished horntail
Holzwespen	Siricidae	woodwasps, horntails
Holzzecken*	*Dermacentor* spp.	eastern wood ticks
Honduras-Lanzenotter	*Bothriechis marchi/ Bothrops marchi*	Honduran palm viper, March's palm pit viper
Honiganzeiger	Indicatoridae	honeyguides
Honigbeutler	*Tarsipes spenserae*	honey possum
Honigbiene, Europäische Honigbiene, Gemeine Honigbiene	*Apis mellifera mellifera*	honey bee, hive bee
Honigdachs	*Mellivora capensis*	honey badger, rattel
Honigesser, Honigfresser	Meliphagidae	honeyeaters (incl. *Ephthianura*, *Ashbyia*)
Honigfadenfisch	*Colisa sota*	honey gourami
Honigkauri	*Cypraea helvola*	honey cowrie
Honigwespen	Masaridae	shining wasps, masarid wasps
Hopfenblattlaus, Hopfenlaus	*Phorodon humuli*	hop aphid, damson-hop aphid

Hopfenerdfloh, Hanferdfloh	*Psylliodes attenuata*	hop flea beetle, European hop flea beetle
Hopfeneule	*Hypena rostralis*	buttoned snout (moth)
Hopfenspinner, Hopfenmotte, Hopfenwurzelbohrer	*Hepialus humuli*	ghost swift, ghost moth
Hopfenwanze	*Calocoris fulvomaculatus*	hop capsid bug, needle-nosed hop bug, shy bug
Hopfenzikade	*Evacanthus interruptus*	hop leafhopper
Hopfkuckuck	*Cuculus saturatus*	Oriental cuckoo
Hopkins Rose	*Hopkinsia rosacea*	Hopkins' rose
Hoplichthyiden	Hoplichthyidae	spiny flatheads, ghost flatheads
Horn-Stachelschnecke	*Bolinus cornutus*	horned murex
Horn-Zahnschnecke	*Dentalium corneum* (Scaphopoda)	horned tusk, horned tuskshell
Hörnchen	Sciuridae	squirrels & chipmunks & marmots & prairie dogs
Hörnchen-Kletterbeutler	*Gymnobelideus leadbeateri*	Leadbeater's possum
Hörnchenschnecken, Kopflappen-Sternschnecken	Polyceridae	polyceras
Hörnchenverwandte	Sciuromorpha	squirrel-like rodents
Hornfliege, Kuhfliege	*Haematobia irritans*	horn-fly, tropical buffalo fly
Hornhaie, Stierkopfhaie	*Heterodontus* spp.	hornsharks
Hornhecht	*Belone belone*	garfish, garpike (FAO)
Hornhechtartige, Ährenfischverwandte	Atheriniformes	silversides & skippers & flying fishes and others
Hornhechte	Belonidae	needlefishes
Hornige Baumkoralle, Bäumchenkoralle	*Dendrophyllia cornigera*	horny tree coral*
Hornisse	*Vespa crabro*	hornet, brown hornet, European hornet
Hornissenraubfliege	*Asilus crabroniformis*	robberfly (Central European sanddune robberfly)
Hornissenschwärmer, Bienenglasflügler	*Aegeria apiformis/ Sesia apiformis*	poplar hornet clearwing, hornet moth
Hornkorallen, Rindenkorallen	Gorgonaria/ Gorgonacea	gorgonians, gorgonian corals, horny corals
Hornkorallen u.a., Seefächer u.a.	*Eunicella* spp.	horny corals a.o., sea fans
Hornmilben & Moosmilben	Orbatida	orbatid mites (beetle mites & moss mites)
Hornraben	Bucorvidae	ground-hornbills
Hornschnecken	Buccinidae	whelks
Hornschnecken, Nadelschnecken	Cerithiidae	ceriths, cerithids, hornsnails (horn shells), needle whelks
Hornschwämme, Netzfaserschwämme	Cornacuspongiae	horny sponges

Hornviper	*Cerastes cerastes*	horned viper, African desert horned viper
Horsfieldlerche	*Mirafra javanica*	eastern singing bush lark
Horsfields Flughörnchen	Iomys horsfieldi	Horsfield's flying squirrel
Hosenbienen	*Dasypoda* spp.	hairy-legged bees*
Hubbs Schnabelwal	*Mesoplodon carlhubbsi*	arch-beaked whale, Hubbs' beaked whale
Hubrichts Schleichensalamander	*Phaeognathus hubrichti*	Red Hills salamander
Hübsche Gitterschnecke	*Trigonostoma pulchra/ Cancellaria pulchra*	beautiful nutmeg
Hübsche Mitra	*Vexillum pulchellum*	beautiful mitre, beautiful miter
Hübsche Schraubenschnecke	*Terebra spectabilis*	graceful auger
Hübsche Walzenschnecke	*Cymbiolacca pulchra*	beautiful volute
Hübsche Zahnschnecke	*Pictodentalium formosum* (Scaphopoda)	beautiful tusk
Hübscher Taschenkrebs*	*Cancer gracilis*	graceful rock crab, graceful crab
Hübschgesichtwallaby, Schönwallaby	*Macropus parryi*	whiptail wallaby, pretty-face wallaby
Hübschkauri	*Cypraea pulchra/ Luria pulchra*	lovely cowrie
Huchen	*Hucho hucho*	Danube salmon, huchen
Hudsonmeise	*Parus hudsonicus*	boreal chickadee
Hufeisen-Azurjungfer	*Coenagrion puella*	common coenagrion, azure damselfly
Hufeisennasen	Rhinolophidae	horseshoe bats
Hufeisennasen, Eigentliche Hufeisennasen	*Rhinolophus* spp.	horseshoe bats
Hufeisennatter	*Coluber hippocrepis*	horseshoe snake, horseshoe whip snake
Hufeisenwürmer, Phoroniden	Phoronidea	phoronids
Hufschnecken	Hipponicidae	hoof limpets, horsehoof limpets, bonnet limpets
Hüftwasserläufer	*Mesovelia furcata*	pondweed bug, water treader
Hüftwasserläufer	Mesoveliidae	water treaders
Hügelhuhn	*Arborophila torqueola*	common hill partridge
Hühner-Kopflaus	*Cuclotogaster heterographus/ Gallipeurus heterographus*	chicken head louse
Hühner-Körperlaus, Körperlaus	*Menacanthus stramineus/ Eomenacanthus stramineus*	chicken body louse
Hühneresser, Hühnerfresser	*Spilotes pullatus pullatus*	tropical chicken snake
Hühnerfederling, Schaftlaus	*Menopon gallinae*	shaft louse, chicken shaft louse
Hühnerfloh, Europäischer Hühnerfloh	*Ceratophyllus gallinae*	European chicken flea, common chicken flea, hen flea
Hühnerkammfloh, Hühnerfloh	*Echidnophaga gallinacea/ Sarcopsylla gallinacea*	sticktight flea

Hühnermilbe, Rote Vogelmilbe	*Dermanyssus gallinae*	red mite of poultry, poultry red mite, red chicken mite, chicken mite
Hühnernestmilbe, Vogelnestmilbe	*Androlaelaps casalis*	poultry litter mite, cosmopolitan nest mite
Hühnervögel	Galliformes	gallinaceous birds, fowl-like birds
Hulman, Hanumanlangur	*Presbytis entellus*	hanuman langur
Hulock, Weißbrauengibbon	*Hylobates hoolock*	white-browned gibbon, Hoolock gibbon
Humboldtpinguin	*Spheniscus humboldti*	Humboldt penguin
Hummelfledermaus, Schweinsschnauzen-Fledermaus, Kittis Kleinstfledermaus	*Craseonycteris thonglongyai*	Kitti's hog-nosed bat (*smallest mammal!*)
Hummelgarnele	*Gnathophyllum americanum*	bumblebee shrimp
Hummelgarnelen	Gnathophyllidae	bumblebee shrimps
Hummelmilbe	*Parasitus fucorum*	bumblebee mite
Hummeln	Bombidae/Bombinae	bumble bees
Hummelnematode	*Sphaerularia bombi*	bumblebee nematode
Hummelschwärmer	*Hemaris fuciformis*	broad-bordered bee hawkmoth
Hummelschweber (Hummelfliegen) & Wollschweber & Trauerschweber	Bombyliidae	beeflies
Hummer	Nephropidae	clawed lobsters
Hummer> Amerikanischer Hummer	*Homarus americanus*	northern lobster, American clawed lobster
Hummer> Europäischer Hummer	*Homarus gammarus*	common lobster, European clawed lobster, Maine lobster
Hummerkrabbe (Hbz.), Rosenberg-Garnele, Rosenberg Süßwassergarnele	*Macrobrachium rosenbergii*	Indo-Pacific freshwater prawn, giant river shrimp, giant river prawn, blue lobster (tradename)
Hund> Afrikanischer Wildhund	*Lycaon pictus*	African wild dog, African hunting dog, African painted wolf
Hund> Haushund	*Canis familiaris*	domestic dog
Hündchen-Fledermaus*	*Peropteryx* spp.	doglike bats
Hunde-Haarbalgmilbe	*Demodex canis*	dog follicle mite
Hunde-Hakenwurm	*Ancylostoma caninum*	canine hookworm
Hunde-Spirocercose-Wurm, Ösophagus-Rollschwanz	*Spirocerca lupi/ Spirocerca sanguinolenta*	esophageal worm (of dog)
Hundebandwurm, Dreigliedriger Hundebandwurm	*Echinococcus granulosus*	dwarf dog tapeworm, dog tapeworm, hydatid tapeworm
Hundebandwurm u.a., Quesenbandwurm	*Multiceps multiceps/ Taenia multiceps*	dog tapeworm a.o. (>gid/sturdy in sheep)
Hundefloh	*Ctenocephalides canis*	dog flea
Hundegel, Rollegel, Achtäugiger Schlundegel	*Herpobdella octoculata/ Erpobdella octoculata*	eight-eyed leech
Hundehaarling	*Trichodectes canis*	dog biting louse
Hundehakenwurm, Fuchshakenwurm	*Uncinaria stenocephala*	dog hookworm

Hundeherzwurm	*Dirofilaria immitis*	heartworm, dog heartworm
Hundelaus	*Linognathus setosus*	dog sucking louse
Hundemangusten	*Bdeogale* spp.	black-legged mongooses
Hundepelzmilbe	*Cheyletiella yasguri*	dog fur mite
Hundertfüßer, Chilopoden	Chilopoda	centipedes, chilopodians
Hundespulwurm	*Toxocara canis*	canine ascarid
Hundezecke	*Ixodes canisuga*	dog tick, British dog tick
Hunds-Fechterschnecke, Hunds-Flügelschnecke	*Strombus canarium*	yellow conch, dog conch
Hundsbarbe	*Barbus meridionalis*	Mediterranean barbel
Hundsfisch> Europäischer Hundsfisch, Ungarischer Hundsfisch	*Umbra krameri*	European mudminnow
Hundsfische	Umbridae	mudminnows
Hundskopf-Wassertrugnatter	*Cerberus rhynchops*	dog-faced water snake, bockadam
Hundskopfboa, Gartenboa	*Corallus enydris*	garden tree boa, Cook's tree boa, Amazon tree boa
Hundskopfboas	*Corallus* spp.	tree boas
Hundslachs, Chum-Lachs, Keta-Lachs, Ketalachs	*Oncorhynchus keta*	chum salmon
Hundsrobben	Phocidae	true seals, earless seals, hair seals
Hundsschnapper	*Lutjanus jocu*	dog snapper
Hundszahn-Umberfisch	*Otolithes ruber*	long-tooth croaker, tiger-tooth croaker
Hundszungen	Cynoglossidae	tonguefishes
Hungerwespen	Evaniidae	ensign wasps
Hunters Leierantilope	*Damaliscus hunteri*	Hunter's antelope, Hunter's hartebeest, hirola, sassaby
Hüpferlinge	*Cyclops* spp.	cyclops
Hüpferlinge (Schwimmer)	Cyclopidae	cyclopids
Hüpfkäfer	Throscidae	false metallic wood-boring beetles, throscid beetles
Hüpfmäuse	*Zapus* spp.	jumping mice
Hüpfmäuse> Eigentliche Hüpfmäuse	Zapodidae	birch mice & jumping mice
Hüpfspinnen, Springspinnen	Salticidae	jumping spiders
Husarenaffe	*Erythrocebus patas*	Patas monkey, red guenon, red monkey
Hutkoralle	*Parahalomitra robusta*	basket coral
Hüttengärtner	*Amblyornis inornatus*	vogelkop gardener bowerbird
Hyänen	Hyaenidae	hyenas
Hyazinthara	*Anodorhynchus hyacinthinus*	hyacinth macaw
Hydrokorallen, Stylasteriden	Stylasteridae	stylasterine hydrocorals
Hydrozoen	Hydroida	hydrozoans, hydroids, hydra-like animals

Iberiensteinbock, Spanischer Steinbock	*Capra pyrenaica*	Spanish ibex
Iberienwühlmaus	*Microtus lusitanicus*	Iberian vole, Spanish vole
Iberische Barbe	*Barbus comizo*	Iberian barbel
Iberische Egelschnecke, Gewächshausschnegel	*Limax valentianus/ Lehmannia valentiana*	glasshouse slug, Iberian slug
Iberische Gebirgseidechse	*Lacerta monticola*	Iberian rock lizard
Iberische Geburtshelferkröte, Spanische Geburtshelferkröte	*Alytes cisternasii*	Iberian midwife toad
Iberische Mauereidechse, Spanische Mauereidechse	*Podarcis hispanica*	Iberian wall lizard
Iberische Smaragdeidechse, Schreibers Smaragdeidechse	*Lacerta schreiberi*	Schreiber's green lizard
Iberische Süßwasserkrabbe	*Potamon ibericum*	Bieberstein's freshwater crab
Iberische Viper, Pyrenäenotter	*Vipera seoanei*	Iberian adder, Pyrenean viper
Iberischer Wasserfrosch	*Rana perezi/ Rana ridibunda perezi*	Coruna frog
Ibisschnabel	Ibidorhynchidae	ibisbill
Ibisvögel	Threskiornithidae	ibises, spoonbills
Igel	Erinaceidae	hedgehogs & gymnures
Igel-Wegschnecke, Igelschnecke, Kleine Wegschnecke	*Arion intermedius*	hedgehog slug, hedgehog arion
Igelfisch> Gewöhnlicher Igelfisch, Gemeiner Igelfisch, Kosmopolit-Igelfisch, Gepunkteter Igelfisch	*Diodon hystrix*	common porcupinefish, spotted porcupinefish, spot-fin porcupinefish (FAO)
Igelfische, Zweizähner	Diodontidae	porcupinefishes (burrfishes)
Igelfloh	*Archaeopsylla erinacei*	hedgehog flea
Igelratten	*Proechimys* spp.	spiny rats, casiraguas
Igelrochen	*Urogymnus asperrimus*	porcupine ray (FAO), thorny ray
Igelrochen	*Urogymnus* spp.	thorny rays
Igelwurm (Rüsselwurm)	*Bonellia viridis* (Echiura)	
Igelwürmer, Stachelschwänze, Echiuriden	Echiura	spoon worms, echiuroid worms
Igelzecke	*Ixodes hexagonus*	hedgehog tick
Iltis> Europäischer Iltis, Waldiltis	*Mustela putorius*	European polecat
Impala, Schwarzfersenantilope	*Aepyceros melampus*	impala
Imperator-Garnele, Verwandlungs-Partnergarnele	*Periclimenes imperator*	emperor shrimp
Imperatorfische, Großaugen-Brassen	*Gymnocranius* spp.	large-eye breams
Indianerblässhuhn	*Fulica americana*	American coot
Indianerfische	Pataecidae	
Indianermeise	*Parus bicolor*	tufted titmouse

Indien-Kantschil, Fleckenkantschil	*Tragulus meminna*	Indian spotted chevrotain
Indigofink	*Passerina cyanea*	indigo bunting
Indigonatter	*Drymarchon corais couperi*	indigo snake
Indigoschlangen	*Drymarchon* spp.	indigo snakes
Indische Bambusotter	*Trimeresurus gramineus*	Indian green tree viper
Indische Buschratte, Elliot Buschratte	*Golunda ellioti*	Indian bush rat, coffee rat
Indische Dachschildkröte	*Kachuga tecta*	Indian tent turtle, Indian roofed turtle
Indische Dornschildkröte	*Cyclemys mouhoti/ Pyxidea mouhotii*	Indian leaf turtle, jagged-shell turtle, Indian keel-backed terrapin, keeled box turtle
Indische Eierschlange	*Elachistodon westermanni*	Indian egg-eating snake
Indische Erdschildkröte	*Melanochelys* spp.	Indian terrapins
Indische Fadenmakrele	*Alectis indicus*	Indian mirrorfish, Indian threadfish (FAO)
Indische Hauptmannsgarnele	*Fenneropenaeus indicus/ Penaeus indicus*	Indian prawn
Indische Honigbiene	*Apis mellifera indica*	Indian honey bee
Indische Klappen-Weichschildkröte	*Lissemys punctata*	Indo-Gangetic flapshell, Indian flapshell turtle
Indische Königsmakrele	*Scomberomorus commerson*	Commerson's mackerel, narrow-barred Spanish mackerel (FAO), barred mackerel
Indische Lackschildlaus	*Laccifer lacca*	Indian lac insect
Indische Makrele, Indische Zwergmakrele	*Rastrelliger kanagurta*	Indian mackerel
Indische Nasenotter	*Agkistrodon hypnale/ Hypnale hypnale*	Indian humpnose viper, hump-nosed viper
Indische Ornament-Vogelspinne, Tiger-Vogelspinne	*Poecilotheria regalis*	Indian ornamental tarantula
Indische Pellona	*Pellona ditchela*	Indian pellona
Indische Sandagame	*Psammophilus dorsalis*	Indian sand agama
Indische Sandboa	*Eryx johnii johnii*	John's sand boa
Indische Schönechse	*Calotes versicolor*	common bloodsucker, variable agama, "chameleon"
Indische Schwebmakrele	*Ariomma indica*	Indian driftfish
Indische Seefledermaus	*Halieutaea indica*	Indian handfish
Indische Spindelschnecke	*Tibia curta*	Indian tibia
Indische Stachelmakrele	*Decapterus russelli*	Indian scad
Indische Trappe, Hindutrappe	*Ardeotis nigriceps*	great Indian bustard
Indische Turmschnecke	*Lophiotoma indica*	Indian turret
Indische Walzenschnecke	*Melo melo*	baler snail (baler shell), Indian volute
Indische Warzenschlange, Kleine Warzenschlange	*Chersydrus granulatus/ Acrochordus granulatus*	Indian wart snake
Indische Wühlschlangen	*Uropeltis* spp.	Indian earth snakes
Indischer Bambushai	*Chiloscyllium indicum*	Indian carpet shark, slender bambooshark (FAO)

Indischer Barrakuda, Indomalayischer Barrakuda	*Sphyraena jello*	banded barracuda, pickhandle barracuda (FAO), Indian barracuda
Indischer Charsa	*Martes gwatkinsi*	Nilgiri marten
Indischer Dornschwanz	*Uromastyx hardwickii*	Indian spiny-tailed lizard, Indian mastigure
Indischer Eberfisch	*Antigonia rubescens*	Indian boarfish
Indischer Elefant, Asiatischer Elefant	*Elephas maximus*	Asiatic elephant, Indian elephant
Indischer Elefantenzahn, Kostbarer Elefantenzahn	*Dentalium pretiosum/ Antalis pretiosum* (Scaphopoda)	Indian money tusk, wampum tuskshell
Indischer Fähnchen-Messerfisch, Tausenddollarfisch	*Notopterus chitala/ Chitala chitala*	clown knifefish (FAO), Indian featherback
Indischer Glasbarsch	*Chanda ranga*	Indian glassfish
Indischer Glaskärfling	*Horaichthys setnai*	Malabar ricefish
Indischer Glaswels	*Kryptopterus bicirrhis*	glass catfish (FAO), ghost catfish
Indischer Hering, Großer Wolfshering	*Chirocentrus dorab*	wolf-herring, dorab wolf-herring (FAO)
Indischer Hutaffe	*Macaca radiata*	bonnet macaque
Indischer Krait, Gewöhnlicher Bungar	*Bungarus caeruleus*	Indian krait
Indischer Mangrovenwaran, Pazifik-Waran	*Varanus indicus*	Pacific monitor, mangrove monitor, Indian mangrove monitor
Indischer Riesenflughund	*Pteropus giganteus*	Indian flying fox
Indischer Sambar, Pferdehirsch	*Cervus unicolor*	sambar
Indischer Schwarzfrosch	*Melanobatrachus indicus*	black microhylid
Indischer Schweinswal	*Neophocaena phocaenoides*	finless porpoise
Indischer Stachelwels, Indischer Goldstreifenwels, Indischer Streifenwels	*Mystus vittatus/ Hypselobagrus tengara*	striped catfish, striped dwarf catfish (FAO)
Indischer Weißrücken-Anemonenfisch, Weißrücken-Clownfisch	*Amphiprion akallopisos*	skunk clownfish (FAO), skunk anemonefish
Indisches Panzernashorn	*Rhinoceros unicornis*	greater Indian rhinoceros, great Indian rhinoceros
Indochinesisches Elefantenohr	*Rhodactis indosinensis*	hairy mushroom anemone
Indonesische Kobra	*Naja sputatrix*	Indonesian cobra
Indonesische Schildkrötenkopf-Seeschlange	*Emydocephalus annulatus*	ringed turtlehead sea snake, annulated sea snake
Indonesischer Schlangenkopffisch	*Channa micropeltes*	Indonesian snakehead
Indopazifische Lederanemone	*Heteractis crispa*	leather anemone
Indopazifische Leuchtfische	Acropomatidae	temperate ocean-basses, lanternfishes
Indopazifische Makrele	*Scomber australasicus*	spotted chub mackerel, blue mackerel (FAO)
Indopazifische Reiterkrabbe	*Ocypode ceratophthalma*	Indo-Pacific ghost crab, horn-eyed ghost crab
Indopazifische Spanische Makrele	*Scomberomorus guttatus*	Indo-Pacific Spanish mackerel, Indo-Pacific king mackerel (FAO)

Indopazifischer Buckeldelphin, Chinesischer Weisser Delphin	*Sousa chinensis*	Chinese humpback dolphin, Indo-Pacific hump-backed dolphin, white dolphin
Indopazifischer Ebarme	*Psettodes erumei*	adalah, Indian halibut, Indian spiny turbot (FAO)
Indopazifischer Fächerfisch, Indopazifischer Segelfisch	*Istiophorus platypterus*	Indo-Pacific sailfish
Indopazifischer Flugfisch, Indopazifischer Fliegender Fisch	*Hirundichthys affinis*	fourwing flyingfish
Indopazifischer Fuchshai	*Alopias pelagicus*	pelagic thresher
Indopazifischer Segelfisch, Indopazifischer Fächerfisch	*Istiophorus platypterus*	Indo-Pacific sailfish
Indopazifischer Stechrochen	*Dasyatis brevicaudata*	short-tail stingray (FAO), giant short-tail stingray of Australia
Indopazifischer Teufelsrochen, Zwergteufelsrochen	*Mobula diabola*	devil ray
Indopazifisches Gorgonenhaupt	*Astroba nuda*	Indopacific basket star
Indri	*Indri indri*	indri
Indus-Delphin, Indusdelphin	*Platanista indi*	Indus River dolphin, bhulan
Inermiiden	Inermiidae	bonnetmouths
Inger, Atlantischer Inger	*Myxine glutinosa*	hagfish
Ingerartige, Inger, Schleimaale	Myxiniformes (Myxinida)	hagfishes
Inkatäubchen	*Scardafella inca*	Inca dove
Innennest-Raubameise, „Diebsameise"	*Solenopsis fugax/ Diplorhoptrum fugax*	European fire ant
Insektenfressende Röhrennasen-Fledermäuse	*Murina* spp.	tube-nosed insectivorous bats
Insektenfressende Waldmaus	*Deomys ferrugineus*	Congo forest mouse, link rat
Insel-Flugfuchs	*Pteropus hypomelanus*	small flying fox, island flying fox
Insel-Glasschleiche, Küstenglasschleiche	*Ophisaurus compressus*	island glass lizard (U.S.)
Insel-Lanzenotter	*Bothrops insularis*	jararaca ilhoa, Queimada Island bothrops, golden lancehead
Insel-Nachtechse	*Xantusia riversiana/ Klauberina riversiana*	island night lizard
Inselmäuse	*Macrotarsomys* spp.	Madagascan rats
Inselratte	*Nesomys rufus*	Madagascan rat
Ioras	Aegithininae	ioras
Ipanema-Fledermaus	*Pygoderma bilabiatum*	Ipanema bat
Irawadi-Delphin, Irrawadydelphin	*Orcaella brevirostris*	Irrawaddy River dolphin, snub-fin dolphin
Iriomoto-Katze	*Felis iriomotensis/ Mayailurus iriomotensis*	Iriomote cat
Irisblattwespe	*Rhadinoceraea micans*	iris sawfly
Irrawadydelphin, Irrawadi-Delphin	*Orcaella brevirostris*	Irrawaddy River dolphin, snub-fin dolphin

Irusmuschel	*Irus irus/* *Venerupis irus*	irus clam
Isabellantilope, Riedbock	*Redunca redunca*	Bohar reedbuck
Isabellfarbene Kauri	*Cypraea isabella/* *Luria isabella*	isabelline cowrie, dirty-yellow cowrie
Isabellsteinschmätzer	*Oenanthe isabellina*	isabelline wheatear
Isabellwürger	*Lanius isabellinus*	isabelline shrike
Isidoradler	*Spizaetus isidori*	black & chestnut eagle
Island-Kammmuschel, Isländische Kammmuschel	*Chlamys islandica*	Iceland scallop
Island-Katzenhai, Island-Tiefen-Katzenhai (Madeira-Katzenhai)	*Apristurus laurussonii* (*Apristurus maderensis*)	Iceland catshark (FAO) (Madeira catshark)
Isländische Herzmuschel	*Clinocardium ciliatum*	hairy cockle, Iceland cockle
Isländische Kammmuschel, Island-Kammmuschel	*Chlamys islandica*	Iceland scallop
Isländische Mondschnecke, Island-Mondschnecke	*Bulbus islandicus/* *Amauropsis islandica*	Iceland moonsnail
Islandmuschel	*Arctica islandica*	Icelandic cyprine, Iceland cyprina, ocean quahog
Israelische Katzennatter	*Telescopus dhara*	Israelian cat snake
Israelischer Scheibenzüngler	*Discoglossus nigriventer*	Israel painted frog
Isthmus-Zwergboa*	*Ungaliophis continentalis*	Isthmian dwarf boa
Italienische Barbe	*Barbus plebejus*	Italian barbel
Italienische Honigbiene	*Apis mellifera ligustica*	Italian honey bee
Italienische Kleinwühlmaus, Savi-Kleinwühlmaus	*Microtus savii/* *Pitymys savii*	Savi's pine vole
Italienischer Kammmolch, Alpenkammmolch	*Triturus carniflex*	alpine crested newt, Italian warty newt
Italienischer Schleuderzungensalamander	*Hydromantes italicus*	Italian cave salamander
Italienischer Skorpion	*Euscorpius italicus*	Italian scorpion
Italienischer Springfrosch	*Rana latastei*	Italian agile frog
Italienischer Taschenkrebs, Gemeine Krabbe, Gelbe Krabbe	*Eriphia verrucosa*	warty xanthid crab, Italian box crab, yellow box crab
Italienischer Wassermolch	*Triturus italicus*	Italian newt
Ivells Seeanemone*	*Edwardsia ivelli*	Ivell's sea anemone

Jacobita, Commerson-Delphin	*Cephalorhynchus commersonii*	Commerson's dolphin, Tonina dolphin, piebald dolphin, black and white dolphin
Jagdelster	*Cissa chinensis*	green magpie
Jagdgarnele*	*Eohippolysmata ensirostris*	hunter shrimp
Jagdraubkäfer, Fellkäfer	Korynetidae/ Corynetidae (>Cleridae)	ham beetles
Jagdsalmler	*Hoplias malabaricus*	tigerfish, tararira, trahira (FAO) haimara
Jagdspinnen	*Dolomedes* spp.	fishing spiders
Jagdspinnen, Riesenkrabbenspinnen	Heteropodidae (Eusparassidae, Spavassidae)	giant crab spiders, huntsman spiders
Jagdwels	*Wallagonia attu*	mulley
Jägerliest, Lachender Hans	*Dacelo novaeguineae*	laughing kookaburra
Jaguar	*Panthera onca*	jaguar
Jaguarundi, Wieselkatze	*Felis yagouaroundi*	jaguarundi
Jakobs-Pilgermuschel, Jakobsmuschel	*Pecten jacobaeus*	St.James's scallop, great scallop
Jakobskreuzkrautbär, Karminbär	*Tyria jacobaeae*	cinnabar moth
Jakobsmuschel, Jakobs-Pilgermuschel	*Pecten jacobaeus*	St.James's scallop, great scallop
Jamaika Feigen-Fledermaus	*Ariteus flavescens*	Jamaican fig-eating bat
Jamaika-Baumratte	*Geocapromys brownii*	Jamaican hutia (Indian coney)
Jamaika-Boa	*Epicrates subflavus*	Jamaican boa, yellow snake
Jamaika-Schmuckschildkröte, Antillen-Schmuckschildkröte	*Pseudemys terrapen/ Trachemys terrapen*	Jamaican slider
Jamaika-Stechrochen, Jamaika-Rund-Stechrochen, Gelber Stechrochen*	*Urolophus jamaicensis*	yellow stingray
Januarkärpfling, Kiesfisch	*Phalloptychus januarius*	barred millionsfish
Japanische „Sonne und Mond"-Muschel	*Amusium japonicum*	Japanese sun-and-moon scallop
Japanische Archenmuschel	*Anadara subcrenata*	mogal clam
Japanische Bachschildkröte	*Mauremys japonica*	Japanese turtle
Japanische Bastardmakrele	*Trachurus japonicus*	Japanese jack mackerel
Japanische Bernsteinmakrele, Japanische Seriola	*Seriola quinqueradiata*	Japanese amberjack
Japanische Fleckfiebermilbe (Tsutsugamushi-Fieber-Erntemilbe)	*Leptotrombidium akamushi/ Trombicula akamushi*	scrub typhus chigger mite, Japanese scrub typhus chigger mite*
Japanische Flunder	*Limanda herzensteini*	Japanese dab, yellow-striped flounder
Japanische Grubenotter, Mamushi	*Agkistrodon blomhoffi/ Gloydius blomhoffi*	Japanese mamushi
Japanische Languste	*Panulirus japonicus*	Japanese lobster
Japanische Makrele	*Scomberomorus niphonius*	Japanese Spanish mackerel

Japanische Perlmuschel	*Pinctada martensi*	Japanese pearl-oyster, Marten's pearl oyster
Japanische Reusenschnecke	*Nassarius fraterculus*	Japanese nassa
Japanische Riesenkrabbe	*Macrocheira kaempferi*	giant spider crab
Japanische Sardelle	*Engraulis japonicus*	Japanese anchovy
Japanische Sardine	*Sardinops melanostricta*	Japanese pilchard
Japanische Seegurke	*Stichopus japonicus*	Japanese sea cucumber
Japanische Stachelschnecke	*Chicoreus asianus*	Asian murex
Japanische Teichmuschel	*Ruditapes philippinarum/ Tapes philippinarum/ Venerupis philippinarum/ Tapes japonica*	Japanese littleneck, short-necked clam, Japanese clam, Manilla clam
Japanische Venusmuschel	*Meretrix lusoria*	Japanese hard clam
Japanischer Aal	*Anguilla japonica*	Japanese eel
Japanischer Barsch	*Lateolabrax japonicus*	Japanese seabass
Japanischer Bergmolch	*Echinotriton andersoni*	Anderson's newt
Japanischer Feuerbauchmolch	*Cynops pyrrhogaster*	Japanese fire-bellied newt, Japanese firebelly newt
Japanischer Flugfisch, Japanischer Fliegender Fisch	*Cheilopogon agoo/ Cypselurus agoo*	Japanese flyingfish
Japanischer Flugkalmar	*Todarodes pacificus*	Japanese flying squid
Japanischer Halbschnäbler	*Hemiramphus sajori*	Japanese halfbeak
Japanischer Hase	*Lepus brachyurus*	Japanese hare
Japanischer Heilbutt	*Hippoglossoides dubius*	Pacific false halibut, flathead flounder (FAO)
Japanischer Inger	*Bdellostoma burgeri*	inshore hagfish
Japanischer Käfer	*Popillia japonica*	Japanese beetle
Japanischer Kardinalfisch	*Synagrops japonicus*	Japanese splitfin
Japanischer Marder	*Martes melampus*	Japanese marten
Japanischer Riesenkalmar	*Architeuthis japonica*	Japanese giant squid
Japanischer Riesensalamander	*Andrias japonicus*	Japanese giant salamander
Japanischer Sandfisch	*Arctoscopus japonicus*	Japanese sandfish, sailfin sandfish (FAO)
Japanischer Schläfer	*Glirulus japonicus*	Japanese dormouse
Japanischer Schnabelwal, Ginkgozahn-Schnabelwal	*Mesoplodon ginkgodens*	ginkgo-toothed beaked whale
Japanisches Glasauge, Goldlachs	*Glossanodon semifasciatus/ Argentina semifasciata*	Pacific argentine, deepsea smelt (FAO)
Japankärpfling, Japanischer Goldhecht, Medaka	*Oryzias latipes*	medaka
Jararaca	*Bothrops jararaca*	jararaca
Jararacussu	*Bothrops jararacussu*	jararacussú
Java-Kuhnasenrochen	*Rhinoptera javanica*	flapnose ray, Javanese cownose ray (FAO)
Java-Vogelspinne	*Selenocosmia javanensis*	Javanese whistling spider
Javanashorn	*Rhinoceros sondaicus*	Javan rhinoceros
Javaneraffe, Langschwanzmakak	*Macaca fascicularis/ Macaca irus*	crab-eating macaque
Javanische Höckernatter, Java-Hökernatter	*Xenodermus javanicus*	xenodermine snake

Javanische Warzenschlange	*Acrochordus javanicus*	elephant-trunk snake
Javanisches Schuppentier	*Manis javanica*	Malayan pangolin
Jeffrey-Grundel	*Buenia jeffreysii*	Jeffreys's goby
Jemen-Chamäleon	*Chamaeleo calyptratus*	Yemen chameleon, cone-headed chameleon, veiled chameleon
Jendajasittich	*Aratinga jandaya*	Jandaya conure
Jenkins Deckelschnecke, Neuseeländische Deckelschnecke	*Potamopyrgus jenkinsi/ Potamopyrgus antipodarum*	Jenkins' spiresnail, Jenkins's spire snail (Jenkins' spire shell), New Zealand spiresnail
Jentinkducker	*Cephalophus jentinki*	Jentink's duiker
Jerichonektarvogel	*Nectarinia osea*	northern orange-tufted sunbird, Palestine sunbird
Johannisbeerblasenlaus	*Cryptomyzus ribis*	red currant blister aphid, currant aphid
Johannisbeerblattgallmücke	*Dasineura tetensi/ Perrisia tetensi*	black currant leaf midge
Johannisbeergallmilbe	*Eriophyes ribis/ Cecidophyopsis ribis*	currant gall mite, big bud gall mite
Johannisbeerglasflügler	*Synanthedon tipuliformis/ Aegeria tipuliformis*	currant clearwing moth, currant borer
Johannisbeermotte	*Incurvaria capitella*	currant bud moth
Johannisbeerzapfenspanner	*Eupithecia assimilata*	currant pug moth
Johannisbrotmotte	*Ectomyelois ceratoniae*	carob moth
Johannisechse	*Ablepharus kitaibelii*	juniper skink, snake-eyed skink
Johanniskäfer, Johanniswürmchen, Gemeiner Leuchtkäfer, Kleiner Leuchtkäfer, Glühwürmchen	*Lamprohiza splendidula*	small lightning beetle
Jonahkrabbe	*Cancer borealis*	jonah crab
Jordans Waldsalamander, Appalachen-Waldsalamander	*Plethodon jordani*	Jordan's salamander, red-cheeked salamander, Appalachian woodland salamander
Josefinenlori	*Charmosyna josefinae*	Josephine's lory
Josefinische Mondschnecke	*Neverita josephina*	Josephine's moonsnail
Jota-Eule	*Autographa jota*	plain golden Y
Joubins Krake, Zwergkrake	*Octopus joubini*	Atlantic pygmy octopus, dwarf octopus, Joubin's octopus
Juan-Fernández-Seebär	*Arctocephalus philippii*	Juan Fernández fur seal
Juan-Fernández-Languste	*Jasus frontalis*	Australian spiny lobster, Juan Fernández rock lobster
Juchtenkäfer, Eremit	*Osmoderma eremita*	hermit beetle
Judenfisch *siehe*> Riesen-Zackenbarsch	*Epinephelus itajara*	giant grouper (FAO) (jewfish)
Jugendliche Venusmuschel*	*Periglypta puerpera*	youthful venus
Julikäfer	*Anomala dubia*	margined vine chafer
Jumbokalmar, Riesenkalmar	*Dosidicus gigas*	jumbo flying squid

Jungfern-Seescheide, Weiße Seescheide	*Ascidia virginea*	virgin sea-squirt*
Jungfernfische, Riffbarsche	*Abudefduf* spp.	sergeant damselfishes
Jungfernkegel	*Conus virgo*	virgin cone
Jungfernkranich	*Anthropoides virgo*	demoiselle crane
Jungfernsohn, Großes Jungfernkind	*Archiearis parthenias*	orange underwing moth
Jungfrau-Nerite	*Neritina virginea*	virgin nerite
Jungfräuliche Stachelschnecke	*Chicoreus virgineus*	virgin murex
Junikäfer (Gemeiner Brachkäfer, Sonnwendkäfer)	*Amphimallon solstitialis/ Rhizotragus solstitialis*	summer chafer
Junko	*Junco hyemalis*	dark-eyed junco
Juno-Walzenschnecke	*Scaphella junonia*	junonia, Juno's volute
Juraviper, Aspisviper	*Vipera aspis*	asp viper, aspic viper
Juwelen-Fahnenbarsch	*Anthias squamipinnis*	orange sea perch, sea goldie (FAO), lyretail coralfish
Juwelenbarsch	*Cephalopholis miniatus*	blue-spotted rockcod, coral trout, coral hind (FAO)

Kabeljau (*Ostsee/Jungform*: Dorsch)	*Gadus morhua*	cod, Atlantic cod (*young*: codling)
Kabinettkäfer, Teppichkäfer	*Anthrenus scrophulariae*	carpet beetle
Kabylenkleiber	*Sitta ledanti*	Algerian nuthatch
Käfer	Coleoptera	beetles
Käfermilbe	*Parasitus coleoptratorum*	beetle mite
Käfermilben	Parasitidae	beetle mites
Käferzikaden	Tettigometridae	tettigometrids
Kaffeebohnenkäfer	*Araecerus fasciculatus*	coffee bean weevil
Kaffernadler	*Aquila verreauxii*	Verreaux's eagle
Kaffernbüffel, Afrikanischer Büffel	*Syncerus caffer*	African buffalo
Kaffernhornrabe	*Bucorvus cafer*	ground hornbill
Kaffernsegler	*Apus caffer*	white-rumped swift
Kagu	Rhynochetidae	kagu
Kagu	*Rhynochetos jubatus*	kagu
Kahlgesicht-Mönchsaffe	*Pithecia irrorata*	bald-faced saki
Kahlhecht, Schlammfisch, Amerikanischer Schlammfisch	*Amia calva*	bowfin
Kahlhechte, Schlammfische	Amiiformes	modern bowfins
Kahlhechte, Schlammfische	Amiidae	bowfins
Kahlkopfatzel	*Sarcops calvus*	bald starling
Kahlkopfpapagei	*Gypopsitta vulturina*	vulturine parrot
Kahlschnabel	Cochleariidae	boatbill heron
Kahnfahrer	*Scapholebris mucronata*	
Kahnfüßer, Grabfüßer, Scaphopoden	Solenoconchae/ Scaphopoda	scaphopods, scaphopodians (tooth shells, tusk shells, spade-footed mollusks)
Kahnkäfer	Scaphididae	shining fungus beetles
Kahnschnecke	*Cymbium olla*	Olla volute
Kahnspinner, Kleiner Kahnspinner, Buchenwickler, Buchenkahneule, Jägerhütchen	*Hylophila prasinana/ Pseudoips fagana/ Bena prasinana/ Bena fagana*	green silverlines, scarce silverlines
Kaimanfisch, Alligatorfisch	*Lepisosteus tristoechus*	Cuban gar
Kaiseradler	*Aquila heliaca*	imperial eagle
Kaiseramazone	*Amazona imperialis*	imperial parrot, imperial amazon
Kaiserbarsch, Alfonsino, Nordischer Schleimkopf	*Beryx decadactylus*	beryx, alfonsino (FAO), red bream
Kaiserbarsch> Mittelmeer-Kaiserbarsch	*Hoplostethus mediterraneus*	rough-fish, roughfish, rosy soldierfish, Mediterranean slimehead (FAO)
Kaiserbuntbarsch	*Aulonocara nyassae*	emperor cichlid
Kaiserfasan	*Lophura imperialis*	imperial pheasant
Kaiserfische	Pomacanthidae/Pomacanthinae	angelfishes
Kaisergans	*Anser canagicus/ Philacte canagicus*	emperor goose

Kaisergranat, Kaiserhummer, Kronenhummer, Schlankhummer, Tiefseehummer	*Nephrops norvegicus*	Norway lobster, Norway clawed lobster, Dublin Bay lobster, Dublin Bay prawn (scampi, langoustine)
Kaiserkegel, Königliche Kegelschnecke	*Conus imperialis*	imperial cone
Kaiserliche Helmschnecke	*Cassis madagascariensis*	cameo helmet, queen helmet, emperor helmet
Kaiserliche Wendeltreppe	*Epitonium imperialis*	imperial wentletrap
Kaisermantel	*Argynnis paphia*	silver-washed fritillary
Kaiserpinguin	*Aptenodytes forsteri*	emperor penguin
Kaiserschnapper	*Lutjanus sebae*	emperor snapper, red emperor, emperor red snapper (FAO)
Kaiserschnurrbarttamarin	*Saguinus imperator*	emperor tamarin
Kaiserskorpion, Afrikanischer Riesenskorpion, Afrikanischer Waldskorpion	*Pandinus imperator*	common emperor scorpion
Kaiserspecht	*Campephilus imperialis*	imperial woodpecker
Kakadus	Cacatuidae	cockatoos (incl. corellas)
Kakao-Motte, Kakaomotte, Tabakmotte, Speichermotte, Heumotte	*Ephestia elutella*	cocoa moth, chocolate moth, tobacco moth, flour moth, warehouse moth, cinereous knot-horn
Kakapo, Eulenpapagei	*Strigops habroptilus*	kakapo, owl-parrot
Kaktus-Seescheide	*Boltenia echinata*	cactus sea-squirt
Kaktusmaus	*Peromyscus eremicus*	cactus mouse
Kaktusmotte	*Cactoblastis cactorum*	cactus moth
Kaktusschwamm, Stachelschwamm	*Acanthella acuta*	cactus sponge
Kaktuszaunkönig	*Campylorhynchus brunneicapillus*	cactus wren
Kalan (Hbz.), Seeotter	*Enhydra lutris*	sea otter
Kalanderlerche	*Melanocorypha calandra*	calandra lark
Kalifornische Alligatorschleiche	*Gerrhonotus multicarinatus multicarinatus/ Elgaria multicarinata multicarinata*	California legless lizard
Kalifornische Fadenschnecke	*Flabellina iodinea*	Spanish shawl
Kalifornische Felsenlanguste, Kalifornische Languste	*Panulirus interruptus*	California spiny lobster, California rock lobster
Kalifornische Filigrankoralle	*Stylaster californicus*	Californian hydrocoral
Kalifornische Froschschnecke	*Crossata californica*	California frogsnail
Kalifornische Geißelgarnele	*Farfantepenaeus californiensis/ Penaeus californiensis*	yellowleg shrimp, yellow-leg shrimp
Kalifornische Kegelschnecke	*Conus californicus*	California cone
Kalifornische Languste, Kalifornische Felsenlanguste	*Panulirus interruptus*	California spiny lobster, California rock lobster
Kalifornische Miesmuschel	*Mytilus californianus*	California mussel (common mussel)

Kalifornische Ringelschleiche	*Anniella pulchra*	California legless lizard
Kalifornische Rötelmaus	*Clethrionomys occidentalis*	red-backed mouse
Kalifornische Sardine, Pazifische Sardine	*Sardinops caeruleus*	Californian pilchard
Kalifornische Schafskrabbe	*Loxorhynchus grandis*	Californian sheep crab
Kalifornische Schildlaus, San-José-Schildlaus	*Quadraspidiotus perniciosus*	San Jose scale
Kalifornische Schmierlaus	*Pseudococcus obscurus*	Californian mealybug
Kalifornische Scholle	*Eopsetta jordani*	petrale sole
Kalifornische Trogmuschel	*Mactrotoma californica*	California surfclam, Californian mactra
Kalifornischer Bärenkrebs	*Scyllarides astori*	Californian slipper lobster
Kalifornischer Barrakuda, Pazifik-Barrakuda	*Sphyraena argentea*	Pacific barracuda (FAO), California barracuda
Kalifornischer Blattfingergecko	*Phyllodactylus xanti*	American leaf-toed gecko
Kalifornischer Bleikabelbohrer	*Scobicia declivis*	short-circuit beetle
Kalifornischer Eidechsenfisch	*Synodus lucioceps*	California lizardfish
Kalifornischer Eselhase	*Lepus californicus*	black-tailed jack rabbit
Kalifornischer Halbschnabelhecht, Kalifornischer Halbschnäbler	*Hyporhamphus rosae*	California halfbeak
Kalifornischer Holzbock	*Ixodes pacificus*	California black-eyed tick, western black-legged tick
Kalifornischer Kondor	*Gymnogyps californianus*	California condor
Kalifornischer Krake	*Octopus californicus*	California octopus
Kalifornischer Molch	*Taricha torosa*	California newt
Kalifornischer Querzahnmolch	*Ambystoma californiense*	California tiger salamander
Kalifornischer Rochen	*Raja inornata*	California ray
Kalifornischer Rundhering	*Etrumeus acuminatus*	California round herring
Kalifornischer Sandskorpion	*Paruroctonus mesaensis*	giant sand scorpion
Kalifornischer Seelöwe	*Zalophus californianus*	Californian sea lion
Kalifornischer Stierkopfhai	*Heterodontus francisci*	hornshark (FAO), California hornshark
Kalifornischer Strandfloh, Kalifornischer Sandhüpfer	*Orchestoidea californiana*	California beach flea
Kalifornischer Taschenkrebs, Pazifischer Taschenkrebs	*Cancer magister*	Dungeness crab, Californian crab, Pacific crab
Kalifornischer Wurmsalamander	*Batrachoseps attenuatus*	California slender salamander
Kalifornischer Zitterrochen, Pazifik-Zitterrochen	*Torpedo californica*	Pacific electric ray
Kalkschwämme	Calcarea	calcareous sponges
Kalkstein-Höhlensalamander	*Hydromantes brunus*	limestone salamander
Kalmar> Gemeiner Kalmar, Roter Gemeiner Kalmar	*Loligo vulgaris*	common squid
Kalmare	Loliginidae	inshore squids
Kalmare	Teuthoidea (Teuthida)	squids

Kalong, Fliegender Hund, Riesen-Flugfuchs	*Pteropus vampyrus*	kalong, flying dog, large flying fox
Kambodscha-Saugbarbe	*Garra cambodgiensis*	stonelapping minnow
Kamel> Einhöckriges Kamel, Dromedar	*Camelus dromedarius*	dromedary, one-humped camel
Kamel> Zweihöckriges Kamel, Trampeltier	*Camelus bactrianus*	Bactrian camel, two-humped camel
Kamelhalsfliegen	Raphidiidae/Raphidioptera	snakeflies
Kamelspinner	*Ptilodon capucina*	coxcomb prominent
Kamelspinner, Uferweiden-Zahnspinner, Zickzackspinner	*Eligmodonta ziczac/ Notodonta ziczac*	pebble prominent
Kamerunfluss-Delphin, Kamerun-Buckeldelphin, Atlantischer Buckeldelphin	*Sousa teuszi*	humpback dolphin, Atlantic hump-backed dolphin, West African hump-backed dolphin
Kamm-Anolis	*Anolis cristatellus*	crested anole
Kamm-Samtmuschel	*Glycymeris pectinata*	comb bittersweet
Kamm-Venusmuschel	*Pitar dione*	royal comb venus, royal pitar
Kammblässhuhn	*Fulica cristata*	crested coot
Kammchamäleon	*Chamaeleo cristatus*	crested chameleon
Kammfinger, Gundis	Ctenodactylidae	gundis
Kammfische	Macristiidae	
Kammkiemer, Fiederkiemer	Protobranchiata	protobranch bivalves
Kammleguane	*Brachylophus* spp.	Fiji iguanas, banded iguanas
Kammloses Himalaya-Stachelschwein*	*Hystrix hodgsoni*	crestless Himalayan porcupine
Kammmolch	*Triturus cristatus*	warty newt, crested newt, European crested newt
Kammmünder	Ctenostomata	ctenostomates
Kammmuschelartige Samtmuschel	*Glycymeris pectiniformis*	scalloplike bittersweet
Kammmuscheln, Kamm-Muscheln	*Chlamys* spp.	scallops
Kammmuscheln, Pilgermuscheln	Pectinidae	scallops
Kammquallen, Rippenquallen, Ctenophoren	Ctenophora	ctenophores (sea gooseberries, sea combs, comb jellies, sea walnuts)
Kammratten	Ctenomyidae	tuco-tucos, ctenomyids
Kammschwanzbeutelmaus	*Dasycercus cristicauda*	mulgara
Kammspinnen, Wanderspinnen	Ctenidae	wandering spiders, running spiders
Kammstachelratten	*Echimys* spp.	arboreal spiny rats
Kammwürmer, Kammborstenwürmer	Pectinariidae	comb worms, trumpet worms, ice-cream-cone worms, pectinariids
Kammzahn-Katzenhai	*Proscyllium habereri*	graceful catshark
Kammzahn-Katzenhaie	Proscylliidae	finback catsharks
Kammzehen-Springmaus	*Paradipus ctenodactylus*	comb-toed jerboa

Kampfadler	*Polemaetus bellicosus/ Hieraaetus bellicosus*	martial eagle
Kampffische, Paradiesfische	Macropodinae	macropods
Kampfläufer	*Philomachus pugnax*	ruff
Kampfuchs> Brasilianischer Kampfuchs	*Lycalopex vetulus/ Pseudalopex vetulus*	hoary fox, Brasilian fox
Kampfwachteln, Laufhühnchen	Turnicidae	button quails, hemipodes
Kamtschatka-Krabbe, Alaska-Königskrabbe, Königskrabbe (Kronenkrebs, Kamtschatkakrebs)	*Paralithodes camtschaticus*	king crab, red king crab, Alaskan king crab, Alaskan king stone crab (Japanese crab, Kamchatka crab, Russian crab)
Kanadagans	*Branta canadensis*	Canada goose
Kanadakleiber	*Sitta canadensis*	red-breasted nuthatch
Kanadakranich	*Grus canadensis*	sandhill crane
Kanadaluchs (Hbz. Silberluchs)	*Felis lynx canadensis*	Canadian lynx, silver lynx
Kanadareiher	*Ardea herodias*	great blue heron
Kanadaschnepfe	*Scolopax minor*	American woodcock
Kanadische Alse, Blaurückenhering	*Alosa aestivalis*	blueback herring
Kanadischer Biber	*Castor canadensis*	North American beaver
Kanadischer Zander	*Stizostedion canadense*	sauger
Kanalkäfer, Kamelläufer	*Amara* spp.	sun beetles
Kanarengirlitz, Wilder Kanarienvogel	*Serinus canaria*	canary, common canary, island canary
Kanarenschmätzer	*Saxicola dacotiae*	Canary Island chat
Kanarenskink	*Chalcides viridanus*	Canarian cylindrical skink
Kanarienvogel-Lungenmilbe	*Sternostoma tracheacolum*	canary lung mite
Kanarienvogel> Wilder Kanarienvogel, Kanarengirlitz	*Serinus canaria*	canary, common canary, island canary
Kanarische Grundel	*Vanneaugobius canariensis*	Canary goby
Kanburi-Lanzenotter	*Trimeresurus kanburiensis*	Kanburian pit viper, Kanburi pit viper, tiger pit viper
Kandelaber-Wellhorn	*Busycon candelabrum*	splendid whelk, candelabrum whelk
Kängurumäuse	*Microdipodops* spp.	kangaroo mice
Kängururatten, Taschenspringer	*Dipodomys* spp.	kangaroo rats, California kangaroo rats
Kängurus	Macropodidae & Potoroidae	kangaroos
Kaninchen> Altweltliches Wildkaninchen, Europäisches Wildkaninchen (Hbz. Wildkanin)	*Oryctolagus cuniculus*	Old World rabbit, domestic rabbit
Kaninchen-Haarmilbe, Kaninchenmilbe	*Listrophorus gibbus*	rabbit fur mite a.o.
Kaninchenerdhacker	*Geositta cunicularia*	common miner
Kaninchenfische	Siganidae/Teuthidae	rabbitfishes
Kaninchenfische	*Siganus* spp.	rabbitfishes

Kaninchenfloh	*Spilopsyllus cuniculi*	rabbit flea, European rabbit flea
Kaninchenkängurus	*Potorous* spp.	potoroos
Kaninchenkauz	*Athene cunicularia*	burrowing owl
Kaninchenlaus	*Haemodipsus ventricosus*	rabbit louse
Kaninchennasenbeutler, Ohrenbeuteldachse	*Macrotis* spp.	bilbies, rabbit-eared bandicoots
Kaninchenpelzmilbe	*Cheyletiella parasitivorax*	rabbit fur mite a.o.
Kaninchenratte*	*Reithrodon physodes*	coney rat
Kaninchenratten	*Conilurus* spp.	tree rats, rabbit rats
Kannibalenfrösche	*Lechriodus* spp.	cannibal frogs
Kanonenkugel-Qualle	*Stomolophus meleagris*	cannonball jellyfish
Kanten-Triton*	*Cymatium femorale*	angular triton
Kantengarnelen, Kanten-Tiefseegarnelen	*Heterocarpus* spp.	nylon shrimps
Kantige Sattelschrecke	*Uromenus rugosicollis*	rough-backed bushcricket
Kantige Wendeltreppe	*Epitonium angulatum*	angulate wentletrap
Kantschile	*Tragulus* spp.	Asiatic chevrotains, mouse deer
Kap Lopez, Fahnenhechtling, Bunter Prachtkärpfling	*Aphyosemion australe*	lyretail, Cape Lopez lyretail
Kap-Blessmull	*Georychus capensis*	Cape mole-rat
Kap-Drachenkopf	*Sebastes capensis*	Jacopever, Cape redfish, false jacopero (FAO)
Kap-Feilennatter	*Mehelya capensis*	Cape file snake
Kap-Hummer, Südafrikanischer Hummer	*Homarinus capensis*	Cape lobster
Kap-Igel	*Atelerix frontalis/ Erinaceus frontalis*	South African hedgehog
Kap-Kalmar	*Loligo reynaudi*	Cape Hope squid, chokker squid
Kap-Knurrhahn	*Chelidonichthys capensis*	Cape gurnard
Kap-Krallenfrosch	*Xenopus gilli*	Cape platanna, Cape clawed toad
Kap-Languste, Afrikanische Languste	*Jasus lalandei*	Cape rock crawfish, Cape rock lobster
Kap-Plattgürtelechse	*Platysaurus capensis*	Cape red-tailed flat lizard
Kap-Wolfsnatter	*Lycophidion capense*	African wolf snake
Kapfuchs, Kamafuchs, Silberrückenfuchs	*Vulpes chama*	Cape fox
Kapgeier	*Gyps coprotheres*	Cape vulture, Cape griffon
Kapgoldmull	*Chrysochloris asiatica*	Cape golden mole
Kaphase, Wüstenhase	*Lepus capensis*	Cape hare, brown hare
Kapitäns-Kegelschnecke	*Conus capitaneus*	captain cone
Kapitänsfisch, Fingerfisch	*Polynemus quadrifilis*	threadfin, five-rayed threadfin, giant African threadfin (FAO)
Kapkobra	*Naja nivea*	Cape cobra, yellow cobra
Kapohreule	*Asio capensis*	marsh owl, Algerian marsh owl

Kapotter, Fingerotter (Hbz. Afrikanischer Otter)	*Aonyx capensis*	African clawless otter
Kapottern, Fingerottern (Hbz. Afrikanische Otter)	*Aonyx* spp.	clawless otters
Kappenammer	*Emberiza melanocephala*	black-headed bunting
Kappengibbon	*Hylobates pileatus*	capped gibbon, pileated gibbon
Kappensäger	*Mergus cucullatus*	hooded merganser
Kappensteinschmätzer	*Oenanthe monacha*	hooded wheatear
Kappentierchen	*Colpoda cucullus*	
Kapscharbe	*Phalacrocorax capensis*	Cape cormorant
Kapsturmvogel	*Daption capense*	pintado petrel
Kaptäubchen	*Oena capensis*	Namaqua dove
Kapuzenbärchen	*Nola cucullatella*	short-cloaked moth
Kapuzenkäfer, Kapuzinerkäfer, Bohrkäfer, Holzbohrkäfer	Bostrichidae/ Bostrychidae	horned powder-post beetles, branch borers & twig borers, bostrichids (wood borers)
Kapuzenkugelkäfer	*Mezium affine*	shiny spider beetle
Kapuzennatter	*Macroprotodon cucullatus*	hooded snake, false smooth snake
Kapuzenspinnen	Ricinulei	ricinuleids
Kapuzenzeisig	*Carduelis cucullata/ Spinus cucullatus*	red siskin
Kapuziner, Weißschulterkapuziner, Weißschulteraffe	*Cebus capucinus*	white-throated capuchin
Kapuzineraffen, Kapuziner, Rollaffen	*Cebus* spp.	capuchins, ring-tailed monkeys
Kapuzinerartige	Cebidae	New World monkeys
Kapverdensturmvogel	*Pterodroma feae*	gon-gon
Karakal, Wüstenluchs	*Felis caracal*	caracal
Karakara, Schopfkarakara	*Polyborus plancus*	common caracara, crested caracara
Karausche	*Carassius carassius*	Crucian carp
Kardeneule, Steppenkräuterhügel- Sonneneule	*Heliothis viriplaca/ Heliothis dipsacea*	marbled clover
Kardenfliege*	*Phytomyza ramosa*	teasel fly
Kardinal, Pandorafalter	*Pandoriana pandora/ Argynnis pandora*	cardinal
Kardinal-Kegelschnecke	*Conus cardinalis*	cardinal cone
Kardinal-Winkelspinne	*Tegenaria parietina*	cardinal spider
Kardinalbarsche	*Apogon* spp.	cardinalfish, cardinal fish cardinals
Kardinalbarsche, Kardinalfische	Apogonidae	cardinal fishes
Kardinäle	Cardinalinae	cardinals
Kardinalfisch, „Arbeiterneon"	*Tanichthys albonubes*	white cloud mountain minnow (FAO), white cloud,
Kardinalspecht	*Dendropicos fuscescens*	cardinal woodpecker

Karettschildkröte, Karette	*Eretmochelys imbricata*	hawksbill turtle, hawksbill sea turtle
Karettschildkröte> Unechte Karettschildkröte, Unechte Karette	*Caretta caretta*	loggerhead sea turtle, loggerhead
Karfunkelsalmler	*Hemigrammus pulcher*	garnet tetra (FAO), pretty tetra
Karibenkaiserfisch, Felsenschönheit	*Holacanthus tricolor*	rock beauty
Karibik-Kaisergranat	*Metanephrops binghami*	Caribbean lobsterette
Karibik-Pilgermuschel	*Argopecten irradians*	bay scallop, Atlantic bay scallop
Karibik-Spitzkopfkugelfisch	*Canthigaster rostrata*	sharpnose puffer
Karibische Asbestkoralle*	*Briareum asbestinum*	corky sea fingers
Karibische Languste, Amerikanische Languste	*Panulirus argus*	West Indies spiny lobster, Caribbean spiny lobster, Caribbean spiny crawfish
Karibische Mönchsrobbe	*Monachus tropicalis*	Caribbean monk seal
Karibische Sandmuschel*	*Asaphis deflorata*	gaudy asaphis
Karibische Spinnenkrabbe, Dreieckskrabbe	*Stenorhynchus seticornis*	yellowline arrow crab, Caribbean arrow crab
Karibischer Ammenhai, Atlantischer Ammenhai	*Ginglymostoma cirratum*	nurse shark
Karibischer Bärenkrebs, „Spanischer" Bärenkrebs	*Scyllarides aequinoctialis*	"Spanish" lobster, "Spanish" slipper lobster
Karibischer Büschelbarsch, Karibischer Korallenwächter	*Amblycirrhitus pinos*	redspotted hawkfish
Karibischer Felsenbarsch	*Epinephelus adscensionis*	rock hind
Karibischer Küstendelphin, Guayana-Delphin	*Sotalia guianensis*	Guayana river dolphin, Guiana coastal dolphin, white dolphin
Karibischer Soldatenfisch	*Myripristis jacobus*	blackbar soldierfish
Karibischer Zitterrochen, Kleiner Zitterrochen	*Narcine brasiliensis*	Brazilian electric ray
Karibisches Gorgonenhaupt	*Astrophyton muricatum*	Caribbean basket star
Karierte Fasanenschnecke	*Eulithidium affine affine/ Tricolia affinis affinis*	checkered pheasant snail
Karierte Kauri	*Cypraea tessellata/ Luria tessellata*	checkerboard cowrie, checkered cowrie
Karminbär, Jakobskreuzkrautbär	*Tyria jacobaeae*	cinnabar moth
Karmingimpel	*Carpodacus erythrinus*	scarlet rosefinch, common rosefinch
Karminroter Kissenstern	*Porania pulvillus*	red cushion star
Karminroter Seereisig	*Lophogorgia miniata*	
Karminspinnmilbe	*Tetranychus cinnabarinus*	carmine spider mite
Karnivoren, Raubtiere	Carnivora	carnivores
Karotten-Kegel, Mohrrüben-Kegelschnecke*	*Conus daucus*	carrot cone
Karpatenmolch	*Triturus montandoni*	Montandon's newt, Carpathian newt
Karpatenneunauge (Donauneunauge)	*Eudontomyzon danfordi*	Carpathian lamprey (FAO), Carpathian brook lamprey (Hungarian lamprey/ Danubian lampern)

Karpfen

Karpfen, Flusskarpfen	*Cyprinus carpio*	carp, common carp (FAO), European carp
Karpfenartige, Karpfenfische	Cypriniformes	carps & characins & minnows & suckers & loaches
Karpfenfische	Cyprinidae	minnows or carps
Karpfenlaus	*Argulus foliaceus*	carp louse
Karpfenläuse, Fischläuse, Kiemenschwänze	Branchiura/ Argulida	fish lice
Karpfenschwanzmilbe, Essigmilbe	*Histiogaster carpio*	vinegar mite*, carptail mite*
Kartoffelbohrer, Rübenbohrer, Schachtelhalmeule, Markeule	*Hydraecia micacea*	rosy rustic (moth)
Kartoffelerdfloh	*Psylliodes affinis*	potato flea-beetle
Kartoffelkäfer (Koloradokäfer)	*Leptinotarsa decemlineata*	Colorado potato beetle, Colorado beetle, potato beetle
Kartoffelkellerlaus, Kellerlaus, Breitröhrige Kartoffelknollenlaus	*Rhopalosiphoninus latysiphon*	bulb and potato aphid
Kartoffelkrätzeälchen (Nematodenfäule der Kartoffel)	*Ditylenchus destructor*	potato rot nematode
Kartoffelmotte	*Phthorimaea operculella*	potato moth, potato tuber moth (potato tuberworm)
Kartoffelzystenälchen, Gelbe Kartoffelnematode	*Heterodera rostochiensis/ Globodera rostochiensis*	golden nematode (golden nematode disease of potato)
Kaschelot, Cachalot, Großer Pottwal	*Physeter macrocephalus (catodon)*	sperm whale, great sperm whale, cachalot
Kaschmir-Wühlmaus, True-Wühlmaus	*Hyperacrius fertilis*	Kashmir vole, True's vole
Kaschmir-Wühlmäuse, Punjab-Wühlmäuse	*Hyperacrius* spp.	Kashmir voles, Punjab voles
Käsefliege	*Piophila casei*	cheese-skipper (cheese maggots)
Käsefliegen	Piophilidae	cheese-skippers
Käsemilbe	*Tyrophagus casei/ Tyrolichus casei*	cheese mite
Kaskadenfrösche	*Amolops* spp.	sucker frogs
Kaspi-Ringelrobbe	*Phoca caspica*	Caspian seal
Kaspikönigshuhn	*Tetraogallus caspius*	Caspian snowcock
Kaspische Bachschildkröte	*Mauremys caspica*	Caspian turtle
Kaspische Smaragdeidechse, Streifensmaragdeidechse	*Lacerta strigata*	Caspian green lizard, Caucasus emerald lizard
Kaspische Wasserschildkröte	*Clemmys caspica*	Caspian pond turtle
Kaspischer Geradfingergecko	*Alsophylax pipiens*	Caspian even-fingered gecko, Caspian straight-fingered gecko
Kastanien-Astarte, Glatte Astarte	*Astarte castanea*	smooth astarte
Kastanien-Froschschnecke*	*Bufonaria bufo/Bursa bufo*	chestnut frogsnail

Kastanien-Scheinolive	*Ancilla castanea*	chestnut ancilla
Kastanienbohrer, Blausieb	*Zeuzera pyrina*	leopard moth
Kastanienkauri	*Cypraea spadicea/ Zonaria spadicea*	chestnut cowrie
Kastanienmaus	*Niviventer fulvescens*	chestnut rat
Kastanienspecht	*Celeus castaneus*	chestnut-coloured woodpecker
Kastenkrabben, Taschenkrebse u.a.	Homolidae	carrier crabs (box crabs a.o.)
Kastilienspitzmaus	*Sorex granarius*	Spanish shrew
Kasuare & Emu	Casuariidae	cassowaries & emu
Katholiken-Frosch	*Notaden bennettii*	crucifix toad
Katipo	*Latrodectus mactans hasselti*	katipo widow
Katta	*Lemur catta*	ring-tailed lemur
Kattfisch, Katfisch, Seewolf, Gestreifter Seewolf	*Anarhichas lupus*	Atlantic wolffish, wolffish (FAO), cat fish, catfish
Katze> Hauskatze	*Felis catus/ Felis silvestris f. catus*	domestic cat
Katze> Wildkatze	*Felis silvestris*	wild cat
Katzen	Felidae	cats
Katzen-Haarbalgmilbe	*Demodex cati*	cat follicle mite
Katzen-Kreuzwels, Weißer Katzen-Kreuzwels, Weißer Meereswels*	*Galeichthys feliceps*	white seacatfish, white baggar (FAO)
Katzen-Kreuzwels*	*Arius felis*	hardhead sea catfish
Katzenaugennatter	*Leptodeira annulata*	banded cat-eyed snake
Katzenaugennattern	*Leptodeira* spp.	cat-eyed snakes
Katzenbandwurm u.a., Dickhalsiger Bandwurm	*Hydatigera taeniaeformis/ Taenia taeniaeformis*	common cat tapeworm
Katzenbär	*Ailurus fulgens*	lesser panda, red panda
Katzenegel, Katzenleberegel	*Opisthorchis felineus*	cat liver fluke, Siberian liver fluke
Katzenfloh	*Ctenocephalides felis*	cat flea
Katzenfrosch	*Neobatrachus pictus*	meeoing frog
Katzengecko	*Aeluroscalabotes felinus*	cat gecko
Katzenhaarling	*Felicola subrostratus*	cat louse
Katzenhaie	Scyliorhinidae	catsharks, cat sharks
Katzenkauri	*Cypraea felina*	cat cowrie, kitten cowrie
Katzenleberegel, Katzenegel	*Opisthorchis felineus*	cat liver fluke, Siberian liver fluke
Katzenmakis	Cheirogaleidae	mouse lemurs
Katzennatter> Europäische Katzennatter	*Telescopus fallax*	cat snake, European cat snake
Katzenspulwurm	*Toxocara cati*	felline ascarid
Katzenvogel	*Dumetella carolinensis*	catbird
Katzenzunge, Katzenzungen-Tellmuschel	*Tellina linguaefelis*	cat's-tongue tellin
Katzenzungen-Auster	*Spondylus linguaefelis*	cat's-tongue oyster
Katzenzungen-Tellmuschel, Katzenzunge	*Tellina linguaefelis*	cat's-tongue tellin

Kaukasischer Schlammtaucher	*Pelodytes caucasicus*	Caucasian parsley frog
Kaukasus-Salamander	*Mertensiella caucasica/ Mertensia caucasica*	Caucasian salamander
Kaukasusagame	*Agama caucasica/ Stellio caucasica*	northern rock agama, Caucasian agama
Kaukasusbirkhuhn	*Tetrao mlokosiewiczi*	Caucasian black grouse
Kaukasuskönigshuhn	*Tetraogallus caucasicus*	Caucasian snowcock
Kaukasusotter	*Vipera kaznakovi*	Caucasus viper, Caucasian viper (IUCN), red viper
Kaulbarsch	*Gymnocephalus cernuus*	ruffe (FAO), pope
Kaulquappentriton, Braunweißband-Triton*	*Gyrineum gyrinum*	tadpole triton
Kaulquappenwelse	Amphiliidae	loach catfishes
Kauri> Falsche Kauri	*Erronea errones*	mistaken cowrie
Kaurischnecken, Kauris, Porzellanschnecken, Porzellanen	*Cypraea* spp.	cowries (*sg.* cowrie or cowry)
Kaurischnecken, Kauris, Porzellanschnecken, Porzellanen	Cypraeidae	cowries
Kea	*Nestor noabilis*	kea
Kegel-Seescheide	*Aplidium conicum*	conical sea-squirt
Kegelchen	Euconulidae	
Kegelige Strandschnecke, Marschenschnecke	*Assiminea grayana*	dun sentinel
Kegelköpfe, Schwertschrecken	Conocephalidae	meadow grasshoppers
Kegelrobbe	*Halichoerus grypus*	gray seal
Kegelschnecken	Conidae	cone snails, cones (cone shells)
Kegelschnecken	*Conus* spp.	cone snails, cones (cone shells)
Kehllappen-Weichschildkröte, Nackendorn-Weichschildkröte	*Trionyx steindachneri/ Palea steindachneri*	wattle-necked softshell turtle
Kehlphallusfische, Priapiumfische, Phallostethiden (Zwergährenfische u.a.)	Phallostethidae	priapium fishes
Kehlstreifpinguin	*Pygoscelis antarctica*	bearded penguin
Kehlzähner	Astronesthidae	snaggletooths
Keilfleck-Bärbling	*Rasbora heteromorpha*	harlequin fish, harlequin rasbora (FAO)
Keilfleck-Buntbarsch	*Uaru amphiacanthoides*	uaru, chocolate cichlid, triangle cichlid (FAO)
Keilfleck-Falterfisch	*Chaetodon falcula*	sickle butterflyfish, blackwedged butterflyfish (FAO)
Keilflecklibelle	*Aeshna isoceles*	Norfolk hawker (Br.)
Keilkopf-Glattstirnkaiman	*Paleosuchus trigonatus*	Schneider's smooth-fronted caiman
Keilschleiche	*Sphenops sepsoides*	wedge-snouted skink
Keilschwanz-Regenpfeifer	*Charadrius vociferus*	killdeer

Deutsch	Wissenschaftlich	Englisch
Keilschwanzadler	*Aquila audax*	wedge-tailed eagle
Kelaarts Langkrallenspitzmaus	*Feroculus feroculus*	Kelaart's long-clawed shrew
Kelchkoralle	*Astroides calycularis*	
Kelchwürmer, Nicktiere, Kamptozooen	Entoprocta	kamptozoans, entoprocts
Kelee-Alse	*Hilsa kelee*	kelee shad
Kellerassel	*Androniscus dentiger*	cellar woodlouse, rosy woodlouse
Kellerassel> Raue Kellerassel	*Porcellio scaber*	common rough woodlouse, garden woodlouse, slater, scabby sow bug
Kellerkäfer, Grotten-Dunkelkäfer	*Laemostenus terricola/ Pristonychus terricola*	cellar beetle*, European cellar beetle
Kellerlaus, Kartoffelkellerlaus, Breitröhrige Kartoffelknollenlaus	*Rhopalosiphoninus latysiphon*	bulb and potato aphid
Kellerspanner, Höhlenspanner	*Triphosa dubitata*	cave moth*, cellar moth*
Kellerspinne	*Segestria senoculata*	snake's back spider, cellar spider*
Kemps Bastardschildkröte, Atlantik-Bastardschildkröte, Atlantische Bastardschildkröte	*Lepidochelys kempii*	Kemp's ridley (sea turtle), Atlantic ridley turtle
Kerb-Seepocke	*Balanus perforatus*	perforated barnacle
Kerbameise, Buchtenkopf-Waldameise	*Formica exsecta/ Coptoformica exsecta*	narrow-headed ant
Kerben-Astarte*	*Astarte crenata*	crenulate astarte
Kermeslaus	*Kermes vermilio*	Mediterranean kermes coccid
Kernbeißer	*Coccothraustes coccothraustes*	hawfinch
Kernkäfer, Kernholzkäfer	Platypodidae	pin-hole borers
Kerzenstint, Kerzenfisch, Eulachon	*Thaleichthys pacificus*	candlefish, eulachon (FAO)
Kescherspinne, Ogerspinne	*Deinopis longipes*	net-casting spider
Kescherspinnen, Ogerspinnen	Deinopidae/Dinopidae	ogre-faced spiders
Keta-Lachs, Ketalachs, Hundslachs, Chum-Lachs	*Oncorhynchus keta*	chum salmon
Kettenhecht	*Esox niger*	chain pickerel
Kettenkärpfling	*Fundulus catenatus*	northern studfish
Kettenklapperschlange, Massasauga	*Sistrurus catenatus*	massasauga
Kettenviper	*Daboia russelli/Vipera russelli*	Russell's viper
Keulen-Eichelwurm	*Balanoglossus clavigerus* (Enteropneusta)	Mediterranean acorn worm
Keulen-Synascidie, Keulensynascidie, Glaskeulen-Seescheide, Gelbe Seescheide	*Clavelina lepadiformis*	light-bulb sea-squirt
Keulenkäfer	Clavigeridae	clavigerid beetles
Keulenpolyp	*Cordylophora caspia*	freshwater hydroid a.o.
Keulenpolypen	*Clava* spp.	club hydroids, club polyps

Keulenrochen, Nagelrochen	*Raja clavata*	thornback skate, thornback ray (FAO), roker
Keulenschwanzgeckos	*Nephrurus* spp.	knob-tailed geckos, knobtail geckos
Keulensynascidien, Glaskeulen-Seescheiden	*Clavelina* spp.	light-bulb tunicates, light-bulb sea-squirts
Keulenwespe> Gemeine Keulenwespe	*Sapyga clavicornis*	common European sapygid wasp
Keulhornblattwespe	*Trichiosoma tibiale*	hawthorn sawfly
Keulige Schließmundschnecke	*Clausilia pumila*	clublike door snail*
Khapra-Käfer, Khaprakäfer	*Trogoderma granarium*	khapra beetle
Kiang	*Equus kiang*	kiang
Kichererbsen-Kauri*	*Cypraea cicercula*	chick-pea cowrie
Kiebitz	*Vanellus vanellus*	lapwing
Kiebitzregenpfeifer	*Pluvialis squatarola*	grey plover
Kieferegel	*Gnathobdelliformes*	jawed leeches
Kieferfüßer	Maxillopoda	maxillopods
Kieferläuse, Läuslinge	Mallophaga	chewing lice, biting lice, bird lice
Kieferlose, Agnathen	Agnatha	jawless fishes, agnathans
Kiefermäuler, Kiefermündchen, Gnathostomuliden	Gnathostomulida	gnathostomulids
Kiefermünder	Gnathostomata	jawed vertebrates, jaw-mouthed animals, gnathostomatans
Kiefernbock, Bäckerbock, Gefleckter Langhornbock	*Monochamus galloprovincialis*	pine sawyer beetle
Kiefernbuschhornblattwespe	*Diprion pini*	pine sawfly
Kieferneule, Forleule	*Panolis flammea*	pine beau, pine beauty moth
Kiefernhäher	*Nucifraga columbiana*	Clark's nutcracker
Kiefernharz-Gallmücke	*Cecidomyia pini*	pine resin midge
Kiefernharzwickler	*Cydia coniferana*	pine resin moth
Kiefernholzwespe, Gemeine Holzwespe	*Sirex juvencus*	polished horntail
Kiefernknospen-Triebmotte	*Exoteleia dodecella*	European pine bud moth*
Kiefernknospenwickler	*Blastesthia turionella*	pine bud moth
Kiefernkreuzschnabel	*Loxia pytyopsittacus*	parrot crossbill
Kiefernkulturrüssler	*Pissodes castaneus*	small banded pine weevil
Kiefernmarkkäfer, Markkäfer, Großer Waldgärtner	*Tomicus piniperda*	large pine shoot beetle
Kiefernmoos*	*Diphasia pinastrum*	sea pine hydroid
Kiefernnadelwickler, Nadelholzwickler	*Archips oporana*	pineneedle tortrix moth
Kiefernnatter	*Pituophis melanoleucus*	pine snake
Kiefernnattern	*Pituophis* spp.	pine snakes, gopher snakes
Kiefernprozessionsspinner, Fichtenprozessionsspinner	*Thaumetopoea pinivora*	pine processionary moth
Kiefernrindenwanze	*Aradus cinnamomeus*	pine flatbug

Kiefernsaateule	*Agrotis vestigialis*	archer's dart moth
Kiefernschwärmer, Kleiner Fichtenrüssler	*Hyloicus pinastri/ Sphinx pinastri*	pine hawkmoth
Kiefernsittich, Arasittich	*Rhynchopsitta pachyrhyncha*	thick-billed parrot
Kiefernspinner	*Dendrolimus pini*	pine lappet, pine-tree lappet, European pine moth
Kieferntangare, Louisianatangare	*Piranga ludoviciana*	western tanager
Kieferntriebwickler	*Evetria buoliana/ Rhyacionia buoliana*	gemmed shoot moth, pine-sprout tortrix, European pine shoot moth
Kiefernwühlmaus	*Microtus pinetorum/ Pitymys pinetorum*	American pine vole
Kiefernzapfenrüssler	*Pissodes validirostris*	pinecone weevil
Kiefernzapfenwanze	*Gastrodes grossipes*	pinecone bug
Kiefernzapfenwickler	*Cydia conicolana*	pine cone moth
Kiefernzeisig	*Carduelis pinus*	pine siskin
Kiefernzweig-Gallmilbe*	*Trisetacus pini*	pine twig-knot mite, pine gall mite, pine twig gall mite
Kieferwurm	*Eunice harassii*	
Kielfußschnecke, Schwimmende Mütze	*Carinaria lamarcki*	Lamarck's nautilus
Kielrückennattern	*Natrix* spp.	grass snakes
Kielschuppen-Wassernatter	*Atretium schistosum*	olive keelback
Kielschwänze	*Tropidurus* spp.	lava lizards
Kielschwanznatter	*Helicops carinicaudus*	Wied's keelback
Kiemenegel, Krebsegel	*Branchiobdella parasitica*	fish-gill leech a.o.
Kiemenegel, Krebsegel	Branchiobdellidae	branchiobdellid leeches
Kiemenfuß, Sommer-Kieferfuß, Großer Rückenschaler	*Triops cancriformis*	tadpole shrimp, freshwater tadpole shrimp, rice apus
Kiemenfüßer, Blattfußkrebse	Phyllopoda/Branchiopoda	phyllopods, branchiopods
Kiemenfußkrebse	*Chirocephalus* spp.	fairy shrimps a.o.
Kiemenringelwurm	*Scoloplos armiger*	
Kiemenschlitzaale, Sumpfaale	Synbranchidae	swamp-eels
Kiemenschwänze, Karpfenläuse, Fischläuse	Branchiura/ Argulida	fish lice
Kiemenwurm, Krebsegel	*Branchiura sowerbyi*	
Kiesbank-Grashüpfer	*Chorthippus pullus*	gravel grasshopper
Kieselkrabbe*	*Myra fugax*	pebble crab
Kieselschwämme (Demospongien)	Silicospongiae	siliceous sponges, demosponges
Kiesfisch, Januarkärpfling	*Phalloptychus januarius*	barred millionsfish
Kilka, Tyulka-Sardelle	*Clupeonella cultriventris*	clupeonella, kilka, Black Sea sprat (FAO), Tyulka sprat

Killerwal, Schwertwal, Orca	Orcinus orca	killer whale, orca
Killifische> Eierlegende Zahnkarpfen	Cyprinodontidae	pupfishes
Kings Skink	Egernia kingii	King's skink
Kinkhorn, Knotiges Tritonshorn, Trompetenschnecke	Charonia lampas/ Tritonium nodiferum	knobbed triton
Kinnblatt-Fledermäuse, Nackrücken-Fledermäuse	Mormoopidae	moustached bats, naked-backed bats, ghost-faced bats
Kirks Agame	Agama kirkii	Kirk's rock agama
Kirschenspanner	Lycia hirtaria	brindled beauty moth
Kirschfliege, Kirschfruchtfliege	Rhagoletis cerasi	European cherry fruit fly
Kirschfruchtstecher	Rhynchites auratus	cherry fruit rhynchites weevil
Kirtland-Schlange	Clonophis kirtlandi	Kirtland's snake
Kissensterne	Oreasteridae	cushion stars
Kitfuchs, Großohrkitfuchs (Hbz. Kitfuchs)	Vulpes macrotis	kit fox
Kittis Kleinstfledermaus, Hummelfledermaus, Schweinsschnauzen-Fledermaus	Craseonycteris thonglongyai	Kitti's hog-nosed bat (*smallest mammal!*)
Kiwis	Apterygiformes/Apterygidae	kiwis
Klaffende Feilenmuschel	Limaria hians	gaping fileclam
Kläfferkauz	Ninox connivens/ Hieracoglaux connivens	barking owl, barking hawk-owl
Klaffmuscheln	Mya spp.	gaper clams
Klaffmuscheln	Myidae	gaper clams
Klammeraffen	Ateles spp.	spider monkeys
Klammerauster	Pycnodonta frons/ Lopha frons/Ostrea frons/ Dendrostrea frons	coon oyster
Klappenassel, Krallenfüßige Meerassel	Idotea chelipes	clawfooted marine isopod
Klappenmütze, Klappmütze	Cystophora cristata	hooded seal (bladder-nose)
Klappenstern	Pteraster militaris	winged seastar
Klappergrasmücke	Sylvia curruca	lesser whitethroat
Klappermoos*	Sarsia tubulosa	clapper hydroid, clapper hydromedusa
Klapperschlangen> Echte Klapperschlangen	Crotalus spp.	rattlesnakes, rattlers
Klappnasen, Mausschwanz-Fledermäuse	Rhinopoma spp.	mouse-tailed bats, long-tailed bats
Klappschildkröten	Kinosternon spp.	mud turtles
Klauenkäfer, Hakenkäfer	Dryopidae	long-toed water beetles
Klebfadenwebspinnen	Ecribellatae	viscid band spiders
Klee-Blutströpfchen, Klee-Widderchen	Zygaena lonicerae	narrow-bordered five-spot burnet
Kleeblattgallmücke	Dasineura trifolii	clover leaf midge
Kleebläuling	Everes argiades	short-tailed blue
Kleeborkenkäfer, Kleewurzelborkenkäfer	Hylastinus obscurus	large broom bark beetle

Kleeluzernerüssler, Luzerne-Dickmaulrüssler	*Otiorhynchus ligustici*	alfalfa snout beetle
Kleemilbe, Klee-Spinnmilbe	*Bryobia praetiosa*	clover mite
Kleesamenwespe	*Eurytoma gibbus/ Bruchophagus gibbus*	lucerne chalcid wasp
Kleespinner	*Pachygastria trifolii/ Lasiocampa trifolii*	grass eggar
Kleestengelbohrer*	*Languria mozardi*	clover stem borer
Kleezystenälchen	*Heterodera trifolii*	clover cyst nematode
Kleiber	*Sitta europaea*	Eurasian nuthatch
Kleiber, Spechtmeisen	Sittidae	nuthatches
Kleiderlaus	*Pediculus humanus* (*P. humanus humanus/ P. humanus corporis*)	body louse, "cootie", "seam squirrel", clothes louse
Kleidermotte	*Tineola bisselliella*	common clothes moth, destroyer clothes moth
Kleidervögel	Drepanididae	Hawaiian honeycreepers
Klein-Flügeltaschen-Fledermaus	*Balantiopteryx* spp.	least sac-winged bats
Kleinabendsegler, Kleiner Abendsegler	*Nyctalus leisleri*	Leisler's bat
Kleinasiatischer Braunfrosch	*Rana macrocnemis*	Brusa frog
Kleinasiatischer Salamander, Lykischer Salamander	*Mertensiella luschani/ Salamandra luschani*	Luschan's salamander, Lycian salamander
Kleinasiatisches Hörnchen	*Sciurus anomalus*	Persian squirrel (Asia Minor)
Kleinäugige Sternkoralle	*Stephanocoenia michelinii*	blushing star coral
Kleinäugiger Hammerhai, Großer Hammerhai	*Sphyrna tudes*	giant hammerhead shark
Kleinäugiger Reismehlkäfer	*Palorus subdepressus*	depressed flour beetle
Kleinäugiger Rochen, Hellfleckiger Rochen	*Raja microocellata*	small-eyed ray (FAO), painted ray
Kleine Achatschnecke, Kleine Glattschnecke	*Cionella lubricella/ Cochlicopa lubricella*	lesser slippery moss snail, thin pillar snail (U.S.)
Kleine Alpenschrecke	*Anonconotus alpinus*	small Alpine bushcricket
Kleine Araberkauri	*Cypraea arabicula*	little Arabian cowrie
Kleine Bambusratte	*Cannomys badius*	lesser bamboo rat
Kleine Bandeule, Heckenkräuterflur-Bandeule	*Noctua orbona/ Triphaena orbona*	lunar yellow underwing (moth)
Kleine Bandikutratte	*Bandicota bengalensis*	lesser bandicoot rat
Kleine Bärenkauri	*Cypraea ursellus*	little bear cowrie
Kleine Bartfledermaus	*Myotis mystacina*	whiskered bat
Kleine Baumratte	*Chiruromys vates*	lesser tree mouse
Kleine Becherqualle, Achtstrahlige Becherqualle	*Haliclystus octoradiatus*	octaradiate stalked jellyfish*
Kleine Bergente	*Aythya affinis*	lesser scaup
Kleine Bernsteinmakrele, Augenstreifen-Königsfisch	*Seriola rivoliana*	Almaco jack (FAO), longfin yellowtail
Kleine Bienenschwebfliege	*Eristalis arbustorum*	lesser drone fly
Kleine Binsenjungfer	*Lestes virens*	lesser emerald damselfly
Kleine Bodenrenke	*Coregonus pidschian*	humpback whitefish (FAO), humpbacked whitefish
Kleine Braunschlange	*Elapognathus minor*	little brown snake
Kleine Brombeerschildlaus	*Aphis ruborum*	permanent blackberry aphid
Kleine Brunnenschnecke	*Bythiospeum acicula*	lesser springsnail*

Kleine Buntschrecke	*Poecilimon elegans*	lesser bushcricket
Kleine Büschel-Seescheide	*Clavelina nana*	dwarf light-bulb sea-squirt
Kleine Celebes-Spitzmausratte*	*Melasmothrix naso*	Celebes lesser shrew rat
Kleine Dasselfliege	*Hypoderma lineatum*	lesser ox warble fly, lesser ox botfly, common cattle grub (U.S.)
Kleine Elfenbeinseepocke	*Balanus improvisus*	bay barnacle, little ivory barnacle
Kleine Feldlerche	*Alauda gulgula*	Oriental skylark
Kleine Felsen-Kantengarnele	*Sicyonia dorsalis*	lesser rock shrimp
Kleine Felsgarnele, Steingarnele	*Palaemon elegans*	rockpool prawn
Kleine Fichtenblattwespe	*Pristiphora abietina*	gregarious spruce sawfly
Kleine Gabelschwanzmakrele	*Seriola fasciata*	lesser amberjack
Kleine Gelbcorvina	*Pseudosciaena polyactis*	lesser yellow croaker
Kleine Gelbe Fledermäuse	*Rhogeessa* spp.	Rogeessa bats, little yellow bats
Kleine Glanzschnecke	*Aegopinella pura*	clear glass snail, delicate glass snail
Kleine Goldschrecke	*Chrysochraon brachypterus/ Euthystira brachyptera*	small gold grasshopper
Kleine Großohr-Fledermäuse	*Micronycteris* spp.	little big-eared bats
Kleine Halmfliege	*Thaumatomyia notata*	small cluster-fly
Kleine Hamsterratte	*Beamys hindei*	long-tailed pouched rat, lesser pouched rat, lesser hamster-rat
Kleine Hausspinne*	*Oonops domesticus*	tiny house spider
Kleine Herzmuschel, Dreieckige Herzmuschel	*Parvicardium exiguum*	little cockle
Kleine Himbeerblattlaus, Kleine Himbeerlaus	*Aphis idaei*	small raspberry aphid
Kleine Höckerschrecke, Kleiner Band-Grashüpfer	*Arcyptera microptera*	small banded grasshopper
Kleine Holzwürmer	Teredilia	lesser wood-boring beetles
Kleine Hornisse, Sächsische Wespe	*Dolichovespula saxonica*	Saxon wasp*
Kleine Hufeisennase, Kleinhufeisennase	*Rhinolophus hipposideros*	lesser horseshoe bat
Kleine Johannisbeertriebblattlaus, Kleine Johannisbeerlaus	*Aphis schneideri*	permanent currant aphid
Kleine Kaktuskoralle	*Isophyllia multiflora*	lesser cactus coral
Kleine Kätzcheneule	*Orthosia cruda*	small quaker moth
Kleine Knopf-Kammuschel*	*Pecten imbricata*	little knobbly scallop
Kleine Kohlfliege, Wurzelfliege, Radieschenfliege	*Delia radicum/ Hylemia brassicae/ Chortophila brassicae/ Paregle radicum*	cabbage fly, cabbage maggot, radish fly
Kleine Königslibelle	*Anax parthenope*	lesser emperor dragonfly
Kleine Maräne, Zwergmaräne	*Coregonus albula*	vendace
Kleine Mittelmeergarnele	*Plesionika edwardsii*	soldier striped shrimp
Kleine Mondschnecke	*Euspira nana/ Neverita nana*	tiny moonsnail

Kleine Moosjungfer	*Leucorrhinia dubia*	white-faced darter, white-faced dragonfly
Kleine Mosaikjungfer	*Brachytron pratense/ Brachytron hafniense*	lesser hairy dragonfly, hairy hawker
Kleine Moschusschildkröte, Zwerg-Moschusschildkröte	*Sternotherus minor/ Kinosternon minor*	loggerhead musk turtle
Kleine Nacktrücken-Fledermaus	*Pteronotus davyi*	lesser naked-backed bat
Kleine Nacktsohlen-Rennmäuse	*Taterillus* spp.	small naked-soled gerbils
Kleine Nelkenkoralle*	*Caryophyllia cornuformis*	lesser cupcoral, lesser horncoral
Kleine Netzmuräne	*Gymnothorax permistus*	black-blotched moray
Kleine Noddiseeschwalbe, Schlankschnabelnoddi	*Anous tenuirostris*	lesser noddy
Kleine Ohrwürmer	Labiidae	little earwigs
Kleine Pappelglucke	*Poecilocampa populi*	December moth
Kleine Pazifik-Auster, Pazifische Plattauster	*Ostrea lurida*	native Pacific oyster, Olympia flat oyster, Olympic oyster
Kleine Pazifikratte, Kleine Burmaratte, Polynesische Ratte	*Rattus exulans*	Polynesian rat
Kleine Pechlibelle	*Ischnura pumilio*	lesser ischnura, scarce blue-tailed damselfly
Kleine Pelomeduse	*Pelusios nanus*	African dwarf mud turtle
Kleine Pfeffermuschel, Weiße Pfeffermuschel	*Abra alba*	white abra (white furrow-shell)
Kleine Pfeffermuscheln	*Abra* spp.	abras (lesser European abras)
Kleine Pflaumenblattlaus	*Brachycaudus helichrysi*	leaf-curling plum aphid
Kleine Pilgermuschel, Bunte Kammmuschel	*Aequipecten opercularis*	queen scallop
Kleine Riesenmuschel, Längliche Riesenmuschel	*Tridacna maxima*	elongated clam, elongate clam, small giant clam
Kleine Röhrichteule	*Archanara neurica*	white-mantled wainscot
Kleine Rosenschildlaus	*Aulacaspis rosae*	rose scale
Kleine Rot-Vogelspinne*	*Paraphysa manicata*	dwarf rose tarantula
Kleine Rote Waldameise, Kahlrückige Rote Waldameise	*Formica polyctena*	small red wood ant
Kleine Salpe	*Salpa fusiformis*	lesser salp*
Kleine Schildmotte	*Heterogenea asella*	triangle (moth)
Kleine Schlangennadel	*Nerophis ophidion*	straight-nosed pipefish
Kleine Schließmundschnecke, Zierliche Schließmundschnecke	*Clausilia parvula*	dwarf door snail
Kleine Schmerle	*Lepidocephalus thermalis*	lesser loach
Kleine Schwarze Lärchenblattwespe	*Pristiphora laricis*	small larch sawfly
Kleine Schwebrenke, Schnepel, Schnäpel, Edelmaräne	*Coregonus oxyrhynchus/ Coregonus oxyrinchus*	houting

Kleine Schwertmuschel, Durchsichtige Messerscheide, Durchscheinende Messerscheide	*Phalax pellucidus*	translucent piddock*
Kleine Seenadel	*Syngnathus rostellatus*	lesser pipefish, Nilsson's pipefish (FAO)
Kleine Seespinne	*Maja verrucosa/ Maia verrucosa/ Maja crispata*	small spider crab, lesser spider crab
Kleine Sepie, Schlammsepie	*Sepia elegans*	elegant cuttlefish
Kleine Sepiette	*Sepietta minor*	lentil bobtail squid
Kleine Stachelbeertriebblattlaus	*Aphis grossulariae*	gooseberry aphid
Kleine Stechfliege, Herbststechfliege	*Haematobosca stimulans/ Haematobia stimulans/ Siphona stimulans*	cattle biting fly (Texas fly)
Kleine Steingarnele, Kleine Felsgarnele	*Palaemon elegans*	rockpool prawn
Kleine Sternkoralle	*Favia fragum*	golfball coral
Kleine Strauchschrecke	*Yersinella raymondi*	Raymond's bushcricket
Kleine Streifenseepocke	*Balanus amphitrite*	little striped barnacle
Kleine Stubenfliege	*Fannia canicularis*	lesser house fly
Kleine Trogmuschel	*Spisula elliptica*	elliptical surfclam (elliptical trough shell)
Kleine Turmschnecke	*Ena obscura*	lesser bulin
Kleine Wachsmotte	*Achroia grisella*	lesser wax moth
Kleine Walddeckelschnecke	*Cochlostoma septemspirale*	
Kleine Walnussblattlaus	*Chromaphis juglandicola*	small walnut aphid
Kleine Warzenschlange, Indische Warzenschlange	*Acrochordus granulatus/ Chersydrus granulatus*	lesser wart snake, Indian wart snake
Kleine Weißnasenmeerkatze	*Cercopithecus petaurista*	lesser white-nosed guenon, lesser spot-nosed monkey
Kleine Wühlmaus, Kleinwühlmaus, Kurzohrmaus, Kleinäugige Wühlmaus	*Microtus subterraneus/ Pitymys subterraneus*	European pine vole
Kleine Zibetkatze, Rasse	*Viverricula indica*	lesser oriental civet, rasse
Kleine Zitterrochen	Narcinidae/ Narkidae	electric rays, numbfishes
Kleiner Röhrennasen-Flughund	*Paranyctimene raptor*	lesser tube-nosed fruit bat
Kleiner Abendsegler, Kleinabendsegler	*Nyctalus leisleri*	Leisler's bat
Kleiner Ährenfisch, Boyers Ährenfisch	*Atherina boyeri*	big-scale sand smelt (FAO), big-scaled sandsmelt, Boyer's sand smelt
Kleiner Armmolch	*Siren intermedia*	lesser siren
Kleiner Asselspinner, Große Schildmotte	*Cochlidion limacodes/ Apoda avellana*	festoon
Kleiner Australkrebs, Yabbie	*Cherax destructor*	yabbie
Kleiner Band-Grashüpfer, Kleine Höckerschrecke	*Arcyptera microptera*	small banded grasshopper
Kleiner Band-Schlangenstern*	*Ophiactis balli*	small banded brittlestar

Kleiner Bärenkrebs, Grillenkrebs	*Scyllarus arctus*	small European locust lobster, small European slipper lobster, lesser slipper lobster
Kleiner Blutsauger, Kammzahnvampir	*Diphylla ecaudata*	hairy-legged vampire bat
Kleiner Bombardierkäfer	*Brachinus explodens*	lesser bombardier beetle
Kleiner Brauner Rüsselkäfer	*Hylobius pinastri*	small fir-tree weevil
Kleiner Bunter Eschenbastkäfer	*Leperisinus varius/ Leperisinus fraxini*	ash bark beetle, lesser ash bark beetle
Kleiner Darmegel, Zwerg-Darmegel	*Metagonimus yokogawai*	Yokogawa fluke
Kleiner Eichelwurm	*Glossobalanus minutus* (Enteropneusta)	lesser acorn worm
Kleiner Eishai, Lemargo	*Somniosus rostratus*	little sleeper shark
Kleiner Eisvogel	*Ladoga camilla/ Limenitis camilla*	white admiral
Kleiner Feuerfalter	*Lycaena phlaeas*	small copper
Kleiner Fleckenkiwi, Zwergkiwi	*Apteryx owenii*	little spotted kiwi
Kleiner Fuchs	*Aglais urticae*	small tortoiseshell
Kleiner Fuchsbandwurm	*Echinococcus multilocularis*	lesser fox tapeworm, alveolar hydatid tapeworm
Kleiner Gabelschwanz	*Harpyia hermelina/ Cerura bifida/ Furcula bifida*	sallow kitten
Kleiner Gelbschenkel	*Tringa flavipes*	lesser yellowlegs
Kleiner Gewächshaus-Tausendfüßer*	*Cylindroiulus britannicus*	lesser glasshouse millepede
Kleiner Heidegrashüpfer, Ramburs Grashüpfer	*Stenobothrus stigmaticus*	lesser mottled grasshopper
Kleiner Herzigel, Herzigel	*Echinocardium cordatum*	common heart-urchin, sea potato, heart urchin
Kleiner Heufalter	*Coenonympha pamphilus*	small heath
Kleiner Igelrochen	*Raja erinacea/ Leucoraja erinacea*	little skate(FAO), hedgehog skate
Kleiner Kampffisch	*Betta imbellis*	peaceful betta, crescent betta (FAO)
Kleiner Kohlweißling, Rübenweißling	*Pieris rapae/Artogeia rapae*	small white, cabbage butterfly, imported cabbageworm (U.S.)
Kleiner Kolben-Wasserkäfer, Schwarzbeiniger Stachel-Wasserkäfer, Stachelwasserkäfer, Großer Teichkäfer	*Hydrochara caraboides*	lesser silver water beetle
Kleiner Lanzennasen-Fruchtvampir	*Phyllostomus discolor*	pale spear-nosed bat, lesser spear-nosed bat
Kleiner Laternenfisch	*Photoblepharon palpebratus*	flashlight fish
Kleiner Laubholzzangenbock	*Rhagium mordax*	oak longhorn beetle
Kleiner Leberegel (Lanzettegel)	*Dicrocoelium dendriticum/ Dicrocoelium lanceolatum/ Fasciola lanceolatum*	lancet fluke, common lancet fluke
Kleiner Leistenkopfplattkäfer	*Cryptolestes pusillus*	flat grain beetle
Kleiner Maivogel	*Euphydryas maturna/ Hypodryas maturna*	scarce fritillary

Kleiner Nandu, Darwinstrauß	*Pterocnemia pennata*	lesser rhea
Kleiner Obstbaumsplintkäfer	*Scolytus rugulosus*	fruit bark beetle, fruit bark borer, shothole borer
Kleiner Ohrwurm, Zwergohrwurm	*Labia minor*	lesser earwig, little earwig
Kleiner Ölfisch	*Comephorus dybowskii*	lesser Baikal oilfish
Kleiner Pappelblattkäfer*	*Phyllodecta laticollis*	small poplar leaf beetle
Kleiner Pappelbock, Espenbock	*Saperda populnea*	small poplar borer, lesser poplar borer, small poplar longhorn beetle
Kleiner Pappelglasflügler, Bremsenglasflügler	*Paranthrene tabaniformis*	lesser poplar hornet clearwing, lesser hornet moth
Kleiner Perlmutterfalter	*Issoria lathonia/ Argynnis lathonia*	Queen of Spain fritillary
Kleiner Pferdepfriemenschwanz	*Probstmayria vivipara*	small pinworm (equine)
Kleiner Puppenräuber, Kleiner Kletterlaufkäfer	*Calosoma inquisitor*	oakwood ground beetle
Kleiner Raschkäfer, Kleiner Uferläufer	*Elaphrus riparius*	lesser wetland ground beetle*
Kleiner Raubwasserfloh, Großäugiger Wasserfloh	*Polyphemus pediculus*	big-eyed water flea*
Kleiner Rindermagennematode, Kleiner Rindermagenwurm	*Trichostrongylus axei*	small stomach worm (of cattle and horses), stomach hair worm
Kleiner Rotaugenfrosch	*Agalychnis saltator*	misfit leaf frog
Kleiner Roter Drachenkopf	*Scorpaena notata/ Scorpaena ustulata*	lesser red scorpionfish, small red scorpionfish (FAO)
Kleiner Rückenschaler, Schuppenschwanz, Langschwänziger Flossenkrebs, Frühjahrs-Kieferfuß	*Lepidurus apus/ Lepidurus productus*	
Kleiner Scheibenbauch	*Liparis montagui*	Montagu's sea snail, Montagus seasnail (FAO)
Kleiner Schildigel, Zwergseeigel, Zwerg-Seeigel	*Echinocyamus pusillus*	green sea urchin, green urchin, pea urchin
Kleiner Schillerfalter, Espen-Schillerfalter	*Apatura ilia*	lesser purple emperor
Kleiner Schlammläufer	*Limnodromus griseus*	short-billed dowitcher
Kleiner Schlingerhai	*Centrophorus uyato*	little gulper shark
Kleiner Schneckenegel	*Glossiphonia heteroclita*	small snail leech
Kleiner Schwarzer Dornhai, Schwarzer Hundfisch	*Etmopterus spinax*	velvet-belly, velvet belly
Kleiner Schwarzer Erdfloh*	*Phyllotreta aerea*	small black flea beetle
Kleiner Schwarzspitzenhai	*Carcharhinus limbatus*	blacktip shark
Kleiner Schwertwal, Unechter Schwertwal (Kleiner Mörder)	*Pseudorca crassidens*	false killer whale, false pilot whale
Kleiner Sonnenröschen-Bläuling, Dunkelbrauner Bläuling	*Aricia agestis*	brown argus
Kleiner Strauch-Hydroid	*Plumularia setacea*	little seabristle
Kleiner Sturmtaucher	*Puffinus assimilis*	little shearwater

Kleiner Süßwasserschwamm, Blasenzellenschwamm	*Ephydatia muelleri*	lesser freshwater sponge
Kleiner Tabakkäfer	*Lasioderma serricorne*	cigarette beetle (tobacco beetle)
Kleiner Tannenborkenkäfer	*Cryphalus piceae*	white spruce beetle
Kleiner Taubeneckkopf	*Campanulotes bidentatus compar*	small pigeon louse
Kleiner Taubleguan	*Holbrookia maculata*	lesser earless lizard
Kleiner Teichfrosch, Kleiner Grünfrosch, Kleiner Wasserfrosch, Tümpelfrosch	*Rana lessonae*	pool frog (little waterfrog)
Kleiner Teufelsrochen, Ostatlantischer Teufelsrochen, Meeresteufel	*Mobula mobular*	devil fish (FAO), devilfish, devil ray
Kleiner Tigerfisch	*Hydrocynus lineatus*	lesser tigerfish
Kleiner Tümmler, Kleintümmler, Schweinswal, Braunfisch	*Phocoena phocoena*	harbor porpoise, common harbor porpoise
Kleiner Uferläufer, Kleiner Raschkäfer	*Elaphrus riparius*	lesser wetland ground beetle*
Kleiner Ulmensplintkäfer	*Scolytus multistriatus*	small elm bark beetle, small European elm bark beetle
Kleiner Waldgärtner, Rotbrauner Waldgärtner	*Blastophagus minor/ Tomicus minor*	minor pith borer
Kleiner Wasserfrosch, Kleiner Teichfrosch, Kleiner Grünfrosch, Tümpelfrosch	*Rana lessonae*	pool frog (little waterfrog)
Kleiner Weinschwärmer	*Deilephila porcellus*	small elephant hawkmoth
Kleiner Zitterrochen	*Narke dipterygia*	numbray
Kleiner Zitterrochen, Karibischer Zitterrochen	*Narcine brasiliensis*	Brazilian electric ray
Kleinere Kolumbianische Riesenvogelspinne	*Xenesthis immanis*	Colombian lesserblack tarantula
Kleines Eichenkarmin, Kleines Eichenkarmin-Ordensband, Kleines Karminrotes Eichenhain-Ordensband	*Catocala promissa*	light crimson underwing
Kleines Elefantenohr	*Rhodactis inchoata*	Tonga blue mushroom anemone, blue Tonga mushroom anemone
Kleines Granatauge	*Erythromma viridulum*	lesser red-eyed damselfly (Continental)
Kleines Johanniswürmchen	*Phausis splendidula*	lesser glow-worm
Kleines Nachtpfauenauge	*Saturnia pavonia/ Eudia pavonia*	emperor moth
Kleines Ochsenauge	*Hyponephele lycaon/ Maniola lycaon*	dusky meadow brown
Kleines Petermännchen, Zwergpetermännchen, Viperqueise	*Echiichthys vipera/ Trachinus vipera*	lesser weever
Kleines Posthörnchen	*Gyraulus parvus*	lesser ramshorn, ash gyro
Kleines Seemoos, Zwergmoos	*Dynamena pumila/ Dynamena cavolini/ Sertularia pumila*	sea oak, minute garland hydroid, minute hydroid

Kleines Sumpfhuhn	*Porzana parva*	little crake
Kleinfleck-Ginsterkatze	*Genetta genetta*	small-spotted genet
Kleinfleckkatze, Salzkatze	*Felis geoffroyi*	Geoffroy's cat
Kleinflügelfisch*, Kleinflügel-Flugfisch	*Oxyporhamphus micropterus*	smallwing flyingfish
Kleingefleckter Katzenhai, Kleiner Katzenhai	*Scyliorhinus canicula/ Scyllium canicula*	lesser spotted dogfish, smallspotted dogfish, rough hound, smallspotted catshark (FAO)
Kleingrison	*Galictis cuja*	little grisón
Kleinhufeisennase, Kleine Hufeisennase	*Rhinolophus hipposideros*	lesser horseshoe bat
Kleinkantschil	*Tragulus javanicus*	lesser Malay chevrotain
Kleinkärpflinge	*Cyprinodontiformes*	killifishes
Kleinköpfiger Ansauger	*Apletodon microcephalus/ Lepadogaster microcephalus*	small-headed clingfish (FAO), small-headed sucker
Kleinlibellen	*Zygoptera*	damselflies
Kleinmäuliger Kalifornischer Seestint	*Hypomesus pretiosus*	surf smelt
Kleinmausohr	*Myotis blythi*	lesser mouse-eared bat
Kleinmünder	*Bathylagidae*	deep-sea smelts
Kleinmündiger Lippfisch, Kleinmäuliger Lippfisch	*Centrolabrus exoletus*	rock cock, rock cook (FAO), small-mouthed wrasse
Kleinohrgalago	*Otolemur garnetti*	small-eared galago
Kleinohrspitzmäuse	*Cryptotis* spp.	small-eared shrews
Kleinsäugerläuse	Hoplopleuridae	small mammal sucking lice
Kleinspecht	*Picoides minor/ Dendrocopos minor*	lesser spotted woodpecker
Kleinstböckchen	*Neotragus pygmaeus*	royal antelope
Kleinste Erbsenmuschel	*Pisidium tenuilineatum*	miniscule pea mussel*
Kleinste Rosenblattwespe	*Blennocampa phyllocolpa*	least leaf-rolling rose sawfly
Kleinster Perlmutterfalter, Hainveilchen-Perlmutterfalter, Magerrasen-Perlmuttfalter	*Clossiana dia/ Boloria dia*	violet fritillary, Weaver's fritillary
Kleistermotte	*Endrosis sarcitrella*	white-shouldered house moth
Kleopatrafalter, Kleopatra	*Gonepteryx cleopatra*	cleopatra
Kletteneule, Markeule	*Gortyna flavago*	frosted orange (moth)
Kletter-Seeigel, Kletterseeigel	*Psammechinus microtuberculatus*	
Kletterfisch	*Anabas testudineus*	climbing perch (FAO), climbing gourami, walking fish
Kletterfische & Buschfische	Anabantidae	climbing gouramies
Kletterholothurie, Kletterseewalze	*Cucumaria planci*	climbing sea cucumber*
Kletterkammseestern	*Astropecten spinulosus*	spiny comb star
Kletternattern	*Elaphe* spp.	rat snakes, ratsnakes
Kletterspitzmaus	*Sylvisorex megalura*	climbing shrew
Kletterwaldsänger	*Mniotilta varia*	black-and-white warbler
Kletterwelse	Astroblepidae	climbing catfishes
Kliesche (Scharbe)	*Limanda limanda*	dab, common dab

Klippenassel, Strandassel	*Ligia oceanica*	great sea-slater, sea slater (quay-louse)
Klippenasseln, Strandasseln	*Ligia* spp.	slaters, sea-slaters, rock lice
Klippenasseln, Strandasseln	Ligiidae	sea-slaters, rock lice
Klippenbarsch, Felsenbarsch	*Ctenolabrus rupestris*	goldsinny, goldsinny wrasse
Klippenkleiber	*Sitta tephronota*	great rock nuthatch
Klippentyrann	*Muscisaxicola albifrons*	white-fronted ground tyrant
Klippkleber> Jungfräuliche Napfschnecke, Weiße Schildkröten-Napfschnecke	*Acmaea virginea/ Tectura virginea*	white tortoiseshell limpet
Klippschliefer	*Procavia capensis*	hyrax, rock dassie
Klippspringer	*Oreotragus oreotragus*	klippspringer
Kloakentiere	Monotremata (Prototheria)	monotremes (prototherians)
Klon-Anemone*	*Anthopleura elegantissima*	clonal anemone, aggregating anemone
Klopfkäfer, Nagekäfer, Pochkäfer	Anobiidae	furniture beetles & drug-store beetles & death-watch beetles
Klumpenschwamm	*Haliclona limbata*	clumpy tube sponge*
Klunkern-Gallmilbe	*Aceria fraxinivora*	ash gall mite
Klunzingers Ponyfisch	*Leiognathus klunzingeri*	pony fish
Knackerlerche	*Rhamphocoris clotbey*	thick-billed lark
Knäkente	*Anas querquedula*	garganey
Knallkrebschen, Mittelmeer-Knallgarnele, Mittelmeer-Pistolenkrebs	*Alpheus dentipes*	Mediterranean snapping prawn
Knallkrebschen, Pistolenkrebse	Alpheidae	snapping prawns, snapping shrimps
Knallkrebschen, Pistolenkrebse	*Alpheus* spp.	snapping prawns, snapping shrimps, pistol shrimps
Knarrschrecken	Catantopidae	catantopid grasshoppers
Knirpsspitzmaus	*Sorex minutissimus*	least shrew
Knoblauchfliege	*Suillia univittata*	garlic fly*
Knoblauchkröte	*Pelobates fuscus*	common spadefoot (garlic toad)
Knoblauchkrötenähnlicher Frosch	*Neobatrachus pelobatoides*	humming frog
Knochenhecht, Gemeiner Knochenhecht, Schlanker Knochenhecht	*Lepisosteus osseus*	longnose gar
Knochenhechte, Kaimanfische	Lepisosteiformes/Lepisosteidae	gars
Knochenmehlkäfer (Keratinkäfer)	*Anthrenus flavipes/ Anthrenus vorax*	furniture carpet beetle
Knochenzüngler, Arawana	*Osteoglossum bicirrhosum*	arawana
Knochenzüngler	Osteoglossidae	bonytongues, osteoglossids
Knochenzüngler, Knochenzünglerartige	Osteoglossiformes	osteoglossiforms
Knochs Mohrenfalter	*Erebia epiphron*	mountain ringlet (butterfly)
Knollen-Kalkschwamm	*Leuconia aspera*	knobby calcareous sponge*

Knollenqualle, Spiegelei-Qualle	*Cotylorhiza tuberculata*	fried-egg jellyfish
Knopfhornblattwespen	Cimbicidae	cimbicids
Knopfkäfer, Derodontiden	Derodontidae	tooth-necked fungus beetles
Knopfkoralle	*Montastrea annularis*	boulder star coral
Knorpel-Weichschildkröte, Asiatische Weichschildkröte	*Amyda cartilaginea/ Trionyx cartilagineus*	Asiatic softshell turtle, black-rayed softshell turtle
Knorpelegel, Plattegel	Glossiphoniidae	glossiphoniid leeches
Knorpelfische	Chondrichthyes	cartilaginous fishes, chondrichthians
Knorrige Gitterschnecke	*Cancellaria nodulifera*	knobbed nutmeg
Knorrige Kreiselschnecke	*Gibbula magus*	turban topsnail
Knospenstrahler	Blastoida	blastoids
Knötchenwurm	*Oesophagostomum radiatum*	nodular worm
Knötchenwürmer	*Oesophagostomum* spp.	nodular worms
Knotenameisen, Stachelameisen	Myrmicidae	harvester ants & others
Knotenhaarlaus	*Chaetosiphon fragaefolii*	strawberry aphid
Knotenwurm	*Onchocerca volvulus*	blinding nodular worm
Knötericheule	*Dypterygia scabriuscula*	bird's wing
Knotige Archenmuschel	*Arca nodulosa*	nodular ark (nodular ark shell)
Knotige Froschschnecke, Riesenfroschschnecke	*Bursa bubo*	giant frogsnail
Knotige Helmschnecke, Knotenschnecke, Knotenschelle	*Galeodea rugosa*	rugose helmet snail, Mediterranean rugose bonnet
Knotige Herzmuschel	*Acanthocardia tuberculata*	tuberculate cockle, rough cockle
Knotige Hirnkoralle	*Diploria clivosa*	knobby brain coral
Knotige Koralle, Felderkoralle	*Madracis decactis*	ten-ray star coral
Knotige Porenkoralle	*Porites astreoides*	mustard hill coral, knobby porous coral
Knotige Schlüssellochschnecke	*Fissurella nodosa*	knobby keyhole limpet, knobbed keyhole limpet
Knotige Sternkoralle*	*Solenastrea hyades*	knobby star coral
Knotiger Walzenstern	*Protoreaster nodosus*	wartstar*
Knubbelanemone	*Entacmaea quadricolor*	four-colored anemone, bubble-tip anemone, bulb-tip anemone, bulb-tentacle sea anemone, maroon anemone
Knüllpapierkoralle	*Pectinia lactuca*	carnation coral
Knurrender Gurami	*Trichopsis vittatus*	talking gourami (FAO), croaking gourami
Knurrhähne	Triglidae	searobins, gurnards
Knutt	*Calidris canutus*	knot
Koalabär, Beutelbär	*Phascolarctos cinereus*	koala
Kob-Antilope, Moorantilope	*Kobus kob*	kob
Kobaltblaue Vogelspinne*	*Haplopelma lividum*	cobalt blue tarantula
Kobaltflügelsittich	*Brotogeris cyanoptera*	cobalt-winged parakeet
Koboldhai, Japanischer Nasenhai	*Scapanorhynchus owstoni/ Mitsukurina owstoni*	goblin shark

Koboldhaie, Nasenhaie	Scapanorhynchidae/ Mitsukurinidae	goblin sharks
Koboldkärpfling, Moskitofisch, Gambuse	Gambusia affinis	mosquito fish, mosquitofish (FAO)
Koboldmakis	Tarsius spp.	tarsiers
Kobra> Königskobra	Ophiophagus hannah	king cobra, hamadryad
Kobras	Naja spp.	cobras
Köcherfliegen, Haarflügler	Trichoptera	caddis flies
Köcherwurm	Amphictene koreni/ Lagis koreni/ Pectinaria koreni	quiver worm*
Köderkalmar	Gonatus fabricii	boreoatlantic armhook squid
Kofferfische	Ostraciidae/ Ostraciontidae	boxfishes (cowfishes & trunkfishes)
Kohl-Juwelendose	Chama brassica	cabbage jewel box
Kohl-Mottenschildlaus	Aleyrodes proletella	cabbage whitefly
Kohlblattminierfliege*	Phytomyza rufipes	cabbage leaf miner (fly)
Kohldrehherzmücke, Kohldrehherzgallmücke	Contarinia nasturtii	swede midge, cabbage midge
Kohlenfisch	Anoplopoma fimbria	sablefish
Kohlenskink	Eumeces anthracinus	coal skink
Köhler, Seelachs, Blaufisch	Pollachius virens	saithe (FAO), pollock, Atlantic pollock, coley, coalfish
Kohlerdflöhe	Phyllotreta spp.	flea beetles, turnip flea beetles
Köhlerschildkröte	Geochelone carbonaria/ Testudo carbonaria/ Chelonoidis carbonaria	red-footed tortoise, South American red-footed tortoise, coal tortoise
Kohleule, Herzeule	Mamestra brassicae	cabbage moth
Kohlgallenrüssler	Ceutorhynchus pleurostigma	turnip gall weevil, cabbage gall weevil
Kohlmeise	Parus major	great tit
Kohlmotte, Kohlschabe, Schleiermotte	Plutella maculipennis/ Plutella xylostella	cabbage moth, diamondback moth
Kohlrübenblattwespe, Rübenblattwespe	Athalia rosae	turnip sawfly
Kohlschabe, Kohlmotte, Schleiermotte	Plutella maculipennis/ Plutella xylostella	cabbage moth, diamondback moth
Kohlschnake	Tipula oleracea	cabbage cranefly, brown daddy-long-legs (Br.)
Kohlschnecke, Kohl-Stachelschnecke	Hexaplex brassica/ Phyllonotus brassica	cabbage murex
Kohlschotengallmücke	Dasineura brassicae	cabbage pod midge
Kohlschotenrüssler	Ceutorhynchus assimilis	cabbage seedpod weevil
Kohlwanze	Eurydema oleraceum	brassica bug
Kohlweißlingsraupenwespe	Apanteles glomeratus	common apanteles, common parasitic wasp
Kohlzünsler, Meerrettichzünsler	Evergestis forficalis	garden pebble moth
Kohtao-Wühle	Ichthyophis kohtaoensis	Koh Tao Island caecilian

Kojote	*Canis latrans*	coyote
Kokardenspecht	*Picoides borealis*	red-cockaded woodpecker
Kokosnusskrebs, Palmendieb	*Birgus latro*	coconut crab, robber crab
Kolbenente	*Netta rufina*	red-crested pochard
Kolbenpolypen, Kölbchenpolypen	*Coryne* spp.	flowerhead polyps
Kolibrirose	*Sagartia troglodytes/ Actinothoe troglodytes*	cave-burrowing anemone*
Kolibris	Trochiliformes/ Trochilinae (Trochilidae)	hummingbirds
Kolios, Thunmakrele, Mittelmeermakrele	*Scomber japonicus/ Scomber colias*	chub mackerel (FAO), Spanish mackerel, Pacific mackerel
Kolkrabe	*Corvus corax*	common raven
Kolonistenkäfer	Colonidae	colonid beetles
Kolumbianische Erdschildkröte	*Rhinoclemmys melanosterna*	Colombian wood turtle
Kolumbianische Riesenkröte	*Bufo blombergi*	Colombian giant toad
Kolumbianische Riesenvogelspinne	*Pamphobeteus ornatus*	Colombian pinkbloom tarantula
Kolumbianische Waldmaus	*Chilomys instans*	Colombian forest mouse
Kolumbianische Wegschnecke	*Ariolimax columbianus*	Pacific banana slug, giant yellow slug
Kolumbianischer Hornfrosch	*Ceratophrys calcarata*	Colombian horned frog
Kolumbianisches Wiesel, Kolumbienwiesel	*Mustela felipei*	Colombian weasel
Kolumbuskrabbe	*Planes minutus*	gulfweed crab, Gulf weed crab, Columbus crab
Kometenstern	*Linckia multifora*	comet star*, comet seastar
Komma-Schildlaus, Gemeine Kommaschildlaus, Obstbaumkommaschildlaus	*Lepidosaphes ulmi*	oystershell scale, mussel scale
Kommaeule	*Mythimna comma*	shoulder-striped wainscot
Kommafalter	*Hesperia comma*	silver-spotted skipper
Kommasalmler	*Moenkhausia comma*	comma tetra
Komodo-Ratte	*Komodomys rintjanus*	Komodo rat
Komodo-Waran, Komodowaran	*Varanus komodoensis*	Komodo dragon, ora
Komoren-Eule	*Otus pauliani*	Comoro scops-owl
Komoren-Quastenflosser, Gombessa	*Latimeria chalumnae*	gombessa
Kompassmuschel	*Amusium pleuronectes*	Asian moon scallop
Kompassqualle	*Chrysaora hysoscella*	compass jellyfish
Kongo-Glaswels	*Eutropiella debauwi*	African glass catfish
Kongo-Otter, Kongo-Fingerotter	*Aonyx congica*	Congo clawless otter
Kongo-Pelomeduse, Gekielte Pelomeduse	*Pelusios carinatus*	African keeled mud turtle
Kongo-Zwergbuntbarsch, Blauer Kongocichlide	*Nanochromis parilus/ Nanochromis nudiceps*	nudiceps, Congo dwarf cichlid*
Kongopfau	*Afropavo congensis*	Congo peafowl

Königin-Drückerfisch	*Balistes vetula*	queen triggerfish (FAO), old-wife
Königin-Engelsfisch	*Holocanthus ciliaris*	queen angelfish
Königin-Randschnecke	*Glabella pseudofaba*	queen marginella
Königinnennatter, Königinnatter, Königinschlange	*Regina septemvittata*	queen snake
Königinnennattern	*Regina* spp.	crayfish snakes, queen snakes
Königinschlange	*Natrix septemvittata*	queen snake
Königliche Felsen-Kantengarnele	*Sicyonia typica*	kinglet rock shrimp
Königliche Kegelschnecke, Kaiserkegel	*Conus imperialis*	imperial cone
Königliche Stachelauster	*Spondylus regius*	royal thorny oyster
Königliche Warzenkoralle	*Balanophyllia regia*	royal star coral
Königs-Corvina	*Cynoscion regalis*	grey weakfish
Königs-Lammzunge, Königs-Lammbutt	*Arnoglossus imperialis*	imperial scaldfish
Königs-Rotgarnele	*Pleoticus robustus*	royal red shrimp
Königs-Schlangenmakrele	*Rexea solandri*	gemfish
Königs-Umberfisch	*Menticirrhus saxatilis*	northern kingfish
Königsamazone	*Amazona guildingii*	St. Vincent parrot, St. Vincent amazon
Königsbärbling, Malabarbärbling	*Danio aequipinnatus/ Brachydanio aequipinnatus*	giant danio
Königsbarsch	*Rachycentron canadum*	cobia (prodigal son)
Königsbartmännchen, Südafrikanischer Kingklip	*Genypterus capensis*	kingklip
Königsbussard	*Buteo regalis*	ferruginous fawk
Königscichlide, Kribensis	*Pelvicachromis pulcher*	kribensis
Königsfadenfisch	*Pentanemus quinquarius*	royal threadfin
Königsgeier	*Sarcorhamphus papa*	king vulture
Königsgeißelgarnele	*Melicertus latisuculatus/ Penaeus latisuculatus*	western king prawn
Königsgliederkoralle, Goldener Seefächer	*Isis hippuris*	sea fan a.o., golden sea fan
Königshelm	*Cassis tuberosa*	Caribbean helmet, king helmet
Königsholothurie, Königsseegurke	*Stichopus regalis*	royal sea cucumber
Königskobra	*Ophiophagus hannah*	king cobra, hamadryad
Königskrabbe (Kronenkrebs, Kamtschatkakrebs), Alaska-Königskrabbe, Kamtschatka-Krabbe	*Paralithodes camtschaticus*	king crab, red king crab, Alaskan king crab, Alaskan king stone crab (Japanese crab, Kamchatka crab, Russian crab)
Königslachs, Quinnat	*Oncorhynchus tschawytcha*	chinook salmon (FAO), chinook, king salmon
Königslanguste, Grüne Languste	*Panulirus regius*	royal spiny crawfish

Königsmakrele	*Scomberomorus cavalla*	king mackerel (FAO), kingfish
Königsmakrelen, Pelamiden	Scomberomoridae (Cybiidae)	Spanish mackerels
Königsmantel	*Cryptopecten pallium*	royal cloak scallop
Königsmeise	*Parus spilonotus*	Chinese yellow tit
Königsnattern	*Lampropeltis* spp.	kingsnakes
Königsnektarvogel	*Nectarinia regia*	regal sunbird
Königsparadiesvogel	*Cicinnurus regius*	king bird-of-paradise
Königspinguin	*Aptenodytes patagonicus*	king penguin
Königspython	*Python regius*	ball python, royal python
Königsriesenhörnchen	*Ratufa affinis*	cream-colored giant squirrel
Königssalmler	*Inpaichthys kerri*	blue emperor
Königsschlange	*Boa constrictor*	boa constrictor
Königsschnapper	*Aprion virescens*	king snapper, green jobfish (FAO), streaker
Königsschnecke*	*Hexaplex regius*	royal murex, regal murex
Königsseegurke, Königsholothurie	*Stichopus regalis*	royal sea urchin
Königsseeschwalbe	*Sterna maxima*	royal tern
Königsseestern	*Asterina cepheus*	royal seastar*
Königssittich	*Alisterus scapularis*	Australian king parrot
Königstyrann	*Tyrannus tyrannus*	eastern kingbird
Konvex-Pantoffelschnecke	*Crepidula convexa*	convex slippersnail
Konvexer Laufkäfer, Kurzgewölbter Laufkäfer	*Carabus convexus*	convex ground beetle
Kopfbinden-Zwergnatter	*Eirenis modestus*	Asia Minor dwarf snake
Köpfchenpolypen	*Tubularia* spp.	tubularian hydroids
Kopffüßer, Cephalopoden	Cephalopoda	cephalopods
Kopfhornschröter, Baumschröter	*Sinodendron cylindricum*	rhinoceros beetle, "small European rhinoceros beetle"
Kopflaus	*Pediculus capitis* (*P. humanus capitis*)	head louse
Kopfringler	Capitellidae	capitellid worms
Kopfschildschnecken, Kopfschildträger	Cephalaspidea/Kephalaspidea	bubble snails
Kopfsteher	Chilodontidae	headstanders
Kopfsteher, Engmaulsalmler	Anostomidae	anostomids
Koppe, Groppe, West-Groppe, Kaulkopf, Mühlkoppe	*Cottus gobio*	Miller's thumb, bullhead (FAO)
Koprakäfer, Rotbeiniger Kolbenkäfer, Rotbeiniger Schinkenkäfer	*Necrobia rufipes*	red-legged ham beetle, copra beetle
Kopramilbe	*Tyrophagus putrescentiae*	mold mite (U.S.), mould mite (Br.)
Korallen-Kammmuschel	*Lyropecten corallinoides*	coral scallop
Korallen-Kaninchenfisch	*Siganus corallinus*	coral rabbitfish

Korallen-Königsnatter	*Lampropeltis zonata*	coral kingsnake, California mountain kingsnake
Korallen-Moostierchen	*Lichenopora radiata*	coralline bryozoan*
Korallen-Rollschlange	*Anilius scytale*	coral pipesnake, false coral snake
Korallen-Seespinne	*Mithrax hispidus*	coral clinging crab
Korallenbarsche & Riffbarsche & Jungfernfische	Pomacentridae	damselfishes
Korallenfinger	*Pelodryas caerulea/ Hyla caerulea/Litoria caerulea*	White's treefrog
Korallengrundel	*Odondebuenia balearica*	coralline goby
Korallenkrabben	*Trapezia* spp.	coral crabs
Korallenkrabben	Trapeziidae	coral crabs
Korallenmoos	*Hydrallmania falcata*	sickle hydroid, sickle coralline, coral moss
Korallenmöve	*Larus audouinii*	Audouin's gull
Korallenschlange> Gewöhnliche Korallenschlange, „Kobra-Korallenschlange"	*Micrurus corallinus*	common coral snake
Korallenschlangen> Amerikanische Korallenschlangen, Echte Korallenottern	*Micrurus* spp.	common coral snakes
Korallenwelse	Plotosidae	eeltail catfishes, eel catfishes
Korallenwurm	*Salmacina dysteri*	coral worm
Körbchenmuschel, Korbmuschel	*Corbula gibba*	common corbula, common basket clam
Körbchenspinne, Strauchradspinne	*Araneus redii/ Agalenatea redii*	bush orbweaver*, basketweaver*
Korbmuschel, Körbchenmuschel	*Corbula gibba*	common corbula, common basket clam
Korbmuscheln	Corbulidae	box clams, little basket clams
Koreanische Miesmuschel	*Mytilus crassitesta*	Korean mussel
Korkenzieheranemone, Fadenanemone	*Macrodactyla doreensis*	corkscrew anemone, long tentacle anemone (L.T.A.), red base anemone
Korkschwamm, Einsiedler-Korkschwamm, Häuschenschwamm	*Suberites domuncula*	sulphur sponge, sulfur sponge
Korkschwamm, Feigenschwamm	*Suberites ficus/ Ficulina ficus*	fig sponge
Kormoran	*Phalacrocorax carbo*	great cormorant
Kormorane	Phalacrocoracidae	cormorants, shags
Körnige Froschschnecke	*Bursa granularis*	granulate frogsnail
Körnige Meerassel	*Idotea granulosa*	granular marine isopod*
Kornkäfer, Schwarzer Kornwurm	*Sitophilus granarius*	grain weevil
Kornmotte	*Nemapogon granellus*	European grain moth
Kornnatter	*Elaphe guttata*	corn snake
Kornschnecken	Chondrinidae	
Kornweihe	*Circus cyaneus*	marsh hawk, hen harrier, northern harrier

Deutsch	Wissenschaftlich	Englisch
Körperlaus, Hühner-Körperlaus	*Menacanthus stramineus/ Eomenacanthus stramineus*	chicken body louse
Korsenkleiber	*Sitta whiteheadi*	Corsican nuthatch
Korsischer Gebirgsmolch	*Euproctus montanus*	Corsian brook salamander, Corsican mountain newt
Koslows Zwergspringmäuse	*Salpingotus* spp.	three-toed dwarf jerboas
Kostbarer Elefantenzahn, Indischer Elefantenzahn	*Antalis pretiosu*m (Scaphopoda)	Indian money tusk, wampum tuskshell
Kotfresser	*Onthophagus* spp.	dung beetles
Kotpillenwurm	*Heteromastus filiformis*	
Kotwanze, Staubwanze, Maskierter Strolch	*Reduvius personatus*	masked hunter bug (fly bug)
Kotwespe	*Mellinus arvensis*	field digger wasp
Kouprey	*Bos sauveli*	kouprey, gray ox
Krabbenanemone	*Triactis producta*	crab anemone*
Krabbenbussard	*Buteogallus anthracinus*	common black hawk
Krabbenesser	*Lobodon carcinophagus*	crabeater seal
Krabbenspinnen	*Thomisidae*	crab spiders
Krabbentaucher	*Alle alle*	little auk, dovekie
Kragen-Teppichhaie	Parascyllidae	collared carpet sharks
Kragenbär	*Selenarctos thibetanus/ Ursus thibetanus*	Asiatic black bear, Himalayan bear, Tibetan bear
Kragenechse	*Chlamydosaurus kingi*	frilled lizard, Australian frilled lizard, frill-necked lizard
Kragenente	*Histrionicus histrionicus*	harlequin duck
Kragenfaultier	*Bradypus torquatus*	maned sloth, Brazilian three-toed sloth
Kragenflughunde	*Myonycteris* spp.	little collared fruit bats
Kragenhai, Krausenhai	*Chlamydoselachus anguineus*	frill shark, frilled shark
Kragenhaie, Krausenhaie	Chlamydoselachidae	frill sharks, frilled sharks
Kragenhuhn	*Bonasa umbellus*	ruffed grouse
Kragenparadiesvogel	*Lophorina superba*	superb bird-of-paradise
Kragentiere, Hemichordaten	Hemichordata/ Branchiotremata	hemichordates
Kragentrappe	*Chlamydotis undulata*	Houbara bustard
Krähe> Saatkrähe	*Corvus frugilegus*	rook
Krähenscharbe	*Phalacrocorax aristotelis*	shag
Krähenschnaken	*Nephrotoma* spp.	spotted craneflies
Krähenstirnvogel	*Psarocolius decumanus*	crested oropendola
Krainer Honigbiene	*Apis mellifera carnica*	Carniola honey bee
Krake> Gemeiner Krake, Gemeiner Octopus, Polyp	*Octopus vulgaris*	common octopus, common Atlantic octopus, common European octopus
Krallenfingermolche	*Onychodactylus* spp.	clawed salamanders
Krallenfrösche	Pipidae (*Xenopus* spp. u.a.)	clawed frogs
Krallengeckos, Wüstengeckos	*Coleonyx* spp.	banded geckos

Krallenkalmar, Hakenkalmar	*Onychoteuthis banksii*	common clubhook squid, clawed squid, clawed calamary squid
Kranich	*Grus grus*	common crane
Kraniche	Gruidae	cranes
Kranichvögel, Kranichverwandte	Gruiformes	cranes & rails and allies
Kranzfühler, Fühlerkranztiere, Tentaculaten	Tentaculata	tentaculates (bryozoans, phoronids, brachiopods)
Kranzquallen, Tiefseequallen	Coronata	coronate medusas
Kraterschwamm	*Dysidea arenaria*	crater sponge*
Krätzemilben (Räudemilben)	Sarcoptidae	itch mites, scabies mites, scab mites
Kratzer	Acanthocephala	spiny-headed worms, thorny-headed worms, acanthocephalans
Krätzmilbe	*Sarcoptes scabiei*	scabies mite, itch mite
Krause Bohrmuschel, Raue Bohrmuschel	*Zirfaea crispata*	great piddock, oval piddock
Kräusel-Riss-Schnecke*	*Anatoma crispata*	crispate scissurelle
Kräuselfadenwebspinnen	Cribellatae	hackled band spiders
Kräuselhaar-Vogelspinne*	*Brachypelma albopilosum*	curlyhair tarantula
Kräuselradnetzspinnen	Uloboridae	hackled orbweavers
Kräuselspinnen, Eigentliche Kräuselspinnen	Dictynidae	dictynid spiders
Krausenhai, Kragenhai	*Chlamydoselachus anguineus*	frill shark, frilled shark (FAO)
Krausenhaie, Kragenhaie	Chlamydoselachidae	frill sharks, frilled sharks
Krauskopfpelikan	*Pelecanus crispus*	Dalmatian pelican
Kräuterdieb	*Ptinus fur*	white-marbled spider beetle, white-marked spider beetle
Krebsegel, Kiemenegel	*Branchiobdella parasitica*	fish-gill leech a.o.
Krebsfrosch	*Rana areolata*	crawfish frog
Krebsspinne*	*Gasteracantha cancriformis*	spinybacked spider
Krebssuppe, Zimteule, Zackeneule	*Scoliopterix libatrix*	herald (noctuid moth)
Krebstrugnatter	*Fordonia leucobalia*	white-bellied mangrove snake, whitebelly mangrove snake
Krefft-Spitzkopfschildkröte	*Emydura kreffti*	Krefft's river turtle, Krefft's tortoise
Kreiselkoralle, Ovale Nelkenkoralle	*Caryophyllia smithii*	Devonshire cupcoral
Kreismünder, Rundmäuler, Rundmünder	Cyclostomata	cyclostomes
Kreisrunde Erbsenmuschel	*Pisidium lilljeborgii*	Lilljeborg pea mussel, Lilljeborg peaclam
Kreisspinne, Konische Radspinne	*Cyclosa conica*	trashline orbweaver
Kreisspinnen	*Cyclosa* spp.	trashline orbweavers
Kreiswirbler	Gymnolaemata/ Stelmatopoda	gymnolaemates, "naked throat" bryozoans
Kreta-Stachelmaus	*Acomys minous*	Cretan spiny mouse

Kreuzbrustschildkröten	*Staurotypus* spp.	musk turtles (IUCN), cross-breasted turtles
Kreuzdelphin, Sanduhrdelphin	*Lagenorhynchus cruciger*	hourglass dolphin, southern white-sided dolphin
Kreuzdorn-Petersilien-Blattlaus	*Dysaphis apiifolia*	hawthorn-parsley aphid
Kreuzflügel, Rosskastanien-Frostspanner	*Alsophila aescularia*	March moth
Kreuzkrabbe	*Charybdis feriatus*	striped swimming crab
Kreuzkröte	*Bufo calamita*	natterjack toad, natterjack, British toad
Kreuzotter	*Vipera berus*	adder, common viper, common European viper
Kreuzqualle*	*Mitrocoma cellularia*	cross jellyfish
Kreuzschrecke	*Oedaleus decorus*	Mediterranean whitecross grasshopper*
Kreuzspinne> Gemeine Kreuzspinne, Gartenkreuzspinne	*Araneus diadematus*	cross orbweaver, European garden spider, cross spider
Kreuzspinnen	*Araneus* spp.	orbweavers (angulate & roundshouldered orbweavers)
Kreuzspinnen, Radnetzspinnen	Araneidae	orbweavers, orb-weaving spiders (broad-bodied orbweavers)
Kreuzwelse	*Arius* spp.	sea catfishes, salmon catfishes
Kribensis, Königscichlide	*Pelvicachromis pulcher*	kribensis
Krickente	*Anas crecca*	teal, green-winged teal
Kriebelmücken	Simuliidae	black flies, blackflies, buffalo gnats
Kriebelmücken	*Simulium* spp.	blackflies
Kriechtiere, Reptilien	Reptilia	reptiles
Krimeidechse	*Podarcis taurica*	Crimean wall lizard, Crimean lizard
Kristall-Geißelgarnele	*Farfantepenaeus brevirostris/ Penaeus brevirostris*	crystal shrimp
Kristallgrundel	*Crystallogobius linearis*	crystal goby
Kristallkoralle	*Galaxea fascicularis*	crystal coral*
Kroatische Gebirgseidechse	*Lacerta horvathi*	Horvath's rock lizard
Krokodile, Panzerechsen	Crocodilia	crocodiles
Krokodilfische u.a.	Pinguipedidae/ Parapercidae	sandperches
Krokodilfische u.a.	Percophidae/ Percophididae/ Bembropsidae	duckbills, flatheads
Krokodilhai, Falscher Sandtiger	*Pseudocarcharias kamoharai*	crocodile shark
Krokodilhecht, Riesen-Hornhecht, Riesen-Krokodilhecht	*Tylosurus crocodilus/ Strongylura crocodila*	hound needlefish (FAO), houndfish, crocodile longtom (Austr.), crocodile needlefish
Krokodilkaiman	*Caiman crocodilus*	spectacled caiman

Krokodilschleichen, Alligatorschleichen	*Gerrhonotus* spp.	alligator lizards
Krokodilschwanz-Höckerechse	*Shinisaurus crocodilurus*	Chinese xenosaur, Chinese crocodile lizard
Krokodilschwanzechse	*Crocodilurus lacertinus*	crocodile tegu, dragon lizard
Krokodilteju	*Dracaena guianensis*	caiman lizard
Krokus-Riesenmuschel	*Tridacna crocea*	crocus giant clam
Krönchen-Seepferd	*Hippocampus kuda*	golden seahorse, yellow seahorse, spotted seahorse (FAO)
Kronen-Erdfloh*	*Phyllotreta diademata*	crown flea beetle
Kronen-Kalkschwamm	*Sycon coronatum*	crown vase sponge*
Kronen-Kegelschnecke	*Conus regius*	crown cone
Kronen-Krötenechse	*Phrynosoma coronatum*	Coast horned lizard
Kronenadler	*Spizaetus coronatus*	crowned eagle
Kronenducker	*Sylicapra grimmia*	common duiker, gray duiker
Kronenflughuhn	*Pterocles coronatus*	crowned sandgrouse
Kronengecko	*Diplodactylus stenodactylus*	crowned gecko, long-fingered gecko
Kronenhummer, Kaisergranat, Kaiserhummer, Schlankhummer, Tiefseehummer	*Nephrops norvegicus*	Norway lobster, Norway clawed lobster, Dublin Bay lobster, Dublin Bay prawn (scampi, langoustine)
Kronenkranich	*Balearica pavonina*	crowned crane
Kronenkraniche	Balearicinae	crowned-cranes
Kronenlaubfrosch	*Anotheca spinosa*	spiny-headed treefrog
Kronenmaki	*Lemur coronatus/ Petterus coronatus/ Eulemur coronatus*	crowned lemur
Kronenmeerkatze	*Cercopithecus pogonias*	crowned guenon, crowned monkey
Kronenqualle*	*Nausithoe punctata*	crown jellyfish
Kronenschlangen	*Aspidomorphus* spp.	crowned snakes
Kronentyrann	*Onychorhynchus coronatus*	royal flycatcher
Kronwaldsänger	*Dendroica coronata*	yellow-rumped warbler
Kropfgazelle	*Gazella subgutturosa*	goitred gazelle
Krötenechsen	*Phrynosoma* spp.	horned lizards, "horny toads"
Krötenfisch, Gestreifter Anglerfisch	*Antennarius hispidus*	anglerfish, fishing frog, shaggy angler (FAO)
Krötenfliege	*Lucilia bufonivora*	toad blowfly*
Krötenfrösche	Pelobatidae	spadefoot toads
Krötengrundel	*Mesogobius batrachocephalus*	knout goby
Krötenkopf-Agamen	*Phrynocephalus* spp.	toadhead agamas, toadhead lizards, toad-headed lizards
Krötenkopf-Schildkröten	*Phrynops* spp.	toad-headed turtles
Krötenkopf-Schildkröte> Gewöhnliche Krötenkopf-Schildkröte, Gewöhnliche Froschkopf-Schildkröte	*Phrynops nasutus/ Batrachemys nasuta*	common South American toad-headed turtle, toad-headed side-necked turtle

Krötenottern, Nachtottern	*Causus* spp.	night adders
Krötenschlangensterne	Phrynophiurida	phrynophiurids
Kroyers Geißelgarnele, Glatthorn-Garnele	*Xiphopenaeus kroyeri*	seabob, Atlantic sea bob
Krugfische, Spitzkopfkugelfische	Canthigasteridae	sharpnose pufferfishes
Krummfühlerwanzen	Alydidae	broad-headed bugs
Krummhals-Triton*	*Cymatium caudatum*	bent-neck triton
Krummschnabel-Spottdrossel	*Toxostoma curvirostre*	curve-billed thrasher
Krummschnauzige Schlangennadel, Große Wurm-Seenadel	*Nerophis lumbriciformis*	worm pipefish
Krusten-Seescheiden	Polyclinidae	sea biscuits
Krustenanemonen	Zoantharia	zoanthids
Krustenechsen, Giftechsen	Helodermatidae	beaded lizards
Krustenkoralle	*Madracis pharensis*	crusty pencil coral
Krustenlederschwamm	*Ircinia fasciculata*	stinker sponge
Krustige Feuerkoralle	*Millepora squarrosa*	crustal fire coral
Krustiges Moostierchen	*Calpensia nobilis*	crusty bryozoan
Kruzifixwels	*Arius proops*	crucifix sea catfish (FAO), salmon catfish
Kuba-Anolis, Braunanolis	*Anolis sagrei*	brown anole, Cuban anole
Kuba-Nachtechse	*Cricosaura typica*	Cuban night lizard
Kuba-Schlankboa	*Epicrates angulifer*	Cuban tree boa, Cuban boa, maja
Kuba-Schlitzrüssler	*Solenodon cubanus*	Cuban solenodon
Kubafink	*Tiaris canora*	melodious grassquit
Kubanische Schabe	*Panchlora nivea*	Cuban cockroach
Kubanische Zwergboa	*Tropidophis melanurus*	Cuban Island ground boa
Küchenschabe, Bäckerschabe	*Blatta orientalis*	Oriental cockroach, common cockroach
Kuckuck	*Cucullus canorus*	cuckoo
Kuckucke	Cuculidae	cuckoos, roadrunners, anis
Kuckucke> Amerikanische Kuckucke, Regenkuckucke	Coccyzidae	American cuckoos, New World cuckoos
Kuckucke> Eigentliche Kuckucke, Altwelt-Kuckucke	Cuculinae	Old World cuckoos
Kuckucks-Knurrhahn, Seekuckuck	*Aspitrigla cuculus*	red gurnard, cuckoo gurnard
Kuckucksbienen, Wespenbienen	*Nomada* spp.	cuckoo bees
Kuckuckslippfisch	*Labrus bimaculatus/ Labrus mixtus*	cuckoo wrasse
Kuckucksrochen	*Raja naevus*	cuckoo ray (FAO), butterfly skate
Kuckucksvögel	Cuculiformes	cuckoos & turacos & allies
Kuckuckswels	*Synodontis petricola*	spotted squeaker
Kuckuckswürger, Stachelbürzler	Campephagidae	cuckoo-shrikes

Kudu> Großer Kudu	*Tragelaphus strepsiceros*	greater kudu
Kudu> Kleiner Kudu	*Tragelaphus imberbis*	lesser kudu
Kugelassel, Gemeine Rollassel	*Armadillidium vulgare*	common woodlouse, common pillbug, sow bug
Kugelbauchmilbe, Heumilbe	*Pyemotes tritici/ Pyemotes ventricosus*	hay itch mite, grain itch mite, grain mite, straw itch mite
Kugelechse, Halsbandleguan	*Crotaphytus collaris*	collared lizard
Kugelfingergeckos	*Sphaerodactylus* spp.	least geckos
Kugelfisch, Nördlicher Kugelfisch	*Sphoeroides maculatus/ Sphaeroides maculatus*	northern puffer (FAO), swellfish
Kugelfische	Tetraodontidae	puffers
Kugelfische	*Tetraodon* spp.	pufferfishes
Kugelfischverwandte	Tetraodontiformes/Plectognathi	plectognath fishes
Kugelfliegen, Spinnenfliegen	Acroceridae	small-headed flies
Kugelgürteltier, Brasilianisches Dreibinden-Gürteltier	*Tolypeutes tricinctus*	Brazilian three-banded armadillo
Kugelgürteltiere	*Tolypeutes* spp.	three-banded armadillos
Kugelige Erbsenmuschel	*Pisidium pseudosphaerium*	pseudospherical pea mussel*
Kugelige Strandkrabbe	*Cyclograpsus integer*	globose shore crab
Kugelkäfer	Sphaeridae	minute bog beetles
Kugelkäfer, Buckelkäfer	*Gibbium psylloides*	smooth spider beetle (bowl beetle)
Kugelkauri*	*Cypraea globulus*	globose cowrie
Kugelkopf-Kammzähner	*Istiblennius periophthalmus*	red-dotted blenny
Kugelkopffliegen, Augenfliegen	Pipunculidae	big-headed flies
Kugelkrabben	Leucosiidae	purse crabs, nut crabs, pebble crabs
Kugelmuschel> Gemeine Kugelmuschel, Hornfarbene Kugelmuschel, Linsenmuschel	*Sphaerium corneum*	horny orb mussel, European fingernailclam
Kugelmuscheln	Sphaeriidae	orb mussels (orb shells, sphere shells)
Kugelmuscheln	*Sphaerium* spp.	orb mussels, fingernailclams
Kugelschildläuse	Hemicoccinae	gall-like coccids
Kugelschnecke, Gemeine Kugelschnecke	*Akera bullata*	common bubble snail
Kugelschnecken, Blasenschnecken	Ampullariidae	bubble snails (bubble shells)
Kugelspinnen	*Theridion* spp.	comb-footed spider
Kugelspinnen, Haubennetzspinnen	Theridiidae	comb-footed spiders, cobweb and widow spiders, scaffolding-web spiders
Kugelspringer	Sminthuridae	globular springtails
Kuhantilope	*Alcelaphus buselaphus*	red hartebeest
Kuhfliege, Hornfliege	*Haematobia irritans*	horn-fly, tropical buffalo fly

Kuhkopffisch, Kuhkopf-Doktorfisch, Hornloser Einhornfisch	*Naso lituratus/ Callicanthus lituratus*	smooth-headed unicornfish, orangespine unicornfish (FAO), green unicornfish
Kuhnasenrochen	*Rhinoptera bonasus*	cownose ray
Kuhnasenrochen	Rhinopteridae	cownose rays
Kuhreiher	*Ardeola ibis/ Bubulcus ibis*	cattle egret, buff-backed heron
Kuhrochen, Afrikanischer Adlerrochen	*Pteromylaeus bovinus/ Myliobatis bovina*	bull ray
Kukrinattern	*Oligodon* spp.	kukri snakes
Kumazeen	Cumacea	cumaceans
Kümmelmotte, Kümmelpfeifer, Möhrenschabe	*Depressaria nervosa*	carrot and parsnip flat-body moth
Kundekäfer	*Callosobruchus chinensis*	cowpea weevil
Kupfer-Goldwespe	*Chrysis cuprea*	copper wasp
Kupfer-Pinzettfisch	*Chelmon rostratus*	copper-banded butterflyfish, copperband butterflyfish (FAO), long-nosed butterflyfish, beaked coralfish
Kupferglucke	*Gastropacha quercifolia*	lappet
Kupferhai, Bronzehai	*Carcharhinus brachyurus*	copper shark (FAO), bronze whaler, narrowtooth shark
Kupferkopf	*Agkistrodon contortrix*	copperhead
Kupfersalmler	*Hasemania nana*	silver-tipped tetra
Kupferstecher, Sechszähniger Fichtenborkenkäfer	*Pityogenes chalcographus*	six-dentated bark beetle
Kuppelgaumen-Bulldogg-Fledermäuse	*Promops* spp.	domed-palate mastiff bats
Kürbisspinne	*Araniella cucurbitina/ Araneus cucurbitinus*	gourd spider, pumpkin spider
Kürbiswanze	*Anasa tristis*	squash bug (U.S.)
Kurol	Leptosomidae/Leptosomatidae	cuckoo-roller
Kurol	*Leptosomus discolor*	courol, cuckoo-roller
Kurter	Kurtidae	nurseryfishes
Kurzbeinige Philippinenkröte	*Pelophryne brevipes*	short-legged toadlet
Kurzdornige Rote Knotenameise, Kurzdornige Rote Gartenameise	*Myrmica laevinodis*	shortsting red myrmicine ant*
Kurze Messermuscheln, Kurze Scheidenmuscheln	Solecurtidae	short razor clams
Kurze Zwergmakrele	*Rastrelliger brachysoma*	short mackerel
Kurzfangsperber	*Accipiter brevipes*	Levant sparrowhawk
Kurzflossen-Dornrückenaal	*Notacanthus bonapartei*	short-finned spiny-eel
Kurzflossen-Grindwal, Indo-Pazifischer Grindwal	*Globicephala seiboldii, Globicephala macrorhynchus*	Indo-Pacific pilot whale, short-finned pilot whale, blackfish
Kurzflossen-Haarschwanz, Espada	*Aphanopus carbo*	black scabbardfish (FAO), espada
Kurzflossen-Mako	*Isurus oxyrinchus/ Isurus oxyrhynchus/ Isurus glaucus*	shortfin mako (FAO), mako shark, mako
Kurzflossenbarbe	*Barbus brevipinnis*	shortfin barb

Kurzflügel-Leuchtkäfer, Kurzflügeliger Leuchtkäfer	Phosphaenus hemipterus	short-wing lightning beetle
Kurzflügler, Raubkäfer	Staphylinidae	rove beetles
Kurzflüglige Beißschrecke	Metrioptera brachyptera	bog bushcricket
Kurzflüglige Schwertschrecke	Conocephalus dorsalis	short-winged cone-head, short-winged conehead
Kurzfuß-Inselratte	Brachytarsomys albicauda	shortfoot Madagascan rat*
Kurzfuß-Stelzenralle	Mesitornis variegata	white-breasted mesite
Kurzfuß-Wasserläufer	Tringa guttifer	Nordmann's greenshank
Kurzhaar-Hummel	Bombus subterraneus	short-haired bumble bee
Kurzhaar-Wasserratte	Paraleptomys wilhelmina	short-haired hydromyine
Kurzhorn-Krötenechse	Phrynosoma douglasi	short-horned lizard
Kurzkammleguan	Brachylophus fasciatus	Fiji banded iguana
Kurzkiefer-Umberfisch	Pseudotolithus brachygnathus	law croaker
Kurzkopf-Gleithörnchenbeutler	Petaurus breviceps	sugar glider
Kurzkopf-Weichschildkröte	Chitra indica	narrow-headed softshell turtle
Kurzkopffrosch	Breviceps gibbosus	rain frog, Cape rain frog (IUCN), South-African short-headed frog
Kurzkopffrösche	Breviceps spp.	short-headed frogs
Kurzkopfige Gespensterkrabbe	Inachus dorsettensis	scorpion spider crab
Kurzköpfige Rinderlaus, Kurznasige Rinderlaus	Haematopinus eurysternus	shortnosed cattle louse
Kurzlappenqualle*	Bolinopsis microptera	short-lobed comb jelly
Kurznasen-Dornhai	Centroscymnus cryptacanthus	shortnose velvet dogfish
Kurznasen-Einhornfisch, Schärpen-Nasendoktor	Naso brevirostris	spotted unicornfish (FAO), short-snouted unicornfish
Kurznasen-Flughund, Schwarzflügel-Flughund	Thoopterus nigrescens	short-nosed fruit bat, swift fruit bat
Kurznasen-Flughunde	Cynopterus spp.	short-nosed bats, dog-faced fruit bats
Kurznasen-Knochenhecht	Lepisosteus platystomus	shortnose gar
Kurznasen-Seefledermaus, Amerikanische Kurznasen-Seefledermaus	Ogcocephalus nasutus	shortsnout batfish, shortnose batfish (FAO)
Kurznasenbeutler	Isoodon spp./Thylacis spp.	short-nosed bandicoots
Kurznasenstör	Acipenser brevirostrum	shortnose sturgeon
Kurzohrfuchs	Atelocynus microtis	small-eared dog
Kurzohrmaus, Kleinäugige Wühlmaus, Kleine Wühlmaus, Kleinwühlmaus	Microtus subterraneus/ Pitymys subterraneus	European pine vole
Kurzohrrüsselspringer	Macroscelides proboscideus	short-eared elephant shrew
Kurzschnabel-Makrelenhecht	Cololabis saira	Pacific saury (FAO), mackerel-pike
Kurzschnabel-Seenadel	Syngnathus abaster	short-snouted pipefish, shore pipefish
Kurzschnabelalbatros	Diomedea albatrus	short-tailed albatross
Kurzschnabeldelphin, Frasers Delphin, Borneo-Delphin	Lagenodelphis hosei	shortsnouted whitebelly dolphin, Fraser's dolphin, Sarawak dolphin, Bornean dolphin

Kurzschnabelflamingo	*Phoenicoparrus jamesi*	Puna flamingo
Kurzschnabelgans	*Anser brachyrhynchus*	pink-footed goose
Kurzschnabeligel	*Tachyglossus aculeatus*	echidna, short-nosed echidna, spiny anteater
Kurzschnauzen-Marlin, Kurzschnäuziger Speerfisch	*Tetrapturus angustirostris*	shortbill spearfish
Kurzschnauziger Schlangenaal, Stumpfnasen-Schlangenaal	*Echelus myrus*	blunt-nosed snake eel
Kurzschnauziges Seepferdchen	*Hippocampus hippocampus*	short-snouted sea horse
Kurzschwanz-Blattnasen	*Carollia* spp.	short-tailed leaf-nosed bats
Kurzschwanz-Chinchilla	*Chinchilla brevicaudata*	short-tailed chinchilla
Kurzschwanz-Hamsterratten	*Saccostomus* spp.	African pouched rats
Kurzschwanz-Maulwurfsratte	*Nesokia indica*	pest rat, short-tailed bandicoot rat
Kurzschwanz-Notothenia	*Patagonotothen brevicauda*	Patagonian rockcod
Kurzschwanz-Rennmäuse	*Diplodillus* spp.	short-tailed gerbils
Kurzschwanz-Rochen, Blonde	*Raja brachyura*	blonde ray (FAO), blond ray
Kurzschwanz-Spitzmausratte*	*Neohydromys fuscus*	short-tailed shrew mouse
Kurzschwanz-Stachelschwein	*Hystrix brachyura/ Acanthion brachyura*	Malayan porcupine
Kurzschwanz-Sturmtaucher	*Puffinus tenuirostris*	short-tailed shearwater
Kurzschwanzbussard	*Buteo brachyurus*	short-tailed hawk
Kurzschwänzige Zwerghamster	*Phodopus* spp.	small desert hamsters, dwarf hamsters
Kurzschwanzkänguru, Quokka	*Setonix brachyurus*	quokka
Kurzschwanzkrebse, Echte Krabben	*Brachyura*	crabs
Kurzschwanzleguane	*Stenocercus* spp.	whorltail iguanas
Kurzschwanzspitzmaus	*Blarina brevicauda*	short-tailed shrew
Kurzschwanzwaran	*Varanus brevicauda*	short-tailed monitor, short-tailed pygmy monitor
Kurzspringer	Hypogastruridae	hypogastrurid springtails
Kurzstachel-Schlangenstern	*Ophioderma brevispina*	short-spined brittlestar
Kurzzahnschlangen*	*Ultrocalamus* spp.	short-fanged snake
Kurzzehenlerche	*Calandrella brachydactyla*	short-toed lark
Kusimansen	*Crossarchus* spp.	cusimanses
Kuskuse	*Phalanger* spp.	cuscuses
Küssende Guramis	Helostomatidae	kissing gourami
Küssender Gurami, Küsser	*Helostoma temmincki*	kissing gourami
Küsten-Pfeilwurm	*Sagitta setosa* (Chaetognatha)	coastal arrow worm
Küsten-Sandlaufkäfer	*Cicindela maritima*	coastal tiger beetle
Küsten-Strauchschrecke	*Pholidoptera littoralis*	littoral bushcricket
Küsten-Waldschabe, Heideschabe	*Ectobius panzeri*	lesser cockroach

Küstenglasschleiche, Insel-Glasschleiche	*Ophisaurus compressus*	island glass lizard (U.S.)
Küstenhüpfer, Sandhüpfer	*Orchestia gammarellus*	shore-hopper, beach-flea, common shore-skipper, common scud
Küstenreiher	*Egretta gularis*	western reef heron
Küstensauger, Remora	*Remora remora*	common remora (FAO), shark sucker
Küstenschnecken	Ellobiidae	coastal snails*
Küstenseeschwalbe	*Sterna paradisaea*	Arctic tern
Küstenwühlmaus	*Microtus breweri*	beach meadow vole, beach vole
Kusu-Grasratten	*Arvicanthis* spp.	unstriped grass mice, kusu rats
Kusus	*Trichosurus* spp.	brush-tailed possums
Kykladen-Eidechse, Ägäische Mauereidechse	*Podarcis erhardii*	Erhard's wall lizard, Aegean wall lizard

La-Plata-Delphin, Franciscana	*Pontoporia blainvillei*	Franciscana dolphin, La Plata dolphin
La-Plata-Erdfresser	*Gymnogeophagus meridionalis*	La-Plata eartheater
Labkrautblattkäfer	*Timarcha tenebricosa*	bloody-nosed beetle
Labkrautschwärmer	*Hyles gallii/Celerio galii*	bedstraw hawkmoth
Labyrinthkoralle, Gefurchte Hirnkoralle, Feine Atlantische Hirnkoralle	*Diploria labyrinthiformis*	labyrinthine brain coral, grooved brain coral
Labyrinthspinne	*Agelena labyrinthica*	grass funnel-weaver, maze spider*
Lachender Hans, Jägerliest	*Dacelo novaeguineae*	laughing kookaburra
Lachfalke	*Herpetotheres cachinnans*	laughing falcon
Lachmöve	*Larus ridibundus*	black-headed gull
Lachs> Atlantischer Lachs, Salm (*Junglachse im Meer*: Blanklachs)	*Salmo salar*	Atlantic salmon (*lake pop. in U.S./Canada*: ouananiche, lake Atlantic salmon, landlocked salmon, Sebago salmon)
Lachse	Salmonidae	salmonids
Lachse> Atlantische Lachse & Forellen	*Salmo* spp.	Atlantic trouts & Atlantic salmons
Lachseeschwalbe	*Gelochelidon nilotica*	gull-billed tern
Lachsforelle, Meerforelle	*Salmo trutta trutta*	sea trout
Lackschildläuse	Lacciferidae (Tachardiidae)	lac insects
Lackvögel	Dasyornithinae	bristlebirds
Lactarius	*Lactarius lactarius*	false trevally (FAO), white fish
Lafrentz-Hautwühle	*Dermophis oaxacae*	Oaxacan caecilian
Lagerhauskäfer	*Trogoderma glabrum*	warehouse beetle
Lagunen-Flohkrebs	*Gammarus insensibilis*	lagoon sand shrimp
Lagunen-Grundel	*Knipowitschia panizzae*	lagoon goby
Lagunen-Herzmuschel	*Cerastoderma glaucum*	lagoon cockle
Lagunen-Sandwurm*	*Armandia cirrhosa*	lagoon sandworm
Lagunen-Winkerkrabbe	*Uca anullipes*	lagoon fiddler crab
Lakeland Down-Kleinmaus	*Leggadina lakedownensis*	Lakeland Downs mouse
Lama	*Lama glama*	llama
Lamagazelle, Stelzengazelle, Dibatag	*Ammodorcas clarkei*	dibatag
Lamarck-Kielwurm	*Pomatoceros lamarcki*	Lamarck's keelworm
Lamarcks Kaktuskoralle, Raue Kaktuskoralle	*Mycetophyllia ferox*	rough cactus coral
Lamarcks Kauri	*Cypraea lamarckii*	Lamarck's cowrie
Lamarcks Salatblattkoralle	*Agaricia lamarcki*	Lamarck's sheet coral
Lamellen-Korallengast, Lamellenschnecke	*Coralliophila meyendorffi*	lamellose coralsnail
Lamellenkiemer, Blattkiemer	Eulamellibranchia	eulamellibranch bivalves
Lammzunge, Lammbutt	*Arnoglossus laterna*	scaldfish
Lampenratte, Lampara de agua (Hbz.), Schwimmbeutler, Yapok	*Chironectes minimus*	water opossum, yapok
Land-Einsiedlerkrebse	*Coenobita* spp.	land hermit crabs
Land-Einsiedlerkrebse	Coenobitidae	land hermit crabs

Landasseln	Oniscidea	oniscideans (pillbugs, woodlice, sowbugs, slaters)
Landegel	Haemadipsidae	terrestrial leeches, haemadipsid leeches
Landkärtchen	*Araschnia levana*	map butterfly
Landkarten-Kegelschnecke, Landkartenkegel	*Gastridium geographus/ Conus geographus*	geography cone, geographic cone
Landkartenkauri, Landkartenschnecke	*Cypraea mappa*	map cowrie
Landkartenschildkröte, Landkarten-Höckerschildkröte	*Graptemys geographica*	map turtle
Landkartenschildkröten	*Graptemys* spp.	map turtles
Landkrabben	Gecarcinidae	land crabs
Landraubtiere	Fissipedia	terrestrial carnivores
Landschildkröten	Testudinidae	tortoises, true tortoises
Landwirbeltiere, Tetrapoden	Tetrapoda	terrestrial vertebrates, tetrapods
Langarmiger Krake	*Octopus macropus*	Atlantic white-spotted octopus, long-armed octopus, grass octopus, white-spotted octopus (FAO)
Langarmiger Schlangenstern	*Acrocnida brachiata*	longarm brittlestar
Langarmkrabbe	*Parthenope angulifrons/ Lambrus angulifrons*	long-armed crab
Langarmkrabben, Ellbogenkrabben*	Parthenopidae	elbow crabs
Langbein-Fledermaus	*Macrophyllum macrophyllum*	long-legged bat
Langbeinfliegen	Dolichopodidae	long-headed flies
Langbeinige Spinnenkrabbe, Gespensterkrabbe	*Macropodia rostrata*	long-legged spider crab
Langbeinige Winkerkrabbe	*Uca stenodactyla*	long-legged fiddler crab
Langdornige Rote Knotenameise, Langdornige Rote Gartenameise	*Myrmica ruginodis*	longsting red myrmicine ant*
Langdornige Schmierlaus, Mehlige Gewächshausschildlaus	*Pseudococcus longispinus*	longtailed mealybug
Lange Herzmuschel	*Laevicardium oblongum*	oblong cockle
Lange Pfeffermuschel	*Abra prismatica*	prismatic abra, elongate abra (elongate furrow-shell)
Lange Riffmuschel	*Trapezium oblongum*	oblong trapezium
Lange Spindelschnecke	*Fusinus longissimus*	long spindle
Langer Elefantenzahn	*Antalis longitrorsum* (Scaphopoda)	elongate tusk
Langer Grünling, Langer Terpug	*Ophiodon elongatus*	lingcod
Langer Schnurwurm, Engländischer Langwurm	*Lineus longissimus* (Nemertini)	giant bootlace worm, sea longworm
Langfinger, Herzzüngler	*Cardioglossa* spp.	long-fingered frogs
Langfinger-Winkerkrabbe	*Uca speciosa*	longfinger fiddler
Langflossen-Buntbarsch*	*Oreochromis macrochir*	longfin tilapia

Langflossen-Gabeldorsch	*Phycis chesteri*	longfin hake
Langflossen-Grindwal, Gewöhnlicher Grindwal	*Globicephala melaena*	northern pilot whale, Atlantic pilot whale, blackfish, pothead, long-finned pilot whale
Langflossen-Hai, Weißspitzenhai, Hochsee-Weißflossenhai, Weißspitzen-Hochseehai	*Carcharhinus longimanus/ Carcharhinus maou*	oceanic whitetip shark, whitetip shark, whitetip oceanic shark
Langflossen-Halbschnäbler	*Hemiramphus saltator*	longfin halfbeak
Langflossen-Mako	*Isurus paucus*	longfin mako
Langflossen-Saibling*	*Salvelinus svetovidovi*	long-finned char
Langflossen-Stachelmakrele	*Decapterus maruadsi*	round scad, long-fin scad, Japanese scad (FAO)
Langflossen-Zigarrenfisch*	*Cubiceps gracilis*	longfin cigarfish
Langflossenbarsch*	*Taractichthys longipinnis*	long-finned bream
Langflossenhecht*	*Dinolestes lewini*	long-finned pike
Langflügel-Fledermäuse	*Miniopterus* spp.	long-winged bats, bent-winged bats, long-fingered bats
Langflüglige Schwertschrecke	*Conocephalus discolor*	long-winged cone-head, long-winged conehead
Langfühler-Dornschrecke	*Tetrix tenuicornis/ Tetrix natans*	longhorned groundhopper
Langfühlermotten, Langhornmotten	Adelidae/ Incurvariidae	longhorn moths, bright moths
Langfuß-Fledermaus	*Myotis capaccinii*	long-fingered bat
Langfuß-Wasserratte*	*Leptomys elegans*	long-footed hydromyine
Langfußratten	*Malacomys* spp.	big-eared swamp rats, long-eared marsh rats, long-footed rats
Langhaargundi, Sahara-Kammfinger, Sahara-Gundi	*Massoutiera mzabi*	Mzab gundi, fringe-eared gundi
Langhals-Schmuckschildkröte	*Deirochelys reticularia*	chicken turtle
Langhalsaale	Derichthyidae	longneck eels
Langhorn-Blattminiermotten	Lyonetiidae	lyonetiid moths
Langhorn-Kofferfisch	*Lactoria cornuta*	longhorn cowfish
Langhorn-Ohrwürmer	Labiduridae	long-horned earwigs, striped earwigs
Langhorn-Porzellankrebs, Schwarzer Porzellankrebs, Schwarzes Porzellankrebschen	*Pisidia longicornis/ Porcellana longicornis*	long-clawed porcelain crab, common porcelain crab, minute porcelain crab
Langhorn-Seeskorpion	*Myoxocephalus octodecemspinosus*	longhorn sculpin
Langhorn-Wüstenheuschrecken	Tanaoceridae	desert long-horned grasshoppers
Langhornmotten, Langfühlermotten	Adelidae/ Incurvariidae	longhorn moths, bright moths
Langkäfer	Brentidae/Brenthidae	straight-snouted weevils
Langkamm-Sägefisch	*Pristis zijsron*	longcomb sawfish
Langkopf-Umberfisch	*Pseudotolithus typus*	longneck croaker
Langköpfige Rinderlaus	*Linognathus vituli*	longnosed cattle louse
Langkopfwespen	*Dolichovespula* spp.	long-headed wasps

Langkrallen-Spitzmäuse	*Solisorex* spp.	long-clawed shrews
Längliche Riesenmuschel, Kleine Riesenmuschel	*Tridacna maxima*	elongated clam, elongate clam, small giant clam
Länglicher Herzigel	*Lovenia elongata*	elongate heart urchin
Länglicher Segelkalmar*	*Histioteuthis elongata*	elongate jewel squid
Langlippenkäfer	Telegeusidae	long-lipped beetles
Langmaul-Schlangenaal, Mittelmeer-Schlangenaal	*Ophisurus serpens*	long-jawed snake eel
Langnasen-Dornhai	*Centroscymnus crepidater*	longnose velvet dogfish
Langnasen-Falkenfisch	*Oxycirrhites typus*	long-nosed hawkfish
Langnasen-Fledermaus	*Choeronycteris mexicana*	Mexican long-nosed bat, hog-nosed bat
Langnasen-Leopardleguan	*Gambelia wislizenii*	longnose leopard lizard
Langnasen-Sägehai	*Pristiophorus cirratus/ Squalus anisodon*	longnose sawshark
Langnasen-Seefledermaus	*Ogcocephalus vespertilio*	longsnout batfish
Langnasenbeutler	*Perameles* spp.	long-nosed bandicoots
Langnasenhai, Großer Schwarzspitzenhai	*Carcharhinus brevipinna/ Carcharhinus maculipinnis*	spinner shark, long-nose grey shark
Langohren-Fledermäuse	*Plecotus* spp.	lump-nosed bats, long-eared bats, lappet-eared bats
Längsband-Schleimfisch	*Parablennius rouxi*	striped blenny
Längsbandkärpfling	*Fundulus notatus*	blackstripe topminnow
Längsbandorfe	*Notropis hypselopterus*	sailfin shiner
Längsbandsalmler	*Nannostomus beckfordi/ Nannostomus aripirangensis*	golden pencilfish
Langschnabel-Messerfisch	*Rhamphichthys rostratus*	bandfish
Langschnabel-Spinnenjäger	*Arachnothera robusta*	long-billed spiderhunter
Langschnabeligel	*Zaglossus bruijni*	long-nosed echidna, long-beaked echidna, New Guinea long-nosed spiny anteater
Langschnabelpieper	*Anthus similis*	long-billed pipit
Langschnabelsittich	*Enicognathus leptorhynchus*	slender-billed conure
Langschnabelwachtel	*Rhizothera longirostris*	long-billed wood partridge
Langschnauzen-Einhornfisch, Blauklingen-Nasendoktor	*Naso unicornis*	long-snouted unicornfish, bluespine unicornfish (FAO)
Langschnauzen-Eisfisch	*Channichthys rhinoceratus*	long-snouted icefish, unicorn icefish (FAO)
Langschnauzen-Kaninchenkänguru	*Potorous tridactylus*	long-nosed potoroo
Langschnauzen-Lippfisch, Schnauzenlippfisch	*Symphodus rostratus*	long-snouted wrasse, cannadelle
Langschnauzen-Pinzettfisch, Langmaul-Pinzettfisch	*Forcipiger longirostris*	long-snouted forceps fish
Langschnauzen-Seeratte	*Trachyrhynchus trachyrhynchus*	roughsnout grenadier (FAO), long-snouted grenadier
Langschnauzenmanguste	*Rhynchogale melleri*	Meller's mongoose
Langschnäuziger Seebader	*Halimochirurgus centriscoides*	longsnout spikefish
Langschnäuziger Speerfisch	*Tetrapturus pfluegeri*	longbill spearfish (FAO), long-billed spearfish
Langschnauziges Seepferdchen	*Hippocampus guttulatus/ Hippocampus ramulosus*	seahorse, European seahorse, long-snouted seahorse (FAO)

Langschopfmaina	*Acridotheres grandis*	great mynah
Langschwanz-Borstenigel, Spitzmaus-Borstenigel	*Microgale* spp.	shrew-tenrecs, long-tailed tenrecs
Langschwanz-Chinchilla	*Chinchilla lanigera*	long-tailed chinchilla
Langschwanz-Flughund	*Notopteris macdonaldi*	long-tailed fruit bats
Langschwanz-Mennigvogel	*Pericrocotus ethologus*	long-tailed minivet
Langschwanz-Schmetterlingsrochen	*Gymnura poecilura*	long-tailed butterfly ray
Langschwanz-Schuppentier	*Manis tetradactyla*	phatagin
Langschwanz-Thun	*Thunnus tonggol*	longtail tuna
Langschwanz-Wühlmaus	*Microtus longicaudus*	long-tailed vole, long-tailed meadow mouse
Langschwanzdrossling	*Turdoides caudatus*	common babbler
Langschwanzeidechsen	*Takydromus* spp.	grass lizards
Langschwanzgrundel	*Knipowitschia longecaudata*	long-tail goby
Langschwänzige Indische Baummäuse	*Vandeleuria* spp.	long-tailed climbing mice
Langschwänziger Bläuling, Großer Wander-Bläuling	*Lampides boeticus*	long-tailed blue
Langschwänziger Gelbsalamander, Langschwanzsalamander	*Eurycea longicauda*	long-tailed salamander
Langschwanzkatze, Margay	*Felis wiedii*	margay
Langschwanzmakak, Javaneraffe	*Macaca fascicularis/ Macaca irus*	crab-eating macaque
Langschwanzratten, Neuguinea-Ratten,	*Macruromys* spp.	New Guinean rats
Langschwanzskunk	*Mephitis macroura*	hooded skunk
Langschwanzwiesel	*Mustela frenata*	long-tailed weasel
Längsgepunktete Zwergnatter	*Eirenis lineomaculatus*	striped dwarf snake
Langstachel-Husar, Karibischer Eichhörnchenfisch	*Holocentrus rufus*	longspine squirrelfish
Langstachel-Schlangenstern*	*Ophiothrix angulata*	Atlantic long-spined brittlestar
Langstacheliger Diademseeigel	*Diadema setosum*	hatpin urchin, longspined sea urchin
Langstachelschnecke*	*Astralium phoebium*	longspine starsnail
Langstachliger Seeigel, Mittelmeer-Diademseeigel	*Centrostephanus longispinus*	Mediterranean hatpin urchin
Langstirn-Maulbrüter	*Tachysurus* spp.	barbels, sea barbels
Langstirnwels	*Auchenoglanis occidentalis*	bubu (FAO), giraffe catfish
Langtaster-Wasserkäfer	Hydraenidae	minute moss beetles
Languren	*Presbytis* spp.	leaf monkeys, langurs
Langusten	Palinuridae	spiny lobsters
Langusten u.a.	*Panulirus* spp.	spiny crawfish
Langzahn-Sägerochen, Langzahn-Sägefisch	*Pristis microdon* (*Pristis perotteti/ Pristis zephyreus*)	largetooth sawfish
Langzehen-Querzahnmolch	*Ambystoma macrodactylum*	long-toed salamander
Langzehen-Strandläufer	*Calidris subminuta*	long-toed stint
Langzungen-Fledermäuse u.a.	*Glossophaga* spp.	long-tongued bats
Langzungen-Flughunde	*Macroglossus* spp.	long-tongued fruit bats

Lannerfalke	*Falco biarmicus*	lanner falcon
Lansbergs Lanzenotter	*Porthidium lansbergi*	Lansberg's hognose viper
Lanzenfisch, Langnasen-Lanzenfisch	*Alepisaurus ferox*	lancetfish, longnose lancetfish (FAO)
Lanzenfische u.a.	Kraemeriidae	sand darts, sandfishes, sand gobies
Lanzenfische u.a.	Alepisauridae	lancetfishes
Lanzenfliegen u.a.	Lonchaeidae	lonchaeids
Lanzenfliegen u.a.	Lonchoptridae	spear-winged flies
Lanzennasen	*Phyllostomus* spp.	spear-nosed bats
Lanzenotter	*Bothrops asper/ Bothrops andianus asper*	terciopelo, fer-de-lance, barba amarilla
Lanzenotter> Gewöhnliche Lanzenotter, Gemeine Lanzenotter	*Bothrops atrox*	common lancehead, barba amarilla
Lanzenottern	*Lachesis* spp. (see also: Bothrops/Porthidium)	American lance-headed vipers
Lanzenratte	*Hoplomys gymnurus*	armored rat, thick-spined rat
Lanzenratten, Stachelratten	Echimyidae	spiny rats
Lanzenseeigel	Cidaridae	cidarids
Lanzenseeigel	Cidaroida	cidaroids
Lanzenskinke	*Acontias* spp.	lance skinks
Lanzettfischchen	*Branchiostoma lanceolatum*	lancelet
Lanzettfischchen	*Branchiostoma* spp.	lancelets
Lanzettförmiger Schnabelwels, Gemeiner Nadelwels	*Farlowella acus*	twig catfish
Lappenchamäleon	*Chamaeleo dilepis*	flap-necked chameleon, flapneck chameleon
Lappenente	*Biziura lobata*	musk duck
Lappenfisch*	*Icosteus aenigmaticus*	ragfish
Lappenmünder, Lippenmünder	Cheilostomata	cheilostomates
Lappenpittas	Philepittidae	asities (asitys)
Lappenqualle	*Bolina hydatina*	
Lappenrippenquallen	Lobata	lobate comb-jellies, lobates
Lappenrüssler, Dickmaulrüssler	*Otiorhynchus* spp.	snout beetles, snout weevils
Lappenspanner	*Lobophora halterata*	seraphim
Lappenstar	*Creatophora cinerea*	wattled starling
Lappentaucher	Podicipediformes/ Podicipedidae	grebes
Lapplandmeise	*Parus cinctus*	Siberian tit
Lar, Weißhandgibbon	*Hylobates lar*	common gibbon, white-handed gibbon
Lärchenblasenfuß	*Taeniothrips laricivorus*	larch thrips
Lärchengallenwickler, Lärchenrindenwickler	*Cydia zebeana*	larch gall moth*, larch bark moth*
Lärchengespinst-Blattwespe	*Cephalcia lariciphila*	web-spinning larch sawfly, larch webspinner
Lärchenknospen-Gallmücke	*Dasineura laricis*	larch bud midge

Lärchenminiermotte	*Coleophora laricella*	larch casebearer, larch leaf miner
Lärchenpilzkoralle, Salatkoralle	*Agaricia agaricites*	lettuce coral
Lärchenspanner, Beerenkrautspanner, Heidelbeerspanner, Tannenspanner, Pflaumenspanner	*Ectropis crepuscularia/ Ectropis bistortata*	larch looper, blueberry lopper, fir looper, plum looper
Largha-Seehund	*Phoca largha*	spotted seal, larga seal
Larvenfische	Cromeriidae	cromeriids
Larvenroller	*Paguma larvata*	masked palm civet
Larvensifaka	*Propithecus verreauxi*	Verreaux's sifaka
Lasso-Spinne, Amerikanische Bola-Spinne	*Mastophora bisaccata*	American bolas spider
Lasso-Spinnen, Bola-Spinnen	*Mastophora* spp.	bolas spiders
Lasurmeise	*Parus cyanus*	azure tit
Laternenangler> Kinnbart-Laternenangler*	*Linophryne arborifera*	lanternfish a.o.
Laternenfisch	*Myctophum punctatum*	lanternfish
Laternenfische	Myctophiformes/Myctophidae	lanternfishes (& blackchins)
Laternenfische, Blitzlichtfische	Anomalopidae	lanterneye fishes, flashlight fishes
Laternenmuscheln	Laternulidae	lantern clams
Laternensalmler, Schlusslichtsalmler, Fleckensalmler	*Hemigrammus ocellifer*	beacon fish, head-and-taillight tetra (FAO)
Laternenträger, Leuchtzikaden	Fulgoridae	lanternflies, lantern flies, fulgorid planthoppers
Laternenzüngler	Neoscopelidae	blackchins
Latiaxis-Schnecken	Coralliophilidae	coralsnails, latiaxis snails
Latrinenfliege	*Fannia scalaris*	latrine fly
Lattichfliege, Salatfliege	*Pegohylemyia gnava/ Botanophila gnava*	lettuce seed fly
Lau	*Chondrostoma genei*	South European nase, casca (*Ital.*), lasca (FAO)
Laubanolis	*Anolis chlorocyanus*	Hispaniola green anole
Laube, Weißer Ukelei	*Alburnus albidus*	Italian bleak
Laubenvögel	Ptilonorhynchidae	bowerbirds
Laubfrosch> Europäischer Laubfrosch	*Hyla arborea*	European treefrog, common treefrog, Central European treefrog
Laubfrösche	*Hyla* spp.	common treefrogs
Laubfrösche	Hylidae	tree frogs, true tree frogs
Laubheuschrecken, Singschrecken	Tettigoniidae	long-horned grasshoppers (incl. katydids & bushcrickets)
Laubholz-Flechtenspinner	*Eilema lurideola*	common footman
Laubholz-Säbelschrecke	*Barbitistes serricauda*	sawtailed bushcricket
Laubnutzholz-Borkenkäfer, Buchennutzholz-Borkenkäfer, Buchen-Nutzholzambrosiakäfer	*Xyloterus domesticus*	broad-leaved wood ambrosia beetle, broad-leaf wood ambrosia beetle

Laucharassari	*Aulacorhynchus prasinus*	emerald toucanet
Lauchgrüner Skink	*Prasinohaema virens*	green-blood skink
Lauchmotte	*Acrolepia assectella*	leek moth
Lauchschrecke	*Mecostethus alliaceus/ Parapleurus alliaceus*	leek grasshopper
Lauer, Weinzwirner, Blutrote Zikade	*Tibicina haematodes/ Tibicen haematodes*	vineyard cicada
Laufflöter	Orthonychidae	logrunners, chowchilla
Laufhühnchen	*Turnix sylvatica*	little button-quail, Andalusian hemipode
Laufkäfer	Carabidae	ground beetles
Laufkuckucke, Spornkuckucke	Centropodidae	coucals
Laufmilben, Erntemilben, Herbstmilben	Trombiculidae	chigger mites, harvest mites
Laufspinnen	Philodromidae	running crab spiders, philodromids, philodromid spiders
Laufspringer	Entomobryidae	entomobryid springtails
Laufvögel, Strauße, Straußenvögel	Struthioniformes	ostriches
Läuse (Tierläuse)	Anoplura (Siphunculata)	sucking lice
Lausfliegen	Hippoboscidae	louseflies (forest flies & sheep keds)
Läuslinge, Kieferläuse	Mallophaga	chewing lice, biting lice, bird lice
Lavendelparadiesvogel	*Paradisaea decora*	Goldie's bird-of-paradise
Laxierfische, Schnauzenbrassen	Centracanthidae	picarels
Layard-Schnabelwal	*Mesoplodon layardi*	strap-toothed beaked whale, Layard's beaked whale
Laysan-Mönchsrobbe	*Monachus schauinslandi*	Hawaiian monk seal
Lazarusklappe, Lazarus-Schmuckkästchen, Stachelige Hufmuschel, Stachelige Gienmuschel	*Chama lazarus*	Lazarus jewel box
Lazulifink	*Passerina amoena*	lazuli bunting
Lear-Ara	*Anodorhynchus leari*	Lear's macaw, indigo macaw
Lebendgebärende Brotulas	Bythitidae	bythitids, viviparous brotulas
Lebendgebärende Kröten	*Nectophrynoides* spp.	live-bearing toads
Lebendgebärende Zahnkarpfen	Poeciliidae	poeciliids
Lebendgebärender Schnurwurm	*Prosorhochmus claparedi* (Nemertini)	
Leberegel> Afrikanischer Leberegel (Afrikanischer Lanzettegel)	*Dicrocoelium hospes*	African lancet fluke
Leberegel> Asiatischer Leberegel*	*Clonorchis sinensis*	Asian liver fluke
Leberegel> Großer Amerikanischer Leberegel*	*Fascioloides magna*	large American liver fluke

Leberegel> Großer Leberegel	*Fasciola hepatica*	sheep liver fluke
Leberegel> Kleiner Leberegel (Lanzettegel)	*Dicrocoelium dendriticum/ Dicrocoelium lanceolatum/ Fasciola lanceolatum*	lancet fluke, common lancet fluke
Leberegel> Riesenleberegel	*Fasciola gigantica*	giant liver fluke
Leberegelschnecke, Kleine Sumpfschnecke	*Galba truncatula*	dwarf pond snail, dwarf mud snail
Leder-Koffermuschel, Ledrige Dreiecksmuschel	*Hecuba scortum/ Donax scortum*	leather donax
Lederanemone	*Urticina coriacea*	leathery anemone, stubby rose anemone
Lederjacken, Einstachler	*Aluteridae*	leatherjackets
Lederlaufkäfer, Leder-Laufkäfer, Lederkäfer	*Carabus coriaceus*	leatherback ground beetle*
Lederrücken*	*Scomberoides lysan*	doublespotted queenfish (FAO), leatherback
Lederschildkröte	*Dermochelys coriacea*	leatherback sea turtle, leatherback, leathery turtle, luth turtle
Lederschwamm	*Cacospongia scalaris*	leather sponge
Lederseeigel	Echinothuridae	leather urchins*
Lederstern*	*Dermasterias imbricata*	leather star
Lederwanze, Saumwanze	*Coreus marginatus/ Mesocerus marginatus*	squash bug a.o.
Lederwanzen, Randwanzen	Coreidae	leaf-footed bugs, coreid bugs (squash bugs, squashbugs)
Lederzecken, Saumzecken	Argasidae	soft ticks, softbacked ticks
Ledrige Dreiecksmuschel, Leder-Koffermuschel	*Hecuba scortum/ Donax scortum*	leather donax
Lefevres Seegurke	*Aslia lefevrei*	brown sea cucumber
Leguane	Iguanidae	iguanas
Lehmwespen & Pillenwespen	Eumenidae	mason wasps, potter wasps
Leier-Herzmuschel	*Lyrocardium lyratum*	lyre cockle
Leierantilope	*Damaliscus lunatus*	topi, tsessebi
Leierfisch> Gemeiner Leierfisch, Gewöhnlicher Leierfisch, Europäischer Leierfisch, Gestreifter Leierfisch	*Callionymus lyra*	common dragonet
Leierfische (Spinnenfische)	Callionymidae	dragonets
Leierherzigel	*Brissopsis lyrifera*	lyriform heart-urchin, fiddle heart-urchin
Leierhirsch, Thamin	*Cervus eldi*	thamin, brow-antlered deer, Eld's deer
Leierkopfagame	*Lyriocephalus scutatus*	lyrehead lizard
Leiernatter	*Trimorphodon biscutatus*	lyre snake
Leierschwänze	Menuridae	lyrebirds
Leierschwanzwida	*Euplectes jacksoni*	Jackson's whydah

Leimschleuderspinne, Speispinne	*Scytodes thoracica*	spitting spider
Leimschleuderspinnen, Speispinnen	Scytodidae	spitting spiders
Leistenkrokodil	*Crocodylus porosus*	saltwater crocodile, estuarine crocodile
Leistenschnabeltukan	*Andigena laminirostris*	plate-billed mountain toucan
Leiterbock	*Saperda scalaris*	scalar longhorn beetle
Leman-Felchen	*Coregonus hiemalis*	Lake Geneva whitefish
Lemargo, Kleiner Eishai	*Somniosus rostratus*	little sleeper shark
Lemminge> Echte Lemminge	*Lemmus* spp.	true lemmings
Lemminge> Steppenlemminge	*Lagurus* spp.	sagebrush voles, steppe lemmings
Lemmingmäuse	*Synaptomys* spp.	bog lemmings
Lemuren-Greiffrosch	*Phyllomedusa lemur*	lemur leaf frog
Lemurenartige	Lemuridae	lemurs
Leng, Lengfisch	*Molva molva*	ling (FAO), European ling
Lenok	*Brachymystax lenok*	lenok
Leopard	*Panthera pardus*	leopard
Leopard-Buschfisch	*Ctenopoma acutirostre*	leopard bushfish
Leopard-Panzerwels	*Corydoras julii*	leopard corydoras
Leopard-Stechrochen	*Himantura uarnak*	honeycomb stingray (FAO), thornycomb stingray
Leopardbärbling	*Brachydanio frankei*	leopard danio
Leoparden-Drückerfisch	*Balistoides conspicillum*	big-spotted triggerfish, clown triggerfish (FAO)
Leoparden-Kegelschnecke	*Conus leopardus*	leopard cone
Leopardenbarsch	*Plectropomus leopardus*	leopard coral trout, leopard grouper, leopard coralgrouper (FAO)
Leopardenhai, Kalifornischer Leopardhai	*Triakis semifasciatus*	leopard shark
Leopardenhai, Zebrahai	*Stegostoma fasciatum*	zebra shark (FAO), leopard shark
Leopardfrosch	*Rana pipiens*	leopard frog, northern leopard frog, grass frog
Leopardgecko, Panthergecko	*Eublepharus macularius*	leopard gecko
Leopardgrundel	*Thorogobius ephippiatus/ Gobius forsteri*	leopard-spotted goby
Leopardkatze, Bengalkatze	*Felis bengalensis*	leopard cat
Leopardleguane	*Gambelia* spp.	leopard lizards
Leopardnatter	*Elaphe situla*	leopard snake
Leptochilichthyiden	Leptochilichthyidae	leptochilichthyids
Leptoniden	Leptonidae	coin shells
Lerchen	Alaudidae	larks
Lerchenlaufhühnchen	*Ortyxelos meiffrenii*	quail plover
Lerchenstärling	*Sturnella magna*	eastern meadowlark
Lesueur-Bürstenkänguru	*Bettongia lesueur*	boodie
Leuchtanemone	*Bartholomea lucida*	luminant anemone
Leuchtende Seefeder	*Pennatula phosphorea*	luminescent sea-pen

Leuchtender Erdläufer	Geophilus electricus	luminous centipede
Leuchtender Flugkalmar	Symplectoteuthis luminosa/ Eucleoteuthis luminosa	luminous flying squid, striped squid
Leuchtender Gabeldorsch	Steindachneria argentea	luminous hake
Leuchtfische	Photichthyidae/Phosichthyidae	lightfishes
Leuchtheringe	Searsiidae/Platytroctidae	tubeshoulders
Leuchtkäfer (Glühwürmchen, Johanniswürmchen)	Lampyridae	glowworms, fireflies, lightning "bugs"
Leuchtkäfer> Gemeiner Leuchtkäfer, Kleiner Leuchtkäfer, Johanniskäfer, Johanniswürmchen, Glühwürmchen	Lamprohiza splendidula	small lightning beetle
Leuchtkäferkalmar	Watasenia scintillans	lightningbug squid*
Leuchtkrebse & Krill	Euphausiidae	euphausiids, krill
Leuchtqualle	Pelagia noctiluca	phosphorescent jellyfish, purple jellyfish, pink jellyfish, night-light jellyfish
Leuchtschwanz-Meerkatze	Cercopithecus solatus	sun-tailed monkey
Leuchtstern-Kardinalfisch	Apogon lachneri	whitestar cardinalfish
Levante-Otter, Levanteotter	Daboia lebetina/ Vipera lebetina/ Macrovipera lebetina	blunt-nosed viper, bluntnose viper, levantine viper
Levante-Wühlmaus, Mittelmeer-Feldmaus	Microtus guentheri	Gunther's vole, Mediterranean vole
Leyte-Erdschildkröte	Heosemys leytensis	Leyte pond turtle
Libanesische Bergotter	Vipera bornmuelleri	Lebanon viper
Liberia-Kusimanse, Liberianische Manguste	Liberiictis kuhni	Liberian mongoose
Libyscher Steppenfuchs (Hbz.), Blassfuchs	Vulpes pallida	pale fox
Lichtensteins Hartebeest	Alcelaphus lichtensteini/ Sigmoceros lichtensteini	Lichtenstein's hartebeest
Lichtensteins Nachtotter	Causus lichtensteini	Lichtenstein's night adder
Lidmücken, Netzmücken	Blephariceridae/ Blepharoceridae/ Blepharoceratidae	net-winged midges
Liebliche Stachelschnecke	Chicoreus venustulus	lovely murex
Lienardos Scheinolive*	Ancilla lienardi	Lienardo's ancilla
Lieste, Jägerlieste	Dacelonidae	dacelonid kingfishers
Ligusterschwärmer	Sphinx ligustri	privet hawkmoth
Lila Aalgrundel	Gobioides broussoneti	violet goby
Lila Schwamm*	Haliclona permollis	purple encrusting sponge, purple sponge
Lila-Tüpfelgrundel	Mogurnda striata/ Mogurnda adspersa	trout gudgeon, purple-spotted gudgeon (FAO)
Lilagold-Feuerfalter	Palaeochrysophanus hippothoe/ Lycaena hippothoe	purple-edged copper
Lilastreifen-Quallen	Pelagia spp.	purplestriped jellyfishes
Lilienhähnchen	Lirioceris lilii	lily beetle
Lilienmoos*	Diphasia rosacea	lily hydroid

Limande, Echte Rotzunge	*Microstomus kitt*	lemon sole
Lincolnammer	*Melospiza lincolnii*	Lincoln's sparrow
Linden-Gallmücke	*Didymomyia tiliacea*	lime leaf gall midge
Linden-Sichelspinner	*Sabra harpagula*	scarce hook-tip (moth)
Lindenbock	*Oplosia fennica*	linden borer (longhorn beetle)
Lindeneule, Aprileule, Grüne Eicheneule	*Dichonia aprilina/ Griposia aprilina*	Merveille-du-Jour
Lindengallmilbe	*Eriophyes tiliae*	lime nail-gall mite (>lime nail gall)
Lindenschwärmer	*Mimas tiliae*	lime hawkmoth
Liniendornwels	*Pseudodoras niger/ Doras niger/ Oxydoras niger*	ripsaw catfish (FAO), black-shielded catfish
Linienkärpfling	*Jenynsia lineata*	onesided livebearer
Linienschwärmer	*Hyles lineata*	striped hawkmoth
Linierter Nutzholzborkenkäfer, Nadelholz-Ambrosiakäfer	*Xyloterus lineatus*	lineate bark beetle, striped ambrosia beetle, conifer ambrosia beetle
Liniierte Anemone*	*Fagesia lineata*	lined anemone
Liniierter Grashüpfer, Panzers Grashüpfer	*Stenobothrus lineatus*	stripe-winged grasshopper, lined grasshopper
Linkshänder-Einsiedlerkrebse	Diogenidae	left-handed hermit crabs
Linnés Zwergschlange	*Calamaria linnaei*	Linné's reed snake
Linsang> Afrika-Linsang, Pojana	*Poiana richardsoni*	African linsang, oyan
Linsang> Bänderlinsang	*Prionodon linsang*	banded linsang
Linsang> Fleckenlinsang	*Prionodon pardicolor*	spotted linsang
Linsenfloh	*Chydorus sphaericus*	
Linsenförmige Tellerschnecke	*Hippeutis complanatus*	flat ramshorn
Linsenkäfer	*Bruchus lentis*	lentil weevil
Lippen-Triton*	*Cymatium labiosum*	lip triton, wide-lipped triton
Lippenbär	*Melursus ursinus/Ursus ursinus*	sloth bear
Lippenfleck-Sandbarsch, Gelbfleck-Krokodilfisch	*Parapercis xanthozona*	spottyfin weever, whitebar weever, yellowbar sandperch (FAO)
Lippenmünder, Lappenmünder	Cheilostomata	cheilostomates
Lippennatter	*Fimbrios klossi*	bearded snake
Lippfische	Labridae	wrasses
Listers Tellmuschel	*Tellina listeri*	speckled tellin
Listige Manguste	*Dologale dybowskii*	African tropical savannah mongoose
Listspinne, Raubspinne	*Pisaura mirabilis*	fantastic fishing spider*
Lisztaffe, Lisztäffchen	*Saguinus oedipus*	cotton-top tamarin
Lithographische Kegelschnecke	*Conus lithoglyphus*	lithograph cone
Litschi-Wasserbock, Litschi-Moorantilope	*Kobus leche*	lechwe
Lobeliennektarvogel	*Nectarinia johnstoni*	red-tufted malachite sunbird
Löcherkrake	*Tremoctopus violaceus*	common blanket octopus, common umbrella octopus

Lodde	*Mallotus villosus*	capelin
Löffelente	*Anas clypeata*	northern shoveler
Löffelhund	*Otocyon megalotis*	bat-eared fox
Löffelstör	*Polyodon spathula*	paddlefish, Mississippi paddlefish (FAO)
Löffelstöre, Vielzähner, Schaufelrüssler	Polyodontidae	paddlefishes, spoonbills
Löffler	*Platalea leucorodia*	white spoonbill
Logbarsch, Schweinsbarsch (Manitou-Springbarsch)	*Percina caprodes caprodes*	logperch
Lönnbergmöve	*Larus relictus*	relict gull
Lorbeertaube	*Columba junoniae*	laurel pigeon
Loretosalmler	*Hyphessobrycon loretoensis*	Loreto tetra
Loriciferen, Korsetttierchen, Panzertierchen	Loricifera	corset bearers, loriciferans
Loris (Vögel)	Loriidae	lories, lorikeets
Loris (Primaten)	Lorisidae	lorises & pottos & galagos
Lotsenfisch	*Naucrates ductor*	pilot fish
Louisiana-Sumpfkrebs, Louisiana-Flusskrebs, Roter Sumpfkrebs	*Procambarus clarkii*	Louisiana red crayfish, red swamp crayfish, Louisiana swamp crayfish, red crayfish
Louisianawürger	*Lanius ludovicianus*	loggerhead shrike
Löwe	*Panthera leo*	lion
Löwenpranke	*Nodipecten nodosus/ Lyropecten nodosa*	lions-paw scallop, lion's paw
Lowes Alfonsino	*Beryx splendens*	Lowe's beryx, splendid alfonsino (FAO)
Lowes Meerkatze	*Cercopithecus lowei*	Lowe's monkey
Luchs> Nordluchs (Hbz. Luchs)	*Felis lynx/ Lynx lynx*	lynx
Luchsfliegen, Stilettfliegen	Therevidae	stiletto flies
Luchsspinnen, Scharfaugenspinnen	Oxyopidae	lynx spiders
Luftröhrenwürmer	*Syngamus* spp.	gapeworms
Lummensturmvögel, Tauchersturmvögel	Pelecanoididae	diving petrels
Lump, Lumpfisch, Seehase	*Cyclopterus lumpus*	lumpsucker (FAO), lumpfish, hen-fish, henfish, sea hen
Lumpfische, Seehasen	Cyclopteridae	lumpfishes (lumpsuckers)
Lungenegel, Ostasiatischer Lungenegel	*Paragonimus westermani*	human lung fluke
Lungenenzian-Ameisen-Bläuling, Kleiner Moorbläuling	*Maculinea alcon*	alcon blue
Lungenfische	Dipnoi	lungfishes
Lungenlose Salamander	Plethodontidae	lungless salamanders

Lungenquallen, Blumenkohlquallen	*Rhizostoma* spp.	cabbage bleb, marigold, blubber, football jellyfish
Lurche, Amphibien	Amphibia	amphibians
Lurchfische	Ceratodontidae	Australian lungfish
Lusitanien-Kleinwühlmaus	*Pitymys lusitanicus*	Lusitanian pine vole
Lusitanischer Kuhnasenrochen	*Rhinoptera marginata*	Lusitanian cownose ray
Lusitanischer Schlingerhai	*Centrophorus lusitanicus*	lowfin gulper shark
Luzerneälchen, Stengelälchen, Stockälchen (Rübenkopfälchen)	*Ditylenchus dipsaci*	lucerne stem nematode, stem-and-bulb eelworm, stem and bulb nematode, stem nematode, bulb nematode (potato tuber eelworm)
Luzerneblatt-Gallmücke	*Jaapiella medicaginis*	lucerne leaf midge
Luzerneblattnager	*Phytonomus variabilis/ Hypera postica*	lucerne weevil
Luzerneblüten-Gallmücke	*Contarinia medicaginis*	lucerne flower midge
Luzernefloh	*Sminthurus viridis*	lucerne flea
Luzernen-Blattschneiderbiene, Alfalfa-Biene	*Megachile rotundata*	alfalfa leafcutter bee
Luzernenmarienkäfer	*Subcoccinella vigintiquatuorpunctata*	twentyfour-spot ladybird, 24-spot ladybird
Luzerneule, Braune Tageule	*Ectypa glyphica/ Euclidia glyphica*	burnet companion
Luzon-Nasenratten	*Rhynchomys* spp.	Luzon shrewlike rats
Luzon-Ratten	*Carpomys* spp.	Luzon rats
Luzon-Spitzmausratte*	*Celaenomys silaceus*	Luzon shrew-rat
Luzon-Streifenratte	*Chrotomys whiteheadi*	Luzon striped rat
Luzon-Waldratte	*Batomys dentatus*	Luzon forest rat
Lynxkauri	*Cypraea lynx/Lyncina lynx*	lynx cowrie, bobcat cowrie
Lyra-Grundel	*Evorthodus lyricus/ Mugilostoma gobio*	lyre goby
Lyraskunk (Hbz.), Fleckenskunks	*Spilogale* spp.	spotted skunks

Macklots Python	*Liasis mackloti/ Morelia mackloti*	water python, Macklot's python
Madagaskar-Boa	*Acrantophis madagascariensis/ Boa madagascariensis*	Madagascan boa
Madagaskar-Braunnatter	*Leioheterodon modestus*	Madagascar brown snake
Madagaskar-Hundskopfboa	*Sanzinia madagascariensis*	Madagascar tree boa
Madagaskar-Inselratten u.a.	*Brachyuromys* spp.	Madagascan rats a.o.
Madagaskar-Leguan u.a.	*Chalarodon madagascariensis*	Madagascar iguana
Madagaskar-Leguane u.a.	*Oplurus* spp.	Madagascar swifts
Madagaskar-Spornschildkröte, Madagassische Schnabelbrustschildkröte	*Asterochelys yniphora/ Geochelone yniphora/ Testudo yniphora*	Madagascan spurred tortoise, angonoka
Madagaskar-Strahlenschildkröte, Madagassische Strahlenschildkröte	*Asterochelys radiata/ Geochelone radiata/ Testudo radiata*	radiated tortoise
Madagaskar-Zwergchamäleon	*Brookesia tuberculata*	Mount d'Ambre leaf chameleon
Madagaskarseeadler	*Haliaeetus vociferoides*	Madagascar fish eagle
Madagaskarweber	*Foudia madagascariensis*	Madagascan red fody
Madagassische Flachrückenschildkröte	*Pyxis planicauda*	Madagascar flat-shelled tortoise (IUCN), flat-backed tortoise
Madagassische Haftscheiben-Fledermaus	*Myzopoda aurita*	Old World sucker-footed bat
Madagassische Schienenschildkröte	*Erymnochelys madagascariensis*	Madagascan big-headed side-necked turtle, Madagascan big-headed turtle (IUCN)
Madagassische Strahlenschildkröte	*Geochelone radiata/ Testudo radiata/ Asterochelys radiata*	radiated tortoise
Madagassisches Erdchamäleon, Madagassisches Zwergchamäleon	*Brookesia minima*	minute leaf chameleon, dwarf chameleon
Madagaskar-Spornschildkröte, Madagassische Schnabelbrustschildkröte	*Geochelone yniphora/ Testudo yniphora/ Asterochelys yniphora*	Madagascan spurred tortoise, northern Madagascar spur tortoise, angonoka, Madagascar tortoise (IUCN)
Madeira-Grundel	*Mauligobius maderensis*	Madeiran goby
Madeira-Mauereidechse	*Podarcis dugesii*	Madeira wall lizard
Madeira-Sardinelle	*Sardinella madarensis*	Madeiran sardinella (FAO), short-body sardinella, short-bodied sardine
Madeira-Schabe	*Leucophaea maderae*	Madeira cockroach
Madeirasturmvogel	*Pterodroma madeira*	freira
Madeirawellenläufer	*Oceanodroma castro*	Madeiran petrel
Madenhacker	*Buphaginae*	oxpeckers
Madenkuckucke, Guira-Kuckuck	*Crotophaginae*	anis, Guira cuckoo
Madenwurm (Springwurm, Pfriemenschwanz)	*Enterobius vermicularis*	pinworm (of man), seatworm
Mädesüß-Perlmutterfalter, Violetter Silberfalter	*Brenthis ino*	lesser marbled fritillary
Magdalena-Ratte	*Xenomys nelsoni*	Magdalena rat

Magellan-Dampfschiffente	*Tachyeres pteneres*	flightless steamer duck
Magellan-Miesmuschel	*Aulacomya ater*	Magellan mussel, black-ribbed mussel
Magellan-Triton	*Fusitriton magellanicum*	Magellanic triton
Magellandrossel	*Turdus falcklandii*	austral thrush
Magellangans	*Chloephaga picta*	Magellan goose
Magellanpinguin	*Spheniscus magellanicus*	Magellanic penguin
Magenbrüterfrosch	*Rheobatrachus silus*	gastric-brooding frog, platypus frog
Magendasseln, Magenbremsen (Dasselfliegen)	Gasterophilidae	botflies, bot flies (horse botflies)
Magendasseln & Rachendasseln & Nasenbremsen & Biesfliegen	Oestridae	warble flies
Magenwürmer u.a.	*Physaloptera* spp.	stomach worms
Magenzystenwürmer	*Gnathostoma* spp.	stomach cyst worms, stomach cyst nematodes (of swine)
Magerrasen-Hauhechelspanner	*Aplasta ononaria*	rest harrow
Magerrasen-Perlmuttfalter, Hainveilchen-Perlmutterfalter, Kleinster Perlmutterfalter	*Boloria dia/ Clossiana dia*	violet fritillary, Weaver's fritillary
Mahagoni-Muschel*	*Nuttallia nuttallii*	California mahogany-clam, mahogany clam
Mahagoni-Schnapper	*Lutjanus mahogoni*	mahogany snapper
Mähnenhirsch	*Cervus timorensis*	Sunda sambar
Mähnenratte	*Lophiomys imhausi*	maned rat, crested rat
Mähnenrobbe	*Otaria flavescens*	southern sea lion, South American sea lion
Mähnenspringer	*Ammotragus lervia*	barbary sheep, aoudad
Mähnenwolf	*Chrysocyon brachyurus*	maned wolf
Maifisch, Alse, Gewöhnliche Alse	*Alosa alosa*	Allis shad
Maiglöckchenlaus	*Aulacorthum speyeri*	lily-of-the-valley aphid
Maikäfer, Feldmaikäfer	*Melolontha melolontha*	common cockchafer, maybug
Mairenke, Schemaja	*Chalcalburnus chalcoides*	Danubian bleak, Danube bleak (FAO), shemaya
Mais-Schnellkäfer*	*Melanotus communis*	corn wireworm, corn click beetle
Maisblattlaus	*Rhopalosiphum maidis*	corn-leaf aphid, corn aphid, cereal leaf aphid
Maiseule, Maiseulenfalter*, Maiskolbenbohrer*, Maismotte*	*Helicoverpa zea/ Heliothis zea*	corn earworm, bollworm, tomato fruitworm
Maiskäfer	*Sitophilus zeamais*	corn weevil
Maiszünsler, Hirsezünsler	*Ostrinia nubilalis/ Pyrausta nubilalis*	European corn borer

Makaken	*Macaca* spp.	macaques
Makibären	*Bassaricyon* spp.	olingos
Makifrösche, Greiffrösche, Affenfrösche	*Phyllomedusa* spp.	leaf frogs
Makrele, Europäische Makrele, Atlantische Makrele	*Scomber scombrus*	Atlantic mackerel (FAO), common mackerel
Makrelen (und Thunfische)	Scombridae	mackerels (incl. tunas & bonitos)
Makrelenhaie	Isuridae	mackerel sharks
Makrelenhecht, Atlantischer Makrelenhecht	*Scomberesox saurus*	saury, Atlantic saury (FAO), saury pike, skipper
Makrelenhechte	Scomberesocidae	sauries
Makropode, Paradiesfisch	*Macropodus opercularis*	paradise fish
Malabar-Zackenbarsch, Malabar-Riffbarsch	*Epinephelus malabaricus/ Epinephelus salmonoides*	Malabar grouper (FAO), Malabar reefcod
Malabarbärbling, Königsbärbling	*Danio aequipinnatus/ Brachydanio aequipinnatus*	giant danio
Malabarhornvogel	*Anthracoceros coronatus*	Malabar pied hornbill
Malaien-Gleitflieger, Temminck-Gleitflieger	*Cynocephalus variegatus*	Malayan flying lemur
Malaien-Mokassinschlange, Malaiische Grubenotter	*Agkistrodon rhodostoma/ Calloselasma rhodostoma*	Malayan pit viper
Malaien-Sumpfschildkröte	*Malayemys subtrijuga*	Malayan snail-eating turtle
Malaienadler	*Ictinaetus malayensis*	Indian black eagle
Malaienbär, Sonnenbär	*Ursus malayanus/ Helarctos malayanus*	sun bear, Malayan sun bear
Malaienwiesel	*Mustela nudipes*	Malayan weasel
Malaiische Baumratte	*Pithecheir melanurus*	monkey-footed rat, Malayan tree rat
Malaiische Dornschildkröte	*Cyclemys dentata*	Asian leaf turtle
Malaiische Weichschildkröte, Malaien-Weichschildkröte	*Dogania subplana/ Trionyx subplanus*	Malayan softshell turtle
Malaiischer Bungar, Malaiischer Krait	*Bungarus candidus*	Malayan krait, blue krait
Malaiischer Pinselstachler, Pinselstachler	*Trichys fasciculata*	long-tailed porcupine
Malariamücken, Fiebermücken, Gabelmücken	*Anopheles* spp.	malaria mosquitoes
Malediven-Kegelschnecke	*Conus maldivus*	Maldive cone
Malaienmausspecht	*Sasia abnormis*	rufous piculet
Malermuschel	*Unio pictorum/ Pollicepes pictorum*	painter's mussel
Malmignatte, Schwarze Witwe, Europäische Schwarze Witwe	*Latrodectus mactans tredecimguttatus*	European black widow, southern black widow
Malta-Eidechse	*Podarcis filfolensis*	Maltese wall lizard
Malven-Dickkopffalter	*Carcharodus alceae*	mallow skipper
Malven-Würfelfleckfalter, Kleiner Malvendickkopf	*Pyrgus malvae*	grizzled skipper

Mamba> Blattgrüne Mamba, Gewöhnliche Mamba	*Dendroaspis angusticeps*	eastern green mamba
Mamba> Grüne Mamba	*Dendroaspis viridis*	western green mamba
Mamba> Schwarze Mamba	*Dendroaspis polylepis*	black mamba
Mambas	*Dendroaspis* spp.	mambas
Mampalon	*Cynogale bennettii*	otter civet
Mamushi, Japanische Grubenotter	*Agkistrodon blomhoffi/ Gloydius blomhoffi*	Japanese mamushi
Mandarinente	*Aix galericulata*	mandarin duck, mandarin
Mandarinfische	*Synchiropus* spp.	mandarinfishes
Mandrill	*Mandrillus sphinx/Papio sphinx*	mandrill
Mandschurenkranich	*Grus japonensis*	red-crowned crane, Japanese crane, Manchurian crane
Mangaben	*Cercocebus* spp.	mangabeys
Mangolderdfloh, Rübenerdfloh	*Chaetocnema concinna*	mangold flea beetle, mangel flea beetle, beet flea beetle
Mangrovekrabbe, Mangroven-Baumkrabbe	*Aratus pisonii*	mangrove tree crab, mangrove crab
Mangroven-Lanzenotter	*Trimeresurus purpureomaculatus*	mangrove viper
Mangroven-Nachtbaumnatter	*Boiga dendrophila*	mangrove snake
Mangroven-Winkerkrabbe	*Uca crassipes*	mangrove fiddler crab
Mangroven-Wurzelkrabbe	*Goniopsis cruentata*	Mangrove root crab
Mangrovenbarsch, Mangroven-Schnapper	*Lutjanus argentimaculatus*	river snapper, mangrove red snapper (FAO)
Mangrovenqualle	*Cassiopeia xamachana*	mangrove upsidedown jellyfish
Mangrovenskink	*Emoia atrocostata*	marine skink
Mangrovereiher	*Butorides striatus*	green-backed heron, green heron
Manilagrundel	*Ophiocara aporos*	snake-headed gudgeon
Manipur-Ratte	*Hadromys humei*	Manipur bush rat
Manitoba-Kröte	*Bufo hemiophrys*	Canadian toad
Manna-Schildlaus	*Trabutina mannipara*	tamarisk manna scale
Manta, Riesenmanta, Großer Teufelsrochen	*Manta birostris*	manta, Atlantic manta, giant devil ray, giant manta (FAO)
Mantas, Teufelsrochen	Mobulidae	mantas (manta rays & devil rays)
Mantel-Kammmuschel	*Gloripallium pallium*	mantle scallop
Mantel-Mangabe	*Cercocebus albigena*	gray-cheeked mangabey, white-cheeked mangabey
Mantelaffe, Manteläffchen	*Saguinus bicolor*	bare-faced tamarin, pied tamarin
Mantelaktinie	*Adamsia palliata/ Adamsia carciniopados*	cloak anemone (hermit crab anemone a.o.)
Mantelbrüllaffe	*Alouatta palliata*	mantled howler
Mantelmöve	*Larus marinus*	great black-backed gull
Mantelpavian	*Papio hamadryas*	hamadryas baboon, sacred baboon
Manteltiere, Tunicaten (Urochordaten)	Tunicata	tunicates
Manul	*Felis manul*	Pallas's cat

Maorischlüpfer, Neuseelandpittas	Xenicidae/ Acanthisittidae	New Zealand wrens
Marabu	Leptoptilos crumeniferus	marabou stork
Maränen, Felchen u.a.	Leucichthys spp.	ciscoes
Maränen, Renken u.a.	Coregonus spp.	whitefishes, lake whitefishes
Maras	Dolichotis spp.	Patagonian cavies, Patagonian "hares", maras
Marbeleule, Binsengraseule, Finkeneule, Waldmoorrasen-Türkeneule, Rotgefranste Weißpunkteule	Mythimna turca	double-lined wainscot, double line moth
Marder-Kegelschnecke	Conus mustelinus	weasel cone
Marderhund, Enok (Hbz. Seefuchs)	Nyctereutes procyonoides	raccoon dog
Margareta-Ratte	Margaretamys spp.	Margareta's rats
Marienbarsch	Tilapia mariae	West African tilapia
Marienkäfer	Coccinellidae	ladybirds, ladybird beetles, lady beetles, "ladybugs"
Marienkäferspinne*	Eresus niger	ladybird spider
Marienprachtkäfer, Großer Kiefernprachtkäfer	Chalcophora mariana	European sculptured pine borer
Mariskensänger	Acrocephalus melanopogon	moustached warbler
Markeule, Schachtelhalmeule, Kartoffelbohrer, Rübenbohrer	Hydraecia micacea	rosy rustic (moth)
Markkäfer, Kiefernmarkkäfer, Großer Waldgärtner	Tomicus piniperda	large pine shoot beetle
Markusfliege	Bibio marci	St.Mark's fly
Marmelente	Marmoronetta angustirostris	marbled teal
Marmor-Bohnenmuschel*	Modiolaria tumida	marbled crenella
Marmor-Eidechsenfisch	Saurida gracilis	graceful lizardfish
Marmor-Käferschnecke, Marmorchiton	Chiton marmoratus	marbled chiton
Marmor-Seewalze	Bohadschia marmorata	marbled holothurian
Marmorbarsch	Notothenia rossii	marbled notothenia, marbled rockcod (FAO)
Marmorbrassen	Lithognathus mormyrus/ Pagellus mormyrus	marmora, striped seabream (FAO)
Marmorfische	Aplodactylidae	marblefishes
Marmorgesichttamarin	Saguinus inustus	mottle-faced tamarin
Marmorgrundel, Marmor-Schläfergrundel	Oxyeleotris marmorata/ Eleotris marmorata	marble goby (FAO)
Marmorierte Bohnenmuschel	Musculus marmoratus	spotted mussel, marbled mussel, marbled musculus
Marmorierte Grundel u.a.	Pomatoschistus marmoratus	marbled goby (FAO)
Marmorierte Grundel u.a.	Proterorhinus marmoratus/ Gobius marmoratus	tubenose goby (FAO)

Marmorierte Kreuzspinne	*Araneus marmoreus*	marbled orbweaver, marbled spider
Marmorierte Schwimmkrabbe	*Liocarcinus marmoreus*	marbled swimming crab
Marmorierter Bachsalamander	*Leurognathus marmoratus*	marbled salamander
Marmorierter Beilbauchfisch	*Carnegiella strigata*	marbled hatchetfish
Marmorierter Blattfingergecko	*Phyllodactylus marmoratus*	Australian marbled gecko
Marmorierter Buntleguan	*Polychrus marmoratus*	many-colored bush anole
Marmorierter Kaninchenfisch	*Siganus argenteus*	roman-nose spinefoot, streamlined spinefoot (FAO)
Marmorierter Salamander	*Aneides ferreus*	clouded salamander
Marmorierter Schleimfisch	*Coryphoblennius galerita*	Montagu's blenny
Marmorierter Zackenbarsch	*Epinephelus microdon*	mottled grouper, camouflage grouper (FAO)
Marmorierter Zitterrochen, Marmor-Zitterrochen, Marmelzitterrochen	*Torpedo marmorata*	marbled crampfish, marbled torpedo ray, marbled electric ray (FAO)
Marmorierter Zwergkalmar, Mittelländischer Zwergkalmar	*Alloteuthis media*	marbled little squid
Marmorkarpfen, Edler Tolstolob	*Hypophthalmichthys nobilis/ Aristichthys nobilis*	bighead carp
Marmorkatze	*Felis marmorata/ Pardofelis marmorata*	marbled cat
Marmorkegel	*Conus marmoreus*	marble cone, marbled cone
Marmormolch	*Triturus marmoratus*	marbled newt
Marmorquerzahnmolch	*Ambystoma opacum*	marbled salamander
Marmosetten, Seidenäffchen	*Callithrix* spp.	short-tusked marmosets, titis
Marnock-Frosch, Felsen-Zwitscherfrosch*	*Syrrhophus marnockii/ Eleutherodactylus marnockii*	cliff frog, cliff chirping frog
Marokkanische Wanderheuschrecke	*Dociotaurus maroccanus*	Moroccan locust
Marokko-Brassen, Marokko-Zahnbrassen	*Dentex maroccanus*	Morocco dentex
Maronenlangur	*Presbytis rubicunda*	maroon leaf monkey
Maronenstirnsittich	*Rhynchopsitta terrisi*	maroon-fronted parrot
Maroni, Schlüsselloch-Buntbarsch	*Cleithracara maronii*	keyhole cichlid
Marschenschnecke, Kegelige Strandschnecke	*Assiminea grayana*	dun sentinel
Marschzecke*	*Dermacentor reticulatus*	marsh tick
Martinique Baum-Vogelspinne	*Avicularia versicolor*	Antilles pinktoe tarantula
Martinique-Lanzenotter	*Bothrops lanceolatus*	fer-de-lance, Martinique lancehead
Märzveilchenfalter, Bergmatten-Perlmutterfalter, Hundsveilchen-Perlmutterfalter	*Fabriciana adippe/ Argynnis adippe*	high brown fritillary
Maß-Hecht	*Esox vermiculatus*	grass pickerel
Maß-Schnecken	Modulidae	modulus snails

Masken-Doktorfisch	*Naso vlamingi*	bignose unicornfish (FAO), zebra unicornfish, Vlaming's unicornfish
Masken-Papageifisch	*Scarus tricolor*	tricolour parrotfish (FAO)
Masken-Sägebarsch	*Serranus phoebe*	tattler
Masken-Springaffe	*Callicebus personatus*	masked titi
Maskenammer	*Emberiza spodocephala*	black-faced bunting
Maskenbienen	*Hylaeus* spp.	yellow-faced bees
Maskeneule	*Phodilus badius*	bay owl
Maskenfisch, Maskenwimpelfisch, Halfterfisch	*Zanclus cornutus/ Zanclus canescens/ Chaetodon canescens*	Moorish idol
Maskengrasmücke	*Sylvia rueppelli*	Rueppell's warbler
Maskenköpfchen, Schwarzköpfchen	*Agapornis personatus*	masked lovebird
Maskenkrabbe, Antennenkrebs	*Corystes cassivelaunus*	masked crab, helmet crab
Maskenkrabben	Corystidae	helmet crabs
Maskenläuse, Maskenblattläuse	Thelaxidae	thelaxid plantlice
Maskenleguan	*Leiocephalus personatus*	Haitian curlytail lizard
Maskenmonarch	*Monarcha melanopsis*	pearly-winged monarch
Maskenschnecke	*Isognomostoma isognomostoma*	
Maskenschwalbenstar	*Artamus personatus*	masked wood swallow
Maskentitya	*Tityra semifasciata*	masked titya
Maskentölpel	*Sula dactylatra*	blue-faced booby
Maskenwürger	*Lanius nubicus*	masked shrike
Maskierter Strolch, Staubwanze, Kotwanze	*Reduvius personatus*	masked hunter bug (fly bug)
Massasauga, Kettenklapperschlange	*Sistrurus catenatus*	massasauga
Masu-Lachs	*Oncorhynchus masou*	masu salmon, cherry salmon (FAO)
Matamata, Fransenschildkröte	*Chelus fimbriatus*	matamata
Matten-Krustenanemone	*Zoanthus pulchellus*	mat anemone
Mattscheckiger Braun-Dickkopffalter	*Thymelicus acteon*	Lulworth skipper
Mattschwarzer Getreideschimmelkäfer	*Alphitobius laevigatus*	black fungus beetle (lesser mealworm beetle)
Mauerassel, Gemeine Mauerassel	*Oniscus asellus*	common woodlouse, common sowbug, grey garden woodlouse
Mauerbienen	*Osmia* spp.	mason bees
Mauereidechse	*Podarcis muralis/ Lacerta muralis*	common wall lizard
Mauerfuchs	*Lasiommata megera*	wall, wall brown
Mauergecko> Gewöhnlicher Mauergecko	*Tarentola mauritanica*	common wall gecko, Moorish gecko
Mauergeckos	*Tarentola* spp.	wall geckos
Mauerläufer	Tichodrominae	wallcreeper
Mauerläufer	*Tichodroma muraria*	wallcreeper
Mauermilbe	*Balaustium murorum*	wall mite*

Mauersegler	*Apus apus*	swift
Mauerseglerlausfliege	*Crataerina pallida*	swift parasitic fly, swallow parasitic fly
Mauerspinne	*Dictyna civica*	dictynid spider
Maulbeerspinner, Echter Seidenspinner	*Bombyx mori*	silkworm moth
Maulbrütender Kampffisch	*Betta pugnax*	mouthbrooding fighting fish, Penang betta (FAO)
Maulbrüterwelse, Meereswelse, Kreuzwelse	Tachysuridae/Ariidae	sea catfishes
Maulfüßer, Fangschreckenkrebse	Stomatopoda	mantis shrimps
Maulwurf-Querzahnmolch	*Ambystoma talpoideum*	mole salamander
Maulwurf> Europäischer Maulwurf	*Talpa europaea*	European mole
Maulwürfe	Talipidae	moles & shrew moles & desmans
Maulwürfe> Altweltmaulwürfe	*Talpa* spp.	Old World moles
Maulwurfsgrille, Werre	*Gryllotalpa gryllotalpa*	mole cricket
Maulwurfsgrillen	Gryllotalpidae	mole crickets
Maulwurfskauri	*Cypraea talpa/Talparia talpa*	mole cowrie
Maulwurfskink	*Eumeces egregius*	mole skink
Maulwurfskrebse	Upogebiidae	mud shrimps, mudprawns
Maulwurfsmäuse*	*Notiomys* spp.	long-clawed mice, mole mice
Maulwurfsnatter	*Pseudaspis cana*	mole snake, molslang
Mauretanische Languste	*Palinurus mauritanicus*	pink spiny lobster
Maurische Bachschildkröte	*Mauremys leprosa*	Maurish turtle, Mediterranean turtle
Maurische Landschildkröte	*Testudo graeca*	spur-thighed tortoise, Mediterranean spur-thighed tortoise, common tortoise, Greek tortoise
Maurische Netzwühle, Maurische Ringelwühle (Ringelschleiche)	*Blanus cinereus*	Mediterranean worm lizard
Maurischer Skorpion	*Scorpio maurus*	Maurish scorpion
Mauritische Bullia	*Bullia mauritiana*	Mauritian bullia
Mauritiusfalke	*Falco punctatus*	Mauritius kestrel
Mauritiussittich	*Psittacula echo*	Mauritius parakeet
Mausartiger Zwerghamster, Mäuseartiger Zwerghamster	*Calomyscus bailwardi*	mouselike hamster
Mäuschen, Mausspanner	*Minoa murinata*	drab looper (moth)
Mäuse> Echte Mäuse	*Mus* spp.	mice
Mäuseartige, Echte Mäuse, Langschwanzmäuse	Muridae	mice family
Mäusebussard	*Buteo buteo*	common buzzard
Mäusefloh	*Leptinus testaceus*	mouse flea
Mäusekegel	*Conus mus*	mouse cone
Mäuseverwandte	Myomorpha	ratlike rodents
Mauskauri, Mausschnecke	*Cypraea mus*	mouse cowrie
Mausmaki, Gewöhnlicher Mausmaki	*Microcebus murinus*	mouse lemur, grey mouse lemur
Mausohr-Fledermäuse	*Myotis* spp.	little brown bats

Mausohrspanner	*Idaea aversata*	riband wave
Mausschläfer	*Myomimus* spp.	mouse-tailed dormice, mouselike dormice
Mausschwanz-Fledermäuse, Klappnasen	*Rhinopoma* spp.	mouse-tailed bats, long-tailed bats
Mausspanner, Mäuschen	*Minoa murinata*	drab looper (moth)
Mausvögel, Buschkletterer	Coliiformes/Coliidae	mousebirds, colies
Mauswiesel	*Mustela nivalis*	least weasel
Maximilianpapagei	*Pionus maximiliani*	scaly-headed parrot
McMahon-Viper	*Eristicophis macmahoni*	McMahon's viper, leaf-nosed viper
Mearns Luzon-Ratte	*Tryphomys adustus*	Mearns' Luzon rat
Medaka, Japankärpfling, Japanischer Goldhecht	*Oryzias latipes*	medaka
Medinawurm, Guineawurm, Drachenwurm	*Dracunculus medinensis*	fiery serpent, medina worm, guinea worm
Medusenhäupter	Gorgonocephalidae	medusa-head stars, basket stars
Medusenwurm, Spaghetti-Wurm	*Loimia medusa*	medusa worm*
Meeraal, Gemeiner Meeraal, Seeaal, Congeraal	*Conger conger*	conger eel, European conger
Meeraale	Congridae	conger eels, congers
Meeräsche> Gemeine Meeräsche, Flachköpfige Meeräsche, Großkopfmeeräsche	*Mugil cephalus*	striped gray mullet, striped mullet, common grey mullet, flat-headed grey mullet, flathead mullet (FAO)
Meeräschen	Mugilidae	mullets, grey mullets
Meerbarbe, Gewöhnliche Meerbarbe, Rote Meerbarbe	*Mullus barbatus*	red mullet (FAO), plain red mullet
Meerbarben	Mullidae	goatfishes
Meerbarbenkönig	*Apogon imberbis*	cardinalfish, cardinal fish (FAO)
Meerbart*	*Nemertesia antennina*	sea beard
Meerdrossel, Grüner Lippfisch, Amsellippfisch	*Labrus viridis*	green wrasse
Meerechse, Galapagos-Meerechse	*Ambylyrhynchus cristatus*	marine iguana, Galapagos marine iguana
Meerengel, Engelhaie	Squatinidae	angelsharks (sand devils)
Meeres-Einsiedlerkrebse, Rechtshänder-Einsiedlerkrebse	Paguridae	right-handed hermit crabs
Meeres-Rollassel	*Sphaeroma hookeri*	marine pillbug a.o.
Meeresdatteln, Seedatteln	*Lithophaga* spp.	datemussels, date mussels
Meeresmilben	Halacaridae	sea mites

Meerespfirsiche	*Halocynthia* spp.	sea peaches
Meeresrose	*Hydatina physis*	brown-line paperbubble, green-lined paperbubble, green paperbubble
Meeresschildkröten	Cheloniidae	sea turtles
Meeresspinnen*	Desidae	marine spiders
Meeresstachelbeeren, Seestachelbeeren	*Pleurobrachia* spp.	sea gooseberries, cat's eyes
Meereswelse, Kreuzwelse, Maulbrüterwelse	Ariidae/ Tachysuridae	sea catfishes
Meerfeige (Goldenes Meerbiskuit)	*Polyclinum aurantium/ Aplidium ficus*	golden sea biscuit
Meerforelle, Lachsforelle	*Salmo trutta trutta*	sea trout
Meergürkchen, Wimper-Kalkschwamm, Wimpern-Kalkschwamm	*Scypha ciliata/ Sycon ciliatum*	little vase sponge
Meerhand, Tote Manneshand, Nordische Korkkoralle (Lederkoralle)	*Alcyonium digitatum*	dead-man's fingers, sea-fingers
Meerjunker, Pfauenfederfisch	*Coris julis*	rainbow wrasse
Meerkatzen	*Cercopithecus* spp.	guenons
Meerkatzenverwandte	Cercopithecidae	Old World monkeys
Meermandel, Englisches Pastetchen, Archenkammmuschel, Gemeine Samtmuschel, Mandelmuschel	*Glycymeris glycymeris*	dog cockle, orbicular ark (comb-shell)
Meernematode	*Enoplus meridionalis*	
Meerneunauge	*Petromyzon marinus*	sea lamprey
Meerohren, Seeohren, Abalones	*Haliotis* spp.	abalones (U.S.), ormers (Br.)
Meerorange, Apfelsinenschwamm, Kugelschwamm	*Tethya aurantia/ Tethya lyncurium*	orange puffball sponge, golfball sponge
Meerpfau	*Thalassoma pavo*	Turkish wrasse, peacock wrasse, sea-peacock, rainbow wrasse, ornate wrasse (FAO)
Meerrabe, Seerabe, Rabenfisch	*Sciaena umbra/ Johnius umbra*	brown meagre
Meerrose	*Pentapora foliacea*	rose 'coral', rose-coral bryozoan, sea rose
Meersau-Haie, Meersäue, Rauhaie, Schweinshaie	Oxynotidae	prickly dogfishes, oxynotids
Meerschlei, Pfauenlippfisch	*Crenilabrus pavo/ Symphodus tinca*	peacock wrasse
Meerschweinchen	*Cavia* spp.	cavies, guinea pigs
Meerschweinchen	Caviidae	cavies & Patagonian "hares"

Meerstrandläufer	Calidris maritima	purple sandpiper
Meertanne, Tannenmoos, Moostanne	Abietinaria abietina	sea fir
Meertäschchen*	Scypha compressa	purse sponge
Meertrauben*	Molgula spp.	sea grapes
Meerwarze	Verruca stroemia	sea wart*
Meerzahn	Antalis dentalis (Scaphopoda)	European tusk
Meerzander	Stizostedion marina	sea zander
Meerzitrone, Warzige Sternschnecke	Archidoris pseudoargus/ Archidoris tuberculata	sea lemon
Mehely-Hufeisennase	Myotis mehelyi/ Rhinolophus mehelyi	Mehely's horseshoe bat
Mehlige Apfellaus, Mehlige Apfelblattlaus	Dysaphis plantaginea	rosy apple aphid
Mehlige Birnblattlaus	Dysaphis pyri	mealy pear aphid, pear-bedstraw aphid
Mehlige Gewächshausschildlaus, Langdornige Schmierlaus	Pseudococcus longispinus	longtailed mealybug
Mehlige Kohlblattlaus	Brevicoryne brassicae	cabbage aphid, mealy cabbage aphid
Mehlige Zwetschgenblattlaus, Pflaumenblattlaus, Hopfenblattlaus, Hopfenlaus	Hyalopterus pruni	damson-hop aphid, hop aphid
Mehlkäfer	Tenebrio molitor	yellow mealworm beetle
Mehlkäfer (Larve: Mehlwurm)	Tenebrio spp.	mealworm beetle
Mehlmilbe (>Bäckerkrätze)	Acarus siro/ Tyroglyphus farinae	flour mite, grain mite
Mehlmotte	Ephestia kuehniella/ Anagasta kuehniella	Mediterranean flour moth
Mehlschwalbe	Delichon urbica	common house martin
Mehlzünsler	Pyralis farinalis	meal moth
Mehrstreifen-Skink	Eumeces multivirgatus	many-lined skink
Mehrteilige Seescheiden	Aplousobranchia	aplousobranchs
Meisen	Paridae (> Parus spp.)	titmice, chickadees (Am.)
Meisensauger	Neottiophilidae	neottiophilids (nestling bird suckers)
Meisenwaldsänger	Parula americana	northern parula
Melanoniden	Melanonidae	melanonids, pelagic cods
Meldenwanzen, Rübenwanzen	Piesmatidae	beet bugs, ash-gray leaf bugs, piesmatids
Mellers Chamäleon	Chamaeleo melleri	Meller's chameleon
Melomys-Ratten	Melomys spp.	mosaic-tailed rats, banana rats
Melonen-Seeigel	Echinus melo	melon urchin
Melonenkopf	Peponocephala electra	many-toothed blackfish, melon-headed whale, small melon-headed blackfish

Melonenqualle, Gurkenqualle	*Beroe cucumis*	melon jellyfish, melon comb jelly
Melonenquallen, Gurkenquallen	*Beroe* spp.	beroids, melon jellyfishes
Melonenquallen, Mützenquallen	Beroida	beroids, melon jellyfishes
Melvera-Fregattmakrele, Unechter Bonito	*Auxis rochei*	bullet tuna (FAO), bullet mackerel
Melyriden	Melyridae	soft-winged flower beetles
Membran-Geißelgarnele	*Solenocera membranacea*	mud shrimp
Menarana Madagaskar-Natter	*Leioheterodon madagascariensis*	Madagaskar Menarana snake
Mendesantilope	*Addax nasomaculatus*	addax
Mensch	*Homo sapiens sapiens*	people, human beings, humans
Mensch> Heidelbergmensch	*Homo heidelbergensis* (*Homo erectus heidelbergensis*)	Heidelberg man
Mensch> Neanderthalmensch	*Homo sapiens neanderthalensis*	Neanderthals, Neanderthal people, Neanderthal man
Menschenaffen	Pongidae	great apes, pongids
Menschenartige	Hominoidea	apes, anthropoid apes
Menschendasselfliege	*Dermatobia hominis*	human bot fly, torsalo
Menschenfloh	*Pulex irritans*	human flea
Menschenflöhe	Pulicidae	common fleas, pulicid fleas
Menschenherz, Ochsenherz	*Glossus humanus*	ox heart, heart cockle
Menschenläuse	Pediculidae	human lice
Mentawai-Gibbon, Zwergsiamang, Biloh	*Hylobates klossii/ Symphalangus klossi*	Kloss's gibbon, dwarf siamang, Mentawai gibbon, beeloh
Mentawailangur	*Presbytis potenziani*	Mentawai leaf monkey
Merlin	*Falco columbarius*	merlin
Mertens Anemone, Riesenanemone, Nessellose Riesenanemone	*Stichodactyla gigantea/ Stichodactyla mertensi/ Stoichactis giganteum*	giant carpet anemone, giant anemone
Mertens Waran	*Varanus mertensi*	Mertens' water monitor
Mesquite-Stachelschuppenleguan	*Sceloporus grammicus*	mesquite lizard
Messeraale> Echte Messeraale, Gestreifte Messeraale	Gymnotidae	naked-back knifefishes
Messerfisch, Butterfisch	*Pholis gunnellus*	butterfish, gunnel
Messerfische, Eigentliche Messerfische	Notopteridae	featherfin knifefishes, Old World knifefishes
Messerfuß	*Pelobates cultripes*	western European spadefoot, Iberian spadefoot
Messergarnele	*Hymenopenaeus triarthrus/ Haliporoides triarthrus*	knife shrimp
Messerkieferfische	Oplegnathidae	knifejaws
Messerzahn-Tiefen-Dornhai	*Scymnodon ringens*	knifetooth dogfish

Messerzahnaale	Muraenescocidae	pike congers, pike eels
Messingbarbe	Barbus semifasciolatus	green barb, China barb, Chinese barb (FAO)
Messingeule	Diachrysia chrysitis	burnished brass
Messingfarbener Weidenblattkäfer	Phytodecta vitellinae	brassy willow leaf beetle
Messingkäfer	Niptus hololeucus	golden spider beetle
Metall-Panzerwels	Corydoras aeneus	bronze corydoras (FAO), aeneus catfish
Metallblauer Spitzmausrüssler	Apion pomonae	vetch seed weevil, tare seed weevil
Metallfarbener Distelbock	Agapanthia violacea	metallic thistle longhorn beetle*
Metallkärpfling	Girardinus metallicus	girardinus, metallic livebearer (FAO)
Mexikanerzeisig	Carduelis psaltria/ Spinus psaltria	lesser goldfinch
Mexikanische Blond-Vogelspinne	Aphonopelma chalcodes	Mexican blond tarantula
Mexikanische Flecken-Erdschildkröte, Rückenflecken-Erdschildkröte	Rhinoclemmys rubida	Mexican spotted wood turtle
Mexikanische Gopherschildkröte	Gopherus flavomarginatus	Mexican gopher tortoise, Bolson tortoise (IUCN)
Mexikanische Hakennasen-Natter	Gyalopion canum	western hooknose snake
Mexikanische Hautwühlen	Dermophis spp.	Mexican caecilians
Mexikanische Mokassinschlange	Agkistrodon bilineatus	cantil, mocassin
Mexikanische Moschusschildkröte	Kinosternon integrum	Mexican musk turtle
Mexikanische Pygmäensalamander	Thorius spp.	pygmy salamanders
Mexikanische Schlangenechse	Anelytropsis papillosus	Mexican blind lizard
Mexikanische Schlankblindschlange	Leptotyphlops humilis	Mexican blindsnake
Mexikanische Trichterohren-Fledermaus	Natalus stramineus	Mexican funnel-eared bat
Mexikanische Vulkanmaus	Neotomodon alstoni	Mexican volcano mouse
Mexikanische Winkerkrabbe, Kalifornische Winkerkrabbe	Uca crenulata	Mexican fiddler, California fiddler
Mexikanische Zwergklapperschlange	Sistrurus ravus	Mexican pigmy rattlesnake
Mexikanischer Getreidekäfer	Pharaxonota kirschi	Mexican grain beetle
Mexikanischer Karmingimpel, Hausgimpel	Carpodacus mexicanus	house finch, California linnet
Mexikanischer Klippenfrosch	Hylactophryne augusti/ Eleutherodactylus augusti	common robber frog, barking frog
Mexikanischer Präriehund	Cynomys mexicanus	Mexican prairie dog
Mexiko-Salamander, Mexikosalamander	Pseudoeurycea spp.	false brook salamanders

Meyers Falterfisch, Gebänderter Falterfisch	*Chaetodon meyeri*	Meyer's butterflyfish, maypole butterflyfish, scrawled butterflyfish (FAO)
Mi-Eule	*Callistege mi*	Mother Shipton
Micromalthiden	Micromalthidae	telephone-pole beetles
Midasohr, Eselohr des Midas	*Ellobium aurismidae*	Midas ear cassidula (snail)
Miesmuscheln	Mytiloidea/ Mytilidae	mussels
Mietenlaus, Tulpenblattlaus	*Rhopalosiphoninus tulipaellus*	tulip aphid, iris aphid
Mikroporenkoralle	*Montipora tuberculosa*	microporous coral
Milben & Zecken	Acari/Acarina	mites & ticks
Milchfische	Chanidae	milkfish
Milchfischverwandte, Sandfische	Gonorhynchiformes	milkfishes & relatives
Milchfleck, Weißband-Mohrenfalter	*Erebia ligea*	arran brown (ringlet butterfly)
Milchweiße Archenmuschel	*Striarca lactea*	milky-white ark
Millard-Ratte	*Dacnomys millardi*	large-toothed giant rat, Millard's rat
Millardia-Ratten, Millard-Mäuse	*Millardia* spp.	Asian soft-furred rats, cutch rats
Millers Grundel	*Millerigobius macrocephalus*	Miller's goby
Millionenfisch, Guppy	*Lebistes reticulatus/ Poecilia reticulata*	guppy
Milos-Eidechse	*Podarcis milensis*	Milos wall lizard
Mimikry-Hechtschleimfisch	*Hemiemblemaria similus*	wrasse blenny
Mindanao-Ratte u.a.	*Limnomys sibuanus*	Mindanao rat (Mount Apo)
Mindanao-Waldratte	*Batomys salomonseni*	Mindanao forest rat
Mindoro-Büffel, Tamarau	*Bubalus mindorensis*	tamaraw, taumarau
Mindoro-Krokodil	*Crocodylus mindorensis*	Philippine crocodile, Mindoro crocodile
Mindoro-Ratte	*Anonymomys mindorensis*	Mindoro rat
Miniatur-Archenmuschel	*Acar bailyi/ Barbatia bailyi*	miniature ark
Minierfliegen	Agromyzidae	leaf miner flies
Miniersackmotten	Incurvariidae	yucca moths & fairy moth & others
Miniervogelspinnen, Falltürspinnen	Ctenizidae	trapdoor spiders
Minzen-Blattkäfer	*Chrysolina menthastri*	mint leaf beetle
Minzen-Erdfloh*	*Longitarsus ferrugineus*	mint flea beetle
Minzenbär, Weiße Tigermotte, Punktierter Fleckleib-Bär	*Spilosoma lubricipeda/ Spilosoma menthastri*	white ermine moth
Mirakelbarsche, Rundköpfe	Plesiopidae	longfins, roundheads
Mississippi-Alligator	*Alligator mississippiensis*	American alligator
Mississippi-Barsch	*Morone mississippiensis*	yellow bass
Mississippi-Knochenhecht	*Atractosteus spatula/ Lepisosteus spatula*	alligator gar

Mississippi-Landkartenschildkröte, Mississippi-Höckerschildkröte	*Graptemys kohnii*	Mississippi map turtle
Mistbiene (Rattenschwanzlarve), Schlammfliege, Große Bienenschwebfliege	*Eristalis tenax*	drone fly (rattailed maggot)
Misteldrossel	*Turdus viscivorus*	mistle thrush
Mistelfresser, Blütenpicker	Dicaeidae	flowerpeckers
Mistkäfer	*Geotrupes stercorarius*	common dor beetle
Mistkäfer	Geotrupidae	dung beetles
Mistkäfer, Rosskäfer	*Geotrupes* spp.	dor beetles
Mistwurm, Dungwurm, Kompostwurm	*Eisenia fetida/ Eisenia foetida*	brandling, manure worm
Mittagsfliege	*Mesembrina meridiana*	noon-fly, noonfly
Mittel-Indische Dachschildkröte	*Kachuga tentoria*	Mid-Indian tent turtle
Mittelafrikanische Klappen-Weichschildkröten	*Cyclanorbis* spp.	sub-Saharan flapshells
Mittelamerikanische Baummäuse u.a.	*Tylomys* spp.	Central American climbing rats
Mittelamerikanische Zwergboas	*Ungaliophis* spp.	Central American dwarf boas, banana boas
Mittelamerikanisches Aguti	*Dasyprocta punctata*	Central American agouti
Mittelamerikanisches Katzenfrett	*Bassariscus sumichrasti*	Central American cacomistle
Mittelasiatische Kobra	*Naja oxiana*	Central Asiatic cobra
Mittelhamster	*Mesocricetus* spp.	golden hamsters
Mittelkudu, Bergnyala	*Tragelaphus buxtoni*	mountain nyala
Mittelmeer-Abalone, Mittelländisches Seeohr	*Haliotis lamellosa*	Mediterranean ormer, common ormer
Mittelmeer-Ackerschnecke, Kastanienschnecke*	*Deroceras panormitanum/ Deroceras carnanae*	longneck fieldslug (U.S.), chestnut slug (Br.), Carnana's slug, Sicilian slug
Mittelmeer-Barrakuda, Quergestreifter Pfeilhecht	*Sphyraena sphyraena*	Mediterranean barracuda, European barracuda (FAO)
Mittelmeer-Bastardmakrele, Mittelmeer-Stöcker	*Trachurus mediterraneus*	Mediterranean horse-mackerel
Mittelmeer-Chiton, Bunte Käferschnecke	*Chiton olivaceus*	variable chiton, Mediterranean chiton
Mittelmeer-Diademseeigel, Langstachliger Seeigel	*Centrostephanus longispinus*	Mediterranean hatpin urchin
Mittelmeer-Dornschrecke	*Paratettix meridionalis*	Mediterranean groundhopper
Mittelmeer-Feldgrille	*Gryllus bimaculatus*	two-spotted cricket, Mediterranean field cricket
Mittelmeer-Felsenkrabbe	*Brachynotus sexdentatus*	Mediterranean crab, Mediterranean rock crab
Mittelmeer-Fruchtfliege	*Ceratitis capitata*	Mediterranean fruit fly
Mittelmeer-Gabeldorsch, Südliche Meerschleie, Kleiner Gabeldorsch	*Phycis phycis*	forkbeard (FAO), lesser hake

Mittelmeer-Gespenstkrabbe, Anemonen-Gespenstkrabbe	*Inachus phalangium*	Leach's spider crab, Mediterranean spider crab
Mittelmeer-Glatthai, Schwarzgefleckter Glatthai, Schwarzpunkt-Glatthai	*Mustelus punctulatus/ Mustelus mediterraneus*	blackspotted smooth-hound
Mittelmeer-Gorgonenhaupt	*Astrospartus mediterraneus*	Mediterranean basket star
Mittelmeer-Gurkenqualle, Mittelmeer Melonenqualle	*Beroe ovata*	ovate comb jelly, ovate beroid
Mittelmeer-Haarstern	*Antedon mediterranea*	orange-red feather star
Mittelmeer-Kleinwühlmaus	*Pitymys duodecimcostatus*	Mediterranean pine vole
Mittelmeer-Knallgarnele, Mittelmeer-Pistolenkrebs, Knallkrebschen	*Alpheus dentipes*	Mediterranean snapping prawn
Mittelmeer-Kugelkrabbe, Mittelmeer-Nusskrabbe	*Ilia nucleus*	leucosian nut crab, Mediterranean nut crab
Mittelmeer-Laubfrosch	*Hyla meridionalis*	stripeless treefrog, Mediterranean treefrog
Mittelmeer-Leng, Spanischer Leng	*Molva dipterygia macrophthalma/ Molva macrophthalma/ Molva elongata*	Mediterranean ling, Spanish ling (FAO)
Mittelmeer-Lippfisch	*Crenilabrus mediterraneus/ Symphodus mediterraneus*	axillary wrasse
Mittelmeer-Maulwurfskrebs	*Upogebia pusilla*	Mediterranean mudprawn, Mediterranean mud lobster
Mittelmeer-Miesmuschel, Blaubartmuschel, Seemuschel	*Mytilus galloprovincialis*	Mediterranean mussel
Mittelmeer-Mönchsrobbe	*Monachus monachus*	Mediterranean monk seal
Mittelmeer-Muräne	*Muraena helena*	Mediterranean moray (FAO), European moray
Mittelmeer-Nacktsandaal, Mittelmeer-Sandaal	*Gymnammodytes cicerelus/ Ammodytes cicerelus*	Mediterranean sand lance, Mediterranean sandeel (FAO), Mediterranean smooth sandeel
Mittelmeer-Pantoffelschnecke	*Crepidula gibbosa*	Mediterranean slippersnail, Mediterranean slipper limpet
Mittelmeer-Putzergarnele	*Lysmata seticaudata*	Mediterranean cleaner shrimp, Mediterranean rock shrimp, Monaco cleaner shrimp
Mittelmeer-Raurochen	*Raja radula*	Mediterranean rough ray, rough ray (FAO)
Mittelmeer-Schamkrabbe, Hahnenkammkrabbe	*Calappa granulata*	Mediterranean shame-faced crab
Mittelmeer-Schlangenaal, Langmaul-Schlangenaal	*Ophisurus serpens*	long-jawed snake eel
Mittelmeer-Seequappe	*Gaidropsarus mediterraneus*	shore rockling
Mittelmeer-Sepiole, Zwerg-Sepia, Zwergtintenfisch, Kleine Sprutte	*Sepiola rondeleti*	Mediterranean dwarf cuttlefish, lesser cuttlefish, dwarf bobtail squid (FAO)
Mittelmeer-Silberdorsch	*Gadiculus argenteus thori*	Mediterranean silvery pout
Mittelmeer-Speerfisch, Langschwänziger Speerfisch	*Tetrapturus belone*	Mediterranean spearfish (FAO), Mediterranean marlin
Mittelmeer-Steinschmätzer	*Oenanthe hispanica*	black-eared wheatear
Mittelmeer-Sternrochen	*Raja asterias*	starry ray
Mittelmeer-Stöcker, Mittelmeer-Bastardmakrele	*Trachurus mediterraneus*	Mediterranean horse-mackerel

Mittelmeer-Strandkrabbe	*Carcinus aestuarii/ Carcinus mediterraneus*	Mediterranean green crab
Mittelmeer-Sturmtaucher	*Puffinus yelkouan*	Yelkouan shearwater
Mittelmeer-Tellmuschel	*Angulus planatus*	Mediterranean tellin
Mittelmeer-Würfelqualle, Mittelmeer Seewespe	*Carybdea marsupialis*	Mediterranean seawasp
Mittelmeer-Zylinderrose	*Cerianthus membranaceus*	Mediterranean cerianthid
Mittelmeer-Zylinderschwamm, Rosa Röhrenschwamm	*Haliclona mediterranea*	Mediterranean tube sponge
Mittelmeerhufeisennase, Mittelmeer-Hufeisennase	*Rhinolophus euryale*	Mediterranean horseshoe bat
Mittelmeerkärpfling, Salinenkärpfling (Zebrakärpfling)	*Aphanius fasciatus*	Mediterranean toothcarp
Mittelmeerkegel	*Conus mediterraneus*	Mediterranean cone
Mittelmeersandaal, Mittelmeer-Nacktsandaal	*Ammodytes cicerelus/ Gymnammodytes cicerelus*	Mediterranean sandlance, Mediterranean smooth sandeel, Mediterranean sandeel (FAO)
Mittelmeertermite, Erdholztermite	*Reticulitermes lucifugus*	common European white ant, common European termite
Mittelsäger	*Mergus serrator*	red-breasted merganser
Mittelspecht	*Dendrocopos medius/ Picoides medius*	middle spotted woodpecker
Mitteltiere, Gewebetiere, „Vielzeller"	Metazoa	metazoans
Mitternachtswels, Zamorawels	*Auchenipterichthys thoracatus*	Zamora catfish
Mittlere Achatschnecke, Mittlere Glattschnecke	*Cochlicopa repentina/ Cionella repentina*	intermediate moss snail, intermediate pillar snail
Mittlere Hornisse	*Dolichovespula media*	medium wasp*
Mittlerer Gleithörnchenbeutler	*Petaurus norfolcensis*	squirrel glider
Mittlerer Katzenmaki, Fettschwanzmaki	*Cheirogaleus medius*	lesser dwarf lemur, fat-tailed dwarf lemur
Mittlerer Rindermagennematode, Mittlerer Rindermagenwurm	*Ostertagia ostertagi*	medium stomach worm (of cattle), brown stomach worm
Mittlerer Weinschwärmer	*Deilephila elpenor*	elephant hawkmoth
Mittleres Jungfernkind	*Archiearis notha*	light orange underwing
Mitu	*Mitu mitu*	Alagoas curassow, razor-billed curassow
Moa, Riesenmoa	*Dinornis maximus*	giant moa
Moabsperling	*Passer moabiticus*	Dead Sea sparrow
Moas	Dinornithiformes	moas
Möbelschabe, Braunbandschabe	*Supella supellectilium/ Supella longipalpa*	brown-banded cockroach, brownbanded cockroach, furniture cockroach
Moderholz> Gemeines Moderholz, Graue Moderholzeule	*Xylena exsoleta*	sword-grass moth
Moderkäfer	Lathridiidae	plaster beetles, minute brown scavenger beetles
Moderlieschen	*Leucaspius delineatus*	moderlieschen, belica (FAO), sunbleak

Moderne Knochenfische	Teleostei	teleosts, teleost fishes
Möhrenblattlaus, Mehlige Möhrenlaus	*Semiaphis dauci*	carrot aphid
Möhrenblattsauger	*Trioza apicalis*	carrot sucker (psyllid)
Mohrenfalter	*Erebia* spp.	ringlets (butterflies)
Möhrenfliege, Rüeblifliege	*Psila rosae*	carrot rust fly, carrot fly
Mohrenkaiman	*Melanosuchus niger*	black caiman
Mohrenmakak	*Macaca maura*	moor macaque
Mohrenmaki	*Lemur macao/Petterus macaco*	black lemur
Möhrenschabe, Kümmelmotte, Kümmelpfeifer	*Depressaria nervosa*	carrot and parsnip flat-body moth
Mohrrüben-Kegelschnecke*, Karotten-Kegel	*Conus daucus*	carrot cone
Moilanatter	*Malpolon moilensis*	moila snake
Mojarras	Gerreidae/Gerridae	pursemouths, mojarras
Mojave-Fransenzehenleguan	*Uma scoparia*	Mojave fringe-toed lizard
Mojave-Klapperschlange	*Crotalus scutulatus*	Mojave rattlesnake
Mokassinschlange, Wassermokassinschlange	*Agkistrodon piscivorus*	cottonmouth, water moccasin
Molchfische	Lepidosirenidae	South American lungfish
Molinasittich	*Pyrrhura molinae*	green-cheeked conure
Moloney-Flachkopf-Fledermaus	*Mimetillus moloneyi*	Moloney's flat-headed bat, narrow-winged bat
Molukkenkakadu	*Cacatua moluccensis*	salmon-crested cockatoo
Mombasa-Vogelspinne*	*Pterinochilus murinus*	Mombasa golden starburst tarantula
Monameerkatze	*Cercopithecus mona*	mona monkey
Monarchen	Monarchidae	monarchs, magpie-larks
Monarchfalter	*Danaus plexippus*	monarch, milkweed
Monasteria-Mikroporenkoralle	*Montipora monasteriata*	Monasteria microporous coral*
Mönch, Klosterfrau	*Panthea coenobita*	pine arches
Mönchsfisch	*Chromis chromis*	damsel fish, damselfish (FAO)
Mönchsgeier	*Aegypius monachus*	black vulture
Mönchsgrasmücke	*Sylvia atricapilla*	blackcap
Mönchsittich	*Myiopsitta monachus*	monk parakeet
Mönchsrobben	*Monachus* spp.	monk seals
Moncktons Schwimmratte	*Crossomys moncktoni*	earless water rat
Mond-Azurjungfer	*Coenagrion lunulatum*	Irish damselfly, lunular damselfly*
Mond-Falterfisch	*Chaetodon lunula*	raccoon butterflyfish (FAO), red-striped butterflyfish
Mondährenfisch, Gezeiten-Ährenfisch	*Menidia menidia*	Atlantic silverside
Mondauge	*Hiodon tergisus*	mooneye
Mondfisch	*Mola mola*	ocean sunfish
Mondfisch> Schlanker Mondfisch, Stutzmondfisch, Schwimmender Kopf, Langer Mondfisch	*Ranzania laevis* (*Ranzania truncata/ Ranzania typus*)	sunfish, truncated sunfish, slender sunfish (FAO)

Mondfische (Sonnenfische), Klumpfische	Molidae	ocean sunfishes
Mondfische*	Menidae	moonfish
Mondfleckspanner	*Selenia tetralunaria*	purple thorn
Mondhornkäfer	*Copris lunaris*	tumblebug, English scarab
Mondmuscheln	Lucinidae	lucines (hatchet shells)
Mondnatter	*Oxyrhopus trigeminus*	Brazilian false coral snake
Mondplaty, Platy, Spiegelkärpfling	*Xiphophorus maculatus*	southern platyfish
Mondschein-Anemone*	*Anthopleura artemisia*	moonglow anemone
Mondscheinfadenfisch, Mondschein-Gurami	*Trichogaster microlepis*	moonlight gourami (FAO), moonbeam gourami
Mondsichelbarsch	*Variola albimarginata*	white-margined lunartail rockcod, crescent-tailed grouper, white-edged lyretail (FAO)
Mondsicheljunker	*Thalassoma lunare*	moon wrasse (FAO), green wrasse, lyretail wrasse
Mondspanner	*Selenia lunularia*	lunar thorn
Mondspinner	*Actias luna*	luna moth
Mondspinner	*Actias* spp.	luna moths
Mondvogel, Mondfleck	*Phalera bucephala*	buff-tip moth
Mongoleigazelle	*Procapra gutturosa*	zeren, Mongolian gazelle
Mongolen-Rennmaus	*Meriones unguiculatus*	Mongolian gerbil, clawed jird
Mongolenregenpfeifer	*Charadrius mongolus*	lesser sand plover
Mongolischer Wüstenrenner	*Eremias argus*	Mongolia racerunner
Mongozmaki	*Lemur mongoz/ Petterus mongoz/ Eulemur mongoz*	mongoose lemur
Monrovia-Doktorfisch	*Acanthurus monroviae*	Monrovian surgeonfish
Monrovia-Hechtling, Querbandhechtling, Rotkehlhechtling	*Epiplatys dageti monroviae*	firemouth epiplatys
Monterey-Makrele	*Scomberomorus concolor*	Monterey Spanish mackerel
Montezuma-Schwertträger	*Xiphophorus montezumae*	Montezuma swordtail
Montezumawachtel	*Cyrtonyx montezumae*	Montezuma's quail
Moor-Jagdspinne	*Dolomedes plantarius*	fen raft spider, great raft spider
Moor-Wegschnecke	*Arion brunneus*	bog slug, bog arion
Moorameise	*Formica candica*	bog ant
Moorantilope, Kob-Antilope	*Kobus kob*	kob
Moorbläuling, Lungenenzian-Ameisen-Bläuling	*Maculinea alcon*	alcon blue
Moorbunteule	*Anarta cordigera*	small dark yellow underwing
Moorente	*Aythya nyroca*	ferruginous duck, white-eyed pochard
Moorfrosch	*Rana arvalis*	moor frog
Moorschneehuhn	*Lagopus lagopus*	willow grouse

Moorspitzmaus, Nördliche Wasserspitzmaus	*Sorex palustris*	northern water shrew
Moorspringer	*Isotomurus palustris*	marsh springtail
Moorweichkäfer, Dascilliden	Dascillidae	soft-bodied plant beetles
Moosblasenschnecke	*Aplexa hypnorum*	moss bladder snail
Moosgrüner Eulenspinner	*Polyploca ridens*	frosted green (moth)
Mooshummel	*Bombus muscorum*	moss carder bee
Moosjungfern	*Leucorrhinia* spp.	white-faced darters
Moosknopfkäfer	*Atomaria linearis*	pygmy mangold beetle
Moosmilben & Hornmilben	Orbatida	orbatid mites (beetle mites & moss mites)
Moosmücken	Cylindrotomidae	moss craneflies*
Moosschimmelkäfer	Dasyceridae	minute scavenger beetles
Moosskorpion	*Neobisium carcinoides*	
Moostierchen, Bryozoen	Ectoprocta/ Polyzoa (Bryozoa)	moss animals, bryozoans
Mopsfledermaus	*Barbastella barbastellus*	barbastelle
Mopsfledermäuse	*Barbastella* spp.	barbastelles
Mördermuschel, Große Riesenmuschel	*Tridacna gigas*	giant clam, killer clam
Mordspanner	*Crocallis elinguaria*	scalloped oak
Morgenammer	*Zonotrichia capensis*	rufous-collared sparrow
Mornellregenpfeifer	*Charadrius morinellus*	dotterel
Mörtelbiene	*Megachile parietina/ Chalicodoma parietina/ Chalicodoma muraria*	wall bee, mason bee
Mörtelgrabwespen	*Sceliphron* spp.	mud daubers
Morwongs	Cheilodactylidae	fingerfins, morwongs
Mosaikfadenfisch	*Trichogaster leeri*	pearl gourami
Mosaikkegel	*Conus tessulatus*	tessellate cone
Mosaikschwanz-Riesenratten	*Uromys* spp.	giant naked-tailed rats
Mosambik-Buntbarsch	*Tilapia mossambica/ Oreochromis mossambicus*	Mozambique tilapia (FAO), Mozambique mouthbreeder
Mosambik-Speikobra	*Naja mosambica*	Mozambique spitting cobra
Moschusbock	*Aromia moschata*	musk beetle
Moschusente	*Cairina moschata*	muscovy duck
Moschuskrake, Moschuspolyp	*Eledone moschata/ Ozeana moschata*	white octopus
Moschusochse	*Ovibos moschatus*	muskox
Moschusrattenkänguru	*Hypsiprymnodon moschatus*	musky "rat"-kangaroo
Moschusschildkröte> Gewöhnliche Moschusschildkröte	*Kinosternon odoratum/ Sternotherus odoratus*	common musk turtle, stinkpot
Moschusschildkröten	*Sternotherus* spp.	musk turtles
Moschusspitzmaus, Asiatische Moschusspitzmaus	*Suncus murinus*	musk shrew, Indian musk shrew
Moschusspitzmäuse, Dickschwanzspitzmäuse	*Suncus* spp.	musk shrews, pygmy shrews, dwarf shrews
Moschustiere, Moschushirsche	*Moschus* spp.	musk deer
Moskitofisch, Koboldkärpfling, Gambuse	*Gambusia affinis*	mosquito fish, mosquitofish (FAO)

Moskitos, Stechmücken	Culicidae	mosquitoes, gnats (Br.)
Mosor-Eidechse	*Lacerta mosorensis*	Mosor rock lizard
Motmots, Sägeracken	Momotidae	motmots
Motten	Heterocera	moths
Mottenmilbe	*Pyemotes herfsi*	moth mite*
Mottenschildläuse (Mottenläuse, Weiße Fliegen)	Aleyrodidae	whiteflies
Mount Apo Ratte	*Tarsomys apoensis*	Mount Apo rat
Mount Lyell-Höhlensalamander	*Hydromantes platycephalus*	Mount Lyell salamander
Mount Roraima Maus	*Podoxymys roraimae*	Mount Roraima mouse
Möwen	Larinae	gulls
Möwen & Seeschwalben	Laridae	gulls & terns
Möwenvögel & Watvögel & Alken	Charadriiformes	gulls & shorebirds & auks
Mozambik-Buntbarsch	*Oreochromis mossambicus*	Mozambique tilapia
Mozambikgirlitz	*Serinus mozambicus*	yellow-fronted canary
Mücken & Schnaken	Nematocera (Diptera)	mosquitoes
Mückenfänger	Polioptilidae	gnatcatchers, gnatwrens
Mückenfresser, Mückenfänger	Conopophagidae	gnateaters
Mückenhafte	Bittacidae	hanging scorpionflies, hangingflies
Muellers Erdviper	*Micrelaps muelleri*	Mueller's ground-viper
Muffelwild, Mufflon	*Ovis musimon*	mouflon
Mühlenberg-Schildkröte, Amerikanische Hochmoor-Schildkröte	*Clemmys muhlenbergii*	bog turtle, Muhlenberg's turtle
Mühlsteinsalmler, Riesenpacu, Gamitana-Scheibensalmler	*Piaractus brachypomus/ Colossoma bidens*	pirapatinga (FAO), cachama
Mulga-Schlange	*Pseudechis australis*	Australian mulga snake
Mull-Lemminge	*Ellobius* spp.	mole-voles, mole-lemmings
Müllers Schlange	*Rhinoplocephalus bicolor*	Müller's snake
Müllers Zwergbutt, Haarbutt	*Zeugopterus punctatus*	topknot
Mulmbock	*Ergates faber*	carpenter longhorn
Mulmkäfer	Aderidae/ Hylophilidae	aderid beetles, hylophilid beetles
Mulmkäfer	Cerophytidae	rare click beetles
Mungos> Echte Mungos	*Herpestes* spp.	mongooses
Muntjak> Riesenmuntjak	*Megamuntiacus vuquagensis*	giant muntjac
Muntjaks	*Muntiacus* spp.	muntjacs, barking deer
Münzen-Eule	*Polychrysia moneta*	golden plusia
Münzennatter	*Coluber nummifer*	coin snake

Muränen, Muränenaale	Muraenidae/ Heteromyridae	moray eels, morays
Muriki	*Brachyteles arachnoides*	woolly spider monkey, muriqui
Murmel (Hbz.)> Alpenmurmeltier & Bobak	*Marmota marmota* & *Marmota bobak*	alpine marmot & bobac marmot
Murmeltiere	*Marmota* spp.	marmots
Murray Steinkrabbe, Subantarktische Steinkrabbe	*Lithodes murrayi*	Murray king crab, Subantarctic stone crab
Muscheldarmkrebs	*Mytilicola intestinalis*	mussel intestinal crab
Muschelkrebse, Ostracoden	Ostracoda	ostracods (shell-covered crustaceans), seed shrimps
Muschelmilben	*Unionicola* spp.	mussel mites
Muscheln	Bivalvia/ Pelecypoda/ Lamellibranchiata	bivalves, pelecypods, "hatchet-footed animals" (clams: *sedimentary*, mussels: *freely exposed*)
Muschelsammlerin, Sandröhrenwurm, Bäumchenröhrenwurm	*Lanice conchilega*	sand mason
Muschelschaler	Conchostraca	clam shrimps
Muschelwächter, Erbsenkrabben	Pinnotheridae	pea crabs, commensal crabs
Muschelwächterkrabbe, Muschelwächter, Erbsenkrabbe	*Pinnotheres pisum*	pea crab, Linnaeus's pea crab
Museumskäfer	*Anthrenus museorum*	museum beetle
Museumsmilbe	*Thyreophagus entomophagus*	museum mite, flour mite
Muskelflosser, Fleischflosser	Sarcopterygii/ Choanichthyes	fleshy-finned fishes, sarcopterygians, sarcopts
Muskellunge	*Esox masquinongy*	muskellunge
Mussurana	*Clelia clelia*	mussurana
Mützenlangur	*Presbytis comata*/ *Presbytis aygula*	grizzled leaf monkey, Javan leaf monkey, Sunda leaf monkey, Sunda Islands leaf monkey
Mützenqualle	*Beroe mitrata*	sea mitre
Mützenschnecken, Haubenschnecken	Calyptraeidae/ Crepidulidae	cup-and-saucer limpets, slipper limpets, slipper shells

Nachtaffe	*Aotus trivirgatus*	douroucouli, night monkey
Nachtaktive Hydromeduse	*Olindias phosphorica*	iridescent limnomedusian
Nachtbaumnattern	*Boiga* spp.	tree snakes
Nachtechsen	Xantusiidae	night lizards
Nachtgarnelen*	Processidae	night shrimps
Nachtigall	*Luscinia megarhynchos*	nightingale
Nachtigall-Grashüpfer	*Chorthippus biguttulus*	bow-winged grasshopper
Nachtkerzenschwärmer	*Proserpinus proserpina*	willowherb hawkmoth
Nachtottern, Krötenottern	*Causus* spp.	night adders
Nachtreiher	*Nycticorax nycticorax*	night heron
Nachtsalmler	*Prochilodus insignis/ Semiprochilodus theraponura*	night characid, kissing prochilodus
Nachtsalmler, Schwanzstreifensalmler	*Semaprochilodus taeniurus*	colored prochilodus
Nachtschattenfresser, Seidenwürger	*Hypocolius ampelinus*	grey hypocolius
Nachtschlangen	*Hypsiglena* spp.	night snakes
Nachtschwalben	Caprimulgidae	nightjars & nighthawks
Nachtschwalbenschwanz, Holunderspanner	*Ourapteryx sambucaria*	swallow-tailed moth
Nachtskink u.a.	*Scincopus fasciatus*	Peters' banded skink
Nachtskink u.a.	*Egernia striata*	elliptical-eye skink
Nachtstint	*Spirinchus starksi*	night smelt
Nackendorn- Weichschildkröte, Kehllappen- Weichschildkröte	*Trionyx steindachneri/ Palea steindachneri*	wattle-necked softshell turtle
Nackenstachler	*Acanthosaura lepidogaster*	brown pricklenape
Nacktamöben	Gymnamoebia	naked amebas
Nacktaugenkakadu	*Cacatua sanguinea*	little corella
Nacktbrustkänguru	*Caloprymnus campestris*	desert "rat"-kangaroo
Nackte Stachelauster	*Spondylus anacanthus*	nude thorny oyster
Nackter Seeschmetterling*	*Clione limacina*	naked sea butterfly
Nacktfingergecko	*Gymnodactylus geckoides*	naked-toed gecko
Nacktfingergeckos	*Gymnodactylus* spp.	naked-toed geckos
Nacktfingergeckos> Bogenfingergeckos	*Cyrtodactylus* spp.	bow-fingered geckos
Nacktfledermäuse	*Cheiromeles* spp.	naked bats, hairless bats
Nacktfliegen	Psilidae	rust flies
Nacktmull	*Heterocephalus glaber*	naked mole-rat, sand puppy
Nacktnasenwombat	*Vombatus ursinus*	common wombat, coarse-haired wombat
Nacktrücken-Fledermäuse	*Pteronotus* spp.	naked-backed bats
Nacktrücken-Fledermäuse, Kinnblatt-Fledermäuse	Mormoopidae	moustached bats, naked-backed bats, ghost-faced bats
Nacktrücken-Flughund	*Dobsonia* spp.	bare-backed fruit bats
Nacktsandaal, Nackt-Sandaal	*Gymnammodytes semisquamatus*	smooth sandeel
Nacktsandaal> Mittelmeersandaal, Mittelmeer-Nacktsandaal	*Ammodytes cicerellus/ Gymnammodytes cicerellus*	sandlance, Mediterranean sandeel (FAO), smooth sandlance

Nacktschwanz-Beutelratte	*Metachirus nudicaudatus/ Philander nudicaudatus*	brown "four-eyed" opossum
Nacktschwanz-Gürteltiere	*Cabassous* spp.	naked-tailed armadillos
Nacktschwanzratten* (Solomon-Inseln)	*Solomys* spp.	naked-tailed rats
Nacktsohlen-Rennmäuse	*Tatera* spp.	large naked-soled gerbils
Nacktzunge	*Gymnachirus melas*	naked sole
Nadel-Hornhecht, Agujon	*Tylosurus acus*	agujon needlefish
Nadelbaum-Spinnmilbe	*Oligonychus ununguis*	spruce spider mite, conifer spinning mite
Nadelfisch, Fierasfer	*Carapus acus/Fierasfer acus*	pearlfish
Nadelholz-Ambrosiakäfer, Linierter Nutzholzborkenkäfer	*Xyloterus lineatus*	lineate bark beetle, striped ambrosia beetle, conifer ambrosia beetle
Nadelholz-Säbelschrecke	*Barbitistes constrictus*	eastern sawtailed bushcricket
Nadelholz-Wellenbock	*Semanotus undatus*	European pine longhorn beetle
Nadelkissensterne u.a.	*Choriaster* spp.	pincushion starfish a.o.
Nadelkoralle	*Seriatopora caliendrum*	
Nadelnknickende Kieferngallmücke	*Contarinia baeri/ Cecidomyia baeri*	pine needle gall midge
Nadelschnecke> Gemeine Nadelschnecke, Gemeine Hornschnecke, Gemeine Seenadel	*Cerithium vulgatum/ Gourmya vulgata*	European cerith, common cerith
Nadelschnecken	Acmidae/Aciculidae	point snails
Nadelschnecken, Hornschnecken	Cerithiidae	ceriths, cerithids, hornsnails (horn shells), needle whelks
Nadelwels> Gemeiner Nadelwels, Lanzettförmiger Schnabelwels	*Farlowella acus*	twig catfish
Nagekäfer> Gemeiner Nagekäfer, Gewöhnlicher Nagekäfer, Holzwurm	*Anobium punctatum*	furniture beetle, woodworm (*larva*)
Nagelfleck	*Aglia tau*	tau emperor
Nagelhai, Brombeerhai, Stachelhai, Alligatorhai	*Echinorhinus brucus*	bramble shark
Nagelhaie, Alligatorenhaie, Alligatorhaie	Echinorhinidae	bramble sharks
Nagelkängurus	*Onychogalea* spp.	nail-tailed wallabies
Nagelrochen, Keulenrochen	*Raja clavata*	thornback skate, thornback ray (FAO), roker
Nagermilbe, Tropische Rattenmilbe	*Ornithonyssus bacoti/ Liponyssus bacoti*	tropical rat mite
Nagermilben	Macronyssidae	rodent mites
Nagetiere	Rodentia	rodents, gnawing mammals (except rabbits)
Nagetierläuse*	Polyplacidae	rodent sucking lice

Nagpur-Weichschildkröte, Vorderindische Weichschildkröte	Trionyx leithii/ Aspideretes leithii	Nagpur softshell turtle, Leith's softshell turtle
Namaqua-Sandrennmaus, Kurzschwanz-Sandrennmaus	Desmodillus auricularis	Cape short-eared gerbil, short-tailed gerbil, Namaqua gerbil
Namaqua-Zwergpuffotter	Bitis schneideri	Namaqua dwarf adder
Namib-Rennmäuse	Gerbillurus spp.	southern pygmy gerbils, hairy-footed gerbils
Namib-Sandgecko (Wüstengecko u.a.)	Palmatogekko rangei	Namib sand gecko, web-footed gecko
Nandaysittich	Nandayus nenday	black-hooded conure
Nanderbarsche, Nanderfische, Blattfische	Nandidae	leaffishes
Nandu	Rhea americana	greater rhea
Nandu> Kleiner Nandu, Darwinstrauß	Pterocnemia pennata	lesser rhea
Nandus	Rheiformes/Rheidae	rheas
Napfschildläuse	Coccidae	soft scales & wax scales & tortoise scales (scale insects)
Narkomedusen	Narcomedusae	narcomedusas
Narwal	Monodon monoceros	narwhal, unicorn whale
Narzissenzwiebelmilbe	Steneotarsonemus laticeps	bulb scale mite
Nase	Chondrostoma nasus	nase
Nasen-Erdschildkröte, Ecuador-Erdschildkröte	Rhinoclemmys nasuta	large-nosed wood turtle
Nasenaffe	Nasalis larvatus	proboscis monkey
Nasenbär, Südamerikanischer Nasenbär, Gewöhnlicher Nasenbär	Nasua nasua	coatimundi, common coati
Nasenbär> Kleiner Nasenbär	Nasuella olivacea	mountain coati
Nasenbären	Nasua spp.	coatis, coatimundis
Nasenbeutler, Bandikuts	Peramelidae	bandicoots
Nasendassel	Gasterophilus haemorrhoidalis	nose botfly, lip botfly, rectal botfly
Nasendasselfliege, Schafbiesfliege, Schafbremse	Oestrus ovis	sheep nostril-fly, sheep bot fly, sheep nose bot fly
Nasenfische, Einhornfische	Nasidae/Nasinae	unicornfishes
Nasenfledermaus	Rhynchonycteris naso	proboscis bat, sharp-nosed bat
Nasenfrösche	Rhinoderma spp.	mouth-brooding frogs, mouth-breeding frogs
Nasenfrösche	Rhinodermatidae	mouth-brooding frogs
Nasenhaie, Koboldhaie	Mitsukurinidae/ Scapanorhynchidae	goblin sharks
Nasenkakadu	Cacatua tenuirostris	long-billed corella
Nasenkröte	Rhinophrynus dorsalis	Darwin's frog, Mexican burrowing toad
Nasenkröten	Rhinophrynidae	burrowing toads
Nasenmuränen	Rhinomuraena spp.	ribbon eels

Nasennatter	*Rhinocheilus lecontei*	longnosed snake
Nasenschrecke, Gewöhnliche Nasenschrecke, Turmschrecke	*Acrida hungarica/ Acrida ungarica*	snouted grasshopper, long-headed grasshopper
Nasentermiten	Rhinotermitidae	rhinotermites, subterranean termites
Nasentierchen	*Didinium nasutum*	
Nashorn> Breitmaulnashorn, Breitlippennashorn, „Weißes Nashorn"	*Ceratotherium simum*	white rhinoceros, square-lipped rhinoceros, grass rhinoceros
Nashorn> Indisches Panzernashorn	*Rhinoceros unicornis*	greater Indian rhinoceros, great Indian rhinoceros
Nashorn> Javanashorn	*Rhinoceros sondaicus*	Javan rhinoceros
Nashorn> Spitzmaulnashorn, Spitzlippennashorn, "Schwarzes" Nashorn	*Diceros bicornis*	black rhinoceros, hooked-lipped rhinoceros, browse rhinoceros
Nashorn> Sumatranashorn	*Dicerorhinus sumatrensis*	Sumatran rhinoceros, hairy rhinoceros
Nashornagamen	*Ceratophora* spp.	horned agamas
Nashornalk	*Cerorhinca monocerata*	rhinoceros auklet, horn-billed puffin
Nashörner, Rhinozerosse	Rhinocerotidae	rhinoceroses, rhinos
Nashornkäfer	*Oryctes nasicornis*	European rhinoceros beetle
Nashornkäfer	*Dynastes* spp.	unicorn beetles, hercules beetles
Nashornleguan	*Cyclura cornuta*	rhinoceros iguana
Nashornpelikan	*Pelecanus erythrorhynchos*	American white pelican
Nashornviper	*Bitis nasicornis*	river jack, rhinoceros viper
Nashornvögel & Hornvögel & Tokos	Bucerotidae	hornbills
Nassau Zackenbarsch	*Epinephelus striatus*	Nassau grouper
Natal -Gelenkschildkröte	*Kinixys natalensis*	Natal hingeback tortoise
Natal-Languste	*Panulirus delagoae*	Natal spiny lobster
Natalrötel	*Cossypha natalensis*	red-capped robin chat
Natterers Sägesalmler, Roter Piranha	*Serrasalmus nattereri*	convex-headed piranha, Natterer's piranha, red piranha (FAO), red-bellied piranha
Nattern	Colubridae	colubrine snakes, common snakes
Nattern-Plattschwanz- Seeschlange	*Laticauda colubrina*	yellow-lipped sea krait, yellow-lipped sea snake, banded sea snake
Naumanndrossel	*Turdus naumanni*	dusky thrush
Nautilusverwandte	Nautiloidea	nautilus (*pl.* nautili)
Neanderthalmensch	*Homo sapiens neanderthalensis*	Neanderthals, Neanderthal people, Neanderthal man
Neapolitanische Triton*, Riesen-Triton	*Cymatium parthenopeum*	giant triton, Neapolitan triton
Nebelhai	*Squatina nebulosa*	clouded angelshark

Nebelparder	*Neofelis nebulosa*	clouded leopard
Nebelwaran	*Varanus nebulosus*	clouded monitor
Nebliger Schildkäfer	*Cassida nebulosa*	beet tortoise beetle, clouded shield beetle
Nektarvögel	Nectariniidae	sunbirds, spiderhunters
Nelkeneule	*Hadena compta*	dianthus moth, campion moth
Nelkenkorallen	*Caryophyllia* spp.	cupcorals
Nelkenwurm, Nelkenbandwurm	*Caryophyllaeus laticeps*	
Nelsons Baumratte	*Nelsonia neotomodon*	Nelson's wood rat
Nelsons Dosenschildkröte	*Terrapene nelsoni*	Nelson's box turtle, spotted box turtle
Nemathelminthen, Aschelminthen, Schlauchwürmer, Rundwürmer	Nemathelminthes/ Aschelminthes	nemathelminths, aschelminths
Nemouriden	Nemouridae	winter stoneflies & spring stoneflies
Neon-Grundel, Neongrundel, Blaue Putzgrundel	*Gobiosoma oceanops/ Elacatinus oceanops*	neon goby
Neonfisch	*Paracheirodon innesi/ Hyphessobrycon innesi*	neon tetra
Neostethiden (Zwergährenfische u.a.)	Neostethidae	neostethids
Neotropische Berghörnchen	*Syntheosciurus* spp.	neotropical montane squirrels
Neotropische Flughunde	*Artibeus* spp.	neotropical fruit bats
Neotropisches Zwerghörnchen	*Sciurillus pusillus*	neotropical pygmy squirrel
Neptunsbecher	*Poterion neptuni*	Neptune's cup sponge
Neptunschleier, Netzkorallen, Neptunsmanschetten	*Retepora* spp./ *Reteporella* spp./ *Sertella* spp.	lace 'corals'
Neptunshorn> Gemeines Neptunshorn, Gemeine Spindelschnecke	*Fusus antiqua/ Neptunea antiqua*	ancient whelk, ancient neptune, common spindle snail, neptune snail, red whelk, buckie
Neptunsschwamm	*Aplysina fistularis*	yellow tube sponge, candle sponge, sulphur sponge (dead man's fingers)
Nereiden	Nereidae	nereids
Nerz> Amerikanischer Nerz, Mink	*Mustela vison*	American mink
Nerz> Europäischer Nerz	*Mustela lutreola*	European mink
Nessel-Grünrüssler	*Phyllobius urticae*	nettle weevil, stinging nettle weevil
Nesselbär, Weißer Fleckleib-Bär	*Spilosoma urticae*	water ermine moth
Nesseleulen, Brennnesseleulen	*Abrostola* spp.	dark spectacles
Nesselquallen	*Chrysaora* spp.	sea nettles

Nesselschnabeleule, Zünslereule	*Hypena proboscidalis*	common snout (moth)
Nesseltiere, Hohltiere, Cnidarien, Coelenteraten	Cnidaria/Coelenterata	cnidarians, coelenterates
Nesselzünsler	*Eurrhypara hortulata*	small magpie (moth)
Nestermotte	*Niditinea fuscipunctella*	brown-dotted clothes moth, poultry house moth
Nestkäfer, Erdaaskäfer	Catopidae/Leptodiridae	small carrion beetles
Netz-Feuerkoralle	*Millepora dichotoma*	ramified fire coral
Netz-Helmschnecke*	*Cypraecassis testiculus*	reticulate cowrie-helmet
Netz-Kegelschnecke, Gewebte Kegelschnecke, Weberkegel	*Darioconus textile/ Conus textile*	textile cone
Netz-Kissenstern	*Oreaster reticulatus*	reticulate cushion star
Netz-Prachtschmerle	*Botia lohachata*	Pakistani loach
Netzanemone*	*Actinauge verrillii*	reticulate anemone
Netzaugenfische	Ipnopidae	ipnopids
Netzfliegen	Nemestrinidae	tangle-veined flies
Netzfliegen> Hornfliegen	Sciomyzidae	marsh flies
Netzhörnchen, Hornschnecke, Mäusedreck, Kleine Gitterschnecke	*Bittium reticulatum*	needle whelk (needle shell)
Netzkärpfling, Panamakärpfling	*Phallichthys pittieri*	iridescent widow, orange-dorsal live bearer, merry widow livebearer (FAO)
Netzkegel	*Conus retifer*	netted cone
Netzkorallen, Neptunschleier, Neptunsmanschetten	*Retepora* spp./ *Reteporella* spp./ *Sertella* spp.	lace 'corals'
Netzpython	*Python reticulatus*	reticulated python (diamond python, Java rock python)
Netzreusenschnecke, Gemeine Netzreuse	*Hinia reticulata*	netted dog whelk
Netzstern, Seefledermaus	*Patiria miniata*	bat star
Netzwanzen (Blasen-, Buckel- oder Gitterwanzen)	Tingidae	lace bugs
Neuguinea Elseya-Schildkröte, Neuguinea-Schnapper	*Elseya dentata novaeguineae/ Elseya novaeguineae*	New Guinea snapping turtle
Neuguinea-Beutelmäuse	*Murexia* spp.	short-tailed marsupial "mice"
Neuguinea-Hüpfmaus	*Lorentzimys nouhuysi*	New Guinea jumping mouse
Neuguinea-Filander	*Thylogale brunii*	dusky pademelon
Neuguinea-Krokodil	*Crocodylus novaeguineae*	New Guinea crocodile
Neuguinea-Nasenbeutler	*Peroryctes* spp.	New Guinean bandicoots
Neuguinea-Ratten, Langschwanzratten	*Macruromys* spp.	New Guinean rats
Neuguinea-Riesenratte*	*Anisomys imitator*	powerful-toothed rat, New Guinea giant rat (squirrel-toothed rat), uneven-toothed rat
Neuguinea-Schlangenhalsschildkröte	*Chelodina novaeguineae*	New Guinea snake-necked turtle

Neuguinea-Zwergpython	*Morelia boa*	barred python
Neuhollandhabicht	*Accipiter novaehollandiae*	white goshawk
Neukaledonisches Perlboot	*Nautilus macromphalus*	New Caledonian nautilus, bellybutton nautilus (FAO)
Neumundtiere, Neumünder, Zweitmünder	Deuterostomia	deuterostomes
Neunaugen, Neunaugenartige	Petromyzontida	lampreys
Neunbinden-Gürteltier	*Dasypus novemcinctus*	nine-banded armadillo
Neunstachliger Stichling, Zwergstichling	*Pungitius pungitius*	nine-spined stickleback
Neuntöter, Rotrückenwürger	*Lanius collurio*	red-backed shrike
Neuseeland-Eisfisch	*Pseudophycis bacchus/ Physiculus bacchus*	red cod, New Zealand red cod, red codling (FAO)
Neuseeland-Fledermaus	*Mystacina tuberculata*	New Zealand short-tailed bat
Neuseeland-Flussbarsch	*Parapercis colias*	New Zealand blue cod
Neuseeland-Grenadiere	Acanthisittidae	New Zealand wrens
Neuseeland-Miesmuschel, Große Streifen-Miesmuschel*	*Perna canaliculus*	New Zealand mussel, channel mussel
Neuseeland-Plattauster	*Ostrea lutaria*	New Zealand dredge oyster
Neuseeland-Stachelmakrele	*Pseudocaranx dentex*	guelly jack, white kingfish, New Zealand trevally, white trevally (FAO)
Neuseeländische Jakobsmuschel	*Pecten novaezealandiae*	New Zealand scallop
Neuseeländische Languste	*Jasus edwardsii*	red rock lobster
Neuseeländische Lappenvögel	Callaeatidae/Callaeidae	New Zealand wattlebirds
Neuseeländische Rinderzecke	*Haemaphysalis longicornis*	New Zealand cattle tick
Neuseeländische Urfrösche	*Leiopelma* spp.	New Zealand frogs
Neuseeländische Urfrösche	Leiopelmidae/Leiopelmatidae	ribbed frogs, leiopelmids (tailed frogs & New Zealand frogs)
Neuseeländischer Grenadier	*Macruronus novaezealandiae*	blue grenadier
Neuseeländischer Kaisergranat	*Metanephrops challengeri*	New Zealand scampi, deep water scampi
Neuseelandlachse	Retropinnidae	New Zealand smelts
Neuvögel	Neornithes	neornithes, true birds
Neuwelt-Gleithörnchen	*Glaucomys* spp.	New World flying squirrels
Neuwelt-Hakenwurm	*Necator americanum*	New World hookworm
Neuwelt-Trogons	Trogonini	New World trogons
Neuweltaffen, Breitnasenaffen	Platyrrhina	New World monkeys (South American monkeys & marmosets)
Neuweltgeier	Cathartidae	New World vultures
Neuweltliche Zwerghörnchen	*Microsciurus* spp.	neotropical dwarf squirrels
Neuwieds Lanzenotter, Jararaca pintada	*Bothrops neuwiedi*	jararaca pintada, Neuwied's lancehead
Nickerls Scheckenfalter, Ehrenpreis-Scheckenfalter	*Melitaea aurelia*	Nickerl's fritillary
Niedrighöckerige Kaktuskoralle	*Mycetophyllia danaana*	lowridge cactus coral

Nieren-Riesen-Palisadenwurm	*Strongylus gigas/ Dioctophyma renale*	renal blood-red nematode
Nierenschwamm, Lederschwamm	*Chondrosia reniformis*	chicken liver sponge
Nierenwurm	*Dioctophyme renale*	kidney worm
Nigeria Blaumaulmeerkatze	*Cercopithecus sclateri*	Sclater's monkey
Nikobaren-Spindelschnecke	*Fusinus nicobaricus*	Nicobar spindle
Nil-Buntbarsch, Nilbarsch	*Oreochromis niloticus/ Tilapia nilotica*	Nile tilapia (FAO), Nile mouthbreeder
Nil-Flösselhecht	*Polypterus bichir*	Nile bichir
Nil-Flughund	*Rousettus aegyptiacus*	Egyptian rousette
Nil-Larvenfisch	*Cromeria nilotica*	naked shellear
Nilbarsch (Viktoriabarsch)	*Lates niloticus*	Nile perch (Sangara)
Nilbarsche, Snooks	Centropomidae	snooks
Nilgans	*Alopochon aegyptiacus*	Egyptian goose
Nilgauantilope	*Boselaphus tragocamelus*	nilgai, bluebuck
Nilgiri-Tahr	*Hemitragus hylocrius*	Nilgiri tahr
Nilgirilangur	*Presbytis johnii*	Nilgiri langur
Nilhechte	Mormyriformes	mormyrids
Nilhechte, Elefantenfische	Mormyridae	elephantfishes
Nilkrokodil	*Crocodylus niloticus*	Nile crocodile
Nilwaran	*Varanus niloticus*	Nile monitor
Nilwels	*Bagrus docmac*	nilotic catfish
Nipponibis	*Nipponia nippon*	crested ibis
Nistfliegen	Milichiidae	milichiids
Noddiseeschwalbe	*Anous stolidus*	noddy tern, noddy
Nonne	*Lymantria monacha*	black arches
Nonnenkranich	*Grus leucogeranus*	great white crane, Siberian crane, Siberian white crane
Nonnensteinschmätzer	*Oenanthe pleschanka*	pied wheatear
Noppenrand-Anemone, Noppenrand-Meerblume, Pizza-Anemone	*Cryptodendrum adhaesivum*	pizza anemone, nap-edged anemone
Nord-Antillen-Schmuckschildkröte	*Trachemys decussata*	West Indian slider, North Antillean slider
Nord-Australische Elseya-Schildkröte	*Elseya dentata*	northern Australian snapping turtle, northern snapping tortoise
Nord-Blattnasennatter	*Langaha nasuta*	northern leafnose snake
Nordafrikanischer Spießbock, Säbelantilope	*Oryx dammah*	scimitar oryx, scimitar-horned oryx
Nordafrikanisches Erdhörnchen, Atlashörnchen	*Atlantoxerus getulus*	barbary ground squirrel
Nordamerikanische Brasse	*Stenotomus chrysops*	scup
Nordamerikanische Flussheringe	*Pomolobus* spp.	alewifes
Nordamerikanische Glattnasen	*Lasiurus* spp.	hairy-tailed bats
Nordamerikanische Graufüchse	*Urocyon* spp.	gray foxes

Nordamerikanische Kleinohrspitzmaus	*Cryptotis parva*	least shrew
Nordamerikanische Pfeifente	*Anas americana*	American wigeon, baldpate
Nordamerikanische Quellkärpflinge	Empetrichthyinae	springfishes & poolfishes
Nordamerikanische Rinderzecke	*Boophilus annulatus*	North American cattle tick
Nordamerikanische Rohrdommel	*Botaurus lentiginosus*	American bittern
Nordamerikanische Rothörnchen, Chikarees	*Tamiasciurus* spp.	American red squirrels, chickarees
Nordamerikanische Tannenmäuse, Heidewühlmäuse	*Phenacomys* spp.	heather voles, tree voles
Nordamerikanische Trockenobstmotte	*Vitula edmandsae*	dried-fruit moth
Nordamerikanische Wassernattern	*Nerodia* spp.	water snakes
Nordamerikanischer Festland-Graufuchs (Hbz. Grisfuchs)	*Urocyon cinereoargenteus*	mainland gray fox, North American gray fox
Nordamerikanischer Fichtentriebwickler	*Choristoneura fumiferana*	spruce budworm (tortrix)
Nordamerikanischer Flusshering	*Alosa pseudoharengus*	alewife (FAO), river herring
Nordamerikanischer Insel-Graufuchs	*Urocyon littoralis*	island gray fox
Nordamerikanischer Kalmar, Langflossen-Schelfkalmar	*Loligo pealeii*	longfin inshore squid, Atlantic long-fin squid, Atlantic long-finned squid
Nordamerikanischer Seehecht, Silberhecht, Silber-Wittling	*Merluccius bilinearis*	silver hake (FAO), silver whiting
Nordamerikanischer Streifenbarsch, Felsenbarsch	*Morone saxatilis/ Roccus saxatilis*	striped bass, striper, striped sea bass (FAO)
Nordamerikanisches Felchen	*Coregonus clupeaformis*	whitefish, common whitefish, lake whitefish (FAO), humpback
Nordamerikanisches Katzenfrett	*Bassariscus astutus*	ringtail
Nordatlantik-Grenadier, Rauer Grenadier, Raukopf Panzerratte	*Macrourus berglax*	rough rattail, rough-head grenadier, onion-eye grenadier (FAO)
Nordatlantik-Strandkrabbe, Strandkrabbe	*Carcinus maenas*	green shore crab, green crab, North Atlantic shore crab
Nordatlantische Zwerg-Tellmuschel	*Tellina agilis*	northern dwarf-tellin, dwarf tellin
Nordatlantischer Weißrochen, Nördlicher Weißrochen	*Raja lintea/ Dipturus linteus*	sailray (FAO), sharpnose skate
Nordaustralische Hüpfmaus	*Notomys aquilo*	northern hopping mouse
Nordaustralische Schlangenhalsschildkröte	*Chelodina rugosa*	northern snakeneck, northern Australian snake-necked turtle, northern long-necked tortoise

Nordeuropäische Elefantenzahnschnecke	Dentalium entale/ Antalis entale/ Antalis entalis (Scaphopoda)	North European elephant tusk, common elephant tusk
Nordfledermaus, Nordische Fledermaus	Eptesicus nilssoni	northern bat
Nordische Apfelwanze, Grüne Apfelwanze	Plesiocoris rugicollis	apple capsid bug
Nordische Fledermaus, Nordfledermaus	Eptesicus nilssoni	northern bat
Nordische Gebirgsschrecke	Melanoplus frigidus	northern migratory grasshopper
Nordische Keulensynascidie	Clavelina borealis	northern light-bulb sea-squirt
Nordische Schiffsbohrmuschel, Nordischer Schiffsbohrwurm	Teredo norvegica/ Nototeredo norvagicus	Norway shipworm, Norwegian shipworm
Nordische Seequappe	Ciliata septentrionalis	northern rockling
Nordischer Kalmar, Forbes' Kalmar	Loligo forbesii	long-finned squid, northern squid
Nordischer Kammstern	Astropecten irregularis	sand star
Nordischer Meerbrassen, Graubarsch, Seekarpfen	Pagellus bogaraveo	red seabream, common seabream, blackspot seabream (FAO)
Nordischer Sandaal	Ammodytes marinus	Raitt's sandeel, lesser sandeel (FAO)
Nordischer Schiffsbohrwurm	Nototeredo norvagicus/ Teredo norvegica	Norway shipworm, Norwegian shipworm
Nordischer Schleimkopf, Kaiserbarsch, Alfonsino	Beryx decadactylus	beryx, alfonsino (FAO), red bream
Nordkaper	Eubalaena glacialis/ Balaena glacialis	northern right whale, black right whale, Pacific right whale
Nordkröte	Bufo boreas	western toad
Nördliche Alligatorschleiche	Gerrhonotus coerulens	northern alligator lizard
Nördliche Amerikanische Schwarze Witwe	Latrodectus variolus	northern black widow spider
Nördliche Astarte	Astarte borealis	boreal astarte
Nördliche Mondschnecke	Euspira heros/Lunatia heros	northern moonsnail, common northern mononsnail, sand collar moon snail
Nördliche Nussmuschel	Nuculana pernula	northern nutclam, Müller's nutclam
Nördliche Quahog-Muschel	Mercenaria mercenaria	northern quahog, quahog (hard clam)
Nördliche Rosa-Garnele, Rosa Golfgarnele, Nördliche Rosa Geißelgarnele	Farfantepenaeus duorarum/ Penaeus duorarum	pink shrimp, northern pink shrimp
Nördliche Schlangenkopfgrundel	Ophiocara porocephala	northern mud gudgeon
Nördliche Schotenmuschel	Solemya borealis	boreal awningclam
Nördliche Steinkrabbe	Lithodes maja	northern stone crab, Norway crab, devil's crab (prickly crab), stone king crab
Nördliche Sternkoralle	Astrangia oculata	northern star coral

Nördliche Taschenratte	*Thomomys talpoides*	northern pocket gopher
Nördliche Thracia	*Thracia septentrionalis*	northern thracia
Nördliche Tiefseegarnele, Grönland-Shrimp	*Pandalus borealis*	northern shrimp, pink shrimp, northern pink shrimp
Nördliche Weiße Geißelgarnele, Atlantische Weiße Garnele	*Litopenaeus setiferus/ Penaeus setiferus*	white shrimp, lake shrimp, northern white shrimp
Nördlicher Bootsmannsfisch	*Porichthys notatus*	plainfin midshipman
Nördlicher Falterfisch	*Chaetodon ocellatus*	spotfin butterflyfish
Nördlicher Glattdelphin	*Lissodelphis borealis*	Pacific right whale dolphin, northern right-whale dolphin
Nördlicher Grillenfrosch	*Acris crepitans crepitans*	northern cricket frog
Nördlicher Kielnagelgalago	*Euoticus pallidus*	pallid needle-clawed galago
Nördlicher Krill, Norwegischer Krill	*Meganyctiphanes norvegica*	Norwegian krill
Nördlicher Kurzflossenkalmar	*Illex illecebrosus*	northern shortfin squid (FAO), common shortfin squid, northern squid, boreal squid
Nördlicher Seebär	*Callorhinus ursinus*	northern fur seal
Nördlicher Ziegenfisch	*Mullus auratus*	golden goatfish
Nördlicher Zweistreifiger Gelbsalamander	*Eurycea bislineata*	two-lined salamander
Nördliches Flechtenbärchen	*Eilema sericea*	northern footman
Nördliches Wurzelgallenälchen	*Meloidogyne hapla*	northern root-knot nematode
Nordluchs (Hbz. Luchs)	*Felis lynx/ Lynx lynx*	lynx
Nordmannstannen-Trieblaus, Tannentrieblaus, Weißtannen-Trieblaus	*Dreyfusia nordmanniana/ Adelges nordmanniana*	silver fir migratory adelges
Nordmeer-Seestern*	*Asterias vulgaris*	northern seastar
Nordopossum	*Didelphis virginiana*	Virginian opossum
Nordpazifik-Glattwal	*Eubalaena japonica/ Balaena glacialis japonica*	North Pacific right whale
Nordplaty	*Xiphophorus couchianus*	Monterrey platyfish
Nordsee-Einsiedlerkrebs, Bernhardskrebs, Bernhardseinsiedler, Gemeiner Einsiedler	*Pagurus bernhardus/ Eupagurus bernhardus*	large hermit crab, common hermit crab, soldier crab, soldier hermit crab, Bernhard's hermit crab
Nordsee-Zylinderrose	*Cerianthus lloydii*	North Sea cerianthid
Nordseegarnele, Granat, Porre	*Crangon crangon*	brown shrimp, common shrimp, common European shrimp
Nordwestamerikanischer Querzahnmolch*	*Ambystoma gracile*	northwestern salamander
Nordwestatlantischer Menhaden	*Brevoortia tyrannus*	Atlantic menhaden (FAO), bunker
Nordwestatlantische Sardelle	*Anchoa mitchilli*	bay anchovy
Norfolk-Brillenvogel	*Zosterops albogularis*	Norfolk Island white-eye, white-breasted white-eye, white-chested white-eye
Norwegische Garnele	*Pontophilus norvegicus*	Norwegian shrimp
Norwegische Grundel	*Pomatoschistus norvegicus*	Norway goby (FAO), Norwegian goby

Norwegische Herzmuschel	*Laevicardium crassum*	Norway cockle, Norwegian cockle
Norwegische Wespe	*Dolichovespula norwegica*	Norwegian wasp, Norway wasp
Norwegischer Krill, Nördlicher Krill	*Meganyctiphanes norvegica*	Norwegian krill
Norwegischer Rotbarsch	*Sebastes viviparus*	Norway haddock, lesser redfish
Norwegischer Zwergbutt	*Phrynorhombus norvegicus*	Norwegian topknot
Notaden	*Notaden* spp.	Australian shovelfoots
Nubierspecht	*Campethera nubica*	Nubian woodpecker
Nubischer Ziegenmelker	*Caprimulgus nubicus*	Nubian nightjar
Nudelfische	Salangidae	noodlefishes, icefishes
Nufar-Seebrasse	*Cheimerius nufar*	soldier, santer seabream (FAO)
Nukleus-Kauri	*Cypraea nucleus*	nucleus cowrie
Nukras-Eidechsen	*Nucras* spp.	sandveld lizards
Nulleneule, O-Eule, Eichen-Nulleneule, Eichenhochwald-Doppelkreiseule	*Dicycla oo*	heart moth
Nussmuscheln	*Nucula* spp.	nut clams, nutclams (nut shells)
Nussmuscheln	Nuculacea/ Nuculidae	nut clams, nutclams (nut shells)
Nuttings Schwamm	*Leucilla nuttingi*	Nutting's sponge
Nyala	*Tragelaphus angasi*	nyala
Nymphensittich	*Nymphicus hollandicus*	cockatiel

Obstbaumminiermotte, Schlangenminiermotte	*Lyonetia clerkella*	apple leaf miner, Clerk's snowy bentwing
Obstbaumspinnmilbe, Rote Spinne	*Panonychus ulmi/ Metatetranychus ulmi*	fruittree red spider mite, European red mite
Obstbaumwickler, Weißdornwickler	*Archips crataegana*	fruit tree tortrix (moth), brown oak tortrix, oak red-barred twist
Obstbaumzweigabstecher, Triebstecher	*Rhynchites caeruleus*	apple twig cutter
Obstfliegen, Taufliegen, Essigfliegen	Drosophilidae	"fruit flies", pomace flies, small fruit flies, ferment flies
Obstgartenspanner	*Chloroclystis rectangulata*	green pug
Ochsenauge, Indopazifischer Tarpon, Indopazifischer Tarpun	*Megalops cyprinoides*	Indopacific tarpon
Ochsenauge> Großes Ochsenauge	*Maniola jurtina*	meadow brown
Ochsenauge> Kleines Ochsenauge	*Maniola lycaon/ Hyponephele lycaon*	
Ochsenaugen-Makrele	*Selar boops*	oxeye scad
Ochsenfrosch	*Rana catesbeiana*	American bullfrog
Ochsenherz, Menschenherz	*Glossus humanus*	ox heart, heart cockle
Ockergelber Blattspanner	*Camptogramma bilineata*	yellow shell
Ockerstern	*Pisater ochraceus*	ochre seastar, ochre star
Odinshühnchen	*Phalaropus lobatus*	red-necked phalarope
Oenpelli-Rautenpython	*Morelia oenpelliensis*	Oenpelli python
Ofenfischchen	*Lepismodes inquilinus/ Thermobia domestica*	firebrat
Ogerspinne, Kescherspinne	*Deinopis longipes*	net-casting spider
Ogerspinnen, Kescherspinnen	Deinopidae/Dinopidae	ogre-faced spiders
Ogilbyducker	*Cephalophus ogilbyi*	Ogilby's duiker
Ogoue Zwergmeerkatze	*Miopithecus ogouensis*	northern talapoin
Ohr-Walzenschnecke	*Aulica aulica/Cymbiola aulica*	princely volute
Ohrbüschel-Seidenäffchen*	*Callithrix aurita*	buffy tufted-ear marmoset
Öhrchenmuschel*	*Anadara notabilis*	eared ark (eared ark shell)
Ohrenfische, Schlankfische	Kneriidae	kneriids
Ohrengeier	*Aegypius tracheliotus*	lappet-faced vulture
Ohrenkrötenkopf, Bärtige Krötenkopfagame	*Phrynocephalus mystaceus*	secret toadhead agama
Ohrenlerche	*Eremophila alpestris*	shore lark, shore horned lark
Ohrenqualle	*Aurelia aurita*	moon jelly, common jellyfish
Ohrenratten, Afrikanische Lamellenzahnratten	*Otomys* spp.	African swamp rats, groove-toothed rats
Ohrensardine, Große Sardine, Sardinelle	*Sardinella aurita*	gilt sardine, Spanish sardine, round sardinella (FAO)
Ohrenscharbe	*Phalacrocorax auritus*	double-crested cormorant
Ohrentaucher	*Podiceps auritus*	Slavonian grebe

Ohrfleck-Brasse	*Archosargus rhomboidalis*	western Atlantic seabream
Ohrfleck-Heringsmakrele, Gefleckte Stachelmakrele	*Decapterus punctatus*	round scad
Ohrwurm> Gemeiner Ohrwurm	*Forficula auricularia*	common earwig, European earwig
Ohrwürmer	Dermaptera	earwigs
Okapi	*Okapia johnstoni*	okapi
Okavango-Pelomeduse	*Pelusios bechuanicus*	Okavango mud turtle
Oktokorallen	Octocorallia	octocorallians, octocorals
Okuliergallmücke, Okuliermade	*Thomasiniana oculiperda/ Resseliella oculiperda*	red bud borer midge (larva)
Okzitanischer Skorpion, Gelber Mittelmeerskorpion, Gemeiner Mittelmeerskorpion	*Buthus occitanus*	common yellow scorpion
Oleanderschildlaus	*Aspidiotus hederae/ Aspidiotus nerii*	oleander scale
Oleanderschwärmer	*Daphnis nerii*	oleander hawkmoth
Ölfisch	*Ruvettus pretiosus*	oilfish
Ölfische, Baikal-Ölfische	Comephoridae	Baikal oilfishes
Olivastrild	*Amandava formosa*	green munia
Olivbartvogel	*Cryptolybia olivacea*	green barbet
Olivbraune Seeschlange, Braune Seeschlange	*Aipysurus laevis*	olive-brown sea snake
Oliven-Braunschlange*	*Demansia olivacea*	olive whipsnake
Olivenfliege	*Bactrocera oleae*	olivefruit fly
Olivenspötter	*Hippolais olivetorum*	olive-tree warbler
Olivgrüne Flunder	*Paralichthys olivaceus*	olive flounder, bastard halibut (FAO)
Olivgrüner Schlangenstern	*Ophiorachna incrassata*	green brittlestar, olivegreen brittlestar*
Olivgrüner Snook	*Centropomus undecimalis*	snook, common snook
Ölkäfer	Meloinae	oil beetles
Ölkäfer (Maiwurm)	*Meloë* spp.	oil beetles, blister beetles
Ölkrug, Große Krötenschnecke, Taschenschnecke, Großes Ochsenauge	*Gyraulus gigantea/ Ranella gigantea*	giant frogsnail, oil-vessel triton
Olme	Proteidae	olms & mudpuppies
Ölpalmenhörnchen	*Protoxerus* spp.	oil palm squirrels
Olymp-Querzahnmolch	*Rhyacotriton olympicus*	Olympic salamander
Onager, Asiatischer Halbesel, Kulan, Khur	*Equus hemionus*	kulan, khur, onager
Onyxkauri	*Cypraea onyx*	onyx cowrie
Opal-Kreiselschnecke	*Cantharidus opalus*	opal jewel topsnail
Opalauge	*Girella nigricans*	opaleye
Opalisierender Kalmar, Pazifischer Opalkalmar	*Loligo opalescens*	opalescent inshore squid, opalescent squid (American "market squid"), common Pacific squid

Opalwurm	*Arabella iricolor*	opal worm
Opalwürmer	Nephtyidae	catworms
Opalwürmer	*Nephtys* spp.	catworms, shimmy worms
Ophidiiden, Bartmännchen	Ophidiidae	cuskeels, cusk-eels
Opossumkäfer	Monommatidae	opossum beetles
Opossummäuse	*Caenolestes* spp.	common "shrew" opossums
Opossums> Amerikanische Opossums	*Didelphis* spp.	large American opossums
Opuntienspinne	*Cyrtophora citricola*	fig-cactus spider
Orang-Utan	*Pongo pygmaeus*	orangutan
Orange Flechtenbärchen	*Eilema sororcula*	orange footman
Orange Hufschnecke	*Hipponix subrufus*	orange hoofsnail
Orange Warzenkoralle*	*Tubastraea coccinea*	orange cupcoral
Orange-Anemonenfisch, Clown-Anemonenfisch, Schwarzgeränderter Orangenringelfisch	*Amphiprion percula*	orange clownfish (FAO), clown anemonefish
Orangefarbene Krustenanemone	*Epizoanthus scotinus*	orange zoanthid, orange encrusting anemone
Orangefarbener Feuerkäfer	*Schizotus pectinicornis*	orange-coloured fire beetle*
Orangefleck-Doktorfisch	*Acanthurus pyroferus*	chocolate surgeonfish
Orangefleck-Lippfisch, Spiegelfleck-Lippfisch	*Coris angulata*	clown wrasse (FAO), clown coris, red-blotched rainbowfish
Orangeflecken-Feilenfisch	*Oxymonacanthus longirostris*	beaked leatherjacket, harlequin filefish (FAO)
Orangeflecken-Grundel	*Fusigobius longispinis*	butterfly goby
Orangegrüner Junker, Zweifleck-Junker	*Halichoeres biocellatus*	red-lined wrasse
Orangekärpfling	*Poecilia heterandria/ Limia heterandria*	dwarf molly
Orangemündige Pazifische Purpurschnecke	*Drupa grossularia*	finger drupe
Orangenflossen- Anemonenfisch	*Amphiprion chrysopterus*	orangefin anemonefish (FAO), orange-finned anemonefish
Orangenfuß-Seegurke*	*Cucumaria frondosa*	orange-footed sea cucumber, "pudding"
Orangenkauri, Goldene Kauri	*Cypraea auriantium*	golden cowrie
Orangenringelfisch, Orangeringelfisch, Falscher Clown-Anemonenfisch	*Amphiprion ocellaris*	clown anemonefish (FAO), false clown anemonefish
Orangenschalen- Moostierchen*	*Tubicellepora magnicostata*	orange peel bryozoan
Orangenschildlaus, Australische Wollschildlaus	*Icerya purchasi*	cottony cushion scale, fluted scale
Oranger Bohrschwamm	*Cliona delitrix*	orange boring sponge
Oranger Fingerschwamm	*Clathria procera*	orange antler sponge*
Oranger Strahlenschwamm	*Spirastrella cunctatrix*	
Orangerote Gorgonie	*Lophogorgia ceratophyta*	
Orangerote Seefeder*	*Swiftia exserta*	orange sea fan, soft red sea fan

Orangerote Weizengallmücke, Rote Weizengallmücke	*Sitodiplosis mosellana*	orange wheat blossom midge
Orangeroter Bohrschwamm	*Spirastrella purpurea*	orange-red boring sponge*
Orangeroter Feilenfisch	*Aluterus schoepfi*	orange filefish
Orangeschwarzer Giftfrosch	*Phyllobates vittatus*	orange-and-black poison frog, orange-and-black poison-dart frog, Golfodulcean poison frog
Orchideenfliege	*Eurytoma orchidearum*	orchidfly, orchid wasp, cattleya 'fly'
Orchideenschildlaus	*Diaspis boisduvalii*	orchid scale
Oregon-Triton	*Fusitriton oregonensis*	Oregon triton
Oreos	Oreosomatidae	oreos
Orfe, Aland	*Leuciscus idus*	ide (FAO), orfe
Orfe> Gewöhnliche Orfe, Gemeine Orfe	*Notropis cornutus/ Luxilus cornutus*	common shiner
Orgelkoralle	*Tubipora musica*	organ-pipe coral, pipe-organ coral
Orgelkorallen	*Tubipora* spp.	organ-pipe corals, pipe-organ corals
Orientalische Süßlippe, Orient-Süßlippe	*Plectorhinchus orientalis*	oriental sweetlip
Orientalisches Wildschaf	*Ovis orientalis*	Asiatic mouflon
Orientturteltaube	*Streptopelia orientalis*	rufous turtle dove
Orinoko-Krokodil	*Crocodylus intermedius*	Orinoco crocodile
Ornament-Baumvogelspinne	*Poecilotheria ornata*	fringed ornamental tarantula
Ornament-Wobbegong, Gezeichneter Wobbegong	*Orectolobus ornatus*	ornate wobbegong (FAO), banded wobbegong
Ornamentschlange	*Denisonia maculata*	ornamental snake
Ornamentschlangen	*Denisonia* spp.	ornamental snakes
Ornatlanguste	*Panulirus ornatus*	ornate spiny crawfish
Orpheusgrasmücke	*Sylvia hortensis*	orphean warbler
Orpheusspötter	*Hippolais polyglotta*	melodious warbler
Ortolan	*Emberiza hortulana*	ortolan bunting
Oryxweber	*Euplectes orix*	red bishop
Oskar, Roter Oskar, Pfauenaugenbuntbarsch	*Astronotus ocellatus*	oscar (FAO), oscar's cichlid, velvet cichlid
Ösophagus-Rollschwanz, Hunde-Spirocercose-Wurm	*Spirocerca lupi/ Spirocerca sanguinolenta*	esophageal worm (of dog)
Ösophagus-Wurm	*Gongylonema pulchrum*	gullet worm, zigzag worm
Ost-Amerikanischer Hundsfisch, Zwerghundsfisch	*Umbra pygmaea*	eastern mudminnow
Ost-Spitzkopfschildkröte	*Emydura signata*	Brisbane short-necked turtle, Eastern Australian short-necked tortoise
Ostafrikanische Bergotter	*Atheris hindii*	East African mountain viper
Ostafrikanische Erdviper	*Adenorhinos barbouri*	worm-eating viper
Ostafrikanische Gelbbauch-Pelomeduse	*Pelusios castanoides*	East African yellow-bellied mud turtle

Ostafrikanischer Lungenfisch	*Protopterus aethiopicus*	East African lungfish
Ostafrikanischer Rotstummelaffe	*Procolobus rufomitratus/ Piliocolobus rufomitratus*	eastern red colobus
Ostafrikanisches Dreihornchamäleon	*Chamaeleo jacksoni*	Jackson's chameleon, Jackson's three-horned chameleon
Ostamerikanischer Maulwurf	*Scalopus aquaticus*	eastern American mole
Ostasiatischer Frühabendsegler	*Ia io*	great evening bat
Ostasiatischer Schlammpeitzger	*Misgurnus anguillicaudatus*	Japanese weatherfish
Ostatlantische Gabeldorsche	*Phycis* spp.	Eastern Atlantic hakes
Ostatlantische Gitterschnecke, Gemeine Muskatnuss	*Cancellaria reticulata*	common nutmeg, common East-Atlantic nutmeg
Ostatlantische Königsmakrele, Westafrikanische Königsmakrele	*Scomberomorus tritor*	West African Spanish mackerel
Ostatlantischer Marmorrochen, Scheckenrochen, Bänderrochen	*Raja undulata*	undulate ray (FAO), painted ray
Ostatlantischer Sägerochen, Sägefisch	*Pristis pristis*	sawfish, common sawfish (FAO)
Ostatlantischer Spitzrochen	*Raja oxyrhynchus/ Raja oxyrinchus*	long-nosed skate, longnosed skate (FAO)
Ostatlantischer Tiefseedorsch	*Mora mora*	hakeling
Ostaustralische Languste	*Jasus verreauxi*	green rock lobster
Ostaustralischer Zirpfrosch	*Crinia signifera*	common eastern froglet
Osterluzeifalter	*Zerynthia polyxena*	southern festoon
Österreichische Wespe	*Vespula austriaca*	Austrian cuckoo wasp
Ostigel, Weißbrustigel	*Erinaceus concolor*	eastern hedgehog
Ostkaukasischer Steinbock	*Capra cylindricornis*	East-Caucasian tur
Ostkreischeule	*Otus asio*	eastern screech owl
Ostküsten-Seezunge	*Austroglossus pectoralis*	East Coast sole (South Africa)
Östliche Glasschleiche	*Ophisaurus ventralis*	eastern glass lizard (U.S.)
Östliche Harlekingarnele	*Hymenocera picta*	eastern harlequin shrimp
Östliche Kettennatter	*Lampropeltis getulus*	common kingsnake
Östliche Weißohr-Riesenratte, Goliathratte	*Hyomys goliath*	eastern white-eared giant rat
Östlicher Barschlachs	*Percopsis omiscomaycus*	trout-perch
Östlicher Blaupfeil	*Orthetrum albistylum*	eastern European skimmer
Östlicher Helmkopfbasilisk	*Laemanctus longipes*	eastern casquehead iguana
Östlicher Kielnagelgalago	*Galago matschiei/ Galago inustus*	eastern needle-clawed bush baby, spectacled galago
Östlicher Schaufelfuß	*Scaphiopus holbrookii*	eastern spadefoot, Holbrook's spadefoot
Östlicher Zaunleguan	*Sceloporus undulatus*	fence lizard, eastern fence lizard
Östliches Fuchshörnchen	*Sciurus niger*	eastern fox squirrel
Östliches Heupferd	*Tettigonia caudata*	eastern green bushcricket
Östliches Zwergchamäleon	*Bradypodion ventrale*	eastern dwarf chameleon
Ostpazifisch-Mittelamerikanischer Manta	*Manta hamiltoni*	Eastern Pacific manta

Ostpazifische Königsmakrele	*Scomberomorus sierra*	Pacific sierra
Ostpazifischer Spitznasen-Grundhai	*Rhizoprionodon longuris*	Pacific sharpnose shark
Ostschermaus	*Arvicola terrestris*	European water vole, northern water vole
Ostsee-Meerassel, Baltische Meerassel	*Idotea baltica*	Baltic isopod, Baltic sea "centipede"
Ostseegarnele	*Palaemon adspersus/ Palaemon squilla/ Leander adspersus*	Baltic prawn
Ostsibirischer Dorsch	*Arctogadus borisovi*	toothed cod
Othniiden, Falsche Tigerkäfer	Othniidae	false tiger beetles
Ottermuschel	*Lutraria lutraria*	common otter clam
Ottermuscheln	Lutrariidae	otter clams
Otterspitzmäuse	Potamogalidae	otter shrews
Ouachita-Landkartenschildkröte	*Graptemys ouachitensis*	Ouachita map turtle
Ovale Trogmuschel, Dickschalige Trogmuschel, Dickwandige Trogmuschel	*Spisula solida*	thick surfclam (thick trough shell)
Ovale Venusmuschel	*Timoclea ovata/ Chione ovata/ Venus ovata*	oval venus
Ovaler Dickmaulrüssler, Erdbeerwurzelrüssler	*Otiorhynchus ovatus*	strawberry root weevil
Owens Chamäleon	*Chamaeleo oweni*	Owen's three-horned chameleon, Owen's chameleon
Ozean-Flugfisch	*Exocoetus obtusirostris*	oceanic two-wing flyingfish
Ozeanischer Flohkrebs*	*Gammarus oceanicus*	scud
Ozeanmaräne*	*Caulolatilus princeps*	ocean whitefish
Ozelot	*Felis pardalis*	ocelot
Ozelotkatze, Oncille	*Felis tigrina*	little spotted cat

Paarhufer	Artiodactyla	even-toed ungulates, cloven-hoofed animals, artiodactyls
Paarzweig-Moostierchen	*Eucrate loricata*	
Pacu	*Myleus pacu*	pacu
Paddelwürmer, Ruderwürmer	*Phyllodoce* spp.	paddleworms
Paddelwürmer, Ruderwürmer	Phyllodocidae	paddleworms
Paganellgrundel, Felsgrundel, Felsengrundel	*Gobius paganellus*	rock goby
Pagoden-Strandschnecke	*Tectarius pagodus*	pagoda prickly winkle
Pagodenschnecke	*Pagodulina pagodula principalis*	pagoda snail
Pagodenstar	*Sturnus pagodarum*	black-headed starling
Pahmi (Hbz.), Chinesischer Sonnendachs	*Melogale moschata*	Chinese ferret badger
Pajarello-Zecke*	*Ornithodoros coriaceus*	pajarello tick
Pakarana	*Dinomys branickii*	pacarana
Pakaranas	Dinomyidae	pacaranas
Pakas	*Agouti* spp./*Cuniculus* spp.	pacas
Palästinaviper	*Vipera palaestinae/ Daboia palaestinae*	Palestinian viper, Palestine viper
Pallasammer	*Emberiza pallasi*	Pallas's reed bunting
Palmenblasenfuß	*Thrips palmi*	palm thrips
Palmenbohrer	*Dinapate wrighti*	palm borer
Palmendieb, Kokosnusskrebs	*Birgus latro*	coconut crab, palm crab, robber crab, purse crab
Palmenflughund	*Eidolon helvum*	straw-colored fruit bat
Palmenroller	*Paradoxurus* spp.	palm civets, musangs, toddy cats
Palmenschwamm*	*Isodictya palmata*	common palmate sponge
Palmenweber	*Ploceus bojeri*	golden palm weaver
Palmgeier	*Gypohierax angolensis*	palm-nut vulture
Palmschwätzer (Palmenschmätzer)	*Dulus dominicus*	palmchat
Palmtaube	*Streptopelia senegalensis*	palm dove
Palometa	*Trachinotus goodei*	palometa
Palpenkäfer, Zwergkäfer	Pselaphidae	short-winged mold beetles
Palpenspinner, Schnauzenspinner	*Pterostoma palpina*	pale prominent
Palpigraden	Palpigradi	microwhipscorpions, palpigrades
Pampasfuchs, Azarafuchs (Hbz. Provincia-Fuchs, Pampasfuchs)	*Pseudalopex gymnocercus*	pampas fox
Pampashirsch	*Odocoileus bezoarticus/ Ozotoceros bezoarticus*	pampas deer
Pampaskatze	*Felis colocolo/ Lynchailurus pajeros*	pampas cat
Panama-Fechterschnecke, Ostpazifische Flügelschnecke	*Strombus gracilior*	Eastern Pacific fighting conch, Panama fighting conch
Panama-Randschnecke*	*Persicula accola*	twinned marginella
Panama-Zwergboa	*Ungaliophis panamensis*	Panamanian dwarf boa

Panda>		
Großer Panda,	*Ailuropoda melanoleuca*	giant panda
Riesenpanda,		
Bambusbär		
Panda> Katzenbär	*Ailurus fulgens*	lesser panda, red panda
Pandorafalter, Kardinal	*Pandoriana pandora*	cardinal
Pandoramotte	*Coloradia pandora*	pandora moth
Pandoramuscheln, Büchsenmuscheln	Pandoridae	pandoras, pandora's boxes
Panga-Meerbrasse	*Pterogymnus lanarius*	panga
Pansenegel	*Paramphistomum* spp.	stomach flukes, rumen flukes
Pansenegel> Gemeiner Pansenegel	*Paramphistomum cervi*	common stomach fluke, rumen fluke
Pantherbarsch	*Plectropomus truncatus*	squaretail grouper
Pantherchamäleon	*Chamaeleo pardalis*	panther chameleon
Panthergecko, Leopardgecko	*Eublepharus macularius*	leopard gecko
Pantherkröte	*Bufo regularis*	panther toad
Pantherschildkröte	*Geochelone pardalis pardalis*	leopard tortoise
Panthervögel	Pardalotinae	pardalotes
Pantherwurm	*Hesione pantherina*	panther worm
Pantoffelkoralle*	*Herpolitha limax*	slipper coral
Pantoffeltierchen	*Paramecium caudatum*	slipper animalcule
Pantoffeltierchen	*Paramecium* spp.	slipper animalcules
Panzer-Gürtelschweif	*Cordylus cataphractus*	armadillo girdle-tailed lizard, armadillo lizard
Panzerchamäleon	*Leandria perarmata*	armored chameleon
Panzerechsen, Krokodile	Crocodilia	crocodiles
Panzergarnelen*	Glyphocrangonidae	armored shrimps
Panzergeißler, Dinoflagellaten	Dinoflagellata	dinoflagellates
Panzergroppen	Agonidae	poachers
Panzerhahn, Panzerknurrhahn	*Peristedion cataphractum*	armed gurnard, armored gurnard, African armoured searobin (FAO)
Panzerknurrhähne, Panzerhähne	Peristediidae	armored gurnards
Panzerkopf-Laubfrosch	*Triprion spatulatus*	shovel-headed treefrog
Panzerkopf-Laubfrösche	*Triprion* spp.	shovel-headed treefrogs
Panzerkopf-Laubfrösche u.a.	*Aparasphenodon* spp.	casque-headed frogs
Panzerköpfe	Pentacerotidae	armourheads, armorheads
Panzerkopfwelse	Cranoglanididae	armorhead catfishes
Panzerkrokodil	*Crocodylus cataphractus*	African slender-snouted crocodile, long-snouted crocodile
Panzerspitzmäuse	*Scutisorex* spp.	armored shrews, hero shrews
Panzerwangen, Drachenkopffischverwandte	Scorpaeniformes	sculpins & sea robins
Panzerwelse, Schwielenwelse	Callichthyidae	callichthyid armored catfishes
Papageien	Psittacidae (Psittaciformes)	parrots (& parakeets)
Papageien-Fahnenbarsch, Tiefenrötling	*Callanthias ruber*	parrot seaperch (FAO), barberfish
Papageienfische, Papageifische	Scaridae/Callyodontidae	parrotfishes

Papageienkärpfling, Veränderlicher Spiegelkärpfling	Xiphophorus variatus	variegated platy, variatus swordtail, variable platyfish (FAO)
Papageifisch u.a., Seepapagei	Sparisoma cretense	parrot fish (Med./Atlantic)
Papageischnäbel, Papageimeisen	Panuridae/ Paradoxornithidae	parrotbills
Papageischnabel-Flachschildkröte, Areolen-Flachschildkröte	Homopus areolatus	parrot-beaked tortoise, beaked Cape tortoise
Papageischnabelgimpel	Pseudonestor xanthophrys	Maui parrotbill
Papageitaucher	Fratercula arctica	puffin, Atlantic puffin
Papier-Kammmuschel	Amusium papyraceum	paper scallop
Papierboot	Argonauta argo	greater argonaut, common paper nautilus
Papierboote	Argonauta spp.	paper nautiluses
Papierschupper	Grammicolepididae	tinselfishes, grammicolepidids
Papillenfische, Rotmäulige Walkopffische	Rondeletiidae	redmouth whalefishes
Papillenschwamm	Polymastia mammillaris	papillate sponge*, teat sponge
Pappelblattgallmücke*	Harmandia loewi	poplar leaf gall midge
Pappelblattkäfer, Roter Pappelblattkäfer	Melasoma populi/ Chrysomela populi	red poplar leaf-beetle, poplar leaf beetle, poplar beetle
Pappelblattminiermotte	Phyllocnistis suffusella	poplar leaf miner a.o.
Pappelblattrippen-Gallenlaus	Pemphigus filaginis/ Pemphigus populinigrae	poplar-cudweed aphid
Pappelblattroller	Byctiscus populi	poplar leaf roller weevil
Pappeleulenspinner, Wollrückenspinner	Tethea or	poplar lutestring
Pappelhain-Eulenspinner, Achtzigeule, Augen-Eulenspinner	Tethea ocularis	figure of eighty
Pappelkarmin	Catocala elocata	poplar underwing
Pappelporzellanspinner, Pappelzahnspinner	Pheosia tremula	swallow prominent
Pappelschwärmer	Laothoe populi	poplar hawkmoth
Pappelspanner	Biston strataria	oak beauty, oak brindled beauty
Pappelspinner, Weidenspinner, Atlasspinner, Atlas	Leucoma salicis	satin moth
Papstfink	Passerina ciris	painted bunting
Papstkrone	Mitra papalis	papal mitre, papal miter
Papua-Schildkröte, Papua-Weichschildkröte	Carettochelys insculpta	pig-nosed turtle, pitted-shell turtle, New Guinea plateless turtle, pignosed softshell turtle
Papuaadler	Harpyopsis novaeguineae	New Guinea harpy eagle
Papuaweih	Aviceda subcristata	crested baza
Paradies-Baumschlangen	Chrysopelea paradisi	paradise tree snake
Paradiesfisch, Makropode	Macropodus opercularis	paradise fish
Paradiesfische	Macropodus spp.	paradisefishes

Paradiesglanzvogel	*Galbula dea*	paradise jacamar
Paradiesvögel	Paradisaeidae	birds-of-paradise
Paradox-Harlekinfrosch	*Pseudis paradoxa*	paradox frog, swimming frog
Paraguay-Kaiman	*Caiman yacare*	Paraguayan caiman
Paraguay-Maulbrüter, Paraguay-Erdfresser	*Gymnogeophagus balzanii*	Argentine humphead
Parahyba-Vogelspinne, Brasilien-Riesenvogelspinne	*Lasiodora parahybana*	Brazilian salmon tarantula
Parakärpfling	*Poecilia parae*	two-spot livebearer
Parascorpididen, Jutjawfische	Parascorpididae	jutjaw fish
Parasitenwelse, Schmarotzerwelse, Schmerlenwelse	Trichomycteridae	pencil catfishes, parasitic catfishes
Parasitische Holzwespen	Orussidae	parasitic woodwasps
Parasitische Krustenanemone	*Parazoanthus parasiticus*	parasitic colonial anemone, sponge zoanthid
Parasitischer Aal, Stumpfnase, Stumpfnasenaal	*Simenchelys parasitica*	snubnosed eel (FAO), snubnose parasitic eel
Parazen	*Parazen pacificus*	parazen
Pärchenegel (Blutegel)	*Schistosoma* spp.	blood flukes
Pardelroller	*Nandinia binotata*	African palm civet
Parkers Schlangenhalsschildkröte	*Chelodina parkeri*	Parker's side-necked turtle
Parkettkäfer	*Lyctus linearis*	true powderpost beetle, European powderpost beetle
Parmawallaby	*Macropus parma*	parma wallaby
Parry-Ziesel, Langschwanzziesel (Hbz. Suslik)	*Spermophilus undulatus*	Arctic ground squirrel
Parson-Spinne	*Herpyllus ecclesiasticus*	Parson spider
Parsons Chamäleon	*Chamaeleo parsoni*	Parson's chameleon
Partnergarnelen	*Periclimenes* spp.	cleaner shrimps, grass shrimps, coral shrimps, anemone shrimps
Pastetenmuscheln, Samtmuscheln	*Glycymeris* spp.	dog cockles (Br.), bittersweet clams (U.S.)
Patagonische Chinchillamäuse	*Euneomys* spp.	Patagonian chinchilla mice
Patagonische Erdleguane	*Diplolaemus* spp.	Patagonia iguanas
Patagonischer Grenadier	*Macruronus magellanicus*	Patagonian grenadier
Patagonischer Seehecht, Argentinischer Seehecht	*Merluccius hubbsi*	Argentine hake (FAO), Southwest Atlantic hake
Patagonischer Zahnfisch	*Eleginops maclovinus*	Patagonian mullet, Patagonian blenny (FAO)
Pazifik-Boa	*Candoia carinata*	tree boa, Solomon's ground boa
Pazifik-Brandungskrebs	*Emerita analoga*	Pacific sand crab
Pazifik-Dorsch, Pazifischer Kabeljau	*Gadus macrocephalus*	Pacific cod (FAO), gray cod, grayfish
Pazifik-Halbschnabelhecht	*Hyporhamphus acutus*	Pacific halfbeak
Pazifik-Hüpfmaus	*Zapus trinotatus*	Pacific jumping mouse

Pazifik-Korallenschlangen	*Micropochis* spp.	Pacific coral snakes
Pazifik-Kronenschnecke	*Melongena patula*	Pacific crown conch
Pazifik-Laubfrosch	*Hyla regilla*	Pacific treefrog
Pazifik-Meerengel, Pazifischer Engelhai	*Squatina californica*	Pacific angelshark
Pazifik-Pottwal, Kleinpottwal	*Kogia simus*	dwarf sperm whale
Pazifik-Purpurseeigel	*Strongylocentrotus purpuratus*	purple sea urchin
Pazifik-Riesenbarsch	*Stereolepis gigas*	giant sea bass
Pazifik-Samtmuschel	*Glycymeris subobsoleta*	Pacific Coast bittersweet, West Coast bittersweet
Pazifik-Sandgarnele	*Crangon franciscorum*	California bay shrimp, Pacific sand shrimp
Pazifik-Seehecht, Nordpazifischer Seehecht	*Merluccius productus*	Pacific hake, North Pacific hake (FAO)
Pazifik-Stachelauster, Pazifische Stachelauster, Panama-Stachelauster	*Spondylus princeps*	Pacific thorny oyster
Pazifik-Waran, Indischer Mangrovenwaran	*Varanus indicus*	Pacific monitor, mangrove monitor, Indian mangrove monitor
Pazifik-Wasserschildkröte	*Clemmys marmorata*	Pacific pond turtle, western pond turtle
Pazifik-Zitterrochen, Kalifornischer Zitterrochen	*Torpedo californica*	Pacific electric ray
Pazifikgecko	*Gehyra mutilata*	common house gecko
Pazifikküsten-Zecke*	*Dermacentor occidentalis*	Pacific Coast tick
Pazifische Anchoveta	*Cetengraulis mysticetus*	Pacific anchoveta
Pazifische Auster, Riesenauster	*Crassostrea gigas*	Pacific oyster, giant Pacific oyster, Japanese oyster
Pazifische Bastardmakrele, Pazifik-Stöcker	*Trachurus symmetricus*	Pacific jack mackerel
Pazifische Bastardschildkröte	*Lepidochelys olivacea*	olive ridley (sea turtle), Pacific ridley turtle
Pazifische Blaukrabbe, Pazifische Blaue Schwimmkrabbe, Große Pazifische Schwimmkrabbe	*Portunus pelagicus*	blue swimming crab, sand crab, pelagic swimming crab
Pazifische Braune Nesselqualle, Braune Nesselqualle	*Chrysaora melanaster*	brown jellyfish
Pazifische Falsche Dreiecksmuschel*	*Heterodonax pacificus*	Pacific false-bean clam
Pazifische Feilenmuschel	*Lima vulgaris*	Pacific fileclam
Pazifische Felsenauster	*Crassostrea rhizophorae*	Pacific cupped oyster, mangrove cupped oyster
Pazifische Karettschildkröte	*Eretmochelys imbricata bissa*	Pacific hawksbill turtle
Pazifische Klaffmuschel	*Tresus nuttallii*	Pacific gaper clam, "horseneck clam"
Pazifische Lachs-Tellmuschel	*Tellina nuculoides/ Tellina salmonea*	salmon tellin
Pazifische Lachse	*Oncorhynchus* spp.	Pacific salmon
Pazifische Löwenpranke	*Lyropecten subnodosa*	Pacific lion's paw
Pazifische Messermuschel	*Siliqua patula*	Pacific razor clam

Pazifische Mondschnecke	*Lunatia lewisi*	Lewis' moonsnail, western moon shell
Pazifische Nussmuschel	*Nucula exigua*	iridescent nutclam, Pacific crenulate nutclam
Pazifische Panopea	*Panopea abrupta/ Panopea generosa*	Pacific geoduck, geoduck (*pronounce*: "gouy-duck")
Pazifische Reiterkrabbe	*Ocypode gaudichaudii*	Pacific ghost crab
Pazifische Riesen-Felskammmuschel	*Hinnites giganteus*	giant rock scallop
Pazifische Riesen-Herzmuschel	*Laevicardium elatum*	giant eggcockle, giant Pacific egg cockle
Pazifische Riesen-Kammmuschel	*Patinopecten caurinus*	giant Pacific scallop
Pazifische Riesen-Löwenpranke	*Nodipecten subnodosus*	giant lions-paw
Pazifische Riesenanemone	*Metridium giganteum*	gigantic anemone, white-plumed anemone
Pazifische Riesenfassschnecke	*Tonna olearium*	giant Pacific tun
Pazifische Rossie	*Rossia pacifica*	North Pacific bobtail squid, Pacific bob-tailed squid
Pazifische Rotpunkt-Schwimmkrabbe	*Portunus sanguinolentus*	blood-spotted swimming crab
Pazifische Sandmuschel	*Asaphis violascens*	Pacific asaphis, violet asaphis
Pazifische Scholle	*Lepidopsetta bilineata*	rock sole
Pazifische Schwebkrabbe	*Pachygrapsus marinus*	drifter crab
Pazifische Seespinne	*Hyas lyratus*	Pacific lyre crab
Pazifische Seestachelbeere	*Pleurobrachia bachei*	Pacific sea gooseberry, cat's eye
Pazifische Seezunge	*Achirus lineatus*	lined sole
Pazifische Thonine	*Euthynnus affinis*	kawakawa
Pazifische Vasenschnecke	*Vasum turbinellum*	Pacific top vase
Pazifische Wasserspitzmaus	*Sorex bendirii*	Pacific water shrew
Pazifische Weiße Garnele	*Litopenaeus occidentalis/ Penaeus occidentalis*	western white shrimp, Central American white shrimp
Pazifischer Butterfisch, Pazifischer Pompano	*Peprilus simillimus*	Pacific pompano
Pazifischer Fadenhering	*Opisthonema libertate*	deep-body thread herring, Pacific thread herring (FAO)
Pazifischer Fleckenbarsch	*Epinephelus analogus*	spotted grouper
Pazifischer Goldregenpfeifer	*Pluvialis fulva*	Pacific golden plover
Pazifischer Heilbutt	*Hippoglossus stenolepis*	Pacific halibut
Pazifischer Hering	*Clupea pallasi*	Pacific herring
Pazifischer Heringshai	*Lamna ditropis*	Pacific porbeagle, salmon shark (FAO)
Pazifischer Kleinhering	*Sardinella zunasi/ Harengula zunasi*	Japanese sardinella
Pazifischer Menhaden	*Ethmidium maculatum*	Pacific menhaden
Pazifischer Opalkalmar, Opalisierender Kalmar	*Loligo opalescens*	opalescent inshore squid, opalescent squid (American "market squid"), common Pacific squid
Pazifischer Peitschenrochen, Pazifischer Stechrochen	*Dasyatis akajei*	whip stingray, whip ray, red stingray (FAO)

Pazifischer Pompano, Pazifischer Butterfisch	*Peprilus simillimus*	Pacific pompano
Pazifischer Riesen-Querzahnmolch	*Dicamptodon tenebrosus*	Pacific giant salamander
Pazifischer Riesenkrake	*Octopus dofleini*	North Pacific giant octopus (FAO), giant Pacific octopus, common Pacific octopus
Pazifischer Rochen	*Raja stellulata*	thorny skate, starry skate (FAO)
Pazifischer Rotbarsch, Pazifik-Goldbarsch	*Sebastes alutus*	Pacific ocean perch
Pazifischer Roter Seeigel	*Strongylocentrotus franciscanus*	red sea urchin
Pazifischer Sandaal	*Ammodytes hexapterus*	Pacific sandlance
Pazifischer Seehecht, Chilenischer Seehecht	*Merluccius gayi*	Chilean hake, Peruvian hake, South Pacific hake (FAO)
Pazifischer Seewolf	*Anarrhichthys ocellatus*	Pacific wolf-eel, wolf-eel (FAO)
Pazifischer Silberbeil	*Argyropelecus affinis*	Pacific hatchetfish
Pazifischer Stechrochen, Pazifischer Peitschenrochen	*Dasyatis akajei*	whip stingray, whip ray, red stingray (FAO)
Pazifischer Taschenkrebs	*Cancer antennarius*	Pacific rock crab
Pazifischer Taschenkrebs> Kalifornischer Taschenkrebs	*Cancer magister*	Dungeness crab, Californian crab, Pacific crab
Pazifischer Tomcod	*Microgadus proximus*	Pacific tomcod
Pazifischer Wasserpieper	*Anthus rubescens*	buff-bellied pipit
Pazifischer Weißseiten-Delphin, Weißstreifendelphin	*Lagenorhynchus obliquidens*	Pacific white-sided dolphin
Pazifisches Glasauge	*Argentina sialis*	North-Pacific argentine (FAO)
Pazifisches Weinhorn*	*Neptunea vinosa*	wine whelk
Pearsons Flughörnchen	*Belomys pearsoni*	hairy-footed flying squirrel
Pediliden	Pedilidae	false antloving flower beetles
Peitschennatter, Peitschenschlange, Rote Kutschenpeitschen-Natter	*Masticophis flagellum flagellum/ Coluber flagellum*	coachwhip, San Joaquin coachwhip
Peitschennattern	*Masticophis* spp.	whipsnakes
Peitschenwürmer	*Trichuris* spp.	whipworms
Pelamide	*Sarda sarda*	Atlantic bonito
Pelamiden, Königsmakrelen	Scomberomoridae (Cybiidae)	Spanish mackerels
Peledmaräne	*Coregonus peled*	northern whitefish, big powan, peled (FAO)
Pelikanaal	*Eurypharynx pelecanoides*	pelican eel
Pelikanaale> Echte Pelikanaale, Großmäuler	Eurypharyngidae/ Eupharyngidae	pelican eels, gulpers
Pelikane	Pelecanidae	pelicans
Pelomedusen	*Pelomedusa* spp.	side-necked turtles
Pelomedusen-Schildkröten	Pelomedusidae	Afro-American side-necked turtles, Afro-American sidenecks
Peloponnes-Eidechse	*Podarcis peloponnesciaca*	Peloponnese wall lizard
Peloponnesische Kieleidechse	*Algyroides moreoticus*	Greek keeled lizard

Pelz-Seezunge, Schnurrbart-Zunge	*Monochirus hispidus*	whiskered sole
Pelzbienen	*Anthophora* spp.	flower bees
Pelzflatterer, Riesengleitflieger, Riesengleiter	Dermoptera	flying lemurs, gliding lemurs, colugos, dermopterans
Pelzflohkäfer, Mausflohkäfer	Leptinidae	leptinids (mammal-nest beetles & beaver parasites, beaver beetles, rodent beetles)
Pelzgroppen, Pelzbarsche	Caracanthidae	coral crouchers, orbicular velvetfishes
Pelzkäfer, Gefleckter Pelzkäfer	*Attagenus pellio*	fur beetle (black carpet beetle)
Pelzmilben	Cheyletiellidae	fur mites
Pelzmotte	*Tinea pellionella*	case-bearing clothes moth, case-making clothes moth
Pendelspanner, Weißgrauer Ringfleckspanner, Blasser Ringelfleck-Gürtelpuppenspanner	*Cyclophora pendularia/ Cyclophora orbicularia*	dingy mocha
Pennant-Rotstummelaffe	*Piliocolobus pennanti*	Pennant's red colobus
Pennants Schwimmkrabbe	*Portumnus latipes*	Pennant's swimming crab
Pennantsittich	*Platycerus elegans*	crimson rosella
Pennsylvania-Klappschildkröte	*Kinosternon subrubrum subrubrum*	mud turtle, common mud turtle, eastern mud turtle
Pennsylvanisches Glühwürmchen	*Photinus pennsylvanicus*	Pennsylvania firefly
Pentland-Steißhuhn, Andensteißhuhn	*Nothoprocta pentlandii*	Andean tinamou
Père-Davids-Wühlmäuse	*Eothenomys* spp.	Père David's voles, Pratt's voles
Perenty, Großwaran	*Varanus giganteus*	perenty, perentie
Pergamentwurm	*Chaetopterus variopedatus*	parchment tubeworm, parchment worm
Pergamentwürmer	Chaetopteridae	parchment tubeworms, chaetopterid worms
Perl-Preußenfisch	*Dascyllus melanurus*	black-tailed humbug
Perlaugen	Scopelarchidae	pearleyes
Perlbarsche*	Glaucosomatidae	pearl perches
Perlbinde, Schlüsselblumen-Würfelfalter, Brauner Würfelfalter, Frühlingsscheckenfalter	*Hamearis lucina*	Duke of Burgundy
Perlboote, Schiffsboote	*Nautilus* spp.	chambered nautiluses, pearly nautiluses
Perleidechse	*Lacerta lepida*	ocellated lizard, ocellated green lizard, eyed lizard, jewelled lizard
Perlfisch*	*Echiodon drummondi*	pearlfish
Perlgrasfalter	*Coenonympha arcania*	pearly heath

Perlhühner	Numidinae	guineafowls
Perlhuhnwels, Engelwels	*Synodontis angelicus*	polka-dot African catfish, angel squeaker (FAO)
Perlige Florfliege*	*Chrysopa perla*	pearly green lacewing
Perlkauz	*Glaucidium perlatum*	pearl-spotted owlet
Perlkreiselschnecken	*Margarites* spp.	margarites, pearly topsnails (pearly top shells)
Perlmuscheln, Vogelmuscheln	Pteriidae	wing oysters & pearl oysters
Perlmuttbärbling	*Rasbora vaterifloris*	pearly rasbora
Perlmutter-Erdfresser, Brasilperlmutterfisch	*Geophagus brasiliensis*	pearl cichlid
Perlmutterbuntbarsch	*Tilapia lepidura/ Oreochromis lepidura*	nacreous mouthbreeder*
Perlmutterkärpfling	*Aphanius dispar*	pearly killifish*
Perlnatter, Gesprenkelte Bodenschlange	*Drymobius margaritiferus margaritiferus*	speckled racer
Perlodiden	Perlodidae	perlodid stoneflies
Perlolive	*Oliva reticularis olorinella*	pearl olive
Perlsteißhuhn	*Eudromia elegans*	elegant crested-tinamou
Perlsuslik (Hbz.), Perlziesel	*Spermophilus suslicus*	spotted suslik
Perlwachtel	*Margaroperdix madagarensis*	Madagascar partridge
Perlziesel (Hbz. Perlsuslik)	*Spermophilus suslicus*	spotted suslik
Permit	*Trachinotus falcatus*	permit
Peron-Seeschlange	*Acalyptophis peronii*	Peron's sea snake
Perrys Schwarzer Dornhai	*Etmopterus perryi*	dwarf dogshark, dwarf lanternshark
Persische Bandschnecke	*Pleuroploca persica*	Persian horseconch
Persische Zecke	*Argas persicus*	fowl tick, Persian poultry tick (bluebug, abode tick, tampan)
Persischer Kammmolch, Balkankammmolch	*Triturus karelini*	Balkan crested newt
Persisches Wüstenhuhn	*Ammoperdis griseogularis*	see-see partridge
Peru-Opossummaus	*Lestoros inca*	Peruvian "shrew" opossum
Peru-Sardelle, Anchoveta	*Engraulis ringens*	anchoveta (FAO), Peruvian anchovy
Peruanische Fechterschnecke, Peruanische Flügelschnecke	*Strombus peruvianus*	Peruvian conch
Peruanischer Felsbarsch	*Paralabrax humeralis*	Peruvian rock bass
Peruanischer Geigenrochen	*Rhinobatos planiceps*	Peruvian guitarfish, Pacific guitarfish (FAO)
Peruanischer Umberfisch	*Cynoscion analis*	Peruvian weakfish
Perwitzki (Hbz.), Tigeriltis	*Vormela peregusna*	marbled polecat
Pestfloh, Tropischer Rattenfloh	*Xenopsylla cheopis*	Oriental rat flea
Petermännchen, Großes Petermännchen	*Trachinus draco*	great weever (FAO), greater weever
Peters Einstreifenmaus	*Hybomys univittatus*	Peter's striped mouse
Peters Spießblattnase	*Phylloderma stenops*	Peter's spear-nosed bat
Peters Wollgesicht-Fledermaus	*Chrotopterus auritus*	Peter's woolly false vampire bat
Petersducker, Harveyducker, Schönsteißducker	*Cephalophus callipygus*	Peters' duiker

Petersfisch, Heringskönig	*Zeus faber*	Dory, John Dory
Petersfischartige> Petersfische & Eberfische	Zeiformes	dories (John Dory) & others
Petersfische	Zeidae	dories
Petroleumfliege	*Psilopa petrolei/* *Helaeomyia petrolei*	petroleum fly
Petschanik (Hbz.), Sandziesel	*Spermophilus fulvus*	large-toothed souslik
Petschorapieper	*Anthus gustavi*	pechora pipit
Pfahlwurm, Schiffsbohrmuschel, Schiffsbohrwurm	*Teredo navalis*	naval shipworm, common shipworm, great shipworm
Pfahlwürmer, Schiffsbohrwürmer	Teredinidae	shipworms
Pfannkuchen-Koralle*	*Dichocoenia stellaris*	pancake star coral
Pfannkuchenfisch*	*Halieutichthys aculeatus*	pancake batfish
Pfau	*Pavo cristatus*	common peafowl (peacock/peahen), Indian peafowl
Pfauen-Butt, Pfauenaugenbutt	*Bothus lunatus*	peacock flounder
Pfauen-Fangschreckenkrebs, Bunter Fangschreckenkrebs	*Odontodactylus scyllarus*	peacock mantis shrimp
Pfauen-Lippfisch	*Symphodus tinca*	painted wrasse
Pfauenaugen- Sumpfschildkröten	*Morenia* spp.	eyed turtles
Pfauenaugen- Weichschildkröte	*Aspideretes hurum/* *Trionyx hurum*	Indian peacock softshell turtle, peacock softshell turtle
Pfauenaugenbarsch, Pfauenaugensonnenbarsch	*Centrarchus macropterus*	flier
Pfauenaugenbuntbarsch, Roter Oskar, Oskar	*Astronotus ocellatus*	oscar (FAO), oscar's cichlid, velvet cichlid
Pfauenaugenbutt, Pfauen-Butt	*Bothus lunatus*	peacock flounder
Pfauenfederfisch, Meerjunker.	*Coris julis*	rainbow wrasse
Pfauenfederwurm, Pfauenwurm	*Sabella pavonina*	peacock worm, peacock feather-duster worm
Pfauenkaiserfisch, Herzogen-Kaiserfisch	*Pygoplites diacanthus*	royal angelfish (FAO), blue-banded angelfish, regal angelfish
Pfauenlippfisch, Meerschlei	*Crenilabrus pavo/* *Symphodus tinca*	peacock wrasse
Pfauenschleimfisch	*Lipophrys pavo/* *Blennius pavo/* *Salaria pavo*	peacock blenny
Pfefferfresser	*Selenidera culik*	Guianan toucanet
Pfefferminzgarnele	*Lysmata wurdemanni*	peppermint shrimp, Carribbean cleaner shrimp, veined shrimp
Pfeffermuräne	*Siderea picta*	peppered moray (FAO), painted moray
Pfeffermuscheln	Scorbiculariidae	furrow clams (furrow shells)
Pfeifen-Knurrhahn, Leierknurrhahn, Pfeifenfisch	*Trigla lyra*	piper, piper gurnard (FAO)

Pfeifente	Anas penelope	wigeon
Pfeiffrösche> Echte Pfeiffrösche	Leptodactylus spp.	white-lipped frogs, nest-building frogs
Pfeifgänse (Baumenten)	Dendrocygnidae	whistling-ducks
Pfeifgecko	Ptenopus garrulus	barking gecko, talkative gecko
Pfeifhasen	Ochotona spp.	pikas, mouse hares, conies (coneys)
Pfeifhasen	Ochotonidae	pikas
Pfeifratten	Parotomys spp.	Karroo rats, whistling rats
Pfeileule	Acronicta psi	grey dagger
Pfeilflossenkalmar	Illex oxygonius	sharptail shortfin squid, arrow-finned squid
Pfeilgiftfrösche, Baumsteigerfrösche, Farbfrösche	Dendrobatidae	dendrobatids, poison dart frogs, poison arrow frogs, arrow-poison frogs
Pfeilhechte, Barrakudas	Sphyraena spp.	barracudas
Pfeilkalmar	Todarodes sagittatus/ Ommastrephes sagittatus	flying squid, European flying squid (FAO), sagittate squid, sea-arrow, red squid
Pfeilnatter	Coluber jugularis	green whip snake, large whip snake
Pfeilotter, Krötenotter	Causus rhombeatus	rhombic night adder, common night adder
Pfeilschwanzkrebse	Xiphosura	horseshoe crabs
Pfeilwürmer, Borstenkiefer, Chaetognathen	Chaetognatha	arrow worms, chaetognathans
Pfeilzahn-Heilbutt, Asiatischer Pfeilzahnheilbutt	Atheresthes evermanni	Kamchatka halibut
Pfeilzahnaale	Dysommidae	arrowtooth eels, mustard eels
Pferd> Hauspferd	Equus caballus	horse
Pferdeaktinie, Purpurrose	Actinia equina	beadlet anemone, red sea anemone, plum anemone
Pferdeantilope, Roan	Hippotragus equinus	roan antelope
Pferdebandwürmer	Anoplocephala spp.	horse tapeworms, equine tapeworms
Pferdebauchhöhlenfilarie	Setaria equina	equine abdominal worm
Pferdebohnenkäfer, Ackerbohnenkäfer	Bruchus rufimanus	broadbean weevil, bean weevil
Pferdebremsen & Rinderbremsen	Tabanus spp.	horseflies
Pferdedarmegel u.a.	Gastrodiscus aegyptiacus	equine intestinal fluke
Pferdeegel, Vielfraßegel	Haemopis sanguisuga	European horse leech
Pferdefuß-Räudemilbe*, OchsenschwanzRäudemilbe*	Chorioptes bovis	chorioptic mange mite, horse foot mange mite, oxtail mange mite, symbiotic mange mite
Pferdehaarling	Bovicola equi/Werneckiella equi	horse-biting louse
Pferdehirsch, Indischer Sambar	Cervus unicolor	sambar

Pferdehufmuschel	*Hippopus hippopus*	bear's paw clam, bear paw, horseshoe clam, horse's hoof, strawberry clam
Pferdelaus	*Haematopinus asini macrocephalus*	horse-sucking louse
Pferdelausfliege	*Hippobosca equina*	forest-fly
Pferdemagenbremse	*Gasterophilus intestinalis*	horse botfly, common horse botfly, common botfly
Pferdemagenwurm u.a.	*Habronema muscae*	equine stomach worm a.o.
Pferdemakrele, Pferde-Makrele, Pferde-Stachelmakrele, Caballa	*Caranx hippos*	crevalle jack (FAO), Samson fish
Pferdeschwamm	*Hippospongia communis*	horse sponge
Pferdespringer	*Allactaga* spp.	four- and five-toed jerboas
Pferdestern, Pferde-Kissenstern	*Hippasteria phrygiana*	horse star, rigid cushion star
Pfirsichmotte	*Anarsia lineatella*	peach twig borer
Pflanze-Tier-Dasselfliege*	*Gasterophilus pecorum*	plant-animal botfly
Pflanzenkäfer	Alleculidae	comb-clawed beetles, comb-clawed bark beetles
Pflanzenmäher	Phytotomidae	plantcutters
Pflanzensauger	Homoptera	homopterans (cicadas & aphids & scale insects)
Pflasterstein-Sandlaufkäfer	*Cicindela marginipennis*	cobblestone tiger beetle
Pflasterzahnnattern	*Salvadora* spp.	patchnose snakes, patch-nosed snakes
Pflaumenbeutelgallmilbe, Pflaumenblatt-Beutelgallmilbe	*Eriophyes similis*	plum pouch-gall mite
Pflaumenböckchen	*Tetrops praeusta*	little longhorn beetle
Pflaumengespinstmotte	*Hyponomeuta padellus/ Yponomeuta padellus*	common hawthorn ermel, small ermine moth
Pflaumenknospenwickler	*Hedya pruniana*	plum tortrix (moth)
Pflaumenkopfsittich	*Psittacula cyanocephala*	plum-headed parakeet
Pflaumenmilbe	*Lepidoglyphus destructor/ Glycyphagus destructor/ Lepidoglyphus cadaverum*	cosmopolitan food mite, 'grocers' itch mite
Pflaumenrostmilbe	*Aculus fockeui/ Aculus cornutus*	plum rust mite, peach silver mite
Pflaumenschnecke*, Pflaumen-Randschnecke*	*Prunum prunum*	plum marginella
Pflaumenspanner, Lärchenspanner, Beerenkrautspanner, Heidelbeerspanner, Tannenspanner	*Ectropis crepuscularia/ Ectropis bistortata*	larch looper, blueberry looper, fir looper, plum looper
Pflaumenspanner, Schlehenspanner	*Angerona prunaria*	orange moth
Pflaumenwickler, Pflaumenmade	*Laspeyresia funebrana/ Cydia funebrana*	plum fruit moth, plum moth, red plum maggot
Pflaumenzipfelfalter	*Strymonidia pruni*	black hairstreak
Pflugnasenchimären, Elefantenchimären	Callorhynchidae	plownose chimaeras, elephantfishes

Pfriemenmücken	Anisopodidae	window midges, wood gnats
Pfriemenschnecke	Terebra subulata	subulate auger, chocolate spotted auger
Pfriemenschnecken, Schraubenschnecken	Terebridae	auger snails, augers (auger shells)
Pfriemenschwanz, Madenwurm, Springwurm	Enterobius vermicularis	pinworm (of man), seatworm
Pfriemenschwanz des Pferdes	Oxyuris equi	horse pinworm, equine pinworm
Pfriemenschwänze	Heterakis spp.	pinworms a.o.
Pfriemschnäbel	Toxorhamphini	longbills, honeyeaters
Pfuhlschnepfe	Limosa lapponica	bar-tailed godwit
Phantom-Glaswels	Kryptopterus macrocephalus	poor man's glass catfish, striped glass catfish (FAO)
Pharaoameise, Pharao-Ameise	Monomorium pharaonis	Pharaoh ant, Pharaoh's ant, little red ant
Pharaonenziegenmelker	Caprimulgus aegyptius	Egyptian nightjar
Phayres Langur	Presbytis phayrei	Phayre's leaf monkey
Phengodiden	Phengodidae	glowworm beetles
Philippinen-Gleitflieger	Cynocephalus volans	Philippine flying lemur
Philippinen-Koboldmaki	Tarsius syrichta	Philippine tarsier
Philippinenfrosch	Rana cancrivora	Philippine frog
Philippinenkröten	Pelophryne spp.	Philippine toads
Philippinensambar	Cervus mariannus	Philippine sambar
Philippinische Ratten u.a.	Apomys spp.	Philippine rats a.o.
Philippinische Segelechse	Hydrosaurus pustulatus	Philippine sail-fin lizard
Philippinische Sumpfratten	Crunomys spp.	Philippine swamp rats
Philippinische Teichschildkröte	Geoemyda leytensis	Philippine pond turtle
Philippinischer Rückenstreifen-Anemonenfisch, Philippinen-Weißrücken-Clownfisch	Amphiprion sandaracinos	white-backed anemonefish, yellow clownfish (FAO)
Philippinisches Täubchen	Pyrene phiippinarum	Philippine dovesnail
Phoebe	Sayornis phoebe	eastern phoebe
Piano-Säbelzahnschleimfisch	Plagiotremus tapeinosoma	scale-eating blenny
Piapia	Ptilostomus afer	piapiac
Picassofisch, Gemeiner Picassodrückerfisch	Rhinecanthus aculeatus	Picasso fish, humuhumu, blackbar triggerfish (FAO)
Pieperwaldsänger	Seiurus aurocapillus	ovenbird
Pierwurm, Sandpier, Sandpierwurm, Köderwurm	Arenicola marina	European lug worm, blow lug
Pikarel	Spicara smaris	picarel, zerro
Pikarels, Schnauzenbrassen	Spicara spp.	picarels
Pilgermuscheln, Kammmuscheln	Pectinidae	scallops
Pillenkäfer	Byrrhidae	pill beetles

Pilotbarsche, Steuerbarsche	Kyphosidae	sea chubs
Pilz-Seescheide	*Polycitor crystallinus*	fungal sea-squirt*
Pilzfliegen	Clythiidae/Platypezidae	flat-footed flies
Pilzkäfer, Behaarter Stäublingskäfer	*Mycetaea hirta*	hairy fungus beetle
Pilzkäfer, Stäublingskäfer, Pilzfresser (Puffpilzkäfer)	Endomychidae	handsome fungus beetles
Pilzkoralle	*Fungia fungites*	mushroom coral
Pilzkorallen	*Fungia* spp.	mushroom corals
Pilzmücke*	*Henria psalliotae*	mushroom midge
Pilzmücken	Mycetophilidae	fungus gnats
Pilzmyzelmilbe, Chamignon-Weichhautmilbe	*Tarsonemus myceliophagus*	mushroom mite
Pilzplattkäfer	Biphyllidae	false skin beetles
Pilzschnegel	*Malacolimax tenellus*	slender slug
Pinguin-Flügelmuschel	*Pteria penguin*	penguin wing-oyster
Pinguine	Sphenisciformes/Spheniscidae	penguins
Pinguinsalmler	*Thayeria obliqua*	penguinfish (FAO), honey stick tetra
Pinien-Prozessionsspinner	*Thaumetopoea pityocampa*	pine processionary moth
Pinkpink-Cistensänger	*Cisticola textrix*	tink-tink cisticola
Pinselfüßer	*Polyxenus lagurus*	brushfoot millepede*
Pinselfüßer	Pselaphognatha/ Penicillata (Polyxenidae)	pselaphognaths
Pinselkäfer	Trichiinae	trichiine beetles
Pinselkäfer	*Trichius fasciatus*	bee chafer
Pinselohrschwein, Buschschwein, Flussschwein	*Potamochoerus porcus*	African bush pig, red river hog
Pinselqualle*	*Polyorchis penicillatus*	penicillate jellyfish
Pinselschwanz-Baummäuse	*Chiropodomys* spp.	pencil-tailed tree mice
Pinselschwanz-Bilche	*Graphiurus* spp.	African dormice
Pinselschwanzbeutler	*Phascogale* spp.	brush-tailed marsupial "mice", tuans
Pinselstachler, Malaiischer Pinselstachler	*Trichys fasciculata*	long-tailed porcupine
Pinzettfische	*Forcipiger* spp.	forceps fishes
Pipras, Schnurrvögel, Manakins	Pipridae	manakins
Pirambeba	*Serrasalmus brandtii*	pirambeba, white piranha (FAO)
Piranha, Piraya, Karibenfisch	*Serrasalmus piraya*	piranha
Piranhas, Pirahas, Sägesalmler	*Serrasalmus* spp.	piranhas
Piratenbarsch	*Aphredoderus sayanus*	pirate perch
Piratenspinne	*Pirata piraticus*	pirate spider
Pirol	*Oriolus oriolus*	golden oriole
Pirole & Feigenpirol	Oriolidae	orioles, cuckooshrikes
Piroltrupial	*Icterus pustulatus*	streak-backed oriole
Pistaziensamenwespe	*Megastigmus pistaciae*	pistachio seed wasp

Pistolenkrebse, Knallkrebschen	Alpheidae	snapping prawns, snapping shrimps
Pistolenkrebse, Knallkrebschen	*Alpheus* spp.	snapping prawns, snapping shrimps, pistol shrimps
Pittas	Pittidae	pittas
Pityusen-Eidechse	*Podarcis pityusensis*	Ibiza wall lizard
Pizza-Anemone, Noppenrand-Anemone, Noppenrand-Meerblume	*Cryptodendrum adhaesivum*	pizza anemone, nap-edged anemone
Planktonwurm	*Tomopteris helgolandica*	plankton worm
Plantagenhörnchen, Plantagen-Schönhörnchen	*Callosciurus notatus*	plantation squirrel
Platanengallmilbe	*Aculops acericola*	sycamore gall mite
Platin-Beilbauchfisch	*Thoracocharax securis*	giant hatchetfish (FAO), greater hatchetfish, greater silver hatchfish
Plattbauch	*Libellula depressa*	broad-bodied libellula, broad-bodied chaser
Plattbauchspinnen, Glattbauchspinnen	Gnaphosidae/ Drassodidae	hunting spiders, ground spiders
Plättchen-Seeschlange, Gelbbauch-Seeschlange	*Pelamis platurus*	yellow-bellied sea snake, pelagic sea snake
Platte Pfeffermuschel	*Abra tenuis*	flat European abra (flat furrow-shell)
Platte Rippenquallen	Platyctenidea	platyctenids
Platte Sandmuschel	*Gari ornata*	ornate sunsetclam
Platte Tellmuschel, Zarte Tellmuschel, Plattmuschel	*Tellina tenuis/ Angulus tenuis*	thin tellin, plain tellin, petal tellin
Platten-Feuerkoralle, Brettartige Feuerkoralle	*Millepora platyphylla*	sheet fire coral
Plattenkiemer (Haie & Rochen)	Elasmobranchii	sharks & rays & skates
Plattenkoralle	*Acropora clathrata*	
Platterbsenspanner	*Scotopteryx chenopodiata*	shaded broad-bar (moth)
Plattfische	Pleuronectiformes	flatfishes
Plattfüßer (Sohlenfliegen, Tummelfliegen)	Platypezidae	flat-footed flies
Plattgürtelechsen	*Platysaurus* spp.	flat lizards
Plattkäfer, Schmalkäfer	Cucujidae	flat bark beetles
Plattmuschel, Platte Tellmuschel, Zarte Tellmuschel	*Tellina tenuis/ Angulus tenuis*	thin tellin, plain tellin, petal tellin
Plattrückenschildkröte	*Notochelys platynota*	Malayan flatshell turtle, Malayan flat-shelled turtle
Plattschildkröte	*Platemys* spp.	flat-shelled turtles
Plattschmerlen, Karpfenschmerlen	Homalopteridae/Balitoridae	river loaches, hillstream loaches
Plattschwanz-Seeschlange> Gewöhnlicher Plattschwanz	*Laticauda laticaudata*	common banded sea snake, black-banded sea snake
Plattschwanz-Seeschlangen, Plattschwänze	*Laticauda* spp.	banded sea snakes, sea kraits
Plattschwänzchen	*Egernia depressa*	pigmy spinytail skink
Plattwanzen (Hauswanzen)	Cimicidae	bedbugs

Plattwürmer, Plathelminthen	Plathelminthes	flatworms, platyhelminths (Platyhelminthes)
Platy, Mondplaty, Spiegelkärpfling	Xiphophorus maculatus	southern platyfish
Plazentatiere	Placentalia/ Eutheria	placentals, eutherians
Pleskes Wüstenrenner, Transkaukasischer Wüstenrenner	Eremias pleskei	Pleske's racerunner
Plötze, Rotauge	Rutilus rutilus	roach (FAO), Balkan roach
Pluma-Brasse	Calamus pennatula	pluma
Plumplori, Großer Plumplori	Nycticebus coucang	slow loris, culan
Plumpschrecke	Isophya kraussi	speckled bushcricket
Plüschkopfente	Somateria fischeri	spectacled eider
Plüschkopftangare	Catamblyrhynchinae	plush-capped finch, plushcap
Pockenschildläuse, Pockenläuse	Asterolecaniidae	pit scales
Podas Waldschabe	Ectobius silvestris	Poda's cockroach
Podolische Blindmaus	Spalax polonicus	Podolian blind mole-rat
Poiretischer Rippenmolch	Pleurodeles poireti	Poiret's newt
Polarbirkenzeisig	Carduelis hornemanni	Arctic redpoll, hoary redpoll
Polardorsch	Boreogadus saida	polar cod (FAO/U.K.), Arctic cod (U.S./Canada)
Polardorsch, Arktisdorsch (Grönland-Dorsch)	Arctogadus glacialis	Arctic cod (FAO/U.K.), polar cod (U.S./Canada)
Polargroppe*	Cottunculus microps	polar sculpin
Polarhase	Lepus arcticus	arctic hare
Polarmöve	Larus glaucoides	Iceland gull
Polarrötelmaus	Clethrionomys rutilus	northern red-backed vole, ruddy vole
Polierfliegen, Faulfliegen	Lauxaniidae	lauxaniids
Pollack, Heller Seelachs, Steinköhler	Pollachius pollachius	pollack (green pollack, pollack lythe)
Pollenia-Schmeißfliege	Pollenia rudis	cluster fly
Polstermilbe, Hausmilbe	Glycyphagus domesticus	furniture mite
Polsterschwamm	Mycale ovulum	padded sponge*
Polymnia, Erdbeerwurm*	Eupolymnia nebulosa	strawberry worm
Polynesische Ratte, Kleine Pazifikratte, Kleine Burmaratte	Rattus exulans	Polynesian rat
Polypenlaus	Trichodina pediculus	
Pompano-Goldmakrele	Coryphaena equisetis	pompano dolphinfish
Pontische Alse, Donauhering	Alosa pontica	Pontic shad (FAO), Black Sea shad
Ponyfische	Leiognathus spp.	ponyfishes
Ponyfische, Schlupfmäuler	Leiognathidae	ponyfishes, slimys, slipmouths, soapies

Poorwill, Winternachtschwalbe	*Phalaenoptilus nuttallii*	poorwill, common poorwill
Popes Lanzenotter	*Trimeresurus popeorum*	Pope's tree viper
Porige Warzenkoralle	*Balanophyllia floridana*	porous star coral, porous cupcoral
Porphyr-Erdeule	*Lycophotia porphyrea*	true lovers knot moth
Porphyrsalamander	*Gyrinophilus porphyriticus*	spring salamander
Porphyrwalze, Zelt-Olive	*Oliva porphyria*	tent olive
Porre, Nordseegarnele, Granat	*Crangon crangon*	brown shrimp, common shrimp, common European shrimp
Portugieserhai, Portugiesischer Dornhai	*Centroscymnus coelolepis*	Portuguese dogfish (FAO), Portuguese shark
Portugiesische Auster, Greifmuschel	*Crassostrea angulata/ Gryphaea angulata*	Portuguese oyster
Portugiesische Galeere, Blasenqualle, Seeblase	*Physalia physalis*	Portuguese man-of-war, Portuguese man-o-war, blue-bottle
Porzellan-Einsiedlerkrebse	*Porcellanopagurus* spp.	porcelain hermit crabs
Porzellankrebse	Porcellanidae	porcelain crabs
Posthörnchen	*Spirula spirula*	ram's horn squid, common spirula
Posthörnchen, Postillion, Wandergelbling	*Colias croceus*	clouded yellow
Posthörnchenwürmer	Spirorbidae	spiral tubeworms
Posthörnchenwürmer	*Spirorbis* spp.	spiral tubeworms
Posthornschnecke	*Planorbarius corneus*	horn-colored ram's horn, great ramshorn (trumpet shell)
Potto	*Perodicticus potto*	potto, potto gibbon
Pottwal> Großer Pottwal, Cachalot, Kaschelot	*Physeter macrocephalus (catodon)*	sperm whale, great sperm whale, cachalot
Pottwal> Pazifik-Pottwal, Kleinpottwal	*Kogia simus*	dwarf sperm whale
Pottwal> Zwergpottwal	*Kogia breviceps*	pygmy sperm whale, lesser cachalot
Pourtales Langarmkrabbe	*Parthenope pourtalesii*	Pourtales' long-armed crab
Powis Spindelschnecke	*Tibia powisi*	Powis's tibia
Pracht-Erdschildkröte	*Rhinoclemmys pulcherrima pulcherrima*	painted wood turtle, Mexican reed turtle, Mexican wood turtle
Pracht-Giftfrosch	*Phyllobates lugubris*	lovely poison-dart frog, lovely poison frog
Pracht-Höckerschildkröte	*Graptemys oculifera*	ringed map turtle
Pracht-Kieleidechse	*Algyroides nigropunctatus*	blue-throated keeled lizard
Pracht-Krötenechse	*Phrynosoma solare*	regal horned lizard
Pracht-Krötenfisch	*Sanopus splendidus*	coral toadfish
Pracht-Olive	*Oliva splendidula*	splendid olive
Pracht-Samtschnecke	*Elysia splendida/ Thuridilla hopei*	splendid velvet snail, splendid elysia

Pracht-Zornnatter	*Coluber elegantissimus*	beautiful whip snake
Prachtadler	*Spizaetus ornatus*	ornate hawk eagle
Prachtanemone	*Heteractis magnifica*	magnificent anemone, magnificent sea anemone
Prachtbarbe	*Barbus conchonius/ Puntius conchonius*	rosy barb
Prachtbuntbarsch, Roter Buntbarsch	*Hemichromis bimaculatus/ Hemichromis guttatus*	jewel fish, jewelfish (FAO), red jewel fish, red cichlid
Prachtbuschfisch, Orange-Buschfisch	*Ctenopoma ansorgii*	ornate climbing perch, ornate ctenopoma (FAO)
Prachteiderente	*Somateria spectabilis*	king eider
Prachtfeuerschwamm	*Latrunculia magnifica*	magnificent fire sponge*
Prachtfinken	Estrildidae	estrildine finches, waxbills
Prachtfregattvogel	*Fregata magnificens*	magnificent frigatebird
Prachtglanzbarbe, Dreibandbarbe	*Puntius arulius*	longfin barb, arulius barb (FAO)
Prachtglanzstar	*Lamprotornis splendidus*	splendid glossy starling
Prächtiger Kanalkäfer	*Amara aulica*	magnificent sun beetle
Prächtiger Morwong	*Cheilodactylus variegatus*	pintadilla
Prachtkäfer	Buprestidae	metallic wood boring beetles, metallic wood borers, splendour beetles, buprestids
Prachtkopfsteher	*Anostomus anostomus*	headstander, striped anostomus, striped headstander (FAO)
Prachtlibellen	Calopterygidae/ Agrionidae	demoiselles, broad-winged damselflies
Prachtlibellen	*Calopteryx* spp.	demoiselles
Prachtmaulbrüter	*Tilapia galilaea/ Sarotherodon galilaeus*	mango tilapia (FAO), St. Peter's fish
Prachtnektarvogel	*Nectarinia superba*	superb sunbird
Prachtparadiesvogel	*Ptilornis magnificus*	magnificent riflebird
Prachtsalmler, Segelflossensalmler	Crenuchidae	crenuchids
Prachtschmerlen	*Botia* spp.	clown loaches
Prachtstaffelschwanz	*Malurus cyaneus*	blue wren
Prachttaucher	*Gavia arctica*	black-throated diver, Arctic loon
Prärie-Flusskrebs	*Procambarus gracilis*	prairie crayfish
Prärie-Königsnatter	*Lampropeltis calligaster*	prairie kingsnake
Prärie-Maulwurfsgrille	*Gryllotalpa major*	prairie mole-cricket
Prärie-Skink	*Eumeces septentrionalis*	prairie skink
Präriebussard	*Buteo swainsoni*	Swainson's hawk
Präriefalke	*Falco mexicanus*	prairie falcon
Präriehase, Weißschwanzeselhase	*Lepus townsendii*	white-tailed jack rabbit
Präriehuhn	*Tympanuchus cupido*	prairie chicken
Präriehunde	*Cynomys* spp.	prairie dogs
Prärieklapperschlange	*Crotalus viridis*	western rattlesnake
Präriekröte	*Bufo cognatus*	Great Plains toad

Präriéläufer	Bartramia longicauda	upland sandpiper
Präriemöve	Larus pipixcan	Franklin's gull
Präriewühlmaus	Microtus ochrogaster	prairie vole
Preuß-Bartmeerkatze	Cercopithecus preussi	Preuss's monkey, Preuss's guenon
Preußenfisch	Dascyllus aruanus	black-and-white damselfish, zebra humbug, white-tailed humbug, whitetail dascyllus (FAO)
Preußenfische	Dascyllus spp.	humbugs
Preuss-Rotstummelaffe, Kamerum-Rotstummelaffe	Piliocolobus preussi	Preuss's red colobus
Prevost-Schönhörnchen	Callosciurus prevosti	Prevost's squirrel
Priapswürmer, Priapuliden	Priapulida	priapulans
Prices Klapperschlange	Crotalus pricei	twin-spotted rattlesnake
Prideaux-Einsiedlerkrebs, Anemonen-Einsiedler	Pagurus prideaux/ Eupagurus prideaux	Prideaux's hermit crab, smaller hermit crab
Prigogine-Eule	Phodilus prigoginei	Congo bay-owl
Primaten, Herrentiere	Primates	primates
Primeleule	Noctua comes	lesser yellow underwing
Prinzessjungferchen	Pomacentrus vaiuli	princess damselfish, ocellate damselfish (FAO)
Prometheus-Maus, Prometheusmaus	Prometheomys schaposchnikowi	long-clawed mole-vole
Propeller-Archenmuschel	Trisidos tortuosa	propeller ark
Prophalangopsiden	Prophalangopsidae	hump-winged crickets
Prostomiden	Prostomidae	jugular-horned beetles
Proteavögel	Promeropidae/Promeropinae	sugarbirds
Protozoen, Urtierchen, Urtiere, „Einzeller"	Protozoa	protozoans, "first animals"
Provencegrasmücke	Sylvia undata	Dartford warbler
Prozessionsspinner	Thaumetopoeidae	processionary moths
Przewalski-Gazelle	Procapra przewalskii	Przewalski's gazelle
Przewalski-Pferd	Equus przewalski	Przewalski's horse
Przewalski-Rennmaus	Brachiones przewalskii	Przewalski's gerbil
Pseudomys-Mäuse	Pseudomys spp.	Australian native mice
Pseudoskorpione, Afterskorpione	Pseudoscorpiones/Chelonethi	pseudoscorpions, false scorpions
Ptilodactyliden	Ptilodactylidae	toed-winged beetles
Puderspecht	Muelleripicus pulverulentus	great slaty woodpecker
Pudus	Pudu spp.	pudus
Puerto Rico-Amazone	Amazona vittata	Puerto Rican parrot, Puerto Rican amazon
Puerto Rico-Schmuckschildkröte	Trachemys stejnegeri	Puerto Rican slider, Central Antillean slider (IUCN)
Puerto-Rico-Boa	Epicrates inornatus	Puerto Rican boa, culebra grande
Puerto-Rico-Pfeiffrosch	Eleutherodactylus coqui	Puerto Rican coqui
Puffotter> Gewöhnliche Puffotter	Bitis arietans (Bitis lachesis)	puff adder
Puffottern	Bitis spp.	puff adders
Puku-Antilope	Kobus vardoni	puku

Puma, Silberlöwe	*Felis concolor/ Puma concolor*	cougar, puma, mountain lion
Pümpwurm, Sandkoralle	*Sabellaria spinulosa*	spiny feather-duster worm
Punamaus	*Punomys lemminus*	Puna mouse
Punaré	*Trichomys apereoides*	punaré
Punataucher	*Podiceps taczanowskii/ Dytes taczanowskii*	Puna grebe
Punkt-Strich-Ziegenfisch*	*Parupeneus barberinus*	dash-dot goatfish
Punkt-Umberfisch	*Leiostomus xanthurus*	spot, spot croaker
Punktfleckspanner, Eichenbuschspanner	*Cosymbia punctaria/ Cyclophora punctaria*	maiden's blush moth
Punktierte Zartschrecke	*Leptophyes punctatissima*	speckled bushcricket
Punktierter Buntbarsch, Punktierter Indischer Buntbarsch	*Etroplus maculatus*	orange chromide
Punktierter Fadenfisch	*Trichogaster trichopterus trichopterus*	three-spot gourami
Punktierter Kopfsteher	*Chilodus punctatus punctatus*	spotted headstander
Punktierter Panzerwels, Marmorierter Panzerwels	*Corydoras paleatus*	peppered corydoras
Punktierter Pfeilgiftfrosch	*Dendrobates histrionicus*	red-and-black poison arrow frog, harlequin poison frog
Punktierter Schilderwels	*Hypostomus punctatus*	spotted hypostomus
Punktierter Welklaub-Kleinspanner	*Idaea diiutaria*	silky wave moth
Punktkäfer	Clambidae	fringe-winged beetles, minute beetles
Punktschnecke	*Punctum pygmaeum*	pygmy snail, dwarf snail
Punkttupfen-Anemonenkrabbe	*Neopetrolisthes maculatus*	dotted anemone crab
Puppen-Kreiselschnecke	*Margarites pupillus*	puppet margarite (puppet top shell)
Puppenräuber	*Calosoma* spp.	caterpillar hunters
Puppenschnecke> Gemeine Puppenschnecke, Zylindrische Puppenschnecke	*Leiostyla cylindracea*	common chrysalis snail
Puppenschnecken	*Pupilla* spp.	moss snails, column snails (U.S.)
Puppenschnecken	Pupillidae	moss snails (column snails, snaggletooths, vertigos)
Purpur-Anemone	*Halcampoides purpureus*	purple anemone*
Purpur-Laufkäfer, Goldleiste	*Carabus violaceus*	violet ground beetle
Purpur-Semele	*Semele purpurascens*	purplish semele
Purpur-Vogelspinne	*Avicularia purpurea*	Ecuadorian purple tarantula
Purpurbär, Stachelbeerbär	*Rhyparia purpurata*	gooseberry tiger (moth)
Purpurforelle	*Salmo clarki*	cutthroat trout
Purpurgimpel	*Carpodacus purpureus*	purple finch
Purpurgrackel	*Quiscalus quiscala*	common grackle
Purpurhähnchen	*Leptopoecile sophiae*	Severtzov's tit warbler
Purpurhuhn	*Porphyrio porphyrio*	purple gallinule

Purpurkopfbarbe	*Barbus nigrofasciatus/ Puntius nigrofasciatus*	black ruby barb
Purpurne Zwergolive	*Olivella biplicata*	purple dwarf olive, two-plaited dwarf olive
Purpurnektarvogel	*Nectarinia asiatica*	purple sunbird
Purpurpfeifdrossel	*Myiophoneus caerulens*	Himalayan whistling thrush
Purpurreiher	*Ardea purpurea*	purple heron
Purpurrose, Pferdeaktinie, Erdbeerrose	*Actinia equina*	beadlet anemone, beadlet, red sea anemone
Purpurrückenkolibri	*Aglaeactis aliciae*	purple-backed sunbeam
Purpurschnecke u.a.	*Trunculariopsis trunculus/ Hexaplex trunculus*	trunk murex, trunculus murex
Purpurschwalbe	*Progne subis*	purple martin
Purpurseeigel	*Strongylocentrotus droebachiensis*	green sea urchin, northern urchin, northern sea urchin
Purpurstärling	*Euphagus cyanocephalus*	Brewer's blackbird
Purpurstern, Purpurseestern, Roter Seestern, Blutstern, Roter Mittelmeerseestern	*Echinaster sepositus*	red Mediterranean sea star
Purpurtyrann	*Pyrocephalus rubinus*	Vermilion flycatcher
Pustel-Reusenschnecke	*Nassarius papillosus*	papillose nassa, pimpled nassa
Pustelschwein	*Sus verrucosus*	Javan pig
Pustelspanner, Gelbgrüner Eichenmittelwaldspanner	*Comibaena bajularia/ Comibaena pustulata*	blotched emerald
Putzende Partnergarnele	*Leandrites cyrtorhynchus*	cleaning anemone shrimp
Putzerfisch> Falscher Putzerfisch, Putzernachahmer	*Aspidontus taeniatus*	mimic blenny, false cleanerfish (FAO)
Putzerfische, Putzerlippfische	*Labroides* spp.	cleaner wrasses
Putzerlippfisch	*Labroides dimidiatus*	cleaner wrasse
Putzernachahmer, Falscher Putzerfisch	*Aspidontus taeniatus*	mimic blenny, false cleanerfish (FAO)
Pygospio-Wurm	*Pygospio elegans*	pygospio worm
Pyjama-Kardinalbarsch, Wimpel-Kardinalbarsch	*Apogon nematopterus*	pyjama cardinal fish
Pyralis-Glühwürmchen	*Photinus pyralis*	pyralis firefly
Pyramiden-Kofferfisch, Buckelkofferfisch	*Tetrosomus gibbosus*	trunkfish
Pyramiden-Kreiselschnecke	*Tectus pyramis*	pyramid topsnail, pyramid top
Pyramideneule	*Amphipyra pyramidea*	copper underwing
Pyramidenschnecke	*Pyramidula rupestris*	rock snail
Pyramidenschnecken	Pyramidulidae	rock snails
Pyrenäen-Gebirgsmolch	*Euproctus asper*	Pyrenean brook salamander, Pyrenean mountain newt
Pyrenäen-Gebirgsschrecke	*Cophopodisma pyrenaea*	magnificent Pyrenean grasshopper*
Pyrenäen-Kleinwühlmaus	*Microtus pyrenaicus*	Pyrenean pine vole
Pyrenäendesman	*Galemys pyrenaicus*	Pyrenean desman
Pyrenäenotter, Iberische Viper	*Vipera seoanei*	Iberian adder, Pyrenean viper

Quallen> Echte Quallen, Schirmquallen, Scheibenquallen, Scyphozoen	Scyphozoa	cup animals, scyphozoans
Quallenfische, Driftfische, Galeerenfische	Nomeidae	driftfishes
Quallenflohkrebse	*Hyperia* spp.	jellyfish amphipods
Quappe, Rutte, Trüsche	*Lota lota*	burbot
Quappenegel, Ruttenegel	*Cystobranchus mammillatus*	burbot leech
Quastenflosser	Crossopterygii	lobe-finned fishes, crossopterygians
Quastenflosser	Latimeridae	coelacanths
Quastenstachler	*Atherurus* spp.	brush-tailed porcupine
Quellblasenschnecke, Quell-Blasenschnecke, Quellenblasenschnecke	*Physa fontinalis*	bladder snail, common bladder snail
Quellensalamander	*Gyrinophilus* spp.	spring salamanders
Querbandhechtling, Monrovia-Hechtling, Rotkehlhechtling	*Epiplatys dageti monroviae*	firemouth epiplatys
Quermundbarben	*Capoeta* spp.	butterfly barbs
Querstreifige Venusmuschel*	*Chione cancellata*	cross-barred venus
Querzahnmolche	Ambystomatidae	ambystomids, mole salamanders
Quesenbandwurm, Hundebandwurm u.a.	*Multiceps multiceps/* *Taenia multiceps*	dog tapeworm a.o. (>gid/sturdy in sheep)
Quetzal	*Pharomachrus mocinno*	resplendent quetzal
Quinnat, Königslachs	*Oncorhynchus tschawytcha*	chinook salmon (FAO), chinook, king salmon

Rabe> Kolkrabe	*Corvus corax*	raven, common raven
Rabengeier	*Coragyps atratus*	black vulture
Rabenkakadu	*Calyptorhynchus magnificus*	red-tailed cockatoo
Rabenvögel	Corvidae	crows & magpies & jays & nutcrackers
Rachenbremsen	Cephenomyiinae	nostril flies
Rachendassel	*Gasterophilus nasalis*	throat botfly
Racken	Coraciidae	rollers
Rackenvögel	Coraciiformes	kingfishers & bee-eaters & hoopoes & rollers & hornbills
Radde-Viper, Armenische Bergotter	*Vipera raddei*	Radde's rock viper, Armenian viper
Rädertiere, Rotatorien	Rotatoria	rotifers
Radgarnele	*Marsupenaeus japonicus/ Penaeus japonicus*	Kuruma shrimp
Radians-Kaurischnecke*	*Trivia radians/ Pusula radians*	radians trivia
Radieschenfliege, Kleine Kohlfliege, Wurzelfliege	*Delia radicum/ Hylemia brassicae/ Chortophila brassicae/ Paregle radicum*	cabbage fly, cabbage maggot, radish fly
Radiolarien, Strahlentierchen	Radiolaria	radiolarians
Radnetzspinnen, Kreuzspinnen	Araneidae	orbweavers, orb-weaving spiders (broad-bodied orbweavers)
Raggiparadiesvogel	*Paradisaea raggiana*	Raggiana bird-of-paradise
Rainfarnblattkäfer	*Galeruca tanaceti*	tansy beetle
„Rainwater" Kärpfling	*Lucania parva*	rainwater killifish
Rallen	Rallidae	rails, gallinules, coots
Rallenkranich, Riesenralle	*Aramus guarauna* (Aramidae)	limpkin
Rallenreiher	*Ardeola ralloides*	squacco heron
Ramburs Grashüpfer, Kleiner Heidegrashüpfer	*Stenobothrus stigmaticus*	lesser mottled grasshopper
Ramsnasen-Einhornfisch, Wulst-Nasendoktor	*Naso tuberosus*	humpnose unicornfish
Rändel-Käferschnecke, Aschgraue Käferschnecke	*Lepidochitona cinerea*	cinereous chiton
Randplatten-Kammstern	*Astropecten articulatus*	plated-margined seastar, plate-margined comb star
Rankenfüßler, Rankenfüßer, Cirripeden, Cirripedier	Cirripedia	barnacles, cirripedes
Rankenwürmer, Fadenkiemer	Cirratulidae	cirratulid worms, cirratulids
Ranzenkrebse	Peracarida	peracarids
Rapfen	*Aspius aspius*	asp
Rappenantilope	*Hippotragus niger*	sable antelope
Rapserdfloh	*Psylliodes chrysocephala*	cabbage stem flea beetle, rape flea beetle
Rapsglanzkäfer	*Meligethes aeneus*	pollen beetle
Rapsweißling, Grünader-Weißling	*Pieris napi*	green-veined white
Rasenameise	*Tetramorium caespitum*	turf ant, pavement ant, black pavement ant

Rasenkoralle, Rasige Röhrchenkoralle	*Cladocora caespitosa*	caespitose tube coral
Rasiermesserfisch, Schnepfenmesserfisch	*Aeoliscus strigatus*	razorfish (FAO), shrimpfish
Raspel-Tellmuschel	*Tellina scobinata*	rasp tellin
Rasse, Kleine Zibetkatze	*Viverricula indica*	lesser oriental civet, rasse
Rasselfrosch	*Crinia glauerti*	rattle froglet
Rasselnder Laubfrosch	*Hyla crepitans*	rattle-voiced treefrog, emerald-eyed treefrog
Rathbunscher Brunnenmolch, Texas-Blindsalamander	*Typhlomolge rathbuni*	Texas blind salamander
Ratten> Eigentliche Ratten	*Rattus* spp.	rats
Ratten-Zwergbandwurm	*Vampirolepis fraterna*	rat tapeworm, dwarf rat tapeworm
Rattenegel, Schweinegel	*Gastrodiscoides hominis*	pig intestinal fluke, rat fluke
Rattenfloh, Europäischer Rattenfloh	*Nosopsyllus fasciatus*	northern rat flea
Rattenschwänze, Grenadierfische	Macrouridae	rattails, grenadiers
Raubfliegen (Mordfliegen)	Asilidae	robberflies & grass flies
Raubkäfer, Kurzflügler	Staphylinidae	rove beetles
Raubmilben	Chelyletidae	quill mites, predatory mites
Raubmöwen	Stercorariidae	skuas & jaegers
Raubseeschwalbe	*Hydroprogne caspia/ Sterna caspia*	Caspian tern
Raubspinne, Listspinne	*Pisaura mirabilis*	fantastic fishing spider*
Raubspinnen, Jagdspinnen	Pisauridae	nursery-web spiders, fisher spiders, fishing spiders
Raubtiere, Karnivoren	Carnivora	carnivores
Raubwanze	*Coranus subapterus*	heath assassin bug
Raubwanzen (Schreitwanzen)	Reduviidae	assassin bugs, conenose bugs, ambush bugs & thread-legged bugs
Raubwelse, Luftatmende Welse	*Clarias* spp.	labyrinth catfishes, air-breathing catfishes
Raubwelse, Kiemensackwelse	Clariidae	airbreathing catfishes, clarid catfishes
Raubwürger	*Lanius excubitor*	great grey shrike, northern shrike
Rauch-Warzenseescheide	*Phallusia fumigata*	fumigate sea-squirt*
Rauchschwalbe	*Hirundo rustica*	swallow, barn swallow
Räudemilben	Psoroptidae	mange mites, scab mites
Raue Atlantische Messermuschel	*Siliqua squama*	rough razor clam, squamate razor clam
Raue Bastardmakrele	*Trachurus lathami*	rough scad
Raue Bohrmuschel, Krause Bohrmuschel	*Zirfaea crispata*	great piddock, oval piddock
Raue Feilenmuschel	*Lima scabra/ Ctenoides scabra*	rough lima, rough fileclam (rough file shell, Atlantic rough file shell)
Raue Felsschnecke	*Thais rugosa*	rough rocksnail (rough rock shell)

Raue Kaktuskoralle, Lamarcks Kaktuskoralle	*Mycetophyllia ferox*	rough cactus coral
Raue Kellerassel, Kellerassel	*Porcellio scaber*	common rough woodlouse, garden woodlouse, slater, scabby sow bug
Raue Kliesche, Gelbflossenzunge*	*Limanda aspera*	yellowfin sole
Raue Krötenkopf-Schildkröte, Warzen-Krötenkopf-Schildkröte	*Phrynops tuberculatus*	tuberculate toad-headed turtle
Raue Nadelschnecke	*Rhinoclavis asper*	rough cerith
Raue Scholle, Raue Scharbe, Doggerscharbe	*Hippoglossoides platessoides*	long rough dab
Raue Seescheide	*Styela partita/Styela rustica*	rough sea-squirt
Raue Sternschnecke	*Onchidoris muricata*	rough doris
Raue Strandschnecke, Spitze Strandschnecke, Dunkle Strandschnecke	*Littorina saxatilis*	rough periwinkle
Rauer Dornhai	*Cirrhigaleus asper/ Squalus asper*	roughskin spurdog (FAO), roughskin spiny dogfish
Rauer Dornhai> Schlingerhai	*Centrophorus granulosus*	gulper shark (FAO), rough shark
Rauer Grenadier, Raukopf Panzerratte, Nordatlantik-Grenadier	*Macrourus berglax*	rough rattail, rough-headed grenadier, rough-head grenadier, onion-eye grenadier (FAO)
Rauer Pomfret	*Taractes asper*	rough pomfret
Rauer Stechrochen, Rauer Stechrochen, Brucko	*Dasyatis centroura*	roughtail stingray
Rauer Süßwasserschwamm	*Trochospongilla horrida*	rough freshwater sponge
Raufuß-Klappschildkröte	*Kinosternon hirtipes hirtipes*	Mexican rough-footed mud turtle, Mexican mud turtle
Raufuß-Springmaus	*Dipus sagitta*	rough-legged jerboa, northern three-toed jerboa
Raufußbussard	*Buteo lagopus*	rough-legged buzzard
Raufußhühner	Tetraoninae	grouse
Raufußkauz	*Aegolius funereus*	Tengmalm's owl, boreal owl, Richardson's owl (N.Am.)
Rauhaar-Schlitznase	*Nycteris hispida*	hairy slit-faced bat
Rauhals-Geißelgarnele	*Trachypenaeus constrictus*	roughneck shrimp
Rauhaut-Fledermaus	*Pipistrellus nathusii*	Nathusius' pipistrelle
Rauhäutige Grasnatter, Raue Grasnatter	*Opheodrys aestivus*	rough green snake
Raukopf Panzerratte, Nordatlantik-Grenadier, Rauer Grenadier	*Macrourus berglax*	rough rattail, rough-headed grenadier, rough-head grenadier, onion-eye grenadier (FAO)
Raunackenwaran	*Varanus rudicollis*	rough-necked monitor, tree lizard, harlequin monitor
Raupenfliegen (Schmarotzerfliegen)	Tachinidae	tachinids, parasitic flies

Raurochen, Australoasiatischer Raurochen	Raja nasuta	rough skate (Australasian)
Raurücken-Geißelgarnele*	Trachypenaeus similis/ Trachysalambria curvirostris	roughback shrimp
Raurücken-Seefledermaus	Ogcocephalus parvus	roughback batfish
Rauschalige Lastträgerschnecke	Xenophora corrugata	rough carriersnail
Rauschuppen-Buschviper	Atheris hispidus	rough-scaled tree viper
Rauschuppen-Eidechse	Ichnotropis spp.	rough-scaled lizard
Rauschuppen-Sandboa	Eryx conicus	rough-scaled sand boa
Rauschuppenboa	Trachyboa boulengeri	rough-scaled boa
Rauschuppige Ostaustralische Giftnatter	Tropidechis carinatus	rough-scaled snake
Rauschuppige Kapeidechse	Ichnotropis capensis	Cape rough-scaled lizard
Rauschwanzgrundel*	Evermannichthys metzelaari	roughtail goby
Rauschwänzige Rollassel	Sphaeroma rugicaudata	roughtail marine pillbug
Rautenflecksalmler	Hemigrammus caudovittatus	Buenos Aires tetra
Rautenkrokodil, Kuba-Krokodil	Crocodylus rhombifer	Cuban crocodile
Rautenpapageifisch	Sparisoma viride	stoplight parrotfish
Rautenpython	Morelia spilota spilota/ Morelia argus	diamond python
Rauzahndelphin	Steno bredanensis	rough-toothed dolphin
Rebenpockenmilbe	Eriophyes vitis/ Colomerus vitis	grapeleaf blister mite
Rebensattelschrecke, Steppen-Sattelschrecke, Gemeine Sattelschrecke	Ephippiger ephippiger	tizi, common saddle-backed bushcricket
Rebenstecher, Zigarrenwickler	Byctiscus betulae	hazel leaf roller weevil
Rebhuhn	Perdix perdix	grey partridge
Reblaus	Viteus vitifoliae/ Viteus vitifolii	vine louse, grape phylloxera
Rechtshänder-Einsiedlerkrebse	Paguridae	right-handed hermit crabs
Regen-Grabläufer*	Pterostichus cupreus	rain beetle
Regenbogen-Blattkäfer	Chrysolina cerealis	rainbow leaf beetle
Regenbogen-Buntbarsch	Herotilapia multispinosa	rainbow cichlid
Regenbogen-Drachenkopf	Scorpaenodes xyris	rainbow scorpionfish
Regenbogen-Rundhering	Dussumieria acuta	rainbow sardine (FAO), round herring
Regenbogen-Stachelmakrele	Elagatis bipinnulata	rainbow runner
Regenbogenboa	Epicrates cenchria cenchria	Brazilian rainbow boa
Regenbogenfische	Melanotaeniidae	rainbow fishes, rainbowfishes
Regenbogenfische	Melanotaenia spp.	rainbowfishes
Regenbogenforelle	Salmo gairdneri/ Oncorhynchus mykiss	rainbow trout
Regenbogennatter	Abastor erythrogrammus/ Farancia erythrogramma	rainbow snake
Regenbogenpitta	Pitta iris	rainbow pitta
Regenbogenschlange	Xenopeltis unicolor	sunbeam snake
Regenbogenspint	Merops ornatus	Australian bee eater

Regenbogenstint, Atlantik-Regenbogenstint	*Osmerus mordax*	Atlantic rainbow smelt (FAO), lake smelt
Regenbrachvogel	*Numenius phaeopus*	whimbrel
Regenbremse	*Haematopota pluvialis*	cleg-fly, cleg
Regenbremsen, Blinde Fliegen	*Haematopota* spp./ *Chrysozona* spp.	clegs, stouts
Regenkuckucke, Amerikanische Kuckucke	Coccyzidae	American cuckoos, New World cuckoos
Regenpfeifer	Charadriidae	plovers, lapwings
Regenwurm> Gemeiner Regenwurm, Tauwurm	*Lumbricus terrestris*	common earthworm, earthworm, lob worm, dew worm, squirreltail worm, twachel (Br.)
Regenwürmer	Lumbricidae	earthworms
Reh> Europäisches Reh	*Capreolus capreolus*	roe deer
Rehantilope, Rehböckchen	*Pelea capreolus*	rhebok, gray rhebok
Rehdasselfliege	*Hypoderma diana*	deer warble fly a.o.
Rehrachenbremse	*Cephenomyia stimulator*	roe deer nostril fly
Reife Trogmuschel	*Mactrellona exoleta*	mature surfclam
Reiher (inkl. Dommeln)	Ardeidae	herons & egrets & bitterns
Reiherente	*Aythya fuligula*	tufted duck
Reiherläufer	Dromadidae	crab-plover
Reinwardtarassari	*Selenidera reinwardtii*	golden-collarded toucanet
Reisfeldratte	*Rattus argentiventer*	rice field rat
Reisfische, Reiskärpflinge, Japan-Kärpflinge	Oryziatidae/ Oryziinae	medakas, ricefishes
Reiskäfer	*Sitophilus oryzae*	rice weevil
Reismehlkäfer	*Tribolium* spp.	flour beetles
Reismotte	*Corcyra cephalonica*	rice moth, raisin honey
Reisratten	*Oryzomys* spp.	rice rats
Reisstengelälchen	*Ditylenchus angustus*	rice nematode (ufra disease of rice)
Reisvogel	*Padda oryzivora*	Java sparrow
Reiswühler	*Oryzorictes* spp.	rice tenrecs
Reiterkrabben	*Ocypode* spp.	ghost crabs
Remora, Küstensauger	*Remora remora*	common remora (FAO), shark sucker
Renken, Maränen	Coregoninae	coregonines, lake whitefishes
Renken, Maränen	*Coregonus* spp.	whitefishes, lake whitefishes
Rennechsen, Renntejus	*Cnemidophorus* spp.	whiptails
Rennfrösche	*Kassina* spp.	running frogs
Rennkrabbe, Felsenkrabbe	*Pachygrapsus marmoratus*	marbled shore crab, marbled rock crab
Rennkuckuck	*Geococcyx velox*	lesser roadrunner
Rennmäuse> Eigentliche Rennmäuse	*Gerbillus* spp.	northern pygmy gerbils

Renntaucher	*Aechmophorus occidentalis*	western grebe
Renntejus, Rennechsen	*Cnemidophorus* spp.	whiptails
Rennvogel	*Cursorius cursor*	cream-coloured coursor
Rennvögel & Brachschwalbenartige	Glareolidae	pratincoles, coursers
Rentier, Ren	*Rangifer tarandus*	caribou (North America), reindeer (Europe)
Rentier-Hautbremse, Rentierdasselfliege	*Oedemagena tarandi*	reindeer warble fly
Rentier-Rachenbremse	*Cephenomyia trompe*	reindeer nostril fly
Reseda-Weißling, Resedafalter	*Pontia daplidice*	bath white
Rettich-Kalkschwamm, Borstiger Kalkschwamm	*Sycon raphanus*	radish sponge*, bristly vase sponge*
Rettichschnecke	*Rapa rapa*	bubble turnip, papery rapa
Reusenhai, Riesenhai	*Cetorhinus maximus*	basking shark
Reusenschnecken	*Nassarius* spp.	dog whelks, nassas
Rheinmücke, Augustmücke	*Oligoneuriella rhenana*	Rhine-river mayfly
Rhesusaffe, Rhesusmakak	*Macaca mulatta*	rhesus monkey
Rhododendron-Weißfliege*	*Dialeurodes chittendeni*	rhododendron whitefly
Rhombenkrabbe	*Goneplax rhomboides*	angular crab, square crab (mud runner)
Rhombenkrabben	Goneplacidae	goneplacid crabs, angular crabs
Rhône-Streber	*Zingel asper*	asper, Rhône streber (FAO)
Riedbock, Isabellantilope	*Redunca redunca*	Bohar reedbuck
Riedböcke	*Redunca* spp.	reedbucks
Riedfrösche, Afrikanische Riedfrösche	*Hyperolius* spp.	African reed frogs
Riefenschnabelani	*Crotophaga sulcirostris*	groove-billed ani
Riemenfisch, Bandfisch	*Regalecus glesne*	oarfish
Riemenfische, Bandfische	Regalecidae	oarfishes
Riemennatter	*Imantodes cenchoa*	blunt-headed tree snake
Riemenwurm, Riemenbandwurm	*Ligula intestinalis*	
Riesen-Schlangenhalsschildkröte	*Chelodina expansa*	giant snake-necked turtle, broad-shell snakeneck, broad-shelled tortoise
Riesen-Schlüssellochschnecke, Riesen-Lochschnecke	*Megathura crenulata*	giant keyhole limpet
Riesen-Segelflossenkärpfling*	*Poecilia velifera*	giant sailfin molly, sail-fin molly (FAO)
Riesen-Elenantilope, Derbyantilope	*Taurotragus derbianus*	Derby eland, giant eland
Riesen-Fechterschnecke, Riesen-Flügelschnecke	*Strombus gigas*	queen conch, pink conch
Riesen-Flugfuchs, Fliegender Hund, Kalong	*Pteropus vampyrus*	kalong, flying dog, large flying fox

Riesen-Goldrose, Karibische Goldrose	*Condylactis gigantea*	giant Caribbean anemone, pink-tipped anemone, "condy" anemone, Atlantic anemone
Riesen-Greiffrosch	*Phyllomedusa bicolor*	giant monkey frog
Riesen-Herzmuschel	*Plagiocardium pseudolima*	giant cockle
Riesen-Holzameise, Riesenholzameise	*Camponotus herculeanus*	giant carpenter ant
Riesen-Hornhecht, Riesen-Krokodilhecht, Krokodilhecht	*Tylosurus crocodilus/ Strongylura crocodila*	hound needlefish (FAO), houndfish, crocodile longtom (Austr.), crocodile needlefish
Riesen-Kängururatte	*Dipodomys ingens*	giant kangaroo rat
Riesen-Kreiselschnecke	*Trochus niloticus*	Nile topsnail, Nile trochus, commercial trochus (pearly top shell)
Riesen-Kronenschnecke	*Pugilina morio*	giant hairy melongena, giant melongena
Riesen-Kugelfisch	*Arothron stellatus*	star puffer
Riesen-Maulbeerschnecke	*Morum grande*	giant morum
Riesen-Napfschnecke	*Patella laticostata/ Patella neglecta*	giant limpet, neglected limpet
Riesen-Nasenbeutler	*Peroryctes broadbenti*	giant bandicoot
Riesen-Pferdeschnecke	*Pleuroploca gigantea*	horse conch
Riesen-Rossie	*Semirossia equalis*	greater bobtail squid
Riesen-Schuppentier	*Manis gigantea*	large African pangolin
Riesen-Schwebegarnele*	*Callianassa gigas*	giant ghost shrimp
Riesen-Seepocke u.a.	*Megabalanus psittacus*	giant Chilean barnacle
Riesen-Seepocke u.a.	*Balanus nubilis*	giant acorn barnacle, giant barnacle
Riesen-Seewalze	*Thelenota anax*	giant sea cucumber*, giant holothurian*
Riesen-Smaragdeidechse	*Lacerta trilineata*	Balkan green lizard, Balkan emerald lizard
Riesen-Süßlippe	*Plectorhinchus obscurus*	giant sweetlip
Riesen-Treppenschnecke	*Pugilina colossea/ Hemifusus colosseus*	colossal false fusus, giant stair shell
Riesen-Vogelspinne, Goliath-Vogelspinne	*Theraphosa blondi*	goliath birdeater tarantula
Riesen-Waldspinne, Gefleckte Seidenspinne	*Nephila maculata*	giant wood spider
Riesen-Zackenbarsch (Judenfisch)	*Epinephelus itajara*	giant grouper (FAO) (jewfish)
Riesenabendsegler	*Nyctalus lasiopterus*	greater noctule
Riesenanemone, Mertens Anemone, Nessellose Riesenanemone	*Stichodactyla gigantea/ Stichodactyla mertensi/ Stoichactis giganteum*	giant carpet anemone, giant anemone
Riesenangler	*Ceratias holboelli*	deep-sea angler
Riesenanolis	*Anolis roosevelti*	giant anole, Culebra Island giant anole
Riesenauster, Pazifische Auster	*Crassostrea gigas*	Pacific oyster, giant Pacific oyster, Japanese oyster
Riesenbachling	*Rivulus harti*	Hart's rivulus (FAO), leaping guabine

Riesenbarbe	*Catlocarpio siamensis*	giant barb (giant "carp")
Riesenbastkäfer	*Dendroctonus micans*	European spruce beetle, great spruce bark beetle
Riesenbaummaus, Riesenklettermaus	*Megadendromus nikolausi*	giant climbing mouse
Riesenborkenratten	*Phloeomys* spp.	slender-tailed cloud rats
Riesenborster	*Eunice gigantea*	giant eunice
Riesenchamäleon	*Chamaeleo oustaleti*	Oustalet's giant chameleon, Oustalet's chameleon
Riesendarmegel	*Fasciolopsis buski*	giant intestinal fluke
Riesenegel	*Haementeria ghilianii*	giant turtle leech, giant leech
Riesenerdschildkröte	*Heosemys grandis*	giant Asian pond turtle, giant spined terrapin
Riesenfadenfisch	*Eleutheronema tetradactylum*	four-finger threadfin
Riesenfroschschnecke	*Tutufa bubo*	giant frogsnail
Riesengalago	*Galago crassicaudatus/ Otolemur crassicaudatus*	thick-tailed bush baby, greater galago, greater bush baby
Riesengeigenrochen, Großer Gitarrenrochen, Schulterfleck-Geigenrochen, Ulavi	*Rhynchobatus djiddensis*	whitespotted wedgefish, spotted guitarfish, shovelnose, sand shark, giant guitarfish (FAO)
Riesengleitbeutler	*Schoinobates volans*	greater gliding possum
Riesengleiter, Riesengleitflieger, Pelzflatterer	Dermoptera	flying lemurs, gliding lemurs, colugos, dermopterans
Riesengleithörnchen	*Petaurista* spp.	giant flying squirrels
Riesengrundel, Große Meergrundel	*Gobius cobitis*	giant goby
Riesengürtelschweif	*Cordylus giganteus*	sungazer, giant girdled lizard (IUCN), giant zonure, giant spinytail lizard
Riesengürteltier	*Priodontes maximus*	giant armadillo
Riesenhai, Reusenhai	*Cetorhinus maximus*	basking shark
Riesenhaie	Cetorhinidae	basking sharks
Riesenhamsterratten	*Cricetomys* spp.	African giant pouched rats
Riesenholzwespe	*Urocerus gigas*	giant wood wasp, giant horntail, greater horntail
Riesenhörnchen	*Ratufa* spp.	Oriental giant squirrels
Riesenhüpferling	*Macrocyclops fuscus*	
Riesenkäfer	Dynastinae	rhinoceros beetles & hercules beetles and others
Riesenkalmar u.a., Jumbokalmar	*Dosidicus gigas*	jumbo flying squid, jumbo squid
Riesenkalmare	Architeuthidae	giant squids
Riesenkalmare	*Architeuthis* spp.	giant squids
Riesenkolibri	*Patagona gigas*	giant hummingbird
Riesenkrabbenspinnen, Jagdspinnen	Heteropodidae (Eusparassidae, Sparassidae)	giant crab spiders, huntsman spiders
Riesenkratzer	*Macracanthorhynchus hirudinaceus* (Acanthocephala)	thorny headed worm

Riesenkugler	Sphaerotheriidae	giant pill millepedes
Riesenlandkrabben, Karibische Landkrabben	*Cardisoma* spp.	giant land crabs, great land crabs
Riesenläufer, Skolopender	Scolopendromorpha	scolopendromorphs
Riesenleberegel	*Fasciola gigantica*	giant liver fluke
Riesenmanta, Manta, Großer Teufelsrochen	*Manta birostris*	manta, giant manta (FAO) Atlantic manta, giant devil ray
Riesenmaulhai, Großmaulhai	*Megachasma pelagios*	megamouth shark
Riesenmaulhaie, Großmaulhaie	Megachasmidae	megamouth sharks
Riesenmeereswels	*Netuma thalassinus/ Arius thalassinus*	giant catfish
Riesenmoa	*Dinornis maximus*	giant moa
Riesenmuntjak	*Megamuntiacus vuquagensis*	giant muntjac
Riesenmuräne	*Gymnothorax javanicus*	giant moray
Riesenmuscheln	*Tridacna* spp.	giant clams
Riesennektarvogel	*Nectarinia thomensis*	Sao Thome giant sunbird
Riesenohr-Springmaus	*Euchoreutes naso*	long-eared jerboa
Riesenotter	*Pteronura brasiliensis*	giant otter
Riesenpacu, Gamitana-Scheibensalmler, Mühlsteinsalmler	*Piaractus brachypomus/ Colossoma bidens*	pirapatinga (FAO), cachama
Riesenpanda, Großer Panda, Bambusbär	*Ailuropoda melanoleuca*	giant panda
Riesenrotschwanz	*Phoenicurus erythrogaster*	Güldenstädt's redstart
Riesensalamander	Cryptobranchidae	giant salamanders
Riesensalpe	*Pegea bicaudata*	giant salp
Riesensamtmuschel	*Glycymeris americana/ Glycymeris gigantea*	giant bittersweet, American bittersweet
Riesenschaben	Blaberidae	giant cockroaches
Riesenschlangen	Boidae & Pythonidae	boas & pythons
Riesenschnake	*Tipula maxima*	giant cranefly
Riesensteinfliegen	Pteronarcidae	giant stoneflies, salmonflies
Riesenstern	*Pisaster giganteus*	giant seastar
Riesensturmvogel	*Macronectes giganteus*	giant petrel
Riesentafelente	*Aythya valisineria*	canvasback
Riesentaschenratten	*Orthogeomys* spp.	taltuzas
Riesentermite	*Mastotermes darwiniensis*	Darwin termite
Riesentermiten	Mastotermitidae	Darwin termites
Riesentigerfisch	*Hydrocynus goliath*	giant tigerfish
Riesentukan	*Ramphastos toco*	toco toucan
Riesenturako	*Corythaeola cristata*	great blue turaco
Riesenunke	*Bombina maxima*	Yunnan firebelly toad
Riesenwaldschwein	*Hylochoerus meinertzhageni*	giant forest hog
Riesenwanzen, Riesenwasserwanzen	Belostomatidae	giant water bugs (toe biters)
Riesenwels	*Pangasianodon gigas*	Mekong catfish
Riesenwelse, Schlankwelse	Pangasiidae	pangasid catfishes

Riesenwildschaf, Argali	*Ovis ammon*	argali
Riesenwurmschnecke	*Serpulorbis arenarius*	giant wormsnail
Riesenzitterrochen	*Narcine entemedor*	giant electric ray
Riff-Falterfisch	*Chaetodon sedentarius*	reef butterflyfish
Riff-Hornhecht	*Strongylura incisa*	reef needlefish
Riff-Ruderschlange	*Hydrophis ornatus*	reef sea snake, spotted sea snake
Riff-Steinfisch	*Synanceia verrucosa*	reef stonefish, poison toadfish, stonefish (FAO)
Riff-Weißspitzenhai, Silberspitzenhai	*Carcharhinus albimarginatus*	silvertip shark
Riffbarsche, Jungfernfische	*Abudefduf* spp.	sergeant damselfishes
Riffdach-Schlangenstern	*Ophiocoma scolopendrina*	reefflat brittlestar*
Riffdach-Seeigel	*Echinometra mathaei*	reefflat sea urchin*
Riffhörnchenfische	*Myripristis* spp.	soldierfishes
Riffhummer	*Enoplometopus* spp.	reef lobsters
Riffkorallen, Steinkorallen	Madreporaria/ Scleractinia	stony corals, madreporarian corals, scleractinians
Rifflanguste	*Panulirus penicillatus*	reef spiny crawfish, reef rock lobster
Rindenkäfer	Colydiidae	cylindrical bark beetles
Rindenkäfer, Rindenglanzkäfer	Rhizophagidae	root-eating beetles
Rindenkorallen, Hornkorallen	Gorgonaria	gorgonians, gorgonian corals, horny corals
Rindenskorpione*	*Centruroides* spp.	bark scorpions
Rindenwanzen	Aradidae	flatbugs, flat bugs
Rinder-Haarbalgmilbe	*Demodex bovis*	cattle follicle mite
Rinderbandwurm, Rinder-Menschenbandwurm	*Taenia saginata*	beef tapeworm
Rinderbremse	*Tabanus bovinus*	large horsefly
Rinderdasselfliege (Hautdasselfliege, Große Dasselfliege)	*Hypoderma bovis*	ox warble fly a.o., northern cattle grub (U.S.)
Rinderfinnen-Bandwurm, Unbewaffneter Menschenbandwurm	*Taeniarhynchus saginatatus*	
Rindergemse, Gnuziege, Takin	*Budorcas taxicolor*	takin
Rinderhaarling	*Bovicola bovis*	cattle-biting louse
Rinderschweißfliege, Schweißfliege	*Morellia simplex*	cattle sweat fly, sweat fly
Ring-Kaiserfisch, Ringelkaiserfisch	*Pomacanthus annularis*	bluering angelfish (FAO), blue-ringed angelfish
Ringälchen (der Weinrebe)	*Macroposthonia xenoplax*	ring nematode (on grapevine)
Ringdrossel	*Turdus torquatus*	ring ouzel
Ringel-Bäumchenpolyp	*Eudendrium annulatum*	annulate stickhydroid
Ringel-Schnurwurm	*Tubulanus annulatus* (Nemertini)	
Ringelanemone	*Bartholomea annulata*	ringed anemone, curley-cue anemone

Ringelbrassen	*Diplodus annularis*	annular seabream (FAO), annular bream, annular gilthead
Ringelgans	*Branta bernicla*	brent goose
Ringelhechtling	*Epiplatys annulatus*	rocket panchax
Ringelmungo, Ringelschwanzmungo	*Galidia elegans*	Malagasy ring-tailed mongoose
Ringelnatter	*Natrix natrix*	grass snake
Ringelnemertine	*Tubulanus annulatus*	football jersey worm
Ringelquerzahnmolch	*Ambystoma annulatum*	ringed salamander
Ringelrose, Gürtelrose	*Actinia cari*	green sea anemone
Ringelschnake	*Culiseta annulata/ Theobaldia annulata*	banded house mosquito, banded mosquito, ring-footed gnat
Ringelschnecke*, Ringel-Randschnecke*	*Prunum cincta*	encircled marginella
Ringelschwanz-Stumpfnasenaffe, Pageh-Stumpfnasenaffe	*Simias concolor/ Nasalis concolor*	pig-tailed snub-nosed monkey, pig-tailed snub-nosed langur
Ringelschwanz-Kletterbeutler	*Pseudocheirus* spp.	ring-tailed possums
Ringelspinner, Gemeiner Ringelspinner, Obsthain-Ringelspinner	*Malacosoma neustria*	lackey, European lackey moth, common lackey
Ringeltaube	*Columba palumbus*	woodpigeon
Ringelwidderchen, Weißfleckenwidderchen, Weißfleckwidderchen	*Syntomis phegea*	yellow-belted burnet
Ringelwühle	*Siphonops annulatus*	ringed caecilian
Ringelwühlen	*Blanus* spp.	Mediterranean worm lizards
Ringelwürmer, Gliederwürmer, Borstenfüßer, Anneliden	Annelida	segmented worms, annelids
Ringhalskobra	*Hemachatus haemachatus*	ringhals, ringneck spitting cobra
Ringschnabelente	*Aythya collaris*	ring-necked duck
Ringschnabelmöve	*Larus delawarensis*	ring-billed gull
Ringspanner	*Cyclophora annulata*	mocha
Rio Magdalena-Schienenschildkröte	*Podocnemis lewyana*	Rio Magdalena River turtle, Magdalena River turtle (IUCN)
Rio-de-la-Plata Miesmuschel	*Mytilus platensis*	River Plate mussel
Rio-Meta-Salmler	*Hyphessobrycon metae*	Rio Meta tetra
Rio-Reisratte	*Phaenomys ferrugineus*	Rio de Janeiro rice rat
Rippelstreifen-Falterfisch, Rotsaum-Falterfisch, Dreistreifen-Falterfisch	*Chaetodon trifasciatus*	melon butterflyfish
Rippen-Käferschnecke	*Lepidopleurus cajetanus*	ribbed chiton*
Rippen-Napfschnecke*, Rostrote Napfschnecke*	*Patella ferruginea*	ribbed Mediterranean limpet
Rippen-Rissoschnecke	*Rissoa costata*	costate risso snail
Rippen-Sattelauster, Gerippte Sattelauster	*Monia patelliformis*	ribbed saddle oyster
Rippenquallen, Kammquallen, Ctenophoren	Ctenophora	ctenophores (sea gooseberries, sea combs, comb jellies, sea walnuts)

Riss-Schnecken	Scissurellidae	scissurelles
Rissos Delphin, Rundkopfdelphin	*Grampus griseus*	Risso's dolphin, gray grampus, white-headed grampus
Rissos Glattkopffisch	*Alepocephalus rostratus*	Risso's smooth-head
Rissos Steinkrabbe	*Xantho pilipes*	Risso's crab, lesser furrowed crab
Ritter (Edelfalter, Schwalbenschwänze)	Papilionidae	swallowtails & apollos
Ritteranolis	*Anolis equestris*	knight anole
Ritterhelm	*Syrinx proboscidiferus*	elefant's trumpet snail*
Riu-Kiu-Kaninchen	*Pentalagus furnessi*	Ryukyu rabbit, Amami rabbit
Riu-Kiu-Stachelratte	*Tokudaia osimensis*	Ryukyu spiny rat
Rivolikolibri, Dickschnabelkolibri	*Eugenes fulgens*	magnificent hummingbird
Robben, Flossenfüßer	Pinnipedia	marine carnivores (seals, sealions, walruses)
Robben-Bachsalamander	*Desmognathus monticola*	seal salamander
Roberts Nacktschwanzmaus	*Gymnuromys roberti*	voalavoanala
Rochen> Echte Rochen	Rajidae, Batoidea	rays & skates
Rochenartige	Rajiformes	skates & guitarfishes
Rochenegel	*Pontobdella muricata*	ray leech*
Rocky-Mountains-Holzzecke	*Dermacentor andersoni*	Rocky-Mountains wood tick
Rocky-Mountains-Wühlmaus	*Microtus montanus*	Rocky-Mountains vole, mountain meadow mouse
Roesels Beißschrecke	*Metrioptera roeselii*	Roesel's bushcricket
Rohrammer	*Emberiza schoeniclus*	reed bunting
Rohrbohrer	*Phragmataecia castaneae*	reed leopard moth
Röhrchenkoralle, Röhrenkoralle	*Cladocora arbuscula*	tube coral
Rohrdommel	*Botaurus stellaris*	bittern
Rohrdommel> Nordamerikanische Rohrdommel	*Botaurus lentiginosus*	American bittern
Röhrenaale	Heterocongridae/ Heterocongrinae	garden eels
Röhrenfächer-Moostierchen	*Tubulipora flabellaris*	
Röhrenholothurie, Röhrenseegurke	*Holothuria tubulosa*	Mediterranean trepang
Röhrenkoralle, Röhrchenkoralle	*Cladocora arbuscula*	tube coral
Röhrenläuse, Röhrenblattläuse	Aphididae	aphids, "plant lice"
Röhrenmaul-Pinzettfisch, Gelber Maskenpinzettfisch	*Forcipiger flavissimus*	longnose butterflyfish (FAO), long-beaked butterflyfish
Röhrenmünder	*Solenostomus* spp.	ghost pipefishes
Röhrenmünder, Geisterpfeifenfische	Solenostomidae	ghost pipefishes
Röhrennasen	Procellariiformes	tubenoses, tube-nosed swimmers: albatrosses & shearwaters & petrels
Röhrennasen-Flughunde	*Nyctimene* spp.	tube-nosed fruit bats
Röhrenpolyp	*Tubularia larynx*	ringed tubularian, ringed tubularia
Röhrenschildläuse	Ortheziidae	ensign coccids

Röhrenschnäbler	Aulorhynchidae	tubesnouts
Röhrenschnäbler	*Aulorhynchus flavidus*	tubesnout
Röhrenschwamm	*Reniera cinerea*	redbeard sponge
Röhrenschwamm u.a.	*Siphonochalina siphonella*	tube sponge
Röhrenspinnen u.a., Sackspinnen	Clubionidae	sac spiders, two-clawed hunting spiders, foliage spiders, clubiones, clubionids
Röhrenspinnen u.a.	Eresidae	eresid spiders*
Röhrenvogelspinnen	Hexathelidae	Australian funnelweb spiders
Röhrenwürmer	Serpulidae	serpulids, serpulid tubeworms, serpulid worms
Rohrkäfer, Schilfkäfer	*Donacia* spp.	reed beetles
Rohrkatze (Hbz. Dschungelkatze)	*Felis chaus*	jungle cat
Rohrkolbeneule	*Archanara sparganii*	Webb's wainscot
Rohrmäuse	*Zygodontomys* spp.	cane mice
Rohrratten	*Thryonomys* spp.	cane rats
Rohrrüssler, Rüsselspringer	Macroscelididae	elephant shrews
Rohrsänger	Acrocephalinae	leaf-warblers
Rohrschwirl	*Locustella luscinioides*	Savi's warbler
Rohrweihe	*Circus aeruginosus*	marsh harrier
Rollandtaucher	*Rollandia rolland*	white-tufted grebe
Rollasseln, Kugelasseln	Armadillidiidae	woodlice, pillbugs, sow bugs
Rollegel, Hundegel, Achtäugiger Schlundegel	*Herpobdella octoculata/ Erpobdella octoculata*	eight-eyed leech
Rollenschnecke	*Latirus carinifer*	yellow latirus, trochlear latirus
Rollschlangen	Aniliidae	pipe snakes, pipesnakes
Rollschnecken, Walzenschnecken, Faltenschnecken	Volutidae	volutes
Rollschwanzleguan	*Leiocephalus carinatus*	curly-tailed lizard
Rollwespe> Gemeine Rollwespe, Rotschenkelige Rollwespe	*Tiphia femorata*	common tiphiid wasp
Rollwespen	Tiphiidae	tiphiid wasps
Römischer Maulwurf	*Talpa romana*	Roman mole
Rosa Anemonenfisch, Halsband-Anemonenfisch, Falscher Weißrücken-Clownfisch	*Amphiprion perideraion*	pink anemonefish (FAO), false skunk-striped anemonefish
Rosa Garnele	*Parapenaeus longirostris & Parapenaeus politus*	rose shrimp, deepwater rose shrimp
Rosa Golfgarnele, Nördliche Rosa Geißelgarnele, Nördliche Rosa-Garnele	*Farfantepenaeus duorarum/ Penaeus duorarum*	pink shrimp, northern pink shrimp
Rosa Kaurischnecke, Rosa Kauri	*Trivia suffusa/Niveria suffusa*	pink trivia
Rosa Kingklip	*Genypterus blacodes*	pink cusk-eel, pink ling (FAO)
Rosa Krallenhummer	*Nephropsis rosea*	rosy lobsterette
Rosa Riesenanemone	*Radianthus ritteri/ Heteractis magnifica*	purple-base anemone

Rosa Tiefseegarnele	*Pandalus montagui*	Aesop shrimp, Aesop prawn, pink shrimp
Rosa Zahnbrassen	*Dentex gibbosus*	pink dentex
Rosaflamingo	*Phoenicopterus ruber*	greater flamingo
Rosagesprenkelte Garnele	*Penaeopsis serrata*	pinkspeckled shrimp
Rosakakadu	*Eolophus roseicapillus*	galah
Rosalöffler	*Ajaja ajaja*	roseate spoonbill
Rosamund-Eischnecke*	*Ovula costellata*	pink-mouthed egg cowrie, pinkmouth ovula
Rosamundiger Röhrenpolyp*	*Tubularia crocea*	pinkmouth tubularian, pink-mouth hydroid
Rosapelikan	*Pelecanus onocrotalus*	white pelican
Rosazungen-Skink	*Tiliqua gerrardii*	pinktongue skink, pink-tongued skink
Rosen-Eulenspinner, Roseneule	*Thyatira batis*	peach blossom
Rosenanemone*	*Sagartia elegans*	rosy anemone
Rosenbauch-Schneegimpel	*Leucosticte arctoa*	rosy finch
Rosenberg-Garnele, Rosenberg Süßwassergarnele, Hummerkrabbe (Hbz.)	*Macrobrachium rosenbergii*	Indo-Pacific freshwater prawn, giant river shrimp, giant river prawn, blue lobster (tradename)
Rosenblasenfuß	*Thrips fuscipennis*	rose thrips
Rosenblattlaus, Große Rosenblattlaus, Große Rosenlaus	*Macrosiphum rosae*	rose aphid, "greenfly"
Rosenblattschneiderbiene, Gemeine Blattschneiderbiene	*Megachile centuncularis/ Megachile versicolor*	common leafcutter bee, common leafcutting bee, rose leaf-cutting bee
Rosenblattwespe	*Blennocampa pusilla*	leaf-rolling rose sawfly
Rosenboa	*Lichanura roseofusca*	rosy boa
Rosenboas	*Lichanura* spp.	rosy boas
Rosenbrust-Kernknacker	*Pheucticus ludovicianus*	rose-breasted grosbeak
Rosenbrustsittich, Bartsittich	*Psittacula alexandri*	moustached parakeet
Rosengallmücke	*Dasineura rhodophaga*	rose midge
Rosenkäfer, Goldkäfer	*Cetonia aurata*	rose chafer
Rosenkäfer*	*Macrodactylus subspinosus*	rose chafer
Rosenköpfchen	*Agapornis roseicollis*	peach-faced lovebird
Rosenkopfsittich	*Psittacula roseata*	blossom-headed parakeet
Rosenkoralle*	*Manicina areolata areolata*	rose coral
Rosenminiermotte	*Stigmella anomalella*	rose leaf-miner
Rosenmotte	*Miltochrista miniata*	rosy footman
Rosenmöwe	*Rhodostethia rosea*	Ross's gull
Rosenmund-Maulbeerschnecke	*Morum lamarcki*	rose-mouth morum
Rosenmund-Kreiselschnecke	*Oxystele sinensis*	rosy-base topsnail
Rosenmuschel*, Rosen-Tellmuschel*	*Tellina lineata*	rose-petal tellin, rose tellin, rose petal
Rosensamenwespe	*Megastigmus aculeatus*	rose seed wasp
Rosenseeschwalbe	*Sterna dougallii*	roseate tern
Rosenstar	*Sturnus roseus*	rose-coloured starling

Rosentaube	*Columba mayeri/ Nesoenas mayeri*	pink pigeon
Rosenwickler	*Archips rosana*	rose tortrix (moth), rose twist
Rosenwurm*	*Eisenia rosea*	rosy worm
Rosenzikade	*Typhlocyba rosae/ Edwardsiana rosae*	rose leafhopper
Rosenzweigschnecke, Rosenzweig-Stachelschnecke	*Murex palmarosae/ Chicoreus palmarosae*	rose-branch murex, rose-branched murex
Roses Gespenstfrosch	*Heleophryne rosei*	Skeleton Gorge ghost frog, Table Mountain ghost frog
Rosige Apfelfaltenlaus	*Dysaphis devecta*	rosy leaf-curling aphid
Rosinenmotte	*Cadra calidella*	raisin moth
Ross-Robbe	*Ommatophoca rosii*	Ross seal
Rossameisen, Riesenameisen	*Camponotus* spp.	carpenter ants
Rosshaarwurm, Wasserkalb, Gemeiner Wasserdrahtwurm	*Gordius aquaticus* (Nematomorpha)	gordian worm, common horsehair worm, common hair worm
Rosskastanien-Frostspanner, Kreuzflügel	*Alsophila aescularia*	March moth
Rostbär, Zimtbär	*Phragmatobia fuliginosa*	ruby tiger (moth)
Rostbauchguan	*Penelope purpurascens*	crested guan
Rostbinde	*Hipparchia semele*	grayling
Rostbinden-Samtfalter	*Arethusana arethusa/ Satyrus arethusa*	false grayling
Rostbürzel-Steinschmätzer	*Oenanthe xanthoprymna*	red-tailed wheatear
Rostfarbiger Dickkopf	*Ochlodes venata*	large skipper
Rostgans	*Tadorna ferruginea*	ruddy shelduck
Rostgelber Magerrasen-Kleinspanner	*Idaea serpentata*	ochraceous wave moth
Rostiger Tausendfüßler*	*Trigoniulus lumbricinus*	rusty millipede
Rostkatze	*Felis rubiginosa/ Prionailurus rubiginosus*	rusty-spotted cat
Rostkopfwaran, Australischer Mangrovenwaran	*Varanus semiremex*	Australian mangrove monitor, rusty monitor
Rostpanzerwels	*Corydoras rabauti/ Corydoras myersi*	Myer's catfish
Rostroter Kalkschwamm	*Leucetta primigenia*	ferruginius sponge*
Rostrotes Spitzhörnchen	*Tupaia glis*	tree shrew
Rostschwanzmonal	*Lophophorus impeyanus*	Himalayan monal pheasant
Roststärling	*Euphagus carolinus*	rusty blackbird
Rot-Weiß-Blau-Wurm*	*Proceraea fasciata*	red-white-and-blue worm
Rotauge, Plötze	*Rutilus rutilus*	roach (FAO), Balkan roach
Rotaugen-Fransenlipper	*Labeo cylindricus*	redeye labeo
Rotaugen-Kuhstärling	*Molothrus aeneus*	bronzed cowbird
Rotaugenfrosch, Australischer Rotaugenfrosch	*Litoria chloris*	Australian red-eyed treefrog
Rotaugenfrosch, Rotaugen-Laubfrosch	*Agalychnis callidryas*	red-eyed treefrog
Rotaugensalmler	*Moenkhausia sanctaefilomenae*	redeye tetra (FAO), red-eyed moenkhausia
Rotaugenvireo	*Vireo olivaceus*	red-eyed vireo

Rotband-Bärenkrebs	*Arctides regalis*	red-banded slipper lobster
Rotband-Seebrassen, Rotgebänderter Meerbrassen	*Pagrus auriga*	red-banded seabream
Rotbarsch, Goldbarsch (Großer Rotbarsch)	*Sebastes marinus*	redfish, red-fish, Norway haddock, rosefish, ocean perch (FAO)
Rotbärtige Sklavenameise	*Formica rufibarbis*	red-barbed ant
Rotbärtiger Mönchsaffe, Zottelschweifaffe	*Pithecia monachus*	monk saki
Rotbauch-Braunschlange	*Storeria occipitomaculata*	red-bellied snake
Rotbauch-Buntbarsch, Zilles Buntbarsch	*Tilapia zilli*	redbelly tilapia (FAO), Zilli's cichlid
Rotbauch-Papageifisch	*Scarus gibbus*	blunt-headed parrotfish, heavybeak parrotfish (FAO)
Rotbauch-Schmuckschildkröte	*Pseudemys rubriventris rubriventris*	red-bellied turtle, American red-bellied turtle
Rotbauch-Schwarzotter	*Pseudechis porphyriacus*	red-bellied black snake
Rotbauch-Wassernatter	*Nerodia erythrogaster/ Natrix erythrogaster*	red-bellied water snake, plainbelly water snake, plain-bellied water snake
Rotbauchfilander	*Thylogale billardierii*	red-bellied pademelon
Rotbauchiger Laubschnellkäfer	*Athous haemorrhoidalis/ Athous obscurus*	garden click beetle
Rotbauchleguan	*Sceloporus variabilis*	rose-bellied lizard
Rotbauchmaki	*Lemur rubriventer/ Petterus rubriventer*	red-bellied lemur
Rotbauchmeerkatze	*Cercopithecus erythrogaster*	red-bellied guenon, Nigerian white-throat monkey, white-throated monkey
Rotbauchmolch	*Taricha rivularis*	redbelly newt
Rotbauchschildkröte	*Emydura subglobosa*	red-bellied short-necked turtle
Rotbauchtamarin	*Saguinus labiatus*	white-lipped tamarin
Rotbauchunke	*Bombina bombina*	fire-bellied toad
Rotbein-Vogelspinne	*Brachypelma emilia*	Mexican redleg tarantula
Rotbeinfilander	*Thylogale stigmatica*	red-legged pademelon
Rotbeinige Baumwanze	*Pentatoma rufipes*	forest bug
Rotbeinige Erdmilbe, Schwarze Sandmilbe	*Halotydeus destructor*	redlegged earth mite, black sand mite
Rotbeinige Spinnenameise	*Smicromyrme rufipes/ Mutilla rufipes*	small velvet ant
Rotbeinige Zecke*	*Rhipicephalus evertsi*	red-legged tick
Rotbeiniger Kolbenkäfer, Rotbeiniger Schinkenkäfer, Koprakäfer	*Necrobia rufipes*	red-legged ham beetle, copra beetle
Rotbeinwitwe*, Rote Witwe*	*Latrodectus bishopi*	red-legged widow, red widow spider
Rotbinden-Schläfergrundel	*Amblyeleotris aurora*	pinkbar goby
Rotbrassen, Roter Meerbrassen	*Pagellus erythrinus*	pandora, common pandora (FAO)
Rotbraune Archenmuschel	*Barbatia cancellaria*	red-brown ark (red-brown ark shell)
Rotbrauner Leistenkopfplattkäfer	*Cryptolestes ferrugineus*	rusty grain beetle, rust-red grain beetle
Rotbrauner Fruchtstecher	*Coenorhinus aequatus/ Rhynchites aequatus*	apple fruit rhynchites

Rotbrauner Keulenkäfer	*Claviger testaceus*	redbrown clavigerid
Rotbrauner Laubkäfer	*Serica brunnea*	brown chafer (beetle)
Rotbrauner Reismehlkäfer	*Tribolium castaneum*	red flour beetle
Rotbraunes Krusten-Moostierchen	*Schizobrachiella sanguinea*	sanguine crustal bryozoan*
Rotbrust-Glanzköpfchen	*Nectarinia senegalensis*	scarlet-chested sunbird
Rotbrust-Sonnenbarsch, Großohriger Sonnenfisch	*Lepomis auritus*	redbreast sunfish (FAO), red-breasted sunfish
Rotbugamazone, Blaustirnamazone	*Amazona aestiva*	blue-fronted amazon
Rotdeckenkäfer	Lycidae	net-winged beetles
Rotdrossel	*Turdus iliacus*	redwing
Rotducker	*Cephalophus natalensis*	red forest duiker
Rote Bohne, Baltische Tellmuschel, Baltische Plattmuschel	*Macoma baltica*	Baltic macoma
Rote Chile-Vogelspinne*	*Phrixotrichus cala*	Chilean rose tarantula
Rote Diamantklapperschlange	*Crotalus ruber*	red diamond rattlesnake
Rote Edelkoralle, Edelkoralle	*Corallium rubrum*	red coral, precious red coral
Rote Eischnecke, Rotes Vogelei	*Primovula carnea/ Pseudosimnia carnea*	red dwarf ovula
Rote Fichtengallenlaus, Frühe Fichten-Kleingallenlaus	*Adelges laricis*	red larch gall adelgid, larch adelges, larch woolly aphid
Rote Fledermaus	*Lasiurus borealis*	red bat
Rote Fuchsschwanz-Gallmücke	*Dasineura alopecuri*	foxtail midge
Rote Garnele, Rote Tiefseegarnele	*Aristaeomorpha foliacea*	giant gamba prawn, giant red shrimp, royal red prawn
Rote Käferschnecke	*Tonicella rubra*	northern red chiton, red northern chiton
Rote Keulenschrecke	*Gomphocerus rufus*	rufous grasshopper
Rote Kiefernbuschhornblattwespe	*Neodiprion sertifer*	fox-coloured sawfly, lesser pine sawfly, small pine sawfly, European pine sawfly (U.S.)
Rote Knotenameise, Rotgelbe Knotenameise	*Myrmica rubra*	red myrmicine ant, red ant
Rote Knotenameisen, Gemeine Knotenameisen	*Myrmica* spp.	red myrmicine ants
Rote Königsnatter, Dreiecksnatter	*Lampropeltis triangulum*	milk snake, eastern milk snake
Rote Korkkoralle	*Alcyonium glomeratum*	red sea-fingers, red fingers
Rote Krötenkopf-Schildkröte	*Phrynops rufipes*	red-footed sideneck turtle, red toad-headed turtle
Rote Mauerbiene	*Osmia rufa*	red mason bee
Rote Messergarnele	*Haliporoides sibogae*	pink prawn, royal red prawn
Rote Pandora	*Pagellus bellottii*	red pandora
Rote Riesenaktinie	*Gyrostoma helianthus*	giant red anemone*
Rote Riesengarnele, Atlantische Rote Riesengarnele	*Plesiopenaeus edwardsianus*	scarlet gamba prawn, scarlet shrimp

Rote Seegurke	*Cucumaria miniata*	red sea cucumber
Rote Seescheide, Roter Seepfirsich*	*Halocynthia papillosa*	red sea-squirt, red sea peach*
Rote Seespinne	*Lissa chiraga*	red masked crab*
„Rote Spinne", Gemeine Spinnmilbe, Bohnenspinnmilbe	*Tetranychus urticae*	twospotted spider mite
„Rote Spinne", Obstbaumspinnmilbe	*Panonychus ulmi/ Metatetranychus ulmi*	fruittree red spider mite, European red mite
Rote Spinnenschnecke, Orange Spinnenschnecke	*Lambis crocata*	orange spider conch
Rote Spottdrossel	*Toxostoma rufum*	brown thrasher
Rote Stern-Turbanschnecke	*Lithopoma gibberosum*	red starsnail, red turbansnail, red turban
Rote Strandschnecke*	*Littorina coccinea*	scarlet periwinkle
Rote Stumpfnasenbrasse	*Chrysoblephus gibbiceps*	red stumpnose
Rote Tellmuschel	*Angulus incarnatus/ Tellina incarnatus*	red tellin
Rote Tiefseekrabbe	*Chaceon maritae*	West African geryonid crab, red crab
Rote Venusmuschel	*Callista erycina*	red callista, red venus
Rote Vogelmilbe, Hühnermilbe	*Dermanyssus gallinae*	red mite of poultry, poultry red mite, red chicken mite, chicken mite
Rote Waldameise, Große Rote Waldameise	*Formica rufa*	wood ant
Rote Walzenschlange, Rotschwanz-Rollschlange	*Cylindrophis rufus*	pipe snake
Rote Wespe	*Vespula rufa*	red wasp
Rote Witwe*, Rotbeinwitwe*	*Latrodectus bishopi*	red-legged widow, red widow spider
Rote Wurmschleiche	*Amphisbaena alba*	red worm lizard
Rote Ziegenantilope, Rotgoral	*Nemorhaedus baileyi*	red goral
Rote Zitrusschildlaus, Rote Citrusschildlaus	*Aonidiella aurantii*	California red scale
Rote Zwerg-Avicularia	*Avicularia pulchra*	red dwarf tarantula*
Rötelfalke	*Falco naumanni*	lesser kestrel
Rötelgrundammer	*Pipilo erythrophthalmus*	rufous-sided towhee
Rötelmaus, Waldwühlmaus	*Clethrionomys glareolus*	bank vole
Rötelmäuse	*Clethrionomys* spp.	red-backed mice, bank voles
Rötelpelikan	*Pelecanus rufescens*	pink-backed pelican
Rötelschwalbe	*Hirundo daurica*	red-rumped swallow
Roter Acouchi	*Myoprocta exilis* (Husson '78)	red acouchi
Roter Bandfisch	*Cepola macrophthalma/ Cepola rubescens*	red bandfish
Roter Bäumchenpolyp	*Eudendrium carneum*	red stickhydroid
Roter Blasenfuß	*Aptinothrips rufus*	red glass thrips
Roter Bohrschwamm	*Cliona lampa*	red boring sponge
Roter Brüllaffe	*Alouatta seniculus*	red howler

Roter Buntbarsch, Prachtbuntbarsch	*Hemichromis bimaculatus/ Hemichromis guttatus*	jewelfish (FAO), red jewel fish, red cichlid
Roter Clownjunker, Roter Clown-Lippfisch	*Coris gaimard gaimard*	yellowtail coris (FAO), clown wrasse
Roter Drachenkopf	*Parascorpaena aurita*	red stingfish
Roter Eichelwurm	*Saccoglossus cambrensis* (Enteropneusta)	red acorn worm
Roter Eichenkugelrüssler, Eichenblattroller, Tönnchenwickler	*Attelabus nitens*	oak leaf roller, red oak roller
Roter Eichhörnchenfisch, Roter Soldatenfisch, Kurzstacheliger Soldatenfisch	*Myripristis murdjan/ Myripristis axillares*	pinecone soldierfish (FAO), crimson soldierfish, white-edge soldierfish
Roter Fahnenbarsch, Rötling	*Anthias anthias*	swallowtail seaperch (FAO), marine goldfish
Roter Fingerschwamm	*Rhaphidophlus cervicornis*	red antler sponge
Roter Flughund	*Stenoderma rufum*	red fruit bat
Roter Gabeldorsch	*Urophycis chuss*	red hake (FAO), squirrel hake
Roter Geweihschwamm*	*Haliclona rubens*	red sponge
Roter Herzigel	*Meoma ventricosa*	sea pussy, cake urchin, red heart urchin*
Roter Husar, Roter Soldatenfisch, Rotgestreifter Soldatenfisch, Gewöhnlicher Eichhornfisch	*Holocentrus rubrum/ Sargocentron rubrum*	red soldierfish
Roter Kalkröhrenwurm, Blutroter Röhrenwurm	*Protulopsis intestinum/ Protula intestinum*	bloody tubeworm*
Roter Kalmar, Breitschwanz- Kurzflossenkalmar	*Illex coindetii*	southern shortfin squid, broadtail shortfin squid
Roter Kammstern, Großer Kammstern, Mittelmeer-Kammstern	*Astropecten aurantiacus*	red sand star, red comb star
Roter Kiefernspanner*	*Thera firmata*	red pine carpet moth
Roter Knospenwickler	*Spilonota ocellana*	eyespotted bud moth
Roter Knurrhahn (Seeschwalbenfisch)	*Trigla lucerna*	tub gurnard (FAO), saphirine gurnard
Roter Kölbchenpolyp	*Coryne pusilla*	red flowerhead polyp
Roter Krustenschwamm	*Monanchora arbuscula*	red-white marbled sponge
Roter Langur	*Presbytis melalophos*	banded leaf monkey
Roter Lanzenseeigel	*Stylocidaris affinis*	red cidarid
Roter Luftröhrenwurm, Rotwurm, Gabelwurm	*Syngamus trachea*	gapeworm (of poultry)
Roter Magenwurm	*Hyostrongylus rubidus*	red stomach worm (of swine)
Roter Moosschwamm*	*Microciona prolifera*	red bread sponge, red moss sponge, red sponge
Roter Neon, „Falscher" Neon	*Cheirodon axelrodi/ Paracheirodon axelrodi*	cardinal tetra
Roter Palisadenwurm*	*Strongylus vulgaris*	bloodworm

Roter Piranha, Natterers Sägesalmler	*Serrasalmus nattereri*	convex-headed piranha, Natterer's piranha, red piranha (FAO), red-bellied piranha
Roter Regenbogenfisch	*Glossolepis incisus*	red rainbowfish
Roter Scheckenfalter	*Melitaea didyma*	spotted fritillary
Roter Scheinhummer	*Pleuroncodes monodon*	red squat lobster
Roter Schnapper, Nördlicher Schnapper	*Lutjanus campechanus*	red snapper, northern red snapper (FAO)
Roter Schnurwurm	*Lineus ruber* (Nemertini)	red bootlace
Roter Soldatenfisch, Roter Husar, Rotgestreifter Soldatenfisch, Gewöhnlicher Eichhornfisch	*Sargocentron rubrum/ Holocentrus ruber*	red soldierfish, redcoat (FAO)
Roter Stör	*Acipenser fulvescens*	lake sturgeon
Roter Stummelaffe	*Colobus badius/ Piliocolobus badius*	red colobus, western red colobus
Roter Sumpfkrebs, Louisiana-Sumpfkrebs, Louisiana-Flusskrebs	*Procambarus clarkii*	Louisiana red crayfish, red swamp crayfish, Louisiana swamp crayfish, red crayfish
Roter Uakari	*Cacajao rubicundus*	red uakari
Roter von Rio	*Hyphessobrycon flammeus*	flame tetra
Roter Vulkanschwamm	*Acarnus erithacus*	red volcano sponge
Roter „Walfisch"	*Barbourisia rufa*	red whalefish
Roter Waldregenwurm	*Lumbricus rubellus*	red earthworm, red worm
Roter Wiesensalamander, Rotsalamander	*Pseudotriton ruber*	red salamander
Roter Würfelfalter	*Spialia sertorius*	red underwing skipper
Roter Zackenbarsch	*Epinephelus morio*	red grouper
Roter Zwergfeuerfisch, Gewöhnlicher Zwergfeuerfisch	*Dendrochirus brachypterus*	dwarf lionfish
Rotes Ordensband	*Catocala nupta*	red underwing
Rotes Rattenkänguru, Großes Rattenkänguru	*Aepyprymnus rufescens*	rufous "rat"-kangaroo
Rotes Riesenkänguru	*Macropus rufus*	red kangaroo
Rotes Vogelei, Rote Eischnecke	*Primovula carnea/ Pseudosimnia carnea*	red dwarf ovula
Rotes Wendeltürmchen	*Turbonilla crenata*	red pyramidsnail
Rotfeder	*Scardinius erythrophthalmus*	rudd
Rotfeuerfische	*Pterois* spp.	firefishes, turkeyfishes
Rotflankenducker, Blaurückenducker	*Cephalophus rufilatus*	red-flanked duiker
Rotfleckiger Ansauger, Rotflecken-Ansauger	*Lepadogaster candollei*	Connemara clingfish (FAO), Connemara sucker
Rotflossen-Hornhecht	*Strongylura notata*	redfin needlefish
Rotflossen-Stachelwels	*Mystus nemurus*	Asian redtail catfish
Rotflossenhecht	*Esox americanus*	redfin, grass pickerel, redfin pickerel (FAO)
Rotflossenorfe	*Notropis lutrensis*	red shiner
Rotflossensalmler	*Aphyocharax rubripinnis/ Aphyocharax anisitsi*	bloodfin
Rotflügel-Brachschwalbe	*Glareola pratincola*	collared pratincole

Rotflügelgimpel	*Rhodopechys sanguinea*	crimson-winged finch
Rotflügelige Schnarrschrecke, Schnarrheuschrecke	*Psophus stridulus*	rattle grasshopper
Rotflügelige Ödlandschrecke	*Oedipoda germanica*	red-winged grasshopper
Rotfuchs	*Vulpes vulpes*	red fox
Rotfuß-Erdhörnchen, Streifen-Borstenhörnchen	*Euxerus erythropus*	striped ground squirrel
Rotfußdrossel	*Turdus plumbeus*	red-legged thrush
Rotfußfalke	*Falco vespertinus*	red-footed falcon
Rotfüßige Vogelspinne	*Brachypelma smithi*	Mexican redknee tarantula
Rotfüßiger Schnellkäfer	*Melanotus rufipes*	redfooted click beetle*
Rotfußtölpel	*Sula sula*	red-footed booby
Rotfußvogelspinne, Weißfußvogelspinne	*Avicularia metallica*	whitetoe tarantula
Rotgazelle	*Gazella rufina*	red gazelle
Rotgefranste Weißpunkteule, Marbeleule, Binsengraseule, Finkeneule, Waldmoorrasen-Türkeneule	*Mythimna turca*	double-lined wainscot, double line moth
Rotgelber Seeigel	*Echinus acutus*	red-yellow sea urchin*
Rotgelenks-Winkerkrabbe, Brackwasser-Winkerkrabbe	*Uca minax*	redjointed fiddler, brackish-water fiddler
Rotgerippte Latirus*	*Latirus craticulatus*	red-ripped latirus
Rotgesichtsmakak	*Macaca fuscata*	Japanese macaque
Rotgestreifte Mondschnecke	*Natica rubromaculata*	red-striped moonsnail
Rotgoral, Rote Ziegenantilope	*Nemorhaedus baileyi*	red goral
Rothaarbock	*Pyrrhidium sanguineum*	scarlet-coated longhorn beetle
Rothals-Ziegenmelker	*Caprimulgus ruficollis*	red-necked nightjar
Rothalsfilander	*Thylogale thetis*	red-necked pademelon
Rothalsgans	*Branta rufucoliis*	red-breasted goose
Rothalsiger Schinkenkäfer	*Necrobia ruficollis*	red-shouldered ham beetle
Rothalsschildkröte	*Chinemys kwangtungensis*	Chinese red-necked pond turtle
Rothalstaucher	*Podiceps griseigena*	red-necked grebe
Rothandbrüllaffe	*Alouatta belzebul*	black and red howler
Rothandtamarin	*Saguinus midas*	red-handed tamarin
Rothirsch, Edelhirsch	*Cervus elaphus*	red deer, wapiti, elk (U.S.)
Rothokko	*Nothocrax urumutum*	nocturnal curassow
Rothörnchen (Hbz. Feh)	*Tamiasciurus hudsonicus*	eastern red squirrel, red squirrel
Rothschild-Goliathratte	*Mallomys rothschildi*	giant tree rat
Rothuhn	*Alectoris rufa*	red-legged partridge
Rothund	*Cuon alpinus*	dhole, red dog, Asiatic wild dog
Rotkaninchen	*Pronolagus* spp.	red rabbits, red rock rabbits
Rotkardinal	*Cardinalis cardinalis*	common cardinal
Rotkehl-Anolis	*Anolis carolinensis*	green anole
Rotkehl-Faulvogel	*Nonnula rubecula*	rusty-breasted nunlet

Rotkehl-Hüttensänger	*Sialia sialis*	eastern bluebird
Rotkehl-Rennechse	*Cnemidophorus hyperythrus*	orange-throated whiptail
Rotkehl-Strandläufer	*Calidris ruficollis*	red-necked stint
Rotkehlchen	*Erithacus rubecula*	robin (*Europe*)
Rotkehlfälkchen	*Microhierax caerulescens*	collared falconet
Rotkehlfrankolin	*Francolinus afer*	red-necked spurfowl
Rotkehlpieper	*Anthus cervinus*	red-throated pipit
Rotklee-Spitzmausrüssler	*Apion apricans*	red clover seed weevil
Rotkopf-Bungar, Rotkopf-Krait	*Bungarus flaviceps*	red-headed krait
Rotkopf-Plattschildkröte	*Platemys platycephala*	twist-neck turtle, twist-necked turtle
Rotkopfente	*Aythya americana*	redhead
Rotkopfpapagei	*Geoffroyus geoffroyi*	red-cheeked parrot
Rotkopfsalmler	*Hemigrammus bleheri*	firehead tetra
Rotkopfspecht	*Melanerpes erythrocephalus*	red-headed woodpecker
Rotkopfwürger	*Lanius senator*	woodchat shrike
Rotlappenkiebitz	*Hoplopterus indicus*	red-wattled plover
Rotleibiger Grashüpfer	*Omocestus haemorrhoidalis*	orange-tipped grasshopper
Rötliche Bänder-Olive	*Oliva rufula*	reddish olive
Rötliche Bernsteinschnecke	*Oxyloma sarsii*	slender amber snail, reddish ambersnail*
Rötliche Laubschnecke	*Perforatella incarnata*	
Rötliche Wurzeleule	*Thalpophila matura*	straw underwing
Rötlicher Schlangenaal	*Ophisurus rufus*	rufous snake eel
Rötling	*Phoxinus eos/ Chrosomus erythrogaster*	northern redbelly dace
Rotlippenschlange	*Crotaphopeltis hotamboeia*	herald snake
Rotlippenschlangen	*Crotaphopeltis* spp.	herald snakes, cat-eyed snakes, tropical water snakes
Rotlippige Olivenschnecke	*Oliva rubrolabiata*	red-lipped olive
Rotlori	*Eos bornea*	red lory
Rotluchs (Hbz. Luchskatze)	*Felis rufus/ Lynx rufus*	bobcat
Rotmaul-Ährenfisch	*Labidesthes siccula*	brook silverside
Rotmaulsalmler	*Hemigrammus rhodostomus*	red-nosed tetra, rummy-nose tetra (FAO), red-nosed characin, rummy-nosed characin
Rotmaulsalmler> Falscher Rotmaulsalmler	*Petitella georgiae*	false rummy-nose tetra
Rotmeer-Doktorfisch	*Acanthurus sohal*	Red Sea surgeonfish, Red Sea clown surgeon, Sohal surgeonfish (FAO)
Rotmeer-Falterfisch, Rotmeer-Schmetterlingsfisch, Tabak-Falterfisch	*Chaetodon fasciatus*	Red Sea raccoon butterflyfish, diagonal butterflyfish (FAO)
Rotmeer-Hering	*Herklotsichthys punctatus*	spotted herring, spotback herring (FAO)
Rotmeer-Kaninchenfisch	*Siganus rivulatus*	marbled rabbitfish, marbled spinefoot (FAO)
Rotmeer-Zunge	*Cynoglossus sinusarabici*	Red Sea tongue-sole, Red Sea tonguesole

Rotmilan	*Milvus milvus*	red kite
Rotmund-Elefantenohr	*Rhodactis rhodostoma*	red-mouth mushroom anemone
Rotmund-Froschschnecke	*Tutufa rubeta*	red-mouthed frogsnail, ruddy frogsnail
Rotmund-Leistenschnecke, Rotmundige Steinschnecke	*Thais haemastoma*	redmouthed rocksnail (red-mouthed rock shell)
Rotmund-Olive, Goldmund-Olive	*Oliva miniacea*	gold-mouthed olive, orange-mouthed olive, red-mouth olive, red-mouthed olive
Rotnackenarassari	*Pteroglossus bitorquatus*	red-necked araçari
Rotnasen-Grüntaube	*Treron calva*	African green pigeon
Rotnasenmaus	*Wiedomys pyrrhorhinos*	red-nosed mouse
Rotnasenratte, Afrikanische Rotnasenratte	*Oenomys hypoxanthus*	rufous-nosed rat, rusty-nosed rat
Rotohrbülbül	*Pycnonotus jocosus*	red-whiskered bulbul
Rotohrfrosch	*Rana erythraea*	red-eared frog
Rotohrmeerkatze	*Cercopithecus erythrotis*	russet-eared guenon, red-eared monkey, red-eared nose-spotted guenon
Rotpelzige Sandbiene	*Andrena fulva/Andrena armata*	tawny burrowing bee
Rotpunktgarnele	*Farfantepenaeus brasiliensis/ Penaeus brasiliensis*	pinkspotted shrimp, redspotted shrimp
Rotrand-Fledermausfisch, Rotsaum-Segelflosser	*Platax pinnatus*	red-margined batfish
Rotrandbär	*Diacrisia sannio*	clouded buff moth
Rotrücken-Totenkopfaffe	*Saimiri oerstedii*	red-backed squirrel monkey, Central American squirrel monkey
Rotrücken-Waldsalamander	*Plethodon cinereus*	redback salamander, red-backed salamander
Rotrückenmeise	*Parus rufescens*	chestnut-backed chickadee
Rotrückenwürger, Neuntöter	*Lanius collurio*	red-backed shrike
Rotrückige Klappen-Weichschildkröte	*Cycloderma aubryi*	Aubry's soft-shelled turtle, Aubry's flapshell turtle
Rotrückiger Irrwisch	*Alydus calcaratus*	redbacked bug, redbacked broadheaded bug
Rotscheitel-Waldsänger	*Dendroica discolor/ Agreocantor discolor*	prairie warbler
Rotschenkel	*Tringa totanus*	redshank
Rotschenkel-Kleideraffe	*Pygathrix nemaeus*	red-shanked douc langur
Rotschenkelfrosch	*Rana aurora*	red-legged frog
Rotschenkelhörnchen	*Funisciurus* spp.	African striped squirrels, rope squirrels
Rotschnabel-Madenhacker	*Buphagus erythrorhynchus*	red-billed oxpecker
Rotschnabel-Tropikvogel	*Phaethon aetherus*	red-billed tropic bird
Rotschulterbussard	*Buteo lineatus*	red-shouldered hawk
Rotschulterstärling	*Agelaius phoeniceus*	red-winged blackbird
Rotschwanz, Buchenrotschwanz, Buchen-Streckfuß	*Elkneria pudibunda/ Calliteara pudibunda/ Dasychira pudibunda/ Olene pudibunda*	pale tussock, red-tail moth
Rotschwanz-Ährenfisch	*Bedotia geayi*	Madagascar rainbowfish
Rotschwanz-Pflanzenmäher	*Phytotoma rara*	Chilean plantcutter

Rotschwanzbärbling	*Rasbora borapetensis*	redtailed rasbora, blackline rasbora (FAO)
Rotschwanzbussard	*Buteo jamaicensis*	red-tailed hawk
Rotschwanzgarnele	*Fenneropenaeus penicillatus/ Penaeus penicillatus*	redtail prawn
Rotschwanzkärpfling	*Lucania goodei/ Chriopeops goodei*	bluefin killifish (FAO), blue-fin killy
Rotschwanzlerche	*Ammomanes phoenicurus*	rufous-tailed desertlark, rufous-tailed lark
Rotschwanzmeerkatze	*Cercopithecus ascanius*	black-cheeked white-nosed monkey, red-tailed monkey
Rotschwanzschnapper	*Lutjanus synagris*	lane snapper
Rotschwarze Schaumzikade, Blutzikade	*Cercopis vulnerata*	red-and-black froghopper
Rotsteißkakadu	*Cacatua haematuropygia*	red-vented cockatoo
Rotstirngazelle	*Gazella rufifrons*	red-fronted gazelle
Rotstirngirlitz	*Serinus pusillus*	red-fronted serin
Rotstreifenbärbling	*Rasbora pauciperforata*	red-striped rasbora, red-line rasbora
Rotstummelaffen	*Piliocolobus* spp.	red colobuses
Rotvioletter Seestern, Purpurstern	*Ophidiaster ophidianus*	purple sea star*
Rotwangen-Schmuckschildkröte	*Trachemys scripta elegans/ Pseudemys scripta elegans*	red-eared turtle, red-eared slider
Rotwangen-Zwergpapagei	*Opopsitta diophthalma*	double-eyed fig parrot
Rotwangenhörnchen	*Dremomys* spp.	red-cheeked squirrels
Rotweiß-Spinne*	*Enoplognatha ovata*	red-and-white spider
Rotwolf	*Canis rufus*	red wolf
Rotwurm, Gabelwurm, Roter Luftröhrenwurm	*Syngamus trachea*	gapeworm (of poultry)
Rotzahn, Rotzahn-Drückerfisch, Atlantischer Drückerfisch, Nördlicher Drückerfisch, Gabelschwanz Drückerfisch, Blauer Drücker	*Odonus niger/ Balistes erythrodon/ Odonus erythrodon/ Xenodon niger*	redtooth triggerfish, redfang, redtoothed triggerfish (FAO)
Rotzahnspitzmäuse, Waldspitzmäuse	*Sorex* spp.	red-toothed shrews, long-tailed shrews
Rotzipfelkäfer	*Malachius bipustulatus*	red-tipped flower beetle
Rotzunge, Hundszunge, Zungenbutt	*Glyptocephalus cynoglossus*	witch
Rotzunge> Echte Rotzunge, Limande	*Microstomus kitt*	lemon sole
Rousselot-Baummaus	*Dendroprionomys rousseloti*	Congo tree mouse
Rübenblattwespe, Kohlrübenblattwespe	*Athalia rosae*	turnip sawfly
Rübenderbrüssler	*Bothynoderes punctiventris/ Cleonus punctiventris*	beet root weevil
Rübenerdfloh, Mangolderdfloh	*Chaetocnema concinna*	mangold flea beetle, mangel flea beetle, beet flea beetle
Rübenfliege, Runkelfliege	*Pegomya betae*	beet fly, beet leafminer, (mangold fly, spinach leafminer)

Rübenmotte	*Scorbipalpa ocellatella*	turnip moth*, sugarbeet moth*
Rübenschwanzgecko	*Thecadactylus rapicauda*	wood slave, turniptail gecko
Rübenwanze	*Piesma quadrata*	beet leaf bug
Rübenweißling, Kleiner Kohlweißling	*Pieris rapae/ Artogeia rapae*	small white, cabbage butterfly, imported cabbageworm (U.S.)
Rübenzünsler, Wiesenzünsler	*Loxostege sticticalis/ Margaritia sticticalis*	beet webworm
Rübenzystenälchen	*Heterodera schachtii*	beet eelworm, sugar-beet eelworm, beet cyst nematode
Rubin-Schlangenstern	*Ophioderma rubicundum*	ruby brittlestar*
Rubingoldhähnchen	*Regulus calendula*	ruby-crowned kinglet
Rubinkehlkolibri	*Archilochus colubris*	ruby-throated hummingbird
Rubinlori	*Vini kuhlii*	Kuhl's lory
Rückenfleck-Preußenfisch	*Dascyllus carneus*	two-striped damselfish, two-bar humbug
Rückenflecken-Weichschildkröte, Gefleckte Klappen-Weichschildkröte	*Cyclanorbis elegans*	Nubian soft-shelled turtle, Nubian flapshell turtle
Rückenschaler	Notostraca	tadpole shrimps, shield shrimps
Rückenschwimmender Kongowels	*Synodontis nigriventris*	blackbellied upside-down catfish
Rückenschwimmer	*Notonecta* spp.	backswimmers (water boatmen)
Rückenschwimmer	Notonectidae	backswimmers (water boatmen)
Rückenstreifen-Pelomeduse	*Pelusios gabonensis*	African forest turtle, stripe-backed sidenecked turtle, Gabon turtle
Rückenstreifenmaus*	*Muriculus imberbis*	stripe-backed mouse, Wurch mouse
Rückenstreifenwiesel	*Mustela strigidorsa*	back-striped weasel
Rückenstreifkänguru, Aalstrichwallaby	*Macropus dorsalis*	black-striped wallaby
Rückfallfieberzecke (trop. Lederzecke)	*Ornithodoros turicata*	relapsing fever tick
Rucksackschnecke	*Testacella haliotidea*	earshell slug, shelled slug
Rucksackschnecken	Testacellidae	shelled slugs, worm-eating slugs
Ruderfische, „Straßenkehrer"	Lethrinidae	emperors, emperor breams
Ruderfrösche, Flugfrösche	*Rhacophorus* spp.	flying frogs, rhacophors
Ruderfrösche, Flugfrösche	Rhacophoridae	flying frogs, Old World tree frogs, rhacophorids
Ruderfüßer, Ruderfüßler	Pelecaniformes	totipalmate swimmers: pelicans and allies
Ruderfußkrebse, Ruderfüßer	Copepoda	copepods
Ruderkrabbe, Gemeine Schwimmkrabbe	*Portunus holsatus/ Liocarcinus holsatus/ Macropinus holsatus*	common swimming crab, flying swimming crab

Ruderschlange	*Hydrophis spiralis*	yellow sea snake, banded sea snake
Ruderschlangen	*Hydrophis* spp.	Asian sea snakes, Australian sea snakes
Ruderwanzen (Wasserzikaden)	Corixidae	water boatmen
Ruderwürmer, Paddelwürmer	*Phyllodoce* spp.	paddleworms
Rüeblifliege, Möhrenfliege	*Psila rosae*	carrot rust fly, carrot fly
Ruhm des Meeres	*Leptoconus gloriamaris/ Conus gloriamaris*	glory-of-the-sea cone
Ruhramöben (Amöbiasis)	*Entamoeba histolytica*	dysentary ameba (amebic dysentery/amebiasis)
Ruineneidechse	*Podarcis sicula*	Italian wall lizard, ruin lizard
Rumänischer Hamster	*Mesocricetus newtoni*	Romanian hamster
Rümmlers Mosaikschwanzratte	*Pogonomelomys* spp.	Rümmler's mosaic-tailed rats
Rund-Stechrochen	Urolophidae	round stingrays, stingarees (Austr.)
Rund-Stechrochen, Runder Stechrochen, Kugelrochen	*Taeniura grabata*	round stingray
Rundblattnasen	*Hipposideros* spp.	Old World leaf-nosed bats
Rundblattnasen	Hipposideridae	Old World leaf-nosed bats (incl. trident-nosed bats)
Runde Nelkenkoralle	*Caryophyllia inormata*	circular cupcoral
Runde Pilgermuschel	*Argopecten circularis*	circular scallop
Rundfelchen*	*Prosopium cylindriceum*	round whitefish (FAO), menominee
Rundfelchen*	*Prosopium* spp.	round whitefishes
Rundhering, Gemeiner Rundhering	*Etrumerus teres*	round herring
Rundkopf-Grenadier, Rundkopf-Panzerratte, Langschwanz	*Coryphaenoides rupestris*	roundhead rattail, roundnose grenadier (FAO), rock grenadier
Rundkopfdelphin, Rissos Delphin	*Grampus griseus*	Risso's dolphin, gray grampus, white-headed grampus
Rundköpfe, Mirakelbarsche	Plesiopidae	longfins, roundheads
Rundköpfiger Reismehlkäfer	*Latheticus oryzae*	longheaded flour beetle
Rundkrabben u.a.	Atelecyclidae & Xanthidae	horse crabs, circular crabs & xanthid crabs, mud crabs
Rundliche Sumpfdeckelschnecke	*Viviparus intertextus*	rotund mysterysnail
Rundmäuler, Kreismünder	Cyclostomata	cyclostomes
Rundmündige Windelschnecke, Blanke Windelschnecke	*Vertigo genesii*	round-mouthed whorl snail
Rundohrenfledermäuse	*Tonatia* spp.	round-eared bats
Rundschuppen-Speerfisch	*Tetrapturus georgei*	round-scaled marlin, roundscale spearfish (FAO)
Rundschwänzige Krötenechse	*Phrynosoma modestum*	roundtail horned lizard, round-tailed horned lizard
Rundschwanzsperber	*Accipiter cooperii*	Cooper's hawk

Rundstirnmotten (Wippmotten)	Glyphipterygidae	glyphipterygid moths
Rundwürmer (sensu strictu), Fadenwürmer, Nematoden	Nematoda	roundworms, nematodes
Runzelfrösche	Platymantis spp.	Fijian ground frogs
Runzelgesicht-Rundkrabbe*	Atelecyclus septemdentatus	old-man's face crab
Runzelige Gitterschnecke	Trigonostoma rugosum	rugose nutmeg
Runzelige Schwimmkrabbe	Liocarcinus corrugatus	wrinkled swimming crab
Runzelige Seespinne	Herbstia condyliata	rugose spider crab
Runzelkäfer	Rhysodidae	wrinkled bark beetles
Runzelkorallen	Rugosa	rugose corals
Rüppelfuchs, Sandfuchs	Vulpes rueppelli	sand fox
Rüppell's Thunfisch, Ungestreifte Pelamide	Orcynopsis unicolor/ Gymnosarda unicolor	plain bonito, scaleless tuna, pigtooth tuna,, dogtooth tuna (FAO), Rüppell's bonito
Rüppellseeschwalbe	Sterna bengalensis	lesser crested tern
Ruß-Gleithörnchen	Pteromyscus pulverulentus	smoky flying squirrel
Rußalbatros	Phoebetria palpebrata	light-mantled sooty albatross
Rußbraune Schabe	Periplaneta fuliginosa	smoky brown cockroach, smokybrown cockroach
Rußkakadu	Calyptorhynchus funereus	black cockatoo
Rußköpfchen	Agapornis nigrigenis	black-cheeked lovebird
Rußseeschwalbe	Sterna fuscata	sooty tern
Rüsselchimären, Langnasenchimären	Rhinochimaeridae	longnose chimaeras, spookfishes
Rüsselegel	Rhynchobdelliformes	proboscis leeches
Rüsselhündchen	Rhynchocyon spp.	checkered elephant shrews, forest elephant shrews
Rüsselkäfer	Curculionidae	snout beetles, weevils (true weevils)
Rüsselkrebschen	Bosmina longirostris	lesser bosminid waterflea*
Rüsselkrebse	Bosminidae	bosminid waterfleas
Rüsselqualle, Rüsselmeduse	Geryonia proboscidalis	trunked jellyfish
Rüsselspringer	Macroscelidea	African elephant shrews
Rüsselspringer, Rohrrüssler	Macroscelididae	elephant shrews
Rüsseltiere	Proboscidea	elephants & relatives
Russischer Bär	Euplagia quadripunctaria/ Callimorpha quadripunctaria	Jersey tiger (moth), Russian tiger (moth)
Russischer Zwerghamster	Phodopus sungorus sungorus	Russian striped hairy-footed hamster
Rüsternblasenlaus	Byrsocrypta ulmi	fig gall aphid
Rutte, Trüsche, Quappe	Lota lota	burbot
Ruttenegel, Quappenegel	Cystobranchus mammillatus	burbot leech
Ryukyu-Ratte	Diplothrix legata/ Rattus legata	Ryukyu Island rat

Saateule, Erdeule, Wintersaateule, Graswurzeleule	*Agrotis segetum*	turnip moth (common cutworm)
Saatgans	*Anser fabalis*	bean goose
Saatkrähe	*Corvus frugilegus*	rook
Saatschnellkäfer	*Agriotes lineatus*	lined click beetle, striped click beetle
Säbelantilope, Nordafrikanischer Spießbock	*Oryx dammah*	scimitar oryx, scimitar-horned oryx
Säbeldornschrecke	*Tetrix subulata/* *Acrydium subulatum*	slender groundhopper
Säbeldornschrecke, Gemeine Dornschrecke	*Tetrix undulata/* *Tetrix vittata*	common groundhopper
Säbelfische, Schwarzfische	Anoplopomatidae	sablefishes
Säbelkrabbe	*Platychirograpsus spectabilis*	saber crab
Säbelschnäbler	*Recurvirostra avosetta*	avocet
Säbelzahnfische	Evermannellidae	sabertooth fishes, sabretoothed fishes
Säbler, Weißbrauensäbler, Australische Scheitelsäbler	Pomatostomidae	Australo-Papuan babblers
Sachalin-Stör	*Acipenser mikadoi*	Sakhalin sturgeon
Sachalin-Trogmuschel	*Mactra sachalinensis*	hen clam
Sackbrassen, Gemeiner Seebrassen	*Pagrus pagrus/* *Sparus pagrus pagrus/* *Pagrus vulgaris*	common seabream (FAO), red porgy, Couch's seabream
Sackkiemer, Kiemenschlauchwelse	Heteropneustidae	airsac catfishes
Sackkiemer	*Heteropneustes fossilis*	airsac catfish
Sackmäuler, Schlinger	Saccopharyngidae	swallowers
Sackmotten, Sackträgermotten> Futteralmotten	Coleophoridae	casebearers, casebearer moths
Sackmotten, Sackträgermotten> Sackspinner	Psychidae (incl. Talaeporiidae)	bagworms, bagworm moths
Sackschnecken, Schlundsackschnecken, Schlauchschnecken	Sacoglossa/Saccoglossa	sacoglossans
Sackspinnen, Röhrenspinnen	Clubionidae	sac spiders, two-clawed hunting spiders, foliage spiders, clubiones, clubionids
Sackträger> Gemeiner Sackträger	*Psyche casta*	common bagworm (moth)
Sackträgermotten, Sackmotten> Futteralmotten	Coleophoridae	casebearers, casebearer moths
Sackträgermotten, Sackmotten> Sackspinner	Psychidae (incl. Talaeporiidae)	bagworms, bagworm moths
Sacramento-Stint	*Spirinchus theleichthys*	Sacramento smelt
Safraneule, Safran-Wintereule, Eichenbuschwald-Safraneule	*Jodia croceago/* *Xanthia croceago*	orange upperwing moth
Saftkäfer	Nosodendridae/ Nosodendronidae	wounded-tree beetles

Saftkugler	*Glomeris* spp.	pill millepedes, European pill millepedes
Saftkugler, Kugelasseln, Kugeltausendfüßer	Glomeridae	pill millepedes, glomerid millepedes
Saftschlürfermotten	Phyllocnistidae	eyecap moths
Sägebarsch, Längsgestreifter Schriftbarsch, Ziegenbarsch	*Serranus cabrilla*	comber
Sägebarsche, Zackenbarsche	Serranidae	seabasses & rockcods (groupers)
Sägebäuche	Trachichthyidae	slimeheads, roughies
Sägebock, Gerberbock	*Prionus coriarius*	prionus longhorn beetle (greater British longhorn)
Sägefisch, Ostatlantischer Sägerochen	*Pristis pristis*	sawfish
Sägefisch> Gemeiner Sägefisch, Westlicher Sägefisch, Schmalzahn-Sägerochen	*Pristis pectinata*	smalltooth sawfish (FAO), greater sawfish
Sägefische, Sägerochen	Pristiformes/Pristidae	sawfishes
Sägegarnele, Große Felsgarnele	*Palaemon serratus/ Leander serratus*	common prawn
Sägehaie	Pristiophoridae	sawsharks, saw sharks
Sägekäfer	Heteroceridae	variegated mud-loving beetles
Sägekauz	*Aegolius acadicus*	saw-whet owl, northern saw-whet owl
Sägeracken, Motmots	Momotidae	motmots
Sägerochen, Sägefische	Pristiformes/Pristidae	sawfishes
Sägesalmler	Serrasalminae/Serrasalminae	pacus & silver dollars & piranhas
Sägesalmler, Pirahas, Piranhas	*Serrasalmus* spp.	piranhas
Sägeschnabel-Eremit	*Ramphodon naevius*	saw-billed hermit
Sägeschwanz, Fleckhai	*Galeus melastomus/ Pristiurus melanostomus*	blackmouth catshark (FAO), black-mouthed dogfish
Sägeschwanz-Katzenhai	*Galeus murinus*	mouse catshark
Sägezahn-Steckmuschel	*Atrina serrata*	sawtooth penshell, saw-toothed penshell
Sägezahnaale	Serrivomeridae	sawtooth eels
Sägezahnkrabbe*	*Parthenope serrata*	saw-toothed crab
Sägezahnmuschel, Gestutzte Dreiecksmuschel	*Donax trunculus*	truncate donax, truncated wedge clam
Sägezahnmuscheln, Sägezähnchen, Stumpfmuscheln, Dreiecksmuscheln, Dreieckmuschel	Donacidae	wedge clams, donax clams (wedge shells)
Sahara-Gundi, Langhaargundi, Sahara-Kammfinger	*Massoutiera mzabi*	Mzab gundi, fringe-eared gundi
Saharaohrenlerche	*Eremophila bilopha*	Temminck's horned lark
Saharasteinschmätzer	*Oenanthe leucopyga*	white-crowned black wheatear
Saiblinge	*Salvelinus* spp.	chars, charrs
Saiga, Saigaantilope	*Saiga tatarica*	saiga

Saitenwürmer	Nematomorpha	horsehair worms, hairworms, gordian worms, threadworms, nematomorphans, nematomorphs
Sakis	*Pithecia* spp.	sakis
Salamander & Molche> Echte Salamander & Molche	Salamandridae	salamanders & newts
Salamanderfische*	Lepidogalaxiidae	salamanderfishes
Salami-Garnele*	*Periclimenes brevicarpalis*	pepperoni shrimp
Salatblattkoralle, Folienkoralle	*Agaricia tenuifolia*	lettuce-leaf coral*
Salatblattkorallen	*Agaricia* spp.	lettuce corals
Salatfliege, Lattichfliege	*Botanophila gnava*/ *Phorbia gnava*	lettuce seed fly
Salatkoralle, Lärchenpilzkoralle	*Agaricia agaricites*	lettuce coral
Salatwurzellaus	*Pemphigus bursarius*	lettuce root aphid, poplar-lettuce aphid, lettuce purse-gall aphid
Salinenkärpfling, Mittelmeerkärpfling (Zebrakärpfling)	*Aphanius fasciatus*	Mediterranean toothcarp
Salinenkrebs, Salinenkrebschen, Salzkrebschen	*Artemia salina*	brine shrimp
Salm, Atlantischer Lachs	*Salmo salar*	Atlantic salmon
Salmler	Characiformes	characins: tetras & piranhas
Salmler (Echte Amerikanische Salmler)	Characidae	characins
Salomonen-Riesenskink	*Corucia zebrata*	giant Solomon Island skink
Salomons-Perlboot	*Nautilus scrobiculatus*	Salomon's nautilus
Salpen, Thaliaceen	Thaliacea	salps, thaliceans
Salpen> Eigentliche Salpen	Salpida	salps
Salvadors Waran, Papua-Waran	*Varanus salvadorii*	Salvador's monitor, Papuan monitor, crocodile monitor, tree "crocodile"
Salvin-Kreuzbrustschildkröte	*Staurotypus salvinii*	Chiapas giant musk turtle, giant musk turtle (IUCN), Chiapas cross-breasted turtle
Salzfliegen, Sumpffliegen	Ephydridae	shoreflies, shore flies
Salzkrautbilch	*Selevinia betpakdalensis*	desert dormouse
Salzkrebschen, Salinenkrebs, Salinenkrebschen	*Artemia salina*	brine shrimp
Samar-Fechterschnecke, Samar-Flügelschnecke	*Strombus dentatus*	Samar conch, toothed conch
Samenfüßer	Chordeumatidae	chordeumatid millepedes
Samenkäfer	Bruchidae	seed beetles (pea and bean weevils), pulse beetles
Samenmotte	*Hofmannophila pseudospretella*	brown house-moth, brown house moth
Samenwespen	*Megastigmus* spp.	seed wasps

Samenzünsler	*Aphomia gularis/ Paralispa gularis*	stored nut moth, brush-winged honey, Japanese grain moth
Sammetmilbe, Samtmilbe	*Trombidium holosericeum*	velvet mite
Samoa-Palolo	*Palola viridis/ Eunice viridis*	Pacific palolo worm, Samoan palolo worm
Samtente	*Melanitta fusca*	velvet scoter
Samtgarnele	*Metapenaeopsis goodei*	velvet shrimp
Samtige Entenmuschel	*Scalpellum scalpellum*	velvet goose barnacle
Samtjungferchen, Schwarzer Preußenfisch	*Dascyllus trimaculatus*	domino damselfish, white-spotted damselfish, threespot dascyllus (FAO)
Samtkopf-Grasmücke	*Sylvia melanocephala*	Sardinian warbler
Samtkorallenfisch, Samtanemonenfisch	*Premnas biaculeatus/ Amphiprion biaculeatus*	spine-cheek anemonefish
Samtkrabbe, Wollige Schwimmkrabbe	*Necora puber/ Liocarcinus puber/ Macropipus puber*	velvet swimming crab, velvet fiddler, devil crab
Samtlanguste	*Palinurellus gundlachi*	furry lobster
Samtlangusten	Synaxiidae	furry lobsters
Samtmilbe, Sammetmilbe	*Trombidium holosericeum*	velvet mite
Samtmilben	Trombidiidae	velvet mites, trombidiid mites
Samtmuscheln	*Glycymeris* spp.	dog cockles (Br.), bittersweet clams (U.S.), bittersweets
Samtmuscheln	Glycymeridae/ Glycymerididae	dog cockles (Br.), bittersweet clams (U.S.), bittersweets
Samtschnecke*	*Monacha granulata*	silky snail, Ashford's hairy snail
San-José-Schildlaus, Kalifornische Schildlaus	*Quadraspidiotus perniciosus*	San Jose scale
Sand-Einsiedler, Strandeinsiedler, Kleiner Einsiedlerkrebs	*Diogenes pugilator*	Roux's hermit crab
Sand-Käferschnecke	*Lepidopleurus intermedius*	intermediate chiton, sand chiton
Sand-Zackenbarsch, Sandbarsch	*Diplectrum formosum*	sand perch
Sandaal, Kleiner Sandaal, Tobiasfisch	*Ammodytes tobianus/ Ammodytes lancea*	sandeel, lesser sandeel, small sandeel (FAO)
Sandaale	Ammodytidae/ *Ammodytes* spp.	sand lances, sandlances, sandeels
Sandagamen	*Psammophilus* spp.	sand agamas
Sandalenauge	*Sandalops melancholicus*	sandal-eye squid
Sandalenkoralle, Pantoffelkoralle	*Calceola sandalina*	sandal coral
Sandbankhai, Atlantischer Braunhai, Großrückenflossenhai	*Carcharhinus plumbeus/ Carcharhinus milberti*	sandbar shark
Sandbankwurm*	*Ophelia denticulata*	sand bar worm
Sandbarsch, Sand-Zackenbarsch	*Diplectrum formosum*	sand perch
Sandbarsche	Mugiloididae	sandsmelts, sandperches

Sandbienen	*Andrena* spp.	mining bees, burrowing bees
Sandbienen	Andrenidae	mining bees, burrowing bees
Sandbutt	*Scophthalmus aquosus*	windowpane (FAO), windowpane flounder
Sanddollar> Gewöhnlicher Sanddollar	*Clypeaster humilis*	common sand dollar
Sanddollars, Schildseeigel	Clypeasteroida	sand dollars, true sand dollars
Sandechse u.a.	*Aporosaura anchietae*	Namib sanddiver
Sandechsen	*Pedioplanes* spp.	sand lizards
Sanderling	*Calidris alba*	sanderling
Sandfelchen, Große Bodenrenke	*Coregonus fera*	broad whitefish
Sandfisch	*Gonorynchus gonorynchus*	beaked salmon (FAO), sand eel
Sandfische u.a.	Trichodontidae	sandfishes
Sandfische u.a.	Gonorhynchidae	beaked sandfishes
Sandflachkopf	*Platycephalus indicus*	Indo-Pacific flathead
Sandfloh	*Tunga penetrans*	chigoe flea, jigger flea, sand flea
Sandflughuhn	*Pterocles orientalis*	black-bellied sandgrouse
Sandflunder	*Rhombosolea plebeia*	sand flounder
Sandfuchs, Rüppelfuchs	*Vulpes rueppelli*	sand fox
Sandgarnele	*Crangon septemspinosa*	sand shrimp
Sandgarnelen	Crangonidae	crangonid shrimp
Sandgecko	*Chondrodactylus angulifer*	whorled sand gecko
Sandgoldrose, Goldrose, Goldfarbige Seerose	*Condylactis aurantiaca*	golden anemone, golden sand anemone
Sandgräber	Bathyergidae	African mole-rats
Sandgräberfische	Creediidae/Limnichthyidae	sandburrowers
Sandgrundel, Sandküling	*Pomatoschistus minutus*	sand goby
Sandhai, Echter Sandhai, Sandtiger	*Eugomphodus taurus/ Carcharias taurus/ Odontaspis taurus*	sand shark, sand tiger shark (FAO), sandtiger shark, gray nurse shark
Sandhaie	Odontaspididae/ Carchariidae	sand sharks, sand tigers, ragged-tooth sharks
Sandhüpfer, Küstenhüpfer	*Orchestia gammarellus*	shore-hopper, beach-flea, common shore-skipper, common scud
Sandkatze	*Felis margarita*	sand cat
Sandknotenwespe	*Cerceris arenaria*	sand-tailed digger wasp
Sandkrebse	Hippidae	sand crabs, mole crabs
Sandküling, Sandgrundel	*Pomatoschistus minutus*	sand goby
Sandläufer	*Psammodromus* spp.	sand lizards
Sandlaufkäfer (Tigerkäfer)	Cicindelidae	tiger beetles
Sandlaus, Schaflaus	*Lepikentron ovis/Bovicola ovis*	sheep biting louse
Sandlerche	*Ammomanes cincturus*	bar-tailed desert lark
Sandmesserfisch	*Gymnorhamphichthys hypostomus*	cuchillo (Venezuela)

Sandmücken	Phlebotomidae	sandflies
Sandmücken	*Phlebotomus* spp.	sandflies
Sandmuschel, Sandklaffmuschel, Strandauster, Große Sandklaffmuschel	*Mya arenaria/ Arenomya arenaria*	sand gaper, soft-shelled clam, softshell clam, large-neck clam, steamer
Sandmuscheln	Psammobiidae	sunsetclams
Sandohrwurm, Dünenohrwurm	*Labidura riparia*	tawny earwig, giant earwig
Sandotter	*Vipera ammodytes*	sand viper, nose-horned viper
Sandpierwurm, Sandpier, Pierwurm, Köderwurm	*Arenicola marina*	European lug worm, blow lug
Sandrasselotter	*Echis carinatus*	saw-scaled viper, saw-scaled adder
Sandregenpfeifer	*Charadrius hiaticula*	ringed plover
Sandrennmäuse, Sandrennratten (Wüstenmäuse)	*Meriones* spp.	jirds
Sandrennnattern	*Psammophis* spp.	sand snakes
Sandrochen	*Raja circularis*	sandy ray
Sandröhren-Haarwurm	*Spio setosa*	sand chimney worm
Sandröhrenwurm, Bäumchenröhrenwurm, Muschelsammlerin	*Lanice conchilega*	sand mason
Sandroller, Amerikanischer Barschlachs, Gewöhnlicher Barschlachs	*Percopsis transmontana*	sand roller
Sandrosen-Anemone	*Urticina columbiana*	sand-rose anemone
Sandschnurfüßer	*Schizophyllum sabulosum*	sand millepede
Sandschuppenwurm	*Sthenelais boa*	boa worm, boa-shaped scaleworm
Sandseegurke, Spindel-Thyone	*Thyone fusus*	sand-gherkin*
Sandskinke	*Scincus* spp.	sand skinks
Sandsterngucker	Dactyloscopidae	sand stargazers
Sandstrandläufer	*Calidris pusilla*	semipalmated sandpiper
Sandtampan	*Ornithodoros savignyi*	eyed tampan, sand tampan
Sandtaucher	Trichonotidae	sand divers, sanddivers
Sandtaucher* (Eidechsenfisch)	*Synodus intermedius*	sand diver
Sandteufel	*Squatina dumeril*	sand devil
Sanduhrdelphin, Kreuzdelphin	*Lagenorhynchus cruciger*	hourglass dolphin, southern white-sided dolphin
Sandwallaby, Flinkes Känguru	*Macropus agilis*	agile wallaby
Sandwespen	Sphecinae	sand wasps, thread-waisted wasps
Sandwürmer	Arenicolidae	lug worms, lugworms
Sandziesel (Hbz. Petschanik)	*Spermophilus fulvus*	large-toothed souslik

Sandzunge, Warzen-Seezunge	*Pegusa lascaris/ Solea lascaris*	sand sole
Sänger, Eigentliche Fliegenschnäpper	Muscicapidae	Old World flycatchers
Sängerin, Buchenmotte	*Chimabacche fagella*	beech moth*
Sängervireo	*Vireo gilvus*	warbling vireo
Sansibar-Rotstummelaffe	*Piliocolobus kirkii*	Zanzibar red colobus, Kirk's colobus
Sansibarstummelaffe	*Colobus kirkii/ Piliocolobus kirkii*	Kirk's colobus, Zanzibar red colobus
Santa-Catalina-Klapperschlange	*Crotalus catalinensis*	Santa Catalina rattlesnake
Saphirkrebse	*Sapphirina* spp.	sapphirines
Saphirkrebse	Sapphirinidae	sapphirines
Saphirlori	*Vini peruviana*	Tahitian lory
Sardelle, Europäische Sardelle, Anchovis	*Engraulis encrasicholus*	anchovy, European anchovy
Sardellen	Engraulidae	anchovies
Sardengrasmücke	*Sylvia sarda*	Marmora's warbler
Sardine, Pilchard	*Sardina pilchardus*	European sardine, European pilchard (FAO), sardine (if *small*), pilchard (if *large*)
Sardinelle, Ohrensardine, Große Sardine	*Sardinella aurita*	gilt sardine, Spanish sardine, round sardinella (FAO)
Sardinen	*Sardina* spp.	sardines
Sardinenmaräne*, Sardinen-Cisco	*Coregonus sardinella*	sardine cisco
Sardinischer Schleuderzungensalamander	*Hydromantes genei/ Speleomantes genei*	Sardinian cave salamander, brown cave salamander
Sardinischer Gebirgsmolch, Hechtkopf-Gebirgsmolch	*Euproctus platycephalus*	Sardinian brook salamander, Sardinian mountain newt
Sardischer Scheibenzüngler	*Discoglossus sardus*	Tyrrhenian painted frog
Sargasso-Fisch	*Histrio histrio*	sargassum fish
Sargassum-Anemone	*Anemonia sargassensis*	sargassum anemone
Sargassum-Drückerfisch, Atlantischer Wangenlinien-Drückerfisch	*Xanthichthys ringens*	sargassum triggerfish
Sargassum-Krabbe	*Portunus sayi*	sargassum swimming crab, sargassum crab
Sargassum-Nacktschnecke	*Scyllaea pelagica*	sargassum nudibranch
Sargassumgarnele	*Latreutes parvulus*	sargassum shrimp
Saruskranich	*Grus antigone*	sarus crane
Satansaffe	*Chiropotes satans*	red-backed saki
Satelliteule	*Eupsilia transversa*	satellite
Satere-Marmosette	*Callithrix saterei*	Satere marmoset
Satrap	*Regulus satrapa*	golden-crowned kinglet
Sattel-Anemonenfisch, Sattelfleck-Anemonenfisch	*Amphiprion polymnus/ Amphiprion laticlavius*	saddleback anemonefish, saddleback clownfish (FAO)
Sattelfleck-Falterfisch	*Chaetodon ephippium*	saddleback butterflyfish, saddle butterflyfish (FAO), black-blotched butterflyfish

Sattelfleckenhai 838

Sattelfleckenhai, Gebänderter Bambushai, Brauner Lippenhai	*Chiloscyllium punctatum*	brown carpet shark, brownbanded bambooshark (FAO)
Sattelkröte	*Brachycephalus ephippium*	Spix's saddleback toad, golden frog, gold frog
Sattelkröten	Brachycephalidae	saddleback toads
Sattelkröten	*Brachycephalus* spp.	saddleback toads
Sattelmücke	*Haplodiplosis equestris*	saddle gall midge
Sattelmuschel, Zwiebelmuschel, Zwiebelschale, Europäische Sattelauster	*Anomia ephippium*	European saddle oyster (European jingle shell)
Sattelmuschel> Gewöhnliche Sattelmuschel	*Anomia simplex*	common saddle oyster (common jingle shell)
Sattelmuscheln, Zwiebelmuscheln	*Anomia* spp.	saddle oysters (jingle shells)
Sattelmuscheln, Zwiebelmuscheln	Anomiidae	saddle oysters (jingle shells)
Sattelrobbe	*Phoca groenlandica*	harp seal
Sattelschrecke> Gemeine Sattelschrecke, Rebensattelschrecke, Steppen-Sattelschrecke	*Ephippiger ephippiger*	common saddle-backed bushcricket, tizi
Sattelschrecken	Ephippigeridae	saddle-backed bushcrickets
Sattelvogel	*Creadion carunculatus*	saddleback
Sauerampfer-Gallrüssler	*Apion frumentarium/ Apion haematodes*	sheep's sorrel gall weevil
Sauerampferfeuchthalden-Goldfalter	*Heodes alciphron/ Loweia alciphron*	purple-shot copper
Sauerkirschenlaus, Schwarze Kirschenlaus	*Myzus cerasi*	black cherry aphid
Sauger	Catostomidae	suckers
Sauger	*Catostomus commersoni*	white sucker (FAO), freshwater mullet
Säugetiere, Säuger	Mammalia	mammals
Saugfischverwandte, Schildfische, Spinnenfischartige	Gobiescociformes	clingfishes
Saugfüßer	Polyzoniidae	polyzoniid millepedes
Sauglippen-Buntbarsch	*Cheilochromis euchilus*	euchilus
Saugschirmqualle	*Cassiopeia andromedra*	sucker upsidedown jellyfish
Saugschmerlen, Algenfresser	Gyrinocheilidae	algae eaters
Saugwelse, Turkestanische Zwergwelse	Sisoridae	sisorid catfishes
Saugwürmer, Trematoden, Egel	Trematoda	flukes, trematodes
Säulenkoralle*	*Dendrogyra cylindrus*	pillar coral
Saumwanze, Lederwanze	*Coreus marginatus/ Mesocerus marginatus*	squash bug
Saumzecken, Lederzecken	Argasidae	soft ticks, softbacked ticks
Saussure-Langnasenfledermäuse	*Leptonycteris* spp.	Saussure's long-nosed bats

Savannen-Schienenschildkröte	*Podocnemis vogli*	savanna side-necked turtle, Orinoco turtle
Savannenadler	*Aquila rapax*	tawny eagle
Savannenbussard	*Buteogallus meridionalis*	savannah hawk
Savetta, Südwesteuropäischer Näsling	*Chondrostoma toxostoma*	soiffe (*Fr.*), soffie (*Fr.*)
Savignys Diademseeigel	*Diadema savignyi*	Savigny's hatpin urchin
Saxaulhäher	*Podoces panderi*	Pander's ground jay
Saxesens Holzbohrer, Kleiner Holzbohrer	*Xyleborus saxeseni*	lesser shot-hole borer, fruit-tree pinhole borer
Scelotes-Skinke	*Scelotes* spp.	burrowing skinks
Schabe> Deutsche Schabe	*Blattella germanica*	German cockroach (croton-bug, shiner, steamfly)
Schaben	Blattidae/Blattodea	cockroaches: Oriental and American cockroaches
Schabenartige Steinfliegen	Peltoperlidae	roachlike stoneflies, forestflies
Schabracken-Panzerwels	*Corydoras barbatus*	banded corydoras (FAO), giant corydoras
Schabrackenhyäne	*Hyaena brunnea*	brown hyena
Schabrackenschakal	*Canis mesomelas*	black-backed jackal
Schabrackenspitzmaus	*Sorex coronatus/Sorex gemellus*	Millet's shrew
Schachbrett, Damenbrett	*Melanargia galathea*	marbled white
Schachbrett-Junker	*Halichoeres hortulanus*	checkerboard wrasse
Schachbrett-Prachtschmerle, Zwerg-Prachtschmerle	*Botia sidthimunki*	dwarf loach
Schachtelhalmeule, Markeule, Kartoffelbohrer, Rübenbohrer	*Hydraecia micacea*	rosy rustic (moth)
Schachwürger	*Lanius schach*	black-headed shrike
Schädellose	Acrania	acranians
Schaf, Hausschaf	*Ovis aries*	domestic sheep
Schafbiesfliege, Schafbremse, Nasendasselfliege	*Oestrus ovis*	sheep nostril-fly, sheep bot fly, sheep nose bot fly
Schafgarbenspanner	*Aspitates gilvaria*	straw belle
Schaflaus, Sandlaus	*Bovicola ovis/Lepikentron ovis*	sheep-biting louse
Schaflaus> Schaf-Gesichtslaus	*Linognathus ovillus*	sheep face louse, sheep sucking louse
Schaflaus, Schaflausfliege	*Melophagus ovinus*	sheep ked (fly)
Schafs-Goldfliege	*Lucilia sericata*	sheep maggot fly, sheep blowfly, greenbottle
Schafs-Knötchenwurm	*Oesophagostomum columbianum*	sheep nodular worm
Schafsfinnen-Bandwurm	*Taenia ovis*	sheep-and-goat tapeworm
Schafsfrosch	*Hypopachus variolosus*	sheep frog
Schafskopf-Brasse, Sträflings-Brasse	*Archosargus probatocephalus*	sheepshead
Schafstelze	*Motacilla flava*	yellow wagtail
Schafswollschwamm*	*Hippospongia lachne*	sheep's wool sponge
Schaftlaus, Hühnerfederling	*Menopon gallinae*	shaft louse, chicken shaft louse

Schafzecke	*Dermacentor marginatus*	European sheep tick
Schalenhäuter	Ostracodermata	ostracoderms
Schalenlose, Kiemenfußkrebse, Kiemenfüße	Anostraca	fairy shrimps, anostracans
Schalenweichtiere	Conchifera	conchiferans
Schalottenfliege, Bohnenfliege, Kammschienen-Wurzelfliege	*Delia platura/ Hylemia platura/ Phorbia platura*	seed-corn fly, seed-corn maggot, bean-seed fly, shallot fly
Schalottenlaus, Zwiebellaus	*Myzus ascalonicus*	shallot aphid
Schamkrabbe	*Calappa flammea*	shame-faced crab, flame box crab
Schamkrabben	Calappidae	box crabs a.o.
Schamlaus, Filzlaus	*Pthirus pubis/ Phthirus pubis*	crab louse, "crab", pubic louse
Schan, Schleimlerche, Atlantischer Schleimfisch	*Lipophrys pholis*	shanny
Scharfaugenspinnen, Luchsspinnen	Oxyopidae	lynx spiders
Scharfkantige Turmschnecke*	*Opeas pyrgula*	sharp awlsnail
Scharfkinn-Flugfisch	*Fodiator acutus*	sharpchin flyingfish
Scharlach-Stachelauster	*Spondylus tenellus*	scarlet thorny oyster
Scharlachgesicht, Uakari	*Cacajao calvus*	white uakari, bald uakari
Scharlachlibelle, Späte Adonislibelle	*Ceriagrion tenellum*	small red damselfly
Scharlachnatter	*Cemophora coccinea*	scarlet snake
Scharlachnektarvogel	*Aethopyga siparaja*	yellow-backed sunbird
Scharlachroter Feuerkäfer (Feuerfliege)	*Pyrochroa coccinea*	scarlet fire beetle, cardinal beetle
Scharlachsichler	*Eudocimus ruber*	scarlet ibis
Scharlachtangare	*Piranga olivacea*	scarlet tanager
Scharnierschildkröten	*Cuora* spp.	box turtles
Scharnierschnabelgarnele*	*Rhynchocinetes rugulosus*	hinged-beak prawn
Schärpen-Nasendoktor, Kurznasen-Einhornfisch	*Naso brevirostris*	short-nose unicornfish, short-snouted unicornfish, spotted unicornfish (FAO)
Scharreidechsen	*Meroles* spp.	desert lizards
Schattenmönch, Grauer Mönch	*Cucullia umbratica*	shark moth, common 'shark'
Schaufelfadenfisch	*Trichogaster pectoralis*	snake-skinned gourami, snakeskin gourami (FAO)
Schaufelfüße	*Scaphiopus* spp.	spadefoot toads
Schaufelkärpflinge	Adrianichthyidae	adrianichthyids (killifish-like medakas)
Schaufelnasen-Hammerhai	*Sphyrna tiburo*	bonnet shark, shovel head, bonnethead (FAO)
Schaufelnasennatter	*Chionactis* spp.	shovelnose snakes
Schaufelstör, Schaufelnasenstör	*Acipenser platorynchus/ Scaphirhynchus platorhynchus*	shovelnose sturgeon
Schaumzikaden (Schaumzirpen)	Cercopidae	froghoppers, spittlebugs

Scheckenfalter> Würfelfalter	Riodinidae	metalmarks
Scheckenfalter	*Melitaea* spp. u.a.	fritillaries
Scheckenfalter> Gemeiner Scheckenfalter, Wegerich-Scheckenfalter	*Melitaea cinxia*	Glanville fritillary
Scheckenrochen, Bänderrochen	*Raja undulata*	undulate ray (FAO), painted ray
Scheckente	*Polysticta stelleri*	Steller's eider
Scheckhorn-Distelbock	*Agapanthia villosoviridescens*	thistle longhorn beetle*
Scheefsnut, Flügelbutt	*Lepidorhombus whiffiagonis*	megrim (FAO), sail-fluke, whiff
Scheelaugennatter	*Helicops angulatus*	mountain keelback
Scheibenanemonen	*Discosoma* spp.	disc anemones, disc corallimorphs
Scheibenbarsch	*Enneacanthus chaetodon/ Mesogonistius chaetodon*	blackbanded sunfish (FAO), chaetodon
Scheibenbäuche	Liparididae	snailfishes
Scheibenhals-Schnellläufer	*Stenolophus lecontei*	seed-corn beetle
Scheibennetzspinnen	*Oecobius* spp.	saucerweb spiders*
Scheibennetzspinnen & Zeltdachspinnen	Oecobiidae	flatmesh weavers
Scheibenquallen, Schirmquallen, Scyphozoen (echte Quallen)	Scyphozoa	cup animals, scyphozoans
Scheibensalmler, Silberdollar	*Metynnis argenteus*	silver dollar a.o.
Scheibenschirme	Opistoteuthidae	flapjack octopuses
Scheibensterne	Asterinidae	cushion stars
Scheibenzüngler	Discoglossidae	disk-tongued toads
Scheidenmuscheln	Solenidae	razor clams (razor shells)
Scheidenschnäbel	Chionididae	sheathbills
Scheidewegschnecken, Kaurischnecken	Triviidae	false cowries, bean cowries, sea buttons
Scheidlers Laufkäfer, Gleichgestreifter Laufkäfer	*Carabus scheidleri*	Scheidler's ground beetle
Scheinbock	*Nacerda melanura/ Nacerdes melanura*	wharf borer
Scheinbockkäfer, Scheinböcke (Engdeckenkäfer)	Oedemeridae	false blister beetles, pollen-feeding beetles
Scheinhummer, Furchenkrebse	Galatheidae	squat lobsters
Scheinschnapper	*Nemipterus* spp.	threadfin bream
Scheinschnapper (inkl. Fadenflosser & Monokelbrassen)	Nemipteridae	threadfin breams (butterfly breams, spinecheeks, monocle breams & dwarf breams)
Scheitelaffe (Hbz.), Bärenstummelaffe, Südlicher Guereza	*Colobus polykomos*	king colobus, western pied colobus (western black-and-white colobus)
Scheitelaffe (Hbz.), Guereza, Mantelaffe	*Colobus guereza/ Colobus abyssinicus*	guereza, guereza colobus (eastern black-and-white colobus)

Schelladler	*Aquila clanga*	spotted eagle
Schellente	*Bucephala clangula*	common goldeneye
Schellfisch	*Melanogrammus aeglefinus*	haddock (chat, jumbo)
Scheltopusik	*Ophisaurus apodus*	European glass lizard, armored glass lizard
Scheltostscheck	*Elopichthys bambusa*	sheltostshek, yellowcheek (FAO)
Schemaja, Mairenke	*Chalcalburnus chalcoides*	Danubian bleak, Danube bleak (FAO), shemaya
Schenkelfliegen	Megamerinidae	megamerinid flies
Scheren-Seedattel*	*Lithophaga aristata*	scissor datemussel
Scherenasseln	Tanaidacea (Anisopoda)	tanaidaceans, tanaids
Scherenfußgarnelen*	Psalidopodidae	scissorfoot shrimps
Scherenhörnler, Chelizeraten	Chelicerata	chelicerates
Scherenschnäbel	Rynchopidae	skimmers
Scherenschwanz, Holz-Floßkrebs	*Chelura terebrans*	wood-boring amphipod
Scherenschwanzsalmler	*Moenkhausia intermedia*	
Scherenschwanzschwalbe	*Psalidoprocne obscura*	fantee saw-wing
Scherentyrann	*Tyrannus forficata*	scissor-tailed flycatcher
Scherg, Sternhausen	*Acipenser stellatus*	starry sturgeon, stellate sturgeon (IUCN), sevruga
Schermaus> Ostschermaus	*Arvicola terrestris*	European water vole, northern water vole
Schermaus> Westschermaus	*Arvicola sapidus*	southwestern water vole, southern water vole
Schermäuse	*Arvicola* spp.	water voles, bank voles
Schermesserfisch	*Xyrichthys novacula*	cleaver wrasse
Scheufliegen, Sumpffliegen, Dunkelfliegen	Heleomyzidae	heleomyzid flies
Schiefe Erbsenmuschel	*Pisidium subtruncatum*	shortended pea mussel, shortended peaclam
Schiefe Grasschnecke	*Vallonia excentrica*	eccentric grass snail
Schiefe Samtschnecke*	*Velutina plicatilis*	oblique velutina, oblique lamellaria
Schieferbussard	*Leucopternis schistacea*	slate-coloured hawk
Schieferdrossel	*Zoothera sibirica*	Siberian thrush
Schieferfalke	*Falco concolor*	sooty falcon
Schienenechsen	Teiidae	whiptails & racerunners
Schienenkäfer	Eucnemidae	false click beetles
Schienenschildkröten	*Podocnemis* spp.	South American river turtles, podocnemids
Schiffsbohrmuschel, Schiffsbohrwurm, Pfahlwurm	*Teredo navalis*	naval shipworm, common shipworm, great shipworm
Schiffsboote, Perlboote	*Nautilus* spp.	chambered nautiluses, pearly nautiluses

Schiffshalter, Saugfisch	*Echeneis naucrates*	whitefin sharksucker, Indian remora, suckerfish, live sharksucker (FAO)
Schiffshalter, Saugfische	Echeneidae (Echeneididae)	remoras (sharksuckers)
Schiffskielgarnele, Bärengarnele, Bärenschiffskielgarnele	*Penaeus monodon*	giant tiger prawn
Schiffswerftkäfer, Eichenwerftkäfer	*Lymexylon navale*	ship timberworm
Schild-Moostierchen*	*Scrupocellaria scabra*	shielded bryozoan
Schild-Napfschnecke	*Patella scutellaris*	scutellate limpet
Schildfische, Schildbäuche	Gobiesocidae	clingfishes
Schildfuß> Echter Schildfuß	*Scutopus ventrolineatus* (Aplacophora)	bellylined glistenworm*
Schildkäfer	Cassidinae	tortoise beetles, shield beetles
Schildkobra	*Aspidelaps scutatus*	shield-nose snake
Schildkobras	*Aspidelaps* spp.	shield-nose snakes
Schildkröten	Testudines/ Chelonia	turtles
Schildkröten-Kugelfisch	*Sphoeroides testudineus*	checkered puffer
Schildkröten-Seepocke, Schildkrötenpocke, Schildseepocke	*Chelonibia testudinaria*	turtle barnacle
Schildkrötenegel	*Haementeria costata*	turtle leech
Schildkrötenfrosch	*Myobatrachus gouldii*	turtle frog
Schildkrötenkäfer	Chelonariidae	turtle beetles
Schildkrötenköpfige Seeschlange, Schildkrötenkopf-Seeschlange	*Emydocephalus ijimae*	turtlehead sea snake
Schildkrötenköpfige Seeschlangen, Schildkrötenkopf-Seeschlangen	*Emydocephalus* spp.	turtlehead sea snakes
Schildkrötenkrabbe*	*Cryptolithodes sitchensis*	umbrella crab, umbrella-backed crab, turtle crab
Schildkrötenmilben	Uropodidae	tortoise mites
Schildkrötenmotte	*Incurvaria koerneriella*	tortoise moth*
Schildkrötenschnecke	*Susania testudinaria*	tortoise snail*
Schildkrötenschnecke, Klippkleber	*Tectura testudinalis/ Tectura tessulata/ Acmaea testudinalis/ Collisella tessulata*	plant limpet, tortoiseshell limpet, common tortoiseshell limpet
Schildläuse	Coccoidea	scale insects & mealy bugs
Schildrabe	*Corvus albus*	pied crow
Schildschnabel	*Buceros vigil/ Rhinoplax vigil*	helmeted hornbill
Schildschnecke, Elefantenschnecke	*Scutus antipodes/ Scutus anatinus*	elephant slug, duckbill shell, shield shell
Schildschwänze	Uropeltidae	shield-tailed snakes, shieldtail earth snakes, shieldtails, roughtails

Schildseepocke, Schildkröten-Seepocke, Schildkrötenpocke	*Chelonibia testudinaria*	turtle barnacle
Schildturako	*Musophaga violacea*	violet turaco
Schildwanzen	Scutelleridae	shield-backed bugs
Schildweber	*Malimbus scutatus*	red-vented malimbe
Schildzahnhai	*Odontaspis ferox/ Carcharias ferox*	smalltooth sand tiger (FAO), fierce shark, bumpytail ragged-tooth shark, shovelnose shark
Schildzecken (Holzböcke)	Ixodidae	hard ticks, hardbacked ticks
Schilf-Flechtenbärchen	*Pelosia obtusa*	small dotted footman
Schilf-Springspinne	*Marpissa radiata*	reed slender spider*, reed spider*
Schilfgallenfliege, Zigarrenfliege	*Lipara lucens*	reed gall fly
Schilfradspinne	*Larinioides cornutus/ Araneus cornutus*	furrow orbweaver
Schilfrohrsänger	*Acrocephalus schoenobaenus*	sedge warbler
Schillerbärbling	*Brachydanio albolineatus*	pearl danio
Schillerfalter	*Apatura* spp.	emperors
Schillernde Schwimmkrabbe	*Portunus gibbesii*	iridescent swimming crab
Schillernder Goldwurm	*Pectinaria auricoma*	golden trumpet worm
Schillernder Seeringelwurm	*Nereis diversicolor/ Hediste diversicolor*	estuary ragworm
Schimmel-Getreideblattkäfer	*Ahasverus advena*	foreign grain beetle
Schimmelkäfer	Cryptophagidae	silken fungus beetles
Schimmelkäfer, Faulholzkäfer	Corylophidae/Orthoperidae	minute fungus beetles
Schimpanse	*Pan troglodytes*	savanna chimpanzee
Schindelauster*	*Ostrea imbricata*	imbricate oyster
Schirmkrabbe	*Dorippe lanata/ Medorippe lanata*	demon-faced porter crab
Schirmquallen, Scheibenquallen, Scyphozoen (echte Quallen)	Scyphozoa	cup animals, scyphozoans
Schirmschnecken	Umbraculidae	umbrella snails, umbrella "slugs" (umbrella shells)
Schläfer	*Nimbochromis livingstonii*	livingstoni
Schläfer> Siebenschläfer	*Glis glis*	fat dormouse, edible dormouse
Schläfergrundeln	Eleotridae	sleepers
Schlagender Fangschreckenkrebs*	*Gonodactylus falcatus*	clubbing mantis shrimp
Schlagschwirl	*Locustella fluviatilis*	river warbler
Schlammbarsch	*Acantharchus pomotis*	mud sunfish
Schlammfisch, Amerikanischer Schlammfisch, Kahlhecht	*Amia calva*	bowfin
Schlammfliege, Große Bienenschwebfliege, Mistbiene (Rattenschwanzlarve)	*Eristalis tenax*	drone fly (rattailed maggot)

Schlammfliegen	Megaloptera	megalopterans: dobsonflies, fishflies, alderflies (neuropterans)
Schlammigel*, Schlamm-Herzigel*	Moira atropus	mud heart urchin
Schlammkarpfen	Cirrhinus molitorella	mud carp
Schlammkriecher	Terebralia palustris	mud creeper
Schlammnatter	Farancia abacura	mud snake
Schlammpeitzger	Misgurnus fossilis	weatherfish
Schlammröhrenwurm> Gemeiner Schlammröhrenwurm, Gemeiner Bachröhrenwurm, Bachwurm	Tubifex tubifex	river worm, sludge-worm
Schlammröhrenwürmer	Tubifex spp.	sludge-worms
Schlammröhrenwürmer	Tubificidae	sludge-worms
Schlammsalamander	Pseudotriton montanus	mud salamander
Schlammschildkröten	Kinosternidae	American mud and musk turtles
Schlammschnecke> Gewöhnliche Schlammschnecke, Gemeine Schlammschnecke, Wandernde Schlammschnecke	Lymnaea peregra/ Radix peregra	wandering pondsnail, common pondsnail
Schlammschnecken	Lymnaeidae	pondsnails, pond snails
Schlammschwimmer, Feuchtkäfer	Hygrobia spp.	screech beetle
Schlammsepie, Kleine Sepie	Sepia elegans	elegant cuttlefish
Schlammspringer	Periophthalmidae	mudskippers, mudhoppers, climbing-fish
Schlammspringer	Periophthalmus spp.	mudskippers (FAO), mudhoppers, climbing-fish
Schlammtaucher	Pelodytes spp.	parsley frogs, mud frogs, mud divers
Schlangen	Serpentes/Ophidia (Squamata)	snakes, serpents, ophidians
Schlangen-Gürtelechsen	Chamaesaura spp.	sweepslangs
Schlangen-Schleimfisch*	Leptoclinus maculatus	snake blenny
Schlangenaale & Wurmaale & Garnelenaale	Ophichthyidae/Ophichthinae	snake eels & worm eels
Schlangenadler	Circaetus gallicus	short-toed eagle
Schlangenauge	Ophisops elegans	snake-eyed lizard
Schlangenaugen-Echse	Ophisops spp.	snake-eyed lizards
Schlangenaugenskinke, Natternaugen-Skinke	Ablepharus spp.	snake-eyed skinks, ocellated skinks
Schlangenhaarrose	Actinothoe clavata	snakehair anemone*
Schlangenhalsschildkröten	Chelidae	Austro-American side-necked turtles, snake-necked turtles
Schlangenhalsvögel	Anhingidae	anhingas, darters

Schlangenhaut-Nixenschnecke, Schlangenhaut-Nixe	Nerita exuvia	snake-skin nerite
Schlangenköpfe, Schlangenkopffische	Channidae/Ophiocephalidae	snakeheads
Schlangenkopffische	Channa spp.	snakeheads
Schlangenkopfgrundel	Gobius lota/ Gobius ophiocephalus/ Zosterisessor ophiocephalus	grass goby
Schlangenmakrelen	Gempylidae	snake mackerels
Schlangenmilbe	Ophionyssus natricis	snake mite
Schlangenminiermotte, Obstbaumminiermotte	Lyonetia clerkella	apple leaf miner, Clerk's snowy bentwing
Schlangenpolyp*	Isaurus tuberculatus	snake polyps
Schlangenschleichen	Dibamidae	blind lizards
Schlangenschleichen	Dibamus spp.	Asian blind lizards
Schlangenschnecke, Wurmschnecke	Tenagodus obtusus	obtuse wormsnail
Schlangenschnecken	Siliquariidae	pod snails, wormsnails (pod shells, wormshells)
Schlangenskinke	Ophiomorus spp.	greater snake skinks
Schlangensterne, Ophiuroiden	Ophiuroidea	brittle stars, serpent stars; basket stars (Gorgonocephalus, Astrophyton)
Schlangenweihe	Spilornis cheela	crested serpent eagle
Schlank-Ilisha	Ilisha elongata	elongate ilisha
Schlank-Korallenschlangen	Leptomicrurus spp.	slender coral snakes, threaded coral snakes
Schlank-Seestern*	Leptasterias tenera	slender seastar
Schlank-Wurmaal	Moringua ferruginea	slender worm-eel
Schlankagame	Agama agilis/Trapelus agilis	brilliant agama
Schlankblindschlangen	Leptotyphlopidae	slender blind snakes
Schlankboas	Epicrates spp.	rainbow boas
Schlankdelphin, Fleckendelphin	Stenella attenuata	bridled dolphin, pantropical spotted dolphin, white-spotted dolphin
Schlanke Becherqualle, Tethys-Becherqualle	Craterolophus tethys	Tethys jelly*
Schlanke Bernsteinschnecke, Dunker's Bernsteinschnecke	Oxyloma elegans/ Oxyloma dunkeri	Pfeiffer's ambersnail, Dunker's ambersnail
Schlanke Birnenschnecke, Schrifttäubchen	Mitrella scripta/ Mitrella flaminea	music dovesnail
Schlanke Fässchenschnecke	Orcula gularis	
Schlanke Glasschleiche	Ophisaurus attenuatus	slender glass lizard
Schlanke Herzmuschel	Laevicardium attenuatum	attenuated cockle
Schlanke Käferschnecke	Stenoplax floridana	Florida slender chiton
Schlanke Sandrennnatter	Psammophis schokari	Forskal's sand snake
Schlanke Schwebegarnele	Schistomysis spiritus	slender ghost shrimp
Schlanke Seefeder	Stylatula elongata	slender sea-pen
Schlanke Seegurke	Holothuria impatiens	slender sea cucumber
Schlanke Sumpfschnecke	Stagnicola turricula	slender pondsnail
Schlanker Bäumchenpolyp	Eudendrium tenue	slender stickhydroid

Schlanker Feilenfisch	*Monoacanthus tuckeri*	slender filefish
Schlanker Kammstern	*Astropecten bispinosus*	slender sand star*
Schlanker Korallenklimmer	*Paracirrhites forsteri*	freckled hawkfish, Forster's hawkfish, blackside hawkfish (FAO)
Schlanker Seestern*	*Luidia clathrata*	slender seastar
Schlanker Tüpfeltausendfüßer	*Boreoiulus tenuis/ Blaniulus tenuis*	slender snake millepede
Schlanker Wieselhai	*Hemipristis elongata*	snaggletooth shark
Schlankfliegen	Leptogasteridae	grass flies, leptogasterid flies
Schlankgibbon, Ungka	*Hylobates agilis*	dark-handed gibbon, agile gibbon
Schlankgrundel	*Gobius geniporus*	slender goby
Schlankhaie, Bärtelige Hundshaie	Leptochariidae	barbelled houndsharks
Schlankhörnchen	*Sundasciurus tenuis*	slender squirrel
Schlankhörnige Spinnenkrabbe	*Macropodia tenuirostris*	slender spider crab
Schlanklibellen	Coenagrionidae	narrow-winged damselflies
Schlanklori	*Loris tardigradus*	slender loris
Schlanknatter	*Leptophis ahaetulla*	parrot snake
Schlanknatter (Dahl'sche Natter/ Steignatter)	*Coluber najadum*	light-green whip snake, Dahl's whip snake
Schlankrüssler	Nemonychidae/ Rhinomaceridae	pine-flower snout beetles
Schlankschnabelnoddi, Kleine Noddiseeschwalbe	*Anous tenuirostris*	lesser noddy
Schlankthun	*Allothunnus fallai*	slender tuna
Schlankwelse	Amblycipitidae	torrent catfishes
Schlauch-Ascidie, Durchsichtige Seeurne	*Ciona intestinalis*	sea vase
Schlauchwürmer, Rundwürmer, Aschelminthen, Nemathelminthen	Aschelminthes/ Nemathelminthes	aschelminths, nemathelminths
Schlegels Lanzenotter, Schlegelsche Lanzenotter	*Bothrops schlegelii/ Bothriechis schlegelii*	Schlegel's viper, eyelash viper, horned palm viper, eyelash palm pit viper
Schlehdornspanner	*Eulithis prunata*	phoenix moth
Schlehen-Grünflügelspanner	*Hemithea aestivaria*	common emerald
Schlehenfrostspinner, Grauer Wollrückenspanner	*Apocheima pilosaria/ Phigalia pilosaria*	pale brindled beauty
Schlehenspanner, Pflaumenspanner	*Angerona prunaria*	orange moth
Schlehenspinner, Bürstenbinder	*Orgyia antiqua/ Orgyia recens*	vapourer moth, common vapourer, rusty tussock moth
Schlei, Schleie, Schleihe	*Tinca tinca*	tench
Schleichen	Anguidae	lateral fold lizards, anguids
Schleichspinne*	*Harpactea hombergi*	sneak spider
Schleiereule	*Tyto alba*	barn owl
Schleiereulen	Tytonidae	barn owls & grass owls

Schleiermeduse	*Rhopalonema velatum*	veiled medusa*
Schleimfische	Blenniidae	blennies, combtooth blennies
Schleimgroppe*	*Cottus cognatus*	slimy sculpin
Schleimköpfe	Berycidae	berycids, alfonsinos
Schleimköpfe, Schleimkopfartige	Beryciformes	squirrel fishes (primitive acanthopterygians)
Schleimlerche, Atlantischer Schleimfisch, Schan	*Lipophrys pholis*	shanny
Schleppensylphe	*Sappho sparganura*	red-tailed comet
Schleuderzungensalamander, Höhlensalamander	*Hydromantes* spp.	web-toed salamander
Schlicht-Stachelschnecke*	*Murex trapa*	rare spined murex
Schlichtmeise	*Parus inornatus*	plain titmouse
Schlichtmungo	*Salanoia concolor*	Malagasy brown-tailed mongoose, salano
Schlichtziesel, Europäisches Ziesel	*Citellus citellus/ Spermophilus citellus*	European ground squirrel, European suslik, European souslik
Schlick-Winkerkrabbe	*Uca pugnax*	Atlantic marsh fiddler, mud fiddler
Schlickaale	Ilyophidae/Ilyophinae	arrowtooth eels, mustard eels
Schlickkrebs	*Neopanope sayi*	mud crab
Schlickkrebs, Wattenkrebs, Wattkrebs	*Corophium volutator*	European mud scud, mud dwelling amphipod
Schlicklanguste	*Panulirus polyphagus*	mud spiny crawfish
Schlicksabelle, Schlickröhrenwurm	*Myxicola infundibulum*	slime feather duster
Schlickstern*	*Ctenodiscus crispatus*	mud star
Schlingerhai, Rauer Dornhai	*Centrophorus granulosus*	gulper shark (FAO), rough shark
Schlitz-Wurmschnecke, Schuppige Schlangenschnecke	*Siliquaria squamata/ Tenagodus squamatus*	slit wormsnail
Schlitzaugen-Grundhai	*Loxodon macrorhinus*	sliteye shark
Schlitzbandschnecken, Schlitzkegelschnecken, Schlitzkegel	Pleurotomariidae	slitsnails (slit shells)
Schlitznasen, Hohlnasen	*Nycteris* spp.	slit-faced bats, hollow-faced bats, hispid bats
Schlitzrüssler	Solenodontidae	solenodons
Schluchtenzaunkönig	*Catherpes mexicanus*	canyon wren
Schlundegel	Pharyngobdelliformes	pharyngobdelliform leeches
Schlundkegel-Glattfuß	*Nematomenia banyulensis* (Aplacophora)	
Schlundsackschnecken, Sackschnecken, Schlauchschnecken	Sacoglossa/Saccoglossa	sacoglossans
Schlupfwespen, Echte Schlupfwespen	Ichneumonidae	ichneumon flies, ichneumons
Schlüsselloch-Buntbarsch, Maroni	*Cleithracara maronii*	keyhole cichlid
Schlüsselloch-Sanddollars	Mellitidae	keyhole urchins

Schlusslichtsalmler, Fleckensalmler, Laternensalmler	Hemigrammus ocellifer	beacon fish, head-and-tail light tetra (FAO)
Schmalbeinige Wolfspinnen*	Pardosa spp.	thin-legged wolf spiders
Schmalblättriges Moostierchen	Securiflustra securifrons	narrow-leaved hornwrack
Schmalbrust-Schlangenhalsschildkröte	Chelodina oblonga	oblong tortoise, narrow-breasted snake-necked turtle
Schmale Teichmuschel, Abgeplattete Teichmuschel	Pseudoanodonta complanata	compressed river mussel
Schmale Windelschnecke	Vertigo angustior	narrow whorl snail, narrow-mouthed whorl snail
Schmaler Nilhecht	Marcusenius longianalis/ Brienomyrus longianalis	slender Nile pike*, mpanda
Schmaler Prachtkäfer	Agrilus angustulus	slender oak borer
Schmalfußbeutelmäuse	Sminthopsis spp.	dunnarts, narrow-footed marsupial "mice"
Schmalkäfer, Plattkäfer	Cucujidae	flat bark beetles
Schmalkiel-Klappschildkröte	Kinosternon angustipons	narrow-bridged mud turtle
Schmalkopf-Querzahnmolch	Ambystoma texanum	small-mouthed salamander, smallmouth salamander
Schmalnasen-Glatthai, Patagonischer Glatthai	Mustelus schmitti	narrownose smooth-hound (FAO), Patagonian smooth-hound
Schmalschnabeldelphin, Spinnerdelphin	Stenella longirostris	pantropical spinner dolphin, spinning dolphin, long-snouted dolphin
Schmalschnäuzige Seenadel, Grasnadel	Syngnathus typhle	deep-snouted pipefish, broad-snouted pipefish, deep-nosed pipefish, broadnosed pipefish (FAO)
Schmalschwanz-Paradieselster	Astrapia mayeri	ribbon-tailed bird-of-paradise
Schmalschwanz-Paradieswitwe	Vidua paradisaea/ Steganura paradisea	paradise whydah
Schmalstreifenmungo	Mungotictis decemlineata	Malagasy narrow-striped mongoose
Schmalzahn-Sägerochen, Gemeiner Sägefisch, Westlicher Sägefisch	Pristis pectinata	smalltooth sawfish (FAO), greater sawfish
Schmarotzerhummeln	Psithyrus spp.	cuckoo bees
Schmarotzerkrake, Höckeriger Seepolyp	Ocythoe tuberculata	tuberculate pelagic octopus
Schmarotzerraubmöve	Stercorarius parasiticus	Arctic skua
Schmarotzerrose, Einsiedlerrose	Peachia parasitica	parasitic anemone a.o., hermit crab anemone
Schmarotzerschlauch	Rhopalomenia aglaopheniae (Aplacophora)	"parasitic" solenogaster*
Schmarotzerwespen	Sapygidae	sapygid wasps, parasitic wasps
Schmätzer & Erdsänger & Rotschwänze etc.	Saxicolini	chats
Schmeißfliegen	Calliphoridae	blowflies
Schmerle> Bachschmerle, Bartgrundel	Noemacheilus barbatulus/ Barbatula barbatula	stone loach
Schmerlengrundeln	Rhyacichthyidae	loach gobies
Schmerlenwelse	Trychomycteridae	
Schmetterlinge & Motten> Schuppenflügler	Lepidoptera	butterflies & moths> lepidopterans

Schmetterlings-Dreiecksmuschel	*Donax variabilis*	variable coquina, coquina clam, pompano (coquina shell, butterfly shell)
Schmetterlings-Fechterschnecke, Schmetterlings-Flügelschnecke	*Strombus pipus*	butterfly conch
Schmetterlings-Mondschnecke	*Natica alapapilionis*	butterfly moonsnail
Schmetterlings-Thunfisch	*Gasterochisma melampus*	butterfly kingfish
Schmetterlings-Zwergbuntbarsch, Schmetterlingsbuntbarsch	*Microgeophagus ramirezi/ Apistogramma ramirezi/ Papiliochromis ramirezi*	Ramirez's dwarf cichlid, Ramirezi, Ram cichlid (FAO)
Schmetterlingsagame	*Liopepis belliana*	butterfly lizard
Schmetterlingsbuntbarsche	*Microgeophagus* spp.	butterfly cichlids
Schmetterlingsbuntbarsch, Schmetterlings-Zwergbuntbarsch	*Microgeophagus ramirezi/ Apistogramma ramirezi/ Papiliochromis ramirezi*	Ramirez's dwarf cichlid, Ramirezi, Ram cichlid (FAO)
Schmetterlingsfische	Pantodontidae	butterflyfishes
Schmetterlingsfisch, Afrikanischer Schmetterlingsfisch	*Pantodon buchholzi*	butterfly fish, butterflyfish, freshwater butterflyfish (FAO)
Schmetterlingsfische, Falterfische	*Chaetodon* spp.	butterflyfishes
Schmetterlingsflügel	*Paphia alapapilionis*	butterfly venus, butterfly-wing venus
Schmetterlingshafte	Ascalaphidae	owlflies
Schmetterlingskrabbe*	*Cryptolithodes typicus*	butterfly crab
Schmetterlingsmücken	Psychodidae	mothflies, owl midges, sewage farm flies, waltzing midges (incl. sandflies)
Schmetterlingsrochen	*Gymnura* spp.	butterfly rays
Schmetterlingsrochen, Falter-Stechrochen	Gymnuridae	butterfly rays
Schmetterlingstintenfische	*Stoloteuthis* spp.	butterfly-squids
Schmetterlingstintenfisch	*Stoloteuthis leucoptera*	butterfly bobtail squid
Schmierläuse, Wollläuse	Pseudococcidae	mealybugs
Schmirgelschnecke*	*Patella granatina*	sandpaper limpet
Schmuck-Baumschlangen, Schmuckbaumnattern	*Chrysopelea* spp.	flying snakes, Asian parrot snakes
Schmuck-Dosenschildkröte	*Terrapene ornata ornata*	western box turtle, ornate box turtle (IUCN)
Schmuck-Höckerschildkröte	*Graptemys pulchra*	Alabama map turtle
Schmuck-Hornfrosch	*Ceratophrys ornata*	ornate horned frog, ornate horned toad, escuerzo
Schmuck-Krötenkopfagame	*Phrynocephalus ornatus*	striped toadhead agama
Schmuck-Mitra	*Vexillum coccineum*	ornate mitre, ornate miter
Schmuck-Zwergkopf-Seeschlange	*Microcephalophis gracilis/ Hydrophis gracilis*	graceful small-headed sea snake, graceful sea snake
Schmuckanemone	*Nemanthus nitidus*	jewel anemone
Schmuckbärbling	*Rasbora elegans*	twospot rasbora

Schmuckfliegen	Otitidae	picture-winged flies
Schmuckflossen-Flösselhecht, Schönflössler	Polypterus ornatipennis	ornate bichir
Schmuckkärpfling	Poecilia ornata/ Limia ornata	ornate molly, ornate limia
Schmuckkästchen-Reusenschnecke, Schmuckkästchen	Nassarius arcularius	casket nassa, cake nassa
Schmucklanguste, Bunte Languste	Panulirus versicolor	painted crawfish, painted spiny crawfish, painted rock lobster
Schmuckottern	Calliophis spp.	Oriental coral snakes
Schmucksalmler	Hyphessobrycon ornatus	ornate tetra
Schmuckseeschwalbe	Sterna elegans	elegant tern
Schmuckspanner	Scopula ornata	ornate wave
Schmuckvögel, Kotingas	Cotingidae	cotingas, plantcutters, sharpbill
Schmuckwanze, Kohlschmuckwanze, Schwarzrückige Gemüsewanze	Eurydema ornatum	ornate cabbage bug
Schmutzgeier	Neophron percnopterus	Egyptian vulture
Schnabelbarsch, Tiefenbarsch	Sebastes mentella	deepwater redfish (FAO), beaked redfish
Schnabelblennide, Spitzkopfschleimfisch	Tripterygion tripteronotum	triple-fin blenny
Schnabeldelphin, Ganges-Delphin, Susu	Platanista gangetica	Ganges River dolphin, su-su
Schnabeldornhai	Deania calceus	shovelnose shark, birdbeak dogfish (FAO)
Schnabelfische	Gibberichthyidae	gibberfishes
Schnabelfliegen	Mecoptera	scorpion flies, mecopterans
Schnabeligel, Ameisenigel	Tachyglossidae	echidnas, spiny anteaters
Schnabelmilben	Bdellidae	snout mites
Schnabelnattern	Rhamphiophis spp.	beaked snakes
Schnabeltier	Ornithorhynchus anatinus	platypus, duck-billed platypus
Schnabeltiere	Ornithorhychidae	platypuses and relatives
Schnabelwal> Indopazifischer Schnabelwal	Indopacetus pacificus	Indo-Pacific beaked whale
Schnabelwale, Spitzschnauzenwale	Ziphiidae	beaked whales
Schnaken & Mücken	Nematocera (Diptera)	mosquitoes
Schnaken, Stelzmücken, Erdschnaken, Pferdeschnaken (Bachmücken, Langbeinmücken, Riesenschnaken)	Tipulidae	crane flies, crane-flies, daddy-long-legs (Br.)
Schnäpel, Schnepel, Edelmaräne, Kleine Schwebrenke	Coregonus oxyrhynchus/ Coregonus oxyrinchus	houting

Schnapper	*Lutjanus* spp.	snapper
Schnapper	Lutjanidae	snappers
Schnäppertyrann	*Myiarchus crinitus*	great crested flycatcher
Schnäpperwaldsänger	*Setophaga ruticilla*	American redstart
Schnäpperwürger, Kleinschnäpper	Platysteiridae	wattle-eyes, puffback-flycatchers
Schnappküling, Schwimmgrundel	*Gobiusculus flavescens*	two-spotted goby (FAO), two-spot goby
Schnappschildkröte	*Chelydra serpentina*	snapping turtle, American snapping turtle
Schnarrheuschrecke, Rotflügelige Schnarrschrecke	*Psophus stridulus*	rattle grasshopper
Schnatterente	*Anas strepera*	gadwall
Schnauzenbrassen, Pikarels	*Spicara* spp.	picarels
Schnauzenfalter	Libytheidae	snout butterflies
Schnauzenspinner, Palpenspinner	*Pterostoma palpina*	pale prominent
Schneckenkanker	Ischyropsalididae	
Schneckenmoos*	*Sertularella rugosa*	snail trefoil hydroid
Schneckennatter	*Dipsas variegata variegata*	snail-eating snake, thirst snake
Schneckennattern, Dickkopfnattern	*Dipsas* spp.	snail-eating snakes
Schneckentöter*	*Pherbellia knutsoni*	snail-killing fly
Schneckenweih	*Rostrhamus sociabilis*	Everglades kite
Schneckenwurm	*Ophelia limacina*	snail opheliid*
Schnee-Eule	*Nyctea scandiaca*	snowy owl
Schnee-Kaurischnecke, Schneekauri	*Trivia nix/Niveria nix*	snowy trivia
Schneeammer	*Plectrophenax nivalis*	snow bunting
Schneeballkäfer	*Pyrrhalta viburni*	viburnum beetle
Schneefink	*Montifringilla nivalis*	snow finch
Schneefloh	*Isotoma nivalis*	snow flea, snow springtail
Schneegans	*Anser caerulescens*	snow goose
Schneehase (Hbz.), Eurasischer Schneehase	*Lepus timidus*	blue hare, mountain hare
Schneekrabbe, Nordische Eismeerkrabbe, Arktische Seespinne	*Chionoecetes opilio*	Atlantic snow spider crab, Atlantic snow crab, queen crab
Schneekrabben	*Chionoecetes* spp.	snow crabs
Schneekranich, Schreikranich	*Grus americana*	whooping crane
Schneeleopard	*Panthera uncia/ Uncia uncia*	snow leopard
Schneemaus	*Microtus nivalis*	snow vole
Schneeschaf	*Ovis nivicola*	snow sheep, Siberian bighorn
Schneeschuhhase (Hbz. Schneehase)	*Lepus americanus*	snowshoe hare, varying hare
Schneetaube	*Columba leuconota*	snow pigeon
Schneeziege	*Oreamnos americanus*	mountain goat

Schneider	*Alburnoides bipunctatus*	riffle minnow, schneider (FAO)
Schneider, Echte Weberknechte	Phalangiidae	daddy-long-legs
Schneiderbarbe	*Puntius viviparus/ Barbus viviparus*	bowstripe barb
Schnelle Felsenkrabbe	*Grapsus albolineatus*	swift-footed crab
Schneller Wüstenrenner	*Eremias velox*	rapid racerunner
Schnellkäfer (Schmiede, Schuster)	Elateridae	click beetles
Schnellläufer	*Harpalus* spp.	black-lustred ground beetles
Schnepfen, Schnepfenvögel (inkl. Wasserläufer)	Scolopacidae	sandpipers & snipes & woodcocks
Schnepfenaal	*Nemichthys scolopaceus*	snipe eel
Schnepfenaale	Nemichthyidae	snipe eels
Schnepfenfisch	*Macroramphosus scolopax*	snipefish
Schnepfenfische	Macroramphosidae	snipefishes
Schnepfenfliege	*Rhagio scolopacea*	snipe fly
Schnepfenfliegen	Rhagionidae	snipe flies
Schnepfenmesserfische	Centriscidae	shrimpfishes
Schnepfenmesserfisch, Rasiermesserfisch	*Aeoliscus strigatus*	shrimpfish, razorfish (FAO)
Schnepfenschnabel-Stachelschnecke, Schnepfenschnabel	*Murex haustellum/ Haustellum haustellum*	snipe's bill murex
Schnurfüßer	Julidae	juliform millepedes
Schnurrbart-Seeskorpion*	*Triglops murrayi*	moustache sculpin
Schnurrbart-Zunge, Pelz-Seezunge	*Monochirus hispidus*	whiskered sole
Schnurrbärtige Fledermaus	*Pteronotus parnelii*	mustache bat, mustached bats
Schnurrbarttamarin	*Saguinus mystax*	moustached tamarin
Schnurwürmer	Nemertini/ Nemertea/ Rhynchocoela	nemertines, nemerteans, proboscis worms, rhynchocoelans, ribbon worms (*broad/flat*), bootlace worms (*long*)
Schokoladen-Demoiselle*	*Pomacentrus variabilis*	cocoa damselfish
Schokoladen-Fledermaus	*Chalinolobus morio*	chocolate bat
Schokoladengurami	*Sphaerichthys osphromenoides*	chocolate gourami
Schokoladenhai	*Dalatias licha/ Scymnorhinus licha/ Squalus licha*	kitefin shark (FAO), seal shark, darkie charlie
Scholle, Goldbutt	*Pleuronectes platessus*	plaice, European plaice (FAO)
Schollen	Pleuronectidae	righteye flounders
Schomburgkhirsch	*Cervus schomburgki*	Schomburgk's deer
Schönbär, Grüner Bär, Spanische Flagge, Spanische Fahne	*Callimorpha dominula/ Panaxia dominula*	scarlet tiger (moth)
Schöne Erbsenmuschel	*Pisidium pulchellum*	beautiful pea mussel
Schöne Grabanemone*	*Edwardsia elegans*	elegant burrowing anemone

Schöne Landdeckelschnecke	*Pomatias elegans*	red-mouthed snail, round-mouthed snail
Schöne Samtmuschel	*Glycymeris formosa*	beautiful bittersweet
Schönechsen	*Calotes* spp.	garden lizards, varied lizards, "bloodsuckers"
Schöner Herzzüngler	*Cardioglossa pulchra*	black long-fingered frog
Schöner Stachelleguan	*Sceloporus graciosus*	sagebrush lizard
Schönflossenbarbe	*Epalzeorhynchus kallopterus*	flying fox
Schönflossenbärbling	*Rasbora kalochroma*	clown rasbora
Schönflossensalmler, Zitronensalmler	*Hyphessobrycon pulchripinnis*	lemon tetra
Schönflössler, Schmuckflossen-Flösselhecht	*Polypterus ornatipennis*	ornate bichir
Schönhörnchen, Eigentliche Schönhörnchen	*Callosciurus* spp.	beautiful squirrels, tricolored squirrels
Schönschrecke, Italienische Schönschrecke	*Calliptamus italicus*	Italian locust
Schönsteißducker, Petersducker, Harveyducker	*Cephalophus callipygus*	Peters' duiker
Schönwallaby, Hübschgesichtwallaby	*Macropus parryi*	whiptail wallaby, pretty-face wallaby
Schoolmaster	*Lutjanus apodus*	schoolmaster, schoolmaster snapper (FAO)
Schopf-Mangabe	*Cercocebus aterrimus/ Lophocebus aterrimus*	black-crested mangabey, black mangabey, crested mangabey
Schopfadler	*Spizaetus occipitalis*	long-crested eagle
Schopfalk	*Aethia cristatella*	crested auklet
Schopfducker	*Cephalophus* spp.	duikers
Schopffasan, Wallichfasan	*Catreus wallichii*	cheer pheasant
Schopffische, Schopfköpfe, Einhornfische	Lophotidae	crestfishes
Schopfgibbon, Weißwangengibbon	*Hylobates concolor*	crested gibbon
Schopfhähnchen	*Leptopoecile elegans*	crested tit warbler
Schopfhirsch	*Elaphodus cephalophus*	tufted deer
Schopfhuhn, Hoatzin	Opisthocomidae	hoatzin
Schopfkleidervogel	*Palmeria dolei*	crested honeycreeper
Schopflangur	*Presbytis pileata*	capped leaf monkey
Schopfmakak, Schopfaffe	*Macaca nigra*	Celebes ape
Schopfstirnmotten	Tischeriidae	apple leaf miners & others
Schopfstummelaffe, Grüner Stummelaffe	*Procolobus verus*	olive colobus
Schopfwachtel	*Lophortyx californica/ Callipepla californica*	California quail
Schopfwürmer u.a.	Terebdellidae	terebdellid worms
Schopfwürmer u.a.	Onuphidae	beachworms
Schornsteinsegler	*Chaetura pelagica*	chimney swift

Schotenförmige Schwertmuschel, Schotenförmige Messerscheide, Taschenmesser-Muschel, Schotenmuschel	Ensis siliqua	pod razor clam, giant razor clam
Schotenschnecke, Schlangenschnecke, Aalwurmschnecke	Siliquaria anguina	pod snail, wormsnail (pod shell, wormshell)
Schottenhaube	Phalium granulatum	Scotch bonnet
Schotter-Schlangenstern*	Ophiopsila annulosa	gravel brittlestar
Schottischer Kreuzschnabel	Loxia scotica	Scottish crossbill
Schrägbinden-Ziersalmler	Nannostomus espei	barred pencilfish
Schrägsalmler, Schrägschwimmer	Thayeria boehlkei	blackline penguinfish
Schrägsteher, Spitzmaulziersalmler	Nannostomus eques	tube-mouthed pencilfish
Schrägstreifiger Ampferspanner	Timandra griseata	blood-vein
Schrätzer	Gymnocephalus schraetzer	striped ruffe, schraetzer (FAO), Danube ruffe
Schraubensabelle	Spirographis spallanzanii/ Sabella spallanzanii	fanworm, feather-duster worm a.o.
Schraubenschnecken, Pfriemenschnecken	Terebridae	auger snails, augers (auger shells)
Schraubenwurmfliege, Altwelt-Schraubenwurmfliege	Chrysomya bezziana	screwworm fly, Old World screwworm fly, screwworm
Schraubenwurmfliege, Neuwelt-Schraubenwurmfliege	Cochliomyia hominivorax	screwworm fly, New World screwworm fly, screwworm
Schraubenziege, Markhor	Capra falconeri	markhor
Schreckensklapperschlange, Schauerklapperschlange, Tropische Klapperschlange, Südamerikanische Klapperschlange, Cascabel	Crotalus durissus	neotropical rattlesnake, cascabel
Schreckliche Seegurke	Stichopus horrens	horrific sea cucumber*
Schrecklicher Giftfrosch	Phyllobates terribilis	golden poison frog
Schreiadler	Aquila pomarina	lesser spotted eagle
Schreibers Langflügelfledermaus	Miniopterus schreibersi	Schreiber's bat
Schreibers Smaragdeidechse, Iberische Smaragdeidechse	Lacerta schreiberi	Schreiber's green lizard
Schreifrosch	Rana clamitans	green frog, common spring frog
Schreikranich, Schneekranich	Grus americana	whooping crane
Schreiseeadler	Haliaeetus vocifer	African fish eagle
Schreitvögel, Stelzvögel	Ciconiiformes	herons & storks & ibises and allies
Schrift-Olive	Oliva sayana	lettered olive
Schrift-Venusmuschel	Circe scripta	script venus
Schriftbarsch	Serranus scriba	painted comber (FAO), banded sea-perch
Schrifttäubchen, Getupftes Täubchen	Pyrene scripta	dotted dovesnail, music dovesnail

Schröter, Hirschkäfer	Lucanidae	stag beetles
Schrumpelhorn*, Schrumpeliges Neptunshorn	*Neptunea lyrata decemcostata*	wrinkle whelk, wrinkled neptune
Schrumpfmuschel*	*Corbula contracta*	contracted corbula, contracted box clam
Schuhschnabel	*Balaeniceps rex* (Balaenicipitidae)	shoebill, whale-headed stork
Schuppen-Drachenfische	Stomiidae/Stomiatidae	scaly dragonfishes, barbeled dragonfishes
Schuppen-Flügelmuschel	*Pteria longisquamosa*	scaly wing-oyster
Schuppen-Riesenmuschel, Schuppige Riesenmuschel	*Tridacna squamosa*	fluted giant clam, giant fluted clam, fluted clam, scaly clam
Schuppenameisen	Formicinae	carpenter ants & others
Schuppenameisen, Drüsenameisen	Dolichoderinae	dolichoderine ants
Schuppenflügler (Schmetterlinge & Motten)	Lepidoptera	lepidopterans (butterflies & moths)
Schuppengeckos	*Geckolepis* spp.	large-scaled geckos
Schuppengrasmücke	*Sylvia melanothorax*	Cyprus warbler
Schuppenhalstaube, Bandtaube	*Columba fasciata*	band-tailed pigeon
Schuppenkalmar	*Lepidoteuthis grimaldi*	scaly squid*
Schuppenkriechtiere	Lepidosauria	lepidosaurs
Schuppenkriechtiere> Eigentliche Schuppenkriechtiere	Squamata	squamata (incl. lizards & amphisbaenians & snakes)
Schuppenlose Drachenfische	Melanostomiidae/ Melanostomiatidae	scaleless dragonfishes
Schuppenräuber	Catoprioninae	
Schuppenräuber, Wimpelpiranha	*Catoprion mento*	wimple piranha
Schuppenschwanzkusu	*Wyulda squamicaudata*	scaly-tailed possum
Schuppentiere	*Manis* spp. (Pholidota)	scaly anteaters, pangolins
Schuppenwürmer u.a.	Polynoidae	polynoids, scaleworms a.o.
Schuppenwurmschnecke	*Serpulorbis imbricata*	scaly wormsnail (scaly wormshell)
Schuppige Archenmuschel	*Arca imbricata*	mossy ark (mossy ark shell)
Schuppige Heuschrecke	*Pseudomogolistes squamiger*	scaly cricket
Schuppige Kammmuschel	*Chlamys squamosa*	squamose scallop
Schuppige Stachelauster	*Spondylus squamosus*	scaly thorny oyster
Schuppiger Peitschenrochen	*Himantura imbricata*	scaly whipray
Schuppiger Schlangenstern	*Amphipholus squamata*	scaly brittlestar
Schuppiger Zwerg-Schlangenstern	*Axiognathus squamatus*	dwarf brittlestar
Schuppiges Meerkalb	*Vitularia salebrosa*	rugged seacalf
Schusterbock, Einfarbiger Langhornbock	*Monochamus sutor*	larch sawyer beetle
Schusterfrosch	*Neobatrachus sutor*	shoemaker frog
Schützenfisch	*Toxotes jaculator*	archerfish
Schützenfische	Toxotidae	archerfishes
Schwalben	Hirundinidae	swallows, martins
Schwalbenfaulvogel	*Chelidoptera tenebrosa*	swallow-wing puffbird
Schwalbenlausfliege	*Stenepteryx hirundinis*	swallow parasitic fly a.o.
Schwalbenmöve	*Larus sabini*	Sabine's gull

Schwalbenschwanz	*Papilio machaon*	swallowtail
Schwalbensittich	*Lathamus discolor*	swift parrot
Schwalbenstare	Artamidae	woodswallows, currawongs
Schwalbentangare	Tersininae	swallow-tanager
Schwalbentangare	*Tersina viridis*	swallow-tanager
Schwalbenwanze	*Oeciacus hirundinis*	martin bug, swallow bug
Schwalbenweih	*Elanoides forficatus*	swallow-tailed kite
Schwämmchengrundel*	*Evermannichthys spongicola*	sponge goby
Schwammfliegen	Sisyridae	brown lacewings
Schwammgallwespe, Eichenschwammgallwespe	*Biorrhiza pallida/ Biorhiza pallida*	oak-apple gall wasp (>oak apple)
Schwammkäfer, Baumschwammkäfer	Ciidae/Cisidae	minute tree-fungus beetles
Schwammkugelgallwespe (>Schwammkugelgalle)	*Andricus kollari*	marble gall wasp (>marble gall/oak nut)
Schwammkugelkäfer	Liodidae/Anisotomidae	liodid beetles
Schwammspinner	*Lymantria dispar*	gipsy moth
Schwan (Schmetterling)	*Euproctis similis/ Porthesia similis/ Sphrageidus similis*	yellow-tail, gold-tail
Schwäne (Vögel)	*Cygnus* spp. (siehe dort!)	swans
Schwäne	Cygninae	swans
Schwanefelds Barbe, Brassenbarbe	*Puntius schwanefeldi*	tinfoil barb, goldfoil barb (FAO)
Schwanen-Fechterschnecke, Schwanen-Flügelschnecke	*Strombus epidromis*	swan conch
Schwanenhalsmuscheln, Schwanenhals-Nussmuscheln	Nuculanidae	swan-neck nut clams (swan-neck shells), elongate nut clams
Schwanenmuschel, Große Teichmuschel	*Anodonta cygnea*	swan mussel
Schwanzfleck-Stachelmakrele, Atlantische Goldmakrele*	*Chloroscombrus chrysurus*	Atlantic bumper
Schwanzfleckkärpfling	*Phalloceros caudimaculatus*	dusky millions fish
Schwanzflossen-Messeraale, Peitschenmesseraale	Apteronotidae	ghost knifefishes
Schwanzfrosch	*Ascaphus truei*	tailed frog
Schwanzfrösche	Ascaphidae	tailed frogs
Schwanzlurche	Urodela/Caudata	urodeles (salamanders & newts)
Schwanzmeise	*Aegithalos caudatus*	long-tailed tit
Schwanzmeisen	Aegithalidae	long-tailed tits, bushtits
Schwanzstrichsalmler	*Hemigrammus unilineatus*	feather fin
Schwanzstrichwaran, Streifenschwanz-Waran	*Varanus caudolineatus*	streak-tailed monitor, stripetail monitor
Schwanztupfensalmler	*Moenkhausia oligolepis*	glass tetra
Schwärmer	Sphingidae	hawkmoths, sphinx moths
Schwarmsalmler	Hydrocinidae	hydrocinids
Schwarz-Weiß-Delphine	*Cephalorhynchus* spp.	southern dolphins, piebald dolphins
Schwarzaugen-Dardanus	*Dardanus lagopodes*	black-eyed hermit crab
Schwarzaugen-Grundel*	*Coryphopterus nicholsi*	blackeye goby

Schwarzband-Sägesalmler	*Serrasalmus spilopleura*	dark-banded piranha
Schwarzbandbarbe, Seitenstrichbarbe	*Barbus lateristriga/ Puntius lateristriga*	T-barb, spanner barb (FAO)
Schwarzbandkärpfling	*Poecilia nigrofasciata/ Limia nigrofasciata*	black-barred molly, blackbarred limia (FAO)
Schwarzbandsalmler	*Hyphessobrycon scholzei*	black-line tetra
Schwarzbär	*Ursus americanus*	American black bear
Schwarzbarsch	*Micropterus dolomieui*	smallmouth bass
Schwarzbauch-Erdschildkröte	*Melanochelys trijuga trijuga*	Indian black turtle, hard-shelled terrapin
Schwarzbauch-Seeteufel	*Lophius budegassa*	black-bellied angler
Schwarzbäuchiger Bachsalamander	*Desmognathus quadramaculatus*	black-bellied salamander
Schwarzbinden-Soldatenfisch	*Myripristis adustus*	shadowfin soldierfish (FAO), blacktip soldierfish, blue squirrelfish
Schwarzbrauenalbatros	*Diomedea melanophris*	black-browed albatross
Schwarzbrauner Reismehlkäfer, Schwarzer Reismehlkäfer	*Tribolium madens*	European black flour beetle
Schwarzbrauner Rippenmoderkäfer	*Lathridius nodifer*	black-brown plaster beetle*, black-brown fungus beetle
Schwarzbrust-Schneehöschen	*Eriocnemis nigrivestris*	black-breasted puffleg
Schwarzbrust-Seenadel, Harlekin Seenadel	*Corythoichthys nigripectus*	gilded pipefish
Schwarzbüscheläffchen	*Callithrix penicillata*	black-plumed marmoset
Schwarzbussard	*Buteo albonotatus*	zone-tailed hawk
Schwarzducker	*Cephalophus niger*	black duiker
Schwarze Angler, Schwarzangler, Tiefseeteufel	Melanocetidae	devil-anglers
Schwarze Birnenblattwespe	*Pristiphora abbreviata*	black pearleaf sawfly
Schwarze Bohnenlaus, Schwarze Rübenlaus, Schwarze Bohnenblattlaus, Schwarze Rübenblattlaus	*Aphis fabae*	black bean aphid, "blackfly"
Schwarze Bohnenmuschel	*Musculus niger*	black mussel, black musculus, little black mussel
Schwarze Dickkopf-Schildkröte	*Siebenrockiella crassicollis*	black marsh turtle, Siamese temple turtle
Schwarze Drachenfische	Idiacanthidae	sawtail-fishes
Schwarze Drahtkoralle*	*Stichopathes lutkeni*	black wire coral
Schwarze Erdschildkröte, Bauchstreifen-Erdschildkröte	*Rhinoclemmys funerea*	black wood turtle
Schwarze Erdwanze	*Aethus nigritus*	black groundbug*
Schwarze Finsterspinne	*Amaurobius ferox*	black lace-weaver
Schwarze Fliege, Gewächshausblasenfuß	*Heliothrips haemorrhoidalis*	glasshouse thrips, greenhouse thrips
Schwarze Gartenameise, Schwarzgraue Wegameise, Schwarzbraune Wegameise	*Lasius niger*	black ant, common black ant, garden ant
Schwarze Garteneule, Knötericheule, Flohkrauteule, Blumeneule	*Melanchra persicariae*	dot moth

Schwarze Hammermuschel, Häufige Hammermuschel	*Malleus malleus*	common hammer-oyster
Schwarze Heidelibelle	*Sympetrum danae*	black sympetrum
Schwarze Kiefernholzwespe, Schwarze Fichtenholzwespe, Tannenholzwespe	*Xeris spectrum*	spectrum wood wasp
Schwarze Kielnacktschnecke	*Milax nigricans*	black slug
Schwarze Kirschenblattwespe	*Caliroa cerasi*	pear sawfly, pear slug sawfly, pear and cherry sawfly, pear and cherry slugworm, *larva*: pearslug
Schwarze Kirschenlaus, Sauerkirschenlaus	*Myzus cerasi*	black cherry aphid
Schwarze Koralle	*Antipathes subpinnata*	black coral
Schwarze Koralle> Echte Schwarze Koralle	*Rumphella antipathes*	black sea rod, black sea whip
Schwarze Kröte	*Bufo exsul*	black toad
Schwarze Landkrabbe, Rote Landkrabbe	*Gecarcinus lateralis/ Gigantinus lateralis*	blackback land crab, black land crab, red land crab
Schwarze Makrele, Dunkle Stachelmakrele	*Caranx lugubris*	black jack (FAO), black kingfish
Schwarze Mamba	*Dendroaspis polylepis*	black mamba
Schwarze Marienkäfer*	*Chilocorus* spp.	black ladybird beetles
Schwarze Notothenia	*Notothenia coriiceps*	black rockcod, broad-headed notothenia, yellowbelly rockcod (FAO)
Schwarze Ohrwürmer	Chelisochidae	black earwigs
Schwarze Ölbaumschildlaus	*Saissetia oleae*	black scale
Schwarze Pazifische Turbanschnecke	*Tegula funebralis*	black tegula, black topsnail
Schwarze Pelomeduse	*Pelusios niger*	West African black forest turtle, black sidenecked turtle, African black terrapin
Schwarze Pfirsichblattlaus	*Brachycaudus persicae*	black peach aphid
Schwarze Pflaumensägewespe	*Hoplocampa minuta*	black plum sawfly
Schwarze Ringelschleiche	*Anniella pulchra nigra*	black legless lizard
Schwarze Rübenlaus, Schwarze Rübenblattlaus, Schwarze Bohnenlaus, Schwarze Bohnenblattlaus	*Aphis fabae*	black bean aphid, "blackfly"
Schwarze Sandmilbe, Rotbeinige Erdmilbe	*Halotydeus destructor*	redlegged earth mite, black sand mite
Schwarze Schlinger	Chiasmodontidae	swallowers
Schwarze Seegurke	*Holothuria atra*	black sea cucumber
Schwarze Seerute	*Plexaura homomalla*	black sea rod
Schwarze Seescheide	*Phallusia nigra*	black sea-squirt
Schwarze Stachelameise, Schwarze Bulldogameise*	*Myrmecia forficata*	black bulldog ant
Schwarze Stachelbeerblattwespe	*Pristiphora pallipes*	small gooseberry sawfly
Schwarze Stachelschnecke, Trauerschnecke	*Muricanthus nigritus*	black murex
Schwarze Thai-Vogelspinne*	*Haplopelma minax*	Thailand black tarantula
Schwarze Thonine	*Euthynnus lineatus*	black skipjack

Schwarze Tigerotter	*Notechis ater*	black tiger snake
Schwarze Weichschildkröte	*Trionyx ater*	black softshell turtle, Cuatro Cienegas softshell turtle
Schwarze Weichschildkröte, Dunkle Weichschildkröte, Tempel-Weichschildkröte	*Trionyx nigricans/ Aspideretes nigricans*	black softshell turtle, sacred black softshell turtle, Bostami turtle, dark softshell turtle
Schwarze Witwe, Europäische Schwarze Witwe, Malmignatte	*Latrodectus mactans tredecimguttatus*	European black widow, southern black widow (hourglass spider, shoe button spider, po-ko-moo spider)
Schwarze Wüstenkobra	*Walterinnesia aegyptia*	black desert cobra, desert black snake
Schwarzer Anemonenfisch, Schwarzroter Anemonenfisch	*Amphiprion melanopus*	black anemonefish, dusky anemonefish, fire clownfish (FAO)
Schwarzer Apollo	*Parnassius mnemosyne*	clouded apollo, black apollo
Schwarzer Bär	*Arctia villica*	cream-spot tiger (moth)
Schwarzer Baumsalamander	*Aneides flavipunctatus*	black salamander
Schwarzer Brüllaffe	*Alouatta caraya*	black howler
Schwarzer Buntbarsch	*Cichlasoma arnoldi*	black cichlid
Schwarzer Crappie, Silberbarsch	*Pomoxis nigromaculatus/ Centrarchus hexacanthus*	black crappie
Schwarzer Dreischwanzbarsch	*Lobotes surinamensis*	tripletail
Schwarzer Drückerfisch	*Melichthys indicus*	blackfinned triggerfish, Indian triggerfish (FAO)
Schwarzer Fichten-Dickmaulrüssler	*Otiorhynchus niger*	black spruce weevil*
Schwarzer Flachserdfloh	*Longitarsus parvulus*	black flea beetle, flax flea beetle, linseed flea beetle
Schwarzer Flaggensalmler, Schwarzer Neon	*Hyphessobrycon herbertaxelrodi*	black neon
Schwarzer Flughund*	*Penthetor lucasii*	dusky fruit bat
Schwarzer Getreidenager	*Tenebroides mauritanicus*	cadelle, cadelle beetle
Schwarzer Grabläufer	*Pterostichus niger*	black cloaker
Schwarzer Guereza	*Colobus satanas*	black colobus
Schwarzer Heilbutt, Grönland-Heilbutt	*Reinhardtius hippoglossoides*	Greenland halibut (FAO), Greenland turbot, black halibut
Schwarzer Karpfen	*Mylopharyngodon piceus*	black carp
Schwarzer Katzenwels, Schwarzer Zwergwels	*Ictalurus melas*	black bullhead
Schwarzer Klammeraffe	*Ateles paniscus*	black spider monkey
Schwarzer Kohlerdfloh	*Phyllotreta atra*	black turnip flea beetle
Schwarzer Kohltriebrüssler	*Ceutorhynchus picitarsis*	rape winter stem weevil
Schwarzer Kornwurm, Kornkäfer	*Sitophilus granarius*	grain weevil
Schwarzer Lederschwamm	*Ircinia muscarum*	black sponge*
Schwarzer Marlin	*Makaira indica*	black marlin
Schwarzer Maulbrüter	*Oreochromis placidus*	black tilapia
Schwarzer Meereswels*	*Galeichthys ater*	black seacatfish

Schwarzer Moderkäfer, Schwarzer Moderkurzflügler	*Ocypus olens/ Staphylinus olens*	devil's coach-horse
Schwarzer Muntjak	*Muntiacus crinifrons*	black muntjac
Schwarzer Neon, Schwarzer Flaggensalmler	*Hyphessobrycon herbertaxelrodi*	black neon
Schwarzer Pierwurm	*Arenicola defoliens*	black lugworm
Schwarzer Piranha	*Serrasalmus niger*	black piranha
Schwarzer Pomfret	*Formio niger*	black pomfret
Schwarzer Porzellankrebs, Schwarzes Porzellankrebschen, Langhorn-Porzellankrebs	*Porcellana longicornis/ Pisidia longicornis*	long-clawed porcelain crab, common porcelain crab, minute porcelain crab
Schwarzer Preußenfisch, Samtjungferchen	*Dascyllus trimaculatus*	domino damselfish, white-spotted damselfish, threespot dascyllus (FAO)
Schwarzer Reismehlkäfer, Schwarzbrauner Reismehlkäfer	*Tribolium madens*	European black flour beetle
Schwarzer Sägebarsch	*Centropristis striata*	black seabass (FAO), black sea bass
Schwarzer Schlangenstern	*Ophiocomina nigra*	black brittlestar, black serpent-star
Schwarzer Schlankskink	*Scincella melanosticta*	black ground skink
Schwarzer Schnapper	*Macolor niger*	black beauty, mottled snapper, black and white snapper (FAO)
Schwarzer Schnegel	*Limax cinereoniger*	ash-black slug
Schwarzer Schnurwurm, Schwarze Nemertine	*Cerebratulus marginatus* (Nemertini)	black ribbon worm
Schwarzer Schwammball	*Ircinia strobilina*	loggerhead sponge, "cake sponge"
Schwarzer Seehecht, Schwarzer Zahnfisch	*Dissostichus eleginoides*	Patagonian toothfish
Schwarzer Seeigel	*Arbacia lixula*	black urchin
Schwarzer Tagkurzflügler	*Aleochara curtula*	black diurnal rove beetle*
Schwarzer Tausendfüßer	*Tachypodoiulus niger*	white-legged black millepede, black millepede
Schwarzer Tiefseeaal	*Cyema atrum*	deepsea arrow-eel
Schwarzer Weißbindendrücker, Dunkler Ringens	*Melichthys niger/ Balistes radula*	black durgon
Schwarzer Zackenbarsch	*Promicrops lanceolatus*	giant grouper (FAO), giant seabass
Schwarzer Zitterrochen	*Torpedo nobiliana*	Atlantic electric ray, Atlantic torpedo, electric ray (FAO)
Schwarzes Bergkänguru, Schwarzes Wallaruh	*Macropus bernardus*	black wallaroo, Bernard's wallaroo
Schwarzes C, C-Eule	*Xestia c-nigrum*	setaceous Hebrew character
Schwarzes Ordensband	*Mormo maura*	old lady
Schwarzeule	*Euxoa nigricans*	garden dart moth
Schwarzfersenantilope, Impala	*Aepyceros melampus*	impala
Schwarzfisch	*Centrolophus niger*	blackfish
Schwarzfische u.a.	Centrolophidae	medusafishes, barrelfishes
Schwarzfische u.a., Säbelfische	Anoplopomatidae	sablefishes

Schwarzfleckiger Grashüpfer	*Stenobothrus nigromaculatus*	black-spotted grasshopper
Schwarzfleckzunge*	*Symphurus nigrescens*	spotted tongue-sole
Schwarzflossen-Anemonenfisch, Malediven-Anemonenfisch	*Amphiprion nigripes*	black-finned anemonefish, Maldives anemonefish (FAO)
Schwarzflossen-Angler*	*Lophius gastrophysus*	blackfin goosefish
Schwarzflossen-Fächerfisch, Sternhimmelfisch	*Cynolebias nigripinnis*	Argentine pearlfish
Schwarzflossen-Haiwels	*Pangasius pangasius*	Pungas catfish
Schwarzflossen-Mojarra, Schwarzflossen-Silberling	*Eucinostomus melanopterus*	flagfin mojarra
Schwarzflossen-Schnapper	*Lutjanus buccanella*	blackfin snapper
Schwarzflossen-Thunfisch, Schwarzflossenthun	*Thunnus atlanticus*	blackfin tuna
Schwarzflossenhai, Schwarzflossen-Riffhai, Schwarzspitzen-Riffhai	*Carcharhinus melanopterus*	blacktip reef shark
Schwarzflügel-Brachschwalbe	*Glareola nordmanni*	black-winged pratincole
Schwarzflügel-Perlmuschel	*Pinctada penguin*	blackwing pearl-oyster
Schwarzflügelfisch*, Rondelets Flugfisch	*Hirundichthys rondeletii/ Exonautes rondeleti/ Exocoetus rondeleti*	blackwing flyingfish (FAO), subtropical flyingfish
Schwarzfrösche	*Melanobatrachus* spp.	black frogs, Malaysian treefrogs
Schwarzfuß-Napfschnecke	*Patella depressa*	black-footed limpet
Schwarzfußiltis	*Mustela nigripes*	black-footed ferret
Schwarzfußkatze	*Felis nigripes*	black-footed cat
Schwarzgebänderte Korallenschlange	*Micrurus nigrocinctus*	black-banded coral snake
Schwarzgebänderte Mitra, Schwarzgebänderte Bischofsmütze	*Vexillum vittatum*	black-striped mitre, black-striped miter
Schwarzgebänderter Buntbarsch	*Cichlasoma biocellatum/ Cichlasoma octofasciatum*	Jack Dempsey
Schwarzgefleckter Glatthai, Schwarzpunkt-Glatthai, Mittelmeer-Glatthai	*Mustelus punctulatus/ Mustelus mediterraneus*	blackspotted smooth-hound
Schwarzgefleckter Kugelfisch, Schwarzflecken-Kugelfisch	*Arothron nigropunctatus*	blackspotted puffer (FAO), black-spotted blowfish
Schwarzgefleckter Wassermolch	*Notophthalmus meridionalis*	black-spotted newt
Schwarzgeringelte Korallenschlange	*Micrurus mipartitus*	black-ringed coral snake
Schwarzgesicht-Raupenfänger	*Coracina novaehollandiae*	black-faced cuckoo shrike
Schwarzgraue Hilfsameise	*Serviformica fusca/ Formica fusca*	silky ant
Schwarzgrundel, Schwarzküling	*Gobius niger*	black goby
Schwarzgrüne Buschviper	*Atheris nitschei*	sedge viper
Schwarzgrüne Meerkatze, Sumpfmeerkatze	*Allenopithecus nigroviridis*	Allen's monkey, Allen's swamp monkey (blackish-green guenon)
Schwarzhai, Schwarzer Hai	*Carcharhinus obscurus/ Carcharhinus macrurus*	dusky shark (FAO), black whaler

Schwarzhalsbandleguan, Wüsten-Halsbandleguan	*Crotaphytus insularis*	black-collared lizard, desert collared lizard
Schwarzhalskobra, Speikobra, Afrikanische Speikobra	*Naja nigricollis*	spitting cobra, black-necked spitting cobra, blackneck spitting cobra
Schwarzhalskranich	*Grus nigricollis*	black-necked crane
Schwarzhalstaucher	*Podiceps nigricollis*	black-necked grebe
Schwarzhöckerschildkröte	*Graptemys nigrinoda*	black-knobbed map turtle
Schwarzkäfer> Dunkelkäfer	Tenebrionidae	darkling beetles, flour beetles, mealworm beetles
Schwarzkäfer> Düsterkäfer	Melandryidae (Serropalpidae)	false darkling beetles
Schwarzkehl-Honiganzeiger	*Indicator indicator*	black-throated honeyguide
Schwarzkehlchen	*Saxicola torquata*	common stonechat
Schwarzkinndelphin, Süddelphin, Peales Delphin	*Lagenorhynchus australis*	Peale's dolphin, Peale's black-chinned dolphin, blackchin dolphin, southern dolphin
Schwarzkinnkolibri	*Archilochus alexandri*	black-chinned hummingbird
Schwarzkirschfruchtfliege	*Rhagoletis fausta*	black cherry fruit fly
Schwarzklaffschnabel	*Anastomus lamelligerus*	African open-bill stork
Schwarzkolbiger Braun-Dickkopffalter	*Thymelicus lineolus*	Essex skipper
Schwarzkopf-Ruderente	*Oxyura jamaicensis*	ruddy duck
Schwarzkopf-Schleimfisch	*Blennius niceps/ Lipophrys nigriceps*	black-headed blenny
Schwarzkopf-Schnurwurm, Schwarzkopf-Nemertine	*Tetrastemma melanocephalum* (Nemertini)	
Schwarzkopf-Tragopan	*Tragopan melanocephalus*	western tragopan
Schwarzkopf-Zwergnatter	*Rhynchocalamus melanocephalus*	black-headed dwarf snake
Schwarzköpfchen, Maskenköpfchen	*Agapornis personatus*	masked lovebird
Schwarzköpfiger Flughund	*Chironax melanocephalus*	black-capped fruit bat
Schwarzkopfmaki	*Lemur fulvus/ Petterus fulvus*	brown lemur
Schwarzkopfmeise	*Parus atricapillus*	black-capped chickadee
Schwarzkopfmöve	*Larus melanocephalus*	Mediterranean gull
Schwarzkopfnattern	*Tantilla* spp.	blackhead snakes, black-headed snakes
Schwarzkopfpythons	*Aspidites* spp.	black-headed pythons
Schwarzkopftaucher	*Podiceps dominicus*	least grebe
Schwarzkopfuakari	*Cacajao melanocephalus*	black-headed uakari
Schwarzleguan	*Ctenosaura pectinata*	spiny-tailed iguana, pectinate ctenosaur
Schwarzleguane	*Ctenosaura* spp.	spiny-tailed iguanas, greater spinytail iguanas
Schwarzlerche	*Melanocorypha yeltoniensis*	black lark
Schwärzliche Braunschlange, Schwarze Braunschlange	*Demansia atra*	black whipsnake
Schwärzliche Mitra	*Mitra nigra*	black mitre (Br.), black miter (U.S.)
Schwärzliche Schläfergrundel	*Eleotris fusca*	dusky sleeper
Schwarzlori	*Chalcopsitta atra*	black lory
Schwarzmakrele*	*Scombrolabrax heterolepis*	black mackerel

Schwarzmakrelen	Scombrolabracidae	black mackerels, scombrolabracids
Schwarzmantel-Salmler*	Moenkhausia takasei	black-jacket tetra
Schwarzmaul-Umberfisch	Atrobucca nibe	black-mouth croaker, black croaker, longfin kob
Schwarzmeer-Kilka	Clupeonella cultriventris cultriventris	Black Sea sardelle, Sea of Azov sardelle
Schwarzmeer-Wittling	Merlangius merlangus euxinus	Black Sea whiting
Schwarzmeise	Parus niger	southern black tit
Schwarzmilan	Milvus migrans	black kite
Schwarzmündige Nabelschnecke	Mammilla melanostoma	black-mouth moonsnail
Schwarznasenhai	Carcharhinus acronotus	blacknose shark
Schwarznatter	Coluber constrictor	racer
Schwarzohr-Laubenvogel	Ailuroedus melanotis	spotted catbird
Schwarzohrpapagei	Pionus menstruus	blue-headed parrot
Schwarzottern	Pseudechis spp.	Australian black snakes
Schwarzperlhuhn	Agelastes niger	black guineafowl
Schwarzpunkt-Stechrochen	Taeniura meyeni	round ribbontail ray
Schwarzrand-Seezunge	Solea kleinii/ Synaptura kleinii	Klein's sole
Schwarzrote Vogelspinne	Brachypelma vagans	Mexican redrump tarantula
Schwarzrücken-Falterfisch	Chaetodon melanotus	striped butterflyfish, blackback butterflyfish (FAO)
Schwarzrücken-Mönchsaffe	Pithecia albicans	white saki
Schwarzrücken-Steinschmätzer	Oenanthe lugens	mourning wheatear
Schwarzrückenducker	Cephalophus dorsalis	bay duiker
Schwarzrückentamarin	Saguinus nigricollis	black and red tamarin
Schwarzrückiger Pfeiffrosch	Leptodactylus melanonotus	Sabinal frog
Schwarzschenkel-Kleideraffe	Pygathrix nigripes	black-shanked douc langur
Schwarzschnabel-Sturmtaucher	Puffinus puffinus	Manx shearwater
Schwarzschnabelkuckuck	Coccyzus erythrophthalmus	black-billed cuckoo
Schwarzschnabeltukan	Andigena nigrirostris	black-billed mountain toucan
Schwarzschwan	Cygnus atratus	black swan
Schwarzschwanz	Cercomela melanura	blackstart
Schwarzschwanz-Grenadier, Hohlmaul-Panzerratte	Coelorhynchus coelorhynchus	hollowsnout rattail, hollowsnout grenadier (FAO) black-spotted grenadier
Schwarzschwanz-Klapperschlange	Crotalus molossus	black-tailed rattlesnake, blacktail rattlesnake
Schwarzschwanz-Lippfisch	Crenilabrus melanocercus	black-lip wrasse
Schwarzschwanz-Mückenfänger	Polioptila melanura	black-tailed gnatcatcher
Schwarzschwanz-Präriehund	Cynomys ludovicianus	black-tailed prairie dog, Plains prairie dog
Schwarzschwanzgrundel	Ptereleotris evides	scissor-tailed goby, blackfin gudgeon, blackfin dartfish (FAO)
Schwarzschwanzlesbia	Lesbia victoriae	black-tailed trainbearer
Schwarzschwingen-Beilbauchfisch	Carnegiella marthae	black-winged hatchetfish

Schwarzspecht	*Dryocopus martius*	black woodpecker
Schwarzspitzen-Blaupfeil, Großer Blaupfeil	*Orthetrum cancellatum*	black-tailed skimmer
Schwarzspitzen-Riffhai, Schwarzflossenhai, Schwarzflossen-Riffhai	*Carcharhinus melanopterus*	blacktip reef shark
Schwarzspitzen-Schiffsbohrwurm*	*Lyrodus pedicellatus*	blacktip shipworm
Schwarzspitzen-Stachelmakrele	*Caranx sem/ Caranx heberi*	blacktip trevally (FAO), blacktip kingfish
Schwarzspitzenmurex	*Murex nigrispinosus*	black-spined murex
Schwarzstirnducker	*Cephalophus nigrifrons*	black-fronted duiker
Schwarzstirnwürger	*Lanius minor*	lesser grey shrike
Schwarzstorch	*Ciconia nigra*	black stork
Schwarzstreifen-Harnischwels	*Panaque nigrolineatus*	royal panaque
Schwarzviper, Türkische Viper	*Vipera barani*	black viper, Turkish viper
Schwarzwanze	*Thyreocoris scarabaeoides*	negro bug
Schwarzwanzen	Thyreocoridae	negro bugs
Schwarzwedelhirsch, Maultierhirsch	*Odocoileus hemionus*	mule deer
Schwarzweiße Erdwanze	*Sehirus bicolor*	pied shieldbug
Schwarzweiße Kobra, Schwarzweiße Hutschlange	*Naja melanoleuca*	forest cobra, black-and-white lipped cobra
Schwarzweißer Birken-Blattspanner	*Rheumaptera hasta*	argent and sable moth
Schwarzweißer Schlankcichlide	*Julidochromis transcriptus*	black-and-white julie, masked julie (FAO)
Schwarzzipfeliger Weichkäfer	*Rhagonycha fulva*	black-tipped soldier beetle
Schwebegarnelen	Callianassidae	ghost shrimps
Schwebegarnelen	*Mysis* spp.	opossum shrimps
Schwebende Schiffsbohrmuschel, Schiffsbohrwurm	*Teredo megotara*	drifting shipworm
Schwebende Entenmuschel	*Lepas fascicularis*	float goose barnacle, buoy-making barnacle, short-stalked goose barnacle
Schwebfliegen	Syrphidae	hoverflies, hover flies, syrphid flies, flower flies
Schwefelvögelchen	*Heodes tityrus/Loweia tityrus*	sooty copper
Schwein> Wildschwein	*Sus scrofa*	wild boar, pig
Schweine-Haarbalgmilbe	*Demodex phylloides*	hog follicle mite, pig follicle mite, pig head mange mite
Schweine-Spulwurm	*Ascaris suum*	pig roundworm
Schweinebandwurm, Schweinefinnenbandwurm (Bewaffneter Menschenbandwurm)	*Taenia solium*	pork tapeworm
Schweineegel, Rattenegel	*Gastrodiscoides hominis*	pig intestinal fluke, rat fluke
Schweinelaus	*Haematopinus suis*	hog louse

Schweinsaffe	*Macaca nemestrina*	pigtail macaque
Schweinsbarsch (Manitou-Springbarsch), Logbarsch	*Percina caprodes caprodes*	logperch
Schweinsdachs	*Arctonyx collaris*	hog badger
Schweinsdrückerfisch, Mittelmeerdrücker	*Balistes capriscus/ Balistes carolinensis*	triggerfish, grey triggerfish (FAO), clown triggerfish
Schweinsfisch, Schweinsgrunzer	*Anisotremus virginicus*	porkfish
Schweinsfisch, Westatlantischer Lippfisch	*Lachnolaimus maximus*	hogfish (FAO), hogsnapper
Schweinsfische, Congiopodiden	Congiopodidae	horsefishes
Schweinsfrosch	*Rana grylio*	pig frog
Schweinsfuß	*Chaeropus ecaudatus*	pig-footed bandicoot
Schweinshaie, Meersau-Haie, Meersäue, Rauhaie	Oxynotidae	prickly dogfishes, oxynotids
Schweinshirsch	*Axis porcinus*	hog deer
Schweinsnasenskunks	*Conepatus* spp.	hog-nosed skunks
Schweinsschnauzen-Fledermaus, Hummelfledermaus, Kittis Kleinstfledermaus	*Craseonycteris thonglongyai*	Kitti's hog-nosed bat (*smallest mammal!*)
Schweinswal, Braunfisch, Kleintümmler	*Phocoena phocoena*	harbor porpoise, common harbor porpoise
Schweißfliege, Rinderschweißfliege	*Morellia simplex*	cattle sweat fly, sweat fly
Schweizerische Glanzschnecke	*Oxychilus helveticus*	glossy glass snail, Swiss glass-snail (U.S.)
Schwellhaie	*Cephaloscyllium* spp.	swell sharks
Schwere Kronenschnecke	*Volema myristica*	nutmeg melongena, heavy crown shell
Schwere Sandrennmaus	*Psammomys obesus*	fat sand rat
Schwertfisch	*Xiphias gladius*	swordfish
Schwertfische	Xiphiidae	swordfishes, billfishes
Schwertgarnele	*Palaemon xiphias/ Leander xiphias*	glass prawn
Schwertgarnele> Steingarnele, Große Felsgarnele	*Palaemon squilla/ Leander squilla*	cup shrimp
Schwertnasen	*Lonchorhina* spp.	sword-nosed bats
Schwertschnabel	*Ensifera ensifera*	sword-billed hummingbird
Schwertschrecken, Kegelköpfe	Conocephalidae	meadow grasshoppers
Schwertschwanzmolch	*Cynops ensicauda*	sword-tailed newt
Schwertstör, Chinesischer Schwertstör	*Psephurus gladius*	Chinese paddlefish
Schwertträger	*Xiphophorus helleri*	green swordtail
Schwertwal, Killerwal, Orca	*Orcinus orca*	killer whale, orca
Schwertwespen	Xiphydriidae	wood wasps
Schwielen-Panzerwels	*Callichthys callichthys*	armored catfish, hassar, cascarudo (FAO)
Schwielen-Seescheide	*Ascidia callosa*	callused sea-squirt

Schwielen-Solariella*	*Solariella varicosa*	varicose solarelle
Schwielensalamander	*Chiropterotriton* spp.	splayfoot salamanders
Schwimmanemone	*Stomphia coccinea*	swimming anemone
Schwimmbeutler, Yapok (Hbz. Lampara de agua = Lampenratte)	*Chironectes minimus*	water opossum, yapok
Schwimmgrundel, Schnappküling, Zweistreifen-Grundel	*Gobius flavescens/ Gobius ruthensparri/ Gobiusculus flavescens*	two-spotted goby (FAO), two-spot goby
Schwimmkäfer	Dytiscidae	predaceous diving beetles, carnivorous water beetles
Schwimmkrabben u.a.	Portunidae (Polybiinae)	swimming crabs a.o.
Schwimmratten	*Hydromys* spp.	water rats, beaver rats
Schwimmschnecken, Kahnschnecken, Nixenschnecken	Neritidae	nerites, neritids
Schwimmwanze	*Ilyocoris cimicoides*	saucer bug
Schwimmwanzen	Naucoridae	creeping water bugs, saucer bugs
Schwimmwühlen	Typhlonectidae	aquatic caecilians
Schwimmwühlen	*Typhlonectes* spp.	aquatic caecilians
Schwingfliegen	Sepsidae	black scavenger flies (spiny-legged flies)
Schwirrammer	*Spizella passerina*	chipping sparrow
Schwirrflügler, Seglervögel, Seglerartige (Segler & Kolibris)	Apodiformes/ Micropodiformes	swifts & hummingbirds
Sclaters Igel, Somalia-Igel	*Atelerix sclateri/ Erinaceus sclateri*	Somali hedgehog
Scotia-See-Eisfisch	*Chaenocephalus aceratus*	Scotia Sea icefish, blackfin icefish (FAO)
Scytaliniden	Scytalinidae	graveldivers
Sechsaugenspinnen, Dunkelspinnen, Walzenspinnen	Dysderidae	dysderids
Sechsaugenzunge*	*Dicologlossa hexophthalma*	six-eyed sole
Sechsfleckenbarsch	*Cephalopholis sexmaculatus*	sixspot grouper, six-barred grouper, sixblotch hind (FAO)
Sechskiemen-Stechrochen	Hexatrygonidae	sixgill stingrays
Sechskiemer-Grauhai, Großer Grauhai	*Hexanchus griseus*	bluntnosed shark, six-gilled shark, sixgill shark, grey shark, bluntnose sixgill shark (FAO)
Sechspunkt-Jagdspinne	*Dolomedes triton*	sixspotted fishing spider
Sechsstachel-Doktorfisch	*Naso hexacanthus*	six-spine surgeonfish, black unicornfish, sleek unicornfish (FAO)
Sechsstreifen-Glasflügler*	*Bembecia scopigera*	six-belted clearwing
Sechsstreifen-Rennechse	*Cnemidophorus sexlineatus*	racerunner, six-lined racerunner
Sechsstreifige Langschwanzeidechse	*Takydromus sexlineatus*	Asian grass lizard
See-Ei	*Tripneustes ventricosus*	sea egg

See-Elefant> Nördlicher See-Elefant	*Mirounga angustirostris*	northern elephant seal
See-Elefant> Südlicher See-Elefant	*Mirounga leonina*	southern elephant seal
See-Erbsenmuschel	*Pisidium conventus*	Alpine peaclam
See-Erdbeere*	*Gersemia rubiformis*	red soft coral, sea strawberry
See-Milzkraut*	*Salacia alba*	sea spleenwort
Seeadler	*Haliaeetus albicilla*	white-tailed eagle
Seeanemonen	Actiniaria	sea anemones
Seebarsch, Wolfsbarsch	*Dicentrarchus labrax/ Roccus labrax/ Morone labrax*	bass, sea bass, European sea bass (FAO)
Seebarsche, Wolfsbarsche, Streifenbarsche	Percichthyidae/ Moronidae	temperate perches, temperate basses
Seebirne*, Meerbirne* (Seepfirsich*, Meerpfirsich*)	*Halocynthia pyriformis*	sea pear (a sea peach)
Seeblase, Blasenqualle, Portugiesische Galeere	*Physalia physalis*	Portuguese man-of-war, Portuguese man-o-war, blue-bottle
Seebrassen> Gemeiner Seebrassen, Sackbrassen	*Pagrus pagrus/ Sparus pagrus pagrus/ Pagrus vulgaris*	common seabream (FAO), red porgy, Couch's seabream
Seebulle, Seebull, Ulk (Seeskorpion)	*Myoxocephalus bubalis/ Cottus bubalis/ Taurulus bubalis*	long-spined sculpin, long-spined sea scorpion, longspined bullhead (FAO)
Seedahlie	*Urticina felina/ Tealia felina*	feline dahlia anemone, feline sea dahlia
Seedahlie, Dickhörnige Seerose	*Taelia felina*	dahlia anemone
Seedattel, Steindattel, Meeresdattel	*Lithophaga lithophaga* (see: *Petrophaga lithographica*)	datemussel, common date mussel
Seedatteln, Meeresdatteln	*Lithophaga* spp.	datemussels, date mussels
Seedrachen, Seekatzen, Chimären	Holocephali	chimaeras, ratfishes, rabbit fishes
Seefächer u.a., Hornkorallen u.a.	*Eunicella* spp.	horny corals a.o., sea fans
Seefächer u.a., Fächerkorallen	*Gorgonia* spp.	sea fans a.o.
Seefedern	Pennatularia	sea pens, pennatulaceans
Seefedern, Federgorgonien	*Pseudopterogorgia* spp.	sea plumes, feather gorgonian
Seefledermaus, Netzstern	*Patiria miniata*	bat star
Seefledermäuse	Ogcocephalidae	seabats, batfishes
Seeforelle	*Salmo trutta lacustris*	lake trout
Seefrosch	*Rana ridibunda*	marsh frog, lake frog
Seefuchs (Hbz.), Marderhund, Enok	*Nyctereutes procyonoides*	raccoon dog
Seegänseblümchen	Concentricycloidea	sea daisies, concentricycloids, concentricycloideans

Seegrasgarnele	*Hippolyte inermis/ Hippolyte varians*	seaweed chameleon prawn
Seegurken, Seewalzen	Holothuriidae	sea cucumbers
Seegurken-Schwimmkrabbe	*Lissocarcinus orbicularis*	sea cucumber swimming crab
Seehase (Lump, Lumpfisch)	*Cyclopterus lumpus*	lumpsucker (FAO), lumpfish, hen-fish, henfish, sea hen
Seehecht, Europäischer Seehecht, Hechtdorsch	*Merluccius merluccius*	hake, European hake (FAO), North Atlantic hake
Seehechte	Merlucciidae	hakes, merluccid hakes
Seehund	*Phoca vitulina*	harbor seal, common seal
Seeigel, Echinoiden	Echinoidea	sea urchins, echinoids
Seeigel> Echte Seeigel	Echinidae	sea urchins
Seekarpfen, Graubarsch, Nordischer Meerbrassen	*Pagellus bogaraveo*	red seabream, common seabream, blackspot seabream (FAO)
Seekatze, Seeratte, Spöke	*Chimaera monstrosa*	rabbit fish (FAO), European ratfish, chimaera
Seekatzen	Chimaeridae	shortnose chimaeras, ratfishes
Seekuckuck, Kuckucks-Knurrhahn	*Aspitrigla cuculus*	red gurnard, cuckoo gurnard
Seekuhaale	*Apteronotus* spp.	apteronotid eels
Seekühe	Sirenia	sea cows & manatees & dugongs, sirenians
Seelachs, Köhler, Blaufisch	*Pollachius virens*	saithe (FAO), pollock, Atlantic pollock, coley, coalfish
Seeleber*	*Eudistoma hepaticum*	sea liver
Seeleopard	*Hydrurga leptonyx*	leopard seal
Seelilien, Crinoiden (inkl. Haarsterne=Federsterne)	Crinoidea	sea lilies, crinoids (incl. feather stars)
Seemandel, Offene Seemandel, Offene Blasenschnecke	*Philine aperta*	paper-bubble, European paperbubble, open-shelled paperbubble*
Seemandeln, Mandelschnecken	Philinidae	paperbubbles a.o.
Seemannshand, Diebeshand, Mittelmeer-Korkkoralle	*Alcyonium palmatum*	Mediterranean sea-fingers
Seemannsliebchen, Seemaßliebchen, Sonnenrose	*Cereus pedunculatus*	daisy anemone
Seemaus> Gemeine Seemaus, Filzwurm, Wollige Seemaus, Seeraupe	*Aphrodita aculeata*	sea mouse, European sea mouse
Seemäuse, Filzwürmer	*Aphrodita* spp.	sea mice
Seemäuse, Seeraupen	Aphroditidae	sea mice (scaleworms a.o.)
Seemoos, Zypressenmoos	*Sertularia cupressina*	sea cypress, whiteweed hydroid

Seemotten	Pegasiformes (Pegasidae)	seamoths
Seenadeln & Seepferdchen	Syngnathidae	pipefishes & seahorses
Seenadelverwandte, Seepferdchenverwandte, Büschelkiemenartige	Syngnathiformes (Syngnathoidei)	sea horses & pipefishes and allies
Seenelke	*Metridium senile*	clonal plumose anemone, frilled anemone, plumose sea anemone, brown sea anemone, plumose anemone
Seenessel	*Chrysaora quinquecirrha*	sea nettle
Seenuss	*Mertensia ovum*	sea nut, Arctic sea gooseberry
Seeotter (Hbz. Kalan)	*Enhydra lutris*	sea otter
Seepapagei, Papageifisch	*Sparisoma cretense*	parrot fish
Seepeitsche	*Funicula quadrangularis*	sea whip a.o.
Seepeitsche u.a., Peitschen-Gorgonie u.a.	*Leptogorgia virgulata*	whip coral a.o., sea whip a.o.
Seepeitsche u.a., Peitschen-Gorgonie u.a.	*Junceela juncea*	sea whip a.o.
Seepfirsiche	*Halocynthia* spp.	sea peaches
Seerabe, Rabenfisch, Meerrabe	*Sciaena umbra/ Johnius umbra*	brown meagre
Seerabe> Atlantik-Seerabe	*Hemitripterus americanus*	Atlantic sea raven
Seeratte, Seekatze, Spöke	*Chimaera monstrosa*	rabbit fish (FAO), European ratfish, chimaera
Seeregenpfeifer	*Charadrius alexandrinus*	Kentish plover
Seerinde	*Membranipora membranacea*	sea-mat
Seeringelwurm> Gemeiner Seeringelwurm, Schwimmender Seeringelwurm	*Nereis pelagica*	pelagic clam worm
Seeringelwürmer	*Nereis* spp.	nereids
Seerosen-Blattkäfer, Erdbeerkäfer	*Galerucella nymphaeae*	waterlily beetle, pond-lily leaf-beetle
Seerosenzünsler	*Nymphula nymphaeata*	brown China-mark moth, brown China-mark
Seerute, Biegsame Seerute	*Plexaura flexuosa*	flexible sea rod
Seesaibling, Wandersaibling	*Salvelinus alpinus*	char, charr (FAO), Arctic char, Arctic charr
Seesaibling> Amerikanischer Seesaibling, Stutzersaibling	*Salvelinus namaycush*	American lake trout, Great Lake trout, lake trout (FAO)
Seesandwurm	*Thalassema neptuni* (Echiura)	Gaertner's spoon worm
Seescheide> Gemeine Seescheide, Gemeine Ascidie	*Ascidiella scabra*	common sea-squirt
Seescheiden, Ascidien	Ascideacea	sea-squirts, sea squirts, ascidians
Seeschlangen	Hydrophiidae	sea snakes
Seeschmetterling	*Blennius ocellaris*	butterfly blenny
Seeschmetterlinge, Beschalte Flossenfüßer (Flügelschnecken)	Thecosomata	sea butterflies, shelled pteropods

Seeschmetterlinge, Flügelrossfische, Flügelrösschen	Pegasidae	seamoths
Seeschwalben	Sterninae	terns
Seeskorpion, Seeteufel	*Cottus scorpius/ Myoxocephalus scorpius*	Father Lasher, shorthorn sculpin (FAO), bull rout, bull-rout, short-spined seascorpion
Seeskorpione	Eurypterida	sea scorpions, eurypterids
Seesonne, Sonnenstern, Europäischer Quastenstern	*Solaster papposus*	common sun star, spiny sun star
Seespinnen	Majidae	spider crabs
Seestachelbeeren, Meeresstachelbeeren	*Pleurobrachia* spp.	sea gooseberries, cat's eyes
Seestern> Gemeiner Seestern	*Asterias rubens*	common starfish, common European seastar
Seesterne	Asteroidea	seastars, starfishes
Seestichling, Meerstichling	*Spinachia spinachia*	fifteen-spined stickleback
Seestiefmütterchen	*Renilla* spp.	sea pansies
Seetamariske	*Tamarisca tamarisca*	sea tamarisk
Seetangfische*	Chironemidae	kelpfishes
Seetaucher	Gaviiformes/Gaviidae	divers, loons (U.S.)
Seeteufel, Seeskorpion	*Cottus scorpius/ Myoxocephalus scorpius*	Father Lasher, shorthorn sculpin (FAO), bull rout, bull-rout, short-spined seascorpion
Seewalzen, Seegurken, Holothurien	Holothuroidea	sea cucumbers, holothurians
Seewespe u.a.	*Carybdea alata*	sea wasp a.o.
Seewespe u.a., Australische Würfelqualle	*Chironex fleckeri*	Australian box jellyfish, box jelly, deadly sea wasp
Seewespe u.a.	*Chiropsalmus quadrigatus*	sea wasp a.o.
Seewolf, Gestreifter Seewolf, Kattfisch, Katfisch	*Anarhichas lupus*	Atlantic wolffish, wolffish (FAO), cat fish, catfish
Seezunge, Gemeine Seezunge	*Solea vulgaris/ Solea solea*	common sole (Dover sole)
Seezungen	Soleidae	soles
Segel-Feilenfisch	*Stephanolepis cirrhifer*	thread-sail filefish
Segelechsen	*Hydrosaurus* spp.	sail-fin lizards
Segelfalter	*Iphiclides podalirius*	scarce swallowtail, kite swallowtail
Segelfische	Istiophoridae	sailfishes, billfishes (incl. spearfishes & marlins)
Segelfische	Tetrarogidae	waspfishes
Segelflossen-Doktorfisch	*Zebrasoma veliferum*	Pacific sailfin tang, sail-finned surgeonfish, sailfin tang (FAO)
Segelflossensalmler, Prachtsalmler	Crenuchidae	crenuchids
Segelflosser, Skalar	*Pterophyllum scalare*	freshwater angelfish (FAO), longfin angel fish, black angelfish, scalare
Segelkalmar	*Histioteuthis bonelli*	sail-finned jewel squid*

Segellibellen	Libellulidae	common skimmers
Segelqualle	*Velella velella*	by-the-wind sailor, little sail
Segelträger	Veliferidae	velifers
Seggeneule	*Mythimna impura*	sedge wainscot
Seggenrohrsänger	*Acrocephalus paludicola*	aquatic warbler
Seggenzaunkönig	*Cistothorus platensis*	short-billed marsh wren, sedge wren
Segler	Apodidae	swifts
Seidenbienen	*Colletes* spp.	plasterer bees
Seidenbienen	Colletidae	plasterer bees & yellow-faced bees, plumed bees, colletid bees
Seidenhaarschnecke	*Trichia plebeia/ Trichia sericea/ Trichia liberta*	velvet hairysnail
Seidenhai	*Carcharhinus falciformis/ Carcharhinus menisorrah*	silky shark
Seidenkäfer	Scraptiidae	scraptiid beetles
Seidenknopfgallwespe (>Grüne Pustelgalle)	*Neuroterus numismalis*	oak leaf blister-gall cynipid wasp, oakleaf silkbutton-spanglegall cynipid wasp (>silk button spangle gall)
Seidenlaubenvogel	*Ptilonorhynchus violaceus*	satin bowerbird
Seidenreiher	*Egretta garzetta*	little egret
Seidensänger	*Cettia cetti*	Cetti's warbler
Seidenschnapper	*Lutjanus vivanus*	silk snapper
Seidenschwanz	*Bombycilla garrulus*	bohemian waxwing
Seidenschwänze (inkl. Seidenschnäpper)	Bombycillidae	waxwings
Seidenspinnen	Nephilidae	silk spiders
Seidenspinnen	*Nephila* spp.	golden-silk spiders, golden orb spiders, golden orbweavers
Seidenspinner, Echte Spinner	Bombycidae	silkworm moths
Seidenwürger, Nachtschattenfresser	*Hypocolius ampelinus*	grey hypocolius
Seidige Seegurke*	*Chiridota laevis*	silky sea cucumber
Seidige Turbanschnecke	*Tegula fasciata*	silky tegula
Seifenfische, Streifenbarsche	Grammistidae	soapfishes
Seilspinnen	*Episinus* spp.	
Seitenfleck-Schlankskink	*Scincella laterimaculata*	spotted ground skink
Seitenfleckbärbling	*Luciosoma spilopleura* (Hbz. *Luciosoma setigerum*)	apollo shark, apollo sharkminnow (FAO) (side-spot barb)
Seitenfleckenleguan	*Uta stansburiana*	side-blotched lizard
Seitenfleckkärpfling	*Poeciliopsis gracilis*	porthole livebearer
Seitenstrichbärbling	*Rasbora lateristriata*	elegant rasbora, yellow rasbora (FAO)
Seitenwinder-Klapperschlange, Gehörnte Klapperschlange	*Crotalus cerastes*	sidewinder

Seiwal	*Balaenoptera borealis*	sei whale
Sekretär	*Sagittarius serpentarius*	secretary bird
Sekretäre	Sagittariidae	secretary bird
Sektorspinne	*Zygiella x-notata*	sector spider
Sektorspinnen	*Zygiella* spp.	sector spiders
Seladoneule	*Moma alpinum*	scarce Merveille du jour
Selenopiden	Selenopidae	selenopid crab spiders
Selleriefliege	*Euleia heraclei*	celery fly
Seltsame Miesmuschel*	*Mytilus trossulus*	foolish mussel
Seltsame Solariella*	*Solariella obscura*	obscure solarelle
Semling, Afterbarbe	*Barbus meridionalis petenyi*	southern barbel, Danubian barbel
Senegal-Garnele, Südliche Rosa Geißelgarnele	*Farfantepenaeus notialis/ Penaeus notialis*	southern pink shrimp, candied shrimp
Senegal-Gundi, Senegalkammfinger	*Felovia vae*	felou gundi, Senegal gundi
Senegal-Umberfisch	*Pseudotolithus senegalensis*	cassava croaker
Senegal-Weichschildkröte	*Cyclanorbis senegalensis*	Senegal soft-shelled turtle, Senegalese flapshell turtle, Senegal flapshell turtle
Senegalamarant	*Lagonosticta senegala*	red-billed fire finch
Senegalesische Mondschnecke	*Polinices grossularius*	Senegal moonsnail, Senegalese moonsnail
Senegalesischer Seehecht	*Merluccius senegalensis*	Senegalese hake (FAO), black hake
Senegalgalago, Steppengalago	*Galago senegalensis*	Senegal bush baby, lesser bush baby, Senegal galago
Senegaltaschagra	*Tachagra senegala*	black-crowned tachagra
Senfweißling, Tintenfleck-Weißling, Tintenfleck	*Leptidea sinapis*	wood white butterfly, wood white
Sensenfische, Bandfische	Trachipteridae	ribbonfishes
Sepie> Gemeine Sepie, Gemeiner Tintenfisch, Gemeine Tintenschnecke	*Sepia officinalis*	common cuttlefish
Seraus	*Capricornis* spp.	serow
Sergeant Major, Hauptfeldwebelfisch, Fünfstreifen-Zebrafisch	*Abudefduf saxatlis*	sergeant major
Seriemas	Cariamidae	seriemas
Serra-Makrele	*Scomberomorus brasiliensis*	Serra Spanish mackerel
Serval	*Felis serval/ Leptailurus serval*	serval
Seychellen Baumfrosch	*Tachynemis seychellensis*	Seychelle Islands tree frog
Seychellen Pelomeduse	*Pelusios seychellensis*	Seychelles mud turtle
Seychellen-Freischwanzfledermaus	*Coleura seychellensis*	Arabian sheath-tailed bat
Seychellen-Riesenschildkröte	*Testudo gigantea/ Geochelone gigantea/ Megalochelys gigantea*	Seychelles giant tortoise, Aldabran giant tortoise, Aldabra giant tortoise (IUCN)
Seychellenfrösche	*Sooglossus* spp.	Seychelles frogs
Seychellenrohrsänger	*Bebrornis sechellensis*	Seychelles brush warbler
Shamrock-Spinne*	*Araneus trifolium*	shamrock orbweaver, shamrock spider

Shanghai-Weichschildkröte	*Rafetus swinhoei/ Trionyx swinhoei*	Swinhoe's softshell turtle
Shark Bay-Maus, Shark Bay Pseudomys-Maus	*Pseudomys praeconis*	Shark Bay mouse, shaggy mouse, shaggy-haired mouse
Sharpe-Greisbock	*Raphicerus sharpei*	Sharpe's grysbok
Shasta-Höhlensalamander	*Hydromantes shastae*	Shasta salamander
Shaw-Mayer-Maus, Ellermans Maus	*Mayermys ellermani*	Shaw-Mayer's mouse, one-toothed shrew-mouse
Shepherd-Wal, Tasmanischer Schnabelwal	*Tasmacetus shepherdi*	Shepherd's beaked whale, Tasman whale
Sherrysalmler, Perez-Salmler, Kirschflecksalmler	*Hyphessobrycon erythrostigma/ Hyphessobrycon rubrostigma*	bleeding-heart tetra
Shiba-Geißelgarnele	*Metapenaeus joyneri*	shiba shrimp
Siam-Krokodil	*Crocodylus siamensis*	Siamese crocodile
Siamang	*Hylobates syndactylus/ Symphalangus syndactulus*	siamang
Siamang> Zwergsiamang, Biloh, Mentawai-Gibbon	*Hylobates klossii/ Symphalangus klossi*	Kloss's gibbon, dwarf siamang, Mentawai gibbon, beeloh
Siamesischer Kampffisch	*Betta splendens*	Siamese fighting fish
Sibirische Azurjungfer	*Coenagrion hylas*	Siberian damselfly
Sibirische Keulenschrecke	*Gomphocerus sibiricus/ Aeropus sibiricus*	club-legged grasshopper, Sibirian grasshopper*
Sibirische Winterlibelle	*Sympecma annulata/ Sympecma paedisca*	Siberian winter damselfly
Sibirischer Froschzahnmolch	*Ranodon sibiricus*	Siberian salamander, Siberian land salamander
Sibirischer Hausen	*Huso dauricus*	freshwater kaluga
Sibirischer Stör	*Acipenser baerii*	Siberian sturgeon
Sibirisches Feuerwiesel, Kolonok	*Mustela sibirica*	Siberian weasel
Sibirisches Streifenhörnchen, Burunduk, Eurasisches Erdhörnchen, Östliches Streifenhörnchen (Hbz. Burunduk)	*Tamias sibiricus/ Eutamias sibiricus*	Siberian chipmunk
Sichel-Seegurke	*Cucumaria elongata*	sickled sea cucumber*
Sichelente	*Anas falcata*	falcated duck, falcated teal
Sichelfleck-Zwergpanzerwels	*Corydoras hastatus*	dwarf catfish, dwarf corydoras (FAO)
Sichelflosser	Drepaneidae	sicklefishes
Sichelflügel-Fledermaus*	*Phyllops* spp.	falcate-winged bats
Sichelflügler	Drepanidae	hooktip moths
Sichelflügler> Gemeiner Sichelflügler, Birkensichler, Weiden-Sichelspinner	*Drepana falcataria*	pebble hook-tip
Sichelhopfe	Rhinopomastidae	scimitarbills
Sichelkaiserfisch, Arabischer Kaiserfisch	*Pomacanthus maculosus*	yellow-blotch angelfish, Red Sea angelfish, blue moon angelfish, bride of the sea, yellowbar angelfish (FAO)

Sichelschrecken	Phaneropteridae	bush katydids & round-headed katydids
Sichelstrandläufer	*Calidris ferruginea*	curlew sandpiper
Sichelwanzen	Nabidae	damsel bugs
Sichler	*Plegadis falcinellus*	glossy ibis
Sichling (Sichelfisch), Ziege	*Pelecus cultratus*	ziege (FAO), chekhon, sabre carp, sichel
Siebenarmiger Seestern, Schmalarmiger Großplattenstern, Großplattenseestern	*Luidia ciliaris*	seven-armed starfish, seven-rayed starfish
Siebenbinden-Gürteltier	*Dasypus septemcinctus*	seven-banded armadillo
Siebenkiemer-Grauhai, Spitzkopf-Siebenkiemer, Perlonhai	*Heptranchias perlo*	seven-gilled shark, sharpnose sevengill shark (FAO)
Siebenkiemiger Kammzahnhai	*Notorynchus cepedianus/ Notorynchus maculatus*	sevengill shark, broadnose sevengill shark (FAO), spotted sevengill shark
Siebenliniengarnele	*Sabinea septemcarinata*	sevenline shrimp
Siebenpunkt	*Coccinella septempunctata*	seven-spot ladybird, sevenspot ladybird, 7-spot ladybird
Siebenrock-Schlangenhalsschildkröte, Siebenrocks Schlangenhalsschildkröte	*Chelodina siebenrocki*	Siebenrock's snake-necked turtle
Siebenschläfer	*Glis glis*	fat dormouse, edible dormouse
Siebenspitz-Schmuckkrabbe*	*Macrocoeloma septemspinosum*	thorny decorator crab
Siebkiemer, Verwachsenkiemer	Septibranchia	septibranch bivalves, septibranchs
Siedelweber	*Philetairus socius*	sociable weaver
Siedleragame	*Agama agama*	common agama
Siegelringnatter	*Nerodia sipedon*	northern water snake
Sifakas	*Propithecus* spp.	sifakas
Sigalioniden	Sigalionidae	sigalionids, scaleworms a.o.
Signalbarbe	*Labiobarbus festivus*	signal barb (FAO), sailfin barb
Signalkrebs	*Pacifastacus leniusculus*	signal crayfish
Sika-Hirsch, Sikahirsch	*Cervus nippon*	sika deer
Silber-Beilbauchfisch, Gemeiner Silberbeilbauchfisch	*Gasteropelecus sternicla*	common hatchetfish, silver hatchetfish, river hatchetfish (FAO)
Silber-Fechterschnecke, Silber-Flügelschnecke	*Strombus lentiginosus*	silver conch
Silber-Gabelmakrele	*Trachinotus blochii*	silver pompano
Silber-Waldsalamander	*Plethodon glutinosus*	slimy salamander
Silber-Wespenspinne	*Argiope argentata*	silver argiope, silver garden spider
Silber-Wittling, Silberhecht, Nordamerikanischer Seehecht	*Merluccius bilinearis*	silver hake (FAO), silver whiting
Silberäffchen	*Callithrix argentata*	silvery marmoset

Silberbarsch, Schwarzer Crappie	*Pomoxis nigromaculatus/ Centrarchus hexacanthus*	black crappie
Silberbeile, Faltbauchfische	*Argyropelecus* spp.	hatchetfishes
Silberbrachsen, Silberbrassen	*Pterycombus brama*	silver pomfret, silver bramid, Atlantic fanfish (FAO)
Silberdachs	*Taxidea taxus*	American badger
Silberdollar, Scheibensalmler	*Metynnis argenteus*	silver dollar a.o.
Silberdorsch	*Gadiculus argenteus*	silvery pout
Silberfasan	*Lophura nycthemera*	silver pheasant
Silberfischchen	*Lepisma saccharina*	silverfish
Silberfleckbläuling, Geißkleebläuling	*Plebejus argus*	silver-studded blue (butterfly)
Silberfledermaus	*Lasiurus cinereus*	hoary bat
Silberflossenblatt, Mondfisch	*Monodactylus argenteus*	silver mono, moonfish, diamondfish, fingerfish, kilefish, butter-bream, silver moony (FAO)
Silberflossenblätter, Flossenblätter	Monodactylidae	moonies, moonfishes (fingerfishes), monos
Silberfrösche	*Phrynobatrachus* spp.	silver frogs
Silbergalago	*Otolemur argentatus*	silver galago
Silbergibbon	*Hylobates moloch*	silvery gibbon, Javan gibbon
Silbergrauer Erdbohrer	*Heliophobius argenteocinereus*	silvery mole-rat, sand rat
Silbergrüner Bläuling, Steppenheidebläuling	*Lysandra coridon*	chalkhill blue (butterfly)
Silbergrunzer	*Pomadasys hasta*	silver grunter, silver grunt (FAO), spotted javelin fish, grunter bream
Silberhaarige Fledermaus	*Lasionycteris noctivagans*	silver-haired bat
Silberhäher	*Cyanolyca argrntigula*	silvery-throated jay
Silberhalstaube	*Columba trocaz*	long-toed pigeon
Silberhecht, Silber-Wittling, Nordamerikanischer Seehecht	*Merluccius bilinearis*	silver hake (FAO), silver whiting
Silberkarpfen, Gewöhnlicher Tolstolob	*Hypophthalmichthys molitrix*	silver carp (FAO), tolstol
Silberköpfe	Diretmidae	spinyfins
Silberlachs, Coho-Lachs	*Oncorhynchus kisutch*	coho salmon (FAO), silver salmon
Silberlöwe, Puma	*Felis concolor/ Puma concolor*	cougar, puma, mountain lion
Silbermönch	*Cucullia argentea*	silver shark (moth)
Silbermotten	Argyresthiidae	argents (cypress moths)
Silbermöve	*Larus argentatus*	herring gull
Silbermund-Turbanschnecke	*Turbo argyrostomus*	silver-mouthed turban, silvermouth turban
Silbermundwespe	*Crabro cribrarius*	slender-bodied digger wasp
Silberne Pampel	*Pampus argenteus*	silver pomfret

Silberne Pennahia	*Pennahia argentata*	silver croaker
Silberne Süßlippe, Silbergrunzer	*Pomadasys argenteus*	lined silver grunt, silver grunt (FAO)
Silberquerzahnmolch	*Ambystoma platineum*	silvery salamander
Silberreiher	*Egretta alba*	great white egret
Silberspinner, Silberspinnerchen	*Cilix glauca*	Chinese character
Silberspitzenhai, Riff-Weißspitzenhai	*Carcharhinus albimarginatus*	silvertip shark
Silberstreif-Halbschnabelhecht, Silberstreif-Halbschnäbler	*Hyporhamphus unifasciatus*	silverstripe halfbeak
Silberstreifen-Rundhering	*Spratelloides gracilis*	silver-stripe round herring
Silbertannen-Samenwespe	*Megastigmus pinus*	silver fir seed wasp
Silberwels	*Schilbe mystus*	grasscutter catfish
Silomilbe*	*Tyrophagus longior*	grainstack mite, cucumber mite (seed mite, French 'fly')
Simonys Eidechse	*Gallotia simonyi*	Hierro giant lizard
Sinai Sandläufer	*Psammodromus mucronata*	Sinai psammodromus
Sing-Winkerkrabbe	*Uca musica*	singing fiddler crab
Singammer	*Melospiza melodia*	song sparrow
Singdrossel	*Turdus philomelos*	song thrush
Singkröte, Kalifornische Hochgebirgskröte	*Bufo canorus*	Yosemite toad
Singschrecken, Laubheuschrecken	Tettigoniidae	long-horned grasshoppers (and katydids/bushcrickets)
Singschwan	*Cygnus cygnus*	whooper swan
Singzikaden (Singzirpen)	Cicadidae	cicadas
Sipo	*Chironius carinatus*	sipo
Sipos	*Chironius* spp.	sipos
Sirintaraschwalbe, Weißaugen-Trugschwalbe	*Pseudochelidon sirintarae*	white-eyed river martin
Sitatunga, Sumpfantilope, Wasserkudu	*Tragelaphus spekei*	sitatunga
Sitkafichten-Gallenlaus	*Gilletteella cooleyi*	Douglas fir adelges
Sizilianische Mauereidechse	*Podarcis wagleriana*	Sicilian wall lizard
Skabiosen-Scheckenfalter, Goldener Scheckenfalter	*Eurodryas aurinia/ Melitaea aurinia*	marsh fritillary
Skabiosenschwärmer	*Hemaris tityus*	narrow-bordered bee hawkmoth
Skalar, Segelflosser	*Pterophyllum scalare*	freshwater angelfish (FAO), longfin angel fish, black angelfish, scalare
Skalpellkoralle, Sternkoralle	*Galaxea astreata*	scalpel coral*
Skelett-Vogelspinne*	*Ephebopus murinus*	skeleton tarantula
Skelettspindel, Spinnenkopf, Venuskamm, Venuskamm-Stachelschnecke	*Murex pecten/ Murex tenuispina*	Venus comb murex, Venus comb, thorny woodcock
Skinke, Glattechsen	Scincidae	skinks
Skolopender, Riesenläufer	Scolopendromorpha	scolopendromorphs

Skorpion-Giftechse, Skorpions-Krustenechse	*Heloderma horridum*	Mexican beaded lizard
Skorpione	Scorpiones	scorpions
Skorpiongrundel*	*Lebetus scorpioides*	diminutive goby
Skorpions-Klappschildkröte	*Kinosternon scorpioides scorpioides*	scorpion mud turtle
Skorpions-Krustenechse, Skorpion-Giftechse	*Heloderma horridum*	Mexican beaded lizard
Skorpionschnecke	*Lambis scorpius*	scorpion spider conch
Skorpionsfliegen	Panorpidae	scorpionflies, common scorpionflies
Skorpionwanzen, Skorpionswanzen	Nepidae	waterscorpions
Skua	*Stercorarius skua*	great skua
Smaragd-Kugelschnecke	*Smaragdinella calyculata/ Smaragdinella viridis*	emerald bubble snail, emerald bubble
Smaragd-Orfe	*Notropis atherinoides*	emerald shiner
Smaragdeidechse	*Lacerta viridis*	green lizard, emerald lizard
Smaragdgecko	*Gekko smaragdinus/ Pseudogekko smaragdinus*	Polillo false gecko, emerald gecko
Smaragdgrundel	*Gobionellus smaragdus*	emerald goby
Smaragdlibelle> Gemeine Smaragdlibelle	*Cordulia aenea*	downy emerald
Smaragdschild-Winkerkrabbe	*Uca beebei*	emerald fiddler crab
Smaragdschnäpperdrossel	*Cochoa viridis*	green cochoa
Smaragdschnecke	*Smaragdia viridis*	emerald nerite
Smaragdsittich	*Enicognathus ferrugineus*	austral conure
Smaragdspint	*Merops orientalis*	little green bee-eater
Smaragdvireo	*Vireolanius pulchellus*	green shrike vireo
Smaragdwaran	*Varanus prasinus*	emerald monitor
Smiths Zwergchamäleons	*Bradypodion taeniabronchum*	Smith's dwarf chameleon
So-iny-Meeräsche	*Liza haematochila*	so-iny mullet
Sojazystenälchen	*Heterodera glycine*	soybean cyst nematode
Sokoke-Eule	*Otus ireneae*	Sokoke scops-owl
Soldatenfische	Myripristinae	soldierfishes
Soldatenfische	*Holocentrus* spp.	squirrelfishes
Soldatenkrabbe, Grenadierkrabbe	*Mictyris longicarpus*	soldier crab
Soldatenwels	*Osteogeneiosus militaris*	soldier catfish
Solitär-Filzgallwespe* (Braune Filzgalle)	*Andricus solitarius*	solitary oak leaf gall wasp
Solomon-Inseln-Kupferkopfnatter	*Salomonelaps par*	Solomon Island brown snake
Somalia-Galago, Somaligalago	*Galago gallarum*	Somali galago
Somalia-Rennmaus	*Ammodillus imbellis*	walo, Somali gerbil
Sommer-Kieferfuß, Großer Rückenschaler, Kiemenfuß	*Triops cancriformis*	tadpole shrimp, freshwater tadpole shrimp, rice apus
Sommerflunder	*Paralichthys dentatus*	summer flounder
Sommergoldhähnchen	*Regulus ignicapillus*	firecrest

Sommerkiemenfuß, Sommer-Kiemenfuß, Echter Kiemenfuß	*Branchipus stagnalis/ Branchipus schaefferi*	fairy shrimp
Sommermuschel*	*Schizothaerus nuttalli*	summer clam, gaper
Sömmerringgazelle	*Gazella soemmerringii*	Soemmerring's gazelle
Sommertangare, Feuertangare	*Piranga rubra*	summer tanager
Sompat-Süßlippe	*Pomadasys jubelini*	Atlantic spotted grunt, sompat grunt (FAO)
Sonnen-Krustenanemone, Große Krustenanemone	*Palythoa grandis*	giant colonial anemone, golden sea mat, button polyps, polyp rock, cinnamon polyp
Sonnenanemone	*Stichodactyla helianthus*	sun anemone, Atlantic carpet anemone
Sonnenbär, Malaienbär	*Ursus malayanus/ Helarctos malayanus*	sun bear, Malayan sun bear
Sonnenbarsche, Sonnenfische	Centrarchidae	sunfishes
Sonnenblumenstern	*Pycnopodia helianthoides*	sunflower star
Sonnendachse	*Melogale* spp.	ferret badgers
Sonnengucker	*Phrynocephalus helioscopus*	sunwatcher toadhead agama
Sonnenhörnchen	*Heliosciurus* spp.	sun squirrels
Sonnenmotten	Heliodinidae	heliodinid moths
Sonnenralle	*Eurypyga helias* (Eurypygidae)	sunbittern
Sonnenrose, Seemaßliebchen, Seemannsliebchen	*Cereus pedunculatus*	daisy anemone
Sonnensalmler	*Hyphessobrycon eos*	dawn tetra
Sonnenschnecken	Solariidae/ Architectonicidae	sundials, sundial snails (sun shells, sundial shells)
Sonnensittich	*Aratinga solstitialis*	sun conure
Sonnenstern, Seesonne, Europäischer Quastenstern	*Solaster papposus*	common sun star, spiny sun star
Sonnenstern-Lastträgerschnecke	*Stellaria solaris*	sunburst carriersnail
Sonnensterne	Solasteridae	sun stars
Sonnenstrahl-Messermuschel, Sonnenstrahl-Scheidenmuschel	*Siliqua radiata*	sunset razor clam, sunset siliqua
Sonnentierchen, Heliozoen	Heliozoa	sun animalcules, heliozoans
Sonnenvogel	*Leiothrix lutea*	Pekin robin
Sonora-Klappschildkröte	*Kinosternon sonoriense*	Sonora mud turtle
Sonora-Krötenechse	*Phrynosoma xanti*	Sonoran horned lizard
Sophienkrautspanner	*Lithostege griseata*	grey carpet moth
South-Georgia Eisfisch	*Pseudochaenichthys georgianus*	South Georgia icefish
Sowerby-Zweizahnwal, Zweizahnwal	*Mesoplodon bidens*	Sowerby's whale, Sowerby's North Sea beaked whale
Sowerbys Kielnacktschnecke	*Milax sowerbyi/ Tandonia sowerbyi*	Sowerby's slug, keeled slug
Soziale Faltenwespen	Vesperidae/Vespidae	vespid wasps, social wasps (hornets & yellowjackets & potter wasps & paper wasps)
Spaghetti-Aale	Moringuidae	spaghetti eels

Spaghetti-Wurm, Medusenwurm	*Loimia medusa*	medusa worm*
Spalt-Fächerkoralle*	*Flabellum macandrewi*	splitting fan coral
Spaltenkreuzspinne	*Nuctenea umbratica/ Araneus umbraticus*	crevice spider*
Spaltenschildkröte	*Malacochersus tornieri*	pancake tortoise, African pancake tortoise
Spaltenskorpione	*Hadogenes* spp.	South African rock scorpions
Spaltfüßer	Mysidacea	opossum shrimps
Spaltfußgans	*Anseranas semipalmata*	magpie goose
Spaltfußgänse	Anseranatidae	magpie goose
Spaltzahn-Flughörnchen	*Aeretes melanopterus*	groove-toothed flying squirrel
Spanfisch, Bandfisch	*Trachipterus trachipterus*	ribbonfish, ribbon fish (FAO), Mediterranean dealfish
Spanfisch> Nordischer Spanfisch, Nordischer Bandfisch	*Trachipterus arcticus*	dealfish, deal-fish
Spanienkärpfling	*Aphanius iberus*	Spanish killifish, Spanish toothcarp (FAO), Iberian toothcarp
Spanische Egelschnecke, Valencianische Egelschnecke	*Lehmannia valentiana*	threeband gardenslug, Iberian slug, Canadian slug
Spanische Flagge, Spanische Fahne, Schönbär, Grüner Bär	*Callimorpha dominula/ Panaxia dominula*	scarlet tiger (moth)
Spanische Fliege	*Lytta vesicatoria*	Spanish fly, blister beetle
Spanische Kieleidechse	*Algyroides marchi*	Spanish keeled lizard
Spanische Makrele, Gefleckte Königsmakrele	*Scomberomorus maculatus*	Atlantic Spanish mackerel
Spanische Mauereidechse, Iberische Mauereidechse	*Podarcis hispanica*	Iberian wall lizard
Spanische Wildkatze	*Felis pardinus/ Lynx pardinus*	Spanish lynx, pardel
Spanischer Frosch	*Rana iberica*	Iberian frog, Spanish frog
Spanischer Grashüpfer	*Dociotaurus hispanicus/ Ramburiella hispanica*	Iberian cross-backed grasshopper
Spanischer Meerbrassen, Achselfleck-Brassen, Achselbrasse	*Pagellus acarne*	Spanish bream, Spanish seabream, axillary bream, axillary seabream (FAO)
Spanischer Rippenmolch	*Pleurodeles waltl*	sharp-ribbed salamander
Spanischer Sandläufer	*Psammodromus hispanicus*	Spanish psammodromus
Spanischer Schweinsfisch	*Bodianus rufus*	Spanish hogfish
Spanischer Steinbock, Iberiensteinbock	*Capra pyrenaica*	Spanish ibex
Spanischer Walzenskink	*Chalcides bedriagai*	Bedriaga's skink
Spanischer Wassermolch	*Triturus boscai*	Bosca's newt
Spanisches Nachtpfauenauge	*Graellsia isabellae*	Spanish moon moth
Spanner	Geometridae	geometer moths
Spargelfliege, Große Spargelfliege	*Platyparea poeciloptera*	asparagus fly (asparagus maggot)

Spargelhähnchen, Spargelkäfer	*Crioceris* spp.	asparagus beetles
Spargelhähnchen, Spargelkäfer	*Crioceris asparagi*	asparagus beetle
Sparrmans Buntbarsch	*Tilapia sparrmanii*	banded tilapia (FAO), peacock cichlid, Sparrman's cichlid, banded bream
Späte Adonislibelle, Scharlachlibelle	*Ceriagrion tenellum*	small red damselfly
Spatelente	*Bucephala islandica*	Barrow's goldeneye
Spatelgroppe*	*Icelus spatula*	spatulate sculpin
Spatelraubmöve	*Stercorarius pomarinus*	pomarine skua
Spatelschwanzpapagei	*Prioniturus discurus*	blue-crowned racket-tailed parrot
Spatelwels	*Sorubim lima*	cucharon, shovel-nosed catfish, duckbill catfish (FAO)
Spatenfisch	*Chaetodipterus faber*	Atlantic spadefish
Spatenfische, Fledermausfische	Ephippidae	batfishes, spadefishes
Spatenmuschel	*Cryptopecten phrygium*	spathate scallop
Spechte	Picidae	woodpeckers (incl. wrynecks, piculets)
Spechtfink	*Camarhynchus pallidus*	woodpecker finch
Spechtvögel, Spechtartige	Piciformes	woodpeckers & barbets & toucans and allies
Speckkäfer & Pelzkäfer	Dermestidae	larder beetles, skin beetles & carpet beetles
Speer-Anemone	*Peachia hastata*	spear anemone
Speer-Azurjungfer	*Coenagrion hastulatum*	northern damselfly, northern blue damselfly (Br.)
Speerälchen	*Paratylenchus curvitatus*	pin nematode, South African pin nematode
Speerfisch	*Anotopterus pharao*	daggertooth
Speerfische	Anotopteridae	daggertooths
Speernase	*Lonchorhina aurita*	Tomes' long-eared bat
Speernasen-Fledermaus	*Mimon* spp.	Gray's spear-nosed bats
Speicherkäfer	*Trogoderma inclusum*	larger cabinet beetle
Speichermotte, Heumotte, Tabakmotte, Kakao-Motte, Kakaomotte	*Ephestia elutella*	tobacco moth, cocoa moth, chocolate moth, flour moth, warehouse moth, cinereous knot-horn
Speikobra, Schwarzhalskobra, Afrikanische Speikobra	*Naja nigricollis*	spitting cobra, black-necked spitting cobra, blackneck spitting cobra
Speisebohnenkäfer	*Acanthoscelides obtectus*	bean weevil
Speispinne, Leimschleuderspinne	*Scytodes thoracica*	spitting spider
Speispinnen u.a.	Sicariidae	sixeyed sicariid spiders
Speispinnen u.a., Leimschleuderspinnen	Scytodidae	spitting spiders
Spekegazelle	*Gazella spekei*	Speke's gazelle
Sperber	*Accipiter nisus*	northern sparrowhawk, sparrow hawk

Sperberbussard	*Kaupifalco monogrammicus*	lizard buzzard
Sperbereule	*Surnia ulula*	hawk owl
Sperbergeier	*Gyps rueppellii*	Ruppell's griffon
Sperbergrasmücke	*Sylvia nisoria*	barred warbler
Sperberwaldfalke	*Micrastur ruficollis*	barred forest falcon
Sperberweihe	*Geranospiza caerulescens*	crane hawk
Sperlinge	Passeridae/Passerinae (> *Passer* spp.)	sparrows & rock-sparrows
Sperlings-Zirpfrosch	*Crinia desertcola*	sparrow froglet
Sperlingskauz	*Glaucidium passerinum*	pygmy owl
Sperlingstäubchen	*Columbina passerina*	common ground dove
Sperlingsvögel	Passeriformes	passerines, passeriforms (perching birds)
Sperlingsweber	Placepasserinae	
Sphagnumfrosch, Torfmoosfrosch	*Kyarranus sphagnicolus*	sphagnum frog
Sphinxeule	*Brachionycha sphinx*	sprawler, common sprawler
Spiegelei-Qualle, Knollenqualle	*Cotylorhiza tuberculata*	fried-egg jelly*
Spiegelfleck-Dickkopffalter, Bruchwald-Dickkopf	*Heteropterus morpheus*	large chequered skipper
Spiegelfleck-Lippfisch, Orangefleck-Lippfisch	*Coris angulata*	clown wrasse (FAO), clown coris, red-blotched rainbowfish
Spiegelflügelfisch*	*Hirundichthys speculiger*	mirrorwing flyingfish
Spiegelkärpfling, Platy, Mondplaty	*Xiphophorus maculatus*	southern platyfish
Spiegelkleiber	Neosittini	sittellas
Spiegelrochen, Europäischer Glattrochen	*Raja batis*	common skate, skate (FAO), common European skate, grey skate, blue skate
Spießbekassine	*Gallinago stenura*	pintail snipe
Spießböcke	*Oryx* spp.	oryx & gemsbock
Spießente	*Anas acuta*	northern pintail
Spießflughuhn	*Pterocles alchata*	pin-tailed sandgrouse
Spießhirsche	*Mazama* spp.	brocket deer
Spindelbarsche	Romanichthyini	Romanian spindleperch
Spindelförmige Schließmundschnecke	*Macrogastra rolphii*	Rolph's door snail
Spindelgallwespe (>Kleine Spindelgalle)	*Andricus nudus*	oak catkin gall wasp
Spindelschmerlen	Psilorhynchidae	mountain minnows
Spindelschnecke	*Tibia fusus*	spindle tibia, shinbone tibia
Spinnen, Webspinnen	Araneae	spiders
Spinnen-Mondschnecke	*Natica arachnoidea*	spider moonsnail
Spinnen-Mörtelgrabwespe	*Sceliphron caementarium*	black and yellow mud dauber, black and yellow mud wasp
Spinnenameisen, Bienenameisen, Ameisenwespen	Mutillidae	velvet-ants

Spinnenassel, Spinnenläufer	*Scutigera coleoptrata*	house centipede
Spinnenasseln	Scutigeridae	centipedes
Spinnenasseln, Spinnenläufer	Scutigeromorpha/ Notostigmophora	scutigeromorphs
Spinnenfische	Bathypteroidae	spiderfishes
Spinnenfresser	Mimetidae	pirate spiders, spider-hunting spiders
Spinnenläufer, Spinnenassel	*Scutigera coleoptrata*	house centipede
Spinnenqueise, Geflecktes Petermännchen	*Trachinus araneus*	spotted weever
Spinnenschildkröte, Madagassische Spinnenschildkröte	*Pyxis arachnoides arachnoides*	spider tortoise, Madagascan spider tortoise
Spinnenschnecke> Gemeine Spinnenschnecke, Große Teufelskralle	*Lambis truncata*	giant spider conch, wild-vine root
Spinnenschnecken, Spinnen-Fechterschnecken	*Lambis* spp.	spider conch snails
Spinnenspringer	Dicyrtomidae	dicyrtomid springtails
Spinnentiere, Arachniden	Arachnida	arachnids
Spinnentöter, Wegwespen	Pompilidae (Psammocharidae)	pompilids, spider wasps, spider-hunting wasps
Spinnerdelphin, Schmalschnabeldelphin	*Stenella longirostris*	pantropical spinner dolphin, spinning dolphin, long-snouted dolphin
Spinnmilbe> Gemeine Spinnmilbe, „Rote Spinne"	*Tetranychus urticae*	twospotted spider mite
Spinnmilben	Tetranychidae	spider mites, red mites
Spiralälchen	*Helicotylenchus dihystera/ Helicotylenchus nannus*	Cobb's spiral nematode, Steiner's spiral nematode (on cowpea/tomato)
Spiralen-Dörnchenkoralle, Spiralige Drahtkoralle	*Cirripathes spiralis/ Cirrhipathes spiralis*	spiraled whip coral
Spiralen-Mondschnecke*	*Neverita heliciodes*	spiral moonsnail
Spiralgallenlaus	*Pemphigus spirothecae*	poplar spiral-gall aphid
Spitzbrassen	*Diplodus puntazzo/ Puntazzo puntazzo*	sharpsnout seabream (FAO), sheepshead bream
Spitze Blasenschnecke	*Physella acuta/ Physa acuta*	pointed bladder snail, European physa
Spitze Reusenschnecke	*Nassarius acutus*	sharp nassa
Spitze Sumpfdeckelschnecke	*Viviparus contectus*	pointed river snail, Lister's river snail
Spitzenfleck	*Libellula fulva*	scarce chaser dragonfly, scarce libellula
Spitzflossen-Ammenhai, Keulenhai	*Nebrius ferrugineus/ Nebrius concolor*	tawny nurse shark (FAO), giant sleepy shark
Spitzhörnchen, Tupaias	Scandentia/Tupaiiidae	tree shrews
Spitzkopf-Flossenfüße	*Lialis* spp.	sharpsnouted snake lizards
Spitzkopf-Sägerochen	*Anoxypristis cuspidata*	pointed sawfish
Spitzkopfeidechse, Dalmatinische Spitzkopfeidechse	*Lacerta oxycephala*	sharp-snouted rock lizard
Spitzkopfgrundel	*Butis butis*	duckbill sleeper
Spitzkopfmaskenkrabbe	*Pisa armata*	Gibb's spider crab

Spitzkopfnatter	Gonyosoma oxycephalum/ Elaphe oxycephala	mangrove ratsnake, red-tailed snake
Spitzkopfotter, Wiesenotter (Karstotter)	Vipera ursinii	meadow viper, Orsini's viper
Spitzkopfpython	Loxocemus bicolor	burrowing python, ground python, New World python
Spitzkopfschildkröten	Emydura spp.	short-necked turtles, short-necked tortoises
Spitzkopfskink, Prärie-Skink	Eumeces obsoletus	Great Plains skink
Spitzkrokodil	Crocodylus acutus	American crocodile
Spitzling, Getreidespitzwanze	Aelia acuminata	bishop's mitre, bishop's mitre bug
Spitzmaul-Ziersalmler, Schrägsteher, Schrägsalmler	Nannostomus eques	tube-mouthed pencilfish
Spitzmaulkärpfling	Mollienesia sphenops/ Poecilia sphenops	marbled molly, molly (FAO)
Spitzmaulnashorn, Spitzlippennashorn, „Schwarzes" Nashorn	Diceros bicornis	black rhinoceros, hooked-lipped rhinoceros, browse rhinoceros
Spitzmaus-Borstenigel, Langschwanz-Borstenigel	Microgale spp.	shrew-tenrecs, long-tailed tenrecs
Spitzmaus-Langzüngler	Glossophaga soricina	shrew-like long-tongued bat
Spitzmausbeutelratten	Monodelphis spp.	short-tailed opossums
Spitzmäuse	Soricidae	shrews
Spitzmausrüssler (Spitzmäuschen)	Apion spp.	flower weevils
Spitzmausrüssler (Spitzmäuschen)	Apioninae	weevils, flower weevils
Spitznasen-Grundhai	Rhizoprionodon acutus	milk shark (FAO), milkshark
Spitznasen-Grundhaie	Rhizoprionodon spp.	milk sharks, milksharks
Spitznasen-Rochen, Westatlantischer Glattrochen	Raja laevis	barndoor skate, sharpnose skate
Spitznasen-Stechrochen*	Himantura gerrardi	sharpnose stingray
Spitznattern, Erzspitznattern	Oxybelis spp.	vine snakes
Spitzrippige Napfschnecke	Patella longicosta	spiked limpet, long-ribbed limpet, star limpet
Spitzsalmlerverwandte	Lebiasinidae	lebiasinids (voladoras, pencil fishes)
Spitzschlammschnecke, Spitzhorn-Schlammschnecke, Spitzhorn, Große Schlammschnecke	Lymnaea stagnalis	great pondsnail, swamp lymnaea
Spitzschnauzenwale, Schnabelwale	Ziphiidae	beaked whales
Spitzschuppenmakrele	Scomberoides tol	needlescaled queenfish
Spitzschwanz-Strandläufer	Calidris acuminata	sharp-tailed sandpiper
Spitzschwanzgrundel	Gobionellus hastatus	sharptail goby
Spitzschwanzwelse	Olyridae	olyrids
Spitzzahn-Flughunde	Harpyionycteris spp.	Harpy fruit bats
Spixara	Cyanopsitta spixii	little blue macaw, Spix's macaw
Splintholzkäfer	Lyctidae	powder-post beetles & shot-hole borers

Splintkäfer	*Scolytus* spp.	bark beetles
Spöke, Seekatze, Seeratte	*Chimaera monstrosa*	rabbit fish (FAO), European ratfish, chimaera
Spöke> Gefleckte Pazifische Seeratte, Amerikanische Spöke, Amerikanische Chimäre	*Hydrolagus colliei*	spotted ratfish
Sporen-Plattschildkröte	*Platemys pallidipectoris*	Chaco side-necked turtle
Sporentierchen, Sporozoen	Sporozoa	spore-former, sporozoans
Sporn-Flachschildkröte	*Homopus femoralis*	Karroo tortoise, Karroo Cape tortoise
Spornammer	*Calcarius lapponicus*	Lapland bunting
Sporngans	*Plectropterus gambensis*	spur-winged goose
Spornkiebitz	*Hoplopterus spinosus*	spur-winged plover
Spornpieper	*Anthus richardi*	Richard's pipit
Spornschildkröte	*Geochelone sulcata*	African spurred tortoise
Spornzikaden, Stirnhöckerzirpen	Delphacidae	delphacid planthoppers
Spottdrossel	*Mimus polyglottos*	northern mockingbird
Spottdrosseln	Mimidae	mockingbirds & thrashers & catbirds
Spring-Lanzenotter	*Bothrops nummifer/ Porthidium nummifer*	jumping viper
Springaffen	*Callicebus* spp.	titi monkeys
Springbeutelmaus	*Antechinomys laniger*	kultarr
Springbock	*Antidorcas marsupialis*	springbuck, springbok
Springfrosch	*Rana dalmatina*	agile frog, spring frog
Springhase	*Pedetes capensis*	springhare, springhaas
Springkraut-Netzspanner, Weißgerippter Haarbuschspanner	*Eustroma reticulata/ Lygris reticulata*	netted carpet (moth)
Springmäuse	Dipodidae	jerboas
Springmeeräsche	*Liza saliens/Mugil saliens*	leaping mullet
Springschwänze, Collembolen	Collembola	springtails, garden fleas
Springspinnen, Hüpfspinnen	Salticidae	jumping spiders
Springtamarin, Callimico	*Callimico goeldii*	Goeldi's marmoset
Springwanzen, Uferwanzen	Saldidae	shore bugs, saldids
Springwimperling	*Halteria grandinella*	
Spritz-Seescheide, Spritz-Ascidie	*Ascidiella aspersa*	rough sea-squirt*
Spritzsalmler	*Copella arnoldi/ Copeina arnoldi*	spraying characin, splashing tetra, splash tetra (FAO)
Spritzwürmer, Sternwürmer, Sipunculiden	Sipunculida	peanut worms, sipunculids, sipunculans
Spross-Seescheide, Spross-Synascidie	*Aplidium proliferum*	prolific sea-squirt
Sprosser	*Luscinia luscinia*	thrush nightingale
Sprotte (Sprott, Brisling, Breitling)	*Sprattus sprattus*	sprat, European sprat (FAO) (brisling)

Spulwurm	*Ascaris lumbricoides*	giant intestinal worm, common intestinal roundworm
Spulwürmer u.a.	*Toxocara* spp.	arrowhead worms
Sri Lanka Ornament-Vogelspinne	*Poecilotheria fasciata*	Sri Lankan ornamental tarantula
St.Lucia-Lanzenotter	*Bothrops caribbaeus*	St.Lucia serpent, Saint Lucia lancehead
Staatsquallen	Siphonophora	siphonophorans
Stäbchenkugler	Trachysphaeridae	trachysphaerid millepedes
Stabförmige Meerassel	*Idotea linearis*	rod-shaped marine isopod*
Stabwanze, Wassernadel	*Ranatra linearis*	water stick insect
Stachel-Erdschildkröte	*Heosemys spinosa*	spiny turtle, common spined terrapin
Stachel-Erdschildkröten	*Heosemys* spp.	spiny turtles, spined turtles
Stachel-Juwelendose*	*Arcinella arcinella*	spiny jewelbox
Stachel-Kissenstern	*Culcita* spp.	pincushion star, pincushion starfish a.o.
Stachel-Knurrhahn	*Lepidotrigla cavillone*	large-scale gurnard
Stachel-Ohrenzecke*	*Otobius megnini*	spinose ear tick
Stachelaale u.a., Eigentliche Dornrückenaale, Dornaale	Notacanthidae	spiny eels
Stachelaale u.a., Pfeilaale	Mastacembelidae	spiny eels
Stachelagame	*Agama planiceps*	Namib rock agama
Stachelameisen, Urameisen	Ponerinae	ponerine ants, primitive ants (bulldog ants)
Stachelameisen, Urameisen	*Ponera* spp.	ponerine ants, primitive ants
Stachelauster, Eselshuf	*Spondylus gaederopus*	European thorny oyster
Stachelaustern	Spondylidae	thorny oysters
Stachelbeer-Seescheide	*Dendrodoa grossularia*	gooseberry sea-squirt
Stachelbeer-Spinnmilbe, Stachelbeermilbe	*Bryobia ribis*	gooseberry mite, gooseberry bryobia, gooseberry red spider mite
Stachelbeerbär, Purpurbär	*Rhyparia purpurata*	gooseberry tiger (moth)
Stachelbeerfliege, Gelbe Stachelbeerblattwespe	*Pteronidea ribesii/ Nematus ribesii*	gooseberry sawfly, common gooseberry sawfly
Stachelbeermilbe, Stachelbeer-Spinnmilbe	*Bryobia ribis*	gooseberry mite, gooseberry bryobia, gooseberry red spider mite
Stachelbeerspanner, Harlekin	*Abraxas grossulariata*	magpie moth, currant moth
Stachelbeinige Bergschrecke	*Antaxius spinibrachius*	spiny-legged bushcricket
Stachelchamäleon	*Brookesia stumpfii*	plated leaf chameleon
Stachelfische, „Dornhaie", Acanthodier	Acanthodii	spiny fishes, acanthodians
Stachelhai, Alligatorhai, Nagelhai, Brombeerhai	*Echinorhinus brucus*	bramble shark
Stachelhaie	Squaliformes	bramble sharks & dogfish sharks and allies
Stachelhals-Plattschildkröte	*Platemys radiolata spixii*	black spine-necked swamp turtle, black spiny-necked swamp turtle

Stachelhäuter, Echinodermen	Echinodermata	echinoderms
Stachelhering, Süßwasserhering	*Denticeps clupeoides*	denticle herring (FAO), spiny herring
Stachelheringe, Süßwasserheringe	Denticipitidae/Denticipidae	denticle herrings
Stachelhornkrake*, Augenbrauenkrake	*Tetracheledone spinicirrus*	spiny-horn octopus, eyebrow octopus
Stachelhummer, Europäische Languste	*Palinurus elephas*	crawfish, common crawfish (Br.), European spiny lobster, spiny lobster, langouste
Stachelige Herzmuschel, Große Herzmuschel	*Acanthocardia aculeata*	spiny cockle
Stachelige Kammmuschel	*Chlamys asperrima*	prickly scallop
Stachelige Königskrabbe	*Paralomis multispina*	spiny king crab
Stachelige Mondmuschel	*Lucinisca muricata*	spinose lucine
Stachelige Nadelkoralle, Stachelige Buschkoralle, Dornige Reihenkoralle, Christusdorn-Koralle	*Seriatopora hystrix*	needle coral, spiny row coral
Stachelige Rattenmilbe	*Laelaps echidnina*	spiny rat mite
Stachelige Scherengarnele	*Stenopus spinosus*	spiny coral shrimp
Stachelige Seefeder	*Pteroides spinosum*	spinose sea-pen, spiny sea-pen
Stachelige Seespinne	*Mithrax spinosissimus*	channel clinging crab, spiny spider crab
Stachelige Steckmuschel, Durchsichtige Steckmuschel	*Pinna rudis*	rude penshell, rough penshell
Stacheliger Fiederzweigpolyp*	*Halecium muricatum*	sea hedgehog hydroid
Stacheliger Knurrhahn	*Lepidotrigla dieuzeidei*	spiny gurnard
Stacheliger Schlangenstern u.a.	*Ophiothrix spiculata* & *Ophiocoma echinata*	spiny brittlestar a.o.
Stacheliger Schmetterlingsrochen	*Gymnura altavela*	spiny butterfly ray
Stacheliger Tiefenrochen, Weißrochen	*Bathyraja spinosissima*/ *Psammobatis spinosissima*	white skate (FAO), spiny deep-water skate
Stachelkäfer	Mordellidae	tumbling flower beetles
Stachelkäfer, Dorn-Blattkäfer	*Hispa atra*	spiny leaf beetle, leaf-mining beetle, wedge-shaped leaf beetle
Stachelkoralle	*Hydnophora rigida*	
Stachelleguane	*Sceloporus* spp.	spiny lizards, swifts
Stachellose Bienen*	*Melipona* spp.	stingless bees
Stachelmakrele	*Caranx elacate*/ *Caranx sexfasciatus*	bigeye trevally (FAO), large-mouth trevally
Stachelmakrelen, Pferdemakrelen	Carangidae	jacks (jack mackerels) & pompanos, kingfishes, horse mackerels
Stachelmäuse	*Acomys* spp.	spiny mice
Stachelnasenbeutler	*Echymipera* spp.	New Guinea spiny bandicoot

Stachelpolyp	*Hydractinia echinata*	snailfur, snail fur
Stachelrand- Gelenkschildkröte	*Kinixys erosa*	serrated hinged tortoise, serrated hingeback tortoise, common hinged tortoise, eroded hingeback tortoise
Stachelrand-Landschildkröte	*Psammobates oculifera*	serrated star tortoise, toothed Cape tortoise
Stachelratten u.a.	*Maxomys* spp.	spiny rats, Rajah rats
Stachelratten u.a., Lanzenratten	Echimyidae	spiny rats
Stachelratten u.a., Taschenratten	Geomyidae	pocket gophers
Stachelrochen, Stechrochen	Dasyatidae	stingrays (butterfly rays)
Stachelrücken	Stichaeidae	pricklebacks
Stachelrücken-Schleimfisch	*Chirolophis ascanii*	Yarrell's blenny
Stachelschnecken	*Murex* spp.	murex snails
Stachelschnecken, Muriciden	Muricidae	murex snails (murex shells, rock shells)
Stachelschwamm, Kaktusschwamm	*Acanthella acuta*	cactus sponge
Stachelschwänze, Igelwürmer, Echiuriden	Echiura	spoon worms, echiuroid worms
Stachelschwanzgecko	*Diplodactylus ciliaris*	spiny-tailed gecko, spinytail gecko
Stachelschwanzleguan	*Hoplocercus spinosus*	weapontail
Stachelschwanzsegler	*Hirundapus caudacutus*	needle-tailed swift
Stachelschwanzwaran	*Varanus acanthurus*	spiny-tailed pygmy monitor, ridge-tailed monitor, ridgetail monitor
Stachelschwein> Gewöhnliches Stachelschwein	*Hystrix cristata*	African porcupine, crested porcupine
Stachelschweine> Altwelt-Stachelschweine	Hystricidae	Old World porcupines
Stachelschweine> Eigentliche Stachelschweine	*Hystrix* spp.	Old World porcupines
Stachelschweinverwandte	Hystricomorpha	porcupinelike rodents
Stachelskink	*Egernia major*	land mullet, major skink
Stachelskinke	*Egernia* spp.	spinytail skinks
Stachelsonnenstern, Warziger Sonnenstern	*Crossaster papposus*	common sun star, spiny sun star, spiny sunstar
Stachelspinnen	*Gasteracantha* spp.	spinybacked spiders, spiny orbweavers
Stachelspinnen	Gasteracanthinae	spiny-bellied spiders, spiny orbweavers
Stachelspitzer Totenkäfer	*Blaps mucronata*	mucronate churchyard beetle, mucronate cellar beetle
Stacheltaschenmäuse	*Liomys* spp.	spiny pocket mice
Stachelwangen- Schläfergrundel	*Eleotris pisonis*	spinycheek sleeper
Stachelwanze	*Leptopus marmoratus*	spiny-legged bug, spiny shore bug
Stachelwanze, Wipfelwanze	*Acanthosoma haemorrhoidale*	hawthorn shieldbug

Stachelwanzen u.a.	Acanthosomatidae	shieldbugs
Stachelwanzen u.a.	Leptopodidae	spiny-legged bugs, spiny shore bugs
Stachelwasserkäfer, Großer Teichkäfer, Kleiner Kolben-Wasserkäfer, Schwarzbeiniger Stachel-Wasserkäfer,	*Hydrochara caraboides*	lesser silver water beetle
Stachelweichtiere	Aculifera/ Amphineura	amphineurans
Stachelwelse	Bagridae	bagrid catfishes
Stachelwelse	*Mystus* spp.	featherbacks
Stachelzwergbarsche	Acanthoclinidae	spiny basslets
Stachelzwergwelse	Scoloplacidae	spiny dwarf catfishes
Staffelschwänze, Australische Sänger	Maluridae	fairywrens, Australian wrens
Stahlblaue Kiefernschonungsgespinst-Blattwespe	*Acantholyda erythrocephala*	pine false webworm
Stahlblauer Sonnenbarsch	*Lepomis macrochirus*	bluegill
Stahlschwalbe	*Hirundo atrocaerulea*	blue swallow
Stallfliege, Hausfliege	*Muscina stabulans*	false stable fly
Stammreptilien	Cotylosauria	stem reptiles, cotylosaurs
Standartenfisch	*Stylephorus chordatus*	tube-eye (FAO), thread-tail
Star	*Sturnus vulgaris*	starling, European starling
Stare	Sturnidae	starlings and allies
Stärlinge	Icteridae	troupials, New World blackbirds, meadowlarks
Starrbrust-Pelomeduse	*Pelomedusa subrufa*	helmeted turtle, African helmeted turtle, marsh turtle
Starweber	*Dinemellia dinemelli*	white-headed buffalo weaver
Staubhafte	Coniopterygidae	white lacewings
Staubpilzkäfer	Sphindidae	dry-fungus beetles
Staubwanze, Kotwanze, Maskierter Strolch	*Reduvius personatus*	masked hunter bug (fly bug)
Stechfliege, Wadenstecher	*Stomoxys calcitrans*	stable fly, dog fly, "biting housefly"
Stechmücke> Gemeine Stechmücke, Hausmücke	*Culex pipiens*	house mosquito, northern common house mosquito, common gnat (Br.), house gnat (Br.)
Stechmücken, Moskitos	Culicidae	mosquitoes, gnats (Br.)
Stechpalmenblattminierfliege	*Phytomyza ilicis*	holly leaf miner (fly)
Stechpalmenwickler	*Rhopobota naevana*	holly tortrix, marbled single-dot bell moth, holly leaf tier, black-headed fireworm

Stechrochen, Gewöhnlicher Stechrochen	*Dasyatis pastinaca*	blue stingray, European stingray, common stingray (FAO)
Stechrochen, Stachelrochen	Dasyatidae	stingrays (butterfly rays)
Stechrochenartige	Myliobatiformes	stingrays
Steckmuschel, Edle Steckmuschel, Große Steckmuschel	*Pinna nobilis*	rough penshell, noble penshell
Steckmuschel-Erbsenkrabbe, Steckmuschelwächter	*Pinnotheres pinnotheres*	Pinna pea crab
Steckmuschel> Gemeine Steckmuschel, Schinkenmuschel	*Pinna squamosa*	common penshell
Steckmuscheln	Pinnidae	pen shells, fan mussels
Steife Steckmuschel	*Atrina rigida*	stiff penshell
Stein-Seeigel, Steinseeigel	*Paracentrotus lividus*	purple sea urchin, stony sea urchin, black urchin
Steinadler	*Aquila chrysaetos*	golden eagle
Steinbarsch	*Ambloplites rupestris*	rock bass (red-eye)
Steinbeißer, Europäischer Steinbeißer, Dorngrundel	*Cobitis taenia*	spined loach (FAO), spotted weatherfish
Steinböckchen	*Raphicerus campestris*	steenbok
Steinbrachsen	Scorpididae	stonebreams
Steinbrechwidderchen, Gewöhnliches Sechsfleck-Blutströpfchen, Erdeichel-Widderchen, Gemeines Blutströpfchen	*Zygaena filipendulae/ Anthrocera filipendulae*	six-spot burnet
Steinbutt	*Scophthalmus maximus/ Psetta maxima*	turbot
Steinbuttverwandte	Scophthalmidae	turbot fishes
Steindachners Rothauben-Erdfresser	*Geophagus steindachneri*	redhump eartheater
Steineichenbaumspanner, Steineichenspanner	*Boarmia roboraria*	great oak beauty
Steinfische, Teufelsfische	Synanceidae/ Synanceinae	stonefishes
Steinfliegen, Uferfliegen	Plecoptera	stoneflies
Steinforelle, Bachforelle	*Salmo trutta fario*	brown trout (river trout, brook trout)
Steingecko	*Diplodactylus vittatus*	wood gecko, stone gecko
Steingressling	*Gobio uranoscopus*	Danubian gudgeon
Steinhuhn	*Alectoris graeca*	rock partridge
Steinhummel	*Bombus lapidarius/ Aombus lapidarius*	red-tailed bumble bee
Steinhuscher	*Origma solitaria*	rock warbler
Steinkauz	*Athene noctua*	little owl
Steinköhler, Pollack, Heller Seelachs	*Pollachius pollachius*	pollack (green pollack, pollack lythe)

Steinkorallen, Riffkorallen	Madreporaria/ Scleractinia	stony corals, madreporarian corals, scleractinians
Steinkrabben, Königskrabben	Lithodidae	stone crabs & king crabs
Steinkrebs	*Potamobius torrentium/ Austropotamobius torrentium/ Astacus torrentium*	stone crayfish, torrent crayfish
Steinläufer	Lithobiomorpha	lithobiomorphs
Steinläufer	Lithobiidae	garden centipedes
Steinlerche	*Ammomanes deserti*	desert lark
Steinmarder	*Martes foina*	beech marten, stone marten
Steinobst-Gespinstblattwespe	*Neurotoma nemoralis*	apple web-spinning sawfly
Steinortolan	*Emberiza buchanani*	grey-necked bunting
Steinpicker	*Agonus cataphractus*	pogge, hooknose (FAO), armed bullhead
Steinrötel	*Monticola saxatilis*	mountain rock thrush
Steinschmätzer	*Oenanthe oenanthe*	northern wheatear
Steinschnecke, Nordische Steinchenschnecke, Nördliche Purpurschnecke, Nordische Purpurschnecke	*Nucella lapillus/ Thais lapillus*	Atlantic dog whelk, northern dog whelk, Atlantic dogwinkle, northern dogwinkle
Steinschwalbe	*Ptyonoprogne fuligula*	rock martin
Steinsperling	*Passer petronia/ Petronia petronia*	rock sparrow
Steinwälzer	*Arenaria interpres*	turnstone
Steinwels	*Noturus gyrinus*	tadpole madtom
Steißhühner	Tinamiformes/ Tinamidae	tinamous
Stejneger-Schnabelwal	*Mesoplodon stejnegeri*	Bering Sea beaked whale, North Pacific beaked whale, saber-toothed whale
Stellers Seelöwe	*Eumetopias jubatus*	northern sea lion, Steller sea lion
Stelzen & Pieper	Motacillidae	wag-tails, pipits
Stelzengazelle, Lamagazelle, Dibatag	*Ammodorcas clarkei*	dibatag
Stelzenkrabben	Palicidae	stilt crabs
Stelzenläufer	*Himantopus himantopus*	black-winged stilt
Stelzenläufer (inkl. Säbelschnäbler)	Recurvirostridae	avocets, stilts
Stelzenläuferleguane	*Plica* spp.	racerunners
Stelzenläuferleguan	*Plica umbra*	harlequin racerunner
Stelzenrallen, Stelzrallen, Madagaskar-Rallen	Mesitornithidae	mesites, monias, roatelos
Stelzenwanzen	Berytidae	stilt bugs
Stelzfliege	*Micropeza corrigiolata*	stilt-legged fly
Stelzfliegen	Micropezidae/ Tylidae	stilt-legged flies
Stelzmücken, Sumpfmücken	Limoniidae	short-palped craneflies

*S*tengelälchen

Stengelälchen, Luzernealchen, Stockälchen (Rübenkopfälchen)	*Ditylenchus dipsaci*	lucerne stem nematode, stem-and-bulb eelworm, stem and bulb nematode, stem nematode, bulb nematode (potato tuber eelworm)
Stentorrohrsänger	*Acrocephalus stentoreus*	clamorous reed warbler
Steppen-Beißschrecke	*Platycleis montana*	steppe bushcricket
Steppen-Grashüpfer	*Chorthippus vagans*	dryland grasshopper (heath grasshopper)
Steppen-Schuppentier	*Manis temmincki*	Cape pangolin
Steppenadler	*Aquila nipalensis*	steppe eagle
Steppenagame	*Trapelus sanguinolentus*	steppe agama
Steppenbirkenmaus	*Sicista subtilis*	southern birch mouse
Steppeneidechse, Steppenrenner	*Eremias arguta/ Ommateremias arguta*	stepperunner, arguta
Steppenflughuhn	*Syrrhaptes paradoxus*	Palla's sandgrouse
Steppenfuchs	*Vulpes corsac*	corsac fox, red fox
Steppengalago, Senegalgalago	*Galago senegalensis*	Senegal bush baby, lesser bush baby, Senegal galago
Steppengrille	*Melanogryllus desertus*	steppe cricket
Steppenheidebläuling, Silbergrüner Bläuling	*Lysandra coridon*	chalkhill blue (butterfly)
Steppeniltis	*Mustela eversmanni*	steppe polecat
Steppenkiebitz	*Chettusia gregaria*	sociable plover
Steppenlemminge	*Lagurus* spp.	sagebrush voles, steppe lemmings
Steppennatter, Dione-Natter	*Elaphe dione*	Dione's snake
Steppenpieper	*Anthus godlewskii*	Blyth's pipit
Steppenrenner, Steppeneidechse	*Eremias arguta/ Ommateremias arguta*	stepperunner, arguta
Steppenschildkröte, Vierzehenschildkröte	*Agrionemys horsfieldi/ Testudo horsfieldii*	Horsfield's tortoise, four-toed tortoise, Central Asian tortoise
Steppenwaran	*Varanus exanthematicus*	Cape monitor, rock monitor, Bosc's monitor, African savannah monitor
Steppenweihe	*Circus macrourus*	pallid harrier
Steppenzebra	*Equus quagga*	quagga
Sterlett, Sterlet	*Acipenser ruthenus*	sterlet (FAO), Siberian sterlet
Stern-Röhrenwurm*	*Pomatostegus stellatus*	star tubeworm
Stern-Seefledermaus	*Halieutaea stellata*	starry handfish
Stern-Seescheide u.a., Sternascidie	*Botryllus schlosseri*	star ascidian, star sea-squirt
Stern-Seescheide u.a., Stern-Synascidie	*Aplidium stellatum*	sea pork (*of the tunic*)
Sternauge*	*Pollichthys mauli*	stareye lightfish
Sternaugen-Einsiedlerkrebs	*Dardanus venosus*	stareye hermit crab
Sternbauchspinne*	*Acanthepeira stellata*	starbellied spider, star-bellied spider
Sternchenanemone*	*Nematostella vectensis*	starlet sea anemone

Sternelfe	*Stellula calliope*	Calliope hummingbird
Sternflecksalmler, Stieglitzsalmler	*Pristella maxillaris/ Pristella riddlei*	X-ray tetra (FAO), X-ray fish, pristella
Sternflunder, Pazifischer Sternflunder	*Platichthys stellatus*	starry flounder
Sterngucker, Himmelsgucker	Uranoscopidae	stargazers
Sternhausen, Scherg	*Acipenser stellatus*	starry sturgeon, stellate sturgeon (IUCN), sevruga
Sternhimmelfisch, Schwarzflossiger Fächerfisch	*Cynolebias nigripinnis*	Argentine pearlfish
Sternkoralle	*Astreopora gracilis*	star coral a.o.*
Sternkoralle, Skalpellkoralle	*Galaxea astreata*	scalpel coral*
Sternlochschwamm, Bruchschwamm	*Spongilla fragilis*	freshwater sponge
Sternmull	*Condylura cristata*	star-nosed mole
Sternmuräne	*Echidna nebulosa*	snowflake moray (FAO), starry moray (Austr.)
Sternporenschwamm	*Asteropus sarassinorum*	starpore sponge
Sternrochen, Nordatlantischer Sternrochen, Atlantischer Sternrochen	*Raja radiata*	starry ray, thorny skate (FAO)
Sternschildkröte	*Geochelone elegans elegans*	Indian star tortoise, starred tortoise
Sternseepocke	*Chthamalus stellatus*	star barnacle
Sterntaucher	*Gavia stellata*	red-throated diver, red-throated loon
Sternwürmer, Spritzwürmer, Sipunculiden	Sipunculida	peanut worms, sipunculids, sipunculans
Stichling> Gemeiner Stichling, Dreistachliger Stichling	*Gasterosteus aculeatus*	three-spined stickleback
Stichlinge	Gasterosteidae	sticklebacks
Stichlingsartige, Stichlingverwandte	Gasterosteiformes	sticklebacks (and sea horses)
Stiefmütterchen-Perlmutterfalter, Mittlerer Perlmuttfalter	*Argynnis niobe/ Fabriciana niobe*	niobe fritillary
Stieglitz	*Carduelis carduelis*	goldfinch, European goldfinch
Stieglitzsalmler, Sternflecksalmler	*Pristella maxillaris/ Pristella riddlei*	X-ray tetra (FAO), X-ray fish, pristella
Stieglitzvögel, Gimpel	Carduelinae	cardueline finches, goldfinches & crossbills
Stiel-Winkerkrabbe	*Uca stylifera*	stalked fiddler crab
Stielaugenfliegen	Diopsidae	stalk-eyed flies
Stielaugengarnelen*	Ogyrididae	long-eyed shrimps, longeye shrimps
Stielqualle	*Craterolophus convolvulus*	stalked jellyfish
Stielquallen, Becherquallen	Stauromedusae	stauromedusas
Stier-Fechterschnecke, Stier-Flügelschnecke	*Strombus taurus*	bull conch
Stierhai, Bullenhai, Gemeiner Grundhai	*Carcharhinus leucas*	bull shark

Stierkäfer	*Typhoeus typhoeus*	minotaur beetle
Stierkopfhaie, Hornhaie	*Heterodontus* spp.	hornsharks
Stierkopfhaie, Hornhaie & Doggenhaie	Heterodontidae	hornsharks, bullhead sharks, Port Jackson sharks
Stilettfliegen, Luchsfliegen	Therevidae	stiletto flies
Stinkdachse	*Mydaus* spp.	stink badgers
Stinkender Ankerschwamm, Riesenschwamm	*Geodia cydonium/* *Geodia gigas*	giant sponge*
Stinknatter	*Elaphe carinata*	stink snake
Stinkschwamm, Teichschwamm (Geweihschwamm)	*Spongilla lacustris*	pond sponge, freshwater sponge
Stint (Spierling, Wanderstint)	*Osmerus eperlanus*	smelt, European smelt (FAO)
Stintdorsch	*Trisopterus esmarki*	Norway pout
Stinte	Osmeridae	smelts
Stirnlappen-Basilisk	*Basiliscus plumifrons*	green basilisk, plumed basilisk, double-crested basilisk
Stockälchen, Stengelälchen, Luzerneälchen, (Rübenkopfälchen)	*Ditylenchus dipsaci*	lucerne stem nematode, stem-and-bulb eelworm, stem and bulb nematode, stem nematode, (potato tuber eelworm)
Stockente	*Anas platyrhynchos*	mallard
Stöcker, Schildmakrele, Bastardmakrele	*Trachurus trachurus*	Atlantic horse mackerel (FAO), 8scad, maasbanker
Stockfisch, Kap-Hecht, Flachwasser-Seehecht	*Merluccius capensis*	Cape hake, stock fish, stockfish, stokvis (S. Africa), shallow-water Cape hake (FAO)
Stockrosen-Spitzmausrüssler	*Apion radiolus*	hollyhock weevil
Stör, Baltischer Stör	*Acipenser sturio*	common sturgeon (IUCN), sturgeon (FAO), Atlantic sturgeon, European sturgeon
Störche	Ciconiidae	storks
Störe & Löffelstöre	Acipenseriformes	sturgeons & sterlets & paddlefishes
Sträflings-Brasse, Schafskopf-Brasse	*Archosargus probatocephalus*	sheepshead
Sträflings-Doktorfisch	*Acanthurus triostegus*	convict tang, convict surgeonfish (FAO)
Strahlen-Dreikielschildkröte	*Geoclemys hamiltonii*	black pond turtle, spotted pond turtle
Strahlen-Plattschildkröte, Brasilianische Plattschildkröte	*Platemys radiolata*	Brazilian radiolated swamp turtle, Brazilian radiated swamp turtle
Strahlenauster*, Strahlenförmige Baumauster*	*Isognomon radiatus*	radial purse-oyster
Strahlende Sonnenuhrschnecke, Strahlende Sonnenuhr	*Philippia radiata*	radial sundial
Strahlenfeuerfisch	*Pterois radiata*	radial firefish (FAO), longhorn firefish, clearfin turkeyfish

Strahlenflosser	Actinopterygii	ray-finned bony fishes, rayfin fishes, actinopterygians, actinopts
Strahlenköpfe	Radiicephalidae	inkfishes
Strahlenkörbchen, Gemeine Trogmuschel	*Mactra cinerea/ Mactra stultorum*	white trough clam
Strahlenmücke	*Dilophus febrilis*	fever fly, blossom fly
Strahlennatter	*Elaphe radiata*	radiated rat snake, copperhead racer
Strahlige Tellmuschel	*Tellina radiata*	sunrise tellin (rising sun)
Strahlige Venusmuschel, Gemeine Venusmuschel	*Chamelea gallina/ Venus gallina*	striped venus, chicken venus
Strandasseln, Klippenasseln	Ligiidae	sea-slaters, rock lice
Strandasseln, Klippenasseln	*Ligia* spp.	slaters, sea-slaters, rock lice
Strandauster, Sandklaffmuschel, Sandmuschel, Große Sandklaffmuschel	*Arenomya arenaria*	sand gaper, soft-shelled clam, softshell clam, large-neck clam, steamer
Strandfliegen u.a.	Canaceridae/Canacidae	beach flies
Strandfliegen u.a.	Helcomyzidae	seabeach flies, helcomyzid flies
Strandfloh, Strandhüpfer	*Talitrus saltator*	sandhopper a.o., greater sandhopper
Strandflöhe, Strandhüpfer	Talitridae	sandhoppers
Strandgräber	*Bathyergus* spp.	dune mole-rats
Strandgrundel, Strandküling	*Pomatoschistus microps*	common goby
Strandhüpfer, Strandfloh	*Talitrus saltator*	sandhopper a.o., greater sandhopper
Strandkrabbe, Nordatlantik-Strandkrabbe	*Carcinus maenas*	green shore crab, green crab, North Atlantic shore crab
Strandküling, Strandgrundel	*Pomatoschistus microps*	common goby
Strandlachs*	*Leptobrama muelleri*	beachsalmon
Strandpieper	*Anthus petrosus*	rock pipit
Strandrose, Hafenrose	*Diadumene luciae*	orange-striped anemone
Strandschnecken	Littorinidae	winkles, periwinkles
Strandseeigel, Strand-Seeigel, Olivgrüner Strandigel	*Psammechinus miliaris*	shore sea urchin, shore urchin, purple-tipped sea urchin
Strandspringer	*Anurida maritima*	seashore springtail
Straßentaube	*Columba livia f. domestica*	domestic pigeon
Strauchkaninchen	*Sylvilagus bachmani*	brush rabbit
Strauchkoralle	*Gerardia savaglia*	bushy crust coral*
Strauchradspinne, Körbchenspinne	*Araneus redii/ Agalenatea redii*	bush orbweaver*, basketweaver*
Strauchratten, Degus	*Octodon* spp.	degus
Strauchs Wüstenrenner	*Eremias strauchi*	Strauch's racerunner
Strauchschrecke> Gewöhnliche Strauchschrecke	*Pholidoptera griseoaptera*	dark bushcricket

Strauß	*Struthio camelus* (Struthionidae)	ostrich
Straußen-Federpolyp, Straußen-Nesselfarn	*Aglaophenia struthionides*	ostrich plume hydroid
Straußenfeder-Weichkorallen	*Xenia* spp.	pulse corals, waving hand polyps
Straußenschnecken	Struthiolariidae	ostrich foot snails
Straußenvögel, Strauße, Laufvögel	Struthioniformes	ostriches
Straußwachtel	*Rollulus roulroul*	crested wood partridge
Streber	*Zingel streber*	streber
Streckerspinnen & Kieferspinnen	Tetragnathidae	thick-jawed spiders, large-jawed orb weavers, four-jawed spiders, elongated orb weavers, horizontal orb-web weavers
Streckfuß, Buchenrotschwanz, Rotschwanz	*Dasychira pudibunda/ Olene pudibunda*	pale tussock, red-tail moth
Streifen-Doktorfisch, Blaustreifen-Doktorfisch	*Acanthurus lineatus*	clown surgeonfish, blue-lined surgeonfish, lined surgeonfish (FAO)
Streifen-Erdfresser	*Gymnogeophagus rhabdotus*	stripefin eartheater
Streifen-Erdhörnchen	*Lariscus* spp.	Malaysian striped ground squirrels
Streifen-Glanzschnecke	*Nesovitrea hammonis*	rayed glass snail
Streifen-Hüpfmäuse, Birkenmäuse	*Sicista* spp.	birch mice
Streifen-Nussmuschel	*Nuculata sulcata*	furrowed nutclam
Streifen-Prachtschmerle, Zebra-Prachtschmerle	*Botia striata*	banded loach
Streifen-Schleimfisch*	*Chasmodes bosquianus*	striped blenny
Streifen-Süßlippe	*Pomadasys stridens*	lined piggy
Streifen-Waldmäuse	*Hybomys* spp.	back-striped mice, hump-nosed mice
Streifenanemone*	*Haliplanella luciae*	striped anemone
Streifenassel*	*Philoscia muscorum*	common striped woodlouse
Streifenbackenhörnchen, Chipmunk (Hbz.)	*Tamias striatus*	eastern American chipmunk
Streifenbarsche	*Roccus* spp.	striped basses
Streifenbasilisk	*Basiliscus vittatus*	brown basilisk
Streifenbeutelmäuse	*Phascolosorex* spp.	black-stripe marsupial "mice"
Streifenbeutler, Streifenphalanger, Streifenkletterbeutler	*Dactylopsila* spp.	striped possums
Streifenbrassen, Brandbrassen	*Spondyliosoma cantharus*	black bream, black seabream (FAO), old wife
Streifenbülbül	*Pycnonotus striatus*	striated green bulbul
Streifendelphin, Blau-Weißer Delphin	*Stenella coeruleoalba*	striped dolphin, blue-white dolphin, euphrosyne dolphin
Streifenfüsilier	*Caesio tile*	bartail fusilier
Streifengans	*Anser indicus*	bar-headed goose
Streifengelbeule	*Xanthia citrago*	striped sallow

Streifengnu	*Connochaetes taurinus*	blue wildebeest (brindled gnu, white-bearded wildebeest)
Streifengrundel	*Gobius vittatus*	striped goby
Streifenhechtling	*Aplocheilus lineatus*	lined panchax
Streifenhörnchen	*Eutamias* spp.	Siberian and Western American chipmunks
Streifenhyäne	*Hyaena hyaena*	striped hyena
Streifenkauz	*Strix varia*	barred owl
Streifenkiwi	*Apteryx australis*	brown kiwi
Streifenkreuzspinne	*Mangora acalypha*	lined orbweaver
Streifenmakrele	*Carangoides ferdau*	horse mackerel, Ferdau's trevally, blue kingfish, blue trevally (FAO)
Streifenohreule	*Otus brucei*	striated Scops owl
Streifenprachtschmerle, Zebraprachtschmerle	*Botia striata*	banded loach
Streifenprinie	*Prinia gracilis*	graceful warbler
Streifenreusenschnecke	*Nassarius consensus*	striate nassa
Streifenroller	*Arctogalidia trivirgata*	small-toothed palm civet, three-striped palm civet
Streifensalamander	*Stereochilus marginatus*	many-lined salamander
Streifenschakal	*Canis adustus*	side-striped jackal
Streifenschwanz-Waran, Schwanzstrichwaran	*Varanus caudolineatus*	streak-tailed monitor, stripetail monitor
Streifenschwanz-Ziegelbarsch	*Malacanthus brevirostris*	banded blanquillo, quakerfish (FAO)
Streifenschwirl	*Locustella certhiola*	Pallas's grasshopper warbler
Streifenskink	*Eumeces fasciatus*	five-lined skink
Streifenskunk	*Mephitis mephitis*	striped skunk
Streifenwaldsänger	*Dendroica striata*	blackpoll warbler
Streifenwiesel, Libysches Streifenwiesel	*Poecilictis libyca*	North African striped weasel, Libyan striped weasel
Strichellori	*Eos reticulata*	blue-streaked lory
Strichelschwirl	*Locustella lanceolata*	lanceolated warbler
Strichfußanolis	*Anolis lineatopus*	stripefoot anole
Striegelmuschel, Gemeine Striegelmuschel	*Solecurtus strigillatus*	scraper clam, rosy razor
Stroben-Rindenlaus	*Pineus strombi*	Weymouth pine adelges
Strömer	*Leuciscus souffia*	souffie, vairone (FAO), telestes (U.S.),
Stromlinien-Panzerwels	*Corydoras arcuatus*	arched corydoras, skunk corydoras (FAO)
Strubbelige Fledermaus	*Centronycteris maximiliani*	shaggy-haired bat
Strudelwürmer, Turbellarien	Turbellaria	turbellarians, free-living flatworms
Strumpfbandfisch	*Lepidopus caudatus*	silver scabbardfish (FAO), ribbonfish, frostfish
Strumpfbandnattern, Bändernattern	*Thamnophis* spp.	American garter snakes, "water snakes"
Stubenfliege, Große Stubenfliege	*Musca domestica*	house fly

Stufige Venusmuschel	*Chione subimbricata*	stepped venus
Stülpnasen-Lanzenotter	*Porthidium nasutum/ Bothrops nasutus*	horned hognose pit viper, rainforest hognose viper
Stülpnasenotter	*Vipera latastei*	snub-nosed viper, Lataste's viper
Stumme Umberfische	*Menticirrhus* spp.	kingcroakers
Stummeldaumen-Fledermäuse	Furipteridae	smoky bats, thumbless bats
Stummelfüßer, Onychophoren	Onychophora	velvet worms, onychophorans
Stummelfußfrösche	*Atelopus* spp.	stubfoot frogs
Stummellerche	*Calandrella rufescens*	lesser short-toed lark
Stummelschwanz-Chamäleons	*Brookesia* spp.	stump-tailed chameleons, leaf chameleons
Stummelschwanz-Zwergtyrann	*Myiornis ecaudatus*	short-tailed pygmy tyrant
Stummelschwanzagutis	*Dasyprocta* spp.	agoutis
Stummelschwanzhörnchen, Biberhörnchen, Bergbiber	*Aplodontia rufa*	sewellel, mountain beaver
Stummelschwanzspitzmaus, Wühlspitzmaus	*Anourosorex squamipes*	mole shrew, Szechwan burrowing shrew
Stumpen-Seescheide, Stumpen-Ascidie	*Ascidia mentula*	stumpy sea-squirt*
Stumpfe Sumpfdeckelschnecke, Sumpfdeckelschnecke	*Viviparus viviparus*	river snail, common river snail
Stumpfe Erbsenmuschel	*Pisidium obtusale*	obtuse pea mussel
Stumpfe Mitra	*Mitra retusa*	blunt mitre
Stumpfe Strandschnecke, Stumpfkegelige Uferschnecke	*Littorina obtusata*	flat periwinkle, yellow periwinkle, northern yellow periwinkle
Stumpfe Tellmuschel	*Arcopagia crassa/ Tellina crassa*	blunt tellin
Stumpfkrokodil	*Osteolaemus tetraspis*	African dwarf crocodile, West African dwarf crocodile
Stumpfmuscheln, Sägezahnmuscheln, Dreiecksmuscheln, Dreieckmuschel, Sägezähnchen	Donacidae	wedge clams, donax clams (wedge shells)
Stumpfnase, Stumpfnasenaal, Parasitischer Aal	*Simenchelys parasitica*	snubnosed eel (FAO), snubnose parasitic eel
Stumpfnasen-Dornrückenaal	*Notacanthus chemnitzii*	snub-nosed spiny-eel
Stumpfnasen-Leopardleguan, Stumpfnasenleguan	*Gambelia silus*	bluntnose leopard lizard, blunt-nosed leopard lizard
Stumpfnasen-Schlangenaal, Kurzschnauziger Schlangenaal	*Echelus myrus*	blunt-nosed snake eel
Stumpfnasen-Stechrochen	*Dasyatis sayi*	bluntnose stingray
Stumpfnasenaale, Parasitische Aale, Schmarotzeraale	Simenchelyidae/ Simenchelyinae	snubnose parasitic eels
Stumpfnasenaffen	*Rhinopithecus* spp.	snub-nosed langurs
Stumpfnasenleguan, Stumpfnasen-Leopardleguan	*Gambelia silus*	bluntnose leopard lizard, blunt-nosed leopard lizard

Stumpfschnecke	*Rumina decollata*	decollate snail
Stumpfschwanzchamäleons	*Rhampholeon* spp.	stumptail chameleons
Stupsnasen-Flugfisch*	*Prognichthys gibbifrons*	bluntnose flyingfish
Sturmmöve	*Larus canus*	common gull
Sturmschwalbe	*Hydrobates pelagicus*	storm petrel
Sturmschwalben	Hydrobatidae	storm-petrels
Sturmvögel	Procellariidae	petrels, shearwaters
Sturzbachente	*Merganetta armata*	torrent duck
Sturzbachtachuri	*Serpophaga cinerea*	torrent tyrannulet
Stutz-Gelenkschildkröte	*Kinixys homeana*	Home's hingeback tortoise
Stutzechse, Tannenzapfenechse	*Trachydosaurus rugosus/ Trachysaurus rugosus*	stump-tailed skink
Stutzkäfer	Histeridae	hister beetles, clown beetles
Stutzkäfer> Gemeiner Stutzkäfer	*Saprinus semistriatus*	common hister beetle (Central European)
Stylopiden	Stylopidae (Strepsiptera)	parasitic beetles
Süd-Anakonda, Gelbe Anakonda, Paraguay-Anakonda	*Eunectes notaeus*	yellow anaconda
Süd-Bachsalamander	*Desmognathus auriculatus*	southern dusky salamander
Süd-Blattnasennatter	*Langaha alluaudi*	southern leafnose snake
Süd-Huftiere	Notoungulata	notoungulates
Süd-Schmierlaus	*Pseudococcus maritimus*	grape mealybug
Süd-Seehechte	Macruronidae	southern hakes
Südafrikanische Buschechse	*Heliobolis* spp.	bushveld lizards
Südafrikanische Lähmezecke*	*Ixodes rubicundus*	South African paralysis tick
Südafrikanische Landschildkröten	*Psammobates* spp.	South African land tortoises
Südafrikanische Langohrmaus	*Malacothrix typica*	gerbil mouse, long-eared mouse
Südafrikanische Korallenschlange	*Aspidelaps lubricus*	South African coral snake
Südafrikanische Riesen-Dreiecksmuschel	*Donax serra*	white mussel, giant South African wedge clam
Südafrikanische Sardelle	*Engraulis capensis*	Southern African anchovy
Südafrikanische Sardine	*Sardinops ocellatus*	South African pilchard
Südafrikanische Turbanschnecke, Südafrikanischer Turban	*Turbo sarmaticus*	South African turban, giant pearlwinkle
Südafrikanischer Galago	*Galago moholi*	southern lesser bush baby, South African galago
Südafrikanischer Hummer, Kap-Hummer	*Homarinus capensis*	Cape lobster
Südafrikanischer Hundshai, Gefleckter Marderhai	*Triakis megalopterus*	spotted gullyshark
Südafrikanischer Spaltenskorpion	*Hadogenes troglodytes*	South African rock scorpion
Südafrikanischer Spießbock	*Oryx gazella*	gemsbock
Südafrikanisches Erdhörnchen	*Geosciurus inauris/ Xerus inauris*	South African ground squirrel
Südalpen-Sattelschrecke	*Ephippiger terrestris*	Alpine saddle-backed bushcricket
Südamerikanische Baumkröte	*Bufo typhonius*	Neotropical leaf toad

Südamerikanische Baumschwammkäfer, Südamerikanischer Pilzkäfer	Erotylidae	pleasing fungus beetles
Südamerikanische Blattohrmäuse	Auliscomys spp.	South American leaf-eared mice
Südamerikanische Braunspinne	Loxosceles laeta	South American brown spider
Südamerikanische Dickkopfnatter	Dipsas indica	Indian snail-eating snake
Südamerikanische Erdschildkröte, Guayana-Erdschildkröte	Rhinoclemmys punctularia	spot-legged turtle, spotted-legged turtle
Südamerikanische Feldmäuse	Akodon spp.	South American field mice, grass mice
Südamerikanische Felsenratte	Aconaemys fuscus	rock rat
Südamerikanische Fischratte u.a.	Anotomys leander	fish-eating rat a.o., aquatic rat a.o.
Südamerikanische Fischratte u.a., Nord-Equador-Fischratte	Neusticomys monticolus	fish-eating rat a.o., aquatic rat a.o.
Südamerikanische Fischratten u.a.	Daptomys spp.	fish-eating rats a.o., aquatic rats a.o.
Südamerikanische Flachkopf-Fledermaus	Neoplatymops mattogrossensis	South American flat-headed bat
Südamerikanische Grabmäuse	Oxymycterus spp.	burrowing mice
Südamerikanische Hornfrösche	Ceratophrys spp.	common horned frogs, South American horned frogs, wide-mouthed toads
Südamerikanische Klettermäuse*	Rhipidomys spp.	South American climbing mice
Südamerikanische Korallenschlange, Kobra-Korallenschlange	Micrurus frontalis	southern coral snake
Südamerikanische Landschildkröten	Chelonoidis spp.	South American tortoises
Südamerikanische Riesenratten	Kunsia spp.	South American giant rats
Südamerikanische Riesenschnecke	Strophochilus oblongus	giant South American snail
Südamerikanische Rotnasenratten*	Bibimys spp.	crimson-nosed rats
Südamerikanische Sardine	Sardinops sagax	Chilean pilchard, Pacific sardine, Peruvian sardine, South American pilchard (FAO)
Südamerikanische Stachelratten*	Neacomys spp.	bristly mice, spiny mice rats
Südamerikanische Sumpfratte*	Holochilus spp.	web-footed rats, marsh rats
Südamerikanische Süßwasserstechrochen	Potamotrygonidae	river stingrays
Südamerikanische Wasserratte	Scapteromys tumidus	water rat
Südamerikanische Wasserratten	Nectomys spp.	Neotropical water rats
Südamerikanischer Flussotter	Lutra provocax	southern river otter

Südamerikanischer Königs-Umberfisch	*Macrodon ancylodon*	king weakfish
Südamerikanischer Lungenfisch	*Lepidosiren paradoxa*	South American lungfish
Südamerikanischer Meerotter	*Lutra felina*	marine otter (sea cat)
Südamerikanischer Ochsenfrosch, Fünffingriger Pfeiffrosch	*Leptodactylus pentadactylus*	South American bullfrog
Sudangoldsperling	*Auripasser luteus*	Sudan golden sparrow
Südatlantische Kammmuschel	*Pecten sulcicostatus*	South Atlantic scallop
Südchinesischer Vielbindenbungar, Vielbindenkrait	*Bungarus multicinctus*	many-banded krait
Süddelphin, Schwarzkinndelphin, Peales Delphin	*Lagenorhynchus australis*	Peale's dolphin, Peale's black-chinned dolphin, blackchin dolphin, southern dolphin
Südeuropäischer Gürtelskolopender	*Scolopendra cingulata*	southern girdled scolopendra*
Südfeldmaus, Epirus-Feldmaus	*Microtus epiroticus*	sibling vole
Südflundern	Achiropsettidae	southern flounders
Südfrösche	Leptodactylidae	southern frogs, tropical frogs, leptodactylids
Südindischer Stachelbilch	*Platacanthomys lasiurus*	spiny dormouse
Südkaper, Südlicher Glattwal	*Eubalaena australis/ Balaena glacialis australis*	southern right whale
Südliche Alligatorschleiche	*Gerrhonotus multicarinatus/ Elgaria multicarinata*	southern alligator lizard
Südliche Beißschrecke	*Platycleis affinis*	tuberous bushcricket, southern bushcricket
Südliche Binsenjungfer	*Lestes barbatus*	southern European emerald damselfly
Südliche Buntschrecke	*Poecilimon ornatus*	ornate bushcricket
Südliche Erdotter	*Atractaspis bibroni*	Bibron's mole viper, Bibron's burrowing viper
Südliche Feuerameise	*Solenopsis xyloni*	southern fire ant
Südliche Flusskrebse	Parastacidae	southern crawfish
Südliche Hakennatter	*Heterodon simus*	southern hognose snake
Südliche Hausstechmücke	*Culex quinquefasciatus*	southern house mosquito
Südliche Heidelibelle	*Sympetrum meridionale*	meridional sympetrum*, southern European sympetrum*
Südliche Kammmuschel	*Chlamys australis*	austral scallop
Südliche Mosaikjungfer	*Aeshna affinis*	southern European hawker
Südliche Quahog-Muschel	*Mercenaria campechiensis*	southern quahog
Südliche Rosa Geißelgarnele, Senegal-Garnele	*Farfantepenaeus notialis/ Penaeus notialis*	southern pink shrimp, candied shrimp
Südliche Sandfische	Leptoscopidae	southern sandfishes
Südliche Schwertschrecke	*Conocephalus conocephalus*	southern cone-head
Südliche Seebären	*Arctocephalus* spp.	southern fur seals
Südliche Seespinne	*Jacquinotia edwardsii*	southern spider crab
Südliche Strauchschrecke	*Pholidoptera fallax*	Fischer's bushcricket
Südliche Taschenratte	*Thomomys homomys*	southern pocket gopher

Südlicher Blauflossenthun	*Thunnus maccoyii*	southern bluefin tuna
Südlicher Blaupfeil	*Orthetrum brunneum*	southern European skimmer
Südlicher Geigenrochen	*Rhinobatos percellens*	fiddlerfish (FAO), chola guitarfish
Südlicher Glattdelphin	*Lissodelphis peronii*	southern right-whale dolphin, Delfin liso
Südlicher Grillenfrosch	*Acris gryllus gryllus*	southern cricket frog
Südlicher Guereza, Bärenstummelaffe (Hbz. Scheitelaffe)	*Colobus polykomos*	king colobus, western pied colobus (western black-and-white colobus)
Südlicher Heufalter	*Colias australis/ Colias alfacariensis*	Berger's clouded yellow
Südlicher Krill, Antarktischer Krill	*Euphausia superba*	whale krill, Antarctic krill
Südlicher Kugelfisch	*Sphoeroides nephelus*	southern puffer
Südlicher Ohrwurm	*Anisolabis annulipes*	ring-legged earwig, spotted earwig
Südlicher Schlamm-Maulwurf (Schildfuß)	*Limifossor fratula* (Aplacophora)	southern mole glistenworm
Südlicher Schnapper	*Lutjanus purpureus*	southern red snapper
Südlicher Seehecht	*Merluccius australis*	southern hake
Südlicher Taubleguan	*Holbrookia subcaudalis*	southern earless lizard
Südlicher Vierzahnwal	*Berardius arnuxii*	southern giant bottlenosed whale, Arnoux's beaked whale, four-toothed whale
Südlicher Warzenbeißer	*Decticus albifrons*	white-faced bushcricket, Mediterranean wart-biter
Südlicher Weißrochen, Ostatlantischer Weißrochen, Spitzrochen, Bandrochen	*Raja alba*	white skate (FAO), bottle-nosed skate
Südlicher Zwergbutt	*Phrynorhombus regius*	Eckström's topknot
Südliches Wurzelgallenälchen	*Meloidogyne marioni/ M. incognita*	southern root-knot nematode
Südopossum	*Didelphis marsupialis*	southern opossum
Südostasiatische Drehhornantilope	*Pseudonovibos spiralis*	linh-duong
Südostasiatische Fleckfiebermilbe (Südostasiatische Tsutsugamushi-Erntemilbe)	*Trombicula deliensis/ Leptotrombidium deliensis*	Southeast-Asian scrub typhus chigger mite*
Südostasiatische Seeschlange	*Enhydrina schistosa*	beaked sea snake
Südostasiatischer Reisfrosch, Südostasiatischer Warzenfrosch	*Rana limnocharis/ Limnonectes limnocharis*	Boie's wart frog
Südöstliche Beißschrecke	*Platycleis stricta*	Italian bushcricket
Südöstlicher Sandlaufkäfer*	*Cicindela arenaria*	southeastern European tiger beetle*
Südostpazifische Süßlippe	*Isacia conceptionis*	Southeast Pacific grunt
Südseeboas	*Candoia* spp.	Pacific boas
Südseedrossel	*Turdus poliocephalus*	island thrush
Südseegrasmücken	Acanthizidae	thornbills, whitefaces
Südseeschnäpper, Südseesänger	Eopsaltriidae	Australasian robins
Südwestasiatischer Gartenschläfer	*Eliomys melanurus*	Southwest-Asian garden dormouse

Südwesteuropäischer Näsling, Savetta	*Chondrostoma toxostoma*	soiffe (*Fr.*), soffie (*Fr.*)
Suira	*Euryzygomatomys spinosus*	guiara
Sulawesi-Riesenratte	*Lenomys meyeri*	trefoil-toothed giant rat, Sulawesi giant rat
Sulawesi-Spitzmausratten	*Tateomys* spp.	Sulawesi shrew rats
Sultansmeise	*Melanochlora sultanea*	sultan tit
Sultanspecht	*Chrysocolaptes lucidus*	greater flame-backed woodpecker
Sumatra-Kaninchen	*Nesolagus netscheri*	Sumatra short-eared rabbit, Sumatran rabbit
Sumatrabarbe, Viergürtelbarbe	*Barbus tetrazona/ Puntius tetrazona*	Sumatra barb (FAO), tiger barb
Sumatranashorn	*Dicerorhinus sumatrensis*	Sumatran rhinoceros, hairy rhinoceros
Sumichrasti-Dämmerungsratte, Dämmerungsratte	*Nyctomys sumichrasti*	vesper rat, Sumichrast's vesper rat
Sumpf-Elritze	*Phoxinus percnurus*	swamp minnow
Sumpf-Erdschildkröten, Amerikanische Erdschildkröten	*Rhinoclemmys* spp.	neotropical wood turtles, American terrapins
Sumpf-Federkiemenschnecke	*Valvata macrostoma*	bog valve snail
Sumpf-Grashüpfer	*Chorthippus montanus*	marsh-meadow grasshopper
Sumpf-Heidelibelle	*Sympetrum depressiusculum*	eastern European sympetrum
Sumpf-Reisratte	*Oryzomys palustris*	marsh rice rat
Sumpf-Schlammschnecke	*Lymnaea palustris*	marsh pondsnail, marsh snail
Sumpf-Strandschnecke*, Sumpf-Uferschnecke*	*Littorina irrorata*	marsh periwinkle
Sumpf-Windelschnecke	*Vertigo antivertigo*	marsh whorl snail
Sumpfammer	*Melospiza georgiana*	swamp sparrow
Sumpfantilope, Wasserkudu, Sitatunga	*Tragelaphus spekei*	sitatunga
Sumpfassel	*Ligidium hypnorum*	moss slater
Sumpfbiber, Nutria	*Myocastor coypus*	nutria, coypu
Sumpfbuschsänger	*Bradypterus baboecalus*	African sedge warbler
Sumpfdeckelschnecken	*Viviparus* spp.	river snails
Sumpfdeckelschnecke, Stumpfe Sumpfdeckelschnecke	*Viviparus viviparus*	river snail, common river snail
Sumpfdeckelschnecken, Flussdeckelschnecken	Viviparidae	river snails
Sumpffliegen, Scheufliegen, Dunkelfliegen	Heleomyzidae	heleomyzid flies
Sumpffrosch	*Rana palustris*	pickerel frog
Sumpffrösche	*Limnodynastes* spp.	swamp frogs
Sumpfgrille	*Pteronemobius heydenii*	marsh cricket
Sumpfhirsch	*Odocoileus dichotomus/ Blastocerus dichotomus*	marsh deer
Sumpfhornklee-Widderchen, Feuchtwiesen-Blutströpfchen	*Zygaena trifolii*	five-spot burnet

Sumpfichneumon, Sumpfmanguste, Wassermanguste	*Atilax paludinosus*	marsh mongoose, water mongoose
Sumpfkäfer	Helodidae	marsh beetles
Sumpfkaninchen	*Sylvilagus palustris*	marsh rabbit
Sumpfkrebs, Galizier	*Astacus leptodactylus*	long-clawed crayfish
Sumpfkrokodil	*Crocodylus palustris*	mugger crocodile, mugger, marsh crocodile, broad-snouted crocodile
Sumpfläufer	*Limicola falcinellus*	broad-billed sandpiper
Sumpfmanguste, Sumpfichneumon, Wassermanguste	*Atilax paludinosus*	marsh mongoose, water mongoose
Sumpfmaus, Nordische Wühlmaus	*Microtus oeconomus*	root vole
Sumpfmeerkatze, Schwarzgrüne Meerkatze	*Allenopithecus nigroviridis*	Allen's monkey, Allen's swamp monkey (blackish-green guenon)
Sumpfmeise	*Parus palustris*	marsh tit
Sumpfohreule	*Asio flammeus*	short-eared owl
Sumpfpflanzenlaus	*Rhopalosiphum nymphaeae*	waterlily aphid
Sumpfräuber	*Dolichonabis limbatus*	marsh damsel bug
Sumpfrohrsänger	*Acrocephalus palustris*	marsh warbler
Sumpfschildkröten	Emydidae	pond terrapins (pond and river turtles), emydid turtles
Sumpfschrecke	*Stethophyma grossum/ Mecostethus grossus*	large marsh grasshopper
Sumpfschwalbe	*Tachycineta bicolor*	tree swallow
Sumpfspitzmaus, Kleine Wasserspitzmaus	*Neomys anomalus*	Miller's water shrew
Sumpfwallaby	*Wallabia bicolor*	swamp wallaby
Sumpfzaunkönig	*Cistothorus palustris*	marsh wren, long-billed marsh wren
Sunda-Gavial	*Tomistoma schlegelii*	false gharial, Malayan gharial
Sunda-Koboldmaki	*Tarsius bancanus*	western tarsier, Horsfield's tarsier
Sundahörnchen	*Sundasciurus* spp.	Sunda tree squirrels
Suniböckchen	*Neotragus moschatus*	suni
Suppenschildkröte	*Chelonia mydas*	green turtle (rock turtle, meat turtle)
Surinam-Kauri	*Propustularia surinamensis*	Suriname cowrie
Surinamperlfisch	*Geophagus surinamensis*	Surinam pearl cichlid
Surinamschabe, Gewächshausschabe	*Pycnoscelis surinamensis*	Surinam cockroach
Süße Napfschnecke	*Patelloida saccharina*	Pacific sugar limpet
Suslik (Hbz.), Parry-Ziesel, Langschwanzziesel	*Spermophilus undulatus*	Arctic ground squirrel
Süßlippen, Weichlipper	Plectorhynchidae/ Gaterinidae	sweetlips
Süßwasser-Grundel	*Gobionellus shufeldti*	freshwater goby
Süßwasser-Muräne	*Gymnothorax tile*	freshwater moray eel

Süßwasser-Sägefisch	*Pristopsis leichhardti*	freshwater sawfish
Süßwasser-Schleimfisch, Fluss-Schleimfisch	*Blennius fluviatilis/ Lipophrys fluviatilis/ Salaria fluviatilis*	freshwater blenny (FAO), river blenny
Süßwasser-Stint	*Hypomesus olidus*	pond smelt
Süßwasserbryozoen, Armwirbler	Phylactolaemata/ Lophopoda	phylactolaemates, "covered throat" bryozoans, freshwater bryozoans
Süßwassergarnelen	Atyidae	atyid shrimps (freshwater shrimps)
Süßwasserkrabben	Potamidae	freshwater crabs
Süßwassermeduse, Süßwasserqualle	*Craspedacusta sowerbyi/ Microhydra sowerbyi (polyp = Microhydra ryderi)*	freshwater jellyfish (Regent's Park medusa)
Süßwassermilben	Hydrachnellae/ Hydrachnidia	freshwater mites
Süßwassernematode u.a.	*Trilobus gracilis*	freshwater nematode a.o.
Swiftfuchs (Hbz. Kitfuchs, Swiftfuchs)	*Vulpes velox*	swift fox
Sydney Felsenauster, Australische Auster	*Saccostrea cuccullata*	Sydney cupped oyster
Symmetrische Hirnkoralle, Gewöhnliche Hirnkoralle	*Diploria strigosa*	symmetrical brain coral
Syrische Honigbiene	*Apis mellifera syriaca*	Syrian honey bee
Syrische Schaufelkröte	*Pelobates syriacus*	eastern European spadefoot, Syrian spadefoot
Syrischer Goldhamster	*Mesocricetus auratus*	golden hamster

Tabakblasenfuß, Zwiebelblasenfuß	*Thrips tabaci*	onion thrips, potato thrips
Tabakmotte, Kakao-Motte, Kakaomotte, Speichermotte, Heumotte	*Ephestia elutella*	tobacco moth, cocoa moth, chocolate moth, flour moth, warehouse moth, cinereous knot-horn
Tabasco-Klappschildkröte	*Kinosternon acutum*	Tabasco mud turtle
Tabasco-Schildkröte	*Dermatemys mawii*	Central American river turtle
Tafelente	*Aythya ferina*	pochard
Tagfalter	Rhopalocera	butterflies & skippers
Taggeckos	*Phelsuma* spp.	day geckos
Taghafte	Hemerobiidae	brown lacewings
Tagkurzflügler	*Aleochara* spp.	small-headed rove beetles
Tagpfauenauge	*Inachis io/ Nymphalis io*	peacock moth, peacock
Tagpfauenauge> Amerikanisches Tagpfauenauge	*Automeris io*	io moth
Tagschläfer	Nyctibiidae	potoos
Taguan	*Petaurista petaurista*	red giant flying squirrel
Tahre	*Hemitragus* spp.	tahrs
Taigazecke	*Ixodes persulcatus*	Taiga tick
Taipan	*Oxyuranus scutulatus/ Oxyuranus scutellatus*	taipan
Takahe	*Porphyrio mantelli*	takahe
Takin, Rindergemse, Gnuziege	*Budorcas taxicolor*	takin
Talang	*Scomberoides commersonianus*	talang queenfish
Talerfisch, Hochrückensalmler	*Ctenobrycon spilurus*	silver tetra
Talgfollikelmilbe*	*Demodex brevis*	lesser follicle mite, sebaceous follicle mite
Tamanduas	*Tamandua* spp.	lesser anteaters, tamanduas
Tamarau, Mindoro-Büffel	*Bubalus mindorensis*	tamaraw, taumarau
Tamariskengrasmücke	*Sylvia mystacea*	Ménétries's warbler
Tammarwallaby, Derbywallaby	*Macropus eugenii*	tammar wallaby, dama wallaby
Tampan-Zecke	*Ornithodoros moubata*	tampan tick, eyeless tampan
Tana River-Rotstummelaffe	*Procolobus rufomitratus rufomitratus*	Tana River red colobus
Tang-Seespinne*	*Libinia erinacea*	seagrass spider crab
Tangaren	Thraupidae	tanagers, Neotropical honeycreepers, seedeaters, flower-piercers
Tangbarsch	*Brachyistius frenatus*	kelp perch
Tangbarsch*	*Paralabrax clathratus*	kelp bass
Tanger-Winkerkrabbe	*Uca tangeri*	Moroccan fiddler crab
Tangfliegen	*Coelopa* spp.	seaweed flies

Tangfliegen	Coelopidae	seaweed flies
Tangkrabbe	*Acanthonyx lunulatus*	surf crab*
Tangrose	*Sagartia rhododactylos*	kelp-rose anemone*
Tannenbäumchen-Röhrenwurm, Weihnachtsbaum-Röhrenwurm	*Spirobranchus giganteus*	spiral-gilled tubeworm, Christmas tree worm
Tannengallläuse	Adelgidae	conifer aphids (pine and spruce aphids)
Tannenhäher	*Nucifraga caryocatactes*	nutcracker
Tannenmaus	*Phenacomys longicaudus*	red tree vole
Tannenmeise	*Parus ater*	coal tit
Tannenmoos, Moostanne, Meertanne	*Abietinaria abietina*	sea fir
Tannennadel-Gallmücke	*Paradiplosis abietis*	fir-needle gall midge
Tannensamenwespe	*Megastigmus suspectus*	fir seed wasp
Tannenstreckfuß	*Dasychira abietis*	fir-and-spruce tussock
Tannentrieblaus, Weißtannen-Trieblaus, Nordmannstannen-Trieblaus	*Dreyfusia nordmanniana*/ *Adelges nordmanniana*	silver fir migratory adelges
Tannenzapfenechse, Stutzechse	*Trachydosaurus rugosus*/ *Trachysaurus rugosus*	stump-tailed skink
Tannenzapfenfische	Monocentridae	pineapple fishes, pinecone fishes
Tanrek> Großer Tanrek	*Tenrec ecaudatus*	tenrec
Tanreks, Borstenigel	Tenrecidae	tenrecs
Tantalus-Meerkatze	*Cercopithecus tantalus*	tantalus monkey
Tanzfliegen (Rennfliegen)	Empididae	dance flies
Tapetenmotte	*Trichophaga tapetzella*	tapestry moth
Tapezierspinne	*Atypus affinis*	purse-web spider
Tapezierspinnen	Atypidae	purse-web spiders
Tapire	*Tapirus* spp.	tapirs
Tapirfisch, Elefanten-Rüsselfisch	*Gnathonemus petersii*	Peter's worm-jawed mormyrid, Peter's elephant nose, elephantnose fish (FAO)
Tapirrüsselfisch	*Mormyrus kannume*	elephantsnout fish
Tarantel	*Lycosa narbonensis*	European tarantula
Tarantel> Apulische Tarantel	*Lycosa tarentula*	Apulian tarantula
Taranteln, Echte Vogelspinnen, Eigentliche Vogelspinnen, Buschspinnen	Theraphosidae/ Aviculariidae	tarantulas & bird spiders
Tardigraden (*sg* Tardigrad *m*), Bärentierchen, Bärtierchen	Tardigrada	tardigrades, water bears
Tarnfarben-Kaninchenfisch	*Siganus sutor*	shoemaker spinefoot
Tarnglattfuß	*Nematomenia corallophila* (Aplacophora)	Mediterranean coral solenogaster*
Tarpune	Megalopidae (Elopiformes)	tarpons
Tarsenspinner, Fußspinner, Embien	Embioptera	webspinners, footspinners, embiids

Taschenflügel-Fledermäuse	*Saccopteryx* spp. u.a.	white-lined bats
Taschengallen-Birkenlaus	*Anuraphis farfarae*	pear-coltsfoot aphid
Taschenkrebs	*Cancer pagurus*	European edible crab
Taschenkrebse	Cancridae	rock crabs, edible crabs
Taschenmäuse	Heteromyidae	pocket mice & kangaroo rats and mice
Taschenmäuse> Eigentliche Taschenmäuse	*Perognathus* spp.	pocket mice
Taschenmesser-Muschel, Schotenförmige Schwertmuschel, Schotenförmige Messerscheide, Schotenmuschel	*Ensis siliqua*	pod razor clam, giant razor clam
Taschenmessermuschel	*Pharus legumen*	
Taschenratten, Nordamerikanische Taschenratten	*Thomomys* spp.	pocket gophers
Taschenspringer, Kängururatten	*Dipodomys* spp.	kangaroo rats, California kangaroo rats
Tasmanischer Beuteldachs	*Perameles gunnii*	Tasmanian long-nosed bandicoot
Tasmanischer Schnabelwal, Shepherd-Wal	*Tasmacetus shepherdi*	Shepherd's beaked whale, Tasman whale
Tasmanischer Sumpffrosch	*Limnodynastes tasmaniensis*	spotted grass frog
Tasmanischer Tiger, Beutelwolf	*Thylacinus cynocephalus*	thylacine, Tasmanian wolf, Tasmanian tiger
Tates Dreispitznäsige Fledermaus, Dreispitz-Fledermaus	*Aselliscus* spp.	Tate's trident-nosed bats
Tates Spitzmausratte	*Tateomys rhinogradoides*	Tate's rat
Tatra-Kleinwühlmaus	*Pitymys tatricus*	Tatra pine vole
Tattersall Sifaka, Goldkamm-Sifaka	*Propithecus tattersalli*	golden-crowned sifaka, Tattersall's sifaka
Taubagamen	*Cophotis* spp.	prehensile-tailed lizards
Tauben	Columbidae	pigeons & doves
Tauben-Fassschnecke	*Tonna perdix*	partridge tun, Pacific partridge tun
Taubenlausfliege	*Pseudolynchia canariensis*	pigeon fly
Taubenschnecken, Täubchenschnecken	Pyrenidae/ Columbellidae	dove snails (dove shells)
Taubenschwänzchen	*Macroglossum stellatarum*	hummingbird hawkmoth
Taubenvögel	Columbiformes	doves & pigeons and allies
Taubenwanze, Geflügelwanze	*Cimex colombarius*	pigeon bug, fowl bug
Taubenzecke	*Argas reflexus*	pigeon tick
Taubleguane	*Holbrookia* spp.	earless lizards
Taubwaran	Lanthanotidae	earless monitor
Taucherschnecke, Holzboot	*Scaphander lignarius/ Bulla lignaria*	woody canoe-bubblesnail
Tauchkäfer*	*Dytiscus* spp.	diving beetles, diving water beetles

Taufliege, Essigfliege, Kleine Essigfliege	*Drosophila melanogaster*	vinegar fly, fruit fly
Taufliegen, Essigfliegen, Obstfliegen	Drosophilidae	"fruit flies", pomace flies, small fruit flies, ferment flies
Taumelkäfer	Gyrinidae	whirligig beetles
Taumelkäfer	*Gyrinus* spp.	whirligig beetles
Taurische Mauereidechse, Taurische Eidechse	*Podarcis taurica*	Balkan wall lizard
Tausenddollarfisch, Indischer Fähnchen-Messerfisch	*Notopterus chitala*	Indian featherback
Tausendfüßler, Tausendfüßer, Myriapoden	Myriapoda	millepedes (Br.), millipedes (U.S.) ("thousand-leggers"), myriapodians
Tausendpunkt-Mondschnecke, Tausendpunkt-Nabelschnecke, Fliegendreck-Mondschnecke	*Natica stercusmuscarum/ Naticarius stercusmuscarum*	fly-speck moonsnail, fly-specked moonsnail
Tausenfüßlerschnecke	*Lambis millepeda*	milleped spider conch
Tautog, Austernfisch	*Tautoga onitis*	tautog
Tayra	*Eira barbara*	tayra
Teich-Federkiemenschnecke, Gemeine Federkiemenschnecke	*Valvata piscinalis piscinalis*	valve snail, European stream valve snail, common valve snail, European stream valvata
Teich-Kugelmuschel, Haubenmuschel, Häubchenmuschel	*Sphaerium lacustre/ Musculium lacustre*	lake orb mussel, lake fingernailclam
Teich-Röhrichteule	*Archanara algae*	rush wainscot
Teichfledermaus	*Myotis dasycneme*	pond bat
Teichfrosch, Wasserfrosch	*Rana esculenta* (*Rana* kl. *esculenta*)	European edible frog
Teichhuhn	*Gallinula chloropus*	moorhen
Teichjungfern	Lestidae	spread-winged damselflies
Teichkäfer, Sumpfkäfer	Limnebiidae	minute moss beetles
Teichläufer	*Hydrometra stagnorum*	water measurer, marsh treader
Teichläufer	Hydrometridae	water measurers, marsh treaders
Teichläufer (Wasserläufer, Schneider)	Gerridae	pond skaters, water striders, pond skippers
Teichmolch, Streifenmolch	*Triturus vulgaris*	smooth newt
Teichmuscheln	*Anodonta* spp.	pond mussels*, floaters
Teichrohrsänger	*Acrocephalus scirpaceus*	reed warbler
Teichschwamm, Stinkschwamm, (Geweihschwamm)	*Spongilla lacustris*	pond sponge, freshwater sponge
Teichwasserläufer	*Tringa stagnatilis*	marsh sandpiper
Teleskop-Kardinalfisch	*Epigonus telescopus*	black cardinal fish, bulls-eye (FAO)

Teleskop-Täubchen	*Pyrene punctata*	punctate dovesnail, telescoped dovesnail
Teleskopfische (Riesenschwänze)	Giganturidae	telescopefishes
Teleskopschnecke	*Telescopium telescopium*	telescope snail, telescope creeper
Teleskopschnecken*	Potamididae	telescope snails, mud whelks
Tellerqualle	*Porpita umbella*	
Tellerschnecke> Gemeine Tellerschnecke, Flache Tellerschnecke	*Planorbis planorbis*	ramshorn, common ramshorn, margined ramshorn, ram's horn (margined trumpet shell)
Tellerschnecken	Planorbidae	ramshorn snails
Tellmuscheln, Plattmuscheln	Tellinidae	tellins, sunset clams (sunset shells)
Temminckkatze, Asiatische Goldkatze	*Profelis temmincki/ Felis temmincki*	Asian golden cat
Temminckstrandläufer	*Calidris temminckii*	Temminck's stint
Tempelfledermaus	*Taphozous melanopogon*	black-bearded tomb bat
Tempelschildkröte	*Hieremys annandalei*	yellow-headed temple turtle
Tempelschildkröten	*Hieremys* spp.	temple turtles
Tennessee-Höhlensalamander	*Gyrinophilus palleucus*	Tennessee cave salamander
Tentaculiferen, Tentakeltragende Rippenquallen (Ctenophora)	Tentaculifera	tentaculiferans, "tentaculates"
Teppichanemone, Haddons Anemone	*Stichodactyla haddoni*	Haddon's anemone, saddle anemone, carpet anemone
Teppichchamäleon	*Chamaeleo lateralis*	jewelled chameleon
Teppichhaie, Ammenhaie, Wobbegongs	Orectolobidae	wobbegongs, carpet sharks
Teppichkäfer, Kabinettkäfer	*Anthrenus scrophulariae*	carpet beetle
Teppichmuschel> Gemeine Teppichmuschel, Kleine Teppichmuschel, Getupfte Teppichmuschel	*Venerupis pullastra/ Venerupis saxatilis/ Venerupis perforans/ Venerupis senegalensis/ Tapes pullastra*	pullet carpetclam, pullet venus (pullet carpet shell)
Teppichmuscheln u.a.	*Tapes* spp.	carpet clams, venus clams
Teppichpython	*Morelia spilota variegata*	carpet python
Terekay-Schildkröte	*Podocnemis unifilis*	yellow-headed sideneck, yellow-spotted sideneck turtle, yellow-spotted Amazon River turtle, yellow-spotted river turtle (IUCN)
Terekwasserläufer	*Xenus cinereus*	Terek sandpiper
Terpug, Atka-Makrele	*Pleurogrammus monopterygius*	Atka mackerel
Terpuge, Atka-Makrelen	*Pleurogrammus* spp.	Atka mackerels
Tethys-Becherqualle, Schlanke Becherqualle	*Craterolophus tethys*	Tethys jelly*
Tetras	Tetragonopterinae	tetras
Teufelsangel, Teufelsangel-Erdfresser	*Geophagus jurupari*	demon eartheater (FAO), eartheating devilfish

Teufelsgarnele*	*Lucifer faxoni*	lucifer shrimp
Teufelskrabbe, Große Seespinne	*Maja squinado/ Maia squinado*	common spider crab, thorn-back spider crab
Teufelsrochen, Mantas	Mobulidae	mantas (manta rays & devil rays)
Texanische Gopherschildkröte	*Gopherus berlandieri*	Texas tortoise, Texas gopher tortoise
Texanische Landkartenschildkröte	*Graptemys versa/ Graptemys pseudogeographica versa*	Texas map turtle
Texas-Alligatorschleiche	*Gerrhonotus liocephalus*	Texas alligator lizard
Texas-Blindsalamander, Rathbunscher Brunnenmolch	*Typhlomolge rathbuni*	Texas blind salamander
Texas-Gelbsalamander, Texas-Höhlensalamander	*Eurycea neotenes*	Texas salamander
Texas-Klapperschlange	*Crotalus atrox*	western diamondback rattlesnake
Texas-Krallengecko	*Coleonyx brevis*	Texas banded gecko
Texas-Krötenechse	*Phrynosoma cornutum*	Texas horned lizard
Texas-Nachtschwalbe	*Chordeiles acutipennis*	lesser nighthawk
Texas-Rennechse	*Cnemidophorus gularis*	Texas spotted whiptail
Texas-Stachelleguan	*Sceloporus olivaceus*	Texas spiny lizard
Texas-Wurmschlange, Texas-Schlankblindschlange, New Mexico-Wurmschlange	*Leptotyphlops dulcis*	Texas blindsnake
Texasrochen	*Raja texana*	roundel skate
Texasspecht	*Picoides scalaris*	ladder-backed woodpecker
Textil-Olive	*Oliva textilina*	textile olive
Textilkäfer	*Anthrenus verbasci*	varied carpet beetle
Textorweber	*Ploceus cucullatus*	village weaver
Teydefink	*Fringilla teydea*	blue chaffinch
Theißblüte	*Palingenia longicauda*	long-tailed mayfly
Thekamöben, Beschalte Amöben	Testacea	testate amebas
Theklalerche	*Galerida theklae*	thekla lark
Thermometerhuhn	*Leipoa ocellata*	mallee fowl
Thersites-Fechterschnecke, Thersites-Flügelschnecke	*Strombus thersites*	Thersites conch
Thomas-Kleinwühlmaus, Balkan-Kurzohrmaus	*Microtus thomasi/ Pitymys thomasi*	Thomas's pine vole
Thomas-Paramomäuse	*Thomasomys* spp.	Thomas's paramo mice
Thomassets Seychellenfrosch	*Nesomantis thomasseti*	Thomasset's Seychelles frog, Seychelle rock frog
Thomsongazelle	*Gazella thomsonii*	Thomson's gazelle
Thonine, Falscher Bonito, Kleiner Thonine	*Euthynnus alletteratus*	little tunny (FAO), little tuna, mackerel tuna, bonito
Thors Lammzunge, Thors Lammbutt	*Arnoglossus thori*	Thor's scaldfish (FAO), Grohman's scaldfish
Thorshühnchen	*Phalaropus fulicarius*	grey phalarope
Thracia-Muscheln	Thraciidae	thraciids (lantern shells)
Thunfisch, Großer Thunfisch, Roter Thunfisch, Roter Thun	*Thunnus thynnus*	tunny, blue-fin tuna, blue-finned tuna, northern bluefin tuna (FAO)

Thunmakrele, Mittelmeermakrele, Kolios	*Scomber japonicus/ Scomber colias*	chub mackerel (FAO), Spanish mackerel, Pacific mackerel
Thymian-Blutströpfchen, Thymian-Rotwidderchen	*Zygaena purpuralis*	transparent burnet
Tibetanische Wasserspitzmaus, Gebirgsbachspitzmaus	*Nectogale elegans*	Tibetan water shrew, web-footed water shrew
Tibetantilope, Tschiru	*Pantholops hodgsonii*	chiru, Tibetan antelope
Tibetfuchs	*Vulpes ferrilata*	Tibetan sand fox
Tibetgazelle	*Procapra picticaudata*	goa, Tibetan gazelle
Tibetmakak	*Macaca thibetana*	Père David's stump-tailed macaque, Tibetan stump-tailed macaque
Tiefen-Katzenhai, Tiefseekatzenhai	*Apristurus profundorum*	deepwater catshark
Tiefen-Stechrochen, Tiefsee-Stechrochen	*Plesiobatis daviesi*	deepwater stingray
Tiefenbarsch, Schnabelbarsch	*Sebastes mentella*	deepwater redfish (FAO), beaked redfish
Tiefenwasser-Einsiedlerkrebse	Parapaguridae	deepwater hermit crabs
Tiefenwasser-Kapseehecht	*Merluccius paradoxus*	deepwater hake, deepwater Cape hake (FAO)
Tiefenwunder	*Bathothauma lyromma*	deepsea squid a.o.
Tiefland-Grablaubfrosch	*Pternohyla fodiens*	lowland burrowing frog
Tieflandanoa	*Bubalus depressicornis*	lowland anoa
Tiefsee-Angler	*Centrophryne spinulosa*	deep-sea anglerfish
Tiefsee-Anglerfische	Ceratiidae	seadevils
Tiefsee-Beilfische, Silberbeilfische	Sternoptychidae	hatchetfishes
Tiefsee-Königskrabbe	*Lithodes couesi*	scarlet king crab, deep-sea crab, deep-sea king crab
Tiefsee-Sternkoralle*	*Deltocyathus calcar*	deepsea star coral
Tiefseeaale, Schwarze Tiefseeaale	Cyematidae/Cyemidae	bobtail snipe eels, arrow eels
Tiefseedorsche	Moridae	deepsea cods, morid cods, moras
Tiefseeflundern	Bembridae	deep-water flatfishes, deepwater flatheads
Tiefseegarnelen	Aristeidae & Pandalidae	gamba prawns, aristeid shrimp & pandalid shrimps
Tiefseehechte	*Bathysaurus* spp.	deep-sea lizardfishes
Tiefseeheringe	Bathyclupeidae	deep-sea herrings
Tiefseequappen	Ateleopodidae	tadpole fishes, jellynose fishes
Tiefseevampire, Vampirtintenschnecken	Vampyromorpha	vampire squids
Tiefseezunge	*Bathysolea profundicola*	deepwater sole
Tiefwasser-Springkrebs	*Mundia rugosa*	rugose squat lobster
Tienschan-Laubsänger	*Phylloscopus humei*	Hume's yellow-browed warbler

Tierläuse	Haematopinidae	wrinkled sucking lice
Tierläuse, Lauskerfe, Läuslinge	Phthiraptera (Mallophaga & Anoplura)	phthirapterans
Tiger	*Panthera tigris*	tiger
Tiger-Kammmuschel	*Chlamys tigerina*	tiger scallop
Tiger-Mondschnecke	*Natica tigrina*	tiger moonsnail
Tiger-Olive	*Oliva tigrina*	tiger olive
Tiger-Vogelspinne, Indische Ornament-Vogelspinne	*Poecilotheria regalis*	Indian ornamental tarantula
Tiger-Zwergvogelspinne, Trinidad Zwergvogelspinne	*Hapalopus incei*	Trinidad olive tarantula
Tigerchamäleon	*Chamaeleo tigris*	tiger chameleon
Tigerfink	*Amandava amandava*	red avadavat
Tigerfische, Sägebarsche	Teraponidae/ Theraponidae	thornfishes, grunters, tiger-perches
Tigerhai	*Galeocerdo cuvieri*	tiger shark
Tigeriltis (Hbz. Perwitzki)	*Vormela peregusna*	marbled polecat
Tigerkärpfling, Gelber Prachtkärpfling	*Aphyosemion gulare*	yellow gularis
Tigerklapperschlange	*Crotalus tigris*	tiger rattlesnake
Tigermakrele	*Scomberomorus semifasciatus*	broadbarred king mackerel
Tigermuschel, Tiger-Kammmuschel	*Palliolum tigerinum*	tiger scallop
Tigerotter	*Notechis scutatus*	Australian tiger snake, mainland tiger snake
Tigerottern, Australische Tigerottern	*Notechis* spp.	Australian tiger snakes, black tiger snakes
Tigerprachtschmerle	*Botia helodes*	tiger loach
Tigerpython	*Python molurus*	Burmese python
Tigerquerzahnmolch	*Ambystoma tigrinum*	tiger salamander
Tigerschnecke	*Maurea tigris*	tiger topsnail, tiger maurea (tiger top shell)
Tigerstreifen-Makifrosch	*Phyllomedusa tomopterna*	barred leaf frog, tiger-striped leaf frog
Tilles Makrele, Südafrikanische Makrele*	*Caranx tille*	tille trevally (FAO), tille kingfish
Timalien	Timaliidae	babblers, *Rhabdornis*
Timor-Python	*Python timorensis*	Timor python
Timor-Waran	*Varanus timorensis*	Timor tree monitor, spotted tree monitor
Timucu	*Strongylura timucu*	timucu
Tintenfisch> Gemeiner Tintenfisch, Gemeine Tintenschnecke, Gemeine Sepie	*Sepia officinalis*	common cuttlefish
Tintenfische	Coleoidea/Dibranchiata	coleoids
Tintenschnecken, Eigentliche Tintenschnecken	Sepioidea (or Sepiida)	cuttlefish & sepiolas
Tischler-Seebrasse	*Argyrozona argyrozona*	carpenter, carpenter seabream (FAO)

Titicaca-Pfeiffrosch	Telmatobius culeus	Lake Titicaca frog
Titicacataucher	Rollandia micropterum	short-winged grebe
Todesotter	Acanthophis antarcticus	death adder
Todies, Todis	Todidae	todies
Toga-Schotenmuschel	Solemya togata	toga awningclam
Tokay, Tokee	Gekko gecko	tokay gecko, tokee
Tollwut-Tarantel*	Lycosa rabida	rabid tarantula, rabid wolf spider
Tölpel	Sulidae	boobies, gannets
Tölpelkopfschwamm	Spheciospongia vesparia	loggerhead sponge
Tolstolob> Edler Tolstolob, Marmorkarpfen	Hypophthalmichthys nobilis/ Aristichthys nobilis	bighead carp
Tolstolob> Gewöhnlicher Tolstolob, Silberkarpfen	Hypophthalmichthys molitrix	silver carp (FAO), tolstol
Tolstoloben	Hypophthalmidae	lookdown catfishes, loweye catfishes
Tomatenfrosch	Dyscophus antongilli	tomato frog
Tomatenmilbe	Aceria lycopersici	tomato erineum mite
Tomatenrostmilbe*	Aculops lycopersici	tomato russet mite
Tomcod, Atlantischer Tomcod	Microgadus tomcod	Atlantic tomcod
Tonkeanamakak	Macaca tonkeana	Tonkean macaque
Tonkinlangur	Trachypithecus francoisi/ Presbytis francoisi	Tonkin leaf monkey, François' leaf monkey
Tonkinstumpfnase	Pygathrix avunculus	Tonkin snub-nosed monkey
Tönnchenschnecken	Pupillacea spp.	
Tonnenfisch*	Hyperoglyphe perciformis	barrelfish
Töpfer-Pelzbiene	Anthophora retusa	potter flower bee
Töpfervögel, Ofenvögel	Furnariidae	ovenbirds
Töpferwespen, Pillenwespen	Eumenes spp.	potter wasps
Topfwurm, Weißer Topfwurm	Enchytraeus albidus	white potworm
Tordalk	Alca torda	razorbill
Torf-Mosaikjungfer	Aeshna juncea	common aeshna, common hawker
Tormahseer	Barbus tor/ Tor tor	mahseer (FAO), tor mahseer
Toros	Isothrix spp.	toros
Torpedo Stachelmakrele	Megalaspis cordyla	torpedo scad
Tote Manneshand, Meerhand, Nordische Korkkoralle (Lederkoralle)	Alcyonium digitatum	dead-man's fingers, sea-fingers
Totengräber	Necrophoridae	burying beetles
Totenkäfer	Blaps spp.	churchyard beetles, cellar beetles
Totenkopfaffe	Saimiri sciureus	common squirrel monkey
Totenkopfaffen	Saimiri spp.	squirrel monkeys
Totenkopfchimäre, Elefantenchimäre	Callorhynchus capensis/ Callorhinchus capensis	elephantfish, silver trumpeter (tradename), Cape elephantfish (FAO)

Totenkopfschabe	*Blaberus cranifer*	death's head cockroach
Totenkopfschwärmer, Totenkopf	*Acherontia atropos*	death's-head hawkmoth
Totenuhr (Bücherlaus, Staublaus)	*Trogium pulsatorium/ Atropos pulsatoria*	larger pale booklouse, larger pale trogiid, lesser death watch beetle (booklouse, dust-louse)
Totenuhr (Gescheckter Nagekäfer)	*Xestobium rufovillosum*	death-watch beetle
Totenuhr, Trotzkopf	*Anobium pertinax/ Dendrobium pertinax*	furniture beetle, furniture borer, common furniture beetle
Tovisittich, Goldkinnsittich	*Brotogeris jugularis*	orange-chinned parakeet
Tracheenmilbe, Bienenmilbe	*Acarapis woodi*	bee mite, honey bee mite
Tracheenmilben	*Acarapis* spp.	tracheal mites
Trampeltier, Zweihöckriges Kamel	*Camelus bactrianus*	Bactrian camel, two-humped camel
Tränenkauri	*Trivia pullicina/Trivia pulex*	teardrop trivia*
Trans-Peco-Königsnatter	*Lampropeltis mexicana*	Trans Peco kingsnake
Trans-Peco-Natter	*Elaphe subocularis*	Trans-Pecos rat snake
Transkaukasische Kletternatter	*Elaphe hohenackeri*	Transcaucasian rat snake, Caucasian snake
Transkaukasischer Wüstenrenner, Pleskes Wüstenrenner	*Eremias pleskei*	Pleske's racerunner
Transkei-Languste	*Panulirus homarus*	scalloped spiny crawfish
Trapez-Bandschnecke, Fuchskopf-Bandschnecke	*Pleuroploca trapezium*	trapeze horseconch, trapezium horse conch
Trapez-Kaiserfisch	*Pomacanthus rhomboides/ Pomacanthus striatus*	old woman angelfish
Trapezeule	*Cosmia trapezina*	dun-bar
Trapezmuscheln	Carditidae	carditas, cardita clams
Trappen	Otididae	bustards
Trappenlaufhühnchen	*Pedionomus torquatus*	plains wanderer
Trappenlaufhühnchen, Steppenläufer	Pedionomidae	plains-wanderer
Traubenkirschenlaus, Haferblattlaus, Haferlaus	*Rhopalosiphum padi*	bird-cherry aphid, oat aphid, wheat aphid
Traubiger Röhrenkalkschwamm	*Leucosolenia botryoides*	organ-pipe sponge
Trauer-Wüstenrenner*	*Eremias lugubris*	mourning racerunner
Trauerente	*Melanitta nigra*	black scoter
Trauermantel	*Nymphalis antiopa*	Camberwell beauty
Trauermantelsalmler	*Gymnocorymbus ternetzi*	black tetra
Trauermeise	*Parus lugubris*	sombre tit
Trauermücken	Lycoriidae	dark-winged fungus gnats, root gnats
Trauerrand-Zackenbarsch	*Epinephelus guttatus*	red hind
Trauerschnäpper	*Ficedula hypoleuca*	pied flycatcher
Trauerschnecke, Schwarze Stachelschnecke	*Muricanthus nigritus*	black murex
Trauerseeschwalbe	*Chlidonias niger*	black tern
Trauerseidenschnäpper	*Phainopepla nitens*	phainopepla

Trauersteinschmätzer	*Oenanthe leucura*	black wheatear
Trauerwaran (Gesprenkelter Waran & Schwarzkopfwaran)	*Varanus tristis*	mournful tree monitor, arid monitor (freckled monitor & black-headed monitor)
Traumkaiserfisch	*Euxiphipops navarchus/ Pomacanthus navarchus*	blue-girdled angelfish
Travancore-Spornschildkröte	*Indotestudo forsteni/ Geochelone travancorica*	Travancore tortoise
Treiberameisen, Wanderameisen	Dorylinae	legionary ants, army ants
Treibhaustausendfüßer*	*Asiomorpha coarctata*	hothouse millepede
Trematoden, Egel, Saugwürmer	Trematoda	flukes, trematodes
Treppennatter	*Elaphe scalaris*	ladder snake
Treppenschnecke, Kleine Treppenschnecke, Treppengiebelchen	*Oenopota turricula/ Lora turricula/ Bela turricula/ Propebela turricula*	small staircase snail*
Trichine	*Trichinella spiralis*	trichina worm
Trichternetz-Vogelspinnen	Dipluridae	funnel-web spiders, funnel-web tarantulas, sheetweb building tarantulas
Trichternetzspinnen, Trichterspinnen	Agelenidae	funnel web weavers, funnel-weavers
Trichterohren-Fledermäuse	Natalidae	funnel-eared bats
Trichterohren-Fledermaus	*Natalus* spp.	funnel-eared bats
Trichterspinnen, Trichternetzspinnen	Agelenidae	funnel web weavers, funnel-weavers
Trichterwurm	*Sabellaria alveolata*	honeycomb worm
Triebstecher, Obstbaumzweigabstecher	*Rhynchites caeruleus*	apple twig cutter
Triel	*Burhinus oedicnemus*	stone curlew
Triele	Burhinidae	stone curlews, thick-knees
Trillerfrosch	*Neobatrachus centralis*	trilling frog
Trilobiten, Dreilapper	Trilobita	trilobites
Trinidad Mahagoni-Vogelspinne*	*Tapinauchenius plumipes*	Trinidad mahogany tarantula
Trinidad Zwergvogelspinne, Tiger-Zwergvogelspinne	*Hapalopus incei*	Trinidad olive tarantula
Trinidad-Baumsteiger	*Colostethus trinitatis/ Mannophryne trinitatis*	Trinidad poison frog
Tristan Languste	*Jasus tristani*	Tristan rock lobster
Tristramstar	*Onychognathus tristramii*	Tristram's grackle
Tritonen	Ranellidae/ Cymatiidae	tritons, rock whelks
Trockenholztermiten	Kalotermitidae	drywood termites & powderpost termites
Trockenrasen-Blütenspanner	*Eupithecia centaureata*	lime-speck pug
Trockenrasen-Erdeule	*Agrotis cinerea*	light feathered rustic (moth)
Trockenrasen-Graswurzeleule	*Apamea lithoxylea*	common light arches
Trogmuscheln	Mactridae	mactras, trough clams (trough shells)
Trogmuscheln	*Spisula* spp.	surfclams a.o. (trough shells)

Trogons	Trogonidae	trogons
Trommelfisch	*Pogonias cromis*	black drum
Trompeten-Becherqualle	*Haliclystus salpinx*	trumpet stalked jellyfish
Trompeten-Becherquallen	*Haliclystus* spp.	trumpet stalked jellyfishes
Trompetenfische	Aulostomidae/Aulostomatidae	trumpetfishes
Trompetentierchen	*Stentor* spp.	trumpet animalcules
Trompetenwurm	*Pectinaria gouldii*	ice cream cone worm, trumpet worm
Trompeterfisch	*Latridopsis ciliaris*	blue moki
Trompeterfische	Latridae	trumpeters
Trompeterparadiesvogel	*Manucodia keraudrenii/ Phonygammus keraudrenii*	trumpet bird
Trompetervögel	Psophiidae	trumpeters
Tropensalamander	*Oedipina* spp.	worm salamanders
Tropfenflughuhn	*Pterocles senegallus*	spotted sandgrouse
Tropfenschildkröte	*Clemmys guttata*	spotted turtle
Tropfentrupial, Schwarzbrusttrupial	*Icterus pectoralis*	spot-breasted oriole
Tropikvögel	Phaethontidae	tropicbirds
Tropisch-Afrikanische Laubfrösche	*Agalychnis* spp.	leaf frogs
Tropische Bettwanze	*Cimex hemipterus*	tropical bedbug
Tropische Buntzecke	*Amblyomma variegatum*	tropical bont tick
Tropische Geflügelmilbe	*Ornithonyssus bursa*	tropical fowl mite
Tropische Klapperschlange, Schreckenklapperschlange, Schauerklapperschlange, Südamerikanische Klapperschlange, Cascabel	*Crotalus durissus*	neotropical rattlesnake, cascabel
Tropische Pferdezecke	*Dermacentor nitens/ Anocentor nitens*	tropical horse tick
Tropische Rattenlaus	*Hoplopleura pacifica*	tropical rat louse
Tropische Rattenmilbe, Nagermilbe	*Ornithonyssus bacoti/ Liponyssus bacoti*	tropical rat mite
Tropische Riesenskolopender u.a.	*Scolopendra gigas*	tropical giant scolopendra a.o.
Tropische Rinderzecke	*Boophilus microplus*	tropical cattle tick
Tropische Speichermotte, Dattelmotte, Mandelmotte	*Ephestia cautella*	almond moth (date moth, tropical warehouse moth)
Tropischer Pelzkäfer	*Attagenus fasciatus*	wardrobe beetle, tropical carpet beetle
Tropisches Wiesel	*Mustela africana*	tropical weasel
Troschels Stachelschnecke	*Murex troscheli*	Troschel's murex
Trottellumme	*Uria aalge*	guillemot, common murre
Trotzkopf, Totenuhr	*Anobium pertinax/ Dendrobium pertinax*	furniture beetle, furniture borer, common furniture beetle
True-Wal, True-Zweizahnwal	*Mesoplodon mirus*	True's beaked whale, True's North Atlantic beaked whale
Trüffelfliege	*Suillia tuberiperda*	truffles fly*
Trüffelkäfer, Schwammkugelkäfer	Leiodidae/ Anisotomidae	round fungus beetles

Trugbaumläufer	Rhabdornithidae	Philippine creepers
Trugkärpfling	Chologaster cornutus	swampfish
Trugkoralle, Hundskoralle (Falsche Koralle)	Myriapora truncata	false coral
Trugmanguste	Paracynictis selousi	gray meerkat, Selous' mongoose
Trugmotten	Eriocraniidae	eriocraniid moths
Trugratten	Octodontidae	octodonts
Trüsche, Quappe, Rutte	Lota lota	burbot
Truthahn> Truthuhn	Meleagris gallopavo	turkey, common turkey
Truthahnflügel	Arca zebra	turkey wing
Truthahngeier	Cathartes aura	turkey vulture
Truthuhn	Meleagris gallopavo	turkey, common turkey
Truthühner	Meleagridinae	turkeys, wild turkeys
Tschirr	Coregonus nasus	broad whitefish
Tschiru, Tibetantilope	Pantholops hodgsonii	chiru, Tibetan antelope
Tsetsefliegen	Glossina spp.	tsetse flies
Tsetsefliegen	Glossinae	tsetse flies
Tuberkelhokko	Crax rubra	great curassow
Tugun	Coregonus tugun	tugun
Tukanbartvogel	Semnornis ramphastinus	toucan barbet
Tukane	Ramphastidae	toucans
Tukotukos	Ctenomys spp.	tuco-tucos
Tulpen-Röhrenlaus	Dysaphis tulipae	tulip bulb aphid
Tumbufliege	Cordylobia anthropophaga	tumbu fly (mango fly)
Tümmler> Golftümmler	Phocoena sinus	Gulf of California porpoise, Chochito, Vaquita gulf porpoise
Tümmler> Großer Tümmler, Großtümmler, Flaschennasendelphin	Tursiops truncatus	bottlenosed dolphin, common bottle-nosed dolphin
Tümmler> Kleiner Tümmler, Kleintümmler, Schweinswal, Braunfisch	Phocoena phocoena	harbor porpoise, common harbor porpoise
Tümmler> Schwarzer Tümmler, Burmeisters Tümmler	Phocoena spinipinnis	black porpoise, Burmeister's porpoise, Marposa espinosa
Tümpelfrosch, Kleiner Wasserfrosch, Kleiner Teichfrosch, Kleiner Grünfrosch	Rana lessonae	pool frog (little waterfrog)
Tümpelwasserfloh	Moina brachiata	
Tupaias, Spitzhörnchen	Tupaiidae	tree shrews
Tupaias, Spitzhörnchen	Tupaia spp.	tree shrews
Tüpfel-Drückerfisch, Blaustreifen-Drückerfisch	Pseudobalistes fuscus	rippled triggerfish, yellow-spotted triggerfish (FAO)

Tüpfel-Kaninchenfisch	*Siganus stellatus*	brown-spotted spinefoot
Tüpfel-Rennechse	*Cnemidophorus lemniscatus*	rainbow lizard
Tüpfel-Ritterfisch	*Equetus punctatus*	spotted jackknife fish
Tüpfelbärbling	*Brachydanio nigrofasciatus*	spotted danio
Tüpfelbeutelmarder	*Dasyurus* spp.	quolls, native "cats", tiger "cats"
Tüpfelbuntbarsch	*Laetacara curviceps*	flag acara (FAO), flag cichlid (U.S.)
Tüpfeldoppelfüßer	*Polydesmus angustus*	flat millepede, flat-backed millepede
Tüpfelgrundel	*Mogurnda mogurnda*	northern trout gudgeon
Tüpfelhyäne	*Crocuta crocuta*	spotted hyena
Tüpfelseifenfisch	*Pogonoperca punctata*	speckled soapfish, clown grouper, spotted soapfish (FAO)
Tüpfelskink	*Eumeces schneideri*	Berber skink
Tüpfelsumpfhuhn	*Porzana porzana*	spotted crake
Tüpfeltausendfüßer	Blaniulidae	snake millepedes, blaniulid millepedes
Turakos & Lärmvögel	Musophagidae	turacos & plaintain-eaters
Turbanschnecken	Turbinidae	turban snails, turbans
Turbellarien, Strudelwürmer	Turbellaria	turbellarians, free-living flatworms
Türkenammer	*Emberiza cineracea*	cinereous bunting
Türkenkleiber	*Sitta krueperi*	Krüper's nuthatch
Türkentaube	*Streptopelia decaocto*	collared dove
Türkische Netzwühle, Türkische Ringelwühle	*Blanus strauchi*	Turkish worm lizard
Türkischer Baumschläfer	*Dryomys laniger*	Turkish dormouse
Türkischer Leistenkopfplattkäfer	*Cryptolestes turcicus*	Turkish grain beetle
Türkischer Maikäfer, Walker	*Polyphylla fullo*	fuller
Türkiselminie	*Elminia longicauda*	blue flycatcher
Türkisgoldbarsch	*Melanochromis auratus/ Pseudotropheus auratus*	golden mbuna (FAO), auratus
Türkisnaschvogel	*Cyanerpes cyaneus*	red-legged honeycreeper
Turmfalke	*Falco tinnunculus*	kestrel, common kestrel
Turmschnecke> Gemeine Turmschnecke	*Turritella communis*	common turretsnail, European turretsnail (common screw shell, common tower shell, great screw shell)
Turmschnecken	Turritellidae	turret snails (tower shells, auger shells, screw shells)
Turteltaube	*Streptopelia turtur*	turtle dove
Tuza	*Zygogeomys trichopus*	tuza
Tyrannen	Tyrannidae	tyrant flycatchers
Tyrrhenische Gebirgseidechse, Bedriagas Gebirgseidechse	*Lacerta bedriagae*	Bedriaga's rock lizard, Tyrrhenian rock lizard

Uakari, Scharlachgesicht	*Cacajao calvus*	white uakari, bald uakari
Ufer-Glanzgraseule	*Apamea unanimis*	small clouded brindle
Ufer-Laufkäfer	*Carabus clathratus*	latticed ground beetle
Uferaas (Weißwurm)	*Ephoron virgo/ Polymitarcis virgo*	virgin mayfly
Uferfeuchtkäfer, Noteriden	Noteridae	burrowing water beetles
Uferfliegen, Steinfliegen	Plecoptera	stoneflies
Ufermaina	*Acridotheres ginginianus*	bank mynah
Uferschlammkäfer	Georyssidae	minute mud-loving beetles
Uferschnepfe	*Limosa limosa*	black-tailed godwit
Uferschwalbe	*Riparia riparia*	sand martin
Uferwanzen, Springwanzen	Saldidae	shore bugs, saldids
Uganda-Rotstummelaffe	*Piliocolobus oustaleti*	Central African red colobus
Uhu	*Bubo bubo*	eagle owl
Ukelei	*Alburnus alburnus*	bleak
Ulk, Seebulle, Seebull (Seeskorpion)	*Cottus bubalis/ Taurulus bubalis*	long-spined sculpin, long-spined sea scorpion, longspined bullhead (FAO)
Ulmen-Zipfelfalter, Weißes W	*Strymonidia w-album/ Strymon w-album/ Satyrium w-album*	white-letter hairstreak
Ulmenblattrollenlaus	*Eriosoma ulmi*	currant root aphid, elm-currant aphid, elm leaf aphid
Ulmenwollschildlaus	*Gossyparia spuria*	European elm scale
Ulreys Salmler, Flaggensalmler	*Hemigrammus ulreyi*	Ulrey's tetra
Ultramarinlori	*Vini ultramina*	ultramarine lory
Uluguru-Schwarzfrosch> Afrikanischer Bananenfrosch	*Hoplophryne uluguruensis*	African banana frog
Umberfisch, Bartumber, Schattenfisch	*Umbrina cirrosa/ Sciaena cirrosa*	shi drum (FAO), corb (U.S./U.K.), sea crow (U.S.), gurbell (U.S.), croaker
Umberfische, Adlerfische, Trommler	Sciaenidae	kobs, drums (croakers)
Umgekehrter Flossenfüßer*	*Limacina retroversa*	retrovert pteropod
Underwood-Langzungenfledermaus	*Hylonycteris underwoodi*	Underwood's long-tongued bat
Unechte Deutsche Schabe	*Blatella lituricollis*	false German cockroach
Unechte Dornhaie, Eishaie	Dalatiidae	sleeper sharks
Unechte Gebärfische	Parabrotulidae	false brotulas
Unechte Karettschildkröte, Unechte Karette	*Caretta caretta*	loggerhead sea turtle, loggerhead
Unechte Muränen	Chlopsidae/ Xenocongridae	false morays, false moray eels
Unechte Spinnmilben, Falsche Spinnmilben	*Brevipalpus* spp.	false spider mites, false red spider mites
Unechte Warane, Südamerikanische Tejus	*Callopistes* spp.	false monitors

Unechte Zwergbarsche	Pseudogrammidae	podges
Unechter Bonito, Melvera-Fregattmakrele	*Auxis rochei*	bullet tuna (FAO), bullet mackerel
Unechter Gründling	*Pseudogobio rivularis/ Abbottina rivularis*	Chinese false gudgeon
Ungebändertes Eichen-Kleinbärchen	*Meganola strigula*	small black arches (moth)
Ungestreifte Pelamide, Rüppell's Thunfisch	*Orcynopsis unicolor/ Gymnosarda unicolor*	plain bonito, scaleless tuna, pigtooth tuna, dogtooth tuna (FAO), Rüppell's bonito
Ungka, Schlankgibbon	*Hylobates agilis*	dark-handed gibbon, agile gibbon
Ungleicher Holzbohrer	*Xyleborus dispar*	shot-hole borer, European shot-hole borer, broad-leaved pinhole borer
Unglückshäher	*Perisoreus infaustus*	Siberian jay
Unken, Feuerkröten	*Bombina* spp.	fire-bellied toads, firebelly toads
Unpaarhufer	Perissodactyla	odd-toed ungulates, perissodactyls
Unscheinbare Seewalze	*Bohadschia tenuissima*	inconspicuous holothurian*
Unterlaufene Reusenschnecke*	*Nassarius vibex*	bruised nassa
Upemba-Pelomeduse	*Pelusios upemba*	Upemba mud turtle
Ur (Hausrind), Auerochse	*Bos taurus/ Bos primigenius*	aurochs (domestic cattle)
Urameisen, Stachelameisen	Ponerinae	ponerine ants, primitive ants (bulldog ants)
Urameisen, Stachelameisen	*Ponera* spp.	ponerine ants, primitive ants
Uräusschlange	*Naja haje*	Egyptian cobra
Urfrösche	Amphicoela	
Urial	*Ovis vignei*	urial
Urmotten	Micropterygidae	mandibulate moths
Urmundtiere, Urmünder, Erstmünder	Protostomia	protostomes
Urson	*Erethizon dorsatum*	North American porcupine
Urvögel	Archaeornithes	ancestral birds, "lizard birds", archaeornithes
Urwaldgalago, Zwerggalago	*Galago demidovii/ Galagoides demidoff*	dwarf bush baby, Demidoff's bush baby, Demidoff's galago
Urwaldkröte	*Cacophryne borbonica*	jungle toad
Urwurzelzähner	Thecodontia	thecodonts, dinosaur ancestors
Usambara-Buschhörnchen	*Paraxerus vexillarius*	Usambara squirrel
Utah-Präriehund	*Cynomys parvidens*	Utah prairie dog

Vadigo	*Campogramma glaycos/ Campogramma vadigo/ Campogramma lirio/ Lichia vadigo*	vadigo
Vahls Wolffisch	*Lycodes vahlii*	Vahl's eelpout
Valenciakärpfling	*Valencia hispanica*	Valencia toothcarp
Vampir> Gemeiner Vampir, Gewöhnlicher Vampir	*Desmodus rotundus*	vampire bat, common vampire bat
Vampire> Echte Vampire	Desmodontidae	vampire bats, common vampire bats
Vampirkalmar	*Vampyroteuthis infernalis*	vampire squid
Vampirtintenschnecken, Tiefseevampire	Vampyromorpha	vampire squids
Vangawürger, Blauwürger	Vangidae	vanga shrikes, helmet-shrikes
Vari	*Varecia variegata*	variegated lemur, ruffed lemur
Variable Seegurke	*Holothuria forskali*	cotton spinner
Variabler Lederschwamm	*Ircinia variabilis*	variable loggerhead sponge
Variabler Schönbock	*Phymatodes testaceus*	tanbark borer (longhorn beetle)
Variabler Weichkäfer	*Cantharis livida*	variable cantharid, variable soldier beetle
Varroamilbe, Bienenbrutmilbe	*Varroa jacobsoni*	varroa mite
Vasapapagei	*Coracopsis vasa*	vasa parrot
Vasenschnecken	Vasidae	vase snails, chank snails (vase shells, chank shells)
Vasenschwamm	*Ircinia campana*	vase sponge
Veilchen-Perlmutterfalter, Silberfleck-Perlmuttfalter, Früher Perlmuttfalter	*Boloria euphrosyne/ Clossiana euphrosyne*	pearl-bordered fritillary
Veilchen-Scheckenfalter, Weißfleckenfalter	*Melitaea cynthia/ Euphydryas cynthia*	Cynthia's fritillary
Veilchenblattrollgallmücke	*Dasineura affinis*	violet leaf midge
Veilchenblauer Wurzelhals-Schnellkäfer	*Limoniscus violaceus*	violet click beetle
Veilchenschnecke, Floßschnecke	*Janthina janthina*	large violet snail, common purple sea snail, violet seasnail, common janthina
Veilchenschnecken	Janthinidae	violet snails
Veilchenschwalbe	*Tachycineta thalassina*	violet-green swallow
Veldkamps Zwergflughund	*Nanonycteris veldkampi*	little flying cow
Venezuela Ornament-Vogelspinne	*Psalmopoeus irminia*	suntiger tarantula
Venezuela-Amazone	*Amazona amazonica*	orange-winged parrot, orange-winged amazon
Venezuela-Erdschildkröte	*Rhinoclemmys diademata*	Venezuelan wood turtle
Venusfächer, Großer Seefächer, Große Fächerkoralle, Große Netzgorgonie	*Gorgonia flabellum/ Gorgonia ventalina/ Rhipidogorgia flabellum*	Venus sea fan, Venus' fan, common sea fan
Venusfisch	*Aphyocypris pooni/ Hemigrammocypris lini*	garnet minnow

Venusgürtel	Cestidea	cestids
Venusgürtel	*Cestus veneris*	Venus's girdle
Venuskamm, Venuskamm-Stachelschnecke, Skelettspindel, Spinnenkopf	*Murex pecten/ Murex tenuispina*	Venus comb murex, Venus comb, thorny woodcock
Venusmuschel, Große Teppichmuschel, Kreuzgemusterte Teppichmuschel	*Venerupis decussata/ Tapes decussata*	cross-cut carpetclam, cross-cut venus (cross-cut carpet shell)
Venusmuschel> Gemeine Venusmuschel, Strahlige Venusmuschel	*Venus gallina/ Chione gallina*	striped venus, chicken venus
Venusmuscheln	Veneridae	venus clams (venus shells)
Veracruz-Salamander	*Lineatriton lineola*	Veracruz worm salamander
Veränderliche Fechterschnecke, Veränderliche Flügelschnecke, Mutabilis	*Strombus mutabilis*	mutable conch, changeable conch
Veränderliche Fechterschnecke, Veränderliche Flügelschnecke	*Strombus variabilis*	variable conch
Veränderliche Hummel	*Bombus humilis*	brown-banded carder bee
Veränderliche Krabbenspinne	*Misumena vatia*	goldenrod crab spider
Veränderlicher Weidenblattkäfer	*Phyllodecta viminalis*	variable willow leaf beetle
Veränderlicher Dornschwanz, Afrikanischer Dornschwanz	*Uromastyx acanthinurus*	African spiny-tailed lizard
Veränderlicher Magerrasen-Kleinspanner	*Idaea degeneraria*	Portland ribbon wave
Veränderliches Blutströpfchen, Veränderliches Widderchen, Veränderliches Rotwidderchen, Wickenwidderchen	*Zygaena ephialtes*	variable burnet
Veränderte Beißschrecke	*Platycleis modesta*	variable bushcricket
Verbogene Tritonschnecke*	*Sassia subdistorta*	distorted rock triton
Verbrannte Archenmuschel	*Anadara uropygimelana*	burnt-end ark (burnt-end ark shell)
Verkannter Grashüpfer	*Chorthippus mollis*	lesser field grasshopper
Verkehrte Schirmschnecke, Goldschwammschnecke	*Tylodina perversa*	
Verkehrtschnecken, Linksnadelschnecken	Triphoridae	left-handed hornsnails
Vernays Baummaus	*Vernaya* spp.	Vernay's climbing mice
Vernedes Zahnschnecke	*Fissidentalium vernedei* (Scaphopoda)	Vernede's tusk
Vervetmeerkatze, Südafrikanische Grünmeerkatze	*Cercopithecus pygerythrus*	vervet monkey
Verwachsenkiemer, Siebkiemer	Septibranchia	septibranch bivalves, septibranchs

Verwandlungs-Partnergarnele, Imperator-Garnele	*Periclimenes imperator*	emperor shrimp
Verwitterte Turmschnecke*	*Tachyrhynchus erosus*	eroded turretsnail
Verzierter Drachenkopf, Schmuck-Drachenkopf	*Scorpaenopsis cirrhosa/ Dendroscorpaena cirrhosa*	hairy stingfish, raggy scorpionfish, weedy stingfish (FAO)
Verzweigte Feuerkoralle u.a.	*Millepora alcicornis*	branching fire coral
Verzweigte Feuerkoralle u.a.	*Millepora tortuosa*	ramified fire coral*
Verzweigte Venusmuschel	*Gafrarium divaricatum*	forked venus
Verzweigter Vasenschwamm	*Callyspongia vaginalis*	tube sponge
Vespermäuse	*Calomys* spp.	vesper mice
Victoria-See-Karpfen	*Rastrineobola argentea*	silver cyprinid
Vielaugen-Krokodilfisch	*Parapercis polyophthalma*	black-tailed weever
Vielbindenkrait, Südchinesischer Vielbindenbungar	*Bungarus multicinctus*	many-banded krait
Vielborster, Borstenwürmer, Polychaeten	Polychaeta	bristle worms, polychaetes, polychetes, polychete worms
Vielfarben-Wespenspinne	*Argiope versicolor*	multicolored argiope
Vielfarbentachuri	*Tachuris rubrigastra*	many-coloured rush tyrant
Vielfarbentodi	*Todus multicolor*	Cuban tody
Vielfarbige Mondschnecke*	*Natica canrena/ Naticarius canrena*	colorful Atlantic moonsnail, colorful moonsnail
Vielfarbige Pelomeduse*	*Pelusios rhodesianus*	variable mud turtle
Vielfarbiger Eidechsenfisch	*Synodus variegatus*	variegated lizardfish
Vielfarbiger Fächerröhrenwurm	*Bispira variegata*	variegated fanworm, multicoloured fanworm*
Vielfarbiger Maulbrüter, Kleiner Maulbrüter	*Pseudocrenilabrus multicolor*	dwarf Egyptian mouthbrooder
Vielfleckiger Wüstenrenner	*Eremias multiocellata*	multi-ocellated racerunner
Vielfraß	*Gulo gulo*	wolverine
Vielstreifen-Kardinalbarsch	*Apogon multitaeniatus*	smallscale cardinal
Vieltrichterkoralle	*Siderastrea radians*	lesser starlet coral
Vielzitzenmäuse	*Mastomys* spp.	multimammate rats
Vierauge	*Anableps anableps*	foureyes, largescale foureyes (FAO)
Vieraugen-Falterfisch, Pfauenaugen-Falterfisch	*Chaetodon capistratus*	foureye butterflyfish
Vieraugenbeutelratten	*Metachirops* spp.	"four-eyed" opossums
Vieraugenfische	Anablepidae	four-eyed fishes
Vierband-Grundel*	*Chromogobius quadrivittatus*	banded goby
Vierbärtelige Seequappe	*Enchelyopus cimbrius/ Rhinonemus cimbrius*	fourbeard rockling (FAO), four-bearded rockling
Viereckige Seemandel	*Philine quadrata*	quadrate paperbubble
Vierfleck-Kreuzspinne	*Araneus quadratus*	fourspotted orbweaver*
Vierfleck-Pfauenaugenschildkröte	*Sacalia quadriocellata*	four-eyed turtle
Vierfleckbutt, Gefleckter Flügelbutt	*Lepidorhombus boscii*	four-spot megrim
Vierflecken-Grundel	*Deltentosteus quadrimaculatus*	fourspot goby
Vierfleckflunder	*Paralichthys oblongus*	four-spot flounder

Vierflecklibelle, Vierfleck	*Libellula quadrimaculata*	four-spotted libellula, four-spotted chaser, four spot
Vierflügel-Flugfisch, Fliegender Fisch	*Cheilopogon heterurus/ Cypselurus heterurus*	Mediterranean flyingfish
Vierhorn-Haarwurm*	*Spio quadricornis*	four-horned spio
Vierhorn-Kofferfisch	*Acanthostracion quadricornis/ Ostracion quadricornis*	scrawled cowfish
Vierhorn-Spitzkopfkrabbe	*Pisa tetraodon*	four-horned spider crab
Vierhornantilope	*Tetracerus quadricornis*	four-horned antelope, chousingha
Vierhornchamäleon	*Chamaeleo quadricornis*	four-horned chameleon
Vierhörniger Seeskorpion	*Myoxocephalus quadricornis/ Oncocottus quadricornis/ Triglopsis quadricornis*	four-horn sculpin, four-horned sculpin, fourhorn sculpin (FAO)
Vierhornkäfer	*Gnathocerus cornutus*	broad-horned flour beetle
Vierhornkrake*	*Pteroctopus tetracirrhus*	fourhorn octopus
Vierkantige Archenmuschel	*Arca tetragona*	tetragonal ark (tetragonal ark shell)
Vierlungenspinnen*	Hypochilidae/Hypochelidae	four-lunged spiders
Vierpunkt-Aaskäfer	*Xylodrepa quadrimaculata/ Dendroxena quadrimaculata*	four-spotted burying beetle
Vierpunkt-Kaurischnecke, Vierpunktkauri	*Niveria quadripunctata/ Trivia quadripunctata*	fourspot trivia, four-spotted trivia
Vierstreifen-Erdhörnchen	*Lariscus hosei*	four-striped ground squirrel
Vierstreifen-Skink	*Eumeces tetragrammus*	four-lined skink
Vierstreifennatter, Vierstreifen-Rattennatter	*Elaphe quatuorlineata*	four-lined snake, yellow rat snake
Vierzahn-Nixenschnecke	*Nerita versicolor*	four-tooth nerite
Vierzahn-Röhrenwurm	*Spirobranchus tetraceros*	four-tooth tubeworm
Vierzähnige Windelschnecke	*Vertigo geyerii*	fourtoothed whorl snail*
Vierzehen-Gelbsalamander, Zwergsalamander	*Eurycea quadridigitata*	dwarf salamander
Vierzehen-Rüsselratte	*Petrodromus tetradactylus*	four-toed elephant shrew, forest elephant shrew
Vierzehensalamander	*Hemidactylium scutatum*	four-toed salamander
Vierzehenschildkröte, Steppenschildkröte	*Agrionemys horsfieldi/ Testudo horsfieldii*	Horsfield's tortoise, four-toed tortoise, Central Asian tortoise
Vierzehenschildkröten	*Agrionemys* spp.	four-toed tortoises
Vierzehnpunktiger Marienkäfer, Geballter Marienkäfer	*Propylea quatuordecimpunctata*	fourteen-spot ladybird
Vietnamesische Scharnierschildkröte, Hinterindische Scharnierschildkröte	*Cuora galbinifrons*	Indochinese box turtle
Vikunja	*Vicugna vicugna*	vicuña
Violettbraune Frühlingseule	*Orthosia incerta*	clouded drab moth
Violette Hornkoralle	*Paramuricea chamaeleon*	violet horny coral
Violette Mondschnecke	*Natica violacea*	violet moonsnail
Violette Pilgermuschel	*Argopecten purpuratus*	purple scallop
Violette Seegurke, Bunte Seegurke	*Pseudocolochirus violaceus*	multicolored sea cucumber*, violet sea cucumber
Violette Spinnenschnecke	*Lambis violacea*	violet spider conch

Violette Strandkrabbe	*Hemigrapsus nudus*	purple shore crab
Violette Streifenqualle*	*Pelagia colorata*	purplestriped jellyfish
Violetter Herzigel, Violetter Herzseeigel, Purpur-Herzigel, Purpurigel	*Spatangus purpureus*	purple heart urchin
Violetter Seeigel, Dunkelvioletter Seeigel	*Sphaerechinus granularis*	purple sea urchin, violet urchin
Violetter Silberfalter, Mädesüß-Perlmutterfalter	*Brenthis ino*	lesser marbled fritillary
Violetter Sonnenstern	*Solaster endeca*	smooth sun star, purple sun star, purple sunstar
Violetter Stechrochen	*Dasyatis violacea*	pelagic stingray (FAO), violet stingray
Violetter Weidenkäfer*	*Helops caerulens*	violet willow beetle
Violettgestreifte Sandmuschel	*Gari fervensis*	Faroe sunsetclam
Violettköpfige Tiefseegarnele	*Aristeus antillensis*	purplehead gamba prawn
Viperfische, Vipernfische	Chauliodontidae	viperfishes
Vipern, Ottern	Viperidae	vipers
Vipernatter	*Natrix maura*	viperine snake, viperine grass snake
Viperqueise, Kleines Petermännchen, Zwergpetermännchen	*Echiichthys vipera/ Trachinus vipera*	lesser weever
Vireos, Laubwürger	Vireonidae	vireos, greenlets, peppershrikes, shrike-vireos
Virginia-Uhu, Amerikanischer Uhu	*Bubo virginianus*	great horned owl
Virginiawachtel	*Colinus virginianus*	northern bobwhite
Viscacha (Hbz.), Feld-Viscacha, Pampas-Viscacha	*Lagostomus maximus*	plains viscacha
Viscacharatten	*Octomys* spp.	viscacha rats
Vogel-Azurjungfer	*Coenagrion ornatum*	ornate damselfly
Vogelbecken-Dinosaurier	Ornithischia	bird-hipped dinosaurs, ornithischian reptiles
Vogelbeersamenwespe	*Megastigmus brevicaudis*	rowan seed wasp
Vogelköpfchen*, Vogelkopf-Moostierchen*	*Bugula avicularia*	bird's head coralline
Vogelmilben	Dermanyssidae	dermanyssid mites, poultry mites
Vogelmuschel, Europäische Vogelmuschel, Flügelmuschel, Europäische Flügelmuschel	*Pteria hirundo*	European wing-oyster (European wing shell)
Vogelmuscheln, Perlmuscheln	Pteriidae	wing oysters & pearl oysters
Vogelnestmilbe, Hühnernestmilbe	*Androlaelaps casalis*	poultry litter mite, cosmopolitan nest mite
Vogelschlange, Graue Baumnatter	*Thelotornis kirtlandii*	bird snake
Vogelschuppenbeinmilbe*	*Knemidokoptes mutans*	scalyleg mite (of fowl)
Vollbartmeerkatze	*Cercopithecus l'hoesti*	L'Hoest's monkey

Vorderindische Pfauenaugen-Sumpfschildkröte	*Morenia petersi*	Indian eyed turtle
Vorderindische Weichschildkröte, Nagpur-Weichschildkröte	*Aspideretes leithii/ Trionyx leithii*	Leith's softshell turtle
Vorderindisches Schuppentier	*Manis crassicaudata*	Indian pangolin
Vorratsmilben	Acaridae/ Tyroglyphidae	acarid mites, house dust mites
Votsotsa	*Hypogeomys antimena*	Malagasy giant rat
Vulkankaninchen	*Romerolagus diazi*	volcano rabbit

Wabenkorallen*	*Favites* spp.	honeycomb corals
Wabenkröte	*Pipa pipa*	Surinam toad
Wabenschwamm*	*Echinoclathria gigantea*	honeycomb sponge
Wabenschwanz-Gurami, Wabenschwanz-Makropode	*Belontia hasselti*	Java combtail
Wabenwasserfloh	*Ceriodaphnia reticulata*	honeycomb waterflea*
Wacholderdrossel	*Turdus pilaris*	fieldfare
Wacholdersamenwespe	*Megastigmus bipunctatus*	juniper seed wasp
Wacholderspanner	*Thera juniperata*	juniper carpet moth
Wachsrose	*Anemonia sulcata/ Anemonia viridis*	snakelocks anemone, opelet anemone
Wachssalmler	*Roeboides* spp.	glass headstander
Wachtel	*Coturnix coturnix*	quail
Wachtelfrankolin	*Francolinus pondicerianus*	Indian grey francolin
Wachtelkönig	*Crex crex*	corncrake
Wächtergarnele (Muschelfreund)	*Pontonia custos*	
Wadenstecher, Stechfliege	*Stomoxys calcitrans*	stable fly, dog fly, "biting housefly"
Waffenfliegen	Stratiomyidae	soldier flies
Waglers Lanzenotter	*Trimeresurus wagleri/ Tropidolaemus wagleri*	Wagler's pit viper, Wagler's palm viper
Wagners Flügeltaschen-Fledermaus	*Cormura brevirostris*	Wagner's sac-winged bat
Wahoo	*Acanthocybium solandri*	wahoo
Wal-Laus	*Cyamus boopis*	whale-louse
Wal-Läuse	*Cyamus* spp.	whale-lice
Wal-Seepocken	*Conchoderma* spp.	whale barnacles
Wald-Sandlaufkäfer	*Cicindela silvatica*	wood tiger beetle
Wald-Stacheltaschenmäuse	*Heteromys* spp.	forest spiny pocket mice
Wald-Wiesenvögelchen	*Coenonympha oedippus*	false ringlet butterfly
Waldameisen & Raubameisen	*Formica* spp.	wood ants & predatory ants
Waldammer	*Emberiza rustica*	rustic bunting
Waldbachschildkröte	*Clemmys insculpta*	wood turtle
Waldbaumläufer	*Certhia familiaris*	treecreeper
Waldbirkenmaus	*Sicista betulina*	northern birch mouse
Waldbock	*Spondylis buprestoides*	European cylinder longhorn beetle
Waldböcke	*Spondylis* spp.	cylinder longhorn beetles, cylinder longhorns
Waldboden-Tausendfüßer*	*Cylindroiulus punctatus*	woodland floor millepede
Waldbrettspiel	*Pararge aegeria*	speckled wood
Walddrossel	*Hylocichla mustelina*	wood thrush
Waldeidechse, Bergeidechse, Mooreidechse	*Lacerta vivipara*	viviparous lizard, European common lizard
Waldfrosch	*Rana sylvatica*	wood frog
Waldfuchs	*Cerdocyon thous*	crab-eating fox
Waldgirlitz	*Serinus scotops*	forest canary
Waldgrille	*Nemobius sylvestris*	wood cricket
Waldhummel	*Bombus sylvarum*	knapweed carder bee, shrill carder bee

Waldhund	*Speothos venaticus*	bush dog
Waldhüpfmaus	*Napaeozapus insignis*	woodland jumping mouse
Waldiltis, Europäischer Iltis	*Mustela putorius*	European polecat
Waldkauz	*Strix aluco*	tawny owl
Waldklapperschlange	*Crotalus horridus*	timber rattlesnake
Waldkobra	*Pseudohaje goldii*	Gold's tree cobra
Waldkobras	*Pseudohaje* spp.	tree cobras
Waldlaubsänger	*Phylloscopus sibilatrix*	wood warbler
Waldlemming	*Myopus schisticolor*	wood lemming
Waldmaikäfer	*Melolontha hippocastani*	field maybeetle
Waldmaus, Feld-Waldmaus	*Apodemus sylvaticus*	wood mouse, long-tailed field mouse
Waldmäuse & Feldmäuse	*Apodemus* spp.	Old World wood and field mice
Waldmurmeltier	*Marmota monax*	woodchuck, ground hog
Waldohreule	*Asio otus*	long-eared owl
Waldohrwurm	*Chelidurella acanthopygia*	forest earwig
Waldpieper	*Anthus hodgsoni*	olive-backed pipit
Waldralle	*Gallirallus sylvestris/ Sylvestrornis sylvestris*	Lord Howe rail, Lord Howe Island woodhen
Waldrapp	*Geronticus eremita*	bald ibis, northern bald ibis
Waldsalamander	*Plethodon* spp.	woodland salamanders
Waldsänger	Parulidae	wood warblers, New World warblers
Waldschaben	*Ectobius* spp.	ectobid cockroaches
Waldschildkröte	*Chelonoidis denticulata/ Testudo denticulata/ Geochelone denticulata*	yellow-footed tortoise, Brazilian giant tortoise, American forest tortoise, Hercules tortoise, South American yellow-footed tortoise (IUCN)
Waldschnepfe	*Scolopax rusticola*	woodcock, European woodcock
Waldschnepfen-Stachelschnecke*	*Murex scolopax*	woodcock murex
Waldskinke	*Sphenomorphus* spp.	forest skinks
Waldspitzmaus	*Sorex araneus*	common shrew, Eurasian common shrew
Waldspitzmäuse, Rotzahnspitzmäuse	*Sorex* spp.	red-toothed shrews, long-tailed shrews
Waldsteiger, Waldsteigerfrösche	*Leptopelis* spp.	forest treefrogs
Waldstorch	*Mycteria americana*	American wood ibis
Waldwasserläufer	*Tringa ochropus*	green sandpiper
Waldwespe	*Dolichovespula sylvestris*	tree wasp, wood wasp
Waldwühlmaus, Rötelmaus	*Clethrionomys glareolus*	bank vole
Wale & Delphine	Cetacea	cetaceans: whales & porpoises & dolphins
Walhai	*Rhincodon typus*	whale shark
Walhaie	Rhinocodontidae	whale sharks
Waliasteinbock, Äthiopischer Steinbock	*Capra walie*	Walia ibex

Walker, Türkischer Maikäfer	*Polyphylla fullo*	fuller
Walkerrochen, Chagrinrochen	*Raja fullonica*	shagreen ray
Walköpfe	Cetomimidae	flabby whalefishes
Wallaruh, Bergkänguru	*Macropus robustus*	wallaroo, common wallaroo, euro
Waller, Wels, Schaiden	*Silurus glanis*	European catfish, wels, sheatfish, wels catfish (FAO)
Wallichfasan, Schopffasan	*Catreus wallichii*	cheer pheasant
Walnuss-Schildlaus	*Aspidiotus juglansregiae*	English walnut scale
Walnussblatt-Gallmilbe	*Aceria tristriatus*	walnut leaf gall mite, Persian walnut leaf blister mite
Walnusspockenmilbe	*Eriophyes erinea/ Eriophyes tristriatus*	walnut blister mite
Walross	*Odobenus rosmarus*	walrus
Walross-Schnecke, Dorn-Bandschnecke	*Opeatostoma pseudodon*	thorn latirus, thorned latirus
Walwelse*	Cetopsidae	whalelike catfishes
Walzen-Ascidie, Falten-Ascidie	*Styela plicata*	barrel-squirt*
Walzen-Seefeder	*Cavernularia obesa*	barrel sea-pen*
Walzenschlangen, Rollschlangen	*Cylindrophis* spp.	pipe snakes
Walzenschnecken, Faltenschnecken, Rollschnecken	Volutidae	volutes
Walzenskink	*Chalcides mauritanicus*	cylindrical skink
Walzenskinks	*Chalcides* spp.	cylindrical skinks
Walzenspinnen u.a.	Solifugae/Solpugida	sun spiders, false spiders, windscorpions, solifuges, solpugids
Walzenspinnen u.a.	Segestriidae	segestriids, tube-web spiders
Wandelbare Reusenschnecke	*Sphaeronassa mutabilis/ Nassarius mutabilis*	mutable nassa
Wander-Gebirgsschrecke*	*Melanoplus sanguinipes*	migratory grasshopper, lesser migratory grasshopper
Wanderalbatros	*Diomedea exulans*	wandering albatross
Wanderameisen, Treiberameisen	Dorylinae	legionary ants, army ants
Wanderdrossel	*Turdus migratorius*	American robin
Wanderfalke	*Falco peregrinus*	peregrine
Wanderfilarie, Afrikanischer Augenwurm	*Loa loa*	African eye worm
Wandergelbling, Posthörnchen, Postillion	*Colias croceus*	clouded yellow
Wanderheuschrecke	*Locusta migratoria*	migratory locust
Wanderlaubsänger	*Phylloscopus borealis*	Arctic warbler

Wandermaräne, Große Maräne, Große Schwebrenke, Lavaret, Bodenrenke	Coregonus lavaretus	freshwater houting, powan, common whitefish (FAO)
Wandermuschel, Dreikantmuschel	Dreissena polymorpha	zebra mussel, many-shaped dreissena
Wandernde Tritonschnecke, Großes Argushorn, Krötenschnecke	Ranella olearium/ Ranella gigantea	wandering triton
Wanderratte	Rattus norvegicus	Norway rat, brown rat
Wandersaibling, Seesaibling	Salvelinus alpinus	char, charr (FAO), Arctic charr
Wanderspinnen, Kammspinnen	Ctenidae	wandering spiders, running spiders
Wanstschrecke	Polysarcus denticauda/ Orphania denticauda	large swordtailed bushcricket
Wanzen	Heteroptera (Hemiptera)	heteropterans, true bugs
Wanzenspinne	Coriarachne depressa	
Warane	Varanidae	monitor lizards
Warmouth-Sonnenbarsch*	Lepomis gulosus	warmouth
Warzen-Anglerfisch	Antennarius phymatodes/ A. maculatus	warty frogfish
Warzen-Ascidie, Weiße Warzenseescheide, Knorpelseescheide	Phallusia mammillata	simple sea-squirt
Warzen-Krötenkopf-Schildkröte, Raue Krötenkopf-Schildkröte	Phrynops tuberculatus	tuberculate toad-headed turtle
Warzen-Napfschnecke	Patella mamillaris	mamillary limpet
Warzen-Seezunge, Sandzunge	Pegusa lascaris/ Solea lascaris	sand sole
Warzenbeißer	Decticus verrucivorus	wart-biter, wart-biter bushcricket
Warzenchamäleon	Chamaeleo verrucosus	giant chameleon, warty chameleon
Warzenkopf	Pityriasinae	Bornean bristlehead
Warzenkoralle> Europäische Warzenkoralle	Balanophyllia europaea	European star coral
Warzenkoralle> Warzige Hornkoralle, Kleiner Seefächer	Eunicella verrucosa	warty coral, pink sea fan
Warzenkoralle> Warzige Sternkoralle	Balanophyllia verrucaria	warty star coral, warty cupcoral
Warzenkrake	Graneledone verrucosa	warty octopus
Warzenmolche, Chinesische Warzenmolche	Paramesotriton spp.	warty newts
Warzenrose, Edelsteinrose	Bunodactis verrucosa	wartlet anemone, gem anemone
Warzenschlangen	Acrochordidae	wart snakes, file snakes
Warzenschwamm*	Polymastia robusta	nipple sponge
Warzenschwein	Phacocoerus aethiopicus	wart hog

Warzenstern, Eisseestern, Eisstern	*Marthasterias glacialis*	spiny starfish
Warzige Asien-Baumkröte, Malabar-Baumkröte	*Pedostibes tuberculosus*	warty Asian tree toad, Malabar tree toad
Warzige Buschkoralle	*Pociliopora verrucosa*	warty bushcoral
Warzige Krustenanemone	*Palythoa mammillosa*	knobby zoanthid
Warzige Schirmschnecke, Mittelmeer-Schirmschnecke	*Umbraculum mediterraneum*	Mediterranean umbrella snail
Warzige Venusmuschel	*Venus verrucosa*	warty venus
Warziger Gallertschirm	*Cranchia scabra*	warty gelatin-squid*
Warziges Papierboot	*Argonauta nodosus*	knobby argonaut
Waschbär> Nördlicher Waschbär	*Procyon lotor*	common raccoon
Waschbären	*Procyon* spp.	raccoons
Waschbärenartige	Procyonidae	raccoons and relatives
Wasser-Borstenigel	*Limnogale mergulus*	aquatic tenrec
Wasseramsel	*Cinclus cinclus*	dipper, white-throated dipper
Wasseramseln	Cinclidae	dippers
Wasseranolis	*Anolis aquaticus*	water anolis
Wasserassel, Gemeine Wasserassel	*Asellus aquaticus*	water louse
Wasserbock, Hirschantilope	*Kobus ellipsiprymnus*	waterbuck
Wasserbüffel, Asiatischer Büffel	*Bubalus bubalis/ Bubalus arnee*	Asian water buffalo, wild water buffalo, carabao
Wasserdrache> Gewöhnlicher Wasserdrache	*Physignathus lesueurii*	eastern water dragon
Wasserdrachen	*Physignathus* spp.	Asian water dragons
Wasserfledermaus	*Myotis daubentoni*	Daubenton's bat
Wasserflöhe	Cladocera	water fleas, cladocerans
Wasserflorfliegen (Schlammfliegen)	Sialidae	alder flies
Wasserfrosch, Teichfrosch	*Rana esculenta* (*Rana* kl. *esculenta*)	European edible frog
Wasserhund, Afrikanischer Hechtsalmler	*Hepsetus odae*	pike characin, kafue pike (FAO)
Wasserkäfer, Kolbenwasserkäfer	Hydrophilidae	water scavenger beetles, herbivorous water beetles
Wasserkalb, Rosshaarwurm, Gemeiner Wasserdrahtwurm	*Gordius aquaticus* (Nematomorpha)	gordian worm, common horsehair worm, common hair worm
Wasserkaninchen	*Sylvilagus aquaticus*	swamp rabbit
Wasserkatze, Blauer Katfisch	*Anarhichas latifrons/ Anarhichas denticulatus*	jelly cat, jelly wolffish, northern wolffish (FAO)
Wasserkobra	*Boulengerina annulata*	banded water cobra
Wasserkobras	*Boulengerina* spp.	water cobras
Wassermanguste, Sumpfmanguste, Sumpfichneumon	*Atilax paludinosus*	marsh mongoose, water mongoose
Wassermokassinschlange	*Agkistrodon piscivorus*	cottonmouth, water moccasin

Wassermoschustier, Afrikanisches Hirschferkel	*Hyemoschus aquaticus*	water chevrotain
Wassermünzenkäfer, Psepheniden	Psephenidae	water-penny beetles
Wasserpfeifer, Quellpfeifer	*Hyla crucifer/ Pseudacris crucifer*	spring peeper
Wasserralle	*Rallus aquaticus*	water rail
Wasserreh, Chinesisches Wasserreh	*Hydropotes inermis*	Chinese water deer
Wasserschlängler, Naiden	Naididae	naidid oligochaetes, naidids
Wasserschleichkatze	*Osbornictis piscivora*	aquatic genet
Wasserschuppenkopf	*Bitia hydroides*	keel-bellied water snake
Wasserschweine	Hydrochaeridae	capybaras
Wasserskorpion	*Nepa cinerea*	water scorpion
Wasserspinne	*Argyroneta aquatica*	European water spider
Wasserspinnen	Argyronetidae	water spiders
Wasserspitzmaus> Große Wasserspitzmaus	*Neomys fodiens*	Old World water shrew
Wasserspitzmaus> Kleine Wasserspitzmaus, Sumpfspitzmaus	*Neomys anomalus*	Miller's water shrew
Wasserspitzmäuse	*Neomys* spp.	Old World water shrews
Wassertreter, Wassertretkäfer	Haliplidae	crawling water beetles
Wasserwanzen	Hydrocorisae/Hydrocorizae	water bugs
Wasserzwerg, Zwergrückenschwimmer	*Plea leachi/Plea minutissima*	pygmy backswimmer
Watt-Winkerkrabbe*	*Uca rapax*	mudflat fiddler
Wattenkrebs, Wattkrebs, Schlickkrebs	*Corophium volutator*	European mud scud, mud dwelling amphipod
Wattschwamm	*Adocia cinerea*	mudflat sponge*
Watvögel & Möwenvögel & Alken	Charadriiformes	shorebirds & gulls & auks
Waxdick	*Acipenser gueldenstaedti*	Danube sturgeon, Russian sturgeon (FAO), osetr
Webbipavian	*Papio cyanocephalus*	yellow baboon
Weberknecht> Gemeiner Weberknecht	*Phalangium opilio*	common harvestman
Weberknechte, Kanker	Opiliones/Phalangida	harvestmen, "daddy longlegs"
Weberschiffchen	*Volva volva*	shuttle volva, shuttlecock volva (shuttle shell), shuttlecock, egg spindle
Weberstar	*Aplonis metallica*	shining starling
Webspinnen, Spinnen	Araneae	spiders
Wechselkröte, Grüne Kröte	*Bufo viridis*	green toad (variegated toad)
Wechselpfäffchen	*Sporophila americana*	wing-barred seedeater
Wechseltierchen, Wurzeltierchen, Rhizopoden, Amöben	Rhizopoda	amoebas, amebas
Weddell-Robbe	*Leptonychotes weddelli*	Weddell seal
Wegekuckuck, Erdkuckuck	*Geococcyx californianus*	roadrunner, greater roadrunner

Wegerich-Scheckenfalter	*Melitaea cinxia*	Glanville fritillary
Wegwespen, Spinnentöter	Pompilidae (Psammocharidae)	pompilids, spider-hunting wasps, spider wasps
Wehrvögel	Anhimidae	screamers
Weichboden-Krustenanemone	*Epizoanthus arenaceus*	gray zoanthid, gray encrusting anemone
Weicher Nagekäfer	*Ernobius mollis*	pine bark anobiid
Weichgürteltiere	*Dasypus* spp.	long-nosed armadillos
Weichhornchamäleon	*Chamaeleo tenuis*	slender chameleon
Weichkäfer	Cantharidae	soldier beetles & sailor beetles
Weichkäfer	*Cantharis* spp.	soldier beetles
Weichkopf-Grenadier	*Malacocephalus laevis*	softhead rattail
"Weichkoralle"	*Paracerianthus lloydi*	
Weichkorallen, Lederkorallen	Alcyonacea/Alcyonaria	soft corals, alcyonaceans
Weichkorallen, Lederkorallen	Alcyoniidae	soft corals, alcyoniids
Weichtiere, Mollusken	Mollusca	mollusks, molluscs
Weichwanzen, Blindwanzen	Miridae	mirids, capsid bugs, plant bugs
Weiden-Saumbandspanner, Espenfrischgehölz-Saumbandspanner, Birken-Braunhalsspanner, Espen-Saumbandspanner	*Epione parallelaria/ Epione vespertaria*	dark bordered beauty (moth)
Weiden-Steinfliegen*	Taeniopterygidae	willowflies
Weidenammer	*Emberiza aureola*	yellow-breasted bunting
Weidenblattrollgallmücke	*Rhabdophaga clausilia*	willow leaf-folding midge
Weidenblattwespe	*Nematus salicis*	willow sawfly
Weidenbohrer	*Cossus cossus*	goat moth
Weidengelbeule	*Xanthia togata*	pink-barred sallow
Weidengelbkehlchen	*Geothlypis trichas*	common yellowthroat
Weidenglucke, Heidelbeer-Glucke	*Phyllodesma ilicifolia*	small lappet (moth)
Weidenholzgallmücke	*Helicomyia saliciperda*	willow shot-hole midge, willow wood midge
Weidenkahneule	*Earias chlorana*	cream-bordered green pea
Weidenkarmin	*Catocala electa*	rosy underwing
Weidenknospenblattwespe	*Euura mucronata*	willow bud sawfly
Weidenlaubsänger, Zilpzalp	*Phylloscopus collybita*	chiffchaff
Weidenmeise	*Parus montanus*	willow tit
Weidenrosen-Gallmücke	*Rhabdophaga rosaria*	willow rosegall midge, willow rosette-gall midge, terminal rosette-gall midge
Weidenrüssler, Erlenrüssler	*Cryptorhynchus lapathi*	poplar-and-willow borer, osier weevil, willow weevil
Weidenruten-Gallmücke	*Rhabdophaga salicis*	willow stem gall midge
Weidenschaumzikade	*Aphrophora salicina*	willow froghopper
Weidenschildlaus	*Chionaspis salicis*	willow scale
Weidenspanner*	*Peribatodes rhomboidaria*	willow beauty
Weidensperling	*Passer hispaniolensis*	Spanish sparrow
Weidenspringrüssler	*Rhynchaenus salicis*	European willow flea weevil

Weidentriebwespe	*Janus luteipes*	willow stem sawfly
Weihnachtsbaum-Röhrenwurm, Tannenbäumchen-Röhrenwurm	*Spirobranchus giganteus*	spiral-gilled tubeworm, Christmas tree worm
Weihnachtsbaumkoralle*	*Studeriotes longiramosa*	Christmas tree coral
Weinberg-Maulwurfsgrille	*Gryllotalpa vineae*	vineyard mole-cricket
Weinhähnchen	*Oecanthus pellucens*	fragile whistling cricket, European tree cricket, Italian cricket
Weinkellermotte, Weinmotte	*Oinophila v-flavum*	yellow V carl
Weinkellerspinne*	*Psilochorus simoni*	wine cellar spider
Weinmotte, Bekreuzter Traubenwickler (Sauerwurm/ Gelbköpfiger Sauerwurm)	*Lobesia botrana*	grape fruit moth, vine moth, European vine moth, grape-berry moth
Weinmotten	Oinophilidae	oinophilid moths
Weinrosenkauri*	*Mauritia eglantina*	eglantine cowrie
Weinzwirner, Lauer, Blutrote Zikade	*Tibicina haematodes/ Tibicen haematodes*	vineyard cicada
Weiß-Orfe	*Notropis albeolus*	white shiner
Weißaale	Myrocongridae	myrocongrids
Weißadereule	*Mythimna pallens*	common wainscot
Weißaugen-Finsterspinne	*Amaurobius fenestralis*	white-eyed spider, window lace-weaver
Weißaugen-Trugschwalbe, Sirintaraschwalbe	*Pseudochelidon sirintarae*	white-eyed river martin
Weißaugenvireo	*Vireo griseus*	white-eyed vireo
Weißbandschwalbe	*Atticora fasciata*	white-banded swallow
Weißbart-Grasmücke	*Sylvia cantillans*	subalpine warbler
Weißbart-Seeschwalbe	*Chlidonias hybridus*	whiskered tern
Weißbärtige Archenmuschel	*Barbatia candida*	white-beard ark (white-beard ark shell)
Weißbartlangur	*Presbytis vetulus/ Presbytis senex*	purple-faced leaf monkey
Weißbartpekari, Bisamschwein	*Tayassu pecari*	white-lipped peccary
Weißbauch-Bürstenfellratte*	*Uranomys ruddi*	white-bellied brush-furred rat
Weißbauch-Schuppentier	*Manis tricuspis*	African tree pangolin
Weißbauchducker, Gabunducker	*Cephalophus leucogaster*	white-bellied duiker
Weißbauchfregattvogel	*Fregata andrewsi*	Christmas Islands frigate bird, Christmas Islands frigatebird
Weißbauchigel	*Atelerix albiventris/ Erinaceus albiventris*	four-toed hedgehog
Weißbauchratte	*Niviventer niviventer*	white-bellied rat
Weißbauchratten	*Niviventer* spp.	white-bellied rats
Weißbauchtölpel	*Sula leucogaster*	brown booby
Weißbein-Garnele	*Litopenaeus vannamei/ Penaeus vannamei*	whiteleg shrimp, Central American shrimp
Weißblumen-Umberfisch	*Nibea albiflora*	white flower croaker (FAO), yellow drum
Weißborsten-Gürteltier, Sechsbinden-Gürteltier	*Euphractus sexcinctus*	six-banded armadillo

Weißbrassen, Großer Geißbrassen	*Diplodus sargus*	white bream, white seabream (FAO)
Weißbrauendrossel	*Turdus obscurus*	eye-browed thrush
Weißbrauengibbon, Hulock	*Hylobates hoolock*	white-browed gibbon, Hoolock gibbon
Weißbrust-Doktorfisch, Weißkehlseebader, Weißkehl-Doktorfisch	*Acanthurus leucosternon*	white-breasted surgeonfish, powder-blue tang, powderblue surgeonfish (FAO)
Weißbrust-Pelomeduse	*Pelusios adansonii*	Adanson's mud turtle
Weißbrust-Perlhuhn	*Agelastes meleagrides*	white-breasted guineafowl
Weißbürzel-Strandläufer	*Calidris fuscicollis*	white-rumped sandpiper
Weißbüscheläffchen	*Callithrix jacchus*	common marmoset
Weißdorn-Karotten-Röhrenlaus	*Dysaphis crataegi*	hawthorn-carrot aphid
Weißdorneule	*Allophyes oxyacanthae*	green-brindled crescent
Weißdornspanner, Gelbspanner	*Opisthograptis luteolata*	brimstone moth
Weißdornspinner	*Trichiura crataegi*	pale eggar
Weißdornspinnmilbe	*Tetranychus viennensis*	hawthorn spider mite
Weißdornwickler, Obstbaumwickler	*Archips crataegana*	fruit tree tortrix (moth), brown oak tortrix, oak red-barred twist
Weißdornzweiggallmücke	*Thomasiniana crataegi/ Resseliella crataegi*	hawthorn stem midge
Weiße Babyschnecke*	*Sinum perspectivum*	baby's ear moonsnail, white baby ear
Weiße Bohrmuschel	*Barnea candida*	white piddock
Weiße Dickkopf-Schwebfliege	*Scaeva pyrastri*	cabbage aphid hover fly
Weiße Eischnecke, Weißes Vogelei	*Pseudosimnia adriatica/ Aperiovula adriatica/ Ovula adriatica*	Adriatic ovula
Weiße Fledermaus	*Ectophylla alba*	white bat
„Weiße Fliege"	*Trialeurodes vaporariorum*	greenhouse whitefly
Weiße Gabelmakrele	*Trachinotus glaucus*	glaucous pompano
Weiße Hammermuschel	*Malleus albus*	white hammer-oyster
Weiße Heideschnecke	*Xerolentha obvia*	
Weiße Hornkoralle, Weiße Gorgonie	*Eunicella singularis*	white horny coral
Weiße Hydra, Weißer Süßwasserpolyp	*Hydra americana*	white hydra
Weiße Käferschnecke, Weißer Chiton, Helle Käferschnecke	*Stenosemus albus/ Ischnochiton albus*	northern white chiton, white chiton, white northern chiton
Weiße Kaurischnecke, Weißkauri	*Trivia candidula/ Niveria candidula*	white trivia
Weiße Koralle	*Isidella elongata*	white coral*
Weiße Mittelmeer-Hornkoralle	*Eunicella stricta*	Mediterranean white horny coral
Weiße Muräne	*Siderea grisea*	geometric moray
Weiße Pfeffermuschel, Kleine Pfeffermuschel	*Abra alba*	white abra (white furrow-shell)
Weiße Schildkrötenschnecke	*Tectura virginea/ Acmaea virginea*	white tortoiseshell limpet

Weiße Seescheide, Jungfern-Seescheide	*Ascidia virginea*	virgin sea-squirt*
Weiße Stumpfnase	*Rhabdosargus globiceps*	white stumpnose
Weiße Tigermotte, Punktierter Fleckleib-Bär, Minzenbär	*Spilosoma lubricipeda/ Spilosoma menthastri*	white ermine moth
Weiße Trogmuschel	*Mactra corallina corallina*	rayed trough clam
Weiße Turmschnecke, Große Turmschnecke, Märzenschnecke	*Zebrina detrita*	
Weißer Atun	*Thyrsitops lepidopoides*	white snake mackerel
Weißer Bäumchenpolyp	*Eudendrium album*	white stickhydroid
Weißer Crappie	*Pomoxis annularis*	white crappie
Weißer Gabeldorsch	*Urophycis tenuis*	white hake
Weißer Gitterkalkschwamm	*Clathrina coriacea*	white lattice sponge*
Weißer Graszünsler	*Crambus perellus*	yellow satin grass-veneer
Weißer Hai, Weißhai, Menschenhai	*Carcharodon carcharias/ Carcharodon rondeletii*	white shark, great white shark (FAO)
Weißer Kalifornia-Umberfisch	*Genyonemus lineatus*	white croaker
Weißer Kartoffelnematode	*Globodera pallida*	white potato cyst nematode, pale potato cyst nematode, potato root eelworm
Weißer Katzenwels	*Ictalurus catus*	fork-tailed catfish
Weißer Marlin, Weißer Speerfisch	*Tetrapturus albidus*	white marlin
Weißer Mond, Weiße Mondschnecke	*Neverita albumen*	egg-white moonsnail
Weißer Oryx, Arabischer Spießbock	*Oryx leucoryx*	Arabian oryx
Weißer Sägebarsch	*Roccus chrysops/ Morone chrysops*	white bass
Weißer Speerfisch, Weißer Marlin	*Tetrapturus albidus*	white marlin
Weißer Stör, Sacramento-Stör	*Acipenser transmontanus*	white sturgeon
Weißer Süßwasserpolyp, Weiße Hydra	*Hydra americana*	white hydra
Weißer Thun, Germon	*Thunnus alalunga*	albacore (FAO), "white" tuna, long-fin tunny, long-finned tuna, Pacific albacore
Weißer Ukelei, Laube	*Alburnus albidus*	Italian bleak
Weißer Waldportier	*Brintesia circe*	great banded grayling, greater wood nymph
Weißer Westindischer Seeigel, Verschiedenfarbiger Seeigel	*Lytechinus variegatus*	variegated urchin
Weißer Zackenbarsch	*Epinephelus aeneus*	white grouper
Weißes C, C-Falter	*Comma c-album/ Polygonia c-album/ Nymphalis c-album*	comma

Weißes L, L-Eule	Comma l-album/ Mythimna l-album	false comma, Compton tortoiseshell
Weißes Vogelei, Weiße Eischnecke	Pseudosimnia adriatica/ Aperiovula adriatica/ Ovula adriatica	Adriatic ovula
Weißes W, Ulmen-Zipfelfalter	Satyrium w-album/ Strymon w-album/ Strymonidia w-album	white-letter hairstreak
Weißflanken-Jackrabbit, Mexikohase	Lepus callotis	white-sided jack rabbit
Weißflecken-Congeraal	Astroconger myriaster	white-spotted conger
Weißflecken-Feilenfisch	Cantherinus macrocerus	white-spotted filefish
Weißflecken-Geigenrochen	Rhinobatos albomaculatus	white-spotted guitarfish
Weißflecken-Kugelfisch, Grauer Puffer	Arothron hispidus	toby, blaasop, toadfish, white-spotted puffer (FAO)
Weißflecken-Randschnecke*	Prunum guttatum	white-spot marginella
Weißflecken-Täubchen	Nitidella ocellata	white-spotted dovesnail
Weißflecken-Ulmeneule	Cosmia diffinis	white-spotted pinion moth
Weißfleckenfalter, Veilchen-Scheckenfalter	Melitaea cynthia/ Euphydryas cynthia	Cynthia's fritillary
Weißfleckenwidderchen, Weißfleckwidderchen, Ringelwidderchen	Syntomis phegea	yellow-belted burnet
Weißfleckige Eichengallwespe	Andricus albopunctatus/ Andricus paradoxus	oak leaf gall wasp
Weißflossiger Gründling, Flachlandgründling	Gobio albipinnatus	Russian whitefin gudgeon, white-finned gudgeon (FAO)
Weißflügel-Seeschwalbe	Chlidonias leucopterus	white-winged black tern
Weißflügelente	Cairina scutulata/ Asarcornis scutulatus	white-winged wood duck
Weißflügelgimpel	Rhodospiza obsoleta	desert finch
Weißflügellalage	Lalage sueurii	white-winged triller
Weißflügellerche	Melanocorypha leucoptera	white-winged lark
Weißflügelsittich	Brotogeris versicolorus	canary-winged parakeet
Weißflügeltaube	Zenaida asiatica	white-winged dove
Weißflügelvampir	Diaemus youngi	white-winged vampire bat
Weißfuchs (Hbz.), Eisfuchs, Polarfuchs	Alopex lagopus	arctic fox
Weißfuß- Eichenlinsengallwespe (>Glatte Linsengalle/ Kleine Blattrandgalle)	Neuroterus albipes/ Neuroterus laeviusculus	oak leaf smooth-gall cynipid wasp, Schenck's gall wasp, smooth spangle gall wasp (>smooth spangle gall)
Weißfuß-Seegurke	Holothuria nobilis	whitefoot sea cucumber*
Weißfußaffe	Saguinus leucopus	white-footed tamarin
Weißfußmaus	Peromyscus leucopus	white-footed mouse
Weißfußmäuse	Peromyscus spp.	white-footed mice, deer mice
Weißfußvogelspinne, Rotfußvogelspinne	Avicularia metallica	whitetoe tarantula
Weißgefleckte Leimkraut-Kapseleule, Abendnelken-Kapseleule, Steppenheidehügel- Nelkeneule	Hadena albimacula	white spot moth, white-spotted coronet

Weißgefleckter Glatthai, Nördlicher Glatthai, Gefleckter Glatthai	*Mustelus asterias*	stellate smooth-hound, starry smooth-hound (FAO)
Weißgeränderte Band-Fechterschnecke*, Geränderte Flügelschnecke*	*Strombus marginatus*	marginate conch
Weißgesichts-Seidenäffchen	*Callithrix geoffroyi*	white-faced marmoset
Weißgesichtsschildkröte, Fitzroy-Schildkröte	*Rheodytes leukops*	Fitzroy's turtle, Fitzroy turtle
Weißgetupfte Bandschnecke	*Leucozonia ocellata*	white-spot latirus, white-spotted latirus
Weißgrauer Breitflügelspanner	*Agriopis leucophaearia*	spring umber, spring usher
Weißhai, Weißer Hai, Menschenhai	*Carcharodon carcharias/ Carcharodon rondeletii*	white shark, great white shark (FAO)
Weißhals-Faulvogel	*Notharchus macrorhynchos*	white-necked puffbird
Weißhandgibbon, Lar	*Hylobates lar*	common gibbon, white-handed gibbon
Weißhaubenkakadu	*Cacatua alba*	white cockatoo
Weißkauri, Weiße Kaurischnecke	*Trivia candidula/ Niveria candidula*	white trivia
Weißkehl-Breitschnabel	*Platyrinchus mystaceus*	white-throated spadebill
Weißkehl-Spinnenjäger	*Arachnothera longirostra*	little spiderhunter
Weißkehl-Spornhuhn	*Galloperdix bicalcarata*	Ceylon spurfowl
Weißkehlammer	*Zonotrichia albicollis*	white-throated sparrow
Weißkehlgirlitz	*Serinus albogularis*	white-throated seedeater
Weißkehlmeerkatze	*Cercopithecus albogularis*	Sykes monkey
Weißkehlsänger	*Irania gutturalis*	irania
Weißkehlseebader, Weißbrust-Doktorfisch, Weißkehl-Doktorfisch	*Acanthurus leucosternon*	white-breasted surgeonfish, powder-blue tang, powderblue surgeonfish (FAO)
Weißkopf-Dasselfliege*	*Chrysomya albiceps*	banded blow fly
Weißkopf-Ruderente	*Oxyura leucocephala*	white-headed duck
Weißkopfmaki	*Lemur albifrons*	white-headed lemur
Weißkopfmöve	*Larus cachinnans*	yellow-legged gull
Weißkopfsaki, Blasskopfsaki	*Pithecia pithecia*	white-faced saki
Weißkopfseeadler	*Haliaeetus leucocephalus*	bald eagle
Weißlachs	*Stenodus leucichthys*	inconnu (FAO), nelma
Weißlicher Kiefernspanner	*Peribatodes secundaria*	white pine beauty*
Weißlinge	Sillaginidae	sillagos (whitlings, smelt-whitlings)
Weißlinge	Pieridae	whites & sulphurs & orange-tips
Weißlinien-Drückerfisch	*Sufflamen bursa/ Hemibalistes bursa*	boomerang triggerfish (FAO), drab triggerfish
Weißlippen-Bambusotter	*Trimeresurus albolabris*	white-lipped tree viper
Weißlippenhirsch	*Cervus albirostris*	Thorold's deer
Weißlippenpython	*Morelia albertisii*	D'Albertis' python
Weißmantelstumpfnase	*Pygathrix brelichi*	Brelich's snub-nosed monkey, Guizhou snub-nosed monkey
Weißmaul-Klappschildkröte	*Kinosternon leucostomum*	white-lipped mud turtle
Weißmaul-Umberfisch	*Micropogonias furnieri*	white-mouth croaker

Weißnacken-Moorantilope	*Kobus megaceros*	Nile lechwe
Weißnacken-Vogelspinne, Blaue Argentinische Vogelspinne	*Aphonopelma saltator*	blue Argentine tarantula*
Weißnackenwiesel	*Poecilogale albinucha*	African striped weasel
Weißnasensaki	*Chiropotes albinasus*	white-nosed saki
Weißohrbülbül	*Pycnonotus leucogenys*	white-cheeked bulbul
Weißpunkt-Buntbarsch	*Tropheus duboisi*	duboisi
Weißpunkt-Kofferfisch, Schwarzer Kofferfisch	*Ostracion meleagris*	whitespotted boxfish
Weißrandfledermaus	*Pipistrellus kuhlii*	Kuhl's pipistrelle
Weißrandiger Grashüpfer	*Chorthippus albomarginatus*	lesser marsh grasshopper
Weißrochen, Pazifischer Weißrochen, Stachliger Tiefenrochen	*Bathyraja spinosissima/ Psammobatis spinosissima*	white skate (FAO), spiny deep-water skate
Weißrücken-Pfeifgans	*Thalassornis leuconotos*	white-backed duck
Weißrückenschwalbe	*Cheramoeca leucosterna*	white-backed swallow
Weißrückenspecht	*Picoides leucotos/ Dendrocopos leucotos*	white-backed woodpecker
Weißscheitel-Scherenschwanz	*Enicurus leschenaulti*	white-crowned forktail
Weißschnauzendelphin	*Lagenorhynchus albirostris*	white-beaked dolphin
Weißschulter-Seidenäffchen	*Callithrix humeralifer*	Santarem marmoset
Weißschulterkapuziner, Weißschulteraffe, Kapuziner	*Cebus capucinus*	white-throated capuchin
Weißschwamm*	*Geodia gibberosa*	white sponge
Weißschwanz-Degenflügel	*Campylopterus ensipennis*	white-tailed sabrewing
Weißschwanz-Doktorfisch	*Acanthurus xanthopterus*	purple surgeonfish, yellowfin surgeonfish (FAO)
Weißschwanz-Erdhummel	*Bombus lucorum*	white-tailed bumble bee
Weißschwanz-Präriehund	*Cynomys leucurus*	white-tailed prairie dog
Weißschwanz-Tropikvogel	*Phaethon lepterus*	white-tailed tropic bird
Weißschwanzbülbül	*Baeopogon indicator*	honeyguide greenbul
Weißschwanzbussard	*Buteo albicaudatus*	white-tailed hawk
Weißschwanzeselhase, Präriehase	*Lepus townsendii*	white-tailed jack rabbit
Weißschwanzgnu	*Connochaetes gnou*	black wildebeest (white-tailed gnu)
Weißschwanzichneumon	*Ichneumia albicauda*	white-tailed mongoose
Weißschwänziger Hamster, Afrika-Hamster	*Mystromys albicaudatus*	white-tailed rat, white-tailed mouse
Weißschwanzkiebitz	*Chettusia leucura*	white-tailed plover
Weißschwinguan	*Penelope albipennis*	white-winged guan
Weißseitendelphin	*Lagenorhynchus acutus*	Atlantic white-sided dolphin
Weißspecht	*Melanerpes candidus*	white woodpecker
Weißspitzen-Riffhai, Weißspitzen-Hundshai, Weißspitzen-Marderhai	*Triaenodon obesus*	blunthead shark, whitetip reef shark (FAO)
Weißspitzenhai, Hochsee-Weißflossenhai, Weißspitzen-Hochseehai, Langflossen-Hai	*Carcharhinus longimanus/ Carcharhinus maou*	whitetip shark, whitetip oceanic shark, oceanic whitetip shark (FAO)
Weißstern-Qualle*	*Staurophora mertensi*	whitecross jellyfish

Weißstirn-Seekuhaal, Weißstirn-Messeraal	*Apteronotus albifrons*	black ghost
Weißstirnamazone	*Amazona albifrons*	white-fronted amazon
Weißstirnkapuziner	*Cebus albifrons*	white-fronted capuchin, brown pale-fronted capuchin
Weißstirnlangur	*Presbytis frontata*	white-fronted leaf monkey
Weißstirnlerche	*Eremopterix nigriceps*	black-crowned sparrow-lark
Weißstirntrappist	*Monasa morphoeus*	white-fronted nunbird
Weißstirnweber	*Amblyospiza albifrons*	grosbeak weaver
Weißstorch	*Ciconia ciconia*	white stork
Weißstreifen-Fledermäuse	*Vampyrops* spp.	white-lined bats
Weißstreifen-Kammmuschel	*Chlamys luculenta*	white-streaked scallop
Weißstreifendelphin, Pazifischer Weißseiten-Delphin	*Lagenorhynchus obliquidens*	Pacific white-sided dolphin
Weißstreifiger Distelbock	*Agapanthia cardui*	whitestripe thistle longhorn beetle*
Weißtannenrüssler	*Pissodes piceae*	white fir weevil
Weißtannentrieblaus	*Mindarus abietinus*	balsam twig aphid
Weißwal, Beluga	*Delphinapterus leucas*	white whale, beluga
Weißwangengans	*Branta leucopsis*	barnacle goose
Weißwangengibbon, Schopfgibbon	*Hylobates concolor*	crested gibbon
Weißwedelhirsch	*Odocoileus virginianus*	white-tailed deer
Weißwollige Fichtenstammlaus	*Pineus pineoides*	small spruce adelges, small spruce woolly aphid
Weitäugiger Butt	*Bothus podas*	wide-eyed flounder
Weitmäulige Purpurschnecke	*Purpura patula*	wide-mouthed purpura
Weitmündige Glasschnecke	*Semilimax semilimax*	wide-mouth glass snail
Weitmundschnecken*	Stomatellidae	false ear snails, widemouth snails (widemouth shells)
Weizen-Furchenwasserkäfer	*Helophorus nubilus*	wheat shoot beetle, wheat mud beetle
Weizenälchen, Weizengallenälchen (Radekrankheit)	*Anguina tritici*	wheat-gall nematode, wheat nematode, wheat eelworm
Weizenwurzelgallenälchen	*Meloidogyne naasi*	root-knot nematode (on wheat)
Wellenastrild	*Estrilda astrild*	common waxbill
Wellenband-Lanzenotter	*Ophryacus undulatus/ Bothrops undulatus*	undulated pit viper, Mexican horned pit viper
Wellenflughuhn	*Pterocles lichtensteinii*	Lichtenstein's sandgrouse
Wellenkoralle	*Leptoseris cucullata*	sunray lettuce coral
Wellenläufer	*Oceanodroma leucorhoa*	Leach's petrel
Wellenlinien-Pilgermuschel	*Aequipecten lineolaris*	wavy-lined scallop
Wellenohrspecht	*Dryocopus galeatus/ Ceophloeus galeatus*	helmeted woodpecker
Wellensittich	*Melopsittacus undulatus*	budgerigar ("budgie", parakeet)
Wellensittich-Schuppenbeinmilbe*	*Knemidokoptes pilae*	scalyleg mite (of budgerigars)
Wellenspanner	*Rheumaptera undulata*	undulate looper*
Wellington-Flugkalmar	*Nototodarus sloani*	Wellington flying squid

Wels, Waller, Schaiden	*Silurus glanis*	European catfish, wels, sheatfish, wels catfish (FAO)
Welse, Echte Welse	Siluridae	sheatfishes
Welse, Welsartige	Siluriformes	catfishes
Wendehals	*Jynx torquilla*	wryneck
Wendehalsfrösche	*Phrynomantis* spp.	snake-neck frogs
Wendelfalten-Olivenschnecke	*Olivancillaria contortuplicata*	twisted plait olive
Wendeltreppen	Scalidae	wentletraps
Wendeltürmchen	*Turbonilla jeffreysii*	staircase pyramid
Wenigborster, Oligochaeten	Oligochaeta	oligochetes
Wenigfüßer, Pauropoden	Pauropoda	pauropods
Werftkäfer	Lymexylonidae	ship-timber beetles
Wermutregenpfeifer	*Charadrius asiaticus*	Caspian plover
Werre, Maulwurfsgrille	*Gryllotalpa gryllotalpa*	mole cricket
Wespenbienen, Kuckucksbienen	*Nomada* spp.	cuckoo bees
Wespenbussard	*Pernis apivorus*	western honey buzzard
Wespenspinne, Zebraspinne	*Argiope bruennichi*	black-and-yellow argiope, black-and-yellow garden spider
West Virginia-Quellensalamander	*Gyrinophilus subterraneus*	West Virginia spring salamander
West-Amerikanische Schwarze Witwe	*Latrodectus hesperus*	western black widow spider
West-Atlantische Seespinne	*Libinia emarginata*	portly spider crab, common spider crab
West-Basilisk	*Basiliscus galeritus*	western basilisk
West-Indische Gorgonien-Porzellanschnecke	*Simnia acicularia/ Cymbovula acicularis*	West Indian simnia
West-Indische Riesenreisratte	*Megalomys* spp.	West Indian giant rice rats
Westafrikanische Erdkobra	*Paranaja* spp.	burrowing cobras
Westafrikanische Erdwühlen	*Geotrypetes* spp.	West African caecilians
Westafrikanische Gebüschratten*, Breitfuß-Gebüschratten*	*Thamnomys* spp.	thicket rats, broad-footed thicket rats
Westafrikanische Ilisha	*Ilisha africana*	West African ilisha
Westafrikanische Klappen-Weichschildkröten	*Cycloderma* spp.	Central African flatshells
Westafrikanische Meerbarbe	*Pseudupeneus prayensis*	West African goatfish
Westafrikanische Nachtotter	*Causus maculatus*	West African night adder
Westafrikanischer Ebarme	*Psettodes bennetti*	spiny turbot
Westafrikanischer Lungenfisch	*Protopterus annectens*	West African lungfish
Westafrikanischer Rotstummelaffe	*Piliocolobus badius*	western red colobus
Westamerikanische Korbmuschel, Gelbe Korbmuschel	*Corbula luteola*	common western corbula, yellow basket clam
Westamerikanische Maulwürfe	*Scapandus* spp.	Western American moles
Westamerikanische Wassermolche	*Taricha* spp.	Pacific newts

Westamerikanischer Kreuzwels, „Minihai"	Arius seemanni/ Arius jordani	Tete Sea catfish (FAO), shark catfish
Westamerikanischer Rauhaut-Molch	Taricha granulosa	rough-skinned newt
Westasiatische Hornviper	Pseudocerastes persicus	Persian false horned viper
Westatlantische Alse	Alosa mediocris	hickory shad
Westatlantische Bohrmuschel*	Barnea truncata	Atlantic mud-piddock
Westatlantische Gabeldorsche	Urophycis spp.	West Atlantic hakes
Westatlantische Holzbohrmuschel*	Xylophaga atlantica	Atlantic woodeater, Atlantic wood piddock
Westatlantische Miesmuschel, Braune Miesmuschel	Perna perna	Mexilhao mussel, South American rock mussel, "brown mussel"
Westatlantische Panopea	Panopea bitruncata	Atlantic geoduck
Westatlantischer Glatthai	Mustelus canis	dusky smooth-hound (FAO), smooth dogfish
Westatlantischer Glattrochen, Spitznasen-Rochen	Raja laevis	barndoor skate (FAO), sharpnose skate
Westatlantischer Lippfisch, Schweinsfisch	Lachnolaimus maximus	hogfish (FAO), hogsnapper
Westatlantischer Sandaal	Ammodytes dubius	northern sandlance
Westatlantischer Scheibenbauch	Liparis atlanticus	Atlantic sea snail
Westatlantischer Teufelsrochen	Mobula hypostoma	West Atlantic devil ray, lesser devil ray (FAO)
Westaustralische Schlangenhalsschildkröte	Chelodina steindachneri	dinner-plate turtle, Steindachner's side-necked turtle
Westblindmaus	Nanospalax leucodon/ Microspalax leucodon	lesser mole-rat
Westigel, Braunbrustigel	Erinaceus europaeus	western hedgehog (European hedgehog)
Westindische Fechterschnecke, Westindische Flügelschnecke	Strombus pugilis	fighting conch, West Indian fighting conch
Westindische Kronenschnecke, Westindische Krone	Melongena melongena	West Indian crown conch
Westindische Stern-Turbanschnecke	Lithopoma tectum	West Indian starsnail
Westindische Wurmschnecke	Vermicularia spirata	West Indian wormsnail, common wormsnail
Westindischer Palolo	Eunice schemacephala	West Indian palolo worm
Westkaukasischer Steinbock	Capra caucasica	West-Caucasian tur
Westküsten-Seezunge	Austroglossus microlepis	West Coast sole (South Africa)
Westliche Baumauster*	Isognomon recognitus	purple purse-oyster, western tree-oyster (purse shell)
Westliche Beißschrecke	Platycleis albopunctata/ Platycleis denticulata	western bushcricket
Westliche Braunschlange	Pseudonaja nuchalis	western brown snake
Westliche Dornschrecke	Tetrix ceperoi	Cepero's groundhopper
Westliche Erdotter	Atractaspis corpulenta	western mole viper
Westliche Flussperlmuschel	Margaritifera falcata	western pearlshell, western freshwater pearl mussel

Westliche Hakennatter	*Heterodon nasicus*	western hognose snake
Westliche Harlekingarnele	*Hymenocera elegans*	western harlequin shrimp
Westliche Hüpfmaus	*Zapus princeps*	western jumping mouse
Westliche Keiljungfer	*Gomphus pulchellus*	western European gomphus
Westliche Kreischeule	*Otus kennicottii*	western screech owl
Westliche Rennechse	*Cnemidophorus tigris*	western whiptail
Westliche Sandboa, Sandschlange	*Eryx jaculus*	Javelin sand boa
Westliche Stülpnasen-Lanzenotter	*Porthidium ophryomegas*	western hog-nosed viper, slender hognose viper
Westliche Taschenratte	*Thomomys mazama*	western pocket gopher
Westliche Weißohr-Riesenratte	*Hyomys dammermani*	western white-eared giant rat
Westlicher Flossenfuß	*Pygopus nigriceps*	western scalyfoot
Westlicher Kielnagelgalago, Südlicher Kielnagelgalago	*Galago elegantulus/ Euoticus elegantulus*	western needle-clawed bush baby, elegant needle-clawed galago
Westlicher Rotrückensalamander	*Plethodon vehiculum*	western red-backed salamander
Westlicher Schaufelfuß	*Scaphiopus hammondii*	western spadefoot, Hammond's spadefoot toad
Westlicher Schlammtaucher	*Pelodytes punctatus*	parsley frog, common parsley frog, mud-diver, spotted mud frog
Westlicher Skink	*Eumeces skiltonianus*	western skink
Westlicher Zaunleguan	*Sceloporus occidentalis*	western fence lizard
Westliches Graues Riesenkänguru	*Macropus fuliginosus*	western gray kangaroo
Westliches Grauhörnchen	*Sciurus griseus*	western gray squirrel
Westliches Strauchwallaby	*Macropus irma*	western brush wallaby
Westmediterrane Süßlippe	*Plectorhinchus mediterraneus*	rubberlip grunt
Westpazifische Brutanemone	*Epiactis prolifera*	proliferating anemone
Westpazifische Pelamide	*Sarda orientalis*	striped bonito
Westschermaus	*Arvicola sapidus*	southwestern water vole, southern water vole
Weynsducker	*Cephalophus weynsi*	Weyns' duiker
Wheeler's Partnergrundel	*Amblyeleotris wheeleri*	Wheeler's goby, gorgeous prawn-goby (FAO)
Wickelbär	*Potos flavus*	kinkajou
Wickler	Tortricidae	tortricids, leaf rollers, leaf tyers
Widderbären, Fleckenschwärmerchen	Syntomidae	wasp moths
Widderbock, Gemeiner Widderbock	*Clytus arietis*	wasp beetle
Wiedehopf	*Upupa epops*	hoopoe
Wiedehopfe	Upupidae	hoopoes
Wiederkäuer, Retroperistaltiker, Ruminantier	Ruminantia	ruminants, "cud chewers"
Wieds Leguan	*Enyalius catenatus*	Wied's fathead anole
Wiegenmuschel	*Donax cuneatus*	cuneate wedge clam, cuneate beanclam, cradle donax

Wiesel	*Mustela* spp.	weasels
Wiesel & Dachse & Skunks & Otter	Mustelidae	weasels & badgers & skunks & otters
Wiesel-Olive	*Oliva mustelina*	weasel olive
Wieselhaie	Hemigaleidae	weasel sharks
Wieselkatze, Jaguarundi	*Felis yagouaroundi*	jaguarundi
Wieselmaki> Großer Wieselmaki	*Lepilemur mustelinus*	greater weasel lemur, sportive lemur
Wieselmaki> Kleiner Wieselmaki	*Lepilemur ruficaudatus*	red-tailed sportive lemur
Wieselmakis	*Lepilemur* spp.	weasel lemurs, sportive lemurs
Wieselmeerschweinchen	*Galea* spp.	yellow-toothed cavies, cuis
Wiesen-Grashüpfer	*Chorthippus dorsatus*	meadow grasshopper
Wiesen-Staubeule, Feuchtwiesen-Staubeule	*Athetis pallustris*	marsh moth
Wiesenälchen	*Pratylenchus pratensis*	De Man's meadow nematode (brown rot of tobacco)
Wiesenameisen & Holzameisen & Wegameisen	*Lasius* spp.	field ants (black ants)
Wiesenbläuling, Hauhechelbläuling	*Polyommatus icarus*	common blue
Wieseneidechse	*Lacerta praticola*	pontic lizard, meadow lizard
Wiesenhummel	*Bombus pratorum*	early bumble bee
Wiesenhüpfmaus	*Zapus hudsonius*	meadow jumping mouse
Wiesenotter (Karstotter), Spitzkopfotter	*Vipera ursinii*	meadow viper, Orsini's viper
Wiesenpieper	*Anthus pratensis*	meadow pipit
Wiesenrauten-Blattspanner	*Perizoma sagittata*	marsh carpet (moth)
Wiesenschaumzikade	*Philaenus spumarius*	common froghopper
Wiesenschnake	*Tipula paludosa*	meadow cranefly, grey daddy-long-legs (Br.)
Wiesenspinner, Herbstspinner	Lemoniidae	lemoniid moths
Wiesenstärling	*Sturnella neglecta*	western meadowlark
Wiesenstrandläufer	*Calidris minutilla*	least sandpiper
Wiesenweihe	*Circus pygargus*	Montague's harrier
Wiesenwühlmaus	*Microtus pennsylvanicus*	meadow vole
Wiesenzünsler, Rübenzünsler	*Loxostege sticticalis/ Margaritia sticticalis*	beet webworm
Wilde Dreiklaue, Florida-Weichschildkröte	*Trionyx ferox/ Apalone ferox*	Florida softshell turtle
Wildesel	*Equus asinus*	donkey, burro
Wildkanin (Hbz.), Altweltliches Wildkaninchen, Europäisches Wildkaninchen	*Oryctolagus cuniculus*	Old World rabbit, domestic rabbit
Wildkatze	*Felis silvestris*	wild cat
Wildschaf> Orientalisches Wildschaf	*Ovis orientalis*	Asiatic mouflon

Wildschwein	Sus scrofa	wild boar, pig
Willards Klapperschlange	Crotalus willardi	ridgenose rattlesnake, ridge-nosed rattlesnake
Williams Krötenkopf-Schildkröte	Phrynops williamsi	William's side-necked turtle
Williams-Gecko	Diplodactylus williamsi	Williams' spinytail gecko, Williams' diplodactyl
Wilsondrossel	Catharus fuscescens	veery
Wilsons Eisfisch	Chaenodraco wilsoni	Wilson icefish, spiny icefish (FAO)
Wilsonwassertreter	Phalaropus tricolor	Wilson's phalarope
Wimpel-Kardinalbarsch, Pyjama-Kardinalbarsch	Apogon nematopterus	pyjama cardinal fish
Wimpelfisch	Heniochus acuminatus	pennant coralfish
Wimpelpiranha, Schuppenräuber	Catoprion mento	wimple piranha
Wimpelträger	Pteridophora alberti	King of Saxony bird-of-paradise
Wimper-Kalkschwamm, Wimpern-Kalkschwamm, Meergürkchen	Scypha ciliata/ Sycon ciliatum	little vase sponge
Wimperbock> Gemeiner Wimperbock, Kiefernzweigbock	Pogonocherus fasciculatus	conifer-wood longhorn beetle*
Wimperfledermaus	Myotis emarginata	Geoffroy's bat
Wimperspitzmäuse	Crocidura spp.	white-toothed shrews
Wimpertierchen, Ciliaten	Ciliata	ciliates
Windelschnecke> Gemeine Windelschnecke, Zwergwindelschnecke	Vertigo pygmaea	dwarf whorl snail, common whorl snail, crested vertigo (U.S.)
Windelschnecken	Vertiginidae	vertigos, whorl snails & chrysalis snails
Windeneulchen	Emmelia trabealis	spotted sulphur
Windeneule	Tyta luctuosa	four-spotted moth
Windenschwärmer	Agrius convolvuli/ Herse convolvuli/ Sphinx convolvuli	convolvulus hawkmoth, morning glory sphinx moth
Windling-Geistchen, Federmotte	Pterophorus monodactylus	common plume, sweet-potato plume moth
Windspielantilopen, Dikdiks	Madoqua spp.	dik-diks
Winkel-Olive	Oliva incrassata	angled olive, angulate olive, giant olive
Winkelkopfagamen u.a.	Acanthosaura spp.	pricklenapes
Winkelkopfagamen u.a.	Gonocephalus spp.	humphead forest dragons
Winkelzahnmolche	Hynobiidae	Asiatic salamanders, hynobiids
Winkelzahnmolche> Echte Winkelzahnmolche	Hynobius spp.	Asian salamanders
Winkerkrabben	Uca spp.	fiddler crabs
Winkerkrabben & Reiterkrabben	Ocypodidae	fiddler crabs & ghost crabs
Winterflunder, Amerikanische Winterflunder	Pseudopleuronectes americanus	winter flounder

Wintergetreidemilbe*	*Penthaleus major*	winter grain mite
Wintergoldhähnchen	*Regulus regulus*	goldcrest
Winterhafte	Boreidae	snow scorpionflies
Winterlibelle, Gemeine Winterlibelle	*Sympecma fusca*	winter damselfly
Wintermücken, Winterschnaken	Trichoceridae (Petauristidae)	winter gnats, winter crane flies
Wintermücken, Winterschnaken	*Trichocera* spp.	winter gnats, winter crane flies
Winternachtschwalbe, Poorwill	*Phalaenoptilus nuttallii*	poorwill, common poorwill
Winterrochen, Westatlantischer Winterrochen	*Raja ocellata*	winter skate
Winterzecke	*Dermacentor albipictus*	winter tick, shingle tick
Wirbellose, Wirbellose Tiere, Invertebraten, Evertebraten	Invertebrata	invertebrates
Wirtelschwanzleguane	*Cyclura* spp.	West Indian rock iguanas, rhinoceros iguanas
Wisent	*Bison bonasus*	European bison, wisent
Wittling, Merlan	*Merlangius merlangus*	whiting
Witwen	*Latrodectus* spp.	widow spiders
Witwen-Drachenkopf	*Sebastes entomelas*	widow rockfish
Witwen-Olivenschnecke	*Oliva vidua*	widow olive, black olive
Witwenmonjita	*Xolmis irupero*	white monjita
Witwenpfeifgans	*Dendrocygna viduata*	white-faced whistling duck
Witwenrose	*Sagartiogeton undulatus*	widow-rose anemone*
Witwenvögel	Viduidae	whydahs
Wobbegongs, Teppichhaie, Ammenhaie	Orectolobidae	wobbegongs, carpet sharks
Wolf	*Canis lupus*	gray wolf
Wolfs Meerkatze, Wolf-Meerkatze	*Cercopithecus wolfi*	Wolf's monkey
Wolfsbarsch, Seebarsch	*Roccus labrax/ Dicentrachus labrax/ Morone labrax*	bass, sea bass, European sea bass (FAO)
Wolfsfische, Seewölfe	Anarhichadidae	wolffishes
Wolfsheringe	Chirocentridae	wolf-herrings, wolf herrings
Wolfsmilch-Ringelspinner, Grasheiden-Ringelspinner, Wolfsmilchspinner	*Malacosoma castrensis*	ground lackey moth
Wolfsmilchschwärmer	*Hyles euphorbiae*	spurge hawkmoth
Wolfsmilchwanzen	Stenocephalidae	spurge bugs
Wolfsnattern	*Lycophidion* spp.	African wolf snakes
Wolfsspinnen, Wolfspinnen	Lycosidae	wolf spiders, ground spiders

Wolfszahnnatter> Gewöhnliche Wolfszahnnatter	*Lycodon aulicus*	wolf snake
Wolfszahnnattern	*Lycodon* spp.	wolf snakes
Woll-Samtschnecke*	*Velutina lanigera*	woolly velutina, woolly lamellaria
Woll-Schildläuse, Wollläuse	Eriococcidae	mealybugs
Wollaffe	*Lagothrix lagotricha*	common woolly monkey
Wollaffen	*Lagothrix* spp.	woolly monkeys
Wollafter, Birkenwollafter	*Eriogaster lanestris*	small eggar
Wollbeutelratten	*Caluromys* spp.	woolly opossums
Wollbiene> Große Wollbiene	*Anthidium manicatum*	wool carder bee
Wollbienen & Harzbienen	*Anthidium* spp.	wool carder bees
Wolleule	*Acronicta leporina*	miller
Wollfledermäuse	*Kerivoula* spp.	painted bats, woolly bats
Wollhaarkäfer	Dasytidae	dasytids, dasytid beetles (soft-winged flower beetles)
Wollhaarratte, Afrikanische Sumpfratte	*Dasymys incomtus*	shaggy swamp rat
Wollhand-Einsiedler	*Pagurus cuanensis*	woolly hermit crab
Wollige Buchenlaus	*Phyllaphis fagi*	beech aphid, woolly beech aphid
Wollkäfer	Lagriidae	long-jointed beetles
Wollkrabbe	*Dromia personata/ Dromia vulgaris*	sponge crab, common sponge crab, sleepy sponge crab, Linnaeus's sponge crab, sleepy crab, little hairy crab
Wollkrabben	*Dromia* spp.	sponge crabs
Wollkrauteule, Brauner Mönch	*Cucullia verbasci*	mullein moth
Wollkrebse	Dromiidae	sponge crabs
Wollläuse, Schmierläuse	Pseudococcidae	mealybugs
Wollmaki	*Avahi laniger*	avahi, wooly lemur
Wollrückenspinner	*Tethea or*	poplar lutestring
Wollschildläuse	Coccoidea	mealybugs
Wollschwamm	*Hippospongia canaliculata*	wool sponge, sheepswool sponge
Wollspinner, Trägspinner, Schadspinner	Lymantriidae	tussock moths & gypsy moths & others
Wrackbarsch, Atlantischer Wrackbarsch	*Polyprion americanum*	wreckfish (FAO), stone bass
Wrackbarsche	Polyprionidae	wreckfishes
Wrights Stachelauster	*Spondylus wrighteanus*	Wright's thorny oyster
Wühlerkakadu	*Cacatua pastinator*	long-billed corella
Wühlmäuse, Feldmäuse	*Microtus* spp.	voles, meadow mice
Wühlspitzmaus, Stummelschwanzspitzmaus	*Anourosorex squamipes*	mole shrew, Szechwan burrowing shrew

Wulst-Nasendoktor, Ramsnasen-Einhornfisch	*Naso tuberosus*	humpnose unicornfish
Wulstlippen-Buntbarsch	*Melanochromis labrosus*	labrosus
Wunderbare Schlitzbandschnecke	*Perotrochus amabilis*	lovely slitsnail
Wunderflosser	Mirapinnidae	mirapinnids (incl. hairyfishes & tapetails)
Wunderkreisel, Wunderbare Japanische Turmschnecke	*Thatcheria mirabilis*	Japanese wonder snail (miraculous Thatcher shell)
Wunderlampe	*Lycoteuthis diadema/ Thaumatolampas diadema*	deepsea luminescent squid a.o.
Wundersylphe	*Loddigesia mirabilis*	marvellous spatuletail
Wundervolle Kammmuschel	*Mirapecten mirificus*	miraculous scallop
Würfel-Rennechse	*Cnemidophorus tesselatus*	checkered whiptail, Colorado checkered whiptail
Würfelfalter, Scheckenfalter	Riodinidae	metalmarks
Würfelnatter	*Natrix tessellata*	dice snake
Würfelquallen	Cubomedusae/Cubozoa	box jellies, sea wasps, fire medusas, cubomedusas
Würgadler	*Morphnus guianensis*	Guiana crested eagle
Würger	Laniidae	shrikes, true shrikes
Würgerkrähen	Cracticidae	butcherbirds
Würgfalke	*Falco cherrug*	saker falcon
Wurmaale	Myrophinae	worm-eels
Wurmaale	*Moringua* spp.	worm-eels
Wurmanemone*	*Scolanthus callimorphus*	worm anemone
Wurmanolis	*Anolis vermiculatus*	Vinales anole
Wurmfische	Microdesminae	wormfishes
Wurmfische & Pfeilgrundeln	Microdesmidae/ Cerdalidae	wormfishes & dartfishes
Wurmholothurie, Kletten-Holothurie, Kletten-Seegurke	*Leptosynapta inhaerens*	common white synapta
Wurmkoralle*	*Stenocyathus vermiformis*	worm coral
Wurmnatter	*Carphophis amoenus*	worm snake
Wurmsalamander	*Batrachoseps* spp.	slender salamanders
Wurmschlange, Blödauge	*Typhlops vermicularis*	European blind snake, Greek blind snake, worm snake
Wurmschleichen, Doppelschleichen	Amphisbaenia	worm lizards, amphisb(a)enids, amphisbenians
Wurmschnecke, Schlangenschnecke	*Tenagodus obtusus*	obtuse wormsnail
Wurmschnecken	Vermetidae	worm snails (worm shells)
Wurmwühlen	Caeciliidae/ Caeciliaidae	common caecilians
Wurzelbohrer	Hepialidae	swift moths (swifts & ghost moths)
Wurzelfäule-Älchen	*Pratylenchus* spp.	root-lesion nematodes

*W*urzelfliege

Wurzelfliege, Radieschenfliege, Kleine Kohlfliege	*Delia radicum/ Hylemia brassicae/ Chortophila brassicae/ Paregle radicum*	cabbage fly, cabbage maggot, radish fly
Wurzelgallenälchen	*Meloidogyne* spp.	root-knot nematodes, root-knot eelworms
Wurzelholzbohrer	*Dinoderus bifoveolatus*	root borer, West African ghoon beetle
Wurzelknötchengallwespe* (>Knötchengalle)	*Andricus quercusradicis*	oak root truffle-gall wasp
Wurzelkrebse	Rhizocephala	rhizocephalans (parasitic "barnacles")
Wurzelmundquallen	Rhizostomeae	rhizostome medusas
Wurzelratten	Rhizomyidae	bamboo rats, East African mole-rats
Wurzelschnecke	*Muricanthus radix/ Hexaplex radix*	root murex
Wurzelspinner, Kleiner Hopfenwurzelbohrer, Queckenwurzelspinner	*Hepialus lupulinus/ Korscheltellus lupulinus*	common swift, garden swift moth
Wüsten-Braunspinne	*Loxosceles deserta*	desert loxosceles, desert brown spider
Wüsten-Halsbandleguan, Schwarzhalsbandleguan	*Crotaphytus insularis*	black-collared lizard, desert collared lizard
Wüsten-Krötenechse	*Phrynosoma platyrhinos*	desert horned lizard
Wüsten-Langohrfledermaus	*Otonycteris hemprichi*	desert long-eared bat
Wüsten-Rennechse	*Cnemidophorus uniparens*	desert grassland whiptail
Wüsten-Sandboa	*Eryx miliaris*	desert sand boa
Wüsten-Schmuck-Dosenschildkröte	*Terrapene ornata luteola*	desert box turtle
Wüsten-Stachelleguan	*Sceloporus magister*	desert spiny lizard
Wüsten-Wurmsalamander	*Batrachoseps aridus*	desert slender salamander
Wüsten-Zwergwaran, Höhlenwaran	*Varanus eremius*	desert pygmy monitor, spinifex monitor
Wüstenagame	*Agama mutabilis/ Trapelus mutabilis*	desert agama
Wüstenasseln	*Hemilepistus* spp.	desert woodlice
Wüstenbussard	*Parabuteo unicinctus*	Harris' hawk
Wüstenchamäleon	*Chamaeleo namaquensis*	desert chameleon, Namaqua chameleon
Wüstenfalke	*Falco peregrinoides*	barbary falcon
Wüstenfuchs, Fennek	*Fennecus zerda/Vulpes zerda*	fennec fox
Wüstengeckos, Krallengeckos	*Coleonyx* spp.	banded geckos
Wüstengimpel	*Bucanetes githagineus*	trumpeter finch
Wüstengrasmücke	*Sylvia nana*	desert warbler
Wüstengrundel	*Chlamydogobius eremius*	desert goby (Austr.)
Wüstenhase, Kaphase	*Lepus capensis*	Cape hare, brown hare
Wüstenkatze, Sandkatze	*Felis margarita*	sand cat
Wüstenkobras	*Walterinnesia* spp.	desert cobras
Wüstenläuferlerche	*Alaemon alaudipes*	hoopoe lark, bifasciated lark
Wüstenleguan	*Diplosaurus dorsalis*	desert iguana

Wüstenluchs, Karakal	*Caracal caracal/ Felis caracal*	caracal
Wüstenprinie	*Scotocerca inquieta*	scrub warbler
Wüstenrabe	*Corvus ruficollis*	brown-necked raven
Wüstenregenpfeifer	*Charadrius leschenaultii*	greater sand plover
Wüstenrenner	*Eremias* spp.	racerunners
Wüstenschabe	*Arenivaga bolliana*	desert cockroach
Wüstenschildkröte, Kalifornische Gopherschildkröte	*Gopherus agassizii*	desert tortoise, western gopher tortoise, California desert tortoise
Wüstensperling	*Passer simplex*	desert sparrow
Wüstenspitzmaus> Graue Wüstenspitzmaus	*Notiosorex crawfordi*	gray shrew
Wüstenspitzmaus> Große Wüstenspitzmaus	*Notiosorex gigas*	desert shrew
Wüstenspringmäuse	*Jaculus* spp.	desert jerboas
Wüstensteinschmätzer	*Oenanthe deserti*	desert wheatear
Wüstenteufel, Dornteufel, Moloch	*Moloch horridus*	moloch, horny devil, thorny devil
Wüstenwaran	*Varanus griseus*	desert monitor (agra monitor)

Yak

Yak	*Bos grunniens/* *Bos mutus*	yak
Yarrow-Stachelleguan	*Sceloporus jarrovi*	Yarrow's spiny lizard
Ypsiloneule	*Agrotis ipsilon/* *Agrotis ypsilon/* *Scotia ypsilon*	dark dart moth, dark sword-grass moth; black cutworm
Yucatan-Klappschildkröte	*Kinosternon creaseri*	Creaser's mud turtle
Yucca-Nachtechse	*Xantusia vigilis*	desert night lizard
Yuccamotten	*Prodoxidae*	yucca moths
Yukatan-Dämmerungsratte	*Otonyctomys hatti*	Yucatan vesper rat
Yünnan-Scharnierschildkröte	*Cuora yunnanensis*	Yunnan box turtle
Yangtze-Stör	*Acipenser dabryanus*	Yangtze sturgeon

Zacken-Elseya-Schildkröte, Breitbrustschildkröte	*Elseya latisternum*	saw-shelled snapping turtle, serrated snapping turtle, East Australian snapping turtle
Zacken-Erdschildkröten	*Geoemyda spengleri*	black-breasted leaf turtle
Zacken-Napfschnecke	*Patella crenata*	crenate limpet
Zackenbarsche	*Epinephelus* spp.	groupers
Zackenbarsche, Sägebarsche	Serranidae	seabasses & rockcods (groupers)
Zackeneule, Zimteule, Krebssuppe	*Scoliopterix libatrix*	herald (noctuid moth)
Zackengrundel*	*Deltentosteus colonianus*	toothed goby
Zackenhirsch, Barasingha, Barasinghahirsch	*Cervus duvauceli*	barasingha, swamp deer
Zackenspanner, Herbstspanner, Herbstlaubspanner	*Ennomos autumnaria*	large thorn
Zagutis	*Plagiodontia* spp.	Hispaniolan hutias
Zäher Orangefarbener Hornschwamm, Zäher Goldschwamm	*Agelas oroides*	orange crater sponge
Zahnarme, Nebengelenktiere	Edentata/Xenarthra	edentates, "toothless" mammals, xenarthrans
Zahnbrassen	*Dentex dentex*	dentex, common dentex (FAO)
Zahnheringe, Mondaugen	Hiodontidae	mooneyes
Zahnleisten-Halbschnäbler	*Hemirhamphodon pogonognathus*	South East Asian livebearing halfbeak, bearded halfbeak
Zahnlose Haarschnecke	*Trichia edentula/ Petasina edentula*	toothless hairysnail, edentate marginsnail
Zahnmoos*	*Sertularella polyzonias*	great tooth hydroid
Zahnschnecken, Elefantenzähne	Dentaliidae (Scaphopoda)	tusks, tuskshells
Zahnspinner	Notodontidae	prominents
Zahntaucher	Hesperornithiformes	western birds
Zahnwachteln, Neuwelt-Wachteln	Odontophorinae	New World quails
Zahnwale	Odontoceti	toothed whales & porpoises & dolphins
Zährte	*Vimba vimba*	East European bream, zährte, Baltic vimba (FAO)
Zaire-Dianameerkatze	*Cercopithecus salango*	Zaire Diana monkey
Zamorawels, Mitternachtswels	*Auchenipterichthys thoracatus*	Zamora catfish
Zander (Hechtbarsch, Schill, Sandbarsch)	*Stizostedion lucioperca/ Sander lucioperca*	pike-perch, zander (FAO)
Zangensterne	Forcipulatida	forcipulatids
Zapfenguan, Bergguan	*Oreophasis derbianus*	horned guan
Zarte Augenkoralle	*Oculina tenella*	delicate ivory bush coral
Zarter Sack-Kalkschwamm	*Leucosolenia fragilis*	fragile calcareous sponge*
Zartfliegen	Tanypezidae	tanypezids

Zartschrecken	*Leptophyes* spp.	bushcrickets a.o.
Zartspinnen	Anyphaenidae	ghost spiders
Zauberhafte Artmuschel	*Dosinia elegans*	elegant dosinia
Zauberhafte Seefeder	*Virgularia mirabilis*	sea rush, elegant sea-pen*
Zauberhafte Tellmuschel*	*Tellina pulcherrima*	beautiful tellin
Zaunadler	*Harpyhaliaetus coronatus*	crowned solitary eagle
Zaunammer	*Emberiza cirlus*	cirl bunting
Zaunanolis	*Anolis distichus*	bark anole
Zauneidechse	*Lacerta agilis*	sand lizard
Zaunkönig	*Troglodytes troglodytes*	wren, winter wren
Zaunkönige	Troglodytinae	wrens
Zaunleguan	*Sceloporus scalaris*	bunch grass lizard
Zaunschrecke	*Sepiana sepium/ Platycleis sepium*	sepia bushcricket
Zebra-Ammenhaie	Stegostomatidae	zebra sharks
Zebra-Buntbarsch, Grünflossen-Buntbarsch	*Cichlasoma nigrofasciatum*	zebra cichlid, convict cichlid (FAO)
Zebra-Grundel	*Zebrus zebrus*	zebra goby
Zebra-Kaiserfisch*	*Genicanthus caudovittatus*	zebra angelfish
Zebra-Mondschnecke, Zebramondschnecke	*Natica undulata/ Notocochlis undulata/ Tanea undulata*	zebra moonsnail (wavy moonsnail)
Zebra-Nixenschnecke	*Nerita pupa*	zebra nerite
Zebra-Prachtschmerle, Streifen-Prachtschmerle	*Botia striata*	banded loach
Zebra-Riffbarsch	*Abudefduf sexfasciatus*	striped damselfish, scissortail sergeant (FAO)
Zebra-Schleimfisch	*Lipophrys basiliscus*	zebra blenny, basilisk blenny
Zebra-Schraubenschnecke	*Terebra strigata*	zebra auger
Zebra-Springspinne, Zebraspinne	*Salticus scenicus*	zebra jumper
Zebra-Stierkopfhai	*Heterodontus zebra*	zebra bullhead shark
Zebra-Strandschnecke	*Littorina ziczac/ Nodilittorina ziczac*	zebra periwinkle, zebra winkle
Zebra-Zwergfeuerfisch	*Dendrochirus zebra*	zebra lionfish
Zebrabärbling	*Brachydanio rerio*	zebra danio (FAO), zebrafish
Zebrabrassen, Bänderbrassen	*Diplodus cervinus*	zebra seabream
Zebraducker	*Cephalophus zebra*	banded duiker
Zebrafink	*Poephila guttata*	spotted-sided finch, zebra finch
Zebrahai, Leopardenhai	*Stegostoma fasciatum*	zebra shark (FAO), leopard shark
Zebrakärpfling, Zebrafundulus, Blaubandkärpfling	*Fundulus heteroclitus*	killifish, mummichog (FAO)
Zebramanguste	*Mungos mungo*	banded mongoose
Zebramondschnecke	*Notocochlis undulata/ Tanea undulata/ Natica undulata*	zebra moonsnail (wavy moonsnail)
Zebraseenadel	*Syngnathus dactyliophorus*	zebra pipefish

Zebraspinne, Wespenspinne	*Argiope bruennichi*	black-and-yellow argiope, black-and-yellow garden spider
Zebraspinnen	Argiopinae	garden spiders
Zecken	Ixodides	ticks
Zederngirlitz	*Serinus syriacus*	Syrian serin
Zedernkäfer*	Callirhipidae	cedar beetles
Zedernseidenschwanz	*Bombycilla cedrorum*	cedar waxwing
Zehnarmige Tintenschnecken, Zehnarmer	Decapoda/ Decabrachia	cuttlefish & squids
Zehnpunkt-Marienkäfer	*Adalia decimpunctata*	ten-spot ladybird
Zehrwespen	Proctotrupidae (Serphidae)	proctotrupids
Zelt-Olive, Porphyrwalze	*Oliva porphyria*	tent olive
Zeltbau-Fledermäuse*	*Uroderma* spp.	tent-building bats
Zeltdachspinnen	*Uroctea* spp.	star-web spiders
Zeltlagermuschel	*Lioconcha castrensis*	chocolate flamed venus, camp pitar-venus
Zeniontiden	Zeniontidae	zeniontids
Zentralafrikanische Langhaarratten	*Stochomys* spp.	Central African target rats
Zentralamerikanische Wassermäuse	*Rheomys* spp.	Central American water mice
Zentrifugal-Sonnenschnecke*	*Spirolaxis centrifuga*	exquisite false-dial
Zerbrechliche Seepeitsche, Helle Seepeitsche	*Junceela fragilis*	fragile sea whip
Zerbrechliche Steckmuschel	*Pinna fragilis/Atrina fragilis*	fragile penshell, fan mussel (fragile fanshell)
Zerbrechliche Tellerkoralle	*Agaricia fragilis*	fragile saucer coral
Zerbrechliche Wurmseegurke	*Oestergrenia adriatica*	
Zerbrechlicher Schlangenstern	*Ophiothrix fragilis*	common brittlestar
Zibetkatzen> Echte Zibetkatzen	*Viverra* spp.	oriental civets
Zickzack-Glockenpolyp	*Laomedea geniculata/ Obelia geniculata*	knotted thread hydroid, zig-zag wine-glass hydroid
Zickzack-Kammmuschel	*Pecten ziczac*	zigzag scallop
Zickzack-Nerite	*Neritina communis/ Theodoxus communis*	zigzag nerite, common nerite
Zickzack-Salamander	*Plethodon dorsalis*	zigzag salamander
Zickzackspinner, Kamelspinner, Uferweiden-Zahnspinner	*Eligmodonta ziczac/ Notodonta ziczac*	pebble prominent
Ziege, Sichling (Sichelfisch)	*Pelecus cultratus*	ziege (FAO), chekhon, sabre carp, sichel
Ziege> Hausziege	*Capra hircus*	domestic goat
Ziegelbarsche	Branchiostegidae	tilefishes
Ziegelfische	Malacanthidae	sand tilefishes
Ziegenantilopen, Gorals	*Nemorhaedus* spp.	gorals
Ziegenhaarling	*Bovicola caprae*	goat-biting louse
Ziegenlaus	*Linognathus stenopsis*	goat sucking louse
Ziegenmelker	*Caprimulgus europaeus*	nightjar

Ziegensittich	Cyanoramphus novaezelandiae	red-fronted parakeet
Zierbarben	Puntius spp.	barbs
Zierblattläuse, Zierläuse	Callaphididae/ Drepanosiphonidae	drepanosiphonid plant-lice
Zierliche Schließmundschnecke	Ruthenica filograna	
Zierliche Schwimmkrabbe	Portunus anceps	delicate swimming crab
Zierliche Strauchschrecke	Rhacocleis germanica	Mediterranean bushcricket
Zierliche Zwergolive	Olivella gracilis	graceful dwarf olive
Ziermotten	Scythridiidae	owlets, owlet moths
Zierschildkröte	Chrysemys picta picta	painted turtle, eastern painted turtle
Zierwurm*	Amphitrite ornata	ornate worm
Ziesel	Citellus spp./ Spermophilus spp.	ground squirrels, susliks, sousliks
Ziesel> Europäisches Ziesel, Schlichtziesel	Citellus citellus/ Spermophilus citellus	European ground squirrel, European suslik, European souslik
Zieselmaus	Spermophilopsis leptodactylus	long-clawed ground squirrel
Ziesteule	Autographa pulchrina	beautiful golden Y
Zigarrenfliege, Schilfgallenfliege	Lipara lucens	reed gall fly
Zigarrenwickler, Rebenstecher	Byctiscus betulae	hazel leaf roller weevil
Zikaden (Zirpen) & Schaumzikaden	Auchenorrhyncha (Homoptera)	cicadas & hoppers (see: spittlebugs, froghoppers)
Zikadenräuber, Rhipiceriden	Rhipiceridae	cicada parasite beetles
Zikadenwespen	Dryinidae	dryinids, dryinid wasps
Zilpzalp, Weidenlaubsänger	Phylloscopus collybita	chiffchaff
Zimmermannsbock	Acanthocinus aedilis	common timberman beetle, timberman
Zimmermannsfrosch*	Rana virgatipes	carpenter frog
Zimokkaschwamm	Spongia zimocca	brown Turkey cup sponge
Zimtbär, Rostbär	Phragmatobia fuliginosa	ruby tiger (moth)
Zimtente	Anas cyanoptera	cinnamon teal
Zimteule, Zackeneule, Krebssuppe	Scoliopterix libatrix	herald (noctuid moth)
Zimtflunder	Pseudorhombus cinnamoneus	cinnamon flounder
Zimtkolibri	Selasphorus rufus	rufous hummingbird
Zimtmuschel	Botula fusca	cinnamon mussel
Zimttao, Buschtinamu	Crypturellus cinnamomeus	thicket tinamou
Zingel	Zingel zingel	zingel
Zinnober-Napfschnecke	Patella miniata/ Patella pulchra	cinnabar limpet
Zipfelfrösche	Megophrys spp.	Asian spadefoot toads, horned-nosed frogs, horned toads
Zipfelkäfer, Warzenkäfer	Malachiidae	malachiid beetles, flower beetles
Zipfelschnecke	Siratus laciniatus	laciniate murex
Zippammer	Emberiza cia	rock bunting

Zirbelkieferwolllaus	*Pineus cembrae*	Swiss stone pine adelges, Arolla pine woolly aphid
Zirpfrösche	*Syrrhophus* spp.	chirping frogs
Zirpkröte	*Ansonia longidigita*	long-fingered stream toad
Zirrenkrake	*Eledone cirrosa/ Octopus cirrhosus/ Ozeana cirrosa*	lesser octopus, curled octopus, horned octopus (FAO)
Zirrenlose Seelilien	Millericrinida	sea lilies without cirri
Zirrentragende Seelilien	Isocrinida	sea lilies with cirri
Zischnattern	*Pseustes* spp.	puffing snakes
Zitronen-Zwergkaiser, Zitronengelber Zwergkaiserfisch	*Centropyge flavissimus*	lemonpeel angelfish
Zitronenbarbe, Gelber Ziegenfisch	*Parupeneus cyclostomus*	yellowsaddle goatfish, yellow-tailed goatfish, goldsaddle goatfish (FAO)
Zitronenfalter	*Gonepteryx rhamni*	brimstone
Zitronenfarbene Austernschildlaus, Austernförmige Schildlaus	*Quadraspidiotus ostreaeformis*	European fruit scale
Zitronengelbe Demoiselle	*Pomacentrus sulfureus*	sulphur damsel (FAO), lemon demoiselle
Zitronengirlitz	*Serinus citrinella*	citril finch
Zitronengrundel	*Gobiodon citrinus*	lemon goby
Zitronenhai, Amerikanischer Zitronenhai	*Negaprion brevirostris*	lemon shark
Zitronenhaie	*Negaprion* spp.	lemon sharks
Zitronensalmler, Schönflossensalmler	*Hyphessobrycon pulchripinnis*	lemon tetra
Zitronenschmierlaus, Citrus-Schmierlaus, Gewächshausschmierlaus	*Pseudococcus citri/ Planococcus citri*	common mealybug, citrus mealybug, citrus scale
Zitronenwaldsänger	*Protonotaria citrea*	prothonotary warbler
Zitrusknospenmilbe	*Eriophyes sheldoni*	citrus bud mite
Zitteraal	*Electrophorus electricus*	electric knifefish
Zitteraale	Electrophoridae	electric knifefishes
Zitterdrossel	*Cinclocerthia ruficauda*	brown trembler
Zitterfliegen	Pallopteridae	pallopterid flies
Zitternde Seerinde*	*Victorella pavida*	trembling sea-mat
Zitterpappel-Blattgallmücke*	*Harmandia tremulae*	aspen leaf gall midge
Zitterpappelblattkäfer	*Phytodecta decemnotata*	aspen leaf beetle
Zitterrochen, Elektrische Rochen	Torpediniformes	electric rays
Zitterspinne	*Pholcus phalangioides*	long-bodied cellar spider, longbodied cellar spider
Zitterspinnen	Pholicidae	cellar spiders
Zitterwels, Elektrischer Wels	*Malapterurus electricus*	electric catfish
Zitterwelse, Elektrische Welse	Malapteruridae	electric catfishes
Zobel	*Martes zibellina*	sable
Zobel, Donau-Brachse	*Abramis sapa*	Danube bream, Danubian bream, white-eye bream (FAO)
Zonierte Mondschnecke	*Tanea euzona*	zoned moonsnail

Zope	*Abramis ballerus*	zope (FAO), blue bream
Zorilla	*Ictonyx striatus*	zorilla, striped polecat
Zottelschweifaffe, Rotbärtiger Mönchsaffe	*Pithecia monachus*	monk saki
Zottenplanarie	*Thysanozoon brochii*	skirt dancer
Zottige Haarschnecke	*Trichia villosa*	villous hairysnail
Zottige Seerinde	*Electra pilosa*	hairy sea-mat
Zuckerkäfer	Passalidae	bess beetles, "bessbugs", peg beetles
Zuckerrohr-Garnele	*Parhippolyte uveae*	sugarcane shrimp
Zuckerrohrzikade	*Perkinsiella saccharicida*	sugarcane delphacid
Zuckerrohrzünsler	*Diatraea saccharalis*	sugar cane borer
Zuckerrübenwurzellaus	*Pemphigus betae*	sugar beet root aphid
Zuckervogel	*Coereba flaveola*	bananaquit
Zuckmücken, Schwarmmücken (Tanzmücken)	Chironomidae	nonbiting midges & gnats
Zügeldelphin	*Stenella clymene*	Atlantic spinner dolphin, clymene dolphin, helmet dolphin
Zügelgrundel	*Coryphopterus glaucofraenum*	bridled goby
Zügelseeschwalbe	*Sterna anaethetus*	bridled tern
Zulia Krötenkopf-Schildkröte	*Phrynops zuliae*	Zulia toad-headed turtle
Zungenbutt, Rotzunge, Hundszunge	*Glyptocephalus cynoglossus*	witch
Zungenkiemer, Weichstrahlenfische	Malacosteidae	loosejaws
Zungenkoralle	*Polyphyllia talpina*	
Zungenlose	Pipidae	tongueless frogs (clawed and Surinam toads)
Zungenwurm> Gemeiner Zungenwurm	*Linguatula serrata*	common tongue worm
Zungenwürmer, Linguatuliden, Pentastomitiden	Pentastomida	tongue worms, linguatulids, pentastomids
Zünsler	Pyralidae	pyralid moths, pyralids (snout moths, grass moths & others)
Zünslereule, Nesselschnabeleule	*Hypena proboscidalis*	common snout (moth)
Zürgelbaum-Schnauzenfalter	*Libythea celtis*	nettle-tree butterfly
Zweiäugiger Plattenegel, Zweiäugiger Plattegel	*Helobdella stagnalis*	
Zweiband-Mondschnecke	*Polinices bifasciatus*	two-banded moonsnail
Zweibinden-Anemonenfisch, Zweiband-Anemonenfisch, Orangeringelfisch, Spitzbinden-Anemonenfisch, Rotmeer-Clownfisch	*Amphiprion bicinctus*	two-banded anemonefish, twoband anemonefish (FAO)
Zweibinden-Brassen	*Diplodus vulgaris*	two-banded bream, common two-banded seabream (FAO)
Zweibinden-Bussard	*Buteo nitidus/Asturina nitida*	gray hawk (U.S.), grey hawk

Zweibinden-Ziersalmler, Zweibindensalmler	*Nannostomus bifasciatus*	whiteside pencilfish
Zweibindiger Pilzschwarzkäfer	*Alphitophagus bifasciatus*	twobanded fungus beetle
Zweifaltenrandige Schließmundschnecke	*Laciniaria biplicata*	two-lipped door snail
Zweifarben-Baumauster*	*Isognomon bicolor*	bicolor purse-oyster
Zweifarben-Demoiselle	*Chromis dimidiata*	bicolor damselfish, half-and-half chromis, chocolatedip chromis (FAO)
Zweifarben-Glanzstar	*Spreo bicolor*	African pied starling
Zweifarbfledermaus	*Vespertilio murinus*	particolored bat
Zweifarbige Beißschrecke	*Metrioptera bicolor*	twocoloured bushcricket
Zweifarbige Fledermäuse, Zweifarbfledermäuse	*Vespertilio* spp.	frosted bats, particolored bats
Zweifarbiger Bratpfannenwels	*Dysichthys coracoideus*	guitarrita
Zweifarbiger Schmalbauchrüssler	*Phyllobius oblongus*	brown leaf weevil
Zweifelhafte Seespinne*	*Libinia dubia*	longnose spider crab, doubtful spider crab
Zweifinger-Faultiere	*Choloepus* spp.	two-toed sloths
Zweifleck, Zweifleck-Libelle	*Epitheca bimaculata/ Libellula bimaculata*	two-spotted dragonfly*
Zweifleck-Buntbarsch	*Cichlasoma bimaculatum*	black acara
Zweifleck-Junker, Orangegrüner Junker	*Halichoeres biocellatus*	red-lined wrasse
Zweifleck-Krake	*Octopus bimaculatus*	two-spotted octopus
Zweifleck-Lippfisch	*Symphodus bailloni*	vielle
Zweiflecken-Ansauger	*Diplecogaster bimaculata*	two-spotted clingfish
Zweifleckiger Grashüpfer	*Chorthippus binotatus*	twospotted grasshopper
Zweifleckiger Laubläufer, Zweigefleckter Eilkäfer	*Notiophilus biguttatus*	burnished ground beetle
Zweifuß-Doppelschleichen	Bipedidae	two-legged worm lizards
Zweigestreifte Quelljungfer	*Cordulegaster boltoni*	golden-ringed dragonfly
Zweigkoralle	*Madracis mirabilis*	yellow pencil coral
Zweikerben-Sanddollar	*Echinodiscus auritus*	two-slit sand dollar
Zweikiel-Haarschnecke	*Trichotropis bicarinata*	two-keel hairysnail, two-keeled hairysnail
Zweipunkt	*Adalia bipunctata*	two-spot ladybird, 2-spot ladybird
Zweipunkt-Dornschrecke	*Tetrix bipunctata*	twospotted groundhopper
Zweipunkt-Eulenspinner	*Ochropacha duplaris*	common lutestring
Zweipunkt-Ohrwurm	*Anechura bipunctata*	two-spotted earwig
Zweipunkt-Tagkurzflügler	*Aleochara bipustulata*	egg-eating rove beetle
Zweipunkt-Wellenstriemenspanner	*Scotopteryx bipunctaria*	chalk carpet moth
Zweipunktbarbe	*Puntius binotatus*	spotted barb
Zweipunktige Grünwanze	*Calocoris norvegicus*	potato capsid bug
Zweipunktiger Eichenprachtkäfer, Zweifleckiger Eichenprachtkäfer, Gefleckter Eichenprachtkäfer	*Agrilus biguttatus*	twospotted oak borer

Zweistreifen-Chamäleon	*Chamaeleo bitaeniatus*	two-striped chameleon, two-lined chameleon
Zweistreifen-Grundel, Schwimmgrundel	*Gobius flavescens/ Gobius ruthensparri*	two-spotted goby
Zweistreifenbarsch	*Apogon binotatus*	barred cardinalfish
Zweitupfen-Raubsalmler	*Exodon paradoxus*	two-blotched tetra, two-blotched characin, bucktooth tetra (FAO)
Zweizahn-Glockenpolyp	*Obelia bidentata*	doubletoothed hydroid
Zweizähnige Laubschnecke	*Perforatella bidentata*	
Zweizähnige Linsenmuschel	*Montacuta bidentata*	bidentate montacutid
Zweizähnige Puppenschnecke	*Pupilla bigranata*	bidentate moss snail
Zweizahnwal, Sowerby-Zweizahnwal	*Mesoplodon bidens*	Sowerby's whale, Sowerby's North Sea beaked whale
Zweizahnwale	*Mesoplodon* spp.	beaked whales
Zweizehen-Aalmolch	*Amphiuma means*	two-toed amphiuma
Zwerg-Darm-Bandwürmer	*Hymenolepis* spp.	dwarf intestinal tapeworms
Zwerg-Darmegel, Kleiner Darmegel	*Metagonimus yokogawai*	Yokogawa fluke
Zwerg-Epaulettenflughunde	*Micropteropus* spp.	dwarf epauleted bats
Zwerg-Gurami	*Trichopsis pumilis*	pygmy gourami
Zwerg-Kieleidechse	*Algyroides fitzingeri*	Fitzinger's algyroides
Zwerg-Koboldmaki	*Tarsius pumilus*	pygmy tarsier
Zwerg-Moschusschildkröte, Kleine Moschusschildkröte	*Sternotherus minor/ Kinosternon minor*	loggerhead musk turtle
Zwerg-Pferdebandwurm	*Paranoplocephala mamillana*	dwarf equine tapeworm
Zwerg-Plumplori, Zwergplumplori	*Nycticebus pygmaeus*	pygmy slow loris
Zwerg-Schlammschnecke	*Lymnaea truncatula*	dwarf pondsnail
Zwerg-Sepia, Mittelmeer-Sepiole, Zwergtintenfisch, Kleine Sprutte	*Sepiola rondeleti*	Mediterranean dwarf cuttlefish, lesser cuttlefish
Zwerg-Turmschnecke*	*Opeas pumilum*	dwarf awlsnail
Zwerg-Veilchenschnecke	*Janthina exigua*	dwarf janthina
Zwergadler	*Hieraaetus pennatus*	booted eagle
Zwergalk	*Aethia pusilla*	least auklet
Zwergameisen	*Plagiolepis* spp.	pygmy ants
Zwergameisenbär	*Cyclopes didactylus*	silky anteater, pygmy anteater
Zwergammer	*Emberiza pusilla*	little bunting
Zwergara	*Ara nobilis*	red-shouldered macaw
Zwergarmmolch, Gestreifter Zwergarmmolch	*Pseudobranchus striatus*	dwarf siren, mud siren
Zwergassel*	*Trichoniscus pusillus*	common pygmy woodlouse
Zwergbachsalamander	*Desmognathus wrighti*	pygmy salamander
Zwergbandwurm	*Vampirolepis nana/ Hymenolepis nana*	dwarf tapeworm
Zwergbärbling	*Rasbora maculata*	dwarf rasbora (FAO), spotted rasbora
Zwergbarsch, Brauner Sägebarsch	*Serranus hepatus*	brown comber
Zwergbeutelratten	*Marmosa* spp.	murine opossums, mouse opossums

Zwergbläuling	*Cupido minimus*	small blue
Zwergboas u.a.	*Charina* spp.	rubber boas
Zwergbuntbarsche	*Apistogramma* spp.	dwarf cichlids
Zwergchamäleons u.a.	*Bradypodion* spp.	dwarf chameleons
Zwergchamäleons u.a.	*Microsaura* spp.	pygmy chameleons
Zwergdarmegel	*Heterophyes* spp.	heterophyid flukes
Zwergdarmegel	Heterophyidae	heterophyid flukes
Zwergdommel	*Ixobrychus minutus*	little bittern
Zwergdorsch	*Trisopterus minutus*	poor cod (FAO), poor-cod
Zwergdrachenflosser	*Corynopoma riisei*	swordtail characin
Zwergdrossel	*Catharus ustulatus*	Swainson's thrush
Zwergelfe, Zwergkolibri	*Mellisuga minima*	vervain hummingbird
Zwergfadenfisch	*Colisa lalia*	dwarf gourami
Zwergfadenwurm u.a.	*Strongyloides ransomi*	intestinal threadworm a.o.
Zwergfelskänguru, Zwergsteinkänguru, Nabarlek	*Peradorcas concinna*	pygmy rock wallaby, nabarlek
Zwergfeuerfische	*Dendrochirus* spp.	lionfishes
Zwergflamingo	*Phoenicopterus minor*	lesser flamingo
Zwergfledermaus	*Pipistrellus pipistrellus*	common pipistrelle
Zwergfledermäuse	*Pipistrellus* spp.	pipistrelles
Zwergflughund	*Aethalops alecto*	pygmy fruit bat
Zwergflusspferd	*Choeropsis liberiensis/ Hexaprotodon liberiensis*	pygmy hippopotamus
Zwergfurchenmolch	*Necturus punctatus*	dwarf waterdog
Zwergfurchenwal, Zwergwal	*Balaenoptera acutorostrata*	minke whale, lesser rorqual
Zwergfüßer	Symphyla	symphylans
Zwerggalago, Urwaldgalago	*Galago demidovii/ Galagoides demidoff*	dwarf bush baby, Demidoff's bush baby, Demidoff's galago
Zwerggalagos	*Galagoides* spp.	dwarf bush baby, dwarf galagos
Zwerggans	*Anser erythropus*	lesser white-fronted goose
Zwerggeißelskorpione	Schizomida/ Schizopeltidia	microwhipscorpions, schizomids
Zwergglattwal	*Caperea marginata*	pygmy right whale
Zwerggleitbeutler, Mausgleitbeutler, Mausflugbeutler	*Acrobates pygmaeus*	pygmy gliding possum, feathertail glider
Zwerggleithörnchen	*Petinomys* spp.	dwarf flying squirrels
Zwerggrindwal, Zwergschwertwal	*Feresa attenuata*	pygmy killer whale, slender blackfish
Zwerggrison	*Lyncodon patagonicus*	Patagonian weasel
Zwerggürteltier	*Zaedyus pichiy/ Euphractus pichiy*	pichi
Zwerghai	*Euprotomicrus bispinatus*	pygmy shark (FAO), slime shark
Zwerghöhlenschnecke	*Lartetia* spp.	
Zwerghörnchen, Asiatische Zwerghörnchen u.a.	*Exilisciurus* spp.	Asian pygmy squirrels (Borneo/Philippines)

Zwerghundsfisch, Ost-Amerikanischer Hundsfisch	*Umbra pygmaea*	eastern mudminnow
Zwergkaninchen	*Sylilagus idahoensis/ Brachylagus idahoensis*	pygmy rabbit
Zwergkärpfling	*Heterandria formosa*	least killifish
Zwergkiwi, Kleiner Fleckenkiwi	*Apteryx owenii*	little spotted kiwi
Zwergklapperschlange> Eigentliche Zwergklapperschlange	*Sistrurus miliarius*	pygmy rattlesnake
Zwergkleiber	*Sitta pygmaea*	pygmy nuthatch
Zwergkopf-Seeschlangen	*Microcephalophis* spp.	small-headed sea snake
Zwergkrabbe	*Rhithropanopeus harrisii*	dwarf crab, dwarf xanthid crab
Zwergkrake, Joubins Krake	*Octopus joubini*	Atlantic pygmy octopus, dwarf octopus, Joubin's octopus
Zwergkugler	Glomeridellidae	lesser pill millipedes
Zwerglarvenfische	Grasseichthyidae	grasseichthyids
Zwergläuse, Zwergblattläuse	Phylloxeridae	phylloxerans (aphids)
Zwerglibelle	*Nehalennia speciosa*	green damsel
Zwergmakis, Mausmakis	*Microcebus* spp.	mouse lemurs
Zwergmangusten, Zwergmungos	*Helogale* spp.	dwarf mongooses
Zwergmaräne, Kleine Maräne	*Coregonus albula*	vendace
Zwergmäuse	*Haeromys* spp.	pygmy tree mice
Zwergmeerschweinchen	*Microcavia* spp.	mountain cavies
Zwergmoos, Kleines Seemoos	*Dynamena pumila/ Dynamena cavolini/ Sertularia pumila*	sea oak, minute garland hydroid
Zwergmotten	Stigmellidae (Nepticulidae)	dwarf eyecap moths, leaf miners, leaf-miners
Zwergmöve	*Larus minutus*	little gull
Zwergmuntjak, Chinesischer Muntjak	*Muntiacus reevesi*	Chinese muntjac, Reeve's muntjac
Zwergohreule	*Otus scops*	Scops owl
Zwergohrwurm, Kleiner Ohrwurm	*Labia minor*	lesser earwig, little earwig
Zwergotter	*Aonyx cinerea*	Oriental small-clawed otter
Zwergpetermännchen, Viperqueise, Kleines Petermännchen	*Echiichthys vipera/ Trachinus vipera*	lesser weever
Zwergpinguin	*Eudyptula minor*	little penguin
Zwergpolyp (Süßwassermeduse, Süßwasserqualle)	*Microhydra ryderi/ Microhydra sowerbyi (Craspedacusta sowerbyi)*	freshwater polyp (freshwater jellyfish, Regent's Park medusa)
Zwergpottwal	*Kogia breviceps*	pygmy sperm whale, lesser cachalot
Zwergpython	*Antaresia perthensis/ Liasis perthensis/ Morelia perthensis*	Perth pygmy python, pygmy python
Zwergradnetzspinne	*Theridiosoma radiosa*	ray spider

Zwergradnetzspinnen	Theridiosomatidae	ray spiders
Zwergrückenschwimmer	Pleidae	pleid water bugs, lesser water-boatmen, pygmy backswimmers
Zwergrückenschwimmer, Wasserzwerg	*Plea leachi/ Plea minutissima*	pygmy backswimmer
Zwergsäger	*Mergus albellus*	smew
Zwergscharbe	*Phalacrocorax pygmeus*	pygmy cormorant
Zwergschilderwelse	*Peckoltia* spp.	tiger clown pleco
Zwergschimpanse, Bonobo	*Pan paniscus*	bonobo, pygmy chimpanzee
Zwergschläfer	*Graphiurus nanus*	pygmy dormouse
Zwergschlangen	*Calamaria* spp.	reed snakes
Zwergschleimfisch, Dalmatinischer Schleimfisch	*Lipophrys dalmatinus*	Dalmatian blenny
Zwergschnäpper	*Ficedula parva*	red-breasted flycatcher
Zwergschnepfe	*Lymnocryptes minimus*	jack snipe
Zwergschwan	*Cygnus columbianus*	tundra swan
Zwergschwertträger	*Xiphophorus pygmaeus*	pygmy swordtail
Zwergsechsaugen(spinnen)	Oonopidae	dwarf sixeyed spiders (minute jumping spiders)
Zwergseeigel, Zwerg-Seeigel, Kleiner Schildigel	*Echinocyamus pusillus*	green sea urchin, green urchin, pea urchin
Zwergseepferdchen	*Hippocampus zosterae*	pygmy seahorse, dwarf seahorse (FAO)
Zwergseeschwalbe	*Sterna albifrons*	little tern
Zwergseezunge, Zwergzunge	*Buglossidium luteum/ Solea lutea*	solenette (FAO), pygmy sole,
Zwergseidenäffchen	*Cebuella pygmaea*	pygmy marmoset
Zwergskolopender, Gewächshaus-Zwergfüßer	*Scutigerella immaculata* (Symphyla)	glasshouse symphylid, garden symphylan, garden centipede
Zwergsonnenbarsch	*Elassoma evergladei*	pygmy sunfish
Zwergsonnenbarsche	Elassomatidae/ Elassomidae	pygmy sunfishes
Zwergspinnen	Erigoninae & Micryphantidae	dwarf spiders
Zwergspitzmaus	*Sorex minutus*	Eurasian pygmy shrew
Zwergspringmaus> Fünfzehen-Zwergspringmaus	*Cardiocranius paradoxus*	five-toed dwarf jerboa
Zwergsteinkänguru, Zwergfelskänguru, Nabarlek	*Peradorcas concinna*	pygmy rock wallaby, nabarlek
Zwergstichling, Neunstachliger Stichling	*Pungitius pungitius*	nine-spined stickleback
Zwergstrandläufer	*Calidris minuta*	little stint
Zwergstrandschnecke	*Littorina littoralis*	dwarf periwinkle
Zwergstrandschnecke, Blaue Strandschnecke, Gewöhnliche Strandschnecke	*Littorina neritoides*	small periwinkle
Zwergsultanshuhn	*Gallinula martinica/ Porphyrula martinica/ Porphyrio martinica*	purple gallinule, American purple gallinule
Zwergsumpfhuhn	*Porzana pusilla*	Baillon's crake

Zwergtaucher	*Tachybaptus ruficollis/ Podiceps ruficollis*	little grebe
Zwergteufelsrochen, Indopazifischer Teufelsrochen	*Mobula diabola*	devil ray
Zwergtigerbarsch	*Serranus tigrinus*	harlequin bass
Zwergtrappe	*Tetrax tetrax*	little bustard
Zwergwabenkröte	*Pipa parva*	Sabana Surinam toad
Zwergwachtel	*Coturnix chinensis*	Indian blue quail
Zwergwal, Zwergfurchenwal	*Balaenoptera acutorostrata*	minke whale, lesser rorqual
Zwergwaldmaus, Kleine Waldmaus	*Apodemus microps*	pygmy field mouse
Zwergwaran	*Varanus storri*	dwarf monitor
Zwergwasserläufer, Uferläufer	Hebridae	hebrids, velvet water bugs, sphagnum bugs
Zwergwasserläufer, Uferläufer	*Hebrus* spp.	hebrids, velvet water bugs, sphagnum bugs
Zwergweberknecht	*Siro duricorius*	mite harvestman
Zwergweberknechte	Cyphophthalmi	mite harvestmen
Zwergwespen	Mymaridae	fairy flies
Zwergwickler	Bucculatricidae	patches, ribbed case-makers, ribbed case-bearers
Zwergwildschwein	*Sus salvanius*	pygmy hog
Zwergwindelschnecke, Gemeine Windelschnecke	*Vertigo pygmaea*	dwarf whorl snail, common whorl snail, crested vertigo (U.S.)
Zwergzackenbarsche	Pseudochromidae	dottybacks
Zwergziersalmler	*Nannostomus marginatus*	dwarf pencilfish
Zwergzikaden	Cicadellidae	leaf hoppers
Zwergzunge, Zwergseezunge	*Solea lutea/ Buglossidium luteum*	pygmy sole, solenette (FAO)
Zwickerkobra	*Naja kaouthia*	monocled cobra
Zwiebelblasenfuß, Tabakblasenfuß	*Thrips tabaci*	onion thrips, potato thrips
Zwiebelfliege	*Delia antiqua/ Phorbia antiqua/ Hylemyia antiqua*	onion fly, onion maggot
Zwiebelmuschel, Zwiebelschale, Sattelmuschel, Europäische Sattelauster	*Anomia ephippium*	European saddle oyster (European jingle shell)
Zwiebelmuscheln, Sattelmuscheln	Anomiidae	saddle oysters (jingle shells)
Zwiebelmuscheln, Sattelmuscheln	*Anomia* spp.	saddle oysters (jingle shells)
Zwischenkärpflinge, Hochlandkärpflinge	Goodeidae	
Zwitscherschrecke	*Tettigonia cantans*	twitching green bushcricket
Zwitter-Käferschnecke	*Lepidochitona raymondi*	Raymond's chiton
Zwölfarmige Schmarotzerrose	*Peachia quinquecapitata*	twelve-tentacle parasitic anemone

Zwölfpunktspargelhähnchen, Zwölfpunkt-Spargelkäfer	*Crioceris duodecimpunctata*	spotted asparagus beetle
Zyklamenmilbe, Alpenveilchenmilbe, Begonienmilbe	*Tarsonemus pallidus/ Phytonemus pallidus*	cyclamen mite, begonia mite
Zylinderrosen	Ceriantharia	tube anemones, cerianthids
Zylinderrosen	*Cerianthus* spp.	cerianthids, cylinder anemones
Zylinderschnecke, Zylinder-Kelchschnecke*	*Cylichna cylindracea*	cylindrical barrel-bubble
Zylinderwindelschnecke	*Truncatellina cylindrica*	cylindrical whorl snail
Zylindrische Porenkoralle	*Porites cylindrica*	cylindrical finger coral, cylindrical porous coral
Zylindrische Seefeder	*Cavernulina cylindrica*	cylindrical sea-pen
Zypern-Schlanknatter	*Coluber cypriensis*	Cyprus whip snake
Zypressen-Federpolyp, Zypressen-Nesselfarn	*Aglaophenia cupressina*	cypress plume hydroid
Zyprische Honigbiene	*Apis mellifera cypria*	Cyprian honey bee, Cyprian bee

Literatur

Abbott RT, Dance SP: *Compendium of Seashells.*
  Odyssey, El Cajon, CA, 1998

Baillie J, Groombridge B: *1996 IUCN Red List of Threatened Animals.*
  IUCN, Gland, Switzerland and Cambridge, UK, 1996

Barnard PC: *Identifying British Insects and Arachnids.*
  Cambridge University Press, Cambridge, 1999

Bernard R: *Vogelnamen (E/G/L).* Aula, Wiesbaden, 1993

Betts CJ: *Checklist of Protected British Species,*
  2nd edn. Betts Env. Biol. 1998, ISBN 1 9000 2303 2

Binot M, Bless R, Boye P, Gruttke H, Pretscher P: *Rote Liste gefährdeter Tiere Deutschlands.* Bundesamt für Naturschutz, Bonn-Bad Godesberg, 1998

Blanke R, Brauser L: *Fellbezeichnungen im Pelzhandel und Pelzgewerbe (sowie die artenschutzrechtlichen Bestimmungen für die Ein- und Ausfuhr).*
  Deutsches Pelz Institut, Bad Homburg/Frankfurt/M., 1995

Borror DJ, Triplehorn CA, Johnson NF: *An Introduction to the Study of Insects,*
  6th edn. Saunders, Philadelphia, 1989

Bosik JJ: *Common Names of Insects and Related Organisms.*
  Entomological Society of America, Washington DC, 1997

Bratton J: *British Red Data Books 3: Invertebrates Other than Insects.*
  JNCC, 1991

Brauns A: *Taschenbuch der Waldinsekten,*
  4. Aufl. Fischer, Stuttgart - Jena, 1991

Breene RG: *Common Names of Arachnids,* 2nd edn.
  ATS, Artesia, New Mexico, 1997

Brohmer - *Fauna von Deutschland,* 18. Aufl. (Schaefer M, ed.)
  Quelle & Meyer, Heidelberg, 1992

Cairns SD et al: *Common and Scientific Names of Aquatic Invertebrates from the United States and Canada: Cnidaria and Ctenophora.*
  American Fisheries Society, Bethesda, MD, 1991

Carter DJ: *Field Guide to Caterpillars.* Collins, London, 1986

Chinery M: *Insects of Britain & Northern Europe,* 3rd edn.
Harper-Collins, London, 1993

Cole TCH: *Wörterbuch der Biologie.* Spektrum Akademischer Verlag,
Heidelberg - Berlin, 1998

Corbet GB, Hill JE: *A World List of Mammalian Species,*
3rd edn. Oxford University Press, Oxford, NY, 1991

Dance SP: *Shells.* Dorling Kindersley, 1992; *Muscheln und Schnecken.*
Meier, Ravensburg, 1994

Emanoil M: *Encyclopedia of Endangered Species.*
Gale Research Inc., Detroit, 1994

Ernst CH, Barbour RW: *Turtles of the World.* Smithsonian Inst. Press,
Washington DC, 1989

European Community: *Multilingual Illustrated Dictionary of Aquatic Animals
and Plants,* 2nd edn. Fishing News Books/Blackwell, Oxford, 1998

Frank N, Ramus E: *A Complete Guide to Scientific and Common Names of
Reptiles and Amphibians of the World.* NG Publ., Pottsville, 1995

Froese R, Pauly D: *FishBase '99.* International Center for Living
Aquatic Resources Management - ICLARM, 1999
www.cgiar.org/iclarm/fishbase/

Godan, D: *Common Names von Schadgastropoden.*
Mitt. Biol. Bundesanst. Berlin-Dahlem. Parey, Berlin/Hamburg, 1974 (159)

Goodwin HA, Holloway CW: *Red Data Book. Mammalia.* International
Union for Conservation of Nature and Natural Resources (IUCN).
Morges, Switzerland, 1978

Grzimek B: *Grzimeks Tierleben, Enzyklopädie des Tierreichs,* 13 Bände.
Kindler, Zürich, 1971; *Grzimek's Animal Life Encyclopedia,* 13 Vols.
Van Nostrand Reinhold, New York, 1972-1975

Harde KW, Severa F: *Der Kosmos-Käferführer. Die Mitteleuropäischen Käfer,*
3. Aufl. Franckh, Stuttgart, 1988

Hayward PJ, Ryland JS: *Handbook of the Marine Fauna of North-West Europe.*
Oxford University Press, Oxford, 1995

Hayward P, Nelson-Smith T, Shields C: *Collins Pocket Guide to the Sea Shore
of Britain & Europe.* Harper-Collins, London, 1996

Honomichl K: *"Jacobs/Renner" Biologie und Ökologie der Insekten,*
3. Aufl. Fischer, Stuttgart, 1998

Howard R, Moore A: *A Complete Checklist of Birds of the World,*
2nd edn. Academic Press, London, 1991

Ingle R: *Crayfish, Lobsters, and Crabs of Europe.*
Chapman/Hall, London, 1997

Jedicke E: *Die Roten Listen: Gefährdete Pflanzen, Tiere, Pflanzengesellschaften und Biotope in Bund und Ländern* (mit CD-ROM). Ulmer, Stuttgart, 1997

Kerney M: *Atlas of the Land and Freshwater Molluscs of Britain and Ireland.* Harley, Colchester, Essex, 1999

King WB: *Red Data Book. Aves.* International Union for Conservation of Nature and Natural Resources (IUCN). Morges, Switzerland, 1979

Kingdon J: *African Mammals.* Academic Press, London, 1997

Kühlmann D, Kilias R, Moritz M, Rauschert M: *Wirbellose Tiere Europas.* Neumann, Radebeul, 1993

Leftwich AW: *A Dictionary of Entomology.* Constable, London, 1977

Leftwich AW: *A Dictionary of Zoology.* Constable, London, 1975

Lieske E, Myers R: *Collins Pocket Guide to Coral Reef Fishes.* Harper-Collins, London, 1994; *Korallenfische der Welt.* Jahr-Verlag, Hamburg, 1994

Lindner G: *Muscheln und Schnecken der Weltmeere,* 5. Aufl. BLV, München, 1999

Mattison C: *Schlangen Enzyklopädie. Alle Arten der Welt.* BLV, München, 1999

Metcalf RL, Metcalf RA: *Destructive and Useful Insects,* 5th edn. McGraw-Hill, New York, 1993

Miller PJ, Loates MJ: *Collins Pocket Guide to Fish of Britain & Europe.* Harper-Collins, London, 1997

Miller PR, Pollard HL: *Multilingual Compendium of Plant Diseases. Viruses and Nematodes.* American Phytopathological Society, St. Paul, IL, 1977

Nelson JS: *Fishes of the World,* 3rd edn. Wiley, New York, 1994

Pennak RW: *Collegiate Dictionary of Zoology.* Ronald Press, 1964

Perrins CM: *The Illustrated Encyclopaedia of Birds.* Headline, London, 1990; *Die Grosse Enzyklopädie der Vögel.* Mosaik, München, 1992

Pfadt RE: *Fundamentals of Applied Entomology,* 3rd edn. Macmillan, New York, 1978

Ragge DR, Reynolds WJ: *The Songs of the Grasshoppers and Crickets of Western Europe.* Harley/NHM, London, 1998

Robins CR et al: *Common and Scientific Names of Fishes from the United States and Canada.* American Fisheries Society, Bethesda, MD, 1991

Roper CFE, Sweeney MJ, Nauen CE: *Cephalopods of the World.* FAO Fisheries Synopsis 125 (3), FAO/UN, Rome, 1984

Schmidt G: *Die deutschen Namen wichtiger Arthropoden.*
Mitt. Biol. Bundesanst. Berlin-Dahlem. Parey, Berlin/Hamburg, 1970 (137)

Seymour PR: *Invertebrates of Economic Importance in Britain. Common and Scientific Names,* 4th edn. Her Majesty's Stationary Office, London, 1989

Sokolov VE: *Dictionary of Animal Names in Five Languages.*
Amphibians & Reptiles (L/R/E/G/F). Russky Yazyk Publ., Moscow, 1988

Sterba G: *Süßwasserfische der Welt,* 2. Aufl. Ulmer, Stuttgart, 1990

Stresemann E: *Exkursionsfauna von Deutschland.* Bd. 1-3.
Volk und Wissen/Fischer, Stuttgart, 1992-1995

Thornback J, Jenkins M: *The IUCN Mammal Red Data Book.*
International Union for Conservation of Nature and Natural Resources (IUCN). Gland, Switzerland, 1982

Tolman T, Lewington R: *Butterflies of Britain & Europe.*
Harper-Collins, London, 1997; *Die Tagfalter Europas und Nordwestafrikas.*
Kosmos, Stuttgart, 1997

Turgeon DD et al: *Common and Scientific Names of Aquatic Invertebrates from the United States and Canada: Mollusks,* 2nd edn. American
Fisheries Society, Bethesda, MD, 1998

Walker's Mammals of the World, 5th edn. (Nowak RM ed.) Johns Hopkins
University Press, Baltimore, 1991

Weidner H: *Bestimmungstabellen der Vorratsschädlinge und des Hausungeziefers Mitteleuropas,* 4. Aufl. Fischer, Stuttgart - Jena, 1982

Wells SM, Pyle RM, Collins NM: *The IUCN Invertebrate Red Data Book.*
International Union for Conservation of Nature and Natural Resources (IUCN). Gland, Switzerland, 1984

Whitehead PJP, Bauchot ML, Hureau JC, Nielsen J, Tortonese E (eds.):
The Fishes of the North Atlantic and Mediterranean,
3 Vols. UNESCO, 1984-1986

Williams AB et al: *Common and Scientific Names of Aquatic Invertebrates from the United States and Canada: Decapod Crustaceans.*
American Fisheries Society, Bethesda, MD, 1989

Wissmann W (Hrsg.): *Wörterbuch der deutschen Tiernamen. Insekten,*
Lieferung 1-6. Akademie-Verlag, Berlin, 1963-1968

Wolters HE: *Die Vogelarten der Erde.* Parey, Hamburg - Berlin, 1975-1982

Wye KR: *The Encyclopaedia of Shells.* Headline, London, 1991

Charles Robert
Darwin
(1809 – 1882)

Britischer Naturforscher und Biologe; erkannte als erster die natürliche Selektion als Triebkraft der Evolution, was er 1859 in *On the Origin of Species* niederschrieb. Damit widerspricht er der Katastrophentheorie Cuviers und den Vorstellungen Lamarcks zur Vererbung erworbener Eigenschaften. Seine Erkenntnisse bilden seither die Grundlage biologischen Denkens. Seine Reise auf der „HMS Beagle" (1831 – 1836) und seine Beobachtungen der Galapagos-Finken bilden einen Meilenstein für die Erkenntnis der Zusammenhänge der biologischen Vielfalt. Hierdurch erhält die systematische Arbeit von Linné, Cuvier, Lamarck und anderen frühen Zoologen und Naturforschern erst ihre wahre Bedeutung ebenso wie Darwins Erkenntnisse die moderne molekularbiologische Systematik inspirieren.

Carolus Linnaeus, Carl von
Linné
(1707 – 1778)

Schwedischer Mediziner, Botaniker und Naturforscher; der wohl bedeutendste Systematiker seiner Zeit; entwickelte die binäre oder binominale Nomenklatur der biologischen Arten, die er erstmals 1735 in *Systema Naturae* vorstellte. Mit der 10. Auflage dieses Werkes (1758) gilt die lateinische Artbenennung nach dem binären System als allgemein anerkannt. In der 12. Auflage der *Systema Naturae* (1766) ordnet er den Menschen (*Homo sapiens*) erstmals dem Tierreich zu. Zur schwedischen Fauna verfasste er 1746 die *Fauna Suecica*. Viele Tiere tragen noch heute die von Linné geprägten Namen, obgleich bei den meisten von ihm benannten Tieren durch Revisionen der Zugehörigkeit, mit Einbeziehung der damals noch unerkannten evolutionären Zusammenhänge, eine Änderung der Gattungsnamen erforderlich geworden ist.

Jean Baptiste Pierre Antoine de Monet, Chevalier de
Lamarck
(1744 – 1829)

Französischer Zoologe; Professor für Naturgeschichte der Niederen Tiere am Musée d'Histoire Naturelle und Jardin des Plantes, Paris; unterschied zwischen Wirbeltieren und Wirbellosen; untergliederte letztere in Mollusken, Anneliden, Crustaceen, Arachniden, „Strahltiere" (Echinodermen) usw. und das zu einer Zeit, zu der Schlangen und Krokodile noch zu den „Insekten" gezählt wurden; benannte viele Wirbellose nach dem Linnéschen System. Lamarck vertrat die Theorie der „Vererbung erworbener Eigenschaften" (Die Giraffe streckt ihren Hals so lange, bis sie die Nahrung erreicht, und vererbt die dadurch erworbene Halslänge an die Nachkommen). Schrieb das siebenbändige Werk *Histoire naturelle des animaux sans vertèbres* (*Naturgeschichte der Invertebraten*), 1815 – 1822.

George Baron de
Cuvier
(1769 – 1832)

Französischer Zoologe; Professor für vergleichende Anatomie und Naturgeschichte am Museé d'Histoire Naturelle und dem Collège de France; Begründer der wissenschaftlichen Paläontologie und vergleichenden Anatomie: untergliederte das Tierreich und benannte viele Tiere, besonders Mollusken und Fische, nach dem Linnéschen System. Cuvier erkannte, dass Fossilien jeweils nur in bestimmten stratigraphischen Schichten vorkommen und vertrat dabei die „Katastrophentheorie", nach der die Arten unveränderlich sind und neue Arten nach epochalem Massenaussterben neu erschaffen wurden. Umfangreiche Sammlungen in den Hallen des Jardin des Plantes in Paris. Schrieb das mehrbändige Werk *Le règne animal (Das Tierreich)*, 1817.

Printed by Printforce, the Netherlands